6th Euro-Asian Pulsed Power Conference (EAPPC 2016)

Held with the 21st International Conference on High-Power Particle Beams (BEAMS 2016) and the 15th International Conference on Megagauss Magnetic Field Generation (MG-XV)

Estoril, Portugal
18 – 22 September 2016

ISBN: 978-1-5108-4607-4

Printed from e-media with permission by:

Curran Associates, Inc.
57 Morehouse Lane
Red Hook, NY 12571

CURRAN ASSOCIATES INC. proceedings .com

Some format issues inherent in the e-media version may also appear in this print version.

Printed by Curran Associates, Inc. (2017)

Association for the Advancement of Pulsed Power (A2P2)
Rua das Calcadas
Vivenda Pereira, 1° andar
2755-277 Cascais
Portugal

Phone: (+351) 21 400 35 40
Fax: (+351) 21 406 08 29

secretariat@a2p2pulsepower.org

Additional copies of this publication are available from:

Curran Associates, Inc.
57 Morehouse Lane
Red Hook, NY 12571 USA
Phone: 845-758-0400
Fax: 845-758-2633
Email: curran@proceedings.com
Web: www.proceedings.com

TABLE OF CONTENTS

EAPPC
WITH BEAMS
& MEGAGAUSS
2016 ESTORIL

EURO-ASIAN PULSED POWER CONFERENCE
& CONFERENCE ON HIGH-POWER PARTICLE BEAMS

PROCEEDINGS

CASCAIS & THE ESTORIL COAST – PORTUGAL
18 TO 22 SEPTEMBER 2016

Organised by

A2P2 POTÊNCIA PULSADA

ENERGY PULSE SYSTEMS DESIGNING ENERGY FOR LIFE

CPL

Institutional Support

TURISMO DE PORTUGAL

estoril MEETINGS

CASCAIS

IEEE NPSS NUCLEAR & PLASMA SCIENCES SOCIETY

ISP

Under the patronage of

ORDEM
DOS ENGENHEIROS
REGIÃO SUL

BEHLKE HIGH-TECH IN HIGH VOLTAGE

DIVERSIFIED TECHNOLOGIES, INC.

ITHPP ALCEN

STANGENES INDUSTRIES INC

R&K

ScandiNova EXCELLENCE IN PULSED POWER

amsTECHNOLOGIES where technologies meet solutions

Logiser

IEEE

ISEL

HVP High Voltage Products. High Voltage Experts.

AMO tronics

OSRAM

Table of Contents

Welcome Letter

Dear Colleagues,

We are delighted to Welcome You at the 6th Euro-Asian Pulsed Power Conference with the 21st International Conference on High-Power Particle Beams and the 15th International Conference on Megagauss Magnetic Field Generation and related topics will be held in Estoril, Portugal from 18 to 22 September 2016 at the Estoril Congress Center.

The conference will provide a forum for the exchange of scientific and technical information between industry, academic institutions and research organizations on the broad range of current and emerging research areas, covering a wide range of scientific knowledge.

In addition to basic research, the conference will feature the technology, the applications and the various commercial opportunities of pulsed power technology, high power particle beams and ultra high magnetic fields and will not only address the present state-of-the-art but will also highlight future trends.

The main subjects to be discussed include pulsed power technology and applications in the biomedical, environmental and food industry, the newest fusion results from the major research centres, high energy density physics, high power microwaves, high-current accelerators, plasma, ion and electron sources, explosive and non-destructive production of high magnetic fields, high energy liners, physics at ultra high fields and many others, such as a growing number of industrial, environmental and biomedical applications that use Pulsed Power Technology.

During the conference a number of important applications and their corresponding commercial opportunities will be discussed such as: the reduction of pollutants from fossil fuels; application of pulsed plasma and electric field techniques for water treatment and food sterilisation, including extraction and cancer treatment; novel pulsed techniques for separating water from oil; magneto-forming techniques including moulding and unconventional welding of metallic sheets, and much more.

We invite you to also visit our exhibitor's area and discuss the latest innovation technologies.

We hope you a have a great time in Estoril and the charming Lisbon Coast and we expect that your networking objectives are fully achieved.

It will be an honor and a pleasure to welcome you all in person at the EAPPC2016 Conference with BEAMS and MEGAGAUSS in Estoril, Portugal.

Looking forward to meeting you all.

Prof. Luís M. S. Redondo
Conference Chair
Pulsed Power Advanced Applications Research Group
from the Lisbon Engineering Superior Institute GIAAPP/ISEL

Programme At A Glance

#	Sunday, 18/9	Monday, 19/9	Tuesday, 20/9	Wednesday, 21/9	Thursday, 22/9
08:30		Opening Ceremony			
09:00					
10:00		Plenary 1	Plenary 2	Plenary 3	Plenary 4
10:30		Oral Session 1 / Oral Session 2	Oral Session 5 / Oral Session 6	Oral Session 9 / Oral Session 10	Oral Session 13 / Oral Session 14
12:30					Closing Ceremony
14:00		Poster Session 1,2,3,4,13,14	Poster Session 5,6,7,8,14	Poster Session 9,10,11,12,13	
15:30	Registration Opens				
16:00		Oral Session 3 / Oral Session 4	Oral Session 7 / Oral Session 8	Oral Session 11 / Oral Session 12	
18:00					
19:00	Welcome Reception		Local Flavour Get Together	Conference Gala Dinner	

Keynote Speakers

Speaker: Dr. Kyle John Peterson
Organisation / Company: Sandia National Laboratories

Kyle Peterson is the manager of the Radiation and Inertial Confinement Fusion (ICF) target design department at Sandia National Laboratories. In this role, he directs the theoretical and computational efforts on the Z facility in the areas of radiation effects sciences, secondary assessment technologies, and ICF. He also is responsible for Sandia's ICF fuel preconditioning and heating scientific program. Before taking on management roles in 2013, he performed research on electrothermal instabilities, magnetohydrodynamic instabilities and turbulence, VISAR based current diagnostics, dynamic hohlraums, z-pinches, and other ICF concepts including radiation driven capsules for the National Ignition Campaign.
Kyle joined Sandia after receiving a Ph.D. in physics from the University of Tennessee in 2003 and a Bachelor's degree in physics from North Park University in 1998.

Speaker: Dr. Michael John Barnes
Organisation / Company: CERN / Technology Department

Michael J. Barnes received the B.Sc. degree in electrical engineering in 1981. Subsequently he carried out an industrially based PhD in power electronics, with research and development undertaken at the General Electric Company of England (GEC) and academic work at the University of Aston in Birmingham, U.K., and received the PhD degree in 1985. He moved to TRIUMF, Canada, in 1989: the main focus of his work at TRIUMF was in the area of pulsed kicker systems for accelerators. Michael joined CERN, Switzerland, in 2006 where he is responsible for several kicker systems, including the high precision injection kicker systems for the Large Hadron Collider (LHC). In addition, he is actively involved in several research projects, including for high power Inductive Adders and semiconductor based Marx generators. He has recently participated in, and is currently taking part in, several international collaborations including MedAustron (Austria), TRIUMF (Canada), the Compact Linear Collider (CLIC), and the Future Circular Collider (FCC). One of Michael's current interests is surface coatings and/or finishes, for insulating surfaces such as alumina, for use in ultra-high vacuum systems with pulsed high-voltage.

Speaker: Dr. Peitian Cong
Organisation / Company: Northwest Institute of Nuclear Technology

Peitian Cong, who was born in 1974, graduated from Xi'an Jiao Tong University (XJTU) with a B. Sci. degree in high voltage engineering in 1996. He received a M. Sci. degree in pulsed power science and technology from NINT in 2000 with a dissertation on a composite insulation for the linear transformer devices (LTDs). He obtained his Ph.D. degree by researching and developing a multi-stage gas switch in 2012 from NINT. In 2010, he spent one year as a visiting researcher with the Plasma Physics Group at Imperial College London.
He is the director of the NINT radiation simulation department specializing in pulsed power driver, high current particle beams and intense radiation. His research interests include close switches, transmission lines and magnetic cores. He is currently focusing on the IVA/LTD pulsed power system and technology.
Dr. Cong is a member of the Chinese Society of Pulsed Power Technology and Application.

Speaker: Dr. Koichi Takaki
Organisation / Company: Iwate University / Department of Electrical Engineering

K. Takaki was born in Fukuoka, Japan in 1963. He received the B.Eng., M.Eng, and Dr.Eng. degrees in Electrical Engineering from Kumamoto University, Kumamoto, Japan in 1986, 1988 and 1995, respectively. In 1989, he worked for the department of Electrical Engineering, Oita National College of Technology. In 1996, he worked on the Faculty of Engineering, Iwate University and was a Professor of Iwate University since 2011. During 2000-2001, he was with McMaster University, Hamilton, ONT, Canada, as a Visiting Scientist. His main research interests include pulsed power technologies such as plasma generation based on solid materials, environmental and agricultural applications. Typically, he has experienced studies of shunting arcs and pulsed-ion technology such as HIPIMS, exhaust gas processing in atmospheric pressure such as decomposition, water remediation by discharge under water, high-voltage stimulation on mushroom for improvement of harvest yield. He is co-authors of 14 books (International: 3 and domestic in Japan: 11 of which two books are appeared in an editor in chief) and 171 papers are published in journals (International: 104 and domestic in Japan: 67).

Pulsed Power at CERN

M.J. Barnes, L. Ducimetière, T. Fowler, B. Goddard
TE-ABT, CERN, Geneva, Switzerland
Mike.barnes@cern.ch

Abstract—**CERN, the European Organization for Nuclear Research, is the world's leading particle physics research laboratory. Founded in 1954, the laboratory sits astride the Franco-Swiss border near Geneva, Switzerland. The laboratory has 22 member states. At CERN, Physicists and Engineers are probing the fundamental structure of the universe. They use the world's largest and most complex scientific instruments to study the basic constituents of matter. Protons and lead ions are collided at close to the speed of light, giving the physicists clues about how the particles interact and providing insights into the fundamental laws of nature. Particle Physicists and their accelerators have for decades required the design, development, construction, control and operation of equipment that utilizes pulsed power and power electronics. In this paper, after a brief introduction to CERN and in particular the Large Hadron Collider, an overview of pulsed power utilized at CERN will be presented. Subsequently an overview of ongoing research and development into pulsed power technology for possible future colliders, including the Future Circular Collider, with significantly higher performance than the LHC, will be given.**

Keywords— Accelerator, analogue modulation, CERN, CLIC, FCC, inductive adder, kicker magnet, kicker system, Marx generator, modulator, pulsed power

I. Introduction

The Big Bang theory of the universe proposes that the universe expanded explosively from an extremely dense and hot state, and continues to expand today. Calculations have dated this Big Bang to approximately 13.7 billion years ago. In the first moments after the Big Bang, the universe was extremely hot and dense. As the universe cooled, conditions became just right to give rise to the building blocks of matter – quarks and electrons. A few millionths of a second later, quarks aggregated to produce protons and neutrons. Within minutes, these protons and neutrons combined into nuclei. As the universe continued to expand and cool, things began to happen more slowly. It took 380,000 years for electrons to be trapped in orbits around nuclei, forming the first atoms. These were mainly helium and hydrogen, which are still by far the most abundant elements in the universe. 1.6 million years later, gravity began to form stars and galaxies from clouds of gas. Heavier atoms such as carbon, oxygen and iron, have since been continuously produced in the hearts of stars.

Stars and galaxies do not tell the whole story. Astronomical and physical calculations suggest that the visible universe is only 4% of what the universe is actually made of. A very large fraction of the universe, 26%, is made of "dark matter". Unlike stars and galaxies, dark matter does not emit any light or electromagnetic radiation of any kind, so it can be detected only through its gravitational effects. "Dark energy" accounts for about 70% of the mass-energy content of the universe: even less is known about it than dark matter. At CERN, Physicists and Engineers use a complex of accelerators to probe the fundamental structure of the universe.

The Large Hadron Collider (LHC) is the world's largest and most powerful particle accelerator. It first started up in September 2008, and remains the latest addition to CERN's accelerator complex. The LHC consists of a 27 km ring of superconducting magnets, to constrain the particles on a circular trajectory, with a number of accelerating structures to boost the energy of the particles. Inside the LHC, two high-energy particle beams travel at close to the speed of light before they are made to collide. The beams travel in opposite directions, in separate beam pipes, kept at ultrahigh vacuum.

The LHC is situated approximately 100 m underground. To achieve the required performance and reduce losses many systems are underground. Hence, given the access time, reliable and modular designs of equipment are essential. In addition, where possible, redundancy is incorporated. Furthermore remote diagnostics are an essential feature. Certain power electronics are necessarily placed close to the accelerators: beam losses can irradiate electronics (digital and power electronics) with negative consequences.

Pulsed power is used in many areas of application at CERN including: (i) Injection of beam into and extraction of beam from an accelerator ring; (ii) RF systems; (iii) Beam focusing system; and (iv) Pulsed ramping of dipole magnets. The following sections give a brief overview of several pulsed power systems currently utilized at CERN. Subsequently a summary of some ongoing research and development, at CERN, into pulsed power technology is given.

II. Overview of Pulsed Power at CERN

A. Injection of Beam into an Accelerator

An accelerator stage has limited dynamic range: a chain of stages is required to reach high energy. Thus beam transfer into (injection) and out of (extraction) an accelerator is required. A combination of septa and kickers are frequently used for injection and extraction. Septa can be electrostatic or magnetic: pulsed septa provide slower field rise and fall times, but stronger field, than kicker magnets. Kicker magnets provide fast field rise and fall times, but relatively weak fields [1].

In a particle accelerator a septum is a device which separates two field regions [2]. Important features of septa are an ideally homogeneous (electric or magnetic) field in one region, for deflecting beam, and a low fringe field (ideally zero field) next to the magnet so as not to affect the circulating beam. Hence a septum provides a space separation of circulating and injected/extracted beam. In contrast a kicker magnet provides time selection [1] of beam to be injected/extracted. A kicker system is used for fast, single-turn, injection and extraction (Fig. 1).

The field produced by a kicker magnet must rise/fall within the time period between the beam batches. In addition, the magnetic field must not significantly deviate from the flat-top of the pulse or from zero between pulses (i.e. very small ripple/excursions). Typical field rise/fall times range from tens to hundreds of nanoseconds and pulse widths range from hundreds of nanoseconds to tens of microseconds. If a kicker exhibits a time-varying structure, in the field pulse shape, this can translate into small offsets with respect to the closed orbit (betatron oscillations). Thus a fast, low ripple, kicker system is generally required.

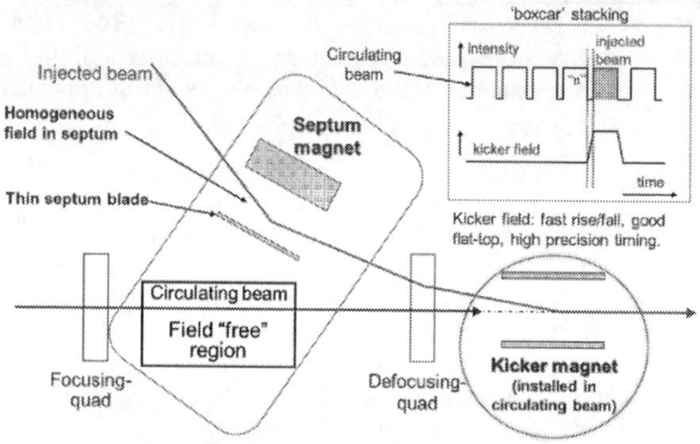

Fig. 1: Fast single-turn injection.

Fig. 1 shows an example of fast single-turn injection in one plane. The injected beam passes through the homogenous field region (aperture) of the septum: circulating beam is in the field free region (i.e. space separation of injected and circulating beam). The septum deflects the injected beam onto the closed orbit at the centre of the kicker magnet; the kicker magnet compensates the remaining angle. A kicker magnet is installed in the accelerator and hence the circulating beam is in the aperture of the kicker. Thus the kicker field must rise from zero to full field in the time interval between the circulating beam and the start of the injected beam (Fig. 1, top right) and fall from full to zero field in the time interval between the end of the injected beam and the subsequent circulating beam.

Line-type modulators, e.g. Pulse Forming Lines (PFLs) and Pulse Forming Networks (PFNs), are currently used in many pulse power systems at CERN. These modulators are used to feed klystrons and kicker magnets in the Large Hadron Collider (LHC) and its injectors [3, 4]: the Proton Synchrotron Booster (PSB), the Proton Synchrotron (PS) and the Super Proton Synchrotron (SPS). Line-type topologies include at least one closing high voltage switch, which can be a spark gap, thyratron or thyristor [5]: a second closing switch can be used to control the pulse duration in the load. The advantages of the line-type modulators are that they are in general relatively simple to design and build, with limited stored energy and good efficiency [5]. Typically, for fast kicker systems, the impedance of the modulator and kicker magnet is matched to avoid reflections: however, this requires that the modulator voltage is twice the magnet voltage. The simplest configuration for a pulse forming circuit is a coaxial cable PFL charged to twice the required pulse voltage. The PFL gives fast and low ripple pulses, but low attenuation is essential, especially with longer pulses, to limit droop and "cable tail" [1]. Attenuation is adversely affected by the use of semiconductor layers to improve voltage rating [1]. Hence, for PFL voltages above 60 kV, SF_6 pressurized polyethylene tape cables are used. A PFL becomes costly, bulky and the droop significant (e.g. ~1%) for pulses exceeding ~3 μs duration. However PFLs may be cost-effective in comparison with solid-state modulator topologies [6]. Where low droop and long duration pulses are required an artificial transmission line, a PFN, made of lumped elements is used [1].

Although a lumped-type magnet has a simple structure, in most cases it cannot be applied to a fast kicker system because of its impedance mismatch and its slow response [1]. The lumped inductance kicker is generally useable only when a rise time above a few hundreds of nanoseconds is required. A resistive termination is generally either in series with the kicker magnet input or else the resistor is omitted. In both cases the lumped inductance kicker magnet only sees (bipolar) voltage during field rise and fall. When the resistor is omitted, magnet current is doubled.

To overcome the long rise time, the first transmission line kicker magnet was developed at CERN in the early 1960's [1]. These magnets consists of few to many "cells" to approximate a broadband coaxial cable. Ferrite C-cores are sandwiched between high voltage (HV) capacitance plates: plates connected to ground are interleaved between the HV plates. The HV and ground plates form a capacitor to ground: one C-core, together with its ground and HV capacitance plates, is termed a cell. Fig. 2 shows an LHC injection kicker magnet, which consists of 33-cells: each cell has 2 ceramic capacitors.

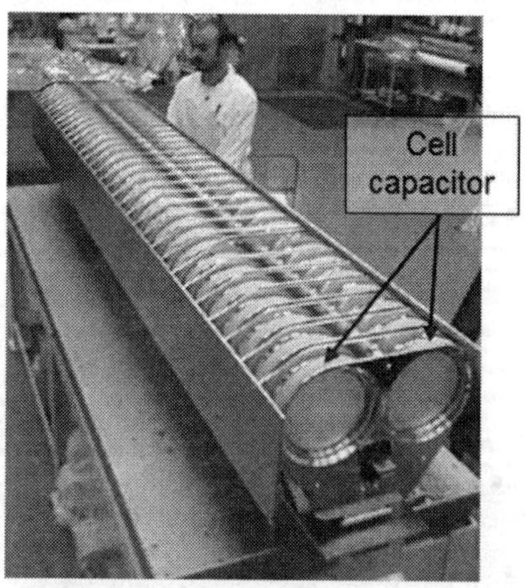

Fig. 2: LHC injection kicker magnet.

To achieve high current and fast field rise time, high voltage is required. Historically thyratrons are used at CERN to switch the high voltage and high current with the required fast rise time. However a thyratron switch is susceptible to erratic (uncontrolled) turn-on, which can result in circulating beam being miskicked and hence cause considerable damage to accelerator components. In accelerators such as the LHC, the PFNs are rapidly charged starting only a few milliseconds before the injection kickers are required: this keeps to a reasonable minimum the time for which high voltage is across the thyratrons [1]. Nevertheless erratic turn-ons of the thyratrons still do, occasionally, occur.

Solid-state modulators differ from the line-type modulators by having a switch that can be opened when conducting full load current and therefore only a portion of the stored energy is delivered to the load during the pulse. Solid-state modulator topologies provide the possibility to modify the pulse duration and thus to reach high efficiency, because the pulse form can be close to an ideal rectangular pulse with very short rise and fall times. Solid-state modulators provide many advantages over line-type modulators, and especially over single switch based modulators. Modulators using thyratrons or spark gaps have limitations with regard to combinations of desired repetition rate, efficiency, pulse width agility, dynamic range, average power and cost, and in some cases even switch availability [7]. In general, solid-state modulators may provide better overall reliability and efficiency, and longer expected life-time over thyratron based modulators [8, 9]. The source impedance of a solid-state modulator can be low and hence, unlike a PFL or PFN, the source voltage does not need to be significantly higher than the load voltage.

A massive improvement program of the LHC injector chain is presently being conducted under the LHC Injectors Upgrade (LIU) project [10]. Upgrades include raising the PSB injection energy from 50 MeV to 160 MeV and the injection energy in the PS from 1.4 GeV to 2 GeV [11]. Thus the PSB requires the construction of new pulse generators to increase the duration of the pulse from 136 μs to a maximum of 620 μs. Each of the five new pulse generators use two 6.5 kV IGBTs in series for operation at ~7 kV and 1000 A. The rise time of the field is ≤ 2 μs, flat-top ripple ≤ ±1 %, and the repetition rate 1.1 Hz. Free-wheeling diodes allow a partial recovery of energy, to the PFN, after pulsing.

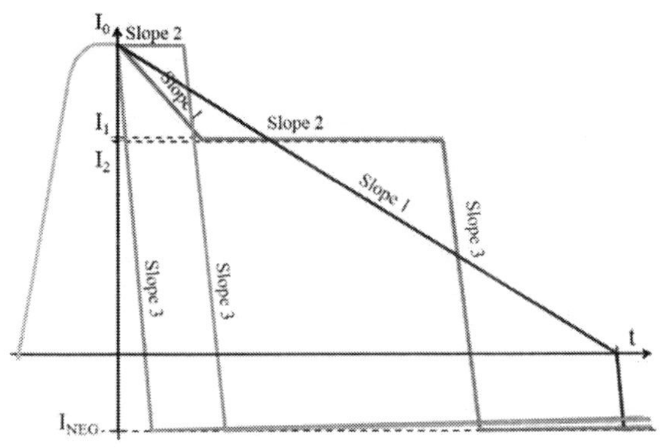

Fig. 3: KSW current waveforms for different users.

As part of the LIU project, 160 MeV H$^-$ ions beam from Linac4 will be distributed to four superposed rings of the PSB and a new multi-turn H$^-$ charge exchange injection system [12] will allow increasing of the beam intensity in the PSB. A series of four horizontal kickers (KSW) will produce a 35 mm closed orbit, with falling amplitude during injection, and uniformly fill the horizontal phase space (transverse painting) and move the circulating beam away from a stripping foil [13]. The KSW kicker magnets will operate with many different waveforms depending on the beam user (Fig. 3). Before beam injection, the current rises to a positive maximum (up to 400 A) and then starts to fall to a negative maximum (−100 A maximum) during the injection with 1, 2 or 3 different linear slopes: the maximum fall-time is 150 μs. The final slope, from negative maximum to 0 A in 1 ms, is identical for all waveforms. All four generators of one ring must work synchronously and deviation from the reference waveform will be less than 1 %. A highly flexible waveform generator has been developed to meet the demands of the KSW and is presently being tested.

Fig. 4 shows a direct drive pulsed septum magnet. This type of magnet is often used under vacuum to minimize the distance between circulating and deflected beam. The coil is generally constructed as a single-turn, so as to minimize magnet self-inductance. The pulsed septum magnet is powered with a half sine wave current, in the range 7 kA to 40 kA, with a half-period duration of typically 3 ms: the peak current density in the septum conductor is up to 300 A/mm^2. For powering a pulsed septum magnet a third harmonic circuit is typically used to obtain a better flat-top current than is given by a basic sinusoidal discharge current [2].

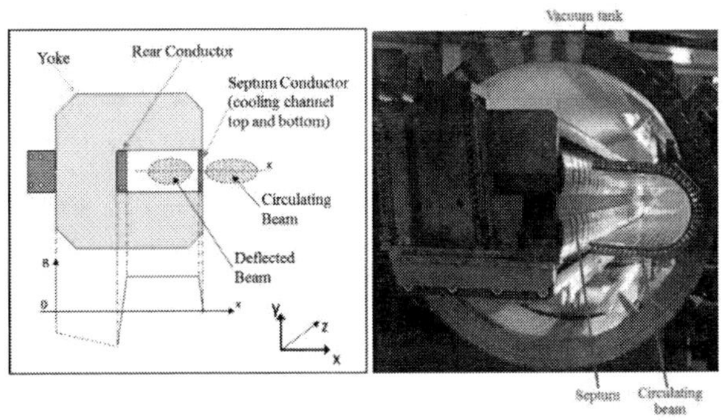

Fig. 4: Direct drive pulsed septum magnet.

B. Radio Frequency Systems

The beams injected into the LHC have an energy of 450 GeV. The maximum design energy of the LHC is 7 TeV per beam. The two counter-rotating beams are accelerated by two identical 400.8 MHz Radio Frequency (RF) systems as the current in the superconducting dipoles is ramped up from 710 A to 11.08 kA, to produce a dipole field of 8.3 T at 7 TeV. Superconducting cavities, operating at 4.5 K, provide the required accelerating field for ramping the beam energy. The RF systems use klystrons: each klystron can deliver up to 330 kW, continuous wave, of RF power at 400.8 MHz, while operating with a cathode voltage of −58 kV and a cathode current of 9.2 A [14]. The klystrons are fitted with a modulation anode, which allows the cathode current to be varied without changing the cathode voltage.

A solid-state klyston modulator recently delivered to CERN, in support of the proposed Compact Linear Collider (CLIC) project, is specified to deliver pulses to −450 kV, 335 A and 1.5 μs flat-top length at 50 Hz repetition rate. The measured pulse flatness is better than 0.4 % and the pulse to pulse stability is better than 0.04% [15].

C. Protection Systems

The nominal LHC beam consists of 2808 bunches per beam, with 1.5 x 10^{11} protons per bunch. Once accelerated to 7 TeV the stored energy per beam is 470 MJ. A beam dump, an 8 m long graphite block, is the only element that can safely absorb this energy. The LHC Beam Dumping System (LBDS) must protect the LHC from damage by reliably and safely extracting and absorbing the circulating beams when requested. The LBDS consists, per beam, of 15 horizontally deflecting extraction kicker magnets (MKD), 15 vertically deflecting septum magnets and 10 dilution kicker magnets (MKB), followed by several hundred meters of beam transfer line before the extracted beam reaches the dump block [16].

The MKD kicker system is the most safety critical and complex element of the LBDS as a fault in the system can lead to significant damage of LHC equipment. The extraction of the beam from the LHC is at a fixed angle of 0.275 mrad, by the MKD magnets, hence the MKD system needs to track the beam energy (between 450 GeV and 7 TeV). The maximum operating voltage of the MKD system is 30 kV to generate a pulsed current of 19 kA with a flat-top of 95 μs. Critical parameters of the MKD pulse are the rise time and the synchronisation with a beam free gap (abort gap) of 3.0 μs, during which the kicker current should rise [16].

In order to achieve high reliability each generator has 2 parallel branches, either of which can carry full load current. At least 14 out of the 15 kicker magnets must operate simultaneously on request to safely extract the beam from the LHC. Any spontaneous firing or fault of one of the pulse generators is detected and the other generators are triggered [17]. To minimize spontaneous self-triggering semi-conductors are used for the power switches: Fast High Current Thyristors (FHCTs), from ABB and Dynex, are utilized.

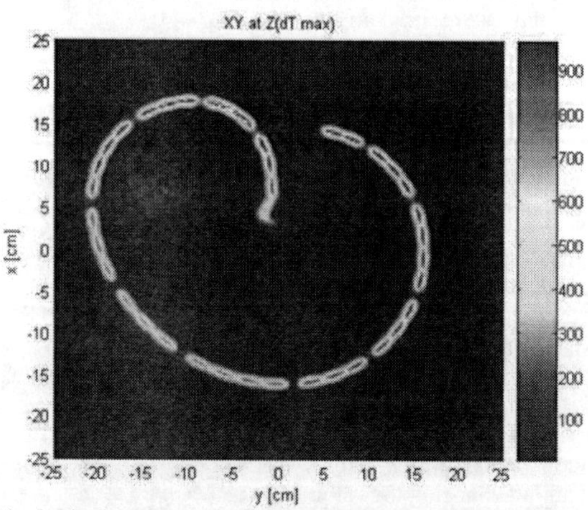

Fig. 5: Lissajous figure of LHC extracted beam on the dump block – right hand side scale shows temperature (°C).

The function of dilution kickers is to sweep the extracted beam in a Lissajous figure on the dump block: the circumference of the figure is approximately 1.2 m. Separate horizontal and vertical MKB systems are used to generate sine and cosine-like current shapes over 90 μs. The peak deflection angle is 0.28 mrad (for 450 GeV to 7 TeV beam). Each MKB generator comprises a capacitor coaxially connected to the magnet inductance. The capacitor is discharged through a solid state closing switch: the discharge pulse is an attenuated sinusoidal oscillation of 25 kA maximum amplitude (at 7 TeV) with a period of 70 μs.

D. Miscellaneous Pulsed Systems

1) AD Horn

Antiprotons for the CERN Antimatter Decelerator (AD) are produced by smashing a beam of protons from the PS onto an iridium target. However, the particles produced by the nuclear interactions are emitted at very wide angles: without a focussing element, all these particles would be lost. A magnetic horn is placed at the exit of the target to focus back a large fraction of the negative particles, including antiprotons, parallel to the beam line and with the right momentum.

The horn is two concentric aluminium conductors, in which the internal one has a double parabolic shape. When a 400 kA current pulse flows through the conductors, a 13 T magnetic field is generated in between the conductors, at a repetition rate of up to 1 pulse per minute. The horn is powered by 18 parallel circuits, each containing a capacitor bank, and an anti-parallel freewheeling diode. The capacitor banks, together with the inductance of the horn and the striplines which connect to the horn, form an LC circuit with a period of ~60 μs. Each anti-parallel diode, which is in series with a resistance, damps the falling edge of the pulse current, reducing the dissipation in the horn. The AD horn and pulse generators date back to the 1970's, and the pulse generators use ignitron switches as both a switch and diode. A project is underway to replace the pulse generators: the new version will use solid-state switches and diodes [18].

2) Solid-State Marx Generator

CERN collaborated with Instituto Superior de Engenharia de Lisboa (ISEL) and Energy Pulse Systems (EPS), Portugal, to develop a high repetition-rate, semiconductor based, Marx generator [19]. The Marx has been designed to power a spark gap system: this system allows the study of breakdown rate phenomena for different materials and surface finishes, for RF structures for the proposed CLIC. The Marx circuit is based on SiC MOSFET (Metal Oxide Semiconductor Field Effect Transistor), with 15 compact stages, capable of running with capacitive type loads during normal operation and also in short-circuit mode at reduced repetition rate. The Marx generates pulses of up to 10 kV with widths from 200 ns to 100 μs, at frequencies up to 1 kHz and pulse voltage rise time of less than 100 ns (Fig. 6) with a load capacitance of 150 pF.

3) Powering of CERN PS magnets

The PS is a key component in CERN's accelerator complex: it usually accelerates either protons delivered by the PSB or heavy ions from a Low Energy Ion Ring. The PS, which presently operates from 1.4 GeV up to 25 GeV, has a circumference of 628 m, contains 277 conventional (room-temperature) electromagnets, including 101 dipoles to bend the beams round the ring. The dipole magnets are connected in series and have a total inductance of 0.9 H and resistance of 0.32 Ω: these magnets must be ramped, together with the beam energy, to a current of up to 6 kA. This requires a voltage of up to 10 kV to be applied to the string of diploes, resulting in a peak power demand of 60 MW during the ramp to the flat-top current [20]. When the current is ramped down to 0 A, the peak power is inverted. To supply the pulses of power required, the PS used a dedicated motor-generator set. The inertia stored an energy of 230 MJ at 1000 rpm, for ramping magnets, to avoid mains power fluctuations. A 6 MW motor ensured the production of the losses.

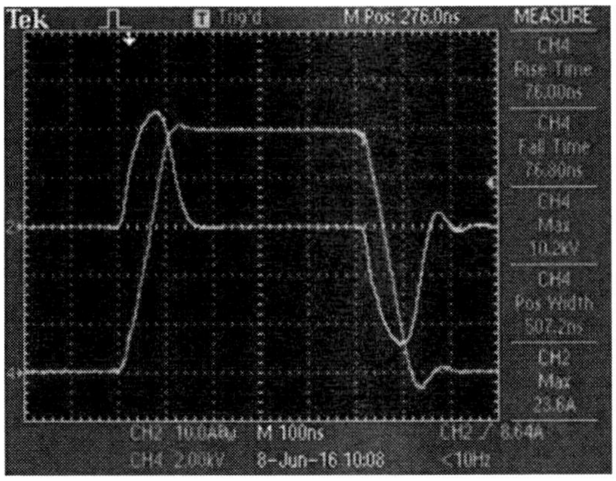

Fig. 6: Measured pulse into a 150 pF load, voltage (green) – 2 kV/div, and current (blue) – 10 A/div, 100 ns/div.

In 2011 a new 60 MW power converter, Power system for PS main magnets (POPS), was commissioned [20]. The POPS system uses 6 capacitor banks, each of almost 0.25 F, as energy storage elements. The system is designed to generate 6 kA peak current with 10 kV peak voltage. The capacitor banks are used to exchange the energy with the PS magnets during converter pulsing operation [21]. As the current in the magnets rises, the voltage across the storage capacitors drops. The energy stored in the magnets is then given back to capacitors at the end of the cycle. Two identical AC/DC converters provide only the losses required by the magnets and the converter itself.

III. R&D for Future colliders at CERN

Originally conceived in the 1980s, the LHC took 25 years to come into being. The LHC is expected to run until 2035. A large-scale upgrade to increase luminosity and thereby exploit the LHC to its full potential is planned: the HL (High Luminosity) LHC is CERN's number-one priority and will increase the number of collisions accumulated in the experiments by a factor of ten from 2024 onwards.

Nevertheless CERN has initiated an exploratory study for a future long-term project centred on a new-generation circular collider with a circumference of 80 km to 100 km. This Future Circular Collider (FCC) programme will focus especially on studies for a hadron collider, similar to the LHC, capable of reaching unprecedented energies in the region of 100 TeV.

The FCC study will be a global venture for particle physics and stems from the recommendation in the European Strategy for Particle Physics, published in May 2013, that a feasibility study be conducted on future fundamental research projects at CERN. The FCC runs in parallel with the CLIC study: CLIC studies have already been under way for a number of years. The aim of the CLIC study is to investigate the potential of a linear collider based on a novel accelerating technology [22].

The goal of the FCC and CLIC studies is to examine the feasibility of various possible machines, to evaluate their costs and to produce a conceptual design report for the FCC and elaborate on the one already produced for CLIC in time for the next European Strategy update circa 2018/2019.

A. Pulsed Power Research and Development

1) Inductive Adders

a) CLIC Damping Rimgs

CLIC would be a high-energy electron-positron collider. It could provide very clean experimental environments and steady production of all particles within the accessible TeV energy range. To achieve high luminosity at the interaction point, it is essential that the beams have very low transverse emittance: the Pre-Damping Ring (PDR) and Damping Ring (DR) damp the beam emittance to extremely low values in all three planes [22].

Stripline kickers are required to inject beam into and extract beam from the PDRs and DRs [23]. Jitter in the magnitude of the kick waveform causes beam jitter at the interaction point [24]. Hence, in particular, the DR extraction kicker must have a very small magnitude of jitter: the 1 GHz specifications call for a 12.5 kV pulse of 900 ns flat-top duration, with a combined ripple and droop of not more than ±0.02 % [22]. An inductive adder (IA) has been selected as a promising means of achieving the specifications for the DR kicker [25]. The IA (Fig. 7) is a solid-state modulator, which can provide relatively short and precise pulses: it is modular and thus the design can be adapted to other requirements.

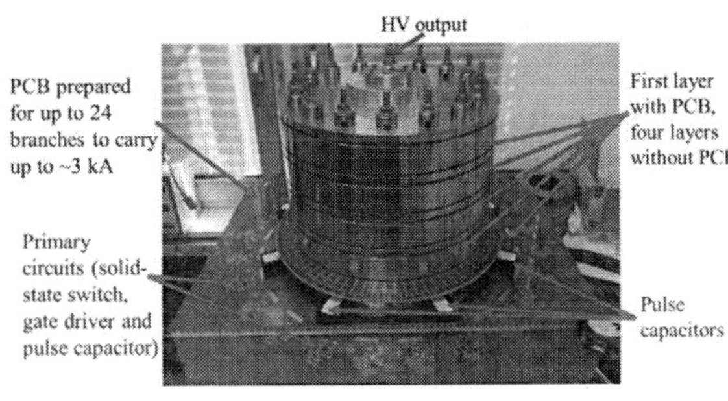

Fig. 7: Prototype inductive adder at CERN.

The solid state switches typically used in the IA are either MOSFETs or IGBTs. These solid-state switches can be opened when conducting full load current, hence only a portion of the stored energy is delivered to the load during the pulse (therefore a PFL or PFN is not required). In addition opening of the solid-state switches potentially limits the duration of fault current in the event of a magnet (load) electrical breakdown. Furthermore, as a result of the low source impedance, source voltage does not need to be doubled.

Considerable research and development has been carried out on the IA for the CLIC DR [26]. A significant advantage of the IA over alternative technologies is the ability to modulate the output [26]. Recent measurements, with a 5-layer prototype, demonstrated a 2.43 kV pulse with a flat-top of ±0.04 % for the required 900 ns (Fig. 8) [27].

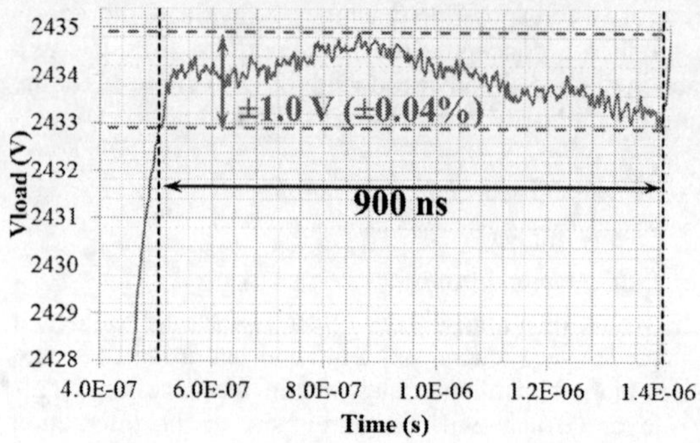

Fig. 8: Measured, modulated, output pulse from a prototype inductive adder.

The CLIC DR kicker specifications also require excellent field homogeneity (±0.01 % specified over a 1 mm radius) and low beam coupling impedance [22]. A stripline kicker has been extensively studied and a prototype built [28]. The prototype striplines will be installed in ALBA, Spain, initially for beam impedance and dc field homogeneity measurements. Following the completion and laboratory testing of the 12.5 kV prototype inductive adders, these may also be installed at ALBA for testing together with the striplines.

b) CERN PS KFA45

The CERN PS KFA45 is a 45-year old PFL based system and it is becoming very challenging to find a commercial replacement for the systems 80 kV SF_6 gas filled coaxial PFL cables: thus an alternative pulse generator design is required. Specifications for the PS KFA45 system include magnet pulse current with 2.6 µs flat-top duration and a magnet impedance of 26.3 Ω: during the flat-top the allowable droop and ripple is ±2%. The nominal single-way delay of the kicker magnet is 35 ns. The required rise and fall times, defined between 1% and 99% of the field pulse, are 46 ns and 96 ns, respectively [29]. The nominal magnet current, with a matched termination, is 1.52 kA: however the possibility of operating the KFA45 with a short-circuit termination, in order to double the pulsed magnetic field, is being studied.

Given the relatively short pulse width an IA may be a good choice for the KFA45 pulse modulator, and detailed studies are underway. In addition research and development is being carried out, in conjunction with ISEL and EPS, on a solid-state Marx generator. High reliability of the system is of great importance and failure modes of the pulse generators must be carefully considered. In addition, the impact of operating the magnet with a short-circuit termination, for generators with low source impedance, will be studied.

c) CLIC Combiner Ring

The specifications for the CLIC Combiner Ring (CR) extraction kickers call for a burst-rate of up to 688 kHz for 140 µs, every 20 ms [22]. The required pulse voltage, of 10 kV into 50 Ω, corresponds to 200 A. The rise and fall time should be no more than 150 ns and the flat-top pulse duration will be between 245 ns to 450 ns, with 96 pulses per burst. Hence the average repetition rate is 4.8 kHz [30].

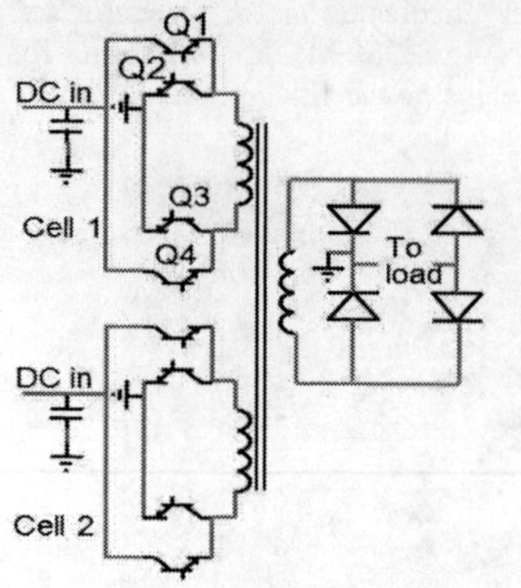

Fig. 9: Novel variant, employing a full bridge inverter, of the typical IA.

The energy transferred during a pulse is approximately 1 J, depending upon the flat-top duration, which corresponds to 96 J per burst. The maximum allowable droop during the burst is ~0.25%, corresponding to 25 V. If this energy were stored in a single capacitor, the 10 kV capacitor would need to store at least 19 kJ – which requires a large capacitance value (384 µF). An alternative approach is to use an inductive adder with appropriate droop correction. Assuming that the magnetic core used for the IA must support the volt-time integral of a burst, this corresponds to a total of ~0.6 V·s. Supposing a 0.5 T saturation flux-density, gives a total cross-sectional area of the core of ~1.2 m^2 [30], which is very large. To reset the cores during the burst, between pulses, requires a reset current of ~100 A, which is similar to the pulse current.

Collaborators at Stanford Linear Accelerator (SLAC), developed a novel variant, employing a full bridge inverter, of the typical IA topology: Fig. 9 shows a 2 cell example. The 2 cells produce alternating positive and negative polarity pulses, but the load output is single polarity. The flux alternates in direction in the magnetic core: hence required cross-sectional area of core material can be reduced by almost 2 orders of magnitude. Tests and measurements on an initial low-voltage prototype show that long pulse trains can be achieved [30].

2) FCC

a) FCC injection

For machine protection reasons the duration of the beam pulse to be injected into the FCC is only 2.25 µs (Table 1). Hence the FCC injection systems will require relatively fast field rise times, in the order of 280 ns, with a flat-top duration of ~2.25 µs [31]. In addition the system must be highly reliable as miskicked beam, e.g. due to an erratic turn-on of a switch, could cause considerable damage to accelerator components: thus thyratrons are not considered for the design.

TABLE I. INJECTION AND EXTRACTION KICKER REQUIREMENTS FOR FCC [31]

Parameter	Injection	Extraction
Kinetic energy [TeV]	3.3	50
Angle [mrad]	0.3	0.13
Pulse length [μs]	2.25	≥ 333
Flat-top tolerance [%]	±0.5	± ~5
Field rise time (0.5-99.5 %) [μs]	0.28	1
Field fall time (99.5-0.5 %) [μs]	0.28	
System impedance [Ω]	5	
Magnet voltage [kV]	3 to 13	
Magnet current [kA]	0.6 to 2.6	0.5 to 8
Repetition rate [Hz]	115	< 0.05

The kicker magnets presently envisaged for the FCC injection are fast transmission line-type magnets similar to the ones used for LHC injection. Each kicker magnet is currently assumed to have a system impedance of 5 Ω, as per the LHC injection kicker magnets. Assuming a magnetic length of 36 m for the magnets, a current of 2.6 kA would be needed, which leads to a required output voltage of the pulse generator of 13 kV. The pulse generator for the injection of a 3.3 TeV beam has to have a fast rise time (< 80 ns) to allow a reasonable magnet filling time. Solid-state switches will be used to ensure reliability: the most appropriate switches, in order to achieve fast and fall rise times, are presently MOSFETs. The relatively low system impedance of 5 Ω requires a low resistance for the primary circuit: hence SiC MOSFETs, with low on-state resistance, are proposed [29]. Both an IA and a Marx are being studied for the FCC injection pulse generator [32].

b) FCC extraction

As per the LHC extraction system, the FCC system will consist of extraction kickers as well as beam dilution kickers, both of which will be part of the FCC beam dump system and will have to reliably abort proton beams with stored energies of up to 8.5 GJ. Both systems have to be precisely synchronized with the particle free abort gab and have to track the beam energy from injection to flat-top level. An unsynchronized beam abort or particles in the abort gap would lead to the so called "asynchronous beam abort". In this situation particles are swept over the accelerator surface into the dump channel until the design trajectory is reached. Whilst in the LHC sophisticated protection devices protect the accelerator components, an asynchronous dump event would be a catastrophic failure case for the FCC. Protection devices for the planned intensities at 50 TeV are a major challenge [33] and currently limited by material parameters.

Table I shows the present specifications for the FCC extraction system: the pulse duration is ≥ 333 μs, for a 100 km circumference. The pulse current is in the range of 0.5 kA (at 3 TeV injection energy) to 8 kA (at 50 TeV), with a rise time of 1 μs. In order to reduce the number of erratic turn-ons, the design for the LBDS excludes thyratron switches. However operational experience of the LHC has shown that the pulse generators, which are based on solid state switches, are prone to erratic turn-on for several reasons: hence, to ensure adequate redundancy in the system, any circulating beam needs to be dumped once any generator integrity is violated. The LHC protection system can thus trigger the LBDS if it detects a potential fault. This fault may be due to, for example, single event burnout of a semiconductor, a high voltage discharge, or even due to self-healing of capacitors.

To mitigate this fault risk for the FCC extraction, a segmented system is under consideration. A relatively long total length would give a reasonable value of current: in addition a high number of active kicker systems results in a shorter length of individual magnets, thus reducing magnet inductance and magnet voltage. Hence each kicker magnet experiences less electric stress and allows for an increased component overrating. In addition the number of modules can be adjusted such that an erratic turn-on of one or two modules would not deflect the beam by more than 1 σ. Hence, in this case, the dump action can be delayed to allow for synchronization with the abort gap [33].

c) Components

Critical components for the Marx generator and IA include semiconductor switches and pulse capacitors. SiC MOSFETs are a very promising technology and samples have been ordered for test and measurement, to verify important parameters for CERN pulsed power applications. Similarly new generations of IGBTs are also promising for pulsed power systems. Studies are also ongoing to improve tolerance to radiation. Furthermore a range of semiconductor switches are being evaluated for future SPS, LHC and FCC applications. These include ring-gate FHCT (e.g. for the SPS beam dump system), SiC FHCTs, and laser triggered thyristors. Laser triggered thyristors have several potential advantages over electrical triggering including elimination of control electronics and thus mitigation of radiation effects and electromagnetic interference issues for these electronics. It is planned to construct a pulse-power facility at CERN to test laser pump triggering of new high voltage semiconductor devices for high power electronics applications [34].

Pulse capacitors, with high capacitance, are required for relatively high operation voltage and to provide high peak current, in a low inductance and high energy-density package. For IA applications NWL Capacitors (USA) have historically been used: Laclanché Capacitors (Switzerland) have recently manufactured capacitors which are being characterized and tested for IA designs [27].

Due to saturation effects of the magnetic core the maximum pulse length of an IA is typically limited to ~3 μs, mainly depending on voltage per layer, magnetic properties of the core and the core cross-sectional area. Tape wound cores are typically used for the IA, however the inter-winding insulation is important too. A selection of cores have recently been purchased and evaluated [35].

The ferrite used under vacuum for CERN kicker magnets is a NiZn ferrite of type 8C11 (Ferroxcube) or CMD5005 (National Magnetics Group): these have a Curie temperature of ~125°C. For some future applications a higher Curie temperature would be beneficial [36], however a disadvantage is that, when operating at elevated temperature, outgassing would be increased – which is detrimental to vacuum.

IV. Summary

The LHC is one of the most complex instruments that has ever been conceived. Much of the equipment is 100 m underground – thus reliability and maintainability are paramount. Pulsed power systems are used throughout the CERN accelerators, including for: (i) injection and extraction of beam; (ii) RF systems; (iii) beam focusing; and (iv) pulsed ramping of dipole magnets. These pulsed power systems cover a wide range of rise time, pulse duration and magnitude of flat-top current and voltage. Many existing pulsed power systems use thyratrons and either PFNs or PFLs. However power semiconductors are starting to make significant inroads into some of the pulsed power applications. Solid-state IAs and Marx generators are being studied as these potentially offer benefits over "traditional" technologies. Pulsed power challenges will be even greater for the next generation of accelerators, and research and development is ongoing to identify suitable components and system topologies.

References

[1] M.J. Barnes, L. Ducimetière, T. Fowler, V. Senaj, L. Sermeus, "Injection and extraction magnets: kicker magnets", in proc. of the CERN-CAS Accelerator School: Magnets, Bruges, Belgium, 16-25 June 2009, edited by D. Brandt, CERN-2010-004, pp 141-166.

[2] M.J. Barnes, J. Borburgh, B. Goddard, M. Hourican, "Injection and extraction magnets: septa", in proc. of the CERN-CAS Accelerator School: Magnets, Bruges, Belgium, 16-25 June 2009, edited by D. Brandt, CERN-2010-004, pp 167-184.

[3] O.S. Brüning (ed.) et al., "LHC design report", Volume I, The LHC Main Ring. CERN, Geneva, 2004.

[4] M. Benedikt (ed.) et al., "LHC design report", Volume III, The LHC Injector Chain. CERN, Geneva, 2004.

[5] K. Takayama (ed.), Briggs, R.J. (ed.), "Induction accelerators". Springer, 2011. ISBN: 978-3-642-13916-1.

[6] K. Whitham, "What is wrong with line type modulators", proc. IECON'03, 29th Vol. 1, pp. 93-95, Roanoke, VA, USA, Nov. 2-6, 2003.

[7] E.G. Cook, "Review of solid state modulators". Linac 2000, proc. XX Int. Linear Accelerator Conf., Monterey, CA, USA, Aug. 21-25, 2000.

[8] A. Parker, "Solid-state technology meets collider challenge". Science & Technology Review, Lawrence Livermore National Laboratory, CA, USA, Sept., 2004. Available on-line: https://str.llnl.gov/str/September04/pdfs/09_04.4.pdf

[9] J.S. Oh et al., "Lifetime issue of a thyratron for a smart modulator in the C-band linear collider", proc. of APAC'04, Gyeongju, Korea, Mar. 22-26, 2004, pp. 767-769.

[10] J. Coupard et al., "LHC injectors upgrade (LIU) project at CERN", proc. of IPAC2016, Busan, Korea, May 2016, MOPOY059.

[11] C. Bracco et al., "LHC injectors upgrade for the HL-LHC", HB2016, 57th ICFA Advanced Beam Dynamics Workshop on High-Intensity and High-Brightness Hadron Beams, Malmö, Sweden, July 3-8, 2016.

[12] W. Weterings et al. "Status of the 160 MeV H- injection into the CERN PSB", proc. of IPAC2012, New Orleans, LA, USA, 20 - 25 May 2012, TUPAN109.

[13] L. Feliciano et al. "A new hardware design for PSB kicker magnets (KSW) for the 35 mm transverse painting in the horizontal plane", proc. of IPAC2015, Richmond, VA, USA, THPF086.

[14] A. Mikkelsen, "Digital measurement system for the LHC klystron high voltage modulator", PhD Thesis, 2014, CERN-THESIS-2014-033 (available on-line).

[15] G.McMonagle, "X band stand alone power source at CERN", CLIC Project Meeting, 8th July 2011, https://indico.cern.ch/event/140986/contributions/159984/attachments/124633/176926/X_Band_Stand_Alone_Power_Source_at_CERN.pdf

[16] J. Uythoven et al., "Calibration measurements of the LHC beam dumping system extraction kicker magnets", proc. of EPAC2006, Edinburgh, Scotland, 26-30 June 2006, TUPLS015.

[17] J. Bonthond, J.H. Dieperink, L. Ducimetière, U. Jansson, E. Vossenberg, "Dual branch high voltage pulse generator for the beam extraction of the large hadron collider", Conference Record of the Twenty-Fifth International Power Modulator Symposium, 2002 and 2002 High-Voltage Workshop, Holywood, CA, USA, June 30 2002-July 3 2002.

[18] G. Gräwer, "AD horn power supply consolidation LS2", unpublished.

[19] L.M. Redondo, A. Kandratsyeu, M.J. Barnes, S. Calatroni and W. Wuensch "Solid-state Marx generator for CLIC breakdown studies", to be pulished in proc. 2016 IEEE Power Modulator and High Voltage Conference, 5-9 July 2016, San Francisco, CA, USA.

[20] F. Boattini, J.P. Burnet, G. Skawinski, "POPS: the 60MW power converter for the PS accelerator: control strategy and performances", CERN-ACC-2015-0098.

[21] F. Bordry, J.P. Burnet, C. Farhni, A. Rufer, "Device for supplying a load with an integrated energy storage", European Patent, EP1868279.

[22] M. Aicheler (ed.) et al., "A multi-TeV linear collider based on CLIC technology: CLIC Conceptual Design Report", CERN-2012-007.

[23] C. Belver-Aguilar, M.J. Barnes, L. Ducimetière, "Review on the effects of characteristic impedance mismatching in a stripline kicker", proc. of IPAC'16, Busan, Korea, 8-13 May 2016, THPMW034.

[24] Y. Papaphilippou et al., "Conceptual design of the CLIC damping rings", proc. of IPAC'12, New Orleans, USA, May 2012, TUPPC086.

[25] J. Holma et al., "Preliminary design of the pulse generator for the CLIC DR extraction system", proc. of PPC'11, Chicago, USA, June 19-23, 2011, pp. 1353-1358.

[26] J. Holma, "A pulse power modulator with extremely flat-top output pulses for the compact linear collider at CERN", Ph.D. Thesis, Aalto University publication series, Doctoral Dissertations 196/2015, Helsinki, Finland, 2015. CERN-THESIS-2014-359 (available on-line).

[27] J. Holma, M.J. Barnes, "Design and initial measurements of a 12.5 kV prototype inductive adder for CLIC DR kickers", to be pulished in proc. 2016 IEEE Power Modulator and High Voltage Conference, 5-9 July 2016, San Francisco, USA.

[28] C. Belver-Aguilar, "Development of stripline kickers for low emittance rings: application to the beam extraction kicker for CLIC damping rings", PhD thesis, Universidad de Valencia, Spain, 2015. CERN-THESIS-2015-175 (available on-line).

[29] M.J. Barnes, L. Ducimetière, J. Holma, T. Kramer, A. Fowler, "Inductive adders for replacing thyratron based modulators at CERN", 17th European Conference on Power Electronics and Applications (EPE'15 ECCE-Europe), 8-10 Sept. 2015.

[30] M. Kemp, "High rep-rate CLIC combiner ring Kkicker", ALERT 2014 Workshop, IFIC-Valencia, Spain, 5-6 May 2014, https://indico.cern.ch/event/306551/contributions/705065/attachments/583693/803464/20_ALERT2014_Kemp.pdf

[31] L.S. Stoel et al., "High energy booster options for a future circular collider at CERN", IPAC'16, Busan, Korea, May 2016, MOPOY007.

[32] M.J. Barnes et al., "FCC-hh injection and extraction kicker topologies and solid state generators", submitted to Physical Review Accelerators and Beams FCC 2016 Conference Edition (unpublished).

[33] W. Bartmann et al., "Dump system concepts for the FCC-hh collider", submitted to Physical Review Accelerators and Beams FCC 2016 Conference Edition (unpublished).

[34] J. Rodziewicz, E. Carlier, B. Goddard, "Prospects for laser triggering of large arrays of semiconductor switches", FCC Week 2016, 11-15 April 2016, Italy.

[35] J. Holma, M.J. Barnes, L. Ducimetière, "Measurements on magnetic cores for inductive adders with ultra-flat output pulses for CLIC DR kickers", proc. of IPAC2016, Busan, Korea, May 2016, THPMW032.

[36] J. Uythoven et al., "Upgrades to the LHC injection and beam dumping systems for the HL-LHC project", proc. of IPAC'14, Dresden, Germany, June 2014, MOPRO032.

Progress on IVA and LTD research at NINT

Peitian Cong[1], Yixiang Hu[1], Weixi Luo[1], Tieping Sun[1], Jiahui Yin[1], Tao Huang[1], Hao Wei[1,2], Zhiguo Wang[1], Jiangtao Zeng[1], Fengju Sun[1], Hailiang Yang[1], Zhengzhong Zeng[1,2], and Aici Qiu[1,2]

1- State Key Lab of Intense Pulsed Radiation Simulation and Effect (IPRSE), NINT, Xi'an, China, 710024.
2- Electrical Engineering college, XJTU, Xi'an, China,710049.

Abstract—**Both IVA and LTD have important applications for flash radiography and driving Z pinches. The induction voltage adder (IVA) is a device to realize voltage addition and current collection for a pulsed power system with series and parallel sub-circuits. As a very compact IVA, the linear transformer driver (LTD) with higher power density is taken as a promising technology for next generation accelerators. This article is mainly on recent development about IVA and LTD work led by the State Key Laboratory of Intense Pulsed Radiation Simulation and Effect (IPRSE) at the Northwest Institute of Nuclear Technology (NINT). In this paper, the IPRSE lab is firstly introduced in brief, and then several IVA issues involving the azimuthal line, secondary impedance and driving pulse timing are addressed before a 3MV/60kA IVA for flash X radiography is introduced. Finally, after elaborating the investigation on some key technologies such as gas switches and trigger system, several newly developed single stage LTDs in the range of 100kA to 800kA are presented.**

Index Terms—**induction voltage adder, linear transformer driver, gas switches, pre-fire rate, jitter.**

I. Introduction

INDUCTION voltage adder (IVA) is a line pulse transformer (LPT) to add driving voltages in series while restricting the summed voltages to a central conductor which is enclosed at a slightly larger radius by a continuous metal cylinder[1,2]. To date the highest voltage and power output belongs to the 20MV, 700kA, 40ns Hermes-III, and late applications include Cygnus, RITS-6 and HRF [3-6]. Evolving from IVA, the linear transformer driver (LTD) has drawn very much attention and efforts because its higher power density, more efficient and better flexibility promises a good feasibility of petawatt-class z-pinch accelerators [7-13].

Compared with the IVA, the advantages of the LTD approach are, therefore, a significant reduction in both overall and component size as well as configuration flexibility. These advantages are gained at the expense of increased complexity. For large facilities this trade-off becomes particularly beneficial.

Supported by NINT, the state key laboratory of intense pulsed radiation simulation and effect (IPRSE) have been devoted to IVA and LTD research and application for over 10 years [14-19]. The IPRSE lab plays a fundamental and supporting role in the radiation simulation and effect research field in China, and its research fields include the radiation simulation, radiation effect, and intense pulsed radiation measurement and diagnostic. One of our main concerns is the performance responses and mechanisms of electronic components and systems in radiation environment, and another part of our concerns is the theories and technologies of intense pulsed radiation, such as X-rays, particle beams and electromagnetic pulses. Since 1970s, a series of intense pulsed power accelerators have been constructed, and some radioactive sources such as Co-60 and Cf-252 were built as well. With these facilities, quantities of research on radiation effects and radiation diagnostics have been carried out [20-24].

This work was supported by the National Natural Science Funds of China under Grants 51577156 and 51307141.

In this paper, we present recent development and application of IVA and LTD technologies undertaken by IPRSE lab. Firstly, several issues about the IVA technology are addressed, for instance, the uniformity of the cavity-driven current, the secondary impedance profile, and the timing of the driving pulses. Then a 3MV/60kA IVA test bed is introduced. Thirdly, some investigations on LTD technologies including gas switches and trigger systems are described. Finally, several newly developed single-stage LTDs with currents 100kA to 800kA are presented.

II. IVA Technology Research and Application

A. *Electro-magnetic model of cavity*

For a cavity, the azimuthal line plays an important role in distributing current uniformly around the coaxial line, which requires the azimuthal line have a low inductance and multi connections with the cathode plate. Thus, a 3D electromagnetic model was built to study the line profile, the injection mode and the stack insulation [25-28]. A full-size cavity in Fig.1 was also established to test the designed capabilities [29-31]. Table I displays for a one-port driven cavity with different injecting structures the simulated current asymmetries and experimental results. For a two-port cavity driven by two pulses respectively, the influence of the synchronization of the two driving pulses on the current uniformity and addition efficiency was investigated (see Fig.2) [32]. Using the model, the structure of a 9-level insulation stack was optimized, and the flashover probability at 1.5MV was minimized to be 8.7×10^{-9}. Fig.3 shows the electrostatic field distributions.

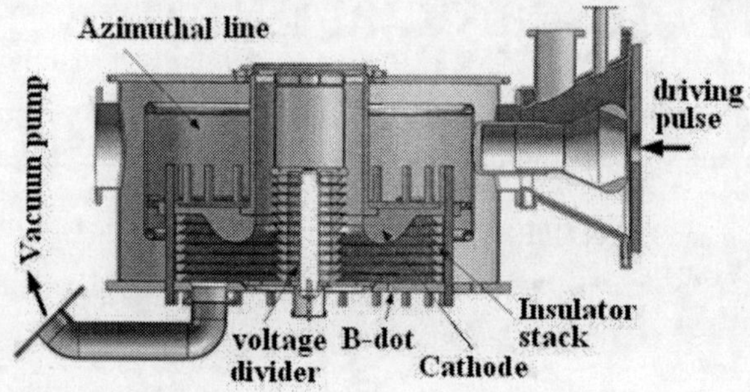

Fig.1 Prototype cavity for test

TABLE I
ASYMETRIC DISTRIBUTION COEFFICIENTS OF THE CURRENT DRIVING

Azimuthal line type	Measured value, δ_m (%)	Calculated value, δ_c (%)
Semi-circle	±9.6	±8.7
Semi-circle with two tabs	±23.4	±21.0
Semi-circle with four tabs	±21.9	±22.1
Whole circle	±24.8	±23.4
No azimuthal line	±58.4	-

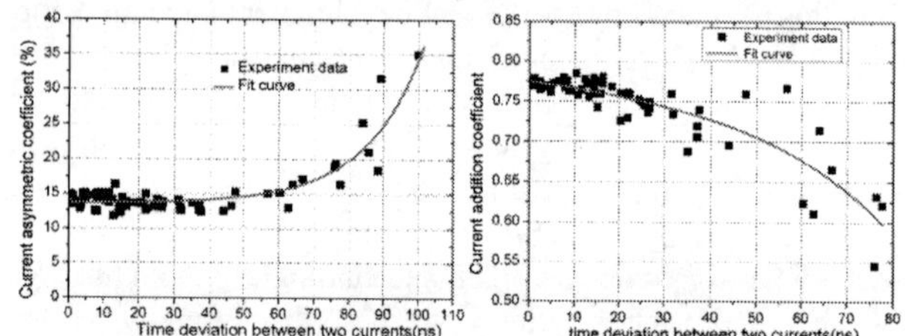

Fig.2 Influence of the timing deviation between two driving pulses on the primary current uniformity (left) and on the secondary current addition efficiency (right)

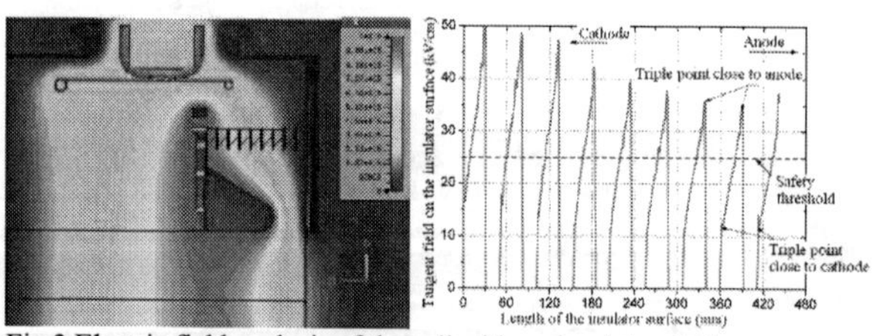

Fig.3 Electric field analysis of the cell with an insulator stack

B. *Circuit model of series cavities*

An IVA accelerator was developed to exchange the difficulty of high voltage pulse formation and switching with the complexity of adding multiple synchronous lower-voltage modules. To design an IVA accelerator with series cavities, a TLCODE model in Fig.4 was developed to evaluate the driving pulses and to determine secondary impedances [32, 34].

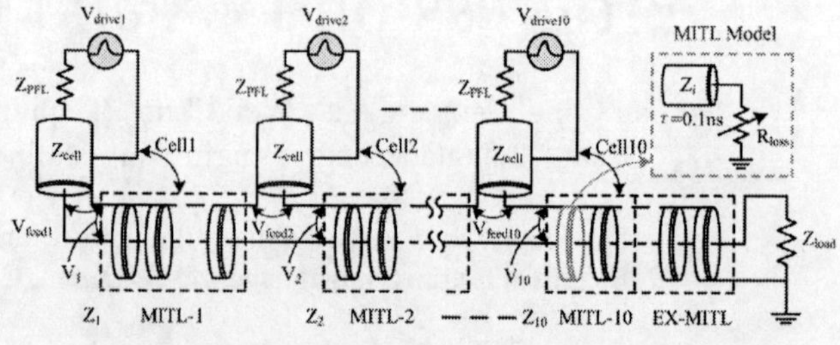

Fig.4 TLCODE model of a magnetically insulated voltage adder (MIVA)

In Fig.5, n means that n cavities were driven simultaneously as one group, and τ_c stands for the electrical length between two neighboring cavities. For a multi-stage IVA, it is called an ideal time sequence operating mode when the time interval of any two driving pulses equals to the electrical length between those two pulse-driven cells. As shown in Fig.5, the simulation for a magnetically insulated voltage adder (MIVA) with 12 cavities explains how the output pulse changes with the cavity grouping option [35], and how the pulses into a cavity vary with the driving time deviation from the ideal timing of $\Delta t=0$.

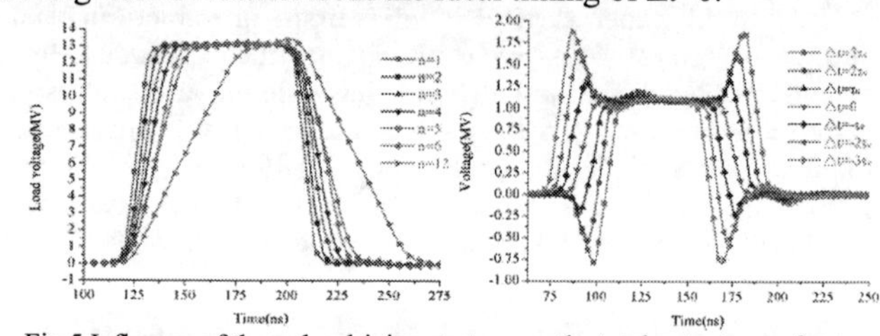

Fig.5 Influence of the pulse driving strategy on the total output waveforms (left) and influence of the driving time deviation from the ideal driving time point on the waveforms of a cavity (right)

The relation between the rise time of output voltage and the jitter of driving pulses was presented. When the number, N of cavities is greater than 30, the rise time of output voltage will increase by 2.563σwith a probability greater than 90%, and σ represents the time jitter of driving pulses. The final rise time of the output voltage, t_{out} is as follows [36],

$$t_{rout} = \sqrt{t_{rin}^2 + (2.563\sigma)^2} \tag{1}$$

where t_{in} is the rise time of input voltage.

For N less than 30, the output rise time should be deemed as a Normal distribution, and its standard deviation σ_{out} could be expressed as [37],

$$\sigma_{out} = -1.53 + 1.53 \cdot \exp(\sigma/6.67) \tag{2}$$

The MITL impedance has close influence on the power transit efficiency between the driver and load. An exponential impedance profile in Fig.6 which has higher voltage addition efficiency than linear and stepped profiles was put forward for the inner stalk design. The impedance is given by [38, 39],

$$Z(x) = \frac{Z_{out}}{N_{ser}} \cdot \exp\left(\frac{x}{L}\ln(N_{ser})\right) \tag{3}$$

where Z_{out} is the output impedance, N_{ser} is the total number of cavities in series, and L is the MITL length.

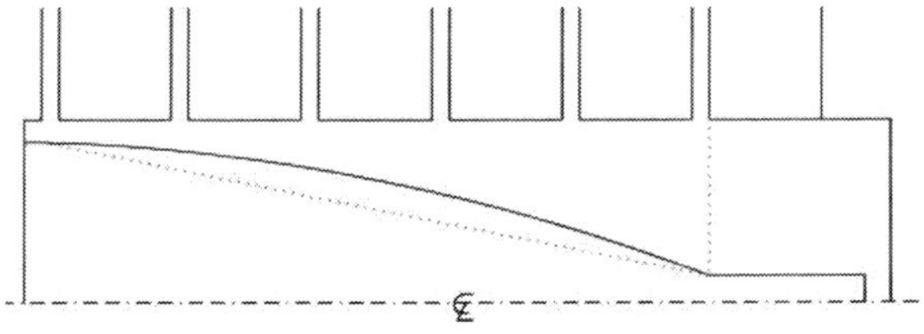

Fig.6 Schematic of a MITL with an exponential profile

C. *IVA accelerator*

An IVA which has three 1.2MV inductively isolated cavities connected in series through a high voltage, vacuum insulating transmission line (VITL) was designed to drive a rod-pinch diode (RPD). The facility in Fig.7 consists of a 3MV Marx, 7.8 ohm pulse forming and transmission lines, impedance transformation water cable, oil manifold, IVA cells and the RPD [40-42]. A 1.2 MV, 70ns induction cell was developed with a successful production of magnetic cores (see Fig.8). The efficiency, I_{eff} of the diode current to the cavity-driving current was no less than 0.9, and the maximum field, E_{bore} on the negatively stressed surface of VITL was 200kV/cm below electron emission threshold.

These magnet cores are made of pre-annealed 25μm thick amorphous metal tapes with a width of 5cm, and 8μm thick capacitor papers are used for tape interlayer insulation. As shown in Fig.9, a convenient method of measuring dynamic magnetization curves accelerated the process of developing magnetic cores[43]. The cores were finally produced to have a packing factor of ~0.6, a saturation flux density B_s of 1.55T, and a ratio of the remanence B_r to B_s was experimentally acquired as 0.5. Each inductive cell contains 840kg pre-annealed amorphous magnetic cores whose radial size parameters were determined by the above mentioned transfer efficiency, I_{eff} and bore field, E_{bore}. With the measured magnetization permeability and the determined geometry parameters, the equivalent inductance and resistance of the cavity with ferromagnetic cores were calculated for modeling the whole IVA circuit. By field analysis and experimental test, the operation fields of the vacuum and magnetic areas were controlled below 70% of the breakdown threshold for the 1.2MV/70ns inductive cavity.

In Fig.10, when the Marx generator was charged at ±35 kV, a pulse of 2.0 MV, 50 kA, and 70 ns (FWHM) was produced to drive the RPD, creating an x-ray source of 1mm size in diameter and 28mGy dose at 1m in 47 ns [44].

Currently, a two-stage MIVA with 2MV/400kA/60ns output parameters is under construction [45].

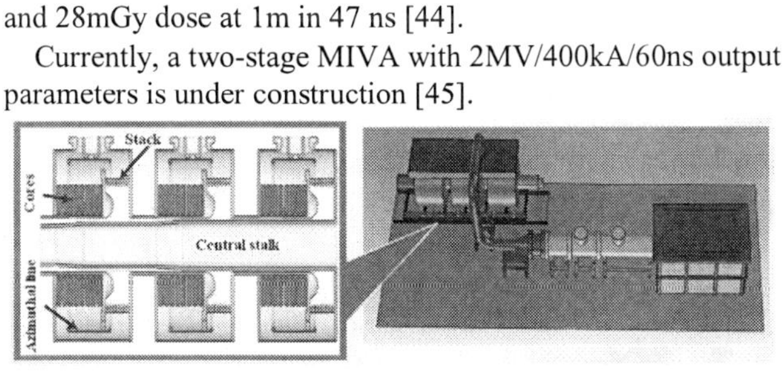

Fig.7 IVA baseline

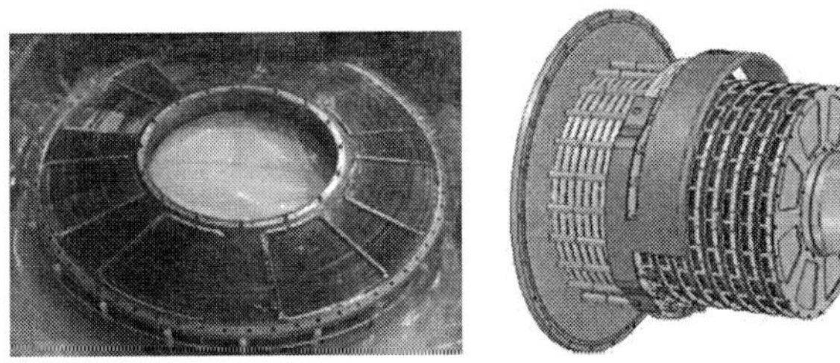

Fig.8 Ferromagnetic core and cavity subassembly

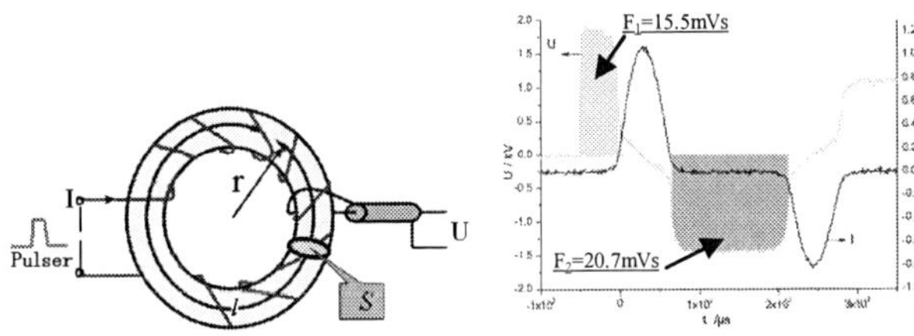

Fig.9 Test schematic and typical results of ferromagnetic cores

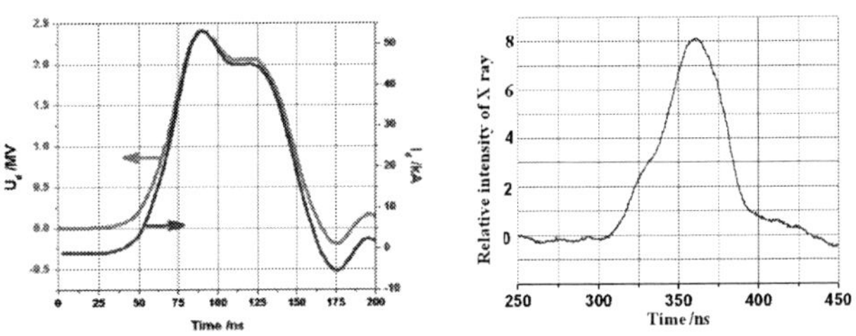

Fig.10 RPD waveforms of voltage, current and X-ray

III. LTD TECHNOLOGY RESEARCH AND APPLICATION

A. *Gas switch*

Since 2009, two multi-stage switches have been developed. As shown in Fig.11, one is a six-stage gas switch with a stackable structure of electrodes and insulating supports, and the other is a four stage switch using corona needles for voltage dividing between gaps [46-56].

Fig.11 Switch appearance (Left: a six-gap switch; right: a four-gap switch)

Based on experimental data of a single gap, a probability analysis method to evaluate the self-breakdown voltage for a multi-stage switch has been put forward. If all series gaps have the same self-breakdown characteristics, the pre-fire probability of the entire switch follows the relation,

$$p_s = 1-(1-p_0)^n \quad (4)$$

where p_s is the pre-fire probability of the switch at voltage u_s, p_0 is the probability of one single gap with a voltage of u_0, n is the number of all gaps in series, and u_0 equals u_s/n when the switch voltage u_s is uniformly distributed into all gaps.

By equation (4), a six-stage switch with a 50% pre-fire probability, p_s demands that six gaps each have a 10.9% self-breakdown probability, p_0. The voltage, u_0 with a self-breakdown probability of 10.9% is solved for the single gap by using a normal distribution function determined by experimental results. So the expected switch voltage, u_s which equals $6u_0$ is obtained. By comparison, the experimental self-breakdown voltage can be evaluated and then might be increased by improving the field distribution.

Fig.12 shows the structure and self-break performance of the finally revised six-gap switch with a stack of electrodes.

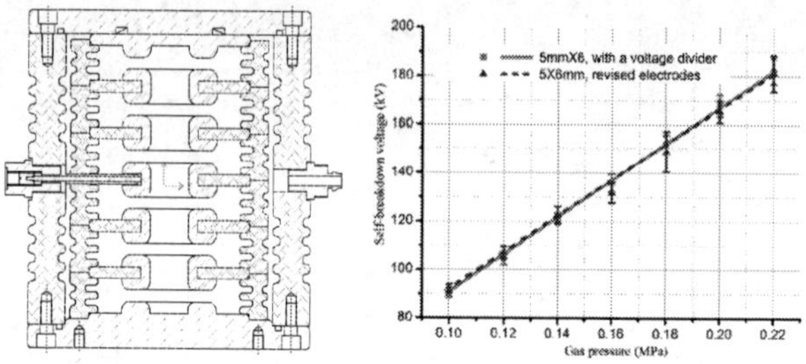

Fig.12 Switch configuration and its self-breakdown performance

A series experiments on different materials displayed in Table III showed that graphite EDM-AF5 demonstrated better self-break performance than stainless steel, tungsten-copper alloy, and copper. As is shown in Fig. 13, the electrode surface roughness is presumably responsible for the consistency of self-break voltage. The life test result in Fig. 13 shows a pre-fire rate of 1.7×10^{-5} for switch V1 with stainless steel electrodes which had 2 pre-fires in an operation of 120'000 shots without failure. Based on the results in Table III, graphite electrodes were then introduced as a replacement to make a new switch V2, and the V2 survived a test of 50'000 shots without pre-fire. The comparison in Table IV promises graphite EDM-AF5 as a potential better choice of electrode material. However, the graphite electrodes need a conditioning process about 10'000 shots which the stainless steel electrodes don't need.

TABLE II

SELFBREAK PERFORMANCE OF A SINGLE GAP SWITCH WITH DIFFERENT MATERIALS (20000 SHOTS FOR EACH MATERIAL)

Material	U_{mean} / kV	σ	U_{min} / kV	P(U<0.9U_{mean})	P(U<0.8U_{mean})
1Cr18Ni9Ti	41.3	4.16 %	34.02	0.186%	0
WCu10	41.1	2.12 %	34.26	0.124%	0
H62	40.6	3.21	26.8	0.55%	0.15%

TABLE III

PERFORMANCES OF TWO 6-GAP SWITCHES USING ELECTRODES OF DIFFERENT MATERIALS (20000 SHOTS EACH)

Material	U_{mean} / kV	σ	U_{min} / kV	P(U<0.9U_{mean})	P(U<0.8U_{mean})
1Cr18Ni9Ti	41.3	4.16 %	34.02	0.186%	0
WCu10	41.1	2.12 %	34.26	0.124%	0
H62	40.6	3.21	26.8	0.55%	0.15%

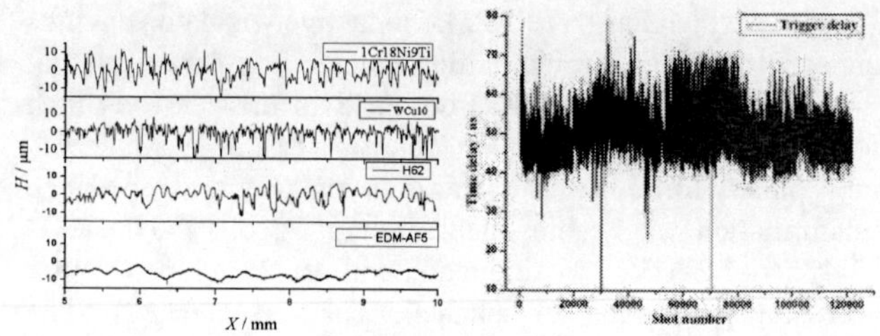

Fig.13 Surface roughness of Electrodes after 20'000 shots (left) and trigger delays of a six-gap switch with 120'0000 shots (right)

TABLE IV

PERFORMANCE COMPARISON OF TWO SWITCHES USING ELECTRODES OF DIFFERENT MATERIALS

	Electrode material	Gas pressure /MPa	Operation ratio	Trigger voltage / kV	Close jitter / ns	Prefire rate	Operation times
V1	1Cr18Ni9Ti	0.3Mpa	65%	70kV	3.4ns	2	120000
V2	EDM-AF5*	0.3Mpa	65%	40kV	3.5ns	0	50000

* This result is after a preliminary conditioning about 10'000 shots

B. Trigger generator

The schematic of a trigger generator which produces a 10ns rise time, 120kV pulse with a jitter of 1ns is shown in Fig.14. The generator in a 40'000 operation test demonstrated well without failure. As displayed in Table V, a pseudospark gap has an order of higher current conduction capability than a hydrogen thyratron. With a pseudospark switch, a 20-channel output generator which provides a 100kV voltage pulse with a 20ns rise time was made. Fig.15 presents the configuration and 10 continuously output pulses of one channel of the generator.

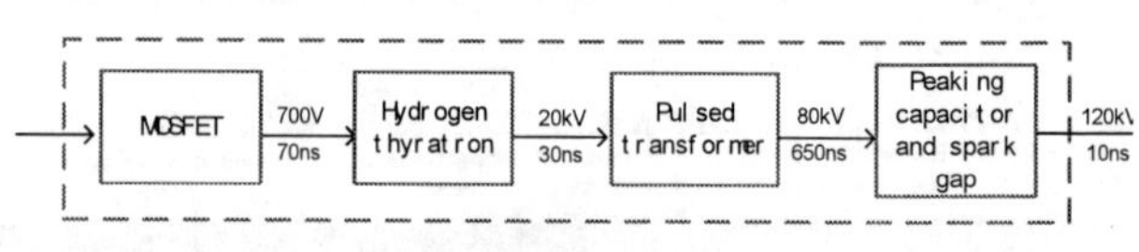

Fig.14 Schematic diagram of the trigger generator setup

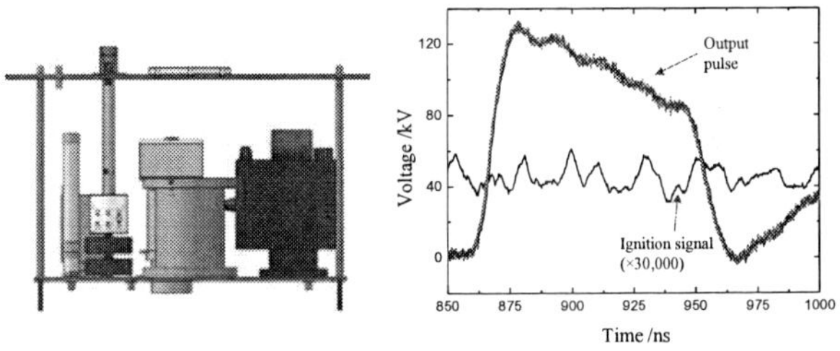

Fig.15 Configuration and 10 output pulses of the trigger generator

TABLE V
PARAMETERS OF HYDROGEN THYRATRON AND PSEUDOSPARK GAP

	Operation voltage	Current	Close jitter	Trigger pulse	Close delay	Heating
Hydrogen thyratron	20-50kV	1-3kA	1ns	>400V	200ns	6.3V/150W
Pseudospark gap	50kV	20kA	1-2ns	5~7kV	200ns	5-6V/30W

C. *LTD testers*

A few LTD testers of 100kA, 300kA, 500kA and 800kA current levels have been developed since 2000. In 2013, a 7-stage LTD driver based on 100kV/300kA single stage was fabricated (See Fig.16). In collaboration with Nuclear Physics and Chemistry Institute of China Engineering Physics Academy, an 800kA, 0.1Hz LTD (LTD800) was produced in 2015 [57]. The LTD800 has 35 drive bricks, and each uses two capacitors and a switch to compose a discharge unit. As shown in Fig.17, one trigger circuit similar to the LTD brick is proposed [58], and the trigger pulse is distributed evenly to the 35 LTD bricks by an azimuthal line, and the trigger unit is triggered through a cable by the generator in Fig.14.

Fig.16 LTD testers: (a) a 7-stage 300kA/100ns LTD and (b) an 800kA/100ns/0.1Hz LTD.

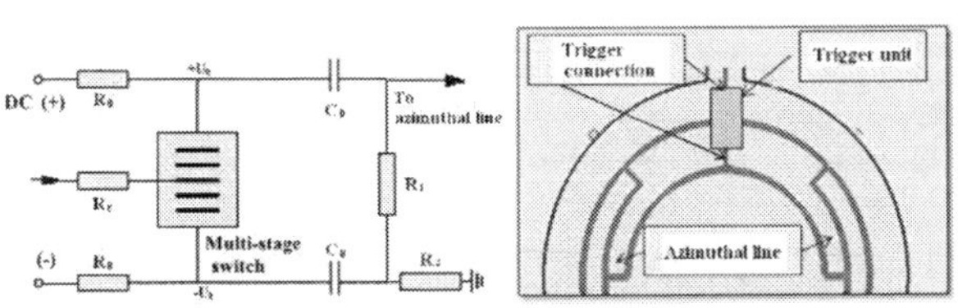

Fig.17 LTD800 trigger (a) circuit and (b) configuration.

IV. SUMMARY

We have briefly described the progress on IVA and LTD technologies at NINT. For an IVA accelerator, some important aspects such as azimuthal line and MITL, are analyzed and discussed to optimize the power efficiency and the current waveform. After reporting investigations on gas switches and trigger systems, two LTD facilities, which output 700kV/300kA/100ns and 800kA/100ns/0.1Hz, are presented respectively.

In the future, our efforts will focus on the current loss of MITL, the pre-fire of gas discharge and the lifetime of gas switch, including both mechanism probing and technology improvement.

ACKNOWLEDGMENT

The authors would like to thank the unmentioned pulsed power teammates of the radiation simulation department in IPRSE Lab for accomplishing the excellent research and hard work.

REFERENCES

[1] G. A. Mesyats. Pulsed power [M]. New York: Kluwer Academic/Plenum Publishers, 2005:10-11

[2] Ian D. Smith, "Induction voltage adders and the induction accelerator family," *Phys. Rev. Spec. Top. Accel. and Beams*, vol.7, no.6, p.064801, 2004.

[3] Juan J. Ramirez, K. R. Prestwich, D. L. Johnson,J. P. Corley, G. J. Denison, J. A. Alexander,T. L. Franklin, P. J. Pankuch, T. W. L. Sanford, T. J. Sheridan, L. L. Torrison and G. A. Zawadzkas, "Performance of the Hermes III Gamma Ray Simulator," in *Proc. 7th IEEE International Pulsed Power Conf.*, June 1989, pp. 26-31.

[4] D. Johnson, V. Bailey, R. Altes, P. Corcoran, I. Smith, S. Cordova, K. Hahn, J. Maenchen, I.Molina, S. Portillo, E. Puetz, M. Sceiford, D. Van De Valde, D. Rose, B. Oliver, D. Welch and D. Droemer, "Status of the 10 M V, 120 kA RITS-6 inductive voltage adder," in *Proc. 2005 Particle Accelerator Conf.*, 2005, pp.314-317.

[5] Weidenheimer D, Corcoran P, Altes R, et al, "Design of driver for the Cygnus X-ray source," in *Proc. 13th IEEE International Pulsed Power Conf.*, 2001, pp. 591-594.

[6] P. Corcoran, V. Carboni, I. Smith, V. Bailey, R. Altes, H. Kishi, J. Pearce, T. da Silva,D. Spelts, R. Stevens, B. Whitney, J. Douglas, J. DeRosa, T. Tatman, K. da Silva, K. Thomas, T. Bryant, M. Bockle, I. Smith, A. Summers, M. Swierkosz and S. Fishlock, "Design of an induction voltage adder based on gas switched pulse forming line," in *Proc. 15th IEEE International Pulsed Power Conf.*, 2005, pp. 308-313.

[7] A A Kim, A N Bastrikov, S N Volkov, V G Durakov, B M Kovalchuk, V A Sinebryukhov, "100 GW fast LTD stage" in *Proc. 10th Symposium on High Current Electronics Conf.*, 2004, pp. 141-144.

[8] A. A. Kim, M. G. Mazarakis, V. A. Sinebryukhov, B. M. Kovalchuk, V. A. Visir, S. N. Volkov, F. Bayol, A. N. Bastrikov, V. G. Durakov, S. V. Frolov, V. M. Alexeenko, D. H. McDaniel, W. E. Fowler, K. LeChien, C. Olson, W. A. Stygar, K. W. Struve, J. Porter, and R. M. Gilgenbach, "Development and tests of fast 1-MA linear transformer driver stages," *Phys. Rev. ST Accel. Beams*, vol. 12, no. 5, p. 050402, May 2009.

[9] W. A. Stygar, W. E. Fowler, K. R. LeChien, F.W. Long, M. G. Mazarakis, G. R. McKee, J. L. McKenney, J. L. Porter, M. E. Savage, B. S. Stoltzfus, D. M. Van De Valde, and J. R. Woodworth, "Architecture of petawattclass Z-pinch accelerators," *Phys. Rev. ST Accel. Beams*, vol.10, no.3, p.030401, Mar. 2007.

[10] Mazarakis M, Fowler W, LeChien K, et al. High-current linear transformer driver development at Sandia National Laboratories [J]. IEEE Trans Plasma Sci, 2010, 38(4): 704-713..

[11] Deng Jianjun ,Wang Meng, Xie Weiping, et al. "Super-power repetitive Z-pinch driver for fusion-fission reactor," *High power laser and particle beams*, vol.26, no.10, 2014, pp.100201.

[12] C L Olson, M G Mazarakis, W E Fowler,et al. "RTL and LTD development for Z-pinch inertial fusion energy (Z-IFE) and high yield," SAND2007-0059
[13] Li Zhenghong;Huang Hongwen;Wang Zhen, Chen Xiaojun, Qi Jianmin, Guo Haibing, Ma Jimin, Xiao Chengjian, Chu Yanyun, Zhou Lin, "Conceptual design of Z-pinch driven fusion-fission hybrid power reactor," *High power laser and particle beams*, vol.26, no.10, 2014, pp.100202.
[14] Bin Kuai, Zhengzhong Zeng, Aici Qiu, Tianxue Liang, Peitian Cong, "A MV-pulsed generator with different pulse-width," *High voltage engineering,* vol.28, no.120, 2002, pp.28-29.
[15] Zhengzhong Zeng, An introduction to practical pulsed power technology (in Chinese), Shannxi Press of Sci. and Tech., 2003.
[16] Zhengzhong Zeng, Bin Kuai, Fengju Sun, Peitian Cong, Aici Qiu, "Criterion of Magnetic Saturation and Simulation of Nonlinear Magnetization for a Linear Multi-core Pulse Transformer," *Plasma Science & Technology,* vol.4, no.5, 2002, pp.1491-1496.
[17] Peitian Cong, Bin Kuai, Aici Qiu, Zhengzhong Zeng, Tianxue Liang, Liangping Wang, Zhong Zhang, Wei Jia, Ning Guo, Hong Chen, "Research of pre-pulse gas switch for Qiangguang-I generator," *High power laser and particle beams,* vol.15, no.7, 2003, pp.725-728.
[18] Fengju Sun, Aici Qiu, Jiangtao Zeng, Zhengzhong Zeng, Jiahui Yin, Bin Kuai, "Relationship of stored energy to core and circuit parameters for pulsed linear transformer driver (in Chinese)," *High power laser and particle beams*, vol.15, no.2, 2003, pp.191-194.
[19] Aici Qiu, Fengju Sun, J "Relationship of stored energy to core and circuit parameters for pulsed linear transformer driver (in Chinese)," *High power laser and particle beams*, vol.15, no.2, 2003, pp.191-194.
[20] Ruibin Li, Guizhen Wang, Wei Chen, Qiang Ma, Yan Liu, Dongsheng Lin, Shanchao Yang, Xiaoyan Bai, Chao Qi, Xiaoming Jin, "Impact of TID on response to pulsed X-ray irradiation in the bipolar operational amplifier," *Science China Technological Sciences*, vol.58, no.2, 2015, pp.390-396.
[21] Chenhui Wang, Wei Chen, Zhibin Yao, Xiaoming Jin, Yan Liu, Shanchao Yang, Zhikuan Wang, "Ionizing/displacement synergistic effects induced by gamma and neutron irradiation in gate-controlled lateral PNP bipolar transistors," *Nuclear Instruments and Methods in Physics Research A*, vol.831, 2016, pp.322-327.
[22] Jin Xiaoming, Yang Shanchao, Li Da, Zhang Wenshou, Wang Chenhui, Li Ruibin, Wang Guizhen, Bai Xiaoyan, Qi Chao, Liu Yan, "Predictive Approach of SEU Occurrence Induced by Neutron in SRAM and EEPROM," *Nuclear Science and Techniques*, vol.26, no.5, 2015, pp. 050501-1-050501-5.
[23] Zujun Wang, Wuying Ma, Shaoyan Huang, Zhibin Yao, Minbo Liu, Baoping He, Jing Liu, Jiangkun Sheng,and Yuan Xue, "Characterization of total ionizing dose damage in COTS pinned photodiode CMOS image sensors," *AIP Advances*, vol.6, no.3, 2016, pp.0325205.
[24] Lili Ding, Simone Gerardin, Marta Bagatin, Dario Bisello, Serena Mattiazzo, "Alessandro PaccagnellaComparison of radiation degradation induced by x-ray and 3-MeV protons in 65-nm CMOS transistors," Chinese *Physics B,* Vol.25, no.9, 2016, pp: 096110.
[25] H. Wei, F. J. Sun, A. C. Qiu, J. T. Zeng, J. H. Yin, T. X. Liang, and Y. X. Hu, "Optimized Design of Azimuthal Transmission Lines for the Cell Driven by Two PFLs in Induction Voltage Adders", IEEE Trans. on Plasma Sci. 8, 2421-2426 (2013).
[26] H. Wei, F. J. Sun, A. C. Qiu, J. T. Zeng, T. X. Liang, J. H. Yin, and Y. X. Hu, "Azimuthal transmission lines for inductive voltage adders with four pfls driving simultaneously or separately," IEEE *Trans. on Plasma Sci.,* vol. 42, no. 10, pp.3057-3063, Oct. 2014.
[27] H. Wei, F. J. Sun, J. H. Yin, A. C. Qiu, "Current density distributions on the cathode plates of inductive voltage cells," IEEE *Trans. on Plasma Sci.,* vol. 42, no. 10, pp.2846-2847, Oct. 2014.
[28] H. Wei, F. J. Sun, J. H. Yin, Y. X. Hu, T. X. Liang, P. T. Cong, and A. C. Qiu, "Numerical simulation of azimuthal uniformity of injection currents in single-point feed induction voltage adders," *Plasma Sci. and Technology,* vol. 17, no. 3, pp.241-246, Oct. 2015.
[29] H. Wei, F. J. Sun, T. X. Liang, J. M. Guo, A. C. Qiu, J. H. Yin, and P. T. Cong, Y.X. Hu, X. F. Jiang, Z.G. Wang, and T.F. Dang, "Experimental study and electromagnetic model of a 1-MV induction voltage cavity," IEEE Trans. on Plasma Sci., vol. 42, no. 10, pp.3057-3063, Oct. 2014.
[30] H. Wei, F. J. Sun, T. X. Liang, J. M. Guo, A. C. Qiu, J. H. Yin, and P. T. Cong, "Experimental validation of an electromagnetic model for a 1-MV induction voltage adder cavity," in *proc. 5th Euro-Asian Pulsed Power Conf.*, Sep. 2014.
[31] H. Wei, F. J. Sun, T. X. Liang, J. H. Yin, T. F. Dang, J. T. Zeng, P. T. Cong, and A. C. Qiu, "Low voltage pulse injection test of a single-stage 1MV prototype induction voltage adder cell," *Review scientific and instruments*, vol. 85, 083596, Aug. 2014.
[32] H. Wei, F. J. Sun, T. X. Liang, P. T. Cong, A. C. Qiu, J. M. Guo, J. T. Zeng, X. F. Jiang, Z.G. Wang, and J. H. Yin, "Test of a single stage 1-MV prototype induction voltage cavity," in *proc. 20th IEEE Int. Pulsed Power Conf.*, Jun. 2015, pp. 843-847.
[33] Yixiang Hu, Fengju Sun, Jiangtao Zeng, Hao Wei, Jiahui Yin, Peitian Cong, and Ai'ci Qiu, "Circuit model of magnetically-insulated induction voltage adders based on the transmission line code," IEEE Transactions on Plasma Science, vol.42, no.8, 2014, pp.2086-2091.
[34] Yixiang Hu, Ai'ci Qiu, Huang Tao, Juanjuan Han, Liangping Wang, Zhengzhong Zeng, Ning Guo, Xinjun Zhang, Weixi Luo, Mo Li, and Yan Li, "Experimental and simulation studies of a 1-m-Long magnetically insulated transmission line with 2-cm anode–cathode gap," IEEE Transactions on Plasma Science, vol.40, no.6, 2012, pp.1743-1748.
[35] Yixiang Hu, Fengju Sun, Jiangtao Zeng, Ai'ci Qiu, Peitian Cong , Jiahui Yin, Jiang Sun, and Hao Wei, "Influences of cell-driving sequences on performances of magnetically- insulated induction voltage adders," IEEE Transactions on Plasma Science, (Accepted, will be published in Oct.).
[36] Yixiang Hu, Fengju Sun, Jiangtao Zeng, Ai'ci Qiu, Peitian Cong , Jiahui Yin, Jiang Sun, and Hao Wei, "Mathematical derivation of cell-driving-jitter effects on the risetime of IVA-output voltages," IEEE Transactions on Plasma Science, (Accepted, will be published in Oct.).
[37] Yixiang Hu, Fengju Sun, Jiangtao Zeng, Ai'ci Qiu, Peitian Cong , Jiahui Yin, Jiang Sun, and Hao Wei, "Effects of cell-driving jitters on the output voltage of magnetically-insulated induction voltage adders," IEEE Transactions on Plasma Science, (Submitted).
[38] Yixiang Hu, Fengju Sun, Jiangtao Zeng, Ai'ci Qiu, Hao Wei, Jiahui Yin, and Peitian Cong, "Exponential impedance-transformer stalks used for magnetically insulated induction voltage adders," IEEE Transactions on Plasma Science, vol.42, no.11, 2014, pp.3598-3560.
[39] Yixiang Hu, Ai'ci Qiu, Tao Huang, Fengju Sun, Peitian Cong, Jiangtao Zeng, Xinjun Zhang, and Tianshi Lei, "Simulation analysis of transmission-line impedance transformers with the gaussian, exponential, and linear impedance profiles for pulsed-power accelerators," IEEE Transactions on Plasma Science, vol.39, no.11, 2011, pp.3227-3232.
[40] Peitian Cong, Guowei Zhang, Hanyu Wu, Jianfeng Sun, Jingya Li, Zhaofeng Su, Hailiang Yang, Fengju Sun, Aici Qiu, "Design of inductive cavity for 3MV induction voltage adder," *High Power Laser and Particle Beams*, vol.23, no.2, 2011, pp.563-568.
[41] Yin Jiahui, Sun Feng ju, Qiu Aici, Yang Hailiang , Zeng Jiang tao, Liang Tianxue, Huang Jianjun, "Self-break water switches study under pulsed voltage in induction voltage adder," *High power laser and particle beams*, vol.21, no.1, 2009, pp.147-151.
[42] Fengju Sun, Aici Qiu, Hailiang Yang , Jiangtao Zeng, Tongyang Gai, Tianxue Liang, Jiahui Yin, Jianfeng Sun, Peitian Cong, Huang Jianjun, Zhaofeng Su, Yi Gao, Zhigang Liu, Xiaofeng Jiang, Jingya Li, Zhong Zhang, Guzhou Song, Mingjing Pei, Shengli Niu, "Pulsed X-ray source based on inductive voltage adder and rod pinch diode for radiography," *High power laser and particle beams*, vol.22, no.4, 2010, pp.936-940.
[43] Peitian Cong, Aici Qiu, Guowei Zhang, Hanyu Wu, Fengjian Sun, Jingya Li, Zaofeng Su, "Development of 1.2MV induction cavity for IVA driver," in *Proc. 18th international conf. on nuclear engineering,* 2009, ICONE18-29459.
[44] Jianfeng Sun, Jiang Sun, Aici Qiu, Pengfei Zhang, Hailiang Yang, Jingya Li, Jiahui Yin, Yang Hu, and Jin Liang, "Effect of plasma density on working characteristics of plasma-filled rod-pinch diode," High Power Laser and Particle Beams, vol.26, no.11, 2014, pp.115009.
[45] Jiahui Yin, Aici Qiu, Fengju Sun, et al. "Performance of a 2.8MV low inductance electrically triggered gas switch," in *Proc.* this conference., 2016.
[46] Peitian Cong, Tieping Sun, ZhengZhong Zeng, et al. "Analysis and enhancing of self-breakdown voltages of a multi-stage gas switch," IEEE *Trans. on Plasma Sci.,* 2013,41(1): 187-191.
[47] Peitian Cong, Aici Qiu, and Tieping Sun. "Comparison of Gaussian and Weibul models used for gas switch self-breakdown probability analysis," *High Power Laser and Particle Beams*, vol.25, no.7, 2013, pp.1831-1834.

[48] Tieping Sun, Peitian Cong, Zhengzhong Zeng. "Multi-gap switch for fast linear transformer driver," *High Power Laser and Particle Beams*, vol.22, no.7, 2010, pp.1673-1676.

[49] Xiaofeng Jiang, Fengju Sun, Tianxue Liang, et al., "Experimental studies on a multi-gap switch for fast linear transformer drivers," *High voltage technology*, vol.35, no.1, 2009, pp103-107.

[50] Peitian Cong, Aici Qiu, Tieping Sun, et al. "Discharge process and closure jitter of six-stage gas switch under pulsed trigger," *High Power Laser and Particle Beams*, vol.25, no.5, 2013, pp.1303-1306.

[51] Tieping Sun, Peitian Cong, Zhengzhong Zeng, et al. "Lifetime of insulator used for multi-gap gas switch,". *High Power Laser and Particle Beams*, vol.24, no.3, 2012, pp.571-575.

[52] Peitian Cong, Tieping Sun, Aici Qiu, et al. "Inductance of six-stage gas switch of fast linear transformer driver," *High Power Laser and Particle Beams*, vol.24, no.8, 2012, pp.2009-2012.

[53] Tao Huang, Weixi Luo, Peitian Cong, et al. "Experimental study on electrode erosion characteristics of spark gas switch under different currents," *High Power Laser and Particle Beams*, vol.26, no4, 2014, pp.045042.

[54] Tieping Sun, Peitian Cong, Weixi Luo. "Analysis and enhancing of self-breakdown voltages of a multi-stage gas switch," High Power Laser and Particle Beams, vol.26, no.11, 2016, pp.115009.

[55] Weixi Luo, Peitian Cong, Tieping Sun, and Tao Huang. Influence of electrode materials on static performance of gas spark switch[J]. *High Power Laser and Particle Beams*, vol.26, no.1, 2016, pp.015022.

[56] Peitian Cong, Weixi Luo, Tieping Sun, et al. "Progress on FLTD gas switch at NINT," *Workshop on pulsed power technology, ICMRE2016*, Chengdu, China, Aug. 2016.

[57] Chuan Liang, Lin Zhou, Fengju Sun, et al. "Investigation of a 0.1Hz, 800kA LTD stage,". *High Power Laser and Particle Beams*, vol.28, no.1, 2016, pp.015005.

[58] Feng ju Sun, Jiangtao Zeng, Tianxue Liang, Hao Wei, Zhiguo Wang, Jiahui Yin, Qiu Aici , "A novel triggering technique based on an internal brick and azimuthal line in cavities for linear transformer drivers," *Modern Applied Physics*, vol.7, no.1, 2016, pp010401.

Peitian Cong who was born in1974, graduated from Xi'an Jiao Tong University (XJTU) with a B. Sci. degree in high voltage engineering in 1996. He received a M. Sci. degree in pulsed power science and technology from NINT in 2000 with a dissertation on a composite insulation for the linear transformer devices (LTDs). He obtained his Ph.D. degree by researching and developing a multi-stage gas switch in 2012 from NINT. In 2010, he spent one year as a visiting researcher with the Plasma Physics Group at Imperial College London.

He is the director of the NINT radiation simulation department specializing in pulsed power driver, high current particle beams and intense radiation. His research interests include close switches, transmission lines and magnetic cores. He is currently focusing on the IVA/LTD pulsed power system and technology..

Yixiang Hu was born in Shannxi Province, China. He received the B.S. and Ph.D. degrees in Electrical Engineering from Xi'an Jiaotong University in 2007 and 2012, respectively.

He is currently with State Key Laboratory of Intense Pulsed Radiation Simulation and Effect, and also with Northwest Institute of Nuclear Technology, Xi'an, China.

Weixi Luo was born in Sichuan Province, China. He received the B.S. degree in pulsed power technology at the Northwest Institute of Nuclear Technology (NINT) in 2011.

He is currently working toward the Ph.D. degree with the Pulsed Power Department at NINT. He is now with State Key Laboratory of Intense Pulsed Radiation Simulation and Effect, and also with Northwest Institute of Nuclear Technology, Xi'an, China.

Tieping Sun was born Shaan'xi, China, in 1979. He received the B. S. degree in industrial electric automation from Xi'an Jiaotong University, Xi'an, in 2002, and the M. S. degree in pulsed power technology from the Northwest Institute of Nuclear Technology (NINT), Xi'an, China, in 2005.

He is currently an associate researcher with the NINT, and his interest includes close switch and fast pulse measurements.

Jiahui Yin was born in Hebei, China. He received the M.S. degree in nuclear technology and applications from the Northwest Institute of Nuclear Technology, Xi'an, China, in 2007. He is currently working toward the Ph.D. degree in the State Key Laboratory of Electrical Insulation and Power Equipment, Xi'an Jiaotong University. His main research interest includes MV triggered switch and trigger generator.

Zhengzhong Zeng was born in Sichuan, P. R. China, in 1956. He received the B.S. degree in nuclear engineering from Tsinghua University, Beijing, China and the Ph.D. degree in electrical engineering from Xi'an Jiaotong University, Xi'an, China.

He has been with Northwest Institute of Nuclear Technology, Xi'an, P. R. China, since 1982. He is now working in the pulsed power field, interested in gas switch, plasma opening switch and Z pinch physics.

Dr. Zeng is the deputy chairman of Chinese Pulsed Power Society (CPPS)

Aici Qiu received a B. Sci. degree in electrical engineering from Xi'an Jiaotong University, Xi'an, China, in 1964.

She is currently a Professor with the Northwest Institute of Nuclear Technology (NINT), and a professor with the High Voltage Division, School of Electrical Engineering, and the State Key Laboratory of Electrical Insulation and Power Equipment, Xi'an Jiaotong University.

Ms. Qiu was honored as an Academician of the academy of Chinese Engineering in 1999.

The Modeling of Thermal and Diffusion Fluxes for Magneto-Inertial Fusion

Sergei V. Ryzhkov
Thermal physics dept.
Bauman Moscow State Technical University
Moscow, Russia
svryzhkov@bmstu.ru

Victor V. Kuzenov
Thermal physics dept.
BMSTU, IPMech RAS, VNIIA
Moscow, Russia
vik.kuzenov@gmail.com

***Abstract*—The magneto-inertial (hybrid) approach to nuclear fusion is considered. Numerical simulation of radiation magneto-gas dynamic processes in the interaction of laser beams with a dense magnetized plasma (target) is performed. A new quasi-monotone numerical method and model for implosion of magnetized target is proposed. Distinctive feature of this problem is presence of initial seed fields (the imposed external pulse magnetic field) and compression of a magnetic flux and target by high energy laser beams or plasma liners. The task for plasma formation and laser induced compression of magnetized target is formulated. Detailed numerical models of laser-driven magnetic flux compression and the pulsed high-temperature processes in two dimensions are developed.**

Keywords—laser driver; magnetized target; mathematical model; heat flux; magneto-inertial fusion

I. INTRODUCTION

Magneto-inertial fusion (MIF) [1-10] is an alternative path to the magnetic and inertial fusion, and magnetic field suppresses the thermal conductivity in a magnetized plasma (the target). Thereby greatly reducing heat losses and increasing the lifetime of the configuration, and hence the thermophysical properties and energy parameters of the reactor.

The mathematical model for determining the heat fluxes and plasma flows to the first chamber wall should be based on the modeling of transport (diffusion, thermal, radiative) processes of atoms, ions, electrons, molecules, photons in a strong magnetic field.

This paper presents a detailed description of magneto-plasmadynamic code that allows to perform modeling of X-rays/first wall interaction. This problem is important in assessing the performance of MIF device as the power plant. This mathematical model can not determine the dynamic response (the evolution of the stress waves and pressures with the strength and compressibility in structural materials) of the first wall on neutron pulse, but the calculations make possible to estimate the interactions between radiation and solid wall, determine the mass of the evaporated material, its temperature, and mechanical impulse.

This work is devoted to the simulation of laser beams interaction with an axisymmetric dense magnetized plasma, e.g. such burning plasma parameters can be produced in magneto-inertial fusion. For this reason the numerical model of the process is based on the multi-component radiative gas dynamics with turbulence model. The radiation transfer is taking into account in multi-group diffusive approach. The main distinguish of this proposed model is the presence of the externally applied magnetic field.

II. MATHEMATICAL MODEL OF LASER-TARGET INTERACTION WITH EXTERNAL MAGNETIC FIELD

The equation of plasma magnetic field generation $\vec{B}$ is:

$$\frac{\partial \vec{B}}{\partial t} = rot\left[\vec{V} \times \vec{B}\right] - \frac{c^2}{4\pi}\frac{t_*}{L_*^2} rot\left(\frac{rot\vec{B}}{\sigma}\right),$$

$$\vec{j} = \sigma\left(\vec{E} + \frac{V_*}{c}\left[\vec{V} \times \vec{B}\right]\right), \qquad (1)$$

where t_* is the characteristic exposure time, L_* is the characteristic size (length of the cylindrical target), V_* is the characteristic velocity, σ is the electrical conductivity coefficient, $\vec{E}$ is the intensity of electric field, $\vec{V}$ is the plasma velocity.

Operator *rot* in the cylindrical coordinate system (r, z, φ):

$$\left(rot\vec{B}\right)_r = \frac{1}{r}\frac{\partial B_z}{\partial \varphi} - \frac{\partial B_\varphi}{\partial z}, \quad \left(rot\vec{B}\right)_\varphi = \frac{\partial B_r}{\partial z} - \frac{\partial B_z}{\partial r},$$

$$\left(rot\vec{B}\right)_z = \frac{1}{r}\frac{\partial r B_\varphi}{\partial r} - \frac{\partial B_r}{\partial \varphi}. \qquad (2)$$

It is known, that plasma dynamical processes are strongly affected by irradiation, if laser plasma temperature achieve 1 eV. In that case, the gas dynamic fields of thermophysical variables may be calculated only with taking into account the radiation fields. In this work the radiation transfer equation is used in the form of multi-group diffusion approach:

$$\frac{1}{J}\frac{\partial\left(Jq_{i\xi}\right)}{\partial \xi} + \frac{1}{J}\frac{\partial\left(Jq_{i\eta}\right)}{\partial \eta} + \chi_i c U_i = 4\chi_i \sigma_i T^4, \qquad (3)$$

$$\frac{c}{3}\frac{\partial U_i}{\partial \xi} + \chi_i q_{i\xi} = 0, \quad \frac{c}{3}\frac{\partial U_i}{\partial \eta} + \chi_i q_{i\eta} = 0,$$

where $U_i(y, z, t)$ is the radiation power density in i-th spectral group, χ_i is the spectral absorption coefficient. Besides of the method mentioned above, the discrete ordinates method has been used in the work, which gets an opportunity to solve the radiation transfer equation on tetrahedral non–structured mesh. Modified alternatively triangular three-layered iterative scheme is applied for the solution of radiation transport equations, where the time step selected via conjugate directions method.

One can use the mechanism of continuum absorption, the opposite to electron bremsstrahlung process under conditions of local thermodynamic equilibrium, to define the plasma absorption coefficient for laser radiation:

$$\chi_\omega = \frac{4,97 g Z_i n_i n_e}{n_c^2 \lambda^2 (kT_e)^{3/2}} \frac{1}{\sqrt{1 - n_e/n_c}},$$

where λ is the laser radiation wavelength (mcm), n_e, n_i are the electron and ion densities correspondingly (cm^{-3}), kT_e is the electron temperature (keV), g is the Gaunt factor, $n_c = 10^{21} \times \lambda^{-2}$ is the critical electron density (cm^{-3}).

The turbulent viscous coefficient $\mu_\Sigma^{\perp,\parallel}$ is calculated by using of the Bussinesk hypothesis, when effective viscous is $\mu_\Sigma^{\perp,\parallel} = \mu_m^{\perp,\parallel} + \mu_t$, where $\mu_m^{\perp,\parallel}$ is a molecular viscosity coefficient and μ_t is a turbulent one, determined from $q-\omega$ turbulence model. The equations of $q-\omega$ turbulence model in curvilinear coordinates ξ,η are:

$$\frac{\partial \rho q}{\partial t} + \frac{d\xi}{dt}\frac{\partial \rho q}{\partial \xi} + \frac{d\eta}{dt}\frac{\partial \rho q}{\partial \eta} + \frac{1}{J}\frac{\partial (J\rho V_\xi q)}{\partial \xi} + \frac{1}{J}\frac{\partial (J\rho V_\eta q)}{\partial \eta} +$$
$$\alpha \frac{\rho q u}{r} = S_q + \frac{\rho q}{2\omega}\left(C_\mu f D - \frac{2}{3}\omega div\vec{V} - \omega^2\right),$$

$$\frac{\partial \rho \omega}{\partial t} + \frac{d\xi}{dt}\frac{\partial \rho \omega}{\partial \xi} + \frac{d\eta}{dt}\frac{\partial \rho \omega}{\partial \eta} + \frac{1}{J}\frac{\partial (J\rho V_\xi \omega)}{\partial \xi} + \frac{1}{J}\frac{\partial (J\rho V_\eta \omega)}{\partial \eta} +$$
$$\alpha \frac{\rho \omega u}{r} = S_\omega + \rho\left(C_1\left(C_\mu D - \frac{2}{3}\omega div\vec{V}\right) - C_2\omega^2\right), \quad (4)$$

$$S_q = \frac{1}{J}\frac{\partial\left\{J\left(\mu_{\Sigma q}^{\perp}\xi_r^2 + \mu_{\Sigma q}^{\parallel}\xi_z^2\right)q_\xi\right.}{\partial \xi} +$$
$$\frac{\left.J\left(\mu_{\Sigma q}^{\perp}\xi_r\eta_r + \mu_{\Sigma q}^{\parallel}\xi_z\eta_z\right)q_\eta\right\}}{\partial \xi} +$$

$$\frac{1}{J}\frac{\partial\left\{J\left(\mu_{\Sigma q}^{\perp}\eta_r\xi_r + \mu_{\Sigma q}^{\parallel}\eta_z\xi_z\right)q_\xi + J\left(\mu_{\Sigma q}^{\perp}\eta_r^2 + \mu_{\Sigma q}^{\parallel}\eta_z^2\right)q_\eta\right\}}{\partial \eta} +$$
$$\alpha\frac{\mu_{\Sigma q}^{\perp}}{r}\left\{\xi_r\frac{\partial q}{\partial \xi} + \eta_r\frac{\partial q}{\partial \eta}\right\},$$

$$S_\omega = \frac{1}{J}\frac{\partial\left\{J\left(\mu_{\Sigma\omega}^{\perp}\xi_r^2 + \mu_{\Sigma\omega}^{\parallel}\xi_z^2\right)\omega_\xi\right.}{\partial \xi} +$$
$$\frac{\left.J\left(\mu_{\Sigma\omega}^{\perp}\xi_r\eta_r + \mu_{\Sigma\omega}^{\parallel}\xi_z\eta_z\right)\omega_\eta\right\}}{\partial \xi} +$$

$$\frac{1}{J}\frac{\partial\left\{J\left(\mu_{\Sigma\omega}^{\perp}\eta_r\xi_r + \mu_{\Sigma\omega}^{\parallel}\eta_z\xi_z\right)\omega_\xi + J\left(\mu_{\Sigma\omega}^{\perp}\eta_r^2 + \mu_{\Sigma\omega}^{\parallel}\eta_z^2\right)\omega_\eta\right\}}{\partial \eta} +$$
$$\alpha\frac{\mu_{\Sigma\omega}^{\perp}}{r}\left\{\xi_r\frac{\partial \omega}{\partial \xi} + \eta_r\frac{\partial \omega}{\partial \eta}\right\},$$

$$\mu_{\Sigma q}^{\perp,\parallel} = \mu_m^{\perp,\parallel} + \mu_t, \quad \mu_{\Sigma\omega}^{\perp,\parallel} = \mu_m^{\perp,\parallel} + 1,3\mu_t, \quad \mu_t = C_\mu f(n)\rho\frac{q^2}{\omega},$$

$$C_1 = 0,045 + 0,405 f(n), \quad C_2 = 0,92, \quad C_\mu = 0,09,$$

$$f(n) = 1 - \exp\left(-0,0065\frac{\rho q n}{\mu_m^{\perp,\parallel}}\right),$$

where q is a pseudo velocity and ω is a pseudo vortex; $f(n)$ is a wall function for correct flux parameters evaluation in laminar sub-layer near a solid streamline surface; n is the distance by normal from given point to nearest surface.

The molecular Prandtle number is $\Pr = 0.72$ and turbulent one is $\Pr_t = \mu_t c_p / \lambda_t = 0.9$ and it is possible to determine the corresponding turbulent heat conductivity coefficient as $\lambda_{e,\Sigma}^{\perp,\parallel} = \lambda_{\perp,\parallel}^e + Cp\mu_t^{\perp,\parallel}/\Pr_t$.

III. Results

A computational domain and target of magneto-inertial fusion consist of the central part and one coaxial layer. They have a cylindrical shape with the following range of values of initial parameters of target and medium. The central part of the target (inner radius R = 0.05 cm) is filled with D-T mixture with density ρ = 5·10^{-2} g/cm^3 and temperature of T = 297 K. It was surrounded by the coaxial layer (outer radius R_c = 0.1 cm) consisting of metal (Al) with density ρ = 2.7 g/cm^3 and temperature of T = 297 K. The computational domain has outer radius r = 0.2 cm. Thermodynamic parameters of external rarefied medium (Ar) are defined: T = 297 K, ρ = 1.29·10^{-3} g/cm^3.

Initial values of intensity of seed magnetic field in the rarefied medium reach the fraction of T. Results of computational modeling [11-13] are presented below. Calculations carried out for the Nd laser radiation for forming rectangular shape (impulse duration = 10 ns). The radiant flux value is $q_{las} = 2{\cdot}10^{14}$ W/cm^2. A thin metallic cylindrical shell material is Al. The target radius is 0.1 cm.

Convective parts of a computer model of magneto-inertial fusion targets are tested on the one-dimensional Riemann problem (for some particular initial data, called the Soda problem) about decay of an unstable fracture configuration. The comparison of exact and approximate solutions showed that the difference is no more than a percent.

The thermal part of a model was tested on some tasks allowing accurate analytical solutions: heating of the continuous medium, filling a flat semi-bounded space $r > 0$, by heat flow through the left-hand fixed boundary $r = 0$.

The spatial distribution of pressure P (atm) and temperature T (kK) is shown in Fig. 1. Plasma velocity $V(x)$ and density ρ (x) versus coordinate r at time t = 2.85 ns are given in Fig. 2. Plots of magnetic pressure P_{mag} (atm) and contact boundary versus r are presented in Fig. 3.

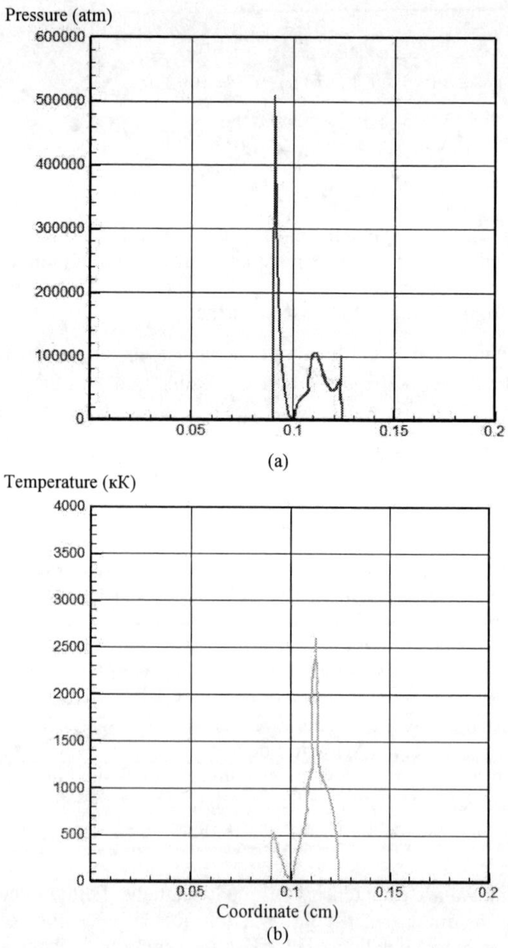

Fig. 1. Spatial dstribution of pressure P (a) and temperature T (b) at 2.85 ns.

IV. Conclusion

In conclusion, the mathematical model of magnetized plasma - laser beams of high energy pulses is developed. It is based on radiation plasma dynamics equations in arbitrary curvilinear coordinates. An improved two-dimensional radiation-hydrodynamics code which simulates plasmas in cylindrical or spherical geometries is designed. It solves single-fluid, two-temperature equations of motion with contributions from diffusion, convection, heat conduction. Electromagnetic processes are described by the Maxwell-Ohm equations in plasma with final conductivity. The transport coefficients in the given system of the equations taking into account magnetized laser plasma. Non-stationary two dimensional radiation magneto-gas dynamic model is developed by authors. The model is based on splitting method in terms of physical processes and spatial directions, that in spatially smooth solution allows to get seventh order of accuracy.

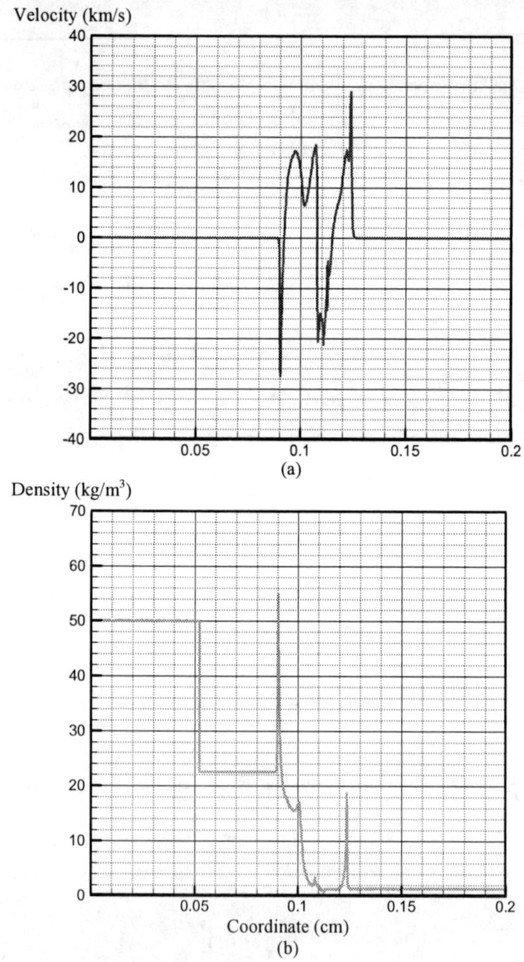

Fig. 2. Distribution of velocity V (a) and density ρ (b) at time t = 2.85 ns.

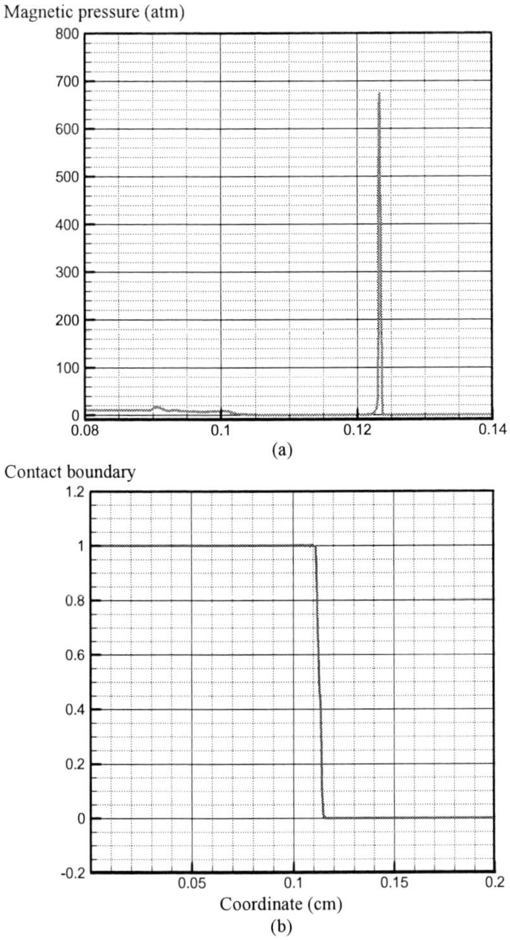

Fig. 3. Spatial distribution of magnetic pressure P_{mag} (a) and contact boundary (b) at time $t = 2.85$ ns.

Modified alternatively triangular three-layered iterative scheme is applied for the solution of radiation transport equations, where the time step selected via conjugate directions method.

Numerical simulation of the magnetized target compression ($q_{las} \leq 10^{14}$ W/cm^2), consisting of a central part and coaxial layer is done. The dynamics of the magnetized plasma expansion into an ambient gas is considered. The key parameters of the laser-target system are calculated. Theoretical results for the case of double-layer target for magneto-inertial fusion are presented. Numerical modeling of the laser target compression process showed the following. The central part of a target is optically transparent for both laser and own broadband plasma radiation during compression. However at the same time heat flow density on the first wall of the reactor chamber at a certain moment of time can reach 10^8 W/cm^2. The magnetic pressure P_{mag} during compression of a target changes in time and is comparable to the static pressure P, reaching values 10^6 atm. The maximum values of plasma pressure and temperature of a magneto-inertial fusion are observed after reflection of a shock wave from a geometrical axis of symmetry at $t = 5$ ns.

Note, that plasma of a magneto-inertial fusion target is a powerful source of magnetic flux and broadband radiation increasing in time (at the first and second stages). The described method is applicable for target compression by high speed plasma jets. Mathematical method developed here may be applied for both impact fast ignition and uniform compression calculations [14-15].

Acknowledgment

This work was supported by the Ministry of Education and Science of the Russian Federation (Project No. 13.79.2014/K).

References

[1] J. T. Cassibry, M. Stanic, S. C. Hsu, F. D. Witherspoon, and S. I. Abarzhi, "Tendency of spherically imploding plasma liners formed by merging plasma jets to evolve toward spherical symmetry," Phys. Plasmas, vol. 19, pp. 052702-9, 2012.

[2] S. C. Hsu, T. J. Awe, S. Brockington, A. Case, J. T. Cassibry, G. Kagan, S. J. Messer, M. Stanic, X. Tang, D. R. Welch, and F. D. Witherspoon, "Spherically imploding plasma liners as a standoff driver for magneto-inertial fusion," IEEE Trans. Plasma Sci., vol. 40, pp. 1287–1298, 2014.

[3] R. B. Baksht, A. G. Rousskikh, A. S. Zhigalin, V. I. Oreshkin, "Metal double-puff Z-pinch," IEEE Trans. Plasma Sci., vol. 41, 182-186, 2013.

[4] P. F. Schmit, P. F. Knapp, S. B. Hansen, M. R. Gomez, K. D. Hahn, D. B. Sinars, K. J. Peterson, S. A. Slutz, A. B. Sefkow, T. J. Awe, R. D. McBride, J. L. Porter, G. K. Robertson, G. A. Rochau, D. C. Rovang, C. L. Ruiz, M. E. Savage, I. C. Smith, W. A. Stygar, R. A. Vesey, "Understanding fuel magnetization and mix using secondary nuclear reactions in magneto-inertial fusion," Phys. Rev. Lett., vol. 113, pp. 155004-6, 2014.

[5] M. E. Cuneo, M. C. Herrmann, D. B. Sinars, S. A. Slutz, W. A. Stygar, R. A. Vesey, A. B. Sefkow, et al., "Magnetically driven implosions for inertial confinement fusion at Sandia National Laboratories," IEEE Trans. Plasma Sci., vol. 40, pp. 3222-3245, 2012.

[6] L. Wu, M. Phillips, S. Messer, et al., "Numerical Simulation of Merging Plasma Jets Using High-Z Gases,“ IEEE Trans. Plasma Sci., vol. 41, pp. 1011-1017, 2013.

[7] V. A. Gasilov, S. V. Zakharov, V.P. Smirnov, "Generation of intense radiation fluxes and megabar pressures in liner systems,“ JETP Lett., vol. 53, no. 2, pp. 85-87, 1991.

[8] S. V. Ryzhkov, "The behavior of a magnetized plasma under the action of laser with high pulse energy,“ Problems of atomic science and technology, no. 4 (7), pp. 105-110, 2010.

[9] I. Yu. Kostyukov and S. V. Ryzhkov, "Magneto-inertial fusion with laser compression of a magnetized spherical target," Plasma Physics Reports, vol. 37, pp. 1092–1098, 2011.

[10] S. V. Ryzhkov, "Current status, problems and prospects of thermonuclear facilities based on the magneto-inertial confinement of hot plasma,“ Bulletin of the Russian Academy of Sciences. Physics, vol. 78, pp. 456–461, 2014.

[11] V. V. Kuzenov, S. V. Ryzhkov, "Numerical modeling of magnetized plasma compressed by the laser beams and plasma jets,“ Problems of Atomic Science and Technology, no. 1 (83), pp. 12-14, 2013.

[12] V. V. Kuzenov, S. V. Ryzhkov, "Evaluation of hydrodynamic instabilities in inertial confinement fusion target in a magnetic field,“ Problems of Atomic Science and Technology, no. 4 (86), pp. 103-107, 2013.

[13] V. V. Kuzenov, S. V. Ryzhkov, "Development of method to modeling physical processes in combined schemes of the magneto-inertial confinement of high temperature plasma,“ Bulletin of the Russian Academy of Sciences. Physics, vol. 80, pp. 598–602, 2016.

[14] A. Yu. Chirkov, S. V. Ryzhkov, "The plasma jet/laser driven compression of compact plasmoids to fusion conditions,“ Journal of Fusion Energy, vol. 31, pp. 7-12, 2012

[15] V. V. Kuzenov, P. A. Frolko, "Standard and combined effects in the concept of magneto-inertial fusion,“ Applied Physics, no. 2, pp. 21-27, 2015

Solid-State Pulsed Power System for a Stripline Kicker*

Dr. Neal Butler, Dr. Marcel P.J. Gaudreau, John Kinross-Wright, Michael Kempkes, Rebecca Simpson
Diversified Technologies, Inc.,
Bedford, MA, USA

***Abstract*—This electronic Diversified Technologies, Inc. (DTI) has designed, built, and demonstrated a prototype pulse amplifier for stripline kicker service capable of less than 5 ns rise and fall times, 5 to 90 ns pulse lengths, peak power greater than 13.7 MW at pulse repetition rates exceeding 100 kHz, and measured jitter under 100 ps. The resulting pulse is precise and repeatable, and will be of great interest to accelerator facilities requiring electromagnetic kickers. The pulse generator is based on the original specifications for the NGLS fast deflector. DTI's planar inductive adder configuration uses compensated-silicon power transistors in low inductance leadless packages with a novel charge-pump gate drive to achieve unmatched performance. The prototyping efforts guided the design of the full unit, however the magnetics and transmission line effects of the system were not revealed until the entire unit was assembled. The unit was brought to LBNL, compared with other researcher's efforts, and was judged very favorably. A number of development prototypes have been constructed and tested, including a successful 18.7 kV, 749 A unit. The modularity of this design will enable configuration of systems to a wide range of potential applications in both kickers and other high speed requirements, including high performance radars, directed energy systems, and excimer lasers.**

Keywords—stripline; kicker; inductive adder; solid-state switch

I. Introduction

Diversified Technologies, Inc. (DTI) recently completed an SBIR effort under a grant from the US Department of Energy to research kicker drivers for the Next Generation Light Source (NGLS) injectors. The NGLS specification required a 10 kV, 200 Ampere pulse into 50 Ω, with a 2 ns rise time (10-90%), a highly repeatable flattop with a pulsewidth from 5-40 ns, and a fall time less than 1 μs (defined as down to 10-4 of the peak value). DTI chose a novel derivative of the classic inductive adder circuit which has demonstrated not only the required pulse rise times required by NGLS, but also the pulses required by the Advanced Light Source (ALS) upgrade at Lawrence Berkeley National Lab (LBNL), and a number of other commercial, medical, and other non-accelerator-based applications.

II. Design

Inductive adders have been used for decades in the accelerator physics community to create ultra-high voltage, short rise time deflection pulses in beams. The inductive adder approach can be thought of as a number of pulse transformers with the primary windings in parallel and the secondary windings in series. Such approaches are not typically used in high-PRF applications, and have not shown the fast rise times required for the NGLS kicker. DTI's improvements to the classic inductive adder design include:

- Multiple cores on a single circuit board
- Secondary return current path on the circuit board
- State-of-the-art 650 V rated compensated MOSFETs
- Low inductance leadless flat packages
- Push-pull double voltage drive
- Low inductance backswing clamp
- High performance low delay charge pump gate drives

Re-arranging the circuitry of the classic inductive adder by placing several cores and drive circuits on a single circuit board allows the secondary return current to flow through the ground plane on the circuit board, which minizes the secondary stray inductance (Fig. 1).

Fig. 1. Dual PCB high speed pulser kicker driver system developed by DTI for the NGLS program. This kicker uses commodity 650 V transistors to switch nearly 17 MW, 700 A in 5 ns.

The best figure of merit power transistors are 650 V rated compensated silicon MOSFETs. These are available in very low inductance leadless flat packs. The relatively low voltage

*Work supported by US Department of Energy contract DE-SC0004255).

rating is compensated for by driving one end of the primary low and the other end high, allowing up to 1 kV added per stage.

At turn-off, the transistor voltage must be limited by a diode backswing clamp. Stray inductance in this circuit allows the voltage to overshoot. This effect can limit the maximum operating voltage, especially with very fast switching times.

The switching performance of transistors can be limited by the gate drive; DTI developed a low delay, high performance, discrete component gate drive which achieves switching performance much faster than the data sheet values.

Fig. 2 shows a simplified schematic of the push-pull circuit. In DTI's final design, four parallel connected MOSFETs drive each side of a primary winding on each stage of the adder. Each individual MOSFET has a diode clamp for overvoltage. All the MOSFETs on each side of each primary share a common high speed gate driver. Isolation for the high side gate drivers is provided using inductive isolation.

III. GATE DRIVER TOPOLOGY

The key innovation is the use of commodity compen-sated silicon MOSFETs with the state-of-the-art figure of merit (on resistance times gate charge) in low-inductance packages at speeds far beyond the data sheet turn-on and turn-off times. This speed was attained with a custom discrete component gate driver developed at DTI.

Typical commercial MOSFETS have an internal dis-tributed gate resistance of a few ohms. The STL57N65M5, for example, has a gate resistance of 1.4 Ω and gate capacitance of 4.2 nF; the RC time constant is nearly 6 ns. Package and stray circuit inductance also increase the gate charging time. To mitigate these problems, a charge pump circuit is used: a 50 V high current, low delay pulse amplifier drives the gate through a series capacitor sized to deliver the correct gate charge. The effective gate circuit RC time constant is reduced by a factor of 5, allowing switching times much faster than the datasheet values.

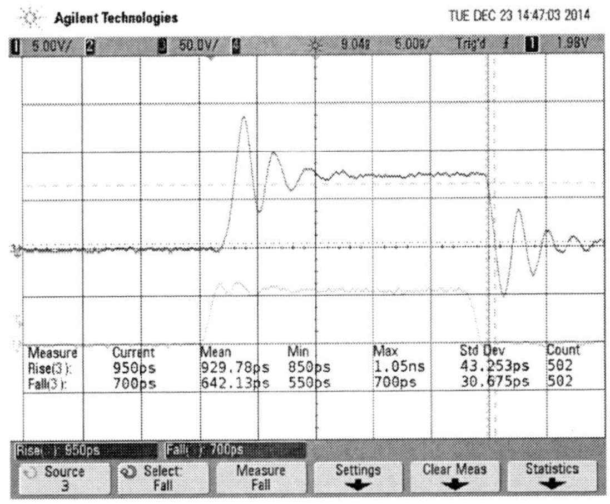

Fig. 3. Performance of the gate driver circuit with GaN transistors in a test board, operating from 75 V into 1.5 Ω and 3.3 nF, or 50 A peak. Note the 950 ps rise time, 700 ps fall time, and 1.5 ns group delay.

The input logic signal is buffered and distributed to the twenty gate drives on a board through a chain of four 74LVC logic gates, with a total delay of about 6 ns. The gate drive circuit and measured output waveform is shown in Fig. 3. The input transistor of the gate driver, the 20 V rated DMG1012T , is driven from both sections of a SN74LVC2G34 buffer. This transistor switches 15 V at 6 A with a propagation delay of less than 2 ns.

The 15 V output from the DMG1012 drives a two tran-sistor totem pole output circuit with IRLML0060 transis-tors. The first stage drives the bottom transistor directly and top transistor through an inverting transformer. The input transistor is normally off, keeping the bottom totem pole transistor on and the power transistors off.

The pulse transformer is a miniature RF balun with re-sponse in the GHz range. This circuit switches 50 V at 35 A

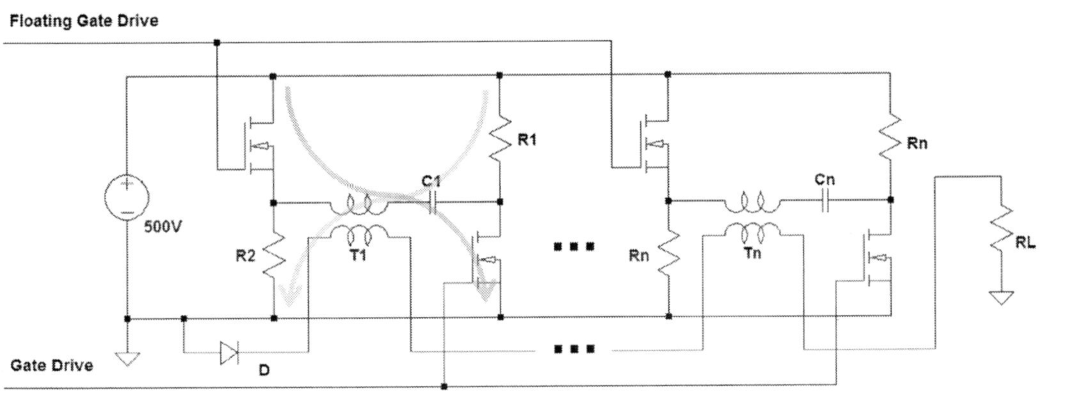

Fig. 2. Electrical schematic of two stages showing main pulse current path in blue and charging current path in green. The floating gate drive is referenced to the source leads of the "flying" transistors. For clarity, snubbers and clamp diodes are not shown. Diode D isolates the transformer core reset voltage from the load.

*Work supported by US Department of Energy contract DE-SC0004255).

with 750 ps rise, 500 ps fall. We also tested a circuit with GaN transistors from EPC. We selected the silicon circuit for ease of manufacturing with acceptable performance.

IV. TRANSISTOR PERFORMANCE

The final layout of the kicker PCBs is shown in Fig. 4. The charge pump gate drive used to push the compen-sated silicon transistors is quite effective. We tested several types of transistors in the circuit: STL57N65M5, STL21N65M6, and IPL65R130C7 were all demonstrated to have excellent rise and fall times. Faster transistors were less rugged than slower ones due to limitations in the backswing clamp circuit.

Fig. 4. Final stripline kicker circuit board.

The backswing clamp diodes and capacitors are mounted directly underneath the power transistors with multiple vias between layers. SiC diodes in the D2-PAK were used. The board and the transistor only accounted for 1 nH of stray inductance, but the diode package measured 2.4 nH. Thus with all losses accounted, at 40 A, the turn-off spike could only clamp at 154 V even with a SiC diode directly across the drain of each MOSFET. This ultimately limited our 650 V MOSFETs to safe operation at a maximum of 450 V. A lower inductance diode is required in future development work.

Fig. 5 shows representative switching performance of STL57N65M5 transistors at 100 A per transistor. Purple trace (channel 3) shows the drain voltage. Note that the turn on occurs in 2.9 ns. The output pulse is 18.6 kV, 370 A into 50 Ω, or 7 MW per pulse. The pulser was configured with two boards in series for this demonstration, and as such only achieved 9 ns rise and 10 ns fall. With a single board, the

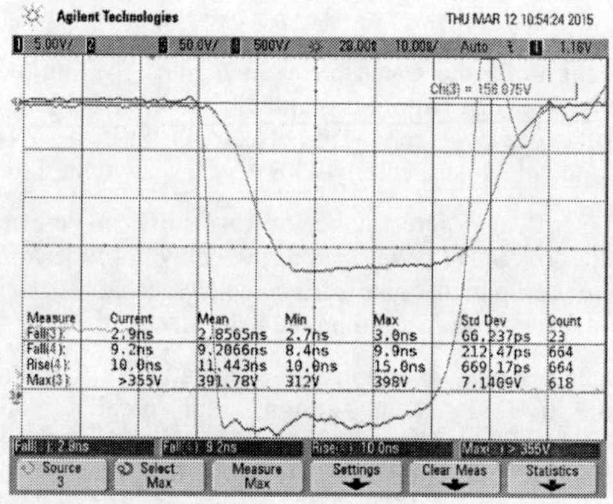

Fig. 5. Performance of two stripline kicker boards in series. Output pulse is 18.6 kV, 370 A into 50 Ω (7 MW).

system is more than twice as fast; it is limited by the inductance of the secondary.

V. CONCLUSION

Given the successes of the pulser demonstrations in the last several months both at LBNL and for various potential commercial customers at DTI, it is DTI's plan to continue development of this core technology. The topology for the pulser coupled with the novel ultra-high-performance gate driver has pushed the inductive adder state-of-the-art. The device physics principles revealed from high speed turn-off in compensated-silicon MOSFETs as it relates to charge storage and clamping capacity of SiC diodes and the theories built around the results from prototype demonstration have been extremely enlightening. Testing of the hardware has pushed the limits of the state-of-the-art in both PCB manufacturing, circuit design, and SPICE simulations. Scaling assumptions made in the initial phases turned out to be prescient in some areas, while completely off-base in others. Prototype after prototype development efforts eventually yielded a pulser that supports full voltage and current at the required PRF and rise and fall times.

Parallel and series operation of two boards and more was successfully demonstrated, and the flexible voltage and repetition rate capability was achieved. Understanding the limitations of our current circuit, we now know how to improve beyond the 6-7 kV limit per board, and the path to sub-5 ns rise and fall times with the full 10 kV per board.

*Work supported by US Department of Energy contract DE-SC0004255).

Design and Operation of Pulsed Power Systems Built to ESS Specifications

Dr. Marcel P.J. Gaudreau, Dr. Ian Roth,
Michael Kempkes, Rebecca Simpson
Diversified Technologies, Inc.
Bedford, MA, USA

Julien Domenge, Jean-Luc Lancelot
SigmaPhi Electronics,
Vannes, France

***Abstract*— Diversified Technologies, Inc. (DTI), in partnership with SigmaPhi Electronics (SPE) has built three long pulse solid-state klystron transmitters to meet spallation source requirements. Two of the three units are installed at CEA Saclay and the National Institute of Nuclear and Particle Physics (IN2P3), where they will be used as test stands for the European Spallation Source (ESS).**

The systems delivered to CEA and IN2P3 demonstrate that the ESS klystron modulator specifications (115 kV, 25 A per klystron, 3.5 ms, 14 Hz) have been achieved in a reliable, manufacturable, and cost-effective design. There are only minor modifications required to support transition of this design to the full ESS Accelerator, with up to 100 klystrons. The systems will accommodate the recently-determined increase in average power (~660 kW), can offer flicker-free operation, are equally-capable of driving klystrons or Multi-Beam IOTs (MBIOTs), and are designed for an expected MTBCF of over ten years, based on operational experience with similar systems.

Keywords—solid-state; modulator; long pulse; klystron; ESS

I. Introduction

Diversified Technologies, Inc. (DTI), in partnership with SigmaPhi Electronics (SPE), has designed and installed advanced, high voltage solid-state modulators for European Spallation Source (ESS)-class klystron pulses (Fig. 1). These klystron modulators use a series-switch driving a pulse transformer, with an advanced, patent-pending regulator to maintain a precise cathode voltage as well as a constant load to the external power grid. The success of the design in meeting the ESS pulse requirements (Table 1) is shown in Fig. 2.

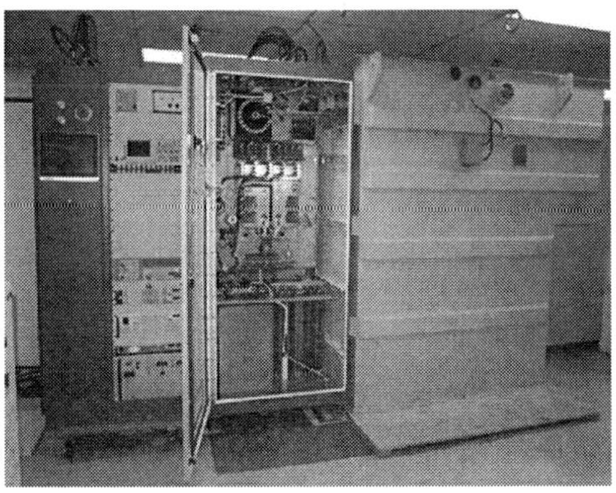

Fig. 1. DTI's prototype solid-state ESS-class klystron modulator, developed under a DOE SBIR grant.

TABLE I. ESS Klystron Modulator Requirements

Specification	
Voltage	-115 kV
Current	25 A per Klystron
Pulse Width	3.5 ms
Frequency	14 Hz (max)
Average Power	160 kW (per Klystron)
Droop	< 1%
Pulse Repeatability	< 0.1%

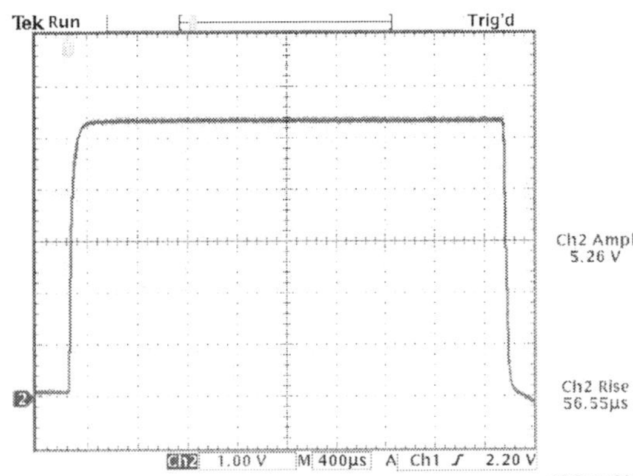

Fig. 2. Modulator pulse at 108 kV, 3.5 ms, 0.07% flattop into a Thales TH2179A klystron during site acceptance testing at IN2P3, 18 December 2015.

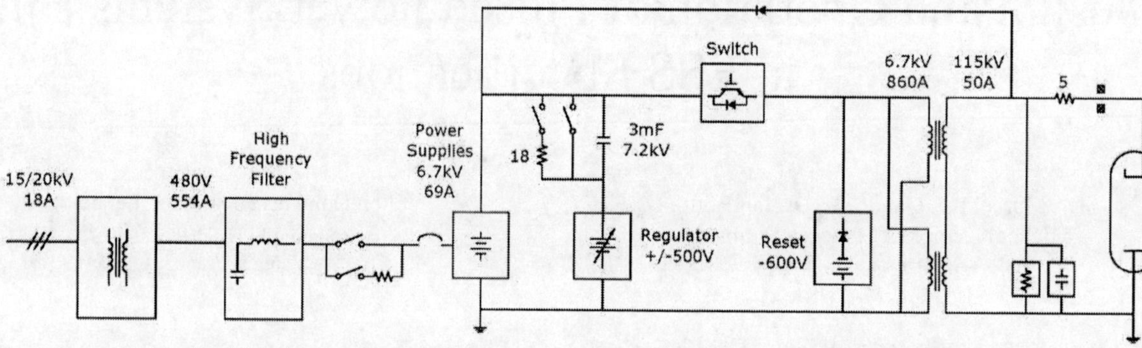

Fig. 3. Simplified schematic of the complete transmitter. The turn-key system included all components up to the klystron.

The DTI/SPE klystron modulator is now a fully proven design, delivering significant advantages in klystron performance through:

- Highly reliable operation, demonstrated in hundreds of systems worldwide, and predicted to significantly exceed ESS requirements
- Flicker- and droop-free operation over a range of operating parameters
- All active electronics in air for easy maintenance.

With the delivery of these initial modulators, the transition to production for the ESS system itself is straightforward.

II. DESIGN

A. Solid-State Switch

The heart of the DTI/SPE modulator design is a high voltage solid-state switch driving a pulse transformer. The switch is made of seven series-connected IGBT modules, and operates at 6.7 kV. This design enables a measured modulator efficiency of 95.7%, primarily due to the fact that the peak power is only switched once per pulse (in contrast to a switching converter, where the peak power is switched at high frequency during the pulse). With a power supply efficiency of 96.9%, this gives an overall wall plug to cathode efficiency of 92.8%.

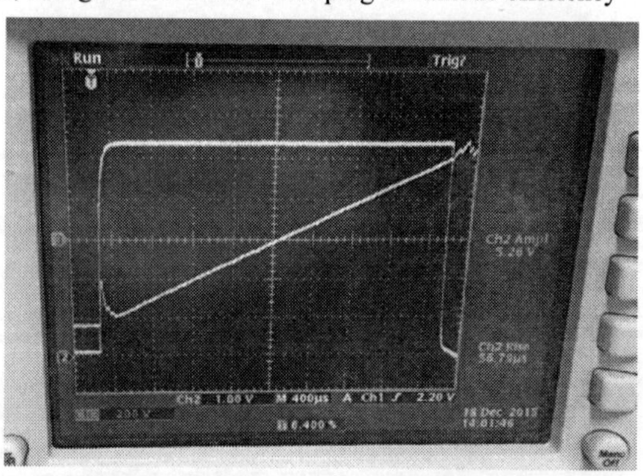

Fig. 4. Regulator voltage during a klystron pulse. Ch 2 (blue) shows pulse voltage, 20 kV/div, 400 μs/div. Ch 3 (pink) shows regulator voltage, 200V/div, 400 μs/div. The pulse settles rapidly and remains extremely flat (< 0.1%) over the long pulse flat top (3.5 ms).

The IGBTs in the switch give N+2 redundancy, meaning two of seven can fail without affecting the ability of the switch to operate at full rated voltage. This is possible because the devices always fail as a short circuit. The series switch also protects the klystron from damage in the event of an arc by opening in less than 800 ns. This rapid opening time limits the dissipated energy from the modulator to 27 mJ, significantly extending the klystron lifetime. As soon as the arc extinguishes, the switch can reclose. Since the arc extinction time is well under 10 ms, this allows the modulator to resume operation before the next pulse.

B. Regulation

A capacitor bank capable of directly meeting the ESS pulse requirements would be unrealistically large and expensive. The DTI/SPE modulator has a much smaller capacitor, which droops by ~15% during a pulse. This droop is eliminated by the switching regulator shown in Fig. 3. The regulator supplies only the droop voltage (~7.5% of the output) rather than the full voltage (Fig. 4). This means that the regulator can be small and efficient. The regulator operates in opposition to the variation in capacitor voltage, and produces both a flat output pulse and a constant load voltage to the DC power supply. As a result, the power supply can operate at constant current and power – and thus does not produce flicker, regardless of the pulse frequency. Because the regulator sinks and sources the same energy during each pulse / charge cycle, the regulator itself is non-dissipative – it uses no net power over a cycle.

The regulator switches ~5% of the peak power via two full bridges in parallel. The IGBTs switch at 100 kHz during pulsing and 5 kHz during charging. Their switching is staggered, achieving an effective switching frequency of 200 kHz during pulsing. The switching transients are filtered by the output filter and the pulse transformer, producing a ripple of only 0.0015% peak-to-peak.

C. Pulse Transformer

The pulse transformer design is similar to that of a heavy-duty power distribution transformer. The cylindrical windings are on two core legs, with the primary windings closer to the core, and a single secondary winding around each primary.

The primaries are connected in parallel, and the secondaries in series.

The low-loss silicon-steel core has a cruciform cross-section with five step sizes, giving a packing fraction of 90.6%. The core cross-sectional area and number of turns were chosen to give a flux swing of 3.4 T for the 115 kV, 3.5 ms pulse. The design is based on well-established criteria for the electric fields. To reduce the electric field at the ends of the windings, there are round field shapers. The transformer tank has voltage and current monitors, and a termination for the high voltage output cable.

D. *Power Supplies*

The high voltage DC supplies are commercial units designed by DTI. Nearly 100 of these have been successfully installed worldwide in large military and civilian radar and accelerator transmitters, operating in both shipboard and land-based systems, where reliability, high performance, and compact footprint are of the utmost importance. Each high voltage DC power supply is rated for 240 kW, with a demonstrated MTBF over 10 years, and regulation much better than 0.02%.

III. ESS Applicability

The systems delivered to CEA and IN2P3 demonstrate that the ESS klystron modulator specifications (115 kV, 25 A per klystron, 3.5 ms, 14 Hz) have been achieved in a reliable, manufacturable, and cost-effective design. There are only minor modifications required to support transition of this design to the full ESS Accelerator, with up to 100 klystrons.

A. *Four Klystron Operation*

The recent decision to move to 4-klystrons per modulator raises the pulse current to 100 A peak, and requires ~ 660 kW average power. Driving four ESS klystrons is straightforward, and within the inherent capabilities of the design. The major modification is achieving the average power. The IN2P3 design was built and tested at 440 kW, with two power supplies. Upgrading this design to 660 kW simply requires a third power supply. DTI has delivered systems combining up to eight of these supplies in a single transmitter, providing approximately twice the ESS average power.

The second issue is maintaining the desired risetime and efficiency of the pulse at 100 A. The regulator grows from two to four bridges in parallel to provide the 100 A current required, while their frequency may be allowed to decrease.

Maintaining the risetime seems challenging, since for a given transformer inductance, the risetime will increase in proportion to the peak current. While it is possible to re-design the pulse transformer for lower inductance, cutting it significantly seems unlikely without making other compromises. Examination of the klystron voltage pulse (Fig. 5), however, shows that the majority of the risetime occurs between 90 and 99% of full voltage – the 0 - 90% occurs very quickly. The total risetime, therefore, can be significantly improved by adding a small voltage 'kicker' at the beginning of the pulse - essentially making 90% of the kicked pulse

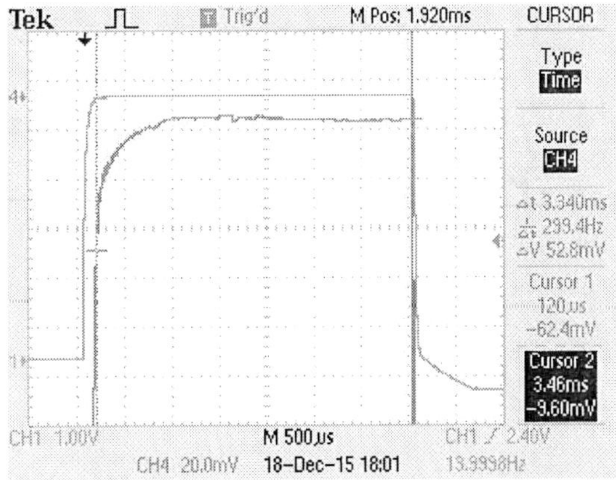

Fig. 5. Output voltage and zoomed-in flattop pulse from a klystron pulse. Ch 1 (orange) shows pulse at 120 kV (20kV/div), 500 μs/div. Ch2 (green) shows a zoomed-in flattop, at 400 V/div, with total ripple < 100 V (~0.07%).

equal to the final cathode voltage. This 'kicked' pulse can be much lower voltage (~12 kV) and much shorter (~20 – 30 μs) than the main pulse, so very little energy is required. DTI has demonstrated a version of this approach on other modulators, with great success.

B. *Power Quality*

The inherent design of the ESS-class Modulator provides flicker-free operation, due to the operation of the regulator. The switching power supplies, however, are 6-pulse designs, which can create harmonics on the mains without filtering (as provided in IN2P3 and CEA). The increase to 660 kW for ESS significantly simplifies harmonic reduction, since three independent power supplies are now required. Winding the MV stepdown transformer with multi-phase secondaries allows the three supplies for each modulator to appear as an 18-pulse rectifier without any other modifications. A single multi-MVA step-down transformer could power multiple modulators.

C. *MBIOT Operation*

ESS is considering the use of Multi-Beam IOTs (MBIOTs) for the High Energy Beam Transport (HEBT) segment of the accelerator, which operate at higher efficiency and lower cathode voltage than the klystrons. The DTI/SPE modulator is equally capable of driving these MBIOTs. While the klystron modulator itself could be used for the MBIOT, dedicated MBIOT modulators would simply use pulse transformers with lower step-up ratios, to minimize costs and maximize supportability. In either configuration, the cathode of the MBIOT would be pulsed in synchronization with the RF drive being applied. The same pulse flattop and ability to provide arc protection to the MBIOT would be preserved.

D. *Reliability / Maintainability*

The DTI/SPE ESS-class Modulator was designed for very high reliability, with an expected MTBCF over ten years, based on operational experience on the majority of assemblies.

The cooling of the electronics is sized for very long life. The individual modules comprising the system are easily replaceable, and readily transportable in the accelerator tunnel.

The exception is the pulse transformer itself, which is similar to a utility distribution transmitter. The failure rate for 33 – 110 kV utility transformers is estimated at 1 failure per decade per 25 transformers [1]. Because the pulse transformer is water-cooled and situated indoors, its reliability will be much higher than a passively cooled transformer subject to weather extremes – potentially as low as one failure every century for 25 pulse transformers (based on typical derating factors). While it might be troublesome to replace a pulse transformer, the probability of ever needing to do so is extremely low.

References

[1] Roos, F. and Lindahl, S., "Distribution system component failure rates and repair times – an overview", Nordic Distribution and Asset Management Conference, 2004.

Thyratron Replacement*

Dr. Ian Roth, Dr. Marcel P.J. Gaudreau, Michael Kempkes, Rebecca Simpson
Diversified Technologies, Inc.,
Bedford, MA, USA

Abstract—**Semiconductor thyristors have long been used as a replacement for thyratrons in low power or long pulse RF systems. To date, however, such thyristor assemblies have not demonstrated the reliability needed for installation in short pulse, high peak power RF stations used with many pulsed electron accelerators. The fast rising current in a thyristor tends to be carried in a small region, rather than across the whole device, and this localized current concentration can cause a short circuit failure. An alternate solid-state device, the insulated-gate bipolar transistor (IGBT), can readily operate at the speed needed for the accelerator, but commercial IGBTs cannot handle the voltage and current required. It is, however, possible to assemble these devices in arrays to reach the required performance levels without sacrificing their inherent speed. Diversified Technologies, Inc. (DTI) has patented and refined the technology required to build these arrays of series-parallel connected switches. DTI is currently developing an affordable, reliable, form-fit-function replacement for the klystron modulator thyratrons at SLAC capable of pulsing at 360 kV, 420 A, 6μs, and 120 Hz.**

Keywords—thyristors; thyratron; solid-state; switch

I. Background

The Stanford Linear Collider (SLC) has used thyratrons in its klystron modulators since its inception in 1963. While the thyratrons function, they need replacement every 10,000 hours at a cost of $13,000 each, plus labor. Furthermore, periodic maintenance is required to adjust their reservoir heater voltage over the thyratron lifetime. As the Stanford Linear Accelerator Center (SLAC) continues to run its accelerator over the next two decades, replacing the thyratrons with a solid-state switch that would last 25 years or more, and does not need maintenance, would provide significant savings – both in the avoided cost of thyratrons as well as the labor in replacing and adjusting them.

SLAC is presently funding the development of a solid-state switch, based on thyristor technology to replace the thyratrons (Fig. 1), meeting the requirements of existing klystron modulators (Table 1). The difficulty is that a fast rising current in a thyristor tends to be carried in a small region, rather than across the whole device, and this localized current concentration can cause a short circuit failure.

II. Design

An alternate solid-state device, the insulated-gate bipolar transistor (IGBT), can readily operate at the speed needed for the accelerator, but commercial IGBTs cannot handle the voltage and current required. It is, however, possible to assemble these devices in arrays to reach the required performance levels without sacrificing their inherent speed. Diversified Technologies, Inc. (DTI) has patented and refined the technology required to build these arrays of series-parallel connected switches. DTI has shipped more than 500 systems leveraging this technology, which have been operating in facilities around the world for many years.

Fig. 1. DTI solid-state replacement for the L-4888 thyratron used at SLAC. The switch, which operates at 48 kV and 6.3 kA, fits in the same location as the legacy thyratron assembly.

TABLE I. ESS Klystron Modulator Requirements Design

Specification	
Voltage	48 kV
Current	6.3 kA
Pulse Width	6 μs
Risetime	1 μs, 10 A/μs
Frequency	120 Hz

*Work supported by DOE under contract DE-SC0011292

In the initial phase of this project, DTI evaluated a number of IGBT switching devices, and selected the one with high current capability, short-circuit tolerance, and lowest cost. We verified that the temperature of this IGBT in operation will be low, giving high reliability. We also tested a series array of 60 devices with the number of faults that they would see in 20 years of operation; there was no sign of damage. We built and tested a 10 × 10 array of IGBTs (Fig. 2), which demonstrated operation in both series and parallel, with peak-to-peak jitter less than 1.5 ns. DTI performed an initial design of the switch assembly, showing that it fits in the modulator cabinet at SLAC. These switches will be delivered to SLAC, and their performance and reliability will be demonstrated in a SLAC modulator.

Fig. 2. 8 kV, 3.2 kA switch plate. Each plate uses 20 devices in series by 16 devices in parallel.

III. MOTIVATION

The market for thyratrons is in decline. As newer solid-state modulators are deployed, and older thyratron systems are taken out of service, the demand for thyratrons has diminished significantly. In response, several vendors have either gone out of business or stopped manufacturing thyratrons, further diminishing their availability. It is not clear how long the supply of thyratrons for these legacy systems will continue, making a solid-state thyratron replacement critical for their continued operation and support.

For this and other reasons, the thyratrons in many klystron modulators represent a major expense in operation – not only from the increasing cost of replacement, but from the labor involved both in such annual replacement and in the need to adjust them for proper operation during their relatively short life.

Most of these systems, including commercial medical accelerators, were designed decades ago, before solid-state switches were available. Even today, the cost of transitioning from thyratrons and pulse-forming networks to solid-state modulators remains prohibitive for many of these systems, despite the declining availability of thyratrons. Consequently, DTI anticipates a wide range of benefits from the development of a cost-effective, drop-in replacement for the thyratrons in operation in commercial and scientific systems around the world.

IV. MECHANICAL DESIGN

For this effort to be practical and cost-effective, the thyratron replacement switch must readily fit in the existing cabinet with minimum modifications required for its installation. The volume available for the new switch is that of the existing thyratron chassis which includes the thyratron and heater supplies. Applying this 17" x 17" x 37" volume to the new switch with 2080 devices, the volume per device would be 5.1 in3. This means that the switch will need to be very tightly packed. There is certainly room for the devices themselves, but providing the required cooling and voltage standoff (preferably with air) presents a challenge.

V. CURRENT LIMITS

The peak current requirement is large, which is not an issue for a thyratron, but does represent a challenge for semiconductor devices. To minimize the size and cost of the modulator, each IGBT should carry the maximum current possible. The IGBT chosen will have the highest current possible consistent with the risetime. The trade-off is that higher current devices have a longer risetimes.

Another limit to the current is the gate voltage; this should be as high as possible, consistent with reliability. The gate voltage also needs to have a fast risetime to minimize the switching losses.

The current rate-of-rise in an IGBT is limited mainly by the inductance in the emitter lead of the device. As dI/dt is increased, the inductive voltage drop on the emitter lead increases, reducing the gate voltage on the device itself. This raises the conduction voltage drop, increasing the power dissipated in the device. Higher power dissipation raises the device temperature, which in turn limits device reliability. Therefore, low inductance in the emitter lead is required for both speed and reliability.

Once the device was selected, 60 devices were connected in series and tested to demonstrate series operation and ruggedness to arcs at a current of 250 A. The switch has are more devices in series than the minimum required, so multiple devices can fail short without affecting the switch operation. The switch has monitors that measure the number of devices that have failed, so that maintenance can be scheduled as necessary. These monitors are voltage dividers, made with the blue resistors next to the toroids.

To determine the current sharing, it is not feasible to make direct measurements, since there are 320 devices in a plate. Instead, we took thermal pictures of the switch after operation (Fig. 3).

VI. TEST RESULTS

DTI built and checked the twelve circuit boards needed for a modulator. The tests were:

- Resistive load, inductance corresponding to normal operation. This test demonstrated that the voltages share evenly in a series stack. It also showed that the current rises to its peak value in under 2 μs (Fig. 4).

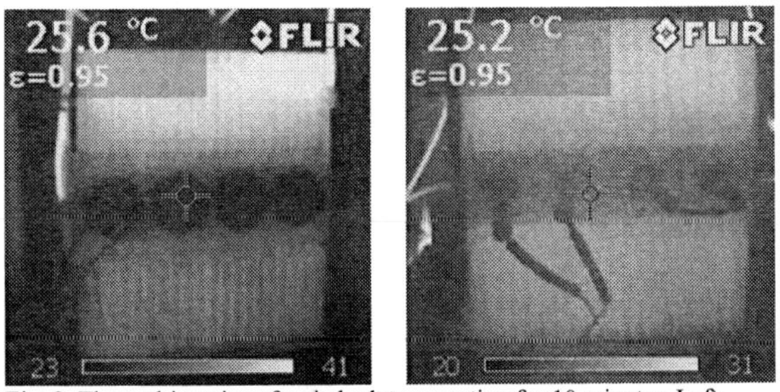

Fig. 3. Thermal imagine of switch plate operating for 10 minutes. Left shows poor current sharing; right shows improved sharing (note scale).

- Arc, inductance corresponding to a klystron arc. We performed 300 tests with the switch stack, which corresponds to the number of klystron arcs the switch would experience in 20 years of operation. Based on other DTI work, the IGBTs are expected to survive many more arcs; our IGBT switches are routinely used for klystron conditioning.
- Arc, inductance corresponding to a cable short. There were 10 tests for the switch stack. This is more than the number of cable shorts the switch would see in 20 years of operation.

VII. Future Plans

DTI will continue testing candidate devices and drives while being mindful of final cost. We estimate that the DTI switch (as presently designed) will provide more than 20X the lifetime of a thyratron, for approximately three times the price, resulting in greatly reduced lifecycle costs – especially after the hours required for thyratron adjustment and replacement are included.

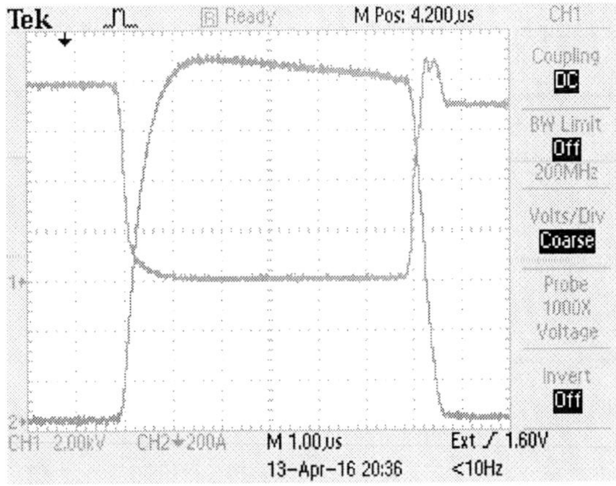

Fig. 4. Plate test at 8 kV (yellow, 2 kV/div), 1.8 kA (blue, 200 A/div).

Following successful development, fabrication, and demonstration of the thyratron replacement switches for SLAC, DTI plans to make the switch assembly commercially available, targeting other legacy systems still using thyratron pulse-forming network pulse modulators.

Acknowledgements

The authors wish to thank SLAC for their collaboration on this project and to acknowledge the funding received from DOE (grant no. DE-SC0011292).

Pulsed Electric Field Application to Aquaculture: Predator Control and Product Extraction

Michael A. Kempkes
Dr. Ian Roth
Diversified Technologies, Inc.
Bedford MA USA
Kempkes@divtecs.com

Dr. Thomas A. Dempster
Arizona State University / AzCATI
Tempe, AZ USA

***Abstract*— Pulsed Electric Field (PEF) processing uses short, high voltage electrical pulses to break cellular membranes. There are at least two distinct applications of PEF to algal growth and processing – extraction of intracellular material and microalgae predator population control. PEF processing has the potential to provide lower costs and higher productivity for the production of biofuels, human and animal food and feed supplements, and high value specialty chemicals from large-scale algae farms, but it is not possible to assess either the applicability or costs of PEF processing in a commercially meaningful manner today. Diversified Technologies, Inc. (DTI) recently received an SBIR grant from the US Department of Agriculture to work with ASU / AzCATI to extend research into two potential PEF applications to algal products – extraction enhancement and predator control.**

Keywords*— *Pulsed Electric Field, algae, biofuels, extraction, predators, pests

I. Introduction

Pulsed Electric Field (PEF) processing is a low temperature, non-thermal, non-chemical, low impact process that induces electroporation of cell membranes in microbes, plant, and animal cells. PEF processing involves the application of short duration (1 – 20 μs), very high voltage pulses that create a high voltage field (from 1 - 50 kV/cm) across a liquid. Depending on the desired effect, this field can open the cell membranes in plant cells (1 – 10 kV/cm) or kill bacteria, molds, and other microorganisms at higher field strengths.

Multiple studies have shown that PEF processing can be successfully applied to both extraction of valuable products from algae, and the control of potentially devastating predators in algal growth ponds and raceways. While these are very different applications of PEF, they both focus on increasing the cost effectiveness of algal products in the marketplace. For extraction, PEF lyses the cell membranes, allowing the intracellular compound to be released into the surrounding solution, as well as allowing solvents to enter the cell itself (Figure 1). In predator control, the differences in size and composition between algal cells and their predators provide the potential to selectively kill predators, while having little impact on the algae itself.

The potential microalgae products market is conservatively estimated to be at least \$500M - 1B annually today, and is growing at a rapid pace. This market consists of two major areas: Biofuels, and 'Everything Else'. 'Everything Else' includes food (human, fish, and pet), nutritional supplements (e.g., Omega-3, spirulina, and astaxanthin), cosmetics, fertilizers, and specialty chemicals [8]. These products vary considerably in value, from specialty chemicals at the apex to fish food near the bottom, with multiple products often available from a single microalgae stream. Estimates for individual market sizes vary wildly, however, because microalgae products are typically a small percentage of the total market for each commodity. This puts the primary emphasis on cost reduction in order to grow market size – sales will follow lower prices [19]. Decreasing the cost of production and extraction has long been recognized by the algal products community as the most significant challenge to the commercialization of microalgae products.

While algal products represent a very small niche today, especially compared to other agricultural markets, recent global investments in developing algal biofuels have significantly decreased the cost of growing, harvesting, and processing algae. This has spurred the growth of algal products in higher value markets. PEF is not known to be part of any commercial algae process chain at this time, however. This is likely due to several factors, including the size of the algal products market today, and the relative novelty of PEF in other applications.

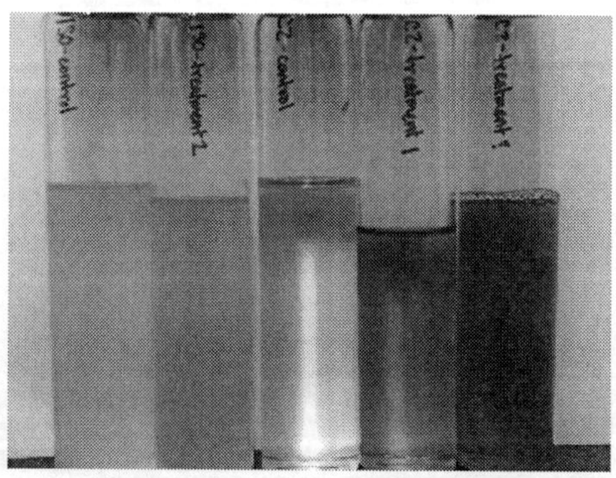

Figure 1. Supernatant from PEF treated Isochrysis galbana (left two) and Chlorella vulgaris (right three). The dark supernatant shows released intra-cellular material after PEF lysing [11].

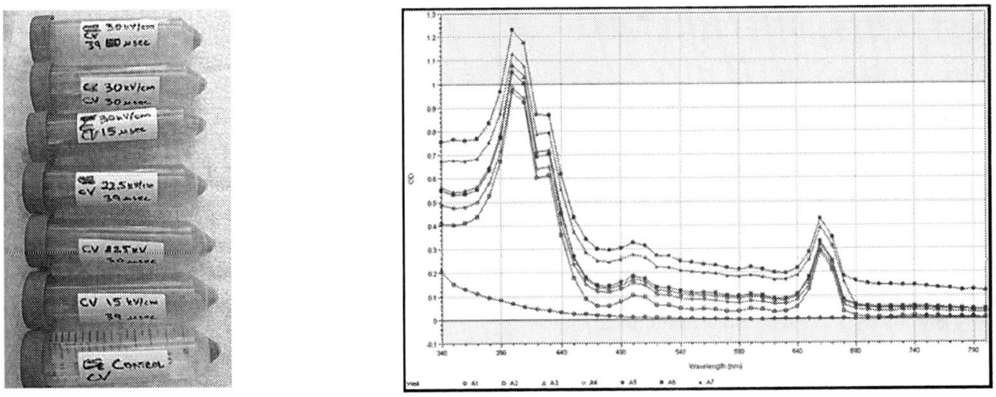

Figure 2. Chlorella vulgaris treated at different field strengths and durations [4]. The control sample (bottom, left) is optically clear after centrifuging and exposure to Triton X for 3 minutes, and the chlorophyll can readily be seen in the biomass pellet (bright green). The PEF treated samples show intracellular material released into the supernatant, and a reduction of the chlorophyll in the biomass pellet (yellow-green). Spectrophotometric scans (R) correlate vertically to the samples (L), showing differentiated levels of cell lysing, with the control at the bottom.

II. PEF Assisted Extraction

There are many steps required to produce economically valuable products from microalgae, with the most expensive being growing the microalgae itself. The second largest cost is typically extracting the desired products from the microalgae [6]. Most algal products today rely on drying and solvent extraction processes (or freeze-drying and separation using supercritical CO_2) to reach a commercial end product. These extraction processes are inherently energy intensive and expensive, limiting the market for microalgae products.

The major impediment to extraction is the algae cell itself, which has evolved specifically to protect the contents of the cell. A rapid, efficient approach to lysing the cell would simplify the process of creating biofuels and extracting other chemicals, lowering the cost of these products. PEF has been shown by multiple researchers to lyse a variety of microalgae species through electroporation, which releases their intra-cellular contents into the surrounding solution (Figures 2 and 3). The primary benefit of lysing the algal cells through PEF is to make those intra-cellular materials, which may include lipids, proteins, and other chemicals, available for downstream extraction, separation, and purification into specific products. PEF will not significantly assist the extraction / separation of compounds found within the cell walls themselves, however.

Although PEF clearly lyses microalgae cells, the majority of published research, unfortunately, does not allow evaluation of the energy required (cost) versus the level of electroporation achieved (benefit). Even less of this data addresses the impact of lysing on downstream processes for the extraction of valuable compounds, and even where data does exist, the results for a given algal species vary widely.

The key to commercial adoption of PEF for extraction of lipids and other compounds from microalgae, will be the overall cost of PEF and downstream extraction processing relative to the alternatives. PEF does appear to have substantial promise, even with the limited data available. One calculated cost was \$2.69 USD for PEF lysing of a volume of *C. Zofingiensis* algae containing one barrel of lipids, at a treatment condition of 9.6 kV/cm, 29 μs into a 3.5 mS/cm concentrated solution [20]. This is over an order of magnitude lower than the cost of drying a concentrated microalgae solution prior to extraction, which was calculated at ~ \$40 USD/barrel under the same assumptions. If the extraction costs are similar after either PEF or drying (which is still a subject of demonstration and confirmation), PEF would be over an order of magnitude cheaper than drying and solvent extraction.

Marine microalgae present an additional complication. The very high conductivity of salt water (~40 – 60 mS/cm, vs. 2 – 4 mS/cm for freshwater) limits both the electric field and duration of the PEF treatment. Boiling can occur even with relatively low PEF doses at these conductivities. PEF processing of saltwater microalgae require that conductivity is reduced by at least an order of magnitude, without adding significant new costs. Using PEF for saltwater algae, at reasonable energy levels, represents a significant challenge.

III. Predator Control

The second potential area of PEF applicability to algal products is predator control in microalgae cultivation. Growing large quantities of microalgae is the single largest cost in the microalgae product stream [1]. Open pond cultivation represents the lowest-cost approach to growing large volumes of microalgae biomass for downstream processing [3]. These ponds, however, are subject to a variety of threats, including microalgae predators.

Predators and grazers are estimated to consume 10 – 30% of the algal biomass in an open raceway pond, and outbreaks can crash a pond overnight [2]. There are a number of potential algal predators, impacting different species of algae. These range from bacteria and amoeba, to multi-cellular animals, such as rotifers. Chemical treatments are not attractive, since they may have deleterious impacts on the microalgae or its intended products. Microalgae predators, such as amoeba, ciliates, flagellates, rotifers, and *Poteriochromonas* can devastate a large microalgae pond in hours, causing extensive losses and production delays for commercial microalgae farms.

Chemical approaches to predator control have shown dismal results, given the range of both predator species and algae [2]. The potential for wind-borne infection by these predators in open raceway ponds, and the frequency of contamination events, has led microalgae producers to accept the high costs of photobioreactors, invest in greenhouses and air filtration systems, change microalgae species, and/or adopt other countermeasures. At commercial scale, these can be cost prohibitive for all but the most valuable products.

Preliminary investigations by researchers on amoeba [22] and rotifers [17] [18] demonstrated that it is possible to apply PEF treatment to microalgae at intensities which are strong enough to kill predators without causing damage to the microalgae cells themselves. Both amoeba and rotifers appear to be susceptible to PEF at < 1 kV/cm field strength, versus the 4 - 40 kV/cm field strengths typically required to kill commercial microalgae strains. The selective destruction of predators, but not microalgae, may be due to the absence of cell walls (amoeba), or their larger size (e.g., rotifers). This distinction in lethality may not hold true for all predators (especially bacterial predators) and microalgae species.

For predator control, the basic parameters required to size the PEF system are the field strength and treatment time required for effective predator kill, and the treated volume required per hour as a percentage of the total pond volume. The low electric field reported for predator lethality makes predator control a potentially very low energy application of PEF, since energy is a function of voltage-squared (e.g., 10% of the field strength requires only 1% of the energy, with all other parameters constant).

With a PEF system continually treating a small fraction of the microalgae cultivation system volume (similar in concept to the operation of a swimming pool filter), it appears possible to prevent the population of predators from ever reaching concentrations required to 'crash' cultivation systems (Figure 8), at very low energy costs. In a 500,000 liter raceway pond, for example, treating 10,000 liters / hour (2% of the volume, or complete treatment every two days) at 1 kV/cm and 20 µs would require approximately 8 kWhrs of electricity, or less than $10 USD per week at $0.08 / kWh.

Even if all of these factors needed to be doubled to achieve the desired results (e.g., 2 kV/cm, 40 µs treatment time, and 20,000 liters / hour), the electricity cost would still be only ~ $150 USD/week – inconsequential compared to the cost of a single pond crash. The PEF equipment required at this scale is very inexpensive and compact, especially compared to systems required for large-scale extraction. The same PEF system could also be used to treat feed water / cultures, to ensure that predators are not introduced via these avenues.

IV. Conclusions

Algal products represent a very small market today, but their potential is enormous. The recent focus on, and of investment in, R&D for algal biofuels has created a new excitement around the cultivation and extraction of products from algae. This investment has led to substantive improvements in cultivating, harvesting, and processing algae. With this activity, the value of algal products for other markets has received close attention, and the recent drop in petroleum prices has shifted substantial activity from biofuels into high value products. These factors, in turn, have spurred a level of commercial activity, with hundreds of companies involved around the world. It remains to be seen if this wave will recede in the face of continued lower petroleum prices (as occurred in the 1980's), or will be continued into the future.

Multiple researchers around the world have demonstrated that PEF processing lyses algal cells, and that this can simplify the extraction of compounds from the algae itself. This research, however, has been limited to a small set of algal species, and yields a wide range of different conclusions as to the energy required to lyse each species. Extending and consolidating this research, with consistently applied and reported PEF processes, would provide significant benefits to algal product manufacturers in assessing the utility of PEF in their operations. It may also open products that are cost-prohibitive under conventional processes to algal cultivation and extraction.

Characterizing the interaction of PEF processing with downstream extraction / separation technologies is in its infancy. Clearly, the elimination of a drying step prior to

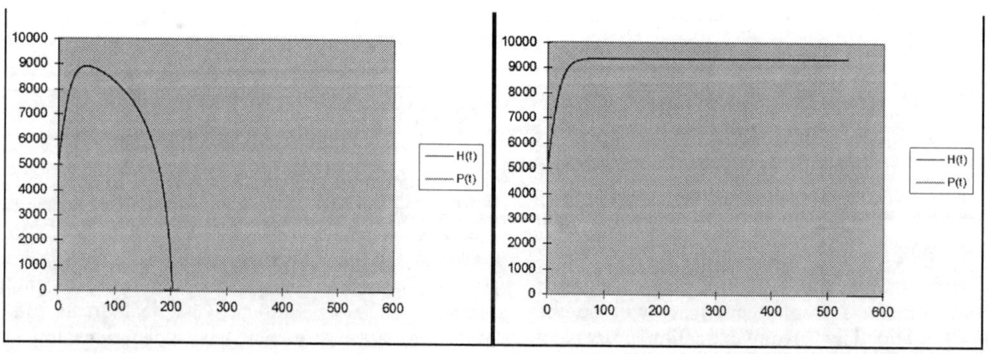

Figure 3. Two simple predator prey models showing microalgae and predator population versus time, with representative growth rates and predation levels. The first shows the crash of a microalgae pond (L) with no predator control, and the second (R) shows the impact of low levels of predator control in the same pond (one complete treatment of the pond volume every 200 time units). All other parameters in this model were identical.

extraction yields significant energy savings. This is highly beneficial in some extraction processes, but irrelevant to others, since some extraction solvents and processes also lyse the cells. This appears to be the major missing element in determining the commercial applicability of PEF to the production of algal products, and DTI/ASU plan to explore this area in detail in Phase II of our USDA SBIR effort in 2017. PEF treatment must reduce costs in the entire extraction and separation process to attract commercial interest from algal producers.

Finally, the use of PEF as a means of predator control remains to be proven in practice, and at large scale. This technique clearly shows promise for some predators, but others (especially microbes) may not be amenable to PEF at non-lethal field strengths for the algae. Clearly, this is another area requiring additional research.

The applicability of PEF to this market has yet to be demonstrated commercially, but it has the promise to lower the cost of algal products significantly. Only after cost savings can be clearly demonstrated will commercial algal products companies adopt PEF in their operations. Finally, all of this depends on the growth and profitability of an algal products market – whether for biofuels, specialty chemicals, or protein/feed. This growth, in turn, depends on the work of a wide range of researchers, across dozens of fields – including PEF.

References

[1] Brennan, L., Owende, P. 2009. Biofuels from Microalgae – A review of technologies for production, processing, and extractions of biofuels and co-products. Renewable and Sustainable Energy Reviews, doi:10.1016/j.rser.2009.10.009

[2] Carney, L., Lane, T. 2014. Parasites in microalgae mass culture. Frontiers in Microbiology, Vol 5, Article 278.

[3] Davis, R. Fishman, D., Frank, E., Wigmosta, M., 2012. Renewable Diesel from Microalgae Lipids: An Integrated Baseline for Cost, Emissions, and Resource Potential from a Harmonized Model. Technical Report ANL/ESD-12-4; NREL/TP-5100-55431; PNNL-21437.

[4] Dempster, T. 2014. Impact of Cell Wall and Cell Membrane Disruption by Pulsed Electric Field Processing on the Liberation of Lipids and Phycopigments. 11th International Bioelectrics Symposium, Columbia, MO.

[5] Frey, W., Eing, C., Goettel, M., Gusbeth, C., Posten, C., and Straessner, R. 2012. Application of Pulsed Electric Field Treatment for Microalgae Processing. Proceedings of the 2012 Euro-Asian Pulse Power Conference.

[6] Gendy, T., El-Temtamy, S. 2013. Commercialization potential aspects of microalgae for biofuels: An overview. Egyptian Journal of Petroleum 22, 43-51/

[7] Grimi, N., Barba, F. 2014. New approaches for the use of non-conventional cell disruption technologies, to extract potential food additives and nutraceuticals from microalgae. Food Engineering Reviews, FERE-D-14-00047.

[8] Henrikson, R. 2010. Real World Experiences in Growing Microalgae. Presented at the Microalgae World Summit, San Diego, CA.

[9] Joannes, C., Coswald, S., Sipaut, J., Dayou, S., Mansa, R., The Potential of Using Pulsed Electric Field (PEF) Technology as the Cell Disruption Method to Extract Lipid from Microalgae for Biodiesel Production, International Journal of Renewable Energy Research, Vol. 5, No. 2, 2015

[10] Joannes, C., Sipaut, C.S., Dayou, J., Yasir, S.M., Mansa, R.F. 2015. Review Paper on Cell Membrane Electroporation of Microalgae using Electric Field Treatment Method for Microalgae Lipid Extraction. Materials Science and Engineering 78 012034.

[11] Kempkes, M. , Roth, I., Dempster, T. 2012. Pulsed Electric Field Processing for Enhanced Algal Oil Extraction. Presented at the 2012 Microalgae Biomass Summit, Denver, CO.

[12] Kempkes, M., Gaudreau, M.P.J., Roth, I. 2015. Pulsed electric field (PEF) method for continuous enhanced extraction of oil and lipids from small aquatic plants. US Patent 9,029,108.

[13] Kretzschmar, M., Nisbet, R.M., Mccauley, E., 1993. A Predator-Prey Model for Zooplankton Grazing on Competing Algal Populations. Theoretical Population Biology, Vol 44, Issue 1.

[14] Lai, Y., Parameswaran, P., Baez, M, Rosov, T., McGowen, J., Rittman, B. 2014. Superior FAME recovery from Scendesmus sp. through pulsed electric field treatment. Presented at the 2014 Microalgae Biomass Symposium, San Diego, CA.

[15] Luengo, E., Condon-Abanto, S., Alvarez, I., Raso, J. 2014. Effect of Pulsed Electric Field Treatment on Permeabiliazation and Extraction of Pigments from Chlorella Vulgaris. The Journal of Membrane Biology, Vol 247, Number 6, June 2014.

[16] Luengo, E., Martinez, J., Coustets, M., Alvarez, I., Teissie, J., Rols, M-P, Raso, J. 2015. A Comparative Study of the Effects of Millisecond- and Microsecond-Pulsed Electric Field Treatments on the Permeabilization and Extraction of Pigments from Chlorella Vulgaris. The Journal of Membrane Biology, DOI 10.1007/s00232-015-9796-7

[17] Rego, D., Costa, L., Pereira, M.T., Redondo, L.M., Cell Membrane Permeabilization Studies of Chlorella sp. by Pulsed Electric Fields, IEEE Transactions on Plasma Science, 0093-3813, 2015.

[18] Rego, D., L.M. Redondo, V. Geraldes, L. Costa, J. Navalho, M.T. Pereira. 2014. Control of predators in industrial scale microalgae cultures with Pulsed Electric Fields, Bioelectrochemistry, doi: 10.1016/j.bioelechem.2014.08.004

[19] Rijffels, R. 2014. Microalgae in Europe. Presented at the 2014 Microalgae Biomass Summit, San Diego, CA.

[20] Roth, I. 2011. Final Report for IIP-1013814, Application of Pulsed Electric Fields to the Extraction of Oil From Microalgae. National Science Foundation SBIR.

[21] Singh, J., Gu, S. 2010. Commercialization potential of microalgae for biofuels production. Renewable and Sustainable Energy Reviews 14 2596-2610.

[22] Vernhes, M.C., Benichou, A., Cabanes, P.A., Teissie, J. 2002. Elimination of free-living amoebae in fresh water with pulsed electric fields. Water Research 36 3429 – 3438.

[23] Zbinden, M., Sturm, B., Nord, R., Carey, W., Moore, D., Shinogle, H., Stagg-Williams, S. 2013. Pulsed Electric Field (PEF) as an Intensification Pretreatment for Greener Solvent Lipid Extraction From Microalgae. Biotechnology and Bioengineering, Vol. 110, No. 6 June 2013

Obtaining Ultrahigh Pressures in Configuration of Multi-Wire Z-θ Pinch on Gamma Facility

V.D. Selemir, N.V. Zaviyalov, V.S. Gordeev, P.B. Repin, A.G. Rep'ev, A.P. Orlov, B.G. Repin

Russian Federal Nuclear Center-VNIIEF,

37, Mira Ave., Sarov, Nizhny Novgorod region, 607188, RUSSIA

Abstract— **This paper presents the results of the numerical experiment of the initial axial magnetic field compression by the cylindrical multi-wire array that was powered by the current pulse from the pulsed power electrophysical facility Gamma [1] (the current amplitude is ~11 MA, rise time is ~60 ns). This facility is designed in RFNC-VNIIEF.**

Keywords— Ultrahigh magnetic pressure, plasma Z-θ pinch, MHD simulation.

I. Introduction

There are ideas of combining the advantages of inertial and magnetic approaches to the problems of the controlled fusion [2]. Rather high magnetic field applied to the system can localize thermonuclear plasma in the space and suppress the conductive thermal outflow and alfa particles diffusion from the fusion area. It was shown theoretically that one can effectively use cylindrical liner systems to create ultra-high (up to 100 MGs) magnetic fields. These systems are radially accelerated by the electrodynamic Ampère force $j_z \cdot B_\varphi$ and simultaneously compress the initial longitudinal magnetic field Bz_0 at the high values of the magnetic Reynolds number $R_m >> 1$ [3]. Such configuration, where the azimuthal and axial components are present, we call Z-θ pinch, as well as the authors of the paper [4]. A shear of the resultant magnetic field is formed in the process of Z-θ pinch implosion. It effectively suppresses the growth of different magneto-hydrodynamic instabilities, which without shear can destroy the initial cylindrical symmetry of the liner system [5].

Hollow supersonic gas jets, solid azimuthal monolithic cylindrical shells, and multi-wire arrays can be used for forming *Z-θ* pinch. Particularly, the initial longitudinal magnetic field Bz_0 was increased 420 times (from 100 kGs up to ≈42 MGs) because of the radial implosion of the hollow gas shell from *Ne* [6] in the experiment on the electrophysical facility Proto-II at the current amplitude of ~7.5 MA. The plasma *Z-θ* pinch based on the cylindrical multi-wire array suppressed the initial field Bz_0=5 kGs up to Bz^{max}~300 kGs [7] in the experiment on Angara 5-1 facility at the current amplitude of ~2MA. In the frames of MagLIF project, it is supposed to compress the initial longitudinal magnetic field by thick-walled *Be* liner up to the value more than 100 MGs [8].

In the current work it is shown that because of the wire array plasma implosion in *Z-θ* pinch configuration on the electro-physical Gamma facility (Fig.1) it is possible to get megabar pressures inside the cylindrical container with an electro-conductive wall. The container is located in the center of a discharge chamber. The resultant pressure of the compressed magnetic field affecting the container surface can be used for different materials compression aimed to investigate their properties [9].

The main goal of the carried out estimated research is predicting parameters of the pressure pulses in the container according to its diameter.

II. Calculations Statement

Numerical simulation of *Z-θ* pinch was carried out using three dimensional resistive magneto-hydrodynamic (RMHD) code FLUX-3D, described in details in [10].

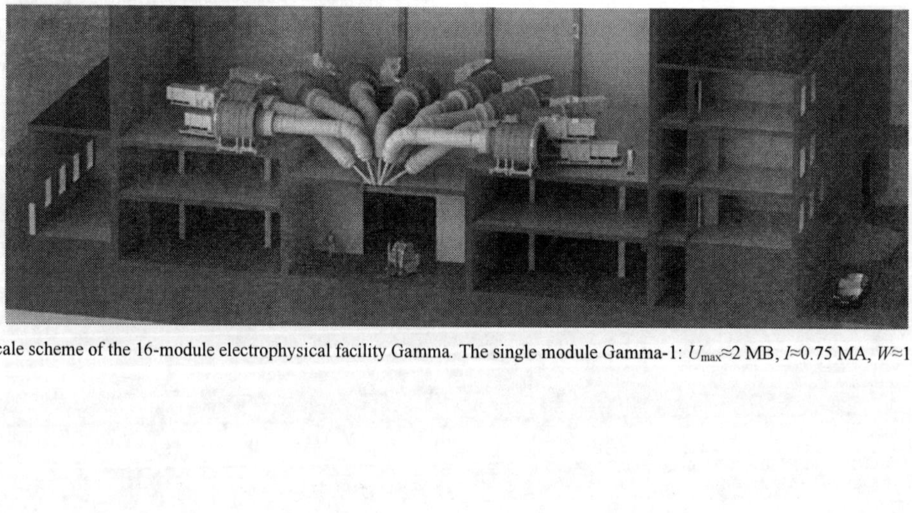

Fig. 1. The full-scale scheme of the 16-module electrophysical facility Gamma. The single module Gamma-1: U_{max}≈2 MB, I≈0.75 MA, W≈1.5 TW.

Physical and mathematical model is based on this code. This model is one-fluid, three-temperature (ions, electrons, and thermal radiation, distribution of the latter is calculated in one-group diffusion approach) and considers the conductive heat transfer in plasma.

While making all the calculations, the electro-explosion process of the thin metal wires (transfer from the solid state into plasma under the current influence) is not calculated but simulated in the frames of phenomenological model of the prolong plasma formation [11].

All the calculations are carried out according to double-loop equivalent electric circuit of the Gamma facility (Fig. 2).

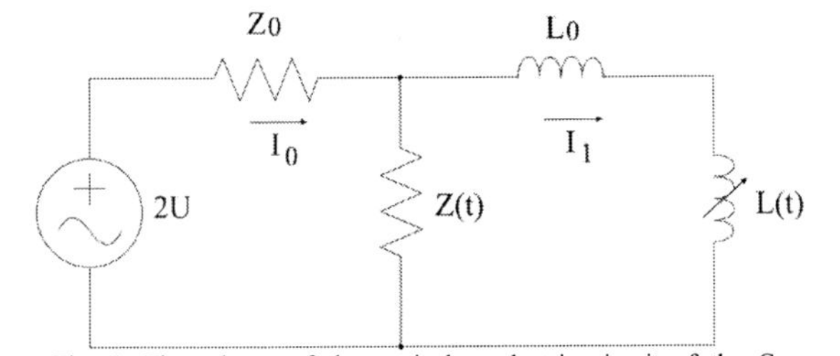

Fig. 2. The scheme of the equivalent electric circuit of the Gamma facility. 2U(t) – voltage generator; Z_0=0.181 Ohm – generator impedance; Z – current dependent resistance simulating electron leakages; L_0 – transmission line inductance; L(t) – load inductance; $L_0+L(t=0)$=15 nH; I_0 and I_1 – currents flowing in double-loop circuit.

The voltage pulse form of the running wave is illustrated in Fig. 3.

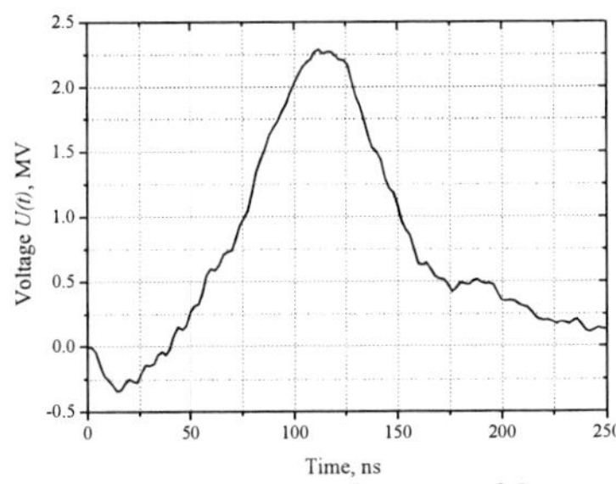

Fig. 3. Voltage pulse form of the running wave of Gamma facility.

The wire array with the height $l = 2$ cm, initial radius $R_0 = 1.6$ cm, and linear mass $m_L = 0.82$ mg/cm was chosen according to the results of the preliminary optimization 0D calculations for magneto-hydrodynamic simulation performance. Hereby, the load chamber has the diameter of 4 cm and height of 2 cm.

The initial configuration for 3D calculation of the multi-wire Z-pinch on the GAMMA facility is presented in Fig.4.

The initial inductance value of the uniform longitudinal field in the load chamber was Bz_0=100 kGs. It was preliminary created by the external source. There were two series of calculations with the diameter of the cylindrical container D_{cont}=2.5 mm and 4 mm. The container height was the third part of the liner height (Fig. 5).

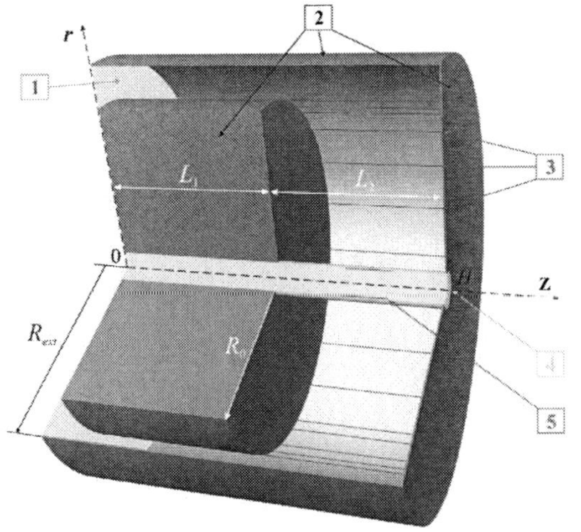

Fig. 4. The scheme of the calculated spatial area. 1 – finite section of the coaxial transmission line; 2 – ideally conducting electrodes of the load chamber; 3 – multi-wire array; 4 – cylindrical dielectric rod with ε=2; 5 - container.

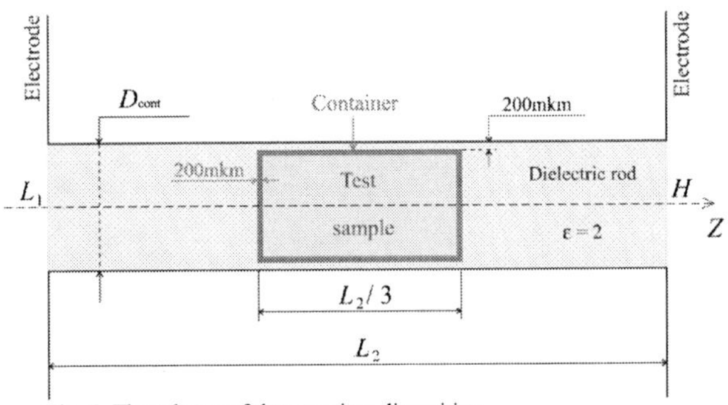

Fig. 5. The scheme of the container disposition.

In both variants, the calculated stationary spatial grid consisted of truncated wedged cells of the size Δz=Δr=100 μm with an opening angle Δφ=0.9 (total number of cells is 17.6 million).

Random azimuthal-uncorrelated perturbations of plasma formation intensity with the amplitude A_0=1.75% were put over the wire array that consisted of 40 tungsten wires. The coefficients value of the substance ablation for the wires is k_{abl}=0.38 mg/(ns·cm^2). One should note that these parameters were chosen according to the paper [1].

III. Results of Numerical 3D Simulation

The calculated time dependencies of plasma formation velocity and production of total mass in the load chamber are shown in Fig. 6. The process of the plasma formation of the wires finishes by the time moment $t \approx 100$ ns.

The spatial distributions of the tungsten plasma density and azimuthal component of the magnetic field in the consecutive time moments at the container usage with a diameter 2.5 mm are provided in Fig. 7

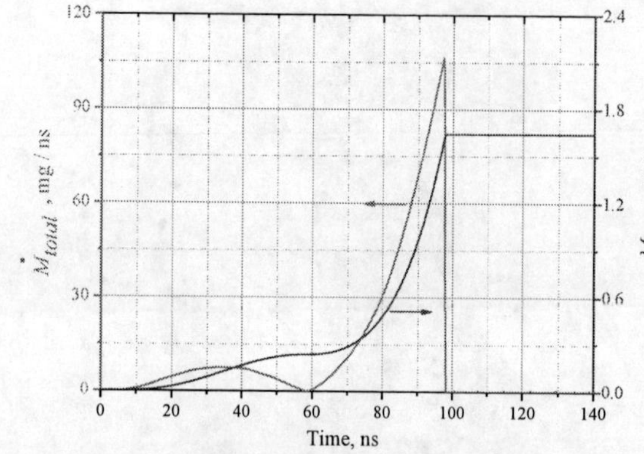

Fig. 6. Time dependence of the plasma formation velocity (left axis) and total mass production (right axis).

The time moment $t \approx 100$ ns characterizes the movement start of the current-plasma shell to the system axis. The load passes approximately half of the way to the system axis at $t \approx 140$ ns. The plasma reaches its maximum radial compression approximately at the time moment $t \approx 158$ ns.

Time dependencies of the current and SXR pulse power for both calculation variants are provided in Fig. 8.

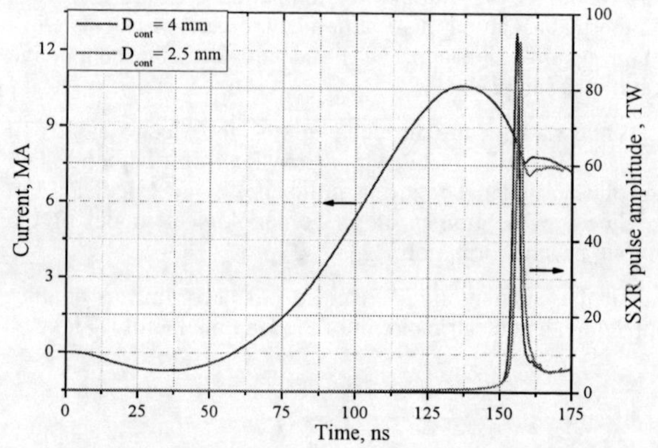

Fig. 8. Time dependence of the current in Z-θ pinch (left axis) and generated SXR pulse (right axis) for the container diameter of 4 mm (—) and 2.5 mm (-----), respectively.

The maximum current in the load reaches the value of ~10.5 MA with characteristic rise time ~60 ns (for the level 0.1-0.9). The SXR pulse power is 90 TW at the container of any diameter. One should note that this value of the SXR power turns out to be a little lower than the value obtained in the work [1] (120 TW) at the same parameters.

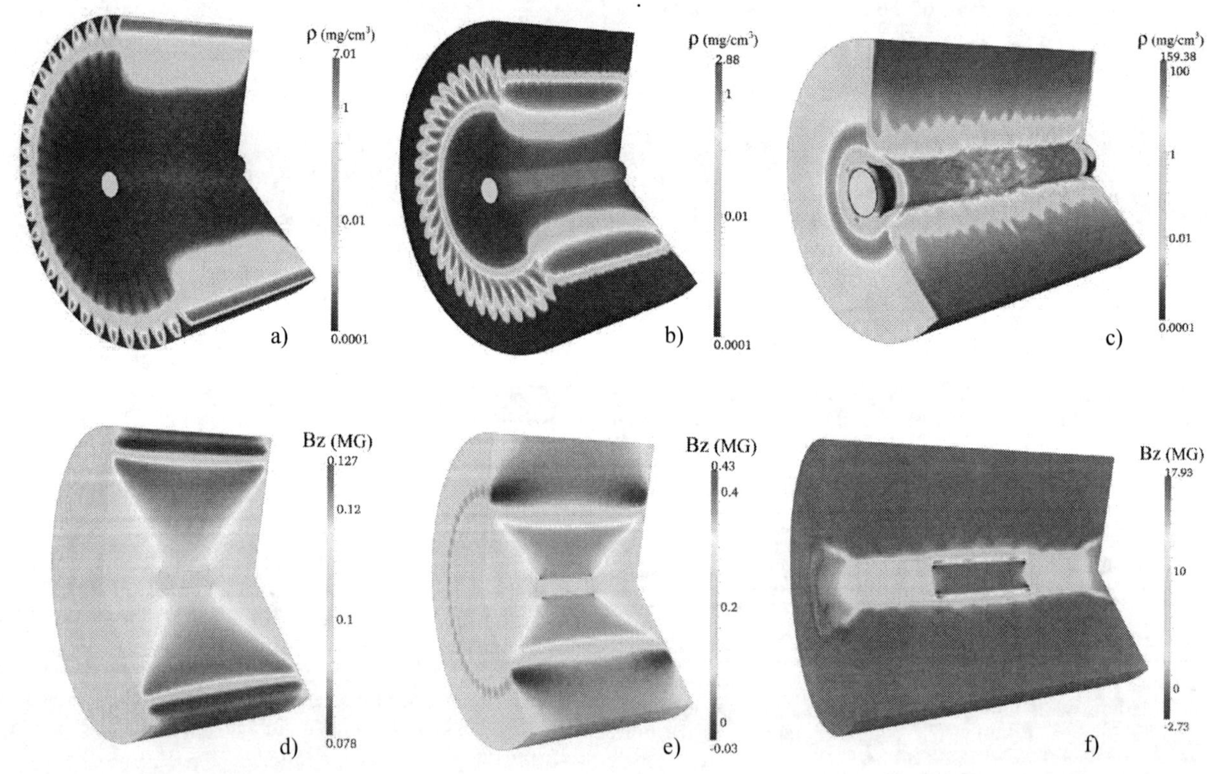

Fig. 7. The spatial distribution of the tungsten plasma density – a), b), c); and axial component of the magnetic field – d),e), f) at consecutive time moments in case of the container with the diameter D_{cont}=2.5 mm. The area: 2 cm $\le z \le$ 4 cm; 0°$\le \varphi \le$ 240° and $0 \le r \le$ 1.6 cm for distributions – a), b), c), d); and $0 \le r \le 0.8$ cm for distributions – c), f).

This result is stipulated by the influence of the external axial magnetic field that decreases the main parameters of the SXR pulse [7] and the existence of the container itself.

Time dependence of the magnetic pressures on the sample wall in its central part for both container sizes are presented in Fig. 9.

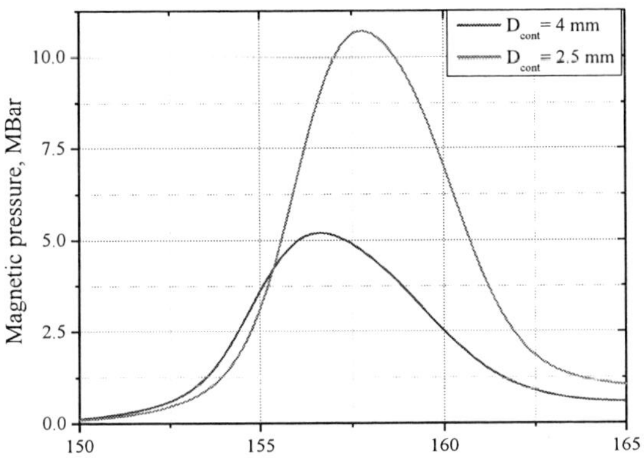

Fig. 9. Time dependence of the magnetic pressure on the sample wall ($z=L_1+0.5 \cdot L_2$) for the container diameter of 4 mm (—) and 2.5 mm (—), respectively.

The maximum pressure is realized in the range of 10.7 Mbar for the container of the diameter 2.5 mm and 5.2 Mbar for the container of 4 mm, respectively. The characteristic length of the pressure influence on the sample is less than 5 ns. The obtained pressures correspond to the magnetic field value of 11.4 MGs and 16.4 MGs, respectively.

It is a rather challenging task to measure directly the axial magnetic field pulses generated on the system axis in the experiment with *Z-θ* pinch. It is possible to use magneto-optical method based on the Faraday effect to solve this diagnostic task. According to this method, the polarized laser radiation is passed through the fiber cable located into the protective shell and put along the axis of the load chamber [12,13]. The calculation of such configurations can be found in the paper [14].

Later on, it is reasonable to make a calculated scaling of the maximum pressures dependence in the studied configuration on the value Bz_0 of the compressed axial magnetic field and also the number of wires in the array.

IV. Conclusion

The numerical simulation of the initial axial magnetic field compression Bz_0 by the cylindrical multi-wire array was powered by the current pulse from the electrophysical facility Gamma designed in RFNC-VNIIEF (the amplitude $I_{max}\approx10.5$ MA, rise time is ~60 ns). It was carried out using three dimensional resistive magneto-hydrodynamic code FLUX-3D and revealed the following:

Multi-wire array with the initial radius $R_0=1.6$ cm, consisted of small number (~40) of thin tungsten wires can effectively make compression of the initial axial magnetic flux.

The amplitude of the generated pressure pulse grows at the decrease of the container diameter. Particularly, it is possible to get short (~5 ns) pressure pulses with the amplitude $P_{max}\approx5.2$ Mbar at $D_{cont}\approx4$ mm and $P_{max}\approx10.7$ Mbar at $D_{cont}\approx2.5$ mm at the initial external magnetic field $Bz_0=100$ kGs as a result of implosion of the multi-wire *Z-θ* pinch on Gamma facility. The obtained pressures are corresponded to the magnetic field values of 11.4 MGs and 16.4 MGs, respectively.

The resultant pressure of the compressed magnetic field affecting the container surface can be used for the research of different materials properties.

V. References

[1] V.S. Gordeev, A.V. Grishin, A.P. Orlov, A.L. Mozgovoj, B.G. Repin, S.L. Glushkov, S.Yu. Puchagin, K.V. Strabykin "Numerical Simulation of Radiating Z-Pinch on «Gamma» Facility". Proceedings of the 41st IEEE Int. Conf. on Plasma Science and 20th Int. Conf. on High-Power Particle Beams, May 25-29, 2014, USA, Washington DC.

[2] Lindemuth I.R. and Kirkpatrick R.C. "Parameter space for magnetized fuel targets in inertial confinement fusion," Nucl.Fusion V.23, No.3, 1983, P.263.

[3] Felber F.S., Liberman M.A. and Velikovich A.L. "Methods for producing ultrahigh magnetic fields," Appl.Phys.Lett. V.46, Issue 11, 1985, P.1042.

[4] Golberg S.M., Liberman M.A., Velikovich A.L. "Plasma compression, heating and fusion in megagauss *Z-θ* pinch systems," Plasma Phys. and Controlled Fusion V.32, No.5, 1990, P. 319.

[5] Bud'ko A.B., Liberman M.A., Velikovich A.L. and Felber F.S. "Suppression of Rayleigh-Taylor and bulk convective instabilities in imploding plasma liners and pinches," Phys. Fluids B V.2, No.6, 1990, P.1159.

[6] Felber F.S., Malley M.M., Wessel F.J., Matzen M.K., et al. "Compression of ultrahigh magnetic fields in a gas-puff Z pinch. Phys.Fluids," V.31, No.7, 1988, pp.2053-2056.

[7] Zukakishvili G. G., Mitrofanov K. N., Grabovskii E. V., and Oleinik G. M. "Measurements of the Axial Magnetic Field during the Implosion of Wire Arrays in the Angara-5-1 Facility", Plasma Physics Reports, Vol. 31, No. 8, 2005, pp. 652–664.

[8] Slutz S.A., Hermann M.C., Vesey R.A., Sefkow A.B., et al. "Pulsed-power-driven cylindrical liner implosions of laser preheated fuel magnetized with an axial field," Physics of Plasmas, V.17, 056303, 2010.

[9] G.V. Boriskov, V.I. Timareva, A.I. Panov, S.S. Sokolov "Simulation of Isentropic Compression Process in a Device Based on Magneto-Cumulative Generator". In: "Megagauss-XI", Proc. of the Eleventh International Conference on Megagauss Magnetic Field Generation and Related Topics, September 10-14, 2006, London, UK, pp. 219-224.

[10] A.P. Orlov and B.G. Repin. Numerical Simulation of Multi-Wire Z-Pinches Within the Framework of 3D Magneto-Hydrodynamics. IEEE Trans. on Plasma Science. 2015. V.43. No.8. P.2515-2519.

[11] P.Yu. Edmund, M.E. Cuneo, M.P. Desjarlais, R.W. Lemke, D.B. Sinars, T.A. Haill, E.M. Waisman, G.R. Bennett, C.A. Jennings, T.A. Mehlhorn, T.A. Brunner, H.L. Hanshaw, J.L. Porter, W.A. Stygar, L.I. Rudakov "Three-dimensional effects in trailing mass in the wire-array Z-pinch", Physics of Plasmas, V.15, 056301, 2008.

[12] A.S.Boriskin, M.I.Dolotenko, A.S.Kravchenko et al. "Diagnostic methods of measuring megavolts voltages and megaamperes currents," Proceedings of 12th Int. Conf. On High-Power Particle Beams, Haifa, Israel, June, 1998. P.503.

[13] F.S.Felber, F.J.Wessel, N.C.Wild, et al. "Ultrahigh magnetic fields produced in a gas-puff Z pinch," J.Appl.Phys., V.64, No.8, 1988, pp.3831-3844.

[14] A.P. Orlov and B.G. Repin. Numerical Experiment on Initial Magnetic Flux Compression by a Multiwire Z-Pinch. IEEE Trans. on Plasma Science. 2015. V.43. No.1. P.360.

*Design and Experimental Study of a kind of land spark source**

Ailong Fan, Yaohong Sun, Xuzhe Xu, Rongyao Fu
Institute of Electrical Engineering
Chinese Academy of Sciences
Beijing, China
fanailong@mail.iee.ac.cn

Ailong Fan, Yaohong Sun, Ping yan, Rongyao Fu
University of Chinese Academy of Sciences, Beijing, China
Ailong Fan
Liaoning institute of science and technology, Benxi, China

***Abstract*—In this paper, a kind of land spark source with a closed cavity transducer is designed, which is used in the mountainous area with a little water. The spark source is mainly composed of the adjustable DC source, charging resistors, the isolation silicon stack, the energy storage capacitor, discharge resistors, the resistor divider voltage measurement circuit, the thyristor and its buffer circuit, discharge cables, the closed cavity transducer including discharge electrodes, etc. Using the spark source and the corresponding experimental platform, the discharge experiment is carried out, and its electrical characteristics are researched. The selected variables are electrode materials, voltage and electrode spacing. We focus on the resistance of the plasma channel, the discharge power and their relationship in the time domain. By comparison, the effective power of the output of electrodes in non-metallic materials is larger than that of metallic materials. With the decrease of the voltage, the longer the duration of low resistance. With the increase of the voltage, the electric energy conversion rate of the discharge plasma is also increased. This study provides the necessary reference for the improvement of the land spark source.**

Keywords—land spark source; discharge experiment; closed cavity transducer; electrode materials

I. RESEARCH BACKGROUND

Currently the land vibroseis mainly includes the tamping source, the air gun source, the electromagnetic driven controlled source, the hydraulic vibrator, the precision active vibrator and the sparker[1]. Sparker is the most used non-explosive source. It uses the pulse capacitor to store the electric energy, and charges can be released quickly in water by high voltage discharge electrodes to produce a strong pressure pulse which couples with the earth to form an earthquake wave. Sparker can be stimulated in the deep well, shoal and waters. It was originally a very low frequency underwater plasma sound sources. When the spark discharge happens in the underwater, a high pressure plasma and a vapor bubble are produced. Then the vapor bubble continues to expand and collapse, producing a great sound, and most of its frequency is between 10 and 200 Hz[2~4] .On the basis of the sound source of the plasma, it is applied to the VSP log and land exploration, and further it can be improved as the electric spark source for land prospecting. In 1966, the American J. W. Miller got a patent on the land electric spark source[5~8]. In the 1960s, the Institute of Electrical Engineering, Chinese Academy of Sciences was in cooperation with the Ministry of oil on development work of marine sparker; In the 1990s, the application of electric spark source was studied in the aspects of vertical seismic profile, inter well seismic, vibration and engineering exploration. In 2005, the first oceanography institute of Chinese oceanic administration conducted the research of high resolution, multi-channel shallow seismic detection system, and achieved good application effect. Netherlands Geographic Resources Company developed 16kJ multi-electrode plasma source, which can be used to study deep-sea shallow strata.

II. SYSTEM DESIGN

The principle of electric spark source is mainly based on the electric and mechanical effect of large current discharge. The discharge electrode needs to be put into a liquid medium. For sparker used in dry areas, few studies have been reported. land spark source described in this article can be applied to dry mountainous environment, in order to provide the necessary technical support for the metal ore seismic exploration.

A. Main Circuit Design

A land spark source main circuit is shown in Figure 1, and it consists of the charging unit and discharge unit. The charging unit is mainly composed of regulating device, a rectifier circuit, charging resistors, isolation silicon stack, etc. Its main function is to produce continuously adjustable DC power supply, and further to charge the storage capacitor. Discharge unit is mainly composed of the energy storage capacitor, discharge resistors, discharge constitute switch, resistor divider voltage measurement circuit, thyristor and its buffer circuit, discharge cables, discharge electrodes. Its main function is to release the charge stored in the capacitor by controlling the opening of the thyristor. The main working principle of the land spark source is: 220V alternating current source is changed into a direct current source (0~6KV) through voltage regulator, rectifier circuit, charging resistor and isolated silicon reactor, then charge the energy storage capacitor C. When the gate electrode of the thyristor received the effective trigger pulse, the thyristor is triggered into the conduction state. Charges in the energy storage capacitor are discharged in the closed cavity transducer through the

*Work supported by R&D of Key Instruments and Technologies for Deep Resources Prospecting (the National R&D Projects for Key Scientific Instruments), Grant No.ZDYZ2012-1-06 and Chinese National Natural Science Foundation of China (51277171, 11411130178).

thyristor, the discharge cable, discharge electrodes, and then a shock wave is generated by coupling with the ground.

The purpose of charging resistance is two: one is to limit the high charging current when the system is charged, and to protect the energy storage capacitor and the rectifier circuit; the other is to limit the short-circuit current, so as not to cause damage to the regulator. The main purpose of setting up the isolated silicon heap is to prevent the high reverse voltage and current from stringing into the charging unit. Essentially, it plays a role in isolation between the charging unit and the discharge unit. Setting the discharge resistance and discharge switch aims to discharge residual charges in the capacitor during the charge and the discharge. The purpose of setting the voltage measuring circuit of the resistance voltage divider is to sample the charging voltage and display in the instrument panel of the chassis. It can be observed in real time and the voltage overshoot can be avoided. The purpose of setting the buffer circuit is to protect the thyristor during its opening and closing, absorbing the instant peak voltage of the thyristor in during the turn-on and turn-off, slow down the speed of the switch, reduce the du/dt, and reduce the switching loss. In addition, it is also conducive to improve the electromagnetic compatibility environment of the machine.

This circuit is simple in structure, easy in maintenance; it has certain advantage in the dry mountain, because the closed cavity transducer is easy to carry and filled-water inside.

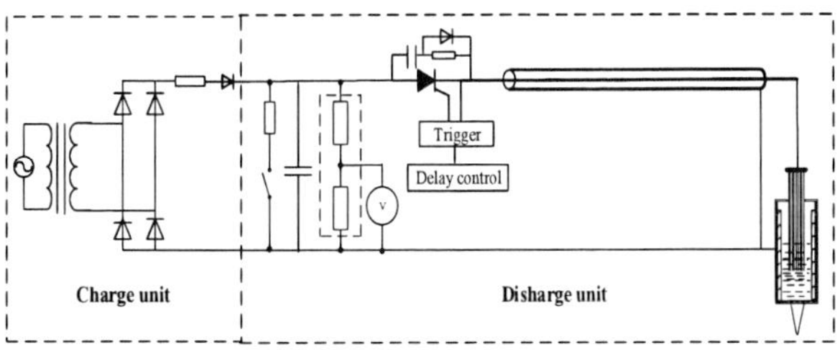

Fig. 1. Schematic diagram of the main circuit

B. Design index

The specific parameters of the land spark source are: Long time stable operation in the output voltage of 6 kV; burst mode operation of the maximum voltage of 7kV; the adjusted continuously output voltage; the peak discharge current of 12kA;1-9999 μ s adjusted delay time; delay input mode by setting up a code switch; 1 μ s pulse delay jitter; pulse width of 100-500 μ s; seismic wave frequency range of 10~300 Hz.

C. Experimental platform

Recording instrument uses TPS 2024B Tektronix oscilloscope, and its bandwidth is 200 MHz and the sampling rate is 2GS/s. Measurement of discharge voltage uses P6015A Tek (bandwidth 75 MHz, partial pressure ratio 1000:1) voltage probe. Measuring thyristor branch current uses coaxial shunt. Measuring transducer discharge current uses Pearson6600 coil whose ratio is 0.001 V/A.

The cathode of the thyristor is connected to the central copper wire of the discharge cable. The network conducting layer of the discharge cable is connected with the cathode of the storage capacitor. The output of the trigger is connected with the control end of the thyristor and the cathode end, and the delay control circuit controls the trigger operation, and then control the thyristor turn-on and turn-off. The closed cavity transducer is embedded into the ground, and the front-end of the discharge electrode is connected with the central copper wire of the discharge cable; the electrode terminal of the discharge electrode is connected with the network layer of the discharge cable. Coil current transformer (current coil) is set on the high voltage line, and the connecting end of the sensor is connected to the CH1 channel of the oscilloscope. The voltage probe is connected to the front-end of the cable beam and the other end is connected to the CH2 channel of the oscilloscope.

The load is discharge cable and the discharge transducer. The discharge transducer is mainly composed of metal cover, buffer spring, upper insulation sleeve, multi-bundle electrodes, metal upper tube, metal cylinder, and lower insulation sleeve. The upper part of the insulating sleeve and the multi-bundle electrodes as a whole are put in the internal section. The discharge cavity is filled with liquid salt, and multi-bundle electrodes are arranged in the lower part of saline liquids. A buffer chamber is between the upper insulating sleeve and the metal cover.

Metal tube design solves the problem of easy breaking of the electrode end and insufficient liquid space. Layered design makes removal and replacement of the electrode more convenient. The split design also makes the measurement more convenient. The design of the inner insulating cylinder can adjust the distance between the positive and negative electrodes, which makes it easy to compare the tests, increasing the flexibility of the field survey adjustment. More importantly, the buffer chamber and the buffer spring is designed to increase the low-frequency component of vibration waves, which can reduce the attenuation of vibration waves in the ground, so as to improve the conversion and utilization efficiency of electric energy.

III. DISCHARGE EXPERIMENT

The discharge experiment was carried out using the land spark source. The effects of different electrode materials and number, different voltage and electrode distance were studied.

A. Discharge experiment of different electrode materials

We chose six different electrode materials: stainless steel, glass, nylon, Teflon, rubber and epoxy. It found when the outer tube used different electrode materials, the discharged state did not change. On the condition of the same discharge voltage, electrode spacing, the measured current waveform and the discharge resistor synthesis waveform are not changed significantly, the typical discharge voltage-current- resistance waveform is shown in Figure 2. Figure 3 and Figure 4 are typical discharge current waveforms using separately an insulating material cylindrical electrode and a metal cylindrical electrode under the 4600V voltage and the electrode spacing of 3.8cm. After conversion, the peak discharge currents in Figure 3 and Figure 4 are respectively about 460A and 485A. The obtained discharge current using the metal material electrode is slightly larger than using the non-metallic material electrode.

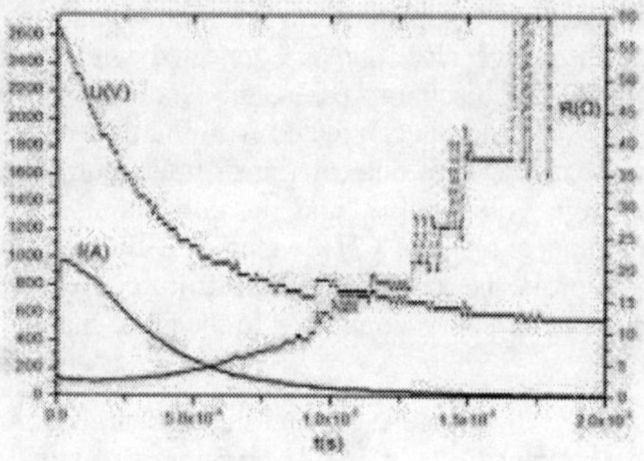

Fig. 2. The typical discharge voltage-current-resistance waveform

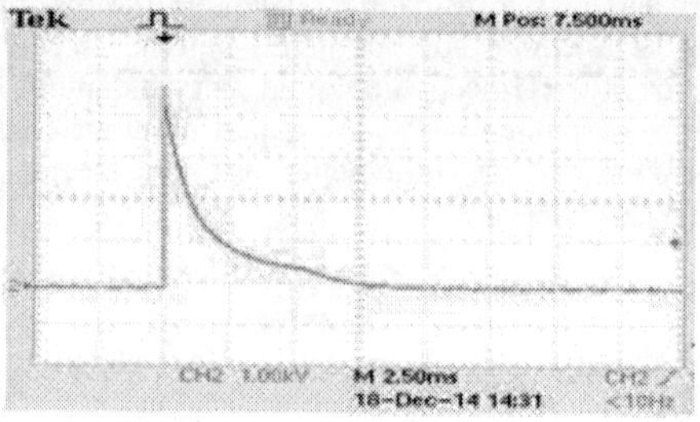

Fig. 3. The discharge current waveform of the insulating tube electrode under the conditions of 3.8cm gap and 4.6kV

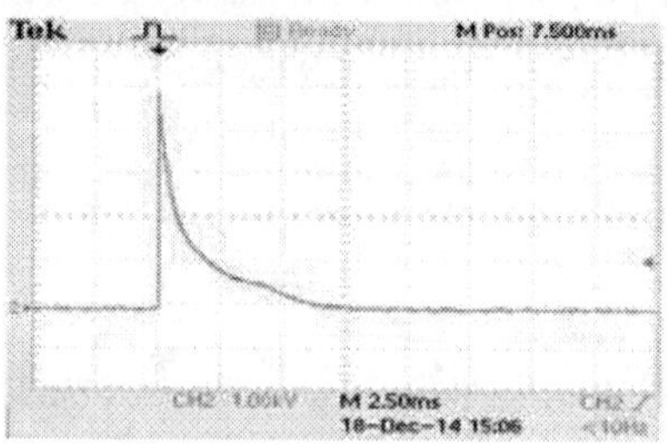

Fig. 4. The discharge current waveform of the metallic outer cylinder under the conditions of 3.8cm gap and 4.6kV

B. *Discharge test under different voltage conditions*

We chose eight voltage levels: 2000V, 2500V, 3000V, 3500V, 3750kV, 4000V, 4500V and 4800V. Figures 5 and 6 are typical time-varying waveforms of discharge resistors under 2000V and 4800V voltage condition and the electrode spacing of 2cm They is obtained by using ORIGIN software according to the corresponding current waveform data. Figures 7 and 8 are typical time-varying waveforms of discharge powers under 2000V and 4800V voltage condition and the electrode spacing of 2cm They is obtained by using ORIGIN software according to the corresponding current waveform data.

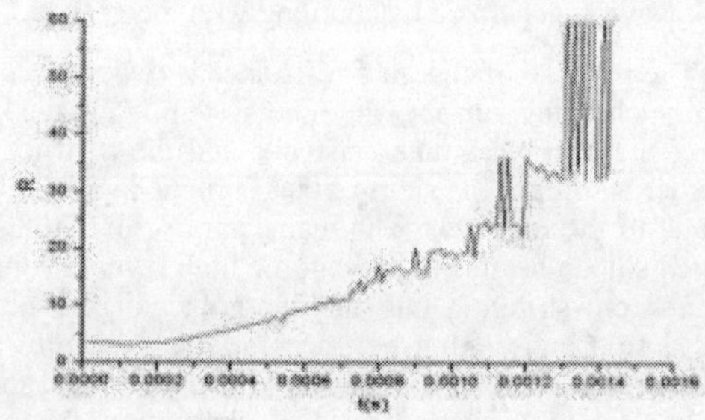

Fig. 5. The curve of discharge resistance under the conditions of 2cm gap and 2kV

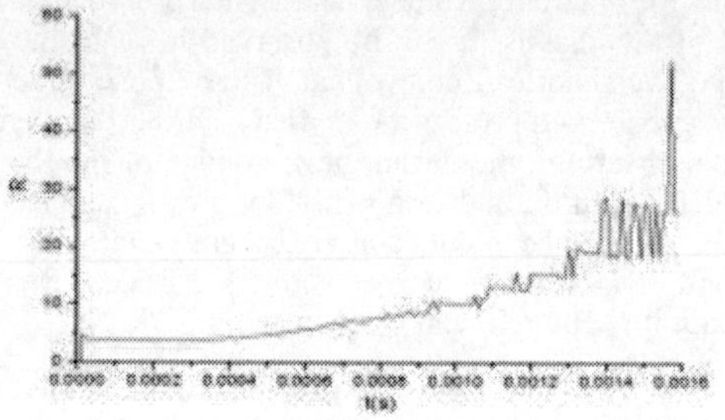

Fig. 6. The curve of discharge resistance under the conditions of 2cm gap and 4.8kV

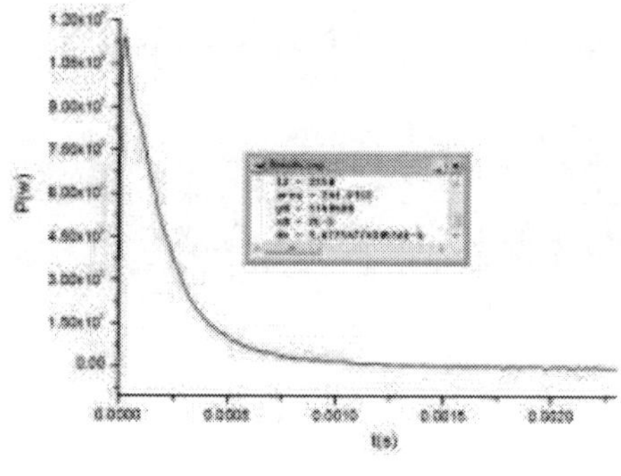

Fig. 7. The curve of discharge power under the conditions of 2cm gap and 2kV

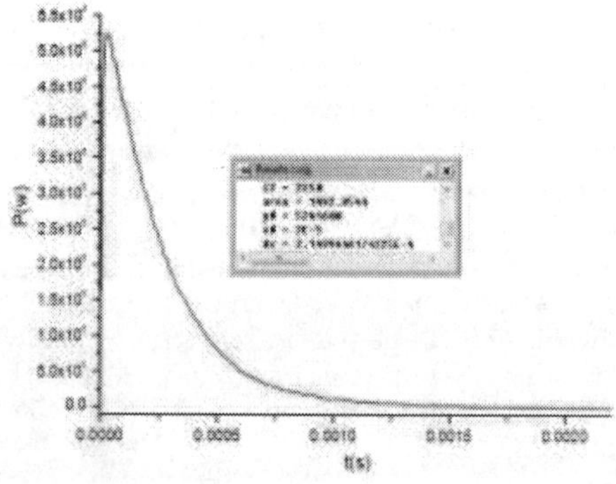

Fig. 8. The curve of discharge power under the conditions of 2cm gap and 4.8kV

As can be seen from the figure, changing trends of the resistance and power of the discharge plasma channel are the same. However, duration of low resistance of the plasma channel in 2000V discharge is greater than that of 4800V. At the early stage of discharge, the resistance of the plasma channel is about 4 Ω. By conversion, under the condition of 2000V discharge, the discharge energy is about 242J, and the discharge efficiency is about 75.6%; under the condition of 4800V discharge, the discharge energy is about 1493J, and the discharge efficiency is about 80.9%. From the above images and other experimental data, it is concluded that with the increase of the voltage, the electric energy conversion rate of the discharge plasma is also increased.

C. discharge experiment under the different electrode spacing

In the experiment, the electrode gap distance is the 1cm at the beginning, then increased by 1cm in each round of the experimental until 6cm. Figures 9, Figures 10 and Figures 11 are the typical discharge current waveforms under the conditions of 1cm, 2cm and 4cm gap distance and the voltage of 3.75kV.

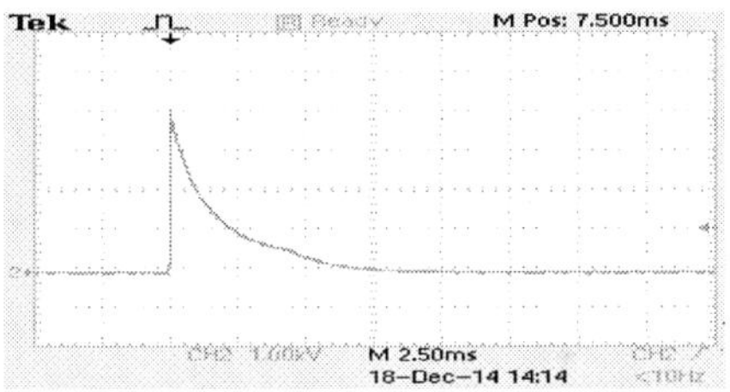

Fig. 9. The typical discharge current waveform under the conditions of 1cm gap and 3.75kV

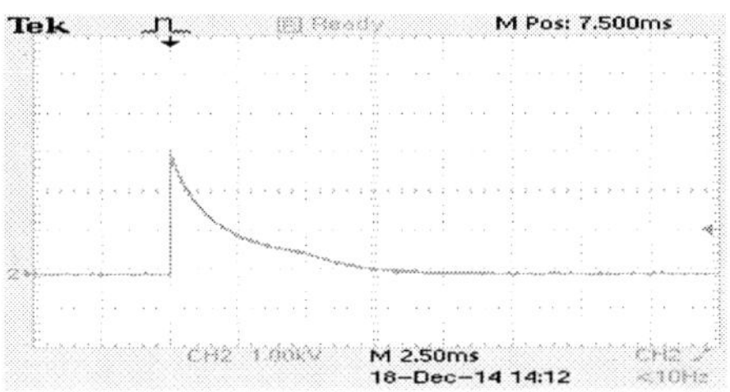

Fig. 10. The typical discharge current waveform under the conditions of 2cm gap and 3.75kV

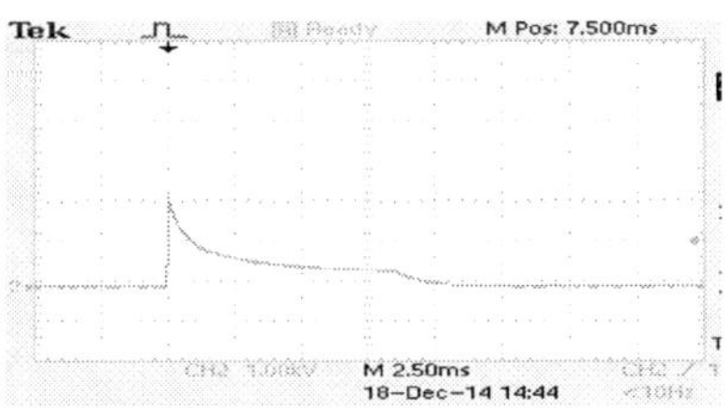

Fig. 11. The typical discharge current waveform under the conditions of 4cm gap and 3.75kV

It can be seen from the above waveforms and other experimental data, in each round of the experiment, under the condition of the same gap distance and discharge voltage, the discharge current waveform is basically the same. When the electrode gap distance is changed, the discharge current changes. When the gap distance of the discharge electrode is greater than 4cm, the current change is small; when the gap distance is less than 4cm, the discharge current increases significantly with the decrease of the electrode distance, but at the same time, the probability of arc discharge is large too.

IV. CONCLUSION

In this paper, a set land spark source system is designed, which can withstand the discharge current of 12kA and can operate stably in the 6kV state. Its pulse width is 100-500 μ s. The working principle of land spark source system is introduced, and the experimental platform is set up. Discharge experiment shows: different electrode material has little effect on the shape of discharge current, and the discharge current of the metal electrode material is slightly greater than that of the insulating material. When using different insulation material, the discharge current is not very different. Under different voltage conditions, the variation trend of the resistance and power of the discharge plasma channel is the same. But with the increase of the voltage, the electric energy conversion rate of the discharge plasma increases too; when the discharge current is below 4cm, the discharge current increases with the decrease of the electrode spacing.

References

[1] W. T. BORNT. A REVIEW OF GEOPHYSICAL INSTRUMENTATION[J]. GEOPHYSICS, 1960, 25(1): 77-91.

[2] Lv Guotao, Wen Mingming, Wu Heng, Feng Qiangqiang. Analysis of acoustic characteristics of 20kj electric spark source[J]. Neijiang science and technology, 2013, 34(1): 49-50

[3] Pei Yanliang,Wang Kuiyang,Liu Chenguang,Li Xishuang,Li Zhengguang.The Sparker Source Charging Technique[J]. Ocean Technology, 2007, 26(3): 73-76

[4] Shi Xiaoxia, Gao Yinghui, Sun Yaohong, Liu Kun, Yan Ping. Development of an underwater sparker with repetitive pulse[J]. Chinese Journal of Power Sources. 2013, 37(4): 646-649

[5] Sun Yaohong, Zuo Gongning. Transferring spark source used in crosshole seismic[J]. Geophysical Prospecting For Petroleum, 2001, 40(3): 57-60

[6] Sun Yaohong ,Zuo Gongning. The Transferring Spark Source Used in Dredging Oil Well[J]. High Voltage Engineering, 2000, 26(4): 31-32

[7] Zhang Yunliang, Sun Yaohong, Wang Yongrong. Application of high power electric spark source in VSP log[J]. Petroleum Instruments, 2006, 20(1): 27-31

[8] Lu Song, Yu Weigang, Chen Jundong. Application of 10kJ ZDF-3 sparker on large distance layer CT[J]. Technical Acoustics, 2013, 32(6): 452-457

Coupling Effect between Accelerating Units of Dielectric Wall Accelerator*

Shen Yi, Liu Yi, Ye Mao, Zhang Huang, Wang Wei, Wang Zhiwen, Yang Chao, Xia Liansheng†, Shi Jinshui, Zhang Linwen, Deng Jianjun

Key Laboratory of Pulsed Power, Institute of Fluid Physics, CAEP, P. O. Box 919-106, Mianyang, China

***Abstract*—The dielectric wall accelerator(DWA), based on technologies of solid-state pulse forming lines, photoconductive semiconductor switches and high gradient insulators, is a new concept pulsed linear accelerator being developed at Institute of Fluid Physics, CAEP. During accelerator test process, the energy gain of accelerated proton beam is much lower than expected. We found that the coupling effect induced connecting circuit is a major factor of causing lower beam energy gain. In this paper, we have confirmed that circuit and accelerating structure coupling between accelerating units are inevitable. The circuit topologies of two kinds of coupling effects are given. The coupling effects between accelerating units are studied, and the coupling efficiency of multiple accelerating units is measured.**

Keywords—dielectric wall accelerator; coupling; accelerating unit; pulsed power technology

I. Introduction

Dielectric wall accelerator (DWA) is a novel compact pulsed linac under development, which is put forward by Lawrence Livermore National Laboratory at the end of last century.[1] The DWA is based on technologies of solid-state pulse forming lines, photoconductive semiconductor switches and high gradient insulators.[2] Which has advantages of high accelerating gradient, compact structure, lower cost. Particularly, it is possible to vary the beam intensity, the beam energy, and the spot size shot by shot in a DWA, which makes it outstanding as a proton accelerator for cancer therapy.[3] A prototype DWA including proton injector, DWA module and measurement system are proposed and developed at Institute of Fluid Physics (IFP), and the layout of a prototype DWA is shown in Figure 1. Proton beams are generated in a 50 Hz ECR ion source and extracted at 40 keV with bunch length of several hundreds of microseconds. The low energy beam transport with two einzel lens is designed to transport the proton bunch through the DWA module.[4] The accelerated proton beams are deflected on a scintillator by using a bends magnet. And the luminescence of scintillator is recorded by an ICCD camera. During accelerator test process, the energy gain of accelerated proton beam is much lower than expected. We found that the coupling effect is a major factor of causing lower beam energy gain.

Each DWA module consists of pulsed power generator and high gradient insulators (HGI), which is called accelerating units. And each accelerating units ought to have operated independently. However, it has a serious circuit coupling by reason of using a public electrode between adjacent modules. Which leads to voltage of each DWA module becomes much lower than expected. Even if each accelerating unit could operate independently, there is still accelerating structure coupling between accelerating units. In this paper, we have confirmed that circuit and accelerating structure coupling between accelerating units are inevitable.

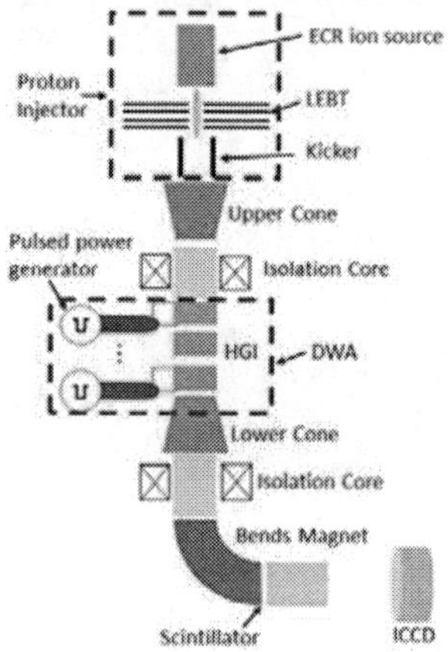

Fig. 1 Layout of a prototype DWA

II. Coupling Effect between Accelerating Units

A. Circuit Coupling between Accelerating Units

The prototype DWA being researched and developed at IFP consists of multiple identical accelerating units including pulsed power generators and HGIs. Each pulsed power generator with typical output voltage pulse of 80~100 kV amplitude, 10 ns duration (FWHM), and 3 ns rise-time loads on HGI. The HGI is seen as accelerating cavity, which is equivalent as a capacitor. The ground and negative electrode of adjacent HGIs both are connected with a pubic electrode as shown in figure 2(a), such as the negative electrode of generator A (GA) is connected with the ground of generator B (GB). When GA drives the HGI A, there is only an interior current, as shown in blue arrows, flows through HGI A if GA and GB have different grounds at far end. However, there is a coupling current, as shown in red arrows, comes from GB's

*This project is supported by National Nature Science Foundation of China (Nos. 11035004, 51407169 and 51507162), NSAF (No. U1530156), Key Laboratory of Inorganic Function Material and Device, Chinese Academy of Sciences (No. KLIFMD-2014-01), Key Projects of Sichuan Provincial Department of Education (No. 16zd1121), and Sichuan Province Science and Technology Innovation Talent project (No. 2016012).

†xialiansheng@caep.cn

ground, even which is larger than interior current. This condition could be described as GA needing to drive a coupling capacitor C_{BA}. This indicates the effect of GB on GA. Equally, there are coupling currents come from GA and GC, which indicate that GB needs to drive two coupling capacitors (C_{AB} and C_{CB}) at the same time. The figure 2(b) is the equivalent circuit of schematic diagram of figure 2(a). This coupling effect is called circuit coupling. The major reason of coupling effect is that each accelerating unit could not operate independently because there is a pubic electrode for different accelerating cavities. The coupling effect leads to effective charge loaded on electrodes reduce, which affects electric field of accelerating cavity. Moreover, it's worth mentioning that we could not block the coupling current at the coaxial cable by using a high-voltage diode, because the flow direction of coupling current and interior current have a same direction.

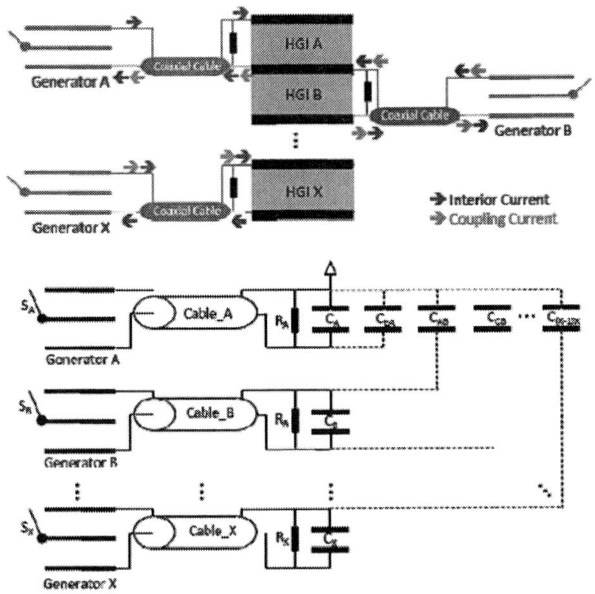

Fig.2 (a) Circuit coupling between accelerating units and (b) its equivalent circuit of schematic diagram. S_A, S_B ... and S_X are switches. R_A, R_B ... and R_X are generator's loads. C_A, C_B ... and C_X are HGIs' equivalent capacitances. C_{AB}, C_{BA} ... and $C_{(X-1)X}$ are generator's equivalent coupling capacitances.

B. Accelerating Structure Coupling between Accelerating Units

Accelerating structure coupling between accelerating units is another coupling effect, which is caused by nonideal charge distribution of accelerating cavity with capacitive structure. First, for only one accelerating cavity, there is decelerating field at the entrance and exit in regions I and III as shown in figure 3(a). Figure 3(b) shows that the electric field distribution of adjacent cavity as only one generator drives accelerating cavity. Another is that the accelerating field decrease for two or more accelerating cavities situation as shown in figure 3(c). The accelerating structure coupling of two or more accelerating cavities could be interpreted as the addition of accelerating field and decelerating field. Actually, it is that the charge redistributes on the accelerating cavity electrode after which the charge interacts with each other. That is to say that each generator not only drives respective cavity, but also drives adjacent and farther cavities. If the load capacity of the generator is limited, the accelerating field decreases. The figure 3(d) is the equivalent circuit of multiple accelerating units.

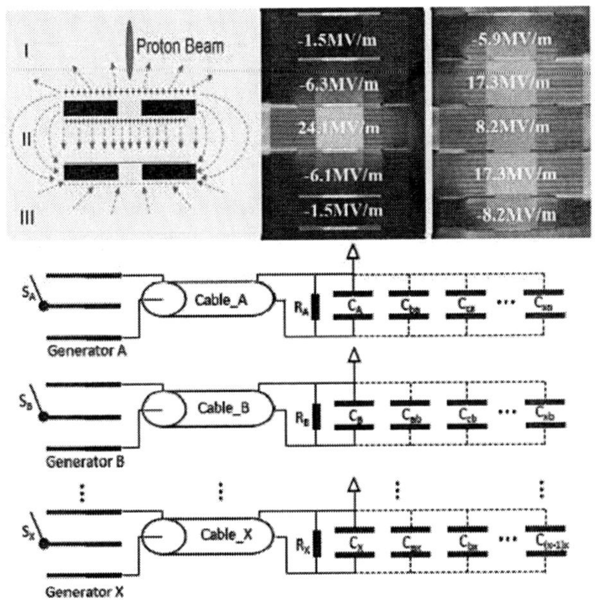

Fig.3 (a) Accelerating structure coupling between accelerating units and (b) its equivalent circuit of multiple accelerating units. $C_{(x-1)x}$ is an equivalent coupling capacitance interpreted the effect of HGI (x-1) on HGI x.

C. Experiments on Coupling of Multiple Accelerating Units

Coupling efficiency of multiple accelerating units have be measured by using a high-voltage probe. As shown in figure 4(a), the accelerating structure consists of eight continuous accelerating cavities. Thus, it has circuit coupling and accelerating structure coupling at the same time. We have measured the voltage of each accelerating cavity from loading only one accelerating cavity to eight accelerating cavities. The voltage of each accelerating cavity has be normalized as shown in figure 4(b). The figure 4(b) shows that the more the accelerating cavities are loaded, the lower the normalized voltage of each accelerating is. It indicates that the coupling effect increases with the increasing of accelerating cavity. And the normalized voltage of middle accelerating cavity is lower than normalized voltages of both ends of accelerating cavities for the continuous accelerating structure.

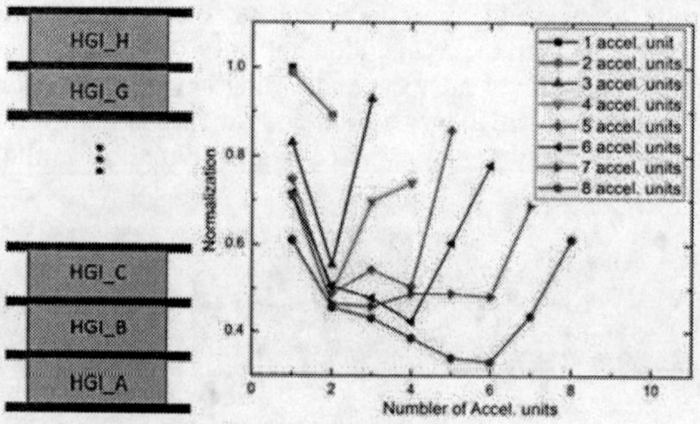

Fig. 4 (a) A continuous accelerating structure and (b) its coupling efficiency including circuit coupling and structure coupling.

III. CONCLUSION

The DWA is being researched and developed at IFP, which consists of multiple identical accelerating units including pulsed power generators and HGIs. Because a public electrode is used in accelerating structure. There is a serious circuit coupling and accelerating structure coupling between accelerating units. In this paper, the circuit topologies of two kinds of coupling effects are given. The coupling effects between accelerating units are studied, and the coupling efficiency of multiple accelerating units is measured. References

References

[1] G. J. Caporaso, Y. J. Chen, and S. E. Sampayan, "The dielectric wall accelerator," *Rev. Accl. Sci. Tech.* Vol. 2, pp. 253-263, 2009.

[2] Y. Shen,W. Wang, Y. Liu, L. Xia, H. Zhang, H. Pan, et al, "A compact 300kV solid-state high-voltage nanosecond generator for dielectric wall accelerator," *Rev. Sci. Instrum.*, Vol. 86, 055110, 2015.

[3] G. J. Caporaso, T. R. Mackie, S. Sampayan, Y. J. Chen, D. Blackfield, J. Harris, et al, "A compact linac for intensity modulated proton therapy based on a dielectric wall accelerator," *Physica Med.*, Vol. 24, pp. 98-101, 2008.

[4] J. Zhu, H. Zhang, S. Peng, H. Ren, L. Xia, W. Wang, et al, "Beam transport design for a 1 MeV prototype dielectric wall accelerator," *Chinese Phys. C*, Vol. 38, 047001, 2014.

[5] K. Yan, "Corona plasma generation," PhD diss., Eindhoven University of Technology, 2001.

[6] Z. Liu, "Multiple-switch pulsed power generation based on a transmission line transformer," PhD diss., Eindhoven University of Technology, 2008.

High Resolution Beam Profile Monitor for High and Low Energy Ion Beams

J. Lopes
Instituto Superior de Engenharia de Lisboa
GIAAPP/ISEL
Lisboa, Portugal

J. Rocha, N. Catarino and M. Peres
Instituto de Plasmas e Fusão Nuclear
Instituto Superior Técnico, Universidade de Lisboa
Lisboa, Portugal

Abstract- **An innovative high resolution beam profile monitor that allows visualization of the ion beam profile over the sample target, for both high and low energies, with target deceleration, is presented. With it, it is possible to evaluate the focus of the ion beam over a high voltage biased target, in the ion implanter installed at the Laboratório de Aceleradores e Tecnologias de Radiação (LATR) at the Campus Tecnológico e Nuclear, of Instituto Superior Técnico (IST). The aim of this work was to develop a high resolution system capable of showing the ion beam profile for energies below 15 keV, using a beam deceleration system, in order to evaluate beam dispersion effect over the implantation area. Important to mention that was also an objective to develop a beam profile system with a shape as flat as possible, so it could be attached to a deceleration target disk without affecting the target geometry, but that could also be used with the target grounded to produce a high energy beam profile.**

Keywords - high resolution, high energy, low energy, beam profile, ion beam

I. INTRODUTION

Low energy, high current ion implantation is important in nowadays technology [1-3]. Most beam profile measurement/monitors systems have a device to display the shape of the beam. Traditionally, beam profile monitoring can be performed by measuring the current through conductors using three different systems: wire scanners, fixed wire multi wire beam detector and Faraday Cups matrix [4-8].

The objective of this work was to develop a system that is suitable for monitoring beams with energies below 15 keV, using a biased target. The most suitable choice for the LATR-IST ion implanter was a modified system based on the wire scanner.

II. SYSTEM DESCRIPTION

The ion implanter installed at LATR-IST is a general purpose Danfysik model 1090, it has the versatility in ion species, beam energy, beam intensity, and allowed sample geometry. However, to produce low energy implantations, below 15 keV with this implanter model, a beam deceleration system is needed, due to long beam path and high beam currents [9].

As part of the deceleration system developed at LATR-IST with the general purpose for low energy implantations of several samples [10], a beam profile monitor system was built to produce a good resolution with a shape as flat as possible, so it could be attached to a deceleration target without affecting the target geometry. With this system should be possible to obtain a beam profile over the target using both a high energy (> 15 keV) beam with target grounded, or a low energy (< 15 keV) beam with target biased.

The developed system consists on the use of a slit scan method that can be described as follows: a flat surface with a slit hole is exposed to an ion beam, a part of the beam will pass through the slit, a conductive wire is placed behind that surface and normal to the slit as shown in Fig. 1, only a small portion of the beam will hit the wire. The resolution of this slit scan beam profile monitor (SSBPM) only depends on the slit and wire dimensions.

Fig. 2 shows a picture of the complete system built inside the target chamber of the high current implanter at the LATR-IST. A 280 mm diameter aluminum target plate was built with a square window of 50×50 mm located 50 mm from the border. Behind the square window and pressed against the target is a 70×100 mm rectangle plate made of tantalum, containing a slit hole of 50×1.5 mm, in a way that the slit can move vertically from the bottom to the top of the window. The target is fixed to the chamber by an insulating support and the slit is fixed to an insulating arm attached to a stepper motor inside the chamber that provides the vertical movement.

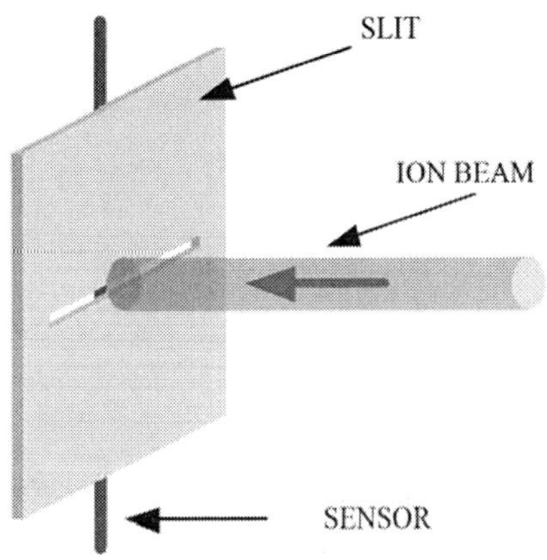

Fig. 1. Slit scan method: a sensor behind a slit is exposed to a small part of the beam.

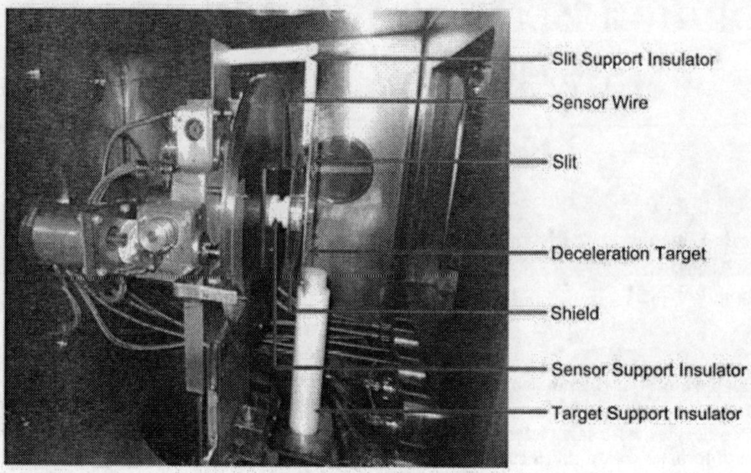

Fig. 2 - Mechanical system implemented.

Few millimeters behind the slit and target, but not attached to it, is a 1 mm width and 80 mm long copper wire (sensor). The sensor has one end connected to an insulating shaft attached to another stepper motor, in a way that the sensor rotates continuously clockwise passing periodically behind the slit. A resistor (R) is connected between the sensor and the target so, when the ion beam hits the sensor through the slit, a current passes through the resistor, so the voltage across R is proportional to the beam current and can be sent to a signal processing circuit:

$$V_R = I_{beam} \times R \tag{1}$$

To avoid capacitive effects due to the excess of positive charges when the ion beam is hitting the target a 180 mm diameter aluminum plate was placed behind the sensor, connected to the target. Without the shield, a permanent negative offset voltage in relation to the target is read by the sensor when the ion beam hits anywhere on the target.

The target and the sensor are connected via a high voltage feedthrough to the outside of the target chamber. Also, the target is connected to a high voltage power supply (HVPS) that sets the deceleration voltage from 0 to 15 kV applied to the target, as shown in Fig. 3.

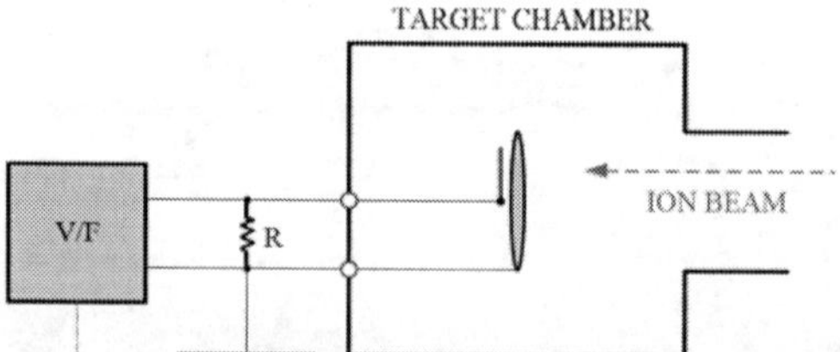

Fig. 3 - Block diagrams for electronic system of the sensor and deceleration target.

The time period for the scan reading is define by an optical switch (X-Switch) connected to the sensor support insulator is set to be ON when the sensor passes behind the slit. Another switch is connected to the slit support insulator that enables the vertical movement of the slit (Y-Switch).

The sensor is connected to a voltage-to-frequency (V/F) converter circuit placed at the target potential that converts V_R, as shown in (1), into an optical frequency signal. The output of the V/F converter is sent through an optical fiber cable to a frequency-to-voltage (F/V) converter at the ground potential. The output of the F/V is connected to a PC through an analog-to-digital converter (ADC). A LabVIEW based program was developed to acquire and save the data from the successive beam scans.

The SSBPM resolution is determined in X direction by the width of the sensor and in Y direction by the width of the slit window. In practice, the resolution of the system is on the order of 1.5mm × 1.5mm, the wire tends to introduce a slight broadening in the measured profile. Thus, the resolution is about 25×16 dpi.

III. EXPERIMENTAL RESULTS

In order to determine the beam profile the slit windows was irradiated by a Argon (Ar^+) ion beam with a 50 μA intensity focused over the slit window, in a vacuum of $2\text{-}5\times10^{-6}$ mbar. The profiles were based on 30 scans with an ADC sample rate of 100 Hz.

Two tests were made in order to obtain beam profiles with energies of 50 keV with the target grounded and 10 keV with the target biased, initial beam energy are 50 keV and 20 keV respectively.

In Fig. 4 the acquired data is presented in 3D charts, where the X and Y axis correspond to the horizontal and vertical dimensions of the slit window in cm, and the Z axis to the intensity of the beam in μA. Fig. 4a) corresponds to the beam profile obtained with the 50 keV beam energy and target grounded. Fig. 4b) corresponds to the beam profile for 10 keV, with initial energy of 20 keV and target biased at 10 kV.

A very sharp peak is observed in the 50 keV beam profile, corresponding to a sharp focus. In the case of 10 keV, the peak broadening is mainly due to the beam dispersion [9].

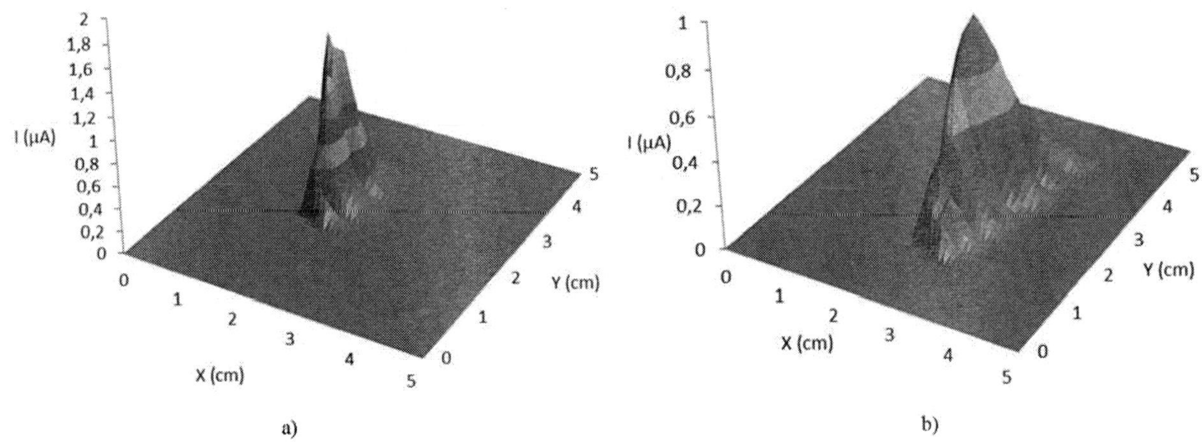

Fig. 4 - Beam profiles: a) 50 keV Ar^+ beam with target grounded; b) 10 keV Ar^+ beam with 20 keV initial energy and target biased at 10 kV.

The differences between the two profiles becomes more clear considering the full width at half maximum (FWHM) of the beam profile. For a better viewing experience, a map plot is presented in Fig. 5 with a cut in the beam profile at the half maximum beam intensity. This map is considered as a reference for the beam dimension over the target. It shows the cross section of the beam profile for the two profiles obtained with the energies of 50 keV (Fig. 5a) and 10 keV (Fig. 5b) respectively.

Through both Figures and analyzing the experimental results obtained by the system, is possible to conclude that the system works correctly for high and low energy.

IV. CONCLUSIONS

The presented system is an important upgrade of the previous system described in [11], which had only a mechanical scan in X direction. The Y direction scan was produced by sweeping the beam over the target, which could leave to erroneous results with target biasing, due to beam trajectory dispersion, depending on beam energy and target potential.

Furthermore, using mechanical scan in both X and Y directions with well-defined scanning area, the resolution of the SSBPM is almost twice the previous system which was about 25×9 dpi.

The developed system proved to be a versatile tool for the study of ion beam behavior over a grounded or biased target. Experimental results reveal differences in shape and dispersion of the focused beam, that otherwise could not be detected.

The SSBPM is mechanically simple and cheap. Despite this, the SSBPM can be used on an ion implantation target chamber like the one at the LATR-IST, with the advantage that the same stepper motors used for rotation and vertical movement of the sample target can be used for the slit scan process.

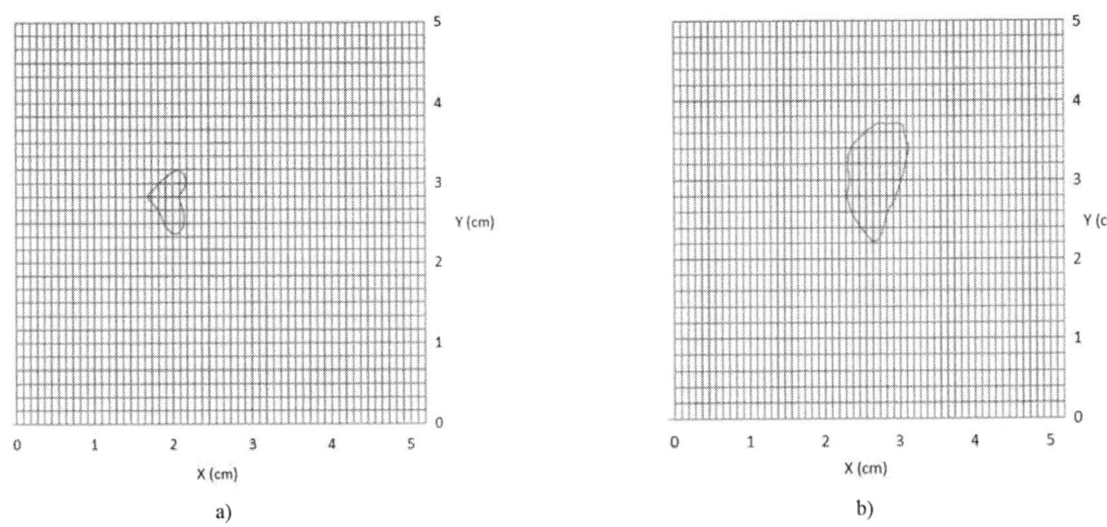

Fig. 5 - Middle section of: a) 50 keV Ar^+ beam with target grounded; b) 10 keV Ar^+ beam with 20 keV initial energy and target biased at 10 kV.

References

[1] J. A. Sánchez-García, L. Vázquez, R. Gago, A. Redondo-Cubero, J. M. Albella and Zs Czigány, "Tuning the surface morphology in self-organized ion beam nanopatterning of Si(001) via metalincorporation: from holes to dots" IopScience - Nanotechnology, vol. 19, number 35, pp. 355306, year 2008. DOI: 10.1088/0957-4484/19/35/355306.

[2] G. Hobler, G. Fehlmann, "A study of ultra-shallow implanted dopant profiles in silicon using BC and MD simulations" Radiation Effects and Defects in Solids, vol. 141,pp 113-125, year 1996.DOI:10.1080/10420159708211562.

[3] G. Hobler, H. H. Vuong, J. Bevk, A. Agarwal, H. J. Gossmann, D. C. Jacobson, M. Foad, A. Murrell, Y. Erokhin, "Modeling of ultra-low energy boron implantation in silicon", Electron Devices Meeting, 1997. IEDM '97. Technical Digest., International. pp. 489-492, year 1997. DOI: 10.1109/IEDM.1997.650430.

[4] J. Lopes, J. Rocha, N. Catarino, L. Redondo, " Characterization of a Single Electrode Focusing Lens for Ion Beam Deceleration", Instrumentation Science & Technology, vol., pp., year 2016. DOI: 10.1080/10739149.2016.1186033.

[5] G. Bassato, A. Andrighetto, N. Conforto, M. Giacchini, J. A.Montano, M. Poggi, J.A. Vásquez, "A beam profiler and emittance meter for the spes project at infn-lnl", Proceedings of ICALEPCS2011, Grenoble, France.

[6] G. Seniutinas, G. Gervinskas, J. Anguita, D. Hakobyan, E. Brasselet, S. Juodkazis, "Nano-proximity direct ion beam writing" Nanofabrication, vol. 2, pp. 54-62, year 2015. DOI:10.1515/nanofab-2015-0006.

[7] T. Lida, Y. Maekawa, R. Taniguchi, M. Byakuno, K. Sumita,"Simple profile monitor for low intensity, low-duty cycle pulse ion beams" Review of Scientific Instruments, vol. 52, pp. 1328-1331 year 1981. DOI: http://dx.doi.org/10.1063/1.1136766.

[8] L. Panitzsch, M. Stalder, and Robert F. Wimmer-Schweingruber em "Direct high-resolution ion beam-profile imaging using a position-sensitive Faraday Cup array" Review of Scientific Instruments, vol. 80, pp. 11302-11308. year 2009. DOI: 10.1063/1.3246787.

[9] Jochen Teichert, Johannes von Borany "Decel Lens System for Low-Energy Ion Implantation with High Dose Uniformity", Ion Implantation Technology, 2000. Conference on, Alpbach, 2000, pp. 388-391. doi: 10.1109/.2000.924169.

[10] J. Lopes, J. Rocha, "Simple General Purpose Ion Beam Deceleration System Using a Single Electrode Lens"World Journal of Engineering and Technology, vol. 3, pp. 127-133, year 2015. DOI: http://dx.doi.org/10.4236/wjet.2015.33014.

[11] J. G. Lopes, J. Rocha, L.M. Redondo F. C. Alegria "High resolution ion beam profile Measurement System", 13th International Conference on Accelerator and Large Experimental Physics Control Systems - Proceedings, de 10 a 14 Outubro, Grenoble - França 2011.

Compact High-Power Microwave Compression Systems for Formation of Long Nanosecond-Duration Pulses with a Rectangular Pulse Shape

Sergey N. Artemenko, Vladimir A. Avgustinovich,
Stanislav A. Gorev, Vladislav S. Igumnov
National Research Tomsk Polytechnic University
snartemenko@mail.ru, vladislavigumnov@gmail.com

***Abstract* — We propose a new approach to designing the geometry of large accumulative systems of compact Microwave Pulse Compressors (MPC's) used to generate ~10-ns rectangular pulses in the S- and X-bands. Our design uses standard structural elements made from single-mode or moderately multimode waveguide sections whose ends are terminated by waveguide tees or their analogs. One arm of each T-junction is connected to a section of the resonant cavity, the second arm is close-circuited, and the third arm connects to an adjacent linear section. The plasma switch, used to Q-spoil the cavity, is mounted in the side arm of another *tee*. The proposed approach leads to develop of very compact systems. Integration of an MPC into an existing RF generator is also straightforward without significantly expanding its footprint.**

We present schemes for several types of MPC's with variable geometry of the accumulative system. Our work shows that a specific architecture of the accumulative system, accounting for the relevant distribution of energy and method for pulse extraction, can be used to build MPC's with discretely controlled output pulse parameters. This paper demonstrates the first results of an experimental study of MPC's with a planar accumulative system.

Keywords — microwave pulse compressor; microwave compression; resonant cavity; plasma switch; H-tee; waveguide

I. Introduction

The 1960's witnessed a new trend in electronics associated with the development of sources of powerful, nanosecond-duration RF pulses. Interest in these microwave sources is based on both fundamental and specific features of nanosecond pulses. Nanosecond-duration pulses are comparable to the characteristic time scales for various physical processes of current scientific interest.

Examples of such processes include RF breakdown in microwave devices, failure of electronic equipment under the influence of pulsed RF radiation, and reflection of short RF pulses from probed objects during radio positioning. Going from microsecond to nanosecond pulses, the delay in discharge formation provides an increase in the dielectric strength of microwave devices. Because the time scale for these transient processes is longer than the pulse lengths discussed here, higher-power RF pulses do not risk breakdown, i.e., these pulses may contain the equivalent energy of longer-duration pulses without incurring any undesirable effects. Shorter pulses increase accuracy and information content during exploration by radio positioning. Thus, powerful RF pulses find applications in plasma-chemical technology [1], where square RF pulses are preferable in most cases. When used in acceleration devices [2,3], short pulses increase the acceleration rate and reduce the dispersion of particle energy. When used in electronic warfare devices, they increase the range and accuracy of failure rate determination. When used for radio positioning, they can increase positional accuracy and simplify processing of the received signal. In addition, compact and inexpensive sources are preferable for obvious reasons [4]. The range of applications is not limited to those mentioned here. All these reasons explain the importance of development of compact sources of powerful nanosecond RF pulses having a rectangular waveform.

II. Active Microwave Pulse Compression System. Problems

Simultaneously with the development of the most powerful albeit rather complicated and expensive relativistic RF generators and amplifiers, passive methods were also studied in an effort to increase the RF pulse amplitude. One of these methods is the relatively simple and effective method of resonant pulse compression.

The microwave pulse compressors used in this method are passive RF amplifiers providing accumulation of energy from a relatively long and weak RF pulse in resonant cavity, followed by fast output of this energy into a load (see Fig. 1). Typically, microsecond pulses are fed into the input port of such amplifiers, while shorter and more powerful nanosecond pulses are generated at the output port.

The output pulse parameters of a resonant MPC depend on the Q-factor and dielectric strength of the resonant cavity as well as on the energy extraction method. The Q-factor and dielectric strength determine the amount of accumulated energy. The energy extraction method and system topology determine the nature and instant of energy output; in any case, the larger the amount of the stored energy, the faster and greater the energy output. Consequently, the efficiency is also improved. Given the relative simplicity of such amplifiers, our prototype devices produced gain coefficients up to ~25 dB, pulse powers up to ~0.1-1 GW, and efficiencies over 60%.

At the same time, several issues important for development of compressors remain unsolved. One of them is the problem of generating long and powerful nanosecond pulses with a rectangular shape in compact MPC's. For X-band and S-band pulses of ~10-100 ns and ~0.1-1 GW, the length of the radiated wave train is a few to tens of meters, which grossly exceeds the length of typical compressors (~1-2 m).

In addition to the difficulty of generating a rectangular shaped pulse, unsolved problems exist concerning the geometry used to form the resonant cavity. As a rule, the storage cavity of the resonator is a cylinder, sphere, or prism. Compressors, being passive amplifiers, depend upon the design of the active source. Integration of fixed-form MPC's with standard sources is problematic, especially if size is a concern.

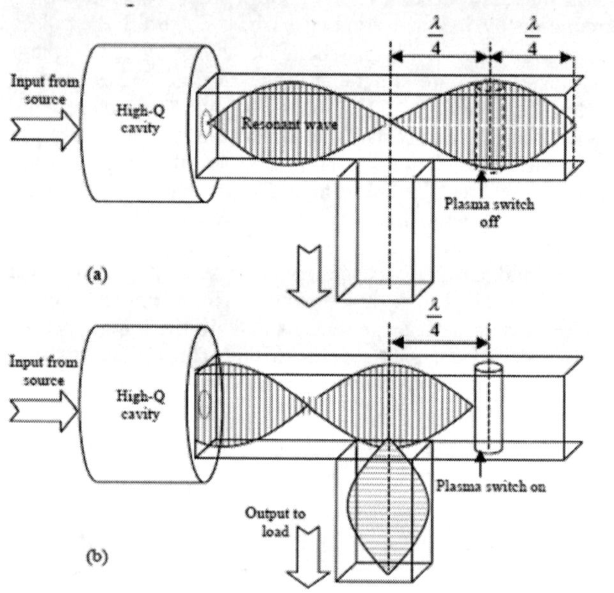

Fig. 1. Scheme of an active high-power MPC with output energy via an interference plasma switch: (a) plasma switch is off; (b) plasma switch is on (picture taken from [5]).

III. Solution of the Problems

To solve the aforementioned problems, a new approach to the design of a resonant storage cavity is proposed. The traditional fixed-form resonant cavity is replaced by a resonant system with a variable geometry. As shown in Fig. 2, our system is constructed using standard structural elements made from single-mode or moderately multimode waveguide sections whose ends are terminated by two waveguide tees or their analogs. One arm of the tee is connected to a section of the resonant cavity, the second arm is close-circuited, and the third arm connects to a tee from an adjacent linear section.

An accumulative system made in this way may be formed utilizing extended linear or three-dimensional structures, or combinations thereof, by alignment of standard elements through open arms of node devices. In addition to solving the problem of generating sharp pulses in a compact compressor, this approach increases the functionality of the compressor. The proposed approach simplifies development of systems that must fit within a small space. Integration of an MPC into an existing RF system is also easier with a concomitant savings in space.

A simple analysis of the resonant system using the dispersion matrix method shows that our system can be implemented with a wave-transmission coefficient that approximates the value obtained with the orthodox configuration shown in Fig. 1.

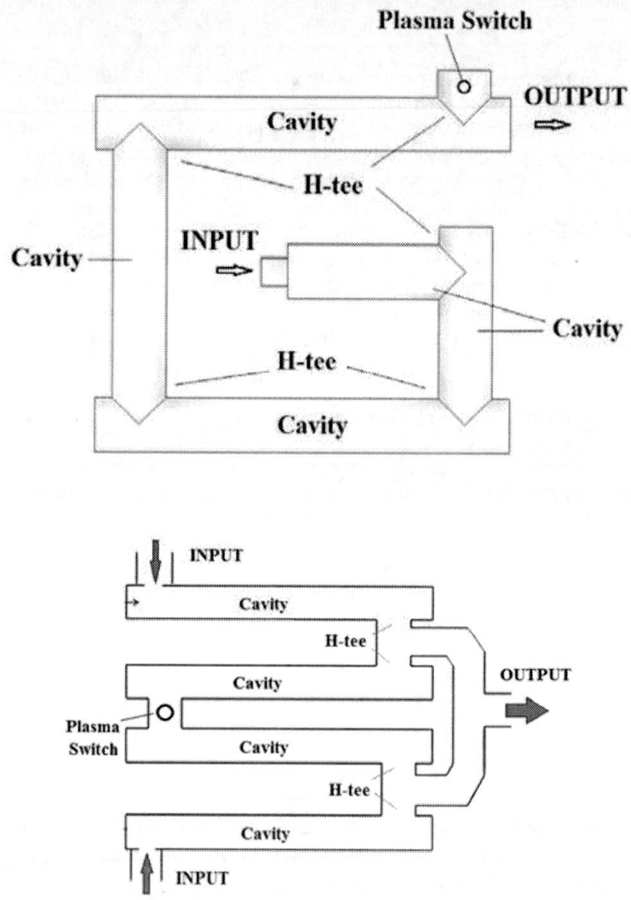

Fig. 2. Examples of the proposed geometry.

IV. Results

As shown in Fig. 3, we built several simple types of S-band MPC's with a compact storage cavity having a variable geometry. Specific architectures of the accumulative system along with placement of the output coupler can be used to make microwave compressors with discretely controlled output parameters. We also show that the proposed approach to the design of storage cavities makes it possible to develop compact, cascade MPC's with power multiplication of the travelling wave from each resonant cavity to the next.

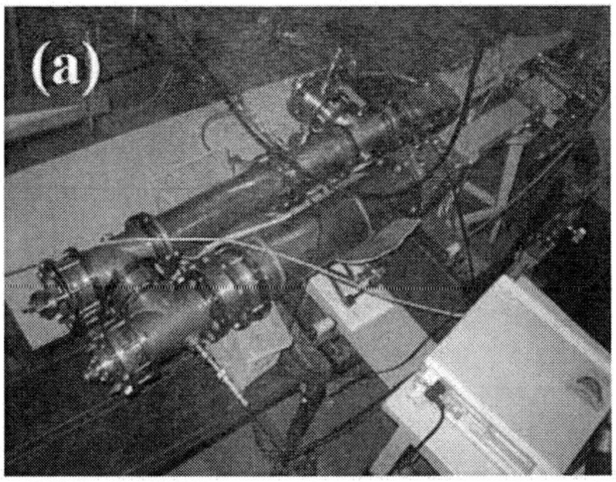

Fig. 3. Photographs of compact, active, high-power MPC's with (a) two- and (b) three-section storage cavities.

The emphasis of the present experiments is the study of MPC's with planar resonant cavities. Representative output from a three-section cavity is shown in Fig. 4; the input pulse parameters are frequency 2.8 GHz, power 2 MW, and pulse width 3.2 μs. These compressors are based on resonant cavities following a *meander* shape. The components are located in one plane and consist of sections of single-mode round waveguide (90 mm internal diameter) and TE-tees. T-junctions are used to connect linear sections as well as to mount the plasma switch [6]. The operating mode is TE11. The intrinsic Q-factor of the resonant cavity is 24,000, while during pulse storage this parameter is approximately 12,500. The output port is connected to a matched load.

Different layouts produce identical pulses if the resonant cavity lengths are identical. We confirmed experimentally that such compressors generate nearly rectangular shaped pulses with duration equal to the double transit time of a traveling wave within the resonant cavity and with power compatible to that generated in a much longer, linear cavity. We defined the necessary conditions under which such pulses are generated.

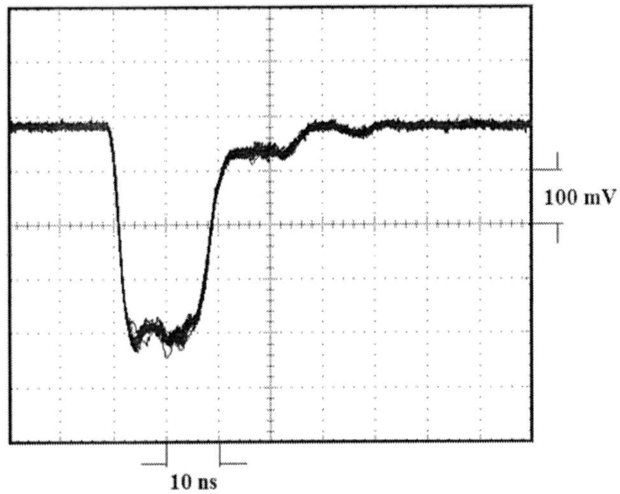

Fig. 4. Typical oscillogram of multiple output pulses from a compressor with a three-section resonant cavity (gain 13 dB, duration 25 ns, output power 40 MW).

V. Prospects

The proposed approach appears also to be effective when the resonant cavity is made from sections of moderately multimode round waveguide. Based on the known equivalence of the field structure in a standard switch and in round waveguide with a TE01 wave, we can assume that the upper cutoff diameter of the section will be less that the critical diameter for a TE02 wave. As shown in Fig. 5, the joining sections can be made from rectangular TE01 waveguides or based on axially symmetric radial feeders. The resulting mode structure will be TE01(p).

The output device for such a compressor can be constructed as an H-tee interference switch using oversized rectangular waveguide [7] or can be based on a packet of standard H-tees with a shared commuting arm that synchronizes the work of the packet [8]. Cascade interference switches will increase the working power [9].

Computational simulations show that such connections allow the resonant cavity to be formed using various planar or three-dimension geometries, including cubes and parallelepi-peds. The wave propagates with very little reflection, which allows generation of long, nanosecond-duration MW pulses of nearly rectangular shape and high power.

As shown in Fig. 6, we developed a prototype device for a compact MPC system operating in the X-band using our folded topology and we are conducting experimental studies at low power. The first results confirm our hypotheses. Excitation of oscillations and simulation of the switch in the output mode demonstrate essential functionality of the system.

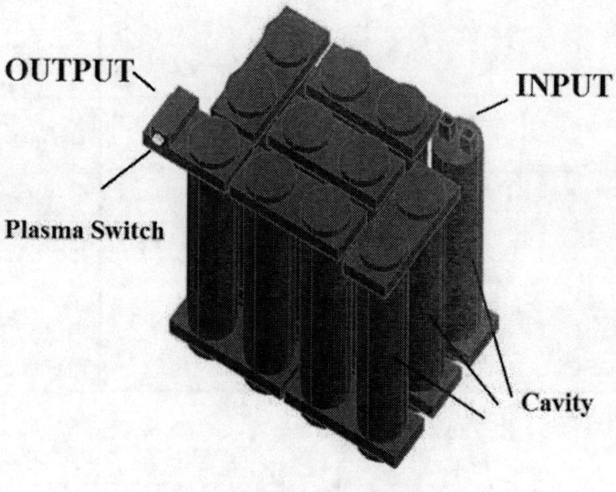

Fig. 5. Scheme of a compact compressor made from sections of moderately multimode round waveguide with a TE01 mode structure.

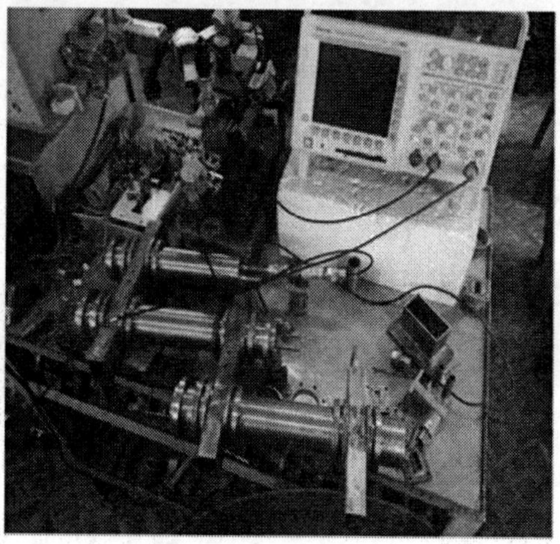
Fig. 6. Photograph of an active MPC with an oversize, three-section cavity.

VI. CONCLUSIONS

This work confirms the possibility of compact, active, microwave pulse compression systems capable of generating long and powerful nanosecond-duration RF pulses with a rectangular shape and sufficiently high gain.

In addition to making the integration of components easier, the use of a variable-geometry resonant volume improves the functionality of the compressor by affording proper connection of various output devices to the tees. Pulses of various waveforms, powers, and durations are more readily generated.

ACKNOWLEDGMENTS

This research was performed with partial financial support from the National Research Tomsk Polytechnic University Competitiveness Program.

The authors wish to thank Vladimir Avgustinovich, Vadim Lis, and Galina Vasiltsova for useful discussions and assistance.

REFERENCES

[1] N. E. Aktaev, "Theoretical approach to modelling the low-barrier chemical reactions initiated by pulsed electron beam," Journal of Physics: Conference Series 552, 012033, pp. 5, 2014.

[2] Z. D. Farkas, H. A. Hogg, G. A. Loew, and P. B. Wilson, "SLED: A Method of Doubling SLAC's Energy," SLAC Pub.#1453, 1974.

[3] J. Benford, J. A. Swegle, and E. Schamiloglu, *High Power Microwaves*, Second Edition, New York: CRC Press, 2007, pp.197–204.

[4] S. H. Gold and G. S. Nusinovich, "Review of high-power microwave source research," Review of Scientific Instruments, Vol. 68, № 11, pp. 3945-3974, 1997.

[5] I. V. Konoplev, P. MacInnes, A. W. Cross, W. He, A. D. R. Phelps, K. Ronald, C. G. Whyte, and C. W. Robertson, "2D and 1D surface photonic band gap structures for accelerator applications," Proceedings of EPAC 2006, Edinburgh, Scotland, pp. 1388-1390.

[6] V. A. Avgustinovich, S. N. Artemenko, et al., "A study of the switching of the microwave compressor switch in a circular waveguide," Instruments and Experimental Techniques, Vol. 52, № 4, pp. 547-550, 2009.

[7] V. A. Avgustinovich, S. N. Artemenko, and A. A. Zhukov, "Microwave-energy extraction from a resonator via oversized interference switch," Technical Physics Letters, Vol. 39, № 5, pp. 492-494, 2013.

[8] V. A. Avgustinovich, S. N. Artemenko, and M. S. Arteev, "Synchronous extraction of microwave energy from cavities through a packet of interference switches," Technical Physics Letters, Vol. 39, № 12, pp. 1040-1043, 2013.

[9] S. N. Artemenko, V. A. Avgustinovich, S. A. Gorev, and V. S. Igumnov, "Analysis of microwave interference switches with distributed power of switched wave and plasma gas-discharge switching," Journal of Physics, Conference Series, Vol. 652, № 1, 012017, 2015.

Rayleigh-Taylor Instability of High-Velocity Condensed-Matter Liners

S.F. Garanin, A.M. Buyko, V.B. Yakubov
Russian Federal Nuclear Center - VNIIEF
Sarov, Russia

***Abstract*—High magnetic pressures applied to a liner can drive it to high implosion velocities, which is of interest for numerous applications, including generation and use of shock waves, generation of high magnetic fields, shockless compression of matter and compression of magnetized plasma. Since the energy density achieved by the liner implosion is determined by the velocity and density of the liner, it is important that the liner has high density, which can be ensured if the liner preserves its condensed state during implosion. Liner acceleration can be accompanied by instabilities distorting the liner's shape and uniformity, which can impair its performance. Among the potentially destructive instabilities one can accentuate the Rayleigh-Taylor (RT) instability, which occurs, when a liner is driven by a magnetic field. Related to the RT instability is the sausage instability caused by the curvature of the streamlines of an azimuthal magnetic field. This work considers the RT instability developing during liner implosion, its manifestation in some experiments, numerical simulations of its development and some mitigation approaches.**

Key words—liner implosion; z-pinch; Rayleigh-Taylor instability; aspect ratio

I. Introduction

Magnetically driven liner implosion (in z-pinch or Θ-pinch geometry) offers unique advantages [1, 2] for its application in high energy density (HED) experiments. As the energy is delivered to the liner by a magnetic field at the speed of light (rather than by the pressure of dense matter, as it is the case in HE and gas gun driven implosion), liners can reach velocities higher than those attainable in HE or gas gun drivers. In this case, magnetic pressure, at least in some directions (for example, in the z direction for the z-pinch liner drive), equilibrates nearly instantaneously, which improves implosion symmetry. Parameters of experimental electric circuits producing drive currents can be controlled in a wide range, so the liner drive rate and the velocity can be varied in a controlled manner in accordance with specifications of the experiment. A magnetic field can drive liners smoothly, without shock waves, making it possible to closely monitor the liner conditions and to study isentropic compression of materials.

Liner implosion is used to produce strong shock waves in targets, to perform work on a magnetic field (and, in so doing, to generate high magnetic fields), to compress magnetized plasma and thus increase its internal energy, for shockless compression of matter, and in other applications.

An application-relevant liner characteristic is the dynamic pressure ρv^2 (ρ and v are the liner density and velocity), because this quantity characterizes the energy density. Therefore under conditions of HED it is important to ensure that the liner has high density, which can be reached if the liner remains condensed during implosion, because vaporization and conversion into plasma decreases its density by several orders.

Liner implosion can be accompanied by instabilities distorting the liner's shape and uniformity, which can impair its performance or even destroy the liner. Among these instabilities one distinguishes those resulting from the interaction between the liner and the current-carrying walls and the Rayleigh-Taylor (RT) instability, which occurs when a low-density matter accelerates a higher-density one, for example, when a magnetic field (zero-density matter) drives a dense liner. The RT instability can also develop if the liner compresses a magnetic field or low-density plasma (on liner deceleration). Related to the RT instability is the sausage instability in an azimuthal magnetic field, which is often used in z-pinch liner implosion.

In this work we will explore the growth of perturbations due to the development of the RT instability during condensed-matter liner implosion, its manifestation in some experiments, numerical simulations of the instability growth and some approaches to its mitigation. In some sense, our work can be complementary to the review talk of A. Velikovich [3] made at the Conference ICOPS-2015.

II. RT Instability and Growth of Perturbations. The Impact of the Sausage Instability

A. Linear stage of the RT and sausage instabilities

For an ideal incompressible fluid disposed in a gravitational field with acceleration g the linear theory [4, 5] suggests that the Fourier component of the surface perturbation with wave number k grows in time exponentially with the increment γ equal

$$\gamma = \sqrt{kg}\,. \tag{1}$$

To study the stability of a liner [6], one should replace the acceleration of gravity with acceleration due to the inertial force directed opposite the acceleration of the fluid. The formula (1) holds true if, instead of an infinite fluid, we consider a magnetically driven liner of thickness d (Fig. 1) such that the liner acceleration g is determined by magnetic pressure $p_B \equiv B^2/8\pi$ (B is the magnetic field):

$$g = p_B / \rho d. \quad (2)$$

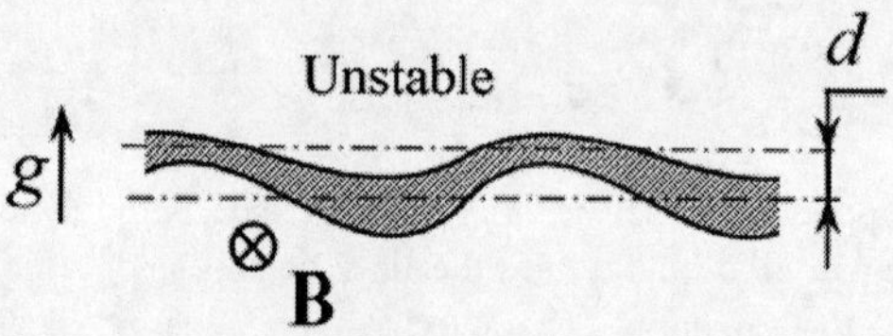

Fig. 1. Rayleigh-Taylor instability of an accelerated liner

If the liner is driven by an azimuthal magnetic field, in addition to the RT instability, some contribution can be made by the sausage instability, associated with the fact that the azimuthal magnetic field depends on the radius as $B \sim 1/r$, and, consequently, the magnetic pressure is proportional to $p_B \sim 1/r^2$. For short wavelengths, $kr >> 1$, the development of the sausage instability reduces [7] to the RT instability and has the increment (1). Here, the liner material plays the role of a heavy fluid, and the magnetic field, of a light one, and in formula (2) instead of acceleration g the quantity

$$g_B = 2p_B / \rho r = B^2 / 4\pi\rho r = c_A^2 / r \quad (3)$$

should be used (c_A is the Alfven velocity).

The increments (1) of the RT and the sausage instability correspond to the case of a distinct matter/magnetic field interface, so that the wavelengths are large compared to the size of the transition region. This appears to be justified for the case of a condensed-matter liner. However, when the liner begins to explode the interface between the magnetic field and the matter becomes gradual. As far as the increments for such instabilities grow with the wave number, the shortest wavelengths can be most destructive. In case of a smooth distribution of the density and the magnetic field for the shortest wavelengths, the increments of the RT or sausage instabilities tend to a constant limit depending on the gradients of corresponding quantities. Calculation of these limiting increments with the joint action of the RT and the sausage instability in an azimuthal magnetic field is considered in [7].

B. Nonlinear development of the RT instability

When amplitudes of perturbations become comparable to their wavelength, the instability development becomes nonlinear and the analysis of arising nonlinear flows becomes more complicated.

However, as applied to turbulent flows developing in instability environments, it turns out to be helpful to look at some special cases of nonlinear instability development. One of such cases is the problem of constant-wavelength periodic perturbations [8, 9] for infinite ratio of heavy/light densities.

In this setup, the problem of instability development at the nonlinear stage reduces on a large time scale to a steady-state case, which simplifies its analysis. Basic features of the steady-state flow in this setup include bubbles rising in a heavy fluid at velocity, which is constant at constant gravity acceleration, and spikes sinking in the frame of reference of the bubbles with constant acceleration g. Experimental studies of periodic perturbations in the magnetic RT instability of initially solid-state liners are documented in [10, 11].

Another type of perturbations, the development of which is universal, are localized perturbations discussed in [12, 13]. The problem of plane localized perturbations [13], i.e. perturbations in the form of a straight line drawn across the surface of a fluid, when the perturbation zone grows with time according to the self-similar law

$$y \sim gt^2,$$

can be of interest for turbulent flows description.

C. Turbulent stage of the RT and sausage instabilities

A characteristic feature of the turbulent stage of the RT (see, for example, Fig. 2) and the sausage instability, just as of the nonlinear stage of periodic and localized perturbations, is the presence of bubbles and spikes. However, as distinct from the periodic perturbations, the bubbles and spikes are located chaotically and the bubbles merge as the instability continues to develop, so the characteristic amplitudes and wavelengths of the perturbations grow. A common property of this stage is the fact that the characteristic wavelengths and amplitudes of the perturbations turn out to be on the order of some part of the path length traveled with acceleration.

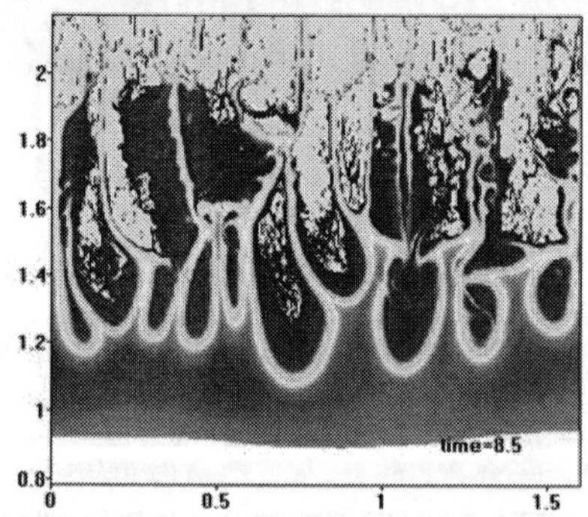

Fig. 2. Distribution of liner density during implosion

Comparison of the accelerations g (2) and g_B (3) determining the mixing zones of the RT and the sausage instability shows that their magnitudes are related as R/d, and for liners with a high aspect ratio, R/d, the role the sausage instability should not be essential.

The turbulence arising in liner implosion in an azimuthal magnetic field (z-pinch geometry) can be two-dimensional (2D), because the magnetic field resists the curvature of magnetic streamlines bending. Therefore, it is reasonable to believe that 2D simulations in the $r-z$ coordinates (of the type of Fig. 2) are able to describe the arising turbulence.

III. Liner Instability Development in Some Experiments

For HED condensed-matter liner applications [1, 7, 14-20] liner velocities should be ~1 cm/μs or higher. This requires megagauss magnetic fields that cause skin layer of the liner to melt or vaporize and convert to plasma [21]. This part of the liner looses its strength and creates conditions for the development of the RT instability.

Growth of perturbations in liner implosion is governed by the complex dynamics, which depends on the instability development, because Joule heat release may increase in the nonuniform liner part. As a result, the way the instabilities develop turns out hard to predict. Even for similar liner drive conditions, the dynamics of RT instability development can differ significantly. For example, in the z-pinch liner implosion experiments conducted on the Shiva-Star facility at the Air Force Research Laboratory (the NTLX series) and on the Atlas facility at the Los Alamos National Laboratory (the HF and the LD series), although the initial liner radii were the same, and the differences in the liner thicknesses, peak currents and current rise times [1, 22] were minor, the radiograms revealed (Fig. 3) moderate or weak instability development in NTLX and HF, while in LD it was very strong.

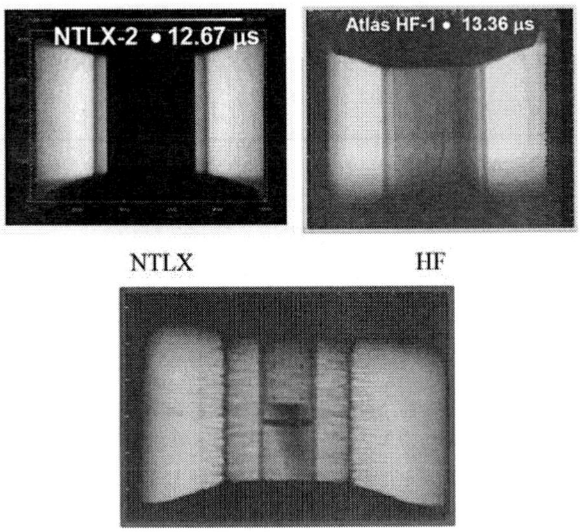

Fig. 3. Results of liner radiography in the NTLX, HF, and LD experiments

We conducted 2D ($r-z$) magnetohydrodynamic (MHD) simulations [23] of these experiments (using the code [24]). The initial radius of the liners was 5 cm. The liner in the NTLX series was 0.1 cm thick, and in HF and LD, 0.13 cm. The liners were driven by azimuthal magnetic fields produced by time-dependent currents with peak currents and current rise times 15.9, 16.9, 19.7 MA and 8.2, 6.1, 6.0 μs for NTLX, HF, and LD consequently.

Initial perturbations in the simulations were assumed to be random. As the mesh size Δ did not allow to track the development of perturbations with wavelengths $\lambda < \Delta$, the simulations did not include the initial phase of perturbation development up to the amplitudes of $a \sim \lambda \sim \Delta$ and assumed that initially $a \sim \lambda \sim \Delta$. One can suppose, however, that the inaccuracy thus introduced is insignificant, because the level of RT instability development is basically controlled by the liner displacement S (such that $a \sim S$) and, consequently, the initial level of perturbations should have virtually no impact on the magnitude of RT instability perturbations.

Simulated density distributions shown in Fig. 4 demonstrate that although the state of the NTLX liner seems to be less perturbed than the state of the LD liner, it still looks noticeably less uniform than in the experiment. As for the HF experiment, the simulations in agreement with the experiment give more stable motion compared to both LD and NTLX.

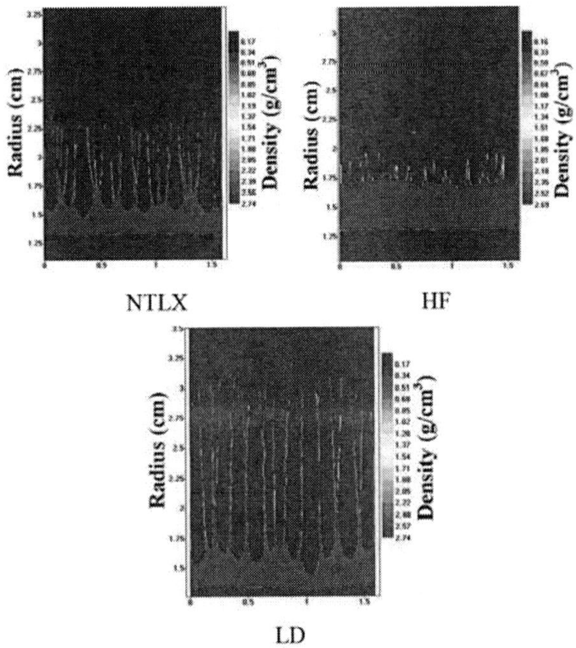

Fig. 4. Liner density obtained in the NTLX, HF, and LD simulations at the inner liner radius R_{in} = 1.3 cm

To quantify the degree of instability development, one can consider perturbations on the inner liner surface, which are much smaller than those on the outer surface, but the size of which characterizes the liner's suitability for experimental purposes. For the experiments under consideration, the amplitude of its perturbations a determined as a half difference between the maximum and minimum radii of the inner surface is $a \approx 0.094$ mm for NTLX, $a \approx 0.265$ mm for LD, and $a \approx 0.09$ mm for HF (at the liner radius R_{in} = 1.3 cm). Thus, by means of the 2D simulations we demonstrated that the experimental conditions in HF and NTLX lead to a smaller perturbation growth than in LD, if the perturbation growth is quantified based on the perturbation development on the inner surface.

Analysis of the correlation between the level of instability development and the mesh size Δ revealed that the results converge as the mesh size decreases for LD and HF, and do not converge for NTLX. Probably, for NTLX, the level of perturbation development on a finer mesh will get lower and more consistent with the experimental radiogram. One can suppose that perturbations for even thinner grids can become so small that they will be invisible in a radiogram, and the NTLX liner in the radiogram will look stable. Unfortunately, further mesh refinement caused significant computational difficulties for the code we used.

IV. Perturbation Stabilization Approaches

One of the approaches to mitigating perturbations is precision liner machining making it possible to minimize the initial amplitude of perturbations. Other stabilization methods include possible variations in the experimental setup that could make the liner motion more stable from the computational and theoretical standpoint. Among these methods one can note reducing the liner's aspect ratio while preserving its mass and applying periodic perturbations of a certain wavelength on the liner, such that the perturbations themselves have no time to destroy the liner during implosion but can efficiently suppress the growth of chaotic perturbations [25]. Additional possible measures for mitigating the RT instability are related to specific properties of the magnetic field as a source of pressure.

Possible RT instability mitigation methods include collision of liners, as a result of which the target liner turns out to be less perturbed than the incident one. Although part of energy is lost in this case, the improvement in the conditions of the target liner compared to the incident one turns out to be beneficial and is practiced on a wide scale [26]. A similar idea is used today in implosion of wire liners, when double (nested) wire arrays are used [27].

The RT instability can also be mitigated by centrifugal acceleration that occurs when the liner is rotated [28].

One should also note a new interesting RT instability mitigation method [29]. Liner experiments on the Z facility at the Sandia National Laboratories, USA, demonstrated that the development of the magnetic RT instability can be suppressed substantially by applying a relatively thick (~70 μm) dielectric coating on the outer surface of the liner.

Let us consider some of these implosion stabilization methods in more detail.

A. *Impact of aspect ratio on liner instability development*

Experiments of the LD series on Atlas (section III) demonstrated strong instability development. This brings up the question of how the liner aspect ratio influences the development of liner instability. For example, what if one reduces the initial radius R of the liner, without changing its mass, compared to the LD experiments? Simulations show that in this case liner velocities can be the same as in LD at the same radius of the measurement unit. However, due to the significant decrease in the aspect ratio R/d (the number of thicknesses traveled by the liner decreases) liner implosion can be expected to be more stable.

Our proposal was to decrease the initial liner radius to 3 cm (it was 5 cm in LD); then, with the liner mass remaining the same, the liner thickness would be $d = 0.23$ cm (instead of 0.13 cm in LD). Let us call this proposed experiment LDR.

In the simulations, the current of the Atlas facility was varied in accordance with its electrical circuit. As shown by the 1D simulations, the velocity of the inner liner surface at an impact radius of 1 cm in this case does not change. This is due to the fact that the LD liner is basically driven at the current fall stage, whereas in LDR the current is close to maximum over a considerable part of the liner drive path.

According 1D simulations the Joule melted zone for LDR is somewhat lower than for LD. Thus, for LDR in addition to a decrease in R/d, liner conditions appear to be less prone to instability development.

The density distribution obtained in the 2D LDR simulation (Fig. 5) shows that the LDR liner appears to be less perturbed than the LD liner (Fig. 4).

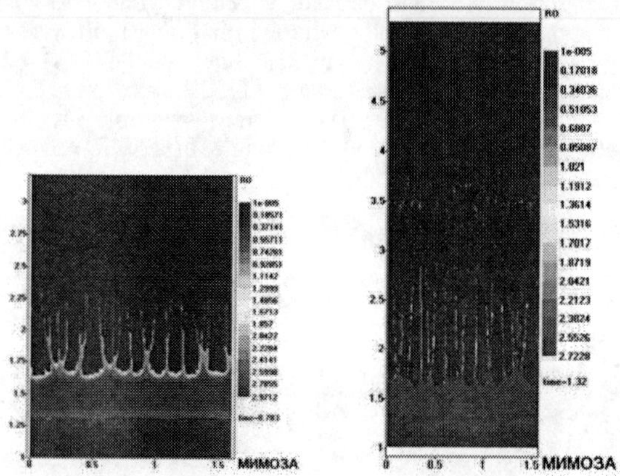

LDR LDP1

Fig. 5. Liner density distribution obtained in the LDR and LDP1 simulations at the inner liner radius $R_{in} = 1.3$ cm

The simulations revealed a significant difference in the degree of the development of the perturbations, the amplitude of which for LD is $a \approx 0.265$ mm and for LDR, $a \approx 0.026$ mm, i.e. an order smaller.

B. *Liner instability development in the presence of periodic perturbations*

Now let us consider the possibility of instability mitigation by liner machining to produce periodic perturbations of a certain wavelength such that the perturbations themselves have no time to destroy the liner during implosion but can efficiently suppress the growth of chaotic perturbations.

The wavelength λ^* and the initial amplitude a_0 of a specified periodic perturbation were chosen as

$$\lambda^* = 0.4 \text{ mm}, \quad a_0 = 0.13 \text{ mm}. \quad (4)$$

To simulate the implosion of a liner with periodic perturbations (LDP simulation) we used the same current waveform as in LD. Initial perturbations represented a sum of the sinusoidal perturbations (4) and random perturbations specified in the cells of the outer surface.

The simulations indicated that the degree of the inner surface perturbation development in the LD and LDP simulations is approximately the same.

However in the LDP simulation, the inner liner surface acquires a substantially higher velocity (≈ 8.8 km/s) than in LD (7.4 km/s), because part of its mass is thrown off with the growth of periodic perturbations. Thus, liner performance in the LD and LDP simulations turns out to be far from equivalent. One can hope that application of periodic perturbations on a thicker liner can stabilize the growth of chaotic perturbations and preserve the velocity of the liner.

In the LDP1 simulation, the liner had a thickness increased to 1.6 mm and, in addition to chaotic perturbations, the same periodic perturbations as in LDP. The results of the LDP1 simulation are shown in Fig. 5. The outer surface perturbations at the final liner radius in LDP1 appear to be somewhat smaller than in LDP, and the inner surface perturbations in LDP1 is about half that ($a = 0.14$ mm) of LD and LDP. Thus, making periodic perturbations on a liner of a greater thickness can stabilize the growth of chaotic perturbations to some extent while preserving its performance.

C. Magnetic stabilization of the RT instability

Ref. [6] considers the RT instability for a magnetically driven liner. For the perturbations, the wave vector of which is parallel to the magnetic field, the growth increment of small perturbations should be smaller than (1). The stabilizing effect of magnetic streamlines tension is the stronger, the shorter is the wavelength. If the perturbation wave vector is directed at some angle θ to the magnetic field, the criterion of complete suppression of the perturbation growth is

$$\lambda < 4\pi d \cos^2\theta . \qquad (5)$$

Paper [30] shows magnetic stabilization of the liner if it is driven by a magnetic field turning to an angle of about 90°.

To test the effect of magnetic RT instability mitigation, a joint LANL/VNIIEF experiment was conducted at Los-Alamos [31]. In the experiment an aluminum liner was driven by an azimuthal magnetic field. The outer surface of the liner on half of its height had azimuthally symmetric sinusoidal perturbations. On the other half of the liner, the perturbations had the same wavelength and amplitude, but the angle θ=45° (helical perturbations). The azimuthally symmetric perturbations should grow under such conditions with an increment (1). The growth of the helical perturbations, however, should be completely suppressed, because the criterion (5) is fulfilled.

Fig. 6 shows two experimental radiograms. In top of them, the liner is at the initial location. The bottom radiogram corresponds to the time, when the liner radius reduces. The azimuthally symmetric perturbations in this radiogram have grown significantly. The other half of the liner (with helical perturbations), however, does not show any perturbation growth. The experimental results obtained validate the effect of magnetic stabilization of the RT instability.

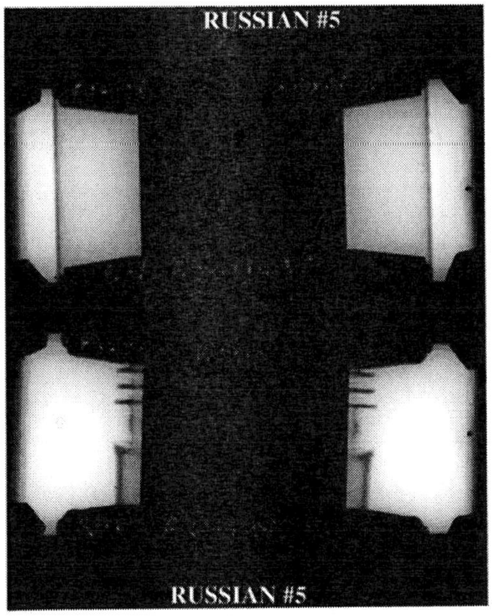

Fig. 6. Radiograms of the liner in the initial position and at the time its radius reduces by about a third

V. CONCLUSION

Literature offers various methods for studying the RT and other hydrodynamic instabilities (like the sausage one). In a case of small perturbations one can use linear analysis. Simplifications are also possible when studying special cases and geometries, like the development of periodic perturbations and local perturbations, when self-similar features are used. However, in real cases, the spectrum of initial perturbations is rather complex, and the fast non-linear development of short-wavelength perturbations results in the turbulization of the mixing zone at the interface between the magnetic field and the liner. In this case, 2D and 3D MHD simulations represent an efficient way of studying the development of instabilities.

Experiments show that even similar drive conditions may lead to disparate liner states with different degrees of liner distortion. For example, in the LD, HF, and NTLX experiments with similar conditions, the radiograms obtained in HF and NTLX demonstrated a comparatively stable motion of the liner, while the radiograms in LD showed that the outer surface of the liner was severely distorted by the growth of instability. 2D numerical simulations of liner implosion in LD, HF, and NTLX suggest that the experimental conditions in HF and NTLX indeed lead to a smaller growth of perturbations compared to LD.

To improve the performance of liner implosion systems, researchers explore the techniques that can help suppress the development of RT instabilities:

- reducing the aspect ratio R/d of the liner without changing its mass,
- liner machining to produce periodic perturbations of a certain wavelength such that the perturbations themselves have no time to destroy the liner during implosion but can efficiently suppress the growth of chaotic perturbations,
- magnetic stabilization, in literature also referred to as magnetic shear,
- collision of liners, as a result of which the target liner turns out to be less perturbed than the incident one,
- the use of liner rotation and centrifugal acceleration induced by it,
- using plastic coatings on metal liners.

References

[1] R.E. Reinovsky, W.E. Anderson, W.L. Atchison, et al., "Pulsed power hydrodynamics: a new application of high magnetic field," Proc. IX Int. Conf. Megagauss Magnetic Field Generation and Related Topics, Moscow-St.-Petersburg, 2002. Eds. V. D. Selemir and L. N. Plyashkevich. Sarov: VNIIEF, pp. 696-705, 2004.

[2] R.E. Reinovsky, W.L. Atchison, C. Rousculp, and A. Kaul, "Pulsed high magnetic fields for exploring the dynamic properties of materials," Proc. XIII Int. Conf. on Megagauss Magnetic Field Generation and Related Topics. 2010. Suzhou, China. Eds. Sun Chengwei and Liu Cangli. Press of National University of Defense Technology, pp. 44-45, 2012.

[3] A.L. Velikovich, "Mitigation of Rayleigh-Taylor instability in high-energy density plasmas. A personal perspective," presented at the 42nd IEEE Int. Conf. on Plasma Science. Belek, Antalya, Turkey, 2015.

[4] G.I. Taylor, "The instability of liquid surfaces when accelerated in a direction perpendicular to their planes," Proc. Roy. Soc. London, vol. A 201, No. 1065, pp. 192-196, 1950.

[5] D.J. Lewis, "The Instability of liquid surfaces when accelerated in a direction perpendicular to their planes. II," Proc. Roy. Soc., vol. A 202, pp. 81-96, 1950.

[6] E.G. Harris, "Rayleigh – Taylor instabilities of a collapsing cylindrical shell in a magnetic field," Phys. Fluids, vol. 5, No. 9, pp. 1057–1062, 1962.

[7] S.F. Garanin. Physical processes in the MAGO/MTF systems. Los Alamos, LA-UR-13-29094, 2013.

[8] P.B. Garabedian, "On steady-state bubbles generated by Taylor instability," Proc. Roy. Soc., vol. A 241, pp. 423–431, 1957.

[9] G. Birkhoff and D. Carter. "Rising plane bubbles," J. Math. Mech., vol. 6, No. 6, p. 769, 1957.

[10] R.E. Reinovsky, W.E. Anderson, W.L. Atchison, et al., "Instability growth in magnetically imploded high-conductivity cylindrical liners with material strength," IEEE Trans. Plasma Sci., vol. 30, No. 5, pp. 1764-1776, 2002.

[11] D.B. Sinars, S.A. Slutz, M.C. Herrmann, et al., "Measurements of magneto-Rayleigh–Taylor instability growth during the implosion of initially solid metal liners," Phys. Plasmas, vol. 18, p. 056301, 2011.

[12] O.I. Volchenko, I.G. Zhidov, E.E. Meshkov, and V.G. Rogachev, "Development of localized perturbations on the unstable boundary of accelerated liquid layer", Pis'ma v ZhTF (Letters to J. Tech. Phys.), vol. 15, No. 1, p. 47-51, 1989 (in Russian).

[13] S.F. Garanin, "Self-similar evolution of Rayleigh-Taylor instability in the corner-point regions," Proc. Fifth International Workshop on Compressible Turbulent Mixing. Eds. R. Young, J. Glimm, and B. Boston. Singapore, New Jersey, London, Hong Kong: World Scientific, pp. 33-39, 1995.

[14] A.M. Buyko, V.V. Zmushko, V.N. Mokhov, et al., "On feasibility to achieve high longitudinal symmetry of cylindrical metal liners compressed by currents from most powerful disk EMG," Digest of Technical Papers 12th IEEE International Pulsed Power Conference, Eds. C. Stallings and H. Kirbie, pp. 1145-1148, 1999.

[15] R.E. Reinovsky, "Pulsed power experiments in hydrodynamics and material properties," Digest of Technical Papers 12th IEEE International Pulsed Power Conference. Eds. C. Stallings and H. Kirbie, pp. 38-43, 1999.

[16] A.M. Buyko, O.M. Burenkov, V.V. Zmushko, et al., "On the feasibility to achieve high pressures with disk EMG driven impacting liners," Digest of Technical Papers, PPPS-2001. Eds. R. Reinovsky and M. Newton. Las Vegas, Nevada, pp. 516-519, 2001.

[17] R.K. Keinigs, W.L. Atchison, W.E. Anderson, et al., "Material science experiments on the Atlas facility," Digest of Technical Papers, PPPS-2001. Eds. R. Reinovsky and M. Newton. Las Vegas, Nevada, pp. 356-360, 2001.

[18] S.F. Garanin, V.I. Mamyshev, and V.B. Yakubov, "The MAGO system: current status", IEEE Trans. Plasma Sci., vol. 34, N 5. Part III. pp. 2273-2278, 2006.

[19] J.H. Degnan, T. Cavazos, D. Clark, et al., "On research on magnetic pressure implosions of long cylindrical liners, suitable for subsequent compression of the field reversed configuration type of compact toroids," Proceedings of Ninth International Conference on Megagauss Magnetic Field Generation and Related Topics. Moscow-St.-Petersburg, 2002. Eds. V. D. Selemir and L. N. Plyashkevich. Sarov: VNIIEF, pp. 730-737, 2004.

[20] D.B. Sinars, "Magneto-inertial fusion research in the United States: A personal perspective," presented at the 42nd IEEE Int. Conf. on Plasma Science. Belek, Antalya, Turkey, 2015.

[21] S.F. Garanin, G.G. Ivanova, D.V. Karmishin, and V.N. Sofronov, "Diffusion of a megagauss field into a metal," J. Appl. Mech. Tech. Phys., vol. 31, No. 2, pp. 153-159, 2005.

[22] W.L. Atchison, R.J. Faehl, I.R. Lindemuth, et al., "Dependence of solid liner stability on drive conditions during magnetic implosion," in Proc. IX Int. Conf. Megagauss Magnetic Field Generation Related Topics. Moscow-St.-Petersburg, 2002. V. D. Selemir and L. N. Plyashkevich, Eds, Sarov, Russia: VNIIEF, pp. 710-717, 2004.

[23] A.M. Buyko, S.F. Garanin, D.V. Karmishin, et al., "Analysis of the liner stability in various experiments," IEEE Trans. Plasma Science, vol. 36, No. 1, pp. 4-9, 2008.

[24] I.D. Sofronov, O.A. Vinokurov, V.V. Zmushko, et al., "The program suite MIMOZA. Multidimensional hydrodynamics simulations," Issues of Mathematical Modeling in Computational Mathematics and Informatics. Transactions. Moscow – Arzamas-16, RFNC-VNIIEF, pp. 94-96, 1994 (in Russian).

[25] A.M. Buyko, S.F. Garanin, V.V. Zmushko, et al., "On stabilization of implosion of condensed liners," J. Appl. Mech. Tech. Phys., vol. 50, No. 3, pp. 361-370, 2009.

[26] A.I. Pavlovsky, A.I. Bykov, M.I. Dolotenko, et al. "Cumulation of ultrahigh magnetic fields". High Energy Densities. Transactions. Sarov: RFNC-VNIIEF, pp. 446-468, 1997 (in Russian).

[27] J. Davis, N.A. Gondarenko, and A.L. Velikovich, "Fast commutator of high current in double wire array Z-pinch loads," Appl. Phys. Lett., vol. 70, pp. 170-172, 1997.

[28] D.L. Book and P.J. Turchi, "Dynamics of rotationally stabilized implosions of compressible liquid shells," Phys. Fluids vol. 22, p. 68, 1979.

[29] K.J. Peterson, T.J. Awe, E.P. Yu, et al., "Electrothermal instability mitigation by using thick dielectric coatings on magnetically imploded conductors," Phys. Rev. Lett., vol. 112, pp. 135002(5), 2014.

[30] S.F. Garanin, S.D. Kuznetsov, V.N. Mokhov, et al., "On feasibility of Rayleigh-Taylor instability magnetic stabilization of liner implosions," Proc. Thirteenth Int. Conf. on Megagauss Magnetic Field Generation and Related Topics. Singapore: World Scientific, pp. 563-567, 2004.

[31] B.G. Anderson, W.E. Anderson, W.L. Atchison, et al., "Liner experiment on verification of Rayleigh-Taylor instability magnetic stabilization effect (joint LANL/VNIIEF experiment Pegasus-2)," Dig. Tech. Papers, Pulsed Power Plasma Science. Eds. R. Reinovsky and M. Newton. Las Vegas, Nevada, pp. 354-355, 2001.

*Analysis of electrical characteristics of the pulse discharge source**

Ailong Fan, Yaohong Sun, Xuzhe Xu, Rongyao Fu
Institute of Electrical Engineering
Chinese Academy of Sciences
Beijing, China
fanailong@mail.iee.ac.cn

Ailong Fan, Yaohong Sun, Ping yan, Rongyao Fu
University of Chinese Academy of Sciences, Beijing, China
Ailong Fan
Liaoning institute of science and technology, Benxi, China

***Abstract*—The discharge characteristics of pulsed discharge source are analyzed in this paper. Firstly, the source discharge circuit is simplified, and the linear analysis of discharge circuit is given. According to the measured data, an approximate mathematical model is proposed. The discharge circuit analysis is carried out when the plasma channel resistance is nonlinear. Secondly, the test case calculation method of the parameters of the discharge characteristic of the source is put forward. Finally, the numerical values of the measured electrical parameters are given. In order to understand the operation law of the source, the effective method to improve the efficiency of the source is explored.**

Keywords—pulsed discharge source; discharge characteristics; plasma channel resistance

I. INTRODUCTION

To the terrestrial seismic exploration, the pulsed discharge source is a kind of better pulse source. It has many characteristics, such as a wide frequency spectrum, repetitive excitation with high efficiency, good repeatability, convenient operation, environmentally friendly, and so on. When the electric energy stored in the capacitor releases into the gap in the water quickly, it can produce many kinds of physical effects in the water, such as force effect, sound effect, light effect and other comprehensive effects. Using the liquid electric effect, the vibration wave is generated, which is the working principle of the pulsed discharge source.

The pulsed discharge source mainly includes the energy storage system, the boosting system, the control system and the power supply system. The high voltage energy generated by the boosting system is stored in the energy storage system. The high voltage energy is controlled by the control system, which is released to form an explosion. This article also uses the capacitor to store the high voltage electric energy which is released through the water medium in a moment. In this, we conducted the simulation analysis to draw relative conclusions of electrical parameters for the pulse discharge source.

*Work supported by R&D of Key Instruments and Technologies for Deep Resources Prospecting (the National R&D Projects for Key Scientific Instruments), Grant No.ZDYZ2012-1-06 and Chinese National Natural Science Foundation of China (51277171, 11411130178).

II. DESIGN ON THE PULSE DISCHARGE EXPERIMENTS SOURCE

The pulse discharge experiments source, as shown in figure 1, includes an adjustable source of direct current, a storage capacitor, a voltage measurement module, a discharge control switch, a discharge cable, and a high-voltage electrode. The adjustable source actually acts as a charging unit. The capacitor, the discharge control switch and the discharge cable form a discharging unit.

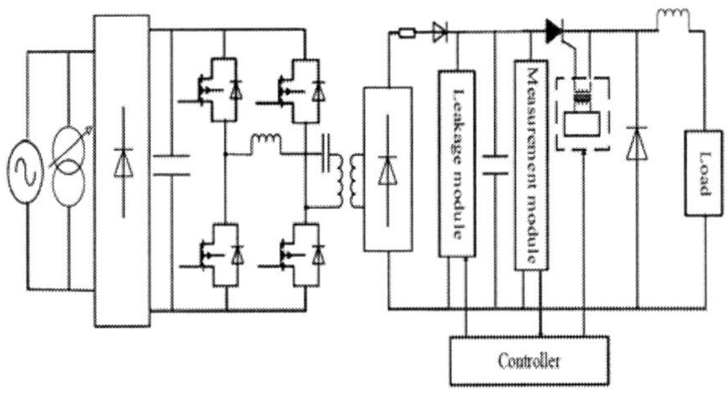

Fig. 1. The electric spark source.

The charging unit adopts a high-frequency resonant full bridge circuit topology structure, which consists of a regulating transformer, the first stage rectifier, a series resonant converter, the second stage rectifiers. The discharging unit is mainly composed of energy storage capacity, the isolation silicon stack, discharge switch, leakage circuit, wave adjusting inductance and continuous current diode. The control module can accurately control the charging and discharging of the capacitor, the release of residual pressure, the detection of voltage and current, and the fault processing.

The connection diagram of experimental devices is shown in Figure 2.

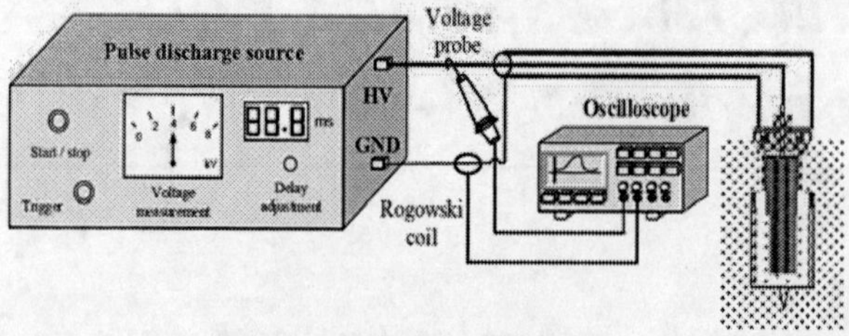

Fig. 2. Experimental platform

Length of each cable is about 25 meters; well depth is 10 meters; the high-voltage electrode discharge is placed in the depth of 8 meter underground.

III. DISCHARGE CIRCUIT ANALYSIS

The discharge circuit is simplified in this section, and the linear analysis of the discharge circuit is given when the resistance of the plasma channel is constant. After the voltage of capacitor reaches the set voltage, the controller sends the trigger pulse to the discharge switch SCR. Charges of the capacitor is released in an instant through the discharge switch SCR, the discharge cable and the discharge electrode. The discharge equivalent circuit is shown in Figure 3.

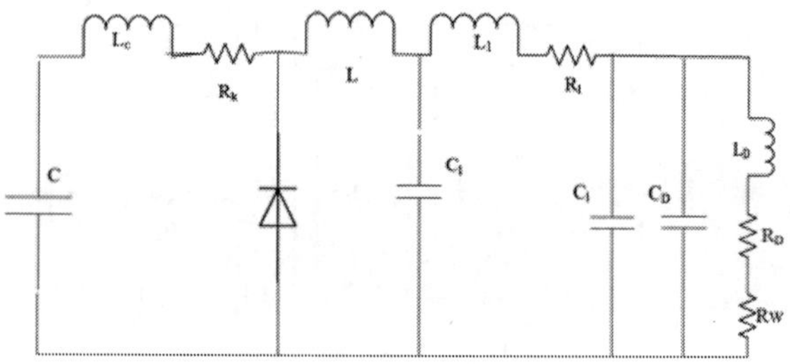

Fig. 3. Discharge equivalent circuit

C - discharge capacity; Lc - equivalent inductance of discharge capacitance; L_i - wave adjusting inductance; R_k - on-state resistance of thyristor or ; L_l - inductance of the discharge cable; C_l - capacitance between discharge cables; R_l - discharge cable resistance; L_D - inductance of the discharge electrode; C_D - capacitance between discharge electrodes; R_D - the resistance of the discharge electrode (the guiding resistance of the plasma channel); Rw - resistance of plasma channel.

Due to the relatively low working frequency of pulse discharge source, the wavelength is much larger than the transmission cable length. The ratio of voltage and current on the cable is constant along the transmission line, which means capacitance between lines C_l of the discharge cable can be ignored. At the same time, C_D of the discharge electrode and equivalent inductance of discharge capacitance Lc is very small. In addition, and the electrode length is very short, so L_D is too very small. Compared with other parts of the circuit, they can be ignored.

At the beginning, the pulse discharge equivalent circuit is a typical RLC discharge circuit. After the voltage is zero for the first time, the diode is switched on, and the current is switched from the RLC circuit to the RLD circuit, and the inductors L and L_l become the independent power supply. Assuming that the plasma channel resistance Rw is constant. Circuit equation are listed as the following

$$\begin{cases} u_c(t) + LC\dfrac{du_c^2(t)}{dt^2} + RC\dfrac{du_c}{dt} = 0 & (t \le t_0) \\ i(t) = C\dfrac{du_c}{dt} & (t \le t_0) \\ L\dfrac{di(t-t_0)}{dt} + Ri(t-t_0) = 0 & (t > t_0) \end{cases}$$

$$\begin{cases} R_0 = R_k + R_l + R_D \\ R = R_0 + R_w \\ L = L_i + L_l \end{cases}$$

$$\begin{cases} u_c(0^+) = kU_0 \\ i(0^+) = 0 \\ \dfrac{du_c(t)}{dt}\Big|_{t=t_0} = 0 \end{cases}$$

According to the circuit equations, the time domain relations of the voltage, current, power and other parameters are as following:

If limit conditions, making

$$\begin{cases} R < \sqrt{\dfrac{L}{C}} \\ \dfrac{kU_0}{d} > 4000V/cm \end{cases}$$

Then the arc discharge will be occurred.

When $t \le t_0$ (t_0 was the time when the voltage decreased to 0 for the first time)

Voltage of capacitor:

$$u_c(t) = U_0 e^{-\frac{R}{2L}t}\left[\cos\left(\sqrt{\frac{1}{LC} - \frac{R^2}{4L^2}}t\right) + R\sqrt{\frac{C}{4L - R^2C}}\sin\left(\sqrt{\frac{1}{LC} - \frac{R^2}{4L^2}}t\right)\right]$$

Loop current:

$$i(t) = \frac{-U_0}{L\omega}e^{-\frac{R}{2L}t}\sin\omega t$$

Output power:

$$p(t) = i^2(t)R_D = \frac{U_0^2 R}{L^2\omega^2}e^{-\frac{R}{L}t}\sin^2\omega t$$

when $i'(t) = \dfrac{U_0\frac{R}{2L}}{L\omega}e^{-\frac{R}{2L}t}\sin\omega t - \dfrac{U_0}{L}e^{-\frac{R}{2L}t}\cos\omega t = 0$,

the maximum time of discharge current and discharge power is obtained:

$$t_m = \frac{\arctan\dfrac{2\omega L}{R}}{\omega}$$

Maximum current:

$$I_{max} = \frac{-U_0}{L\sqrt{\omega^2 + (\frac{R}{2L})^2}} e^{-\frac{R}{2L}\arctan\frac{2\omega L}{R}}$$

Maximum power:

$$P_{max} = \frac{U_0^2 R}{L^2(\omega^2 + \frac{R^2}{4L^2})} e^{-\frac{R}{L}\arctan\frac{2L\omega}{R}}$$

$$u_L(t) = U_0 e^{-\frac{R}{2L}t}\left(\frac{R}{2L\omega}\sin\omega t - \cos\omega t\right)$$

Current steepness:

$$\frac{di}{dt} = \frac{1}{L}u_L(t) = \frac{U_0 e^{-\frac{R}{2L}t}}{L}\left(\frac{R}{2L\omega}\sin\omega t - \cos\omega t\right)$$

Output Energy:

$$W(t) = \int_0^t p(t)dt = \int_0^t \frac{U_0^2 R}{L^2\omega^2} e^{-\frac{R}{L}t}\sin^2\omega t\,dt$$

$$= \frac{U_0^2}{2L\omega^2} e^{-\frac{R}{L}t}\left[\frac{\frac{R}{2L}\cos 2\omega t - \omega\sin 2\omega t}{\frac{R^2}{L^2} + 4\omega^2} - \frac{L}{2R}\right]$$

When uc(t)=0, the time t0 is obtained (when t=t0, the capacitor voltage is reduced to 0 for the first time):

$$t_0 = \frac{\pi - \arctan\sqrt{\frac{4L}{R^2C} - 1}}{\sqrt{\frac{1}{LC} - \frac{R^2}{4L^2}}} = \frac{\pi - \arctan\frac{2\omega L}{R}}{\omega}$$

First half wave output energy:

$$W(t) = \int_0^{t_0} p(t)dt$$

$$= \frac{U_0^2}{2L\omega^2} e^{-\frac{R}{L}\frac{(\pi - \arctan\frac{2L\omega}{R})}{\omega}}\left[\frac{RC}{8} - \frac{2L}{R}\right]$$

Among them,

$$\omega = \sqrt{\frac{1}{LC} - \frac{R^2}{4L^2}}$$

If limit conditions, making

$$\begin{cases} R > 4\sqrt{\frac{L}{C}} \\ \frac{kU_0}{d} < 2000V/cm \end{cases}$$

Then the corona discharge will be occurred.

There, $b = \frac{R_0 + R_D}{2L}$ $\qquad \omega = \sqrt{\frac{1}{LC} - \frac{(R_0 + R_D)^2}{4L^2}}$

$$p_{1,2} = -\frac{(R_0 + R_D)^2}{2L} \pm \sqrt{\frac{(R_0 + R_D)^2}{4L^2} - \frac{1}{LC}}$$

In the same way, we get the following equation.

$$u_C(t) = \frac{U_0}{p_1 - p_2}\left(p_1 e^{p_2 t} - p_2 e^{p_1 t}\right)$$

$$i(t) = \frac{kU_0(e^{p_2 t} - e^{p_1 t})}{L(p_1 - p_2)}$$

$$I_m = \frac{kU_0}{L(p_1 - p_2)}\left[\left(\frac{p_1}{p_2}\right)^{\frac{p_2}{p_1 - p_2}} - \left(\frac{p_1}{p_2}\right)^{\frac{p_1}{p_1 - p_2}}\right]$$

$$p(t) = \frac{k^2U_0^2R_D}{L^2(p_1 - p_2)^2}(e^{p_2 t} - e^{p_1 t})^2$$

$$W(t) = \frac{k^2U_0^2R_D}{L^2(p_1 - p_2)^2}\left[\frac{e^{2p_2 t}}{2p_2} - \frac{e^{2p_1 t}}{2p_1} - \frac{e^{(p_1+p_2)t}}{p_1 + p_2}\right]$$

First half wave output energy:

$$E_0 = \frac{Ck^2U_0^2}{2}$$

IV. SIMULATION AND EXPERIMENT ANALYSIS

A. *Effect of inductance variation on output waveform*

When $u_C(0) = 5kV, C = 160uF, R = 1\Omega$

$L = 1uH, 10uH, 20uH, 40uH, 100uH, 300uH$,

Time domain and frequency domain waveform of load power is shown in the figure 4 and figure 5.

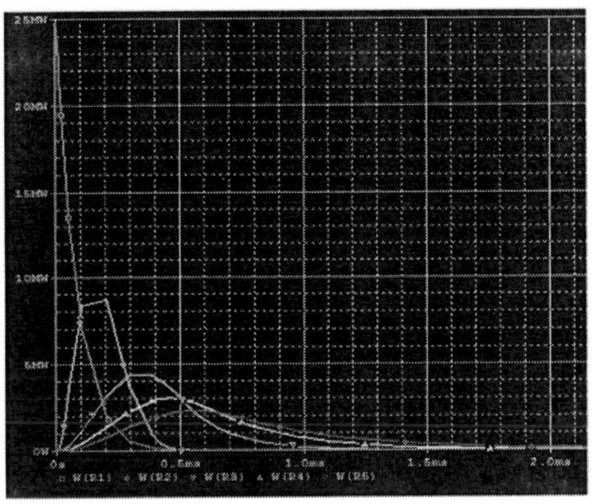

Fig. 4. Current diagram of different inductance parameters

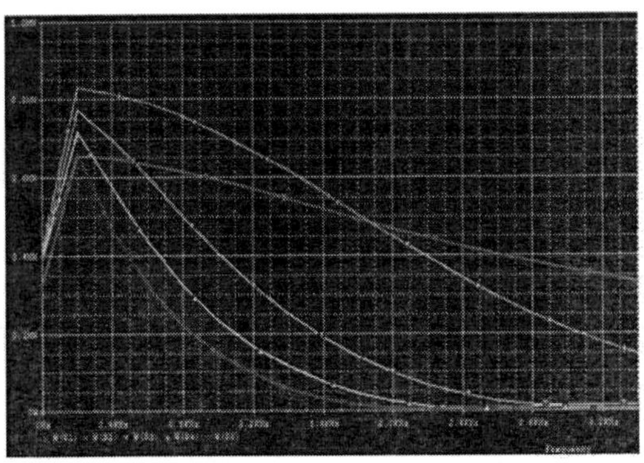

Fig. 5. Amplitude-frequency diagram of different inductance parameters

It can be seen from the figure 4, with the increase of the inductance value, the peak value of power is decreased, and the peak moment of power tended to shift backwards. From the amplitude frequency characteristics, with the increase of the inductance value, the main frequency is nearly constant, and the corresponding peak value is slightly decreased.

B. Effect of capacitance variation on output waveform

When $u_C(0) = 5kV, L = 640uH, R = 1\Omega$

$$C = 160uF, 320uF, 480uF, 640uF,$$

Time domain and frequency domain waveform of load power is shown in the figure 6 and figure 7.

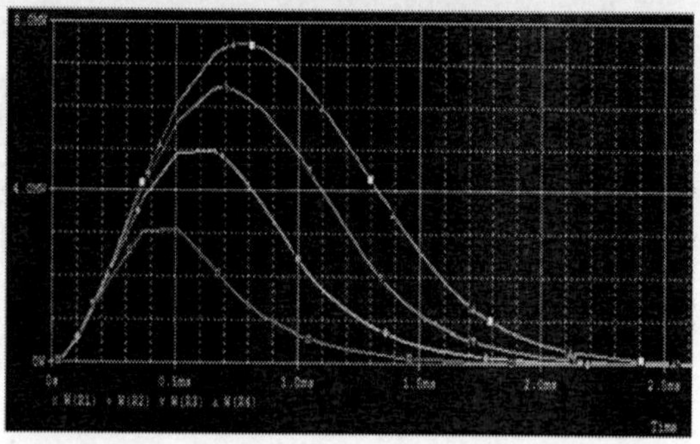

Fig. 6. Current diagram of different capacitance parameters

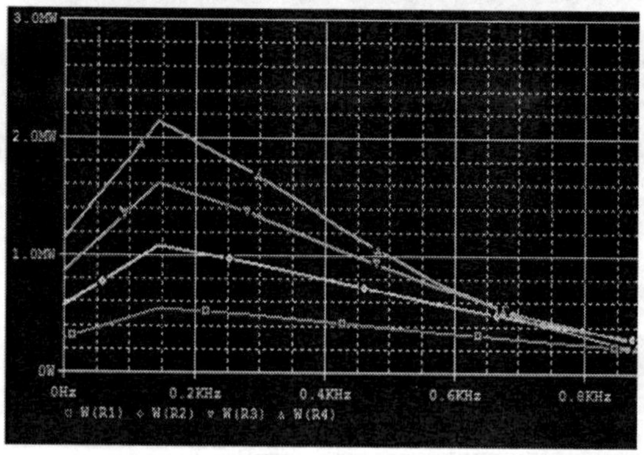

Fig. 7. Amplitude-frequency diagram of different capacitance parameters

It can be seen from the figure 7, the main frequency is not changed when the capacitance value increases exponentially. The increase of the power dissipation is slightly smaller than the increase of the capacitance value, but the low frequency band of the power is relatively large.

Discharge parameters: C = 1mF; L = 20 uH; U0 = 8400V; D = 0.6 cm. The plasma channel resistance curve in the time domain is as shown in the figure 8.

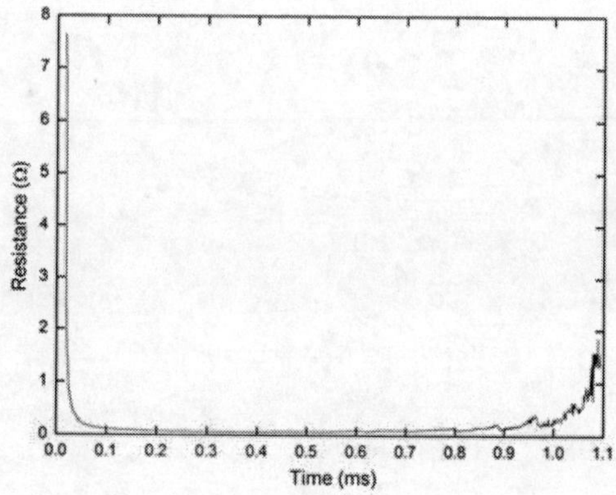

Fig. 8. The plasma channel resistance

V. Conclusion

In this paper, the discharge experimental platform is set up, and the working principle of pulse discharge source is introduced. In especial, the discharge characteristics of pulsed discharge source are analyzed in this paper. The time domain relations of the voltage, current, power and other parameters are obtained.

References

[1] Pei Yanliang, Wang Kuiyang, Liu Chenguang, et al. The Sparker Source Charging Technique[J]. Ocean Technology, 2007, 26(3): 73-76.

[2] Shi Xiaoxia, Gao Yinghui, Sun Yaohong, et al. Development of an underwater sparker with repetitive pulse[J]. Chinese Journal of Power Sources. 2013, 37(4): 646-649.

[3] Sun Yaohong, Zuo Gongning. Transferring spark source used in crosshole seismic[J]. Geophysical Prospecting For Petroleum. 2001, 40(3): 57-60.

[4] Fan Ailong, Sun Yaohong, Yan ping. Development of the metal mine detection technology[C]. Digest of Technical Papers-IEEE International Pulsed Power Conference. October 12, 2015:1-4

[5] Fan Ailong, Sun Yaohong, Yan ping. Experimental study of the underwater plasma based on the phased array vibrator system[C]. Digest of Technical Papers-IEEE International Pulsed Power Conference. October 12, 2015:1-4

[6] Jin Mingjian,Sun Yaohong. The electrical characteristics of underwater pulsed discharge under different experiment parameters[J] . High voltage engineering, 2004, 30(7):46-49

Pulse Generator for Beam Deflector at the SNS Beam Test Facility *

Vladimir V. Peplov, Baoxi Han, Robert B. Saethre, Martin P. Stockli, Robert F. Welton,
Oak Ridge National Laboratory (ORNL),
Spallation Neutron Source (SNS)
1 Bethel Valley Road, Oak Ridge, TN 37831 USA

Abstract— **A new Beam Test Facility (BTF) has been built at the Spallation Neutron Source (SNS). It contains an ion source, Low Energy Beam Transport (LEBT) system, and a new Radio Frequency Quadrupole (RFQ) cavity. To determine the transmission losses of the ion beam through the RFQ, beam currents at RFQ input and output have to be measured. Input current measurements can be performed with an electrostatic deflector within the LEBT to displace the beam to a current pick up sensor located radially outward from the beam axis. A new pulse generator has been designed and built at the SNS to produce up to 6 kV pulses, with a pulse width of 1 to 10 usec at a repetition frequency of 10 Hz into the capacitive load to create the appropriate deflecting electrical field. This paper will review the beam deflector system; describe the new pulse generator and its performance, and present preliminary results of the beam current measurements.**

Keywords—**beam current; pulse generator; deflector**

I. Introduction

The Spallation Neutron Source (SNS) Beam Test Facility (BTF), originally known as Integrated Test Facility (ITF), has been developed and constructed for testing a new spare Radio Frequency Quadrupole (RFQ) cavity which is planned to be installed in the SNS linac as a replacement for the existing RFQ which has suffered performance degradation over a decade of operation. The status of the ITF as of the end of 2015 has been presented in [1]. Later, the ITF was renamed to BTF to be used as a multipurpose test stand for ion beam studies, advanced instrumentation development, and equipment testing. To verify performance of the ion source and RFQ, ion beam current measurements have to be conducted under different operating conditions. An electrostatic beam deflector is used for RFQ input current measurements. This paper reviews BTF layout and describes the deflection system and its main component, a +6 kV pulse generator. Preliminary results of the beam current measurements are presented.

II. Beam Test Facility Configuration

A. *Ion Beam Linear Accelerator*

The configuration of the 2.5 MeV BTF accelerator is a duplicate of the existing low-energy front end part of the SNS linac and was presented in [1]. It consists of the H- Ion Source, Low Energy Beam Transport (LEBT), RFQ cavity, Medium Energy Beam Transport (MEBT) line and the 7.5 kW Beam Dump. The LEBT contains several electrostatic electrodes for extraction, focusing and steering of the ion beam before injection into the RFQ [2]. The MEBT line contains focusing magnets and beam diagnostic tools. A simplified block-diagram of the SNS BTF is shown in Fig. 1.

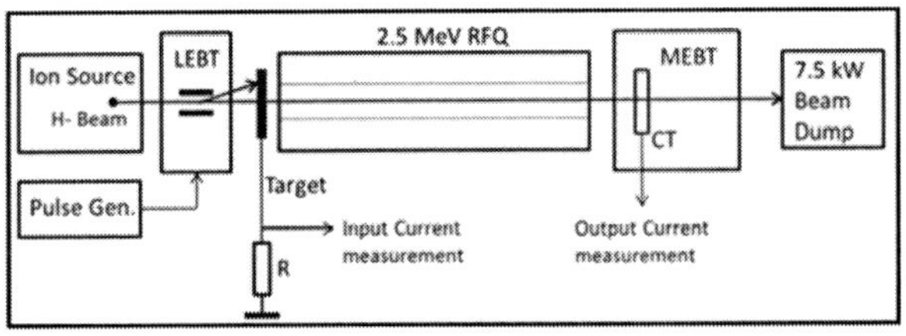

Fig. 1. Simplified block-diagram of the BTF

To evaluate beam transmission through the RFQ, measurements of two currents are required: one at the input and another at the output of the RFQ. The output current is measured by a current transformer (CT) which is located in the MEBT, but there is no space for CT in the LEBT section. Another simple and well-known method to measure beam current is to electrostatically deflect the charged particles to a metal diagnostic plate (target) which is grounded via a fixed-value resistor, and measure a voltage drop across this resistor. In contrast to the SNS front end linac, the BTF has no requirements for a beam chopping system, but the LEBT structure final lens contains four electrodes which serve to focus and align the beam [2]. These electrodes can be effectively used for beam deflection and current measurements.

B. *Beam Deflection System*

The deflecting electrical field is created by applying a high voltage (HV) pulse to one or two sections of the LEBT lens. The pulse generator has to provide fast full beam deflection out of the aperture to the diagnostic plate in order to measure the full beam current. A block-diagram of the deflecting system is presented in Fig. 2.

This manuscript has been authored by UT-Battelle, LLC under Contract No. DE-AC05-00OR22725 with the U.S. Department of Energy. The United States Government retains and the publisher, by accepting the article for publication, acknowledges that the United States Government retains a non-exclusive, paid-up, irrevocable, worldwide license to publish or reproduce the published form of this manuscript, or allow others to do so, for United States Government purposes. The Department of Energy will provide public access to these results of federally sponsored research in accordance with the DOE Public Access Plan (http://energy.gov/downloads/doe-public-access-plan).

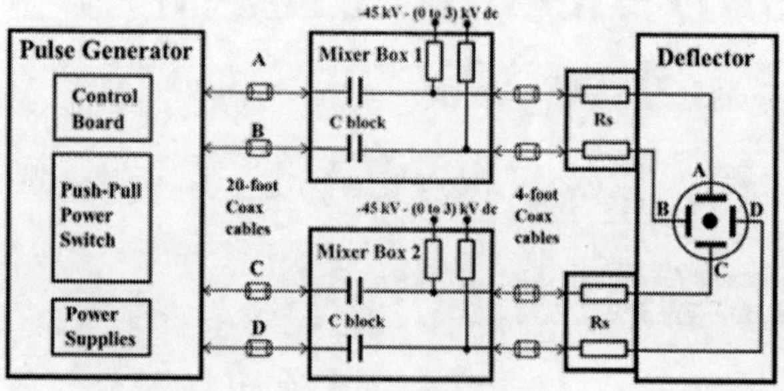

Fig. 2. Block-diagram of the LEBT deflection system

The four outputs of the HV pulse generator are connected to the LEBT electrodes via blocking capacitors installed inside two oil-filled mixer boxes. The mixer boxes are connected to the LEBT electrode penetrations through series resistors located in two side boxes. Electrode focusing and steering voltages of -45 kV and 0 to -3 kV, respectively, are supplied from separate power supplies via high-impedance series resistors located within the mixer boxes. Four 20-foot coaxial cables are used to connect the pulse generator and the mixer boxes and four 4-foot special HV coaxial cables connect outputs of the mixer boxes to the LEBT.

Fig. 3 shows transversal and longitudinal cross-sections of the last part of the LEBT. The direction of beam deflection can be selected by choosing which section or combinations of sections are connected to the pulse generator. In order to create a pulsed electrical field strong enough to fully deflect the beam onto the plate, and to avoid beam losses on the RFQ vanes during transients, two adjacent sections normally are used to kick the beam to the diagonal direction. For example, the HV pulse generator connected to the deflector's sections A and B is shown on the left. An arrow shows the beam deflection from the beamline axis to a point on the diagnostic plate in the upper left quadrant.

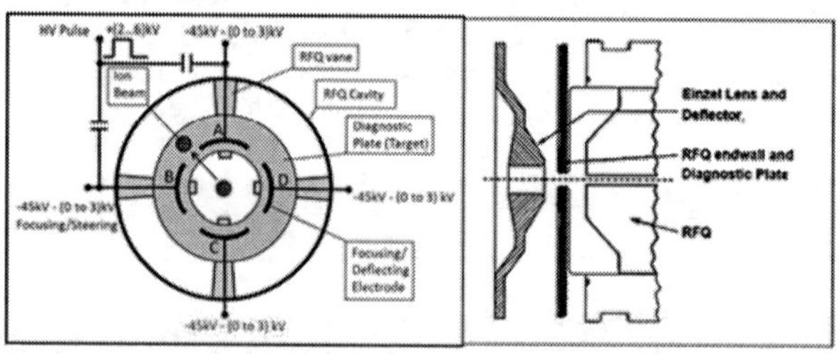

Fig. 3. Transversal and longitudinal cross-sections of the deflector

III. HV Pulse Generator

A. Requirements for the Pulse Generator

The pulse generator has to be well suited for driving the deflection plates for electrostatic modulation of the particle beams and has to produce fast, 6 kV high voltage square waveforms to high impedance low capacitance loads. To minimize damage to the plate, the pulse duration and pulse repetition frequency are limited. Table 1 shows requirements of the BTF deflector pulser.

TABLE I. Pulser Requirements

Input power	110 Vac
Operation mode	Pulsed
Trigger input	TTL pulse
Output voltage	Vout = 0 to +6.0 kV
Number of outputs	N=4, all synchronous
Load (each output)	Rs = 200 ohm in series with C = 100 pF
Pulse Repetition Frequency	PRF max = 10 Hz
Pulse Width	Tpulse = 1 : 10 μs
Rise/Fall time on the load	Trise/Tfall ≤ 250 ns

B. Design of the Pulse Generator

The pulse generator is an air-cooled solid-state half-bridge (totem pole) design, offering equally fast pulse rise and fall times and low power dissipation. The main components of the pulse generator are shown in Figure 4.

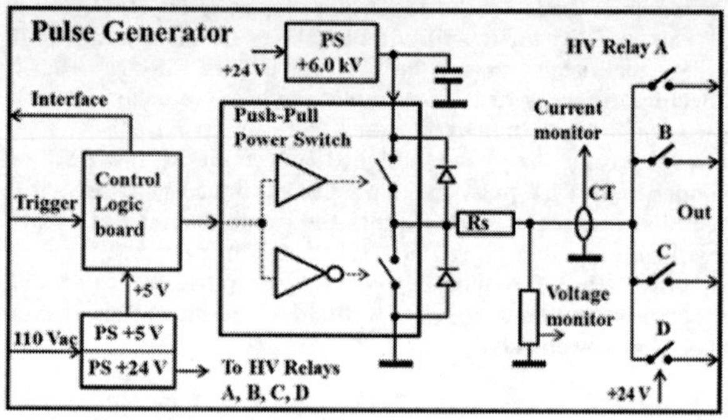

Fig. 4. Simplified block-diagram of the pulse generator

A single output +6.0 kV & 20 W adjustable dc-dc converter is used as a HV power supply to charge a storage capacitor. An 8 kV solid-state push-pull power module, manufactured by BEHLKE, with external antiparallel HV diodes in the high voltage and ground legs, serve as a main switch to charge/discharge the load capacitance. The output pulse amplitude is controlled by the output level of the dc power supply via a multi-turn potentiometer and a digital voltage monitor on the front panel. The output HV pulse is divided into 4 channels. Each output can be connected to its load separately through a 10 kV relay by choosing the appropriate switch on the pulser's front panel. A current transformer is installed at the BEHLKE module output for current monitoring and over-current control. The pulser contains a Control Logic board which was designed for a new version of SNS LEBT chopper and other pulsed power applications and which has been described in detail in [3]. This board accepts TTL electrical or fiber optic input trigger signals to control the HV output pulse width, time delay and pulse repetition frequency. The Control Logic board also provides several protection functions such as over-current, over-heating, pulse repetition frequency limit, and HV switch fault. The pulse generator features a front panel LED status indicator and voltage and current monitors which provide a straightforward means to view the output voltage and current waveforms in real-time, eliminating the need for an external high voltage oscilloscope probe. The pulse generator contains internal +5 V and +24 V ac-dc power supplies and requires only 110V ac for operation. A view of the unit with the upper lid opened is demonstrated in Fig. 5.

Fig. 5. View of the pulse generator

C. Performance of the Pulse Generator

The pulse generator has been tested on the test stand. Voltage and current waveforms were observed and rise/fall times and amplitude of the HV pulses were measured for different configurations: the pulser with no load and the pulser with single or multiple outputs connected to the load(s) via short (4-foot) or long (20-foot) HV coaxial cables. A series resistor Rs=200 ohm and a capacitor C=100 pF to ground have been used as an equivalent load for each channel. Pulse width and output voltage level were varied from 1 μs to dc and 200 to 6000 V.

Measurements with no load connected to the pulser outputs demonstrated that it can produce +6 kV pulses with pulse width from 150 ns up to infinite and switching rise and fall times of less than 25 ns. Any load connected to the pulser's output through a coaxial cable and/or loads connected in parallel add capacitance and thus increases rise/fall times of the output pulse. The rise/fall time increased to 50 ns with one and to 75 ns with two loads connected via 4-foot output cables. In the BTF the location of the pulse generator and two mixer boxes requires the use of 20-foot cables. Two examples of the output voltage and current waveforms are shown in Fig. 6. The upper scope snapshot presents the case when just one load is connected via a 20-foot coaxial cable of 30 ohm impedance, and the traces on the bottom shows waveforms when two outputs are connected. The black trace shows the voltage on the 100 pF capacitor measured by an external HV voltage scope probe, and the gray trace is the waveform measured by the internal voltage divider. For both cases, the set point was 2 kV with a 1μs pulse width. The rise/fall times of the HV pulse with one and two loads are approximately 110 ns and 250 ns, respectively. The distortions during transients on the gray traces are due to reflections from the mismatched impedances of the load and source.

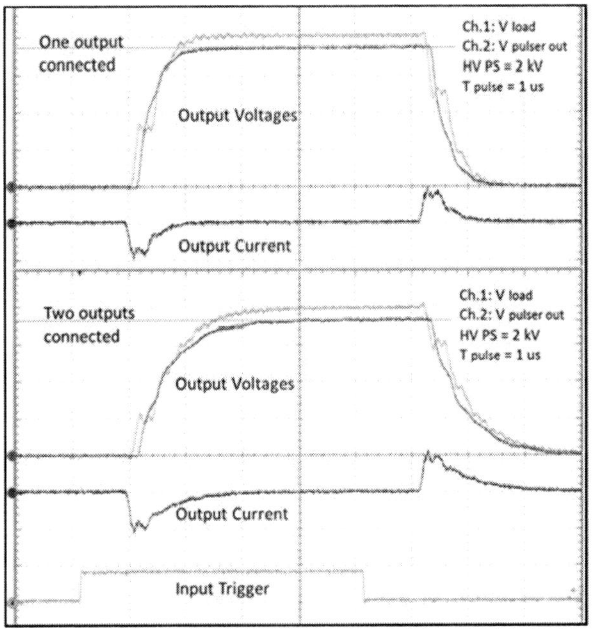

Fig. 6. Example of the voltage and current waveforms with one (top) and two (bottom) outputs connected to the load via 20-foot cables at 2kV & 1μs

IV. Ion Source Beam Current Measurements

In April 2016, the pulse generator was installed in the BTF and the first measurements of the beam current from the ion source at the LEBT output were performed by applying a combination of the dc steering and pulsed transverse voltages across segments of the second LEBT. For the test, the deflecting pulsed voltage applied to segments A and D was increased incrementally while holding the four steering voltages evenly set at -1.5 kV. At lower voltages only part of the beam reaches the diagnostic plate and the value of the measured beam current is lower. Fig. 7 shows the intercepted beam current versus the applied deflection voltage for these measurements. The maximum diagnostic plate current measured was 42 mA at 5.5 kV and did not increased at higher voltages indicating that the full beam is intercepted.

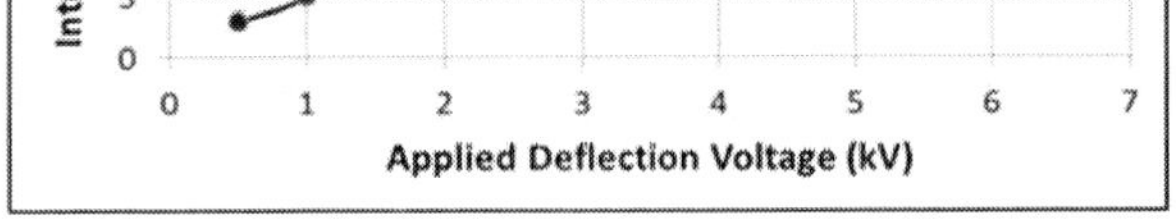

Fig. 7. Measured deflecting beam current vs. applied pulse deflection voltage

Fig. 8 shows an example of the waveforms for the pulse generator output voltage and the beam current on the diagnostic plate for a 6 kV pulse applied to two adjacent electrodes and a deflector pulse width of 10 μs.

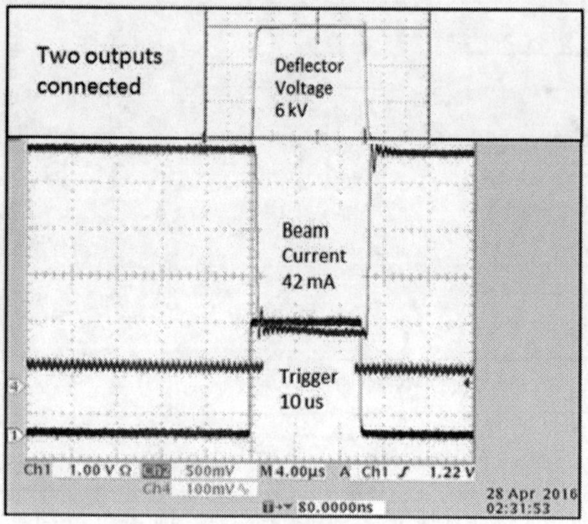

Fig. 8. Voltage and current monitor waveforms at 6kV &10μs

The length of the ion beam pulse from the ion source (macro-pulse) is approximately 1ms, while the pulse width of the deflector HV pulse normally used for current measurements is in the range of (5-10) μs. This permits sampling the ion beam current at different points in the macro-pulse by changing the delay time of the pulse generator trigger pulse relative to the start of the ion beam.

Recently, during summer 2016 studies at the BTF, beam currents of ~60 mA have been measured at the RFQ entrance. This was accomplished by tuning and adjustment of the ion source. This is within expectation for that particular ion source.

V. CONCLUSION

The ion beam study at the SNS Beam Test Facility is ongoing. A low-power high voltage pulse generator for the ion beam deflection system has been designed, built and tested. The generator produces rectangular positive 6 kV pulses of 1 to 10 μs pulse width at 10 Hz repetition rate into 4 capacitive electrostatic electrodes. The pulser is an easy-to-use instrument embedded into the LEBT system for the ion source current tuning and measurements, as well as for RFQ transmission evaluation. Its performance fully complies with the requirements.

Acknowledgment

This material is based upon work supported by the U.S. Department of Energy, Office of Science, Office of Basic Energy Sciences. The authors acknowledge the efforts of all team members participated in making the BTF operational. We would like to thank Ken Fowkes and Syd Murray Jr. for their support assembling the pulse generator and installing and testing it at the BTF.

References

[1] R.F. Welton, A.V. Aleksandrov, V.G. Dudnikov, B.X. Han, Y. Kang, S.N. Murray, T.R. Pennisi, C. Piller, M. Santana and M.P. Stockli1,"The status of the SNS external antenna ion source and spare RFQ test facility", Rev. Sci. Instrum. 87, 02B146 (2016).

[2] J. W. Staples, J. J. Ayers, D. W. Cheng, J. B. Greer, M. D. Hoff, A. Ratti, "The SNS four-phase LEBT chopper", Proceedings of the 1999 Particle Accelerator Conference, New York, 1999.

[3] V. V. Peplov, D. E. Anderson, R. B. Saethre "New pulsed power supply for the Spallation Neutron Source linac LEBT chopper system", Proceedings of the 2015 Pulsed Power Conference, Austin, TX, USA, 2015.

Research on control technology of multi-level battery cascade power supply for high-voltage large-current charging

Kun Liu[1, 2], Ailong Fan[1, 2], Rongyao Fu[1, 2]
1. Institute of Electrical Engineering
Chinese Academy of Sciences
Beijing, China
liukun@mail.iee.ac.cn
gyh@mail.iee.ac.cn

Yinghui Gao[1, 3], Yaohong Sun[1, 2, 3], Ping Yan[1, 2, 3]
2. University of Chinese Academy of Sciences
Beijing, China
3. Key Laboratory of Power Electronics and Electric Drive,
Chinese Academy of Sciences
Beijing, China

***Abstract*—As a high-voltage large-current rapid charging technology, the multi-level battery cascade charging technology has been gradually used in the charging of pulse power equipment, especially the large value capacitor. The control technology of the multi-level battery cascade charging power supply becomes the key to realize the charging effectively. This paper introduced a ten-level battery cascade power supply and researched the control technology, including the design of the optical fiber communication system and optical fiber drive circuit of cascade switches, the research of sequential control method, the design of voltage and current detection circuits and various faults synthetize circuits, the research of voltage feedback control and the realization of switchover charging method used in the multi-group loads. In the experiments of the ten-level battery cascade charging power supply, the average charging power is 92 kJ/s. The waveforms of the experiment results proved that the control system can work effectively.**

Keywords—battery cascade; time sequence control; various faults synthetic; switchover charging; voltage feedback

I. Introduction

As the development of pulse power technology, the energy that pulse power equipment needed increase rapidly, so that the pulse power source module become much larger[1]. The high-voltage large-current charging technology of the primary power supply become one of the research direction in the pulse power technology development[2]. Using series resonant constant current charging power supply (CCPS) is a common method because of its characteristics of simple structure, high output power, easily control and so on[3]. But the higher requirements such as faster charging[4] and larger power are difficult to satisfy by using the CCPS. Developing a new CCPS with higher power is not appropriate because of the limitation of volume, device and thermal management[5].

As a kind of large capacity energy storage device, the battery is used widely as the primary power supply of the pulse power equipment. On condition that there is an effective control technology to connect multi-level of batteries together and make them discharging as a proper sequence, the high-voltage large-current charging for the capacitance load will be realized. Figure 1 is a diagram of multi-level battery cascade charging power supply (BCCPS). There is a switch and a control circuit in every level of the cascade battery, which used to connect the batteries together or break off the cascade circuit. The control signals of the switches are sent out by the top control system which composed by the industrial personal computer (IPC) and digital signal processor (DSP) control circuit. There also is a protect circuit with current detection and faults synthesis functions used in every level. The inductance L is used for the charging current waveform adjustment and after the L is the multi-group of load. The load select circuit and voltage feedback circuit are used for the multi-group control. To divide the load into groups can reduce the peak value of the charging current so that to design the inductance L will be easier, the current stress will be decrease and the cost will be lower.

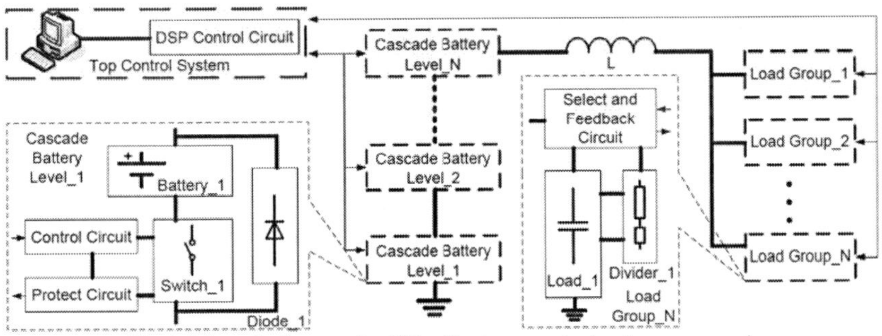

Fig. 1. Schematic diagram of multi-level battery cascade charging power supply

II. OPTICAL FIBER COMMUNICATION AND DRIVE CIRCUIT

The pulse power system is usually applied in the high-voltage large-current occasion, so that there is high electromagnetic interference (EMI) and the weak electric signals are easily to be interfered [6]. Using the optical fiber communication (OFC) system is an effective method to enhance the electromagnetic compatibility (EMC) of the control system of the BCCPS. The time sequence control of the switch device in the BCCPS is also realized by the OFC between the three of the IPC, the DSP control board and the control circuit of every level.

A. Optical fiber communication system

The OFC system include the communication among the IPC, the DSP control board, the control circuit of every level, the load select circuit and the voltage feedback circuit. The schematic diagram of the OFC system is as the figure 2 shows. The upper computer software can send the control signals by IPC with the 232 serial communication protocols and the optical module converts the electric signals to optical signals and transfers them by the optical fiber. The DSP control board receives the control signals and the relevant program will be executed to realize sorts of functions. The DSP also sends kinds of data and the status signals to the IPC by the optical fiber and the upper computer software interface will show them out. The time sequence drive signals and load select control signals that send by DSP control board are convert by HFBR1521 and transferred by glass optical fiber to the control board of every level of battery and every group of load select board. The fault signals of every level of battery and the voltage and current feedback signals are also converted and transferred to DSP by the OFC system.

B. Optical fiber drive circuit of cascade switches

Using the IGBT as the switch in the BCCPS system can connect the batteries together or break them off effectively because of its advantages [7]. The DSP control board sends out the drive pulse signals of the IGBT and the control board receives and converts then to electric signals by HFBR2521. Because the output of HFBR2521 is a negative pulse signal, the optical coupler is used to convert the electric level. The protect circuit will send a level signal which obtained by synthesize the faults signals, then the pulse signal and the level signal will send to an AND gate in the control circuit so that to give out the ENA signal to enable the drive circuit of the IGBT. By this way the IGBT can be controlled to switch off when there are some faults in the BCCPS circuit. Figure 3 is the diagram of the drive pulse transfer and convert progress.

C. Time sequence control strategy

The time when the battery should be connected to the main circuit and the sequence of the multi-level batteries to connect is the key to realize the cascade charging in the BCCPS system. An upper computer software [8] control interface is designed to set up the time sequence of the batteries. Figure 4 is part of the screen capture of the interface in which the sequence can be set up by fill in the text box with the number of the batteries. The time interval between two adjacent batteries can be set up by fill in the text box with an appropriate value. The time sequence control strategy can balance the total working time of the batteries so that they can work as a relatively uniform state. At the same time, the output current and voltage can be regulated by setting different time interval when the parameter of the load or the inductance L changed and the charging time can also be controlled. The time sequence value can be send to DSP by press the Set Up button and the DSP will execute the program to drive the IGBT as the time sequence so that the multi-level battery cascade charging can be realized.

III. CURRENT DETECTION AND VARIOUS FAULTS SYNTHETIC

Although the discharging current of the battery is relatively larger, the large current discharge instantaneously may damage the core of the battery. So it is necessary to detect the currents of the battery so that to ensure the battery can work normally. On the other hand, the IGBT and the Diode are also prone to failure when there is a large current flow through, so that the fault protect circuit is necessary in the BCCPS system. Figure 5 is the diagram of current detection

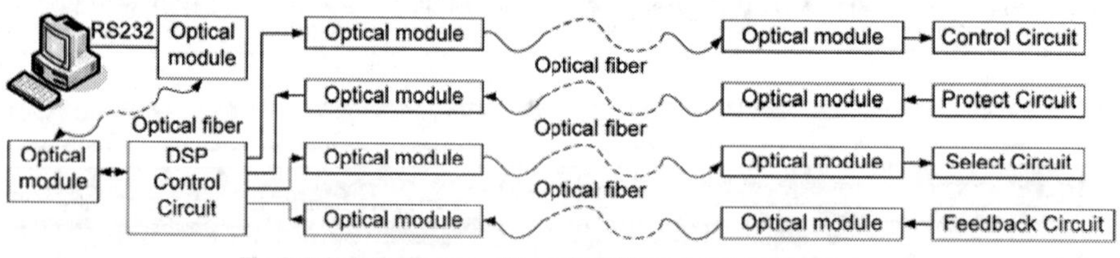

Fig. 2. Schematic diagram of the optical fiber communication system

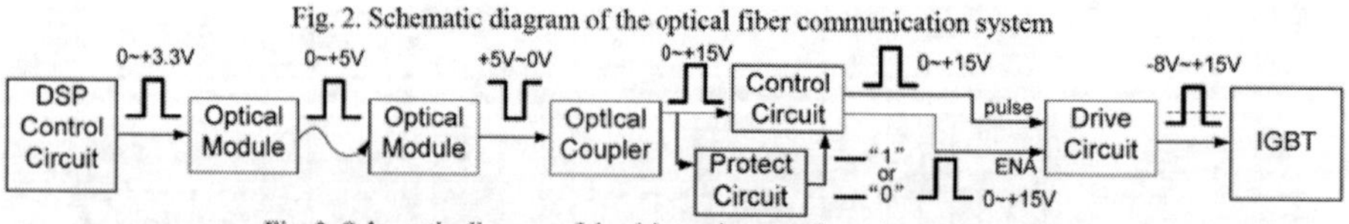

Fig. 3. Schematic diagram of the drive pulse transfer and convert progress

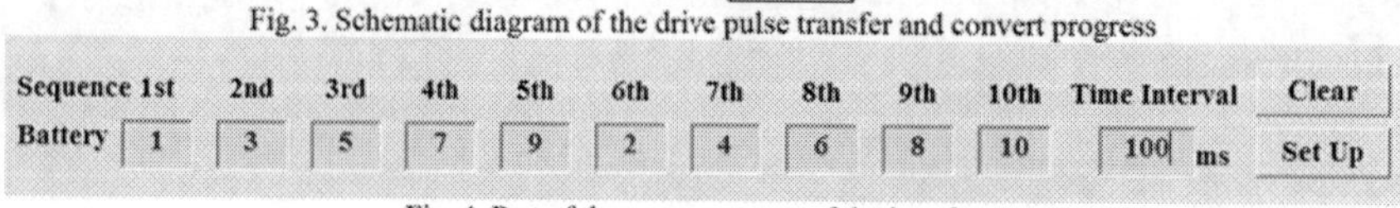

Fig. 4. Part of the screen capture of the interface

and fault protect circuit.

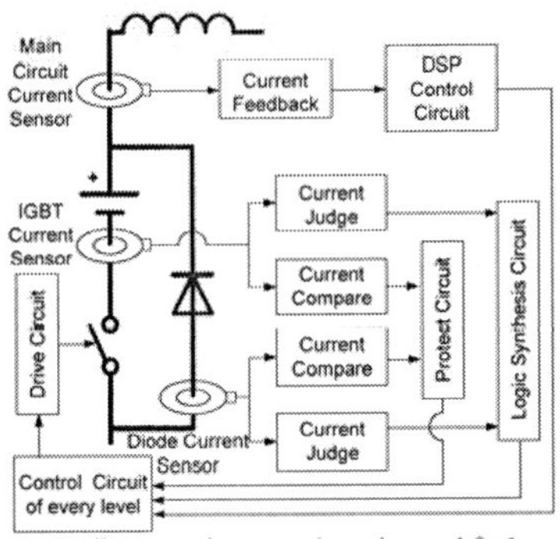

Fig. 5. Schematic diagram of current detection and fault protect circuit

A. Current detection of switch devices and diodes

In the control system of the BCCPS system, the Hall current sensor is used to detect the current which flow through the IGBT, the Diode and the main circuit. The IGBT and Diode currents of every level are detected so that the over current protection of every level can be realized. The current compare circuit is used to compare the maximum current value which has been preset and the detected current value and send out a fault pulse when the detected current is larger than the pre-set value. The detected current of main circuit will be feedback to the DSP and the waveform will be shown on the upper computer control interface.

B. Logical relationship of drive signal and current signal

In the control strategy of BCCPS, the drive signal of the IGBT decides if there is current flow through the IGBT theoretically. Define the following parameters: the drive signal of IGBT is I_F; the current which flows through the IGBT is I_G; the current flows through the diode is I_D and the "1" represents the current flows through the switch and "0" represents there is no current. A table can be obtained which describe the logical relationship of the signals and the faults, just like table 1 shows. By using logic gates device to build protect circuit is an effective method to judge the faults as table 1 shows and ensure the BCCPS to avoid the further enlargement of faults when there is some logical fault before the battery is connect into the circuit.

TABLE I. THE LOGICAL RELATIONSHIP OF THE SIGNALS

Switching Value			Fault or not	Fault Type
I_F	I_G	I_D		
0	0	0	N	/
0	1	0	Y	IGBT Short
0	0	1	N	/
0	1	1	Y	IGBT and Diode Short
1	0	0	Y	IGBT and Diode Open
1	1	0	N	/
1	0	1	Y	IGBT Open
1	1	1	Y	Battery Short

C. Various faults synthetic and protect circuit

The various faults such as the over current fault of IGBT, over current fault of diode, over voltage fault (OV), over heat fault (OH) and logic fault should be synthesized and feedback to the DSP control board. Figure 6 is the various faults synthesize circuit, which contains the logic gates protect circuit that introduced above. The various faults signals are synthesized by the OR gate then the fault signal is sent out. The fault signal is reversed by the NOT gate and synthesized with the pulse signal by the AND gate then the enable signal of the drive circuit is sent out.

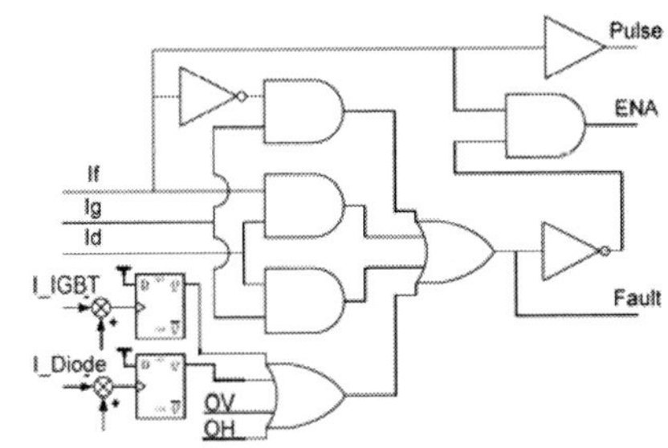

Fig. 6. Various faults synthesize circuit

IV. LOADS SWITCHOVER AND VOLTAGE FEEDBACK

The number of power supply module is increasing in the pulse power system, so that the charging power supply should be expansibility. The multi-level BCCPS has to charging for the load with certain value because of the limitation of the inductance and the output current of the battery so that the loads should be divided into different groups and the grouping control is necessary in the control technology of the BCCPS system. The voltage feedback control is also necessary to realize the accurate charging voltage control of every group of loads. Figure 7 is the diagram of loads switchover and voltage feedback control circuit.

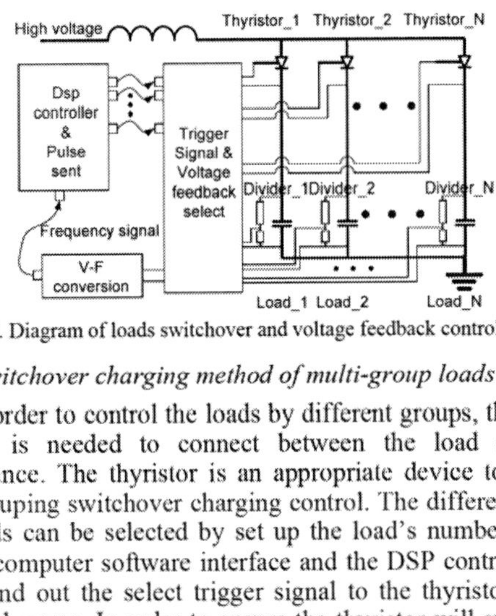

Fig. 7. Diagram of loads switchover and voltage feedback control circuit

A. Switchover charging method of multi-group loads

In order to control the loads by different groups, the select switch is needed to connect between the load and the inductance. The thyristor is an appropriate device to realize the grouping switchover charging control. The different group of loads can be selected by set up the load's number on the upper computer software interface and the DSP control board will send out the select trigger signal to the thyristor of the selected group. In order to ensure the thyristor will switch on when every level of battery is connected into the circuit, the trigger signal should be sent out when there is any pulse to drive the IGBT of every level. Figure 8 is the waveforms of the control signals. In the figure 8 (a), one group of load is select and ten level of battery is cascade so that there are ten trigger pulses as the CH4 shows. Figure 8 (b) is the waveforms of control signals when there are two level of battery charging for two groups of loads, in which it is easy to see that the trigger pulse will a little earlier than the drive signal so that the thyristor will be sure to switch on when the battery is connected in the circuit. Figure 8 (c) is the waveforms when there are ten level of battery and eight groups of loads, in which the pulsed beams of CH2 and CH4 is the trigger pulses of the first and the eighth groups of loads, and the CH1 and CH3 are the drive signal of the IGBT of the first and the tenth level.

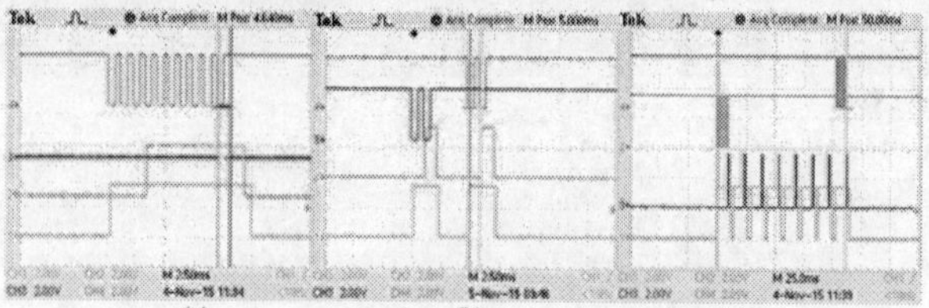

(a) (b) (c)

Fig. 8. Waveforms of the control signals

B. Voltage feedback control

The load voltage is sampled by a resistance voltage divider with a ratio of 1:1000 and the feedback voltage is in the range of 0 - 10V. This voltage is filtered and processed through the voltage-frequency converter and the frequency signal is converted to an optical signal and transferred by the optical fiber. The DSP control board receives the signal and converts it to a digital signal so that the voltage can be calculated and the charging process can be controlled[9]. As the limitation of the timer and counter of the DSP, the feedback voltage signals of the dividers should be also selected by the trigger signal of the thyristor of every group, so that the charging voltage will be controlled accurately.

V. Experiment Results and Conclusion

Ten levels of batteries that each voltage is 50 V are used in the cascade charging experiment to verify the control strategy of the BCCPS. Figure 9 is the experiment waveforms. As figure 9 (a) shows, ten batteries are connected to charge one group of load. CH2 is the waveform of the charging current; CH4 is the waveform of charging voltage which rises linearly; CH3 is the trigger pulse of the selected load which make sure the thyristor is switch on. Figure 9 (b) is the waveform of charge current and load voltage when use ten levels of batteries to charge for eight groups of loads. The current waveforms are almost the same of every group of load.

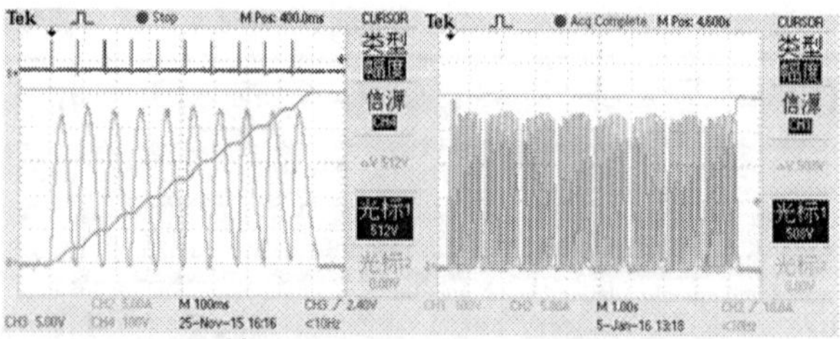

(a) (b)

Fig. 9. Experiment waveforms of 10 level of battery cascade

The high-voltage large-current charging experiment is carried out by using the upper computer software interface and ten levels of batteries with each voltage is 350V. Figure 10 is the screen capture of the interface which shows that the charging voltage of every group of load reached 3 kV and the peak value of the charging current is about 140 A. The value of each capacitance is 22.5mF and the total charging time is 8.8s. The average charging power can be calculated by using the formula (1), and the result is 92kJ/s. All the results of the experiments proved that the control strategy is effective and the BCCPS can realize the high-voltage large-current charging.

$$P=\frac{1}{2}CU^2/t \tag{1}$$

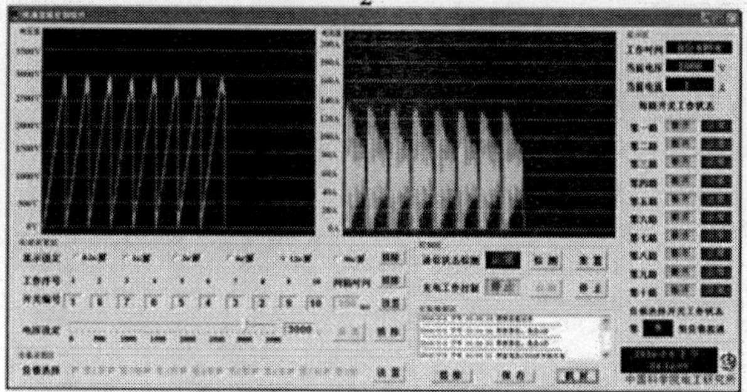

Fig. 10. Screen capture of the interface

Acknowledgment

This work was supported in part by National Natural Science Foundation of China (51307161, 51577178, 51577177.)

References

[1] SH Kim, M Ehsani, YH Kim. "Leg-Rotation Control Method for Capacitor Charging Power Supply used in Pulsed Power System". IEEE Transactions on Dielectrics & Electrical Insulation. 2013, 20(4):1012-1019

[2] Y Zhang, J Wu, Z Li, Y Jin. "Design of compact high-voltage capacitor charging power supply for pulsed power application". IEEE Pulsed Power Conference. 2015:1-5.

[3] Gao Ying-hui, Shi Xiao-xia, Yan Ping. "High power density capacitor charging power supply". High Power Laser and Particle Beams, 2012, 24(4): 943-948.

[4] Liu Kun, Fu Rongyao, Gao Yinghui, Sun Yaohong, Yan Ping. "High-voltage repetition-frequency charging power supply for pulsed laser". IEEE Transactions on Plasma Science, Vol. 43, No. 5. May 1, 2015. pp. 1387-1392

[5] Gao Ying-hui, Sun Yao-hong, Yan Ping, et al. "40kW digital control capacitor charging power supply". High Power Laser and Particle Beams, 2009, 21(8): 1259-1262.

[6] Liu Kun, Gao Yinghui, Wang Jiong, Sun Yaohong, et al. "Control system based on optical fiber communications for the high-frequency high-voltage charging power supplies' parallel operation". EML 2014, 17th International Symposium on Electromagnetic Launch Technology, October 10, 2014. pp. 1-4

[7] SL Holt, CF Lynn, JM Parson, JC Dickens. "Burst Mode Operation of a High Peak Power High Pulse Repetition Rate Capacitor Charging Power Supply". Pulsed Power Conference, 2015.

[8] Kun Liu, Dongdong Zhang, Yinghui Gao, Yaohong Sun, Ping Yan. "Remote control system of nanosecond pulse power supply". Proceedings of the 2014 IEEE International Power Modulator and High Voltage Conference, IPMHVC 2014, 344-347

[9] Kun Liu, Yaohong Sun, Yinghui Gao, et al. High-voltage high-frequency charging power supply based on voltage feedback and phase-shi control. Electromagnetic Launch Technology, 2012 16th Symposium, 2012, Page(s): 1–5.

Study on pulse power supply technology for electrostatic precipitator

Yinghui Gao[1,2], Kun Liu[1,2], Jing Han[1,2]
1 Institute of Electrical Engineering Chinese Academy of Sciences, Beijing, China
gyh@mail.iee.ac.cn

2 Key Laboratory of Power Electronics and Electric Drive, Chinese Academy of Sciences
Yaohong Sun, Ping Yan, Kun Liu
University of Chinese Academy of Sciences, Beijing, China

***Abstract*—The electrostatic precipitator is effective technology in control of boiler flue gas emission concentration. With improvement of boiler flue gas discharge standards, conventional DC electrostatic dust removal technology is difficult to reach the high discharge standard of boiler flue gas. Using DC superimposed pulse power supply for electrostatic precipitator can effectively solve this question. But all the time pulse excitation electrostatic precipitator power supply was limited by pulse power technology development. The paper discusses the electrostatic precipitator power development process, analyze the advantages and disadvantages of three typical pulse power supply topology structure. The pulse excitation electrostatic precipitator using high-voltage switch was studied and its design ideas were put forward. The design was verified by simulation. At last, 10kV DC superposition 10kV pulse power system was established. The experimental results verify the feasibility of the design.**

Keywords*—electrostatic precipitator;pulse power supply; high voltage switch; dust removal efficiency

I. Research Background

Air pollution is a global concerned question, electrostatic precipitator (ESP) technology has become the main means to control the boiler flue gas emissions of the world owing to its high efficiency, gas processing capacity, suitable for high temperature dust, small wind resistance and low operation cost. In Europe, the United States, Japan and other developed countries and regions, the application ratio of ESP is more than 80%. It's also the main dust removal equipment of coal-fired power plants, steel mills, cement plants and other enterprises in China.

From the point of view of atmospheric pressure non-equilibrium plasma physics, electrostatic precipitator is a combination of a huge plasma source and reaction chamber, so the electrostatic precipitator technology involves three main research contents of electrostatic discharge mechanism, electrostatic precipitator power supply and electrostatic precipitator mechanical structure. Electrostatic precipitator power supply is a key technology. Its performance directly affects the effect of dust removal.

II. Development of Electrostatic Precipitator Power Supply

Electrostatic precipitation technology has been originated in the beginning of nineteenth Century. In 1824, the German mathematics teacher named M.Hoheled proved that the electric spark can make the bottle of smoke dissipated and proposed the concept of electric dust removal for the first time. In 1907, A professor of university of California developed the electric dust removal devices for industrial use, and then the dust removal devices gradually used in various industrial fields. Up to now, there are mainly three types of electrostatic precipitator power supply: power frequency electrostatic precipitator power supply, high frequency electrostatic precipitator power supply, pulse excitation electrostatic precipitator power supply.

A. Power frequency electrostatic precipitator power supply

Power frequency electrostatic precipitator power supply is the initial power supply scheme which used in industrial fields. At the beginning of twentieth Century, the mechanical synchronous rectification was used. In 1950s, with the successful development of silicon controlled devices, thyristor controlled high voltage rectifier was used. The topology of single-phase power frequency power supply for electrostatic precipitator is showed in Figure 1. Bigger power fluctuation, longer control delay time and anti corona are the fatal flaws of power frequency ESP power supply. Although the later three-phase rectifier reduces the voltage ripple, but because the three-phase power frequency ESP power supply adopts SCR phase shift control, control response time hasn't be changed in essential. The topology of three phase power frequency ESP power supply is shown in figure 2. Subject to the development of the switch technology, until 1980s, the power frequency high voltage rectifier electrostatic precipitator power supplies were still be mainly used in industrial fields.

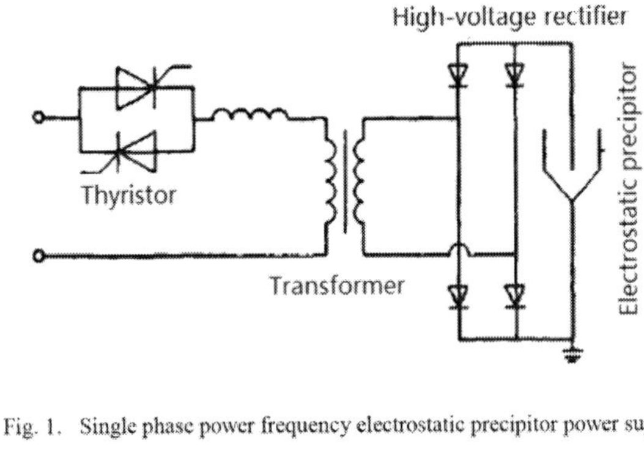

Fig. 1. Single phase power frequency electrostatic precipitor power supply

This work was supported by Chinese National Natural Science Foundation of China (51307161).

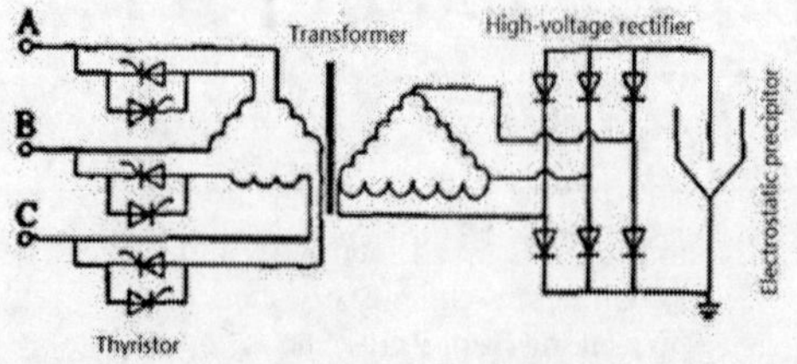

Fig. 2. Three phase power frequency electrostatic precipitor power supply

B. High frequency electrostatic precipitator

At the beginning of 1980s, with the process of high frequency power electronic switching devices, such as high frequency thyristor and transistors, high frequency ESP power supply appears. Based on the maturity of high frequency IGBT technology, the high frequency ESP power supply has been developed rapidly in 1990s. The products of Swedish ALSTOM and Danish SMITH company was its typical representative. The topology of high frequency ESP power supply is showed in figure 3. Due to the high frequency control, the dynamic response speed of the system can be improved rapidly and the ripple of the output voltage can be controlled very small which greatly improved the efficiency of dust removal. On the other hands, high frequency technology has significantly reduced the size of the transformer and hardware costs. At present, high frequency and high voltage rectification ESP power supply have been used extensively in Europe, Japan, the United States and other developed countries, and it is developing towards higher power and higher operating voltage (100kV). However, due to the large number of devices used in high frequency ESP power supply, and its structure is complex, the failure rate of the high frequency ESP power supply is significantly higher than the power frequency electrostatic precipitator.

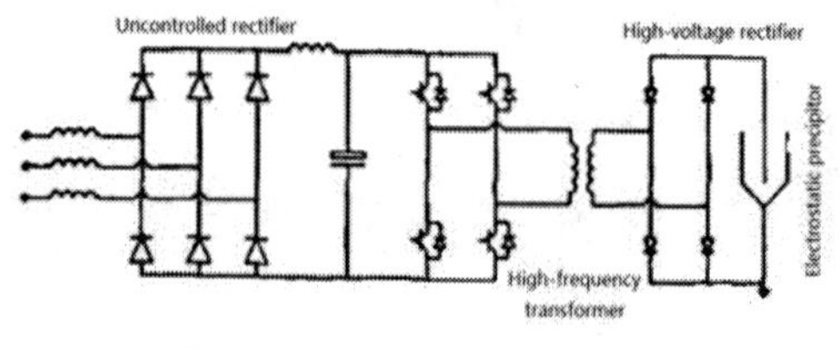

Fig. 3. High-frequency electrostatic precipitor power supply

C. Pulse excitation electrostatic precipitor power supply

Both power frequency and high frequency ESP power supply, their output voltage is DC voltage. The DC electric field is responsible for the generation of electronics, and it is also responsible for dust particle charging and directional movement. With certain electric fields structure, these three functions interact with each other. The output voltage and output current of the power supply can not be adjusted independently. When the spark frequency is high or the counter corona occurs, the power supply voltage will be reduced, so the output current of the power supply is reduced, and then the dust removal efficiency descended. In the process of high resistivity dust, anti corona phenomenon becomes more and more frequent, the reduce of DC voltage frequently will lead to the direct discharge of dust and dust removal function failure. So from 1930s researchers launched a study of pulse excitation ESP power supply.

In 1931, Heinrich R. applied for a US patent about an ESP power supply scheme by adding pulse voltage to the DC voltage, which is used to improve the efficiency of dust removal. In 1952, H. J. WHITE developed a pulsed excitation ESP power supply through a rotary spark gap switch, pulse transformer and blocking diode. For the first time, pulse field test excitation type ESP power supply was carried out a fields test. After a year of trial operation, the data showed that the corona current increases from 20 to 35%, which means that the higher dust removal efficiency. In the following years, many countries have carried out the research on the pulse excitation ESP power supply. Although limiting to the level of switching technology, there was no industrial power products available, but the power supply topology gradually clear. In 1982, Denmark F.L.Smidth mentioned two kinds of topology of pulse excitation ESP power supply in its patent. One was composed of a low voltage switch and a pulse transformer, which is showed in Figure 4. Another was realized by using a high-voltage switch, which is showed in figure 5. In the same year, H. I. Milde who worked in the Ion Physics Department of United States High Voltage Engineering Corporation confirmed that two proposed topology of Denmark F.L.Smidth in its patent can be achieve commercialization, and He pointed out that the biggest difference between the two topologies were the differences of the pulse width and pulse rise time. The pulse width of the pulse transformer topology is generally 100 microseconds and the high voltage switch mode can realize the output pulse width of 1 us. In 1980s, the power supply of pulse transformer was first used in industrial application and was developed with commercial operation. Until now it is the typical representative of Smidth company in Denmark. Their power supply products of 50 ~ 200us pulse width and 25 ~ 400Hz repetition frequency passed the industrial test. In the test high resistivity dust particle migration velocity increased significantly. The size of dust collector is inversely proportional to the migration velocity of charged particles, so higher migration velocity of charged particles can reduce the size of dust collector area. At present, the Danish F.L.Smidth company is still one of the main manufacturers of pulse excitation ESP power supply.

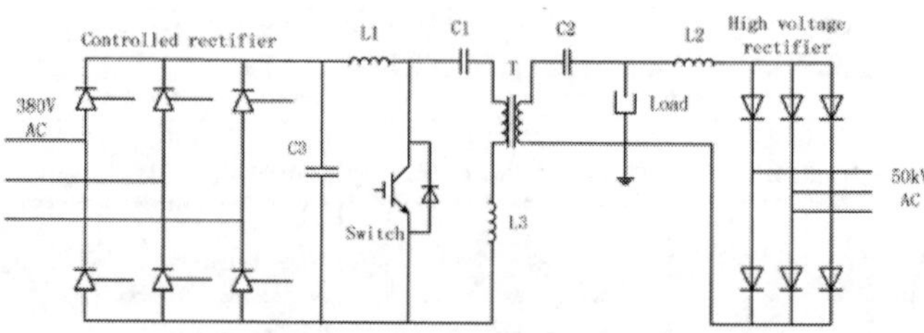

Fig. 4. Topology of pulse excitation electrostatic precipitor power supply using high-voltage transformer

Research progress in the field of electrostatic discharge shows that flow injection breakdown time in ESP is about 10 ~ 20us range. So in order to avoid the occurrence of streamer breakdown, the output pulse width of pulsed excitation ESP power supply should be less than 10us. The topology of high voltage switch mode can output narrower pulse width which makes the advantage of pulse excitation ESP power supply more prominent. But in 1980s, there was only spark gap mechanical switch and non contact ion devices such as mercury thyratron which can meet the requirements of pulse excitation ESP power supply. That seriously restricted the industrial application of pulse excitation ESP power supply.

In 1981 J.F. Shoup and T. Lugar reported they developed a high voltage switch using thyristor. In 1986, a pulse excitation ESP power supply using high voltage thyristor was developed in Italian. In 1999, Posco Co Ltd in South Korea reported that a pulsed excitation ESP power supply using thyristor series switch was developed. Its pulse width is 140us, its repetition frequency is 200Hz and its output voltage is 70kV. At present, The Posco Co Ltd is a typical representative of the pulse excitation ESP power supply using high voltage switch mode. But the output pulse width of their product is closed to the products of Simdth company, it is about 150us.

In order to achieve a better effect of dust collection, at present, researchers study on the power supply of narrower pulse width. In 1997, Heinemnn L. used the magnetic switch to develop the pulse excitatory ESP power supply. Its minimum output pulse width was 10us. In 2002, German company SIEMENS reported that the develop a pulsed excitatory ESP power supply using pseudospark switch. Its output pulse width was 6 ~ 10us, pulse amplitude is 55kV, and pulse peak current is up to 2000A. The dust removal effect contrast test was carried out in a sintering plant between their pulse excitatory power supply, power frequency and high frequency DC ESP power supply. With the power frequency DC power supply the best emission value is 90mg/m^3, with the high frequency DC power supply, emissions is 76mg/m^3, and emissions of pulse excitation power supply is 59mg/m^3. Even compared with high frequency DC power supply, its emission was reduced by 22%. It can be seen that pulse excitation ESP power supply have higher dust removal efficiency, and the use of switching mode pulse excitation ESP power supply can be better to achieve a narrow pulse width, which can strengthen the advantages of pulse excitation ESP power supply.

III. SYSTEMS ANALYSIS AND EXPERIMENTAL VERIFICATION

A. Simulation of pulse excitation electrostatic precipitor power supply using high-voltage switch

Pulse excitation ESP power supply using high-voltage switch not only can achieve a narrow pulse width output, but also easy to achieve the pulse width adjustment, more easy to achieve industrial applications. In this paper, the saber software is used to analyze the system parameters of the power supply. Figure 5 is a simulation circuit. C1 and R3 is the equivalent load, C2 and R6 is the DC filter parameters, L5 and R5 is pulse isolating inductance parameters, L1 is the wave modulation inductance, SW_VCSP is the high voltage switch, C5 and R7 is absorption circuit parameters of the high-voltage switch, C3 is pulse discharge capacitors, L3 and R2 is direct current isolating inductance parameters, C4 and R4 is DC filter parameters. It can be seen from the figure, in the topology, high-voltage switch is the core and key device, system parameters will directly affect the voltage across the switch. Figure 7 (a), (b), (c), (d) are simulated and analyzed the effect of C1, L5, C3, C5 on the voltage of both ends of the switch.

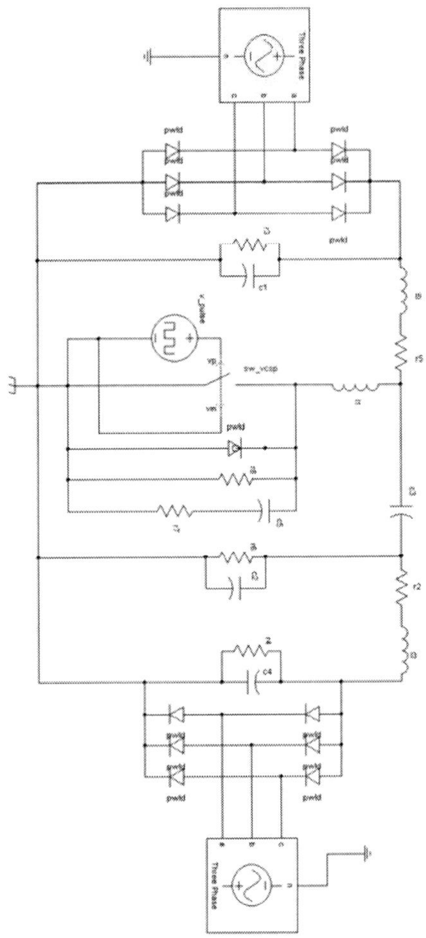

Fig. 5. topology of pulse excitation electrostatic precipitor power supply using high-voltage switch

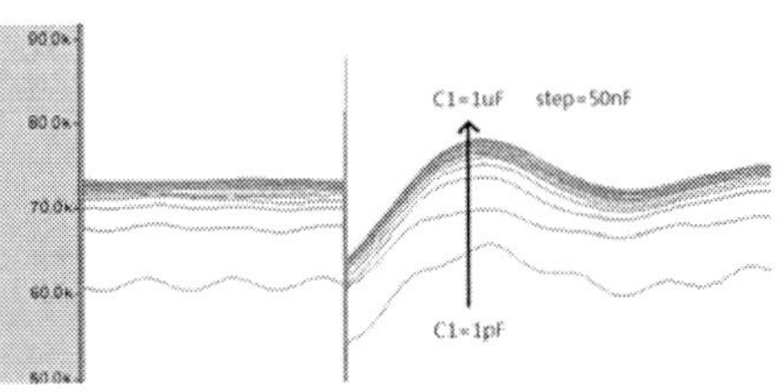

(a) The change trend of voltage at both ends of the switch with C1

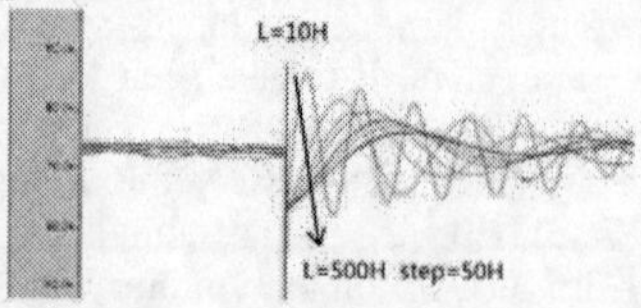

(b) The change trend of voltage at both ends of the switch with L5

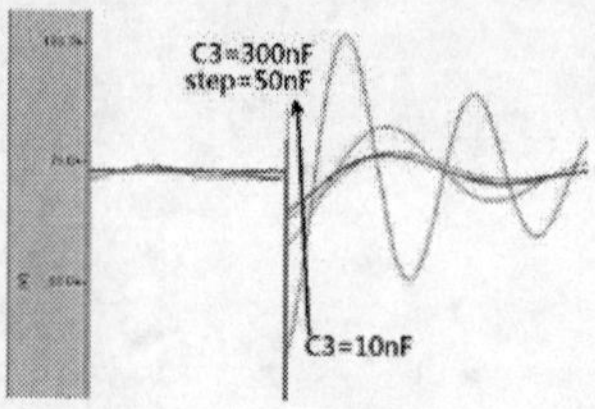

(c) The change trend of voltage at both ends of the switch with L5

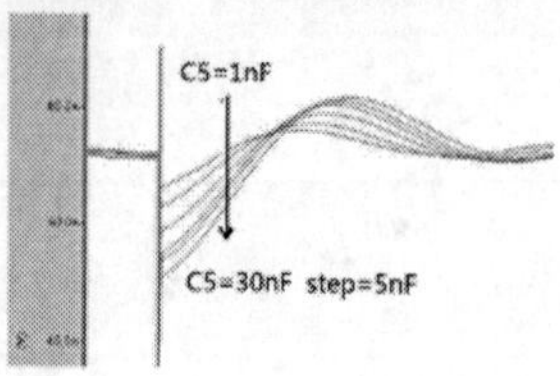

(d) The change trend of voltage at both ends of the switch with C5

Fig. 6. The influence of power parammters on switch

The simulation shows that the oscillation voltage amplitude on both ends of the switch increase with the increase of C1 and C5, and it decreased with increase of L5 and C3. With different system parameters, the oscillation voltage on both ends of the switch has obviously difference. So it also shows that in different power and output characteristic power supply, the design of high-voltage switch should be different.

B. Experimental verification

According to the simulation analysis, the experimental system of 10kV DC superimposed 10kV pulse is established, and the switching voltage, switch current waveform and the load voltage waveform are shown in Figure 7. The peak voltage and oscillation voltage of the switch are well suppressed.

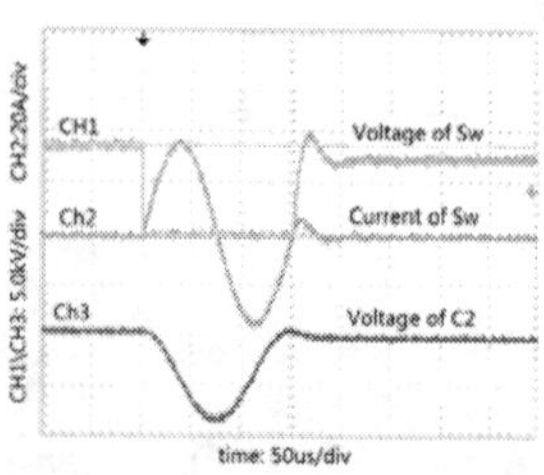

Fig. 7. Current and voltage waveforms of high-voltage switch and voltage waveforms of load

IV. Conclusion

In the pulse excitation ESP power supply adopting high voltage switch, high voltage switch is one of the key technologies, but the system parameters of the power supply may play a crucial role. With the optimization of system parameters, the oscillation voltage at both ends of the switch can be reduced significantly, which can reduce the switch rated voltage and development difficulty of the high voltage switch.

References

[1] White, H. J. Electrostatic precipitator pulse power supply[J]. Electrical Engineering, 1953, 72(3):236-236.

[2] G. W. Penney, P. C. Gelfand, The Trielectrode Electrostatic Precipitator for Collecting High Resistivity Dust, JAPCA, Jan. 1978, Vol. 28,53-55.

[3] S. Masuda, A. Mizuna, H. Nakatani, H. Kawahara, Application of Boxer Charger in Pulsed Electrostatic Precipitators, Conf. record of IEEE/1AS Annual Mtg. 1980, 904-911.

[4] L. Heinemnn, K. Porle,P. Ranstad, Design of a Low Cost Pulsed Power Supply for Electrostatic Precipitators Using Magnetic Switching Technique[C]: ICESP, Budapest, 1996

[5] Heinemann, L.; Porle, K.; Ranstad, P. Design of a high power pulse voltage generator forelectrostatic precipitators using magnetic switching technique[C]. Applied Power Electronics Conference and Exposition, 1997,2:948-952.

[6] Grass, N.; Hartmann, W.; Romheld, M. Microsecond pulsed power supply for electrostaticprecipitators[C]. Industry Applications Conference, 2001,4:2520-2524.

[7] A. C. Mermigkas, I. V. Timoshkin, S. J. MacGregor, et al. Sub-microsecond impulsive corona discharges for electrostatic precipitation applications. IEEE Pulsed Power Conference, 2011: 1194 – 1198.

[8] I. V. Timoshkin, A. C. Mermigkas, S. J. MacGregor, et al. Analysis of particle charging mechanism for optimisation of precipitation efficiency. IEEE Power Modulator and High Voltage Conference, 2012: 114 – 117.

[9] A. C. Mermigkas, I. V. Timoshkin, S. J. MacGregor, et al. Superposition of DC Voltage and Submicrosecond Impulses for Energization of Electrostatic Precipitators. IEEE Plasma Science, 2012, 40 (10): 2388 - 2394.

[10] A. C. Mermigkas, I. V. Timoshkin, S. J. MacGregor, et al. Micro-Electrostatic Precipitation Process for Removal of Fine Particles from Air. 4th Euro-Asian Pulsed Power Conference - 19th International Conference on High-Power Particle Beams, 2012.

[11] C. Mermigkas, I. V. Timoshkin, S. J. MacGregor, et al. Removal of Fine and Ultrafine Particles From Air by Microelectrostatic Precipitation. IEEE Transactions on Plasma Science, 2013, 41 (10): 2842 – 2850.

[12] A. C. Mermigkas, I. V. Timoshkin, S. J. MacGregor, et al. Impulsive Corona Discharges for Fine Particles Precipitation in a Coaxial Topology. IEEE Transactions on Plasma Science, 2014, 42 (10): 3089 – 3094.

[13] S. Masuda, H. Nakao, Control of NOx by positive and negative pulsed corona discharges,Industry Applications, IEEE Transactions on, vol.26,pp. 374-383,1990.

[14] S. Masuda, Pulse corona induced plasma chemical process: a horizon of new plasma chemical technologies, Pure Appl. Chem, vol.60, pp. 727-731,1988.

[15] J. S. Clements, A. Mizuno, W. C. Finney, R. H. Davis, Combined removal of SO2, NOx, and fly ash from simulated flue gas using pulsed streamer corona, Industry Applications, IEEE Transactions on, vol.25,pp. 62-69, 1989.

[16] G. Dinelli, L. Civitano, M. Rea, Industrial experiments on pulse corona simultaneous removal of NOx and SO2 from flue gas, in Proc, 1988 Industry Applications Society Annual Meeting, 1988.,Conference Record of the 1988 IEEE, pp. 1620-1627,

Transient Analysis Modeling of Modular Marx Multilevel Converters

L. Lamy Rocha
Instituto Superior de Engenharia de Lisboa, (GIAAPP/ISEL), INESC-ID, Portugal
luisrocha@deea.isel.pt

J. Fernando Silva
Instituto Superior Técnico - Universidade de Lisboa, INESC-ID, Portugal
fernandos@alfa.ist.utl.pt

L. M. Redondo
Instituto Superior de Engenharia de Lisboa, (GIAAPP/ISEL), Portugal
lmredondo@deea.isel.ipl.pt

Abstract—**This paper presents equivalent models of Modular Marx Multilevel Converter Diode (M^3CD) plus load, for bipolar or unipolar operation. With these models, it's possible to analyze and understand the dynamic behavior of the pulse output voltage and current at the load. Simulation and experimental results, for different voltage levels, and various types of load are presented.**

Keywords-Modeling, Multilevel Converter, Pulsed Power.

I. Introduction

The modeling of a Modular Marx Multilevel Converter Diode (M^3CD) together with its load is presented in this work. The M^3CD can be used as generator of high-voltage positive and/or negative pulses with multi-level voltages lower than the input power supply voltage, as described in [1].

This study can be applied to other multi-level converters based on a half-bridge topology. The half-bridge concept uses two stacks of switching modular cells, one stack in the upper arm and the other in the lower arm. Each modular M^3CD cell is derived from the semiconductor based Marx generator [2] and contains two ON/OFF power semiconductors with anti-parallel diodes (or two bidirectional switches), a diode and a capacitor.

The modular M^3CD converters may be used in the area of electric power systems (in High Voltage DC transmission, HVDC, for example) or in the area of pulsed power, in applications such as generators of high voltage or as a solution to compensate the generated pulse voltage droop, which could be higher than 10%.

The use of multiple levels for voltage generation is advantageous because it allows the use of kV rated semiconductors to generate tens of kV. The rate of change of the pulse voltage is also reduced enabling the increase of the pulse voltage, without significantly increasing the electromagnetic interference.

The operation principle of M^3CD converter contains mainly two steps [3]. The first charges all the capacitors in series (with a fraction of the voltage dc power supply, U_{dc}). The M^3CD states are chosen so that in the second step applies the needed voltage level to the load, and the diode interconnecting the modular cells is forward biased placing the capacitors in parallel to balance their voltages [3].

As the M^3CD converter uses modular cells it is possible to obtain many voltage levels at the output. The herein presented analysis can be extended to a high number of levels and can be generalized for any converter based on a half-bridge topology.

A multilevel topology M^3CD with n levels requires $2(n-1)$ modular cells [1] to bipolar or unipolar voltages (Fig. 1).

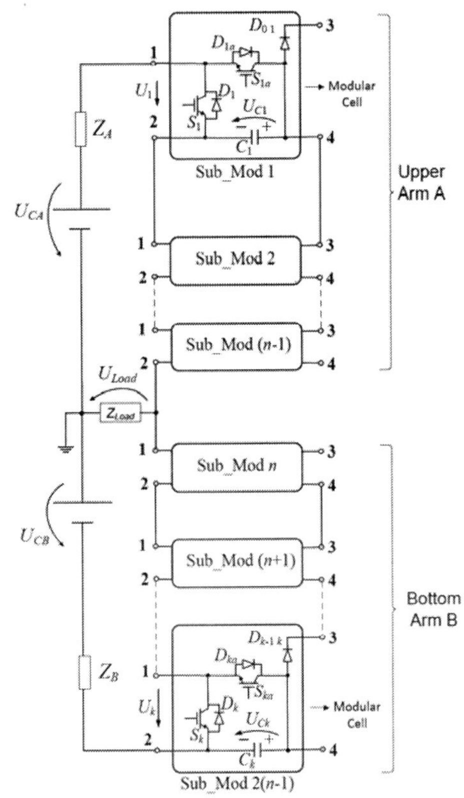

Figure 1. Modulator with n levels using $2(n-1)$ basic cells.

II. Modeling of M³CD with Load for Bipolar Voltage

The equivalent model of a modulator based on half-bridge topology (Fig. 1) with generic load Z_{Load} for bipolar voltage is shown in Fig. 2, where it is considered the power sources (U_{CA} and U_{CB}), the total impedance of the upper arm A (Z_A), the total impedance of the lower arm B (Z_B) and the load impedance (Z_{Load}).

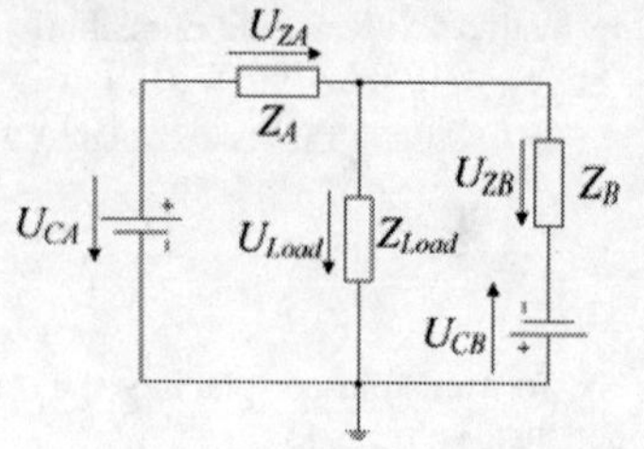

Figure 2. Equivalent model topology half-bridge for bipolar voltage.

To calculate the impedance of each arm, it is considered the equivalent capacitance of the capacitors, which are connected in series for each level, the parasitic resistances (r_1 or r_2) and parasitic inductance (L_1 or L_2), which represent the impedances of the semiconductor and the connections between modular cells.

The impedance value of the upper arm A (Z_A) is given by equation (1), where C_{eqA} are the equivalent capacitance of the capacitors that are connected in series (in upper arm A), and L_1 and r_1 are the parasitic terms of upper arm A components.

$$Z_A = \frac{1}{C_{eqA}s} + L_1 s + r_1 \tag{1}$$

For the bottom arm the impedance value (Z_B) is given by (2), where L_2 and r_2 represent the parasitic terms of bottom arm B and C_{eqB} are the equivalent capacitance of the capacitors that are connected in series (in upper arm B).

$$Z_B = \frac{1}{C_{eqB}s} + L_2 s + r_2 \tag{2}$$

The transfer function of load impedance (Z_{Load}) is obtained depending on the type of load (resistive, resistive capacitive or R //C, inductance L in series with R // C), to exemplify it was considered an inductance L in series with R // C, the impedance value is,

$$Z_{Load} = \frac{RLC_L s^2 + Ls + R}{RC_L s + 1} \tag{3}$$

The parasitic inductors increase the order of the polynomial of the overall system transfer function, while the intrinsic resistors provide little damping. For this reason the output voltage and current responses in capacitive load (parallel RC) are oscillatory with high overshoot.

The models will change accordingly to the voltage level, this is because at each level the number of connected capacitors varies and therefore the impedance of each arm is changed.

Applying Kirchhoff's laws to the equivalent model of Fig. 2 the relationship (4) is received relating the voltage load (U_{Load}) and the voltages sources U_{CA} and U_{CB}.

$$\begin{aligned} U_{Load} &= \frac{Z_B\, Z_{Load}}{Z_B\, Z_{Load} + Z_A\, Z_B + Z_A\, Z_{Load}} U_{CA} + \\ &+ \frac{Z_A\, Z_{Carga}}{Z_B\, Z_{Load} + Z_A\, Z_B + Z_A\, Z_{Load}} (-U_{CB}) \end{aligned} \tag{4}$$

In bipolar topology U_{CA} and U_{CB} voltages are approximately equal, $U_{CA} \approx U_{CB} \approx U_{dc}/2$, yielding equation (5):

$$U_{Load} = \frac{Z_{Load}(Z_B - Z_A)}{Z_B\, Z_{Load} + Z_A\, Z_B + Z_A\, Z_{Load}} \frac{U_{dc}}{2} \tag{5}$$

In order to study this mathematic model (5) it was considered a modulator M³CD with 5-level bipolar ($-U_{dc}/4$; $-U_{dc}/2$; 0; $+U_{dc}/4$; $+U_{dc}/2$) working in the level 3, U_{Load}=0V (Fig. 3), however, this study may be applied in other possible combination (i.e. to another voltage level).

A. *M³CD with five- level working in the level 3*

At this level, two capacitors in upper arm A are connected in series (C_1 and C_2 of Fig. 3), assuming that the capacitance of all capacitors is C then $C_{eqA}=C/2$, S_3 and S_4 are conducting and therefore C_3 and C_4 do not appear in the Z_A impedance.

In bottom arm B, C_5 and C_6 are connected in series ($C_{eqB}=C/2$), switches S_7 and S_8 are conducting and thus C_7 and C_8 do not appear in the Z_B impedance.

The load impedance (Z_{Load}) it is given by equation (3) and corresponds to an inductance L in series with R // C.

Replacing the impedances Z_A (1), Z_B (2) and Z_{Load} (3) in equation (5), the equation (6) is obtained:

$$U_{Load} = \frac{As^5 + Bs^4 + Ds^3 + E\, s^2 + Fs}{Gs^5 + Hs^4 + Is^3 + J\, s^2 + Ks + M} \left(\frac{U_{dc}}{2}\right) \tag{6}$$

Where the numerator polynomial coefficients are given in (7):

$$\begin{aligned} A &= C_{eqA}\, C_{eqB}\, R\, LC_L (L_2 - L_1) \\ B &= C_{eqA}\, C_{eqB} L(L_2 - L_1) + C_{eqA}\, C_{eqB}\, R\, L\, C_L (r_2 - r_1) \\ D &= C_{eqA} C_{eqB} R(L_2 - L_1) + C_{eqA} C_{eqB} L(r_2 - r_1) + RLC_L (C_{eqA} - C_{eqB}) \\ E &= C_{eqA} C_{eqB} R(r_2 - r_1) + L(C_{eqA} - C_{eqB}) \\ F &= R(C_{eqA} - C_{eqB}) \end{aligned} \tag{7}$$

This transfer function has a polynomial of 5th order in the numerator with 18 terms and other polynomial of 5th order in the denominator with 36 terms. Looking at the numerator polynomial coefficients (7), if $r_1 > r_2$ or $L_1 > L_2$ or $C_{eqB} > C_{eqA}$, and using the Hurwitz test, this means the numerator of system has at least one root (a zero) in the right half of the complex plane (non-minimum phase system).

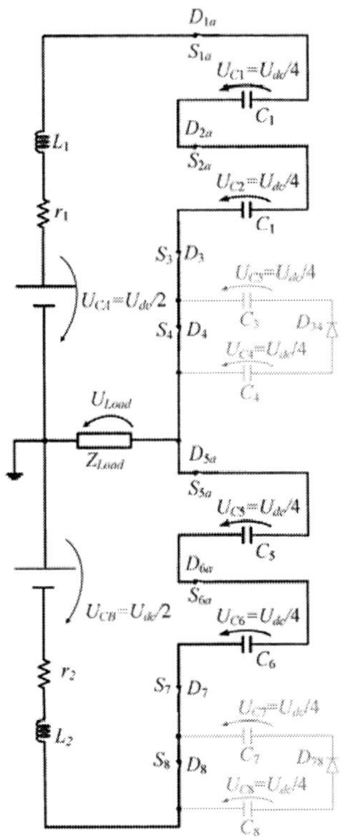

Figure 3. Modulator with five-levels at level 3, U_{Load}=0V.

And the denominator polynomial coefficients are given in (8):

$$
\begin{aligned}
G &= RC_L C_{eqA} C_{eqB}\left(L\,L_1 + L\,L_2 + L_1\,L_2\right)\\
H &= C_{eqA}C_{eqB}\left(L\,L_1 + L\,L_2 + L_1\,L_2\right) +\\
&\quad + C_{eqA}C_{eqB}RC_L\left(r_2 L + r_2 L_1 + r_1 L_2 + r_1 L\right)\\
I &= C_{eqA}C_{eqB}\left(R\,L_1 + R\,L_2 + r_1\,L + r_2\,L + r_2\,L_1 + r_1\,L_2\right) +\\
&\quad + RC_L\left(LC_{eqA} + LC_{eqB} + L_2C_{eqB} + L_1C_{eqA} + r_1 r_2 C_{eqA}C_{eqB}\right)\\
J &= RC_{eqA}C_{eqB}\left(r_1 + r_2\right) + C_{eqA}\left(L + L_1\right) + C_{eqB}\left(L + L_2\right) +\\
&\quad + RC_L\left(r_1 C_{eqA} + r_2 C_{eqB}\right) + r_1 r_2 C_{eqA} C_{eqB}\\
K &= r_1 C_{eqA} + r_2 C_{eqB} + R\left(C_{eqA} + C_{eqB}\right) + RC_L\\
M &= 1
\end{aligned}
\tag{8}
$$

It can be seen that the static gain, using the final value theorem, coincides with the expected value for U_{Load} at level 3 is 0V (9). Considering the input value of the system as $U_{dc}/4$ (or $-U_{dc}/4$) the value of the load voltage tends to U_{Load}=0.

$$\lim_{s\to 0} U_{Load}(s) = 0 \tag{9}$$

The load current is obtained as follows:

$$I_{Load}(s) = \frac{U_{Load}(s)}{Z_{Load}(s)} \tag{10}$$

The obtained models for U_{Load} (6) and I_{Load} (10) can be extended to any level number, although it is necessary to adapt the impedances (Z_A and Z_B) which are dependent on the voltage level. The models (6) and (10) may also be adapted for unipolar voltages (U_{CA}=0 for negative pulses or U_{CB}=0 for positive pulses).

III. Simulation and Experimental Results of M³CD with Load for Bipolar Voltage

To validate the study of the dynamic behavior of the mathematical models obtained in section II, simulations and laboratory tests using a M³CD converter with 5 levels (bipolar voltages) were made. To obtain a modulator with five voltage levels (Fig. 3), eight modular cells are needed (four cells for the upper arm and four cells for bottom arm).

In the first set of simulations and laboratory tests the following parameters were used (Table 1):

TABLE I.

U_{DC}=330V (DC side of the converter) Step value: $U_{DC}/4$=82.5V
$C_{1\text{-}8}$=1μF (capacitors used in cells)
Load with inductance L=1.25 μH in series with $R//C_L$ where R=20MΩ and C_L=0.5nF
Signal description: t_1 =10ms at each intermediate level; width of the minimum and maximum level t_2=t_3=50μs and the interval between pulses t_4 = 500μs.

In Fig. 4 it is observed the bipolar pulse voltage U_{Load} (CH1) with five levels ($+U_{dc}/2$; $+U_{dc}/4$; 0; $-U_{dc}/4$; $-U_{dc}/2$) and the output current I_{Load} (CH2) for an almost complete period.

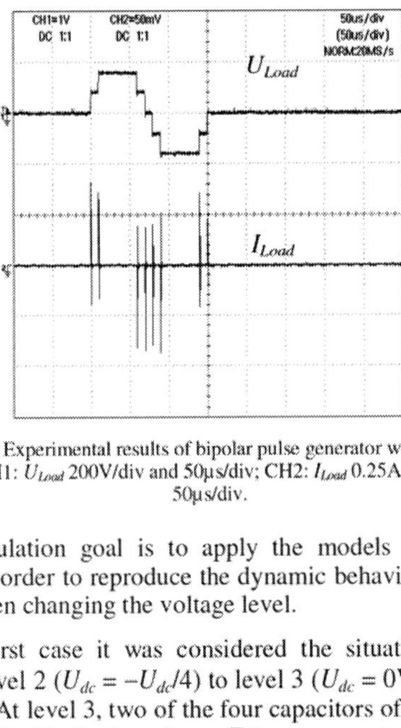

Figure 4. Experimental results of bipolar pulse generator with 5 levels: a) CH1: U_{Load} 200V/div and 50μs/div; CH2: I_{Load} 0.25A/div and 50μs/div.

The simulation goal is to apply the models studied in section II in order to reproduce the dynamic behavior of U_{Load} and I_{Load} when changing the voltage level.

In the first case it was considered the situation of the change of level 2 ($U_{dc} = -U_{dc}/4$) to level 3 (U_{dc} = 0V) (Fig. 4 and Fig. 5). At level 3, two of the four capacitors of upper arm A (associated with the impedance Z_A) are connected in series, in the bottom arm (which are related with impedance Z_B) also two capacitors are connected in series: Equivalent capacitors in upper and bottom arm: C_{eqA}=C/2=0.5μF, C_{eqB}=C/2= 0.5μF.

The parasitic remaining values L_1, r_1, L_2 and r_2, needed to determine the impedances Z_A and Z_B (measured and estimated values considering 10nH/cm) are: Parasitic inductance of upper arm and bottom arm L_1=1.2μH and L_2=1.2μH; Parasitic resistance of upper arm and bottom arm r_1=50Ω (due to switching semiconductors) and r_2=1.2Ω.

Using the values shown in Table 1 and in the previous text in the models obtained in (6) and (10), the Fig. 5a) illustrates a simulation result of dynamic behavior of U_{Load} and I_{Load} when the pulse voltage level goes from level 2 to level 3. Fig. 5b) shows an experimental result (U_{Load} and I_{Load}) zoom of Fig. 4 of going from level 2 to level 3.

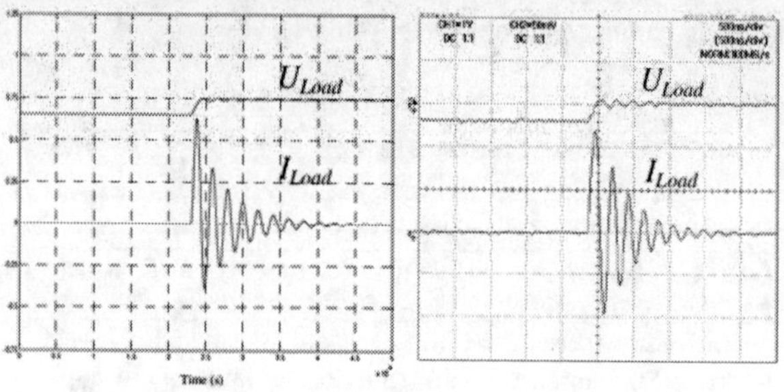

a) b)

Figure 5. Simulation and experimental results of bipolar pulse generator with 5 levels: a) and b) CH1: U_{Load} 200V/div and 500ns/div; CH2: I_{Load} 0,25A/div and 500ns/div.

Fig. 6 shows the simulation and experimental results of switching from level 4 to level 3. A step signal (the set-point) with −82.5V ($-U_{dc}/4$) was applied in this simulation (Fig. 6a). The model used was the same as in Fig. 5 (the initial conditions were not changed). Fig. 6b) shows an experimental result zoom of Fig. 4 switching from level 4 to level 3 for U_{Load} and I_{Load}.

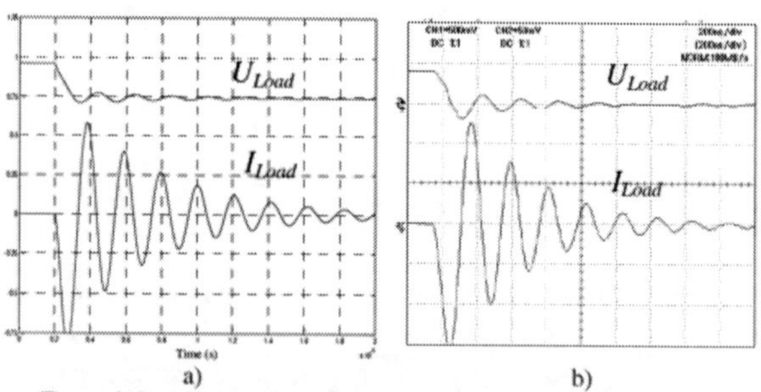

a) b)

Figure 6. Simulation and experimental results of bipolar pulse generator with 5 levels: a) and b) CH1: U_{Load} 100V/div and 200ns/div; CH2: I_{Load} 0,25A/div and 200ns/div.

In a second test, the following parameters were considered: U_{dc}=300V; Load with inductance L=1.1μH in series with $R//C_L$ where R=40MΩ and C_L=10nF; Capacitors with 1μF and charged with $U_{Ck}=U_{dc}/4$=75V; Signal description: t =10μs at each intermediate level and the interval between pulses t_L = 500μs. The switching from level 2 ($U_{dc} = -U_{dc}/4$) to level 3 (U_{dc} = 0V) was considered (Fig. 7). As mentioned in previous case, to obtain the impedance Z_A, in the level 3, two of the four capacitors of upper arm A are connected in series (C_{eqA}=C/2=0.5μF), the same is true for the arm B (C_{eqB}=C/2= 0.5μF).

The other values of Z_A and Z_B correspond to parasitic terms, and the values considered (and estimated) were parasitic resistance of upper arm and bottom arm r_1=50Ω and r_2=1.2Ω, parasitic inductance of upper arm and bottom arm L_1=1.2μH and L_2=1.2μH. Using these values Fig. 7a) from simulation closely matches Fig. 7b) from experimental results.

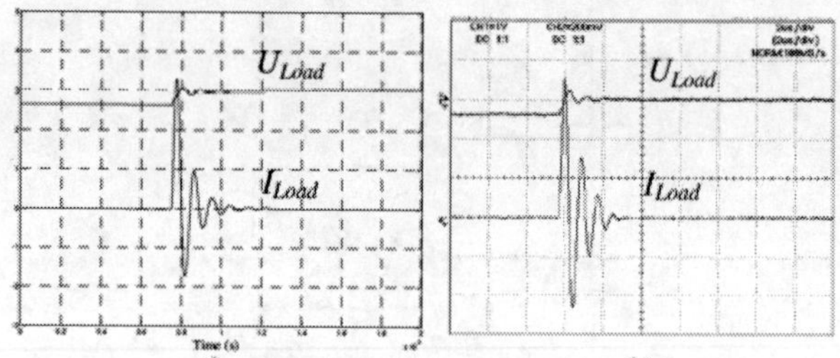

a) b)

Figure 7. Simulation and experimental results of bipolar pulse generator with 5 levels: a) and b) CH1: U_{Load} 200V/div and 2μs/div; CH2: I_{Load} 1A/div and 2μs /div.

IV. CONCLUSIONS

This work proposed a new dynamic model for the half bridge based modular Marx multilevel converters. The main objective was the study of the dynamic behavior and the switching times of the output voltage and current. Results obtained in simulation, using the derived mathematical models, and in the laboratory showed reasonable agreement in the transient response of the voltage and current waveforms namely in overshoot and damping values. Therefore, the usefulness of the theoretical analysis is practically verified. The subtle differences may be due to the difficulty in estimating some of the parameters in the real circuit, namely the values of parasitic inductances and semiconductors equivalent resistances.

V. ACKNOWLEDGEMENTS

This work was supported by the National Funds through the Fundação para a Ciência e a Tecnologia under Grant UID/CEC/50021/2013.

VI. REFERENCES

[1] L. Lamy Rocha, J. F. Silva, and L. M. Redondo, "Multilevel high voltage pulse generation based on a new modular solid-state switch," IEEE Trans. Plasma Sci., vol. 42, no. 10, pp. 2956–2961, Oct. 2014.

[2] LM Redondo, E Margato, J Fernando Silva, A new method to build a high-voltage pulse supply using only semiconductor switches for plasma-immersion ion implantation, Surface and Coatings Technology 136 (1), pp 51-54.

[3] Rocha, L.; Silva, J. Fernando; Redondo, L.; "Seven Levels Unipolar/Bipolar Pulsed Power Generator", Plasma Science, IEEE Transactions on, (published to IEEE Xplorer in February 2016), doi: 10.1109/TPS.2016.2519269.

Broad-band High Pulsed Power Coaxial Applicators

for in Vitro and in Vivo Biological Experiments

R. Vezinet, M. Croizer, T. Chretiennot, A. Catrain
CEA, DAM, GRAMAT; F-46500 Gramat, France

M. Breton, L. M. Mir
UMR 8203 CNRS; F-94805 Villejuif, France

E. Bellard, M. Golzio, M-P. Rols
UMR 5089 CNRS/IPBS; F-31077 Toulouse, France

***Abstract*— We have developed a new concept of broad-band coaxial applicator to expose biological samples to sub-nanosecond electric field pulses at high field amplitude (800 kV/m). A first device makes use of a short-circuited 50 Ohm TEM coaxial structure matched at its extremity by a hollow high voltage resistor. We show that the axial electric field is very homogenous inside the empty volume of the hollow resistor. A cylindrical glass cuvette can be axially installed into this volume; its bottom and its top metallic caps are respectively in contact with the inner and external conductors of the coaxial. The E field is directly related to the voltage and to the length of the cuvette. Two cuvette volumes (168 and 298 µl) are available. A radial V-dot sensor is installed close to the load. This system is presently used with a 25 kV sub-nanosecond generator.**

With the aim to increase the test volume of the coaxial applicator, we have developed a larger one making use of a circular array of resistors in place of the initial hollow resistor. This applicator allow to apply high power pulsed E fields to the content of a dorsal window chamber implanted in the skin of a living mouse. The dorsal window chamber is directly inserted through a rectangular hole in the center of the extremity plate. A lateral hole into the external conductor allows real time observations of the dorsal window content under a microscope.

We give the description of the two devices and show some experimental and numerical results in terms of S11, pulse shape and electric field homogeneity.

Keywords—Subnanosecond pulsed electric field; coaxial chamber; TEM transmission line; exposure device

I. INTRODUCTION

There is a large interest, in the bioelectrical community, for the development of broad-band applicators able to expose, to High Power Pulsed Electric Field, biological samples like Giant Unilamellar Vesicles (GUV), cells, spheroids in suspension or small laboratory animals. The frequency spectrum of the pulses can be very large (DC to few GHz), especially for sub-nano pulses (with very fast rise time and short duration). The electric field amplitude of interest is generally high (1 to few MV/m). The dimensions of the Biological Sample Under Test (BSUT) implie the use of moderate or high voltage generators to reach the required applied voltages (a few kV). Such repetitive generators present typically 50 Ohm output impedance. Thus their load has to be matched.

Finally, suppressing or controlling the parasitical external radiated ElectroMagnetic Field is also a need in terms of EM security for biologists using these applicators at such voltage levels.

In a previous work we developed new high power microwave applicators enabling optimal E-field coupling and homogeneity into biological samples of small BSUT volumes (75 nl to 14 µl) [1]. We present here two new broadband applicators dedicated to larger volumes and high voltages.

The first one is dedicated to *in vitro* experimentation with GUV, cells or spheroids in suspension; the second one has been developed for the needs of a specific *in vivo* experimentation involving living small animals (e.g. mice). Both applicators take advantage of using a hollow load resistor providing a homogeneous E field distribution within the test volume where the BSUT is installed.

II. COAXIAL APPLICATOR FOR IN VITRO EXPERIMENTATIONS

For experiments involving cells in suspension in the sub-nanosecond regime, authors have developed different kinds of coaxial chambers; they present a large frequency bandwidth compared to regular planar TEM transmission lines and they are secure versus high voltage and EM radiations. In [2], the BSUT is installed into a cylindrical cuvette located at the extremity of the coaxial transmission line, but a matching resistor has to be installed inside the liquid. In [3] there is not any matching resistor, but the dimensions of the BSUT (length and diameter) impose constraint in such a way that the liquid in suspension has to present an acceptable matching impedance.

A. Geometry description

In our design [4] (Fig.1) we make use of a hollow 50 Ohm matching resistor (3); the glass cuvette (2) takes place into its center empty volume; the cuvette is closed at its bottom by a stainless steel cap (5) in contact with the positive central conductor of the coaxial transmission line. A removable stainless top cap (4) is connected to the ground return conductor; an air vent (5) avoids compression of the liquid (1) during installation and pulsing. A SMA connector constitutes an open-ended coaxial voltage (7) sensor; the called V-dot delivers a derivative image of the voltage near the load. A second V-dot can optionally be installed in place of the regular top cap to control the internal voltage.

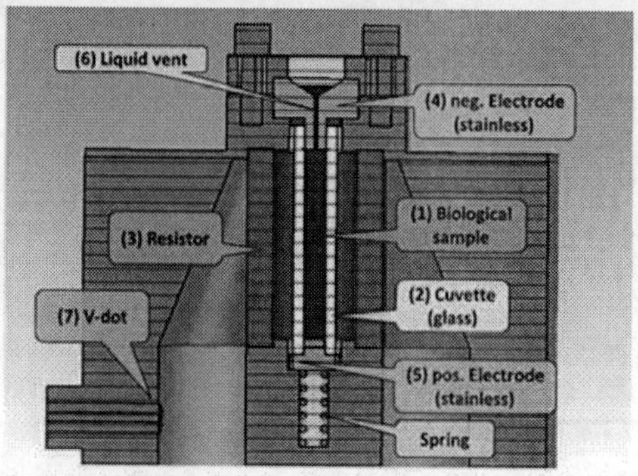

Fig. 1. Details of the coaxial chamber developped for cells suspension in sub-nano second regime.

Considering the intrinsic capacitance of the lateral V-dot and its 50 Ohm transmission line matched load, its sensitivity can be calculated from measured S11 and S21 parameters; making use of the Laplace operator p=jω, the output V-dot voltage V1(p), with R=50 Ohm, can be expressed as follows :

$$V1(p) = Vs(p)\,\frac{1+50\,Cp}{50\,Cp} \qquad (1)$$

Assuming that the transmission line between the input connector of the device and the V-dot position is characterized by its electric length L, its propagation constant β=ω/c and its constant characteristic impedance Zc=50 Ohm, C(p) can be expressed as an equivalent capacitor

$$C_{(p)} = \frac{S21}{50p}\,\frac{1}{(1+S11e^{j2\beta L})e^{-j\beta L} - S21} \qquad (2)$$

V1(t) can be rigorously evaluated using inverse Fourier transform or, if |C(p)| is quite constant:

$$V1(t) \approx K\int_0^t Vs(t)\,dt \quad with\ K = \frac{1}{50|C|} \qquad (3)$$

For |S11|≤ -10 dB (bandwidth range) , we can neglect S11:

$$C \approx \left|\frac{S21}{j\omega}\right|/50 \qquad (4)$$

In this case, the maximum error for K in (3) is 5.4% for |S11| = -10 dB.

The measured value of the K factor, using (2) and (3), is 23.10^{12} for the lateral V-dot.

The coupled E-field homogeneity in the cuvette is evaluated thanks to 3D electromagnetic simulation, using 3D CST software, and assessed by $1-(E_{max}-E_{min})/E_{max}$ where E_{max} and E_{min} are the maximum and minimum E-field strength in the volume; it has been estimated to 84% [4]. The figure 2 represents the maxima E field calculated by CST software on the longitudinal cutting plane.

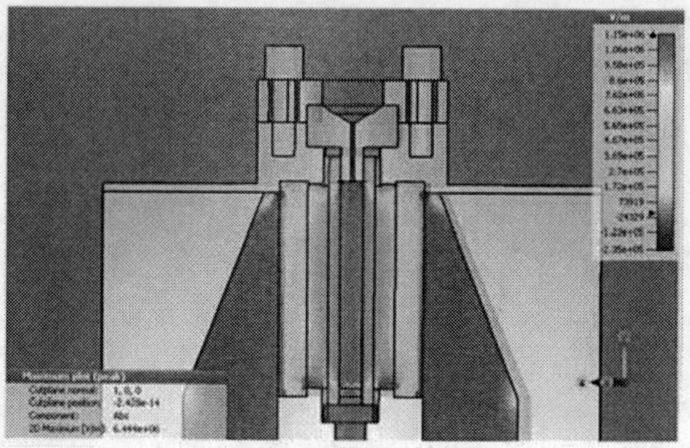

Fig. 2. Maxima E field on the axial cutting plane.

In the present versions, the useful length between electrodes is 23.8 mm; two different diameters (3 and 4 mm internal diameter) of the cuvettes (2) are available. The associated BSUT volumes are respectively 168 and 298 µl. The Fig.3 shows a 4 mm cuvette and the hollow HVR resistor (3).

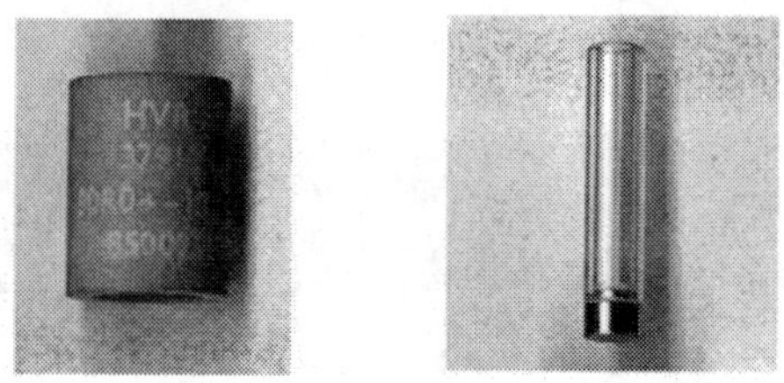

Fig. 3. Hollow HVR resistor and glass cuvette without his top cap.

The Fig.4 gives a picture of the device with its vertical support.

Fig. 4. Picture of the coaxial applicator with his support.

B. Experimental results

The measured S11 parameter is lower than -10 dB from DC to 1300 MHz. The tapered shape for the return conductor in the resistor section (Fig.1) contributes to these performances [4].

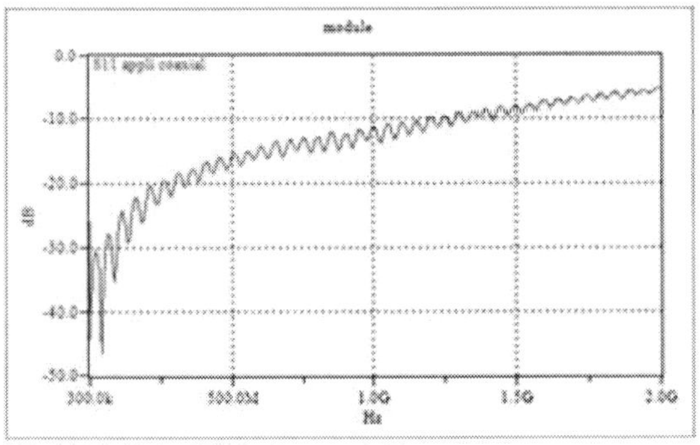

Fig. 5. Measured S11 parameter of the coaxial applicator.

The applicator is driven by a 20 kV repetitive KENTEC PBG5 generator; a spark gap is inserted into the cable connector as a tail cut switch to limit the duration of the pulses [4]. Fig.4 gives the electric field inside the suspension estimated from the V-dot voltage measurement; the maximum E field is 800 kV/m with a rise time of 150 ps and a duration of 530 ps.

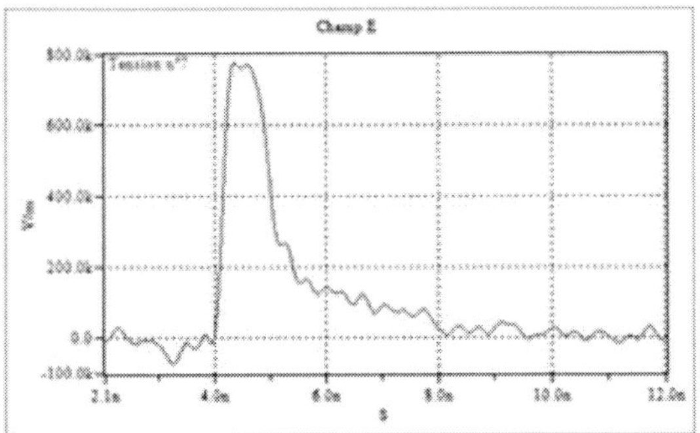

Fig. 6. Measured E field in the biological sample.

We currently use it at CEA Gramat and at UMR 8203 in Villejuif, under repetitive regimes (up to 50000 pulses at 500 Hz).

III. COAXIAL APPLICATOR FOR IN VIVO EXPERIMENTATIONS

Using the same concept, we have developed a larger applicator, allowing to expose a dorsal window chamber implemented on the back of a mouse to high power pulsed E field.

The size of such a dorsal insulated window chamber (2.5x3x1.2 cm, Fig. 7) implies to increase the test volume, inside and consequently around the matching resistor; for that, the previous hollow resistor is replaced by a circular array of resistors and the dimensions of the transmission line are increased: respectively 27.18 and 62.54 mm for the inner coaxial radius.

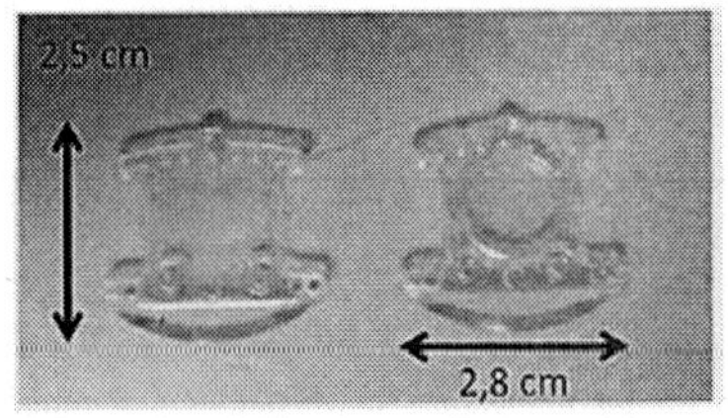

Fig. 7. Views and dimensions of the dorsal window chamber.

Fig. 8 shows the new device. A lateral hole into the external conductor allows real time observations, under microscope, of the mouse dorsal window chamber.

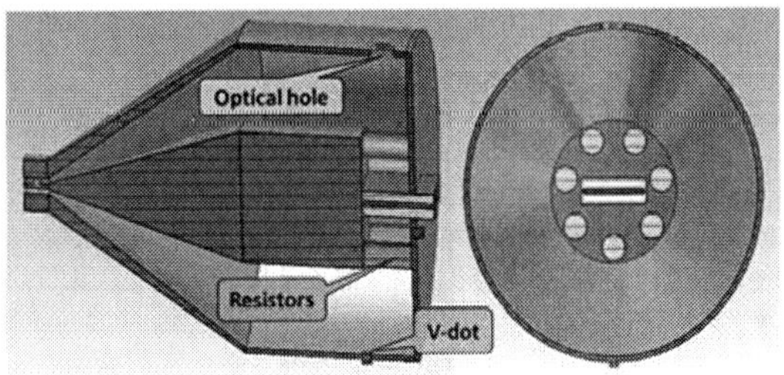

Fig. 8. Geometry details of the mouse dorsal window chamber applicator. The central brick, inserted into the test volume, represent the dorsal window chamber.

The mouse dorsal window chamber is directly inserted through a rectangular hole in the center of the extremity plate; it comes in contact within the central conductor; it has been modeled (Fig.8) by a 2 mm thick bloc of lossy material.

The measured S11 parameter (Fig. 9) is lower than -10 dB from DC to 480 MHz with and without the chamber.

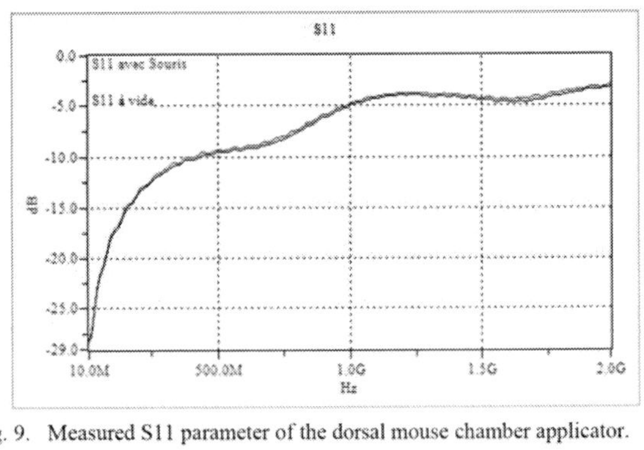

Fig. 9. Measured S11 parameter of the dorsal mouse chamber applicator.

The poor values of S11 for the upper frequencies is attributed to the high frequency behavior of the High Voltage resistors; a 3D simulation with ideal resistors shows that this geometry is able to provide acceptable values of S11 (<10 dB) until 2 GHz. The measured value of the K factor, using (2) and (3), is 18.10^{12} for the lateral V-dot.

The homogeneity of the electric field has been estimated numerically, using the 3D CST software; Fig. 10 shows the maxima of the calculated E field, over a central section of the test volume.

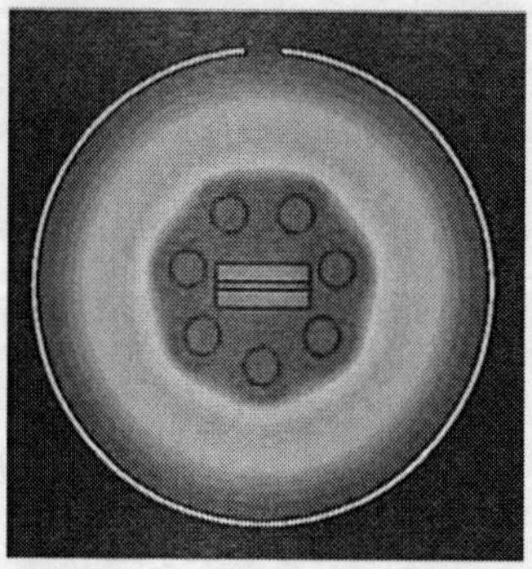

Fig. 10. Maximal E field amplitude on the central test volume cutting plane.

Fig. 11 shows the variations of the maximum of the electric field amplitude along the central axis, into the test volume (l=0-20 mm) and into the near zone outside the window (l=20-30 mm). The homogeneity of the E field has been estimated to 85%. To guaranty this value, the skin of the mouse has to be in contact with the central conductor.

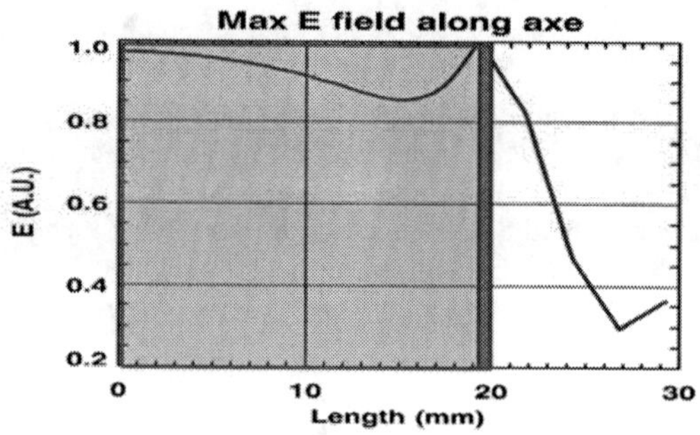

Fig. 11. Maxima E field on the central test volume cuttinp plane.

We currently use it at IPBS in Toulouse in real situation (Fig. 12); the excitation signal is a 150 MHz damped sinusoid with amplitude of 1 kV delivered by an homemade generator based on high voltage MOSFET and resonant transmission line; we are now planning to develop a higher voltage generator using spark gaps switches.

Fig. 12. Applicator with mouse in real situation under the microscope.

Acknowledgment

This work is funded by DGA (Direction Générale de l'Armement) and ANR (Agence Nationale de la Recherche) under the contracts PIERGEN (ANR-12-ASTR-0039-02) and OXILIPI2 (ANR-13-ASTR-0029-02).

References

[1] T. Chretiennot, A. Catrain, R. Vezinet, L. Gibot, F. Pillet, M.-P. Rols "New Efficient High Power Microwave Applicators Enabling Optimal E Field Coupling and Homogeneity into Biological Samples" Proceedings of the 20th IEEE International Pulsed Power Conference (2015), Austin, Texas, 978-1-4799-8403-9/15/

[2] J. T. Camp, S. Xiao, S. J. Beebe, P. F. Blacmore, K. H. Schoenbach "Bioelectric studies with subnanosecond pulsed electric field" Proccedings of the 17th IEEE International Pulsed Power Conference (2009), Washington DC DOI: 10.1109/PPC.2009.5386377

[3] T. Heeren, J. Kolb, S. Xiao, K. H. Schenbach, H. Akiyama "Pulsed power generators and delivery devices for bioelectrical applications" 1-4244-0019-8/06/ IEEE 2006

[4] M. Croizer ; « Applicateurs destinés aux Etudes d'Effets Biologiques des Ondes Électromagnétiques Sub-nano secondes », Ph D thesis, Université de Limoges 2015.

[5] R. Vezinet, M. Croizer, J.C. Diot, A. Catrain, « Device for applying an electromagnetic field to a biological sample » Patent n° WO 2014/086725 A1

Prototype Inductive Adders with Extremely Flat-top Output Pulses for the Compact Linear Collider at CERN

Janne Holma, Michael J. Barnes
CERN, TE/ABT
Geneva, Switzerland
janne.holma@cern.ch

The CLIC study is exploring the scheme for an electron-positron collider with high luminosity and a nominal centre-of-mass energy of 3 TeV. The CLIC pre-damping rings and damping rings will produce, through synchrotron radiation, ultra-low emittance beam with high bunch charge, necessary for the luminosity performance of the collider. To limit the beam emittance blow-up due to oscillations, the pulse generators for the damping ring kickers must provide extremely flat, high-voltage, pulses. The specifications for the damping ring extraction kickers call for a 160 ns or 900 ns duration flat-top pulses of +/-12.5 kV, with a combined ripple and droop of not more than +/-0.02 % (+/-2.5 V) for each pulse: an inductive adder is a very promising approach to meeting the specifications. Two 5-layer, 3.5 kV, prototypes have already been built and tested at CERN and recently, the first five layers of a 20-layer, 12.5 kV, full-scale prototype inductive adder have been assembled. This paper presents flat-top stability and repeatability measurements of the output waveforms of this first 5-layer assembly of a full-scale prototype inductive adder. Methods to measure the output pulse with the required high precision are also discussed. For repeatability measurements, the pulse waveforms have been recorded with an oscilloscope with 16 effective bits, which allows measurements of minimum and maximum envelope curves for a large number of consecutive output waveforms. Both passive and active modulation methods have been applied to improve the relative flat-top stability of the prototype inductive adders to be close to the specification of +/-0.02 %.

Keywords—inductive adder, analogue modulation, flat-top stability, flat-top repeatability, CLIC

I. INTRODUCTION

The Compact Linear Collider (CLIC) would be a high-energy electron-positron collider [1]. It could provide very clean experimental environments and steady production of all particles within the accessible TeV energy range. To achieve high luminosity at the interaction point, it is essential that the beams have very low transverse emittance: the Pre-Damping Ring (PDR) and Damping Ring (DR) damp the beam emittance to extremely low values in all three planes.

Stripline kickers are required to inject beam into and extract beam from the PDRs and DRs [2]. Jitter in the magnitude of the kick waveform causes beam jitter at the interaction point [3]. Hence, in particular, the DR extraction kicker must have a very small magnitude of jitter: the 2 GHz specifications call for a 12.5 kV pulse of 160 ns duration flat-top, with a combined ripple and droop of not more than ±0.02 % [1]. The 1 GHz specifications call for a burst of two 160 ns duration flat-top pulses with 580 ns between the end of the flat-top of the first pulse and the beginning of the flat-top of the second pulse, which can be fulfilled e.g. with a 900 ns, continuous, flat-top duration. The requirements for the voltage and stability during these two 160 ns sections of the flat-top are the same as for 2 GHz design, i.e. ±0.02 % [1]. The flat-top repeatability requirements are also extremely tight, ±0.01%, for both RF system designs [1].

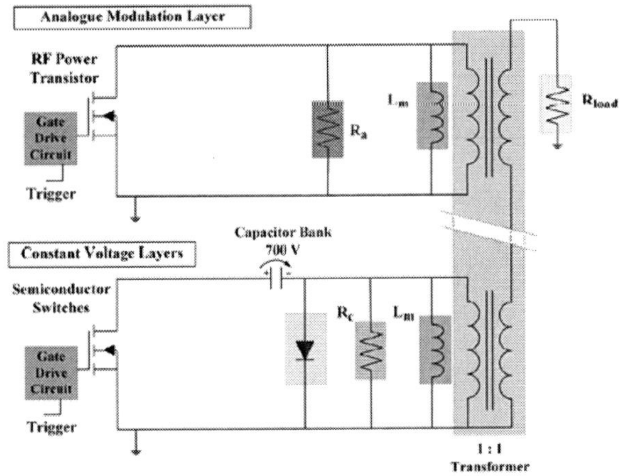

Fig. 1. Schematic of an inductive adder with a single constant voltage layer with core loss resistance R_c and magnetizing inductance L_m.

II. THE INDUCTIVE ADDER

A review of literature of existing pulse generators has been carried out and an inductive adder (Fig. 1) has been selected as the most promising means of achieving the specifications for the DR kickers [4]. The inductive adder is a solid-state modulator, which can provide relatively short and precise pulses. With a careful design of the adder, it may be possible to directly meet the ripple and droop requirements of the PDR kicker and analogue modulation may provide a means to meet the demanding specifications for the DR kicker [5, 6]. The reasoning for choosing the main components of the inductive adder has been given in [6]. Two 5-layer prototype inductive adders have been assembled at CERN. The design parameters and the initial results for these pulse modulators were presented in [7] and [8]. The prototype inductive adders have been equipped with an analogue modulation layer, which can be used to compensate the droop and ripple of the output waveform. Operation of the passive and active modulation layers has been verified with measurements and the results have been presented in detail in [9-11]. Evaluation of magnetic core material and custom-made magnetic cores for the full-scale, 12.5 kV, 250 A, prototype inductive adder was presented in [12]. Two recent papers [12, 13] presented initial measurements on the first five layer full-scale prototype inductive adder and this paper is a continuation of that study. The detailed electrical design of the full-scale prototype inductive adder was presented in [13].

III. Specifications for the 12.5 kV Prototype Inductive Adder

Table I shows the specifications for the 12.5 kV prototype inductive adder for the CLIC DR. In the CLIC DR kicker system, the inductive adder generates pulses for a stripline kicker, which has an odd-mode impedance of 40.9 Ω [2]. The striplines may be terminated with their odd-mode characteristic impedance and therefore the nominal output current of the inductive adder, at 12.5 kV, would be 305 A. The flat-top duration for the output pulse is either 160 ns or 900 ns, depending on the RF system design of the damping rings [1]. The full-scale prototype inductive adder has been designed to supply pulses with up to 900 ns flat-top duration. The maximum allowed total pulse duration, including rise time, flat-top duration and fall time, is approximately 2.2 µs [1]. However, to limit stress on the kicker system, the desired rise and fall times are in the range of 100 ns. The required pulse flat-top stability of ±0.02 %, i.e. ±2.5 V in absolute numbers for ±12.5 kV operation, defines the allowance for the combined droop and ripple for a single pulse. The pulse flat-top repeatability defines the allowed difference of any consecutive pulses and it is specified to be ±0.01 %, which corresponds to ±1.25 V, for ±12.5 kV operation.

IV. MEASUREMENTS ON THE 5-LAYER FULL-SCALE PROTOTYPE

A. Measurement Set-up

Measurements were carried out with five, full-scale, layers of the 20-layer prototype inductive adder. This assembly is shown in Fig. 2. The printed circuit boards (PCBs) of the primary circuit have the same design as for the final 12.5 kV prototype. However, for the tests and measurements reported here, only two current branches were powered in each layer: each current branch consisted of either a single NWL T00216, 12 µF, [14] or a Leclanché PPM-4 170-23.0, 12 µF [15], pulse capacitor, and an APT12057LFLL [16] MOSFET. The nominal maximum output voltage was 3.5 kV, corresponding to 700 V per layer. The output waveform was recorded with a Rohde & Schwarz RTO1004 oscilloscope [17], which has a specified resolution of 14 effective bits in the required bandwidth, 100 MHz.

TABLE I. Specifications for the 12.5 kV inductive adder for CLIC DR extraction kicker

Output voltage (kV)	12.5
Nominal output impedance (Ω)	50
Output current (A)	305
Flat-top duration (ns)	160 or 900 (excl. settling time)
Desired pulse rise/fall time (ns)	100/100
Flat-top stability (%)	±0.02
Flat-top repeatability (%)	±0.01

Fig. 2. Photograph of the first 5 layers of the full-scale 12.5 kV prototype inductive adder assembled at CERN.

B. Measurements without Analogue Modulation

Fig. 3 shows a measured output waveform of the 5-layer full-scale prototype: the output pulse waveform was not modulated. In this measurement, the pulse capacitors of each layer were initially charged to 500 V. The high-voltage output connector of the prototype inductive adder was connected to a 50 Ω load. The maximum output voltage was 2.402 kV. The pulse flat-top duration was 950 ns and the total pulse duration approximately 1.3 µs. In Fig. 3, the single pulse is shown as a green curve, the average of 100 pulses as a red curve and the minimum and maximum envelopes of 100 pulses as a blue and a black curve, respectively. The measured droop of the output waveform is 20 V (0.83 %) over 900 ns. The flat-top repeatability is approximately ±8.0 V (±0.3 %).

C. Measurements with an Active Analogue Modulation Layer

Measurements were also carried out by applying passive and active analogue modulation using the previous version of the prototype inductive adder. The operation principle of an analogue modulation layer is described in detail in [18]. In the measurements shown here, the output of the 5-layer full-scale prototype was connected with a 50 Ω high-voltage coaxial cable to the bottom connector of another 5-layer prototype inductive adder built at CERN (Fig. 4). The second inductive adder was equipped with an analogue modulation layer and

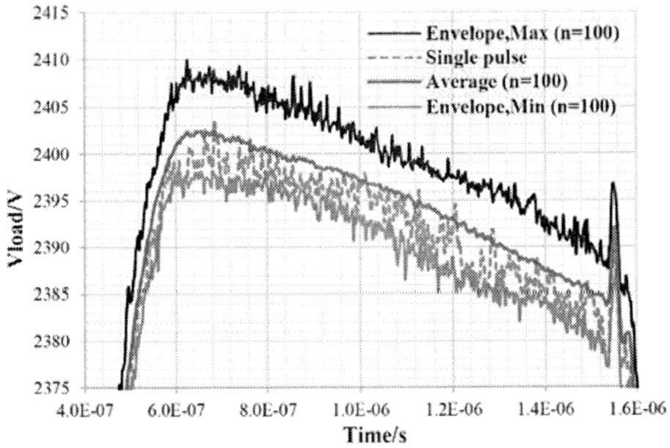

Fig. 3. Measured load voltage V_{load} of the 5-layer full-scale prototype for a single pulse and for an average of 100 pulses: the plot also shows a minimum and a maximum envelope for 100 pulses. The inductive adder was equipped with five constant voltage layers. The initial capacitor voltage was 500 V.

the output of this inductive adder was connected to a 50 Ω load. Effectively, these two inductive adders were in series. The second inductive adder was operating only as an analogue modulation layer, which was used to pulse condition the output pulses of the full-scale 5-layer prototype. The analogue modulation layer consisted of a resistor R_a, which was 7.96 Ω, and an RF power transistor type ARF463AP1G, made by Microsemi [16]. The schematic and design of this analogue modulation layer are described in detail in [11].

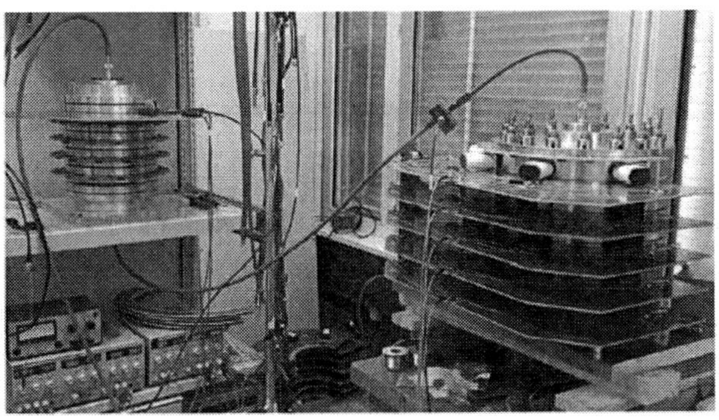

Fig. 4. Photograph of the measurement set-up with the 5-layer full-scale prototype (right) inductive adder and the second (i.e. previous version) 5-layer prototype with an analogue modulation layer (left).

Fig. 5 shows the measured output voltage for the measurement set-up with the full-scale 5-layer prototype, connected in series with the second inductive adder which contains the active analogue modulation layer. In these measurements, the initial pulse capacitor voltage of the constant voltage layers was 555 V. The control signal for the RF power transistor of the analogue modulation layer was a linear ramp and the RF power transistor was biased with a DC power supply, similarly to that described in [11]. The maximum load voltage during the pulse flat-top was 2.435 kV for an average of 100 pulses. The output waveform was recorded in three manner at the same time: as a single pulse, as an ensemble average for 100 pulses and as minimum and

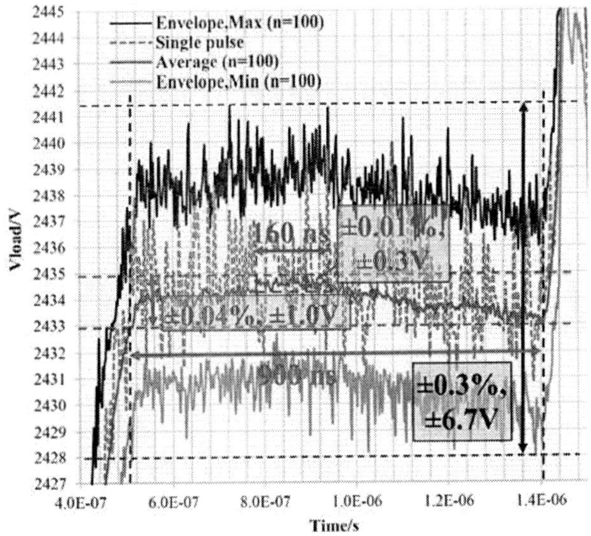

Fig. 5. Measured load voltage V_{load} of the 5-layer full-scale prototype for a single pulse and for an average of 100 pulses: the plot also shows a minimum and a maximum envelope for 100 pulses. The inductive adder was equipped with five constant voltage layers and it was connected in series with a second inductive adder, which contained only an analogue modulation layer. The initial capacitor voltage of each of the five layers was 555 V.

maximum envelopes for the 100 recorded pulses. In Fig. 5, all these pulse waveforms are shown.

In Fig. 5, the recorded flat-top stability, for the average of 100 pulses is 2.0 V peak-to-peak, for 900 ns, at 2.435 kV, which corresponds to ±1.0 V in absolute numbers and ±0.04 % relative flat-top stability. The minimum and maximum envelopes, for 100 pulses, are 13.4 V peak-to-peak, which corresponds to ±0.03 % (±6.7 V). In the waveforms shown in Fig. 5, the output waveform has no clear droop but it has a round "bump", which can be compensated with a properly defined control function, as was demonstrated in [11]. This will be carried out during future measurements with the prototype.

V. Flat-top Stability and Repeatability

In the measurements presented in this paper, the best measured flat-top stability, i.e. the combined ripple and droop of a single pulse, was ±0.01 % (±0.3 V) for a 160 ns flat-top duration and ±0.04 % (±1.0 V) for a 900 ns flat-top duration. These measurements fulfil the relative and absolute flat-top stability of the 2 GHz specifications for the CLIC DR extraction kicker systems. The relative flat-top stability is not yet adequate for the 1 GHz specifications, for which ±0.02 % at 12.5 kV for up to 900 ns flat-top is required: however the measured absolute flat-top stability, ±1.0 V, is adequate provided it does not increase with higher output voltage. The flat-top repeatability was ±0.3 % for measurements both with and without analogue modulation. A point to note is that applying analogue modulation did not degrade the flat-top repeatability.

Several measurements were also carried out in order to improve flat-top repeatability, i.e. by applying common-mode or differential mode noise filters on; (i) the output of the high-voltage power supply, which was used to charge the pulse capacitors of the prototype inductive adder, and (ii) the low-voltage power supply, which feeds the gate drivers. However, these filters did not improve the repeatability.

In addition, measurements were carried out at different output voltages, by charging the capacitors of five constant voltage layers to voltages in the range of 100 V to 500 V, and pulse waveforms were recorded. In these measurements, modulation was not applied. There was not a clear correlation between the output voltage and the measured flat-top repeatability. However it was observed that the repeatability, measured for 100 pulses, was approximately ±0.2 % of the maximum range of a channel of the oscilloscope: this occurred for all three sensitivities selected. Therefore, it is supposed that the measured repeatability was actually determined by noise which was created in the measurement set-up, and not by the inductive adder. However, this requires further investigations.

VI. CONCLUSIONS

In this paper measurements on the first five layers of the full-scale prototype have been presented. For the initial measurements analogue modulation was not applied and there was a clear droop in the output pulse. In the second measurements analogue modulation was applied to the output pulses, in order to improve the flat-top stability of the output waveform. In the measurements presented here, the best measured flat-top stability, i.e. the combined ripple and droop of an ensemble average of 100 pulses, was ±0.04 %, corresponding to ±1.0 V for 900 ns pulse flat-top duration. The relative flat-top stability is not yet adequate for the CLIC DR kicker system. However, the measured absolute flat-top stability, ±1.0 V, is already adequate. The flat-top repeatability was approximately ±0.3 % with and without modulation, and the main source for the asynchronous noise is supposed to be the measurement set-up and not the inductive adder. However, improving repeatability requires further studies and application of different measurement methods, i.e. the pulse cancelling technique described in [11] or balanced measurements with a differential amplifier. An example of the later technique is given in [19].

VII. FUTURE WORK

The PCBs for the remaining 15 layers for the first, 20-layer, 12.5 kV, inductive adder have been ordered. The other main components, i.e. mechanical parts, magnetic cores, pulse capacitors, MOSFETs and gate drivers are already in house. The measurements with a 20-layer prototype will commence in autumn 2016. This prototype adder will be used to demonstrate the feasibility of the CLIC DR extraction kicker systems with the required flat-top stability and repeatability with nominal 12.5 kV output voltage. In the near future, a second full-size 12.5 kV prototype inductive adder will be assembled: the final goal is to demonstrate the feasibility of the CLIC DR kicker system with the required flat-top stability. Finally, the prototype adders will be tested together with the CLIC DR prototype extraction stripline kicker [2] in an accelerator test facility. The inductive adder technology is also seen as a promising technology for use in both existing and future systems at CERN [20, 21].

References

[1] M. Aicheler (ed.) et al., "A multi-TeV Linear Collider based on CLIC technology: CLIC Conceptual Design Report", CERN-2012-007.

[2] C. Belver-Aguilar, M.J. Barnes, and L. Ducimetière, "Review on the effects of characteristic impedance mismatching in a stripline kicker", Proc. IPAC'16, THPMW034, Busan, Korea, May 2016.

[3] Y. Papaphilippou et al., "Conceptual design of the CLIC damping rings", Proc. IPAC'12, TUPPC086, New Orleans, USA, May 2012.

[4] J. Holma, M.J. Barnes, and S.J. Ovaska, "Preliminary design of the pulse generator for the CLIC DR extraction system", Proc. PPC'11, pp. 1353-1358, Chicago, USA, June 2011.

[5] J. Holma, and M.J. Barnes, "Pulse power modulator development for the CLIC damping ring kickers", CLIC-Note-938, CERN, Geneva, Switzerland, April 27, 2012.

[6] J. Holma, and M.J. Barnes, "Preliminary design of an inductive adder for CLIC damping rings", Proc. IPAC'11, THPO032, San Sebastian, Spain, Sept. 2011.

[7] J. Holma, M.J. Barnes, and S.J. Ovaska, "Modelling of parasitic inductances of a high precision inductive adder for CLIC", Proc. IPAC'13, MOPWA033, Shanghai, China, May 2013.

[8] J. Holma, and M.J. Barnes, "Initial measurements on a prototype inductive adder for the CLIC kicker systems", Proc. PPC'13, San Francisco, CA, USA, June 2013.

[9] J. Holma, and M.J. Barnes, "The prototype inductive adder with droop compensation for the CLIC kicker systems", IEEE Trans. Plasma Sci., Vol. 42, No. 10, October 2014, pp2899-2908.

[10] J. Holma, M.J. Barnes, and C. Belver-Aguilar, "Measurements on prototype inductive adders with ultra-flat-top output pulses for CLIC DR kickers", Proc. IPAC'14, MOPRO028, Dresden, Germany, June 2014.

[11] J. Holma, "A Pulse power modulator with extremely flat-top output pulses for the Compact Linear Collider at CERN", Ph.D. Thesis, Aalto University publication series, Doctoral Dissertations 196/2015, Helsinki, Finland, 2015.

[12] J.Holma, M.J. Barnes, and L. Ducimetière, "Measurements on magnetic cores for inductive adders with ultra-flat-top output pulses for CLIC DR kickers", Proc. IPAC'16, THPMW032, Busan, Korea, May 2016.

[13] J.Holma, M.J. Barnes, "Design and initial measurements of a 12.5 kV prototype inductive adder for CLIC DR kickers", Proc. IPMHVC'16, San Francisco, CA, USA, July 5-9, 2016.

[14] NWL Inc., http://www.nwl.com

[15] Leclanché Capacitors, http://www.lcap.ch/

[16] Microsemi, http://www.microsemi.com

[17] Rohde & Schwarz, https://www.rohde-schwarz.com/

[18] E.J. Gower, and J.S. Sullivan, "Analog amplitude modulation of a high voltage, solid state inductive adder, pulse generator using MOSFETs", Conf. Rec. Power Modulator Symposium and High Voltage Workshop, pp. 321-324, Hollywood, CA, USA, June 2002.

[19] M. Paraliev, "Comparison of high resolution "balanced" and "direct conversion" measurements for SwissFEL resonant kicker amplitude", Proc. PPC'15, Austin, TX, USA, May 31 – June 4, 2015.

[20] M.J. Barnes, L. Ducimetière, J. Holma, T. Kramer, and A. Fowler, "Inductive adders for replacing thyratron-based modulators at CERN", Proc. EPE'15,Geneva, Switzerland, Sept. 2015.

[21] T. Kramer, "FCC-hh injection and extraction kicker topologies and solid state generators", FCC Week 2016, Rome, Italy, Apr. 2016.

A Kind of Atmospheric Repetitive Nanosecond Pulse Discharge Source for Applications on Surface Modification and Sterilization

Ming-yang Peng,·Lee Li, Yong Cheng
State Key Laboratory of Advanced Electromagnetic Engineering and Technology, School of Electrical and Electronic Engineering
Huazhong University of Science and Technology, HUST
Wuhan, 430074, China
leeli@hust.edu.cn

Yun Teng
State Grid Jiangsu Electric Power Company Research Institute
Nanjing, 211103, China

Abstract—**This paper designs a repetition frequency nanosecond pulse source using the Tesla transformer and the gas spark gap instead of multistage magnetic compression device, according to which, it introduces a repetitive nanosecond-pulse glow-like discharge reactor using multirod-to-cylinder electrode configuration, which would generate a large-scale of glow-like non-thermal plasmas steadily at atmospheric pressure without any barrier dielectric materials in it. The experimental study is made on the treatment of polyamide 66(PA66) fabrics using this kind of plasma. The experiments of wicking property are made to measure the hydrophilicity changes of the treated PA66 fabrics. The atomic force microscopy (AFM) and X-ray photoelectron spectroscopy (XPS) are adopted respectively, to detect the surface changes in physical microstructure and the variations in the type and quantity of chemical functional groups. The results show that, this kind of glow-like plasma will make the PA66 fabric surface etched obviously, the surface roughness is aggravated, the surface energy is enhanced. On the surface of the treated PA66 fabrics, the quantity of oxygen-containing functional groups increases along with the reduction of the original total C-C bonds and C-N bonds. Through half minute of sterilization by the glow-like plasmas, most of the bacterial colonies on the fabric have been vanished. The effectiveness of this kind of plasma treatment could last 3 days in a sealed environment.**

Keywords—nanosecond pulse; glow-like discharge; polyamide 66 (PA66); surface modification; wicking property; sterilization; aging effect

I. INTRODUCTION (*HEADING 1*)

Making use of the cold plasmas to the surface modification in organic polymer materials has become a newly arisen technique. Compared with other methods in material surface modification, for one thing, this new technique is a relatively environmental drying processing method without need of any chemical reagent or large quantities of water. For another, the treatment on the material surface mainly depends on the collisions between the treated materials and the high-energy electrons, excited or metastable active particles in the plasmas. The modification occurs on the material surface within tens of nanometers, and without a remarkable influence on their internal characters or physical structures [1-4].

The treatment effect achieved by the cold plasmas reflects on both physical and chemical aspects. The experimental results of A. Nakahira et al. showed that, through short time treatment, the cold plasmas could change the roughness of the organic glass surface in a relatively low power, and the superficial area of the material would increase [5]. Using unipolar nanosecond pulses, two typical discharges including homogeneous and filamentary modes were applied to modify the surface of PET (Polyethylene glycol terephthalate) films by Zhang Cheng, Shao Tao et al.[6] Their report showed that the surface roughness of the PET films increased and the contact angle decreased effectively. Meiners et al. researched the influences on the hydrophily of polypropylene films treated by atmospheric air plasmas, which were generated by DBD (dielectric barrier discharge) [7]. The results showed that the hydrophily of the treated polypropylene films improved effectively. With the further measurement, the surface tension of the treated polypropylene films increased by 72 mN/m. Yuri Akishev et al. studied on the surface chemical influences of non-thermal plasmas on the PE, PP, PET and polyester fabrics [8, 9]. It was shown that the polyester material gained a better hydrophily in a short time with the generation of new function groups. According to the research of Lai et al.[10] they founded that C=O, C-O, COOH and C-NH2 appeared on the polymer film after the plasma treatment, what's more, the carbonyl group was the key factor to the improvement of the hydrophily, the contact angle decreased with the increase of the carbonyl.

Up to now, most of the surface modification studies are focused on plasma jet or dielectric barrier discharge (DBD) at atmospheric air. However, the plasmas produced by plasma jet have relatively small active area, and it needs inert gas environment, which makes it inconvenient for application. As for DBD, the dielectric barriers in the DBD may lead to plasma pollution, and the discharge size is limited in the direction of the discharge. According to the recent studies, integrating wire electrodes with repetitive nanosecond pulses had become an effective method to generate large-volume stable glow-like diffuse plasmas at atmospheric pressure in air without dielectric barrier [11, 12], which was a kind of uniform plasma attracting many attentions [13,14].

Given this, this paper has designed a surface modification reactor using a cylindrical discharge chamber, which can generate cylindrical cold plasmas in air environment without any dielectric barriers. To verify the potency of this kind of plasma in material modification, especially the high molecular polymer, we choose the nylon fabric to be the test object. The X-ray photoelectron spectroscopy (XPS) and atomic force microscopy (AFM) are adopted to investigate the nylon surface changes in physical microstructure, the variations in the type and quantity of polar functional groups, before and after the plasma treatment. For the widely application of the high molecular polymer, such as textile and medical industry [15, 16], the improvement in hydrophily and the sterilizing effect of the treated nylon material under different discharge conditions are observed. Also, the timeliness of this kind of plasma treatment is studied. The research results of this paper may provide new idea for the surface modification of polymer materials.

II. EXPERIMENTAL

A. Nylon materials

Fig. 1. Microstructure of the PA66 fabric observed by scanning electron microscope

The nylon material used in this paper is a kind of polyamide 66 (PA66) fiber fabrics. The mesh number of the processed PA66 fabrics is 400 (38 μm pore size). Generally, the PA66 fabric has the poor hydrophily, the surface-to-volume ratio is low, and it is hard to be colored [17]. Previously, the chemical processing was usually employed in surface modification of the PA66 fabric. But the process was complex, the cost was high, the polluted water produced during the process did harm to the environment [1, 18]. On the other hand, in the medical application, the material needed to be sterilized, the traditional methods liked pasteurization was cumbersome, and it wasted a lot of time and energy. Fig. 1 gives the microstructure of the PA66 fabric observed by scanning electron microscope. All the samples are sterilized by alcohol firstly, and then cleaned up by the deionized water. At last, they are dried in vacuum drying device for 1 hour before the plasma treatment.

B. Surface modification reactor

The whole surface modification reactor can be divided into a repetitive nanosecond pulse generator which is designed and made by our laboratory, and a cylinder reactor. As is shown in the dash box of Fig. 2, the nanosecond pulse generator is mainly composed of a repetitive charging circuit, a Tesla transformer and a sharpening switch. The repetitive charging circuit is formed by a DC supply, a storage capacitor C_{main} (2 mF), a charging thyristor S_{ch} and a charging inductor L_{ch}. The output voltage of the DC supply is adjustable. After S_{ch} turns on, C_1 (10 μF) will be resonantly charged by C_{main} through L_{ch}.

After being charged to a defined voltage, C_1 discharges through a controllable fast thyristor S_{main}. Afterwards, high voltage will be obtained across the secondary ceramic capacitor C_2 (0.1 nF) of the Tesla transformer T. A trigatron gap is selected as sharpening switch G. After sharpened by a sharpening switch G, repetition nanosecond pulses will be obtained. G is designed to break down at nearly the maximum voltage across C_2. The trigger electrode of G is connected to the ground through a resistor R_{tr} of 1.0 MΩ. Controlled by the trigger at the repetition frequency, the primary capacitor C_1 (10 μF) is alternant in the state of charge/discharge through a charging thyristor S_{ch} and a controllable fast thyristors S_{main}. The output pulse voltage waveform is shown in Fig. 2(a). The maximum peak voltage is 100 kV, the rise-time is around 25 ns every pulse, the full width at half maximum (FWHM) is ~750 ns.

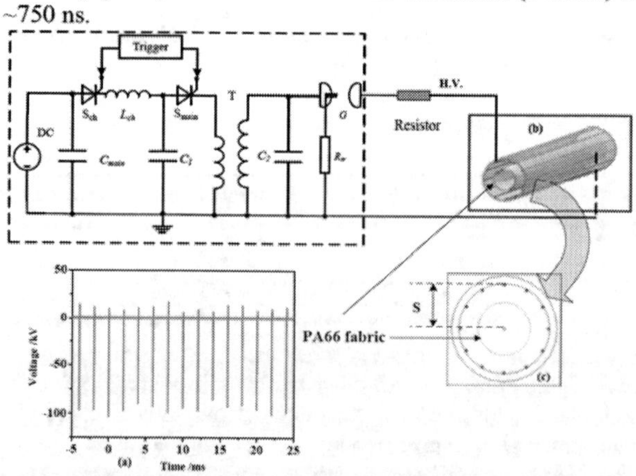

Fig. 2. Schematic diagram of the reactor. (a) The output pulse voltage waveform; (b) The 3D view (The dots of orange and blue stand for the high-voltage terminal and grounding pole, respectively. The blue circles represent

the material waiting to be processed); (c) Side image of the position of the Nylon fabrics being processed.

The discharge chamber adopts a multirod-to-cylinder electrode configuration. The Fig. 2(b) gives the 3D view of the cylindrical electrode configuration, the orange and blue spots stand for the high-voltage (H.V.) and grounding poles, respectively. Additionally, for the purpose of the uniformity and the larger volume of glow-like plasmas [19], the number of high-voltage poles surrounding the grounding pole is set at 12, and the distance between high-voltage pole and the grounding pole is 4.0 cm. All these poles are made of straight bare copper wires. The selected copper wires are 2.0 mm in diameter and 50.0 cm in length. All the wires are parallel to each other. The side view image of the cylindrical electrode configuration is shown in Fig. 2(c). The relative positions of the high-voltage pole, the polyamide fabrics and the grounding can be fixed by some independent insulation supporting structure. Meanwhile, the discharge chamber can be rotatable at a certain speed, which will obtain more uniform treatment results.

The exposure times of Fig. 3(a) and Fig. 3(b) are 1 s and 2 ms, respectively. From the Fig. 3(a), we can see that the atmospheric-air glow-like plasmas fill up the whole discharge chamber, there is no apparent dark space. The exposure time of Fig. 3(b) is equal to the period of a single pulse (500 Hz). It is obvious that, the plasma channels develop between all the H.V. and grounding poles during a sing pulse. By employing longer wire electrodes with the corresponding matching circuit, it is possible to obtain larger volume of glow-like plasmas.

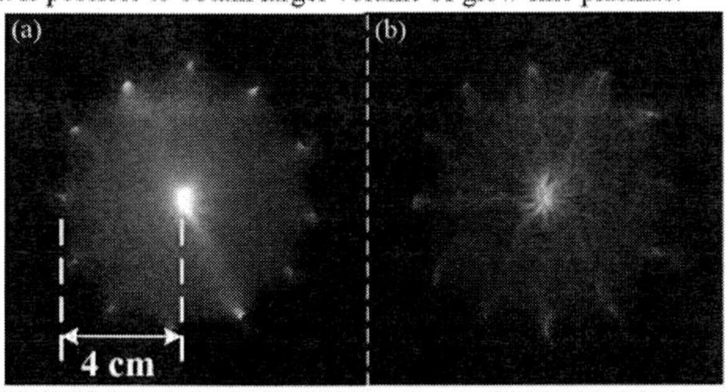

Fig. 3. The side view of the cylindrical electrode configuration in discharge. (a) exposure for 1 s; (b) exposure for 2 ms.

III. Physical Analysis of the Surface

The atomic force microscope is Japanese Shimadzu model SPM9700, which is applied to detect the changes of the surface microcosmic three-dimensional (3-D) structure under different glow-like discharge conditions, before and after the treatment. This AFM instrument can do 3-D scanning of the nylon fabric surface. The lateral resolution is 2.0 nm. This microscope has two micro-cantilevers, which are very sensitive to the weak force. One micro-cantilever is settled, the other one installs a tiny probe. During the test, the tiny probe touches the sample surface, scanning around. The tip atom of the tiny probe has weak repulsive force against the sample surface, the micro-cantilever with a tiny probe moves on the surface wavily. It can obtain the equipotential surface of the atomic force between the tip atom and the sample surface, and then it can get the morphology information of the sample surface.

Making the control group, PA66 fabrics of treated 10 s and treated 60 s scanned by the AFM. The scanning area is $1\mu m \times 1\mu m$, and Fig. 4 gives the scan results.

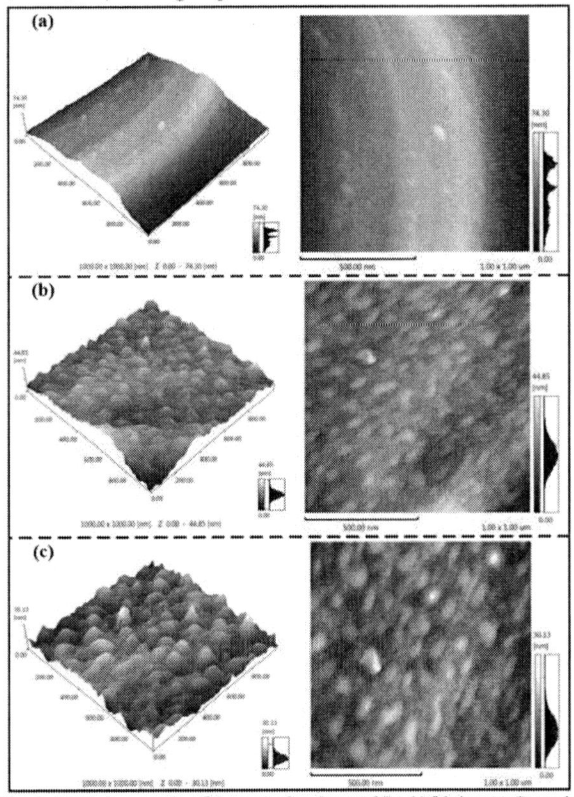

Fig. 4. AFM photographs of the control and treated PA66 fabrics. (a) Control; (b) Treated for 10s; (c) Treated for 60s.

As is shown in Fig. 4(a), untreated PA66 fabric surface seems to be very smooth. It nearly has no protrusions. The surface roughness (Ra) of the control sample is 3.794 nm. Through the Fig. 4(b) and Fig. 4(c), after the glow-like plasma treatment, a lot of protrusions have appeared on the fabric surface. Meanwhile, the Ra increases to 7.221 nm and 8.392 nm, respectively. The more time it is treated, the rougher the surface is. However, the protrusions increasing from 10 s to 60 s are not significant. It illustrates that the modification occurs on the surface of the polyamide fabrics without a remarkable influence on their internal physical structures. This etching phenomenon is an important form and a distinguishing feature of the improvement of the fabric hydrophilic characteristic [20, 21]. The plasma etching results in the surface's becoming rough. And the surface layer is degraded, macromolecular fractures by incorporation of polar functional groups and cross-linking on the surface. Therefore, the hydrophily of the material surface can be improved.

The PA66 fabric is made of a few of filaments, so that there will be some micron-sized apertures between those filaments. Then these apertures will act as the channels for traversing of the plasmas. This approves that, after the high-energy electrons and other active particles in the stable or metastable state arrive at the surface of the PA66 fabric, a few of them act on the surface, but most of the particles will go through the fibrage with several collisions. They will continue to accelerate under the effect of the electric field, and ionize the air of the gap until reach the ground.

Besides, after the treatment, the distribution of the protrusions on the surface of PA66 fabric is relatively uniform. This phenomenon supports the uniformity of the treatment effect achieved by the glow-like plasmas.

IV. Chemical Analysis of the Surface

Surface analyses of the processed PA66 fabrics are made by the X-ray photoelectron spectroscopy (XPS). The instrument is the Japanese Shimadzu-Kratos model AXIS-ULTRA DLD-600W X-ray photoelectron spectrometer with the Bi-anode Al/Mg X-ray source, the spatial resolution is less than 3 μm, the acceptance angle is 90° , the vacuum degree is 10^{-7}~10^{-8} Pa. All XPS-peaks are referred to the C1s signal at a binding energy of approximately 285 eV. Then the data are fitted by Gaussian-Lorentzian curves after removing Shirley background using XPSpeak software.

Fig. 5 shows the spectra characteristics of the PA66 fabrics untreated and treated with different treatment time.

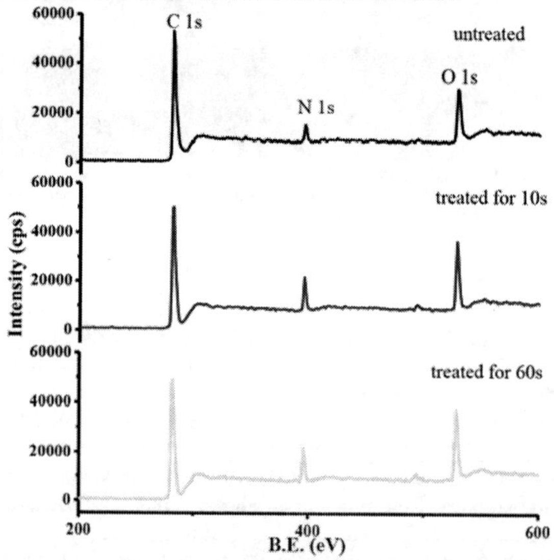

Fig. 5. XPS scan of untreated PA66 fabrics and after diffuse plasma treatment for 10s, 60s.

Table I and Fig. 6 give the details about the relative chemical compositions, atomic ratios determined by XPS for PA66 fabric untreated and treated by glow-like plasmas. The element content is calculated by the ratio of the corresponding integral area of the peaks. Comparing the three XPS spectrum of different conditions, it can be found that the control group has the highest C1s peak. After processing, the C element content of PA66 fabric decreases significantly, at the same time, O element content increases obviously, and N element content has changed accordingly. This is because the plasma treatment can import some polar functional groups onto nylon material surface [22, 23], results in the changes of the microstructure on the fiber surface. As a result of the corresponding chemical bonds' breaking and the formation of new functional groups, it is finally reflected in the C, O and N element content changes in the XPS spectra. With the increase of the treatment time, the range of this kind of element content change is more apparent.

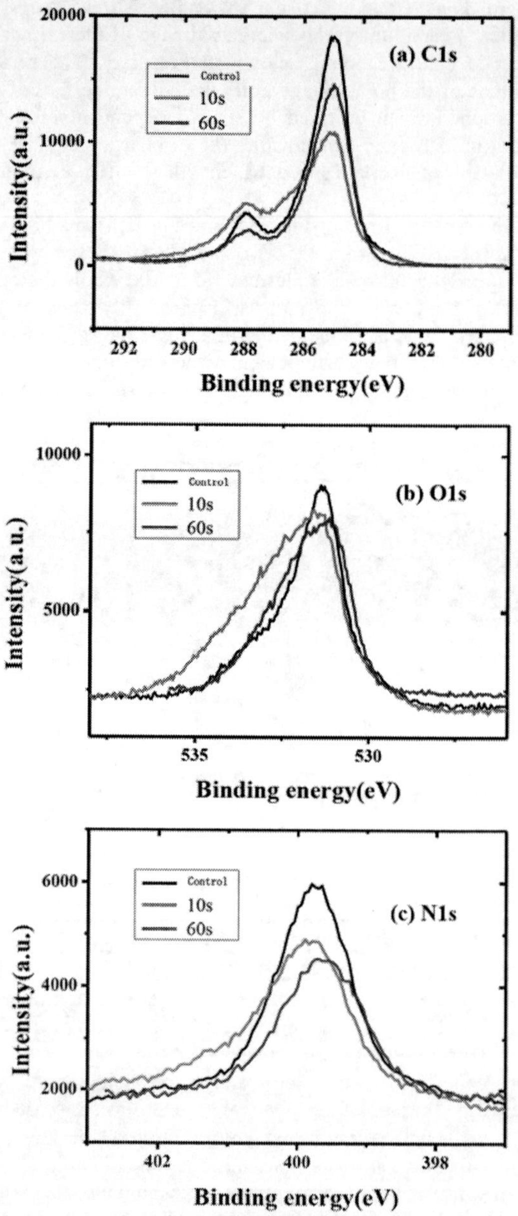

Fig. 6. XPS scan of C/O/N elements of nylon 66 fiber surfaces.

TABLE I
RELATIVE CHEMICAL COMPOSITION AND ATOMIC RATIOS

Sample	Element Content (%)			Ratio		
	C1s	O1s	N1s	O/C	N/C	(O+N) /C
Control	77.93	13.75	8.32	0.18	0.11	0.29
Treated10s	76.63	16.38	6.99	0.21	0.10	0.31
Treated60s	72.40	18.76	8.84	0.26	0.12	0.38

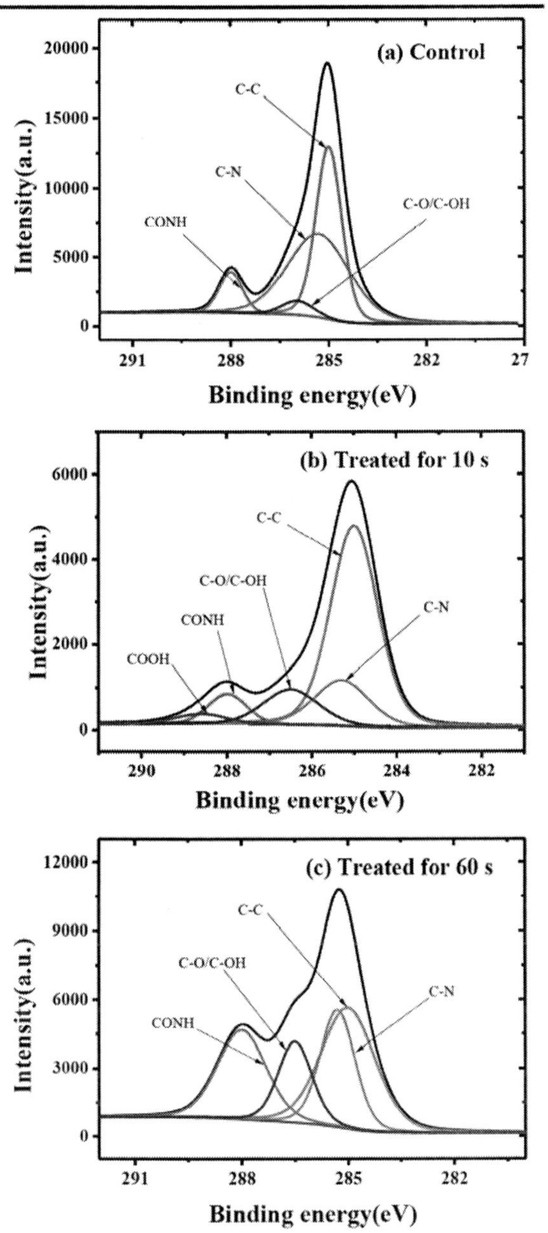

Fig. 7. Synthesis curve of C1s-spectra for the control and diffuse plasma treated PA66 fibers.

To make further analyses about the changes on the types and ratios of the surface functional groups, the C1s and O1s spectra are spilt by software XPSpeak. As is shown in Fig. 7, the C1s peak of untreated samples can be split into four peaks, which represent four carbon-containing groups: C-C (285.0 eV), C-N (285.3 eV), C-O/C-OH (286.5 eV) and CONH (288.0 eV) [24, 25]. The content of carbon groups corresponding to each peak is shown in Table Ⅱ, then Fig. 7 gives the peak splitting diagrams of the samples treated 10 s and 60 s, the corresponding changes of the contents of carbon functional groups are shown in Table Ⅱ line 2 &3.

TABLE Ⅱ
RELATIVE AMOUNT OF DIFFERENT CARBON-CONTAINING GROUPS (%)

Sample	Carbon-containing Groups (%)				
	C-C	C-N	C-O/C-OH	CONH	COOH
Control	40.35	47.05	4.06	8.54	0
Treated10s	60.87	15.87	12.36	7.59	3.30
Treated60s	37.00	23.05	15.54	24.42	0

The Ols XPS spectra under different conditions are shown in Fig. 8, The O1s peak of the samples can be split into two peaks, which represent two oxygen-containing groups: C-O/C-OH (533.3 eV), CONH /COOH (531.6 eV) [26, 27]. Table Ⅲ gives the relative content of oxygen-containing groups corresponding to each peak. It is shown that, when the treatment time is 10 s, the ratios of the two oxygen-containing groups don't change a lot. When the treatment time reaches 60 s, the content of C-O/C-OH increases from 10.04% to 47.10%, the content of CONH/COOH decreases from 89.96% to 52.90%.

It is shown that, the total amount of C-C bonds and C-N bonds on the untreated sample, the treated10s sample and the treated 60s sample are 87.4%, 76.74% and 60.05%, respectively. The quantities of oxygen-containing functional groups increase along with the reduction of the total C-C bonds and C-N bonds in the original polyamides molecule. The glow-like plasmas fracture the original carbon-containing bonds, which is helpful to constitute carboxyl, hydroxyl and double bond of carbon to oxygen using the oxygenium in the air. Those hydrophilic groups will improve the hydrophily of the treated fabrics distinctly. The CONH is the original chemical bond in the polyamides molecules, the COOH and C-OH appears after the treatment. As is shown in Table Ⅱ and Table Ⅲ, when it have reached a certain time with a longer oxidation reaction, the hydroxyl will convert into the carboxyl more easily, but the excessive treatment may lead to that the C-O/C-OH combines closely to the PA66 molecules, the carboxyl converts into other substances or just sheds, which results in the fact that few of carboxyls can be found after processed 60s. We can clearly distinguish the changes of the functional groups varied with the treat time through the peaks of the corresponding peak splitting diagrams in Fig. 7 and Fig. 8. And this phenomenon corresponds to the C, N and O element content changes in Fig. 5.

TABLE III
RELATIVE AMOUNT OF DIFFERENT OXYGEN-CONTAINING GROUPS (%)

Sample	Oxygen-containing Groups (%)	
	C-O/C-OH	CONN/COOH
Control	10.04	89.96
Treated10s	7.64	92.36
Treated60s	47.10	52.90

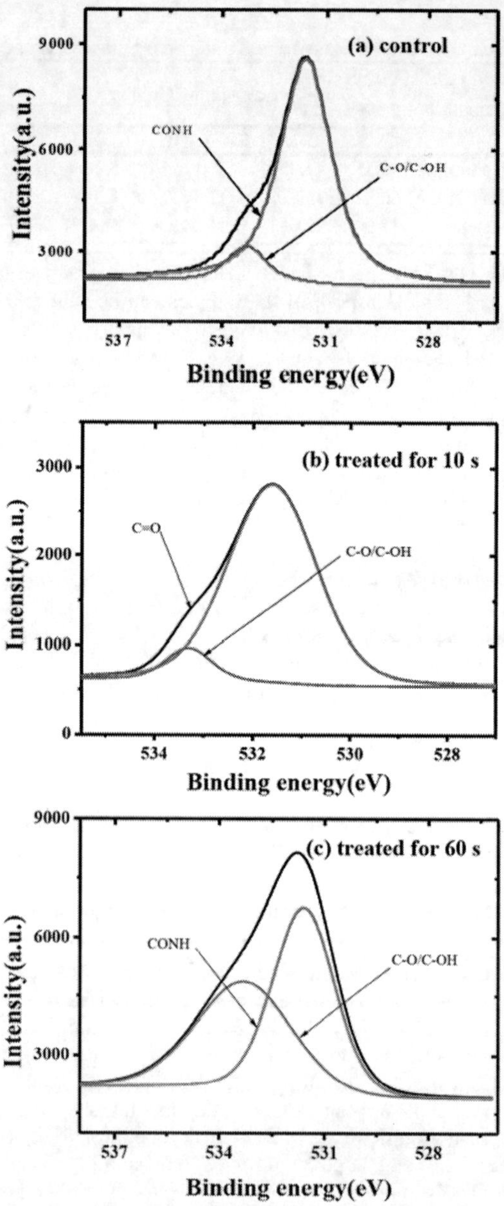

Fig. 8. Synthesis curve of O1s-spectra for the control and diffuse plasma treated PA66 fibers

V. STERILIZATION EFFECT

The high-energy electrons, ions, active particles in the plasmas and the ultraviolet rays during the discharge have inactivation effect on the microorganisms to some degree [28]. This kind of inactivation effect is achieved by their damage on the cytomembrane, cytoderm and the genetic material. To study on the bactericidal performance of this kind of glow-like plasma, the Escherichia coli (E. coli) HT115 is chosen to be the test object. Before the experiment, the wire electrodes and the whole reactor are sterilized strictly, the E. coli has penicillin resistance processing to eliminate the disturbance from the microbial contamination in air.

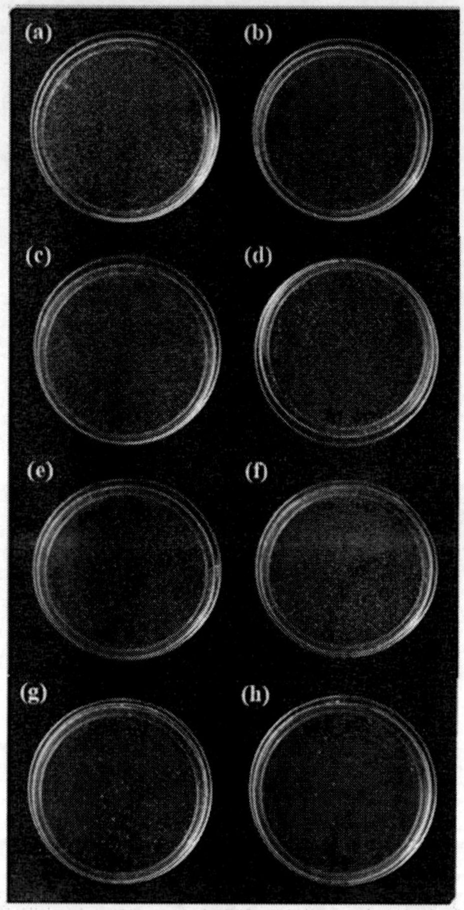

Fig. 9. Sterilizing effect on Escherichia coli by atmospheric air diffuse plasmas(a)control; (b)0.5 s; (c)1 s; (d)2 s; (e)4 s; (f)8 s; (g)16 s; (h)32 s

The experimental procedures are as follows: (1) the samples are soaked in the 70% alcohol, then placed in the high-pressure steam sterilization pot for drying and sterilizing with the temperature is 121 ℃ and the pressure is 0.1 MPa; (2) On the Clean Bench, 10 μL bacteria solution (culturing for 13 hours) is inoculated onto the PA66 fabric surface, a control group and a series of experimental groups are set; (3) Having the samples treated by glow-like plasmas under different experimental conditions; (4) After the treatment, cutting off the region of the PA66 fabric where is inoculated by the E. coli, Making this piece of fabric vibrated fully in 5 mL sterile water to elute the thalli. 100 μL solution is absorbed and coat on the solid medium of penicillin resistance; (5) Having the solid media of control group and experimental groups cultured in the 37 ℃ thermostat for a certain time, then counting the number of E. coli colonies on the solid media.

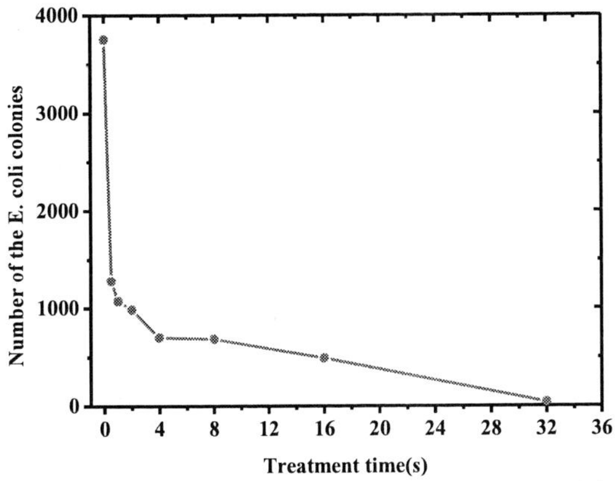

Fig. 10. Number of the E. coli colonies with the treatment time varied.

Fig. 9 is the effect picture of the atmospheric-air glow-like plasmas treated E. coli colonies under different treatment time. Fig. 10 gives changing curve of the number of the E. coli colonies with treatment time varied. Fig. 9(a) is the control group, the Fig. 9(b)~ (h) are sterilizing effect pictures of E. coli colonies treated 0.5~32 s. During the experiment, the pulse repetition frequency (PRF) is 500 Hz. It is clear that, the sterilizing effect achieved by the glow-like plasmas enhances with the treatment time increases. When the treatment time reaches 32 s, there is nearly no colony appearing on the solid medium. Just having 0.5 s treatment of glow-like plasmas, there is approximately only 1/3 of the colonies left on the solid medium compared with the control group. All of these approve that, this kind of atmospheric-air glow-like plasma has a promising sterilizing effect.

Fig. 11 gives the relative numbers of E. coli colonies treated by glow-like plasmas with PRF varied. It is shown that, when the treatment time is constant, the E. coli colonies on the solid medium reduce observably along with the increase of the PRF. What's more, when the number of discharges is constant, the E. coli colonies reduce at first and then increase while the PRF is augmenting. This is due to that, for a certain treatment time, the higher the PRF is, the higher the concentration of the bactericidal-effect active substances in glow-like plasma is, which will has a deeper damage on the E. coli colonies. With regard to a constant number of discharges, the interval between two sequential pulses is long while the PRF is relative low. The short-lived active particles in the plasmas will disappear, and the long-lived active particles will diffuse to some degree, so the concentration of the bactericidal-effect active substances in the reacting chamber isn't high enough to have a valid sterilizing effect on the E. coli on the PA66 fabrics. While the PRF augments to a certain frequency, the active particles reaches a certain concentration to have a good enough sterilizing effect on the E. coli. However, when the PRF continues augmenting to a certain extent, even through the activity of the plasmas is enhanced, the acting time is too short to have a sufficient contact between the active particles and the PA66 fabrics, so the sterilizing effect is weakened. According to this phenomenon, in the practical application, treatment time can be shortened by increasing the PRF, meanwhile, the same sterilizing effect can be achieved by extending treatment time at the optimal PRF.

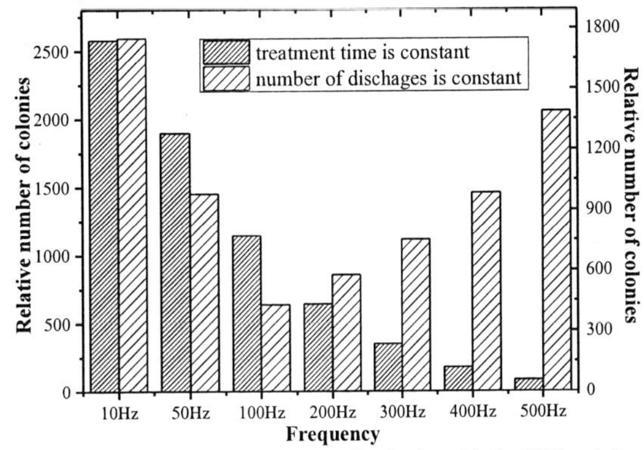

Fig. 11. Relative number of the E. coli colonies with the PRF varied.

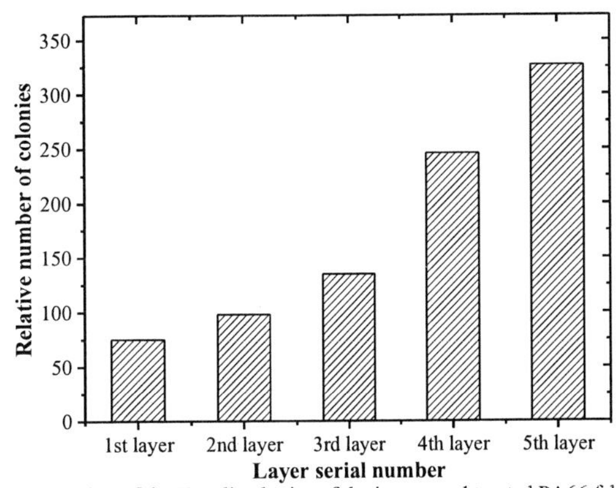

Fig. 12. Number of the E. coli colonies of the juxtaposed treated PA66 fabrics.

Fig. 12 gives the relative numbers of the E. coli colonies of the side-by-side treated PA66 fabrics through the atmospheric-air glow-like plasmas. Before the treatment, the different layers of PA66 fabrics are inoculated by E. coli culture solution of the same concentration. During the discharge, the total 5 layers of PA66 fabrics are placed side-by-side. From the figure, it is shown that, the first layer, i.e. the top layer has the best sterilizing effect, and the bottom layer has the worst sterilizing effect. In other words, the relative number of E. coli colonies increase sequentially from the 1st layer to the 5th layer. This is because that, most of the active particles in the plasmas, which have the bactericidal effect, are absorbed by the top layer, few of them can reach the bottom layer. It should be noted that, extending the treatment time appropriately, the bottom layer also can achieve a good enough treatment effect. The reasons leading this phenomenon may as follows, for one thing, there will be more active particles surrounding the layers of PA66 fabrics while the treatment time increases. And then more active particles can touch the bottom layer. For another, the bactericidal-effect active substances in plasmas may dissolve in the solution on the fabrics, then some properties like acid-base property may be changed so that the E. coli will be inactivated. And this solution may penetrate to the bottom layer through the apertures of the PA66 fabrics, then the living environment of the E. coli is changed, the E. coli is inactivated.

VI. AGING EFFECT

The Table IV is focused on the timeliness of this kind of glow-like plasma treatment. In this experiment, 5 pieces of the PA66 fabrics are processed by the atmospheric-air glow-like plasma respectively with the processing time set at 10 s and PRF set at 500 Hz. One piece of the treated fabrics has been taken out to have the XPS test directly. Two pieces of the rest treated PA66 fabrics are placed in the air for 1 day and 3 days. The last two pieces of the treated PA66 fabrics are stored in sealed polythene bags for 1 day and 3 days, respectively. And then they are tested by XPS. The results of relative chemical compositions and atomic ratios of treated PA66 fabrics stored under different conditions are shown in the Table IV.

TABLE IV
RELATIVE CHEMICAL COMPOSITION AND ATOMIC RATIOS WITH STORAGE TIME AND ENVIRONMENT VARYING

Sample	Element Content (%)			Ratio		
	C1s	O1s	N1s	O/C	N/C	(O+N)/C
Reference	80.11	15.76	4.13	0.20	0.05	0.25
One day(air)	77.08	13.85	9.07	0.18	0.12	0.30
One day(bag)	73.02	17.13	9.85	0.23	0.13	0.36
Three days(air)	77.41	14.17	8.42	0.18	0.11	0.29
Three days(bag)	74.75	16.28	8.96	0.22	0.12	0.34

TABLE V.
RELATIVE AMOUNT OF DIFFERENT CARBON-CONTAINING GROUPS (%) WITH STORAGE TIME AND ENVIRONMENT VARYING

Sample	Carbon-containing Groups (%)			
	C-C	C-N	C-O/C-OH	CONH
One day(air)	42.37	40.20	5.59	11.84
One day(bag)	47.10	27.48	10.19	15.23
Three days(air)	47.74	35.44	5.48	11.35
Three days(bag)	45.60	32.42	9.02	12.96

The reasons why plasmas can improve the hydrophily of PA66 fabrics are not only the introduction of oxygen-containing functional groups, but also the removal of impurities on the surface, which can increase surface energy. When the treated fabrics are placed in the air, the impurities may adsorb on the surface again, to decrease the hydrophily. That is why the timeliness, i.e. the aging effect is an important influence factor in the industrial application of the surface modification [29]. Comparing the data in the Table IV, when the treated fabrics are placed in the same external conditions, the O/C ratio almost remains unchanged in the air over storage time. When stored in sealed polythene bags, the O/C ratio of storing 3 days is only 4.3% less than the one of storing 1 day. All of the data reveal the fact that this kind of glow-like plasma has a relatively good timeliness, meanwhile, the aging effect can be delayed to some degree. What's more, the storing in the sealed polythene bag one has the higher O/C ratio while the storage time is constant. This is due to the small quantity of oxygen and impurities, which restrict the oxidation-reduction reaction and the repetition pollution, respectively. On the other hand, the less the flow disturbance is, the longer time the newly generated polar functional groups will survive. Table V and Table VI give relative amounts of different carbon-containing and oxygen-containing groups with storage time and environment varying.

TABLE VI.
RELATIVE AMOUNT OF DIFFERENT OXYGEN-CONTAINING GROUPS (%) WITH STORAGE TIME AND ENVIRONMENT VARYING

Sample	Oxygen-containing Groups (%)	
	C-O/C-OH	CONN/COOH
One day(air)	8.73	91.27
One day(bag)	16.53	83.47
Three days(air)	13.25	86.75
Three days(bag)	20.02	79.98

VII. CONCLUSIONS

In this paper, large-volume, barrier-free glow-like diffuse plasmas have been obtained using the multi-wire cylinder electrode configuration. All of the experiments are under atmospheric pressure air, the conclusions are as follows:

1) Through AFM analyses, this kind of glow-like plasma will make the PA66 fabric surface etched obviously, the surface roughness is aggravated, the surface energy is enhanced, all of which can be helpful to improve the hydrophily of PA66 fabrics.

2) The XPS analyses show that, after the treatment, the quantity of oxygen-containing functional groups increases along with the reduction of the original total C-C bonds and C-N bonds, and N element content has been changed accordingly. This is due to that the glow-like plasmas fracture the original carbon-containing bonds, which leads to the formations of carboxyl, hydroxyl and double bond of carbon to oxygen using the oxygenium in the air. At the beginning of the treatment, the COOH and C-OH appears after the treatment. The excessive treatment may lead to the shedding of the carboxyl or its conversion into other substances.

3) After almost half minute glow-like plasma treatment for sterilization, most of the E. coli inoculated on the PA66 fabrics is inactivated. For the treatment effect of sterilization, when the number of discharges is constant, the treatment effect doesn't monotonically increase with the augment of the PRF, there exists an optimal frequency. When the treatment

time isn't the dominant factor, extending the treatment time appropriately is an effective method to reach the same or better treatment effect while the PRF can be lowered.

4) Experimental results show that the treatment effect obtained by the glow-like plasmas can maintain a certain time, i.e., the treatment effect has a good enough timeliness. What's more, the aging effect can be delayed by the sealed storing environment. Before further processing, the treated fabrics are suggested to be stored in a sealed or oxygen-free environment.

It should be noted that, the experimental conclusions can be generalized to many kinds of porous polymer fabrics.

Acknowledgment

HUST analytical and testing center is gratefully acknowledged for the assistance in AFM and XPS tests.

References

[1] H. R. Yousefi, M. Ghoranneviss, A. R. Tehrani, and S. Khamseh, "Investigation of glow discharge plasma for surface modification of polypropylene," Surface & Interface Analysis, vol. 35, no. 12, pp. 1015–1017, 2003.

[2] R. D. Mundo, R. D'Agostino, and F. Palumbo, "Long-Lasting Antifog Plasma Modification of Transparent Plastics," Acs Applied Materials & Interfaces, vol. 6, no. 19, pp. 17059-17066, 2014.

[3] N. Abidi, and E. Hequet, "Cotton fabric graft copolymerization using microwave plasma. I. Universal attenuated total reflectance–FTIR study," Journal of Applied Polymer Science, vol. 93, no. 1, pp. 145-154, 2004.

[4] C. X. Wang, and Y. P. Qiu, "Two sided modification of wool fabrics by atmospheric pressure plasma jet: Influence of processing parameters on plasma penetration," Surface & Coatings Technology, vol. 201, no. 14, pp. 6273-6277, 2007.

[5] A. Nakahira, Y. Suzuki, S. Ueno, H. Akamizu, K. Kijima, and S. Nishijima, "Effect of Plasma Treatment on Microstructure and Surface of Glass for Plastic-Based Composite," Science & Engineering of Composite Materials, vol. 8, no. 3, pp. 129-136, 1999.

[6] C. Zhang, T. Shao, K. Long, Y. Yu, J. Wang, D. Zhang, P. Yan, and Y. Zhou, "Surface Treatment of Polyethylene Terephthalate Films Using DBD Excited by Repetitive Unipolar Nanosecond Pulses in Air at Atmospheric Pressure," IEEE Transactions on Plasma Science, vol. 38, no. 6, pp. 1517-1526, 2010.

[7] S. Meiners, J. G. H. Salge, E. Prinz, and F. Förster, "Surface modification of polymer materials by transient gas discharges at atmospheric pressure," Surface & Coatings Technology, vol. 98, no. 1, pp. 1121-1127, 1998.

[8] Y. Akishev, M. Grushin, A. Napartovich, and N. Trushkin, "Novel AC and DC Non-Thermal Plasma Sources for Cold Surface Treatment of Polymer Films and Fabrics at Atmospheric Pressure," Plasmas & Polymers, vol. 7, no. 3, pp. 261-289, 2002.

[9] Y. S. Akishev, M. E. Grushin, A. E. Monich, A. P. Napartovich, and N. I. Trushkin, "One-Atmosphere Argon Dielectric-Barrier Corona Discharge as an Effective Source of Cold Plasma for the Treatment of Polymer Films and Fabrics," High Energy Chemistry, vol. 37, no. 5, pp. 286-291, 2003.

[10] J. Lai, B. Sunderland, J. Xue, S. Yan, W. Zhao, M. Folkard, B. D. Michael, and Y. Wang, "Study on hydrophilicity of polymer surfaces improved by plasma treatment," Applied Surface Science, vol. 252, no. 10, pp. 3375-3379, 2006.

[11] L. Li, L. Liu, Y. L. Liu, Y. Bin, Y. F. Ge, and F. C. Lin, "Analysis and experimental study on formation conditions of large-scale barrier-free diffuse atmospheric pressure air plasmas in repetitive pulse mode," Journal of Applied Physics, vol. 115, no. 2, pp. 023301 - 023301-6, 2014.

[12] Y. Teng, L. Li, Y. L. Liu, L. Liu, and M. Liu, "Generation of large-scale, barrier-free diffuse plasmas in air at atmospheric pressure using array wire electrodes and nanosecond high-voltage pulses," Physics of Plasmas (1994-present), vol. 21, no. 10, 2014.

[13] Z. Zhu, G. Zhang, L. Liu, and Y. Li, "The research on atmospheric pressure glow-like discharge plasma." p. 172.

[14] B. Qi, C. Ren, D. Wang, S. Z. Li, K. Wang, and Y. Zhang, "Uniform glowlike plasma source assisted by preionization of spark in ambient air at atmospheric pressure," Applied Physics Letters, vol. 89, no. 13, pp. 131503 - 131503-3, 2006.

[15] H. Mei, J. Feng, J. Wang, X. Zhang, Y. Li, and Y. Yan, "Synthesis and characterization of nano-HA/PA66 composites," Journal of Materials Science Materials in Medicine, vol. 14, no. 7, pp. 655-660, 2003.

[16] [P. Heikkilä, A. Taipale, M. Lehtimäki, and A. Harlin, "Electrospinning of polyamides with different chain compositions for filtration application," Polymer Engineering & Science, vol. 48, no. 6, pp. 1168–1176, 2008.

[17] A. Bismarck, D. Richter, C. Wuertz, and J. Springer, "Basic and acidic surface oxides on carbon fiber and their influence on the expected adhesion to polyamide," Colloids & Surfaces A Physicochemical & Engineering Aspects, vol. 159, no. 2-3, pp. 341-350, 1999.

[18] Z. Cai, Y. J. Hwang, Y. C. Park, C. Zhang, M. Mccord, and Y. Qiu, "Preliminary investigation of atmospheric pressure plasma-aided desizing for cotton fabrics," Aatcc Review, vol. 2, no. 12, pp. 18-21, 2002.

[19] Y. Teng, L. Li, Y. Cheng, N. Ma, M. Y. Peng, and M. H. Liu, "Optical and electrical investigation of a cylindrical diffuse-discharge chamber," Physics of Plasmas, vol. 22, no. 3, 2015.

[20] T. Wakida, M. Lee, and Y. Sato, "Dyeing properties of oxygen low - temperature plasma - treated wool and nylon 6 fibres with acid and basic dyes," Coloration Technology, vol. 112, no. 9, pp. 233-236, 2008.

[21] J. Yip, K. Chan, K. Sin, and L. Kai, "Study of plasma-etched and laser-irradiated polyamide materials," Material Research Innovations, vol. 6, no. 2, pp. 44-50, 2002.

[22] D. S. Wavhal, and E. R. Fisher, "Membrane Surface Modification by Plasma-Induced Polymerization of Acrylamide for Improved Surface Properties and Reduced Protein Fouling," Langmuir, vol. 19, no. 1, pp. 79-85, 2002.

[23] R. K. Y. Fu, Y. F. Mei, G. J. Wan, G. G. Siu, P. K. Chu, Y. X. Huang, X. B. Tian, S. Q. Yang, and J. Y. Chen, "Surface composition and surface energy of Teflon treated by metal plasma immersion ion implantation," Surface Science, vol. 573, no. 3, pp. 426-432, 2004.

[24] S. H. Kim, H. C. Su, N. E. Lee, H. M. Kim, W. N. Yun, and Y. H. Kim, "Adhesion properties of Cu/Cr films on polyimide substrate treated by dielectric barrier discharge plasma," Surface & Coatings Technology, vol. 193, no. s 1–3, pp. 101-106, 2005.

[25] O. K. Wha, K. S. Hun, and K. E. Ae, "Improved surface characteristics and the conductivity of polyaniline–nylon 6 fabrics by plasma treatment," Journal of Applied Polymer Science, vol. 81, no. 3, pp. 684–694, 2001.

[26] N. Inagaki, S. Tasaka, and H. Kawai, "Surface modification of aromatic polyamide film by oxygen plasma," Journal of Polymer Science Part A Polymer Chemistry, vol. 33, no. 12, pp. 2001-2011, 2003.

[27] N. Inagaki, S. Tasaka, H. Kawai, and Y. Yamada, "Surface modification of aromatic polyamide film by remote oxygen plasma," Journal of Applied Polymer Science, vol. 64, no. 5, pp. 831–840, 1997.

[28] S. Perni, G. Shama, J. L. Hobman, P. A. Lund, C. J. Kershaw, G. A. Hidalgo-Arroyo, C. W. Penn, X. T. Deng, J. L. Walsh, and M. G. Kong, "Probing bactericidal mechanisms induced by cold atmospheric plasmas with Escherichia coli mutants," Applied Physics Letters, vol. 90, no. 7, pp. 073902-073902-3, 2007.

[29] A. Zille, M. M. Fernandes, A. Francesko, T. Tzanov, M. Fernandes, F. R. Oliveira, L. Almeida, T. Amorim, N. Carneiro, and M. F. Esteves, "Size and Aging Effects on Antimicrobial Efficiency of Silver Nanoparticles Coated on Polyamide Fabrics Activated by Atmospheric DBD Plasma," Acs Applied Materials & Interfaces, vol. 7, no. 25, 2015.

An Improved All-metal Circularly Polarized Radial Line Slot Antenna

Shengren Peng, Chengwei Yuan, Ting Shu, Zicheng Zhang, Xuelong Zhao

(College of optoelectronic Science and Engineering, National University of Defense Technology, Changsha 410073, China)

***Abstract*: An improved circularly polarized all-metal radial line slot antenna (RLSA) without slow wave structure for high power microwave (HPM) application is proposed in this study. Comparing with the previously investigated HPM RLSA, the reflection cancelling slot and phase tuning structure are globally optimized on the same disk to achieve broadband design of this improved HPM RLSA, synchronously keeping high radiation efficiency and aperture efficiency. The bandwidth of reflection less than -10dB is increased by more than 100%, which is more adaptable to the operation frequency drift of HPM generator causing by the fluctuation of the diode voltage in the application.**

***Key words*: HPM; RLSA; broadband design; reflection cancelling slot; phase tuning structure; integrated applications.**

I. INTRODUCTION

A high power microwave radial line slot antenna (HPM RLSA) is a planar antenna as the radiation terminal connected to HPM generator for X-band or Ku-band applications [1-4]. It is a type of slotted radial waveguide array. Three disks comprise a twofold radial line feeding waveguide, the top of which is an aperture with slots arrayed spirally for circular polarization, the middle of which is a special disk with slow wave structure diminishing the guide wavelength to suppress the grating lobes from the array, the bottom of which is added a spire-shaped reflection cancelling slot to cancel out the reflection from the array on the top. Electromagnetic power is fed in at the bottom through a coaxial-to-radial line adapter in TEM mode. This generates a radial inward travelling TEM mode in the upper waveguide which couples to the slots array and the radiation is added in phase in the antenna broad side.

It is the first time that RLSA is utilized to radiate HPM in X-band or Ku-band with a high efficiency, compact structure, and notable power handing capacity. However, the reflection cancelling slot on the bottom is located far away from the slots array on the top, leading to the bandwidth of the antenna with large aperture is too narrow, and it becomes worse when it goes into millimeter wave range. Besides, the reflection cancelling slot also awkwardly occupied the lower radial waveguide [2-3], which could make room for the radial line HPM generator [5-7], unfortunately obstructing the optimal integrated design of radial line HPM generate-radiate system with quite compact structure.

To overcome the defects aforesaid, this communication presents the design and optimization of an improved broadband HPM RLSA for integrated applications with radial line HPM generator. And a prototype antenna working in X-band is fabricated and tested to confirm the design.

II. DESIGN OF THE ANTENNA

Fig.1 shows the geometry of the proposed improved HPM RLSA. Comparing with the original HPM RLSA, the three major improvements proposed in the design approach are exhibited as follows:

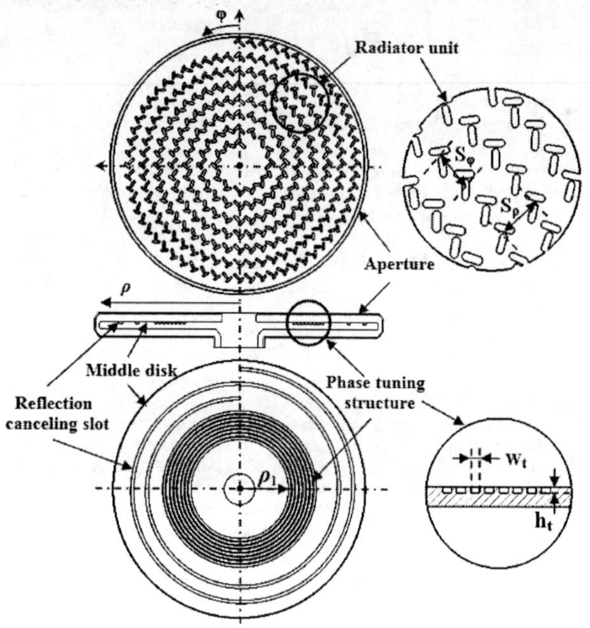

Fig.1 Structure of the improved HPM RLSA

A. Design of the Spiral Slots Array

In the original HPM RLSA, the spacing between adjacent pairs S_ρ along the ρ direction is limited about one guide wavelength. However, to broaden the bandwidth, the slow wave structure must be removed from the upper radial waveguide to make room for the reflection cancelling slot, so the S_ρ is no longer equal to one guide wavelength. In order to keep high aperture efficiency, the value of S_ρ here is chosen between $0.75\lambda_0$ and $0.95\lambda_0$, and the track of the slots array can be expressed as function (1), where r_0 is the radius of head of the array track. The optimization of slot coupling keeps the same way analysed in Ref.2.

$$\rho_a(\varphi)=\left(r_0+\frac{\varphi S_\rho}{2\pi}\right),\quad \varphi\in[0,14\pi] \tag{1}$$

B. Design of the Reflection Cancelling Slot

As analysed in Ref.3, the track of the slots array on the top could give rise to serious reflections, i.e. two degenerate TE_{21} modes, which disturbs the normal antenna operation. Actually, most of the reflections are excited by the slots pairs near the periphery of the antenna aperture, so the reflection cancelling slot should be set here close to the periphery on the upper waveguide. The distance between these special slots pairs mentioned above and the reflection cancelling slot is shorter, the bandwidth of antenna is wider. The function (2) of the track of the spired slot is shown below, where ρ_0 is the radius of head of the track.

$$\rho_c(\varphi)=\left(\rho_0+\frac{\varphi S_\rho}{2\pi}\right),\quad \varphi\in[0,4\pi] \tag{2}$$

C. Design of the Phase Tuning Structure

For realizing high aperture efficiency, the radiator units in the array should be excited in the same phase, ensuring the radiation can be added in phase in the antenna broad side. However, in the improved HPM RLSA, the S_ρ is less than λ_0, accordingly there will be a phase lag between the units near the centre of antenna aperture and the ones near the periphery as the microwave travels radially inward in TEM mode on the upper waveguide. So the phase tuning structure composing of several concentric slots, located in the middle on the disk, is utilized to compensate the phase difference between them.

D. The whole Structure of Antenna

The three improvements discussed above should be consider as a whole in the optimal design of the antenna, as the reflection cancelling slot not only affects the coupling of the slots array, but also has an impact on the phase of the wave travelling inward. A three-dimensional finite integration time domain solver, CST Microwave Studio, was utilized to achieve the final optimization, and the main parameters of the designed antenna are listed in table 1.

TABLE I

Main parameters of the optimized antenna

Height of upper waveguide	H	10 mm
Thickness of top disk	t	5 mm
Width of slot unit ($0 \le \varphi \le 10\pi$)	W_1	5 mm
Width of slot unit ($10 \le \varphi \le 14\pi$)	W_2	4 mm
Slot length ($0 \le \varphi \le 4\pi$)	L_1	14.95 mm
Slot length ($4\pi \le \varphi \le 10\pi$)	L_2	14.69 mm
Slot length ($10 \le \varphi \le 14\pi$)	L_3	14.94 mm
Radius of blend edge of slot unit ($0 \le \varphi \le 10\pi$)	r_1	2.5 mm
Radius of blend edge of slot unit ($10 \le \varphi \le 14\pi$)	r_2	2 mm
Curvature radius of the slot aperture periphery	r_3	1 mm
Spacing between adjacent slot pairs along ρ direction	S_ρ	28.83 mm
Spacing between adjacent slot pairs along φ direction	S_φ	21.5 mm
Number of all slot pairs	n	267
Width of reflection canceling slot	w_c	6.2 mm
Depth of reflection canceling slot	d_c	3 mm
Radius of the head of the reflection cancelling slot track	ρ_0	174.5 mm
Radius of the head of the slot pairs array track	r_0	38 mm
Period of phase tuning structure	p	7.2 mm
Depth of phase tuning structure	h_t	3.27 mm
Gap of phase tuning structure	w_t	4.6 mm
Initial radius of phase tuning structure	ρ_1	95.7 mm

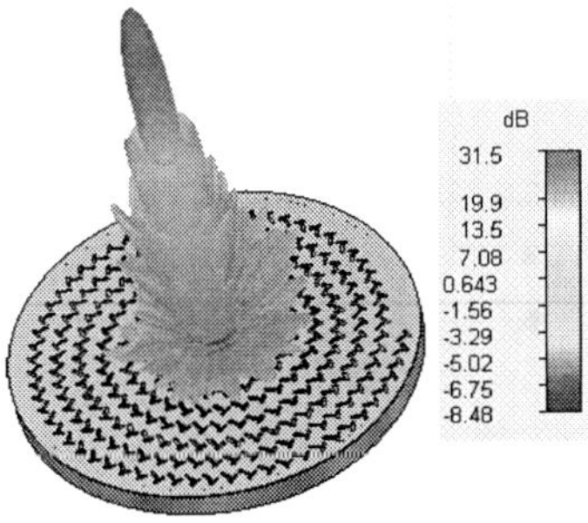

(a) 3D view

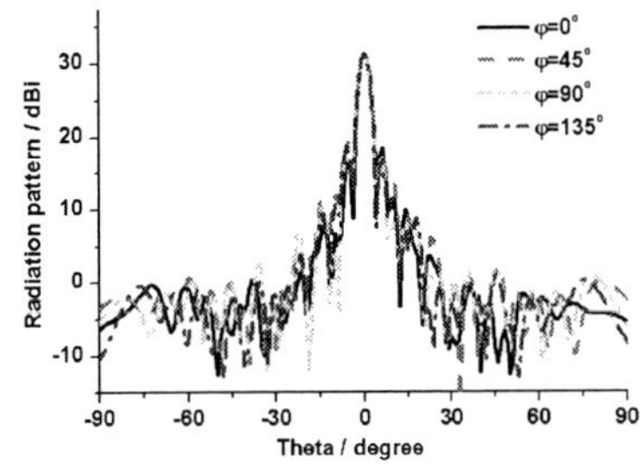

(b) Cartesian view

Fig.2 Simulated radiation pattern of improved HPM RLSA at 9.4 GHz

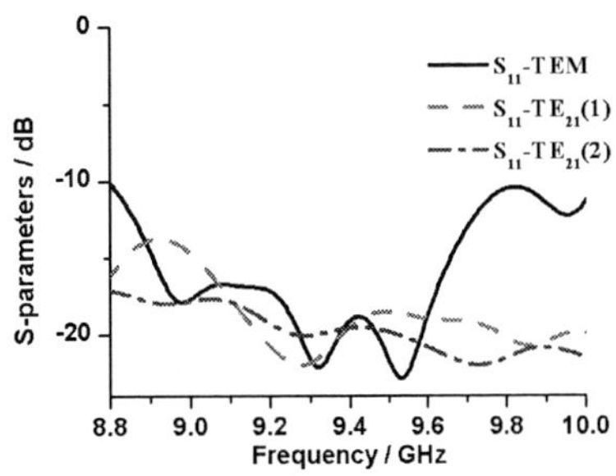

Fig.3 Simulated reflection of improved HPM RLSA

The Fig.2 and Fig.3 indicate that the improved HPM RLSA with aperture diameter of 51 cm presents good simulated performances. At 9.4 GHz, the calculated directivity of the radiation pattern is 31.5 dBi, showing a good quality of rotational symmetry, and accordingly the aperture efficiency reaches 56.34%, the axial ratio of the antenna on broad side is about 1.08, which confirms the feasibility of the remove of slow wave structure. That is to say, three is no need to design special component to diminish the guide wavelength for the reasons of suppressing grating lobes from the array. Over the frequency range of 8.9 to 9.7 GHz (bandwidth of 800 MHz), the reflection is less than -10 dB, which is much improved comparing with original one, whose bandwidth is only about 150 MHz [3]. In the practical application, the operation frequency drift (in the range of about ±70 MHz) of HPM generator often occurs causing by the fluctuation of the diode voltage [8], which is impossibly eliminated, so this proposed improved antenna of wideband design is more adaptable to the situation aforementioned.

III. Experimental Investigations

To verify the feasibility of this improved HPM-RLSA, experimental investigations should be carried out in the anechoic chamber with an antenna test system. A prototype antenna working at 9.4 GHz is fabricated as shown in Fig.4, as this antenna radiates microwave in circularly polarized mode, which is quite difficult to directly tested, the measurement of the antenna gain is performed by the standard linearly polarized antenna comparison method [2, 9]. As illuminated in Fig.1, horizontal direction here is defined along φ=90°, and vertical direction is φ=0°.

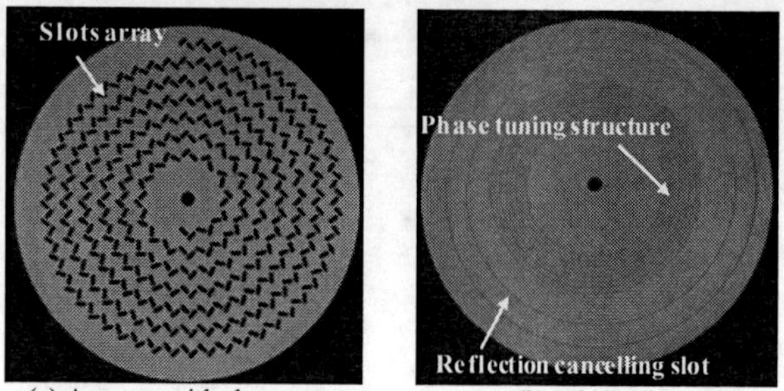

(a) Aperture with slots array (b) Middle disk

Fig.4 Fabrication of improved HPM RLSA

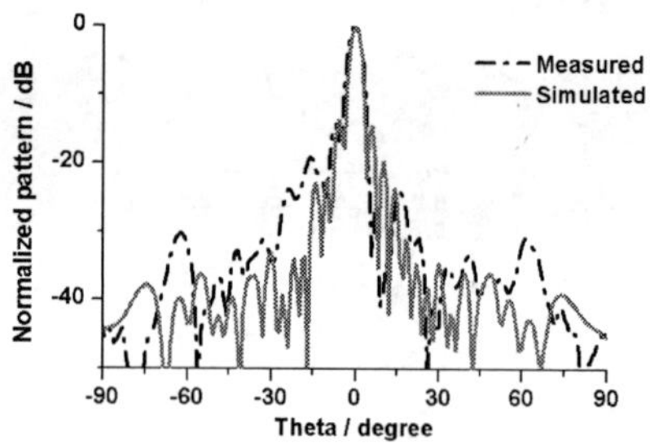

(a) Vertical polarization (φ = 0°) at 9.4 GHz

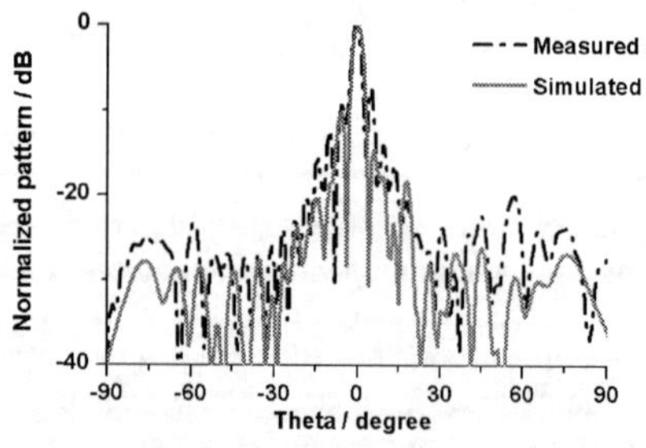

(b) Horizontal polarization (φ = 90°) at 9.4 GHz

Fig.5 Preliminary experimental results of the radiation pattern

TABLE II

Axial Ratio and Gain of the fabricated Antenna

Frequency / GHz	9.36	9.38	9.40	9.42	9.44
Gain / dBi	27.45	28.73	29.92	29.12	28.78

Fig.5 shows the farfield radiation patterns of vertical and horizontal component of the circularly polarized microwave radiated from the antenna, respectively, and the measured results mainly agree with the simulations. The tested axial ratio and gain of the antenna versus frequency are listed in table 2. At the centre frequency of 9.4 GHz, the gain of the antenna reaches 29.92 dBi, and the axial ratio is about 2.21.

To measure the radiation efficiency, a special designed TEM mode adapter is utilized to realize impedance matching between the input port of the antenna and the coaxial line connected to network analyzer. Near the centre frequency, range of 9.35 GHz to 9.5 GHz, the reflection is very low thanks to effective suppression by the reflection cancelling slot, and the radiation efficiency exceed 99% at 9.4 GHz. The measured bandwidth is narrowed by the TEM mode adapter [2] comparing with the simulated results shown in Fig.3.

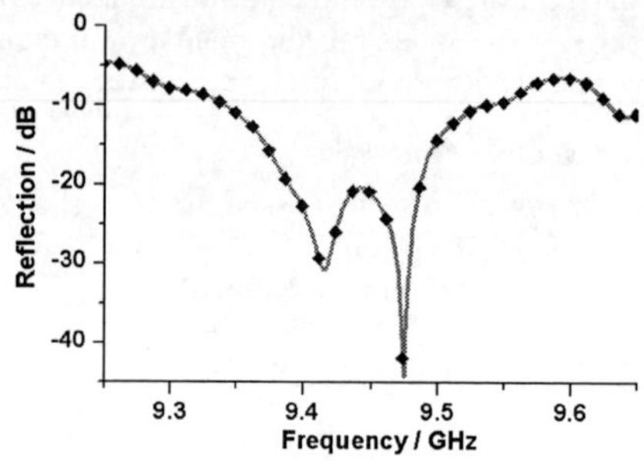

Fig.6 Measured reflectance coefficient at the input port of the TEM mode adapter

IV. Conclusion

In this communication, we proposed an improved broadband HPM RLSA without slow wave structure to realize the integrated design with the radial HPM generator. An antenna prototype is designed and measured. Experimental results prove the feasibility of this antenna, which has a high efficiency and compact structure.

References

[1] S.R. Peng, C.W. Yuan, T. Shu, Z.Q. Li, Q. Zhang, "Kind of high-power microwave radial line slot antenna," *Electronics Letters*, vol. 49, no. 15, 2013.

[2] C.W. Yuan, S.R. Peng, T. Shu, *et al.*, "Designs and experiments of a novel radial line slot antenna for high-power microwave application," *IEEE Trans. Antennas Propag.*, vol.61, no.10, pp.4940-4946, 2013.

[3] S.R. Peng, C.W. Yuan, and T. Shu, "Analysis of a high power microwave radial line slot antenna," *Rev. Sci. Instrum.*, vol.84, p. 074701, 2013.

[4] S.R. Peng, C.W. Yuan, T. Shu, Q. Zhang, "Design of a concentric array radial line slot antenna for high-power microwave application," *IEEE Trans. Plasma Sci.*, 2015, published on line.

[5] M. J. Arman, "High Power Radial Klystron Oscillator," *IEEE Trans. Plasma Sci.*, vol 24, no. 3, pp. 964-969, 1996.

[6] M. J. Arman, "Plasma-filled radial acceletron," *IEEE Trans. Plasma Sci.*, vol 28, no. 3, pp. 767-771, 2000.

[7] F.C. Dang, X.P. Zhang, H.H. Zhong, Y.M. Li, and Z.M. Qi, *Phys. Plasmas*, vol. 21, p.063307, 2014.

[8] D. Zhang, J. Zhang, H.H. Zhong, *et al.*, "Purification of the output modes of overmoded relativistic backward wave oscillators," *Phys. Plasmas*, vol. 21, p.023115, 2014.

[9] C. L. Lin, *Antenna engineering handbook*, Beijing: Electronic industry press, 2002.

Utilizing Ultrahigh Magnetic Fields to Promote the Ignition for Magnetized Plasma*

Xian-jun Yang
Institute of Applied Physics and Computational Mathematics,
Beijing 100094, China
Mail: yang_xianjun@iapcm.ac.cn

Xiao-guang Wang, Guan-qiong Wang
Center for Fusion Energy Science and Technology,
Chinese Academy of Engineering Physics,
Beijing 100088, China
Mail:wang_xiaoguang@iapcm.ac.cn

Abstract——Ultrahigh magnetic field plays an important role in promoting the ignition for high temperature magnetized plasma, which not only can reduce the thermal conductivity, but also can enhance the charged-particle reduction product energy deposition in this kind of plasma. An new scheme is proposed in this paper to investigate the process of ignition and a serial work has been done, which includes generating the ultrahigh magnetic field over 1400 tesla, producing the high temperature plasma target of Field Reversed Configuration and simulating the process of collision-merging of two FRC while an two-dimensional MHD tool has been developed to simulate and analyses the process of ignition.

Keywords— ultrahigh magnetic field; magneto inertial fusion; collision-merging; reversed field configuration; electromagnetic pulse power

I. INTRODUCTION

Magneto-inertial fusion (MIF) is a middle fusion energy path between magnetic confined fusion (MCF) and inertia confined fusion (ICF). The MIF takes advantage of the ultra-high magnetic field to improve energy confinement time by reducing thermal conduction and power deposition of fusion α particles, reducing the required power and precision of implosion dramatically relative to the traditional ICF [1-3]. Since then many MIF schemes have been proposed, such as MAGO [4], MagLIF [5], MagICF, GF device, PJMIF [6]. Magnetized target fusion (MTF) is also one of these MIF schemes, where the reversed-field configuration (FRC) is compressed to fusion ignition condition by imploding solid metal liner.

The MTF concept is first proposed by Los Alamos National Laboratory (LANL) [1]. The integrated experiments for MTF were already conducted [7]. A 1.36 T vacuum field was successfully compressed to 540 T by imploding an aluminum liner in advance. A single FRC is produced, translated into and captured by mirror field in the liner successfully. Furthermore, the aluminum liner was also imploded utilizing electromagnetic pulse power technology. But the obvious compression and heating of FRC was not observed because of the short lifetime of FRC.

It is exciting that the long-lived FRC has been produced in the world's largest compact toroid (CT) facility, C-2, by colliding and merging two high-β plasmoids produced by utilizing the advanced field-reversed θ-pinch technology [8]. The lifetime of FRCs is improved dramatically by this method. Its lifetime exceeds 2 ms [8], even up to 5 ms [9]. The poloidal flux, confinement time and stability of FRCs were improved significantly, contributing to the improved lifetime. But it is difficult to heat the plasma further for C-2. In addition, although the density of FRC for C-2 is far from the requirement of MTF, yet the collision-merging method of formatting FRC is believed to improve the lifetime.

A hopeful MIF scheme is proposed referring both the MTF and C-2 schemes in this paper, described in Sec. II. The four key scientific issues of this scheme were investigated in past years given in Sec III: 1) generating strong magnetic field, 2) producing the initial FRC, 3) formatting plasmoid (with closed flux surfaces) by collision-merging, 4) the feasibility of the ignition. Finally the summary and conclusion are given in Sec. IV.

II. THE FUSION SCHEME

The fusion ignition is achieved through three main steps for the new MIF scheme as Fig. 1 shows. Firstly two symmetric initial FRCs (or plasmoids) are formatted by advanced reversed field θ-pinch discharge technique at two sides, and accelerated up to high velocity by magnetic force due to non-uniform field produced by theta coils. In the second step the initial FRCs are translated into the solid liner at the same velocity with head on. Then these FRCs of high velocity collide and merge into a single FRC (or plasmoid with closed flux surface). The collision-merging FRC is easier to be captured by the mirror field inside the solid liner because of little axial momentum than the single-translated FRC

* This work is supported by National Natural Science Foundation of China (Grant No. 11575029 and 11605013)

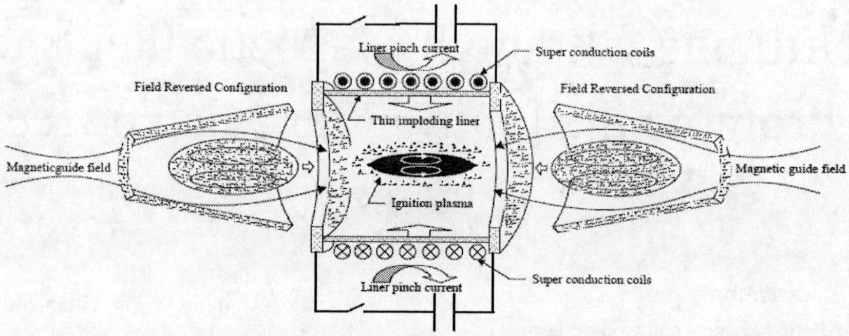

Fig. 1. Scheme for collision-merging FRC based MIF. The initial FRCs are produced in the components at two sides. Then these two symmetric FRCs are translated into and collision-merge in the liner (the middle component). Finally, the collision-merged FRC (or plasmoid with closed flux surfaces) is compressed by imploding the liner.

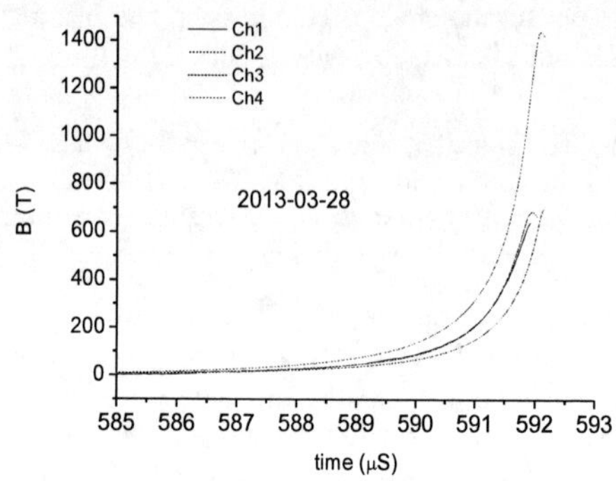

Fig. 2. The magnetic field as a function of time is given from experiment data of different channels.

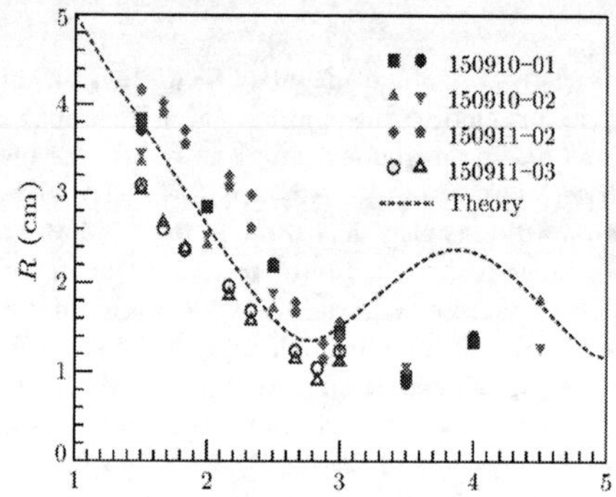

Fig. 3. Evolution of separatrix radius calculated by MPF-2D (blue dash) compared with the experiments. Reprinted with permission from [13]. Copyright 2016 by the Chinese Physical Society and IOP Publishing Ltd.

with high velocity. The lifetime of FRC as well as the density and temperature is improved and expected to meet the requirement of MTF. Finally the solid liner is imploded to compress the collision-merging FRC to fusion ignition condition utilizing the pulse power technology.

III. KEY SCIENTIFIC ISSUES

A. Generating Ultrahigh Magnetic Field

A MC-1 like facility was designed and built to achieve the ultrahigh magnetic field [10]. An array of theta coils is employed to generate the initial magnetic field in a conducting metal liner. Then the liner is driven by explosion of TNT to compress the seed magnetic field, generating the ultrahigh field by the principle of flux conservation. Fig. 2 presents the magnetic field as a function of time from the experiment data of different signal channels. The ultrahigh magnetic field about 1400 T is generated successfully using the MC-1 facility as the Fig. 2 shows. A one-dimensional MHD model has also been developed to simulate and analyses the experiment process of generating ultra-high field [11]. Both the results of experiment and simulation fit very well.The maximum magnetic field B about 1400 T is achieved according to the simulation. Combining with the experiment, where a 1.36 T vacuum field was compressed to 540 T by imploding aluminum liner, it is concluded that the extreme strong field required by MTF could be achieved by compressing seed field with imploding liner.

B. Producing FRC and YingGuang-1 Program

YingGuang-1 device which has been built at China Academy of Engineering Physics in 2013 aims at the research on formation, instabilities and confinement properties of FRC plasma target. The designed value of peak current of theta coil is 1.5 MA, and the maximum of reversed magnetic field can access to 4 T [12]. The FRC was produced successfully at 2015 campaign of YingGuang-1 [13]. In order to understand the experiments better, a new two-dimensional MHD code, MPF-2D [13, 14], has been developed. The single-fluid multi-temperature MHD equations are solved in cylindrical symmetric geometry with the use of arbitrary Lagrangian-Eulerian (ALE) method, which can overcome the computational difficulties due to the large deformation of mesh. The separatrix radius simulated by MPF-2D is compared to the experiments from different shots, as shown in Fig. 3. For each of four shots (150910-01, 150910-02, 150911-02 and 150911-03), at one moment two data points are obtained by measuring the radii of an image along both horizontal and vertical directions respectively, and the image is taken with the high-speed multi-frame visible CCD camera (Fig. 3 in Ref. [13]). The simulation repeated the process of formation of FRC in Yinguang-1 very well as Fig.3 shows.

C. Collision-Merging Process

The present driver capability requires that the plasma target must satisfy some conditions before imploding, i.e., plasma densities n~10^{17} cm^{-3}, plasma temperatures T~300 eV, and the energy confinement time $\tau E > 10$ μs [1]. The collision-merging technology is one of the candidate formation methods of FRC target because it may produce the FRCs with higher temperature [15] and larger poloidal flux [16] than the traditional methods, and these FRCs seem to be suitable for MIF. To understand the physics of the merging process and investigate how to prove the performance of merged FRCs, MPF-2D is used to study the collision-merging numerically. In our simulation, the initial density and temperature are set as n~ 10^{15} cm^{-3} and T~ 1.8 keV, respectively. The main magnetic field is 2.0 T and the translation speed before merging is about 200 km/s. Here, the classical resistivity is applied. The evolution of magnetic field, density and temperature in the different phases including generation, translation and collision-merging are shown in Fig. 4. Three columns represent the contours of flux, density, and temperature, respectively, and six time slices (six rows) are chosen to demonstrate the time evolution.

As can be seen in Fig. 4, the density of the merged plasmoid at field null decreases compared to the single plasmoid with before colliding. Meanwhile, the separatrix radius increases to 3.4 cm from 3.3 cm, and the axial length is also increased. This illustrates that the separatrix volume of the merged plasmoid expands to some extent, so that the density is decreased if the particle numbers is assumed to be conservative. However, the total temperature (electron temperature plus ion temperature) is slightly increased to 2.4 keV from the initial state with 1.8 keV, the reason should be that the part of the kinetic energy of translating plasmoids is converted into plasma thermal energy. Note that two plasmoids do not merge perfectly. In other words, the magnetic reconnection is just accomplished partly and the double-null structure exists in the merged FRC. As mentioned in the previous literatures [15, 17], the complete merging has not been obtained by the numerical simulations, whereas it was usually observed in the experiments. In principle, the merging process is one of the magnetic reconnection phenomena which is significantly influenced by the plasma resistivity and has not been understood well. It implies that the non-fully merged plasmoid in simulations may be caused by the setting of resistivity in the code, here the classical resistivity is employed and the Hall Effect is not considered. In the next five years, YingGuang-1 group plans to do the experiments on the formation of FRCs using collision-merging technology, and verify that this method can generate a suitable target for MTF. In addition, a new device named KMAX (Keda Mirror with Axisymmetricity) has been built in University of Science and Technology of China (USTC).

D. Feasibility of Ignition Condition

The one-dimensional description is appropriate if the elongation of FRC $E=l_s/2r_s$ is larger than two where l_s and r_s are the length of FRC and the radius of separatrix respectively. Usually, the elongation E is 6~7 for MTF. Therefore, the compression of FRC by imploding cylindrical liner is simulated approximately utilizing a one-dimensional MHD code DEIRA, built by Dr. Basko [19]. The rigid-rotor equilibrium model is chosen as the initial profile of FRC. The boundary condition of magnetic field $B_{zb}=I_0\sin(t/2T_p\cdot\pi)/R$ is set as the pulse power source to drive the metal liner where I_0 is the maximum pulse current, T_p is the rising period of pulse current and R is the outer radius of metal liner. The main initial parameters are listed in Tab. 1.

Fig. 5 gives the plots of the radius (a) and imploding velocity (b) as a function of time for the liner (red lines), the vacuum (cyan lines) and plasma (blue lines) regions of FRC during compression. The inserted figure presents the profile of normalized Ti, Bz and ρ at maximum compression. The outer radius of plasma R and the inner radius of liner are 1.24 mm and 2.27 mm respectively at the peak compression. The maximum convergence ratio of FRC is about C=21.5. The maximum velocity at inner surface of liner is about -0.083 cm/μs according to calculation. The maximum temperature T=18.92 keV, external magnetic field B_{ze}=1225 T, maximum mass density $\rho=3.28\times10^{-4}$ g/cm3 (number density 1.21×10^{20} cm^{-3}) at maximum compression according to simulation. The FRC is compressed and heated effectively.

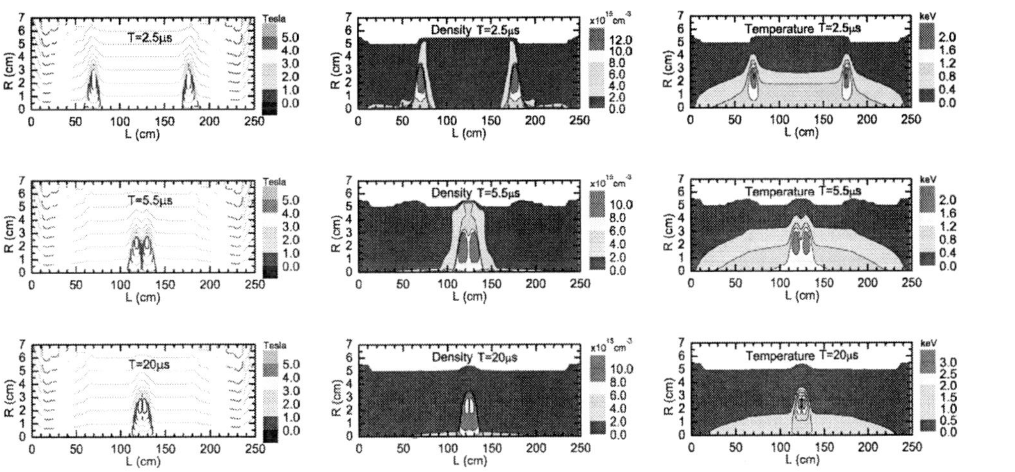

Fig. 4. Time evolution of the flux, density, and temperature showing the dynamics of FRC generation, translation, and merging predicted by MPF-2D. Reprinted with permission from ([18], figure 3). Copyright 2016 by American Institute of Physics.

TABLE I. THE INITIAL PARAMETERS FOR COMPRESSION OF FRC BY IMPLODING ALUMINUM LINER.

Liner			FRC				Pulse Current	
Radius r_w	L_t	ρ	$T_e=T_i$	Density n_m	B_{ze}	r_s	I_0	T_p
4.89 cm	0.11 cm	2.7 g/cm^3	0.13 keV	1.07×10^{17} cm^{-3}	2.7 T	2.5 cm	15 MA	10 μs

The ignition condition is achieved if the power balance condition meets

$$P_\alpha - P_{br} > P_{tc} \quad (1)$$

where Pα is the power of α particles deposited in plasma, P_{br} is the bremsstrahlung radiative loss power and P_{tc} is the loss power of thermal conductivity. $P_\alpha = k_\alpha <\sigma\upsilon>$ where k_α is the deposition ration of α particles. $P_{br}=k_{br}T_e^{1/2}$ where k_{br} is the bremsstrahlung loss coefficient. The ignition condition is about BR≥ 4.5~6×10^5 G·cm with T~10 keV for MTF according to Ref. [2]. The $B_{ze}R$~1.52×10^6 G·cm with T~18.92 keV in the simulation meeting the ignition condition.

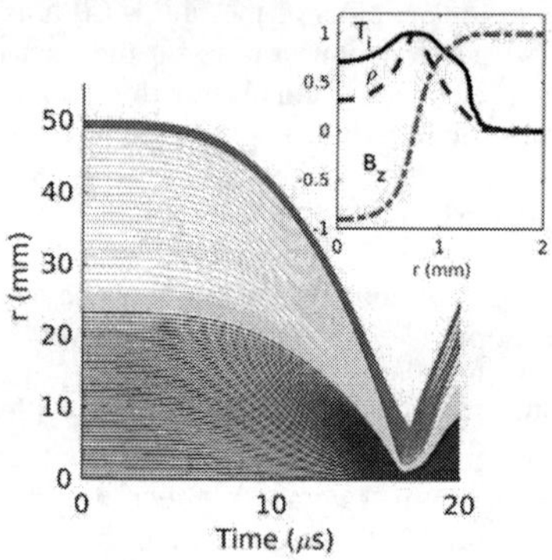

Fig. 5. The radius as a function of time are plotted for liner (red lines), the vacuum (cyan lines) and plasma (blue lines) regions of FRC. The inserted figure gives the profile of normalized T_i, B_z and ρ at maximum compression.

IV. SUMMARY AND CONCLUSION

A new scheme was proposed to demonstrate the ignition of MIF. It is still designed that the FRC is compressed to ignition condition by imploding conducting liner. The FRC (or plasmoid with closed flux surface) target, however, was produced inside the metal liner directly by collide-merging two high-β plasmoids instead of translating singular FRC. It is expected the lifetime of FRC is improved to meet the requirement of MIF by this method. The key scientific issues of this scheme were investigated in past years. Some kind of progress was made on the experimental and theoretical analysis for the new MIF scheme.

Firstly generating ultra-high magnetic field was demonstrated utilizing the like facility. The 1400 T field was achieved in the experiments. And the one-dimensional MHD simulation also repeated the experiment results very well. The device YinGuang-1 was designed and built, producing the required initial FRC successfully. Furthermore, the collision-merging of two plasmoids into a FRC were identified by two-dimensional MHD simulation. The ignition condition could be achieved by compressing the FRC target with imploding liner from the preliminary simulation. Note that the MHD instability and disruption of FRC, which could terminate the life of FRC, is not considered in the simulation.

Although the FRC of high density and temperature could be produced by collision-merging method according to the simulation, it is also not clear whether the lifetime of FRC could be improved effectively in MTF regime. The experimental and theoretical investigation is still needed for this objective. Besides, the other key issues, such as the ignition mechanism and implosion dynamics, is still open, which is planned to be investigated more carefully in future. The integrated experiment and simulation for demonstrating this new scheme is planning to be done in the future.

Acknowledgment

Xian-jun Yang thanks Dr. Qi-zhi Sun, and Lu-lu Li for providing the relevant material for this paper. Use of code DEIRA, built by Dr. Basko (Institute for Theoretical and Experimental Physics), is acknowledged.

Reference

[1] Schoenberg, K.F., et al., Magnetized Target Fusion, A Proof-of-Principle Research Proposal. 1998.

[2] Basko, M.M., A.J. Kemp, and J. Meyer-ter-Vehn, Nuclear Fusion. Vol. 40: pp. 59, 2000.

[3] SIEMON R. E., et al., J. Plasma Fusion Res. SERIES. Vol. 5: pp. 63, 2002.

[4] Garanin, S.F., IEEE Transactions on Plasma Science. Vol. 26: pp. 1230, 1998.

[5] Cuneo, M.E., et al., IEEE Transactions on Plasma Science. Vol. 40: pp. 3222, 2012.

[6] Hsu, S.C., et al., IEEE, Transactions on Plasma Science. Vol. 40: pp. 1287, 2012.

[7] Degnan, J.H., et al., Nuclear Fusion. Vol. 53: pp. UNSP, 2013.

[8] Guo, H.Y., et al., Physics of Plasmas. Vol. 18: pp. 056110, 2011.

[9] Gota, H., et al., Review of Scientific Instruments. Vol. 85: pp. 11D836, 2014.

[10] Knauer, J.P., et al., Physics of Plasmas. Vol. 17: pp. 056318, 2010.

[11] Hohenberger, M., et al., Physics of Plasmas. Vol. 19: pp. 056306, 2012.

[12] Sun, Q.Z., et al., Acta Physica Sinica. Vol. 62: pp. 078407, 2013.

[13] Li, L.L., et al., Chinese Physics Letters. Vol. 33: pp. 045202, 2016.

[14] Li, L.L., H. Zhang, and X.J. Yang, Acta Physica Sinica. Vol. 63: pp. 165202, 2014.

[15] Slough, J., G. Votroubek, and C. Pihl, Nuclear Fusion. Vol. 51: pp. 053008, 2011.

[16] Binderbauer, M.W., et al., Physical Review Letters. Vol. 105: pp. 045003, 2010.

[17] Woodruff, S., A.I.D. Macnab, and N. Mattor, Journal of Fusion Energy. Vol. 27: pp. 128, 2008.

[18] Wang, G.Q., et al., Physics Of Plasmas. Vol. 23: pp. 062701, 2016.

[19] Basko, M.M., DEIRA, A 1-D 3-T hydrodynamic code for simulating ICF targets driven by fast ion beams, Version 4, Institute for Theoretical and Experimental Physics, Moscow, 2001, p. 48

Effects of Pulsed Magnetic Field on Performances of Semiconductor Devices

Yang Yang, Ling Dai, Jinchang Hu, Shuyun Tian, Qin Zhang, Fuchang Lin
State Key Laboratory of Advanced Electromagnetic Engineering and Technology, School of Electrical and Electronic Engineering
Huazhong University of Science and Technology, HUST
wuhan, china
dailing@hust.edu.cn

***Abstract*—The electromagnetic launch system generates strong pulsed magnetic field radiation when discharging, which may cause performance failure or damage to electronic devices, especially semiconductor devices, in low voltage parts of the system. To analyze effects of pulsed magnetic field on semiconductor devices, this paper mainly took diode and triode for research objects. Based on fundamental theory of Electromagnetics and Semiconductor Physics, magnetic injury effect model of typical semiconductor devices, such as diode and triode, was established. The relation equations of magnetic induction density and some parameters, which involving forward conducting voltage and electrode current, was obtained through the derived equations between magnetic induction density with semiconductor mobility and scattering rate. To test and verify the theoretical model, experiments was designed to study effects of pulsed magnetic field on diode and triode. A planar disk-shaped coil was used as the pulsed magnetic field source to simulate the strong magnetic field radiation of the system, and the intensity of magnetic field radiation was up to T magnitude order. The experiment results shows that the diode's conducting voltage regulation is proportional to the square of the magnetic induction intensity of external magnetic field. In addition, common-emitter amplification coefficient of triode decreases with the increasing of magnetic flux density in pulsed magnetic field. And the experimental results match well with the derived equations.**

Keywords—pulsed magnetic field; semiconductor devices; diode; triode; semiconductor mobility; scattering rate; forward conducting voltage; common-emitter amplification coefficient

I. Introduction

Due to limitation of its principle, chemical energy emission is difficult to reach the ultra-high speed. As a consequence, a new generation of ultra-high speed electromagnetic(EM) launch technology appears. EM force serves as the driving force to accelerate a variety of objects, including projectiles, shells, missiles, rockets, satellites, aircraft and etc. During this process, the EM energy is converted to dynamic kinetic energy [1-2]. Compared with conventional emission technology, electromagnetic emission technology has many advantages, such as a large range of emitter quality, excellent work performance and high energy efficiency [3].

The electromagnetic launch system mainly includes the pulse forming network (PFN), the rail and the control unit. Pulsed current up to several megampere flows through PFN and the rail when discharging, therefore, inductive component inside PFN and the rail generate strong pulsed magnetic field radiation [4]. However, the control unit of the system contains a variety of electronic devices, such as diodes, transistors and integrated circuits. The pulsed magnetic radiation may cause performance degradation or even damage to electronic devices, especially semiconductor devices, and furtherly leading to decline of system reliability [5].

Many researchers have studied effects of pulsed magnetic field on electronic devices and electronic systems [6-9]. Schwartz utilized H-type device of Army Research Laboratory (ARL) to simulate the strong magnetic field of railgun launching process. The results showed that both of unshielded electronic watches and electric detonators were not affected, when the peak magnetic induction intensity of the pulsed magnetic field was 0.2T [6]. Ciolini performed a series of static railgun experiments(no projectile movement) to investigate the influence of electromagnetic field on electronic devices such as diode, CMOS integrated circuit and digital watches. During tests, the peak magnetic flux density reached 4.1T and a part of tested devices, such as micro-magnetic fluxgate and digital watch, were out of order [7]. Mandic studied the effect of external magnetic field on the readout of diodes under constant magnetic field environment, and come to a conclusion that the forward voltage of the diode increases with the increasing of magnetic induction intensity [8]. Zhang Guobin measured the reverse saturation current and forward current of diodes based on Si and Ge between 0 and 2 Tesla(T). The results demonstrated that the reverse saturation current of Ge diode is decreased by 100nA when the magnetic field is 2T. What's more, the decrease of the Ge diode is more evident than that of the Si diode [9].

Based on fundamental theory of *Electromagnetism* [10] and *Semiconductor Physics* [11-12], magnetic injury effect model of typical semiconductor devices, such as diode and triode, was established. In addition, the relation equations of magnetic induction density and some parameters, which involving forward conducting voltage and electrode current, was obtained. Finally, the relation curve between magnetic flux density and these parameters was obtained through experiments, in which the peak magnetic induction intensity of pulsed magnetic field reached 2.8T.

II. THEORETICAL ANALYSIS

A. *Effects of Magnetic Field on Semiconductor*

Supposing the semiconductor with a direct current flowing through is placed in a uniform magnetic field. And the electric field intensity is along the x direction at an electric field intensity of ε_x; perpendicular to the direction of the electric field, the magnetic filed is along the z direction at a magnetic induction intensity of B_z. Then, a transverse electric field ε_y is created at the +y or –y direction, which is perpendicular to the direction of the electric field and magnetic field. This phenomenon is known as Hall effect [11].

Placing the semiconductor material in a magnetic field and an electric field, and the motion equation of the carrier can be expressed by [11][12]:

$$m\frac{dv}{dt}=-q(\varepsilon_x+v\times B_z) \tag{1}$$

As is shown in Fig.1, trajectory of carrier in magnetic field is a bight instead of a straight line, due to the effect of lorentz force. In consequence, the carrier scattering rate increases and average freedom time decreases, which will cause the increase of resistivity of semiconductor.

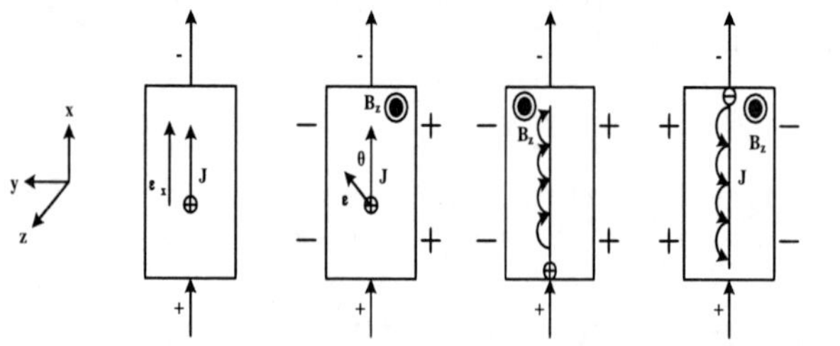

The carrier mobility μ_B and diffusion coefficient D_B in the magnetic field may be revised as formula (2), without considering the statistical theory of Hall effect [8].

$$\begin{cases} \mu_B=\dfrac{\mu_0}{1+\mu_H{}^2B_z{}^2} \\ D_B=\dfrac{D_0}{1+\xi\cdot\mu_H{}^2B_z{}^2} \end{cases} \tag{2}$$

where, μ_0 and D_0 are the carrier mobility and diffusion coefficient under zero magnetic field, respectively; μ_B and D_B are the carrier mobility and diffusion coefficient in magnetic field, respectively; ξ is transverse magnetic resistance coefficient; μ_H is hall mobility.

Identify applicable sponsor/s here. If no sponsors, delete this text box (*sponsors*).

According to Nernst-Einstein relation, diffusion coefficient can be expressed by

$$D=\frac{kT}{q}\mu \tag{3}$$

where, k is boltzmann's constant; T is environment temperature; μ is carrier mobility

$$\mu=\frac{nq}{\delta} \tag{4}$$

where, n is carrier concentration; δ is conductivity.

Thus the change in resistivity ρ of semiconductor may be expressed by

$$\frac{\Delta\rho}{\rho_0}=-\frac{\delta_B-\delta_0}{\delta_B}=\frac{D_0-D_B}{D_B}=\xi\mu_H{}^2B_z{}^2 \tag{5}$$

As the magnetic field strength increases, the resistivity of semiconductor tends to be saturated when $\mu_H B_z$>>1.

According to equation (5), resistivity of semiconductor is proportional to $\mu_H{}^2B_z{}^2$, and the increase in resistivity approaches saturation when $\mu_H B_z$>>1.

B. *Effects of Magnetic Field on Diode*

1) *Forward voltage drop of the diode*

According to the above formula derivations, the relation between variation of forward voltage drop ΔU_d and magnetic flux density of outside magnetic field may be expressed by

$$\Delta U_d\approx I\cdot\rho_0\xi\mu_H{}^2B_z{}^2=U_0\cdot\xi\mu_H{}^2B_z{}^2 \tag{6}$$

where, I is the direct current flowing through; U_0 is conducting voltage of the diode without effects of magnetic field.

Equation (6) shows that voltage of the diode for a given forward current is proportional to $\mu_H{}^2B_z{}^2$, and this increase in voltage drop tends to be saturated when $\mu_H B_z$>>1.

2) *Reverse leakage current of the diode*

The I-V characteristic equation of an ideal p-n junction is

$$I_0=qA(\frac{D_{p0}p_n}{L_{p0}}+\frac{D_{n0}n_p}{L_{n0}})(e^{\frac{qV}{KT}}-1) \tag{7}$$

where, p and n are hole concentration and electron concentration, respectively; L is carrier diffusion length

$$L_{p0}=\sqrt{D_{p0}\tau_{p0}},L_{no}=\sqrt{D_{no}\tau_{no}} \tag{8}$$

Thus the reverse leakage current of the diode may be expressed by

$$I_{dr}=\frac{1}{\sqrt{1+\xi\mu_H{}^2B_z{}^2}}I_{s0}(e^{\frac{qV}{KT}}-1) \tag{9}$$

where, I_{so} is

$$I_{s0} = qA(\frac{p_n}{\sqrt{\tau_p}}\sqrt{D_{p0}} + \frac{n_p}{\sqrt{\tau_n}}\sqrt{D_{no}}) \quad (10)$$

As is shown in Equation (9), reverse leakage current of the diode for a given reverse voltage decreases with the increase of magnetic flux density, and this effect tends to be saturated when $\mu_H B_z >> 1$.

C. Effects of Magnetic Field on Triode

Taking the PNP type transistor as an example, the influence of the magnetic field on the transistor is analyzed. Assuming that the transistor works in an amplified state, the emitter junction is forward biased and the collector junction reverse biased. Then the emitter, collector and base current can be expressed by [13]

$$\begin{cases} I_E \approx qA\frac{D_p}{L_p}\Delta p_E ctnh\frac{W_b}{L_p} \\ I_C \approx qA\frac{D_p}{L_p}\Delta p_E \csc h\frac{W_b}{L_p} \\ I_B \approx qA\frac{D_p}{L_p}\Delta p_E \tan h\frac{W_b}{L_p} \end{cases} \quad (11)$$

where W_b is base region thickness; A is a constant coefficient; Δp_E is excess carrier concentration

$$\Delta p_E = p_n\left(e^{qV_{EB}/kT} - 1\right) \quad (12)$$

Combined with equation (8) and the series expansions of hyperbolic function (one order), equation (11) can be converted to

$$\begin{cases} I_E \approx \frac{qA}{\tau_p}\frac{\Delta p_E}{3W_b}(3D_{p0}\tau_p + W_b^2) \\ I_C \approx \frac{qA}{\tau_p}\frac{\Delta p_E}{6W_b}(6D_{p0}\tau_p - W_b^2) \\ I_B \approx \frac{qAW_b\Delta p_E}{2\tau_p} \end{cases} \quad (13)$$

Taking effects of magnetic field into consideration, the electrode currents may be expressed by

$$\begin{cases} I_{EB} \approx \frac{qA}{\tau_p}\frac{\Delta p_E}{3W_b}(3D_{p0}\tau_p\frac{1}{1+\xi\mu_H^2 B_Z^2} + W_b^2) \\ I_{CB} \approx \frac{qA}{\tau_p}\frac{\Delta p_E}{6W_b}(6D_{p0}\tau_p\frac{1}{1+\xi\mu_H^2 B_Z^2} - W_b^2) \\ I_{BB} \approx \frac{qAW_b\Delta p_E}{2\tau_p} \end{cases} \quad (14)$$

Thus common-emitter amplification coefficient β of the triode can be calculated as

$$\beta_B = \frac{I_{CB}}{I_{BB}} = \frac{(6D_{p0}\tau_p\frac{1}{1+\xi\mu_H^2 B_Z^2} - W_b^2)}{3W_b^2} \approx \frac{1}{1+\xi\mu_H^2 B_Z^2}\frac{2D_{p0}\tau_p}{W_b^2} \quad (15)$$

The formula shows that if other conditions are constant, the electrode currents decreases with the increase of magnetic induction intensity. What's more, the common-emitter amplification coefficient is affected by the magnetic field in the same way.

III. EXPERIMENTS

Experiments was designed to study effects of pulsed magnetic field radiation of EM system on performances of typical semiconductor devices, such as diode and triode. The experimental circuit is shown in Fig.1. A charger was used to charge the capacitor to a given voltage, and the thyristor was the switch for this pulsed forming circuit.

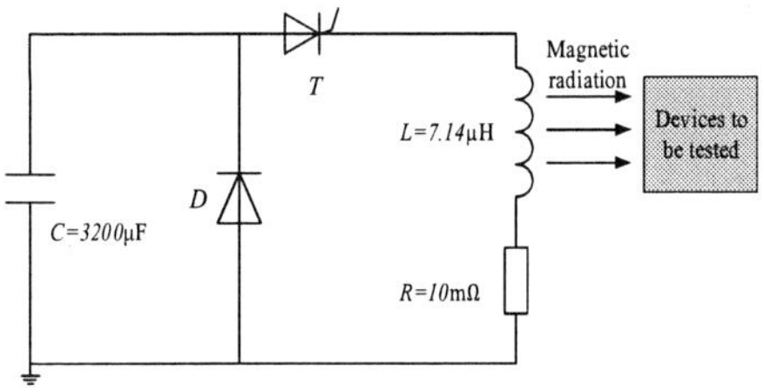

Fig. 1. Diagram of experimental circuit

In the experiment, a pancake coil was used as the pulsed magnetic field source to simulate the strong magnetic field radiation of the system. And according to simulation, the coil generates 784Gs/kA at the testing point, which is 2cm above the center of the coil. The magnetic simulation model of the coil is shown in Fig.2.

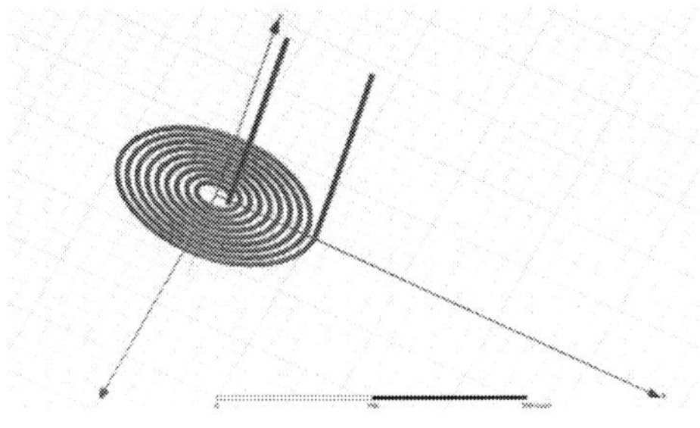

Fig. 2. Magnetic simulation model of the pancake coil

According to previous theoretical analyses and feasibility of the experiment, forward voltage drop (U_d) was selected as testing parameters for diode; common-emitter amplification coefficient(β) for triode.

A. Diode

During the test, the diode was put in the pulsed magnetic field, meanwhile, measuring the forward voltage drop (U_d) of diode when a direct current of 1mA flows through the diode (the product number is IN4007). The experimental circuit is shown in the Fig.3.

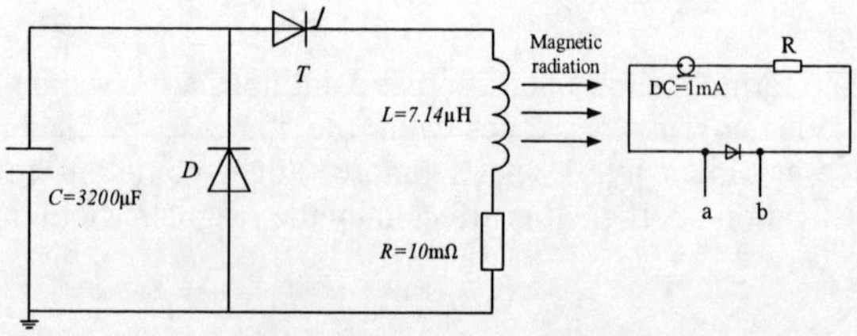

Fig. 3. Diagram of diode's experimental circuit

In the experiments, Pem probe was used to measure the current flowing through the capacitor and inductor, Meanwhile taking advantage of Tektronix oscilloscope probe to measure the diode's conducting voltage.

B. Triode

Similar to diode's experiment, the triode was placed in the pulsed magnetic field, meanwhile, measuring the collector current and base current to calculate the common-emitter amplification coefficient. The experimental circuit is shown in the Fig.4.

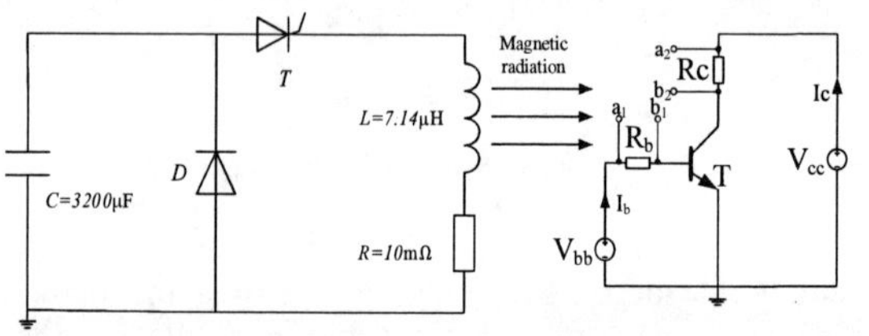

Fig. 4. Diagram of triode's experimental circuit

During the test, measuring the voltage drop of R_b and R_c to obtain base current and collector current.

IV. Results and Analyses

A. Diode

The measurement results of diode's voltage drop for a given forward current in the pulsed magnetic field are listed in TABLE I (According to previous simulation, a current of 1kA will produce a magnetic field of 784Gs at the testing point).

TABLE I. EXPERIMENT DATA

I_L/kA	B_Z/T	U_d/mv	ΔU_d/%
0	0	651	0.00
2.98	0.23	651	0.00
4.6	0.36	651	0.00
6.21	0.49	652	0.15
7.82	0.61	653	0.31
9.49	0.74	655	0.61
11.1	0.87	656	0.77
12.67	0.99	657	0.92
14.25	1.12	659	1.23
16.07	1.26	661	1.54
17.68	1.39	663	1.84
21.04	1.65	667	2.46
24.26	1.90	672	3.23
27.46	2.15	681	4.61
32.04	2.51	691	6.14
33.72	2.64	697	7.07
35.36	2.77	701	7.68

The relation curve between the square of magnetic flux density and diode's conducting voltage regulation was obtained in accordance with the experiment data.

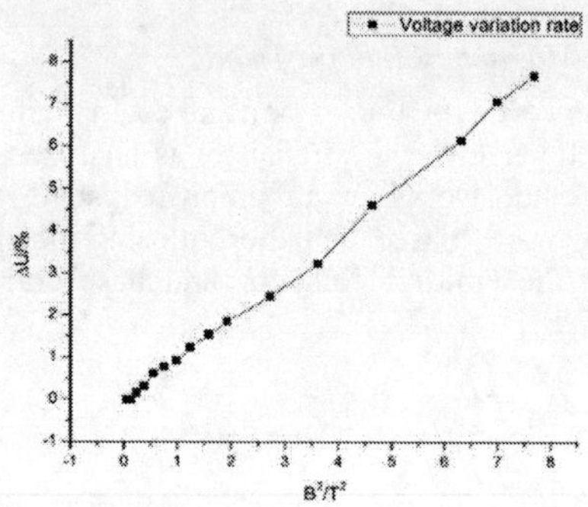

Fig. 5. The relation between the square of magnetic flux density and diode's conducting voltage regulation

It reveals that the magnetic field increases forward voltage drop of the diode. What's more, the conducting voltage regulation is proportional to the square of the magnetic induction intensity of external magnetic field, which is consistent with equation (6).

B. Triode

The measurement results of triode's common-emitter amplification coefficient in the pulsed magnetic field are listed in TABLE II (According to previous simulation, a current of 1kA will produce a magnetic field of 784Gs at the testing point).

TABLE II. EXPERIMENT DATA

I_L/kA	B_z/T	I_b/μA	I_c/mA	β
0	0	43.10	13.08	304
3.15	0.2469	43.10	13.07	303
6.67	0.5229	43.00	12.98	302
10.07	0.7894	43.40	13.00	299
13.54	1.0615	42.90	12.85	300
17.12	1.3422	43.10	12.81	297
20.67	1.6205	42.90	12.65	295
24.19	1.8964	43.20	12.56	291
27.68	2.1701	43.30	12.43	287
29.43	2.3073	43.10	12.21	283
32.93	2.5817	43.10	12.11	281
34.74	2.7236	43.00	12.00	279
36.53	2.8639	43.10	11.85	275

The relation curve between the magnetic flux density and triode's common-emitter amplification coefficient was obtained in accordance with the experiment data.

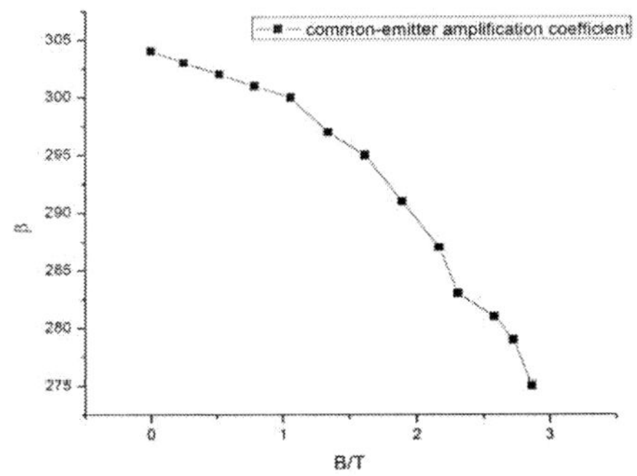

Fig. 6. The relation between magnetic flux density and β

It's easy to know the pulsed magnetic radiation affects triode's output parameter, to be specific, β decreases with the increase of the magnetic induction intensity of external magnetic field.

V. Conclusions

This paper is mainly on the effects of pulsed magnetic field radiation on semiconductor devices, such as diode and triode. Conclusions are summarized as follows:

(1) Theoretical analyses on effects of magnetic field on diode reveals that conducting voltage of the diode for a given forward current is proportional to B_z^2, and this increase in voltage drop tends to be saturated when $\mu_H B_z >> 1$. Experiment results confirm the conducting voltage regulation of the diode is proportional to the square of the magnetic induction intensity of external magnetic field. However, due to limitation of experimental conditions, B_z is not high enough to make $\mu_H B_z >> 1$.

(2) According to the relation curve between common-emitter amplification coefficient and magnetic flux density, β decreases with the increasing of magnetic induction intensity of external magnetic field. In addition, this effect approaches saturation when $\mu_H B_z >> 1$. The experiment results is consistent with equation (15), which shows β decreases with the increase of magnetic induction intensity.

(3) Theoretical analyses and experiment results together indicated that pulsed magnetic field radiation may affect performances of semiconductor devices. Appropriate shielding measures should be taken to reduce the influence and guarantee the reliability of the electromagnetic launch system.

References

[1] Wang Ying. Principle of electrical cannon[M]. Beijing: National Defence Industry Press. 1995 : 25-38.

[2] Richard A.Marshall, Wang Ying. Science and technology of electromagnetic railgun[M]. Beijing: Weapons Industry Press. 2006.

[3] Wang Ying, Ma Fuxue. New concept weapon(The First Edition)[M]. Beijing: Weapon Industry Press, 1997.116~119.

[4] Tang Lingling. Analysis of Electromagnetic Railgun Projectile Fuze's High Magnetic Field Environment and its Shield Design and Utilization[D]. Nanjing University of Science and Technology, 2014.

[5] Ren Xingrong. Research on the Electromagnetic Damage effects and Mechanisms of Semiconductor Devices[D]. Xidian University, 2014.

[6] Barry Schwartz, Fred Herr, Robert Bryan. Fuzing for Use in an Electromagnetic(EM) Gun[C]. HeiT, 2005.

[7] Ciolini R,Schneider M,Tellini B. The Use of Electronic Components in Railgun Projectiles [J]. IEEE Transactions on Magnetics. 2009,45(1):578-583.

[8] Ravotti F, Glaser M, Hartert J, et al. The Effect of Magnetic Field on Readout of Diodes Used as NIEL Counters[J]. Nuclear Science, IEEE Transactions on, 2011, 58(3): 1112-1116.

[9] Zhang Guobin, Shen Kaisheng, et al. Influences of Magnetic Filed on I-V Characteristics of Diodes Based on Silicon and Germanium[J]. Chinese Journal of Electron Devices, 2010, 05: 553-556.

[10] Electromagnetism[M]. Higher Education Press, 2011 : 25-60.

[11] Liu Enke, Zhu Bingsheng, Liu Jinsheng. Semiconductor Physics[J]. Shanghai:Shanghai science and Technology Press. 19B4, 2003, 3: 1988-7.

[12] Brennan K F. The physics of semiconductors[J]. World Scientifc, Singapore, 1999.

[13] Zhang Guobin. Investigation of the Influence of Intense Magnetic Field on Electric Devices[D]. LanZhou University, 2011.

Design of a 6 MW solid-state pulse modulator using Marx generator for the medical linac*

Heuijin Lim*, Dong Hyeok Jeong†, Manwoo Lee, Mujin Lee, Jungyu Yi
Research Center
Dongnam Institute of Radiological & Medical Sciences,
Jwadong-gil 40, Jangan-eup, Gijang-gun Busan, Korea
*heuijin.lim@gmail.com, †physics@dirams.re.kr

Sung Chae Ro
Dong-A HiTECH Co., LTD, 75-8 Mola-3-dong, Sasang-gu, Busan, Korea

***Abstract*— The linacs producing high energy and high power of electron-beam or X-ray beam have been used in medicine, industry, national security, etc. In the linac, the electrons are generated by the electron-gun and accelerated in the accelerating column with the high power RF fields. The high voltage pulses from the pulse modulator are supplied to the RF power source and the electron-gun. The pulse modulator is one of the big and expensive components in the linac. The commercial medical linacs commonly use the pulse modulator based on the thyratron-switched pulse-forming network. In order to improve the power efficiency, achieve the system compactness, and optimize the cost and space, the solid-state pulse modulator based on the Marx generator was proposed. The low power solid-state pulse modulator was developed for the electron-gun operation. The conceptual design and functional results were confirmed. In order to apply it to the RF power source such as a magnetron or a klystron, the 6 MW pulse modulator with the same Marx scheme is proposed. It consists of 40 storage-switch stages and one high voltage pulse transformer, producing the pulse of 50 kV and 120 A required by the magnetron in the medical linac. A storage-switch stage was designed for IGBTs to switch high current of 280 A and 720 V and to use the capacitor of 25 μF which was chosen for the voltage droop of 10 % with the pulse width of 5 μs. The prototype system with eight storage-switch stages was fabricated and tested with a load system. The performance results show that it can be extended to be the 6 MW solid-state pulse modulator. In this paper, we describe the design features, and discuss the results and also the future plan to optimize the solid-state pulse modulator in the medical linac.**

Keywords*— Medical linac, Pulse Modulator, Marx generator, IGBTs

I. INTRODUCTION

More than 30,000 accelerators have been built worldwide over the past 60 years for use in medicine, industry, national security, basic science research, etc. The more than 14,000 particle accelerators have been produced exclusively for medical therapy with electrons, ions, neutrons, or X-rays [1,2]. The majority of the radiation therapies are still performed with linacs producing electron and X-ray beams. The linac accelerates the charged particles using the radio-frequency electric field with a harmonic time dependence provided by the RF power source. Since the RF power sources are operated by the pulse modulators, the market data of radiotherapy machines tell us how many the pulse modulators are used for the cancer treatment application.

The C-band linear accelerator has been designed and constructed to generate the electron and X-ray beams with megavoltage energy in Dongnam Institute of Radiological & Medical Sciences (DIRAMS), Busan, Korea. For the medical application, we proposed a radiotherapy machine including a gantry based on this 6 MeV C-band linac, a rotatable couch for treatment, and a control console [3]. In the linac, the electrons are produced by an electron-gun and gained their energy up to 6 MeV in the accelerating column using 2.5 MW RF power. The electron-gun and the RF power source are operated with the heating power and the high voltage pulse from the pulse modulator. The pulse modulator is commonly constructed with the thyratron-switched pulse-forming network [4] and is the high-cost and large-size one in the linac components. With the evolution of power semiconductor devices, the solid-state pulse modulator was proposed. This kind of modulator has been developed to improve the power efficiency using the various pulsed power topologies [5]. The Marx generator invented by Erwin O. Marx in 1924 [6] has been applied in the pulse modulator using insulated gate bipolar transistors (IGBTs) or metal-oxide-semiconductor field-effect transistors (MOSFETs) instead of the spark gaps. The fast switching speed, high voltage rating, and high current rating of IGBTs and MOSFETs allow this generator to produce high-power square-shaped output pulses at high repetition rate. If this modulator can be manufactured with the components including high voltage switches which easily can be found in the market, can produce the controllable pulse shape and can be constructed with the compact size and the moderate cost, it would be the good candidate to be used in the medical linac.

We developed the solid-state pulse modulator based on the Marx generator to provide the pulsed power to a medical linac electron-gun. This modulator was designed and constructed to generate high voltage pulse up to 25 kV by 35 storage-switch stages. Each stage has a storage capacitor of 500 nF and two IGBTs to control charging and discharging of the capacitor. The results verified the conceptual design and the functional operation with the medical linac electron-gun [7]. With this Marx generator scheme, we designed the high power solid-state pulse modulator for 50 kV and 120 A magnetron operation, consisting of 40 storage-switch stages. For the prototype system, eight storage-switch stages, a bias power board, and a trigger optic distributor were constructed and tested with a resistive load system. This paper focuses on the design of the solid-state pulse modulator in the medical linac and the performance test. And the test results and future plan were discussed.

II. DESIGN

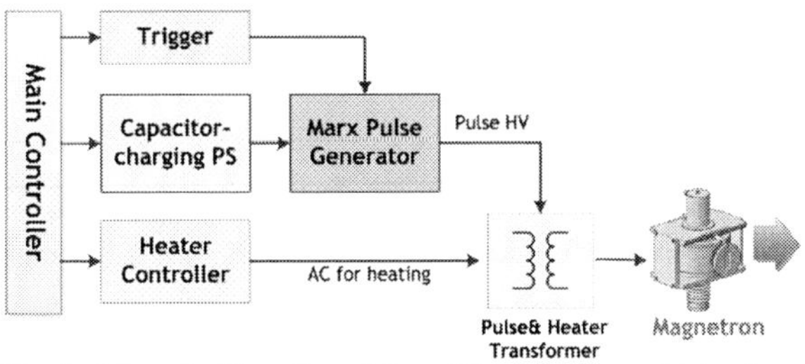

Fig. 1. Schematic illustration of the solid-state pulse modulator for a medical linac magnetron.

In the linac, the electron-gun and the RF power source such as magnetron or klystron require the pulsed power. The high power solid-state pulse modulator for 50 kV and 120 A magnetron was designed. As shown in Fig. 1, the solid-state pulse modulator for a magnetron contains a capacitor-charging power supply, a Marx pulse generator, a pulse & heater transformer, a heater controller, a trigger, and a main controller.

The capacitor-charging power-supply was used for providing the input power and was designed with a series resonant high frequency converter technology [8]. The operating frequency is 50 kHz in LC resonant circuit. It provides the output voltage of 100-1000 V and the peak charging rate of 15 kJ/sec. The Marx pulse generator produces the high voltage pulse by multi-stacked stages with the input voltage from the capacitor-charging power-supply. In order to reduce the number of storage-switch stages and to use the intermediate input voltage for protecting the internal circuits from the high voltage arcing, the input voltage in the Marx pulse generator was chosen as 720 V and will be amplified up to 25 kV with 40 storage-switch stages. And the pulse transformer is added for boosting two times of the pulse voltage from the Marx pulse generator to be satisfied with the magnetron requirement. Due to the magnetizing current of the pulse transformer, the large back-swing can appear at the end of pulse. The tail clipper of series combination with resistors and diodes is proposed to be installed in parallel with the primary winding of the pulse transformer to reduce the back-swing. When the 6 MeV C–band linac was constructed, our first modulator based on the thyratron-switched pulse-forming network included this kind of tail clipper installed after the thyatron switch for the above reason [9]. The heater controller and transformer were designed for 6.3 VAC and 22 A which are normally used for the magnetron heater. The pulse tank will contain the pulse transformer and the heater isolated transformer and be filled with the insulation oil. The main controller with the trigger determines the pulse amplitude of 4-28.8 kV, the pulse width of 1-10 μs, and the repetition rate of 10-300 Hz. It monitors the input power, the output pulse, and the environmental parameters of heater states, temperature, etc. The interlock is also implemented in the main controller. Since the capacitor-charging power-supply and the Marx pulse generator were designed for 25 kV and 240 A, the required pulse voltage for the load system can be adjusted with the pulse transformer. Therefore, this solid-state pulse modulator can be easily used to other applications required up to 6 MW by a matched pulse transformer.

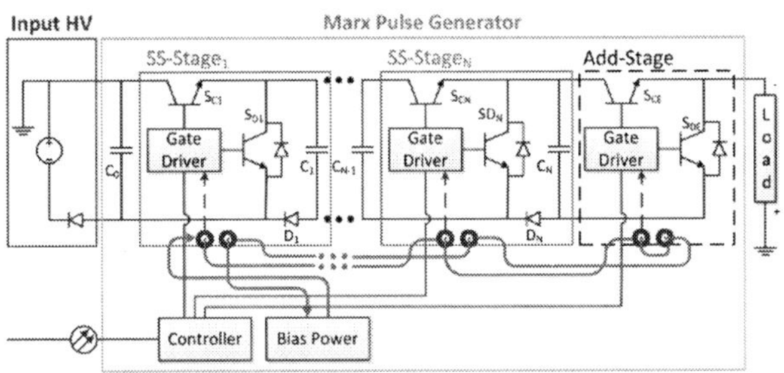

Fig. 2. Circuit diagram for producing the negative high voltage pulse with series semiconductor switches based on Marx concept [7].

The detailed description of our designed Marx pulse generator can be found in Fig. 2. The negative high voltage pulse is generated with the multi-stacked stages. A storage-switch stage consists of an energy storage capacitor C_i, two semiconductor switches for charging S_{Ci} and for discharging S_{Di}, a diode D_i, a gate driver, bias supply circuits for powering the gate drivers, and some auxiliary protection circuits. Since two switches can be controlled independently, they allow to generate the pulse with various shape in discrete steps and to have a fast rise and fall time. If the load system contains the capacitance component, the pulse shape is distorted by the remaining charge of load system. Therefore, additional switches of S_{CE} and S_{DE} are installed after the last stage and discharge the charge of the load system to the ground, resulting in improving the voltage droop. Since the capacitor-charging power supply is based on the opening circuit, the C_0 capacitor is introduced to protect the storage-switch stages for the discharging mode. The gate driver is controlled by the fiber-optic and powered from the bias power board through the isolated transformer [7]. As the input voltage becomes higher and the discharging current becomes lower for the fixed storage capacitance at each stage, the voltage droop ratio becomes lower and better. However, our proposed low input voltage and high discharging current can give us the big droop ratio. In order to solve the voltage droop issue, the capacitor of 25 μF was chosen for the voltage droop of 10 % with the pulse width of 5 μs.

III. Prototype System and Results

Fig. 3. Constructed prototype solid-state pulse modulator. It was tested with the resistive load system.

For the prototype system, eight storage-switch stages, a bias power board and a optic distributor were constructed. In the storage-switch stage, we chose the IXYS model, IXYN120N120C3 (S_{Di}) for the discharging switch and the Fairchild model, FGA30S120P (S_{Ci}) for the charging switch. These IGBTs have the voltage rate up to 1200 V and the fast falling time to allow the high speed switching. Also the freewheeling diodes were additionally attached to these IGBTs. A fast recovery diode of IXYS DSEI30-12A (D_i) with the fast reverse recovery time of 40 ns was chosen to pass the current in the charging mode and immediately to block it in the discharging mode. The film capacitor of UL3_L256K (C_i) manufactured by Electronic Concepts was used for the storage capacitor. The board size of each stage was designed to be 150x260 mm^2. These stages were stacked with the space gap of 100 mm. From this test, we learned that the board size can be reduced by 15 % and the space gap can be also reduced by 10-15 %, achieving the compact size for the fully integrated system.

The test was performed with the resistive load of 24.36 Ω. As shown in Fig. 4, the results of pulse voltage were achieved from -1 kV to -5.76 kV for the different input voltages. They were fitted to the form $V_{OUT}=n \cdot V_{IN}+V_{CONST}$ where V_{OUT} is the pulse voltage, n is the number of stages in the Marx pulse generator, V_{IN} is the input voltage, and V_{CONST} is an additional offset voltage. The fit yields n=8 and V_{CONST} = -167 V with χ^2/N_{data}=0.99 and describes that the number of stages is eight. The current was measured from 46 A to 288 A for the different input voltages (see Fig. 5), resulting that the net impedance of the final load was estimated to be 21.15±0.45 Ω. It confirms that each stage drives the current rate up to 288 A with the input voltage of 720 V and the pulse voltage is amplified with the number of stages.

With an over-current protection circuit, the safe operation was also tested. In the over-current case, the short-circuit immediately discharges the storage capacitors to the ground.

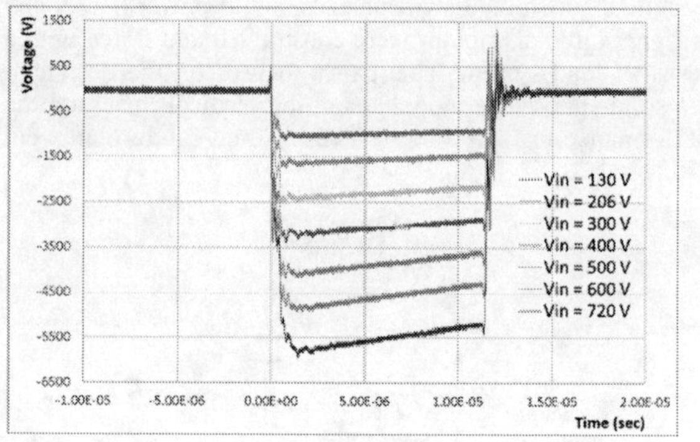

Fig. 4. Test with the resistive load of 24.36 Ω. Different amplitudes of pulse voltage were taken from the input voltages of 130 V to 720 V. The pulse width is 12 μs.

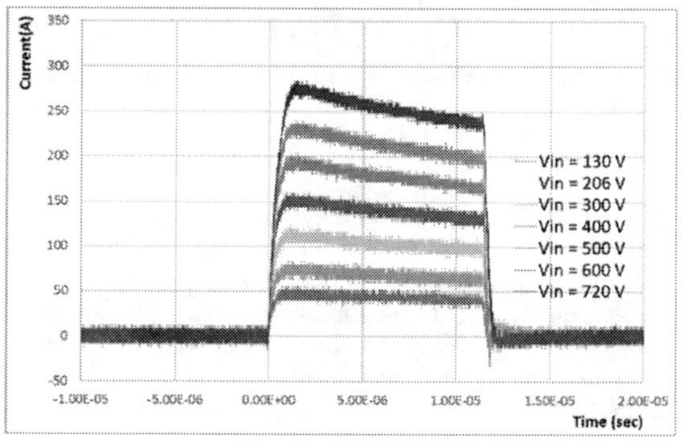

Fig. 5. Test with the resistive load of 24.36 Ω. Different amplitudes of pulse current were taken from the input voltages of 130 V to 720 V. The pulse width is 12 μs.

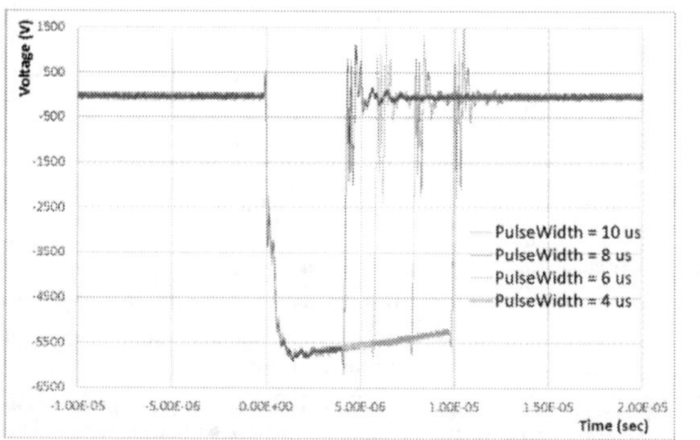

Fig. 6. Test with the resistive load of 24.36 Ω. Generated pulse voltages for the input voltage of 720 V with variable width from 4 μs to 10 μs.

The output pulses were taken with the width on 12 μs in Fig. 4 and Fig. 5. As shown in Fig. 6, the pulse shapes were studied with various pulse widths from 4 μs to 10 μs. For the voltage droop, it was measured to be 6 % at 4 μs and

12 % at 12 μs. It is satisfied with the required voltage droop of our magnetron. Therefore, our system will be optimized without any active circuit of the voltage droop correction.

IV. Conclusion and Future Plan

Fig. 7. Constructed storage-switch stage boards with component assembly.

The 6 MW solid-state pulse modulator with the multi-stacked storage-switch stages based on the Marx generator for the medical linac was proposed and designed according to the same scheme which was already developed for the low power solid-state modulator for the electron-gun. Especially, the stage was designed with two active switches to control charging and discharging the storage capacitor of 25 μF for switching the voltage of 720 V and the current of 288 A with the pulse width of 5 μs. The prototype eight storage-switch stages were constructed and tested with the resistive load of 24.36 Ω. The performance results showed that the voltage droop ratio with the pulse width of 5 μs is less than 10 %, and the storage capacitor of 25 μF and IGBTs were chosen properly. They also verified to amplify the input voltage with the number of stages.

This design is applied for the solid-state pulse modulator with 40 storage-switch stages and optimized with the pulse & heater transformer for the 2.5 MW magnetron operation. As shown in Fig. 7, we are producing these stage boards with components assembly. After passing the board-level quality test, these stages will be integrated and stacked.

As the turns ratio of pulse transformer increases, the number of stages can be reduced. It also leads that the discharging current increases and the voltage droop in the given storage capacitance becomes worse. The current design was optimized with 40 storage-switch stages and two-time boost transformer. However, there is an cost issue to use the expensive pulse transformer of 1:2. We will perform the pulse droop test as increasing the discharging current of IGBTs at each stage and also check the protection limitation as increasing the input voltage of stacked stages with the constructed system. From these studies, we expect to re-adjust the turns ratio of the pulse transformer with the reduced number of stages, achieving the system compactness and minimizing the cost.

Acknowledgment

This research was supported by the Government Research Foundation of DIRAMS grant funded by the Korea government(MSIP) (No. 50495-2016).

References

[1] W. Henning and C. Shank., "Accelerators for America's Future", Report from the US Department of Energy, Jun. 2010.

[2] R. W. Hamm and M. E. Hamm , "Industrial Accelerators and Their Applications", World Scientific, (2012)

[3] H. Lim et al., "Design of a Radiotherapy using the 6 MeV C-band Standing-wave Accelerator", presented at IPAC'16, Busan, Korea, May 2016, paper TUPOY008.

[4] G. N. Glasoe and J. V. Lebacqz, "Pulse Generator", MIT Radiation Laboratory Series, Vol. 5 (McGraw-Hill,1948).

[5] M. H. Rashid, "POWER ELECTRONICS HANDBOOK:Devices, Circuits, and Applications", 3rd ed., Boston, MA:Elsevier, 2011.

[6] E. Marx, "Verfahren zur Schlagprufung von Isolatoren und anderen elektrischen Vorrichtungen", Deustsches Patent No. 455933, 1928.

[7] H. Lim et al., "Solid-state pulse modulator using Marx generator for a medical linac electron-gun," 2016, JINST 11 P04003.

[8] J.S. Oh et al., "DEVELOPMENT AND APPLICATION OF AN INVERTER CHARGING SUPPLY TO A PULSE MODULATOR", in Proc. 21st Linear Accelerator Conference, Gyeongju, Korea, 2002, pp. 205-207.

[9] H. Lim et al., "Construction and Commissioning of the Prototype C-band Linac", presented at the 7th Korea-Japan Joint Meeting on Medical Physics, Busan, Korea, 2014, unpublished.

LARGE AREA ELECTRON BEAM SOURCE WITH ARC PLASMA EMITTER

Kurkuchekov Viktor
Novosibirsk State University
NSU
Novosibirsk, Russia
kurkuchekov.victor@gmail.com

Ivanov Ivan, Kandaurov Igor, Trunev Yuri
Budker Institute of Nuclear Physics
BINP SB RAS
Novosibirsk, Russia

***Abstract* — A 120 keV, up to 250 A, 1 ms pulsed electron beam injector has been developed using an arc-discharge plasma cathode and multi-apertured diode electron optics. The last comprises two flat molybdenum "grids" with many round apertures, drilled in hexagonal order. Such a design is applicable for producing high current large-area beams. The beam parameters can be raised simply by increasing the number of apertures in the diode, if radial uniformity of the emission plasma is provided. In the presented work an experimental comparison has been carried out for two versions of the diode with different size: 241 and 499 apertures arranged within a circle with a diameter of ca. 8 cm and ca. 12 cm, respectively. An X-ray imaging diagnostic was implemented to obtain 2D-picture of the cross-sectional distribution of the beam emission current density.**

***Keywords* – electron beam; plasma cathode; multi-aperture diode; X-ray imaging diagnostic.**

I. Introduction

An electron beam source has been developed at BINP, Novosibirsk, with plasma cathode and multi-apertured electron optical system (EOS). State-of-the-art beam parameters are as follows: energy up to 120 keV, total current up to 250 A, pulse length up to 1 ms. The beam is relevant for simulation of thermal shock loads on plasma-facing components of future tokamaks, plasma heating and stabilisation in linear magnetic systems, processing of materials, etc.

The beam extraction/acceleration features multi-aperture diode optics with cathode and anode performed as flat molybdenum "grids" with multiple round openings, drilled in hexagonal order. Electrons are extracted from the plasma produced by a single on-axis arc-discharge generator. The more detailed description of the beam source design can be foung in [1]. The multi-aperture plasma emission diode is quite suitable for large area beam applications and the beam current increase can be achieved in principle by simply increasing the number of apertures (i.e. effective emission area). However, this approach faces the challenge of the radial uniformity of the emission plasma, in other words, the efficiency of use of the peripheral apertures in EOS with large crossection. Application of an external guiding magnetic field contributes to the problem, leading to a contraction of the arc discharge in near-axis region.

This paper presents an experimental study of the emission current profile for two instances of the multi-aperture EOS with different sizes. The first one has 241 round apertures arranged hexagonally in a circle with diameter of 8.2 cm and the second has 499 apertures within a circle with diameter of 11.8 cm. Effective emission area is ca. 11 sq. cm. and ca. 26 sq. cm., respectively. Diameter of the emission openings is 0.24 cm and 0.26 cm, the diode gap is 1 cm.

In these experiments, the diagnostic tool was implemented for determination of the beam current profile through X-ray imaging of the electron beam footprint on a flat metal target. The diagnostic allows to obtain 2D-picture of the cross-sectional distribution of the beam current density in every single shot. The spatial resolution of this diagnostic has been measured in special tests as one pair of lines per 0.25 cm at 10% contrast level.

II. Experimental setup

The experiments have been carried out using electron beam test facility designed for simulation of pulsed heat loads on construction materials [2]. A schematic of the experimental setup is shown in Fig. 1. The beam injector is mounted inside a cylindrical vacuum chamber. An electron beam passes in a guiding magnetic field to the collector plate located at the distance of 0.85 m from the beam source. The field is produced with external magnetic windings, its value is 10 mT at the beam start and 17 mT at the collector.

The beam collector is designed as a thin metal plate and serves as an X-ray converter. The phosphor screen (Gd_2O_2S:Tb phosphor on a thin plastic film) is attached directly to the back side of the collector plate for imaging the beam X-ray footprint. The phosphor glow is registered by the use of 45° mirror and fast CCD camera. It should be noted that luminescence decay time for Gd_2O_2S:Tb is of ca. 0.6 ms, whereas typical beam duration in the experiments was of ca. 0.2 ms. Thus integrated over time 2D-distribution of the beam current density has been imaged.

For ~ 100 keV beam electrons an angular distribution of X-rays produced in the collector is nearly isotropic. Therefore, to achieve the better image spatial resolution, the phosphor

Russian Science Foundation (project N 14-50-00080) supported the work at the electron beam test facility. The development of X-ray imaging diagnostic supported by RFBR, research project No. 16-38-00739 mol_a.

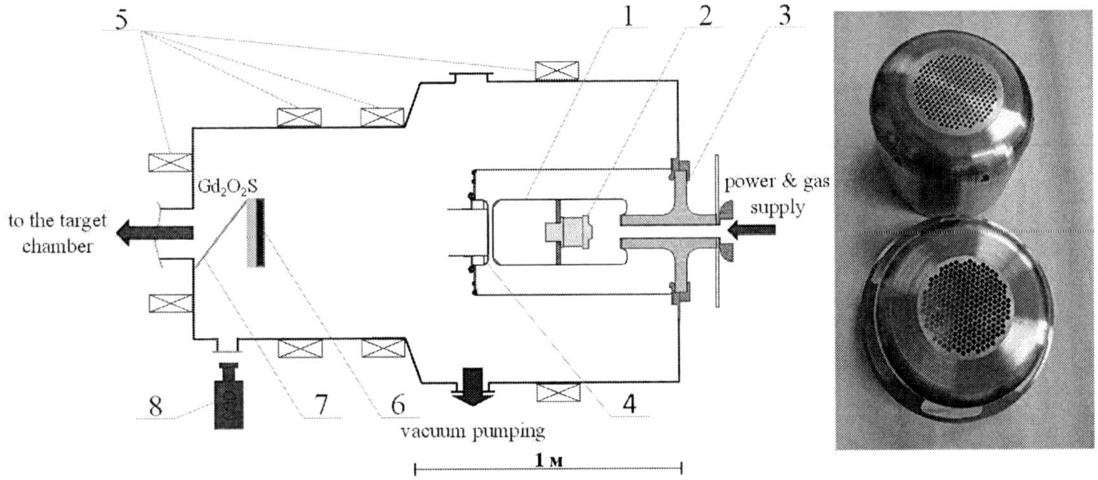

Figure 1 Schematic layout of the experimental facility: 1 – high-voltage electrode (hollow anode of the arc discharge) with emission grid; 2 - arc plasma generator (washer gun); 3 - feedthrough insulator; 4 - extracting grid (diode anode); 5 - magnetic field coils; 6 - beam collector (x-ray converter) with phosphor screen; 7 - 45°- mirror; 8 - fast CCD-camera;

screen should be placed as close as possible to the collector surface, which in turn requires minimal thickness of the collector plate. The thin collector is needed as well to reduce the attenuation of X-rays passing through the collector material onto the phosphor screen. However, the use of a thin collector is hampered by its strong heating under electron beam load. The phosphor screen will be damaged if the back side temperature of the converter exceeds 80° C. The heating of the collector plate has been calculated for different materials (W, Ta and stainless steel) with the beam parameters as follows: $E_b = 100$ keV, $I_b = 100$ A, $\tau_b = 0.15$ ms and the diameter of 7 cm. It is assumed that the absorbed energy is uniformly distributed over the beam footprint. The electrons backscattering factor for all materials were taken into account. The temperature curves vs. collector thickness are shown in the Fig. 2. As it can be seen, an acceptable temperature of the back side of converter is obtained at a thickness of 0.1 cm or more.

Material and thickness of the beam converter also affect the amount of light emitted by the phosphor. The number of photons S emitted by the phosphor screen can be calculated as:

$$S = k \cdot \int_E N(E) \cdot T(E) \cdot A(E) \cdot E dE, \quad (1)$$

where E stands for X-ray photon energy, $N(E)$ is a spectral response of the converter, $T(E)$ is the converter transmittance, $A(E)$ is an absorption coefficient of the phosphor layer and k is the light yield of the phosphor (k = 35 photon/keV). The spectral response of the collector is calculated using the model outlined in [3, 4]. The absorption coefficient is calculated based on the surface density of the phosphor active substance – 55 mg/cm^2. All calculations use the material properties specified in the NIST database [5], the electron beam parameters and the converter materials were the same as for the converter heating calculations.

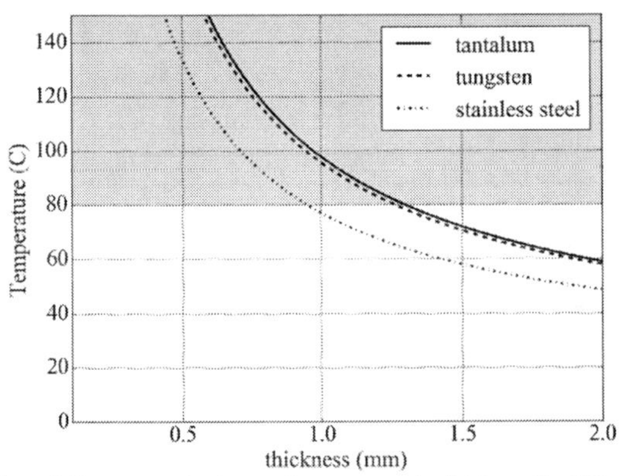

Figure 2 X-ray converter temperature vs converter thickness for different materials.

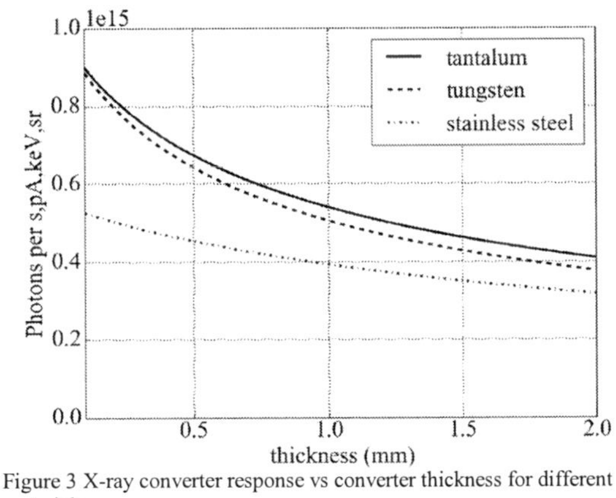

Figure 3 X-ray converter response vs converter thickness for different materials

The number of photons vs. collector thickness are shown in the Fig. 3. With a converter thickness of over 0.1 cm a difference in the emitted light is inessential for all considered materials. Thus stainless steel was chosen for the X-ray converter as inexpensive and easy to process material. The thickness of the converter was taken as 0.1 cm.

III. EXPERIMENTAL RESULTS

A typical X-ray image of the beam (80 keV, 70 A, 0.1 ms) produced with 241 apertured EOS is shown in Fig.4. The beam cross-section is very close to circular, and the image shape and center of mass are stable from shot to shot. The upper part in the Fig. 4 shows the beam current density profile, measured along the line crossing the center of the image. The profile is rather flat and reaches 3.2 A/cm^2 at maximum. An effective beam diameter, defined as a diameter of the circle which encloses 90% of the beam current, is found to be ca. 7 cm. on the target.

Special experiments have been performed on measuring of a sharp edge spread function to determine the spatial resolution of the imaging diagnostics. The edge is formed with a tungsten plate of 0.1 cm thickness, that overlaps part of the beam target and completely shadows the X-rays. The resulting image is shown in Fig. 5. A normalized intensity profile is measured along the line perpendicular to the edge and crossing the center of the beam footprint. On the basis of the profile the modulation transfer function of the diagnostic was calculated [6]. The limiting spacial resolution was found as 4 pair of lines per 1 cm at 10% contrast level.

While the diagnostic measures the current density distribution on the collector at a distance of 0.85 m from the beam injector, it is possible to evaluate the current density distribution directly in multi-aperture diode.

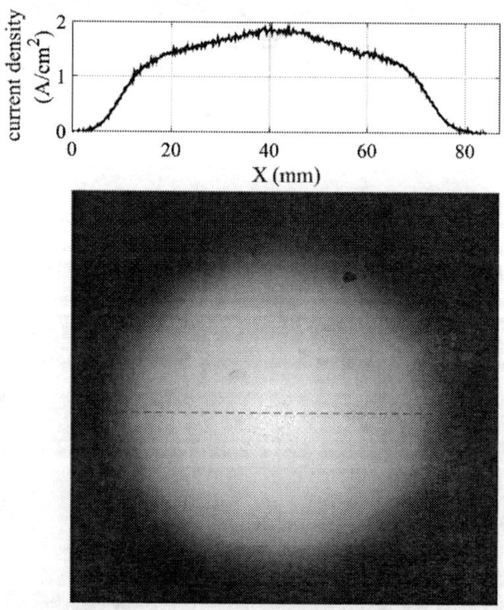

Figure 4 Typical X-ray image of the beam and current density profile measured along dotted line crossing the center of the beam.

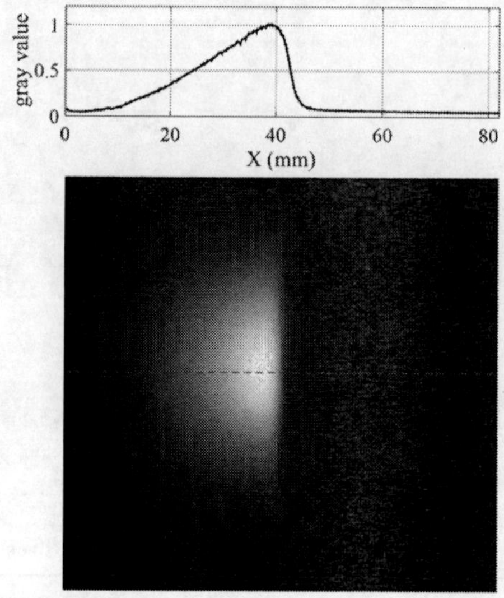

Figure 5 Image of the edge response and intensity profile measured along dotted line crossing the center of the beam.

Originally the beam consists of many separate jets (beamlets). In a guiding magnetic field, the beamlet envelope oscillates radially due to the transversal velocities of electrons. That leads to periodic focusing of the beamlet, ultimately, to its initial radius. Thus, varying the magnitude of the accelerating voltage and magnetic field it is possible to reproduce (on the collector) the initial beam structure and to evaluate the emission yield of every single aperture. An example of such image is shown in Fig.6. As can be seen from the picture, the beam structure completely resembles the EOS structure, all of the 241 apertures are utilized in the beam generation. An emission is

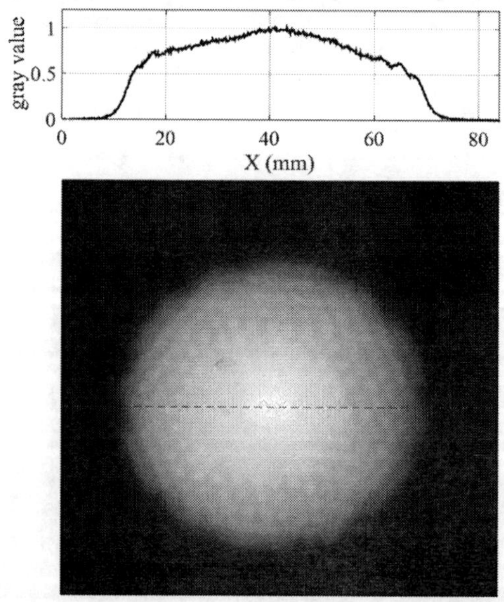

Figure 6 Image of the beam structure for 241-apertured EOS, and intensity profile measured along dotted line crossing the center of the beam.

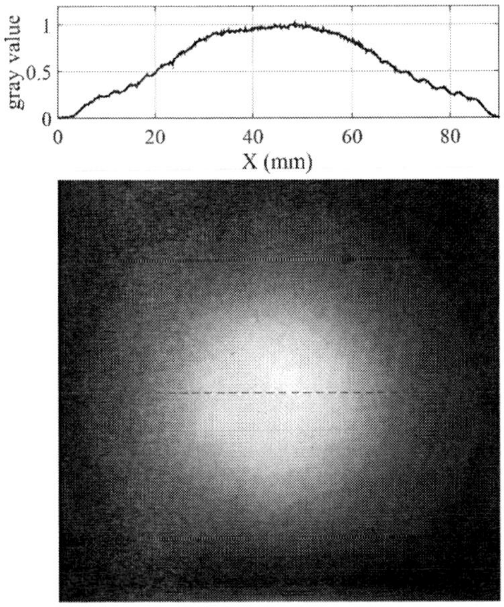

Figure 7 Image of the beam structure for 499-apertured EOS, and intensity profile measured along dotted line crossing the center of the beam.

quite uniform all over the EOS, the current density decreases by 20% from the maximum at a radius of 2.5 cm and falls near twice at the periphery of EOS (at the radius of 4 cm). It should be noted that the discrete structure of the beam on the target disappears with change of the magnetic field and the energy of electrons, which is also confirmed by the melting imprint at the collector.

At the next experimental step the number of apertures was nearly doubled (499 vs 241) and EOS diameter increases ca. 1.4 times (11.8 cm vs 8.3 cm). As a result, the maximum emission current of 250 A at 0.25 ms pulse is achieved. The beam structure in 499-apertured EOS is shown in Fig.7. All of 499 apertures are utilized, as with previous EOS, the current density decreases by 20% from the maximum at a radius of ca. 2.5 cm and falls to ~ 0.25 of maximum in the peripheral apertures (at the radius of 5.5 cm).

IV. Conclusion

An electron beam current density distribution in multi-aperture diode with plasma cathode have been studied with the use of imaging X-ray diagnostic tool. It is demonstrated that the large-area electron beams can be effectively formed by multi-aperture system with a single on-axis arc-discharge plasma generator even in the presence of the guiding magnetic field. All apertures are emitting in the 499-apertured EOS with diameter ca. 12 cm. The beam current density quite uniform to the radius 2.5cm and gradually decreases to the periphery.

V. Reference

[1] V. V. Kurkuchekov, V.T. Astrelin, A.P. Avrorov, A.V. Burdakov, P.V. Bykov, V.I. Davydenko,G.E. Derevyankin, A.A. Ivanov, I.A. Ivanov, I.V. Kandaurov, S.V. Polosatkin, A.F. Rovenskikh, Yu.A. Trunev, V.A. Yarovoy, "Novel Injector of Intense Long Pulse Electron Beam for Linear Plasma Devices," *Fusion Science and Technology, vol.63, No. 1T,* vol. 1T, no. 63, pp. 292-294, 2013.

[2] Takeshi Hirai, Koichiro Ezato, Patrick Majerus, "ITER relevant high heat flux testing on plasma facing surface," *Materials Transactions.,* vol. 46, no. 3, p. 412, 2005.

[3] Horst Ebel , "X-ray Tube Spectra," *X-RAY SPECTROMETRY,* no. 28, p. 255–266, 1999.

[4] Horst Ebel , "FUNDAMENTAL PARAMETER PROGRAMS: ALGORITHMS FOR THE DESCRIPTION OF K, L AND M SPECTRA OF X-RAY TUBES," *Advances in X-Ray Analysis,* no. 49, p. 267 – 273, 2006.

[5] J. H. Hubbell and S. M. Seltzer, "Tables of X-Ray Mass Attenuation Coefficients and Mass Energy-Absorption Coefficients from 1 keV to 20 MeV for Elements Z = 1 to 92 and 48 Additional Substances of Dosimetric Interest," *Available: http://physics.nist.gov/xaamdi. National Institute of Standards and Technology, Gaithersburg, MD.,* 2004.

[6] Steven W. Smith, "The Scientist and Engineer's Guide to Digital Signal Processing," *California Technical Publishing, San Diego, CA,* 1999.

Conceptual electrical design for a 15-TW LTD accelerator*

R. B. Spielman, D. H. Froula, G. Brent,
E. M. Campbell, and M. Shoup
Laboratory for Laser Energetics, University of Rochester
Rochester, NY 14623, USA

D. B. Reisman, M. E. Savage, W. A. Stygar, and
M. L. Wisher
Sandia National Laboratories
Albuquerque, NM 87185, USA

***Abstract*— We describe the conceptual design for a 15-TW pulsed-power accelerator based on the linear-transformer-driver (LTD) architecture. This driver is proposed for high-energy-density-physics (HEDP) experiments is sized to allow multiple experiments per day in a University environment and will enable experiments that are scalable to larger facilities. In this design, many individual energy storage units (bricks), each composed of two capacitors and one switch, directly drive the load without additional pulse compression. Ten modules in parallel drive the load. Each module consists of 16 cavities in series, where each cavity is composed of 22 bricks in parallel. This design stores up to 2.75 MJ and delivers up to 15 TW to the constant impedance water transmission lines in 100 ns and, depending on the load parameters, peak currents of 10 MA to 12.5 MA can be reached.**

Keywords—pulsed power, LTD

I. Introduction

Since 1970, low-impedance pulsed-power accelerators have been based on Marx-generator/water-pulse-forming technology. The world's largest and most powerful pulsed-power accelerator is the Z accelerator [1,2], refurbished from the original facility in 2007 and located at Sandia National Laboratories (SNL) in Albuquerque, New Mexico, USA. Z stores 20 MJ of electrical and delivers as much as 26 MA in 110 ns to a wide range of loads. Z represents the state-of-the-art for Marx-generator, water-pulse-forming accelerators.

We describe the conceptual design for a compact, highly efficient pulsed-power accelerator, based on the LTD architecture. [3,4] We want to demonstrate and validate the LTD architecture proposed by Stygar, et al. [4] for accelerators larger than Z.

Our design has a single stage of pulse compression and has a matched electrical impedance throughout, as described by Stygar, et al. [4], thereby minimizing electrical reflections and maximizing overall electrical coupling efficiency. The LC time constant of the LTD energy store of ~ 100 ns eliminates the need for further pulse compression (with their attendant losses in efficiency). Our LTD design stores 2.75 MJ and efficiently delivers up to 450 kJ of kinetic energy to a dynamic load. The LTD design enables current pulse shaping via independent triggering of the bricks.

Multiple bricks are assembled in parallel in a cavity. The cavities are arranged in series into a single module. Multiple modules, feeding through coaxial water lines, comprise the entire accelerator and drive a single load via a tri-plate magnetically insulated transmission line.

Detailed circuit simulations using SCREAMER [5,6] model the performance of the accelerator with different loads. These simulations model all of the elements of the accelerator.

II. Conceptual Design

A. Overview

This conceptual electrical design has roughly 1/4 the energy and power from SNL's original Z machine (3 MV, 20 MA, 100 ns, 0.12 Ω, 60 TW). [7-10] This design operates at ~2× lower peak current than the original Z and it addresses the following key design goals:

- Use of LTD architecture & pulse shaping:The driver is based on an LTD prime-power architecture. [4] This choice is motivated by the fundamental efficiency and pulse-shaping flexibility of LTD designs, the maturity of SNL's LTD engineering, the reduced maintenance of LTD components, and the lack of water shock.
- Efficiently drive a wide range of loads: For example currents must be large enough to stably implode plasmas to relevant velocities (~10^7 cm/sec). Experiments that exploit the x- ray energy and power produced by wire arrays or gas puffs are also a desired capability.
- High shot rate and manageable debris levels: We desire up to three shots per day to reduce the cost per shot (annual facility cost/shots) and to increase the number of experiments conducted. The low shot cost encourages innovative high-risk experiments.
- Maximize diagnostic flexibility: Provide multiple, equatorial diagnostic access for x- ray diagnostics and laser-beam access. The target chamber will have further diagnostic access at ~15° and ~30° above the equator.

Figure 1 shows the schematic view of our 15-TW accelerator. The LTD-based architecture has ten, radial water-filled modules arranged azimuthally around the load, with each module containing 16 cavities, and with each cavity containing 22 bricks. The total electrical energy stored in all the brick capacitors within the modules is 2 MJ at a charge voltage of 85 kV (2.75 MJ @ 100 kV). The cathode stalk in each module will be supported at the outer radial location and extend to, and be supported by, the coaxial water transmission line at the output of the module. The cathode stalk can be removed when the water is drained from the module and the inner water tank. With the cathode stalk removed, the individual cavities can be removed from the module for maintenance or the entire module can be replaced. Each module couples to a 2-m-long, 1.25-Ω coaxial water transmission line. [11] These water transmission lines connect to a large water tank containing the water convolutes. Each coaxial transmission line transitions, at constant impedance, to a horizontal, tri-plate transmission line. The ten tri-plate water lines merge into a single tri-plate, disk water line just outside the vacuum insulator stack. These water lines provide complete transit- time isolation of the cavities from the load.

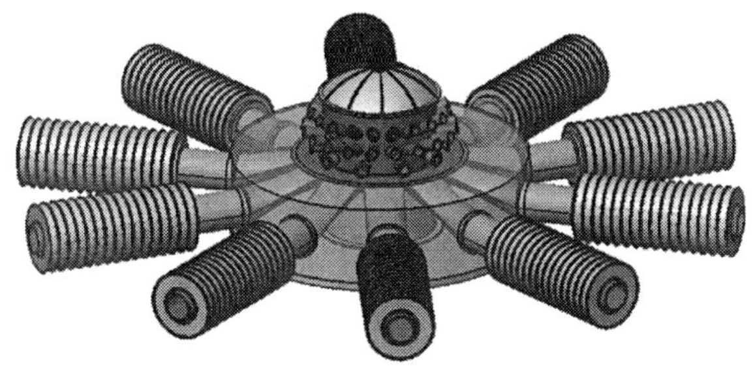

Fig. 1. This schematic of the 15-TW driver shows 10 LTD modules, each having 16 cavities, coupled to coaxial water lines, to the water section, and then to the inner vacuum power feed.

The insulator-stack assembly consists of two, well-graded insulator stacks, each composed of four insulators and three grading rings. Inside the insulator stacks are three, magnetically insulated disk vacuum transmission lines (MITLs)—two anodes and one cathode—arranged in a single, tri- plate configuration.

The two, constant-impedance disk MITLs tie together with a single, post-hole convolute. [7-10] Inside the post-hole convolute (PHC) there is an inner MITL coupling to the load.

B. Brick Design

The basic building block of the LTD driver is a "brick" composed of two capacitors in series and a gas-filled, spark-gap switch. (See Fig. 2.) The baseline accelerator design consists of 3,520 bricks (3,520 switches and 7,040 capacitors). Design and optimization efforts will focus on producing the lowest-possible brick inductance and highest- possible, peak capacitor current within the desired 100-ns rise time. The brick current, upon which our design is based, is presently limited by the 50-kA current limit of the individual capacitors. The individual brick performance effectively defines the overall electrical performance of the driver as a whole through the brick's capacitance, inductance, resistance, peak current, allowed voltage reversal, lifetime, reliability, pre-fire rate, hang-fire rate, size, and jitter.

Our brick design is the baseline Z300 SNL design [12-14]. Our design uses two, 80-nF capacitors arranged in a two-stage Marx configuration (two capacitors and one switch), with the switch placed between the two capacitors and the mid-plane of the switch at ground potential. The two capacitors can each charged to a maximum voltage of +/-100 kV. The capacitors are the low- inductance "double-ended" style with the electrical connections placed on opposing ends of the capacitors. The present switch is a low-inductance, high-pressure, 200-kV spark-gap design. [12-14] The switch is triggered by applying a fast rising ~ 50-kV pulse to the mid-plane of the switch.

Each brick has a matched impedance Z_b where $Z_b = (L / C)^{1/2}$, where L is the total inductance of the brick and C is the total capacitance of the brick. This matched impedance is optimum for power coupling to the load. A resistance of $2 \times Z_b$ would be critically damped.

Energy losses in the brick (~24%) are due to the time-dependent resistance in the gas switch and the effective-series-resistance (ESR) of the capacitors.

Fig. 2. A schematic of an LTD brick showing two, double-ended capacitors and a single, 200-kV low-inductance gas switch. (Figure courtesy of Sandia National Laborities.)

C. Cavity Design

Our oil-filled LTD cavity is composed of 22 bricks arranged radially in parallel. A cavity has the same peak output voltage as a single brick, but has the total current of N_b bricks and an output impedance of Z_b/N_b. Each cavity includes isolation magnetic cores to enable the cavities to be stacked in series. (See Fig. 3.) Common charging connections and a common pulsed triggering system are used for all of the bricks.

The number of bricks in the cavity and, later, the number of cavities in a module was varied to get the desired 1.25-Ω module impedance, cavity current, and output voltage.

The gas switches for each brick within the cavities will be triggered electrically. At *least* two, separate & independent triggering systems per cavity would be required to provide minimal pulse shaping. Multiple, independent trigger systems are needed to achieve the pulse shaping needed for equation-of-state experiments. [15] A typical trigger pulse will have an amplitude of ~ 50 kV with a rise time of ~ 5 ns.

Each cavity contains ferromagnetic magnetic cores that prevent parasitic current losses. (See Fig. 3.) The ferromagnetic cores must have sufficient cross sectional area to support the $V \cdot t$ of the pulse. Acceptable ferromagnetic core materials are 2605CO MetGas™ and FT-3M™ FineMet. [16]

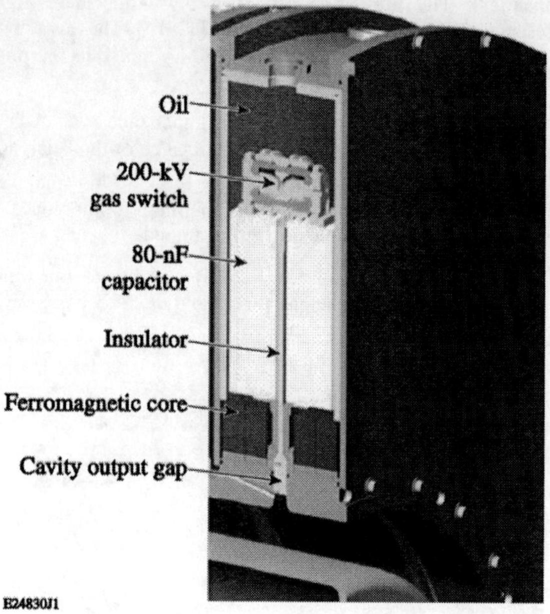

Fig. 3. A schematic of an LTD cavity shows one (of 22) bricks driving the output. The cavity contains ferromagnetic core material used to eliminate parasitic current losses. The cavity is oil filled. (Figure courtesy of Sandia National Laborities.)

D. Module Design

Figure 4 shows the conceptual design of the LTD module that feeds a 1.25-Ω impedance coaxial water transmission line. We used 16 cavities per module to achieve the desired final output voltage (~1.25 MV) and the desired impedance (1.25 Ω). The final module output current is ~1 MA. The power generated by a single module is 1 TW at 85- kV charge voltage and a maximum of 1.5 TW at full 100-kV charge voltage.

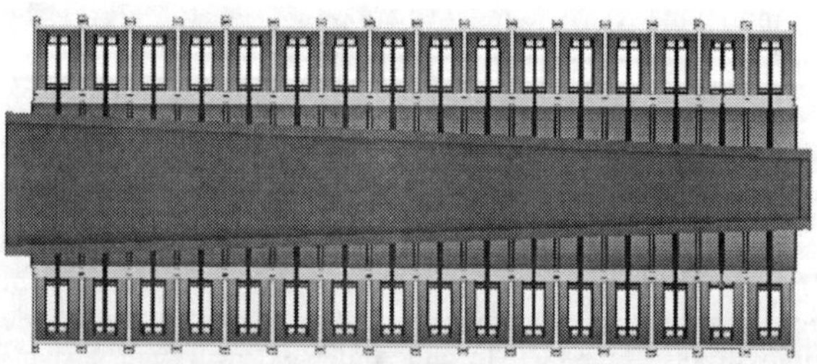

Fig. 4. A schematic of an LTD module shows 16 cavities bricks arranged in series. The module has a tapered cathode that matches the impedance of the cavities as they are added axially. The module is filled with de-ionized water.

Each module has a tapered cathode stalk as the center conductor. The diameter of the cathode stalk varies radially to maintain the correct water, transmission-line impedance. After 16 cavities, the impedance of the water line is a constant 1.25 Ω.

E. Water Transmission-Line Design

The water transmission lines are simple, constant-impedance coaxial designs. These water transmission lines connect the LTD module to the water convolutes. All of the coaxial transmission lines meet at a tank filled with de-ionized water. Inside the stainless-steel water tank, the coaxial transmission lines convolute from a coaxial shape to tri-plate lines that then merge into disk water transmission line and then into a large, monolithic, vacuum-insulator stack. (See Fig. 5.)

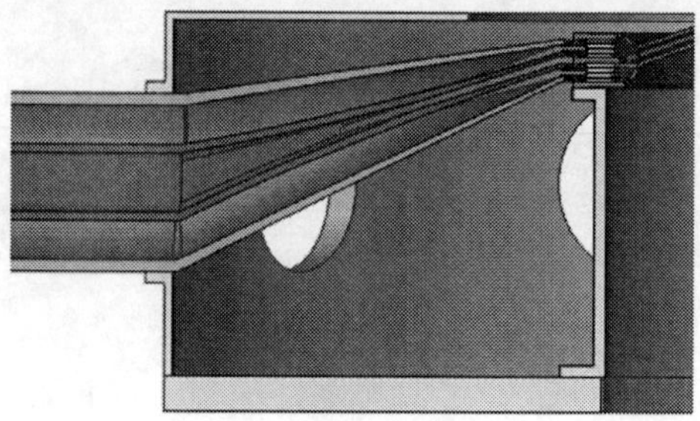

Fig. 5. A close up of the constant-impedance transition from the coaxial water line to a horizontal tri-plate. The module is filled with de-ionized water.

The radial spacing between the outer & inner conductors of the coaxial water lines is determined by the high-voltage breakdown limit in water. We chose the OD of the water coax to be 60 cm and the ID to be 49.74 cm for a 1.25-Ω transmission line. Stygar et al. [17] gives the breakdown requirement for large-area water lines ($A > 10^4$ cm^2):

$$E_p \, \tau_{eff}^{0.33} \leq 0.108 \, , \tag{1}$$

where E_p is the peak value of the average electric field in the water and τ_{eff} is the effective pulse width of the voltage pulse in μs at 63% of peak voltage. The mean peak electric field is ~ 234 kV/cm, the τ_{eff} is ~ 0.089 μs, which yields an acceptable breakdown value of 0.024.

Modeled after the original Z insulator stack [7-10], the outer diameter of the insulator stack will be ~3 m. The overall stack diameter and disk-conductor spacing in the water was determined by the need for a total, 0.125-Ω impedance and limited by water breakdown in the disk section of the water transmission line just outside the insulator stack. Each level of the insulator stack will have a disk water-line impedance of 0.25 Ω. This impedance and the radius of the insulator stack set the A/K gap in the disk water line.

F. Insulator Stack Design

The insulator-stack assembly consists of two insulator stacks, each composed of four Rexolite™ insulators that are ~2.5-cm high. Three hard-anodized, aluminum grading rings separate the insulators and extend from the water section into the vacuum section. The insulators are capacitively graded and must be designed to have a uniform grading to within a few percent.

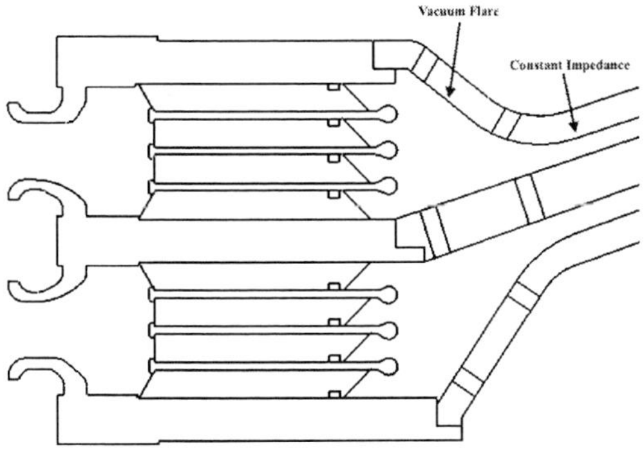

Fig. 6. A close up view of the MITLs near the insulator is shown. Specifically, the vacuum flare region of the vacuum feed shows the transition between the insulator stack and the constant gap of the MITL.

Figure 6 shows the insulator stack that separates the water section of the pulsed-power machine from the vacuum section. Its design must not only take into account the pulsed-power engineering limits but mechanical engineering issues (structural integrity, vacuum quality, and lifetime). The insulator-stack vacuum interface has the lowest limit on electric field in the pulsed-power chain.

The height of the insulator stack is always greater than the constant-impedance water gap because the breakdown field in the vacuum is lower than that of water. This creates an unavoidable impedance mismatch on the water side of the insulator stack and excess inductance in the insulator volume and in the vacuum flare region.

Reductions to the insulator height yield a significant improvement in machine performance by decreasing the impedance mismatch in the water, reducing the inductance in the insulator stack, and lowering the inductance of the local vacuum section. Experiments on the refurbished Z demonstrated successful, insulator operation at 150 kV/cm. [1,2] Our insulator stack is ~0.30 of the area of the A & B levels of the Z insulator This means that our insulator should be able to achieve a slightly higher field stress than Z based on the statistical area effects described by Stygar, et al. [18,19] and on experiments by Savage, et al. [20]. Equation (2) below shows the scaling relationship for a single, 45° Rexolite™ insulator:

$$E_p \left(t_{eff} C\right)^{1/10} \exp\left(-0.24/d\right) = 248 \qquad (2)$$

where E_p is the peak electric field across the insulator, C is the circumference of the insulator surface, and d is the height of the insulator. The effective pulse width t_{eff} is defined as the integral of the electric field divided by the peak field.

Z data suggests that our design for a well-graded, Rexolite™ insulator stack should be capable of handling more than 150 kV/cm.

The calculated peak voltage is 1.5 MV at the vacuum-insulator interface. Based on Eq. 2, the insulator design point is a conservative, peak-mean electric field of 150 kV/cm. This gives an insulator design height of ~10 cm. Figure 6 shows the preliminary insulator design that uses four, 2.54-cm-high insulators (10.16-cm total plastic height) and three standard 6.4-mm-thick grading rings. The total height of the insulator stack is 12 cm. The grading rings are spaced sufficiently far from the anode to keep the local electric fields below the emission threshold of 300 kV/cm for hard-anodized aluminum.

G. *Magnetically Insulated Transmission Lines*

Figure 7 shows the overall vacuum feed design that uses a simple tri-plate MITL arrangement. The MITL mechanical design is based directly on the Z MITLs. [1,2,7-10] The key differences are:

- the minimum gap between the anode and the cathode will be ~1/2 the gap on Z (while *E/cB* will be the same),
- the 18.6° angle of the MITLs will be slightly larger to ensure that the load is accessible to diagnostics on the equatorial plane, and
- the vacuum electron flow in the MITLs will be slightly lower.

Figures 6 & 7 show the anodes plunging rapidly toward the cathode disk just inside the insulator stack to minimize the vacuum-flare inductance. The radial spacing between the anodes and the adjacent insulator grading rings is limited by the need to keep the electric field on the grading rings below 300 kV/cm at all times to prevent electron emission of the grading rings. As the anode disk approaches the cathode disk the electric field on the cathode exceeds the self-emission threshold of ~250 kV/cm and the electrodes emit electrons and the transmission line acts as a magnetically insulated transmission line (MITL). Magnetic insulation is established by a carefully tailored transition from the non-emissive cathode near the insulator to the emissive inner cathode. The cathode will be made from 304L stainless steel. The anodes will be fabricated either from aluminum or stainless steel at large diameters and stainless steel at diameters <1.5 m.

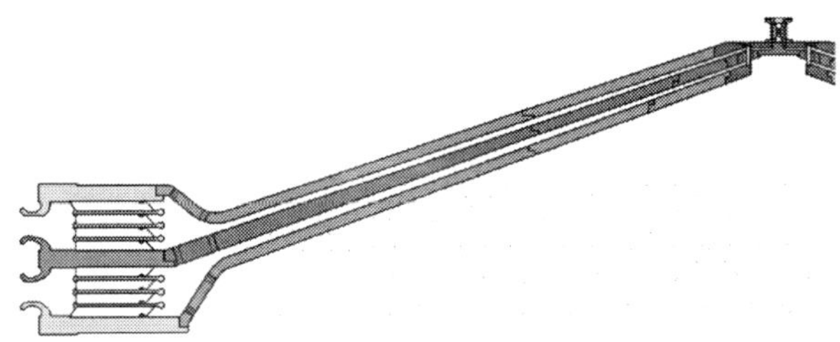

Fig. 7. A schematic of the proposed MITLs showing the transition from the insulator stack, the constant-impedance MITL, the post-hole convolute, and the load.

Figure 7 shows constant impedance MITLs that extend from the vacuum flare to the post-hole convolute. The gap in the MITLs decreases from an initial value of ~ 10 cm to a final gap of ~ 0.5 cm. The MITLs are designed with nearly the same *E/cB* as the SNL Z MITLs. The constant impedance means that the amount of electron current and the spatial extent of the electron sheath flowing in the vacuum will be constant over the entire radial length of the MITL.

The two MITL levels of the accelerator will be coupled to the load via a single, post-hole convolute (Fig. 8). [7-10] In this design, ~12, anode posts penetrate 12, cathode holes to couple the upper and lower anodes. This design generates 12, magnetic- field nulls in the upper disk and 12, magnetic-field nulls on the lower disk. At the post-hole convolute, a constant gap is used (increasing impedance radially inward). Any vacuum electron flow is assumed lost to the magnetic field nulls in the convolute. This worst-case assumption was found to work very well on Z. [8-10]

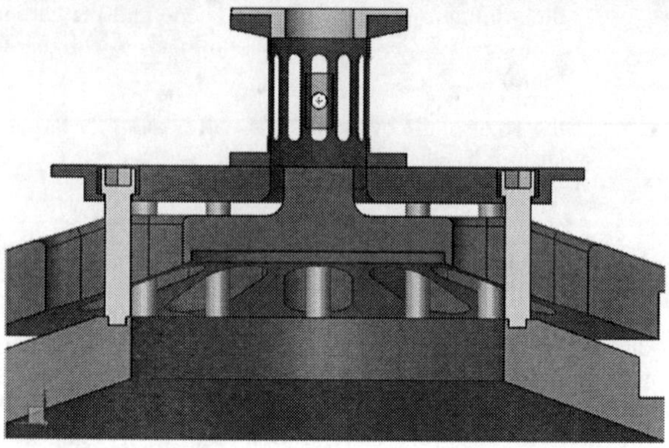

Fig. 8. The tri-plate MITLs are coupled by a post-hole convolute. The convolute is similar to a single level of the Z convolute.

Inside the post-hole convolute there is a short MITL. (See Fig. 8.) This portion of the MITL leading to the load can have many shapes depending on the specific details of the actual load.

III. Circuit Modeling

We describe in this Section the detailed circuit modeling of an LTD module and a wire-array z-pinch driven by the entire machine. All of the circuit modeling is conducted with SCREAMER [5,6].

A. Modeling the Performance of an LTD Module

A single LTD module is composed of 16 individual cavities (arranged in series). Each cavity is composed of 22 bricks in parallel. We have built a SCREAMER model of a module that is terminated in a 1.25-Ω resistance. The charge voltage for the capacitors in the bricks is +/-85 kV. We include in this model the ferromagnetic core losses from all of the cavities. [16]

Figure 9 shows a 1-MA current delivered to the load resistor in ~ 100 ns. Figure 10 shows a peak output voltage of 1.25 MV. The peak power was 1.3 TW and the energy was 170 kJ.

B. Wire Array Z-Pinch Performance

We modeled a standard SNL z-pinch load having an initial radius of 2 cm and a length of 2 cm. The initial mass of the load was 1.5 mg. There was a 6-ns-long impedance mismatch (0.125-Ω to 0.250-Ω taper) in the water section outside the insulator. The total insulator and vacuum inductance was 10.82 nH.

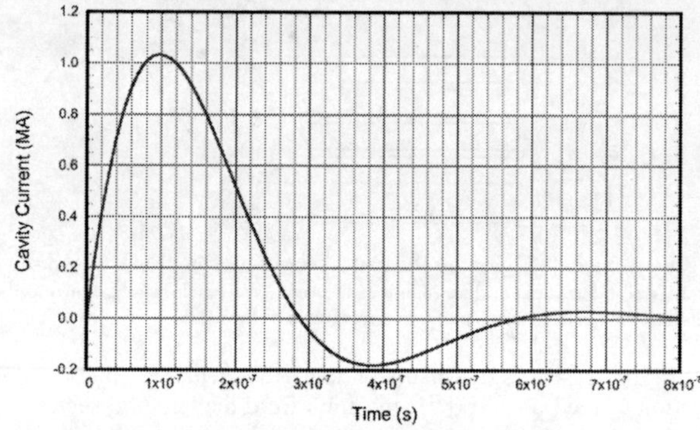

Fig. 9. We plot the current generated by the LTD module into a 1.25-Ω load.

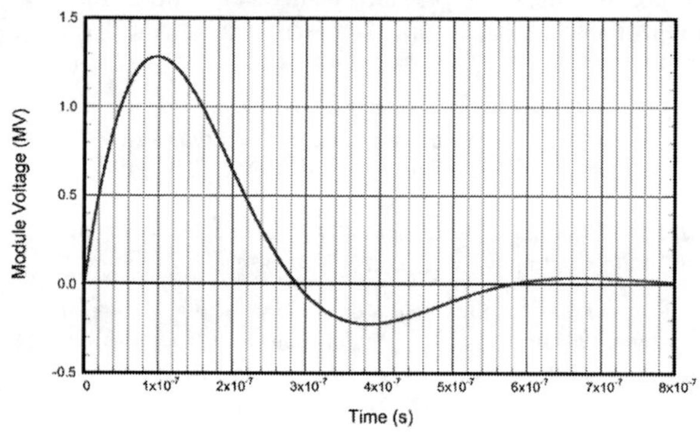

Fig. 10. We plot the module output voltage.

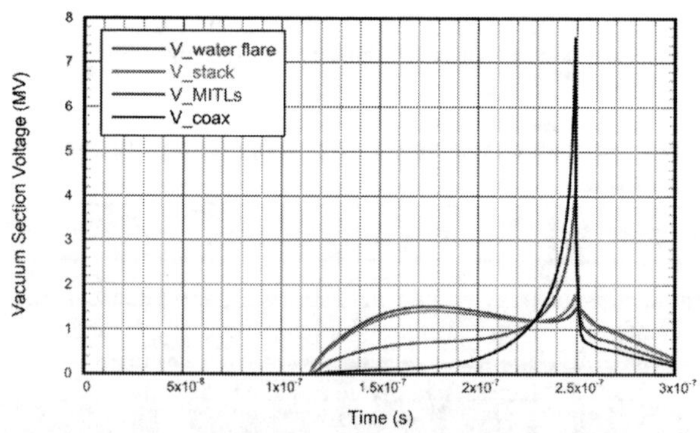

Fig. 11. The voltage waveforms in the vacuum section are plotted.

We calculated a peak current at the load of 10 MA, a peak load kinetic energy of 450 kJ, and a peak implosion velocity of 77 cm/μs. These calculated results are very close to scaled z-pinch performance from Z and have a high confidence of yielding up to 750 kJ and 60 TW of x-ray radiation from a tungsten z pinch. The coupling from the LTD modules to kinetic energy is ~ 25%.

Figure 11 shows the voltage waveforms in the water outside the insulator stack before the water flare (blue curve), at the vacuum side of the insulator stack (red curve), in the MITLs prior to the convolute (green curve), and at the base of the load (black curve). The peak voltage on the insulator stack is 1.42 MV. The peak, mean electric field on the insulators is 145.7 kV/cm.

Figure 12 shows the currents delivered to the vacuum insulator stack (blue curve), the vacuum MITL (red curve), and the z-pinch load (black curve). The loss current is the green curve. Note, the loss current in the calculation reaches only ~50 kA at peak current and reaches 830 kA at the moment of stagnation. There is a slight loss during the set up of magnetic insulation at the beginning of the current pulse. The total loss at peak current is ~0.5% and is energetically insignificant.

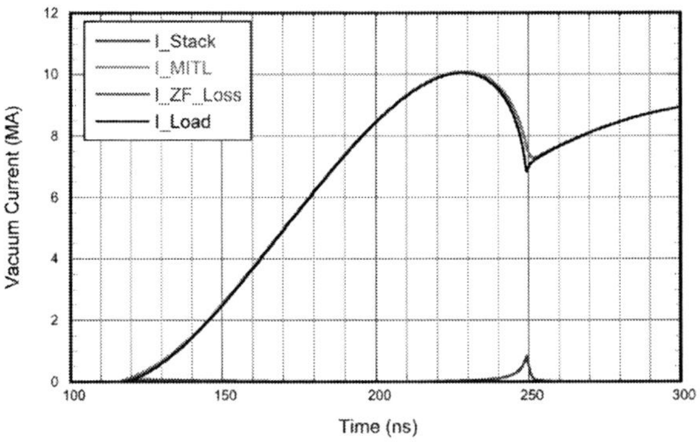

Fig. 12. The current at the insulator stack (blue), in the MITLs (red), and reaching the load (black) is plotted. The loss current (green) shows that the current loss at peak current is at ~0.5% and the current loss at stagnation reaches ~12%.

The energy coupled to the total inductance of the vacuum feed (past the insulator stack) reaches a peak of ~ 1.04 MJ at the time of implosion. The energy in the magnetic field near the load This is the energy that is directly available to the load on a time scale of ~ ns. The kinetic energy delivered to the imploding load reaches 450 kJ.

IV. Conclusions

We have described a detailed conceptual design for a 15-TW LTD accelerator. The water transmission lines, the vacuum insulator stacks, and the magnetically insulated transmission lines are all based on the successful Z and ZR machines.

We have conducted extensive circuit simulations using SCREAMER models that have been validated by similar circuit simulations done for Z. These simulations predict a peak current of 10 MA into a 10.8-nH z-pinch load. The z-pinch load reaches a kinetic energy of ~ 450 kJ and is expected to emit ~ 600 kJ of x rays in a 60-TW pulse.

Such a pulsed-power driver would be the first large-scale demonstration of LTD architecture and serve as "prototype" for future, more capable pulsed-power facilities.

Acknowledgments

The authors wish to acknowledge the extensive work on LTD systems conducted by Acad. B. Koval'chuk and his colleagues at the High Current Electronics Institute (HCEI), Tomsk, Russia and the engineering for LTDs conducted by ITHPP, Thegra, France, We wish to thank W. A. Stygar, M. E. Savage, M. L. Wisher and many others at Sandia National Laboratories for the technical information and unstinting support that enabled many aspects of our designs.

References

[1] M. E. Savage, et al., in Proc. of the 16th IEEE Int. Pulsed Power Conf., edited by E. Schamiloglu and F. Peterkin (2007), p. 979.

[2] M. E. Savage, et al., in Proc. of the 18th IEEE Int. Pulsed Power Conf., edited by R. D. Curry and B. V. Oliver, Chicago, IL (2011), p. 983.

[3] M. G. Mazarakis and R. B. Spielman, in Proc. of the 12th IEEE Pulsed Power Conference, Monterey, CA, (1999), p. 412; A. A. Kim and B. M. Koval'chuk, "High Power Direct Driver for Z-Pinch Loads." Proc. 1st International Conference on Radiation Physics, High Current Electronics, and Modification of Materials, Tomsk Russia, September 2000, p.263; M. G. Mazarakis, R. B. Spielman, K. W. Struve, and F. W. Long, in Proc. of the 20th Int. LINAC Conf., Monterey, CA (2000), p. 497; M. G. Mazarakis, R. B. Spielman, K. W. Struve, and F. W. Long, in Proc. of the 13th IEEE Int. Pulsed Power Conf., edited by R. Reinovsky and M. Newton, Las Vegas, NV (2001), p. 587; M. G. Mazarakis, et al., IEEE Trans. Plasma Sci. **38**, 05373875 (2010).

[4] W. A. Stygar, et al., Phys. Rev. ST Accel. Beams **18**, 110401 (2015); W. A. Stygar, et al., Phys. Rev. ST Accel. Beams **10**, 030401 (2007).

[5] M. L. Kiefer and M. M. Widner, in Proc. of the 5th IEEE Pulsed Power Conf., Arlington, VA, June 10-12 (1985), p. 685.

[6] R. B. Spielman and Y. Gryazin, in Proc. of the 20th IEEE Pulsed Power Conf., Austin, TX (2015), p. 637.

[7] R. B. Spielman, et al., in Proc. of the 10th IEEE Pulsed Power Conf., Albuquerque, NM (1995), p. 396.

[8] R. B. Spielman, et al., in Proc. of the 11th Int. Conf. on High Power Particle Beams, Prague, Czech Republic (1996).

[9] R. B. Spielman, presented at the 39th Ann. Meeting of the Division of Plasma Phys., Pittsburgh, Pennsylvania (1997).

[10] R. B. Spielman, et al., presented at the 1997 Particle Accelerator Conf., Vancouver, Canada (1997).

[11] K. LeChien, et al., in Proc. of the 17th IEEE Pulsed Power Conf., edited by F. Peterkin and R. Curry, Washington, D.C. (2009), p. 1186; M. G. Mazarakis, et al., in Proc. of the 18th IEEE Int. Pulsed Power Conf., edited by R. D. Curry and B. V. Oliver, Chicago, IL (2011), p. 625.

[12] J. R. Woodworth, et al., Phys. Rev. ST Accel. Beams **12**, 060401 (2009).

[13] J. R. Woodworth, et al., Phys. Rev. ST Accel. Beams **13**, 080401 (2010).

[14] B. Stoltzfus, et al., "Design and performance of two 100-GW linear-transformer drivers", to be published.

[15] M. E. Savage, et al., in Proc. of the 18th IEEE Pulsed Power Conf., Chicago, IL (2011), p. 844.

[16] A. A. Kim, et al., Phys. Rev. ST Accel. Beams **12**, 040402 (2009); A. A. Kim et al., Phys. Rev. ST Accel. Beams **13**, 070401 (2010).

[17] W. A. Stygar, et al., Phys. Rev. ST Accel. Beams **09**, 070401 (2006).

[18] W. A. Stygar, et al., Phys. Rev. ST Accel. Beams **08**, 050401 (2005).

[19] W. A. Stygar, et al., Phys. Rev. ST Accel. Beams **07**, 070401 (2004).

[20] M. E. Savage, et al., in Proc. of the 20th IEEE Pulsed Power Conf., Austin, TX (2015), p. 14.

Coherent summation of radiation from four-channel shock-excited rf source operating at 4 GHz and repetition rate of 1000 Hz

M.R. Ulmaskulov, K.A. Sharypov, S.A. Shunailov, V.G. Shpak, M.I. Yalandin, M.S. Pedos and S.N. Rukin.
Institute of Electrophysics UB RAS
Ekaterinburg, Russia
marat@iep.uran.ru

***Abstract*—Results of testing of a generator based on the parallel gyromagnetic nonlinear transmission lines with saturated ferrite are presented. Practically identical and stable rf-modulated high-voltage nanosecond pulses were shaped in each of the four channels. Pulse amplitude reaches -175 kV at a modulation depth of rf-oscillations to 50 % and effective frequency ~4 GHz. Power in 1 s packets with 1000 Hz repetition frequencies was supplied by a solid-state drive. Electric field strength achieved 250 kV/m at a distance 3 m from antennas.**

Keywords— multichannel radiator; nonlinear transmission line; ferrite; coherent summation.

I. Introduction

Multichannel microwave sources [1-7] considerably increase a power flux density of rf-emission at a far-field region of a major maximum of a directional pattern (DP) and sharpen its beam [1-3]. On in-phase summation of rf-fields the flux power density quadratic increases relative to sources quantity. So it is clear the interest in the systems, when each channel power reaches megawatt and high [1-5]. The summation calls for accuracy the microwave sources. On-line control of delays of the output pulses adjusts mutual phasing [5-7], and this permits an integral DP scanning in space.

The four channel rf-source is presented below. The external axial magnetic biasing (H_z) coaxial nonlinear transmission lines NLTL partially filled with ferrite rings were used as sources of high-power microwave [3-10]. The source possesses rf modulation frequency increased to 4 GHz, a rather high output peak power of hundreds of megawatts, and operates at maximum pulse repetition frequency (PRF) of 1000 Hz.

The microwave oscillations come due to the gyromagnetic phenomena. The gyromagnetic resonance takes place under action of couple of orthogonal magnetic fields on the magnetic moments of the ferrite domains. An external bias H_z-field forms the ferrite saturation in an axial direction, while an incident high-voltage pulse creates an azimuthally demagnetizing field. As an incident pulse with sufficient amplitude and rise time propagates through the NLTL, and sufficient external bias presents, the magnetic moments of ferrite domains rotate around. The damped gyromagnetic precession excites microwave oscillations.

This work was supported by research program of IEP UrD RAS 0389-2014-0005 and grant of RFBR №16-08-00058.

II. Experimental Setup

At the work the two section NLTLs were used [3,5-7,9]. The input section operated as a pulse sharpening device and sharpened front sufficient for the second NLTL section operation in the oscillator's mode. Both sections had integrated construction depicted in Fig. 1. The ferrite line implements a continuous unit of NiZn-rings of a type M200VNP ($2R_1$=25 mm; $2R_2$=8 mm; $2r_1$=10 mm; $2r_2$=16mm) in total length 700 mm. A direct current solenoid consisted of five sections and it was symmetrically placed over a case around ferrite. In the implementation NLTL had a condition partition in the sharpening and generating parts by a self-contained power supply. It has been found experimentally that it is optimal to utilize the two first coils for sharpening and delay line tuning of a pulse. The bias field H_z of the remaining NLTL part was optimized in the rang 35-55 A/m to form the rf oscillations. The design of NLTL and biasing field data gave the acceptable delay tuning range (more than rf half-period), maximum rf modulation deepness, and minimized distortion of excited oscillations envelop for a full scale of the delay control.

The installation appearance and the basic components of the rf-oscillator are illustrated in Fig. 2. The full solid state high-voltage driver S500 [11,12] served as a supply source. The driver formed pulse with a rise time ~1 ns, duration (FWHM) of ~5 ns and amplitude ~5 ns V_d≈-500 kV in single pulse mode across a 50 Ohm load. When the system operated at 1000 Hz PRF, the voltage amplitude reduced to V_{d1000}≈-430 kV.

An output driver pulse was split into four parallel channels by a mismatched cross-type power divider (50→4x48 Ohm) presented in Fig. 3,a. The instability of the pulse amplitude (total spread 10%, standard deviation σ~2%) was related to the main voltage ripple across a capacitive filter of primary charging device (PCD).

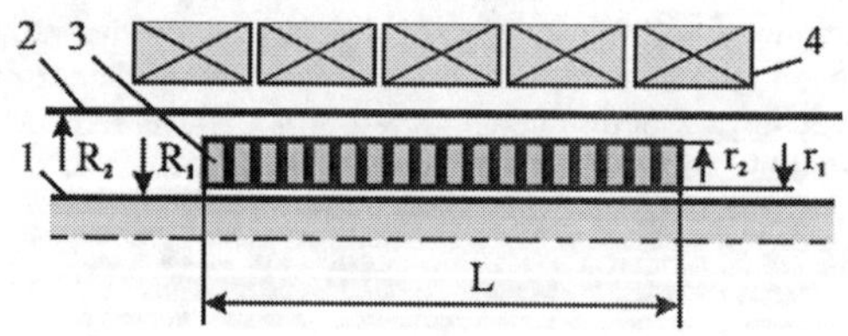

Fig. 1 Schematic of ferrite-loaded coaxial NLTL.1,2- outer and inner electrode of coaxial line, 3-ferrite rings, 4-solenoid.

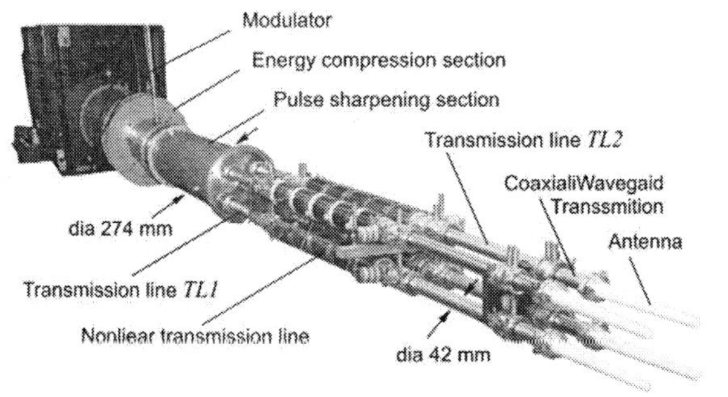

Fig. 2 Appearance of the 4-channel rf oscillator on NLTL.

Depending on the current tasks the each channel was terminated in either the absorbing water load or dielectric conical antenna (Fig. 4). The coaxial and waveguide transitions to water load had a small-angle long conical insulator to minimize reflections. The voltage pulses were recorded in the oil-filled transmission lines (TL1 and TL2) and circular waveguide TL3 by using the capacitive dividers capable to reproduce ~70 ps rise time. The accuracy of amplitude calibration of the dividers was ~5%. The location of the dividers D1, D2 and D3 in the lines provided about 5 ns delay of the reflected signals.

An oil coaxial-to-waveguide transition (CWT) [10,13-15] with oil insulation had invariable diameter 40 mm of an outer conductor (Fig. 4,b,c). At first, the coaxial waveguide inner conductor transformed in strip with identical impedance (48 Ohm), then it tapered and at a constant axial angle 22^0 shorted against the case at a distance 40 mm. CWT converted coaxial line *TEM* wave to TE_{11} of a hollow circular waveguide (CW).

A circular Teflon cone 340 mm long was utilized as an antenna. The beginning cone diameter was identical to waveguide diameter of CWT and tapered at angle ~1.5^0 to diameter 23 mm (Fig. 4,c). The radiator design matched an open end of oil waveguide to the open space with low level signal reflection (<15-20%). CWT design defined the linear polarization of an emitted wave. The vertical and horizontal axle bases of the four channel system were 125 mm.

III. Examination of Individual NLTL

Testing of NLTL electric strength and stability of the output parameters of exited oscillations were held by a single channel at repetition frequency 1000 Hz during 1 second. The

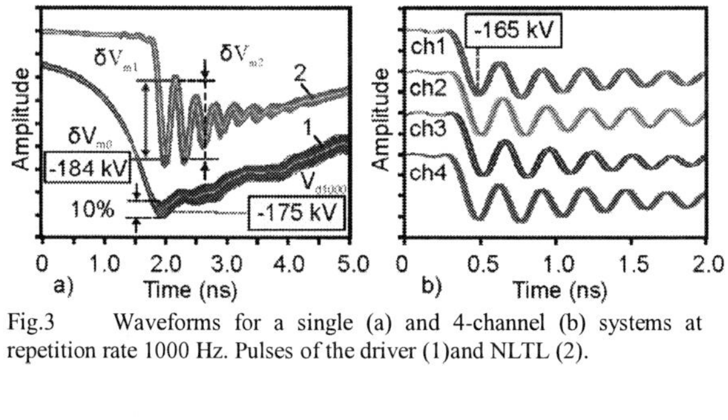

Fig.3 Waveforms for a single (a) and 4-channel (b) systems at repetition rate 1000 Hz. Pulses of the driver (1)and NLTL (2).

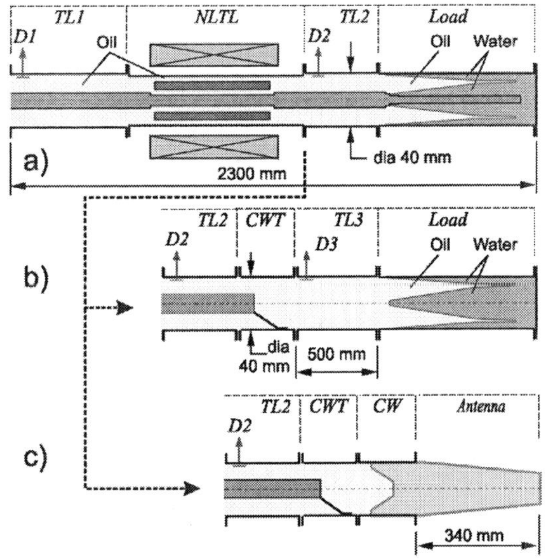

Fig. 4 Schematic of the loads of NLTL: a) coaxial water line; coaxial-waveguide transition terminated in the water waveguide load (b) and dielectric cone antenna (c).

coaxial water load was installed at a NLTL output. Mean input pulse amplitude was V_0≈-175 kV; standard deviation - of σ=3.5 kV. On the parameters and mode the evidence of breakdown has not been found. Waveforms of an output pulse burst are presented in Fig. 3,a. Mean value of a first peak of rf-modulated pulse was $|V_{m0}|$ =184 kV. An electric field strength on a surface of the inner electrode 8 mm of the NLTL achieved ~425 kV/cm, and in the ferrite - ~255 kV/cm. Despite this, short rf-peak duration made it possible avoiding both oil and ferrite breakdowns. Insignificant increase of amplitude is due to the shock wave has been formed at the incident pulse front. The maximum increase was obtained by utilizing of the greater diameter ferrite rings [4].

The NLTL excitation mode was chosen so that the modulation depth of the oscillations was a maximum after the leading peak (V_{m0}). This was achieved by varying the bias field By varying the solenoid bias field H_z the NLTL excitation mode have been adjusted to maximize the modulation depth (δV_{m1}), which followed after the leading peak. The average amplitude and standard deviation (σ_0) of the peak (V_{m0}), second peak amplitude (δV_{m2}) and ratio of them to V_0 were measured. The effective frequencies (f_n) of the three initial peaks were defended as an inversely proportional double time within points at the half-high of rise and fall times of the corresponding peaks. The obtained values are listed in the table 1.

TABLE I.

$\|V_{m0}\|;\|\delta V_{m1}\|;\|\delta V_{m2}\|$ (kV)	$V_{m0};\delta V_{m1};\delta V_{m2}$ (%)	$f_{(n)}$ (GHz)	H_z (kA/m)
184; 89; 81	105; 51; 47	2.7; 3; 3.5	55

An analyze of the mathematical sample of pulses from a packet on an absolute spread of an output pulse amplitude ΔV_0 quantitatively demonstrates possibility for increase in stability of output pulse parameters on reducing of amplitude spread of the incident pulses. Reduction of the sample range of ΔV_0 from ±5% to ±1% reduces instability of the V_{m0} and jitter of duration of an output pulse rise time t_f as well as its standard deviations σ_0 and σ_f. The further range reducing cause no spread reducing, probably because this takes no away effect of all burst pulses on the ferrite medium (heating, variability of remanent magnetization). Because of this time delay jitter Δt between the input and output pulse edges at a level 0.5 and its standard deviations σ_t have no change in all samples. The sample measurement results are listed in a table 2.

TABLE II.

ΔV_0 (%)	Параметры импульсов выборки пачки 1000 Hz 1 секунда			
	$V_{m0};\sigma_0$ (kV)	$t_f;\sigma_f$ (ps)	$\Delta t;\sigma_t$ (ps)	N (pulses)
±5	184; 2.8	128; 1.8	~100; 22	980
±1	184; 1.2	128; 1.5	~100; 22	222

IV. High-PPR Operation of Multi-Channel Oscillators

The four-channel system was tested to investigate the mutual stability of operation of individual NLTLs in excitation of rf oscillations. Analysis of obtained results was carried out for a packet of 100 pulses at frequency 1000 Hz. An NLTL identity was provided by equal quantity of the rings and symmetric location of the solenoids under the rings. All channels were terminated in a coaxial water load.

The external bias field of three output coils of a solenoid of the main section of each NLTL was identical and was lowered to H_{z1}=40 kA/m. This field gave a maximum delay time turning range and more sharpen a leading peak rise time up to t_f=120 ps. An effective frequency increased to ~ 4 GHz, but mean value of an output pulse peak amplitude decreased down to -165 kV Fig. 3,b. The output pulse waveforms coincided within a reproduction accuracy of a divider D2 of each channel.

The channel delay times were some different, because the ferrite rings parameters at the high voltage, short pulse durations and on a frequency ~4 GHz were no controlled by a producer. The time-lock of a pulse appearance had achieved by current turning in the two beginning sections of a solenoid at an output of the NLTLs. In identical delays the calculated fields H_{z2} generated by the coils were 4; 40; 48 and 57 kA/m.

The time instability of generation of rf-modulated pulses depended on H_{z2} in the channels. A cross-coupling comparison of the channels shown the maximum relative spread in time-locking of δt=55 ps (average 15 ps) with standard deviation σ_δ=20 ps had only one channel with most minimum H_{z2}=4 kA/m. The most minimum δt=35 ps (average 8 ps) and σ_δ=1.4 ps were between two channels with sub-equal and high values H_{z2} 40 kA/m и 48 kA/m, which were practically identical to H_z of the NLTL oscillator section. The decrease in instability in increase in a bias field is predictable, because the deeper ferrite saturation by an external field reduces the ferrite medium state uncertainty on action of previous pulse with unrepeatable parameters. It is necessary to note, that in the system with the unseparated sharpening and main (gyromagnetic) NLTLs with more high H_{z2}=55 and unequal to a field in the main NLTL had high σ_δ=9 ps.

V. RF Emission Measurements

A Conversion of a partially modulated unipolar pulse into rf wave train by above described CWT was firstly tested when channel operated at water load Fig. 4,b. The sharpened linearly polarized wave TE_{11} propagated both toward the load and backward direction. A shape of rf-wave in CW was controlled in the E-plane by the divider D3 and is presented in Fig. 5,a. A minimum interval between the peaks corresponded to a period of f_{cw}=4 GHz, and a minimum electric strength at a waveguide wall was E_{cw1}≈25 kV/cm. A waveform control capability makes it possible to decouple an incident pulse reflection from CWT with the driver direction rf-wave, which an electric strength was of the same order E_{cw2}≈15 kV/cm. The lack of H-plane oscillations in a reflection demonstrates an excellent matching of CWT. All CWT of the four-channel radiators were positioned in an identical vertical polarization. A strip horn was utilized as a receiving antenna, which was placed at a distance 3 m from the radiators and at a height 1.25 m from the underlying no absorber terrain, which ensured a time lag of a reflection.

The pulse waveforms in the main DP direction for single and four channels operation are shown in Fig. 5,b. The four radiator pulse was received on in-phase summation of all wave beams. The electric field strength of the single antenna E_{01}≈680 V/cm and four channel system differs approximetly in 3.7 times, which is same less than an expected quadruple increase. The comparison is not quite correct, because the single antenna tested with no other channel antennas and correspondently without of its interference, which took place in the four channel system.

A four channel DP steering was demonstrated by the in-phase and anti-phase modes summation of the wave beams This provided by current turning of the delay sections of the NLTL solenoid (Fig. 6,a). The presence of no a zero level on anti-phase mode can be caused by the inaccuracies fabrication of the waveguide components and mispositioning, as well as difference in the channel waveforms. The direction characteristic of the radiators is defined by the H-plane patches of DPs. The half-power beam width (at a level 0.7, -3 dB), of the four channel is more sharp approximetly in 3 times in comparison with a single antenna.

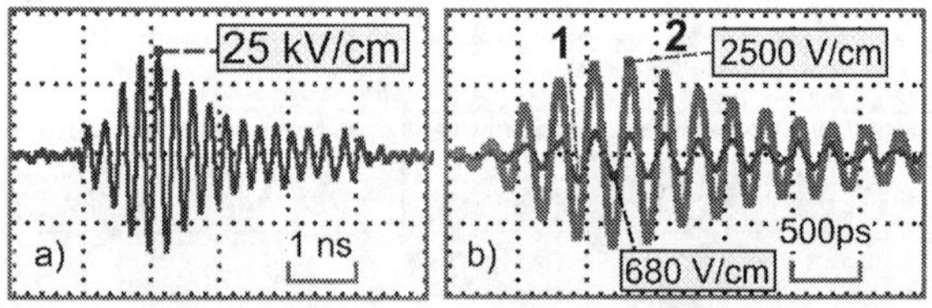

Fig. 5 Waveforms of the wave E-field in a circular waveguide (a) and in a free-space at a distance 3m. 1–single, and 2–four-channel oscillator.

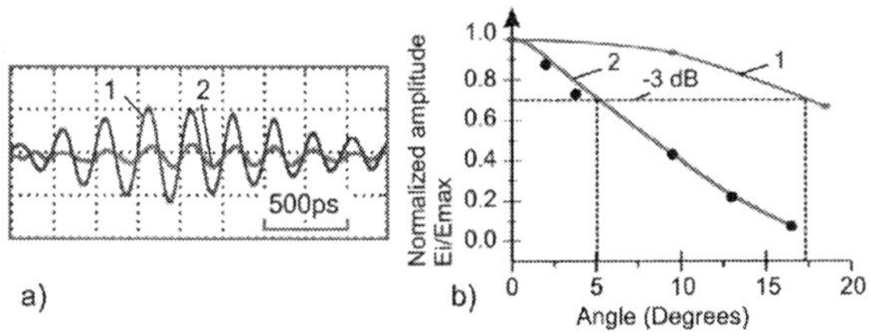

Fig. 6 (a) Waveforms of the wave E-field in the DP maximum at a distance of 3m for in-phase (1) and antiphase (2)summation. (b) DP shapes for a single (1) and four-channel (2) radiators.

It is significant that spectrums (Fast Fourier transform) of pulses at the NLTL outputs, in CWT and in the open space on a coherent summation of the four channel pulses well agree in a high frequency range (Fig. 7). The spectrum maximum values (effective frequency) correspond to ~4 GHz, and it's some shift explains by instability of a pulse amplitude of the driver.

VI. Conclusions

The designed four channel rf-source based on the solid-state driver and NLTLs provided to achieve the rf-emission pulse with electric field up to 2500 V/cm at a distance 3 m. As tis takes place, the microwave energy propagates toward a driver. In principle, the energy can be redirected toward antennas and increase in a power level appreciably. At a driver side the coaxial line section with outer diameter providing cutoff waveguide mode can be utilized. A maximum modulation frequency value of an unidirectional pulse achieved to 4 GHz, with a voltage peak at the NLTL output reached 175 kV. The electric strength provided operating at frequency rate 1000 Hz during 1 second. The integral packing of timing and modulating sections of NLTLs allowed for the electronic scanning of the utilized transmitting antenna. A maximum level of a field of the integral radio pulse on in-phase summation mode have stood out above the single antenna field level in 3.7 times. The result is close to an ideal coherent summation. The flux power density practically increased in 16 times against a single antenna.

Acknowledgment

This work was supported by research program of IEP UD RAS 0389-2014-0005 and by RFBR grant 16-08-00058.

References

[1] I. Yalandin, S.A. Shunailov, M.R. Ul'maskulov, K.A. Sharypov, V.G. Shpak, V.V. Rostov**, I.V. Romanchenko, A.A. El'chaninov, and A.I. Klimov, "Synphase operation of nanosecond relativistic 37_GHz Backward_Wave Oscillators without Electrodynamic Coupling," Technical Physics Letters Vol. 38, No. 10, pp. 917–920, 2012.

[2] K. A. Sharypov, A.A. El'chaninov, G.A. Mesyats, M.S. Pedos, I.V. Romancheko, V.V. Rostov, S.N. Rukin, V.G. Shpak, S.A. Shunailov, M.R. Ul'masculov, and M. I. Yalandin, "Coherent summation of Ka-band microwave beams produced by subgigawatt superradiance backward wave oscillators," J. Appl. Phys. Lett., 103, 134103, pp. 1-4, 2013.

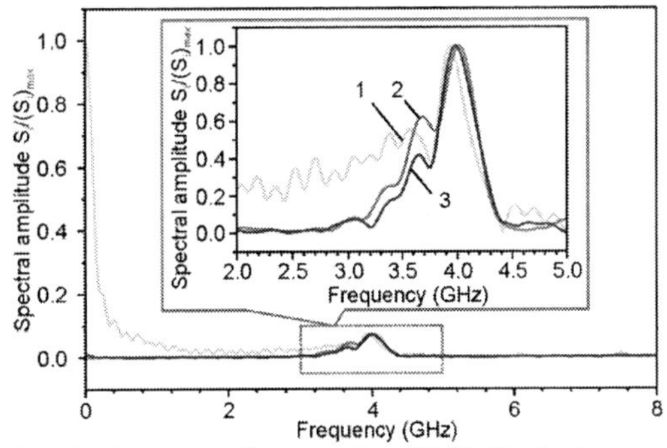

Fig. 7 FFT spectrums of output pulse of NLTL (1), circular waveguide's wave (2), and the wave in a free-space (3).

[3] V.V. Rostov, A.A. El'chaninov, A.I. Klimov, V.Y. Konev, I.V. Ro manchenko, K.A. Sharypov, S.A. Shunailov, M.R. Ul'maskulov, and M.I. Yalandin, "Phase control in parallel channels of shock-excited microwave nanosecond oscillators," IEEE Trans. Plasma Sci. 41, 2735–2741, 2013.

[4] M. R. Ulmaskulov, M. S. Pedos, S. N. Rukin, K. A. Sharypov, V. G. Shpak, S. A. Shunailov, M. I. Yalandin, I.V. Romanchenko, and V. V. Rostov, "High repetition rate multi-channel source of high-power rf-modulated pulses," Rev. of Sci. Instruments 86, 074702, 2015.

[5] I. Romanchenko, V. Rostov, A. Gunin, and V.Y. Konev, "High power. microwave beam steering based on gyromagnetic nonlinear transmission lines," J. Appl. Phys. 117, 214907, 2015.

[6] Johnson, J M; Reale, D V; Krile, J T; Garcia, R S; Cravey, W H; Neuber, A A; Dickens, J C; Mankowski, J J, "Characteristics of a four element gyromagnetic nonlinear transmission line array high power microwave source," Rev. of Sci. Instruments 87,5, 054704, 2016.

[7] J.M. Johnson, D.V. Reale, W.H. Cravey, R.S. Garcia, D.H. Barnett, A.A. Neuber, J.C. Dickens, and J.J. Mankowski, "Material selection of a ferrimagnetic loaded coaxial delay line for phasing gyromagnetic nonlinear transmission lines," AIP Review of Scientific Instruments 86, 084702 pp. 1-5, 2015.

[8] J.-W. Bragg, J. Dickens, and A. Neuber, "Material selection considerations for coaxial, ferrimagnetic-based nonlinear transmission lines," J. Appl. Phys. 113, 064904, 2013.

[9] D. Reale, J.W. Bragg, N. Gonsalves, J. Johnson, A. Neuber, J. Dickens, and J. Mankowski, "Bias-field controlled phasing and power combination of gyromagnetic nonlinear transmission lines," Rev. Sci. Instrum. 85, 054706, 2014

[10] I.V. Romanchenko, V.V. Rostov, V.P. Gubanov, A.S. Stepchenko, Alexander V. Gunin, and Ivan K. Kurkan, "Repetitive sub-gigawatt rf source based on gyromagnetic nonlinear transmission line," Rev. Sci. Instrum. 83, 074705, 2012.

[11] S. Rukin A. Bushlyakov, S. Lyubutin, A. Ponomarev, B. Slovikovsky, S. Timoshenkov, S. Tsyrano , "Solid-state repetitive SOS-based generators providing apeak power of GW-range," Pulsed Power Conference, 2007. Digest of Technical Papers. IEEE International, pp. 698 – 701, 2007 .

[12] A. I. Gusev, M. S. Pedos, S. N. Rukin, S. P. Timoshenkov, and S. N. Tsyranov, "A 6 GW nanosecond solid-state generator based on semiconductor opening switch," Rev. Sci. Instrum. 86, 114706, 2015.

[13] L.Swern, "broadband coaxial-line to wavegate adaptor using step-ridge transformers," Proc.Nat.Electronix conf., 9, p.922, 1953.

[14] A.F.Harvey, "Microwave engineering", Academic press, London and New York, 1963.

[15] V.N.Dulin, M.S. Zhuk, "Spravochnik po elementam radioelectronnih ustroistv"(handbook on the elements of the radioelectonice divaices), "Energy" Moscow, 1977.

Drift characteristics of fast rise-time, moderately magnetized high current electron beam

S. A. Shunailov, K. A. Sharypov, V. G. Shpak, M. R. Ulmaskulov, M. I. Yalandin
Institute of Electrophysics UD RAS
Ekaterinburg, Russia
ssh@iep.uran.ru

V. V. Rostov
Institute of High Current Electronics SD RAS
Tomsk, Russia
rostov@lfe.hcei.tsc.ru

Abstract **— Fast processes associated with acceleration and drift motion of high current, moderately magnetized electron beam (particles energy ~300 keV; current ~3 kA) were studied experimentally when tubular explosive-emission cathode was powered by subnanosecond-rise-time voltage pulse. Kinematic effect causes observed beam front sharpening which is proven by particle-in-cell numerical model. Temporal structure of azimuth beam's fractions recorded with time resolution no worse than 10 ps was analyzed after beam acceleration as well as in the process of further electrons drift in the finite guiding magnetic field.**

Keywords — explosive electron emission; electron beam; acceleration; drift motion, magnetic insulation

I. Introduction

Demonstration of coherent power summation from independent relativistic backward wave oscillators (BWO) [1]-[2] revived attention to the problem of high-current electron beam formation in magnetically isolated coaxial vacuum diodes (MICDs) with explosive electron emission (EEE) cathodes. This is because, in accordance with the concept of microwave generation phase fixing [3]-[6], the beam current front should be stable from pulse to pulse within picoseconds and its rate of rise should be high enough to produce sufficient seed rf signal at the instant when electrons energy gains the band of Cherenkov synchronism with excited wave.

Below we investigate a subnanosecond-time-scale dynamics of the beam acceleration in MICD and its drift motion in finite longitudinal magnetic field. Presented researches append our previous results [7], [8]. However, there is essential distinct determined by new capabilities of recording instruments (oscilloscopes bandwidth gain up to 59 GHz) and involvement of recently developed advanced method of dynamic time domain reflectometry (DTDR) [9] recovering exact characteristics of the MICD.

II. Experiments Arrangment

A. High-Voltage Setup

The device used in our experiments was, in fact, designed to study microwave generation of Ka-band BWOs with output power exceeding 500 MW in superradiance generation mode [2] and approaching to 200 MW in quasi-stationary regime [6], [10]. The difference is in the substitution of slow-wave microwave structure with the collector sensor to register the beam current. For the formation of a high-voltage pulse the sharpener in the form of gyromagnetic non-linear transmitting line (NLTL) with biased ferrite [11] was applied, which shortens the voltage front of the supply source RADAN-303 [12] up to ~300 ps. Sketch drawing of the MICD with tubular graphite cathode is presented in Fig. 1.

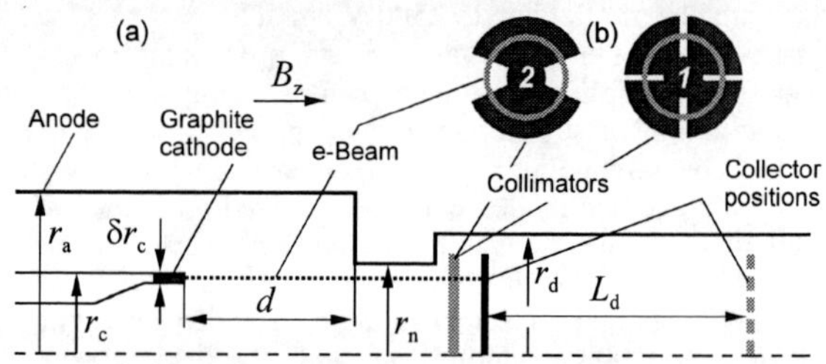

Fig. 1. (a) MICD schematic: r_a=7.5 mm; r_n=4.95 mm; r_c=3.7 mm; δr_c=0.15 mm. (b) Beam's collimators.

B. Beam Current Probes

Collector-type electron beam probes used in experiments are described briefly in [7], [13]. Depending on the gap distance between the beam collimator and the collector plane [Fig. 1, a], the probes were capable to record the current bursts possessing a front no worse than 20 ps ("plane gap" of ~1 mm) or the current pulse having the width a little above 50 ps (FWHM) when the beam's collector resembling a Faraday's cup was used. Residual gas ionization (pressure of ~10^{-2} Tor) and secondary electrons emission from the collector led to the probe's sensor area short circuiting in a certain time if a high beam current hits the collector. Reduction of the current by collimators [Fig. 1, b] made it possible to increase the beam current record length up to 4-5 ns, however, the current propagation factor through collimator's slits becomes approximate in this case. Note that the beam current reduction to ~0.5 kA [*2* in Fig. 1, b] and below 100 A [*1* in Fig. 1, b] minimized the breaking potential of the collector which originally represents the edge of central electrode in low-impedance (6-7 Ohm) coaxial line.

The work was supported by IEP UB RAS Project No. 0389-2014-0005 and, in part, by RFBR Grant No. 17-08-00046.

C. MICD Reflectometry

MICD full current was determined by DTDR [9], which allows to study the impedance dynamics of such nonlinear load when the voltage pulse $U_{in}(t)$ supplied to the cathode is of irregular shape and unstable. Since the reflection from the load occurs on the background of the trailing edge of the pulse $U_{in}(t)$ and the noise, separation of the reflected signal $U_{ref}(t)$ is achieved by means of two identical voltage sensors. If the delay between the signals is equal τ, so a real reflection is find $U_{ref}(t)=U_{in1}(t)-U_{in2}(t-\tau)$ if the path from sensors to the cathode is homogeneous, that is, its impedance $\rho=const$. Measurement error of the impedance ρ (<1%) and the signal delay from the voltage sensors to the cathode edge (<10 ps) sets the precision of the voltage recovery at the cathode $U_c(t)$, the current $I_e(t)$ and hence of the MICD impedance $Z_d(t)$. With that, all these parameters have the absolute time reference.

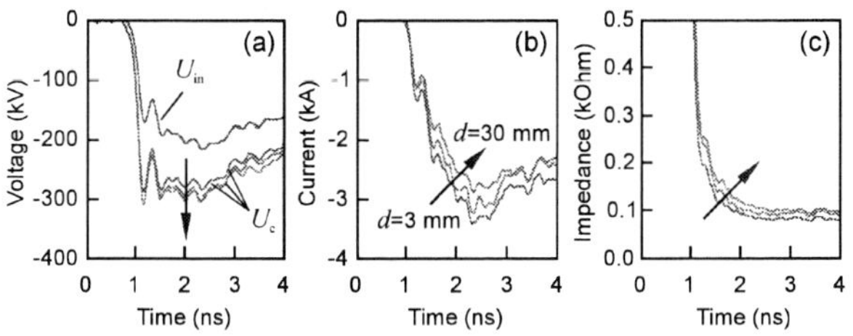

Fig. 2. Input voltage U_{in} identical for all below described experiments and voltage across the cathode U_c (a), MICD current I_e (b) and impedance Z_d (c) for accelerating gaps varied from d=3 mm to d=30 mm. Recorder: Tektronix DPO73304D real-time digital oscilloscope (33 GHz; 100 Gs/s).

III. Effects During Beam Aceeleration

A. MICD parameters Versus Beam Acceleration Distance

DTDR technique allows to precise control of the MICD current (i.e., beam current) when changing the length of the accelerating gap d [Fig. 1, a]. The MICD impedance at the gap variation [Fig. 2, c] has the difference already after 0.1 ns when the beginning of the emission and comes to some steady state value for every d in ~1.5 ns. Used in the analytical solution of the parameters of the magnetized tubular beam [14], the approximation of a homogeneous anode pipe [$r_a = r_n = r_d$, Fig. 1, a] is certainly satisfied when $d > 3r_a$ (in our case r_a=7.5 mm). In particular, when d=30 mm and the emission starts from the graphite after ~0.5 ns the electron current of ~1.6 kA is achieved, which is described by the model [14] in approaches of cathode's infinite emission capacity and small wall thickness of a tubular beam. Following to notions [8], beam wall thickness could be estimated from above as maximum radial distance δ_{max} of electrons from the cathode edge when particles move in crossed electric and magnetic fields.

The value of δ_{max} (in practical units) looks like

$$\delta_{max}(\text{mm}) \approx 1.4 \times E_0(\text{MV/cm}) \times B_z^{-2}(\text{T}) \,, \qquad (1)$$

and for the cathode potential of 270 kV and magnetic field B_z~2.5 T it gives δ_{max}~0.2 mm. Thus, regardless the above value of B_z is not high for the experiment in Fig. 2, it appears $\delta_{max} \approx \delta r_c$ and, hence, analytical approaches in [14] (including cathode's infinite emission capacity) are close to validation in ~500 ps since emission onset time.

B. Beam Front Sharpening

Absolute time reference of $I_e(t)$ and $U_{in}(t)$ DTDR oscillograms [Fig. 2] allows to compare kinematic transformation of the beam front during its acceleration for different gaps d in real experiments (collector measurement) and numerical simulation by particles-in-cell (PIC) method [Fig. 3]. In the PIC-model [15], the cathode is supplied by a running TEM-pulse with the envelope $U_{in}(t)$ as in Fig. 2, a, and the current is given in the form of synchronized DTDR function $I_e(t)$ [Fig. 2. b]. Control calculations show that in this case, within the interval of at least 1 ns the voltage $U_c(t)$ sets at the cathode coinciding with the DTDR- measurement.

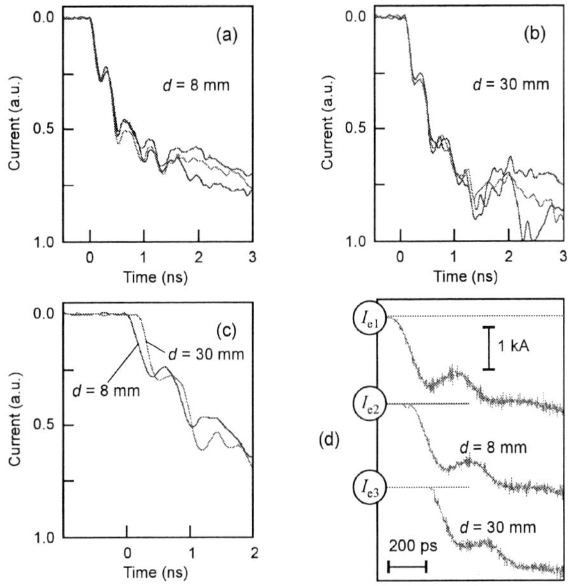

Fig. 3. Oscillograms of slit-collimated beam current for MICD gaps d=8 mm (a) and d=30 mm (b). (c) Comparison of currents fronts for the cases (a) and (b). (d) – PIC simulation of beam's current front transformation as it accelerated from the cathode (I_{e1}) throughout entire gaps d=8 mm (I_{e2}) and d=30 mm (I_{e3}). Recorder: Tektronix DPO73304D real-time digital oscilloscope (33 GHz; 100 Gs/s).

For current registration the 4-slit collimator of the tubular beam was used (*1* in Fig. 1, b; slit width of 0.4 mm). Comparison of Fig. 3, a and Fig. 3, b shows the following. In the interval ~0.5 ns from the start of emission the currents fronts are well reproducible. That is, the emission is uniform at the initial stage of the cathode activation. Another feature is that by increasing d from 8 mm to 30 mm the front of the first current surge escalates to ~100 ps [Fig. 3, c], and this is confirmed by the numerical simulation in Fig. 3, d. Note, that for both gaps the calculated fronts of the beam at the collector

(I_{e2} and I_{e3}) are shorter than the front of the current I_{e1} near the cathode edge. The reason of increasing current sharpening with d growth is due to the approximation of acceleration time to the length of the voltage rise time. Then the later fast electrons catch up with the particles emitted at lower potential on the cathode. Increasing the randomization of current at $t >$ 1 ns [Fig. 3, a; b] is connected, apparently, with the localization dynamics of the EEE on the most protruding agglomerates of the graphite microemitters. This is accompanied by the appearance of discrete plasma structures on the edge of the cathode (see, e.g., Ref. [16] and citation therein). The effect of the kinematic sharpening may also contribute to the increase in the amplitude of the current bursts (micro-bunches) in the greater length of particle acceleration [Fig. 3, b].

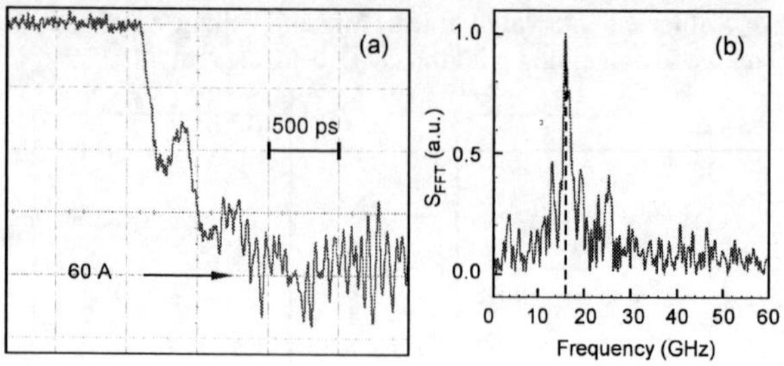

Fig. 4. (a) - UHF modulation of a slit-collimated beam current recorded by collector probe and a 59-GHz-bandwidth oscilloscope. (b) FFT (spectrum) of the current modulation recovered for the signal representing difference of (a) and the same pulse equalized over 70 ps intervals close to modulation period.

IV. Beam in Drift Space

A. Beam's Picosecond Structure

Application of a 59-GHz oscilloscope makes visible in more details for a slit-collimated beam [Fig. 4, a]. Here, the main feature is the ultra-high-frequency (UHF) current modulation. This modulation could not be interpreted as the oscilloscope's noise because of interference along recorder's zero trace is weak (i), the modulation amplitude rises over time (ii), and it remains insignificant for the first half-nanosecond (iii). It can be assumed that the current modulation is associated with the excitation of UHF oscillations in some constructive low-Q cavity at the inlet section of the beam's drift space. Indeed, after the accelerating gap it is a short circular waveguide (radii r_n=4.95 mm), which is from the collector end has a reflector (beam's collimator), and the input differential $r_n \rightarrow r_a$ also gives a reflection growing as we approach the critical frequency of the waveguide. The sharp fronts of the integrated beam current (<100 ps) and its azimuthal fractions can induce at the $r_n \rightarrow r_a$ jump a spontaneous transition radiation in a wide frequency range. Apparently, quasi-critical frequencies should prevail in this radiation. For r_n=4.95 mm the critical frequency of the lower mode of the round waveguide is ~17 GHz. As in the presence of high current beam the loaded frequency is shifted down, shown in Fig. 4, b the spectrum of the beam's modulation with central frequency of ~16 GHz and a half-height-width of about 2 GHz confirms above assumption. The connection of the electron flow with H-type wave is possible for several reasons: the deviation of the beam axis or the inaccuracy of the system alignment, the azimuthal inhomogeneity of the electron current tube, the transverse motion of electrons etc.

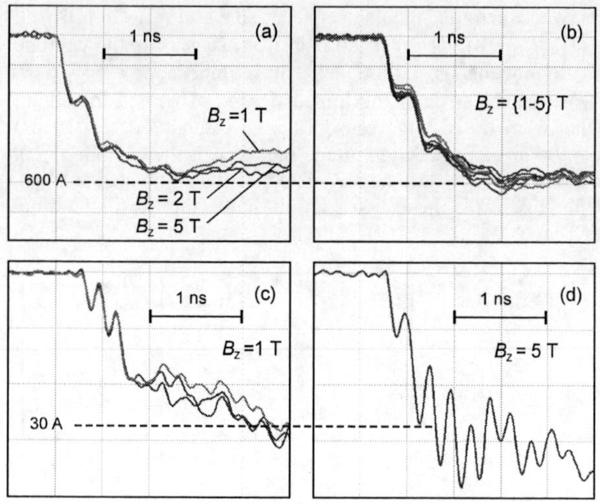

Fig. 5. Current measured using a segment-type collimator (*2*) at the drift space inlet (a), and after the beam passage the distance of ~6 cm (b). (c) Current waveforms obtained with the use of a narrow-slit collimator for drift length 6 cm and B_z=1 T. (d) Typical current waveform obtained using a narrow-slit collimator for the beam drift length 6 cm and B_z=5 T. Recorder: Tektronix DPO73304D real-time digital oscilloscope (33 GHz; 100 Gs/s).

B. Beam Features in Drift Space with Finite B_z-Field

The heterogeneity of the beam structure increases when the drift in guiding B_z-field. Figures 5, a, and b present the current pulses recorded by collector probe with one-quarter collimator [*2* in Fig. 1, b]. At the drift space inlet [L_d=0, Fig. 1, a], for the guiding B-field variation of 1-5 T, the current front [Fig. 5, a] is stable and nearly identical during 0.5 ns, starting from the EEE onset. Later on (t>1.5 ns), the difference is obvious: at B_z=1 T, the current amplitude reduces. This correlates with the behavior of the MICD impedance according to DTDR measurements [8], and could be associated with the EEE expansion toward the cathode cylindrical part adjacent to the edge. As the spatial screening scale [17] is higher at reduced magnetic field, then less EEE centers are active.

When beams' extended drift length (L_d≈6 cm), the recorded current is less reproducible for all magnetic fields [Fig. 5, b], starting from very early times. The reason may be in the azimuthal freedom of micro-bunches encountered at the stage of acceleration. As long as their transversal size depends on the B_z-field (see formula (1)), then the less magnetic field, more chaotic collimation of the micro-bunches having different angular position.

The above mentioned assumption is confirmed [Fig. 5, c and d] by current registration with the collector after narrow-slit collimator [*1* in Fig. 1, b]. According to the formula (1), the beam wall thickness (and probable micro-bunch transversal size) decreases from ~1.5 mm at B_z=1 T to less than 0.1 mm at B_z=5 T. Thus, in the weak B_z field, the beam wall is rather thick and micro-bunch azimuth size exceeds a lot the collimator's slit (0.4 mm). As a result, the current records even after beam's drift [Fig. 5, c] looks rather stable during the first half of nanosecond – just like in the waveforms obtained for the beam at the drift space inlet [Fig. 5, a]. Extended gradual rise of currents in Fig. 5, c is, probably, related with increase in micro-bunches quantity due to emission area expansion over time.

In the contrast, at high *B*-field, the beam clusters after kinematic grouping during drift motion pass throughout collimator slits (if well targeting occurs) perfectly, and produce the train of intensive current bursts with only a few hundreds of picoseconds delay from the EEE onset [Fig. 5, d]. This also confirms the gain of cathode's emission regions in high *B*-field.

V. Conclusion

In conclusion, it is shown experimentally that, for the interval of ~0.5 ns starting from the EEE onset, the subnanosecond beam's current front is well reproducible from pulse to pulse. This front steepness increases during acceleration by a fast-rise-time voltage and becomes more pronounced for extended accelerating gap.

In view of the total achievable beam's current, experiments demonstrate a minor distinct in the MICD emissive properties for the magnetic fields of 2-5 T at time interval of 1 ns. However, we have evidences on essential difference in the beam's microstructure development during drift motion at different magnetic fields.

Discovered UHF modulation of the beam is behind the instant at the current front where the phase establishment in the relativistic BWO occurs. However, this current perturbation may be important for some HPM schemes.

Acknowledgment

The authors are grateful to the representative office of Keysight Technologies in Russia and Ural Federal University's electrophysics department for the possibility of performing tests using the Keysight DSAZ594A signal analyzer and the Tektronix DPO73304D oscilloscope.

References

[1] A.A. El'chaninov, A.I. Klimov, O.B. Koval'chuk, G.A. Mesyats, I.V. Pegel', I.V. Romanchenko, V.V. Rostov, K.A. Sharypov, M.I. Yalandin, "Coherent summation of power of nanosecond relativistic microwave generators", Technical Physics, 2011, Vol. 56, No. 1, pp. 121–126.

[2] K. A. Sharypov, A. A. El'chaninov, G. A. Mesyats, M. S. Pedos, I. V. Romancheko, V. V. Rostov, S. N. Rukin, V. G. Shpak, S. A. Shunailov, M. R. Ul'masculov, and M. I. Yalandin, "Coherent summation of Ka-band microwave beams produced by sub-gigawatt superradiance backward wave oscillators," Appl. Phys. Lett., vol. 103, p. 134103, September 2013.

[3] D. M. Grishin, S. K. Lyubutin, G. A. Mesyats, V. V. Rostov, S. N. Rukin, B. G. Slovikovsky, S. P. Timoshenkov, M. R. Ul'maskulov, K. A. Sharypov, V. G. Shpak, S. A. Shunailov, and M. I. Yalandin, "Generation of gigawatt 10-GHz pulses with stable phase," Tech. Phys. Lett., vol. 34, pp. 822-824, October 2008.

[4] V. V. Rostov, A. A. El'chaninov, I. V. Romanchenko, and M. I. Yalandin, "A coherent two-channel source of Cherenkov superradiance pulses," Appl. Phys. Lett., vol. 100, p. 224102, May 2012.

[5] V. V. Rostov, A. A. El'chaninov, A. I. Klimov, V. Yu. Konev, I. V. Romanchenko, K. A. Sharypov, S. A. Shunailov, M. R. Ul'maskulov, and M. I. Yalandin, "Phase control in parallel channels of shock-excited microwave nanosecond oscillators," IEEE Trans. on Plasma Sci., vol. 41, pp. 2735–2741, October 2013.

[6] V. V. Rostov, A. A. Elchaninov, I. V. Romanchenko, S. A. Shunailov, M. R. Ul'maskulov, K. A. Sharypov, V. G. Shpak, S. N. Rukin, M. I. Yalandin, "Two-channel generator of the 8-mm wavelength range for radiation with subgigawatt power level pulses," Radiophysics and Quantum Electronics, vol. 56, no. 8-9, pp. 475-493, Jan. 2014.

[7] M. I. Yalandin, A. G. Reutova, K. A. Sharypov, V. G. Shpak, S. A. Shunailov, M. R. Ul'maskulov, V. V. Rostov, and G. A. Mesyats, "Stability of injection of a subnanosecond high-current electron beam and dynamic effects within its rise time," IEEE Trans. On Plasma Sci., vol. 38, pp. 2559–2564, October 2010.

[8] S. Ya. Belomyttsev, V. V. Rostov, I. V. Romanchenko, S. A. Shunailov, M. D. Kolomiets, G. A. Mesyats, K. A. Sharypov, V. G. Shpak, M. R. Ulmaskulov, and M. I. Yalandin, "Magnetically insulated coaxial vacuum diode with partial space-charge-limited explosive emission from edge-type cathode," J. Appl. Phys., vol. 119, issue 2, pp. 023304(1-10), Jan. 2016.

[9] K. A. Sharypov, M. R. Ul'masculov, V. G. Shpak, S. A. Shunailov, M. I. Yalandin, G. A. Mesyats, V. V. Rostov, and M. D. Kolomiets, "Current waveform reconstruction from an explosively emissive cathode at a subnanosecond voltage front" Rev. Sci. Instrum., vol. 85, p. 125104, December 2014.

[10] V. V. Rostov, E. M. Totmeninov, and M. I. Yalandin, "High-power relativistic microwave sources based on the backward wave oscillator with a modulating resonant reflector," Tech. Phys., vol. 53, pp. 1471-1478, November 2008.

[11] I. V. Romanchenko, V. V. Rostov, V. P. Gubanov, A. S. Stepchenko, A.V. Gunin, and I. K. Kurkan, "Repetitive sub-gigawatt rf source based on gyromagnetic nonlinear transmission line," Review of Sci. Instrum., vol. 83, no.7, p. 074705, July 2012 .

[12] V. G. Shpak, S. A. Shunailov, M. I. Yalandin, and A. N. Dyadkov, "The RADAN SEF-303A, a small high-current pulsed-power supply," Instrum. Exp. Tech., vol. 36, no. 1, pp. 106-111, January-February 1993.

[13] G. A. Mesyats, M. I. Yalandin, A. G. Reutova, K. A. Sharypov, V. G. Shpak, and S. A. Shunailov, "Picosecond runaway electron beams in air", Plasma Physics Reports, 2012, Vol. 38, No. 1, pp. 29–45.

[14] A. I. Fedosov, E. A. Litvinov, S. Ya. Belomytsev, and S. P. Bugaev, "Characteristics of electron beam formed in diodes with magnetic insulation," Sov. Phys. J. vol. 20, no 10, pp. 1367–1368, Oct. 1977. http://dx.doi.org/10.1007/BF00898924 Translated from Izvestiya Vysshikh Uchebnykh Zavedenii, Fizika, No. 10, pp. 134–135, October, 1977.

[15] V. P. Tarakanov, User's Manual for Code KARAT (Berkeley Research Associates, Inc., Springfield, VA, USA, 1992).

[16] S. D. Korovin, E. A. Litvinov, G. A. Mesyats, V. V. Rostov, S. N. Rukin, V. G. Shpak, M. I. Yalandin / Degradation and recovery of the emission from a graphite cathode in relation to the repetition frequency of nanosecond accelerating pulses // IEEE Transactions on Plasma Science, October 2006, Volume 34, Issue 5, Part 1, pp. 1771-1776.

[17] S. Ya. Belomytsev, S. D. Korovin, and G. A. Mesyats, "Shielding effect in the high-current diodes," Pis'ma v Zh. Tekh. Fiz., vol. 6, no. 18, pp. 1089-1092, 1980.

Modification of high-voltage pulse waveform by the spiral and core-transformer ferrite-filled lines

M.R. Ulmaskulov, S.A. Shunailov, K.A. Sharypov, M.I. Yalandin, and V.G. Shpak.
Institute of Electrophysics UB RAS
Ekaterinburg, Russia
marat@iep.uran.ru

***Abstract*— Results of testing of high-voltage transformer and spiral ferrite filled lines are presented. The devices convert the envelopes of a bell-shape of incident nanosecond pulses with amplitude of about -150 kV. The spiral lines sharpen a front to hundred of picoseconds, shorten pulse duration or form an extended continues pedestal-shape prepulse. A transforming line sharpens a tailing edge of a pulse from 2ns to 1 ns and changes its polarity.**

Keywords—nonlinear transmission lines; ferrite line; high voltage subnanosecond pulse; spirale line.

I. Introduction

In high-current electronics, investigations of charged particles emission and fast electric-discharge processes require application of high-voltage pulses of various shape and different polarity. As a rule, pulses with rise times of the order of hundreds picoseconds and shorter and with an amplitude of tens-hundreds kilovolts should be driven to discharger caps [see, e.g., [1] and citation therein]. Along with the traditional schemes of formation of such high-voltage pulses based on gas discharge switches [1,2], increasing attention is paid for options on the solid-state switchers [3], as well as devices convert a pulse of primary supply source by ferrite-filled gyromagnetic nonlinear transmission lines (NLTL) [see [4,5] and citation therein].

As the properties of ferrite change in an external bias magnetizing field (hereinafter "bias"), then NLTLs with a variable ferrite's magnetization change the driver pulse gradually. Below we describe two options of devices based on NLTLs with biased ferrite cores, operating in traveling wave regime. The first group of devices (the converters) sharpens the driver's pulse front to hundreds of picoseconds or forms the voltage prepulse. The second (the transformer) device changes the pulse polarity and sharpens the pulse tailing edge.

The RADAN-303 small-sized generator [6] was used as a source (driver) of nanosecond voltage pulses. It forms a pulse with a maximum amplitude up to U_0=–200 kV at a matched load of 45 Ω. At a repetition frequency of 100 Hz, the spread of the voltage amplitude was of ~10%. The pulse duration was 4.5 ns (FWHM), and its rise and fall times were about 1.5 ns. The voltage capacitive divider was located at the driver output to control the pulse waveform.

This work was supported by research program of IEP UB RAS 0389-2014-0005 and grant of RFBR №16-08-00230

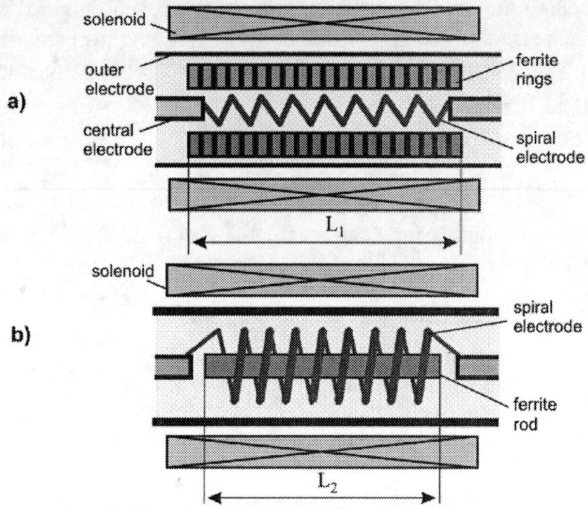

Fig. 1. Schematic of the coaxial ferrite spiral lines. (a) FSL1 – ferrite rings are located in the inter-electrode space. (b) FSL2 - ferrite rode is placed inside the spiral electrode.

II. Principle of Converters Opereation

Pulse waveform converters represent a spiral, high voltage oil-filled lines with ferrite cores (inserts) located in the inter-electrode space of the line (ferrite spiral line 1, FSL1, Fig. 1, a) or inside internal spiral electrode (FSL2, Fig. 1, b).

The axial bias created by external direct-current solenoid is used to saturate the ferrite. Magnetization of the ferrite insert in axial direction can be shifted by changing both the magnitude and direction of the bias (+/-H_Z) [7]. The feature of the spiral line is in generation of an axial component of magnetic field (H_{PZ}) with incident high-voltage pulse propagation. Depending on result of summation (+/-H_Z)+(H_{PZ}), a variable front sharpening takes place in the FSL1 with hollow ferrite insert between the line electrodes (Fig. 1, a). In general, the FSL1 operates like conventional, coaxial ferrite-filled NLTL with extended bias-control capabilities.

When there is no the ferrite between spiral and external electrodes of the line, the last represents a short-circuited turn for axial component of the magnetic field generated by an incident pulse. A certain portion of input pulse energy could be pumped into the turn, and this brings the features in the

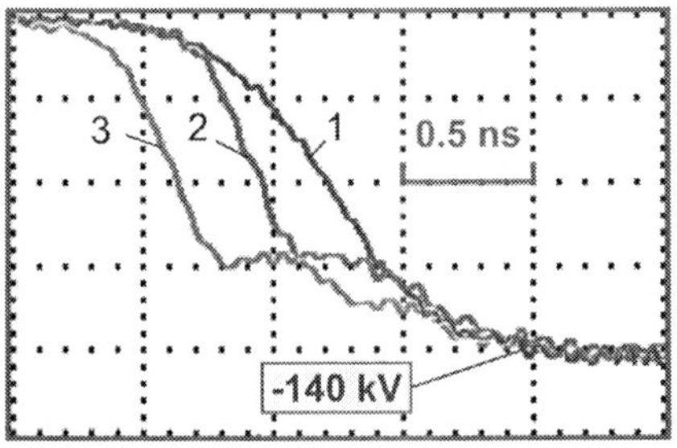

Fig. 2. Waveforms of output pulses of the FSL1 when H_Z=0 (1), +H_Z=25 kA/m (2) and -H_Z=25 kA/m (3).

ferrite magnetization when such a rod-type core insert is inside the spiral electrode (FSL2, Fig. 1, b). When the core is saturated by the bias which has an opposite direction (-H_Z) towards H_{PZ}, then the field H_{PZ} demagnetizes the rod, and the turn pumping is present. This process is responsible for the prepulse shaping. When the rate of magnetization slows down, the pumped energy is released, saturates the ferrate rod completely and makes the voltage pulse rise faster. Further, the input pulse propagates without changes in the shape.

III. Disign and Tests of FSLs

A. Pulse Sharpening and Delay Adjustment

The converter FSL1 presented in Fig. 1, a, is implemented by utilizing of ferrite rings type of M200VNP with diameters 32x20 mm. A total length of a ring assembly is L_1≈165 mm. An outer conductor diameter was 40 mm both for FSL1 and all other tested devices. The central triple spiral electrode was made from a copper wire with a diameter of 2.5 mm. Each spiral has 7 coils. An inner diameter of a spiral is 12 mm, a coil pitch is 21 mm. A matched coaxial oil-filled line with a voltage capacitive divider was used as a load of the FSL1. The load line was terminated by the resistor.

FSL1 output pulses with various bias fields are shown in Fig. 2. In case of H_Z=0 a rise time of an initial pulse was sharpened from ~2 ns to ~1.2 ns at the levels of 0.1-0.9. On identical magnitude but the opposite directions of bias (|+/-H_Z|~25 кA/m) the output pulse fronts are sharpened almost identically to ~0.35 ns, but are formed with a delay with difference by a value of front duration. In field summation +H_Z and H_{PZ} the ferrite is saturated at same time point. In subtraction mode (field -H_Z) a saturation takes place at a higher value of H_{PZ}, that is later at a pulse front. On the waveform it looks like a time shift between fronts by ~0.3 ns. This mode of a FSL1 can be utilized for control of the line delay time.

B. Pulse Envelop Convertion

A central electrode of a FSL2 (Fig. 1, b) was made in the form a triple spiral from a cooper wire with a diameter of 3 mm. A spiral pitch is 21 mm, an inner diameter is 12 mm, a length is L_2=184 мм. A NiZn-ferrite core of type M400NN with a diameter of 8mm and of length 200 mm was inserted into the spiral-form electrode. An impedance of the line with no ferrite core was ~59 Ohm. This line increased a rise time of the test pulse (<100 ps) to 500 ps (Fig. 3,a). A driver pulse with amplitude ≈-150 kV was utilized to test the FSL2. When no bias (H_Z=0) the driver pulse sharpened to 0.7 ns at a level 2/3 of amplitude (trace 4 Fig. 3,b). In the bias band {+H_Z=1 kA/m; |-H_Z| ~10 кA/м} a front of ~0.7-0.5 ns is being formed, and a pulse duration is being sharpened by ~1ns. Most short front was achieved in "counter" bias of |-H_Z|=10 кA/м.

Further increase in the bias |-H_Z|>10 кA/m leaded to the prepulse shaping and front extension. As it is shown in Fig. 3, b, the prepulse parameters are controled within certain range. FSL2 parameters which have practice significance are the following. In prepulse forming mode a delay time exceeds ~0.7 ns at an amplitude reflection coefficient of a driver pulse of ~(0.25-0.3). When the prepulse is not formed, the reflection coefficient does not exceed 0.15.

IV. Ferrite-Line Transformer

A device for conversion the polarity of high-voltage nanosecond pulses presents a transformer with a core as an assembly of NiZn ferrite rings of M200VNP type with diameters of 32x20 mm, which are built into a coaxial oil line (Fig. 4,a). Total length of the rings assembly is L_3=145 mm. The output and the input central electrodes have a magnetic couple by a ferrite core and, including grounding conductors, form the primary and secondary coils of the transformer, which operates in a traveling wave mode (traveling wave ferrite transformer, TWFT). The input and the output central electrodes of a transformer section were identical and contained two cooper wires of 3 mm in diameter. At a ferrite core section the conductors were diametrical disposed at a distance of 5 mm from the axis.

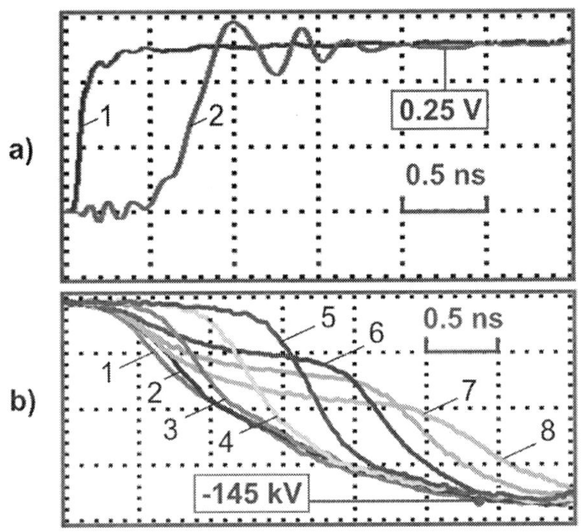

Fig. 3. Waveforms of: (a) input (1) and output (2) pulses of the spiral line FSL2 with no ferrate. (b) Output pulses (1)/(2)/(3) when +H_Z=10/2.5/1 kA/m, (4) H_Z=0 and (5)/(6)/(7)/(8) when -H_Z~10/25/35/50 kA/m.

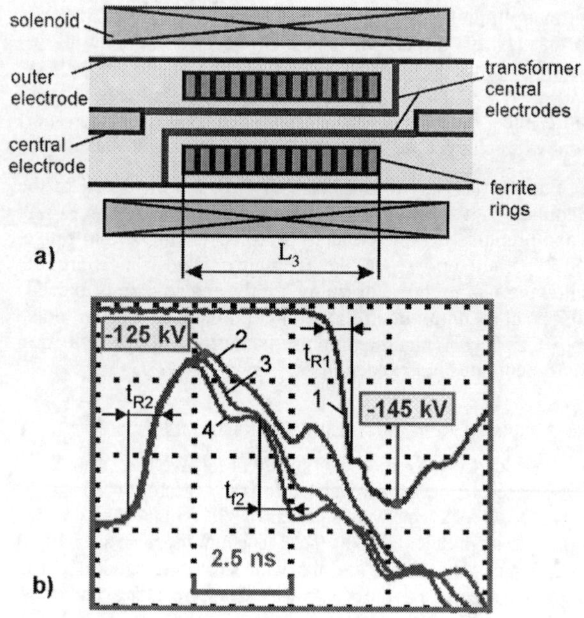

Fig. 4. (a) Schematic of the traveling wave transformer and (b) waveforms of the input (1) and output (2)/(3)/(4) pulses when H_Z=0/20/28 kA/m, respectively.

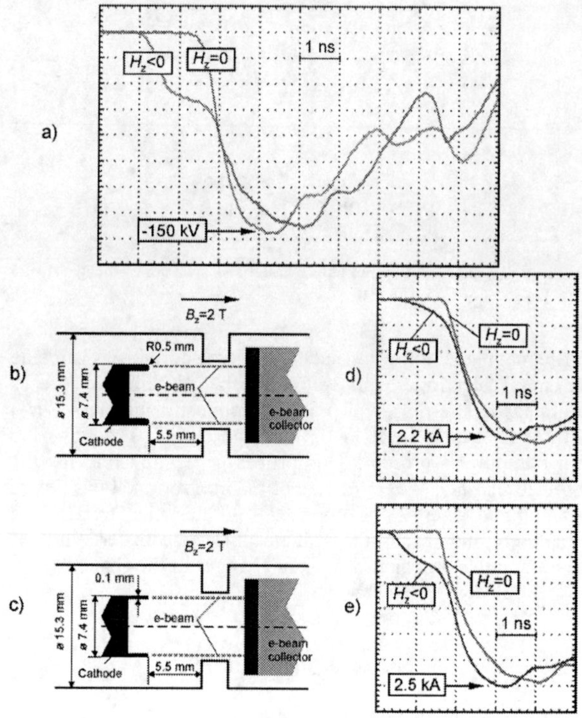

Fig. 5. (a) Voltage pulses formed by FSL2 at different bias fields applied to the EEE graphite cathodes (b,c) having different emission edges. (d,e) Electron beam currents recorded by collector probe at the input of drift space for the cases of different cathodes (b) and (c), and for the voltage pulses in (a). All waveforms (d) and (e) have absolute time reference.

The TWFT was tested by a pulse with the amplitude of~ -150 kV, 3.5 ns (FWHM) and subnanosecond rise time (0.5 ns at a level from 0.1 to 0.75 to waveform inflection, below maximum). This pulse was formed by the use of an additional NLTL, which was installed within the driver and TWFT. Low-distortion pulses with an opposite polarity were formed at the device output Fig. 4, b. The pulse amplitude was decreased by ~15%. When biasing of the core with an external axial field H_Z the ferrite saturated earlier and its nonlinear properties manifested. For example, if H_Z=28 kA/m the output pulse duration and the amplitude are reduced and the fall time is shortened from 2 ns to 1 ns. The rise time is increased to ~0.8 ns. The voltage tail with the amplitude less then 20-25% against the input pulse follows the main pulse, because a sufficient great deal of energy is being stored in a leakage inductance. The tail amplitude is changed in accordance with the degree of ferrite saturation.

V. Example of FSL Converter Application

An example on the effect of a voltage prepulse for the initiation dynamics of the explosive electron emission (EEE) with a graphite cathode of magnetically insulated coaxial diode is made in the work [1]. There, in particular, a role of the prepulse duration and amplitude was shown. In other things being equal, the amplitude factor must dominate, so long as transition of the field emission (FE) into the EEE stage associated with a strong nonlinear dependence of FE current from electric field across the cathode's emission edge. To study these features the voltage pulses formed by FSL2 converter are in interest.

In contrast to high-voltage devices with a sharpening spark discharger [2], prepulse amplitude at the FSL2 output is adjusted, in some ways, independently and can be higher. The example of the effect of the same prepulse on EEE activation at graphite cathode is presented in Fig. 5. In contrast to the case of no bias (H_z=0), a prepulse of a main pulse begins increasing by 1.5 ns earlier against the front, and if field $-H_z$ ~ 35 kA/m takes a pedestal-form (Fig. 5, a). Under these conditions the activation of the EEE depends on a field enhancement on a cathode. With a smooth semi-toroidal edge of a cathode as in Fig. 5, b, the field is enhanced significantly less then on a thin-edge (Fig. 5, b). As a result, from comparison of Fig. 5, d and Fig. 5, e it will be obvious that under a strong field (Fig. 5, e) the EEE comes at hundreds picoseconds earlier and rises so, that electron beam current, measured by a collector probe [8] with picoseconds resolution, follows the shape of voltage prepulse. Analogous conclusion follows from the measurements of the emission current by method of dynamic time-domain reflectometry [9]. This is quite important result in problems of stable in-phase excitation of relativistic microwave generators [10], which shows the possibility of achieving kiloampere electron current before the cathode comes in quickly rising edge of the main accelerating pulse.

VI. Conclusion

It is shown, that the presence of an axial component of a pulse magnetic field, transferring through a ferrate filled spiral line gives additional abilities to control the output pulse parameters by external bias. In consequence, it gives shaping of composite waveform pulses, prepulse forming or the delay time extending of a sharpened front. The location of a ferrite core inside the conductor of the spiral line increases the electrical strength of the device in whole. The traveling wave ferrite transformer allows inverting the high-voltage nanosecond signals with adjustable pulse duration control.

The described devices are useful when precision synchronization of several pulsed power devices, as well as in studies related to explosive emission cathodes and the formation of electron beams.

References

[1] G. A. Mesyats and M. I. Yalandin, "High-power picosecond electronics", Phys. Usp., 2005, 48 (3), 211-229

[2] M.I. Yalandin, V.G. Shpak "Compact high-power subnanosecond repetitive-pulse generators," Instrument and experimental techniques, 2001,V.4, Issue 3, pp 285-310.

[3] A. I. Gusev, M. S. Pedos, S. N. Rukin, S. P. Timoshenkov, and S. N. Tsyranov, "A 6 GW nanosecond solid-state generator based on semiconductor opening switch," Rev. Sci. Instrum. 86, 114706, 2015.

[4] S. Rukin A. Bushlyakov, S. Lyubutin, A. Ponomarev, B. Slovikovsky, S. Timoshenkov, S. Tsyrano , "Solid-state repetitive SOS-based generators providing apeak power of GW-range," Pulsed Power Conference, 2007. Digest of Technical Papers. IEEE International, pp. 698 – 701, 2007

[5] I.V. Romanchenko, V.V. Rostov, V.P. Gubanov, A.S. Stepchenko, Alexander V. Gunin, and Ivan K. Kurkan, "Repetitive sub-gigawatt rf source based on gyromagnetic nonlinear transmission line," Rev. Sci. Instrum. 83, 074705, 2012.

[6] V. G. Shpak, S. A. Shunailov, M. I. Yalandin, and A. N. Dyadkov, "The RADAN SEF-303A, a small high-current pulsed-power supply," Instrum. Exp. Tech., vol. 36, no. 1, pp. 106-111, Jan.-Feb.1993.

[7] L.A. Morugin, G.V Glebovich, "Nanosecond pulse instruments," Moscow. Sov. Radio, 1964.

[8] M. I. Yalandin, A. G. Reutova, K. A. Sharypov, V. G. Shpak, S. A. Shunailov, M. R. Ul'maskulov, V. V. Rostov, and G. A. Mesyats, "Stability of injection of a subnanosecond high-current electron beam and dynamic effects within its rise time," IEEE Trans. On Plasma Sci., vol. 38, no. 10, pp. 2559–2564, Oct. 2010.

[9] K. A. Sharypov, M. R. Ul'masculov, V. G. Shpak, S. A. Shunailov, M. I. Yalandin, G. A. Mesyats, V. V. Rostov, and M. D. Kolomiets, "Current waveform reconstruction from an explosively emissive cathode at a subnanosecond voltage front" Rev. Sci. Instrum., vol. 85, no. 12, pp. 125104(1-6), Dec. 2014.

[10] D. M. Grishin, S. K. Lyubutin, G. A. Mesyats, V. V. Rostov, S. N. Rukin, B. G. Slovikovsky, S. P. Timoshenkov, M. R. Ul'maskulov, K. A. Sharypov, V. G. Shpak, S. A. Shunailov, M. I. Yalandin, "Generation of gigawatt 10-GHz pulses with stable phase," Tech. Phys. Lett., vol. 34, no. 10, pp. 822-824, Oct. 2008.

Two-stage hybrid energy compressor for nanosecond high-voltage pulse

M.R. Ulmaskulov, S.A. Shunailov, K.A. Sharypov, M.I. Yalandin, A.G. Sadykova, and V.G. Shpak.
Institute of Electrophysics UB RAS
Ekaterinburg, Russia
marat@iep.uran.ru

***Abstract*—The results of testing of the high-voltage subnanosecond pulse oscillator with a hybrid, two-stage energy compression scheme, are presented. After the first section made in a form of L-C network with a gas discharger, a ferrite-filled nonlinear gyromagnetic transmitting line (NLTL) was used. The offered solution enabled increasing of initial nanosecond pulse voltage from -150 kV to -340 kV, and the pulse rise time of ~185 ps was obtained. Alternatively, for output amplitude of -240 kV, the pulse front was shortened to ~85 ps. The studies of influence of a transformation high-voltage pulse front on an output signal waveform after NLTL were conducted. He generator with limiting amplitude characteristics was utilized to obtain in air-filled diode ultra-short beams of the runaway electrons with maximum energy approaching to 700 keV**

Keywords—energy compression; gas discharger; nonlinear transmission line; high voltage subnanosecond pulse; runaway electrons;.

I. Introduction

The subnanosecond converters based on energy compression schemes with the L-C (high-impedance–forming-line ring circuit [1]) selection and gas discharger with an adjustable gap (subnanosecond discharger-based converters, SDC) are presented in the works [2-7]. Discharger gap adjustment gives varying of breakdown voltage and, consequently, energy accumulating time before commutation. As a result, in the devices output pulse amplitude can be increased from -150 kV to -290 kV, i.e. in practically ~2 times against converting pulse (K_{UI} –increase factor) [7]. Time instability of discharger gas breakdown defined instability of an output voltage amplitude, which increased in increase in K_{UI}. Therefore, the better repeatability of output pulse parameters achieved on K_{UI} was less the maximum. An output pulse rise time of the converters was ≥200 ps.

The ferrite-filled NLTLs with external bias are more stable pulse formers [8,9,10]. Due to ferrite non linear properties NLTL shapes the shock front, notable, the input nanosecond pulse front was sharpened to hundreds picoseconds and shorter, and the amplitude spike of voltage after front can increase to 30% [9,11]. The gyromagnetic resonance phenomena [11,13,14] comes when sufficient voltage rate. As the fast-rise-time pulse propagates through the NLTL, the ferrite domains magnetic moments come to precession in a crossed magnetic fields: azimuth magnetic field (H_A) of travelling voltage pulse (TEM-wave) and external bias field (H_z). Due to energy dissipation, the precession has a damped character, and microwave oscillations excited in transmitting line damp also. Excitation of microwave oscillations is associated with attaining a maximum primary voltage burst the stop in further voltage pulse sharpening.

By preparing initially sharpened voltage pulse using SDC, the process of oscillations excitation in biased NLTL could be simplified and made more effective. This also means, the faster primary voltage front, the shorter admissible length of the NLTL section is required. Finally, if SDC-formed pulse is short enough, the NLTL's section may form a very limited number of strongly-decay oscillations following to the first voltage burst. These features verification was our goal in tests of developed two-stage pulse compressor.

This work was supported by research program of IEP UrD RAS 0389-2014-0005 and grant of RFBR №16-08-00230.

II. High-Voltage Driver

A RADAN-303 high-voltage nanosecond generator [15] was used as setup driver. It formed voltage pulse with amplitude of up to –200 kV at a matched load of 45 Ω. The half-height width of the pulse was 4.5 ns, and its rise time and fall time were ~1.5 ns. Pulse double forming line (PDFL) of the driver was charged by means of Tesla transformer. Gas discharger of PDFL operated in the control mode by external, low-jitter triggering pulse in order to stabilize breakdown at a maximum PDFL charging voltage. General setup layout is shown in Fig.1 (1). As a load either an absorber (coaxial water-filled line, WL) or a gas diode for generating the runaway electrons were used.

III. First Compression Stage

Design of the first compression stage [2,3,4] is depicted in Fig. 2. This high-voltage former consists of in-series helical line (high-impedance line, HL) and coaxial forming line (FL), which compose the ring circuit. As long as lines impedances Z_{FL}<Z_{HL}, and Z_{FL} is equal to driver's impedance (Z_0), there is time-dependent stepwise voltage rise in the capacitive storage (FL) due to cyclic reflections of traveling incident voltage pulse between junctions "*HL* → driver", "HL → FL" and an open end of FL [2,3]. Depending on adjustable breakdown voltage (and switching instant) of discharger *S1*, the voltage

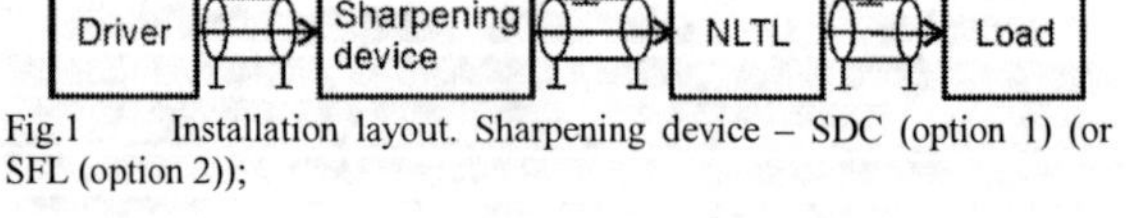

Fig.1 Installation layout. Sharpening device – SDC (option 1) (or SFL (option 2));

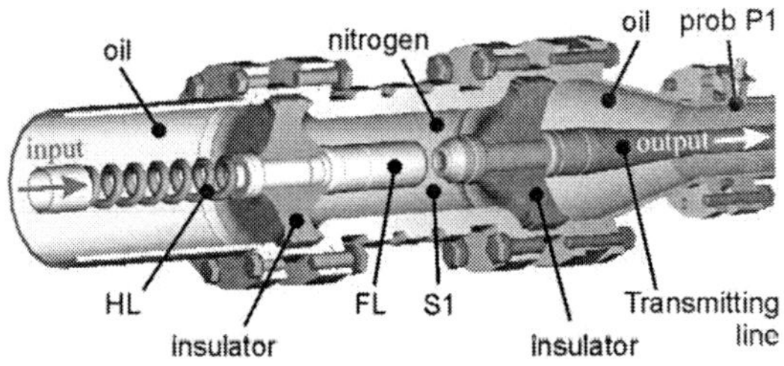

Fig. 2 Design of SDC-based compression stage.

across the load with impedance $Z_L=Z_{FL}=Z_0$ and, hence, the factor K_{UI} could be varied. A nitrogen spark gap pressurized up to 60 atm was used as a switch *S1*. The gap distance (breakdown voltage) is varied by mechanical gear which withstands high potential.

Parameters of the main SDC components were as follow: $Z_{HL}\approx135$ Ohm; $Z_{FL}=50$ Ohm; $t_{HL}\approx1$ ns; and $t_{FL}\approx0.5$ ns. For the ratio of delay times of HL and FL $(t_{HL}/t_{FL})\sim2$, maximum attainable value of K_{UI} was 1.3. With that, an incident driver's pulse U_0=-145 kV is converted at the load (oil line) as shown in Fig. 3, and attains the amplitude of U_I=-190 kV. The maximum rate of voltage rise for the pulse U_I was as high as ~1 MV/ns, and such a pulse we used for further transformation in NLTL.

IV. Tests of NLTLs

The four NLTL releases with oil insulation under pressure 7 atm. were pretested to utilize them in the second stage of pulse energy compressor. All NLTLs were loaded with the NiZn ferrite tape M200VNP with total length *L*=400 mm. The radiuses of the ferrite rings r_2 and r_1 and coaxial electrodes R_2 and R_1 varied (Fig. 4). Correspondingly, impedance of a line with no ferrite (Z_{w0}) and ferrite filling factor of a cross section (k_L) changed. The NLTL parameters are listed in the table 1.

TABLE I.

Line	$2r_2$; $2r_1$ (mm)	$2R_2$; $2R_1$ (mm)	Z_{w0} (Ohm)	k_L
NLTL1	28; 16	32; 8	55	0.55
NLTL2	28; 16	32; 14	32.5	0.63
NLTL3	20; 12	25; 10	37	0.49
NLTL4	16; 10	18; 8	32	0.59

Bias field H_z in ferrite of NLTLs was created by external dc solenoid and varied in order to get the maximum of amplitude and modulation depth of pulse. A special pulse converter based on a spiral, ferrite-loaded line (FSL) [16] was used instead of SDC to study dependence of output pulse parameters of NLTLs on front duration of an output pulse. (*2* in Fig.1). The driver pulse with amplitude $U_0\approx$-150 kV at FSL output had front duration ranging from $t_{f1}\approx1.5$ ns to $t_{f2}\approx0.5$ ns, depending on the bias field adjustment. The coaxial oil delay lines with capacitive voltage probes (P_1 and P_2) were installed before and after testing NLTL, which provided cutout of reflections from recoding signals during 5 ns.

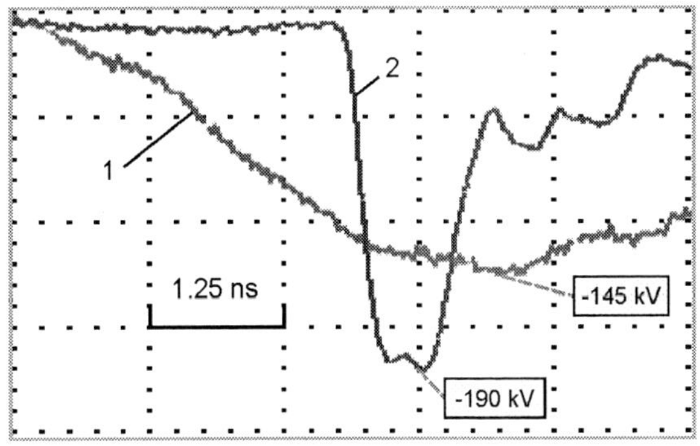

Fig.3 Waveform of output pulse former by SDC.

TABLE II.

Line No	U_{n1};U_{n2} (kV)	t_{n1};t_{n2} (ps)	f_1;f_2 (GHz)
NLTL1	170; 190	255; 240	2.2; 2
NLTL2	160; 170	180; 155	2.7; 2.5
NLTL3	150; 170	145; 130	3.2; 3.1
NLTL4	135; 160	125; 100	4.7; 5

The output pulse waveforms of NLTL1 and NLTL2 with identical rings are shown in Fig. 5, and pulses waveforms of NLTL3 and NLTL4 with less size rings are shown in Fig. 6. At the pictures parameters of the initial and shaped pulses are littered, which value are listed in the table 2. Here too the effective frequencies of exiting oscillations f_1 and f_2 determined by inversely of time between the maximums of

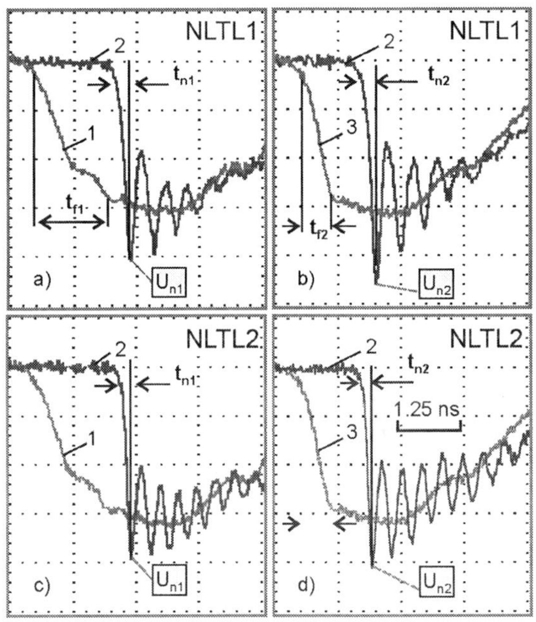

Fig. 5 Waveforms of input pulses (-150 kV) from SFL (1) and output pulses of NLTL1 and NLTL2 (2) for two options of the input pulse rise time: t_f=1.5 ns (a;c) and t_f =0.5 ns (b;d). Vertical scale ~50 kV/div.

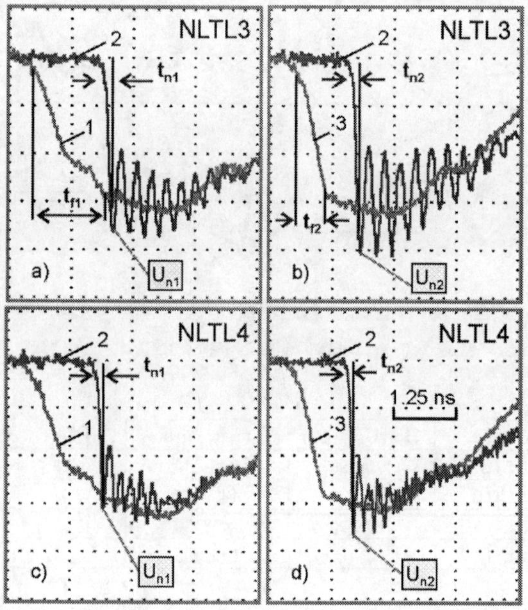

Fig. 6 Waveforms of input pulses (-150 kV) from SFL (1) and output pulses of NLTL3 and NLTL4 (2) for two options of the input pulse rise time: t_f=1.5 ns (a;c) and t_f =0.5 ns (b;d). Vertical scale ~50 kV/div.

two firs voltage peaks are driven.

Increase in amplitude and decrease in rise time of the output leading peak in rise time shortening of input pulse are common quality. So, in t_{f1}≈1.5 ns the saturation of chosen size ferrite rings and shock wave shaping takes place with increasing voltage amplitude at an incident pulse front – when amplitude is out of maximum. In supplying the NLTLs by short rise time pulse (t_{f1}≈0.5 ns) shock wave shaping shifts on the input pulse top with maximum amplitude.

The increase in filling factor as a cost of an increase in R_1 result in impedance reducing and increase in current (and field H_A) in NLTL2 as compared with NLTL1. Therefore a shock wave front is formed earlier and faster. As a result, the output rise time of NLTL2 is shorter, and voltage amplitude of a peak is less. In reducing of the diameters of the ferrite rings and coaxial electrodes of NLTL3 and NLTL4, a less current in the lines is required for ferrite saturation. Therefore, the shock wave is formed earlier and faster too, and the front duration and output amplitude peaks decrease.

While sufficient external bias H_z presents, waveforms in Fig. 5 and Fig. 6 demonstrate that gyromagnetic resonance in NLTL has a different features depending on the front of incident pulse. Gyromagnetic precession of magnetic moments in ferrite damps slower for a shortened front and for higher amplitude of a shock wave. As a result, excited microwave oscillations have increased modulation depth and less decay rate.

V. NLTL As The Second Compression Stage

Two NLTLs were chosen as a result of the carried out research. Most high increase in voltage (K_{U2}≈1.35) had NLTL1, and NLTL4 shaped most short rise time. The lines were tested as a second compression stage after SDC which formed an increased amplitude (-240 kV) and fast rise time (~250 ps) pulse. This incident pulse was short enough: 0.9 ns (FWHM). In the tests, water-filled load was used.

The pulse waveforms from the probes P_1 *and* P_2 are shown in Fig. 7. One can see that the maximum increase in voltage amplitude in the two stage compression scheme including NLTL1 attains U_1=-325 kV. The energy stored in two leading voltage peaks was ~50% relative to the incident pulse. With that, the first peak accumulating ~35% of energy has the rise time of ~185 ps, rate of the voltage rise of ~1.5 MV/ns and FWHM of ~200 ps.

Output pulse amplitude of NLTL4 practically hasn't increased, but a rise time became much shorter ~85 ps (at levels 0.1-0.9) and effective modulation frequency was nearly 5 GHz. The rate of the voltage rise was ~2.3 MV/ns Last value comes close to the physical limit for gyromagnetic precession in ferrites [14].

VI. Electrons Acceleration in Gas-Filled Diode

The installation with two-stage peaker based on SDC and NLTL1 (Fig. 8) was utilized in experiment on generation of ultra short beams of runaway electrons (RE) in air [17]. Agas diode with sharply inhomogeneous distribution of an electric field at a gap of length 2 cm. A diode is similar in design to which described in Ref. [18].

In accordance to existed concepts of RE appearance, electrons continuous acceleration mode and extremely fast interruption of RE emission (see, e.g., Ref. [19] and citation therein), the fast rise time voltage pulse is a key instrument for electrons to runaway in a critical electric field (E_c) and gain at the anode the energy which is close (and not exceeding [20]) to the integral of *eφ(t)* over acceleration time. Here e – elementary charge; *φ(t)* – potential at the cathode. The absence of extended voltage pre-pulse preceding the fast voltage front of two-stage compressor (Fig. 9,a) permits one to avoid advanced development of conventional pulsed breakdown in the gap. More over, by control electric field distortion at the cathode we can properly shift the instant of the E_c attaining toward the voltage rise at the front. If RE acceleration time in the gap is close to t_{max}-t_{re} (Fig. 9,a), than electrons gain a

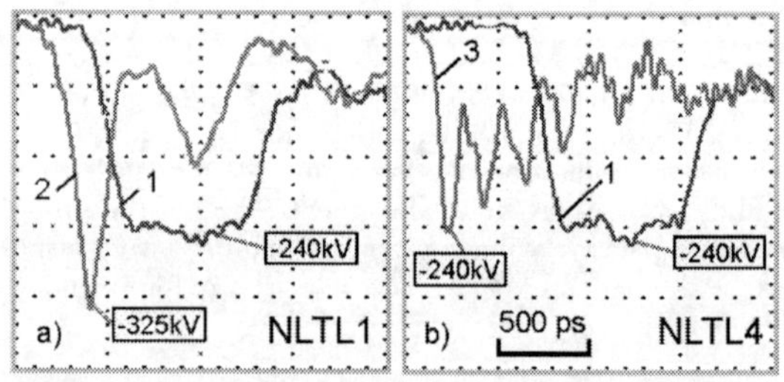

Fig. 7 Output waveforms of 1–SDC, 2–NLTL1 and 3–NLTL4 when a) NTLT1 and b) NTLT4 were used in the hybrid compression scheme.

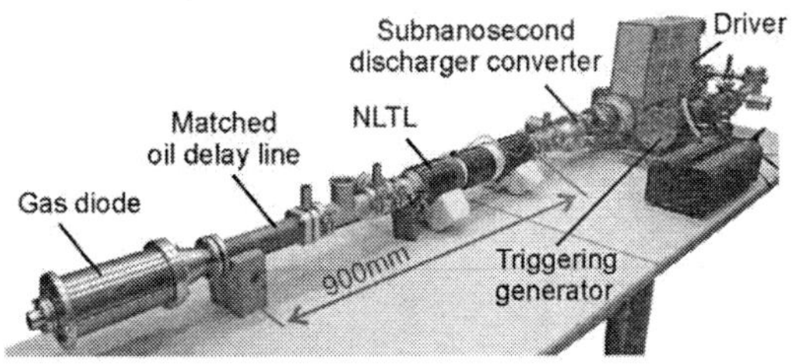

Fig. 8 Appearance of two-stage pulse compression setup.

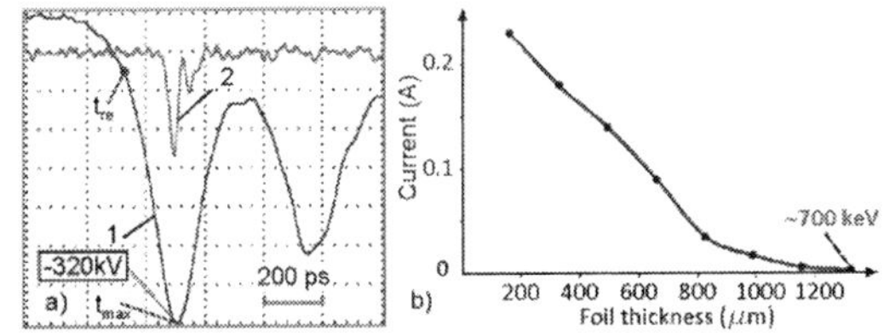

Fig. 9 a) Waveforms of accelerating voltage (1) and current of RE beam (2); b) graph of RE current at the anode as a function of Al- foil filter thickness.

double energy as long as RE current and charge are week and gas diode actually is an idle operation with the voltage duplication at the cathode: φ_{max}=|2U_{max}|. Fig. 9,a shows attenuation of the RE current on varying the thickness of the aluminum foil filter shielded the e-beam collector which installed behind the anode grid. When no filter, typical RE current waveform is shown in Fig. 9,a. As idle amplitude of accelerating pulse at the cathode was close to –650 kV, than approaching of maximum RE beam energy to 700 keV (Fig. 9) looks realistic.

VII. Conclusions

The research results have showed that shortening of initial pulse front has positive influence to output parameters of NLTL, notably the rate of rise time and subnanosecond amplitude spike of voltage increase. Most short rise time ~85 ps at NLTL output was obtained. When pulse rise time duration is ~185 ps voltage amplitude was increased from -240 kV to -325 kV. It can concludes that short front and duration of pulse after SDC increased in efficiency of microwave oscillations. The hybrid two-stage compression scheme made it possible to generate subnanoscong beam of runaway electrons with energy approaching to ~700 keV.

Acknowledgment

This work was supported by research program of IEP UB RAS 0389-2014-0005 and grant of RFBR №16-08-00230

References

[1] L.A. Morugin, G.V Glebovich, "Nanosecond pulse instruments," Moscow. Sov. Radio, 1964.

[2] M.R. Ulmascoulov, V.G. Shpak, S.A.Shunailov, K.A. Sharypov, "Optimization Of High-Voltage Nanosecond Pulse Compression," in proc. of 17th International Conference on High-Power Particle Beams, Xi'an P.R.China, pp.91-94, 2008.

[3] M.R. Ulmaskulov, V.G. Shpak, S.A. Shunailov, K.A. Sharypov, M.I. Yalandin, "Correction of shape and energy compression of high-voltage nanosecond pulses by compact converters with a traveling wave mode," Instrument and experimental techniques, v.51, No4, pp535-539, 2008.

[4] M.R. Ulmaskulov, V.G. Shpak, S.A. Shunailov, K.A. Sharypov, M.I. Yalandin, "High Voltage Subnanosecond Pulse Former Based On High Perfomance Enegy Compression Unit With Low Impedance Forming Line," IET European Pulsed Power Conference (EPPC 2009), 21-25 September CERN Geneva Switzerland, 2009.

[5] M.R. Ulmaskulov, S.A. Shunailov, K.A. Sharypov, V.G. Shpak and M.I. Yalandin "Optimization of High-Voltage Devices for Energy Compression of Nanosecond Pulses," Instruments and Experimental Techniques, V. 53, №1, pp. 100-106, 2010.

[6] M.R. Ul'maskulov, S.A. Shunailov, K.A. Sharypov, V.G. Shpak and M.I. Yalandin, "A compact subnanosecond high-voltage bipolar pulse generator with spark-gaps," Instruments and Experimental Techniques, V. 53, Issue 6, pp825-829, 2010.

[7] M. R. Ulmaskulov, V. G. Shpak, S. A. Shunailov, K. A. Sharypov and M. I. Yalandin, "Formers of High-voltage Subnanosecond Pulses Based on Compression Energy Schemes with Capacitor Storage and Spark Dischargers," JKPS Journal of the Korean Physical Society, December, V. 59, No. 61:pp.3499-3502, 2011.

[8] M I. Yalandin, S.A. Shunailov, M.R. Ul'maskulov, K.A. Sharypov, V.G. Shpak, V.V. Rostov**, I.V. Romanchenko, A.A. El'chaninov, and A.I. Klimov, "Synphase operation of nanosecond relativistic 37_GHz Backward_Wave Oscillators without Electrodynamic Coupling," Technical Physics Letters Vol. 38, No. 10, pp. 917–920, 2012.

[9] M. R. Ulmaskulov, M. S. Pedos, S. N. Rukin, K. A. Sharypov, V. G. Shpak, S. A. Shunailov, M. I. Yalandin, I.V. Romanchenko, and V. V. Rostov, "High repetition rate multi-channel source of high-power rf-modulated pulses," Rev. of Sci. Instruments 86, 074702, 2015.

[10] S.N. Rukin, A. Bushlyakov, S. Lyubutin, A. Ponomarev,B. Slovikovsky, S. Timoshenkov, S. Tsyranovl, "Solid-state repetitive SOS-based generators providing apeak power of GW-range," Pulsed Power Conference, Digest of Technical Papers. IEEE International, pp.698–701, 2007.

[11] J.-W. Bragg, J. Dickens, and A. Neuber, "Material selection considerations for coaxial, ferrimagnetic-based nonlinear transmission lines," J. Appl. Phys. 113, 064904, 2013.

[12] M. R. Ulmaskulov, M. S. Pedos, S. N. Rukin, K. A. Sharypov, V. G. Shpak, S. A. Shunailov, M. I. Yalandin, I.V. Romanchenko, and V. V. Rostov, "High repetition rate multi-channel source of high-power rf-modulated pulses," Rev. of Sci. Instruments 86, 074702, 2015.

[13] I.G. Kataev, Shock electromagnetic waves, Moscow, Sov. Radio, 1963.

[14] A.L. Mikoeliylan, Theory of application of ferrites on microwave frequency, Moscow, State energy press, 1963.

[15] M.I. Yalandin, V.G. Shpak, "Compact high-power subnanosecond repetitive-pulse generators," Instrument and experimental techniques, 2001,V.4, Issue 3, pp 285-310.

[16] M.R. Ulmaskulov, S.A. Shunailov, K.A. Sharypov, M.I. Yalandin, and V.G. Shpak, "Modification of high-voltage pulse waveform by the spiral and core-transformer ferrite-filled lines,", in press in the current Proc.

[17] G. A. Mesyats, M. I. Yalandin, K. A. Sharypov, V. G. Shpak, and S. A. Shunailov, "Generation of a Picosecond Runaway Electron Beam in a Gas Gap with a Nonuniform Field," IEEE Transactions on Plasma Science, Volume 36, no. 5, part 3, pp. 2497-2504, October 2008.

[18] A.V. Gurevich, G.A. Mesyats, K.P. Zybin A.G. Reutova, V.G. Shpak, S.A. Shunailov, M. I. Yalandin, "Laboratory demonstration of runaway electron breakdown of air," Physics Letters A vol. 375, issues 30-31, pp. 2845–2849, July 2011.

[19] G. A. Mesyats, M. I. Yalandin, A. G. Reutova, K. A. Sharypov, V. G. Shpak, and S. A. Shunailov, "Picosecond Runaway Electron Beams in Air", Plasma Physics Reports, 2012, Vol. 38, No. 1, pp. 29–45.

[20] G.A.Mesyats, A.G.Reutova, K.A.Sharypov, V.G.Shpak, S.A.Shunailov, M.I.Yalandin / On the observed energy of runaway electron beams in air // Laser and Particle Beams, 2011, Vol.29, No.4, pp. 425-435

Thermal Stress at the Surface of Thick Conductive Plates Induced by Sinusoidal Current Pulses

Markus J. Löffler, Klaus Mecking, Tobias Lindenblatt
High Voltage and Pulsed Power Technology
Westfaelische Hochschule University of Applied Sciences
Gelsenkirchen, Germany

Markus Schneider
French-German Research Institute of Saint Louis
Saint Louis, France

Abstract—**Transient high magnetic pressures are characteristic for pulsed power applications such as electromagnetic forming, electromagnetic acceleration and other applications involving high electric current pulses. This study presents results of calculations concerning temperatures and thermal stresses at the surface of a thick conductive plate induced by sinusoidal current pulses. Thermal diffusion inside an isotropic material (Cu, Al, Au, Ag, Mo, W) is considered assuming adiabatic conditions. The temperature dependence of the material properties is taken into account. It is shown that at peak magnetic surface pressures between 100 MPa and 600 MPa the surface temperature reaches 0.6±0.1 K/MPa. The thermally induced compressive surface stress reaches 1.7±0.3 MPa/MPa. With regard to the relative performance of the materials considered here, it can be concluded that pure Copper or pure Aluminum are not the best choice for industrial pulsed power applications, as their lifetimes are critical.**

Keywords—skin effect; thermal stress; surface temperature

I. Introduction

In Pulsed Power Technology sinusoidal currents or currents with sinusoidal rising edge are used to generate magnetic flux densities between 1 and 100 T e.g. for high velocity forming [1], for mass acceleration [2], and electromagnetic welding [3]. Moreover, pulses with amplitudes above 10 T are applied to investigate physical properties of materials.

Under such conditions, thermal fatigue caused by local Joule heating limits the lifetime of coils or other conductive structures [4, 5]. The rapid ohmic heating of the conductor's surface leads to thermal quenching of the cold bulk material directly under the surface. In combination with strong magnetically induced mechanical pressures high thermal gradients cause additional stress peaks in the surface region of the conductor which may cause crack formation and growth.

In [6] an analytical formula is given to calculate the maximum surface temperature of a semi-infinite plate assuming a unit step waveform of the magnetic flux density at the plate's surface. Although the formula is useful as a rule of thumb it does not give information about temperature and thermal stress when applying a sinusoidal field pulse. Moreover the formula assumes average material properties independent of temperature. Assuming a sinusoidal current pulse at arbitrary frequency, this paper gives results for Cu, Al, Au, Ag, Mo, and W with temperature dependent material properties enabling an estimate of peak surface temperature and peak thermal surface stress.

II. Theoretical Model

A. Assumptions

A thick plate of thickness d shall be considered, see Fig. 1.

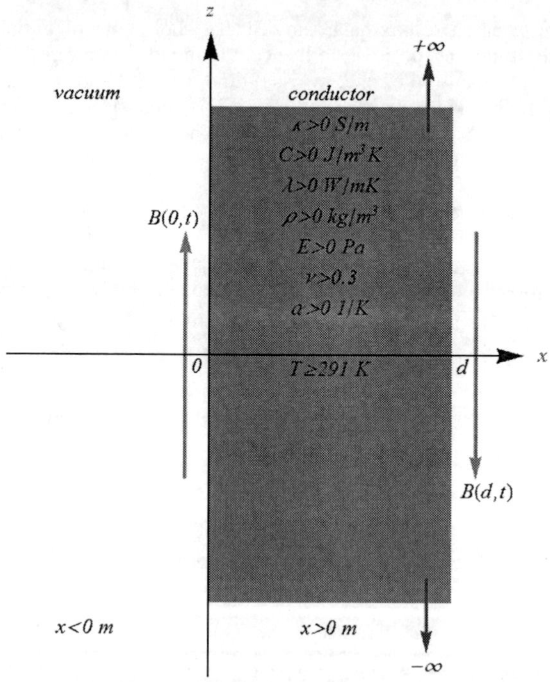

Fig. 1: Cross section of conducting plate of thickness d.

The plate consists of isotropic material with physical properties depending on temperature T: electric conductivity κ, density ρ, thermal conductivity λ, Young's modulus E, Poisson ratio ν, coefficient of linear thermal expansion α, and magnetic permeability $\mu=\mu_0=4\cdot\pi\cdot10^{-7}$ H/m. With specific heat c the volumetric heat capacity yields to $C=\rho\cdot c$.

The surface of the plate is placed in the y-z-area. In y- and in z-direction its dimensions are "infinite". The plate is placed in vacuum. At time t=0 s a pulsed sinusoidal magnetic flux density B with amplitude $\hat{B}$ and angular frequency ω is switched on, see Fig. 2.

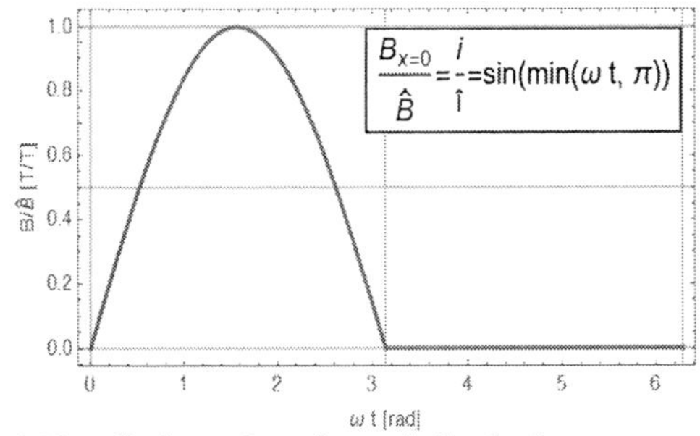

Fig. 2: Normalized wave form of magnetic flux density.

The B-vector points in positive z-direction at x=0 and in negative z-direction at x=d.

For all calculations d is given by

$$d = \frac{200}{\sqrt{\omega\,\kappa_{291}}} S^{-1/2} m^{-1/2} \tag{1}$$

meaning that the plate is considerably thicker than the transient magnetic and thermal diffusion depths.

B. Material properties

Prior to further considerations a data base is established with temperature-depending physical properties of the isotropic materials considered herein [7]. The data are interpolated with polynomials. Fig. 3 gives values for Cu and Al in reference to the values at T=291 K.

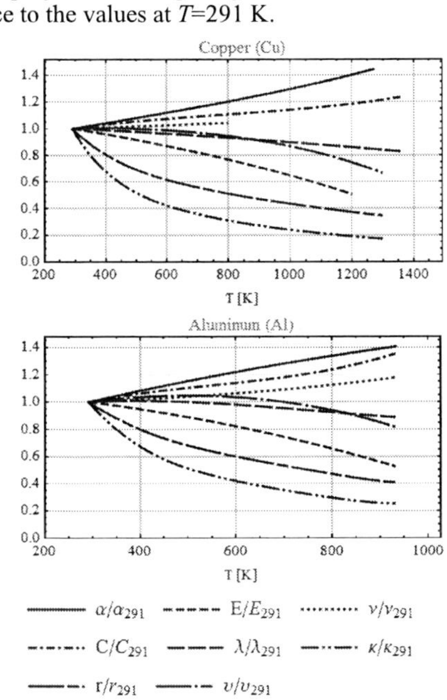

Fig. 3: Material properties of Cu and Al.

r is a parameter as given in [6]:

$$r = \sqrt{\frac{\kappa\,\lambda\,\mu}{C}} \tag{2}$$

g is defined by

$$g = \frac{E\,\alpha}{1-\nu} \tag{3}$$

Linear thermal expansion α, volumetric heat capacity C, Poisson's ratio ν increase with temperature whereas electric conductivity κ, thermal conductivity λ, and Young's modulus E decrease. Other metals show a similar behavior.

C. Calculus

With magnetic flux density B, current density j, electric field E, and local ohmic power density p_Ω Maxwell's equations and Ohm's law

$$\nabla \times \vec{B} = \mu_0 \vec{j},\ \nabla \times \frac{\vec{j}}{\kappa} = -\frac{\partial \vec{B}}{\partial t}, p_\Omega = \frac{j^2}{\kappa} \tag{4}$$

lead to

$$\frac{\partial}{\partial \xi}\left(\frac{1}{\kappa\,\mu_0}\frac{\partial B}{\partial \xi}\right) = \frac{\partial B}{\partial t}, p_\Omega = \frac{\omega}{\mu_0^2 \kappa}\left(\frac{\partial B}{\partial \xi}\right)^2 \tag{5}$$

$$\frac{j}{\sqrt{\omega}} = -\frac{1}{\mu_0}\frac{\partial B}{\partial \xi} \tag{6}$$

where $\quad \alpha + \beta = \chi. \quad (1) \quad (1)$

$$\xi = \sqrt{\omega} x, \tau = \omega\, t \tag{7}$$

with the boundary conditions

$$B(0,\tau) = -B(\sqrt{\omega} d, \tau) = \hat{B} \sin \min(\tau, \pi) \tag{8}$$

$$B(\xi, 0) = 0\, T \tag{9}$$

The PDE describing the diffusion of temperature T is given by

$$\frac{\partial}{\partial \tau}(C\,T) = \frac{\partial}{\partial \xi}\left(\lambda \frac{\partial T}{\partial \xi}\right) + \frac{1}{\mu_0^2 \kappa}\left(\frac{\partial B}{\partial \xi}\right)^2 \tag{10}$$

Assuming negligible heat transfer from the plate's surfaces to the adjacent vacuum, the boundary conditions are given by

$$\frac{\partial}{\partial \xi} T(0,\tau) = \frac{\partial}{\partial \xi} T(\sqrt{\omega}\, d, \tau) = 0\ \frac{K\, s^{1/2}}{m} \tag{11}$$

$$T(\xi, 0) = T_0 = 291\, K \tag{12}$$

The temperature increase is defined by

$$\Delta T = T - T_0 \tag{13}$$

Assuming free expansion of the plate and applying formulas given in [8] the local thermal stress in direction of y and z can be calculated using

$$\sigma(\xi, \tau) = \frac{g}{\sqrt{\omega}\, d}\int_0^{\sqrt{\omega}\, d} \Delta T\, d\xi - g \cdot \Delta T \tag{14}$$

Fig. 4 justifies to set $g \approx g(291\text{ K})$:

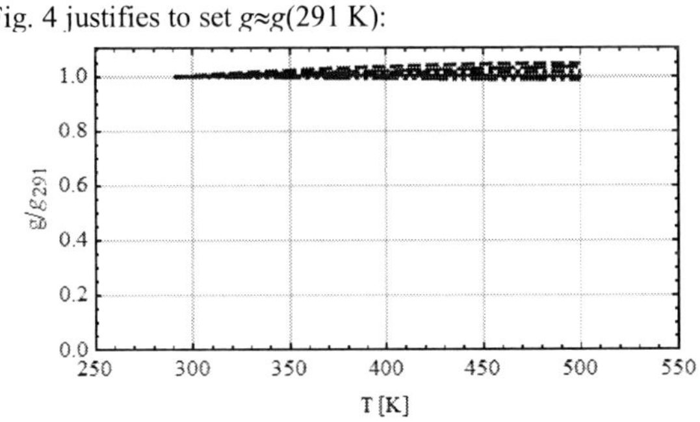

Fig. 4: Nearly constant factor g of Ag, Al, Au, Cu, Mo, W.

$\alpha + \beta = \chi. \quad (1) \quad (1)$

In ξ-direction the stress is zero because magnetic forces are not taken into account. The requirement of finiteness of the integral term at very large d leads to

$$\sigma(\xi, \tau) \approx -g \cdot \Delta T \qquad (16)$$

At the surface the stress is compressive because the colder regions inside the plate "try to hinder" the heated surface to expand. With $0 \leq \tau \leq 2 \cdot \pi$ and with the peak magnetic pressure

$$\hat{p} = \frac{\hat{B}^2}{2\,\mu_0} = 90\ MPa\ (\rightarrow \hat{B} \approx 15\ T) \qquad (17)$$

at the surface of a Cu-plate the diffusion profiles of B, j and ΔT are shown in Fig. 5.

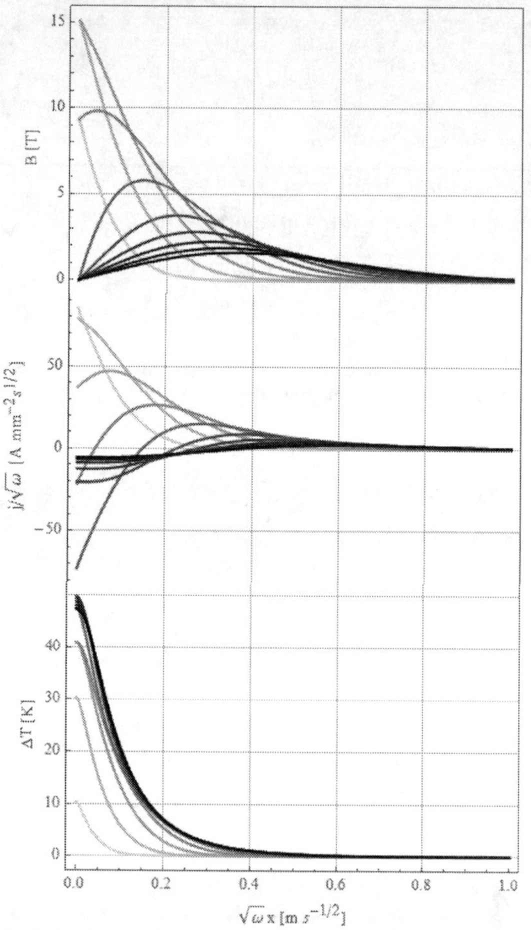

Fig. 5: Spatial distribution of magnetic flux density, current density, temperature increase in Cu at different times. Step size of τ is Δτ=2·π/10.

The curves are drawn darker with advancing τ. The abscissas are restricted to the interesting part of electromagnetic and thermal effects nearby x=0 mm.

At x=0 B follows the sinusoidal field excitation and diffuses into the material. At decreasing surface field (τ>π/2) a field wave travels into the plate. The current density also diffuses into the material. At τ=π B at the surface and overall current i through the plate vanish: j has a negative and a positive part keeping the magnetic flux inside the plate. Due to ohmic losses j tends to zero. ΔT diffuses with increasing amplitude into the material. Due to its smaller diffusion velocity the thermal diffusion depths is only about half the magnetic diffusion depth. Fig. 6 shows time histories of j and ΔT.

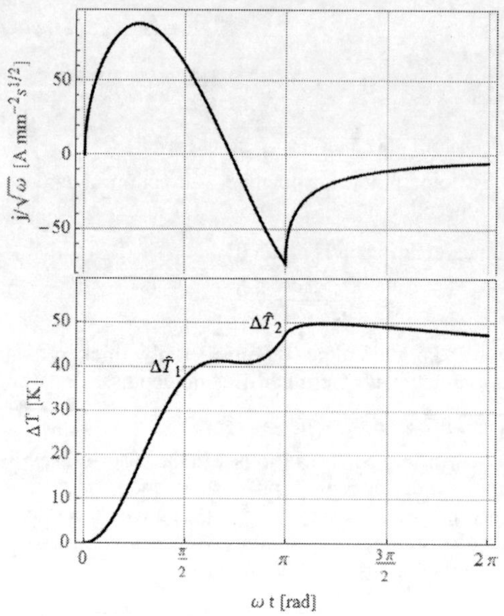

Fig. 6: Characteristics of surface current density and temperature increase.

At τ≈π/4 j reaches a maximum and drops to its negative amplitude when B zeroes at τ=π. For τ>π the surface current density approaches Zero in a more or less exponential decay. ΔT increases according to the two extrema of j. After reaching a first maximum $\Delta\hat{T}_1$ at zeroing j it reaches its second maximum $\Delta\hat{T}_2$ shortly after τ=π. ΔT increases to ≈50 K independently of the frequency.

III. Further Interpretation

For other materials Fig. 7 shows the 2nd maximum $\Delta\hat{T}_2$ at different pressures $\hat{p}_{mag} = 100(100)600\ MPa$.

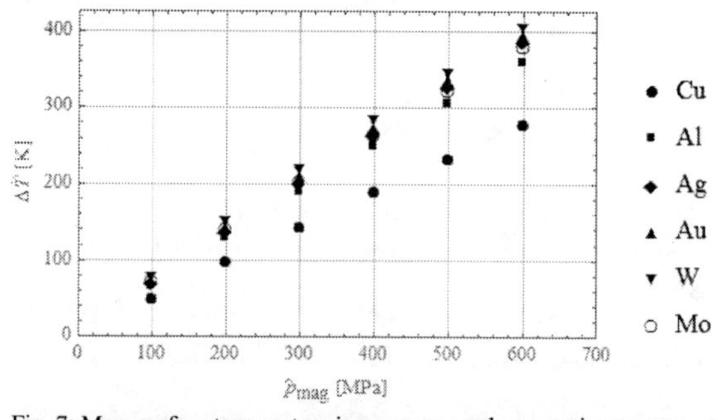

Fig. 7: Max. surface temperature increase vs. peak magnetic pressure.

The temperature maxima increase with increasing magn. pressure. Cu shows the smallest, W the largest temperature increase. Fig. 8 shows the ratios of temperature maxima and peak magnetic pressures.

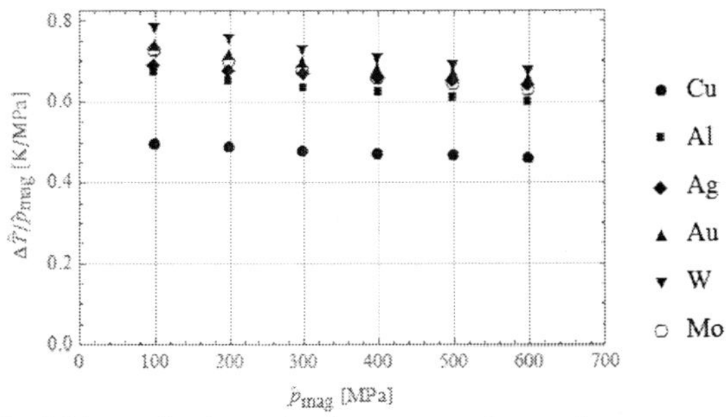

Fig. 8: Max. surface temperature increase vs. peak magnetic pressure.

With increasing peak magnetic pressure the ratios weakly decrease: at higher magnetic peak pressures the associated increase in temperature (Fig. 7) leads to a faster magnetic diffusion due to reduced conductivity (eq. 5).

For the materials investigated so far Fig. 9 shows the average temperature-to-pressure ratios taken from Fig. 8.

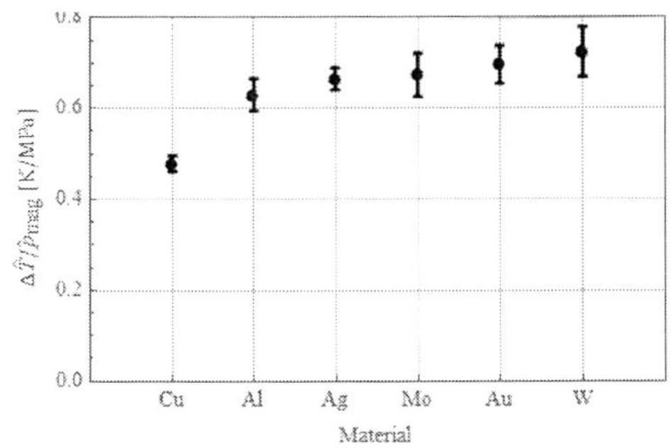

Fig. 9: Surface temperature increase vs. peak magnetic pressure.

At magnetic pressures up to 600 MPa Cu has the lowest (0.48 K/MPa) and W has the highest (0.72 K/MPa) surface temperature rise.

The comparison of maximal thermal stress vs. peak magnetic pressure is shown in Fig. 10:

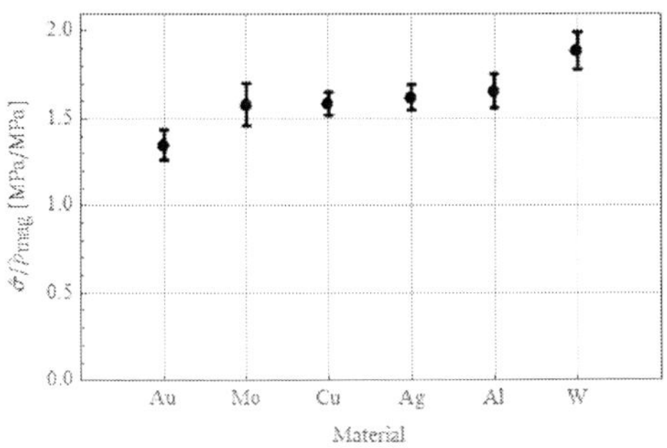

Fig. 10: Max. compressive stress vs. peak magnetic pressure.

Au is stressed least (1.35), W (1.9). Mo, Cu, Ag, and Al are charged with comparable stresses. Note that Cu or Al exhibit a 1.6-times higher thermal stress of 640 MPa, if subjected to magnetic pressures of 400 MPa. At 291 K, the maximal yield strengths of hard-drawn Cu respectively of rolled Al is 470 MPa resp. 145 MPa [9]. Compressive yield strengths of all metals, except those of cold-worked, equal their tensile yield strengths [9]. Hence at peak magnetic surface pressures of 400 MPa ($\rightarrow$32 T) the yield strength of Cu is exceeded by 40% and of Al by 340% respectively. This may lead to lifetime-reducing material fatigue. The design of crucial metal elements for pulsed power applications should take into account the findings of this study, in particular the limits given above for Cu and Al.

IV. Conclusion

If the surface of a thick metal plate (Al, Ag, Au, Cu, Mo, W) is subjected to sinusoidal magnetic pulses it heats up to temperatures between 0.4 and 0.8 K per MPa peak magnetic pressure. Due to the temperature difference between surface and bulk material compressive thermal stress appears at the surface with stress rates between 1.3 and 2.0 MPa per MPa peak magnetic pressure. The findings obtained in this study allow defining design criteria for various metal elements used in pulsed power applications. For instance, the use of Cu or Al at peak pressures of 400 MPa can lead to material fatigue.

References

[1] Li Zhong et al., "Electromagnetic Forming with Solenoid Coil," in, 2015 7th Int. Conf. Meas. Techn. and Mech. Automation, pp.419-422, 2015

[2] R.J. Kaye, "Operational requirements and issues for coilgun electromagnetic launchers," IEEE Trans. on Magn., vol.41, no.1, pp.194-199, Jan. 2005

[3] A.K.Jassim, "Magnetic pulse welding technology," 1st Int. Conf. on Energy, Power and Control, pp.363-373, 2010

[4] K. Svendsen, S.T. Hagen, "Thermo-Mechanical Fatigue Life Estimation of Induction Coils", Int. Sc. Colloquium Modelling for Electromagnetic Processing, Hannover, Oct. 27-29,2008, pp. 127-132

[5] E. Baussan et al., "Target, magnetic horn and safety studies for the CERN to Fr\.b4ejus Super Beam," IOP Publishing Journal of Physics: Conference Series 408 (2013) 012061

[6] M.J. Loffler; M. Schneider, "Temperature Versus Magnetic Pressure at the Surface of a Semi-Infinite Plate," IEEE Trans. on Plasma Science, vol.41, no.10, pp.2815-2818, Oct. 2013

[7] T. Lindenblatt, "Simulation and stress distribution in coaxial lines loaded with pulsed currents," Westf. Hochschule, Bach. Th., Nov. 2013.

[8] T.J. Lu, N.A.Fleck, "The thermal shock resistance of solids," Acta mater. Vol. 46, No. 13, pp. 4755-4768. 1998.

[9] J. Symonds et al.: "Chapter 5: Strength of Materials," Mark's Standard Handbook of Mechanical Engineering, McGraw-Hill Education; 11 edition (December 7, 2006), ISBN-13: 978-0071428675.

RF Formation Using Transmission Lines Built with Ceramic PZT Capacitors of Weak Nonlinearity and a Fast ns Rise Time Pulse Generator*

Lauro P. Silva Neto
José Osvaldo Rossi
Associated Plasma Laboratory
National Institute for Space Research
São José dos Campos, Brazil
silvaneto007@yahoo.com.br

Joaquim J. Barroso
Technological Institute of Aeronautics
São José dos Campos, Brazil

Edl Schamiloglu
Electrical and Computer Engineering Department
University of New Mexico
Albuquerque, USA

***Abstract*— Nowadays RF signals are used in several applications such as in satellite communications, pulse radars, mobile phones, and military communications, etc. Based on a transmission line system that employs ceramic-based capacitors of weak nonlinearity, fixed-value inductors and a fast pulse generator with rise time < 10 ns, RF oscillations have been generated at the line output by means of a filtering process. This is caused essentially by the LC line dispersion, which does not depend on nonlinear effect of the components as in nonlinear transmission lines (NLTLs). In this work, two transmission lines were built by using linear inductors of 3.3 μH and capacitors of weak nonlinearity. All lines were built with commercial ceramic capacitors, where the first one used 1.0 nF capacitor dielectric based on barium titanate (BT) composite while the other used 1.8 nF capacitor dielectrics based on a mixture of lead zirconate and titanate, known as PZT. The results show that using BT capacitors in the line does not generate any oscillation at the line output due to higher losses of BT. On the other, using PZT that has lower losses compared to BT, frequency oscillation of the order of 2 MHz were obtained for de-poled line. The performed tests were made on a single shot and at a repetition pulse of about 100 kHz with the FFT output spectrum measured to confirm RF frequency obtained around 2 MHz.**

Keywords—Soliton; high power; nonlinear capacitor; radiofrequency

I. Introduction

Many institutes and universities around the world have been studied radiofrequency (RF) generation for space applications, because of the focus on communication systems to provide link between different devices such as in mobile phones, television, satellites, battlefield communication disruption and radars [1], [2].

*Work supported by the Southern Office of Aerospace Research and Development (SOARD – USAF) under contract number FA9550-14-1-0133.

One way to produce RF signals could be by using nonlinear transmission lines (NLTLs), which are lumped lines built with nonlinear capacitors C(V) in most cases [3], [4] as shown in Fig. 1. On the other hand, in this work two transmission lines (TL) were built using commercial ceramic capacitors of weak nonlinearity and 3.3 μH linear inductors.

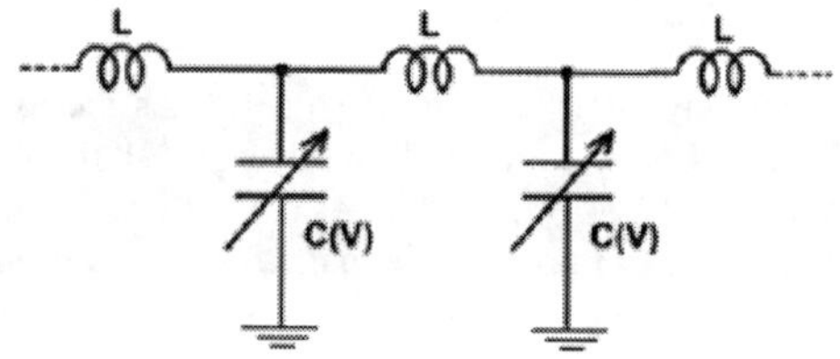

Fig. 1. Transmission line using linear L and nonlinear C.

Nowadays, ceramic materials employed in TLs and NLTLs for RF generation have been studied as BTs (barium titanate) and PZTs (Lead Zirconate Titanate), where the latter is most used as sensors in electronic applications [5]. In this work, it was demonstrated that de-poled PZT ceramic capacitors can be used as linear capacitive elements in a TL to generate RF because of their lower losses compared to BT capacitors. This is corroborated by a previous work [6] that demonstrated that in a low frequency range (100 kHz – 2 MHz) de-poled PZTs have a dielectric loss factor varying between 0.001 and 0.01 while BT above between 0.01 and 0.1. Therefore, this makes PZT more suitable for RF applications using linear TLs than BT as will be demonstrated ahead. Other ceramic dielectric such as PMN (Lead Magnesium Niobate) ceramic has been used as a nonlinear element in transmission lines, but the results found so far are very discouraging as experimental results demonstrated that huge dielectric losses compromise seriously the RF generation [7].

II. Methodology

Two lines were built using linear inductors and capacitors of weak nonlinearity. The first line used commercial-off-the-shelf ceramic capacitors (COTS) of 1.0 nF based on barium titanate (BT) while the second one employed as line capacitors de-poled PZTs, normally used as piezoelectric sensors in industrial applications. Tab. I show the electrical characteristics of the lines built.

Table I. Electrical parameters of the lines tested.

Line	C (nF)	Dielectric	n	Z (Ω)	L (μH)
#1	1.0	BT	30	57	3.3
#2	1.8	PZT	10	41	3.3

Fig. 2 shows the schematics of the lumped lines constructed, which is a LC network. In this case, C is represented by a linear capacitor since BT and PZT dielectrics have weak nonlinearity. As input HV source, the fast pulse generator, model NSG5500 from Teseq Company, was used. Generally, it is employed in electromagnetic compatibility (EMC) and electromagnetic interference (EMI) tests and, according with the maker datasheet, the generator pulse can be produced with the following specifications: pulse rise time (t_{ri}) $\leq$ 5 ns, pulse width (t_w) $\leq$ 50 ns and maximum amplitude (V_{max}) = 5 kV. The experimental arrangement using the fast pulse generator and transmission line is presented in Fig.3.

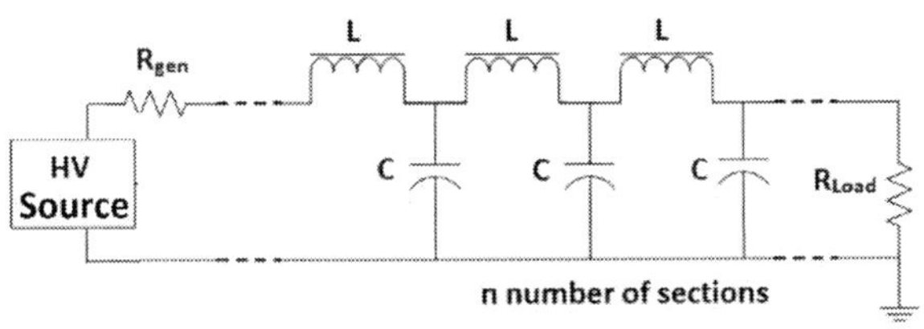

Fig. 2. The circuit used to test the lumped LC transmission line.

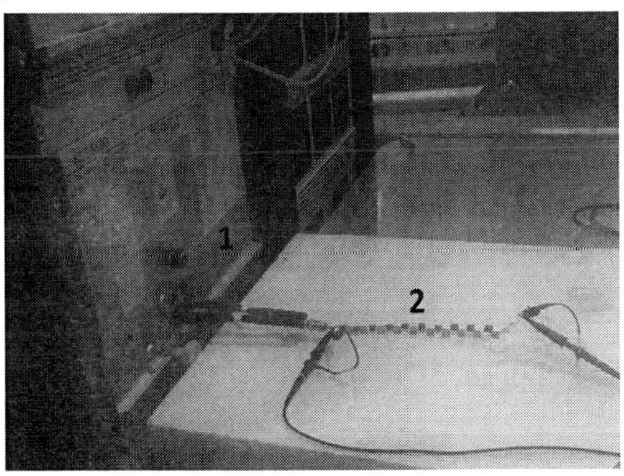

1- Fast Pulse Generator 2- TL

Fig. 3. View of the experimental set-up used showing the PZT line.

To simulate the transmission line, a LT-Spice circuit simulator software was used. In the Fig. 4 it is shown the electrical circuit used to simulate the TL, where R_L and R_C represent the losses in inductor and capacitor, respectively. In Spice simulation, it was assumed constant values for L as well for C for both lines per Tab. I because of the use of the dielectrics with weak nonlinearity including inductance losses, simulation results considered R_L=0.01 Ω for both lines. However, with dielectric losses the assumed R_C was 0.1 Ω for line #2 while R_C was 1.0 Ω for line #1 because of the higher losses of BT compared to PZT. The simulator approximates the Gaussian-type waveform shape of the input pulse generator by a trapezoidal function.

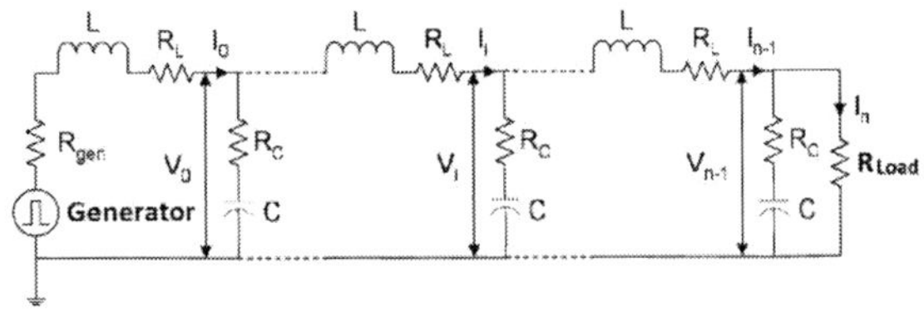

Fig. 4. The Spice circuit model for a linear transmission line.

III. Results

The first results presented are for the transmission line #1 (see Tab. I) with a matched load resistance of about 60 Ω as given by the input and output pulses on same plot in Fig. 5. Observe the pulse width enlargement and the lower peak amplitude, caused by the line dispersion effect and BT dielectric losses, respectively. In Fig. 6, it is observed the input pulse on a reduced x-scale to show the fast pulse rise time obtained around 20 ns. In Fig. 7, a zoom also is given on the experimental output pulse for comparison with simulation result, where one can note some noise produced at the beginning of the pulse in the experimental result. Despite this, one can see that experimental and simulation results are in a good agreement, confirming the lack of oscillations at output due to the BT dielectric losses as demonstrated by the line test.

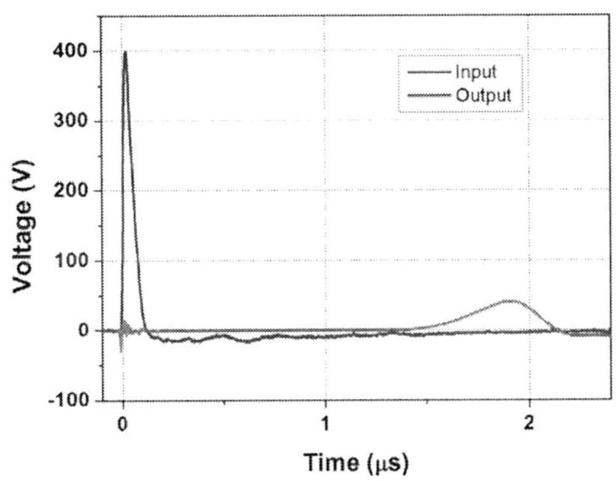

Fig. 5. Output and input pulses obtained for the BT line.

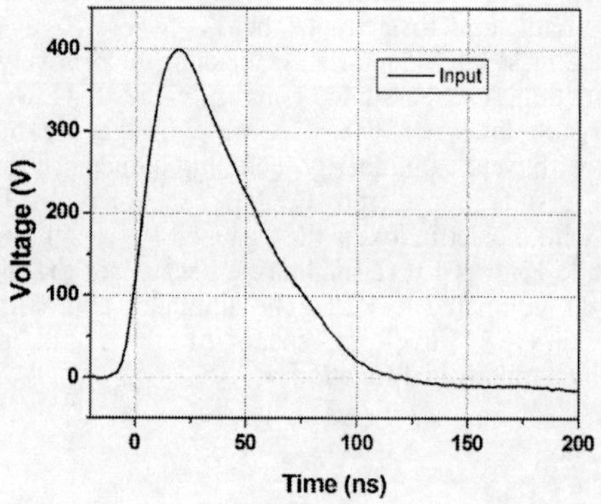

Fig. 6. The input pulse applied into the line #1 shown on a shorter scale.

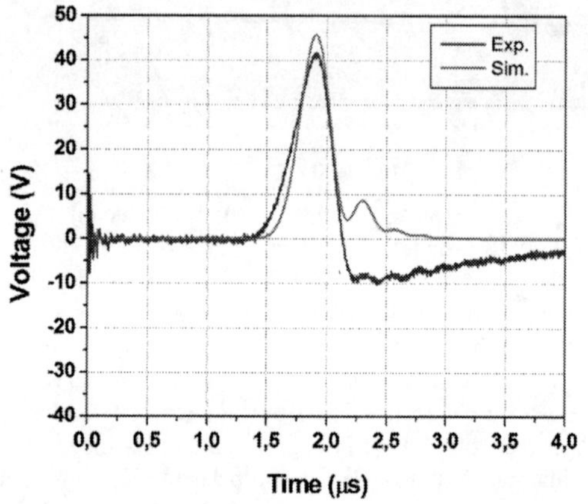

Fig. 7. Output pulses obtained for line #1 shown on a shorter scale.

With the test of line #2 terminated by a matched load (40 Ω), Fig. 8 shows again the input and output pulse on expanded scale and Fig. 9 shows the corresponding applied input pulse shown in a shorter time interval with rise time (t_{ri}) of about 20 ns. As in the previous case, one can note in Fig. 8 the pulse enlargement and the peak amplitude decrease as result of the line dispersion and dielectric losses, respectively. However, now oscillations are seen on the tail of the output pulse in Fig. 8 as the PZT lower dielectric losses are lower. As shown in Fig. 10, by the output pulse (red line) on shorter scale RF oscillations were generated at a frequency of the order of 2.0 MHz which corresponds to the angular frequency $\omega=1/\sqrt{(LC)}$. This also confirmed by a Spice simulation curve in black with a good agreement with the experimental result and by the fast Fourier transform (FFT) spectrum of the output pulse shown in Fig.11. It is clearly seen in the FFT that the plateau is at 2 MHz with an abrupt drop at 4 MHz, which corresponds to the line cutoff frequency. The RF generated on this line is due to the filtering process, where the low loss of the dielectric and the fast t_{ri} has contributed to the RF generation observed. As seen in Fig. 10 by the red line, the RF generated has seven noticeable cycles approximately, where the first cycle has a maximum voltage peak of about 50 V.

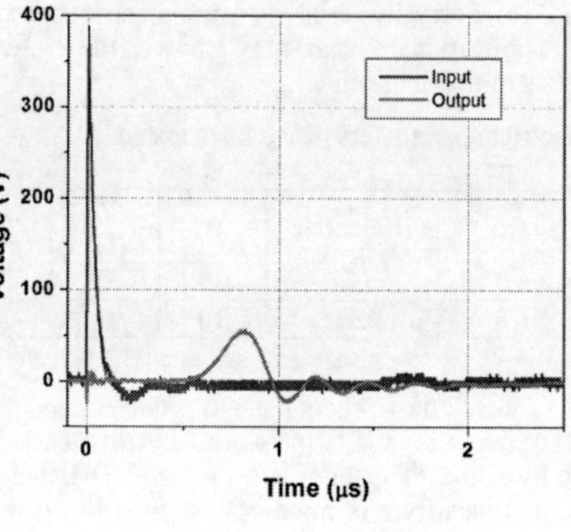

Fig.8 Input and output pulses obtained for the PZT line.

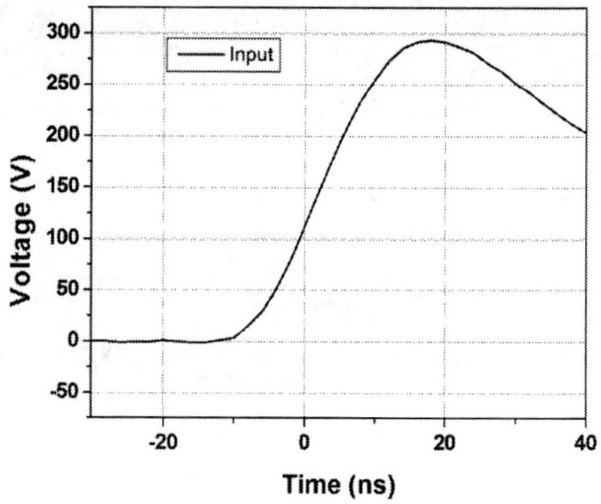

Fig. 9. The input pulse applied into line #2 shown on a shorter scale.

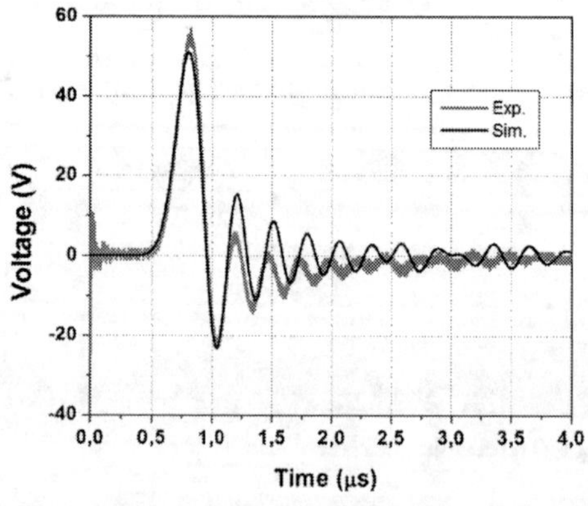

Fig. 10. Output pulses on a shorter scale for line #2 showing RF oscillations.

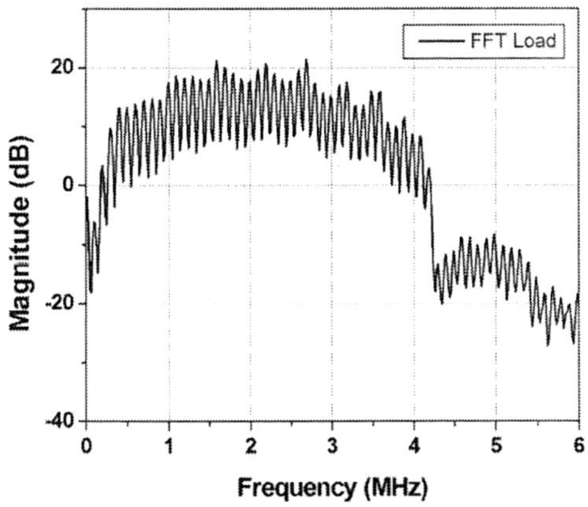

Fig. 11. Measured FFT spectrum for output pulse on line #2.

As pulsed RF is normally used in radars, the PZT NLTL was tested working with the NSG5500 pulse generator at a repetition rate of about 100 kHz as shown in Fig. 12. The output pulse is shown in an expanded time scale up to 40 μs. This result shows that de-poled PZTs of low loss factor and weak nonlinearity are suitable for RF generation using transmission line at high repetition rates.

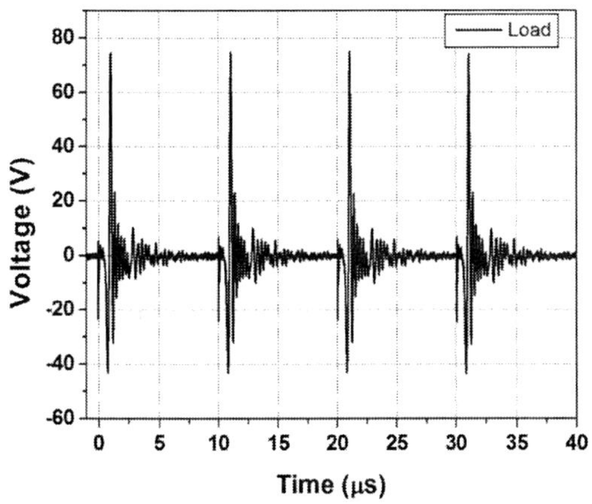

Fig. 12. The sequence of output pulses obtained with line #2 at 100 kHz.

IV. Conclusion

In this work, two transmission lines were built using linear inductors and ceramic capacitors with dielectrics (BT and de-poled PZT) of weak nonlinearity. By applying a 100-ns short pulse with voltage peak of 400 V and rise time of about 20 ns into both lines, it was demonstrated that RF oscillations are generated only in the PZT line of lower dielectric loss. As shown by the simulated and experimental results dielectric loss is an important parameter for RF generation in transmission lines using almost linear dielectrics of weak nonlinearity. The results also show that RF oscillations in this case are generated due to the line dispersion effect as shown by the broadening of the output pulse. However, maximum pulse peak at output is reduced considerably to less than 50/60 V due to line losses in both cases (BT and PZT lines). In short, this work showed an alternative way of generating low power RF using a linear TL built with an almost linear dielectric of low loss factor (< 0.01) and such as the ceramic de-poled PZT.

References

[1] L. P. Silva Neto, J. O. Rossi, and J. J. Barroso. High power RF generation in nonlinear lumped transmission lines using commercial ceramic capacitors as nonlinear elements. In Proc. *of the IEEE International Pulsed Power Conference*, 2015, pp. 1-5.

[2] N. S. Kuek, A.C. Liew, E. Schamiloglu, and J. O. Rossi. Pulsed RF oscillations on a nonlinear capacitive transmission line. *IEEE Transaction on Dielectric and Electrical Insulation*, vol. 20, n. 4, p. 1129-1135, 2013.

[3] P. W. Smith, "Pulsed, high Power, RF generation from nonlinear dielectric ladder networks- performance limits," in Proc. *of the 18th International Pulsed Power Conference*, 2011, pp. 167-172.

[4] J. D. Darling and P. W. Smith "High power pulse burst generation by soliton-type oscillation on nonlinear lumped element transmission lines." in: Proc. of the IEEE *International Pulsed Power Conference*, 2009, pp. 119-123.

[5] L. P. Silva Neto, J. O. Rossi, and J.J. Barroso, A. R. Silva Junior, "Characterization of ceramic dielectrics for Sub-GHz applications in nonlinear transmission lines," *IEEE Transaction on Plasma Science*, vol. 42, n. 10, pp. 3274-3282, 2014.

[6] L. P. Silva Neto and J. O. Rossi, "PZT dielectric ceramic characterization for application in nonlinear transmission lines," Advanced Materials Research, vol. 975, pp. 23-28, 2014.

[7] D. M. French, B. W. Hoff, S. Heidger, and D. Shiffler "Dielectric nonlinear transmission line." in Proc. *of the 18th International Pulsed Power Conference*, 2011 pp. 1-4.

Preliminary Data Analysis and Calculations from PHELIX Crenulation-1 at pRad Experiment

C. L. Rousculp, D. M. Oro, J. T. Bradley, W. A. Reass, A. R. Patten, J. R. Griego, F. Fierro, R. E. Reinovsky, P. J. Turchi, A. Saunders, F. G. Mariam, B. J. Hollander, M. S. Freeman, L. G. Margolin, B. Cheng
Los Alamos National Laboratory
Los Alamos, USA
rousculp@lanl.gov

***Abstract*— The first PHELIX driven Crenulation experiment was fielded at the LANL LANSCE proton radiography (pRad) facility in Dec 2015. This was the first time a thin-walled tin cylinder with single-mode perturbations was shocked to melt-on-release (P > 35 kbar) in converging geometry. The perturbation inversion and growth due to the Richtmyer-Meshkov Instability (RMI) is diagnosed in a 21-image pRad movie. The motivation, design, preliminary data and calculations will be discussed.**

I. Introduction

The RMI instability is of fundamental interest in hydrodynamics and material science. It occurs when a shock wave passes over a crenulated interface between two materials. The case under consideration here is when the shock passes from the more dense material to the less dense material. Figure 1 shows a schematic of the pre and post state of a material with the shock wave incident from the left. Here the less dense material is vacuum. As the shock passes the interface the crenulations invert. The post shock state is characterized by bubbles that carve out material that feed into spikes that stretch into the vacuum. In a solid metal, RMI evolution is greatly determined by the state of the material. When in the liquid state, RMI growth is limited only by surface tension, while in the solid state it is limited by material strength.

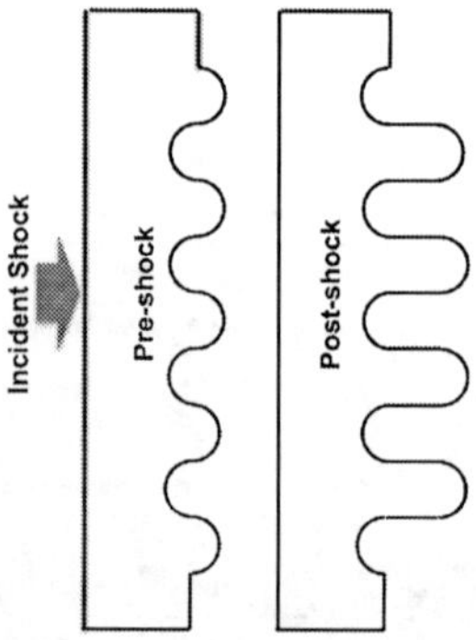

Fig. 1. Schematic of the RMI.

The most sophisticated model of the material state in a continuum hydro-code considers multiple solid phases as well as the liquid phase. Typically a hydro-code will utilize a tabular equation-of-state (EOS) and melt curve. In addition, strength parameters will be different for each of the solid phases. For our experiments, a new tabular EOS has recently been made available. By including beta and gamma solid phase, the shock release state has new limits. For release to a pure liquid state, $P_{shock} > 35$ GPa. For release to pure solid state, $P_{shock} < 20$ GPa. This means that for shock strengths between these limits, the material is in a mixed solid/liquid state [1].

To study the various release states of tin in a controlled manner, a magnetic-drive, Z-pinch, pulsed-power system is used. With a suitably designed liner-on-target system, the pulsed power system has the advantage that the drive (and shock pressure) can be continuously varied to span the range of pressures of interest.

Fig. 2. (a) The PHELIX portable pulsed power system at the LANL pRad laboratory. (b) The PHELIX current step-up transformer. (c) A typical load cassette that is mounted in the middle of the transformer.

The Precision High Energy-density Liner Implosion Experiment (PHELIX) portable capacitor bank has been interfaced with the LANL proton radiography (pRad) laboratory. pRad provides up to twenty-one axial images of a liner-on-target experiment. The images have an inter frame timing as low as 20 ns within a 250 ms experimental time window. PHELIX stores up to 300 kJ, has a peak current of 5 MA with a rise time of 4 us. This allows practical liners between 2 and 6 cm radius with thickness between 0.5 and 2 mm.

The fields produced are up to 0.5 MG for the smallest liners. PHELIX can accelerate these liners between 1.0 and 3.5 km/s. The unique feature of PHELIX is its air-core, toroidal, current step-up transformer. This allows a footprint of less that 20 m^2 [2].

II. The Crenulation Experimental Series

In order to study the evolution of the RMI a load cassette has been designed. Figure 3 shows a cutaway view of the Al return conductor, Al liner (OD = 3.0 cm, dr = 0.6 mm), Al glide planes, a vacuum cap, and tin target (OD = 1.0 cm, dr = 1.0 mm). On the inside of the tin target, three sectors of single-mode, sinusoidal perturbation have been machined. The linear wavelength, λ = 1.1 mm and the amplitude, A, of each sector is such that kA = 0.1, 0.2, 0.3, where $k = 2\pi/\lambda$. Each sector has 2.5 wavelengths of perturbation with smooth ($R_a \sim 1$ μm) between.

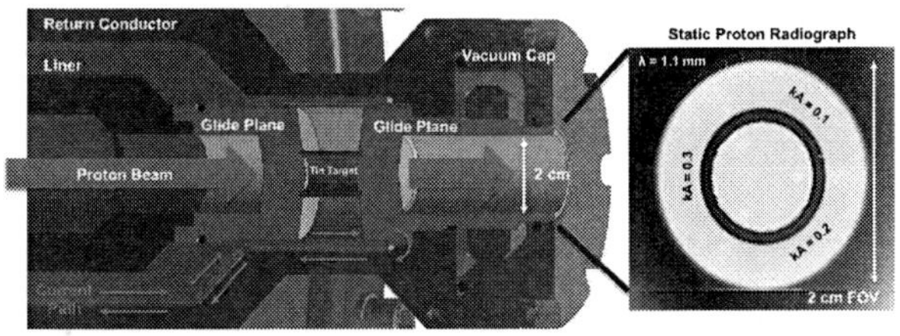

Fig. 3. The Crenulation load cassette and the static proton radiograph.

It was found that by varying the charge voltage on the PHELIX bank between 30 and 80 kV in the tin cylinder could be shocked into the three release states (solid, mixed, liquid) separate experiments. The first experiment, described in this paper, examines the release to the pure liquid state. Figure 4 shows a 1D MHD calculation of the liner, target, return conductor. It also shows a phase-space plot of the Lagrangian cells in the tin target indicating the shock above the 35 GPa threshold for release to pure liquid.

The main technique for the study of RMI growth and validation of models has been single-mode sinusoidal profile of a planer tin surface. If conservation of cylindrical fluid mass is assumed then the amplitude at early time grows linearly as $A(t) \sim A_0(1+vt/R_0)$. Here, R_0 is the initial cylinder radius; v is the surface's radial velocity. Note that it is independent of the mode number of the perturbation [3].

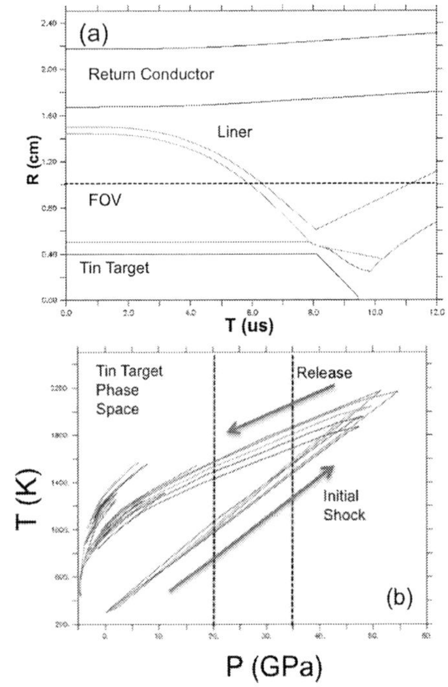

Fig. 4. 1D MHD design calculations. (a) Radius vs time plot of liner, target, RC. The 1 cm dashed line indicates the field of view (FOV) prad. (b) Phase space plot of selected Lagrangian cells in the tin target. All are above the 35 GPa threshold for release to pure fluid.

III. Experimental Results

The 1-D MHD calculations include a coupled circuit model. Shown in Figure 5 is a two-loop system with transformer coupling. The constant values are V_0 = 70 kV, C = 68 μF, R_0 = 29 mΩ, L_0 = 40 nH, L_p = 390 nH, L_s = 24.3 nH, L_1 = 4 nH, R_1 = 0.1 mΩ. The mutual inductance between the windings of the transformer is $M^2 = k^2L_pL_s$, where k ~ 0.93.

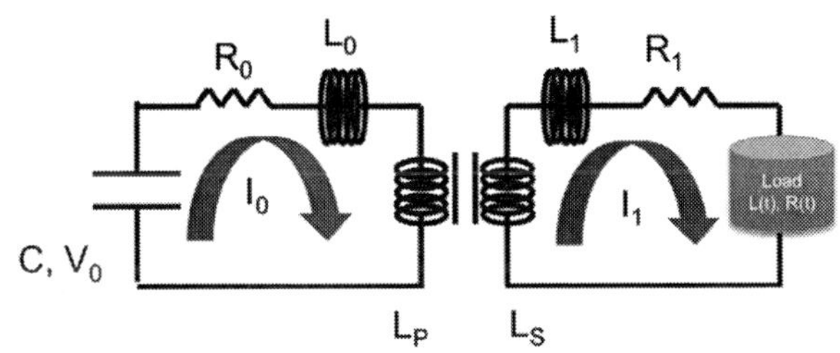

Fig. 5. PHELIX circuit model used in 1D MHD calculation.

Shown in Figure 6 is the load current from both the above mentioned 1D MHD calculations as well as the measured current from a Faraday rotation diagnostic. There is very good agreement during the current rise (0 < t < 4 μs), but the calculation under predicts during the current fall (4 < t < 12 μs). At t = 9 μs, both show a kink due to the liner impacting the target, which causes a sudden change in the inductance in the load cassette.

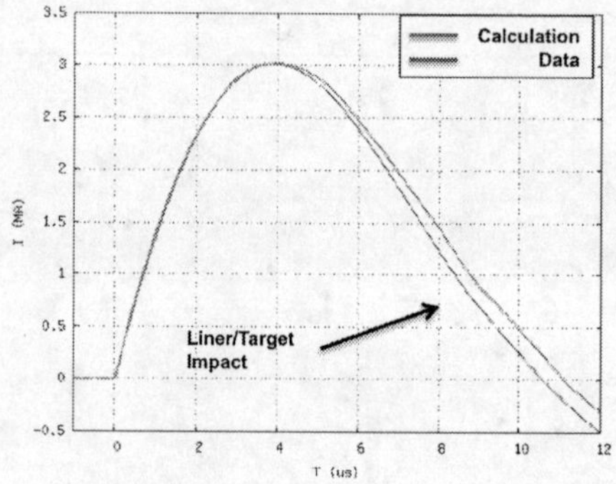

Fig. 6. Load current versus time comparing data and calculation. The kink at ~ 9 μs indicates liner-target impact.

The twenty-one frames of pRad data are shown in Figure 7 The tin target and single-mode perturbations are clearly visible. The three light circles within the target cylinder are evacuation holes. The liner first enters the FOV at T = 1200 ns. Impact with the target occurs at T = 2400 ns. Given the 3.5 mm travel, this gives a liner velocity is ~3 km/s.

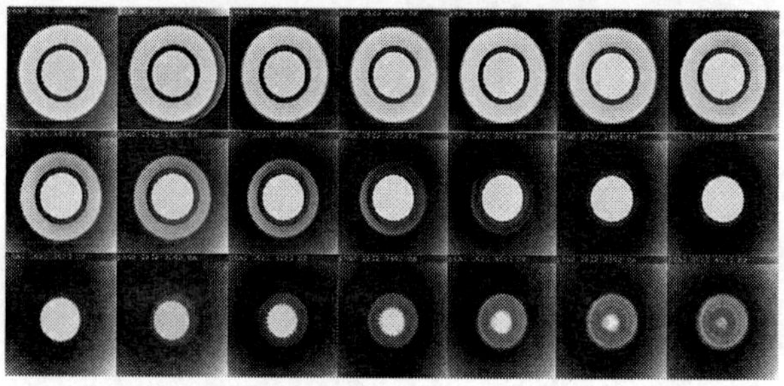

Fig. 7. Twenty-one pRad axial images of the Crenulation-1 experiment. Time runs left-to-right, top-to-bottom.

The RMI growth is seen in frames 16-20 (T = 3000-3800 ns). By fitting a circle with Gaussian indentations to the inner surface of the tin for each frame, an amplitude growth rate can be inferred. The initial IR of the target is used as a fiducial for conversion of pixels to mm. Shown in Figure 8 are images fitted with an interface contour and a plot of the growth rates. Linear growth for all kA values predicted by theory is seen.

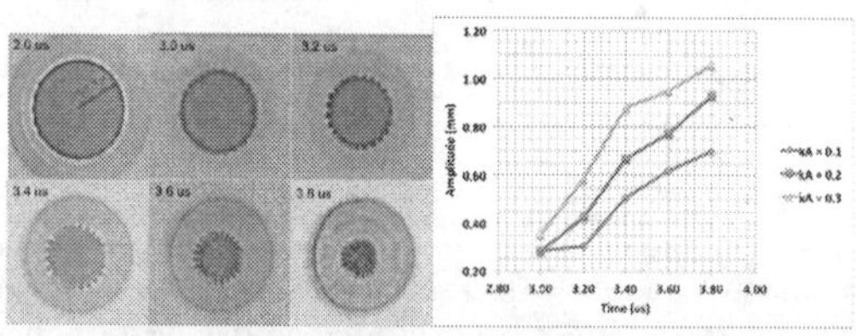

Fig. 8. (Top) Five frames of RMI growth on the IR of the tin target. The black line shows a fit circular function with Gaussian indentations. The initial inner radius of the target (T = 2.0 us) is 75 pixels = 4.0 mm. (Bottom) Amplitude vs time for the different kA.

Finally, 2D calculations have been undertaken with an ALE hydro-code code. Figure 9 shows the evolution of the kA = 0.3. Note calculation t = 0 corresponds to experimental t = 5.2 μs. On the right, calculation growth rate compares very well to the experimental values. Also of note is the low-density cavitation in the bulk of the target (blue regions) that is also seen as density structure in the pRad data.

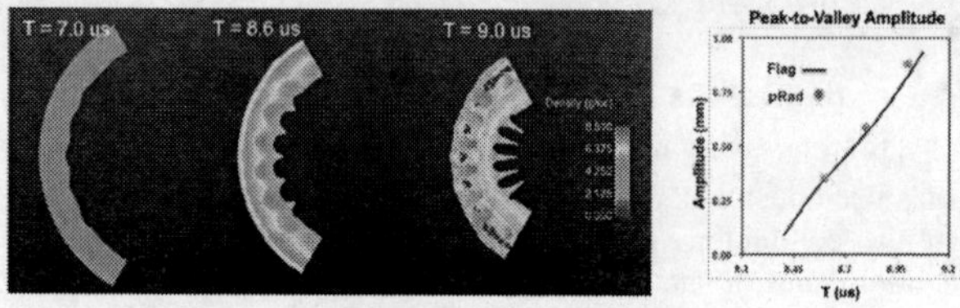

Fig. 9. Hydro code calculation of the kA = 0.3 sector of the tin target with a comparison of the peak-to-valley amplitude to data.

IV. CONCLUSIONS

The Fastest PHELIX driven cylindrical liner to date (v ~ 3 km/s) has been performed. This is three times faster than previous experiments. The thinner (~0.6 mm), smaller diameter (~3.0 cm) liner achieved higher performance with lower capacitor charge voltage. The liner velocity was sufficient to shock a cylindrical tin target to P > 35 GPA. This is the new, higher threshold for pure melt on release. This is the first measurement of RMI spike and bubble growth in converging geometry into an evacuated center from initial single-mode perturbations of three different wavenumber/amplitude products (kA = 0.1, 0.2, 0.3) in tin. Proton radiography image analysis shows peak-to-valley growth rate is linear in time and agrees well with 2D calculations. This supplements previous planer RMI data in tin at similar wavenumber/amplitude product. The crenulation series of experiments will take advantage of PHELIX's continuously variable drive to span the shocked release states of tin (e.g. pure solid, mixed solid/liquid, pure liquid) into vacuum and gas. The next experiment in the crenulation series will have an identical liner and tin target, but will reduce the drive in order to produce a shock to a mixed state (20 < P < 35 GPa) on release.

V. REFERENCES

[1] C. Greef, E. Chisolm, D. George, "SESAME 2161: An Explicit Multiphase Equation of State for Tin," LA-UR-05-9414.

[2] C. L. Rousculp et al., The PHELIX Liner Demonstration Experiment (PLD-1), Proceedings of the 2011 IEEE-PPC, 1282, 2011.

[3] L. G. Margolin and M. J. Andrews, "Models for Crenulation of a Converging Shell," Transactions of the ASME, **136**, 084501-4, 2014.

Estimation Of Warm Dense Matter Formation Stability at Electro-Explosion of Cylindrical Foils on High-Power Electrophsycal Facilities

S. D. Kuznetsov
FSUE RFNC-VNIIEF – Federal Nuclear Center – All-Russian Research Institute of Experimental Physics, Sarov, Russia

***Abstract* — It has been proposed in the papers [1-2] to use high-power electrophysical facilities and generators for obtaining warm dense matter (WDM) at electro-explosions of thin foils constricted between two layers of insulator. This method has already been used, e.g., in [3] at lower energies in a pulsed power system for the study of material properties in the temperature range of around several eV and densities of the order of magnitude of 0.001…0.1 of the solid state one. At that, a question occurs on the spatial homogeneity of the parameters of WDM formed at the foil electro-explosion. The increments of small perturbation growth were calculated that appear due to the material non-homogeneity and technological peculiarities with account for conductivity dependence on the foil temperature and density. Calculations were made for the materials to be used in the experiment. Conditions were obtained at which the discharge instability growth is within the WDM parameters ranges suitable for measuring.**

Keywords—component; formatting; style; styling; insert (key words)

I. Introduction

Systems were numerically studied in the papers [1]-[2] for WDM generation in the process of the cylindrical foil explosion with different power sources. The foil is placed between two layers of insulator, which tamper the explosion. These systems could give good homogeneity of explosion by the foil cross-section and better opportunities for produced plasma diagnostics. The same method of plasma generation is already used for plain foils with the less powerful pulsed power system [3].

It is very important to have uniformity of the foil electrical explosion for better diagnostics with account of initial perturbations due to foil fabrication. The uniformity of the electrical explosion was studied experimentally, for example, in paper [4] with the energy deposited in foil 2-5 kJ/g, which corresponds to the region of the foil melting and close to the electrical explosion. Growth of initial perturbations of temperature nonuniformities could create local overheating and explosion of different foil sections at different time in case of material properties dependence on temperature.

The magnetic Rayleigh-Taylor instability (MRT) plays the leading role in the development of small initial perturbations in the case of cylindrical liners [5], [6], which is connected with the acceleration of the massive fluid by the magnetic field. The increment of the MRT instability can be estimated with the magnetic pressure at the surface of thin liners as

$$\gamma = \sqrt{kg}, \qquad (1)$$

here $g = P_{mag} / \rho d$, $P_{mag} = B^2 / 8\pi$ -the magnetic pressure, ρ , d - the specific density and thickness of a liner.

The MRT-instability in the systems proposed in the papers [1]-[2] starts to develop at the inner surface of the foil, which is placed under the layer of the insulator and also separated from the central current carrying road by the insulating layer. Nevertheless the estimate (1) gives overestimated value of the instability increment as it corresponds to a free liner accelerated by the magnetic pressure at one side and zero pressure at another. It is also essential to take into account the layers of insulator hindering free change of the foil boundaries in the case of the system shown at the fig.1. It is necessary to take into consideration presence of the mater besides the foil boundary defined by the Atwood number of foil and insulator and also to account for the insulator strength, which can be more important for suppression of the short-wave perturbation growth. These perturbations appear on the outer side of the foil with the certain time lag (especially for exploding foils) that is connected with the involvement of all foil mass in the movement. The wave vectors for the perturbations of the outer side with the maximal growth rate are of the order of magnitude $k \sim 1/d$ for the estimate of the MRT-instability according to (1).

The electro-thermal (ET) instability could play the leading role in the systems proposed in [1]. This instability is connected with non uniform heating of the foil material by electric current. Comparison is made below of the electro-thermal instability growth rate with the MRT-instability growth rate according to estimate (1). It is shown that the growth rate of these instabilities are comparable that could lead to the mutual interference of these instabilities. Influence of electro-thermal instability on the foil heating was studied. The problem setting for this study is similar to the approach used in the paper [7] for wire explosion in the vacuum.

II. Physical Scheme of The Load and The Problem Setup for The Growth Rates Calculations of The Exploding Foil.

The physical scheme of the load was described in details in the paper [1] and in the fig.1a. The cross section of the load is presented in the fig.1b.

a) b)

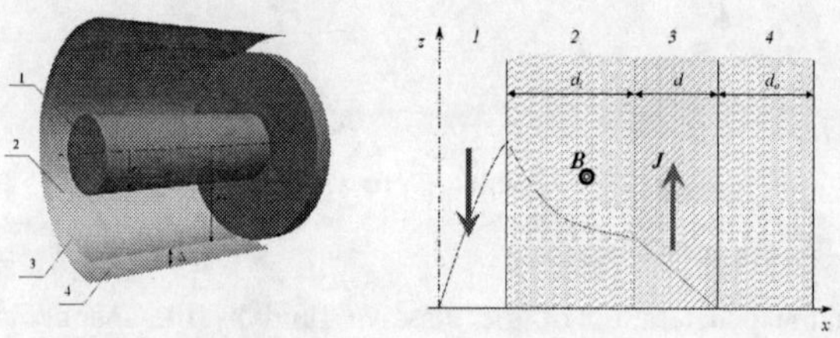

Figure 1 - a) the physical scheme of the load: 1-the central rod; 2-the internal insulator; 3-the foil;4- the external insulator; b) – the currents and the magnetic field distribution in the system cross section.

The computational region in fig. 1b can be divided on the several sub-regions: 1-the central rod; 2-the internal insulator; 3-the foil; 4-the external insulator. The current flowing through the central rod creates the magnetic field in the layers that are sketched in the fig. 1b. The magnetic field in the insulating layer between the central rod and the foil is decreasing by the law ~$1/r$.Inside the foil, which is thinner than the skin layer, the electric current is evenly distributed by the foil cross-section. The magnetic field decreases linearly from the value on the internal surface to zero field at the external surface in assumption that the foil thickness is much smaller than the foil radius.

The small foil thickness and study of perturbations with the large wave numbers enables us to consider the problem in the plane geometry. It is convenient to use the linear current density the current density per unit of the foil circuit length. The relation for undisturbed current with the electric field intensity can be written as

$$E_{0z} = \eta I / 2\pi Rd = \eta_s J_z \,. \qquad (2)$$

The current density per unit length and corresponding resistance are introduced in expression (2).

В формуле (2) введены ток на единицу длины фольги и соответствующее удельное сопротивление для плоской фольги

$$\eta_s = \eta / d \,,\ J_z = I / 2\pi R \,. \qquad (3)$$

The magnetic field is connected with the current by the following equations

$$\nabla \times \mathbf{E} = -\frac{\partial \mathbf{B}}{\partial t},\ \mathbf{E} = \eta_s \mathbf{J} \,. \qquad (4)$$

The magnetic field variation outside the foil material can found through the potential function Ψ

$$\mathbf{\delta B} = -\nabla \Psi,\ \ \Delta \Psi = 0 \,. \qquad (5)$$

The current variation can be found with help of the equation (4)-(5).

$$\delta J_z = -\frac{k_y^2}{k^2}\frac{\delta \eta_s}{\eta_s} J_z \frac{1}{1+\dfrac{\gamma}{\gamma_0}},\ \gamma_0 = 2\eta k / d\mu \qquad (6)$$

γ_0 has the meaning of the characteristic time of the current perturbation smoothing for the perturbation with the wave vector k in the media with the constant conductivity. The estimate gives for the wavelength ~d, which is equal to the foil thickness ~60 μm, in the state with the initial conductivity

$$\gamma_0(\lambda = d) = 2\eta / d^2 \mu \sim 5.59 \cdot 10^6 \text{ c}^{-1}. \qquad (7)$$

The equation of the heat balance for the incompressible foil material is

$$\rho c_v d \frac{\partial T}{\partial t} = \eta_s J_z^2 + \nabla(\lambda d \nabla T) \,. \qquad (8)$$

The solution of this equation for T with the zero order solution and the variation is

$$T(t,\mathbf{r}) = T_0(\mathrm{t},\mathbf{r}) + \delta T(t,\mathbf{r})\,,\ \delta T(t,\mathbf{r}) \sim \exp(\gamma\, \mathrm{t} + \mathrm{i}\,\mathbf{kr})\,. (9)$$

If it is assumed that the variation of conductivity depends on the temperature only then for the growth rate γ in (9) is

$$\gamma = \frac{J_z^2}{\rho c_v d}\frac{\partial \eta_s}{\partial T}\left(1 - 2\frac{k_y^2}{k^2}\frac{1}{1+\dfrac{\gamma}{\gamma_0}}\right) - \chi k^2 \,. \qquad (10)$$

The region of the wave vectors for which the instability can develop is defined from (10) with $\gamma = 0$:

$$A\left(\frac{k_z^2 - k_y^2}{k^2}\right) < k^2,\ \ A = \frac{J_z^2}{\rho c_v d^2}\frac{\partial \eta}{\partial T}\frac{1}{\chi} \,. \qquad (11)$$

Equation (11) is in presentation symmetrical by k_z, k_y to show the existing symmetry of the solution in the wave space. The instabilities can develop in certain directions with all signs of the specific resistance derivative on temperature. The case $A>0$ or $\partial \eta / \partial T > 0$ is for the ordinary metals heating from the room temperature. This mode of instability develops perpendicular to the axis of symmetry and called striation. The case $A<0$ or $\partial \eta / \partial T < 0$ corresponds to the metal plasma existence region. This mode of instability develops parallel to the axis of symmetry and called filamentation. The heat conduction along the foil suppresses development of instabilities for the short wavelength with wave numbers $k \geq \sqrt{|A|}$.

The maximal value of the growth rate is for the perturbation developing orthogonal to the axis of symmetry at the initial stage of the electric explosion with the temperature close to the initial for which $\partial \eta / \partial T > 0$

$$\gamma = \frac{J_z^2}{\rho c_v d^2}\frac{\partial \eta}{\partial T} - \chi k^2 \,. \qquad (12)$$

The upper estimate for the growth rate in (12) can be obtained without the last term or the absence of the heat conduction. The growth rate of the electro-thermal instability without the heat conduction does not depend on the wave number. The uniform growth of the initial perturbation profile takes place.

The numerical estimate of critical wavelength for metals with the known specific resistance can be made with help of the Wiedemann-Franz law $\chi\eta = LT$, here L=2.45·10^{-8} Вт·Ом/К^2.

The critical wavelength is

$$\frac{1}{\lambda(\gamma=0)} = k(\gamma=0) \geq \sqrt{\frac{J_z^2}{L\rho c_v d^2}\frac{\eta\partial\eta}{T\partial T}}. \qquad (13)$$

III. Growth Rate Calculations for the Electro-Thermal Instability in Some Experimental Setups

The numerical estimate according to equation (12) can be made for several systems investigated before to choose more robust in respect to the growth of initial perturbations. The integral growth rate is expressed as

$$\exp(\Gamma);\ \Gamma = \int \gamma dt \qquad (14)$$

The more intensive instability growth could be expected in the systems with higher current density per unit length of the foil circumference due to the quadratic dependence of γ in (12) from J. But parameters of heating such as temperature and pressure are also proportional to the current action and consequently are also have the quadratic dependence. That is, the relative growth of the variation of unperturbed solution could be small in comparison with the unperturbed solution. The integral growth of instabilities is the same for different systems by the amplitude and the time of current growth developed to achieve relatively close parameters of the plasma.

Let investigate in more details growth of perturbation for PHELIX [9] high current installation with the current transformer. This installation can produce in the load current with the maximal value I~5 MA and the character growth time 5 μs. Computations were also made with the current reduced to ~0.7 I. The initial radius of the copper or aluminum foil is 1.5 cm and initial thickness from 62.5 to 250 μm. Thickness of the external insulator with specific density ~1 g/cm^3 in these calculations was 2 cm. Characteristic parameters on time are shown in fig. 2

The growth rate of the initial perturbations was calculated with help of the UP-code with the azimuthal magnetic field [8]. These calculations were made in two approaches.

In the first approach all values necessary to make calculations by (12) were taken from the foil central region. The central region of the foil was chosen because the results of calculation were obtained for the foil more thin than the skin-layer depth and explosion is uniform by the foil thickness. The estimate obtained with this method is certainly applicable at the initial stage of foil expansion when the work made by the foil on the insulators is negligible.

The current derivative on temperature in (12) can be calculated using the derivatives of temperature and specific resistance on time. The integral growth rate $\Gamma = \int \gamma dt$ on time for k_y=0 и χk^2=0 is shown in fig. 3 (curve 3).

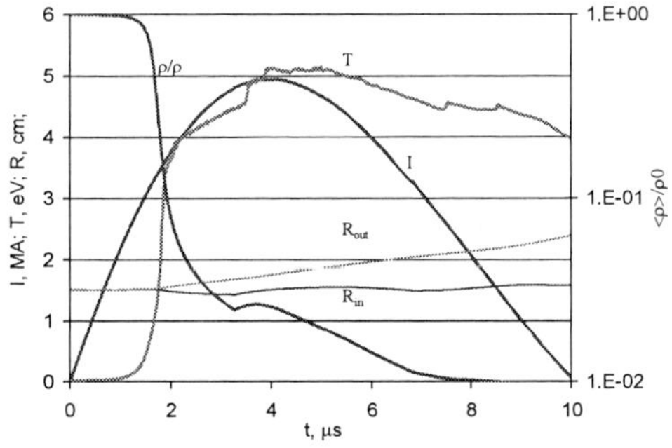

Figure 2 - The current for PHELIX installation with the maximal value 5 MA and the character growth time 5 μs, the external and internal radii of the foil, temperature and density (in the units of initial) of the copper plasma.

In the second approach the growth rate was calculated by variation of the foil thickness in the direct calculations. The growth rate of small perturbations is obtained in the limit of the infinitesimally small initial variation of the thickness. Lets investigate $\delta T(t) = T(t) - T_0(t)$ from (9), here $T(t)$ and $T_0(t)$ are the dependences of the temperature in the nominal computation and the varied initial parameter computation. $\delta T(t)$ from (9) is known for certain time t, then at $t+\Delta t$ $\delta T(t+\Delta t) \sim \delta T(t)\exp(\gamma(t)\Delta t)$. It gives for a finite time interval $\delta T(t) \sim \delta T(t_0)\exp(\int_{t_0}^{t}\gamma(t)\,dt)$. Time dependence is calculated for $e^{\Gamma} - 1 = \exp(\int_{t_0}^{t}\gamma(t)\,dt) - 1 \sim \delta T(t)/\delta T(t_0) - 1$. As the Γ integral buildup from (14) is small at the small times, choice of the initial moment t_0 has small influence on the maximal value of the integral Γ.

The integral of the growth rate Γ from (14) is shown as curve 1 in fig. 3. The curves for Γ calculated in these two approaches used are close enough. The curve 3 that define the upper limit of wavelength of perturbations above which the perturbations grow with account for heat conduction is also shown in fig. 3. It could be seen that the short wave instabilities with the wavelength less than $1/d$ are only effectively suppressed. The MRT-instability growth rate estimate according to (1) is also shown in the fig.3 (curve 4) with the character wavelength equals to the doubled foil thickness. It could be seen from comparison of the increments of the MRT and ET instabilities in fig. 3 that the ET instability growth rate is higher than the growth rate for the more critical modes of the MRT instability.

It is more essential for practical design of the systems to obtain relative values of temperature change for small perturbation in the initial conditions. This relative change can be characterized by the following expression

$$\delta T d_n / T_n \delta d|_{\delta d\to 0} \sim |\ln(T/T_n)/\ln(d/d_n)|_{d\to d_n}.$$

Some curves illustrating this relative change without account for smoothing due to the heat conduction are presented in the fig. 4 for different materials and currents used in calculations. Curves corresponding to different materials and different currents have substantial difference in the times of the maximum growth achievement nevertheless maximal values are close to each other in all computations.

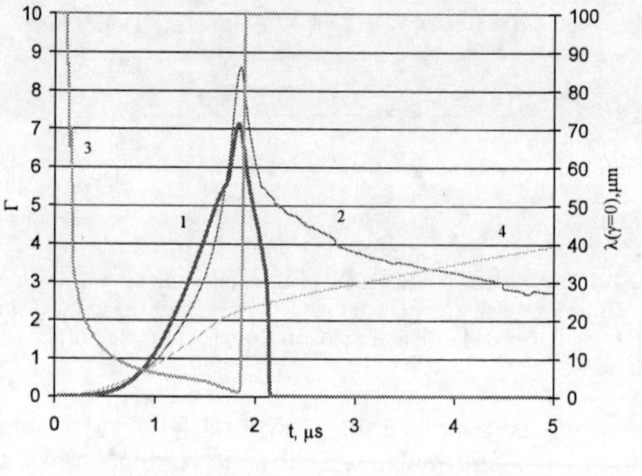

Figure 3 - Time dependences for the copper foil 62.5 μм thick: 1) $\int \gamma dt$ from (12); 2) from computations with the foil thickness variation, 3) the vawe number limiting the instabilities due to the heat conduction, 4) the integral growth rate for the MRT instability with the wavelenght by the formula (1).

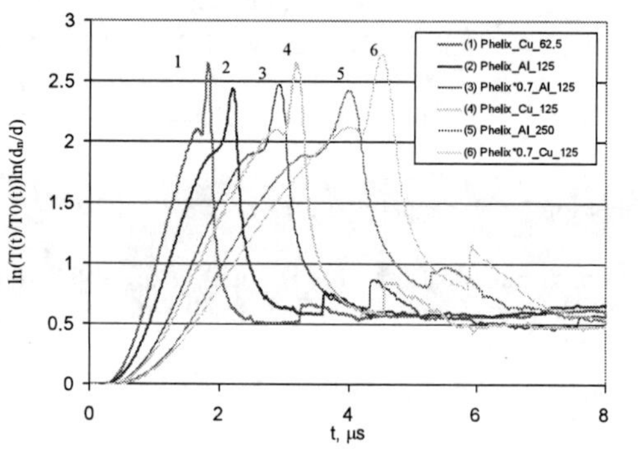

Figure 4 - The logarithmic ratio of the relative temperature change to the nominal temperature from computations for the Al and Cu foils of different thickness with the Phelix current waveform with the maximal values 5 и 3.5 MA.

Relatively close parameters of the metal plasma after the electric explosion and approximately the same growth rate are obtained in spite of difference in the currents. The total growth of the foil temperature for the small initial thickness perturbations is ~12-15$|\ln d/d_n|$, here $|\ln d/d_n|$ is the characteristic of the initial foil uniformity. This value characterizes the required quality of the foil fabrication. The uniformity of the foil thickness, density and conductivity should be on the level of several percents to avoid overheating of some regions which is technically achievable. In this case it is possible to obtain the variation of the temperature along the foil on the level of average temperature. With the uniformity of the foil parameters below per mille it is possible to design system with negligible variations of the temperature along the foil. After the electrical explosion or achievement of the maximal growth and change in the sign of the derivative $\partial\eta/\partial T$ the integral growth rate decreases to the value ~1.6-1.7$|\ln d/d_n|$. The development of instability corresponding to striation stops after the explosion, but instability modes corresponding to filamentation, which are parallel to the system axis of symmetry, are developing.

IV. CONCLUSION

Calculations of the electro-thermal (ET) instability development were made for the system with the foil placed between two insulating layers powered by the current of the Phelix pulsed power system. Growth rates of ET instability in the linear approximation were calculated for the character current of this system. Comparison of growth rate of ET and MRT instabilities was made. It is shown that the growth of ET instability can play the leading role for such systems. Results of the estimates indicate the necessity of the accurate calculation of the conductivity change in calculations of the instabilities in the exploding foils.

References

[1] A.M.Buyko, S.F.Garanin, S.D.Kuznetsov, R.E.Reinovsky, " Numerical simulations of wire and foil electrical explosion under helical EMG current drive for warm dense matter generation", High-Power Pulsed Electrophysics. Proc. of Int. Conf. XIV Khariton's topical scientific readings 2012. Sarov: RFNC VNIIEF, 2013. pp. 95-103.

[2] S.F.Garanin, S.D.Kuznetsov, R.E.Reinovsky, "Study of the feasibility of warm dense matter generation using aluminum and copper foil electric explosion under the Phelix facility current drive", PMTPh, v. 56, 1. pp. 17-23, 2015

[3] V. N. Korobenko and A. D. Rakhel "Electrical resistivity and equation of state measurements on hot expanded aluminum in the metal-nonmetal transition range", Physical Review B 75, 064208, 2007.

[4] R.B. Baksht, A.G. Rousskikh, A.S. Zhigalin, V.I. Oreshkin, A.P. Artyomov "Stratification in Al and Cu foils exploded in vacuum", Physics of Plasmas 22, 103521 (2015)

[5] Harris, E. G., "RayleighTaylor Instabilities of a Collapsing Cylindrical Shell in a Magnetic Field", 1962, Phys. Fluids 5, 1057.

[6] D.D.Ryutov, M.S.Derzon, and M.K.Matzen: "The physics of fast *Z* pinches", Rev. Mod. Phys., Vol. 72, No. 1, January 2000

[7] V. I. Oreshkin, "Thermal instability during an electrical wire explosion", Physics of Plasmas 15, 092103 (2008)

[8] Gavrilov N. F., Ivanova G. G., Selin V. I., Sofronov V. N. "Code UP for solution of 1D problems of continious media movement in 1D complex", VANT, Ser.: Methodic and coding. 1982. N 3(4). P. 11-14

Performance of a 2.8 MV low inductance electrical-triggered gas switch

Jiahui Yin[1,2], Aici. Qiu[1,2]
1) The State Key Laboratory of Electrical Insulation and Power Equipment,
Xi'an Jiaotong University
Xi'an, China
Yinjiahui@nint.ac.cn

Fengju Sun[2], Tianxue Liang[2], Xiaofeng Jiang[2], Zhiguo Wang[2] and Peitian Cong[2]
2) The State Key Laboratory of Intense Pulsed Radiation Simulation and Effect,
Northwest Institute of Nuclear Technology,
Xi'an China

Abstract—**The paper introduces a 2.8-MV, low-inductance, low-jitter, electrical-triggered, gas SF_6 switch. The primary test indicates that the self-breakdown voltage of the switch is in the range of 0.85-2.76 MV as the pressure of the SF_6 changing from 0.1 to 0.9 MPa. When the gap spacing and the charging voltage time are fixed, the voltage is linearly related to the logarithm of pressure. Besides, the charging time affects the breakdown voltage more seriously than Martin's formula in that case. Depend on the pressure of pre-triggered gap and whether trigger pulse is applied or not, the switch has three working modes: self-breakdown, self-triggered and ex-triggered. In self-triggered mode, the jitter of switch is about 3 ns. In ex-triggered mode, the later of the trigger time, the shorter of the trigger delay and the worse of trigger effect. Without the influence of trigger time, the jitter of switch is less than 2 ns. In triggered modes, the switch inductance is usually below 150 nH.**

Keywords—Induction voltage adder, MV gas switch, triggered gas switch, low inductance, low jitter

I. INTRODUCTION OF THE SWITCH

The first induction voltage adders (IVA) accelerator was developed at Northwest Institute of Nuclear Technology (NINT) in 2008, which was named as Jianguang-I. Jianguang-I has a structure similar to Cygnus. It is consisted of a single Marx, a pulse forming line (PFL), a self-breakdown PFL water switch and three induction cells [1-2]. At NINT, a new IVA accelerator will be developed and a 2.8-MV triggered switch was designed for it. The new IVA accelerator has two induction cavities and each cavity is driven by two high voltage pulses in parallel. The voltage of driving pulse is about 1.0 MV, the current is about 180 kA and the pulse width is about 60 ns. To make the accelerator more compact and less expensive, the primary driving pulse is generated based on low inductance Marxes, gas-switched PFLs and no transfer capacitors. Therefore, the rise times of the driving pulses will be compressed only once from hundreds of nanoseconds to less than 30 ns. If so, the PFL switches should be triggered with low jitter and low inductance [3]. The requirements of this switch are shown in Table I.

TABLE I
The Requirements of PFL Switch

Operating voltage	1.0-2.6 MV
Operating time	about 450 ns
Current	up to 200 kA
Jitter	less than 5 ns
Inductance	less than 150 nH
Switch pressure	1 MPa SF_6 (Max.)

Fig. 1 shows the structure of the designed 2.8-MV gas triggered switch. In order to reduce the inductance of the switch, multi channels are necessary when the switch closes, therefore, the switch is triggered by high voltage pulse. The switch consists of a main gap and a pre-triggered gap. The main gap has a V/*n* structure, which consists of three electrodes, manufactured from Type 304 stainless steel. The *n* means the ratio of the voltage across the whole switch to that of trigger disc[4]. The total spacing of the main gap was initially determined to be 10 cm and increased to 11.5 cm later. Because the stray capacitances between the trigger disc, the HV and ground electrodes, as the PFL is pulsed charged, via capacitive voltage dividing, the potential of the disc will increase to hundreds of kV. As the spacing of the disc-ground gap is 2.75 cm and the charging voltage is -2.8 MV, before triggering, the floating potential of the trigger disc is -603 kV, *n* is about 4.64. Meanwhile, the highest electric field, which occurs at the front of the HV electrode, is 87 kV/mm. After triggering by 200 kV pulse, the pre-triggered gap breaks down and the disc potential is changed to 200 kV, the electric field at the edge of trigger disc is increased 6.79 times, from 34 to 231 kV/mm, but the peak electric field at the HV electrode is only increased 1.16 times, from 87 to 101 kV/mm. Then there is followed by the breakdown of the HV-disc and the disc-ground gaps. The pre-triggered gap can also be seen as a trigatron switch. According to self-breakdown and trigger experimental data, in the pre-triggered gap, the HV electrode is spherical, which has a gap spacing of 20 mm and the disc is indented 1 mm into the ground electrode [6].

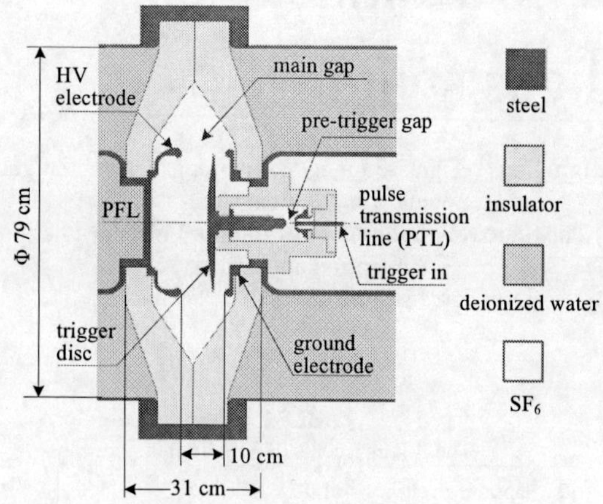

Figure 1. Diagram of the electrical-triggered gas switch

II. THE TEST PLATFORM

The pulse forming line and the triggered switch was tested with the Marx of the "Jianguang-I" accelerator. The equivalent-circuit model of test platform is shown in Fig. 2. The erected capacitance of Marx is 6.67 nF and the erected inductance is about 5.5 μH. The maximum PFL voltage is about 3.0 MV and the 0-100% duration time is 350 ns. By adding inductance between the rows of the Marx, the duration time can be increased to 450 ns. The voltage of PFL and PTL are measured by the differential capacitive voltage monitors and RC integrators, which are shown in Fig.3 with the calibration waveforms. For the PFL voltage monitor, the time constant of integrator is about 5 μs, while the integral time constant for the PTL voltage monitor is about 2 μs. The PFL voltage monitor is located at 13 ns upstream of PFL switch and the PTL voltage monitor is located at 12.5 ns downstream of PFL switch.

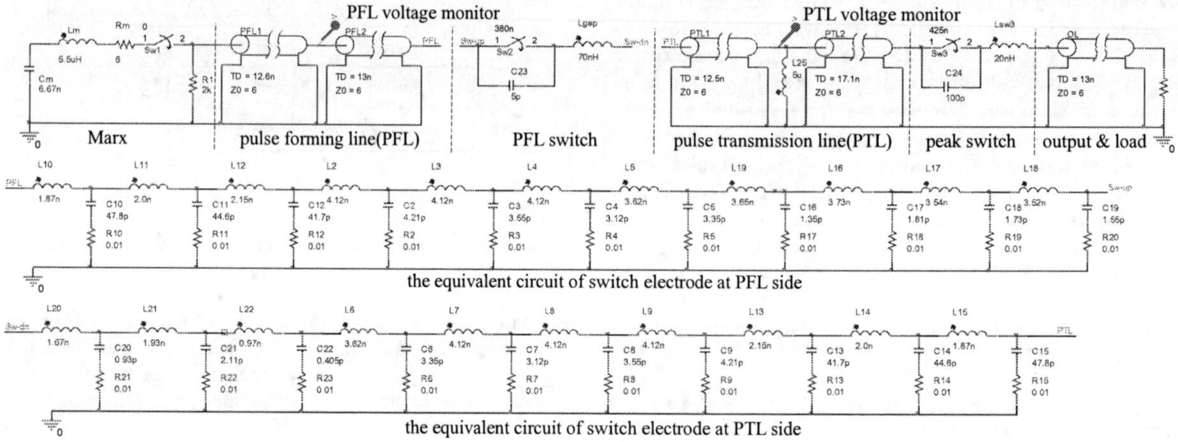

Figure 2. The circuit model of switch test platform

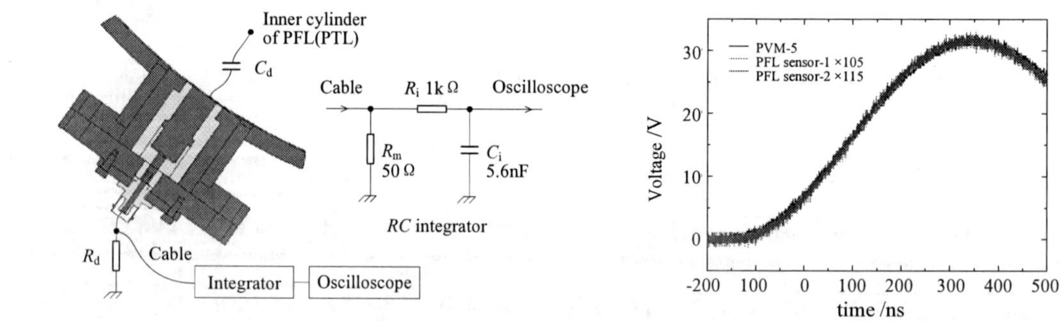

Figure 3. The differential capacitive voltage monitor and the calibration waveforms

III. THE PERFORMANCE OF THE SWITCH

A. Working modes and jitter

Depend on the breakdown sequence of different gaps in the switch and whether trigger pulse is applied or not, switch may work in three modes.

a) Ex-triggered mode: If the trigger pulse is applied, the pre-triggered gap breaks down firstly, and the potential of the trigger disc is changed from negative hundreds of kV to 200 kV, and then the HV-disc and disc-ground gaps break. The switch gets closed completely.

b) Self-triggered mode: When the PFL is pulsed charged, the potential of trigger disc gets increased as well. If the pressure of the pre-triggered gap is much lower than that of the main gap, although no trigger pulse is applied, the pre-triggered gap still breaks down much early. Then the potential of the trigger disc is changed from negative hundreds of kV to ground, just like a trigger pulse is applied. Because of the electric field distortion, the main gap breaks and followed the completely breaks of the switch.

c) Self-breakdown mode: the HV-disc or disc-ground gap breaks down firstly, then the other part of the main gap breaks, and the switch gets closed completely. Occasionally, the pre-triggered gap may break down a little early, but it has little effect on the breakdown of main gap.

When the charging voltage of Marx is ±35kV and all pressures of the main gap are 0.3 MPa, the PFL voltage waveforms in three modes are shown in Fig.4. In self-breakdown mode, the switch breaks not stably and has the highest voltage. When the trigger pulse is applied at the proper time, the switch will work in ex-triggered mode. While if there is no trigger pulse applied and the pressure of pre-triggered gap is below 0.1 MPa, the early breakdown of trigatron gap will lead to the self-triggered mode of switch.

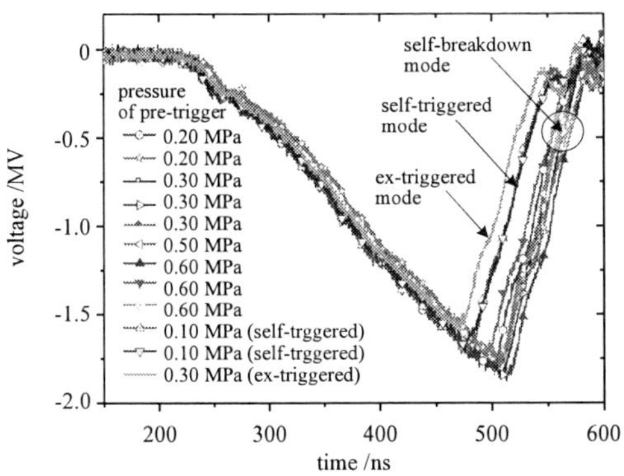

Figure 4. PFL voltage waveforms in three working modes (the pressure of main gap is 0.30 MPa, the pressure of pre-triggered gap is listed in the figure)

B. *Self-Breakdown mode*

In self-breakdown mode, the pressure of the trigatron gap is higher than the main gap pressure, in case that the breakdown of the trigatron switch in advance may affect the breakdown of the main gap. In order to eliminate the effect of pulse duration time and the gap spacing on the breakdown voltage, the 0-100% duration time of voltage was selected 300±10 ns and the gap spacing was all 10 cm. The PFL voltage waveforms are shown in Fig. 5 and the self-breakdown voltage versus the SF_6 pressure of main gap is shown in Fig. 6. The self-breakdown voltage of the switch is ranged from 0.85 MV to 2.76 MV as the pressure of the SF_6 changing from 0.1 to 0.9 MPa.

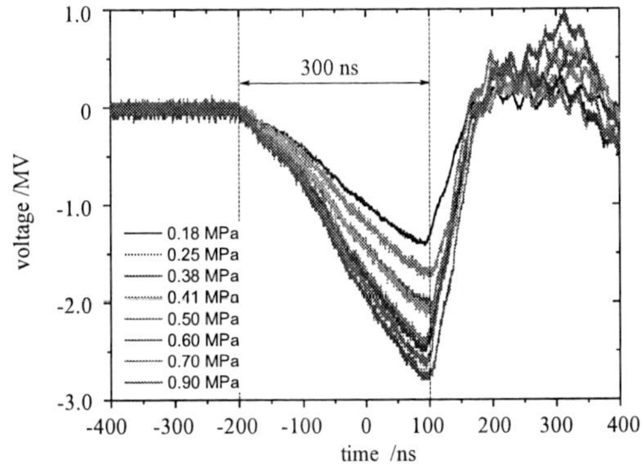

Figure 5. the PFL voltage waveforms with the different pressure of main gap (the gap spacing is 10 cm, the 0-100% duration time of voltage is 300±10 ns)

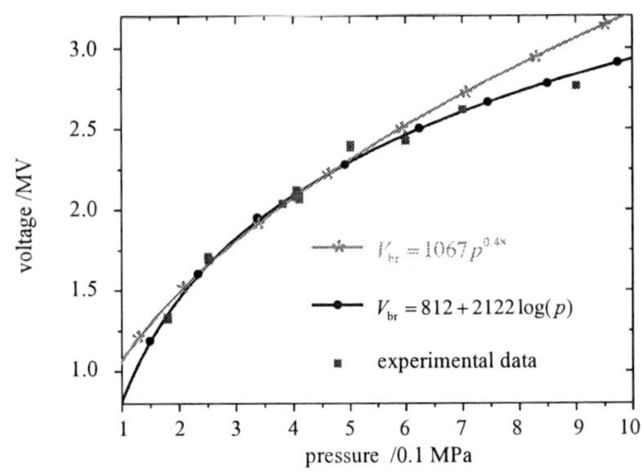

Figure 6. the self-breakdown voltage versus the SF_6 pressure

The breakdown voltage can be fit as

$$V_{br} = 1067 p^{0.48} \tag{1}$$

or

$$V_{br} = 812 + 2122\log(p) \tag{2}$$

where V_{br} is the breakdown voltage [kV], p is SF_6 pressure of main gap [0.1 MPa]. The formula (2) is more accurate, but the formula (1) is more convenient to combine with other variables, such as the spacing d or the duration time t.

The effect of duration time to the breakdown voltage is studied as well. As the pressure of the main and pre-triggered gaps are all 0.5 MPa, the PFL voltage waveforms are shown in Fig. 7 and the self-breakdown voltage versus the duration time is shown in Fig. 8. The breakdown voltage can be fit as

$$V_{br} = 765 t_{0.63}^{-0.53} \tag{3}$$

where $t_{0.63}$ is the time the voltage exceeds 63% of the breakdown voltage [μs]. Compared with the Martin's breakdown formula [7]

$$V_{br} = 72 p^{0.4} d^{5/6} t_{0.89}^{-1/6} \tag{4}$$

where d is the gap spacing [cm] and $t_{0.89}$ is the time the voltage exceeds 89% of the breakdown voltage [μs]. We may find that the duration time affects the breakdown voltage more seriously than Marin's formula in that case.

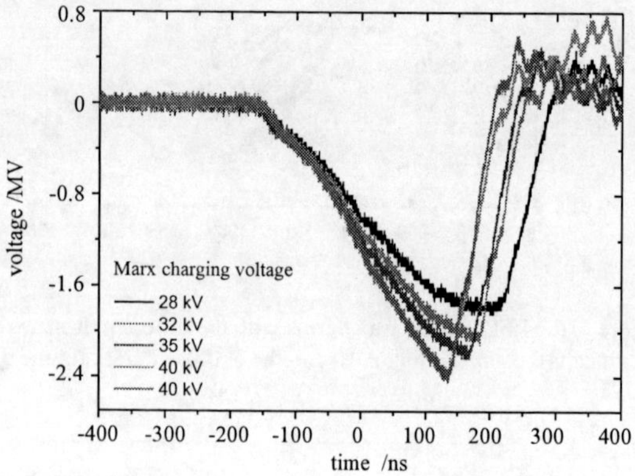

Figure 7. the PFL voltage waveforms with the different duration time (the gap spacing is 10 cm, the pressure of the main and pre-triggered gaps are all 0.5 MPa)

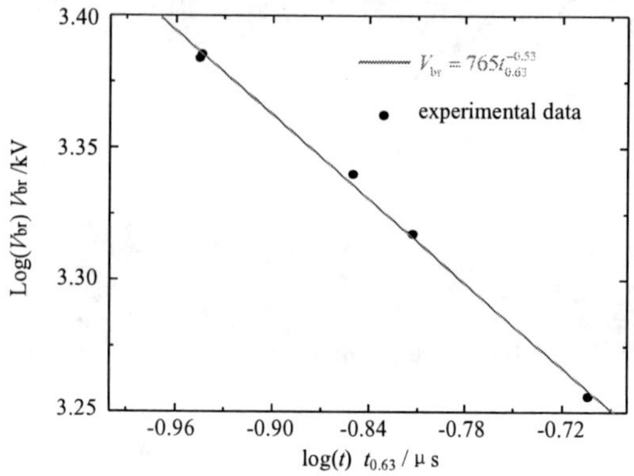

Figure 8. the self-breakdown voltage versus the duration time

C. *Self-triggered mode*

Fig. 9 shows the PFL voltage waveforms in self-triggered mode as the charging voltage of Marx is ±40kV. The peak charging voltage of PFL is about 2.64 MV, and the pressure of the main and pre-triggered gaps are all 0.5 and 0.2 MPa. Obviously in self-triggered mode, the switch breaks more stalely than that in self-breakdown mode. According to the waveforms, the range of breakdown time of the PFL switch is about 8.1 ns and the range of breakdown voltage about 51 kV.

If the pre-triggered gap has different pressures, the breakdown voltage of the switch will be slightly affected by the pressure of pre-triggered gap. Fig. 10 shows the PFL voltage waveforms in self-triggered mode as the pressure of pre-triggered gap is different.

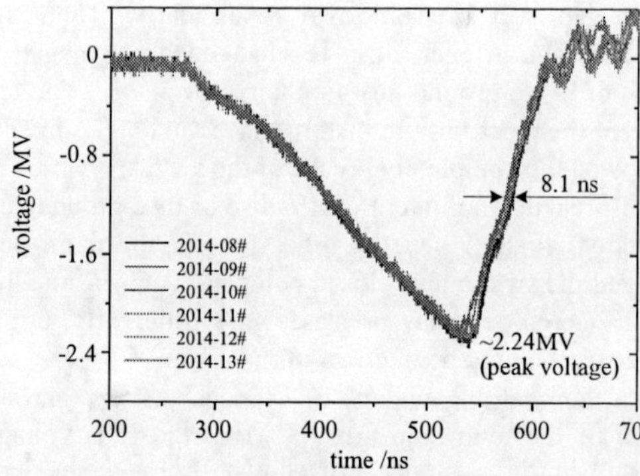

Figure 9. PFL voltage waveforms with the same pressure of pre-triggered gap (the pressure of main gap is 0.50 MPa, the pressure of pre-triggered gap is 0.20 MPa)

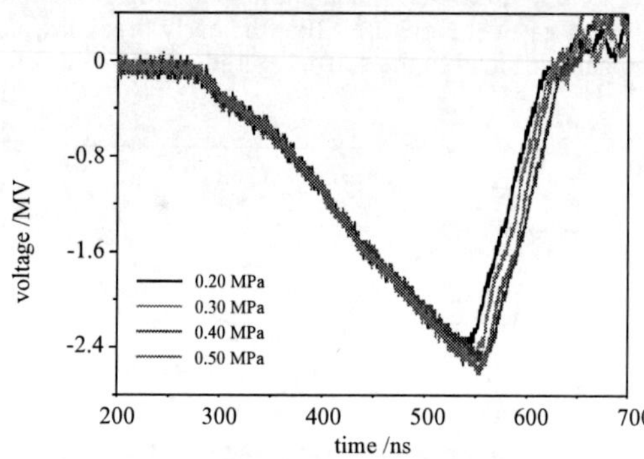

Figure 10. PFL voltage waveforms with the different pressure of pre-triggered gap (corresponding the waveforms from left to right, the pressure of pre-triggered gap is from 0.20 MPa to 0.50 MPa, while the pressure of main gap is 0.50 MPa,)

D. *Ex-triggered mode*

Fig. 11 shows trigger delay versus trigger time, where the gap spacing is 12 cm, the trigger voltage is 220 kV, the pressure of the main and pre-triggered gaps are all 0.5 and 0.4 MPa. The trigger time is defined as the difference between the time of trigger pulse applied to the time of PFL voltage star to increase, while the trigger delay is defined as the difference between the time of switch closes to the time of trigger pulse applied. We can find that the later of the trigger time, the shorter of the trigger delay and the worse of the trigger effect. As the trigger time range from 240 ns to 290 ns, the range of trigger delay is about 0.66 times of the range of the trigger time. While the trigger time is about 282 ns and the jitter of that is 4.5 ns, the trigger delay is about 90.5 ns and the jitter of that is 4.3 ns, as shown in the Table II. As the trigger time is nearly fixed at 282 ns and the jitter of that is 1.5 ns, but the trigger voltage is decreased to 150 kV, the trigger delay time is about 103.8 ns and the jitter of that is 1.4 ns, as shown in the Fig.12.

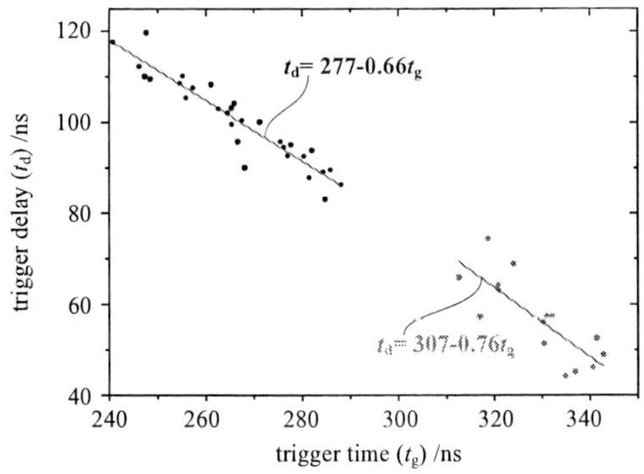

Figure 11. Trigger delay versus trigger time (the gap spacing is 12 cm, the pressure of the main and pre-triggered gaps are all 0.5 and 0.4 MPa)

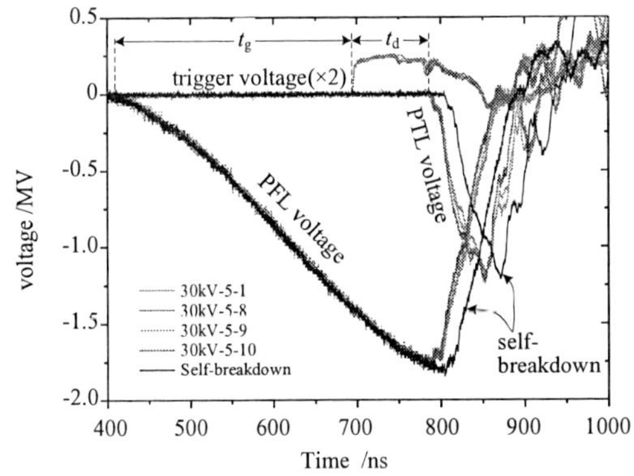

Figure 12. The voltage waveforms of PFL, PTL and trigger cable in ex-triggered mode (the gap spacing is 12 cm, the pressure of the main and pre-triggered gaps are all 0.5 and 0.4 MPa)

TABLE II
TRIGGER TIME AND TRIGGER DELAY IN EX-TRIGGERED MODE

Shot No.	30kV-4-20	30kV-4-21	30kV-4-22	30kV-4-23	30kV-4-27	30kV-4-38	30kV-4-39	30kV-4-48	30kV-4-49
Trigger time t_g /ns	284.7	284.4	288.1	276.2	285.9	277.6	275.5	280.4	281.5
Trigger delay t_d /ns	83.1	89.2	86.4	94.6	89.6	95.1	95.8	92.6	87.9

E. Switch inductance

Fig. 13 shows the voltage waveforms following the PFL switch for the same shots as Fig. 9. The reflected voltage wave from peak switch was assumed, with the inflection point on the rise edge taken as the peak of output pulse. The slowest rise time measured in self-breakdown experiment was 38 ns, corresponding to a switch inductance 205 nH. If a single channel discharge, the electrodes inductance is 70 nH, while the calculated value is 51 nH. Table III shows the rise time and the switch inductance in self-triggered mode. The data indicates that the switch inductance is usually less than 150 nH.

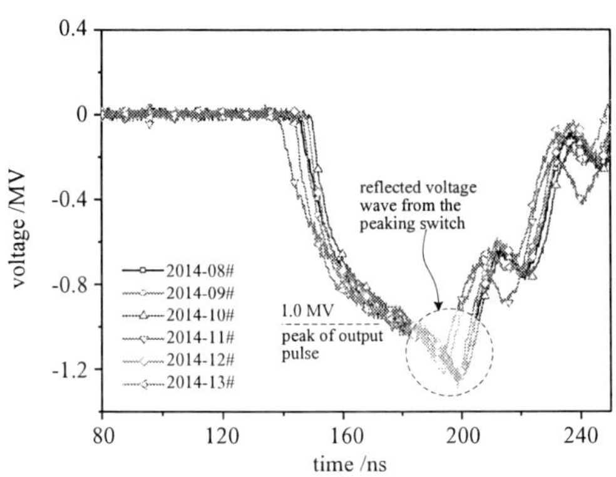

Figure 13. Voltage waveforms after PFL switch in self-triggered mode. (the same shots in Fig. 9)

TABLE III
SWITCH INDUCTANCE AND RISE TIME IN SELF-TRIGGERED EXPERIMENT

Shot No. (2014)	08#	09#	10#	11#	12#	13#
Peaking voltage /MV	1.05	1.02	1.02	0.93	1.01	1.04
Rise time /ns	27.0	23.5	24.2	28.2	22.4	24.7
Total switch inductance /nH	147	128	132	154	122	135
Channels inductance /nH	71	52	56	78	46	59

IV. CONCLUSION

This paper provides the initial design of a 2.8 MV, low-inductance, low-jitter, electrical-triggered gas switch. Based on the analysis and experimental investigation presented in this paper, the following conclusions can be drawn:

(1) The self-breakdown voltage rises from 0.85 MV to 2.77 MV as the pressure of SF_6 increases from 0.1 MPa to 0.9 MPa. When the gap spacing and voltage duration time are fixed, the voltage is linearly related to the logarithm of pressure. The duration time affects the breakdown voltage more seriously than Martin's formula in that case.

(2) Depend on the breakdown sequence of different gaps in the switch and whether trigger pulse is applied or not, the switch has three working modes: self-breakdown, self-triggered and ex-triggered. In self-triggered mode, the jitter of switch is about 3 ns. In ex-triggered mode, the later of the trigger time, the shorter of the trigger delay and the worse of trigger effect. Without the influence of trigger time, the jitter of switch is less than 2 ns.

(3) With a single spark channel discharge, the switch

inductance is about 205 nH, and the rise time of the output pulse is about 38 ns, which leads to the slow of rise time and the reduction of voltage. In triggered modes, multi-channels are formed and the switch inductance is usually less than 150 nH, and the rise time of the main pulse is about 22-28 ns.

Acknowledgment

The authors thank Yuang Zhang, Jiangtao Zeng, Hongyu Jiang, Jianfeng Sun, Zhong Zhang for the many useful suggestions and discussions, and other colleagues in the group of "Jianguang-I" accelerator for their assistance.

References

[1] F J Sun, A C Qiu, H L Yang, et al. "Pulsed X-ray source based on inductive voltage adder and rod pinch diode for radiography". *High Power Laser and Particle Beams*, vol. 22, no. 4, pp.935-940, 2010

[2] D. Weidenheimer, P. Corcoran, R. Altes, et al. "Design of a Driver for the Cygnus X-ray Source" in *Proc. IEEE 13th Int. Pulsed Power Plasma Science Conf.*, 2001, pp. 591-595

[3] Corcoran, P, Carboni, V, Smith, et al. "Design of an induction voltage adder based on gas-switched pulse forming lines" in *Proc. IEEE 15th Int. Pulsed Power Conf.*, 2005, pp. 308-313

[4] S. Mercer, I. Smith, and T. Martin, "A compact, multiple channel 3 MV gas switch", in *Proc. Int. Conf. Energy Storage, Compression Switching,* 1974, pp. 459–462

[5] A. J. McPhee, S. J. MacGregor, S. M. Tumbull. "An investigation of trigatron breakdown by two different mechanisms" in *Proc. IEEE 10th Int. Pulsed Power Conf.* 1995, pp. 775-780

[6] Dang T F, Yin J H, Sun F J, et al. "Influence of electrode structure on breakdown characteristics of trigatron switch" *High Power Laser and Particle Beams*, vol. 27, no. 6, pp. 065004, 2015

[7] Martin T H. "Pulse charged gas breakdown" in Proc. *IEEE 5th Int. Pulsed Power Conf.* 1985, pp. 74-83

Jiahui Yin was born in Hebei, China. He received the M.S. degree in nuclear technology and applications from the Northwest Institute of Nuclear Technology, Xi'an, China, in 2007.
In 2015, he joined the Plasma and Pulsed Power group as an academic visitor for one year at Loughborough University (UK). He is currently working toward the Ph.D. degree in the State Key Laboratory of Electrical Insulation and Power Equipment, Xi'an Jiaotong University. His main research interest includes MV triggered switch and trigger generator.

Aici Qiu received the B.S. degree in electrical engineering from Xi'an Jiaotong University, Xi'an, China, in 1964. She was honored the Academician of the Chinese Academy of Engineering, Beijing, China, in 1999. She is currently a Professor with the State Key Laboratory of Electrical Insulation and Power Equipment, Xi'an Jiaotong University. She is also with State Key Laboratory of Intense Pulsed Radiation Simulation and Effect, Northwest Institute of Nuclear Technology.

Fengju Sun was born in Shandong, China. He received the M.S. degree from the Northwest Institute of Nuclear Technology, Xi'an, China, in 1996, and the Ph.D. degree from Xi'an Jiaotong University, Xi'an, in 2002.
He is currently with the Northwest Institute of Nuclear Technology. His research interests include gas spark switches, pulsed power driver for Z-pinch, and pulsed X-ray radiography source based on induction voltage adder.

Tianxue Liang received the B.S. degree in electrical engineering from Xi'an Jiaotong University, Xi'an, China, in 1998. He received the M.S. degree in electrical engineering from Chongqing University, Chongqing, China, in 2007. He received the Ph.D. degree at the Northwest Institute of Nuclear Technology, Xi'an, China, in 2015.
He is currently with the Northwest Institute of Nuclear Technology. His research interest includes the linear transformer driver technology, gas spark switches.

Xiaofeng Jiang received the M.S. degree in pulsed-power technology from the Northwest Institute of Nuclear Technology, Xi'an, China, in 2009. He is currently with the Northwest Institute of Nuclear Technology. His research interest includes multigap multichannel gas switches.

Zhiguo Wang was born in Shandong, China. He received the M.S. degree from the Northwest Institute of Nuclear Technology, Xi'an, in 2011. He is currently with the Northwest Institute of Nuclear Technology. His research interest includes magnetic cores.

Peitian Cong who was born in1974, graduated from Xi'an Jiao Tong University (XJTU) with a B. Sci. degree in high voltage engineering in 1996. He received a M. Sci. degree in pulsed power science and technology from NINT in 2000 with a dissertation on a composite insulation for the linear transformer devices (LTDs). He obtained his Ph.D. degree by researching and developing a multi-stage gas switch in 2012 from NINT. In 2010, he spent one year as a visiting researcher with the Plasma Physics Group at Imperial College London.
He is the director of the NINT radiation simulation department specializing in pulsed power driver, high current particle beams and intense radiation. His research interests include close switches, transmission lines and magnetic cores. He is currently focusing on the IVA/LTD pulsed power system and technology..

Bacteria inactivation by pulsed electron beam

Camille Lamarche, Gauthier Demol
ITHPP
Thégra, France
clamarche@ithpp-alcen.fr

Flavien Pillet, Houda Baaziz, Marie-Pierre Rols
Cellular Biophysics
IPBS/CNRS
Toulouse, France

Pulsed electron beam is a promising physical technology which enables to sterilize products or packaging. However, the mechanisms involved in bacteria inactivation are still not fully known. In this work, we have investigated by using electron microscopy the effects of pulsed electron beam on the bacteria-envelope. The results reveal a high efficiency of bacteria inactivation but curiously, without any structural modification in the cell-envelope. The explanation of the killing effect remains pending.

Keywords—pulsed electron beam, Bacillus pumilus, inactivation, SEM, TEM

I. Introduction

Food and pharmaceutical industries use different types of sterilization methods: chemical (hydrogen peroxide...), thermal or physical (gamma ray, X-ray or electron beam) technologies. However, these solutions are time consuming, cost effective and sometimes harmful to health (e.g. hydrogen peroxide). Within context, the low energy pulsed electron beam is promising and allows sterilization without using chemical compounds and can be an IN-LINE process. The equipment presented delivers more than 1 kGy in a 10 ns pulse width at a repetition rate of 100 Hz that leads to very high dose rate and makes sterilization possible at less than 25 kGy in the millisecond range. Electron beam was extracted from a cathode and went through an anode, consisted of metallic foil and called the extraction window.

Here, we first present the efficiency of pulsed electron beam on bacteria (vegetative form and spore) to select the best model for the study. Then, we investigated if an adaptive mechanism of spores to electron beam is present. Finally, we tried to explain the mechanisms involved in the killing effect.

II. Materials and Methods

A. Strain selection and growth conditions

The first strain selected is *Bacillus atrophaeus* ATCC 9372 which is the reference for food industries. The second one is *Bacillus pumilus* ATCC 27142. This strain is the reference for ionization treatment [1] [2].

These strains can be found under vegetative (growth phase) and spore (dormant) forms. The last form is resistant to environmental stresses.

Région Midi-Pyrénées and DGA-DGCIS

Vegetative cells were cultivated in Luria broth (LB) medium overnight at 37°C. Cells were centrifuged during 5 minutes at 6 000g and the pellet was resuspended in sterile deionized water. The final concentration of bacteria was 10 times more important than the initial solution.

Spore cells had grown after stationary phase, during 5 days at 37°C. The medium was the following: 8g/L of Difco Nutrient Broth (Difco), 100 mM of $MgSO_4$, 10% of KCl, 1M of NaOH, 1M of $Ca(NO_3)_2$, 10 mM of $MnCl_2$ and 1mM of $FeSO_4$. To eliminate vegetative forms, the solution was heat-shocked at 80°C during 20 minutes. Spores were recovered by centrifugation (5 minutes at 10 000g). Spores were collected on sterile deionized water in order to have 10 times more bacteria than the cultivated solution.

B. Treatment

The voltage and current of the generator used for this experiment were 250 kV and 5 kA, respectively. The pulse duration was 10 ns. For the experiment, two parameters were studied: the pulsed repetition frequency (from 5 to 100 Hz) and the distance from the extraction window (from 2 cm to 9 cm).

C. Determination of inactivation rate

- For treatment on agar, different concentrations of bacteria were spread on Petri dishes. The dishes were treated with the pulsed electron beam and incubated overnight at 37°C. The colonies were counted and the inactivation rate was calculated thanks to the equation (1).
- For dry treatment, 10 µL of solution were put on empty Petri dished and dried under laminar flow. Following the treatment, spots were resuspended in sterile distilled water and different dilutions were applied and spread on Plate Count Agar (PCA) medium overnight at 37°C. The equation (1) was used to determine the inactivation rate.

$$IR = \log_{10}(N0/N) \quad (1)$$

IR inactivation rate

N0 initial microbial concentration

N number of survivors after the treatment

D. *Resistance evaluation*

After an exposition to electron beam at sub-lethal dose, bacteria can acquire a better resistance to the treatment for the same dose [1] [3] [4].

To determine if there is a modification of resistance with our technology, two solutions of *Bacillus pumilus* spores with 10^{10} spores/mL were prepared and a spot of 10µL was deposed on Petri dishes. The first solution called Wild Type (WT) was the stock solution. It was the reference for the efficiency. The other one was cultivated from bacteria which was treated under lethal dose x time(s). After the irradiation, spots were recovered as described on *II. C.* part for 'dry treatment' and survival strain were counted. The decontamination rates were compared. The two Petri dishes were placed at the same time under the electron beam.

The treatment applied for each sample was 10 pulses à 7 centimeters from the extraction window, with a pulsed repetition frequency of 100 Hz.

E. *Scanning Electron microscopy and Transmission Electron Microscopy observation*

P.R. Chalise and S.E. Fiester shown pores formation due to electron beam treatment. A high density of current could induce the formation of irreversible pores around the bacteria [5] [6]. Electron microscopy was used to observe if this phenomenon appears with pulsed electron beam. Bacteria were fixed with 2% glutaraldehyde in 0.1M Sorensen phosphate buffer pH 7.2 just after the recovered. The samples were prepared by the CMEAB platform, Toulouse (France).

For Scanning Electron microscopy (SEM), images were acquired with an electron microscope Quanta™ 250 FEG (FEI, USA) at an accelerating voltage of 10 kV.

For Transmission Electron Microscopy (TEM), sections had a thickness of 70 nm and were placed on a grid. Images were visualized with HT 7700 at 80 kV (Hitachi, USA).

III. RESULTS

A. *Selection of reference strain*

We made treatment on agar placed at 7 centimeters and with a pulsed repetition frequency of 100 Hz to select the most resistant strain. Vegetative form of *Bacillus atrophaeus* was irradiated with 2 pulses. In this condition, the logarithmic reduction is more than 5.7. For the spore form of this bacteria, we need 10 pulses to obtain 4.47 $\log_{10}$ reductions. Consequently, as described in the literature, spore form of bacteria is more resistant than vegetative form. Indeed, the different layers which composed the spore, the presence of a core with condensed DNA, the low concentration of water and other parameters explain the faculty of this form to survive in extreme conditions [7] [8] [9].

The experiment was made on *Bacillus pumilus* spores, which are the reference for irradiation technologies. This strain is effectively more resistant than *Bacillus atrophaeus* spores because after a treatment with 10 pulses, the logarithmic reduction obtained was 4.1 $\log_{10}$.

This is the strain selected for the rest of the study.

B. *Parameters influence*

- The first parameter studied was the pulsed repetition frequency (Hz). For these experiments, *Bacillus pumilus* spores were spotted on empty Petri dishes and placed under the extraction window at 7 centimeters. Inactivation was calculated by colony counting and with the equation (1). The results revealed that for 50Hz and 100Hz, the decontamination rate was the same. However, for 5Hz, the efficiency was better (Fig. 1). These phenomena can be explained because the switch of the generator is self-triggered, consequently, for low frequency, the parameters are different than parameters for high repetition frequency. Consequently, we cannot conclude that the frequency has an influence on the mortality rate of this strain. Because this equipment can be used for industrial applications, the maximum repetition frequency was chosen for the next experiments.

- The second parameter was the distance from the extraction window. For these experiments, *Bacillus pumilus* spores were spotted on empty Petri dishes and placed under the extraction window at 2, 4, 7 and 9centimeters. The repetition frequency was 100Hz. Inactivation was calculated by colony counting and with the equation (1). On Fig. 2, the graph revealed a better efficiency when bacteria were placed close to the extraction window. These phenomena can be explained because of electron beam interaction with molecules present in the air and their path was modified. The interaction with molecules decreased also energy of electrons. These two phenomena induced a decreased of the dose with the distance.

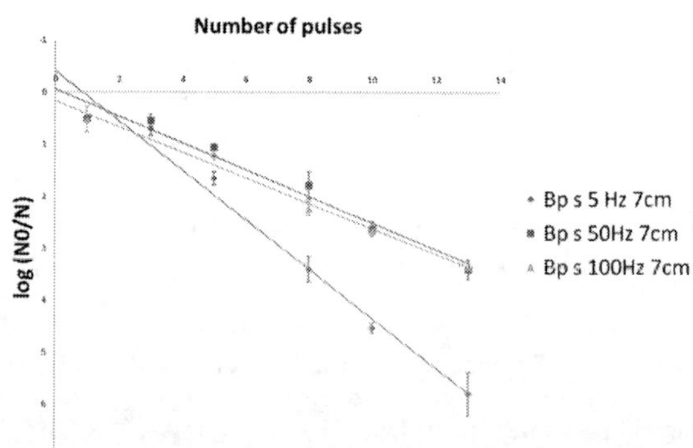

Fig. 1. *Influence of the pulsed repetition frequency on the treatment efficiency on Bacillus pumilus spores.* Effect of different number of pulses when samples were placed at 7centimeters from the extraction window and for different frequency (Hz): 5Hz (diamond), 50Hz (square) and 100Hz (triangle) on the inactivation rate.

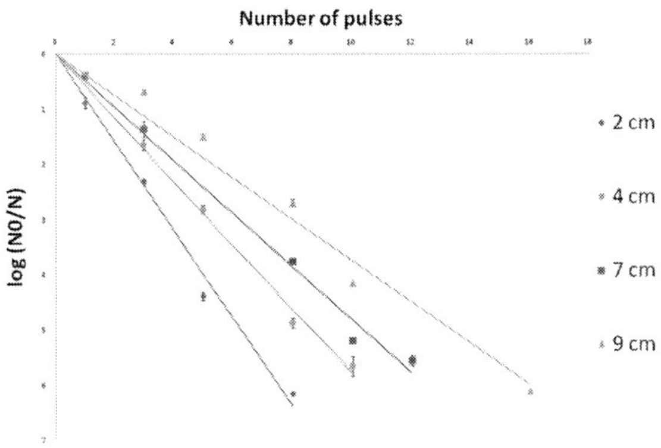

Fig. 2. *Influence of the distance from the extraction window on the treatment efficiency on Bacillus pumilus spores.* Effect of different number of pulses when samples were placed at different distances from the extraction window: 2 centimeters (diamond), 4 centimeters (round), 7 centimeters (square) and 9 centimeters (triangle) on the inactivation rate.

C. Evolution of the resistance after several treatments at sub-lethal dose

The efficiency of the technology was validated. However, it is important to know if any resistance can appear in the case of multi-exposure at sub-lethal dose. Fig.3 presents the logarithmic reduction obtained for the Wild type (WT) in comparison with the reduction of bacteria which had been irradiated at sub-lethal dose before (n+x). The letter x means the number of time that the solution had been exposed at sub-lethal dose before the treatment apply. The WT and the n+x had been placed under the electron beam at the same moment to ensure the same energy deposition between the reference and the assay. The results showed no difference of efficiency after 10 times of treatment. Pulsed electron beams do not induce resistance on *Bacillus pumilus* spores.

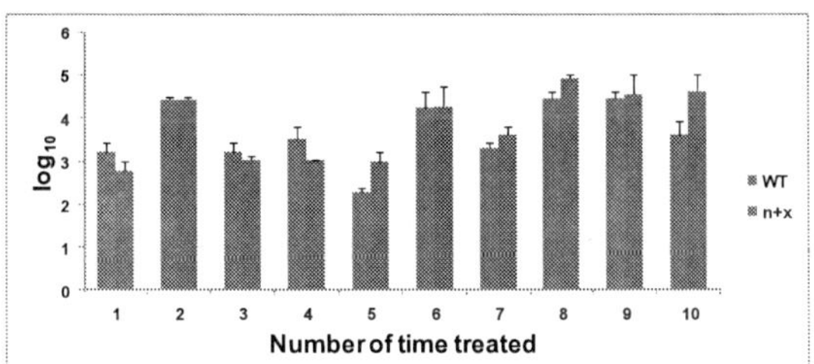

Fig. 3. *Treatment at sub-lethal dose not induced resistance.* Comparaison of the inactivation rate after 10 pulses at 7centimeters and a repetition frequency of 100Hz between bacteria which never see any irradiation treatment (WT) and bacteria which was treated x time(s) at sublethal dose (n+x).

D. Cell wall observation by electronic microscopy

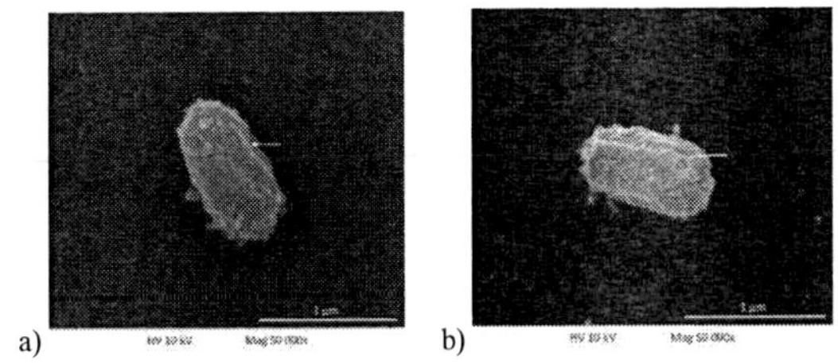

Fig. 4. *Observation of Bacillus pumilus spores wall by SEM.* The comparison of the non-treated spore (a) with the spore irradiated (b) did not reveal difference. Ridges (indicated with white arrows) were present for the two conditions.

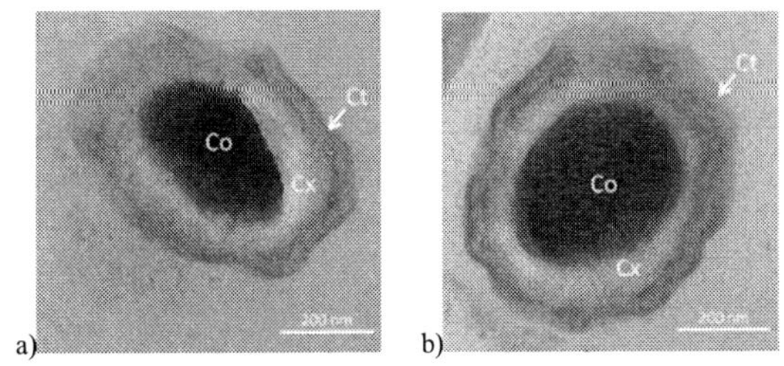

Fig. 5. *Observation of Bacillus pumilus spores wall by TEM.* The comparison of the non-treated spore (a) with the spore irradiated (b) did not reveal difference. The coat (Ct), the cortex (Cx) and the core (Co) are observable. The different layer of the coat are present in all case.

Spores were observed by SEM (Fig. 4) and TEM (Fig. 5). Ridges can be observed at the surface of non-treated and treated bacteria and none difference was observed.. To validate these observations, we performed TEM experiments. Indeed, the structure of the coat (Ct), the cortex (Cx) or the core (Co) can be observed and perfectly distinguished. The protein multilayers of the coat, the cortex and the core were present and do not seem altered after irradiation.

IV. DISCUSSION

Pulsed electron beam, is an efficient irradiation technology to kill bacteria even under spore form. *Bacillus pumilus* appears to be the most resistant bacteria, as previously described by U.S. Pharmacopeia. Moreover, experiments do not reveal an evolution of the resistance after 10 treatments under lethal dose. However, SEM and TEM, do not reveal any changes in the cell wall and organization after irradiation. These results are in contradiction with S.E. Fiester which shown damage on the spore coat by MEB [6]. The hypothesis is that, our technology uses high dose rate and could interact with molecules and cell wall differently. Consequently, to complete the cell wall analysis, Atomic Force Microscopy (AFM) on liquid medium will be used to measure, in living condition and at the nanoscale, the structural organization and the mechanical properties of the cell wall after irradiation [11].

In other way, some scientists supposed that electron beam induced double strand break on DNA [5] [12] [13]. To validate this hypothesis on our technology, Pulsed Field Gel Electrophoresis (PFGE) will be used.

Acknowledgment

I thank all my colleagues from the ITHPP Company for their help to understand the technology and the members of M-P. Rols team for their scientific help, explanation and for their availabilities.

To finish, I thank CMEAB platform for their technical support and explanation for electron microscopy experiments.

References

[1] A. Parisi, and A. D. Antoine, "Increased Radiation Resistance of Vegetative Bacillus Pumilus." *Applied Microbiology* 28, no. 1 (1974): 41–46.

[2] D. A. Cleghorn, J. Dunn, and S. V. Nablo. "Sterilization of Plastic Containers Using Electron Beam Irradiation Directed through the Opening." *Journal of Applied Microbiology* 93, no. 6 (2002): 937–43.

[3] L. Levanduski, and J. Jaczynski. "Increased Resistance of Escherichia Coli O157:H7 to Electron Beam Following Repetitive Irradiation at Sub-Lethal Doses." *International Journal of Food Microbiology* 121, no. 3 (February 10, 2008): 328–34.

[4] A. T. Tesfai, S. K. Beamer, K. E. Matak, and J. Jaczynski, "Radioresistance development of DNA repair deficient Escherichia coli DH5a in ground beef subjected to electron beam at sub-lethal doses", *International Journal of Radiation Biology* 83, no. 6 (2011) : 571-578.

[5] P.R. Chalise, E. Hotta, K.E. Matak, and J. Jaczynski, "Inactivation Kinetics of Escherichia Coli by Pulsed Electron Beam." *Journal of Food Science* 72, no. 7 (2007): M280–85.

[6] S.E. Fiester, S. L. Helfinstine, J.C. Redfearn, R.M. Uribe, and C.J. Woolverton, "Electron Beam Irradiation Dose Dependently Damages the *Bacillus* Spore Coat and Spore Membrane." *International Journal of Microbiology* 2012, (2012): 1-9

[7] M.J. Leggett, "Bacterial spore structures and their protective role in biocide resistance", *Journal of Applied Microbiology*, (2012) 485-498

[8] P. Setlow, "Spores of *Bacillus subtilis*: their resistance to and killing by radiation, heat and chemicals." *Journal of Applied Microbiology* 101, (2006) 514–525.

[9] P. Setlow, "I will survive: protecting and repairing spore DNA." *J Bacteriol* 174, (1992) 2737–2741.

[10] J. Errington, "Regulation of endospore formation in *Bacillus subtilis*." *Nature Reviews Microbioly* 1, (2003) 117–26

[11] E. Dague, D. Alsteens, J-P. Latgé, C. Verbelen, D. Raze, A. R. Baulard, and Y. F. Dufrêne, "Chemical Force Microscopy of Single Live Cells", *Nano Letters* 7, no. 10, (2007): 3026-3030

[12] H. Zhu, JZ Xu, SQ Li, XY Sun, SD Yao, and SL Wang. "Effects of High-Energy-Pulse-Electron Beam Radiation on Biomacromolecules." *Science in China Series B: Chemistry* 51, no. 1 (January 2008): 86–91.

[13] E. Urgiles, J. Wilcox, O. Montes, S. Osman, K. Venkateswaran, M. Cepeda, J. Maxim, L. Braby, and S. D. Pillai. "Electron Beam Irradiation for Microbial Reduction on Spacecraft Components." *In Aerospace Conference*, 1–15. IEEE, 2007.

Calculation of Spectral-Temporal Characteristics of SXR Pulse on GAMMA-4 Facility

V.D.Selemir, N.V.Zaviyalov, V.S.Gordeev, P.B.Repin, A.G.Rep'ev, A.P.Orlov, B.G.Repin, A.L.Mozgovoy, K.V.Strabikin, S.Y.Puchagin

Russian Federal Nuclear Center-VNIIEF, Sarov, Russian Federation

***Abstract*— This paper presents the predicted calculation results of the soft x-ray radiation pulse parameters on the four-module electrophysical GAMMA-4 facility have been creating in RFNC-VNIIEF. The facility operates in the regime of several modules into the single plasma load (Z-pinch).**

Keywords*— *X-ray production, plasma Z-pinch, radiation transfer, multi-group diffusion approximation.

I. Optimality Calculations Based on 0D Model

As a rule, the simplified 0D model is used to obtain the initial value of the output parameters of the complex electrophysical pulsed power facilities that power the dynamic load by the current. This model includes the equations of the equivalent electric circuit of the facility and motion equations of the ideal thin cylindrical shell in the dynamical Z-pinch geometry [1]. Improved double-loop electric circuit of the GAMMA-4 facility is presented in Fig. 1.

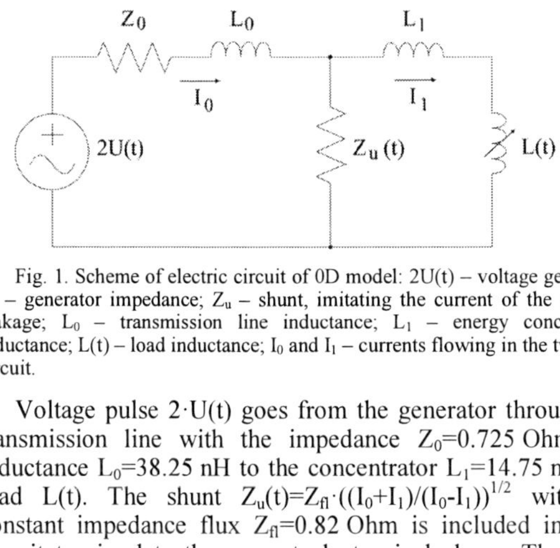

Fig. 1. Scheme of electric circuit of 0D model: 2U(t) – voltage generator; Z_0 – generator impedance; Z_u – shunt, imitating the current of the electron leakage; L_0 – transmission line inductance; L_1 – energy concentrator inductance; L(t) – load inductance; I_0 and I_1 – currents flowing in the two-loop circuit.

Voltage pulse 2·U(t) goes from the generator through the transmission line with the impedance Z_0=0.725 Ohm and inductance L_0=38.25 nH to the concentrator L_1=14.75 nH and load L(t). The shunt $Z_u(t)=Z_{fl}\cdot((I_0+I_1)/(I_0-I_1))^{1/2}$ with the constant impedance flux Z_{fl}=0.82 Ohm is included into the circuit to simulate the current electronic leakage. The pulse voltage form of the running wave is shown in Fig.2.

As a result of optimality 0D calculation series aimed at obtaining the peak kinetic energy, the tungsten load of 1.5 cm height, initial radius of R_0 = 0.9 cm, and optimal linear mass of m_L = 0.18 mg/cm was chosen for further magneto-hydrodynamic calculations. The load was situated in the cylindrical chamber with a return current radius of R_{cam} = 1.3 cm.

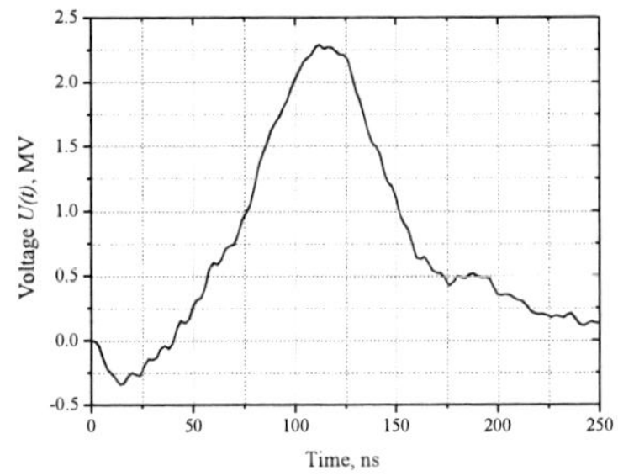

Fig. 2. Running wave voltage pulse form of the Gamma-4 facility

II. Numerical 2D Simulation Results

The calculations of implosion process of the dynamical Z-pinch were carried out with the help of two-dimensional (in Eulerian cylindrical coordinates r-z) radiative magneto-hydrodynamic (RMHD) code FLUX-rz [2]. The code was developed in RFNC-VNIIEF. The calculated model considers the radiation transfer in the multi-group (spectral) diffusive approximation and describes the ablation process of the load in the prolonged plasma formation [3].

Calculations are carried out on the stationary spatial-uniform grid with rectangular cells that have the axial size Δz_i = 100 µm (i = 1…150) and radial size $\Delta r_j \approx$ 29 µm (j = 1…450).

The calculation of the thermal radiation transfer in plasma is realized in 100-group diffusion approximation at the group energy width of 20 eV.

In the process of 2D RMHD simulation, multi-wire one-stage arrays as well as two-stage ones were considered as a load of the GAMMA-4 facility.

The amplitude dependence scaling of SXR pulse and its duration of the quantity of wires was performed for the one-stage array 1.5 cm height consisted of the tungsten wires of 4 µm diameter that were located on the radius R_0 = 0.9 cm. The investigation results are presented in Fig. 3. Let us note that according to the data of SXR pulse characteristics generated in the experiments with multi-wire Z-pinch on the Angara-5-1 facility at the close values of the current pulse parameters [4], the ablation index value of the substance was chosen equal to k_{abl} = 380 µg/(ns·cm^2) in the described scaling. The relative amplitude of the artificial axial perturbations of plasma formation intensity was fixed on the level of $A_0 \sim 20\%$. The given value A_0 was chosen for the load of the linear mass $m_L = 0.18$ mg/cm. This mass was recommended according to the results of 0D calculation series; this corresponds to $N_{wire} \approx 75$ tungsten wires of 4 μm diameter. Change of the perturbation amplitude value was considered at the total quantity variation of the wires in the load according to the law $A_0(N_{wire}) \sim 1/(N_{wire})^{0.5}$ [5].

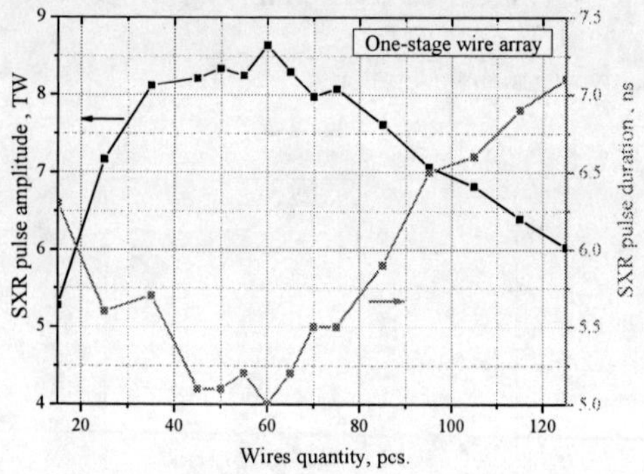

Fig. 3. Dependence of the SXR pulse amplitude (—) and its duration (—) on the wires quantity in case of one-stage wire array.

As resulted from Fig. 3, it is more reasonable to use the load consisted of 60 wires. The peak pulse power of SXR ~8.6 TW with the least duration of ~5 ns is realized at this value.

2D RMHD calculations of the two-stage wire array systems implosion were carried out according to the obtained results. The external array consisted of 40 tungsten wires of 4 μm diameter and was located on the initial radius R_0: the internal array was located on the radius $R_2 = R_0/2$, and the wires quantity (=mass) varied. According to [6], the ablation index k_{abl} for the internal array was defined with regard to the correction for its location radius $\sim (R_0/R_2)^{1.4}$. The results of 2D RMHD calculations are presented in Fig.4.

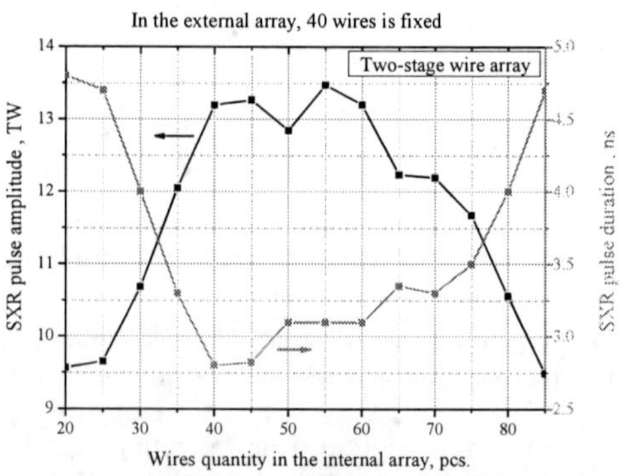

Fig. 4. Dependence of SXR pulse amplitude (—) and its duration (—) in case of two-stage load on the wires quantity in the internal array. In the external array, the wires quantity is fixed and equal to 40.

In Fig. 4 one can see that the optimal parameters of SXR pulse can be achieved when the mass of the internal array is equal to the mass of the external array. This differs from the classical (2:1) two-stage wire array configuration [7]. Based on this result, the authors carried out additional scaling of power and SXR pulse duration varying the quantity of 4 μm wires in such (1:1) loads. The calculated SXR pulse parameters are presented in Fig. 5.

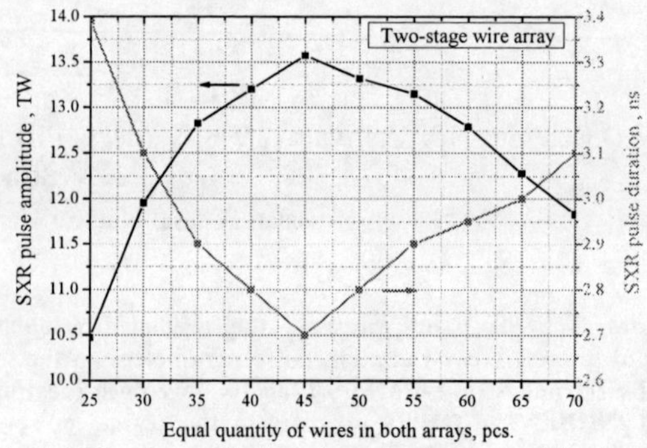

Fig. 5. SXR pulse amplitude (—) and its duration (—) in case of two-stage wire array at equal quantity of wires in both arrays.

The presented analysis of the calculated dependences allows making of a conclusion that in case of two-stage wire array usage, where each stage consists of 45 4 μm wires, it is possible to increase SXR pulse power from 8.6 TW till 13.6 TW and decrease pulse duration from 5 ns to 2.7 ns in comparison with one-stage wire array consisted of similar 60 wires. Such improvement of two-stage multi-wire assembly implosion quality is experimentally proved on many electro-physical facilities [8].

The scaling of SXR pulse basic parameters was performed depending on the internal wire array location for the given two-stage load (with $R_0 = 0.9$ cm). Calculation results are presented in Fig. 6.

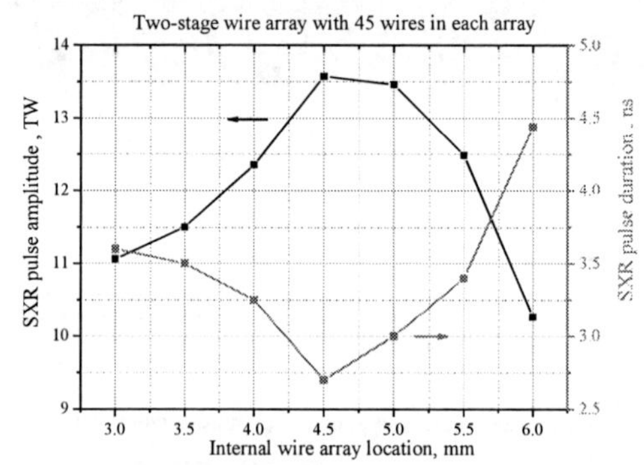

Fig. 6. The calculated values of SXR pulse amplitude (—) and its duration (—) in case of two-stage wire array with 45 wires in each array depending on internal wire array location.

The presented dependencies show that the optimal initial radius of the internal wire array location is the radius $R_0/2$.

III. One-stage and Two-stage Wire Arrays Comparison

In this section we are presenting the comparison results of different characteristics of the implosion process of the best variants of one- and two-stage loads.

Time dependencies of the current and SXR pulses power are shown in Fig. 7. The maximum current in the load in both cases reaches ~3 MA.

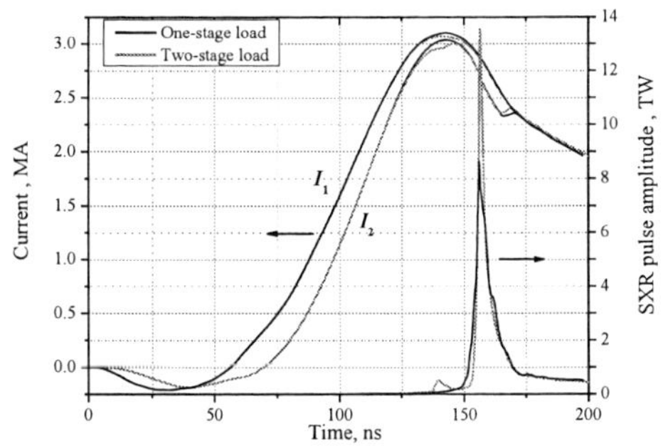

Fig. 7. Time dependence of the currents in the circuit loop and SXR pulses power in case of one-stage (—) and two-stage (-----) loads.

Figure 8 shows time dependence of the load total irradiated energy (at an angle of 4π). It follows from the chart that in both described cases ~75 kJ is irradiated by the moment of SXR pulse end. In addition to the irradiated energy, the temperatures of the irradiating plasma are also identical. It is clearly seen from the approximation of the spectral power radiation distribution by the Planck functions with the temperatures T=160 eV (Fig.10).

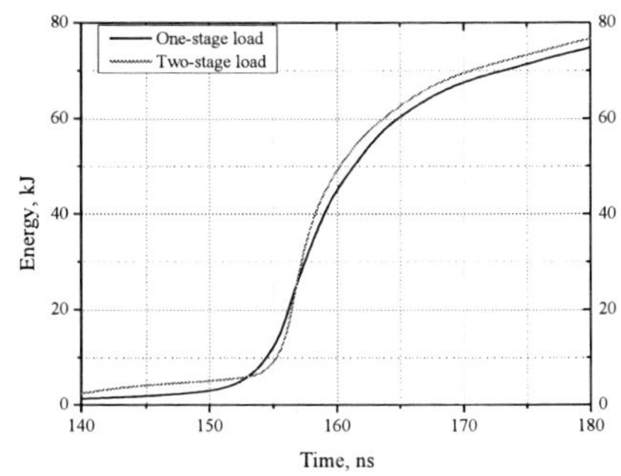

Fig. 8. Time dependence of the irradiating energy in case of one-stage (—) and two-stage (—) loads.

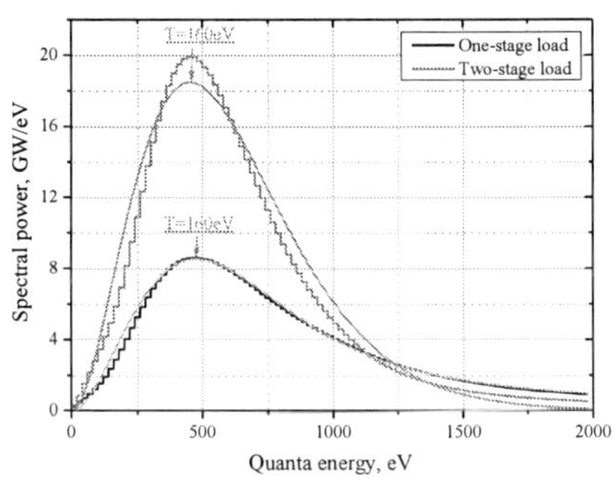

Fig. 9. Quanta energy dependence of radiation spectral power in case of one-stage (—) and two-stage (—) loads and their approximation by the Planck functions with temperature T=160 eV.

In spite of the plasma temperature equality, the area of the irradiating surface in two-stage load turns out to be substantially larger (because of the 1.5 times increase of the total mass if compare with one-stage load). In this case, it leads to increase of the spectral power radiation absolute value.

IV. Results of numerical 3D simulation

Peculiarities of the dynamical Z-pinch implosion, associated with its multi-wire structure, were investigated with 3D (in Eulerian cylindrical coordinates r-φ-z) RMHD code FLUX-3D [9], developed in RFNC-VNIIEF.

The nested multi-wire array implosion was numerically simulated with the optimal parameters defined at 2D RMHD simulation. The internal cascade is located on the radius of 0.9 cm, the external one is on the radius of 0.45 cm; each cascade consists of 45 tungsten wires of 4 μm diameter. An issue about the mutual azimuthal orientation influence of the stages on the generated SXR pulse parameters was investigated. To be exact, in the first calculated variant the wires of the internal cascade were located strictly under the external ones, in the second variant – between them (the external cascade was twisted at the azimuthal angle ΔF=360°/45/2=4°). In both variants, the calculated stationary spatial grid consisted of truncated V-shaped cells of Δz=Δr=100 μm size with an opening angle Δφ=1. Artificial accidental perturbations of A_0=20% amplitude were imposed on the plasma formation intensity in the axial direction only in the wires of the external cascade. However, the samples of the accidental perturbation distribution were not the same (not correlated) between the different wires [9]. The ablation index value of the substance k_{abl} for the wires of the internal and external cascades were the same as at 2D RMHD simulation.

The currents and generated SXR pulses for two variants of two-stage system are shown in Fig.10.

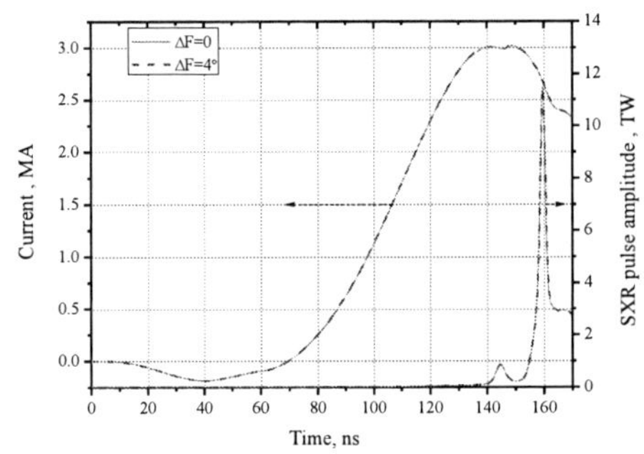

Fig. 10. Time dependence of the currents in the load chamber and SXR pulse power for two-stage wire array in case ΔF=0 (—) and ΔF=4° (- - -).

It follows from the figure that the main calculated characteristics of both systems implosion are practically the same. Thus, the authors will analyze only the variant of two-stage wire array with ΔF=0. Time dependencies of the current and power in time moment t≈145 ns have some peculiarities. They are associated with an interaction process of the current-plasma shell (moving to the axis) of the external cascade with a resting internal multi-wire cascade. Similar peculiarities are experimentally observed on many electro-physical facilities [10] and are explained by the effect of the fast current transmission to the hydrodynamically "transparent" internal cascade.

The spatial density distribution of the tungsten plasma for three time moments are shown in Fig. 11. The moments correspond to: a) the substance ablation termination of the external cascade; b) the substance ablation termination of the internal cascade; c) near the SXR pulse maximum.

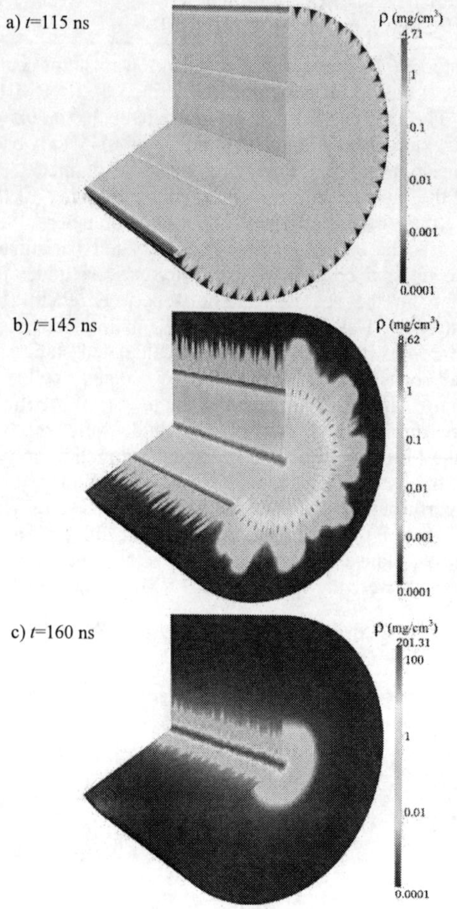

Fig. 11. Spatial distribution of the plasma density in two-stage wire array with ΔF=0 at different time moments.

It is seen in Fig. 11-a that all the space restricted by the internal cascade is uniformly filled by the low-density tungsten plasma carried to the axis by the moment of the external cascade ablation termination. In time moment in Fig. 11-b, the current-plasma shell of the external cascade perturbed as a result of magnetic Rayleigh-Taylor instability growth passes through the internal multi-wire cascade initiating its final electric explosion. The dense (~0.1-0.2 g/cm^3) column of the tungsten plasma is formed in the process of the main SXR pulse generation on the axis of the load chamber (Fig. 11-c).

V. Conclusion

The improved two-loop equivalent electric circuit parameters of the GAMMA-4 facility are determined.

In the frames of 2D RMHD ablative model, the spectral multi-group calculations of the tungsten load implosion powered by the current pulse $I_{max}\approx3$ MA and rise time $\tau\approx60$ ns (at the level of 0.1…0.9) on the GAMMA-4 facility were carried out. The results of the different calculated scalings aimed at obtaining the optimal parameters of the radiating Z-pinch were provided. The comparison of the implosion process efficiency in case of one-stage and two-stage loads usage was made.

Thus, the SXR pulse is generated with a peak power of 8.6 TW, full width at half maximum of 5 ns in case of one-stage wire array (60 tungsten wires of 4 µm diameter) and power of 13.6 TW, length of 2.7 ns in case of two-stage wire array (45+45 tungsten wires of 4 µm diameter) at the peak current of 3 MA in the load. In both cases, the total energy in the pulse is 75 kJ and the emergent radiation spectrum is close to the Planck with the temperature T=160 eV.

It is necessary to note that the presented above calculation parameters (mass distribution) of the two-stage wire array are optimum for GAMMA-4 facility. These parameters differ from the common "standard" nested array, where the mass of internal cascade is twice less than that of the external one. So, coming to the real experiments, it would be useful to foresee the technological possibility of using "standard" (2:1) arrays. Along with calculations and experimental research, this will allow optimizing of SXR pulses generation processes in the two-stage wire array systems.

Performed 3D RMHD calculations of nested wire array of GAMMA-4 facility confirmed efficiency of SXR pulse generation at implosion of an optimum load. The load was selected based on the results of two-dimensional calculated scaling. Besides, these calculations predict absence of the relative angle orientation of the external and internal stages influence on the generated SXR pulses parameters.

VI. References

[1] J. Katzenstein. Optimum coupling of imploding loads to pulse generators. J. Appl. Physics., V.52, No.2, 1981, pp.676-680.

[2] B.G. Repin, A.P. Orlov, P.B. Repin, and V.D. Selemir. Calculation method of radiation spectral transfer in frameworks of two-dimensional magnetohydrodynamic code FLUX-rz. IEEE Trans. on Plasma Science. 2010. V.38. No.8. pp.1822-1827.

[3]V.S. Gordeev, A.V. Grishin, A.P. Orlov, A.L. Mozgovoy, B.G. Repin, et al. On prospects of powerful SXR pulse generation on "Gamma 4" facility. Int. J. Modern Physics: Conference Series, Vol.32 (2014), 1460335.

[4] V.V. Alexandrov, V.A. Barsuk, E.V. Grabovskiy, et al. Plasma Physics. 2009. V.35. №3. pp. 229-250.

[5] M.G. Haines. A Heuristic Model of the Wire Array Z-Pinch. IEEE Trans. Plasma Sci. 1998. Vol.26. No.4. pp. 1275-1281.

[6] K.N. Mitrofanov, E.V. Grabovskiy, V.V. Alexandrov, et al. Implosion investigation of the foam wire assembly on the Angara-5-1. Plasma Physics. 2012. V.38. №12. pp. 1022-1041.

[7] B. Jones, C.A. Coverdale, C. Deeney, D.B. Sinars, and et al. Implosion dynamics and K-shell x-ray generation in large diameter stainless steel wire array Z pinches with various nesting configurations. Phys. Plasmas **15**, (2008), 122703.

[8] Xian-Bin Huang, Shao-Tong Zhou, Jia-Kun Dan, et al. Preliminary experimental results of tungsten wire-array Z-pinches on primary test stand. Phys. Plasmas **22** (2015) 072707.

[9] A.P. Orlov and B.G. Repin. Numerical simulation of multi-wire Z-pinches within the framework of 3D magneto-hydrodynamics. IEEE Trans. on Plasma Science. 2015. V.43. No.8. pp.2515-2519.

[10]T.W.L. Sanford, M.E. Cuneo, D.E. Bliss, C.A. Jennings, and et al. Demonstrated transparent mode in nested wire arrays used for dynamic hohlraum Z pinches. Phys. Plasmas **14**, (2007), 052703.

On the Potential Advantage of a Bipolar Capacitor for Driving Electromagnetic Railguns

Markus Schneider and Gregory Vincent
Electromagnetic Acceleration Group (ERG)
French-German Research Institute of Saint-Louis
Saint-Louis, France

Abstract— **One of the most challenging topics with regard to railgun technology concerns the realization of an adequate power supply. The typical fields of application of powerful (GW-range) railguns require compact and weight-optimized solutions because they should be installed on mobile platforms e.g. ships. While various energy storage methods (kinetic energy, electric field, magnetic field) and combinations of them were suggested and realized, the most mature technique available today is based on the use of capacitors. Currently available compact capacitors are unipolar. Therefore, they are typically used in a crowbar configuration.**

In this study we analyze the pros and cons of an imaginary bipolar capacitor for high power railguns. Using PSPICE, we study the electric circuit and the launch dynamics of such a "ringing" operation of a railgun and compare the performance of the power supply with conventional systems using crowbar diodes. We also put attention to the question, which types of switches are required.

Keywords—power supply, railgun

I. Introduction

Recently, remarkable progress has been made in the field of railgun technology. Most prominently, the US Navy has realized experiments with 32 MJ of kinetic muzzle energy and intends to carry out sea trials in the near future [1]. From an operational point of view, very important aspects of the technology are the volume and the mass of the power supply. Currently, most of the laboratory railguns operating in the GW-range are driven by capacitor banks. While this technology is reliable, in particular in combination with the use of semiconducting switches, many operational scenarios require an installation of the launcher on a mobile platform. Therefore, compact and light-weight power supplies are to be developed.

In principle, the available capacitor technology for GW-range railguns, which is characterized by energy densities between 1 and 3 MJ/m^3 [2] has to be matched against other solutions such as the use of rotating masses or inductive storage. However, in this short contribution we focus on a particular aspect of the capacitor technology. In fact, the high power capacitors used for railgun experiments are unipolar. The question raised here is: would a bipolar capacitor with the same energy density as a unipolar one offer any advantages – in particular with regard to weight and volume?

We tackle that problem in three steps. First, we present generic considerations without specifying the switch to be used in the circuit. Then, we consider a circuit using "realistic" i.e. currently available (semiconductor) switches and make some remarks on the implications of bipolar operation for the launcher. Finally, we illustrate that the use of bipolar capacitors would offer options not available for unipolar versions.

II. Simulation Model

The typical electric circuit for a capacitively driven railgun is sketched in Fig. 1. The capacitor C is protected against a reversal voltage by the diode D. The inductance L limits the peak current and determines the pulse shape. The closing switch S is considered to be generic.

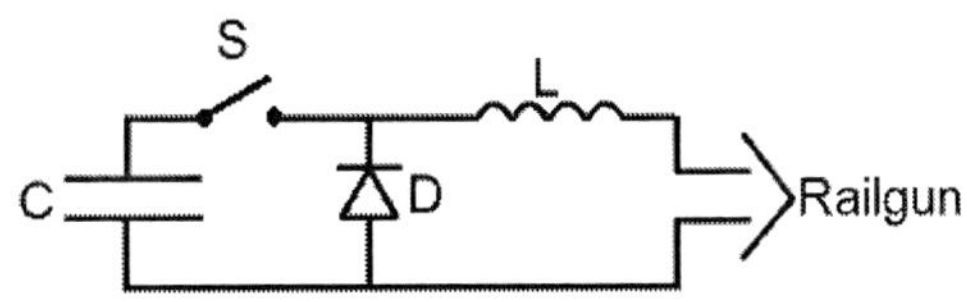

Fig. 1: Sketch of generic electric circuit.

In order to simulate the circuit sketched in Fig. 1 with a Kirchhoff solver, the railgun needs to be implemented. How to do this with PSPICE has been shown earlier [3]. In order to model the launcher, the position and the velocity of the projectile have to be calculated. Fig. 2 shows the circuit version used for our work here. Table 1 lists the circuit parameters. We did not take into account the impedances of the cables and we did not specify the length of the railgun.

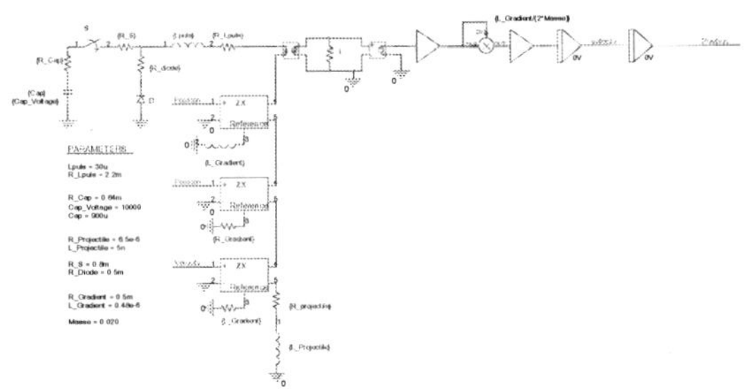

Fig. 2: PSPICE model used for the numerical simulations.

TABLE I. SIMULATION PARAMETERS

Element	*Values*
L	L = 30 μH, R_L = 2.2 mΩ
C	C = 900μF, R_C = 0.64mΩ
D	R_{on} = 0.5mΩ
Rails	L' = 0.48 μH/m, R' = 0.5 mΩ/m
Projectile	L = 5 nH, R = 6.5 μΩ

III. Generic Considerations

We start our considerations by returning to Fig. 1. As the diode D has the job to protect the capacitor C against reversal voltage one could ask the question: would it be advantageous to get rid of it? Assuming a bipolar capacitor of the same volume (and cost) as its unipolar counterpart, what could be gained? Indeed, the railgun force is proportional the square of the current amplitude and therefore not sensitive to changes in its sign.

It was this simple question which triggered the simulations presented here. In this section, we do not specify the closing switch S – apart from its resistance. In the following we compare the crowbar (or unipolar) case depicted in Fig. 1 with the ringing (or bipolar) version without the diode D. Even if the railgun is not a trivial load, we would expect that the performance of our generic power supply depends on the resistances of their components. In the bipolar case C and S are to be taken into account whereas they get disconnected in the unipolar case. In the Figs. 3-5 we summarize the numerical results obtained for different values for the resistances of C, S and D. We have plotted the profiles for velocity and current for three cases. It can be stated that for the simulation parameters chosen here, the performance of the power supply depends indeed on the values of the resistances of C, S and D. A higher end velocity is obtained if the Joule losses are smaller.

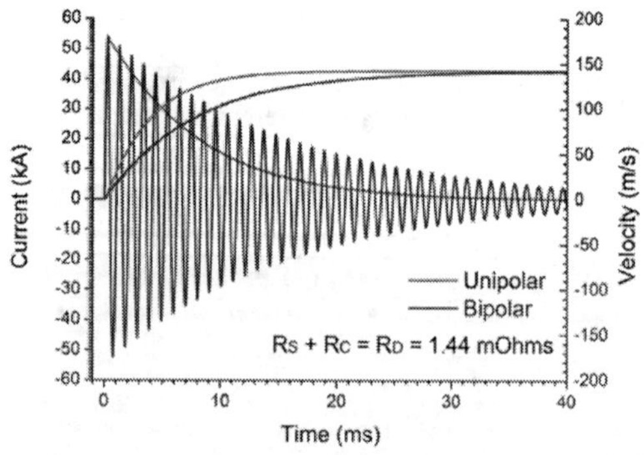

Fig. 3: Simulated profiles of current and velocity .

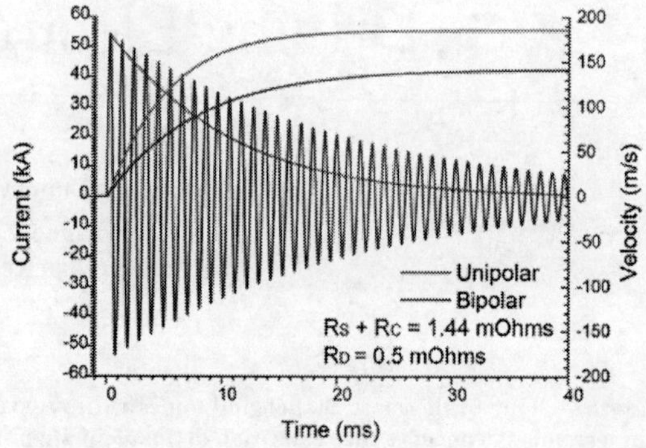

Fig. 4: Simulated profiles of current and velocity.

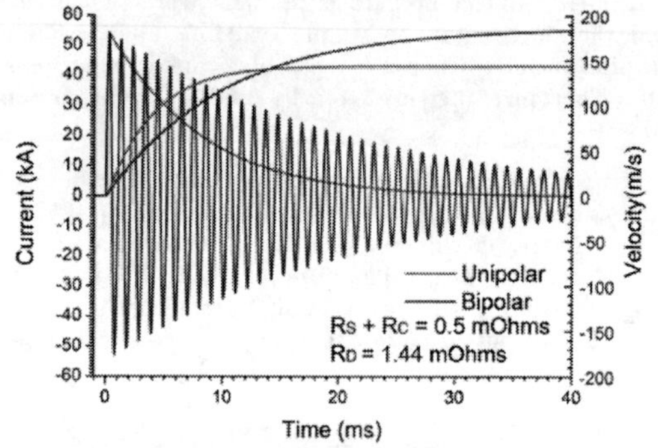

Fig. 5: Simulated profiles of current and velocity.

IV. Realistic Conditions

A. Circuit with realistic switches

The generic considerations of the previous section have shown that the performance of the power supply depends strongly on the realization of the switches. Among the semiconducting closing switches, the thyristor is the most suited for the current pulses required by large railguns. Therefore, we replaced in Fig. 6 (left) the generic switch S by a thyristor T. However, a thyristor is not a symmetric device. The "ringing" circuit (Fig. 6, right) requires an antiparallel diode D which takes the reversal current. It becomes clear that a more compact power supply would only rely on topology but not on the number of elements. In Fig. 7 we compare the performance of both configurations. It can be seen that the crowbar (or unipolar) power supply is clearly superior. Therefore, we conclude that with the assumptions taken here, not even a potential advantage of a bipolar power supply could be demonstrated.

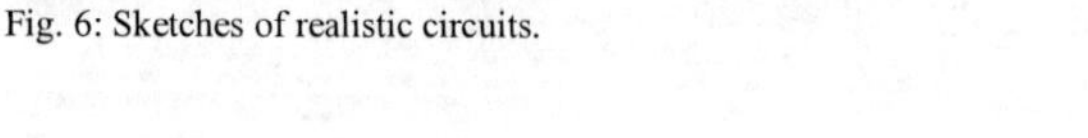

Fig. 6: Sketches of realistic circuits.

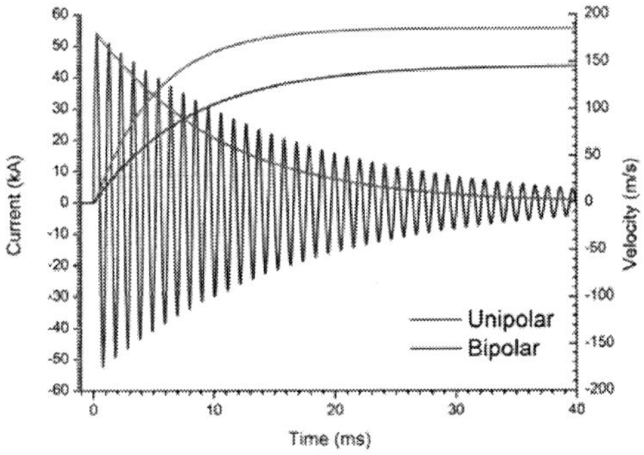

Fig. 7: Simulated profiles of current and velocity.

B. Remarks on railgun operation

While it is not in the scope of this work to include detailed analysis of the consequences of the use of a ringing current for railgun operation, two remarks concerning pivotal technological challenges are appropriate. First, the quality of the sliding electric contact between armature and rails can get worse if a current profile has sections with negative dI/dt. This phenomenon was coined "down-slope transition" [4]. It means that the launch efficiency decreases and the lifetime of the rails is reduced. The second aspect concerns the structural mechanics of the launch. Experiments and numerical simulations have shown that the launch performance depends critically on the structural response of the launcher [5, 6]. During launch experiments, elastic waves are induced in the structures and a proper design has to take into account their behavior. In comparison to a classical operation of a railgun, the oscillating current profile would initiate additional excitations in the housing, which would have to be controlled.

C. Example of application: velocity control

In this last section of our study we would like to point out an interesting option of the oscillating operation of a railgun. In Fig. 8, we show the circuit. In comparison with Fig. 6 the passive diode D is replaced by an active thyristor T. This allows for realizing an elegant velocity control as is shown in Fig. 9. The gate voltage of the "negative" thyristor is removed meaning that the device will block. Consequently, the current flow stops as the capacitor C is charged to its nominal voltage. Note that this option does not exist for the unipolar circuit because it is the negative voltage occurring at its terminals which switches off the "negative" thyristor.

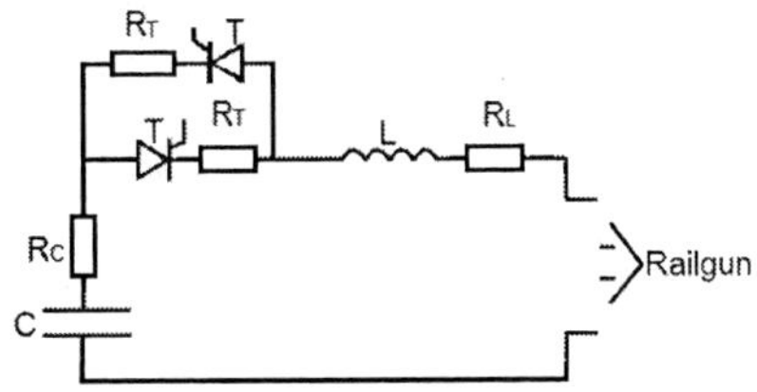

Fig. 8: Circuit sketch for implementation of velocity control.

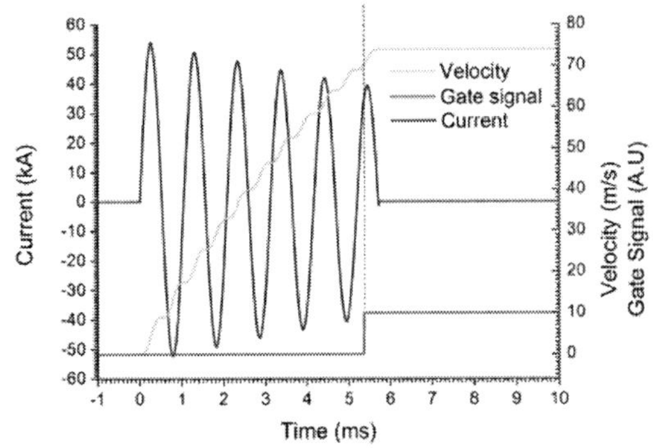

Fig. 9: Numerical results obtained using the circuit of Fig.8.

V. Conclusion

At first sight the use of a bipolar capacitor seems to offer advantages with regard to the compactness of the power supply because the crowbar diode is not needed. However, the numerical circuit analysis presented here reveals that even if both capacitor types would have the same energy density, the unipolar version is to be preferred with regard to compactness. We also indicate that niche applications for a bipolar version exist, such as an elegant velocity control.

References

[1] LaGrone, Sam (2010-12-15). "Electromagnetic railgun sets new world record". Jane's. Archived from the original on 2010-12-17. Retrieved 2014-12-22.

[2] I. R. McNab, "Large scale pulsed power opportunities and challenges," IEEE Trans. Plasma Science, vol. 42, pp. 1118-1127, May 2014.

[3] J. Wey, E. Spahn and M. Lichtenberger, "Railgun modeling with the P-Spice code," in IEEE Transactions on Magnetics, vol. 33, no. 1, pp. 619-624, Jan 1997.

[4] F. Stefani, S. Levinson, S. Satapathy and J. Parker, "Electrodynamic transition in solid armature railguns," in IEEE Transactions on Magnetics, vol. 37, no. 1, pp. 101-105, Jan 2001.

[5] A.J. Johnson, and F.C. Moon, "Elastic waves in electromagnetic launchers," IEEE Trans. Magn., vol. 43(1), pp. 141–144, Jan. 2007

[6] C. Schuppler, L. Tumonis, R. Kacianauskas, and M. Schneider, "Experimental and numerical investigations of vibrations at a railgun with discrete supports," IEEE Trans. Plasma Science, vol. 41, pp. 1508-1513, May 2013.

The Peculiarities of the Application of Magnetic-Pulse Method for Forming Controlled Pressure Pulses to Test Metal Samples

Krivosheev S.I., Magazinov S.G., Alekseev D.I.
Peter the Great Saint-Petersburg Polytechnic University,
Saint-Petersburg, Russia
ksi.mgd@gmail.com

***Abstract* — Experimental study of brittle nonconductive materials with magnetic-pulse method revealed a number of common patterns of fracture process. Loading schemes of metal samples, which exclude the impact of electroplastic effect on the deformation properties of metal, were considered. Numerical simulation results demonstrate the ability to test metal samples with magnetic-pulse method in microsecond duration range. The possibility of using of energy storage time concept to describe the elastic-plastic transition at strain rates up to 10^5 1 / s was shown. Ways of reducing the induced currents in the zone of plastic deformation of samples were suggested.**

Keywords* — *magnetic-pulse loading, quasi-coaxial magnetic-pulse driver, energy storage time, elastic-plastic transition.

I. Introduction

Magnetic pulse method for creating controlled pressure pulses of microsecond duration in the study of the pulse strength of brittle non-conductive materials revealed a number of common patterns of fracture process [1, 2]. Using of thermodynamic approach for analysis of the results allowed revealing specific relations among extreme destructive loading, the duration of this loading and specific material parameter – energy storage time [3]. The special feature of this method is the ability to form stressed states in which there is no pre-stressed modes in material, typical for spall test schemes. Implementation of this feature is especially important for the test of metals, composite and laminate materials, with an explicit spatial anisotropy of the deformation characteristics.

II. Conductive Samples with Defect of Type of Crack

A. Induced currents structure

Typical magnetic pulse loading of samples with macro defects is shown in Fig. 1. Selection of parameters of the generator and a magnetic system for generating controlled pressure pulses of microsecond duration can be implemented on the recommendations of [4]. The pressure amplitude can reach 2 - 10 GPa. Such pressure amplitude is sufficient for testing metallic materials. However, the transition to tests of conductive samples imposes a number of limitations related to the possible impact of the induced current in the sample in the deformation curves of metal.

Work is supported by Russian Federal Targeted Program for 2014-2020 Grant #14.584.21.0019.

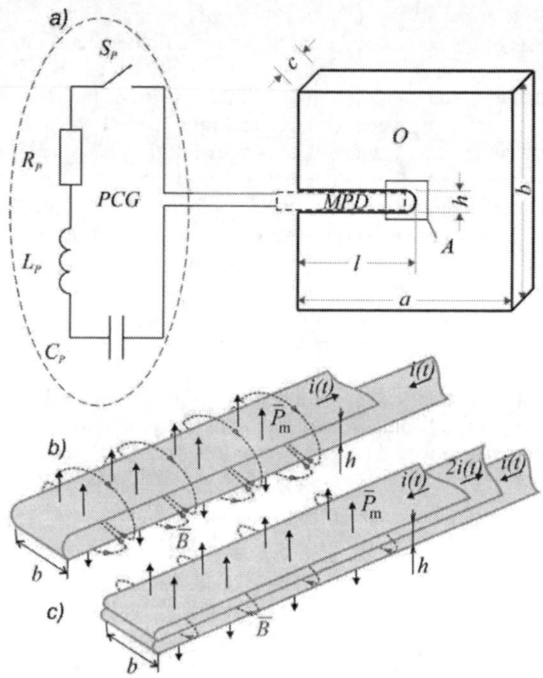

Fig. 1 Current density distribution near the groove top for different magnetic pulse drivers.

The results of modeling of induced currents structure in metallic sample for the simple driver in the form of a system of flat bars and quasi coaxial magnetic pulse driver given in [3] are shown in Fig. 2. Analysis on modeling results describes that quasi coaxial driver provides the formation of stressed state free from the effects of induced currents in the zone of maximum stress. This zone is free from the influence of electroplastic effect (EPE), which determines in accordance with [5, 6] for the critical current density EPE $10^9 A / m^2$, is sufficient to study of deformation in conditions of shock loading when using quasi coaxial magnetic pulse driver system. The results of calculation of current density distribution on the sample during the discharge of the pulse

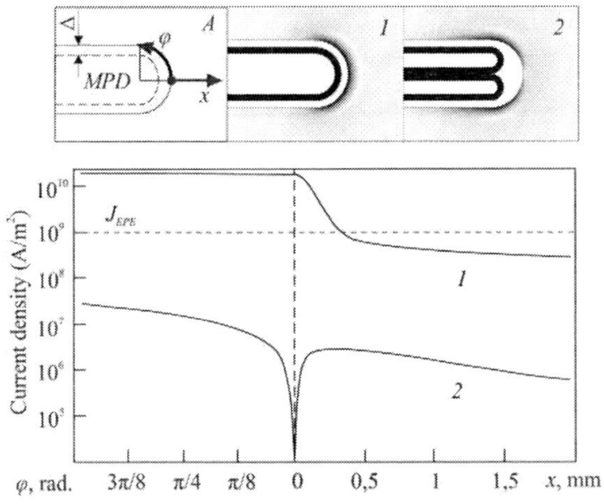

Fig. 2 Current density distribution near the groove top for different magnetic pulse drivers.

current generator, which forms current with linear density 200 kA/sm and duration 10 mks in bars of magnetic pulse driver, is given in Fig. 2. The amplitude of pressure, which impact on the sample, is ~ 0.25 GPa.

B. Elastic-Plastic deformation modeling

Modeling of elastic – plastic deformation of copper bar with macro defect of type of crack in the tree-dimension setting was carried out environment ANSYS Autodyn [7]. The Johnson – Cook plasticity model for OFHC [8] copper was selected as a calculation model of deformation. The reliability of this model for various metals is supported by its implementation in different programming environments, e.g.[7]. According to this model the elastic limit of material varies depending on strain, strain rate and temperature:

$$\sigma = \left[A + B\varepsilon_P^n\right]\left[1 + C\ln v_p^*\right]\left[1 - T^{*m}\right], \qquad (1)$$

where ε_p is an effective plastic strain, v_p^* is a standardized effective plastic strain rate, $T = (T - T_{room})/(T_{melt} - T_{room})$ is homologous temperature, $A = 90$ MPa, $B = 292$ MPa, $n = 0{,}31$, $C = 0{,}025$, $m = 1{,}09$, $T_{melt} = 1082{,}9$ °C are model parameters.

The Johnson-Cook fracture model is applied as the failure criterion:

$$D = \sum \frac{\Delta\varepsilon}{\varepsilon^f}, \qquad (2)$$

$$\varepsilon^f = \left[D_1 + D_2 e^{D_3\sigma^*}\right]\left[1 + D_4 \ln v_p^*\right]\left[1 + D_5 T^*\right], \qquad (3)$$

where ε^f is an effective deformation fracture, v_p is a strain rate $D_1 = 0.54$, $D_2 = 4.89$, $D_3 = -3.03$, $D_4 = 0.014$, $D_5 = 1.12$ are parameters of fracture model. When the parameter D reaches the value 1, the material is destroyed.

The Johnson – Cook plasticity model describes deformation process up to fracture, and the validity of the use of this models is justified by the good correspondence between calculated and experimental deformation curves for different plastically deformable materials at strain rates up to ~ 2000 1 / s. The applicability of this model for description of the high-speed deformation process for steel 09G2S at strain rates up to 1500 1 / s is shown in [9].

The character of the stress state is defined by the characteristic size of the sample b (Fig. 1) and the loading wavelength $\lambda = c_l \cdot T$, where c_l - speed of sound in the material. If λ is much smaller than b shock wave mode is implemented, otherwise, the quasi-static is. Calculations were provided for two samples with different sizes: 1 – $a = 3$ sm, $b = 3$ sm, $c = 1$ mm, $h = 2$ mm, $l = 1{,}5$ sm; 2 – $a = 12$ sm, $b = 12$ sm, $c = 1$ mm, $h = 2$ mm, $l = 6$ sm (Fig. 1). Quasi static mode of loading was implemented in the first sample by impacting distributed pulse of pressure with duration $T = 50$ mks to groove shores. Shock wave mode was implemented in the second sample by impacting distributed pulse of pressure with duration $T = 8$ mks. The results of modeling show that in both cases the maximum stresses are formed in a zone located near the top of the groove.

The deviation the deformation curve from a linear one due to plastic flow leads to a significant reduction in the achievable stress for the case of the shock-wave loading (curves 1, 2, 3 in Fig. 3a). This leads to the failure to ensure a high strain rate, as it shown in Fig. 3. Transition to quasi

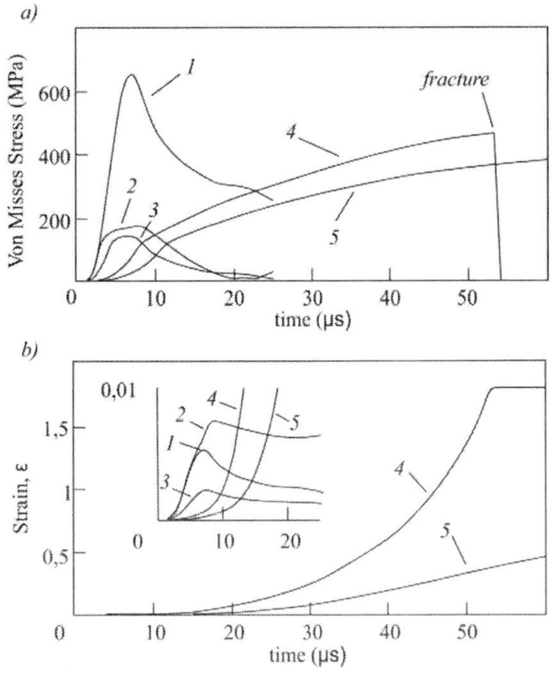

Fig. 3 a) Relation between strees on the groove top and time; b) Relation between relative deformation on the groove top and time. 1 – $P_m = 250$ МПа, $T = 8$ mks (excluding plastic deformation); 2 - $P_m = 250$ МПа, $T = 8$ мкс, $v_p = 1600$ 1/c; 3 - $P_m = 85$ МПа, $T = 8$ мкс, $v_p = 500$ 1/c; 4 - $P_m = 250$ МПа, $T = 50$ мкс, $v_p = 90000$ 1/c; 5 - $P_m = 85$ МПа, $T = 50$ мкс, $v_p = 14000$ 1/c.

stationary mode allows to extend significantly the range of variation of achievable stress, and to provide a large deformation of the material at the top of the groove, until the fracture of sample (Fig. 3, curve 4.).

Analysis of numerical simulation results show that when the sample is impacted with evenly distributed along the shores of the groove pressure created by magnetic pulse driver, it is possible to provide the deformation of the sample at a strain rate of up to 10^5.

C. Experimental setup and procedure

The experimental setup the generator of pulse currents GIT-50.12 (storage capacity $C_p = 12$ µF, charging voltage $U_p = 50$ kV, self-inductance $L_p = 53$ nH) was used to provide initial test of samples according to the scheme shown in Fig. 1.

The material of the sample was copper Cu ETP, sample sizes – $a = 0{,}03$ m, $b = 0{,}03$ m, $h = 0{,}002$ m, $l = 0{,}015$ m aperture diameter 0,01 m (Fig 4.). After charging the generator up to 30 kV, GIT - 50.12 forms pulse current with amplitude ~ 360 kA in magnetic pulse driver. Current has form of strongly damped sinusoid with the attitude of neighboring amplitudes ~ 0.6 and period of 5 mks The corresponding to this current pressure pulse had an amplitude ~ 0.69 GPa in test with magnetic pulse driver with width 11 mm. Comparison of experimentally registered permanent deformation with the results of numerical modeling is given in Fig. 4. and shows the inflated value of the groove deformation $h_m = 3{,}6$ mm, compared with the experimental value $h_e = 2{,}8$ mm. According to the results of numerical modeling the highest mechanical stresses during deformation are concentrated in the top of the groove and plastic deformation strain rate reaches $v_p \sim 15000$ 1 / c, however the Johnson-Cook fracture model described in paragraph B may be insufficiently adequate for a given strain rate.

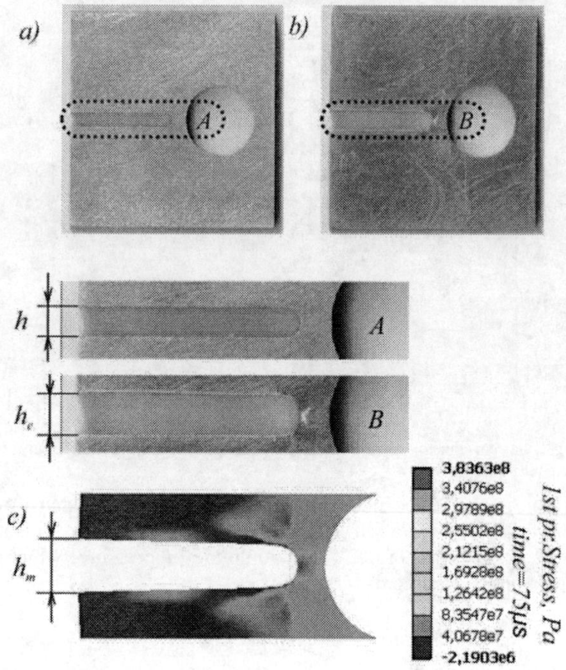

Fig. 4 Comparison of experimental results and numerical modeling. a) The sample before test; b) The sample after test; c) Numerical modeling results – 1st.Principal Stress; time, t=75µs.

III. UNIAXIAL DEFORMATION OF METAL SAMPLE

To expand the range of strain rates and to exclude the influence of form factor of the sample on the deformation characteristics of the material the scheme of loading was developed. This scheme can be seen in Fig. 5.

The studied system consists two round copper plates 1 the radius $R_p = 50$ mm between which the test object - a wire of the test material 4 the radius $R_w = 1$ mm is fixed , the central part of test object has a thinning $R_d = 0.5$ mm radius. Near

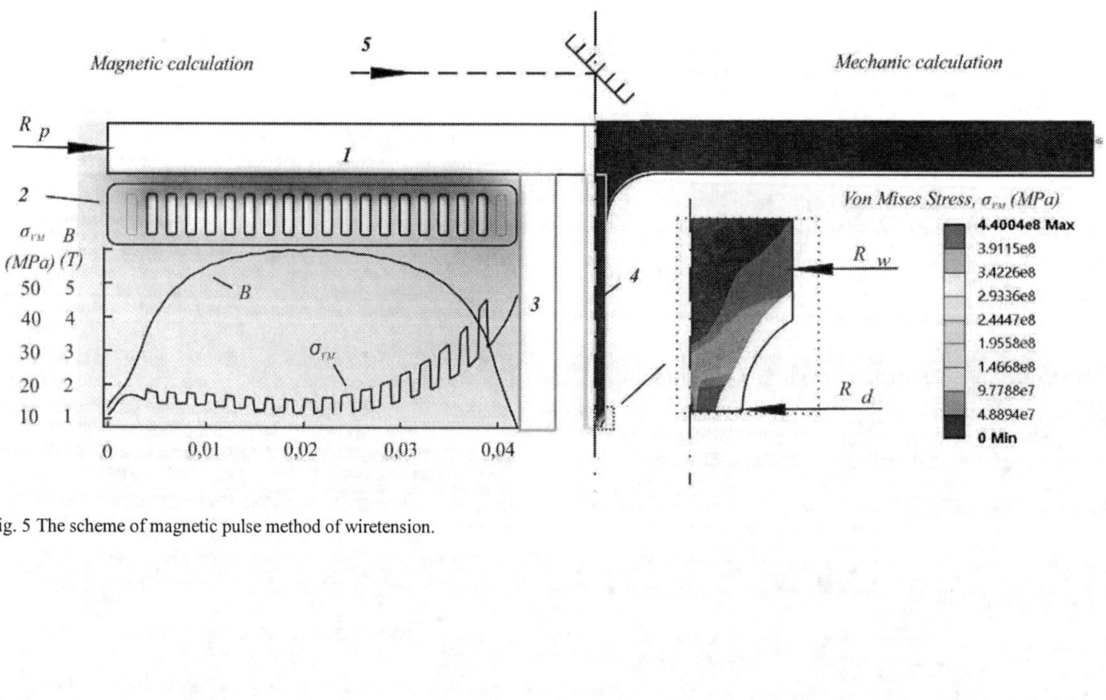

Fig. 5 The scheme of magnetic pulse method of wiretension.

plate there is a magnetic system 2, which creates magnetic pressure to push the copper plate, through which moving the wire is stretched. The test object is placed in a steel shielding tube 3, to reduce the influence of the induced current in it.

The calculation of the magnetic system with considering of existing materials was provided. The magnetic system has a form of solenoid in the shape of an Archimedean spiral with 18 turns. The coils are made of copper Cu ETP bus with size 4x1 mm, the solenoid is filled with rigid insulating compound. This system has an inductance equal to 5.1 mH, and can create magnetic field with induction up to 10 T without failure. Application of laser interferometer in this scheme makes it possible to register the displacement of the top point B during the experiment.

IV. Conclusion

Analysis of modeling results showed the possibility of using magnetic-pulse method of deformation of conductive materials at strain rates up to ~ 100 000 1/sec.

The proposed loading scheme allows combining the process of deformation with a pulse current exposure and investigating the impact of electro plastic effect.

Transition to higher strain rates will provide data for Johnson – Cook model verification in a wider range of influences.

References

[1] S.I. Krivosheev, N.F. Morozov, Y.V. Petrov, and G.A. Shneerson, "Problems of fracture of brittle bodies under pulse loading," Materials Science, vol. 32, is. 3, pp. 286-295, 1996.

[2] S.I. Krivosheev and G.A. Shneerson, "The use of pressure pulses arising from the creation of strong pulsed magnetic fields for the study of the dynamic strength of materials," in 2006 International Conference on Megagauss Magnetic Field Generation and Related Topics, including the International Workshop on High Energy Liners and High Energy Density Applications, MEGAGAUSS 2008. Article number 4530709. P. 407-412. DOI: 10.1109/MEGAGUSS.2006.4530709

[3] S.I. Krivosheev, N.V. Korovkin, V.K. Slastenko, and S.G. Magazinov, "Destruction of brittle materials by microsecond pressure pulses at their formation by magnetic pulse method," International Journal of Mechanics, vol. 9, pp. 293-299, 2015.

[4] S.I. Krivosheev, "Pulsed magnetic technique of material testing under impulsive loading," Technical Physics, vol. 50, is. 3, pp. 334–340, 2005.

[5] O.A. Troitskii, and L.G. Maistrenko, "Electroplastic deformation in metals," Soviet Materials Science, vol. 8, is. 6, pp. 686–689, November 1974.

[6] H. Conrad, "Thermally activated plastic flow of metals and ceramics with an electric field of current," Material Science and Engineering, vol. A322, pp. 100-107, 2002.

[7] ANSYS Academic Research Customer #00420725 (from 2010), CompMechLab SPbSPU, http://www.compmechlab.ru/

[8] G.R. Johnson and W.H. Cook, "Fracture characteristics of three metals subjected to vari-ous strains, strain rates, temperatures and pressures," J. Eng. Frac. Mech., vol. 21, is. 1, pp. 31-48, 1985.

[9] Vl. Vas. Balandin, Vl. Vl. Balandin, A. M. Bragov, L. A. Igumnov, A. Yu. Konstantinov, and A. K. Lomunov, "High-rate deformation and fracture of steel 09G2S," Mechanics of Solids, vol. 49, No 6, pp. 666-672, 2014.

On the Impact of the Elastic-Plastic Flow upon the Process of Destruction of the Solenoid in a Super Strong Pulsed Magnetic Field

Krivosheev S.I., Magazinov S.G., Alekseev D.I.
Peter the Great Saint-Petersburg Polytechnic University,
Saint-Petersburg, Russia
ksi.mgd@gmail.com

***Abstract* — During the interaction process of super strong magnetic fields with solenoid material a specific flow mode of the material forms. For a description of this process magnetohydrodynamic approximation is traditionally used. The formation of plastic shock-waves in material in a rapidly increasing pressure of 100 GPa/μs, can significantly alter the distribution of the physical parameters in the medium and affect the flow modes.**

Keywords* — *super strong magnetic fields, shock-wave, plastic shock-wave

I. INTRODUCTION (*HEADING 1*)

It is characteristic for generation of super strong magnetic fields to have an especially nonstationary character of forming of energy release zone, which is close to or even exceeds material sublimation energy, obtained during numerous experiments. Achieving high energy density is provided by high speed of current rise, which leads to its uneven distribution across the thickness of the conductor. Distribution of induction in the zone of current flow at in the generation of strong and super strong pulse magnetic fields in single-turn coil by the direct discharge of the capacitor bank at least does not decrease radially from the conductor in the direction of the field. Penetration of the diffusion layer and the material transfer is determined by the processes in the wall of the coil and its extension in terms of high energy impact on the inner side. In other words, there are no spatial constraints to increase the area occupied by the magnetic field in single-turn solenoids – an open system. Z-pinch configuration of the electromagnetic field implemented with an electrical explosion of a wire is characterized by the decline of the magnetic induction at the distance from the conductor in the area occupied by the current and the field and the possibility of movement of the material and the diffusion field to the wire axis which is limited with conductor size – a closed system.

II. POSSIBLE FLOW MECHANISMS

Traditionally, the analysis of the processes occurring during the electric explosion of conductors and generation of strong and super strong pulse magnetic fields is carried out within the framework of magnetohydrodynamics (MHD) equations with corresponding equations of state for material and conductivity models. The transition to the use of the MHD approximation for open systems with the wall thickness greater than the depth of the magnetic field penetration may be limited to the process of elastic-plastic material flow, which is not affected by the energy release in the current flow area, but subjected to the action of magnetic and gas-kinetic pressure. Plasticity wave velocity is significantly lower than the speed of sound and the shock wave velocity in the material, so it is possible for this process to impact strongly on the results of the MHD calculations. Consider this situation as an example of experimental studies of generation strong and super strong pulse magnetic fields by direct discharge in miniature single-turn copper solenoids.

Experimental data on generation of magnetic field with the induction of 360 T with rise time of 0.55 μs in a single-turn coil [1] show, that visible coil motion is absent until the maximum induction, and a sharp change in its dimensions begins after exceeding maximum magnetic field (Fig. 1.). Within the two-dimensional simulation of the hydrodynamic flow the experimentally observed delay in the start of movement of the internal radius of the solenoid was explained and the picture of nonlinear field diffusion processes was substantially clarified. Numerical simulation of the interaction process of magnetic field and a conductor with the use of the conductivity model [2, 3] and the three-term equation of state

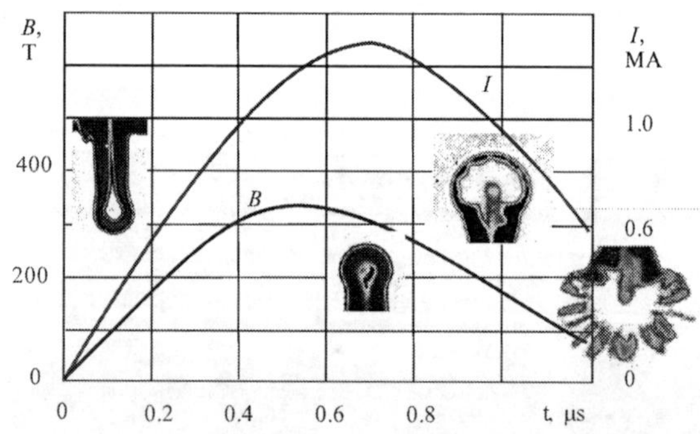

Fig. 1. Typical oscillogram of current I, the induction of the magnetic field in the center of the solenoid B and evolution of solenoid during generation of strong and super strong pulse magnetic field - according to [1]. The location of the X-ray picture of the solenoid corresponds to the time of registration.

of the solenoid material in a two-staged MHD approach was provided. Here, according to the findings of the simulation, the mechanical characteristics of the materiala formed in the process of magnetic field generation of with a pressure amplitude of about 50 GPa do not affect the coil expansion.

A. Induced currents structure

Nonlinear magnetic field diffusion in material with energy up to material sublimation energy is a complicated process. The description of this process requires adequate equations of state for material and appropriate conductivity model. Analytical estimation of the diffusion rate for linearly increasing magnetic field is presented in [5] and is clarified in experimental research [6], by adding additional dimensionless parameter ξ of order unity.

$$\Delta_j = \xi \frac{B(t)}{B_0}\sqrt{\frac{t}{\mu_0\sigma_0}} = \xi\frac{B(t)}{B_0}\Delta. \quad (1)$$

where *B(t)* – relation between the induction of magnetic field on the conductor surface and time, Δ – depth of the skin layer for linear diffusion, B_0 – characteristic induction of magnetic field – dimensionless parameter.

B. Some typical modes of shock wave

It is known [7] that in the field with the induction B of 1000 T the medium speed u_f behind the shock-wave front can be described using the approximation:

$$u_f = a_0 B^{3/2} \quad (2)$$

for copper a_o=0.15.

In view of the Rankine-Hugoniot relations the distance traveled by the shock wave is determined by the following equation:

$$\Delta_f = \int_0^t D dt = \frac{1}{2a_0\mu_0\rho_0}\int_0^t \sqrt{B(t)}dt \quad (3)$$

where D – shock –wave velocity, ρ_0 – material density.

Shock-wave velocity significantly exceeds magnetic field diffusion rate in magnetic fields up to 240 T. However, the implementation of mode when the forming of shock–wave is slower than current diffusion is possible as shown in [5].

Further propagation of the wave front occurs in conjunction with the captured current forming a specific course of flow. Here take place an additional energy release in the shock front due to joule heating of the material. Analysis of experimental data represented in [8] shows shock-wave velocity with additional energy inlet can exceed usual shock-wave velocity, initiated by appropriate magnetic field pressure. Relation which demonstrates shock-wave propagation with external energy inlet is presented in Fig. 3, curve 3. Energy release area forming is determined by the combined action on single-turn coil material of magnetic pressure, Joule heating, field diffusion. An additional factor, which may influence on evolution of energy release area, is transitional character of flow mode forming. It is mentioned in the number of papers [9] that process of shock-wave forming in relation with its amplitude may have the duration from 20 to 100 ns.

Results of experimental shock-wave loading reveal that shock-wave does not have time to form at a distance of 2 mm in copper at pressure amplitudes of up to 20 GPa. Flow mode, in which the velocity of the propagation perturbation is significantly below the speed of sound, is observed in a large number of papers. Since the material in these modes is in a state of deep plastic deformation, this flow mode is classified as elastic-plastic wave mode. Typical view of wave profiles generated during shock loading of copper samples according to [10] is shown in Fig. 2.

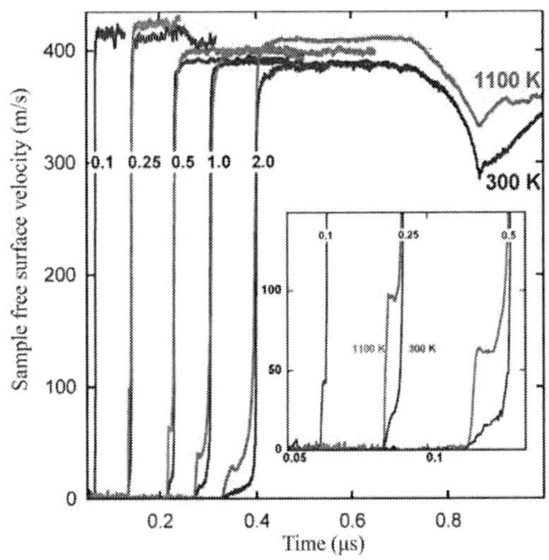

Fig. 2. Evolution of the shock waves in polycrystalline copper at temperatures 300 and 1100K. The numbers on the waveforms correspond to the nominal sample thickness. The inset displays the elastic parts of velocity histories of thinner samples [10].

The relation of the distance traveled by a plastic shock-wave, defined according to [10], is shown in Fig. 3, curve 4.

C. Analysis of the results

Possible trajectories of advancement of wave fronts and front of nonlinear current diffusion into the copper coil wall with thickness 2 mm, during generation of a magnetic field with induction amplitude 360 T are shown in Fig. 3.

As it shown in Fig. 3. the depth of plastic wave front propagation is substantially less than the distance traveled by the shock-wave and the shock-wave with external energy inlet. The output of this wave to the outer wall of the solenoid occurs at the moment of maximum current and is not contrary to the behavior of the solenoid outer wall, observed in the experiment.

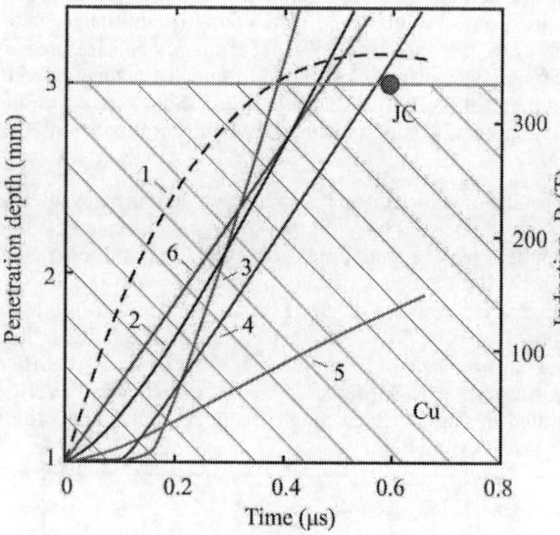

Fig. 3. The induction of the magnetic field – 1; the distance passed by: the shock-wave front– 2; front of shock-wave with external energy inlet – 3; plastic wave front - 4, the wave front of the current diffusion – 5, a sound wave – 6.

From the side of the magnetic field the slow rate of plastic shock-wave advance limits the growth of the area occupied by the flowing current, in which energy of source is released. The modification of parameters in this area is slower than in case of shock-wave and shock-wave with external energy inlet. It may lead to energy density increase. This energy releases in current flow area, limited by propagation of plastic shock-wave.

Simulation of flow process can be provided using the Johnson-Cook model, which describes the material deformation curve behavior in wide range of loading rates The applicability of this model for the description of fast processes is confirmed in several papers, e.g. [12]. Results of numerical simulation in ANSYS environment allow determining release moment of the plastic shock-wave front (Fig. 3 point **JC**). This time moment almost coincides with release moment of the plastic shock-wave, calculation is based on the experimental data [10]. The simulation of flow process of shock impact in the hydrodynamic and elastic-plastic formulation is provided in a number of papers [13, 14]. Results presented in [14] demonstrate the need to consider elastic-plastic material properties in the simulation of shock-wave processes. It should be noted the hydrodynamic approach seem to be justified at pressures typical for process of generation of strong and super strong pulse magnetic fields, because it allows determining the basic patterns of non-linear diffusion and material flow process [15].

III. Conclusion

In cases where the depth of magnetic field penetration into the conductor as a result of non-linear diffusion (nonlinear skin layer) is smaller than the characteristic size of the conductor in both closed and open systems, one can expect a significant impact of flow characteristics, caused by the avalaibility of elastic-plastic material properties, on the specific characteristics of the plasma formed during the explosion of the skin layer, especially in the initial phase of the process.

References

[1] S.I. Krivosheev, V.V. Titkov, and G.A. Shneerson, "Two-dimensional field diffusion and magnetohydrodynamic flow in an electric explosion of a miniature single-turn solenoid in a megagauss magnetic field," Technical Physics, vol. 42, is. 4, pp. 352-366, April 1997.

[2] A.R. Bryant, Megagauss magnetic field generation by explosives and related experiments, Brussel, p. 183, 1966.

[3] S.N. Kolgatin, A.Ya. Polishchuk, G.A. Shneerson, "Numerical simulation of a conductor's skin layer explosion in a superstrong magnetic field," Teplofizika Vysokikh Temperatur, vol. 31, is. 6, pp. 890-896 , 1993.

[4] I.B. Zel'dovich and Yu.P. Raizer, The physics of Shock waves and high-temperature hydrodynamic phenomena, pp. 1966-1967, 1966.

[5] S.I. Krivosheev, V.S. Pomazov, and G.A Shneerson, "Specific features of shock wave formation in superstrong magnetic field," Technical Physics Letters, vol. 37, is. 9, pp. 877-880, September 2011.

[6] A.A. Chaikovsky, V.I. Oreshkin, I.M. Datsko, N.A. Labetskaya, D.V. Rybka, and N.A. Ratakhin, "Experimental study of the nonlinear diffusion of a magnetic field and skin explosion of cylindrical conductors," Physics of Plasmas, vol. 22, is. 11, 1 November 2015.

[7] J.W. Shearer, "Interaction of capacitor-bank-produced megagauss magnetic field with small single-turn coil," Journal of Applied Physics, vol. 40, is. 11, pp. 4490-4497, 1969, doi: 10.1063/1.1657220.

[8] S.I. Krivosheev, "Physical constraint to superstrong magnetic fields by a method of direct discharge" in 12th IEEE International Pulsed Power Conference; Monterey, CA, USA, 27 June 1999 through 30 June 1999; Code 56431 (Conference Paper), vol. 2, pp. 750-753, 1999.

[9] N.V. Vovnenko, B.A. Zimin, and Y.V. Sud'enkov, "Experimental study of thermoelastic stresses in heat-conducting and non-heat-conducting solids under submicrosecond laser heating," Technical Physics, vol. 56, is. 6, pp. 803-808, June 2011.

[10] E.B. Zaretsky, G.I. Kanel, "Response of copper to shock-wave loading at temperatures up to the melting point," Journal of Applied Physics, vol. 114, is. 8, 28 August 2013.

[11] G.R. Johnson and W.H. Cook, "Fracture characteristics of three metals subjected to vari-ous strains, strain rates, temperatures and pressures," J. Eng. Frac. Mech., vol. 21, is. 1, pp. 31-48, 1985.

[12] Vl. Vas. Balandin, Vl. Vl. Balandin, A. M. Bragov, L. A. Igumnov, A. Yu. Konstantinov, and A. K. Lomunov, "High-rate deformation and fracture of steel 09G2S," Mechanics of Solids, vol. 49, No 6, pp. 666-672, 2014.

[13] A.E. Mayer, K.V. Khishchenko, P.R. Levashov, and P.N. Mayer, "Modeling of plasticity and fracture of metals at shock loading,"Journal of Applied Physics, vol. 113, is. 19, article number 193508, 21 May 2013.

[14] T.V. Popova, A.E. Mayer, and K.V. Khishchenko "Access Numerical investigations of shock wave propagation in polymethylmethacrylate" Journal of Physics: Conference Series, vol. 653, is. 1, article number 012045, 11 November 2015.

[15] S.I. Krivosheev, G.A. Shneerson, and N.G. Karlykhanov "Peculiarities of conducting medium MHD flow under skin layer electrical explosion in a megagauss magnetic field," in. Proc. Of IX inter. Conf. On Megagauss Magnetic Field Generation and Related Topics. M.-SPb., Ed by V.D.Selemir and L.N. Plyashkevich//Sarov, VNIIEF, pp. 529-536, July 7-14, 2002.

Single-turn magnet with an additional balanced winding and flux concentrators

Shneerson G.A., Danilin K.A., Nenashev A.P., Parfentev A.A., Pozdeev A.A., Dyogtev D.A., Petrov D.A.
Peter the Great St. Petersburg Polytechnic University
Polytechnicheskay str., St.Petersburg, 195251, Russia
gashneerson@mail.ru

Abstract.— **The balanced winding displaced in the cavity of the magnet provides the possibility for obtaining the stronger magnet field without enhancing of mechanical stresses in conductors. The model of such a system is the single-turn magnet with an additional triple-thread spiral winding. The equilibrium condition in respect to the radial force takes place at definite values of the axial current and pitch of the turns. In order to decrease the transverse field provoking the azimuthal force, the correction system in the form of cylindrical conductors with slits has been used. The discharging of the end part of the additional winding from the axial force is achieved using the flat concentrators of the flux.**

Keywords* — *nondestructive magnet, additional balanced winding, flux concentrators.

Introduction

Recent publications (for example [1]) consider the opportunities of the attaining of the magnetic field with the induction of about 100 T in the nondestructive magnet and point to the difficulties which arise in the process. In spite of the fact, that the researchers used the most elaborate materials, progressive design solutions and power sources with the energy of the order of tens megajoules, the transition from the fields with an induction of 70 - 80 T to the fields of megagauss range lasted longer than 20 years. The further growth of the fields in the magnets with an azimuthal current is possible through the application of larger scale and more expensive devices.

Another way of the obtaining of the fields with an induction above 100 T in the nondestructive magnets can be the development of the magnets with quasi-force free windings [2]. Along with the development of such magnets, it is of interest to consider the possibility of the application of the magnets with a balanced winding as an additional modulus displaced in the cavity of the magnet with a conventional winding [3]. Such a "hybrid" magnet allows the field values to be raised to the level determined by up-to-date materials. In this case the stresses in the additional winding are small as compared with the magnetic pressure on the magnet axis. It is reasonable to evaluate the potentials of such a system and direct the ways of its practical implementation.

1. Initial propositions

Fig.1 shows the magnetic system consisting of the main winding with the azimuthal current 1 generating the field with the induction B_1, and the additional winding, the turns of which are placed under the angle γ towards to the axis z. The current of this winding produces the azimuthal field B_φ and the field with the axial induction $B_2 = B_0 - B_1$, where B_0 is the induction of the resulting field on the – magnet axis.

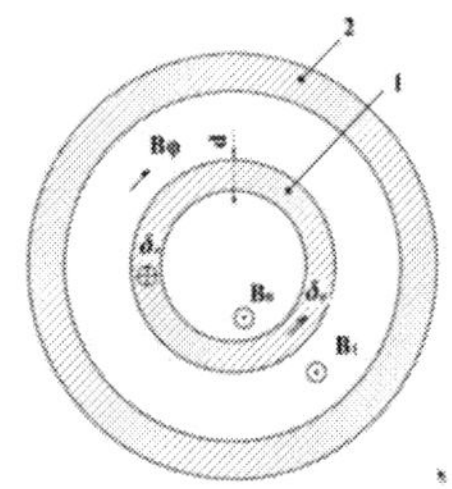

Fig. 1a. Magnet cystem with edditional balanced winding.

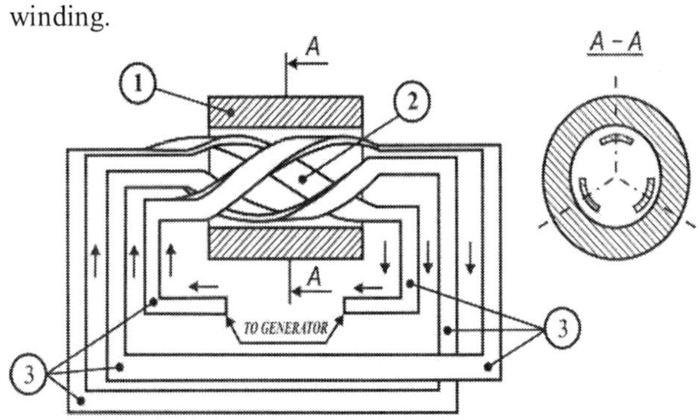

Fig. 1b. The circuit of the hybrid magnet.
1. The single-turn magnet
2. The additional balance winding.
3. Connecting conductors.

The characteristic parameter of this system is the ratio $B_1/B_0 = g$. The paper [3] gives the results of calculations of the mechanical stresses in the thin balanced winding. The calculations are made through the solution of the elasticity theory equations for the idealized system of unlimited length with the continuous current distribution. The assumption on the absence of axial deformations has been used. Such an assumption is also acceptable for the finite length windings with dead-ends. In the calculations it was assumed that the sum of the volume forces was equal zero (the equilibrium condition):

$$\int_{R_i}^{Re} f_r dr = 0, \qquad (1)$$

where f_r is the radial component of the volume electromagnetic force; R_i, R_e are the internal and external radii of the winding. This condition is valid if the angle γ takes the value

$$\gamma = \text{arctg}\frac{B_0 - B_1}{B_\varphi} = \text{arctg}\frac{\sqrt{1-g}}{\sqrt{1+g}}, \qquad (2)$$

The stress estimated from the von Mises formula has a maximal value in the middle of the winding layer:

$$\sigma_M = \frac{b}{2}(1-g)\frac{B_0^2}{2\mu_0} \qquad (3)$$

for the material with the Poisson coefficient 0.3 we have $b \approx 0.4..3$

From these calculations it follows that in the balanced winding the stresses are essentially lower than the field magnetic pressure on the axis $B_0^2/(2\mu_0)$. Thus, for values g = 0.7 and 0.5 we have $2\mu_0\sigma_M/B_0^2 = 0.06$ and 0.1, consequently. The results of the two-dimensional numerical calculations made by using Comsol Multiphysics confirm the analytical evaluations.

2. The application of flux concentrators for the field correction and the compensation of axial forces

In the winding of the finite length the field is not uniform and it is featured by the appearance of the radial components of the magnetic induction. Due to this fact appear the additional forces directed perpendicular to the conductors. That to reduce the field inhomogenietis one can apply the cut screens (flux concentrators).

Fig. 2 shows the calculated model of the magnet with the single-turn magnet, and additional winding and flux concentrators. The concentrator ***C1*** in the shape of a cylinder with a longitudinal section is placed in the clearance between the magnet ***1*** and the balanced winding ***2***. The pulsed axially symmetric field is calculated in the ideal-conductivity approximation.

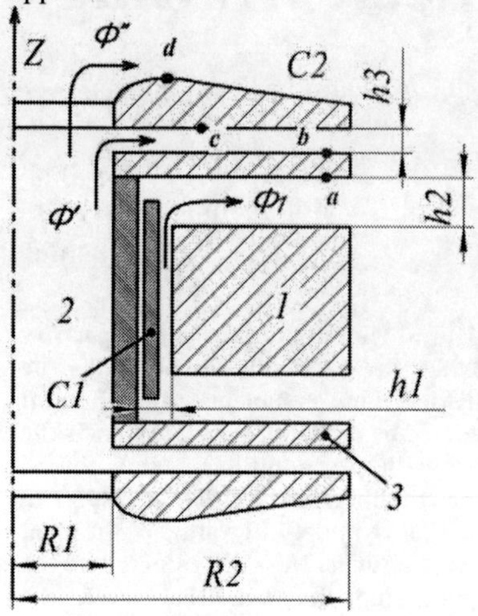

Fig. 2. Magnetic system with balanced winding and flux concentrators.
1. The single turn magnet
2. Additional winding.
3. End part of the additional winding. C1, C2 – the flux concentrators

For the control of the calculated results, the model of the magnet has been built (Fig. 3).The model is fed from the generator of sinusoidal current with a frequency 1 KHz. It is equipped with the gauges for the measuring of the distributions of the induction and flux.

Fig. 3. Single- turned magnet with additional winding.
1.Magnet.
2. End part of the additional winding

Fig. 4 shows as an example the flux distribution in the system at the presence of the screen and at its absence, that indicates the efficiency of the using of this device.

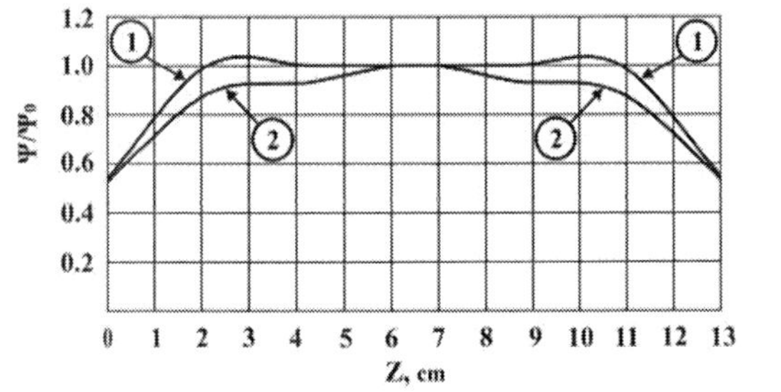

Fig. 4. Distribution of the magnetic flux on the length of the additional winding
1. Without concentrator (cut screen)
2. With the concentrator

3.The application of flux concentrators for the compensation of axial forces

Fig.2 shows the configuration of the end part of the magnet. It also illustrates the distribution of the magnetic field fluxes at a pronounced skin-effect and currents at the boundaries of the conductors. For connecting of the additional winding with the current source it is reasonable to use the plane insulated conductors. Being assembled, they make the plane ring ***3***. In points ***a*** of this ring the induction has both the radial component $B_r(a) \approx \Phi_1/(2\pi r(a)h_2) \approx B_1R_1 / r(a)$, and the azimuthal one $B_\varphi(a) \approx \mu_0 iz/(2\pi r(a)) \approx B_\varphi(R_1)R_1/r$. From the side 1a the ring is subjected to the magnetic pressure both of the poloidal field $B_r^2(a)/(2\mu_0)$ and the azimuthal field $B_\varphi^2(a)/(2\mu_0)$.

The corresponding total force is

$$F_a = \int_{R_1}^{R_2} P_M(a) 2\pi r(a)\, dr(a), \text{ where}$$

$$P_M(a) = \left(B_\varphi^2(a) + B_r^2(a)\right)/(2\mu_0) \approx B_0^2 (R_1/r)^2 (1 - g^2 + h_1^2 g^2 / h_2^2)/(2\mu_0). \quad (4)$$

The plane ring is in the state of the local equilibrium, provided that in the points ***a*** and ***b*** with the same radial coordinates the magnetic pressure is the same:

$$P_M(b) = \left(\frac{\Phi'}{2\pi r(b) h_3}\right)^2 \cdot \frac{1}{2\mu_0} = P_M(a) \cdot \quad (5)$$

The ring ***3*** may be balanced with the help of the additional flux concentrator ***C2***. The end flux concentrator can be made in the form of plane ring with the slot.

In the approximation of the pronounced skin-effect, the currents at pointes ***c*** and ***d*** and the total forces are expressed in a following way:

$$i_c \approx -\int_{R_1}^{R_2} \frac{\Phi' dr(c)}{2\pi\mu_0 h_3 r(c)} \approx -\frac{\Phi'}{2\pi\mu_0 h_3} \ln\rho \quad (6)$$

$$i_d \approx \xi \int_{R_1}^{R_2} \frac{\Phi'' dr}{2\pi r^2 \mu_0} = \xi \frac{\Phi''}{2\pi\mu_0 \rho R_1}\left(1 - \frac{1}{\rho}\right) \quad (7)$$

$$F_c = \int_{R_1}^{R_2} \left(\frac{\Phi'}{2\pi h_3 r}\right)^2 \frac{\pi r}{\mu_0} dr = \frac{(\Phi')^2}{4\pi\mu_0 h_3^2} \ln\rho \quad (8)$$

$$F_d = -\eta \int_{R_1}^{R_2} \left(\frac{\Phi''}{2\pi r^2}\right)^2 \frac{\pi r}{\mu_0} dr = \frac{-\eta (\Phi'')^2}{8\pi\mu_0 R_1^2}\left(1 - \frac{1}{\rho^2}\right), \quad (9)$$

where $\rho = R_2 / R_1$.

In these formulas ξ and η are the dimensionless coefficients taking into account the edge effect, namely, the growth of the current density on the external surface in the vicinity of the orifice with the radius R_1.

The current in the concentrator equals zero at the condition $i_c + i_d = 0$. Hence follows the equality

$$\Phi' = \frac{\xi \Phi'' h_3}{R_1 \ln\rho} \frac{(\rho - 1)}{\rho}, \quad (10)$$

Let's consider the system which satisfies the conditions $h_2 << h_1$, $h_3 << R_1, \Phi' << \Phi''$, $P_M(a) \approx (1 - g^2) B_0^2 / (2\mu_0)$. At these conditions from the eq.(5) it follows that $\Phi' \approx (1 - g^2) 2\pi h_3 R_1 B_0$, moreover, $\Phi'' \approx \pi R_1^2 B_0$. Using these relations and formula (10), we obtain the equation coupling the geometric characteristics with the parameter g:

$$\frac{2\rho}{\rho - 1} \ln\rho = \frac{\xi}{\sqrt{1 - g^2}} \quad (11)$$

Along with that, from the given relations there follows the ratio for the total forces affecting the concentrator:

$$\frac{F_c}{F_d} = \frac{2\xi^2}{\eta \ln\rho} \frac{(\rho - 1)}{(\rho + 1)} \quad (12)$$

The numbers ξ and η in formulas (7) and (9) depend both on the parameter ρ and on the configuration of the external boundary of the concentrator C1 (points d). In the case of the flat boundary, at the condition $\rho >> 1$ we have $\xi \approx 1.34, \eta \approx 3.16$. Variations of the boundary shape allows for changing the numbers ξ and η. Here the force ratio $|F_c|/|F_d|$ changes. In the case it equals unity, the concentrator discharging the end of the magnet, is itself at the equilibrium state. The possibility of the realization of this condition should be the subject of further studies.

Conclusion

The given report suggests the possibility of the constructing of the magnetic system with an additional balanced weakly-loaded winding generating the field stronger than that in the initial system, not increasing the mechanical stresses.

The problems are pointed out, which prevent the attaining of the calculated values of the maximal magnetic field (the structure discreteness, presence of the transverse field), and the ways of their solution are proposed. Considered is the feasibility of an unloading of the end parts of the magnet by the application of the balanced flux concentrators. The developed model permits to prove and correct the calculation data and offer the base for the construction of the magnet pilot sample.

References

[1] S.Zyerlitsin, B.Wustmann, T.Herrmannsdorfer, and J.Wosnitsa. IEEE.Trans. an appl. Supercond. V.22, №3. 2012, 4300603.

[2] G.A Shneerson., M.I Dolotenko., S.I. Krivosheev. Strong and Superstrong Pulsed Magnetic Fields Generation. De Gruyter, Berlin, 2014. p.429

[3] G. A. Shneerson. Minimization of Electromagnetic Forces on Solenoid Wall in External Magnetic Field Technical Physics Letters, 2011, № 19, p.51.

Cumulation of High-current Electron Beams: Theory and Experiment

Sergei V. Anishchenko, Vladimir G. Baryshevsky, Nikolai A. Belous, Alexandra A. Gurinovich, Elizaveta A. Gurinovich, Evgeny A. Gurnevich, Pavel V. Molchanov
Institute for Nuclear Problems, Belarusian State University
Minsk, Belarus
Electrophysical Laboratory, Minsk, Belarus

Abstract—**High-power electron beam cumulation in a relativistic vacuum diode with a ring-type cathode is considered. The basis of this cumulation mechanism is electrostatic repulsion of electrons from the explosive-emission plasma on the inner edge of the cathode. Both simulation and experimental results are presented.**

Keywords—electron beams, explosive electron emission, cathode, current density, plasma properties

I. Introduction

High-current electron beams are used for research in radiation physics [1], generation of high-power microwaves [2,3], collective acceleration of ions [4,5], and pumping gas lasers [6]. Nonlinear phenomena originating from the high-current-beam interaction with self- and external electromagnetic fields figure prominently in all these processes. Self-focusing of high-current electron beams with their own magnetic fields [7,8] provide the charged-particle beam high intensities, thus finding applications in many fields of physics [9–11].

In the experiments [12] with ring-type cathodes (see Fig.1) some peculiarities of electron beam current density distribution were observed. Simulation of electron beam dynamics in planar vacuum diode [13-15] enabled comprehension of the electron beam cumulation effect. The cumulation mechanism could be described as follows: Once the voltage is applied across the cathode-anode gap, the field-emission current is emitted from the cathode surface. Explosive electron emission begins with the formation and expansion of cathode flares with the smaller time delay at higher electric field strength values. Electric field enhancement in the vicinity of the cathode edges initiates fast plasma emission from the inner edge of the cathode. Coulomb repulsion causes the charged particles to rush to the region free from the beam. As a result, the accelerated motion of electrons toward the anode comes alongside the radial motion and the high-current beam density could increase multifold on the axis of the relativistic vacuum diode as compared to the average current density in the cathode-anode gap. Thus, the basis of the considered cumulation mechanism was shown to be electrostatic repulsion of electrons from the explosive-emission plasma on the inner edge of the cathode.

The method for analysis of electron beam imprints used in this paper was developed under support of Federal Targeted Program within the Ministry of Education and Science of Russian Federation through the Project ID RFMEFI57714X0092 entitled Research and Development in Priority Areas of Development of the Russian Scientific and Technological Complex for 2014–2020 under Grant 14.577.21.0092.

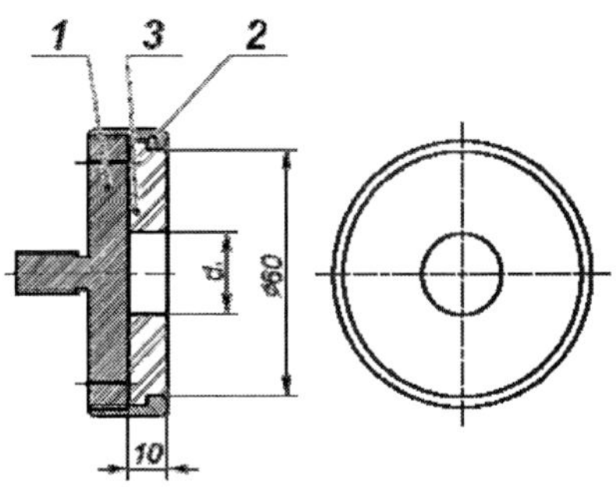

Fig. 1. Cathode unit drawing: 1 – cathode holder (stainless steel), 2 – cathode holder nut (stainless steel), 3 – ring-type cathode emitter (graphite MPG-8 (produced by NIIGraphite, Russia)). Cathode dimensions are in millimeters.

This paper considers high-power electron beam cumulation, which occurs in relativistic vacuum diodes with a ring-type cathode. Both simulation and experimental results are presented. The self-developed 3D computer code [13-15] was used for numerical analysis of electron beam energy density distributions for different cathodes. The same were experimentally evaluated by imprints of the electron beam obtained using a radiochromic film [17].

II. Experimental Setup

Cathodes experimental investigation was carried with pulsed electron accelerator fed from a compact high-voltage Marx generator (12 stages, capacitance 3.125 nF, max. output voltage 600 kV, pulse rise time 30-35 ns, pulse width 120-150 ns, SF_6 insulation) [18]. All the experiments were performed at fixed generator charging voltage. The vacuum level of $\sim 5*10^{-3}$ Pa was provided in the vacuum chamber. The output voltage was measured with a capacitive divider. The diode current and the total generator current were registered using self-integrating Rogowski coils built in the generator flanges. All the sensors were preliminarily calibrated and provided the measurement accuracy not worse than 5%. Graphite ring-type cathodes had outer diameter 60 mm. Inner diameter d_1 had values 10, 20, 23, 26, 30, 34, 38, 42, 46 and 50 mm. The cathode-anode gap was 16 and 20 mm. Anode was made of stainless steel mesh with geometric transparency 0.77.

Distributions of the beam energy density were evaluated by full-sized imprints of the electron beam obtained using a radiochromic film [19]. Radiochromic film was placed 3 mm behind the anode on the cathode faced surface of beam collector (graphite disc of 8 mm thickness). The film enables detection of the total absorbed dose due to the passage of charged particles [20]. The distribution of the absorbed dose (and hence the beam energy density) over the beam cross section was derived from the scanned film images (transmission scanning using the optical filter) and the dose-response calibration curve according to the procedure described in [17]. According to [17,20] the value of the absorbed dose D at a certain point is directly proportional to the energy density of the electron beam (averaged over pulse duration) in this point. In the reported experiments the voltage (current) pulses in different shots are quite identical (see section IV below) and are quite similar to those reported in [17]. Thus, the dose absorbed by the radiochromic film can be converted to the current density in each point of beam cross-section by the simple expression obtained in [17]:

$$j = \alpha D, \quad \text{where } \alpha \approx 1 \frac{\text{A}}{\text{kGy} \cdot \text{cm}^2}. \qquad (1)$$

III. Simulation Results

Modeling of the relativistic vacuum diode operation is carried with the self-developed 3D computer code [12]. Nonuniform explosive electron emission is described by considering the expansion of the cathode plasma emitted from separate explosive emission centers. Self-consistent simulation of particle motion in electric and magnetic fields is performed using the particle-in-cell method [20–22] in a quasi-stationary approximation [23,24]. The diode total current was obtained to be determined by the cathode outer diameter and the cathode-anode gap, while the inner cathode diameter assigned ratio of current in the intense axis flow to the total diode current. Since the beam energy density can be experimentally evaluated by the distribution of the dose absorbed by radiochromic film, just the absorbed dose value was calculated. The intent of simulation was to obtain maximal-to-average current density ratios for different inner cathode diameters as well as to analyze absorbed dose distributions over the anode. The sample of simulated absorbed dose at the anode position for a cathode with outer and inner diameters 60 and 20 mm, respectively, and cathode depth 10 mm is shown in Fig.2.

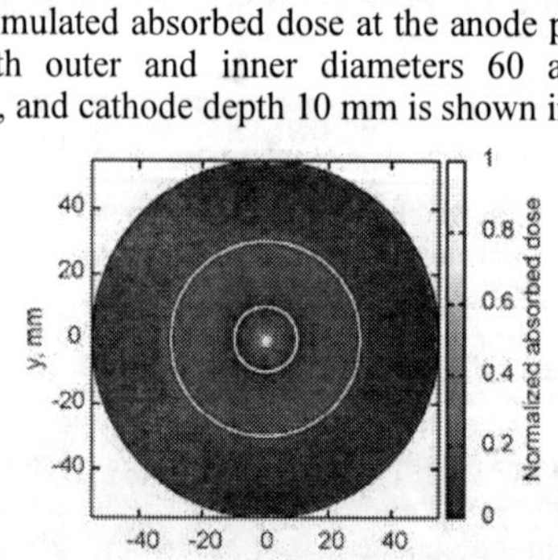

Fig. 2. Simulated normalized absorbed dose at the anode position. Black circles mark outer and inner cathode diameters, which are 60 and 20 mm, respectively.

The voltage pulse with maximal amplitude 360 kV and rise time 30 ns was applied to cathode anode gap 20 mm producing current of 6.5 kA maximal value. According to simulation results the current density distribution at the beam periphery (outside the inner circle on Fig.2) is similar to that for the solid cathode [14]. The simulated current density in the region of the central spot was as large as 1 kA/cm^2, about 3 times exceeding the average current density value. Simulated values of absorbed dose (see Fig. 3) are the statistic quantities due to fluctuation of emission center positions over the cathode surface. The estimated standard deviation is 10-20%.

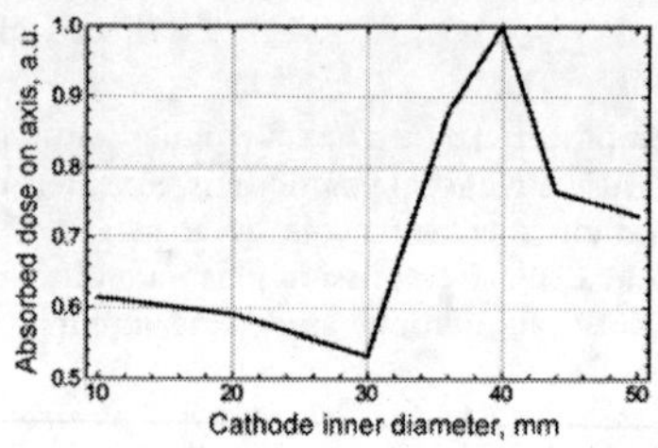

Fig. 3. Simulated absorbed dose on the system axis for different values of cathode inner diameter.

IV. Experimental Results

Typical voltage and current pulses detected in the experiments with different inner cathode diameters for cathode-anode gap 16 mm are presented in Fig.4. They slightly changes with the inner cathode diameter increase. The similar was observed for cathode-anode gap 20 mm.

The average value of the dose absorbed by radiochromic film in conditions of the experiment reported in [17] measured 70—100 kGy. On the other hand the operation range of the used film is 5—150 kGy [17,20]. According to simulation results the current density in the central spot is expected to be several times greater than the average current density value. Therefore, the central spot of the beam imprint could be overexposed. To evaluate current density distribution using the radiochromic film in the anticipated conditions, the central spot was attenuated by two layers of 70μm aluminium foil placed in front of the film. Calibration gives for each foil layer the dose attenuation factor as high as 2.8 times for 20 mm cathode-anode gap and 3.2 times for 16 mm gap. The samples of absorbed dose distributions obtained by electron beam imprints are shown in Figs.5, 6, 7 and 8. White circles in figures mark cathode outer (60 mm) and inner diameters. Average current density j, evaluated by absorbed dose distribution within the area of 60 mm diameter was ~70 A/cm^2 and 100 A/cm^2 for 20 mm and 16 mm gap, respectively. Central spot on each imprint displays higher absorbed dose value then the rest film that means higher current density. Following the procedure, which is briefly described in section II herein and in [17] in more details, all the imprints were analyzed to obtain the absorbed dose (and, hence, the current density) in the central spot for different inner cathode diameters. The central spot on the imprints had no sharp edges, therefore, averaging of calculated dose value over azimuth angle is performed (the averaged absorbed dose as a function of distance from the imprint center r and its linear fit are shown in Fig.9). The maximal value of the averaged absorbed dose at $r=0$ is find using this linear approximation, which perfectly fits the experimental curve at small distances from the center.

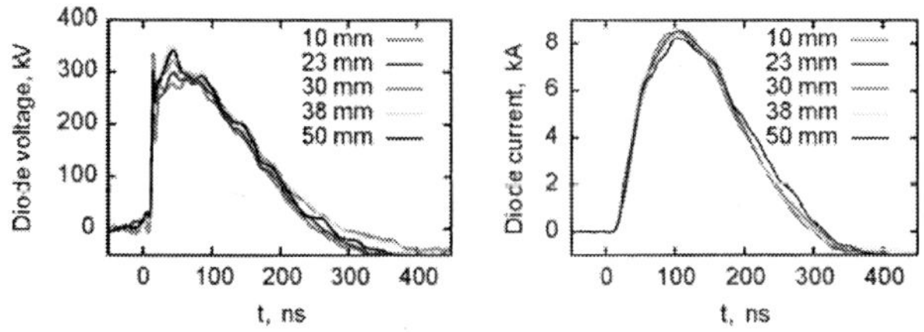

Fig. 4. Typical voltage (left) and current (right) pulses detected in the experiments with different inner cathode diameters for cathode-anode gap 16mm. Figure legend shows inner cathode diameter values.

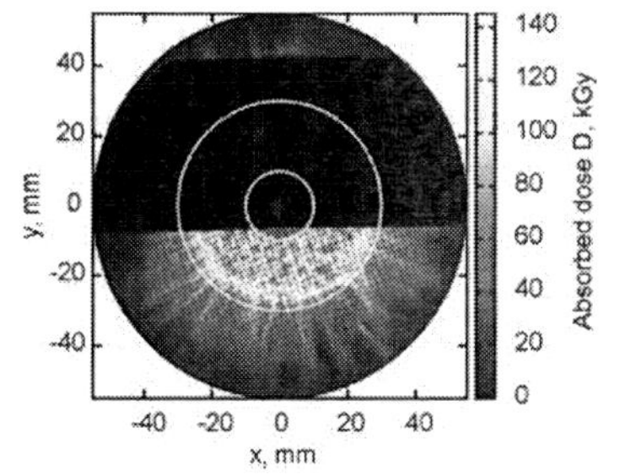

Fig. 5. The distribution of the dose absorbed by the radiochromic film at 3 mm distance behind the anode mesh (beam imprint) for graphite ring-type cathode with outer and inner diameters 60 and 20 mm, respectively. Cathode-anode gap is 16 mm. Dark band indicates position of two layers of 70μm aluminium foil of the same width. Maximal current density in the spot center is evaluated about 0.65 kA/cm^2.

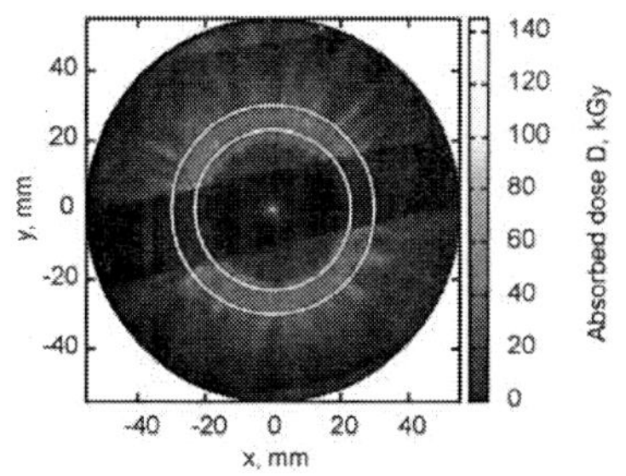

Fig. 6. Beam imprint for graphite ring-type cathode with outer and inner diameters 60 and 46 mm, respectively. Cathode-anode gap is 16 mm. Darker band indicates position of two layers of 70μm aluminium foil, the rest of imprint is covered with one foil layer (foil boundaries are visible near the imprint edges). Maximal current density in the spot center is evaluated about 1.6 kA/cm^2.

The current density value at the diode axis can be evaluated considering (1) and attenuation provided by 70μm aluminium foil. Thus obtained values are included in captions to Fig.5 – Fig.8. The maximal current density 2.2 kA/cm^2, which corresponds to beam energy density 0.7GW/cm^2 was observed for graphite ring-type cathode with outer and inner diameters 60 and 46 mm (see imprint in Fig.8). Absorbed dose on the system axis evaluated for different values of cathode inner diameter is shown in Fig.10 for cathode-anode gap 16 and 20 mm.

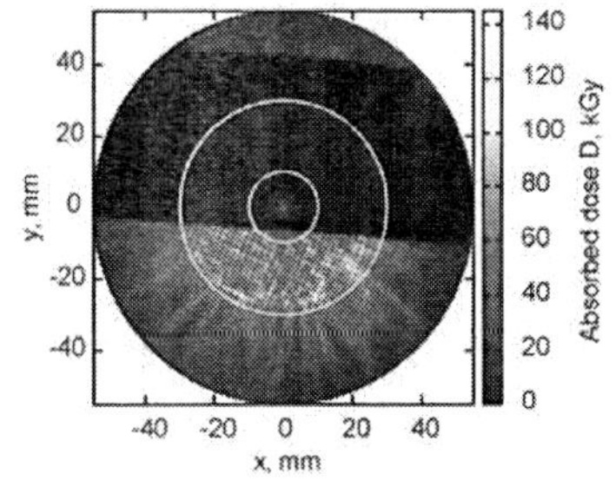

Fig. 7. Beam imprint for graphite ring-type cathode with outer and inner diameters 60 and 20 mm, respectively. Cathode-anode gap is 20 mm. Dark band indicates position of two layers of 70μm aluminium foil of the same width. Maximal current density in the spot center is evaluated about 0.61kA/cm^2.

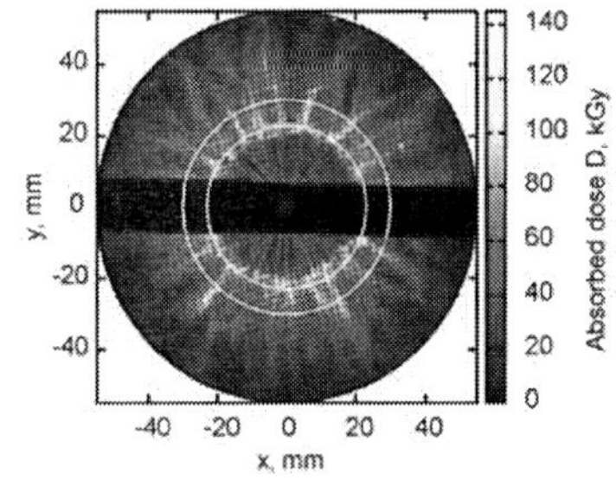

Fig. 8. Beam imprint for graphite ring-type cathode with outer and inner diameters 60 and 46 mm, respectively. Cathode-anode gap is 20 mm. Darker band indicates position of three layers of 70μm aluminium foil of the same width. Maximal current density in the spot center is evaluated about 2.2 kA/cm^2.

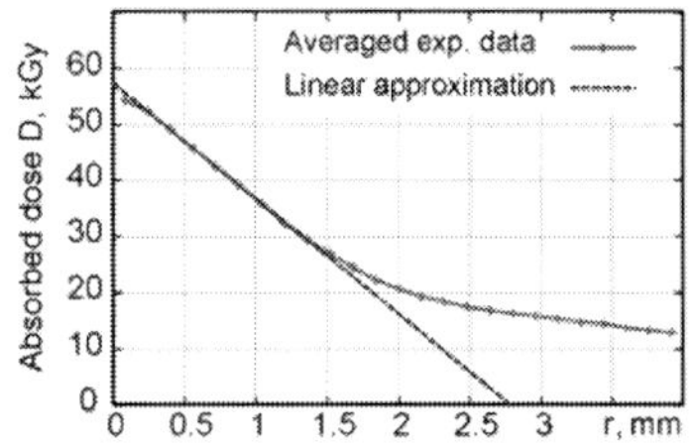

Fig. 9. The averaged over azimuth angle absorbed dose as a function of distance from the imprint center r for ring-type cathode with outer and inner diameters 60 and 34 mm, respectively, and cathode anode gap 16 mm.

According to [16] the cumulation mechanism doesn't depend on whether or not the aperture axis coincides with the cathode axis. To evaluate the effect for eccentric apertures a cathode with 7 apertures was tested (see Fig.11). Centered aperture had 7 mm diameter, centers of the rest six apertures were placed equidistant at the distance 14 mm from the cathode axis and apertures had diameters 1, 3, 5, 7, 9 and 12 mm. Beam imprint (Fig.11) was obtained at 20 mm cathode-anode gap. At smaller aperture diameters (1, 3 and 5 mm) the cumulation effect was scarcely apparent. For eccentric apertures 7, 9 and 12 mm the maximal current density in the spot read approximately 0.36, 0.4 and 0.42

kA/cm^2, respectively. For the centered aperture of 7mm diameter the same is evaluated as 0.43 kA/cm^2. For eccentric appertures the position of the spot was shifted from the aperture center to the cathode axis.

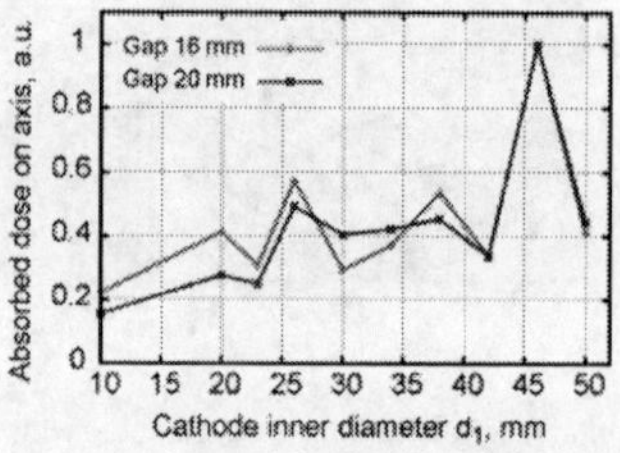

Fig. 10. Absorbed dose on the system axis evaluated for different values of cathode inner diameter for cathode-anode gap 16 and 20 mm.

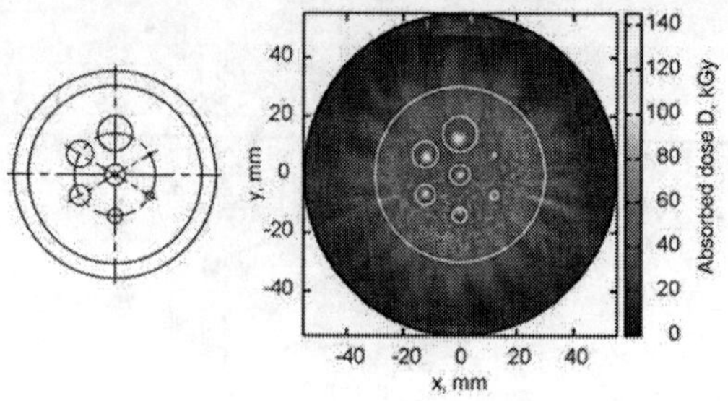

Fig. 11. Beam impprint (right) for the graphite cathode of 60 mm diameter with multiple apertures (drawing left) obtained with one layer of 70μm aluminium foil. at cathode-anode gap 20 mm.

V. Conclusion

High-power electron beam cumulation in a relativistic vacuum diode with a ring-type cathode is considered. The effect is experimentally observed. Dependence of electron beam current density on inner cathode diameter is obtained. Maximal current density in the central spot is shown exceeding the average value about an order. Effect evaluation for a centered aperture in a cathode is provided along with that for eccentric apertures. The undeniable advantage of this cumulation mechanism over a conventional one based on the self-focusing of a high-current beam by its own magnetic field is a very low energy spread of particles in the region of the maximum current density due to the laminar flow of charged particles [16].

References

[1] T. H. Martin, "Design and performance of the Sandia laboratories Hermes II flash X-ray generator", *IEEE Trans. Nucl. Sci.*, pp. 59–63, 1969.

[2] N. F. Kovalev, M. I. Petelin, M. D. Raiser, A. V. Smorgonsky, and A. E. Tsopp, "Generation of high-power electrmagnetic radiation pulses by a flux of relativistic electrons", *JETP Lett.*, vol. 18, pp. 232–235, 1973.

[3] Y. Carmel, J. Ivers, R. E. Kribel, and J. Nation, "Intense coherent cherenkov radiation due to the interaction of a relativistic electron beam with a slow-wave structure", *Phys. Rev. Lett.*, vol. 33. pp. 1278–1282, 1974.

[4] J. Rander, B. Ecker, G. Yonas, D. J. Drickey, "Charged-particle acceleration by intense electron streams", *Phys. Rev. Lett.*, vol. 24. pp. 283–286, 1970.

[5] A. E. Dubinov, I. Yu. Kornilova, and V. D. Selemir, "Collective ion acceleration in systems with a virtual cathode", *Phys. Usp.*, vol. 45, pp. 1109–1129, 2002.

[6] N. G. Basov and V. A. Danilychev, "Condensed- and compressed-gas lasers", *Sov. Phys. Usp.*, vol. 29, pp. 31–56, 1986.

[7] D. L. Morrov, J. D. Phillips, R. M. Stringfield, W. O. Doggett, and W. H. Bennett, "Concentration and guidance of intense relativistic electron beams", *J. Appl. Phys.*, vol. 19., pp. 441–443, 1971.

[8] L.P. Bradley and G.W. Kuswa, "Neutron production and collective ion acceleration in a high-current diode", *Phys. Rev. Lett.*, vol. 29., pp. 1441–1445, 1972.

[9] A. C. Kolb, "Uses of intense electron beams", *IEEE Trans. Nucl. Sci.*, pp. 956–961, 1967.

[10] G. Yonas, "Fusion power with particle beams", *Sci. American*, vol. 239. no. 5. pp. 40–51, 1978.

[11] G. A. Mesyats, *Pulsed Power*, Berlin: Springer, 2005

[12] V.G. Baryshevsky, A.A. Gurinovich, E.A. Gurnevich and P.V. Molchanov, "Experimental Study of an Axial Vircator With Resonant Cavity," *IEEE Trans. Plasma Sci.*, vol. 43, no. 10, pp. 3507-3511, Oct. 2015.

[13] S. V. Anishchenko and A. A. Gurinovich, "Modeling of high-current devices with explosive electron emission", *Computational Science & Discovery*, vol. 7, p. 015007, 2014.

[14] S. Anishchenko and A. Gurinovich, "Simulation of explosive emission and electron beam dynamics in a cathode-anode gap for cathode and anode optimization in an axial vircator", *Book of Abstracts of 5th Euro-Asian Pulsed-Power Conference*, report OA1-6, Kumamoto, Japan, 2014.

[15] S. Anishchenko and A. Gurinovich, "Simulation of electron beam dynamics in a nonmagnetized high-current vacuum diode", arXiv: 1608.04255 [physics.plasm-ph] (2016).

[16] S. V. Anishchenko, V. G. Baryshevsky, N. A. Belous, A. A. Gurinovich, E. A. Gurinovich, E. A. Gurnevich, and P. V. Molchanov, "Cumulation of high-current electron beams: theory and experiment", arXiv:1509.00522 [physics.acc-ph].

[17] V. G. Baryshevsky, N. A. Belous, A. M. Belov, A. A. Gurinovich, E. A. Gurinovich, E. A. Gurnevich, and P. V. Molchanov, "Generation of high-current electron beams in diodes with large-area explosive emission cathodes", *IEEE Trans. Plasma Sci.*, vol. 44., no. 7, pp. 1103-1111, 2016.

[18] E. Bolshakov, D. Getman, V. Eremkin, M. Vasilevsky, V. Vodovozov, and V. Engels, "Compact 600 kV, 585 J, 10 Hz high-voltage generator with SF6 gas insulation," *Voprosy atomnoj nauki i techniki, Electrofizicheskaya apparatura (Problems Atomic Sci. Technol.)*, vol. 5, no. 13, pp. 137–147, 2010.

[19] V. Abramov et al., "Ionizing radiation film chemical dose meter and method of making the dose meter," Russian Federation Patent 2 298 811 C1, May 10, 2007.

[20] D. Goncharov, V. Ezhov, A. Pushkarev, and G. Remnev, "The study on the energy density of high-current pulsed electron beam," (in Russian), *Bull. Tomsk Polytech. Univ.*, vol. 308, no. 6, pp. 76–80, 2005.

[21] A. S. Roshal, Modeling of charged particle beams, Moscow: Atomizdat, 1979.

[22] R. W. Hockney and J. W. Eastwood, Computer simulation using particles, New York: McGraw-Hill, 1981.

[23] C. K. Birdsall and A. B. Langdon, Plasma physics via computer simulation, New York: McGraw-Hill, 1985.

[24] J. W. Poukey, J. R. Freeman, and G. Yonas, "Simulation of relativistic electron beam diodes", *J. Vac. Sci. Technol.* vol. 10. no. 6. pp. 954–958, 1973.

[25] G. T. Golovin, "On the problem of numerical modeling of stationary beams of charged particles", *Zh. Vychisl. Matem. i Matem. Fiz.*, vol. 29., pp. 423–437, 1989.

The problems and development of the magnets with a quasi-force-free winding

Nenashev A.P., Parfentev A.A., Shneerson G.A., Vecherov I.A., Shimanskiy S.A.,Ushakov N.A.

Peter the Great St. Petersburg Polytechnic University

Polytechnicheskay str., St.Petersburg, 195251, Russia

gashneerson@mail.ru

***Abstract* — Calculations based upon two-dimensional models with a continuous current distribution have shown the possibility to construct balanced windings with drastically reduced mechanical stresses. In such windings the reduction of mechanical stresses at the transition from the disbalanced system to the system with equal magnetic pressures from the external and internal side of the winding was experimentally proved. The experiments have shown the deviations from the expected stress values. The reasons are the insulations gaps between the turns, the winding shape from the calculated profile. and induction normal components. We developed the methods for decreasing of these.**

***Keywords* — balanced windings, mechanical stresses, insulations gaps, thin-walled screens.**

1. Introduction

Over several years at St. Petersburg Polytechnical University the investigations have been carried out to develop free-force magnets. The analytical estimates shows that in the winding of such a magnet consisting of N thin balanced layers the residual tension σ_M, calculated by von Mises formula, is close to the value $\eta B_0^2/(2\mu_0 N^2)$, where B_0 is the induction on the magnet axis, and the coefficient $\eta = 0.19$, in the case of a single- layer magnet, and close to 0.21 at $N \geq 2$ [1].

The idealized winding configurations have been calculated for which the equilibrium conditions are fulfilled and normal component of induction is absent. There were considered the systems in which the mechanical stresses didn't exceed the allowed values on the periphery of the magnetic system where the field cannot be analogues to the force-free model. The similar configurations are described in the paper [2] and in a number of other publications reviewed in the book [1]. It has been outlined that the ratio of the external magnet radius with quasi force-free winding in respect to the internal one is significantly less than in the case of the magnet with equally loaded winding and azimuthal current.

The practical realization of the force-free winding leads to the deviations from the original model. One of essential differences is the presence of the gaps between the turns. They reduce the strength of the system. Mechanical stresses grow under the action of the forces appearing at the presence of the normal component of magnetic induction.

The role of admitted factors is convenient to study by numerical calculations taking as an example the magnetic systems of the simplest configuration in the form of single layer magnet of finite length. Besides the results of calculation herewith we also describe the initial experiments aimed to confirm the growth of the strength threshold at the switching-on the azimuthal field providing the balance of the winding in radial direction.

2. Computer models of the magnet with balanced winding

Let us consider the hierarchy of the models describing the stresses in the balanced winding.

Base model. The simplest prototype of the magnet with balanced winding is the solenoid of unlimited length with continuous current distribution. In such an "ideal" winding there are no gaps between turns, radial component of the induction is absent and volume forces possess only the radial component. The stresses in the conductors were analytically calculated earlier in the approximation of the small thickness ($d << R$) at the condition of equality of magnetic pressure of poloidal and azimuthal fields on the both sides of the winding. More accurate calculations were carried out by the method of the finite elements. The highest relative von Mises stress $\eta = \sigma_M \cdot 2\mu_0 / B_0^2$ has been calculated for each ratio of d/R at the different pitch angle γ. The dependences $\eta = f(\gamma)$ are exemplified in Fig.1.

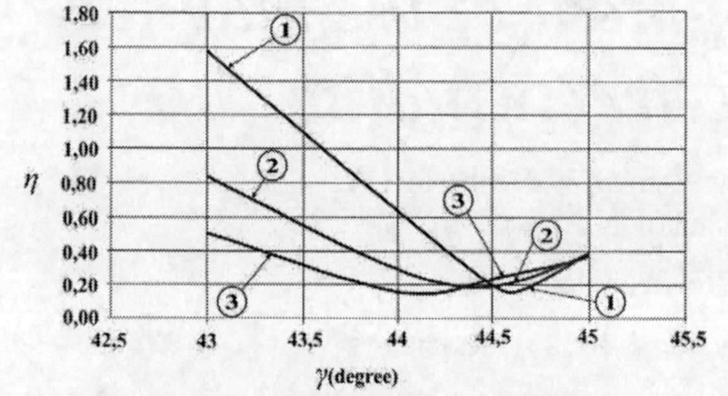

Fig. 1. The examples of the dependences of normalized stresses on the angle γ
1 - d/R =0.077, 2- d/R =0.14, 3 - d/R =0.21

The dependence minimum of normalized von Mises stress $\eta_1 = (\sigma_M)_{min} \cdot 2\mu_0 / B_0^2 = F(d/R)$ shown in Fig. 2.

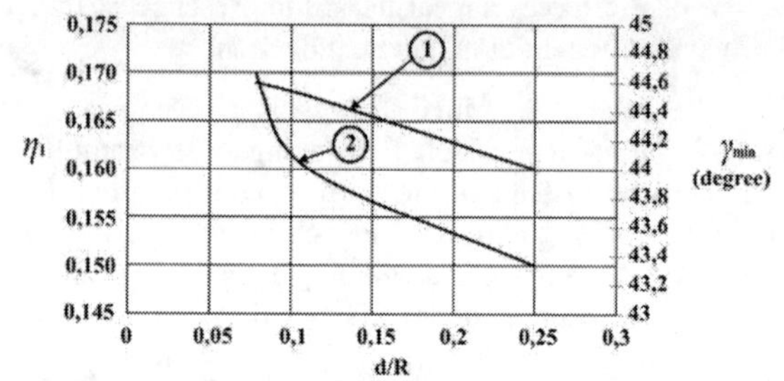

Fig. 2. Dependences of the minimal stress and the angle γ_{min} on the ratio of the winding thickness over the radius, where: 1- γ_{min}, 2 – η_1.

The same figure shows the consequent values of the angle γ_{min}. Further on we shall call the number η_1 as the calculated strength parameter for the winding with a given ratio *d/R*. This parameter weakly depends on the ratio *d/R* at the condition that this ratio is rather small ($d/R \leq 0.1$) For the conditions of the experiment described below, $d/R=0.07$, $\gamma=\pi/4$. In this case the calculated value of the ratio $\eta=\sigma_M \cdot 2\mu_0/B_0^2$ is 0.4

The influence of the discreet structure on the mechanical stresses in a balanced winding. The presence of the gaps between the conductors reduces significantly the system rigidity. Each turn becomes the analog, up to some extent, of the long thin band fixed at the ends. The calculations show that the stresses in the conductors in such a system are much higher than those in the base model. However, the rigidity of the magnet could be increased at the filling of the gaps with the dielectric of the sufficiently high Young modulus and at the application of the external and internal reinforcements [3].

In order to support of the mentioned above, the 3-D calculation has been conducted by using the finite element method for the 3-turn winding in the form of the cupper band of 10 mm-width and 0.7 mm-thickness, displaced at an angle of 45° on the cylinder of 10 mm-radius with the gaps of 2mm between the turns. The action of the transverse field has been neglected, and it has been assumed that the dielectric filled not only the gaps between the turns but also the 1 mm-layers adjacent to the winding from the inside and outside (reinforcements). The condition of the equality of the magnetic pressure on each side of the conductor has been achieved by the applying the modest correcting azimuthal field. It has been produced by the axial current with an amplitude not higher than 10 % of the total current. The calculation has been carried out at the conditions of the regard for the skin-effect.

E, GPa	0	0.038	0.38	3.8	38
η_1	62	3.5	2.7	2.4	0.80
η_2		0.026	0.043	0.14	0.83

The table presents the calculated amplitude values of the normalized von Mises stresses η_1 for the conductors and η_2 for the reinforcements. The value $E=0$ corresponds to the stress in the winding without the dielectric medium. In the considered case it 62 times higher than the magnetic pressure of the field on the magnet axis. The application of the dielectric even with a relatively low Young modulus of 0.38 GPa enables one to reduce the normalized stress down to the value of $\eta_1 \approx 2.7$ When using the dielectric with the Young modulus of 0.38 GPa, the normalized stress decreases down to $\eta_1 \approx 0.8$. The presented data exhibit the opportunity of constructing of the magnet with the mechanical stresses close to those in the base model. The calculated stresses approximately the same as in a thick-walled single-turn magnet. They could be probably reduced additionally at the further optimization of the system: changes in the width and pitch of the winding, and the thickness of the band and reinforcements, as well.

Effects affecting the mechanical stresses in the uniform winding and the ways of their compensation. In the development of the magnet the possible deviations from the balanced configuration are of the "integral' character, i.e. leading to the balance violation of the large part of the winding. In the case, when the balance is violated at the action of the forces directed from the normal to the winding, the correction of the magnet system by switching on the additional poloidal current is possible [4].

The another "integral" effect is the appearance of the axial and azimuthal forces tangential to the boundary of the thin winding. They are determined the crossing magnet flux. The examples are the forces originating in the winding having the form of thin cylinder of limited length. In this winding the induction has radial component normally directed to the vector of the current density. Along with the balanced radial forces, considered above, in this winding the azimuthal and axial forces take also place. The conductors of the winding with currents directed under the angle γ to the axis Z, are affected by axial and azimuthal electromagnetic forces $F_z' = j_\varphi \Phi'_\perp$, $F_\varphi' = j_z \Phi'_\perp$ As one can see, this formula contains the mentioned forces, the magnetic field flux calculated by the unit length of the magnet, and also the components of the linear current density. In the thin balanced winding $j_\varphi \approx j_z$, hence $F_z' = F_\varphi'$. The first of these forces produces the axial stresses σ_z [1]. The second gives arise to the rotational momentum causing the tangential stresses τ. For the middle cross-section of the thin wall magnet of the length exceeding significantly its radius R, we have $\sigma_z = \tau = B_0\Delta\Phi/(2\pi\mu_0 Rd)$, where $\Delta\Phi$ is the flux intersecting the winding [5]. In such a magnet the transverse flux is $\Delta\Phi = 0.5\pi R^2 B_0$, therefore the ratios of the indicated stresses to the magnet pressure attain the values $R/(2d) >> 1$. The efficient way to attenuate the transverse field is the application of the cylindrical sectional screen (flux concentrator) displaced inside or externally the winding. The example of such a device is given in the paper [5]. Fig. 3 shows the three-turn magnet of the length 200 mm and inner radius 10 mm, in the cavity of which the screen is installed.

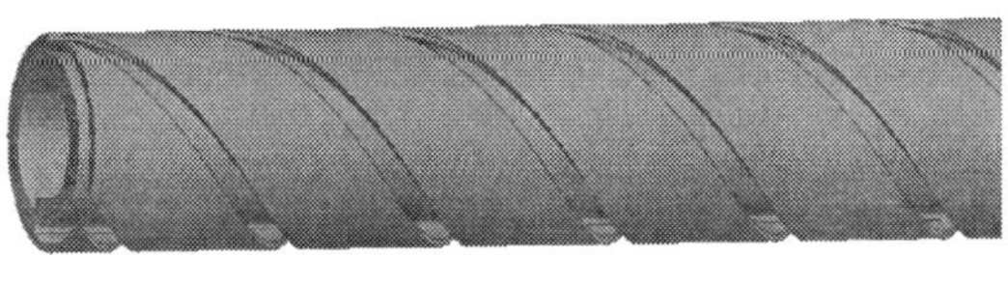

Fig. 3. The screen displaced inside of the winding.

The last is made of the cupper sheet of 1 mm-thickness and has a shape of the fragment of the Archimedes spiral. The screen is separated from the winding by the distance Δ=1mm. The calculations carried out with the use of the ANSYS program with regard for the skin effect have shown that under the sinusoidal change of the field with the period 2 ms, at the presence of the screen, the transverse flux acquires the value $\Delta\Phi = 0.18\pi R^2 B_0$ instead of the value mentioned above. The further reduction of the transverse flux could be obtained by decreasing the gap Δ between the winding and the screen.

3. The experimental sample of the magnet with the spiral winding ant its tests

The action of the considered factors affecting the strength of the balanced winding could be exhibited experimentally. Here we describe the initial stage of the experiments. The aim of experiments is to determine under simple conditions in which extent the processes in the winding correspond to the described models. For this purpose we used the samples of simple configuration: 3-winding magnets made of standard cupper band with the width of 10 mm and thickness of 0.7 mm. The band is displayed in the grooves on the surface of the cylinder of 10-mm diameter with thickness of the wall of 2 mm. The conductors are separated by ribs with height of 3 mm and width of 2 mm. The cylinder is manufactured with high precision from polyamide by the method of additive technology. Unlike quasi force-free magnets with the balanced butts, described above [1, 2], in this case the ends of the winding are tightly fixed and, thus, the conditions of the tight sealing of the ends of the magnet are met, i.e. there is no the axial deformation on the ends (Fig. 4a,b). The applied fixing of the butts is acceptable in the range of fields under study. Note, that the transverse field of the described solenoids does not follow from the constructions faults, but varies in a controlled way.

Two types of the samples have been used for the comparative experiments. In the first of them (Fig. 4a) the conductors with reverse current are displaced in the cavity of the cylinder. In this case the summary axial current and the induction of the azimuthal field on the external boundary of the winding are equal to zero. In this case winding is not balanced.

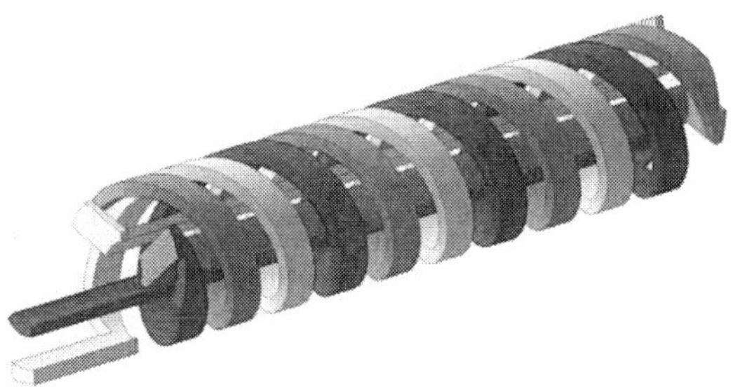

Fig. 4a. Eexperimental sample scheme without balanced winding.

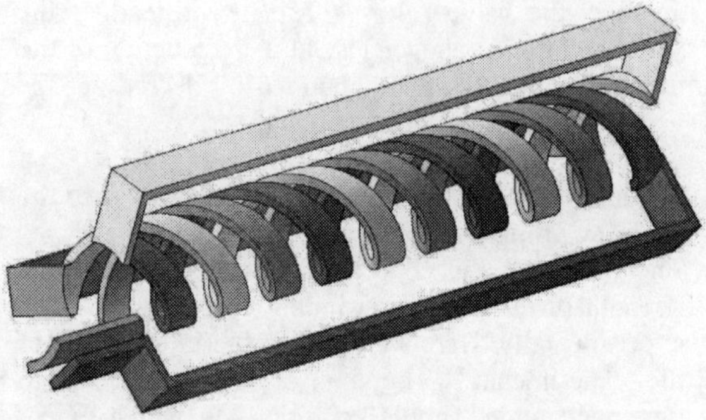

Fig. 4b. Eexperimental sample scheme with balanced winding.

In the second sample the current is closing on the conductors displaced at the large distance from the winding. In this case the axial current component of the winding is not compensated by the current in the cavity of the cylinder. Therefore the azimuthal field is formed with the magnetic pressure equal to the pressure of the poloidale field. It should be noted that the transverse field B_r of both samples is the same. Thus, the experiments with the different samples differed only by the mode of the loading in radial direction. The results of the tests have shown that the samples with unbalanced winding in the field with induction not higher than 0.3 exhibited the residual deformation after a single discharge: the increasing of the winding radius by 10 % approximately. The testing of the samples with external azimuthal field have been carried out at stepwise increasing of the amplitude induction with an interval of 0.5 T. At every step one discharge has been produced followed by measuring the residual deformation (incremental radius growth). In the field with induction amplitude up to 6 T the residual deformation was absent. Thus, for the balanced winding subjected to radial forces the strength threshold has not been exceeded. In the field with induction of 7 T there occurred the disruption of one of ribs between the turns followed by the breakdown. One can assume that this disruption is caused by the forces induced by transverse magnetic flux. In spite of imperfection of the winding (presence of the cuts and transverse field), the first experiments confirmed the sharp increase of the threshold induction value (at least by 6/0.3=20) on switching of an azimuthal field providing the equilibrium of the winding in radial direction. This threshold field is higher than that in the single-turn magnet with the same ratio of the radius to the width, made from cupper with the yield point $\sigma_T = 10^8$ Pa. The threshold induction in such a magnet is $B_T=\sqrt{(2\mu_0\, \sigma\, T\, d/R)} = 4.2$ T.

The application of the disconnected screen makes possible to decrease the transverse flux. Along with the use of the dielectric with the high Young modulus, it gives the possibility to obtain the field with the induction approaching that calculated for the base model.

Conclusion

We have obtained the more accurate data on mechanical stresses in the winding with balanced radial forces. On the example of the simple solenoid with the spiral winding it was shown the effect of the transverse field and discreet current distribution on the stresses. Our experiments confirmed the sharp reduction of the radial deformation at the use of the azimuthal field for the sustaining of the winding balance. The calculations elucidated that the mechanical stresses could be reduced by applying the cut cylinder screens and the isolation having the high Young modulus.

References

[1] Shneerson G.A.,Dolotenko M.I., Krivosheev S.I. Strong and Superstrong Pulsed Magnetic Fields Generation. De Gruyter, Berlin, 2014. p.429

[2] Shneerson G.A , O. S. Koltunov, , S. I. Krivosheev, Yu.E.Adamian, A. N. Berezkin A. P. Nenashev, A. A. Parfentiev . Models of Magnets With the Quasi-Force-Free Winding Manufactured From the Thin Solid Sheets or Transposed Conductors. IEEE TRANSACTIONS ON PLASMA SCIENCE, 2010, № 8 pp.1731-1738. ↓

[3] Shneerson G.A , O. S. Koltunov, A. N. Berezkin, I.. A. Vecherov, S. I. Krivosheev, A. P. Nenashev, A. A. Parfentiev. DEVELOPMENT AND INVESTIGATION OF ONE-LAYER QUASI-FORCE-FREE MAGNETS. Proceedings of 14th International Megagauss Conference, Maui, Hawai, USA, 2012, pp.46-50. ISBN 9781467357197(CD-ROM)

[4] Shneerson G.A , O. S. Koltunov, A. P. Nenashev, A. A. Parfentiev, I.. A. Vecherov, D.A.Degtev, S. I. Krivosheev,. A. N. Berezkin Computer and full-scale models of magnets with a single-layer quasi-force-free winding and evaluations of the possibility of using such magnets in inductive energy stores. THERMAL ENGINEERING, 2013, № 14

[5] Shneerson G.A A. P. Nenashev, A. A. Parfentiev, I.. A. Vecherov, . I. Krivosheev,. N.Yu. Kashirina, M.A.Pilinina. Azimuthal Forces and Method of Their Reduction in Imperfect Winding of Quasi-Force-Free Magnets TECHNICAL PHYSICS LETTERS, 2015, v.41б, № 8, c.26-34

An explosively driven high-power microwave system

V.G.Baryshevsky, N.A.Belous, A.M.Belov, P.T.Bogdanovich, V.A.Evdokimov, A.A.Gurinovich, E.A.Gurnevich, P.V.Molchanov, A.V.Oskin, A.A. Rovba, I.I.Vasiliev
Institute for Nuclear Problems, Belarusian State University
Minsk, Belarus
Electrophysical Laboratory, Minsk, Belarus

Abstract—An explosively driven high power microwave pulsed power system designed, built, and tested in Belarus is presented. The system is composed of a helical flux compression generator (HFCG) seeded from a capacitor bank, a power conditioning system including a pulsed transformer and a fast opening electro-explosive switch with inductive energy storage, and a high-power microwave (HPM) source. The paper includes an overview of system design and parameters. Description of several experiments along with analysis of system performance is provided.

Keywords—flux compression generator; high power microwave; high-voltage pulse transformer

I. Introduction

Explosively driven particle accelerators [1-2] and radiation sources [3-4] are discussed and developed recent decades. High power and increased flexibility of parameters are among the advantages of such devices, though being fundamentally single shot.

An explosively driven high power microwave pulsed power system composed of a helical flux compression generator (HFCG) seeded from a capacitor bank, a power conditioning system including a pulsed transformer and a fast opening electro-explosive switch with inductive energy storage, and a high-power microwave (HPM) source is presented. Overview of system design and description of several experiments is provided along with analysis of system performance.

II. System Design

The design of an explosively driven high power microwave pulsed power system is presented herein.The system is developed with the goal of maximum flexibility in subsystems parameters to enable testing different system configurations at the open air test range.

It is composed of four subsystems as follows: a seed source, a helical flux compression generator (HFCG), a power conditioning unit, and a high-power microwave (HPM) source (see Fig. 1.) Five Rogowski coils enable detecting the current derivatives in the system at points 1, 2, 3, 4 and 5 in Fig. 1. The EEOS voltage and the voltage applied to the cathode can be evaluated when the circuit parameters are known. The microwave emission is detected at a distance of 11.5 m from the output window by two receiving antennas Geozondas 1–4.5 GHz and Tektronix oscilloscopes TDS7704 and TDS7354. All the measuring channels are synchronized.

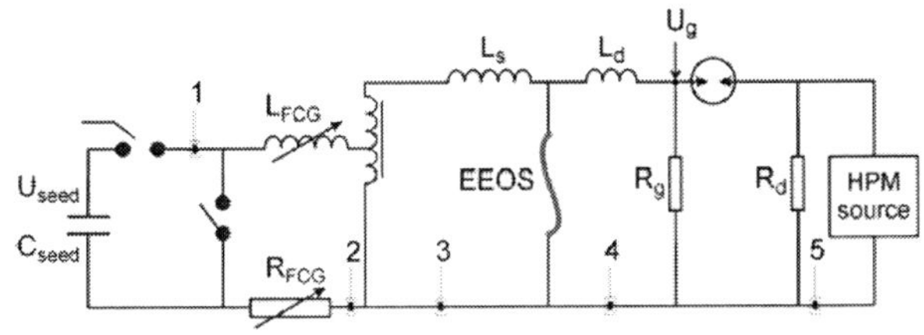

Fig. 1. Block diagram of complete system operation: location of five Rogowski current monitors are marked, R_g= 2.3 kΩ , R_d= 6 kΩ

The seed source, when triggered by a spark gap, discharges into the HFCG coil, thus providing initial flux. Synchronization unit matches maximum of the seed source produced current pulse with the HFCG crowbar closing moment. The explosively driven armature tube inside the HFCG compresses the initial magnetic flux, thus providing current levels measuring hundreds of kiloamperes. The high current output of the HFCG is delivered to a power conditioning unit that transforms the high current output of the HFCG into a higher voltage input for the HPM source. The power conditioning unit is composed of a high-voltage pulse transformer, an energy storage inductor and an electro-explosive fuse as the opening switch. High-voltage pulse produced by the power conditioning unit is applied to a closing switch (a peaking spark gap) inducing a high potential across the cathode-anode gap of vacuum diode, thus producing a high-current electron beam. Interaction of electron beam with the HPM source structure results in high power microwave generation. The open air test range is equipped with an explosion-proof building distant 25 meters from the explosion ground, where an enclosure made of concrete blocks ensures fragmentation protection (Fig.2). During the experiment only HFCG and pulsed transformer are to be destroyed, all other subsystems are designed for multiple use that allows for up to three full system shots daily. Manufacturing of a HFCG and pulsed transformer assembly requires about a week, though. HPM source wheeled bed and specially conditioned car trailer enable its easy transportation to the open air test range.

Fig. 2. HPM source at the open air test range

A. *Seed Source*

The capacitor based stand-alone seed source includes a capacitor bank, a charger, a spark-gap switch, which is controlled via optic fiber, and a current monitor. It directly seeds the helical flux compression generator. The capacitor bank contains four capacitors 10μF/8kJ each, which could be connected either in series or in parallel providing variety of seed source parameters. Both positive and negative chargers are available. A delayed ignition is used to synchronize FCG explosion and seed current maximum. Start signal from the synchronization unit controls spark-gap closing and, therefore, capacitor bank discharge to FCG circuit. The seed source is sheltered in the explosion-proof building. A 25m coaxial cable connects seed source to FCG, which is placed in the concrete enclosure.

B. *Helical Flux Compression Generator*

The single stage helical flux compression generators with a tapered coil geometry and winding are used. A copper armature tube, which has outer/inner diameter either 57/50mm or 41/35 mm, is centered inside the helical winding and electrically connected to it. Detonation is initiated end-on and HMX based explosive provides the detonation velocity about 8mm/μs along the axis and armature tube expansion cone angles within the range 10.4-11.6°. Three types of HFCG coils are designed (see Table I).

TABLE I. HFCG PARAMETERS

HFCG model	armature outer/inner diameter, mm	max/min coil diameter, mm	coil length, mm	number of sections
#1	41/35	90/84	425	7
#2	41/35	104/84	425	6
#3	57/50	122/84	490	6

The initial HFCG inductances are 307, 230 and 530μH, respectively. A Rogowski coil integrated in the HFCG output flange (mark 2 in Fig.1) monitors the current derivative in the HFCG circuit during experiment.

C. *High-Voltage Pulse Transformer*

For impedance matching an explosive generator to higher impedance load (an electron beam diode) two types of high voltage pulse autoconnected transformers are developed: a cable transformer and a compact coaxial one (see Table II). Both are designed to be locked directly to the HFCG output flange. The cable transformer is air insulated, while to mitigate electric breakdown inside the coaxial transformer working volume, transformer oil is applied.

TABLE II. TRANSFORMER PARAMETERS

Parameter \ Transformer type	Cable transformer	Coaxial transformer
Primary winding inductance, μH	0.27	0.29
Primary winding number of turns	5.0	1.5
Secondary winding inductance, μH	12.5	11.0
Secondary winding number of turns	25.0	9
Mutual inductance, μH	1.385	1.6
Secondary winding ac resistance measured at 100kHz, mΩ	90	33
Transformer dimensions (max. diameter/height), mm	315/280	185/210 (200/240)[a]

[a] (dimensions with oil tub)

A high voltage coaxial cable serving as inductive energy storage connects the transformer output to the electro-explosive switch, the junction is also oil insulated.

Fig. 3. High-voltage pulse transformers: cable transformer (left) and coaxial transformer (right) locked to HFCGs of different types

D. *Electro-Explosive Opening Switch*

The electro-explosive opening switch (EEOS) is placed inside a plastic casing, which enables use of high-pressure insulating gas (up to 0.5MPa). The EEOS design allows installing of max. 36 wires, which length can be varied from 0.3m to 0.75 m for straight wires and can be increased up to 1 m for zigzag packing. Copper wires of 100 μm diameter are used in the experiments. Storage inductance measures 2-4 μH.

The switch high voltage output is attached to the peaking gap, which is a part of the HPM source assembly. The peaking gap operates at 0.6 MPa pressure of SF_6.

E. HPM Source

HPM source assembly is composed of an input high-voltage feed through, SF_6 filled tunable peaking gap, a vacuum diode assembly housed in a high-vacuum sectional flanged chamber and a set of output horns. The peaking gap operates at 0.6 MPa pressure of SF_6. Attached vacuum pumps allow chamber evacuation to vacuum level of $\sim 5*10^{-3}$ Pa in 1.5-2 hours. Chamber inner diameter is 300mm. High voltage pulse of up to 500kV of either positive or negative polarity can feed the HPM source assembly, which is designed for testing different structures for generating microwaves, namely: axial vircator [5], triode reflex geometry vircator [6], relativistic split-cavity oscillator [7] and volume free electron laser [8-10]. Installed on the wheeled bed the HPM source can be easily relocated inside the laboratory or transported to the open air test range.

Fig. 4. Electro-explosive openning switch: inside plastis casing (top), for zigzag packing

Fig. 5. HPM Source assembly

III. Experiments

A. HFCG preliminary tests

HFCG preliminary tests were made with mostly inductive load 150 nH (see Fig.6 and Table III). Seed current value was chosen to provide maximal output power.

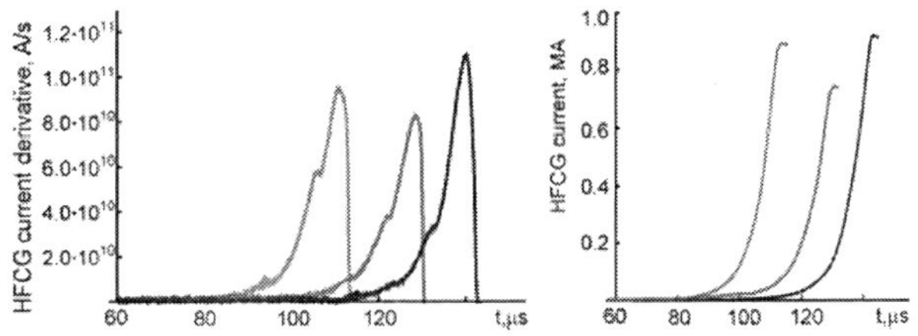

Fig. 6. Current derivative (left) and current in HFCG circuit: #1(red), #2(green) and #3 (blue)

TABLE III. Results of HFCG preliminary tests

HFCG model	Max. Current derivative, A/s	Max. current, kA	Peak power, GW	Seed source capacitance/ charging voltage (μF/kV)
#1	$8.2\ 10^{10}$	740	7.6	20/16
#2	$9.3\ 10^{10}$	890	10.2	10/15
#3	$1.1\ 10^{11}$	920	12.2	10/8

B. Power Conditioning Unit preliminary tests

Results of HFCG with power conditioning unit tests are gathered in Table IV. Storage inductance measured 2μH, EEOS had initial resistance 56 mΩ and included 36 copper wires of 100 μm diameter and 750 mm length. EEOS casing was filled with pressurized N_2 at 0.2 MPa.

TABLE IV. Results of HFCG with Power Conditioning Unit Preliminary Tests

HFCG model	Max. current derivative, A/s	Max. current, kA	Peak power, GW	Transformator type
#1	$2.5\ 10^{11}$	47	40	cable
#2	$3.0\ 10^{11}$	55	55	cable
#3	$3.2\ 10^{11}$	65	65	cable
#3	$4.5\ 10^{11}$	70	80	coaxial

Experiments with cable transformers revealed high probability of isolation breakdown caused by small bending radius of transformer coils. Thus, coaxial transformer was selected for the full-scale system tests.

C. HPM source preliminary tests

To be used for microwave emission two types of HPM sources were tested: axial vircator with resonant cavity [5] and triode reflex geometry vircator [6]. Both sources were driven by a pulsed power supply described in [5,6].

The axial vircator with three cavity resonator was housed in the drift tube of HPM source, superposing the resonator output section with the horn input. The solid cathode Ø71 mm made of dense fine-grained graphite placed with 16.3 mm gap to the stainless steel woven anode mesh with 0.77 geometric transparency was used in the experiments. Microwave output (see Fig.7) with peak power as high as 200 MW at 3.2 GHz frequency was produced at maximal diode voltage 400 kV and amplitude of current pulse 20 kA.

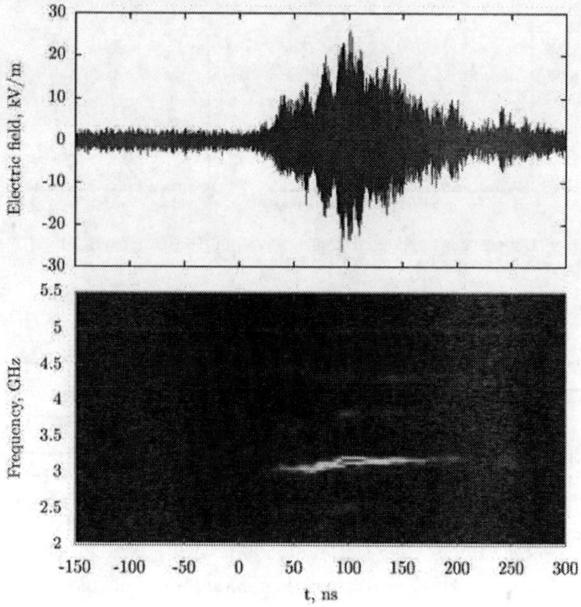

Fig. 7. Detected microwave signal and its spectrogram

Triode reflex geometry vircator [6] was housed in a high-vacuum sectional flanged chamber. The solid cathode Ø60 mm made of dense fine-grained graphite is grounded to the housing. A high-voltage pulse of positive polarity is applied to the centered in the vacuum chamber anode of 200 mm diameter made of stainless steel mesh. Microwave pulse (Fig.8) of peak power about 400 MW at 3.2 GHz frequency was produced by 400 keV electron beam at 17 kA current amplitude.

IV. Conclusion

An explosively driven high power microwave pulsed power system composed of a helical flux compression generator (HFCG) seeded from a capacitor bank, a power conditioning system and a high-power microwave (HPM) source is reviewed. Description of system design and parameters along with the results of subsystem experimental testing is provided. Full-scale system tests are coming soon.

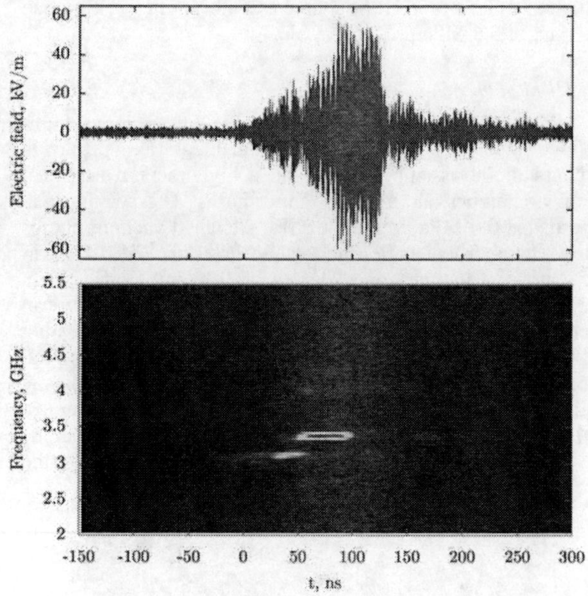

Fig. 8. Detected microwave signal and its spectrogram

References

[1] A. D. Sakharov, Magnetoimplosive Generators, *Sov. Phys. Usp.*, Vol. 9 (1966) pp. 294–299; doi: 10.1070/PU1966v009n02ABEH002876.

[2] A.I. Pavlovskii, G.D. Kuleshov, R.Z. Lyudaev, et al, Pulsed air-cored betatron powered from a magnetocumulative generator, *At. Energ.* Vol. 41 (1976) pp 757-760; doi:10.1007/BF01132663.

[3] K. V. Gorbachev, S. D. Korovin, G. A. Mesyats, et al., High-power microwave pulses generated by a resonance relativistic backward wave oscillator with a power supply system based on explosive magnetocumulative generators,*Tech. Phys. Lett.*, Vol. 31, No. 9 (2005) pp. 775–778; doi:10.1134/1.2061744.

[4] M. A. Elsayed, A. A. Neuber, J. C. Dickens, et al., An explosively driven high-power microwave pulsed power system, *Rev. Sci. Instrum.* 83, 024705 (2012); doi: 10.1063/1.3681443

[5] V. Baryshevsky, A. Gurinovich, E. Gurnevich and P. Molchanov, Experimental Study of an Axial Vircator with Resonant Cavity, *IEEE Trans. Plasma Sci.*, Vol. 43, no.10 (2015) P.3507—3511; doi: 10.1109/TPS.2015.2439332.

[6] V. Baryshevsky, A. Gurinovich, E. Gurnevich and P. Molchanov Experimental Study of a triode reflex geometry vircator, a*rXiv: 1609.02385 [physics.acc-ph].*

[7] V. Baryshevsky, Relativistic Split-Cavity Oscillator, *arXiv:1402.3403 [physics.acc-ph].*

[8] V. Baryshevsky, Spontaneous and induced radiation by electrons/positrons in natural and photonic crystals. Volume free electron lasers (VFELs): From microwave and optical to X-ray range, *NIM B*, Vol. 355 (2015) P.17–23; doi: 10.1016/j.nimb.2015.04.004.

[9] V.G. Baryshevsky, N.A. Belous, A.A. Gurinovich, A.S. Lobko, P.V. Molchanov, V.I. Stolyarsky, Experimental study of a volume free electron laser with a "grid" resonator, *Proceedings of 28th Free Electron Laser Conference FEL2006*, TUPPH012 (2006) P. 331—334.

[10] V.G. Baryshevsky, N.A. Belous, A.A. Gurinovich, E.A. Gurnevich, V.A. Evdokimov, P.V. Molchanov, Experimental studies of volume FELs with a photonic crystal made of foils, P*roceedings of 32nd Free Electron Laser Conference FEL2010* (2010) P. 632—635..

High Performance Triplate TEM Cell

Dr. P. Delmote, F. Bieth
ISL - ERG
French-German Research Institute of Saint-Louis
Saint Louis, France

Abstract—**This paper deals with the design and testing of a compact triplate TEM cell. The device has been optimized to perform experiments with very high electrical fields at high frequencies. The cell consists of an air insulated 50 ohm transmission line, which connects a couple of coaxials to triplate line transducers facing each other. Each transducer offers a 48 mm diameter coaxial input for high power tests. For low power experiments, additional ultra-wideband adaptors allow the connection with 50 ohm N type devices. The global design is based on numerical simulations with a particular attention paid to the improvement of the cell matching. Then, a prototype has been manufactured and characterized in-house. The transmission parameters are at optimum up to 3.1 GHz (S_{21} > -1dB). Once the triplate TEM cell is connected to a fast pulse generator delivering a signal with a 160 kV total amplitude, the measured E-field between two plates can reach more than 9 MV/m peak to peak without breakdown. Considering the dimensions and performances, the developed device could be a good candidate for the study of biological-electromagnetic effects in cell culture dish.**

Keywords—stripline; transmission lines; TEM cell; Pulsed power; high electrical field; electromagnetic biology.

I. INTRODUCTION

A. *Context*

This device has been originally developed as a 50 Ω - 50 Ω transition between a coaxial waveguide and a large biplane microstrip waveguide. The goal was to feed an ultra-wideband antenna with ultra-short electrical signals, for applications like "radar thru the wall" or disruption of electronic devices. The selected antenna uses an antipodal Vivaldi design, made of two metallic arms separated by a dielectric (Fig. 1). The arms use an exponential profile, which provides non resonant properties to the radiator. Microwaves are emitted when the excitation signal travels along the curved arms, from high frequencies to low frequencies at the output. The major drawback of this design consists in losses all over the area forming a biplane microstrip transmission line, and in the fact that the polarization is not stable over the covered bandwidth [1]. Moreover, the design of a compact and efficient transition is a non-straightforward task to achieve. In order to overcome these points, an improved design has been proposed, based on the use of an antenna made of three arms and connected to a large stripline waveguide. These waveguides are made of three metallic layers separated by a dielectric; they offer a low level of losses over a wide bandwidth.

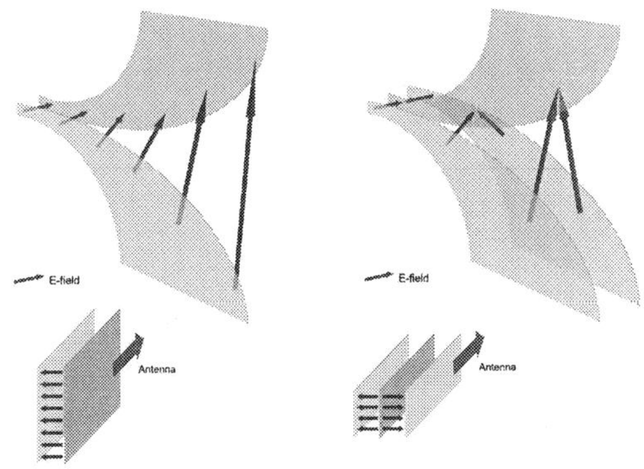

Fig. 1. Vivaldi antipodal antennas in 2 layers and 3 layers configurations.

B. *From waveguide transition to TEM cell*

Most of the time, characterization of waveguide adaptors are made using a couple of devices facing each other. One of the limits of this method relies on the fact that the considered devices emit and receive non guided signals (each of them acts as emitting and receiving antenna), which can affect the measurement of the S parameters. Nevertheless this protocol has been applied and kept from electromagnetic simulations to lab experiments. Once connected to a generator at the input, and to a perfect matched load at the output, this configuration creates a triplate cell, allowing the exposure of specific materials or probes, to high power transient electromagnetic waves.

II. EM SIMULATIONS

As mentioned previously, the main goal was to develop a coax to stripline transition, with some particular specifications regarding compactness, bandwidth and reliability when powered by GW-class electrical discharges. Stripline waveguide has been one of the first forms of transmission lines to be used at microwaves frequencies. The dominant operating mode is a transverse electromagnetic wave (TEM) which is the selected one for the system. There is no lower frequency cut off and higher propagating modes appear at higher frequencies. This mainly depends on the size of the considered waveguide, and more specifically on (1) the width of the inner conductor and (2) the electrical distance between the metallic plates (first cut off frequency for $h = \lambda_0/2 \sqrt{\varepsilon_r}$) (Fig. 2). In our case, the height between the two external plates has been fixed considering the 48 mm diameter of the coaxial input/output of the ISL pulse forming lines, and consists in a compromise between high frequency performances and high resistance towards electric breakdown. The insulation is made of specific polyoxymethylene plastic which offers good mechanical and electromagnetic properties. The dielectric permittivity is in the order of 2.3.

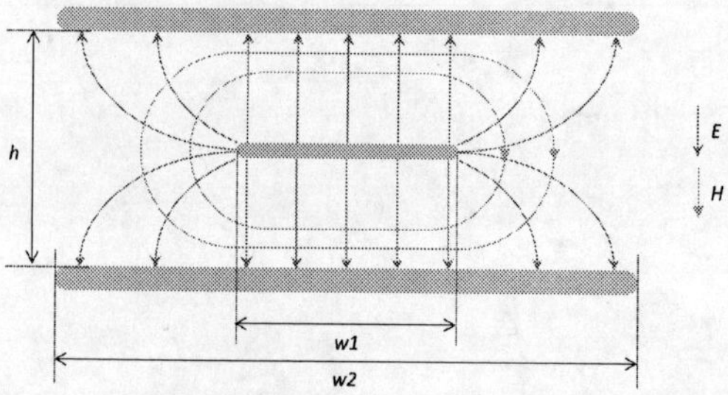

Fig. 2. Typical stripline parameters.

On the stripline side, the waveguide is air-insulated. The width of the outer conductors should be chosen as large as possible in order to minimize losses due to the EM field lines flowing outside of the waveguide. A typical value for this parameter is known as five time the width of the inner conductor. In the case of this prototype, it has been chosen to stay below 200 mm for compactness and weight constraints. The impedance is fixed by adjusting the size of the inner conductor. Some equations can be found in the literature has a first approximation [2]; final sizing has been optimized under CST microwave studio (Fig. 3). As the nominal impedance of 50 Ω is kept on both sides of the transition, the matching only depends on the geometrical parameters. The drawing has been optimized using CST microwave studio with its time domain solver. A specific attention has been paid on the matching and the risk of flash over, while keeping a compact and simple design.

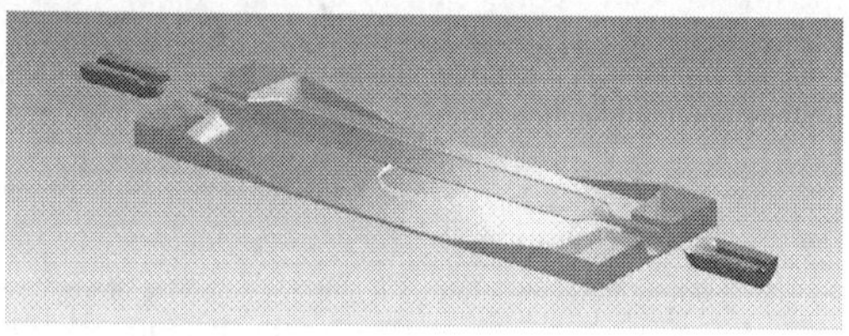

Fig. 3. Exploded view of the proposed design.

III. Manufactured prototype and test setup

In order to characterize the device, a couple of transitions have been manufactured by the machining workshop of ISL. By using additional sheets of metal, they have been easily mounted together, facing each other and forming the TEM cell. It is a modular and robust system, made of brass, stainless steel and plastic. In this configuration, the cell has a 48 mm diameter input and output and can be easily connected to high voltage sources and loads available in the laboratory (Fig. 4).

For the low power experiments, two dedicated adapters are used to ensure the connection with standard N devices. Theses adapters are made following the design rules explained in [3] and they offer high performances in proportion to the needs (losses smaller than -0.5 dB up to 6 GHz).

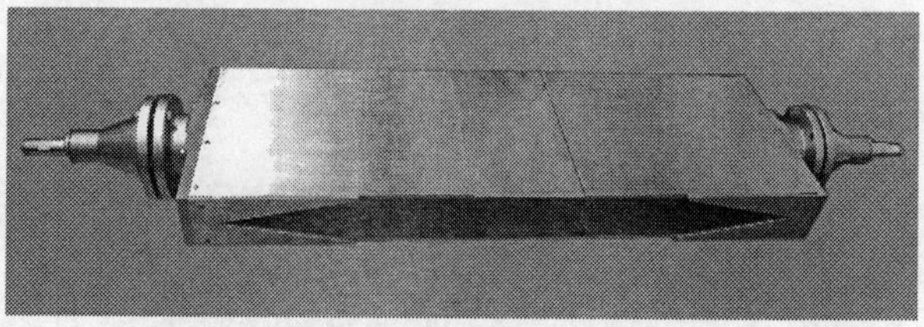

Fig. 4. Manufactured cell equipped with N→48 transitions at both ends.

IV. Results

A. *Low power experiments*

In order to measure the transmission parameter in the frequency domain, the cell equipped with N connectors has been characterized with a vector analyzer (Fig. 5). Measured S_{21} is in good agreement with simulated results, taking into account the use of additional adaptors at both ends (Fig. 6, Fig. 7). The losses remain lower than -1 dB up to 3.1 GHz, and the phase is very stable over this frequency range. At higher frequencies, some ringing occurs at both ends of the cavity formed by the cell, which starts to degrade the overall performances.

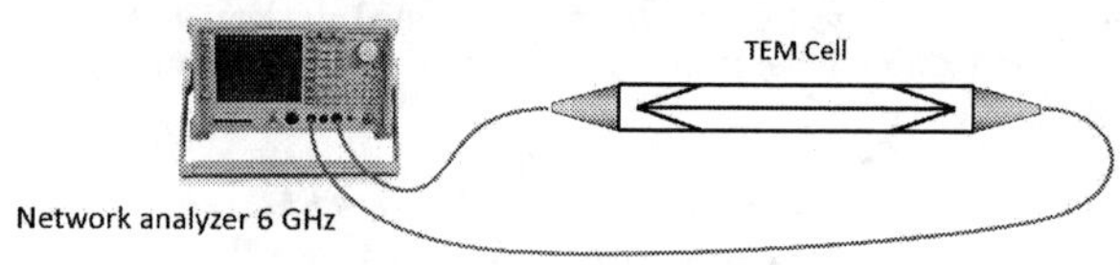

Fig. 5. S21 measurement.

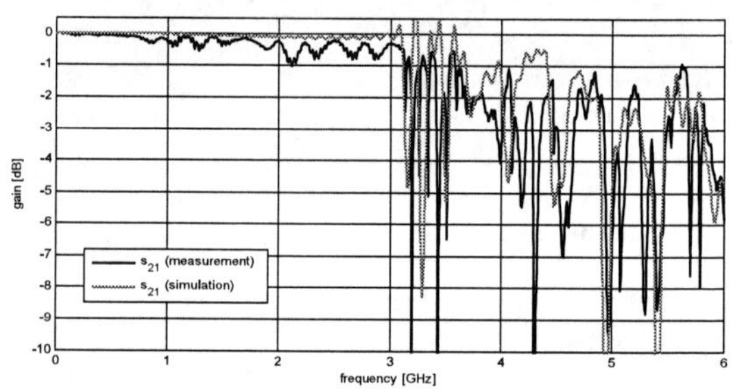

Fig. 6. S21 parameters – amplitude.

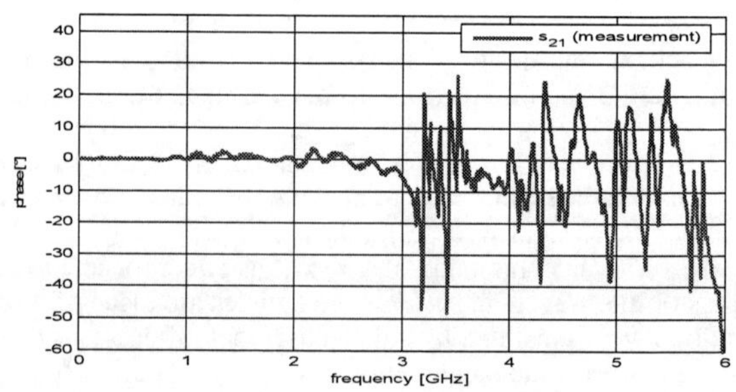

Fig. 7. S21 parameters – phase.

In order to perform time domain measurements, a fast pulser Kentech APG1 has been connected to the input of the cell. Internal electrical fields can be recorded in the middle of the cell, all along a cross section of the waveguide, by inserting a small D-dot probe thru holes drilled inside the external conductor (each position of the probe is separated by 15 mm, position n°1 being at the center) (Fig. 8). Signal are recorded thanks to a fast oscilloscope from Agilent (16 GHz, 80 Gs/s).

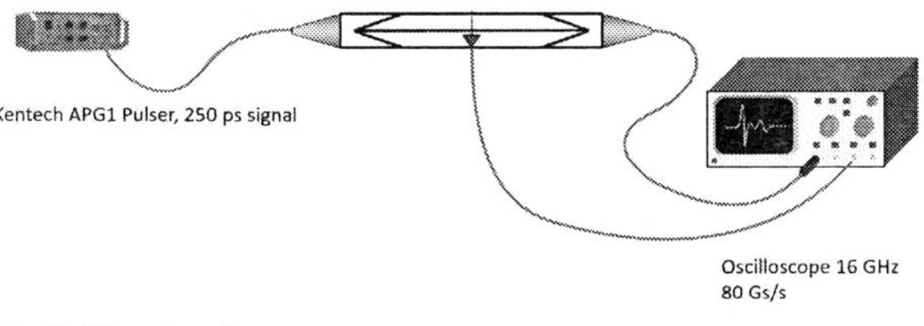

Fig. 8. Time domain measurement.

It can be seen that the amplitudes of the signals, measured at the three first positions, are quite similar (Fig. 9). The impulse shape is preserved, even so with some residual oscillations. Going back to the frequency domain, the frequency content of the signals recorded under the inner conductor (+/- 30 mm from the main axis) is very stable up to 3 GHz (Fig. 10).

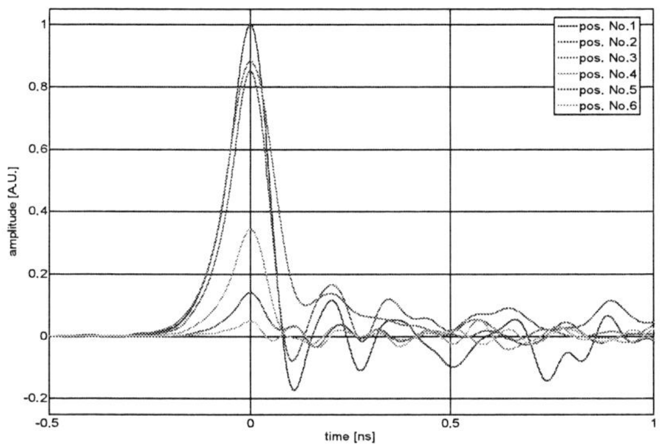

Fig. 9. Traveling signals in time domain, transverse section.

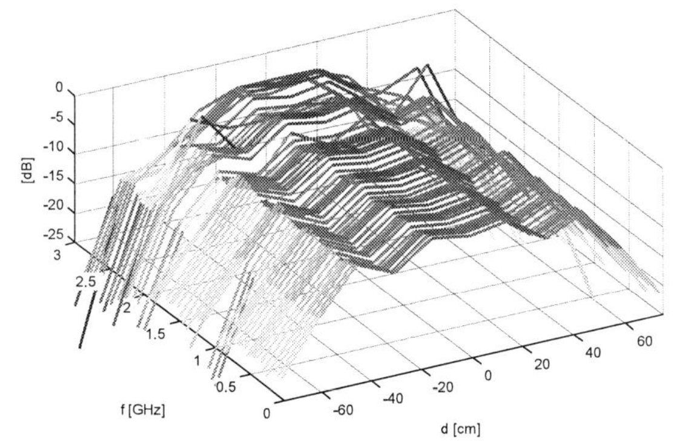

Fig. 10. Traveling signals in frequencies domain, transverse section.

B. High power experiments

To perform high power experiments, the cell has been connected to an ultra-short pulse generator, composed of a Marx generator followed by a pulse forming line. This generator delivers a bipolar electrical impulse with an adjustable rise time, amplitude and duration [4]. Here, the source was set to deliver a 160 kV bipolar impulse with a total duration of 2 ns and rise time up to 200 ps. At the output, a high voltage divider acts both as a load and a measurement device. Signals are recorded before the cell (SMA I-dot probe), in the middle of the cell (D-dot probe) and at the output (VOSS probe) (Fig. 11). At the cell center, the traveling electromagnetic wave can reach more than 9 MV/m even if the test was performed at atmospheric pressure, without SF_6 gaz. The risk of flash over or parasitic discharges is minimized because of the ultra-short duration of the selected signal and the good matching of the cell (Fig. 12).

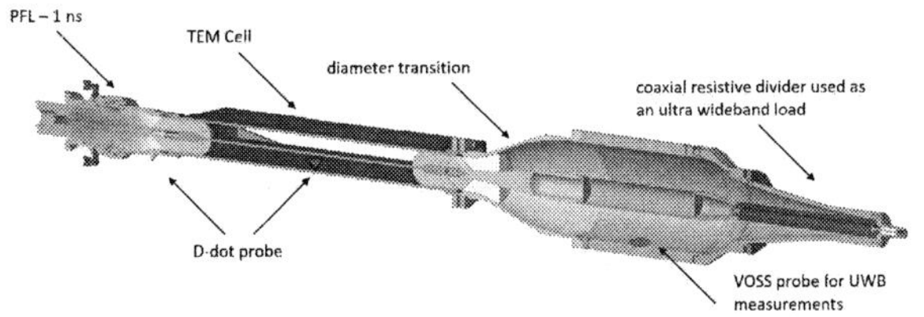

Fig. 11. Setup for high power experiment.

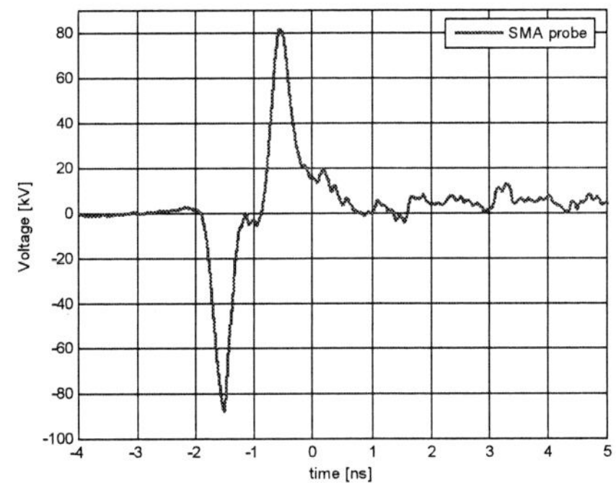

Output voltage of the pulse forming line

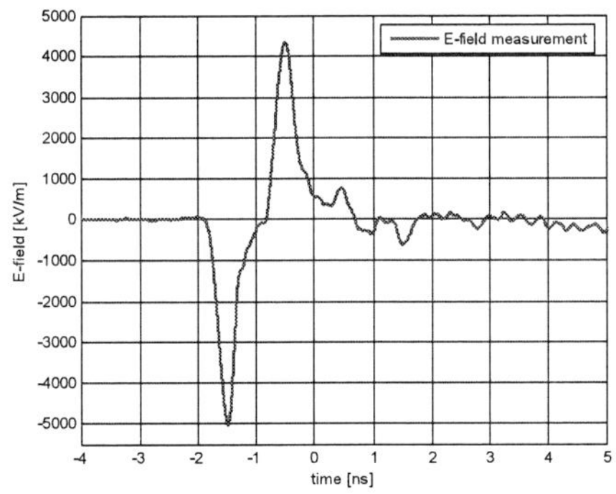

E-field inside the cell

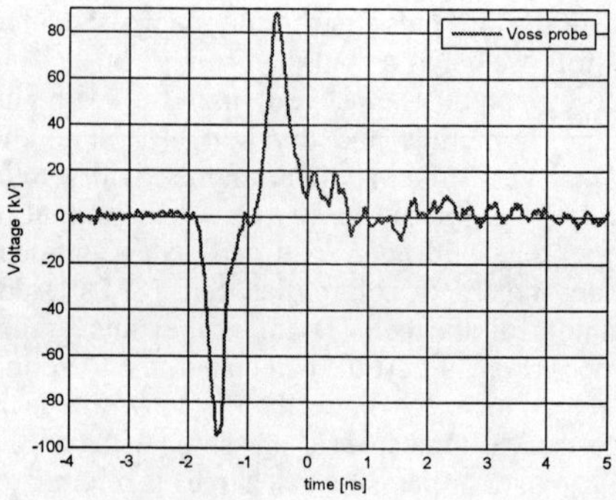

Output voltage of the cell

Fig. 12. Transient signal at different locations.

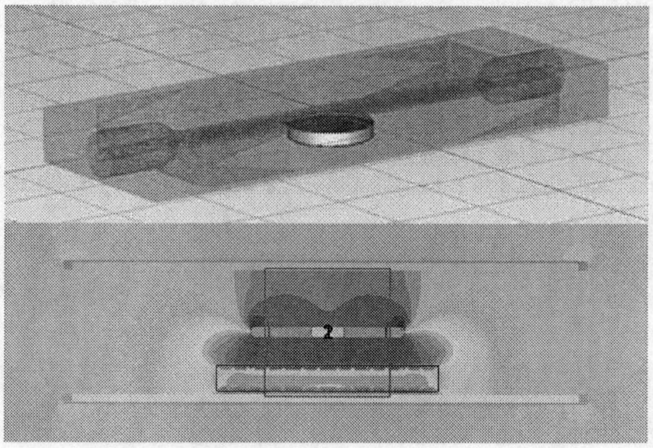

Fig. 13. Petri dish exposed to transient E-field.

V. Conclusion

The proposed cell achieves a very good compromise between size, reliability and performances. The volume available under the inner conductor allows experiments on probes, electronic components or biological material like standard Petri dishes, offering a direct access to the test area (Fig. 13) [5]. Depending on the power supply connected to the developed device, targets can be exposed to electrical fields up to 3 GHz (medium power, continuous waves) or 10 MV/m (short impulses, 100 ps rise time, duration below 5 ns). The use of a bipolar electrical source allows the simulation of radiated waveforms in guided mode (taking into account the waveguide impedance). If necessary, the device can be matched for measurements on specific materials or impedances. Lastly, going back to the initial objective, the coaxial to stripline transition itself delivers an excellent performance and can be used to feed an UWB HPM antenna.

Acknowledgment

The authors would like to thank the ISL Engineering Department and the machining workshop for their participation in this study.

References

[1] Thomas A. Milligan, "Modern Antenna Design", Edited by Wiley, pp510-515.

[2] Paul-François Combes, Raymond Crampagne, "Circuits passifs hyperfréquences", Techniques de l'ingénieur, pp E1 403 10.

[3] J. Wey, F. Bieth, "Very High Voltage Divider with Large Bandwidth", Ultra-Wideband Short-Pulse Electromagnetics 6, Edited by Mokole, Kluwer Academic/Plenum Publishers, 2003.G, pp. 523-536.

[4] P. Delmote, S. Pinguet, F. Bieth, "Performances of a compact, high power WB source with circular polarization", Ultra-Wideband, Short-Pulse Electromagnetics 10, Edited by Springer, pp239-250.

[5] Andrei G. Pakhomov,Damijan Miklavcic,Marko S. Markov, "Advanced Electroporation Techniques in Biology and Medicine", edited by CRC Press, 2010.

Calibration with a Monocone on a Ground Plane (CALMOPLA)

Francois Bieth, Philippe Delmote
French-German Research Institute of Saint-Louis
Saint Louis, France

Abstract—**CALMOPLA (CALibration with a MOnocone on a ground PLAne) is a tool for the calibration of low sensitivity, electric or magnetic field probes. The device works in the microwave range, from several tens of MHz to several GHz. It consists of a small conical monopole (D-dot), positioned at the center of a ground plane and emitting electromagnetic waves. Close to the reference plane, the radiated signals are quasi spherical waves which can be described by well known analytical expressions.**

The time expression of an electromagnetic field at a point in space depends on the distance between the point and the monopole: in the case where the wavelength of the emitted wave is much larger than the size of the monopole, the border between the far field (time second-derivative response) and the near field (time proportional response) is given by the wavelength divided by 2π.

A vector network analyzer is sufficient to cover the frequency range of most of the sensors to be calibrated. Nevertheless, the use of a low noise amplifier and a low frequency power amplifier allow to extend the measurement range to a very low sensitive sensor anywhere on CALMOPLA.

Keywords— CALMOPLA, Monocone, Sensor calibration, UWB.

I. INTRODUCTION

In the field of High Power Microwaves (HPM) and more specifically in the case of Ultra Wideband (UWB) high power microwaves, there is a strong need to characterize large band sensors. In order to measure the frequency response of these sensors, it is very helpful to get access to large band calibration systems. CALMOPLA (CALibration with MOnocone on a ground PLAne), consists of a small conical radiating monopole (D-dot) placed at the center of a ground plane (2 m x 2 m), and emitting quasi-spherical waves. Unlike facilities like EMP Engineering Research Omnidirectional Radiator (EMPEROR) [1], which uses a monocone, a hundred times bigger than CALMOPLA and which emits high power electromagnetic fields for the calibration of guided modes under the cone, the principle of CALMOPLA is to calibrate the sensor in a "free space" configuration.

The analytical behavior of monocone antennas was described by e.g. Schelkunoff [2]. The conical antenna with radius 'a' and half-angle θ_2 is located at the center of a spherical coordinate system (r, θ, ϕ) and supplied by an input voltage V_{in} on its apex (Fig. 1). It can be assumed that the diameter of the coaxial line is small compared to the smallest equivalent wavelength λ: non-propagating modes which could be generated at the interface between the coaxial line and the conical line can be neglected.

From the Maxwell equations and assuming the separation of variables hypothesis, the expression of the electric and magnetic field in spherical coordinates can be given in the antenna region in the guided area and in the free space area (Fig. 1). Assuming that only the TEM (Transverse Electromagnetic) waves propagate under the cone and that the electric component on the metallic surface is zero, equalizing the expression of the electric and the magnetic field at the border of each region provides the expression of the impedance of a wide-angle conical antenna (given by Papas & King [3]):

$$Z_{in} = Z_0 \frac{1-\dfrac{\beta}{\alpha}}{1+\dfrac{\beta}{\alpha}} \quad (1)$$

with $Z_0 = 60\ln\left(\cot\frac{\theta_2}{2}\right)$ characteristic impedance of a conical line and α and β are combinations of Legendre polynomial and spherical Hankel functions of the second kind.

In order to keep the 50 Ω impedance of the coaxial cable all over the conical line, θ_2 is chosen to be 47°. The radius 'a' is 40 mm long and the dimensions of the metal plate (ground

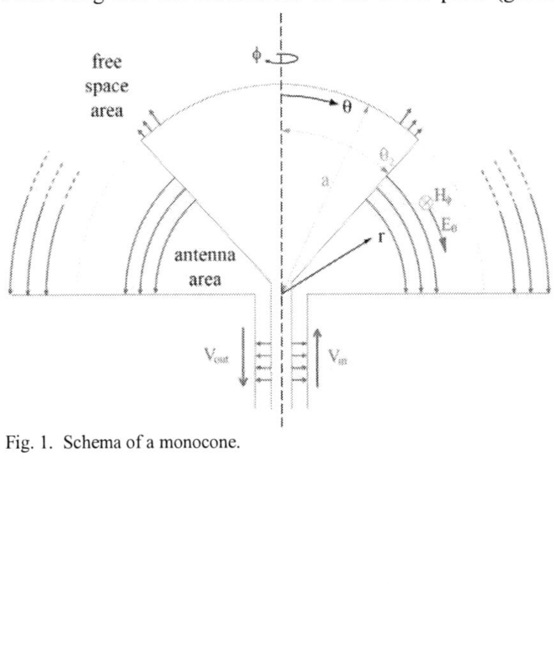

Fig. 1. Schema of a monocone.

plane) are 2 x 2 m^2.

II. Conical Monocone Radiation

It is well known that, up to the cutoff frequency f_c, the conical antenna response of an incident electrical field is a time derivative function. The D-dot sensor response is:

$$V_{out} = R \cdot A_{eq} \cdot \varepsilon_0 \frac{dE_\theta}{dt} \qquad (2)$$

with R the sensor characteristic load impedance, A_{eq} the sensor equivalent area and ε_0 the vacuum permittivity. Considering the measurement of the D-dot sensor (θ_2=47° and a=40 mm) on 50 Ω impedance and a first order derivative response, the cutoff frequency f_c equals roughly to 650 MHz.

Motohisa Kanda [4], after the David M. Kerns studies [5], affirms that, the transmission transient response of the antenna $T_t(f)$ is the time derivative of the reception response $T_r(f)$:

$$\frac{T_t(f)}{T_r(f)} = -\frac{jk\eta_L}{2\pi\eta_0}\frac{e^{-jkr}}{r} \qquad (3)$$

with $jk = j\omega\sqrt{\varepsilon\mu}$, and $\omega = 2\pi f$, η_L antenna admittance, η_0 free space admittance, and ε and μ are the permittivity and permeability, respectively. Since the conical antenna response in reception is a first-order time derivative, the transmission response should be a second-order time derivative.

For antennas much smaller than the wavelength of the considered radiation, the far and near regional boundaries are given by λ/2π (reactive near-field region).

Several measurements of electric fields over the ground plane, with sensors placed at different distances from the center, have been performed to verify the behavior of the frequency response (Fig. 2):

- The near field and far field regions are limited by f_o:

$$f_0 = \frac{c}{2\pi d} \qquad (4)$$

- Like a "small doublet" in the near field region, the amplitude of the electric field decreases according to the cubic distance 'd' and is independent of the frequency,

- In the far field region (by definition, the amplitude decreases as a function of the distance 'd'), the electric field varies according to f^2 until the cutoff frequency f_c of the monopole.

The magnetic field is measured on the ground plane with a magnetic field sensor. The behavior of the magnetic field is similar to that of the electric field in the far field, but the amplitude decreases with the distance 'd' and increases with the square of the frequency.

Finally, on the ground plane of CALMOPLA, the electric and magnetic fields shapes vs frequency are given by Fig. 3. In case of the far field condition, the variation of the electric field as a function of the direction angle θ $M(\theta) = \frac{|E_\theta(r,\theta)|}{|E_\theta(r,\pi/2)|}$ was calculated by Papas & King [6].

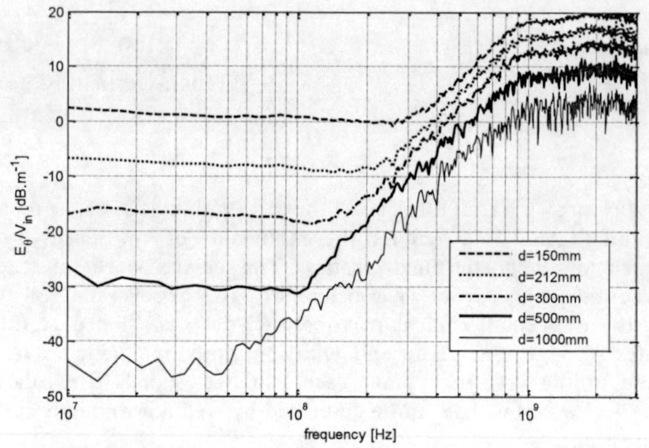

Fig. 2. Measurements of the electric field radiated by the monopole at different distances.

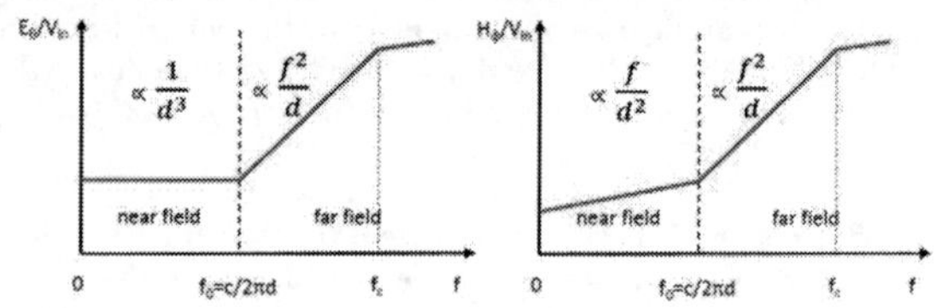

Fig. 3. Electric and magnetic field radiated by CALMOPLA on the ground plane.

Fig. 4. Photograph of CALMOPLA: monocone in the center; inserts can be seen on the diagonals of the ground plane.

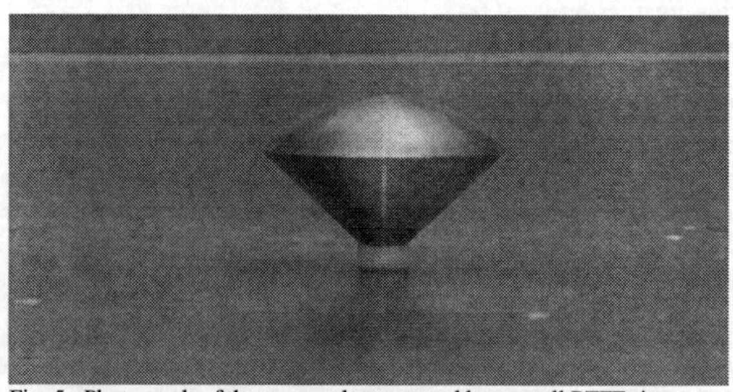

Fig. 5. Photograph of the monopole supported by a small PTFE ring.

III. Principle of Calibration

A. Measuring System

Several inserts are distributed on the ground plane at 150 mm, 212 mm, 300 mm, 500 mm and 1000 mm from the apex of the monocone on the two diagonals of the ground plane (Fig. 4). The monocone is supported by a PTFE (Polytetrafluoroethylene) ring whose dielectric constant is 2.1. The effects of the impedance mismatch are minimized by the small size of the ring (Fig. 5). For each position, the electric (or magnetic) field is calculated by an analytical expression and verified by a reference electric (or magnetic) sensor with a well-known antenna factor. A vector network analyzer (Anritsu® MS4623A) measures the transmission between the monopole and the experimental sensor to be calibrated. Dynamic range reaches 100 dB for 10 Hz IF Bandwidth. Generally, the sensibility level of high frequency and high power sensors is very low. Due to the size of the monocone (40 mm) the ratio between the radiated electric (and/or magnetic) field and the input voltage applied is low. To overcome this problem, a Miteq® Ultrawideband 23 dB LNA (Low Noise Amplifier) is introduced at the output of the sensor to be calibrated. In the case of a derivative probe, a 50 W power amplifier from EMPower® (46 dB power gain on 1-1000 MHz range) is used to increase the level of input voltage of the monocone at low frequencies. Thus, the noise floor of the transmission measurement (Fig. 6) can reach 180 dB.

B. Time Window

In order to improve the quality of the frequency response, a time filter is applied to eliminate the oscillations of the signal. The ground plane of CALMOPLA is not infinite and therefore reflections disturb the frequency response. After performing an inverse Fourier transform in the time domain, a time filter is judiciously chosen (Fig. 7) in order to keep only the intrinsic response of the sensor and to eliminate the reflections due to the environment. These reflections occur mainly at the outer end of the ground plane. Furthermore, the time response of the sensor must be shorter than the gate of the filter and the sensor should be placed as close as possible to the monopole in order to maximize the delay of parasitic signals due to the outer end of the ground plane.

IV. Conclusion

CALMOLA is a helpful and low-cost tool dedicated to the calibration of electric or magnetic field sensors. The potential problem of low level measurements, due to the relatively low power radiated and to the low sensibility of such sensors could be solved by the use of large band amplifiers in the measurement chain. With the assumption of sensors having a shorter response than the first reflections at the outer end of the ground plane, a time window filtering improves the quality of the computed results. As a complementary result, assuming that the position of the sensor to be characterized is well known, CALMOPLA can provide the phase center of a large TEM sensor over the frequency range.

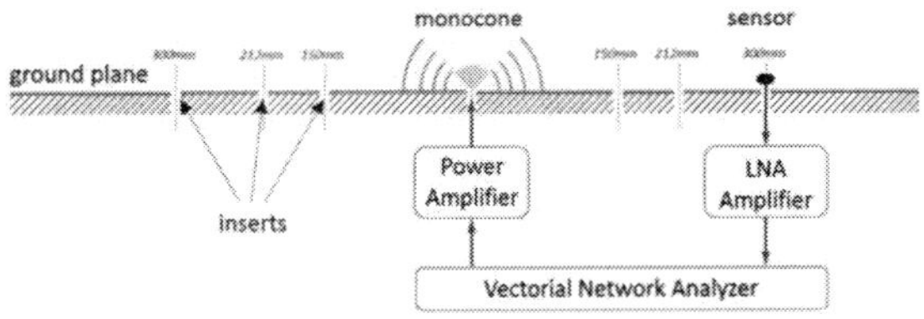

Fig. 6. CALMOPLA Measurement system.

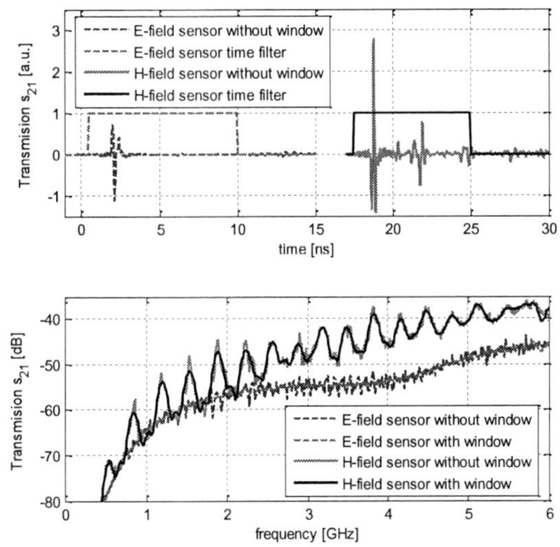

Fig. 7. Two examples of frequency responses with and without time windows at 500mm from the monopole: ground plane electric field and free magnetic field sensors.

References

[1] K. S. Kunz, H. G. Hudson, J. K. Breakall, R. J. King, S. T. Pennock, A. P. Ludwigsen, "Lawrence Livermore National Laboratory Electromagnetic Measurement Facility", IEEE Trans. On Electromagnetic Compatibility, vol. EMC-29, No. 2, pp. 93-103, May 1987.

[2] S. A. Schelkunoff, "Electromagnetic Waves", Van Nostrand, 1943.

[3] C. H. Papas, R. King, "Input Impedance of Wide-Angle Conical Antennas fed by a Coaxial Line", Proc. I.R.E, vol. 37, pp. 1269-1271, 1949.

[4] M. Kanda, "Time-Domain Sensor and Radiator", E.K. Miller, editor, Time-Domain Measurements in Electromagnetics, Van Nostrand Reinhold, New York, 1986.

[5] D. M. Kerns, "Plane-Wave Scattering Matrix Theory of Antennas and Antenna-Antenna Interactions", National Bureau of Standards, Monograph 162, 1981.

[6] C. H. Papas, R. King, "Radiation from Wide-Angle Conical Antennas Fed by Coaxial Line", Proc. I.R.E., vol. 39, No. 1, pp. 49-51, 1951.

A repetitive PFN-Marx generator based on an innovative design for compactness and rise time improvement

F.Lassalle, A.Morell, A.Loyen, T.Chanconie, B.Roques, M.Toury
CEA, DAM, GRAMAT
F-46500 Gramat, France
francis.lassalle@cea.fr

***Abstract*— Repetitive high voltage square pulses are of great importance for producing long-pulse electron beams and high-power microwaves. One of possible technologies for the generation of such pulses is a pulse forming network (PFN) based Marx generator, often combined with a pulse sharpening technique to reduce the rise time to a few nanoseconds (peaking stage). This paper presents an innovative design, named the "zigzag design", for the optimization of the compactness and of the rise time of a 400 kV - 85 ns PFN-Marx. Thanks to this design, the 16 stages of this generator, which deliver an open circuit output voltage of 720 kV, fit in a 650 mm length. For a slightly overmatched load (Z_{load} = 100 Ohm), the output voltage reaches 400 kV with a rise time as less as 5ns. The inductance reduction associated to the innovative zigzag design, which allows this sharp rise time with no need for a peaking stage, is described. The 85 ns plateau duration of the pulse is given by the PFN construction of each stage, which is based on 6 ceramic capacitors (2.1nF – 45kV) connected within a strip line. The 16 PFN stages are housed in a 360 mm diameter gas pressurized vessel. Burst mode operation for a duration of 10 s at a pulse repetition frequency of 100 Hz is reported, for a resistive load and for the electron beam diode of a X-band relativistic BWO [1]. To reach further compactness, the BWO system is integrated on side of the generator vessel and a U shaped gas pressurized line connect both systems through a compact conical vacuum insulator.**

Keywords*— pulsed power technology; high power microwave; pulse generator; PFN-Marx generator

I. Introduction

A. Global scope

This study deals with the development of HV drivers with :

- Generation of a square pulse which main characteristics shown on Fig. 1 are a voltage plateau ranging from 100 kV to 1 MV, a duration ranging from 30 ns to 300 ns and a rise time < 10 ns.
- Operation at a repetition rate ranging from 10 pps to 300 pps.

This work is supported by the DGA (Délégation Générale pour l'Armement)

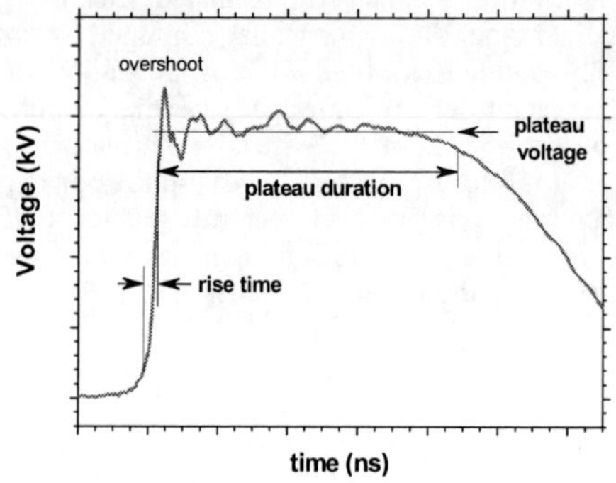

Fig. 1. A typical square pulse and its main characteristics.

- Compactness of the driver, characterized by the ratio (generated voltage / driver length) > 0.5 MV/m and the ratio (stored energy / driver volume) > 1 kJ/m^3.

Such drivers or modulators can be used as pulsed power source for generation of high power microwaves, electromagnetic waves, lasers, electron beams, ion beams, X-rays, plasmas, etc. These applications cover several domains like defense, security, research and industry.

B. Specific objective

These square pulse drivers are of particular interest for military and industrial applications of High Power Microwave (HPM) technologies. These practical applications need such a HPM system that should not only radiate high microwave power with high efficiency but should also be compact, transportable, reliable and work in the repetitive-rate mode.

For this specific study, the objective described in [1] is to demonstrate the possibility to integrate a powerful (sub-GW class) narrow X band HPM system into a generic cubic volume of 650 mm side. The 3D view of Fig. 2 shows this arrangement, which main subsystems are prime power (Batteries), capacitor charger, HV driver, HPM source and antenna. To fulfill the requirements, the HV driver must deliver 400 kV pulses with 80 ns duration, 5 ns rise time and 100 pps repetition rate, be compact (length ≤ 650 mm, diameter ≤ 400 mm) and connected to the microwave source

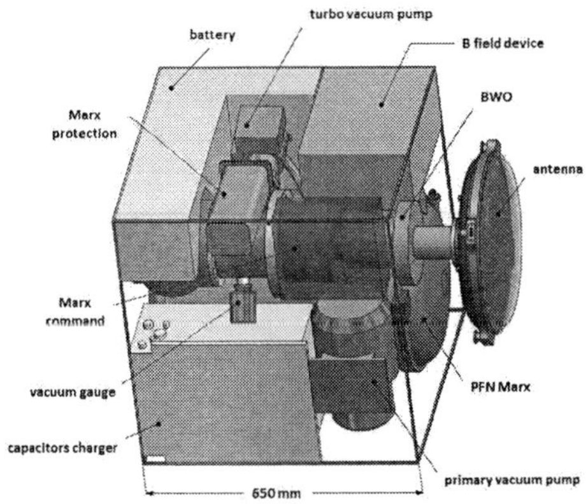

Fig. 2. 3D view of the 300 MW High Power Microwave system, integrated into a cubic volume 650 mm side. The PFN Marx is the grey cylindrical part in the background.

through an U-shaped line. With this connection, the HV driver and the microwave source fit in the diagonal of the cubic volume.

C. Possible technics

In laboratory, several technics are used to generate such repetitive square pulses : HV pulse forming line charged by a transformer [2] or by a Marx generator [3], Marx, IVA or LTD using cables [4], striplines [5] or PFNs [6], [9], etc. This last technic, based on stacked PFNs with a voltage addition using the Marx scheme, is chosen here for its potentialities, in particular in terms of compactness, reliability and quality shape of the pulse.

For any driver which impedance is matched to a resistive load (RLC circuit with "$R = (L/C)^{0.5}$"), the rise time [7] of the pulse is "$\tau_r = 1.2\ L/R$". This rise time is "$\tau_r = 2\ L/R$", for an overmatched load (circuit with "$R = 2\ (L/C)^{0.5}$"). Reaching rise time $\tau_r < 10$ ns is often a challenge, as it needs a very low inductance for the driver.

For most achievements, a peaking stage [8] is used to sharpen the rise time given by the driver. The design of this peaking stage needs special care, as it must use a capacitor and a switch working at the high amplitude level of the output voltage and within a very low inductance allocation. Moreover the insertion of this peaking stage is made to the detriment of system compactness. So an extra objective of this study is to reach a 5 ns rise time without the use of such peaking stage, by designing a very low inductance Marx generator.

II. THE ZIGZAG DESIGN

Some Marx generators described in bibliography already uses a zigzag architecture [6],[9], etc. The innovative design presented here is an optimization of this zigzag concept with two main objectives: reduction of compactness of the Marx generator and improvement of the rise time of the delivered pulse.

A. Marx compactness optimization

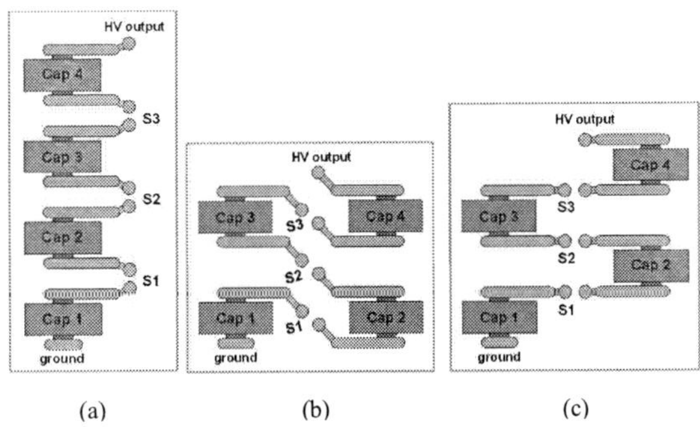

Fig. 3. Design approach for compactness optimization, for an example of a four stages Marx generator. (a) a "straight" Marx, (b) a "zigzag" Marx, (c) the innovative zigzag design proposed for compactness and rise time optimization.

Fig. 3 illustrates the gain in compactness allowed by a zigzag design, for a Marx generator where the switches are arranged along a column. Compared to a regular "straight" design (Fig. 3-a), the zigzag designs of Fig. 3-b and Fig. 3-c allow a gain in a 25% to 50% range along the length of the Marx. The innovative zigzag design of Fig. 3-c is an optimization of this concept, where the required insulation distance between stages is intrinsically obtained as the length of a capacitor and where the length of the connections between capacitors and switches are minimized.

B. Marx inductance optimization

According to the flux expression, the inductance of a circuit travelled by a current I is $L = \vec{B}.d\vec{S} / I$, where $\vec{B}$ is the magnetic field, j is the current density and $I = j.d\vec{S}$ is the current. For any Marx generator, during the discharge, the volume occupied by the magnetic field B is between the Marx circuit (capacitors and switches) and the return current electrode, which for most case is the metallic vessel housing the stages. For many Marx generators, this magnetized volume is not minimized, e.g. the devices where the switches are closed to the axis of the vessel, far from the return current electrode. For this Marx generator, as illustrated on Fig. 4, the switches column is located as close as possible to the vessel (within the limit of a good HV insulation). On the other direction, the zigzag design allows to reduce the length. The reduction of the magnetized volume along these two directions corresponds to a significant reduction of the Marx inductance L_{MARX}.

An extra inductance reduction allowed by this zigzag design is due the phenomena illustrated on Fig. 5.

The current flowing in two successive switches generates antagonistic B-fields within the magnetized volume, and the coupling between these antagonistic fields leads to an inductance reduction. A detailed analysis should be necessary to quantify the contribution of each of the above phenomena (Fig. 4 and Fig. 5) on the reduction of L_{MARX}.

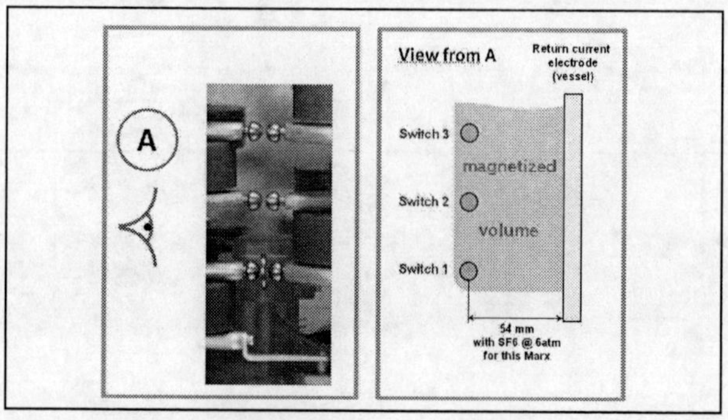

Fig. 4. The inductance of a Marx is related to the volume occupied by the magnetic field B between the Marx circuit (capacitors and switches) and the return current electrode

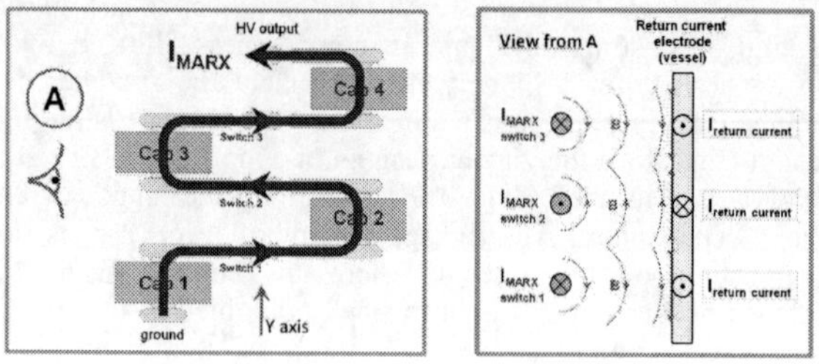

(a) (b)

Fig. 5. With the zigzag design, the extra reduction of Marx inductance is related to the coupling of antagonistics B-fields generated around 2 successives switches.

As a whole, this innovative zigzag design leads to an inductance as small as 25nH / stage. For a same distance between the switches column and the vessel, this inductance would be at least 40nH/stage for a "straight" Marx.

As the rise time of the output pulse is proportional to L/R, this Marx inductance optimization allows to reach a fast rise time (5ns) without the use of a peaking stage.

C. Column of switches

For a reliable timing sequence during Marx erection phase, the switches are arranged along a column. Only the first switch is triggered and the other switches are closed in an avalanche mode. This switching mode is favored by this column arrangement, which allows the photo-ionization of each switch gap by the UV radiation emitted during the closing of the previous switch. A small increase of gap value from switch to switch also contributes to a reliable avalanche mode. For the Marx generator presented on Fig. 8 and for an operation under SF6 gas at 6 atm pressure, the gap gradually increase from 1.3 mm to 2.2 mm from switch #2 to switch #16.

The first switch is electrically triggered, by applying a 30 kV pulse on its intermediate disc electrode. This trigger pulse is delivered by a compact (0.4 liters), cheap and repetitive system based on the pulse transformer of a car motor ignition system. Stray capacitances of stages #2 to #4 are increased to favor the overvoltage and so the breakdown of their switches just after the closing of switch #1. These stray capacitances have been adjusted by reducing the gap between the upstream stage electrodes and the grounded vessel (see Fig. 8).

Each switch electrode is a small hemisphere diameter 10mm, made of Tungsten-Copper alloy to limit its erosion for repetitive operation.

III. PFN DESIGN

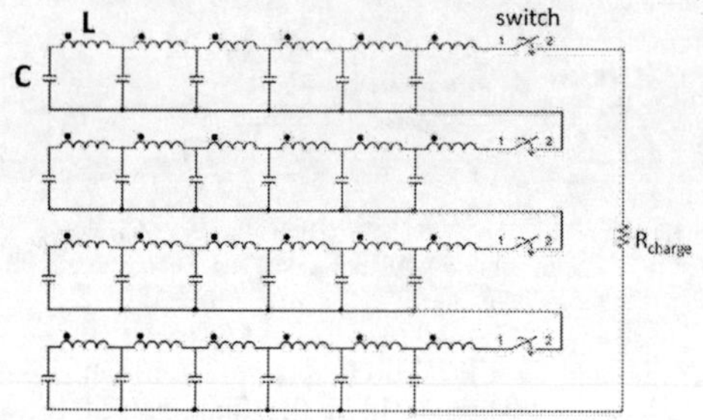

Fig. 6. Principle of a PFN Marx with 4 stages

Beyond the rise time optimization reached with the reduction of the Marx inductance, the voltage of the output pulse is kept constant during the time T_p (i.e. "voltage plateau" or "square" pulse) by using the PFN principle described on Fig. 6. Each stage is a pulse forming network composed of S cells (LC) connected in serie. For a first approximation, the plateau duration is $T_p = 2.S.(LC)^{0.5}$ and for N stages, the Marx impedance is $Z_{MARX} = N.\ (L/C)^{0.5}$ and the voltage amplitude of the plateau is $V_p = N.V_{ch}.(R_{load}/(R_{load} + Z_{MARX}))$.

The detailed design of a PFN stage is shown on Fig. 7. The capacitance C = 2.1 nF is made with a ceramic capacitor, model HP60E40242K from AVX. Each capacitor is 58 mm diameter and 30 mm height. The 6 capacitors are screwed on a strip electrode (16 mm width) and their opposite electrodes are connected with a wire (2.3 mm diameter). The inductance L = 36 nH is given by this strip line geometry. The curved shape is chosen for a compact integration in the cylindrical vessel housing the stages. The hemispherical electrodes of the switch are screwed straight to the strip electrode ends, minimizing the inductance of the switch and allowing its gap adjustment.

IV. THE 400 KV – 85 NS PFN-MARX

A. PFN-Marx design

16 PFN stages are assembled in accordance with the zigzag design to form the Marx generator presented on Fig. 8. The switches column is located closed to the cylindrical vessel, at the distance 54 mm shown on Fig. 4. The PFN stages are distributed alternatively on the right and on the left of this switches column.

The charging circuit is made with 4 columns of stacked inductances, which distribute ground and charging voltage on

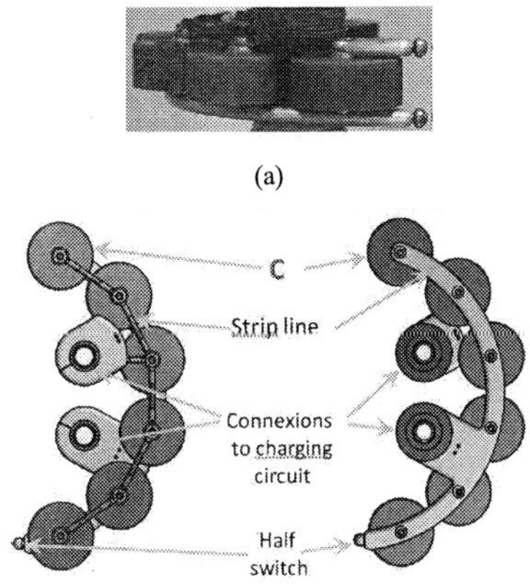

Fig. 7. Each stage is a pulse forming network composed of 6 cells (LC) connected in serie. (a) photo of a PFN stage, (b) top view, (c) bottom view.

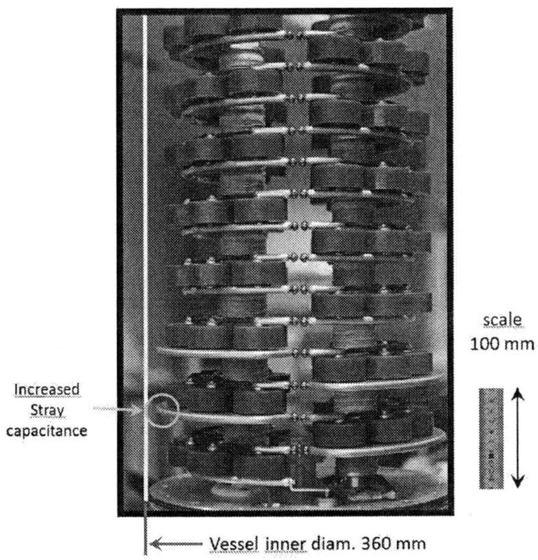

Fig. 8. Photo of the 16 stages PFN-Marx based on the innovative zigzag design.

even and odd stages. Inductive charging is preferred to resistive charging to reduce energy losses and save batteries autonomy for repetitive operation. Each inductance is a 3 layers, 23 turns per layer coil, wound with an insulated wire (single strand 1.2 mm diameter) withstanding a 3 kV dc voltage. The connection of each inductance to the strip electrode of a stage is shown on Fig. 7.

The entire assembling makes a self-supporting structure: the 4 inductances column support the 16 PFN stages and the strip electrodes of these stages support the switches. None additional insulating support is used for the mechanical assembling of the Marx. This self-supporting structure is a significant advantage in terms of reliability, as it minimizes the risks of losses due to surface ramping, but also in terms of easiness of maintenance and weight reduction. Its rigidity allows to work in any position: vertical, horizontal or tilted.

Moreover, with this self-supporting structure, the same gas volume is used for the switches and for the stages insulation. This big gas volume is beneficial for a better dilution of by-products coming from switches electrodes erosion and from gas breakdown by-products. This improves the reliability of the system for repetitive operation.

The complete structure shown on this Fig. 8 weighs 44 kg and fits in the 63 liters cylindrical vessel which inner diameter is 360 mm and total height is 650 mm. These 16 stages stores a 204 J energy for a charging voltage V_{ch} = 45 kV.

B. Simulation and test results on a resistive load

A detailed simulation of this PFN-Marx is performed on PSPICE. Fig. 9 shows the main equivalent scheme, where each blocks details the electric circuit including: R, L, C and transmission line components, dynamic model for switches resistance (Braginsky model [10]), stray capacitances, etc.

Fig. 10 compares a PSPICE simulation with a preliminary test where the load connected at the output of the PFN-Marx is a liquid resistor which characteristics are R_{load} = 68 Ω and L_{load} = 90 nH. The simulations fits with the experimental data which, for a 45 kV charging voltage, show a 360 kV plateau of 80 ns duration.

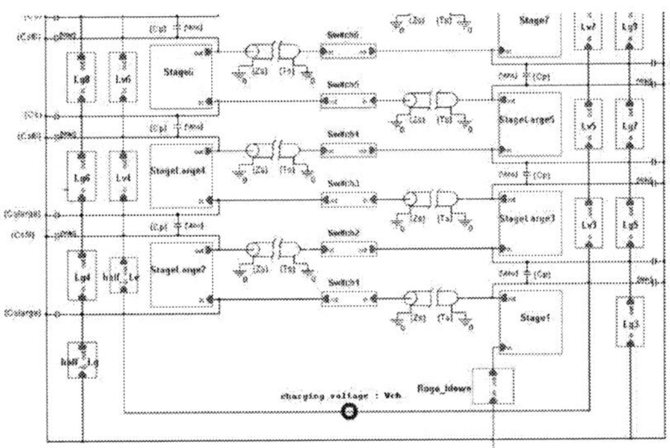

Fig. 9. PSPICE main equivalent scheme of the 6 first stages of the PFN-Marx.

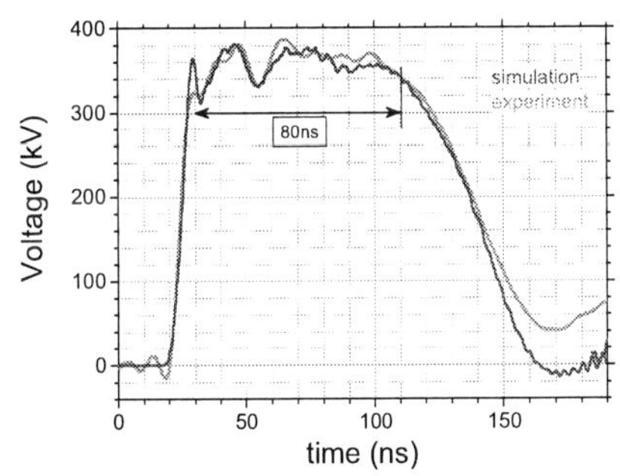

Fig. 10. Comparison experiment - simulation of the PFN-Marx discharge in a resistive and inductive load (68 Ω - 90 nH).

V. Experimental Tests with a Microwave Tube

As described in reference [1], the preliminary qualification of the system shown on Fig. 2 is based on experiments connecting the PFN-Marx to a microwave source (compact X-band BWO, 280 mm in diameter, 550 mm in length) with its antenna.

A. Design of the compact connexion between the PFN Marx and the microwave tube

Fig. 11 shows the U shaped gas pressurized line designed to connect the PFN-Marx to the e-beam diode of the microwave source. Due to its complex shape, the impedance of this line is not constant but is close to 80 Ω, which allows to limit pulse reflections between the Marx (66 Ω) and the e-beam diode (100 Ω). The impedance of the Marx was deliberately chosen to be under-matched to the diode impedance, to reach an output voltage of 380 kV with only 16 stages.

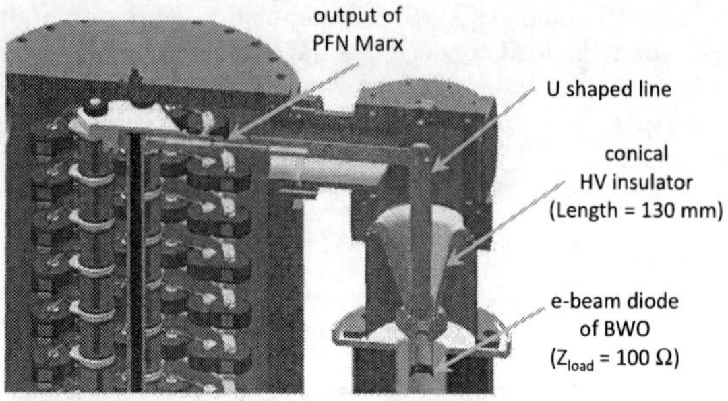

Fig. 11. PFN-Marx connected to a microwave tube (BWO) through an U-shaped coaxial line and a conical HV insulator

The U-shaped line is insulated under the same pressurized gas as the Marx and a conical insulator is used for the connection with the vacuum e-beam diode. This conical insulator, which withstands the 400 kV output voltage, is made of polyamide 6 and is very compact: only 70 mm in outer radius and 134 mm in total length. Its design was optimized through 3D EM simulation (Fig. 12). The stress is limited to 150 kV/cm at its surface and to 80 kV/cm at its triple point. The ovoid shape on the diode holder and the conical shape of the insulator, combined with the B-field topology in this region, avoid the impact of free electrons on the insulator surface.

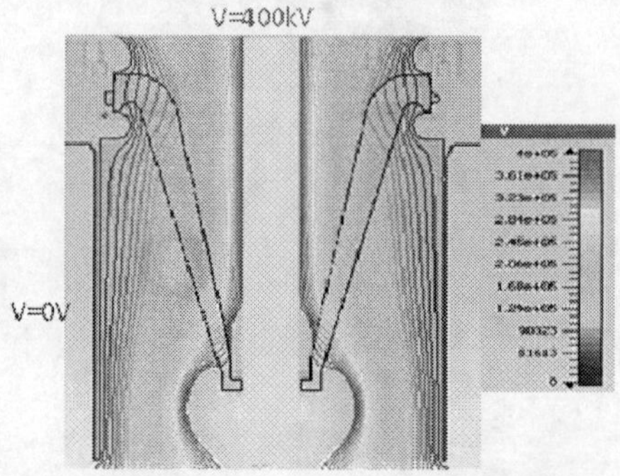

Fig. 12. Electrostatic simulation of the compact conical HV insulator

B. Test results

The system performances have been validated for bursts of 1000 pulses at a repetition rate of 85 pps. The repetition rate is limited to this value because of the capacitor charger available for these tests. A new capacitor charger is needed for higher repetition rates.

Fig. 13 shows the output voltage signals corresponding to the first 100 pulses of a 1000 pulses burst. As expected, the PFN-Marx design gives a plateau of 380 kV amplitude and 85 ns duration. The ± 8% spread in voltage amplitude is fully compatible with the microwave source operation. Thanks to the innovative zigzag design, the 10%-90% rise time is only 5 ns. This sharp rise time is also fully compatible with the system operation, with no need for a peaking stage.

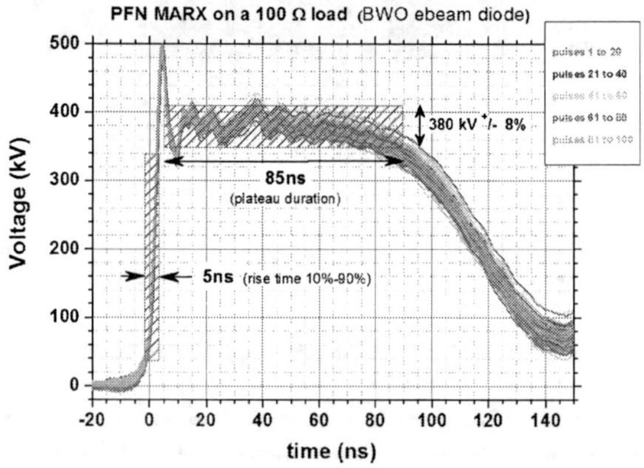

Fig. 13. Voltage profiles of the output pulse for the PFN-Marx discharge on the e-beam diode of a microwave source. First 100 pulses of the 1000 pulses burst (85 pps repetition rate) are shown.

VI. Conclusion

A repetitive PFN-Marx has been developed and successfully validated by achieving several bursts of 1000 square pulses at 85 pps repetition rate, with 380 kV amplitude and 85 ns duration. The innovative zigzag design and other design rules described in this paper allow to reach a high level of compactness and a sharp rise time (5 ns, with no need for a peaking stage). Its clean design (self-supporting structure, simplicity) shows high potentialities, particularly in terms of reliability, modularity, easiness of maintenance and compactness.

Tests will continue to complete the characterization of this prototype and to analyze for example what is the repetition rate when working with SF6 gas, what are the characteristics when working with other gases (dry air), which service procedure is needed and for what life time, etc. Ongoing circuit and 3D EM simulations will also help to analyze the detailed behavior of this prototype and to refine the improvement capabilities. On a technical point of view, many improvements are possible, for example on compactness (higher performances capacitors, dielectric strip line instead of PFN lines, etc.) and on rise time sharpening (switches closer to vessel, conical vessel for a progressive gap increase versus each stage voltage, etc.).

Based on this successful development, main perspectives are to extend the scope of applications of drivers and modulators based on the design rules presented in this paper.

References

[1] R. Vezinet, F. Lassalle, S. Tortel, J.C. Diot, A. Morell , A. Loyen, A. Catrain, Q. Saurin, A. Paupert; "Development of a Compact Narrow-band High Power Microwave System"; 2016 IEEE International Power Modulator and High Voltage Conference.

[2] G.A. Mesyats et al.; "Repetitively pulsed high-current accelerators with transformer charging of forming lines"; Laser and Particle Beams (2003), 21, 197–209.

[3] E.N. Abdulin et al.; "High-power Current Pulse Generator for Inductive-Resistive Load Operation"; proceedings of SHCE, Tomsk (2010); pp.357-360

[4] S M Turnbull, S J MacGregor and J A Harrower; "A PFN Marx generator based on high-voltage transmission lines"; Meas. Sci. Technol. 11 (2000) N51–N55.

[5] S. Li, J-M. Gao, H-W. Yang, and B-L. Qian; "A high-voltage, long-pulse generator based on magnetic pulse compressor and Blumlein-type rolled strip pulse forming line"; Laser and Particle Beams (2015), 33, 511–518.

[6] Wei Li, Zhi-qiang Li, Xiao-liang Sun, and Jun Zhang; "A reliable, compact, and repetitive-rate high power microwave generation system"; Rev. Sci. Instrum. 86, 114704 (2015).

[7] M.G.Mazarakis and R.B.Spielman, "A compact, High-Voltage E-beam Pulser"; proceedings of the 12th IEEE International Pulsed Power Conference, edited by C. Stallings and H. Kirbie (IEEE, Piscataway,NJ, 1999), p. 412.

[8] M. M. Kekez; "Simple sub-50-ps rise-time high voltage generator"; Rev. Sci. Instrum. 62 (12), December 1991; pp.2923-2930

[9] S.V. Tevari et al.; "Development and Analysis of PFN Based Compact Marx Generator Using Finite Integration Technique for an Antenna Load"; IEEE Transactions on Plasma Science, vol.41, no 10, October 2013.

[10] S.I. Braginsky ; 'Theory of the development of a spark channel' ; Soviet. Physics JETP, vol. 34(7), n° 6, déc. 1958, p.1068-1074.

Magneto-Hydrodynamic Code FLUX-RZ Validation Based on Results of Scintillation Measurements of Z-Pinch Thermal Radiation

V.D. Selemir, P.B. Repin, A.G. Rep'ev, A.P. Orlov, B.G. Repin, V.S. Pokrovsky
Russian Federal Nuclear Center-VNIIEF, Sarov, Russian Federation

***Abstract*—This paper demonstrates main possibilities of numerical simulation of spectral measurements in experiments with radiating Z-pinch in the frameworks of two-dimensional magneto-hydrodynamic (MHD) code FLUX-rz. The code is developed in RFNC-VNIIEF.**

Keywords*— *X-ray production, plasma Z-pinch, radiation transfer, multi-group diffusion approximation.

I. Introduction

The calculation-physical model includes radiation transfer in a multi-group diffusion approximation [1]. The process of a substance ablation from the multi-wire array under the influence of the flowing current is described phenomenologically in the regime of prolonged plasma formation [2].

Validation of the calculation method involves measurement results of soft x-ray radiation (SXR) pulses that were obtained using the scintillation probes equipped with different thin metal filters.

Peculiarities of MHD simulation are demonstrated by an example of the experiment on nested wire array implosion. The load was powered from a magneto-cumulative generator with a current pulse I_{max}≈4.2 MA and rise time τ≈400 ns (at the level of 0.1…0.9).

II. Calculations Setting

The load chamber had the ideally conducting walls. It consisted of a cylinder with the radius R_{ext}=4 cm and height H=1.5 cm. The load represented itself a nested array which consisted of 360 tungsten wires of diameter d=7.1 μm and length l=1.5 cm (total mass M_{total}=4.14 mg). The first cascade consisted of 240 wires, which were evenly located on a cylindrical generatrix of the radius R_1=3 cm. The second cascade was formed of 120 wires and located on the radius R_2=2 cm. Supply of the electromagnetic energy was made from one side with coaxial vacuum transmission line.

Numerical simulation of this experiment was made with experimental ("imposed") current pulse. Boundary conditions for azimuth component of the magnetic field were setup in accordance with this current pulse.

Stationary spatially uniform Eulerian grid with 100 μm square cells was used to increase the precision of the obtained results. Calculation of the thermal radiation transfer in plasma was made in 100-group diffusion approximation; energy width of the group was 5 eV.

III. Simulation Results

Pulsed of soft x-ray radiation (Fig. 1) were recorded by three scintillation probes. The probes were located in the radial channels behind different filters. Thicknesses of the filters were: aluminum filter ~3.1 μm, silver filter ~1.05 μm, copper filter ~0.78 μm. Analysis of the experimental data gives the following values of full width at half maximum of SXR pulse behind the filters: aluminum filter $\tau \approx 15$ ns, copper filter $\tau \approx 12.3$ ns, and copper filter $\tau \approx 17.7$ ns.

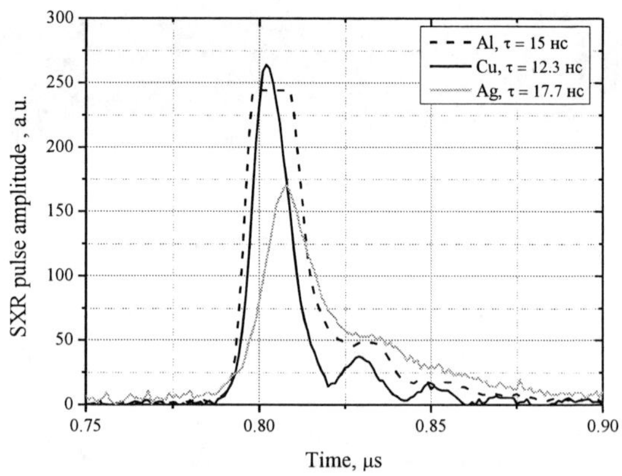

Fig. 1. Signals of SXR pulses recorded by the scintillation probes that were located behind the aluminum (- - -), silver (—), and copper (—) filters.

Calculation "hitting" to the recorded SXR pulses was made simultaneously to all scintillation probes in one RMHD calculation. For this purpose we performed convolution of the radiation intensity, going out from Z-pinch, with the spectral transmission coefficients of a specific filter, taken from Henke tables [3]. "Matching" to the SD-method was performed by varying a specific rate of plasma formation k_{abl} and relative amplitude of random axial disturbances of ablation intensity A_0. In the calculation, the closest experimental values of SXR pulses length were reproduced at k^{l}_{abl} =45 μg/(ns·cm^2) and A_0=3.4%.

Figure 2 shows serial stages of the wire array implosion in this experiment. They were calculated using the parameters mentioned above.

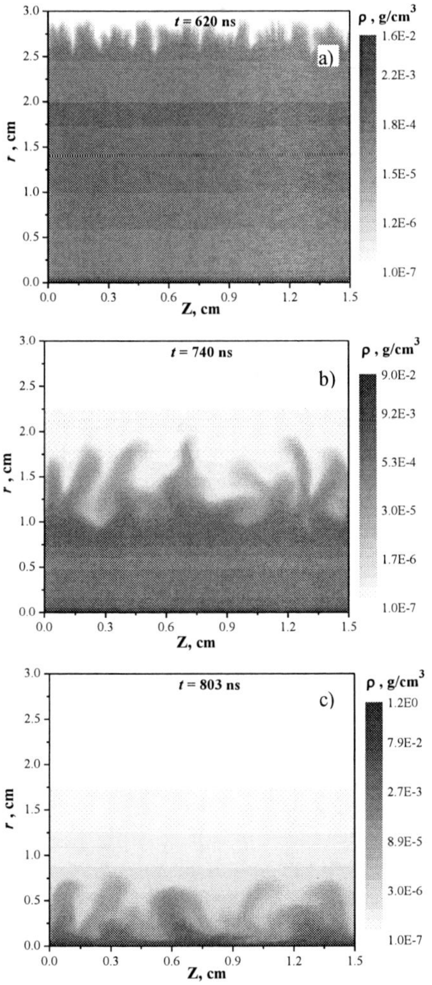

Fig. 2. Spatial distribution of plasma density in time. (a) – termination of ablation of the inner cascade. (b) – load's half way to the axis. (c) – radiation maximum.

These figures clearly demonstrate that the implosion in the experiment takes place in "snow plough" regime. At the moment of ablation phase termination (t_2≈620 ns), the plasma density distribution shows that the substance fills up the whole volume of the load chamber; a column of precursor plasma with the density $\rho \approx 0.016$ g/cm^3 and electrons temperature $T_e \approx 22$ eV is formed on the system axis. In the result of further implosion, the substance, remaining on the upper radii, is "grabbing" to the system axis and carrying along the swept mass. The substance density on the chamber axis reaches the values $\rho \approx 1.2$ g/cm^3 at the time moment t≈803 ns, which corresponds to SXR pulse maximum.

Figure 3 shows calculated SXR pulses (generated in the result of the tungsten plasma implosion) that leave the load chamber through the cylindrical surface of the return current.

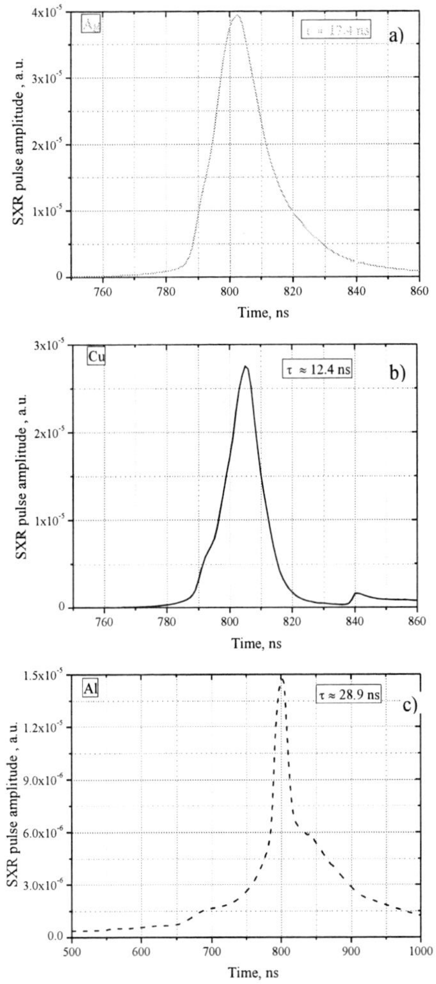

Fig.3. SXR calculation pulses. (a) – behind the silver filter. (b) – behind the copper filter. (c) – behind the aluminum filter.

Comparison with the experimental pulses allows making of the following conclusions. Time position of maximums of all calculated pulses matches well with readings of the scintillation probes. Lengths of the calculated pulses behind the silver and copper filters match the experiment. However, length of the calculated pulse behind the aluminum filter is 2 times higher than the diagnostic measurements. Also, it is necessary to pay attention to existence of significant pedestal before the main pulse that was obtained in the calculation behind the aluminum filter.

Figure 4 shows spectral distributions of the radiation power, going out through the whole cylindrical surface of axially symmetrical Z-pinch. Time moment t≈803 ns corresponds to the intensity maximum of SXR pulse. Time moments t≈794 ns and t≈812 ns correspond to the pulse half-height at its fore and back fronts, respectively. Also, figure 4 shows approximation of the calculation spectra by the Planck functions $f_{Plank}(\varepsilon, T)$ those consider coincidences of the maximums distribution locations and equalities of integral areas.

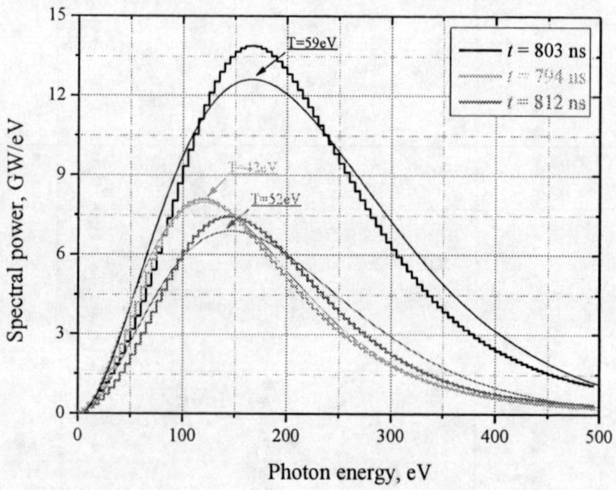

Fig.4. Calculation distributions of the pinch radiation spectrum power at the moment of output radiation maximum; the moments which correspond to the pulse full width at half maximum at its fore and back fronts without radiation losses to the electrodes.

From the presented distributions one can see that maximum temperature of the radiation is realized at the maximum of SXR pulse and equal to T~59 eV. Temperature difference (~10 eV) of the radiation, generated by the plasma in time moments, which correspond to SXR pulse half-height, is a very interesting peculiarity.

IV. Influence of Oxide Films on Filters Transmission

In the previous section we showed that there are significant differences in the shape of experimental and calculated SXR pulses behind the aluminum filter. This should be considered in detail. In the result of detailed analysis of different experiments with irradiating Z-pinches we found that there are, at least, two possible factors significantly influencing on discrepancy between the calculation and experiment. Formation of the oxide film on the aluminum surface belongs to the first factor. It is known that thickness of the oxide film on a recently rolled out aluminum plate is ~2.5 nm. In the air at room temperature, its thickness could increase up to 10-20 nm (see [4,5]). A set of five forged aluminum foils of ~0.62 μm thickness was used as the aluminum filter in the majority of the experiments. The aluminum was obtained several years ago. So, maximum total thickness of estimated oxide in the used filters could be ~200 nm. This is 6% of the total filter thickness. In connection with this, we considered the oxide film with mentioned thickness range in our calculations.

The best results were obtained at total thickness of the oxide film of 3% from the total filter thickness. This value corresponds to the thickness of the oxide surface layer of 9 nm. Figure 5 shows comparison of the calculated radiation pulses for 100% pure aluminum filter with the filter having the oxide film Al_2O_3 of 90 nm total thickness.

From the calculated dependencies one can see that existence of the aluminum oxide significantly influences on the recorded SXR pulse. In addition to disappearance of the pedestal, it should be noted that the radiation pulse full width at half maximum decreased from 28.9 ns to 14.7 ns. This corresponds to the experiment. The pulse amplitude decreased 2.5 times.

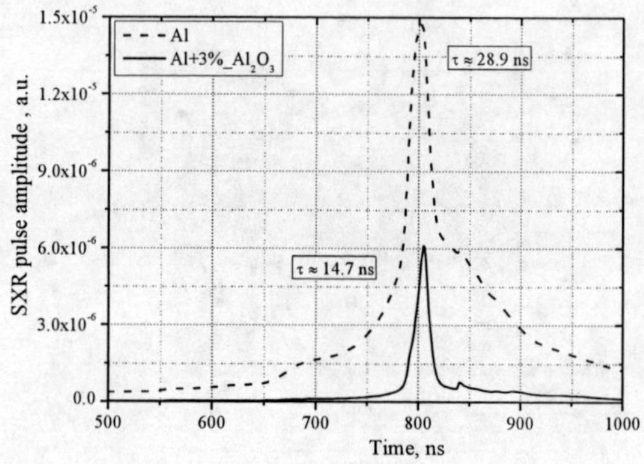

Fig.5. Calculated SXR pulses obtained behind the aluminum filter with oxide film on the filter surface of 90 nm total thickness (—) and without it (- - -).

Similar investigations were performed with different silver and copper films. They show weak influence of these filters on the recorded SXR pulses. The obtained results are the consequence of a spectral transmittance degree change. Spectral transmittance of pure silver and pure copper filters does not differ from spectral transmittance of the filters with dopants. However, one can observe significant decrease of the radiation transmission in the range of low-energy quanta (hv~20-70 eV), irradiated by a plasma precursor, for the aluminum filter with 3% oxide film.

So, consideration of the oxide film on the aluminum filter surface gives good agreement of the calculated and experimental results and should be considered in simulation of radiating Z-pinches.

V. Influence of Film Scintillator Dead Layer

Possible existence of a "dead layer" of used film scintillator is the second factor that significantly influences on the shape of recorded SXR pulse. This layer can be an additional filter. Most probable reasons of such layer appearance are: degradation of chemical compositions in near-surface layer at the scintillator long-term storage and roughness of its surface. Characteristic thickness of such layer depends on the scintillator type, its production method, and energy of incident quanta; according to different estimations, it could be up to 1 μm.

In this experiment, we used 200 μm film scintillator Ej-208, made by American company Eljen Technology [6]. According to the technical data, its chemical composition is presented by $C_{469}H_{517}$ compound. Several calculation series were made for different values of the supposed thickness of the "dead layer". The goal of these calculations was to study the effect of this factor on SXR pulses diagnostic. Since the scintillator is located behind each of three filters, it changes the shapes of all radiation pulses. This complicates determination of its possible thickness because it plays the role of the third free parameter in the calculations. In connection with this, we did not change the found values of specific plasma formation rate k_{abl} and relative amplitude of random axial disturbances of ablation intensity A_0; we varied only the layer depth.

In the calculations, the most optimum "matching" to all three radiation pulses was obtained at the scintillator "dead layer" thickness of 450 nm. Figure 6 shows calculated SXR pulses that were generated with such layer.

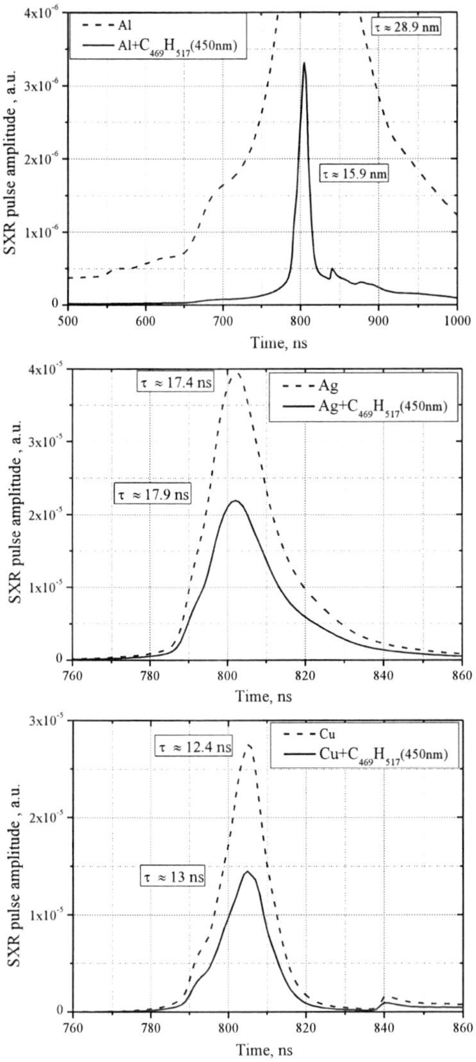

Fig.6. Calculated SXR pulses that were obtained behind different filters with the scintillator's "dead layer" of 450 nm and without it

Full width at half maximum of SXR pulses behind all filters (considering "dead layer" of the scintillator) match the diagnostic measurements. There is no pedestal behind the aluminum filter. A distinctive peculiarity of this simulation results is significant decreasing of the amplitudes of all radiation pulses: 2 times behind the silver and copper filters and 5 times behind the aluminum filter. Such behavior could be explained by the change of spectral transmission factors of the corresponding filters. Additional filter $C_{469}H_{517}$ of 450 nm thickness leads to significant decrease of the quanta transmission with the energies hν > 284 eV for all filters. Additionally, as in the case with oxide film, the "dead layer" of the scintillator significantly decreases the radiation transmission of the aluminum filter in the range of low-energy quanta (hν~20-70 eV). This determines absence of the pedestal and decrease of the SXR pulse length.

Studying effect of this factor on the recorded SXR pulses, we did not reveal any serious contradictions. So, consideration of the scintillator's "dead layer" could be also used for adequate simulation of main parameters of SXR pulse generated in the result of implosion of the cylindrical wire array. Besides, we think that it is expedient to determine experimentally thickness of the "dead layer" of the scintillator that is used in real detectors.

VI. Conclusions

The developed calculation code FLUX-rz allows reproducing of SXR pulses main parameters (implosion time, full width at half maximum, irradiated energy) and derive the temperature of the pinch plasma with good precision. The pulses were recorded by the scintillation probes located behind different filers both in radial and axial diagnostic channels.

The simulation shows that the observed deviations in the form of calculated and experimental SXR pulse (recorded using SD-method behind the aluminum filter) could be associated with existence of the oxide film on the filter surface, or with presence of the scintillator's "dead layer". Additional laboratory investigations will let us know which of two factors is the determining one in this experiment.

References

[1] B.G. Repin, A.P. Orlov, P.B. Repin, and V.D. Selemir. Calculation Method of Radiation Spectral Transfer in Frameworks of Two-Dimensional Magnetohydrodynamic Code FLUX-rz // IEEE Trans. on Plasma Science, 2010, v.38, № 8, pp.1822-1827.

[2] B.G. Repin, A.P. Orlov. Numerical Simulation of Multi-Wire Z-pinches Accounting For Prolonged Plasma Formation // Int. Conf. XII Khariton's Topical Sci. Readings, Dig. Tech. Papers, Sarov, Russia, 2010, pp. 526–531.

[3] B.L. Henke, et al. Low-energy X-ray interaction coefficients: photoabsorption, scattering, and reflection // Atomic data and nuclear data tables, 1982, V.27, pp.1-144.

[4] Yu.L. Bakshaev, S.A. Danko, E.E. Sokolov, and K.V. Chukbar Pulsed bolometer to measure soft x-ray radiation in high-current X-pinch: calculation of limiting parameters and realization // VANT, Series: Thermonuclear fusion, 2011, Iss. 1, pp. 54-62.

[5] http://masterweld.ru/svarka_alyuminiya

[6] Eljen technology. Scintillation products. URL: http://www.eljentechnology.com.

Influence of foil thickness on homogeneity of its electric explosion

A.N. Grigoryev, E.I. Karnaukhov, A.V. Pavlenko

Russian Federal Nuclear Center - Zababakhin All-Russia Research Institute of Technical Physics (RFNC-VNIITF), 456770, Snezhinsk, Chelyabinskaya Oblast, Russia
Fax: (35146) 5-11-01, e-mail: dep5@vniitf.ru

***Abstract* – The paper presents experimental results that demonstrate how thickness of the flat electrically-exploding foil contributes to inhomogeneity and time delay of its explosion. In the experiments, quartz sensors measured the "amplitude-time" pressure profiles in the central and side (explosion moves from edges to the center) regions of the flat foil. Thick foils (14 - 100 µm) made of aluminum, copper, titanium, and nickel alloy 80 NHS were used for experimentation. The collected data show that inhomogeneity and time delay of the foil explosion increase with the foil thickness.**

Keywords* – *electrical foil explosion; pressure; foil thickness; homogeneity; quartz sensor of pressure.

I INTRODUCTION

The authors of [1-3] determined that under electrical explosion of flat foil (EEF), one can observe spatial inhomogeneity ("nonflatness") of pressure distribution over the foil square (explosion moves from the foil edges). Supposition was expressed that this is related to the current scanning– initial distribution of the current density on the foil edges and subsequent increase of the current density in the axial region of the foil.

Nevertheless under electrical explosion of a flat foil, there is a possibility to create flat shock wave with great square. As it is shown in papers [2, 3], it can be achieved by using the foils of low-conductivity metals, such as titanium, lead, steel 08KP, nickel alloy 80NHS. The decrease of nonuniformity of the foil explosion with increase of its specific electrical resistance ρ (all other things being equal) was explained by determining influence of skin effect (distribution of the current density over the foil width), because the typical size of skin layer is $\delta_{sk} \sim \rho^{0,5}$. One can ask a reasonable question: what is the influence of the foil dimensions (thickness, width and length) on explosion inhomogeneity?

One of important factors that influence the pressure amplitude and specific mechanical impulse is the thickness of exploded foil. On the basis of supposition about the influence of skin effect on explosion inhomogeneity we can expect that along with increase of the foil thickness, explosion inhomogeneity will also increase.

This paper presents experimental results that show the influence of the thickness of exploded foil on inhomogeneity and time delay of explosion. Experiments used foils out of copper, aluminum, titanium and nickel alloy 80NHS.

II SET-UP AND RESULTS OF EXPERIMENTS

Electrical explosion of foil was made on a stand which was described in [3, 4], with the following parameters: condenser battery capacity C_o was 23 microfarad, loop inductance $L_c \sim 60$ nanohenry, charge voltage U_o = 22…44,7 kV. In separate experiments (aluminum foil 14 and 100 µm, width b_f = 40 mm) capacity C_o was 79 microfarad. For exploded foils, we used foils out of copper, aluminum, titanium and nickel alloy 80NKhS. At similar length l_f = 100 mm and width b_f = 50 mm (for aluminum 14 and 100 µm – b_f = 40 mm) of the foil and fixed relative energy introduced in the foil w/w_c , nominal foil thickness was varied d_f = 14…100 µm.

Experiments were performed according to a scheme given in figure 1(a). Quartz cylindrical pressure sensors were located on free surface of dielectric substrate (polymethylmethacrylate 3 mm thick), which was glued to exploded foil. The sizes of sensors were (diameter/ thickness): 10 / 3 mm (the majority of experiments) or 8 / 2.5 mm (aluminum 14 and 100 µm). Experiments registered amplitude-time pressure profiles, according to which inhomogeneity coefficient K_n and explosion time delay Δt_P were determined.

Experiments were performed for foils with various thickness that were fabricated out of one metal all other things being equal: fixed length l_f and width b_f of the foil and fixed relative energy w/w_c introduced in the foil (w – thermal energy introduced in the foil, w_c – energy of exploded metal sublimation). Similar to papers [1–3], inhomogeneity of pressure formation under electrical explosion of flat foil was determined with the help of comparison of amplitude-time pressure profiles in central and side (lateral-edge) regions (figure 1(a)).

Inhomogeneity of foil explosion is characterized quantitatively by inhomogeneity coefficient K_n:

$$K_n = \frac{P_{m.c}}{P_{m.ed}} \qquad (1)$$

where $P_{m.c}$ and $P_{m.ed}$ are pressure amplitudes measured in central and side regions on free surface of substrate, to which exploded foil is glued.

Time delay of achieving maximum pressure will be determined by the following:

$$\Delta t_P = t_{m.c} - t_{m.ed}, \qquad (2)$$

where $t_{m.c}$ and $t_{m.ed}$ are times of achieving maximum pressure in central and side regions of exploded foil.

Figures 1 and 2 show the typical amplitude-time profiles under aluminum explosion (figure 1(b)–(c)), nickel alloy 80NKhS (figure 2(a)) and titanium (figure 2(b)).

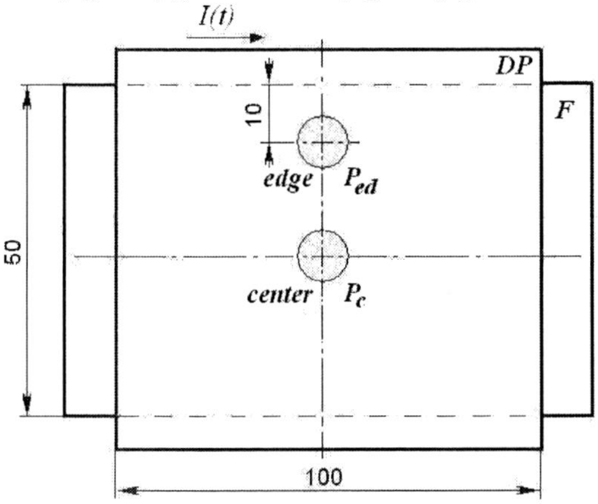

a)

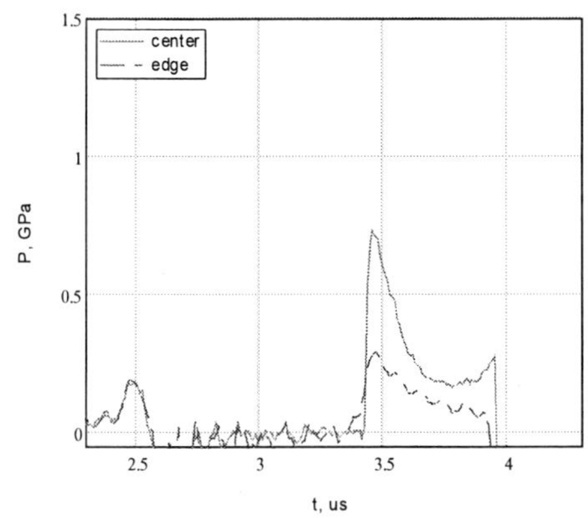

b)

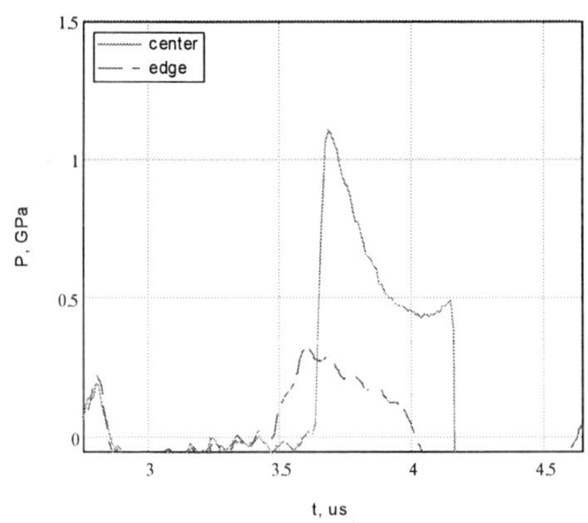

c)

Fig. 1 Location of pressure sensors on substrate in experiments with electrical foil explosion (a) and pressure impulses registered by sensors under electrical explosion of aluminum foil AD1M: with the thickness30 μm (U_o = 27,3 kV, w = 1,14 w_c) (b) иand with the thickness 50 μm (U_o = 36,4 kV, w = 1,18 w_c) (c). Pre-impulse is electromagnetic pick-up from foil explosion on pressure sensor. *DP* is dielectric substrate out of polymethylmethacrylate 3 mm thick, *F* is a foil.

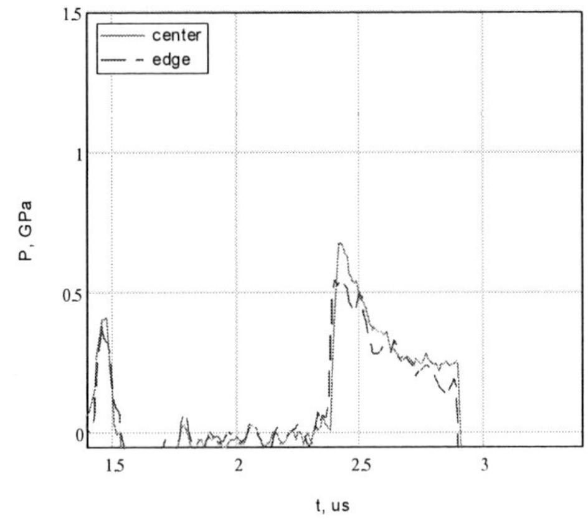

a)

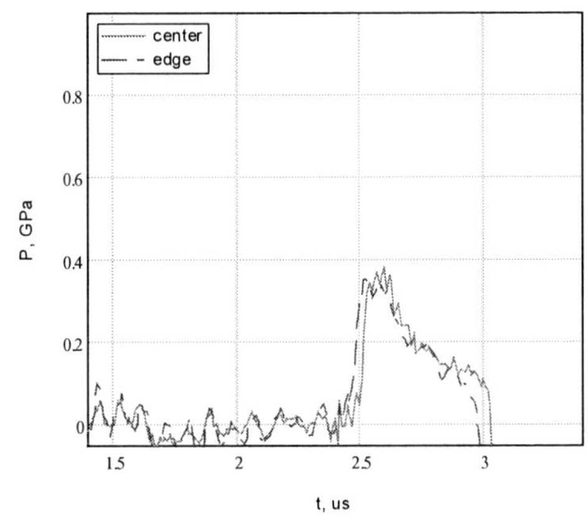

b)

Fig. 2 Pressure pulses registered by pressure sensors under electrical explosion: of 20 μm foil out of nickel alloy 80NHS (U_o = 36,4 kV, w = 1,18 w_c) (a) and 50 μm foil out of titanium VT1-00 (U_o = 44 kV, w = 1,08 w_c) (b)

Fig. 3(a) shows obtained dependences of inhomogeneity coefficient K_n on the thickness of exploded foil d_f. It is necessary to note that in the experiments with aluminum foils, the thickness 14 and 100 μm with $w \sim 1{,}3w_c$ width b_f was 40 mm and 8 / 2,5 mm pressure sensors were used, capacity C_o was79 microfarad. These circumstances can be the reason for increase of registered explosion inhomogeneity.

Fig/ 3(a) shows that along with increase of foil thickness d_f one observes the increase of inhomogeneity coefficient K_n. At the same time, there was sufficient increase of explosion time delay Δt_P from units of nanoseconds (for 16 μm copper foil and 14…30 μm aluminum foil) up to ~ 100…150 ns (for 30 μm copper foil and 50…100 μm aluminum foil) (Fig. 3(b)). These facts speak about governing influence nonuniform distribution of current density (skin effect) over the foil width on explosion inhomogeneity.

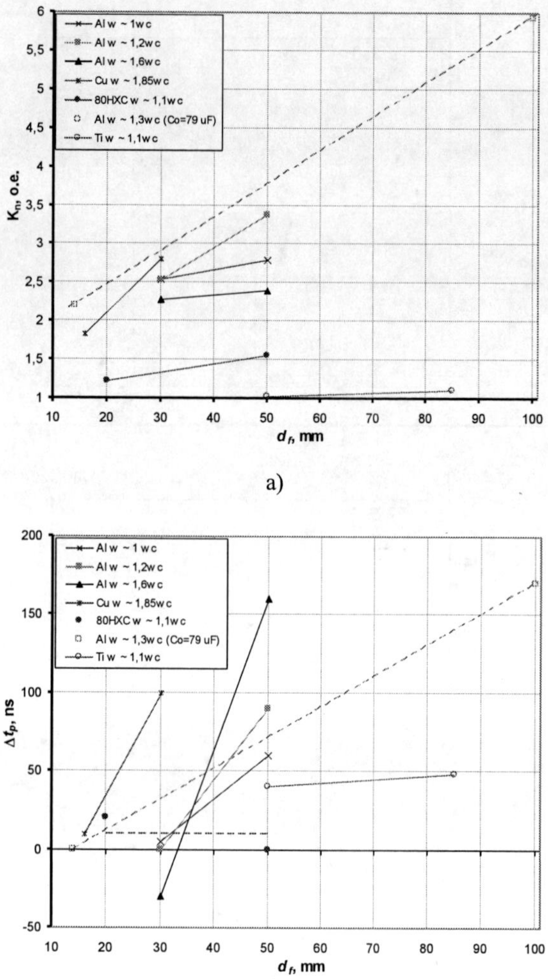

a)

b)

Fig. 3 Dependences of inhomogeneity coefficient K_n (a) and time delay Δt_P (b) on the thickness of exploded foil d_f for the foils out of aluminum, copper, nickel alloy 80NHS and titanium.

In case of electrical explosion of the foil out of low-conductivity metals – titanium and nickel alloy 80NHS, one can observe the best explosion homogeneity both for thin, and thick foil (with inhomogeneity coefficients: titanium – $K_n \approx$ 1,03…1,1, alloy 80NKhS – $K_n \approx$ 1,2…1,6) and complete reproduction of shape for amplitude-time profile of pressure pulse in side and central regions of the foil (Fig. 2(б)).

III CONCLUSION

According to the results of this work, we can make the following conclusions:

- under explosion of all foils without regard for material and the value of energy w/w_c introduced in the foil, we observed the growth of explosion inhomogeneity K_n with increase of exploded foil thickness. Along with this, increase of exploded foil thickness leads to increase of time delay of pressure impulse formation.
- the influence of exploded foil thickness on inhomogeneity coefficient K_n and explosion time delay Δt_P confirm the influence of skin effect (current density distribution over the foil width) on explosion inhomogeneity.

REFERENCES

[1] Pavlenko A.V., Grigoriev A.N., Karnaukhov E.I.. Pis'ma v Zhurnal tekhnicheskoi fiziki 2014, V. 40, issue 2, p. 45-51.

[2] Pavlenko A.V., Grigoriev A.N., Karnaukhov E.I.. Pis'ma v Zhurnal tekhnicheskoi fiziki 2014, V. 40, issue 20, p. 62-68.

[3] Grigoriev A.N., Karnaukhov E.I., Pavlenko A.V. et al. Prikladnaya mekhanika i tekhnicheskaya fizika, 2015, V. 56, # 1, p. 162-170.

[4] Grigoriev A.N., Pavlenko A.V. IEEE Trans. on Dielectric and Electr. Insul. 2007, V. 14, # 4, p. 964–967.

Research in Dynamic Behavior of Materials using GNUV-P Facility

E. I. Karnaukhov, A.N. Grigoryev, A.V. Dulov, A.V. Pavlenko, S.S. Mokrushin

Russian Federal Nuclear Center - Zababakhin All-Russia Research Institute of Technical Physics (RFNC-VNIITF), 456770, Snezhinsk, Chelyabinskaya Oblast, Russia
Fax: (35146) 5-11-01, e-mail: dep5@vniitf.ru

Abstract – **The paper presents the results of research activities aimed to further apply methods of strong magnetic field loading. These studies used GNUV-P facility with the accumulated energy up to 250 kJ. High-strain-rate deformation of Ø30mm cylindrical liners was observed within deformation rates 10^4 through $10^6 s^{-1}$: the total deformation was ~ 100% and the liner compression rate was ~ 0.8 km/s. Acceleration of aluminum current-carrying plates and strikers made of various materials was attained in the speed range from 0.5 to ~ 3 km/s. Velocity profiles for flat plates and cylindrical liners, as well as free-surface velocity profiles for the samples under consideration were registered with PDV-, or VISAR-interferometer. It is shown that GNUV-P facility allows loading pulses sufficient to conduct research into the dynamic properties of materials at the strain rate from 10^4 to 10^6 s^{-1}.**

Keywords – magnetic pulse acceleration of plates; magnetic pulse compression of liners; profile of plates (liners) motion velocity; profile of free surface velocity; strain rate; shock wave

I INTRODUCTION

Studying the dynamic properties of materials is a very important R&D direction at RFNC-VNIITF [1]. Experimental studies form the basis for the development of new and enhancement of available mathematical models of the plastic-viscous-elastic deformation [2]. Currently, the Experimental Physics Division at RFNC-VNIITF uses light-gas guns to investigate dynamic properties of materials [3] and also VISAR [4] and PDV techniques [5] to register the free-surface velocity profiles. Loading devices and interferometers are continually modified for experimentation purposes.

One method that can help to attain high parameters of loading pulses is the use of pulse magnetic fields for acceleration of flat impactors or for compression of cylindrical shells (liners). The principle of the flat plane acceleration is based on interaction of two current-carrying conductors. The Lorentz force created by the mega-ampere current pulse causes the conductors to move. A current-carrying plate and also a striker positioned at its surface can serve as an impactor. For example, flat impactors were accelerated up ~ 8,5 km/s using the CQ-4 facility (the current amplitude was I~4 MA, time of current pulse rise was τ~0,5 μs) [6]. The possibility of flat impactor acceleration up to the ~ 10 km/s velocities was demonstrated on the GEPI facility [7]. The current-carrying plate was accelerated to the velocity higher than 45 km/s on the more powerful Z facility (I~26 MA, τ~0,1 μs) [8].

Contrary to experiments with flat samples, the cylindrical geometry provides additional capabilities for investigation of dynamic properties of materials. Several independent experimental setups are used in this study. The first setup investigates high-strain-rate deformation of the cylindrical liner itself or of the magnetically deformed shell from the material under study. The dielectric substrate can be placed between the liner and the sample. There are no gaps, and this ensures smooth, shock-free deformation of the sample. The sample and the liner have a characteristic diameter of 3 – 10 cm. In this case, pressure pulses with the amplitude ranging from units up to tens of kbar are generated. In the second setup, isentropic compression, the current-conducting liner, and the sample have a less-than-1cm initial diameter. Here, the sample can be loaded by the compression wave with the amplitude more than 10 Mbar with currents from 30 up to 36 MA as on the Atlas facility [9]. In the third setup, if a gap between the liner and the sample is specified such that the liner would have enough time to accelerate up to the velocity above 300 m/s, then during the liner deceleration, the converging cylindrical shock wave with the amplitude sufficient to induce spallation at the internal surface of the hollow cylindrical sample would be formed at the external surface of the cylindrical sample. Paper [9] reports that on the PHELIX facility with the current amplitude I~4 MA and current pulse duration τ~10 μs, the velocity of cylindrical liner compression was ~1 km/s. More powerful facilities SHIVA STAR, Pegasus II, and ATLAS enabled regimes wherein the velocity of metal liners compression achieved ~ 7,5 km/s [10]. Record values of the liner compression velocities are currently obtained with the use of the magnetic explosion generators. In experiments HEL-1, HEL-2, ALT-1, ALT-2, and ALT-3 at currents with the amplitude from 30 up to 100 MA, compression velocities for liners with the diameter from 10 up to 40 cm were registered to fall within 8 – 20 km/s [11]. However, only electro-physical facilities can be used for experiments in laboratory conditions.

This work presents results of the effort aimed to investigate the material loading methods that use strong magnetic fields. Investigations were performed at RFNC-VNIITF on the GNUV-P facility with the accumulated energy up to 250 kJ. We tested the regimes of high-strain-rate deformation of cylindrical liners in the range from 10^4 up to 10^6 s^{-1} at the liner compression velocity of ~ 0,8 km/s and also the acceleration regimes for the flat current-carrying plates and impactors in the range of velocities from 0,5 up to ~ 3 km/s.

The levels of loading pulses generated on the GNUV-P facility permit the dynamic behavior of materials to be studied.

II FACILITY AND MEASUREMENT METHODS

Experiments were performed on the GNUV-P facility [12,13] schematically shown in Figure 1. This facility is a pulse current generator which creates short-circuited loads – a cylindrical liner or a flat current-carrying plate. The capacitor bank consists of 88 capacitors IK50-3 connected in parallel. The total capacity of the capacitor bank is $C_o \approx 249$ μF. Capacitors are grouped in four sections. This arrangement makes possible to use sections independently and to adjust the capacity of accumulator from 60 up to 249 μF in experiments. Each capacitor 1 has a current collector for cable connection. Power cables are connected to capacitor sections from the one side and to corresponding high-voltage electrodes of commutation switches 2 from the second side. Four commutation switches, which use breakdown effect over the surface of a solid dielectric, disconnect the load node 3 from the capacitor bank in the process of charging and with the help of the system that starts the facility 4 they simultaneously connect individual sections to the load node during the facility start-up. The load node is designed such that it is possible to conduct experiments either in the cylindrical (Fig. 2), or flat geometries of the samples (Fig. 3).

The principle of the facility operation in the cylindrical geometry is the following (Fig. 2). Via cores of the current-conducting cables RK50, the flange *1*, and the lead electrode *2*, current *I(t)* from the capacitor bank comes to the high-voltage electrode *3* of the load node. The upper electrode and the fit ring *4* make the electrical contact with the liner *6* with the help of the deformable ring *5*. Via the liner *6*, the lower electrode *7*, and adapters *8* and *9*, the current drains to the grounded flange *12* to which the braiding of power cables is connected. The internal surface of the lead electrode *2* and the external surface of the liner *6* form a gap, wherein the pulse magnetic field is generated to compress the liner. The pass-through insulators *10*, *11*, *13* between the high-voltage and low-voltage electrically insulate metal parts against electrical breakdown in the process of implosion.

In the flat-geometry setup, the electrode unit consisting of the current-carrying bus 2 and the reverse wireway 4 is connected to the load node instead of parts 2-8 (Fig. 3). The metallic impactor *1* (if any) is located on the magnetically-accelerated plate 2. On the one side, the current-carrying plate *2* is connected to the current-carrying flange *3*, and on the other side – to the electrode *4* and to the grounding flange *5*. In order to insulate these electrodes and minimize loading inductance, we use PET film *6* with the 100-μm thickness.

Current pulses are measured with the help of the Rogowski coils [14], and voltage pulses are measured with the help of the resistive potential dividers [15]. Digital oscilloscopes register signals from the current and voltage sensors. And VISAR [4] or PDV [5] is used to register the velocity of liner flight (compression) and the velocity of the sample free-surface. If necessary, we use the pulse X-ray device that is fabricated on the basis of the commercial tube IMA-320D and also the cable pulse transformer [16].

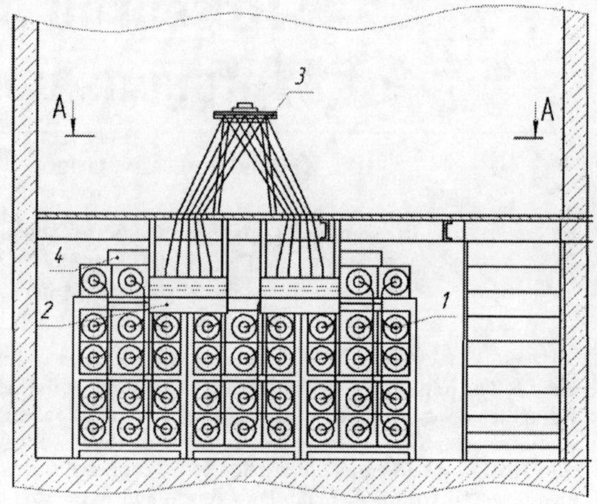

Fig. 1 Scheme of facility GNUV-P

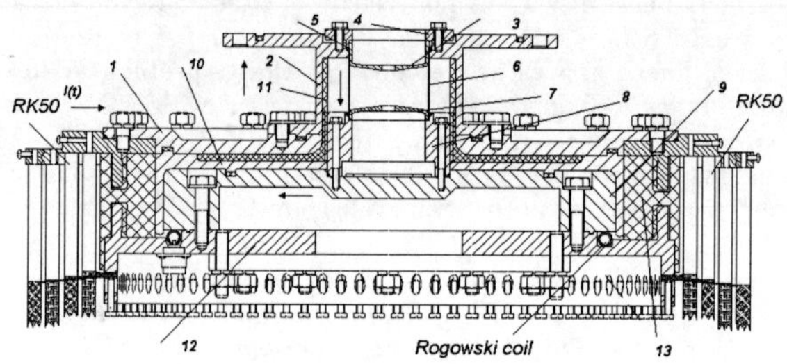

Fig. 2 Design of the load node to investigate the dynamic behavior of cylindrical samples

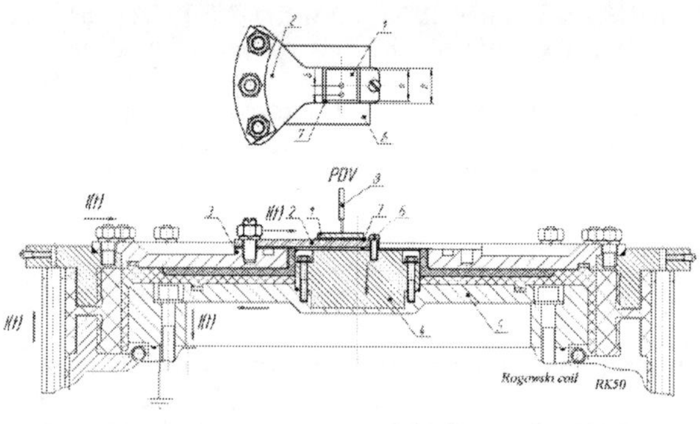

Fig. 3 Design of the load node to investigate the dynamic behavior of flat samples

III EXPERIMENTS IN CYLINDRICAL GEOMETRY

Compression of metal liners can be accompanied by the loss of stability. So, experiments were performed in two series. In the first series, the X-ray method [15] was used to investigate the dynamics and the symmetry of the liner implosion, to determine the parameters of loading pulses. In the second series, the well-developed loading regimes were used to test the samples.

The first-series experiments were performed according to the scheme shown in Figure 2. The experiments used liners made from aluminum AMts with the 30-mm diameter and the 0,3-mm wall thickness. In order to extend the facility lifetime, the charge voltage was Uo=26 kV, the current pulse amplitude was I~2 MA, and current pulse-rise time was about τ ~3,5 μs. The X-ray images of the liner were used to determine the external diameter of the shell at the time of X-ray radiography. The measurement of the liner external-diameter helps to estimate the magnetic pressure $P(t)$ at the liner surface:

$$\mathrm{P(t)} = \frac{\mu_0}{2\pi^2}\left(\frac{I(t)}{D(t)}\right)^2, \qquad (1)$$

where μ_0 is the magnetic constant, $D(t)$ is the liner diameter, and $I(t)$ is the current in the loop.

Figure 4 shows typical X-ray images of the aluminum shell at different times. Figure 5 shows the dependence of magnetically-compressed liner diameters, and Figure 6 shows characteristic dependences of both current and pressure on time calculated according to the formula (1).

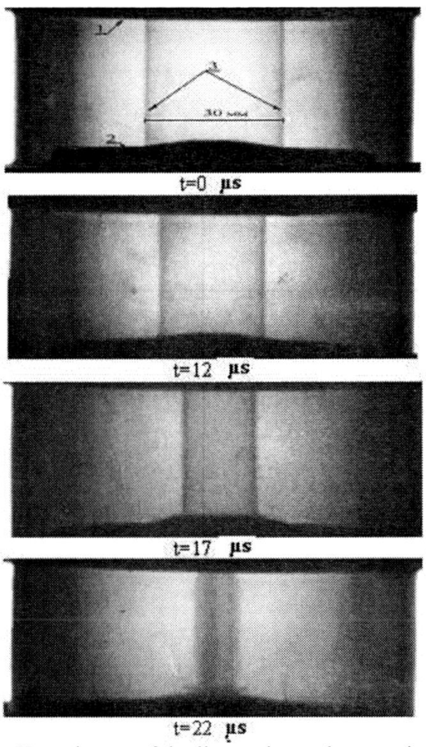

Fig.4 X-ray image of the liner prior to the experiment and at different times under compression by the pulsed magnetic field

In the process of compression up to the 17-mm diameter, the aluminum liner is seen to ideally preserve its symmetry that gets violated by the time of 22 μs. Therefore, the samples with the 17-mm diameter and the 3-mm thickness were used for experimentaion.

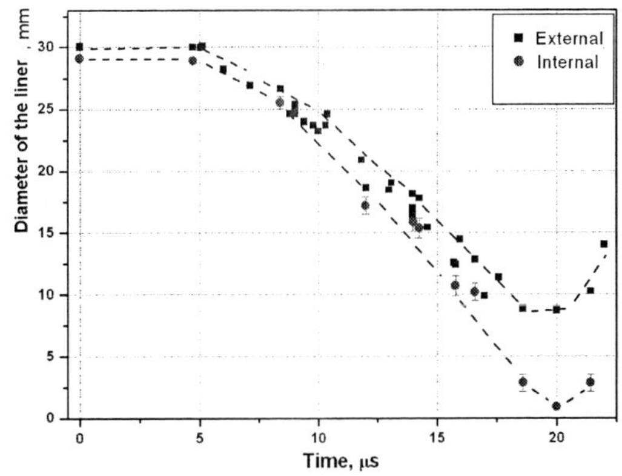

Fig.5 External and internal diameters of the magnetically compressed liner versus time

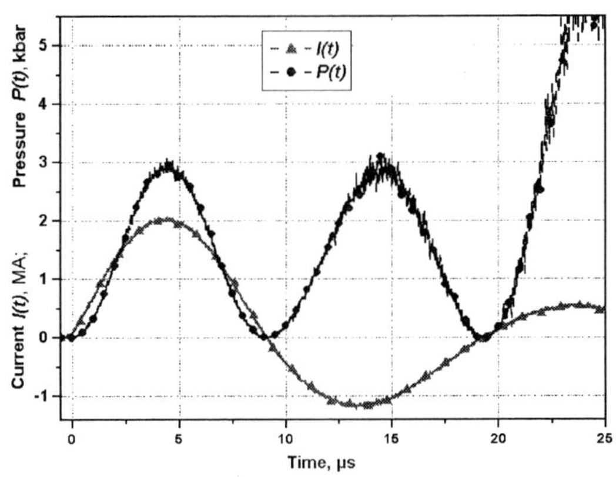

Fig.6 Current $I(t)$ and magnetic pressure $P(t)$ compressing the liner versus time

The second-series experiments tested samples by the shock method shown in Figure 7. As opposed to the first-series experiments, VISAR registered the liner compression velocity. Figure 8 shows how the velocity of the liner accelerated by its magnetic field depends on time and Figure 9 shows a photo with the longitudinal section of the test sample made from the aluminum alloy AD1 after the impact by the liner accelerated up to ~ 600 m/s.

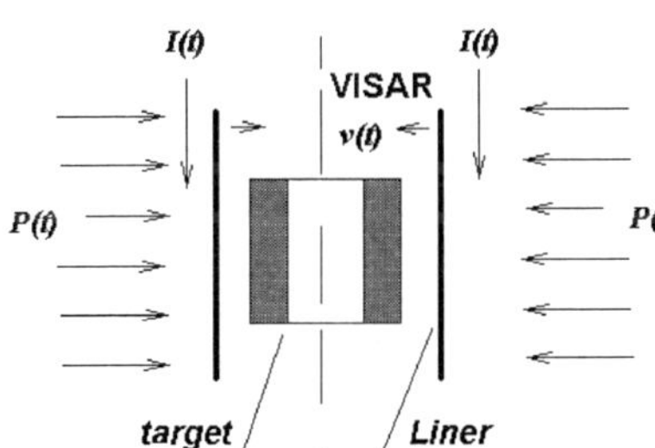

Fig. 7 Experimental setup with the sample

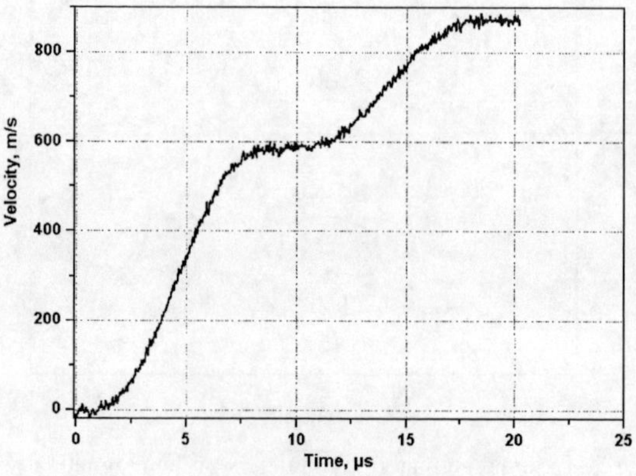

Fig. 8 Liner velocity profile measured with the help of VISAR

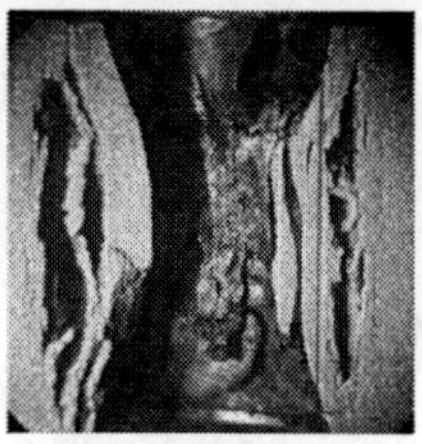

Fig. 9 Macro photo: longitudinal section of the cylindrical sample made from the aluminum alloy AD1 after loading

Impact of the liner accelerated by the pulsed magnetic field up to ~600 m/s caused multiple spallation in the sample. The longitudinal section clearly has several main spallation cracks. The first spall occurred at a distance of ~ 1 mm from the internal surface on the one side and 1,5 mm on the other side. One section displays four main cracks with the height from 12 up to 14 mm; the distance between the main cracks was 0,3 through 0,6 mm. There is no complete symmetry because of nonideal initial geometry of the sample.

IV EXPERIMENTS IN FLAT GEOMETRY

Experiments were performed in two series. In the first series, we checked operability of the flat impactor acceleration schemes and determined the loading modes. The second series included experiments aimed to test samples. The first-series experiments were performed according to the scheme shown in Figure 10, and the second-series experiments – according to schemes shown in Figures 3 and 11. All experiments used a bus of the aluminum alloy Amts 6 with the 0,94-mm thickness as the current-carrying accelerating plate.In experiments, we varied the charge voltage *Uo*, the facility capacity *Co*, the width of the current-carrying plate, and sizes of impactors. All experiments used PDV-interferometer to measure the motion velocity or the free-surface velocity. In the first-series experiments, we measured velocity of current-carrying plates (experiments # 2 – 7) or flat metal impactors (experiments # 9 –11). In the second series, we loaded test samples (St.3, aluminum alloy 1424 and AMts). In experiment #14, the current-carrying plate served as an impactor. Table 1 shows the setup for all experiments.

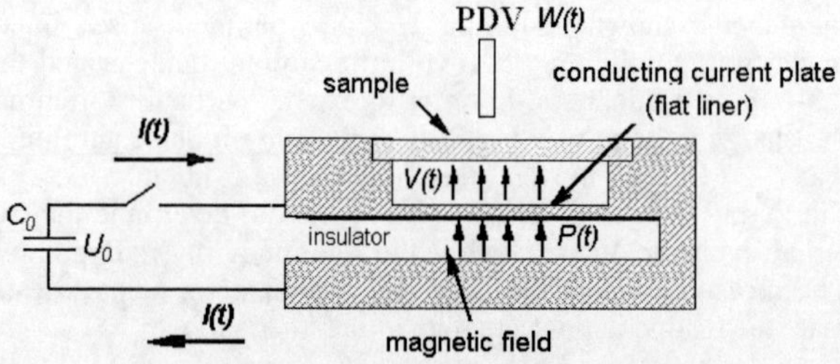

Fig.10 Sample loading scheme that uses the current-carrying plate as the impactor

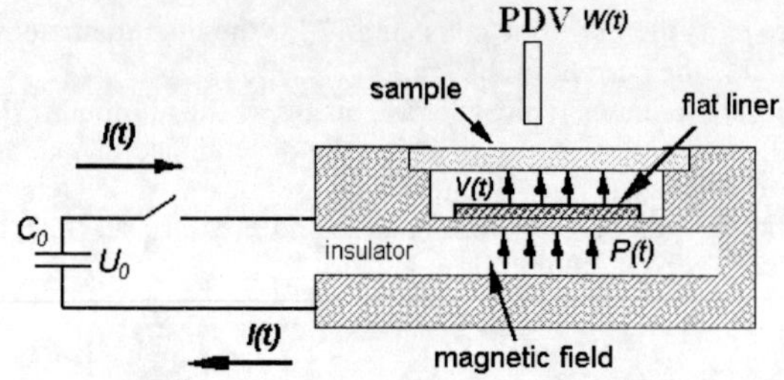

Fig. 11 Sample loading scheme that uses the plate as the impactor

Figure 12 shows the experimental dependence "velocity of the current-carrying plate and flat impactors versus time" and Figure 13 shows the free-surface velocity profiles registered by the PDV-interferometer when we test samples.

The highest velocities were achieved when the current-carrying plates themselves were magnetically accelerated in the range from 1,5 up to 3 km/s. With flat metal impactors, the velocity was in the range from 0,8 up to 1,1 km/s.

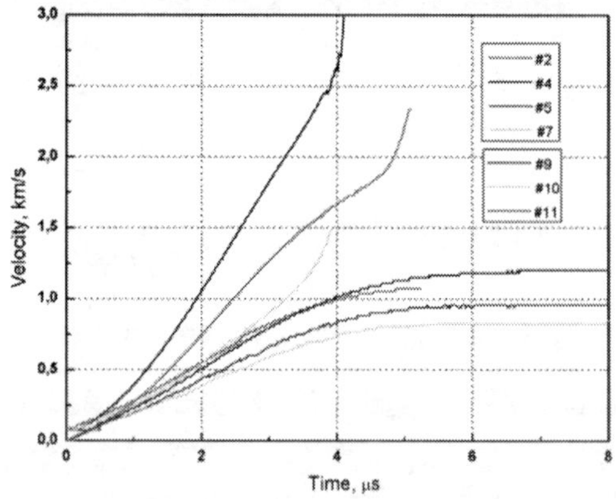

Fig. 12 Velocity of current-carrying plates and flat impactors versus time

Table 1. Experiment setup

Parameters of facility					Parameters of impactor				Parameters of sample			
Experiment #	C_0, µF	U_0, kV	*d*, mm	*b*, mm	material	*a*, mm	*c*, mm	*t*, mm	material	a^*, mm	c^*, mm	t^*, mm
#2	187	22	0,94	7,9	-	-	-	-	-	-	-	-
# 4	187	22	0,94	27,9	-	-	-	-	-	-	-	-
# 5	187	22	0,94	13,4	-	-	-	-	-	-	-	-
# 7	60	22	0,94	7,9	-	-	-	-	-	-	-	-
# 9	187	22	0,94	27,9	St.3	25	25	0,1	-	-	-	-
# 10	187	22	0,94	27,9	St.3	25	25	0,1	-	-	-	-
# 11	187	22	0,94	14,1	St.3	19,8	12,2	0,1	-	-	-	-
Second series of experiments												
#14	120	20	0,9	14,9	Current-carrying plate				Al AMts	13	13	1,9
#16	120	25	0,95	27,7	St.3	20	20	0,4	St.3	20	20	1
#18	120	31	0,94	14	Al 1424	20.5	10	0,42	Al 1424	12	12	1
#19	120	28,5	0,94	14	Al 1424	20.5	10	0,42	Al 1424	11	11	1

C_0– facility capacity; U_0– facility's charge voltage; *d* – thickness of the magnetically accelerated plate; *b* – width of this plate; *a*– impactor length; *c* – impactor width; a^* – sample length; c^*– sample width; t^*– thickness of samples; Al – aluminum alloy; St – steel.

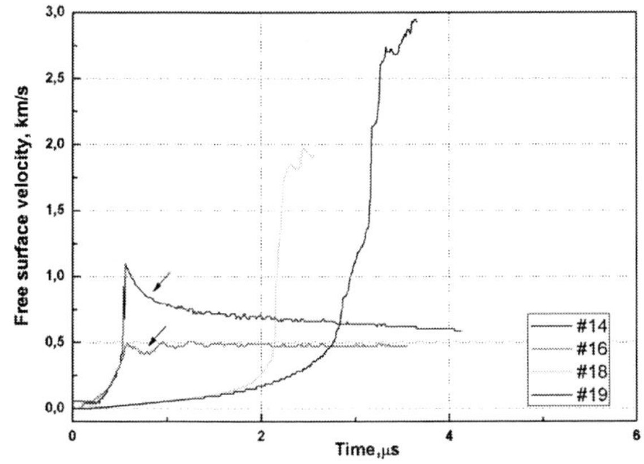

Fig. 13 Experimental dependences of free surface velocity of samples *W(t)* on time (an arrow shows spallation signal)

Loading the sample from aluminum AMts with the 1,9-mm thickness by an impact of the 0,94-mm thick current-carrying aluminum plate (experiment # 14) accelerated up to ~ 1,1 km/s induced the shock wave whose free-surface velocity profile had the shape close to that of the triangular pulse. Note that the sample gets significantly fused when the current-carrying plate is used as the impactor. Thus, it is expedient to use impactors rather than the current-carrying plate for sample loading. |The spall signal is recorded when the 0,4-mm thick steel impactor accelerated up to ~ 0,5 km/s hits against the 1-mm thick steel sample. Spall signals failed to be formed at the velocity of ~2 km/s and ~3 km/s due to skews in the design of the sample's fixturing point, air "pillows", and, perhaps, due to the Foucault current. Further investigations are required to understand the reasons thereof.

V CONCLUSION

The first trial experiments wherein the pulse magnetic fields were used to load the samples were performed on the GNUV-P facility. Modes of the high-strain rate deformation of cylindrical liners with the 30-mm diameter in the strain-rate range from 10^4 up to 10^6 s^{-1} were developed: the total strain was ~ 100% at the velocity of liner compression up to ~0,8 km/s. Acceleration modes for flat aluminum current-carrying plates and impactors from different materials in the velocity range from 0,5 up to ~3 km/s were developed. There is a possibility to expand the range of the impactor velocity up to ~3,5 km/s due to the involvement of the total facility capacity and due to the capacitor bank charge stepped up to 40 kV.

Testing of flat and cylindrical samples demonstrated that attained loading levels are enough for spallation to take place in materials with the density up to 10 g/cm^3.

This method requires further development. The load node design will be developed to use bipolar voltage. This would enable the range of collision velocity to be expanded up to ~ 5-6 km/s.

REFERENCES

[1] Avrorin E.N., Vodolaga B.K., Simonenko V.A., Fortov V.E. Powerful shock waves and extreme states of matter. Uspekhi fizicheskikh nauk, 1993, V.163, #5, p.1-34.

[2] Petrovtsev A.V., Kovalenko G.V., Zadorozhnyi G.A. Zababakhin Scientific Talks: proceedings of VII International conference – Snezhinsk RFNC-VNIITF publishing office, 2003.

[3] Pavlenko A.V., Balabin S.I., Kozelkov O.E., Kazakov D.N. One-stage gas gun for investigation into dynamic properties of structural materials in the range up to 40 GPa. Pribiry I tekhnika eksperimenta, 2013, #4, p.122-124.

[4] Pavlenko A.V., Malyugina S.N., Pereshitov V.V., Lisitsyna I.N. Two-channel interferometric complex VISAR for investigation into material properties under shock-wave loading. Pribiry I tekhnika eksperimenta, 2013, #2, p.127-129.

[5] Mokrushin S. S., Anikin N. B., Malyugina S. N., Pavlenko A. V., Tyaktev A. A. Interferometer with frequency-time multiplexing for investigation into material properties in shock-wave experiments. Pribiry I tekhnika eksperimenta, 2014. – # 4. – p. 107–110.

[6] HU Haibo, YU Jidong, CAO Luoxia, CHEN Yongtao, WANG Guiji, WU Gang. About the Dynamic Response of Material under Intensive Shock Loading in Near Free Surface Layer. The 17th International Topical Conference "Extreme states of substance. Detonation. Shock waves", Sarov, Russia, March 23-27, 2015.

[7] Avrillaud G., Courtois L., Guerre J. et al. GEPI: a Compact Pulsed Power Driver For Isentropic Compression Experiments And For Non Shocked High Velocity Flyer Plates // Proc. 14th IEEE Int. Pulsed Power Conference, Dallas, Texas, USA, June, 15-18, 2003. – P. 913–916.

[8] J.R. Asay. Isentropic compression experiments on the Z accelerator. Proc. APS Shock Compression of Condensed Matter, 1999. – P. 261.

[9] C.L. Rousculp, W.A. Reass, D.M. Oro and ect. The PHELIX pulsed power project: bringing portable magnetic drive proton radiography. The 14th International Topical Conference, Sarov, Russia, March 11-16, 2012.

[10] R.E. Reinovsky. Pulsed power hydrodynamics: a discipline offering high precision data for motivating and validating physics models. 15th IEEE Int. Pulsed Power Conference, Monterey, California, 2005 – P. 910–915

[11] A.M.Buiko. Modeling of interaction of high-speed condensed liners with current-carrying walls. Presentation on international conference XVIII Khariton Scientific Talks on the problem of high energy density physics, April 19 – 22, 2016.

[12] Afanasyev V.N., Galitsky M.V., Pavlenko A.V. et. al.Electro-explosive generator of shock waves. Extreme processes and states: Proceedings of V Zababakhin Scientific Talks. – Snezhinsk, 199. – p 127-128.

[13] Pavlenko A.V., Afanas'ev V.N. Kucherenko Yu.A., et al. Possibilities of Laboratory Complex «GNUV» for Studying the Mechanical Properties of Materials in the Wide Range of Impulse Loading // Proc. 14th IEEE Int/ Pulsed Power Conference, Dallas, Texas, USA, June, 15-18, 2003. - P. 902-904.

[14] Okun' I.Z.. Measurements of discharge currents by Rogowsky coil. Pribiry I tekhnika eksperimenta, 1968, #6, p. 120.

[15] Kuzhekin I.P. test facilities and measurements on high voltage. M.: Energiya, 1980. – p. 101.

[16] A.V. Pavlenko, E.I. Karnaukhov, D.N. Kazakov et. al. Pulse X-ay radiography of metal shells compressed by magnetic filed. Proceedings of VII Khariton Scientific Talks, Sarov, March 2005.

Simulation of Physical Processes at High-Current Discharges in Spherical Chambers with Plasma Focus

V. Yu. Dolinskii, S. F. Garanin, V. I. Mamyshev, N. G. Makeev, Yu. S. Shigaev
All-Russia Research Institute of Experimental Physics,
Sarov, Nizhni Novgorod region, Russia
vasilijdolinskij@gmail.com

***Abstract*–The results of two-dimensional MHD-code development to be used in calculation researches of plasma sheath dynamics in spherical chambers with plasma focus are presented. In the paper the equations of magnetic hydrodynamics are used taking into account magnetic field diffusion, heat conductivity, and plasma radiation. An implicit scheme of magnetic field calculation was used that makes it possible to describe plasma motion in the region with low density next to plasma sheath. The formulas taking into account an anomalous resistance in plasma were used to calculate plasma conductivity. The neutron yield calculation was performed under the assumption of thermonuclear mechanism of neutron generation only. The code parallelization was realized to accelerate the computations. The effects of magnetic field diffusion, heat conductivity, and plasma anomalous resistance on plasma sheath dynamics and voltage at the chamber input was studied. The comparison of calculated dependences and experimental data demonstrated good agreement.**

Key words–plasma focus; current plasma sheath; spherical plasma chamber; neutron yield

I. INTRODUCTION

In the middle of 50-s N.V. Filippov and T.I. Filippova (Institute of Atomic Energy employees) and somewhat later J. Mather (LANL employee) independently advanced, investigated, and published two versions of coaxial gas discharge chambers with metal walls [1,2]. In such chambers a high-voltage discharge occurs in deuterium along the insulator surface; there a current plasma sheath (CPS) is formed, which is then accelerated by the magnetic field of the discharge current. When the sheath descents towards the axis the neutral gas is partially raked together and compressed. As a result of the CPS collapse near the axis dense plasma focus (DPF) is formed with high energy density. In the early sixties N.G. Makeev (VNIIEF employee) advanced and began investigations a new original design of gas-discharge chamber with DPF. It was called SFC [3,4]. The chamber anode and cathode are of spherical shape and form a compact lengthy gas-discharge gap through which the CPS moves smoothly changing the moving direction 180° during a discharge.

Currently SFC high-current discharges of DPF type are being studied in VNIIEF. They are aimed at increasing the neutron yield. These researches are of high scientific and application value. They extend capabilities of using DPF-based neutron sources in nuclear physics, neutron radiography and other applications.

The neutron yield in the discharge of DPF type depends on a large number of factors including: value of initial energy stored in a capacitor bank, fraction of energy delivered to a discharge, composition and pressure of the gas in the chamber, amount of impurities, geometry of electrodes, and parameters of the current circuit.

To investigate CPS dynamics in SFC by calculations and to study the effect of the chamber geometry and current circuit parameters on the value of neutron yield a two-dimensional (in Eulerian cylindrical coordinates) MHD code was developed and realized.

II. DESCRIPTION OF THE COMPUTATIONAL SCHEME OF MHD-EQUATIONS

MHD-calculation involves the following steps:

- Gas-dynamic calculation step at which standard gas dynamic values are determined: ion velocity, density, electron and ion energies;
- Conductivity calculation, distribution of magnetic and electric fields;
- Calculation of energy exchange between electron and ion components and calculation of electron and ion heat conductance, energy losses by radiation, and Joule heating of plasma;
- Calculation of ionization degree, electron and ion temperature, and pressure.

Metal boundaries in the chamber are assumed to be «rigid, cold, ideally conducting walls». The value of boundary magnetic field on the insulator is found basing on the value of current in the current circuit. The circuit current is determined through solving the electrotechnical equation:

$$L_0 \frac{dI}{dt} = U(t) - R \cdot I(t) - U_{PF}(t) \qquad (1)$$

where

$$U(t) = U_0 - \frac{1}{C}\int_0^t I(t)dt ,$$

L_0 is circuit inductance, U_0 is capacitor bank charging voltage, C is capacitance of the capacitor bank, R is circuit resistance, U_{PF} is voltage at the chamber input that is determined basing on the results of MHD-calculations.

To calculate conductivity σ classical formulas were used [5]

$$\frac{1}{\sigma}=\frac{1}{\sigma_{ei}}+\frac{1}{\sigma_{en}} \tag{2}$$

where

$$\sigma_{ei}(s^{-1})=5.44\cdot10^{18}\frac{T_e^{3/2}}{L_e}k\,,$$

$$\sigma_{en}(s^{-1})=\frac{\alpha}{1-\alpha}\max(3.7\cdot10^{15},\sigma_{ei})\,,$$

T_e is electron temperature expressed in keV, L_e is electron Coulomb logarithm, k is conductivity magnetization factor that depends on electron magnetization ($\omega\tau_e$), α is ionization degree.

The following formula [6] was used to take into account anomalous resistance:

$$\sigma_\Sigma=\frac{\sigma}{1+\nu_{eff}\tau_{ei}} \tag{3}$$

where

$$\nu_{eff}\tau_{ei}=0.74\cdot10^{-2}\beta^2\frac{T_e^{3/2}\sqrt{A}}{\alpha\rho L_e}H\,,\ \beta^2=(u/v_{T_i})^2\,,$$

u is current electron velocity, v_{Ti} is thermal velocity of ions, A is atomic weight of ions, ρ is plasma density, H is azimuthal component of magnetic field expressed in MG.

The calculation of neutron yield was performed with the assumption of thermonuclear mechanism of neutron generation only:

$$\frac{dN}{dt}=\int\langle\sigma v\rangle\cdot n^2\cdot2\pi r\,dr\,dz \tag{4}$$

where $\langle\sigma v\rangle$ - is rate of DD or DT reaction (here σ is reaction cross-section).

Gas-dynamic step of calculation is performed in accordance with the explicit scheme. To approximate differential equations a difference "upstream" scheme is applied. Such scheme has got the first order of accuracy regarding time and space. As electron and ion heat conductivities were also calculated basing on explicit schemes an artificial constraint was introduced for the value of energy "flowing out" from the cell within the time step in order to provide calculation stability.

Explicit schemes of calculations impose Courant constraint for the time calculation step value: $\Delta t\le q\Delta l/(c_{ms}+|v|)$, where Δl is cell linear size, $q\approx0.25$–0.75, $c_{ms}^2=c_s^2+c_A^2$, c_s^2 is substance sound velocity squared, $c_A^2=H^2/(4\pi\rho)$ is Alfven velocity squared, $|v|$ is module of ion velocity. For sufficiently low gas densities in the rarefied region behind the sheath ($\rho<\rho_0/10^3$), the c_A^2 value becomes high and the time step decreases significantly. One can get rid of this constraint if the calculation of ion velocity components and the calculation of electric fields are combined and the implicit scheme is used to determine electric fields E_z and E_r as it is done in paper [5]:

$$\frac{\partial H}{\partial t}=\frac{\partial E_z}{\partial r}-\frac{\partial E_r}{\partial z}$$

$$E_z=\kappa\frac{1}{r}\frac{\partial(Hr)}{\partial r}-v_rH\,,\ E_r=-\kappa\frac{\partial H}{\partial z}+v_zH \tag{5}$$

where $\kappa=c^2/(4\pi\sigma)$ is coefficient of magnetic field diffusion, $c=3\cdot10^{10}$ cm/s is velocity of light in vacuum, v_r and v_z are ion velocity components.

Using difference expressions of E_z and E_r (5) and the equation for $\partial H/\partial t$ one can get a three-point diffusion difference system for E_z, E_r but with diffusion coefficient ($\kappa+c_A^2\Delta t$). This system is solved by implicit scheme using sweep method. Iterative convergence is determined basing on the value of H. Such approach makes it possible to reduce considerably the effect of gas density in the rarefied region on the value of time step and conduct calculations with the density value up to 6 orders less than its initial value.

As the difference scheme used in the given paper is of the first order of accuracy in space a large number of cells is required to provide adequate accuracy at calculating the process of plasma focusing near the axis for which more computation time is needed. To decrease computation time on multiprocessors with common memory, parallelization of cycles with the use of program interface OpenMP 3.1 tools was applied in this code. The test calculations performed on personal computers with different architecture of processors demonstrated considerable increase of program parallel version computation rate as compared to a serial version.

III. Calculation Results and Comparison to the Experiment

The calculations were performed for a spherical plasma chamber with 6 cm anode diameter and 12 cm cathode diameter. The chamber geometry is presented in Fig. 1.

The calculation area is of rectangular shape. Computational grid consists of square cells with side size of 0.01 cm. The internal space of the chamber (green) is filled with DT-mixture. The gas density is $\rho_0=4.8\cdot10^{-6}$ g/cm^3. The initial temperature of electrons and ions is $T_{e0}=T_{i0}=11600$ K

(1 eV). The boundary conditions for magnetic field H_0 are set in the cells on the boundary with the insulator (red). To calculate H_0 the value of the circuit current was found from equation (1). The electric circuit parameters for equation (1) were assumed the following: $C = 69.5$ μF, $L = 27$ nH, $R = 3$ mΩ, $U_0 = 21$ kV. The space colored blue shows chamber electrodes. The normal component of velocity on the boundary with electrodes was assumed to be equal to zero. Heat conductivity and radiation losses of energy were not included in the calculation. To exclude the process of magnetic field diffusion the conductivity was considered to be very high (σ_0 = constant = 10^{37} 1/s).

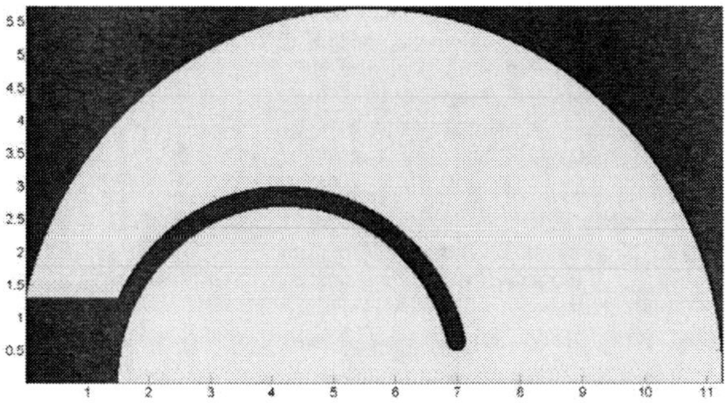

Fig. 1. Geometry of SFC designed for calculations

One of problems in simulating plasma flows in the chamber for plasma focus in Eulerian coordinates is how to conduct calculations in the regions where no substance is left. Such regions occur behind the plasma sheath in the course of its motion towards the chamber axis. To overcome this difficulty one should assume that the density cannot be lower than some minimal value $\rho_{min} = \rho_0/10^n$. For low n = (2–3) the density of gas behind plasma sheath is high enough, thus the current shunting takes place in the region behind the sheath and the process of plasma focusing on the axis becomes less expressed. For large n the computation time step starts decreasing because of the requirement of computation stability.

To study how the quantity of residual gas density affect to the process of plasma focusing a series of calculations was performed with different values of n within the range from 2 to 6.

Basing on the tendencies achieved in the course of calculations we can make a conclusion that at high values of residual density there actually takes place current shunting in the vacuum region and magnetic field screening that does not agree with the experimental data. At low values of residual density ($n \geq 5$) the difference between the calculation results for various values of n becomes insignificant and the agreement with the experiment improves. In the presented calculations the density of residual gas was accepted to be $\rho_{min} = \rho_0/10^5$.

To study the effects of different physical processes on the results of shell dynamics simulation the following series of calculations was conducted: calculation with frozen field; calculation with magnetic field diffusion; calculation with diffusion and heat conductivity. For conductivity and electron heat conductivity coefficient there were used both classical formulas and formulas with the correction for anomalous plasma resistance. For comparison with the experiment we used the measured and calculated time dependence of voltage at the chamber input.

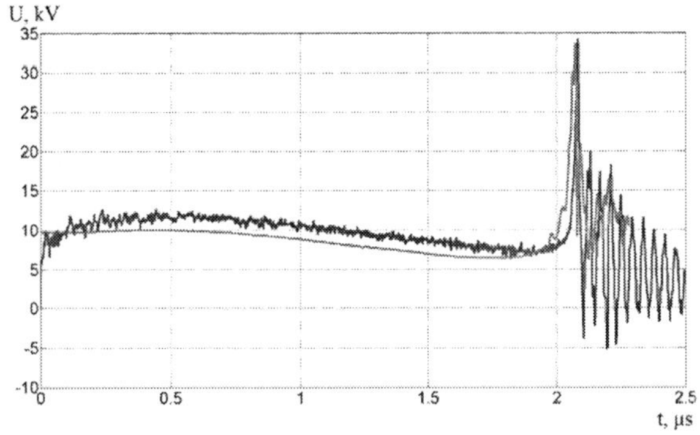

(a)

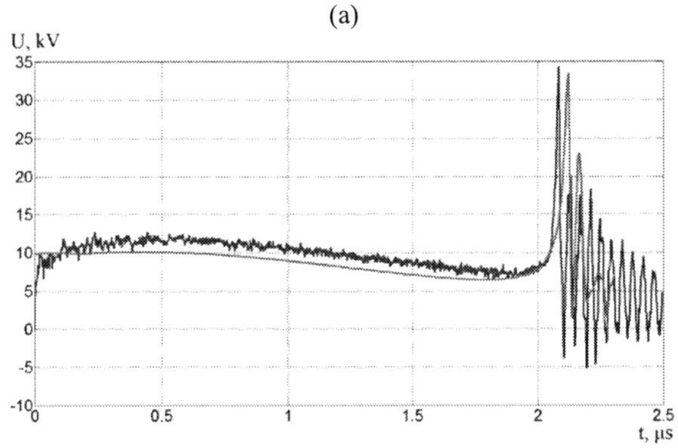

(b)

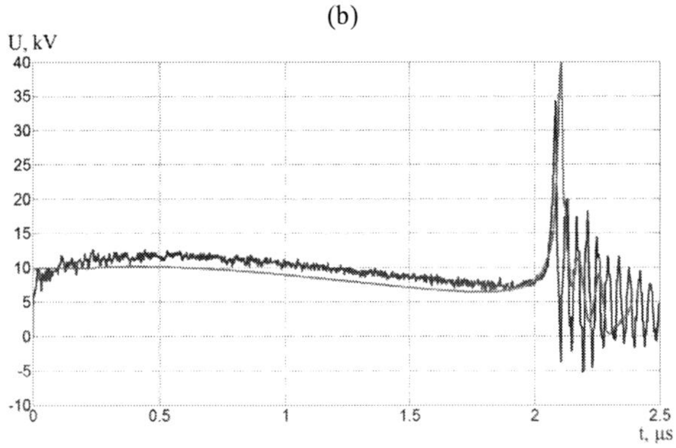

(c)

Fig. 2. Diagrams of chamber input voltage dependence on time. Blue is experiment, red is calculation. (a) is calculation with frozen field, (b) is calculation including magnetic field diffusion, (c) is calculation with diffusion and heat conductivity

Basing on the results of calculations one may note that the inclusion of the process of magnetic field diffusion somewhat changes the implosion time as compared to the calculation with frozen field. Thus, if in case with the frozen field, focusing starts a little earlier than in the experiment (Fig. 2a), then, in the calculation with diffusion (Fig. 2b), focusing occurs ~40 ns later than in the experiment.

Inclusion of heat conductivity makes the time of focusing closer to the experimental one (~20 ns delay), but the calculated amplitude of voltage becomes somewhat overestimated (Fig. 2c).

The calculations demonstrated as well that the inclusion of anomalous corrections [6] in resistance and heat conductivity coefficient insignificantly affects the time of plasma sheath focusing and the voltage amplitude at the time of focusing. Probably the correction for anomalous resistance corresponding to the development of low-hybrid drift instability [7], will affect the calculation results more significantly. The value of anomalous resistance is additionally important because, as it was demonstrated in the paper [5], that to determine neutron yield due to accelerating mechanism of neutron generation, it is necessary to take into account the anomalous corrections for plasma resistance.

The formula (4) evaluated value of neutron yield as $4 \cdot 10^9$ DT-neutrons. The measured in the experiment neutron yield was $1.9 \cdot 10^{11}$ DT-neutrons per pulse. This disagreement between calculation and experiment (~ 50 times) testifies to the fact that thermonuclear mechanism of neutrons generation does not essentially contribute to the neutron yield from plasma focus and confirms conclusions made in paper [5].

IV. CONCLUSION

The two-dimensional MHD model is developed to calculate plasma flows in DPF chambers. A code based on this model was created, tested on model calculations, and then applied to calculations of plasma sheath dynamics in SFC. The comparison with experiment demonstrated adequacy of the developed model and possibility of the code using for calculation research of CPS dynamics in the DPF chambers.

In future the calculation of neutron yield in terms of accelerating mechanism of neutrons generation will be added to the existing model. The improved model will make it possible to perform calculations in optimization of chambers design in order to increase neutron yield. It will also assist in estimating the significance of accelerating and thermonuclear mechanisms of neutron generation in PF at variation of initial and boundary conditions.

References

[1] N.V. Filippov, T.I. Filippova, and V.P. Vinogradov, "Dense high-temperature plasma in a noncylindrical Z-pinch compression," Nucl. Fusion, Suppl., vol. 2, pp. 577-587, 1962.

[2] J.W. Mather, "Formation of a High-Density Deuterium Plasma Focus," Phys. Fluids, vol. 8, No. 2, pp. 366-377, 1965.

[3] N.G. Makeev, V.G. Rumyantsev, and V.V. Maslov, "Physics and technology of spherical gas-discharge chambers with plasma focus," Encyclopedia of low-temperature plasma, V.Ye. Fortov, ed.-in-chief, vol. IX-3. Radiation plasma dynamics, V.A. Gribkov, ed., Moscow: Yanus, 2007, p. 176.

[4] N.G. Makeev, V.G. Rumyantsev, and G.N. Cheremukhin, "Development and research of spherical chambers with plasma focus," in Physics and Technology of Pulsed Ionizing Radiation Sources for Studying Fast Processes, N.G. Makeev, ed., Sarov: VNIIEF, 1996, pp. 281-291 [in Russian].

[5] S.F. Garanin and V.I. Mamyshev, "Two-dimensional MHD simulations of a plasma focus with allowance for the acceleration mechanism for neutron generation," Plasma Phys. Rep., vol. 34, No. 8, pp. 639-649, 2008.

[6] V.V. Vikhrev and S.I. Braginskii, "Z-pinch dynamics," Rev. Plasma Phys., M.A. Leontovich, ed., Moscow: Atomizdat, vol. 10, pp. 243-318, 1980, New York: Consultants Bureau, 1986.

[7] R.C. Davidson and N.T. Gladd, "Anomalous transport properties associated with the lower-hybrid-drift instability," Phys. Fluids, vol. 18, No. 10, pp. 1327-1335, 1975.

Calculated investigation of flat strikers magnetic acceleration in the device based on the magneto-cumulative generator

V.A. Demidov, A.S. Boriskin, Yu.V. Vlasov
Russian Federal Nuclear Center –
All-Russian Research Institute of Experimental Physics (RFNC-VNIIEF)
Sarov, Russia
demidov@ntc.vniief.ru

***Abstract*—The authors provide the task statement and numerical simulation results of the conductive plates acceleration under the influence of magnetic field created by the current of the helical magneto-cumulative generator (MCG) with a stator of 200 mm diameter (MCG-200). Two types of the current pulses are considered: directly from MCG and MCG with an explosive opening switch. The current pulses differ in form, amplitude and characteristic rise time. The parameters of the flat strikers are chosen based on the simulation results. At such parameters their velocity will be 3-5 km/s after the current pulse effect. Hereby, the sides of the strikers opposite to those that are under the magnetic pressure will remain in the solid state and will not be perturbed by the hydrodynamic instabilities. The device project is proposed based on MCG-200 generator that is intended for the equations of state investigation.**

Keywords*—magnetic acceleration; magneto-comulative generator; explosive opening switch; flat striker; numerical simulation.

I. Introduction

One of the methods of defining the equations of states parameters is the analysis of measurement results of wave and mass velocity in the target at the plate strike. The plate is moving with the velocity of several kilometers per second. The acceleration of the plates up to such velocities can be performed by explosive [1] or electromagnetic methods [2, 3]. The advantage of the electromagnetic method of acceleration is absence of the formal limit for the plate final velocity [2].

In the first experiments on wave and mass velocity measurement in the target, it will be necessary accelerate the flat strikers of the size 5×5 cm^2 up to the velocity of 3-5 km/s. To achieve this goal, it is proposed to use the source of the electromagnetic energy based on the helical MCG with a stator of 200 mm diameter (MCG-200). MCG-200 forms the current pulse of 9 MA with characteristic rise time (in "*e*" times at the final operation stage) of 13 μs in the load of 50 nH [4]. The possibility of using the electromagnetic energy source on the basis of MCG-200 generator with the explosive opening switch (EOS) forming the current pulse of 6.5 MA with the characteristic rise time (from 0.1 to 0.9 of its peak value) up to 0.5 μs in the load of 20 nH is also considered [4].

Choosing parameters of the plates of the acceleration unit, it is necessary to take into account that the thickness of the plates should be at least 1 mm to provide the precise physical measurements [2]. Besides, to provide the stable mode, the plates should accelerate at the distance no more than 4 of their thicknesses [2] (according to other data – no more than 10 thicknesses). The plates sides that are opposite to those that are influenced by the magnetic pressure should remain in the solid state by the end of the current pulse influence.

The task statement and the numerical simulation results of the acceleration of the plates under the current pulse influence of the energy source on the basis of MCG-200 generator are provided in the paper. According to the simulation results, the device project for investigations of the equations of state is proposed.

II. Calculated Model

Let us suppose that the acceleration of the conductive plates is produced by the preset current pulse $I(t)$. The current $I(t)$ creates the magnetic field $\boldsymbol{H}(t)$ and the plates move under its influence.

For the flat geometry $H(t) = kI(t)/b$, where b – the plate width, k – coefficient, considering the magnetic field decrease on the plate surface in comparison with $H_f = I/b$ because of the edge effects [3]. When writing down the equation of the plate motion with mass $m = \gamma S\delta$ under the magnetic pressure $p_M = \mu_0 H^2/2$, where $\mu_0 = 4\pi\cdot10^{-7}$ H/m – permeability of vacuum, γ, l, $S = bl$ – density, length, and area of the plate surface that the magnetic pressure influence on, the results are:

$$\dot{v} = \frac{\mu_0}{2}\cdot\frac{k^2 I^2(t)}{b^2\gamma\delta},\ v(0) = 0, \qquad (1)$$

where v – the plate velocity. If in the plates acceleration process the distance between them remain much less than their length and width, one can consider k as constant. In this case, if the current pulse is known, the plate velocity is defined by the current action integral, $F(t) = \int_0^t I^2(\xi)d\xi$, and its shift $\Delta y = y - a(0)/2$ is found by integrating the velocity with respect to time. Here y – distance from the symmetry plane of the acceleration unit till the internal plate surface, $a = 2y$ – distance between the plates (Fig. 1).

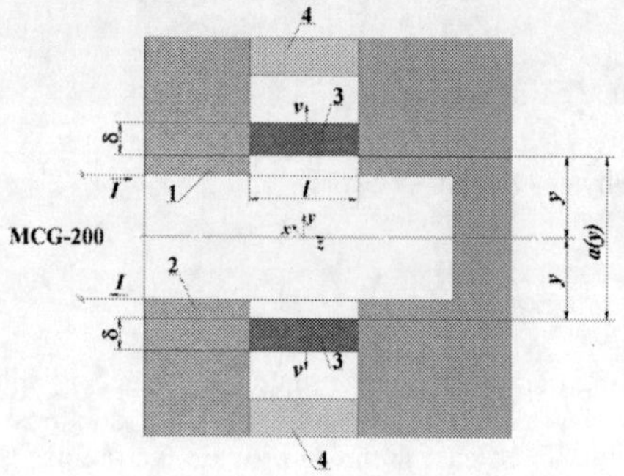

Fig. 1. A sketch of the plates acceleration unit. 1 – direct current guide, 2 – return current guide, 3 – accelerated sections of current guides, 4 – investigated samples, l, δ – length and thickness of the accelerated plates sections, a – distance between the plates

Using (1), one can find the relation between the plate width and thickness at which the plate velocity will achieve the given value v_f by the moment t_f of the magnetic field influence termination on it: $\delta(b) = \mu_0 k^2 F_f / (2\gamma v_f b^2)$. Here $F_f = F(t_f)$. The plate shift Δy_f by the moment t_f is defined as $\Delta y_f = \mu_0 k^2 S_f / (2b^2 \gamma\delta) = v_f S_f / F_f$, where $S_f = \int_0^{t_f} F(t)dt$. As evident from this formula, it is defined by the relation S_f/F_f, at the set final plate velocity; i.e. it is defined only by the form and amplitude of the current pulse and does not depend on the material and plate size.

In the acceleration process, the magnetic field gradually diffuses into the plate and, as a result, its temperature increases and can even reach the vaporization temperature. Correct analysis of the plate acceleration process with the consideration of the magnetic field diffusion process and the conductor heating requires the application of the magneto-hydrodynamic simulation. Some estimation can be done using the formula [5] for the skin depth depending on time at the nonlinear diffusion of the magnetic field: $s(t) \approx H_0(t)\sqrt{\rho_0 t / \mu_0} / H_c$, where $H_0(t)$ – magnetic field on the plate surface, ρ_0 – resistivity of the conductor material at the normal conditions, H_c – critical field. For copper, the value $H_c = 0.35$ MA/cm, for aluminum $H_c = 0.27$ MA/cm [5]. At the moment t_f, the value $s_f = s(t_f)$. Using this formula, let us define at what values of the plate thickness by t_f moment the magnetic field will not reach its opposite surface.

If in the process of the plates acceleration the distance between them became more than their width, it is necessary to consider the magnetic pressure decrease on the plates. The inductance of the generator circuit also increases, and this leads to the current reduction. In this case, the equation for the current in the MCG circuit

$$(L_g + L_t + L_l)\dot{I} + (\dot{L}_g + \dot{L}_l)I + (R_g + R_l)I = 0, \quad I(0) = I_0 \quad (2)$$

is solved together with the equation (1) of the plate motion under the difference of magnetic pressures affecting its internal and external surfaces in the central section (at $x = 0$, Fig. 1). Here L_g, R_g – MCG inductance and resistance, L_t – total inductance of the transmission line and non-accelerated sections of the plates acceleration unit, L_l, R_l – inductance and resistance of the accelerated section, I – current. It is supposed that while the magnetic pressure does not exceed the ultimate strength of the material of the plates, they do not move.

To solve the system of equations (1), (2) it is necessary to know the time dependences of the MCG and acceleration unit inductance and resistance; transmission line inductance and k dependence on the distance between the plates. For the MCG with EOS, the equation (2) is changed by the equations of the current switching scheme.

The coefficient k can be defined from the calculation of the magnetic field distribution at the given geometry of the acceleration unit. For estimations, one can use the formulas for the magnetic field distribution of the thin plate from [5]. In the system from two parallel plates the magnetic field affects both sides of the plates. It is possible to prove that the coefficient k (which characterizes the total effect of the magnetic pressure force on the plate in its central cross-section) dependence on the distance between the plates is defined by the expression: $k(a) = \sqrt{(2/\pi)\mathrm{arctg}(b/(2a))}$.

III. Calculation Results

Let us consider the acceleration of the plates by MCG-200 with EOS current pulse (Fig. 2). Using (1), one can define that the plate should be 5 cm width and made of aluminum of 0.039 cm thickness or copper of 0.012 cm thickness in order to be accelerated up to the given velocity $v_f = 4$ km/s up to the end of the magnetic pressure influence $t_f = 1$ μs. Such thickness is less than required (at least 0.1 cm). The shift by the moment t_f will be 0.09 cm, i.e. the aluminum plate displaces for 2.3 and the copper plate – 7.5 of its thickness. Thus, at the given restrictions it is impossible to accelerate the conductor of ~5×5 cm^2 up to the velocity of 3–5 km/s by the given current pulse. Therefore, it is proposed to use longer MCG-200 current pulse without an EOS.

The experimental dependence of the helical generator MCG-200 current at its operation into the load of 35 nH is presented in Fig. 3. In $t_f = 100$ μs after the MCG operation start, its circuit inductance is ~50 nH, current is 9.3 MA [4], $F_f = 5.78 \cdot 10^8$ A^{2}c, $S_f = 3792$ A^{2}c^2. The plate shift by the moment t_f is 2.6 cm. It is found that the plate thickness should be at least 1 cm and maximum width 3.2 cm to avoid the magnetic field reaching its opposite surface at the copper plate acceleration up to 4 km/s. The plate shift of 1 cm thickness and 3.2 cm width will be 2.6 of its thickness. At the same conditions, the thickness of the aluminum plate should be at least 0.8 cm and maximum width 6.5 cm. The plate shift of 0.8 cm thickness and 6.5 cm width will be 3.3 of its thickness. This does not exceed the limit for the shift mentioned above.

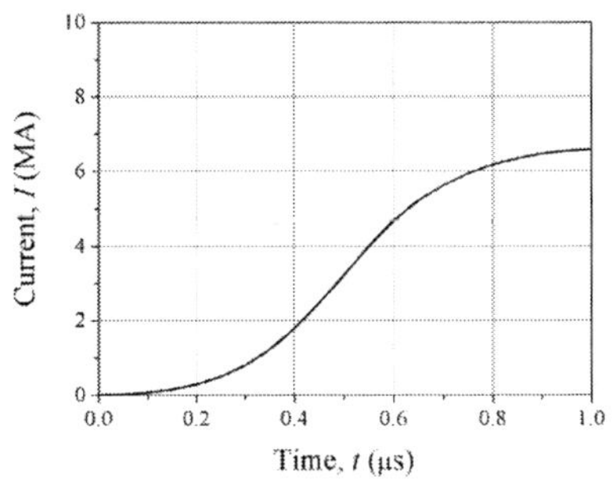

Fig. 2. MCG-200 with EOS at its operation into the load of 20 nH

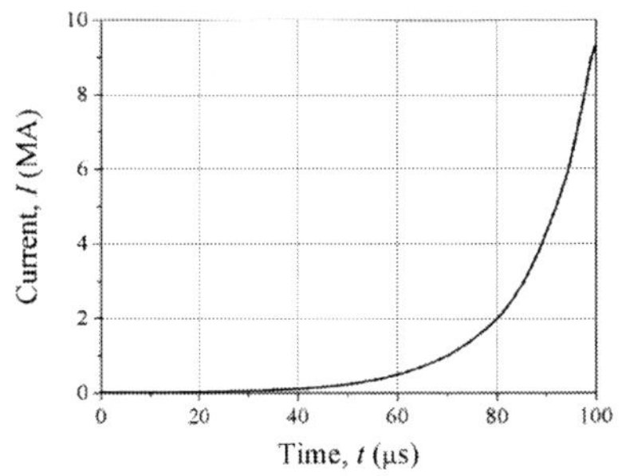

Fig. 3. MCG-200 current at its operation into the load of 35 nH

Thus, if the plate width should be at least 5 cm, only aluminum will be suitable as the material of the plates. Considering that the distance between the plates by the moment t_f will be increased by 5.2 cm in comparison with the initial distance, the authors discover that the value k by the end of the acceleration process will be decreased up to ~0.5 for the plate of 5 cm width. The load inductance will be also increased. This leads to the reduction of MCG current rise rate or even to its decrease by the end of the acceleration process. Therefore, the provided estimations should be specified solving the equation system (1), (2), considering the reduction of the magnetic pressure force effect on the plates and inductance load growth in the process of their acceleration.

Time dependences of the MCG inductance and resistance were used in the calculations. The dependencies were obtained using the method [6]. Inductance L_t of the transmission line from MCG to the acceleration unit is valued by the magnitude ~10 nH if to suppose that it will be of the same form as in the device [7] for the single-turn solenoid powering from MCG - 100 generator, but all its linear dimensions will be increased twice. The acceleration unit inductance of the length l can be calculated using the formula $L = \mu_0 a l K(a/b)/b$ for the inductance calculation of two thin parallel plates depending on the distance between them presented in [5]. The load circuit resistance should not be increased considerably in comparison with the initial resistance, as the main part of the current in the plates acceleration process due to the nonlinear diffusion flows through the cold sections of the accelerated conductors [5]. It was supposed in the calculations that $R_l = 0$.

The ultimate strength value for copper (~0.2 GPa) and aluminum (~0.1 GPa) are taken from the reference book [8].

Calculations considering the magnetic pressure decrease and load inductance increase in the process of the acceleration of the plates were performed for the plates with sizes chosen according to the calculation results with the given current pulse. The length of the accelerated plates section was equal to its width. The initial distance between the plates was 0.35 cm [7]. It was considered that the total length of the non-accelerated sections was twice more than the length of the accelerated section. The calculation was carried out up to the end of MCG operation in 109 μs after its operation start.

The calculation results of the acceleration of aluminum plates of 0.8 cm thickness and 5 cm width are presented in Fig. 4. As it is evident from Fig. 4, MCG current reaches its maximum of 6.9 MA in 105 μs after its operation start. In the process of the acceleration of the plates, the coefficient k decreases from 0.96 to 0.37, load inductance increases from 22 nH to 58 nH. At the end of MCG operation, the calculated plates shift is 6.9 of their thickness, velocity is 3.5 km/s, i.e. it matches the given interval. The magnetic field penetrates into ~0.5 depth of the plate thickness, i.e. it does not reach its opposite surface.

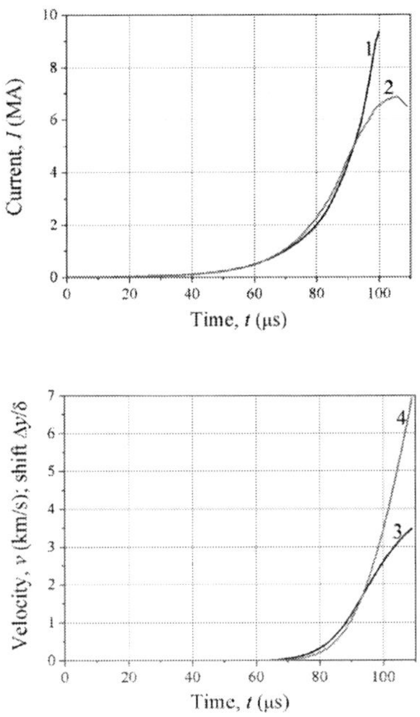

Fig. 4. Calculation results of the aluminum plates acceleration of 0.8 cm thickness and 5 cm width. 1 – current in the experiment [4], 2 – current in the calculation, 3 – plate velocity, 4 – shift in the plate thicknesses

IV. Device Sketch

The device construction for the acceleration of conductive plates by the magnetic pressure can be similar to the used one for the single-turn solenoid powering from MCG-100 [7] (Fig. 5) if to displace the single-turn solenoid by the acceleration unit consisting of two parallel plates. The device (Fig. 6) consists of the helical magneto-cumulative generator MCG-200, plates acceleration unit and transmission line to it.

Fig. 5. The external view of the device for a single-turn solenoid powering based on MCG with a stator of 100 mm diameter. 1 – MCG; 2 – EOS, 3 – coaxial transmission line, 4 – bus transmission line, 5 – single-turn solenoid, 6 – transition from coaxial to bus line

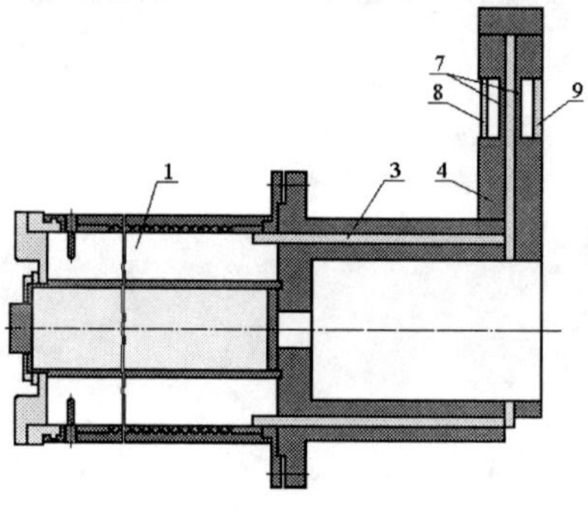

Fig. 6. The device sketch for the conductors acceleration by the magnetic pressure. 7 – the accelerated plates, 8, 9 – the studied samples of different thickness. The rest signs are the same as in Fig. 5

The transmission line contains coaxial and bus sections. At the calculation of the bus section inductance, the formula was used from [5]: $L = \mu_0 alK(a/\bar{b})\ln(b_1/b_2)/(b_1 - b_2)$, where $\bar{b} = (b_1 + b_2)/2$. The total transmission line inductance from MCG to the acceleration unit will be ~10 nH at the gap between the conductors of the transmission line of 3.5 mm [7].

The acceleration unit consists of two plates with the current flowing in opposite directions. To reduce the influence of conductor connecting the direct and return current guides on the magnetic field in the area, where the accelerated sections of the plates are situated, it was proposed that the total length of the non-accelerated sections was twice more than the length of the accelerated one.

If we take the plate of the width b = 50 mm, length $l \sim 3b$ = 150 mm (the length of the accelerated sections $l \sim b$), and the distance between the plates a = 3.5 mm (equal to the gap between the conductors in the transmission lines [7]), the acceleration unit inductance will be ~12 nH. Thus, the initial inductance of the load circuit in the developed device will be ~22 nH. According to the calculations, the load circuit inductance will increase up to ~60 nH in the process of the plates acceleration.

V. Conclusion

Estimations revealed that at the given limits for the plates thickness, shift value, and skin layer thickness it is impossible to accelerate the plates of the area ~5×5 cm^2 up to the velocity of 3–5 km/s by the current pulse of MCG-200 with the opening switch. To achive this goal, it was proposed to use longer current pulse of MCG-200 without the opening switch. In this case, it is possible to accelerate the plates of the given area up to the given velocity if they are produced from aluminum. More powerful MCG will be needed to accelerate copper plates of such area.

The device construction for the acceleration of conductive plates by the magnetic pressure can be similar to the used one in the device based on MCG with a stator of 100 mm diameter for the powering of a single-turn solenoid if to replace the single-turn solenoid in it by the acceleration unit consisting of two parallel plates.

References

[1] R.F. Trunin, "Metalls compressibility at high pressures of shock waves," In High energy density: Digest of scientific papers, Sarov: RFNC-VNIIEF, 1997, pp. 253-270.

[2] A.V. Ivanovsky, "High-speed flat strickers accelerated by the current of magneto-comulative generators," in Proc. Int. Conf. VII Khariton's topical scientific readings. Extreme substance states. Detonation. Shock waves. Dr. A.L. Mikhailov, Ed. Sarov: RFNC-VNIIEF. 2005. pp. 506-511.

[3] P.-L. Hereil, G. Avrillaid, "GEPI: an isentropic compression generator for dynamic material studies," in Proc. Int. Conf. VII Khariton's topical scientific readings. Extreme substance states. Detonation. Shock waves, Dr. A.L. Mikhailov, Ed. Sarov: RFNC-VNIIEF. 2005. pp. 659-665.

[4] V.D. Selemir, V.A. Demidov, "Pulsed Power Explosive Generators for Physical Studies," in Megagauss X, M. von Ortenberg. Ed. Berlin: Humboldt University at Berlin, 2005, pp. 161-168.

[5] H. Knoepfel, "Pulsed High Magnetic Fields," Amsterdam, The Netherlands: NHPC, 1970.

[6] Yu.V. Vlasov, S.N. Golosov, V.A. Demidov, S.A. Kazakov, "Calculational simulation of the explosive magnetic generator – magnetoplasma compressor system," in Megagauss and megaampere pulse technology and applications, V.K. Chernyshev, V.D. Selemir, L.N. Plyashkevich, Eds. Sarov: VNIIEF, 1997, pp. 491-495.

[7] A.S. Boriskin, A.Ya. Brodsky, Yu.V. Vlasov, et al. "Experimental results on single-turn solenoid powering from helical MCG with current opening switch," in Megagauss X, M. von Ortenberg, Ed. Berlin: Humboldt University at Berlin, 2005, pp. 49-53.

[8] Table of physical quantities. Reference book, Academician I.K. Kikoin, Ed. Moscow: Atomizdat, 1976.

Explosive current opening switches

V.A. Demidov
Russian Federal Nuclear Center –
All-Russian Research Institute of Experimental Physics (RFNC-VNIIEF)
Sarov, Russia
demidov@ntc.vniief.ru

***Abstract* — Explosive opening switches (EOS) operate based on the principle of the conductors mechanical destruction of the inductive energy storage system by detonation products of the high explosive charge (HE-charge). At the conductor breakage with the current, the further current flow is connected with the electric arc creation and electromagnetic energy transformation into thermal one. Arc plasma cooling by detonation products causes resistance growth and pulse voltage creation on EOS. Explosive opening switches are applied in general for current pulse sharpening of the magneto-cumulative generators (MCG). Some common experimentally checked schemes of EOS constructions are described in the paper. The investigation results on the flat and cylindrical models and also full-scale experiments with powering from helical and disc MCG are provided.**

Key words—current sharpening; magneto-cumulative generator; explosive opening switch; characteristic current rise time; explosive opening switch resistance.

I. Introduction

Explosive opening switches operate based on the principle of the conductor (foil) mechanical damage of the inductive energy storage system by detonation products (DP) of the high explosive charge (HE charge). The further current flow is connected with the electric arc initiation at the conductor damage by HE-charge. Electromagnetic energy transforms in the arc into thermal one. Arc cooling by detonation products causes a current breakage and pulse voltage creation on EOS. Peak value of EOS resistance and its rise time are defined by velocity of breakages creation in the damaged conductor and extraction velocity of thermal energy from discharges into the arc-suppressing medium.

It is shown in [1] that switching limiting characteristics of EOS are defined by the relation: $EH < Pc$, where E, H – electric and magnetic fields intensity; P – medium pressure; c – sound velocity in the medium. It follows from this relation that EOS effectiveness depends on the HE-charge characteristics as well as medium properties, where the arc is initiated and also on the organization of the conductor mechanical damage and arc plasma cooling.

The conductivity σ_{DP} and electrical strength of HE-charge detonation products refer to the factors that influence the current pulse formation in the load. In [2], the values σ_{DP} = (1-7) Ohm^{-1} cm^{-1} are received for PETN, RDX and HMX. Electric strength of DP of TNT, RDX and their mix structures is 20-70 kV/mm [3-4].

II. Explosive Opening Switches with Ribbed Barrier

Foil breaking in EOS with a ribbed barrier (RB) from dielectric takes place due to different acceleration under the influence of shock waves (SW) of foil sections situated over the barrier grooves and ribs. The opening switch (Fig.1) with a ribbed barrier and HE-charge of toroidal form [5] was used in the first experiments of current pulse sharpening of helical MCG in VNIIEF. The barrier was made of Plexiglas with grooves of 1 mm width, 4 mm depth, and ribs width of 1 mm. HE-charge from TNT/RDX 30/70 in the shape of a truncated toroid had the external diameter of 100 mm, internal diameter of 42 mm. HE-charge initiation was performed in the internal ring by the shock wave from HE-charge of MCG liner.

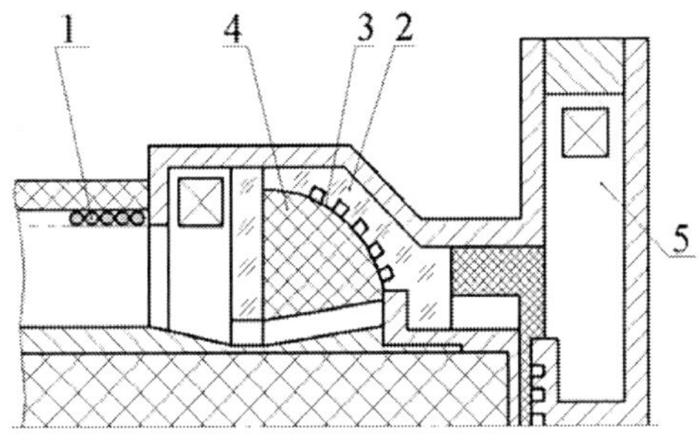

Fig. 1. EOS with toroidal HE-charge. 1 – helical MCG, 2 – ribbed barrier, 3 – foil, 4 – HE-charge, 5 – load

This EOS construction was studied at the powering from helical MCG by the currents up to 3 MA. The load was 18 nH. In the experiments, thickness of aluminum foil was 0.1 – 0.3 mm, width – 170 mm, length – 38 mm.

Fig. 2 shows dependencies of EOS peak resistance (R) and characteristic current rise time τ (current rise time from 0.1 to 0.9 of its maximum value) in the load on foil thickness (δ). It is evident that the main EOS characteristics become rather worse with δ increase. Using copper foil of 0.2 mm thickness, the peak resistance of the opening switch was no more than 5 mΩ and the time τ was ~3 μs. At the increase of the breaking current from 450 kA to 2.7 MA, the values R and τ

changed very little. The peak voltage (~55 kV) was observed in the experiments with foils of 0.1 mm and 0.15 mm thickness.

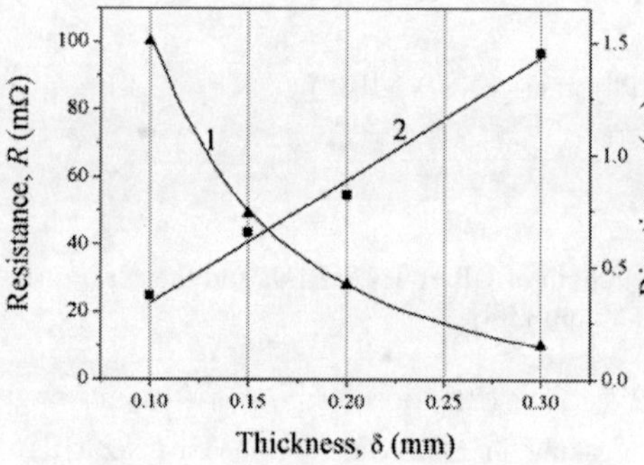

Fig. 2. Dependencies of the opening switch peak resistance (1) and current characteristic rise time in the load (2) on the foil thickness

Low output parameters of EOS with aluminum foil of 0.3 mm thickness and copper foil of 0.2 mm thickness are obviously connected with the destruction peculiarities of comparatively heavy foil on the ribbed barrier made of Plexiglas. At any time moment after SW coming from HE-charge on the foil $U_g > U_r$, where U_g and U_r are foil sections velocity over the groove and rib, respectively. Foil movement velocity over the groove is determined by the material and foil thickness, and also HE-charge type and its thickness. Calculation results made according to [6] for copper foil of 0.2 mm thickness at the HE-charge thickness (TNT/RDX 50/50) 5-13 mm revealed that U_g reaches 3-4.5 mm/μs for the time 0.7-1.5 μs.

Foil movement over the rib is defined by acoustic stiffness of the barrier material $\widetilde{R} = \rho \cdot c$, where ρ – density, c – sound velocity of the RB material. In Fig. 3 there are calculation dependencies $U_r(t)$ obtained in [7] for two materials: Plexiglas ($\widetilde{R} \approx 3$) and steel ($\widetilde{R} \approx 33$). Copper foil had thickness of 0.2 mm, HE-charge thickness from TNT/RDX 50/50 was 5 mm. It is evident from Fig. 3 that the foil velocity over the rib from Plexiglas is approximately twice more than the velocity over the rib from steel. It means that the length of the broken foil section and hereby the resistance of the opening switch with RB from Plexiglas will be less than in case of using the dielectric material with shock-wave steel properties for RB.

Support of $\widetilde{R}$ high value is a necessary condition but not sufficient for effective [7] foil breakage. As a result of unloading wave, there is a cross flow of the material to the groove direction in the rib; at some conditions this can block the foil movement inside the groove. To decrease the cross flow, the RB material should satisfy two conditions: material density should be maximum and the sound velocity – minimal. The experiments with flat RB were carried out. In these experiments, different materials were used to check the influence of the side unloading of the barrier ribs on the foil breakage. The copper foil of 0.2 mm thickness, 10 mm width and 80 mm length was used. HE-charge (PETN) of 5 mm thickness had the width of 30 mm and length of 80 mm. The experiments results are provided in Table I.

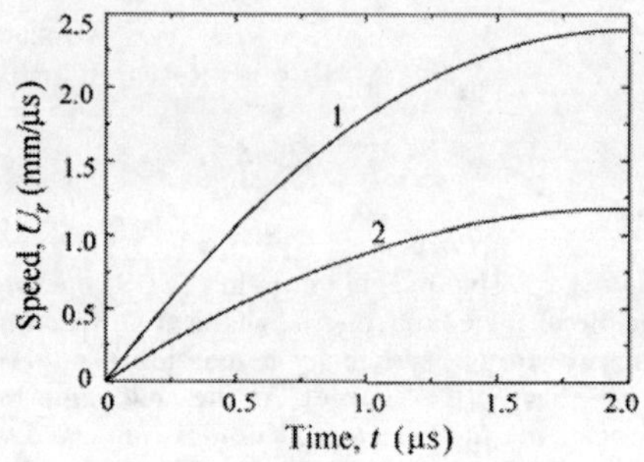

Fig. 3. Calculation dependencies U_r on time at different materials of RB. 1 – Plexiglas, 2 – steel

Table I. RESULTS OF EOS TESTS WITH DIFFERENT MATERIALS OF RIBBED BARRIER

№ of test	RB material	ρ, g/sm^3	c, km/s	τ, μs
1	Plexiglas	1.18	2.5	0.85
2	Ceramics ЦТС-19	7.5	3	0.55
3	Corundum ceramics	3.7	7	2.5

The best material of the tested ones in the experiments is Ceramics ЦТС-19. However, high price of ceramics and technical difficulties of barrier production from this material limits the possibilities of Ceramics ЦТС-19 usage in the opening switches. Thus, the combined barrier [8, 9] made of two available high-tech materials: steel and Plexiglas was experimentally checked. Steel stripes situated on the ribs allow decreasing of the velocity U_r and reducing of the side unloading of the ribs. The switching time τ = 0.6 μs was obtained in the experiment with such barrier at the conditions described above.

The dependence of the opening switch speed with the combined ribbed barrier on the initial conditions was studied. The goal was to get current pulses with characteristic rise time τ = 1 μs and less. The experiments were carried out on the facility consisted of the helical MCG and cylindrical EOS with RB (Fig.4).

Two types of MCG, i.e. with the internal helix diameters of 100 mm and 200 mm were used in the experiments. The first type allows getting of the current up to ~7 MA in the breaking circuit, the second one – up to ~10 MA. The opening switch contains the combined (Plexiglas-steel) barrier with the internal diameter of 100 mm or 200 mm. The groove width is 2.0 mm, rib width is 1.5 mm, rib height is 5.5 mm (steel rings cross-section is 1.5×1.5 mm^2)· In EOS of 100 mm diameter, the HE-charge had the thickness of 5 mm, in EOS of 200 mm diameter it was 10 mm. HE-charge initiation is axial with diversity $\Delta t \le 0.3$ μs. Copper foils were used in the opening switches. The experiments results are provided in Table II.

Table II. EXPERIMENT RESULTS OF HELICAL MCG AND EOS WITH RIBBED BARRIER

Experiment #	D, mm	δ, mm	l, mm	L_k, nH	I_k, MA	L_l, nH	I_l, MA	U_p, kV	τ, µs	τ_R, µs	p, GW/cm^2
1	100	0.3	90	30	5.4	10	3.8	67	0.6	0.3	0.9
2	100	0.3	90	30	9.5	10	6.7	67	1.0	0.4	1.2
3	100	0.2	90	30	5.1	10	3.9	72	0.5	0.2	1.0
4	100	0.3	40	30	6.2	10	4.7	39	0.9	-	1.6
5	100	0.3	190	20	6.9	40	2.0	210	0.43	-	1.5
6	200	0.3	200	30	9.2	10	7.0	110	0.6	0.3	0.5
7	100	0.3	300	20	5.6	17	3.4	190	0.3	0.12	0.7
8	200	0.3	400	50	9.3	21	6.6	340	0.48	0.15	0.7

Table of symbols: D – diameter of foil location, δ – foil thickness, l – foil length, L_k – inductance of the breaking circuit, I_k – peak current in the breaking circuit, L_l – load inductance, I_l – peak current in the load, U_p – peak voltage on the opening switch, τ – characteristic rise time of the current in the load (from $0.1I_l$ up to $0.9I_l$), τ_R – characteristic rise time ("e" times) of the opening switch resistance, p – flux density of the electromagnetic energy through foil

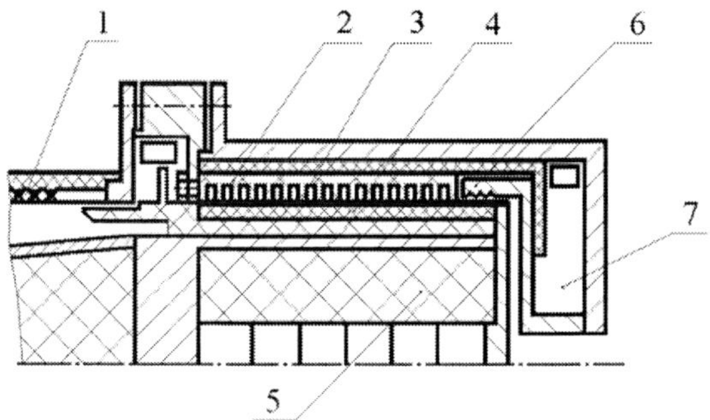

Fig. 4. General view of the helical MCG with the explosive opening switch. 1 – helical MCG, 2 – ribbed barrier, 3 – foil, 4 – HE-charge, 5 – HE-charge, 6 – closing switch, 7 – load

According to the execution conditions, two first experiments differed only in the value of the breaking current. Values of the current liner density in the foil differed ~1.7 times (i = 0.17 MA/cm and 0.3 MA/cm, respectively). The oscillogram of the current derivative received in the experiment #2 is illustrated in Fig. 5. The dependencies $R(t)$ of the experiments are provided in Fig. *6a*, dependencies of the currents in MCG, load, and opening switch of the experiment #2 are provided in Fig. 6*b*. As well as in [10], the curves $R(t)$ are well approximated by exponential dependency $R(t) = R_0 \exp(t/\tau_R)$.

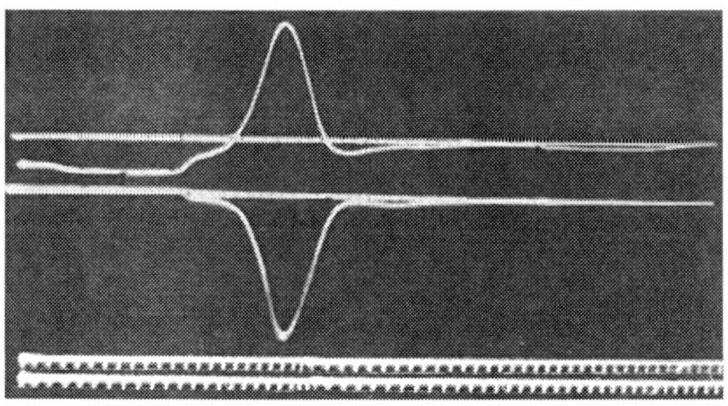

Fig. 5. Oscillogram of the MCG current derivative (the upper beam) and current derivative in the load (lower beam) in the experiment #2. Time marker is in 0.2 µs.

The current characteristic rise time τ in the experiment #2 is ~1.7 times more than in the experiment #1. Thus, in the range i = 0.17-0.3 MA/cm τ is proportional to the current liner density in the foil. In the early experiments [11] with this device at the breaking currents I_k = 3.5-5.5 MA, the weaker current rise time dependency in the load was observed from I_k ($\tau \sim i^{0.25\text{-}0.5}$).

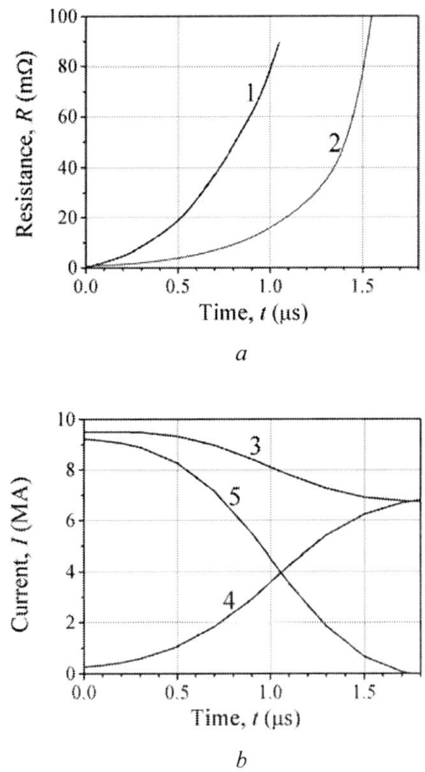

Fig. 6. $R(t)$ dependencies taken from the experiments #1 (1) and #2 (2) (*a*), and also current dependencies in MCG (3), in the load (4), and in the opening switch (5) in the experiment #2 (*b*)

Using the more thin foil leads to some reduction of the load current rise time.

The experiments #4 and #5 were carried out at the high flux density of the electromagnetic energy $p = EH = I_R^2 R/(\pi Dl)$, where E and H – intensities of the electric and magnetic fields in the opening switch; I_R and R – current through the opening switch and its resistance. Just this parameter defines the heat of the conductive layer of the medium in the opening switch and its resistance. High p values can be reached by several methods: raising the breaking current, reducing the foil length or increasing the load inductance. In the experiment #4 with the reduced foil up to 40 mm and in the experiment #5 with the raised load up to 40 nH, p reached the values of 1.5 and 1.6 GW/cm^2. Hereby, the dependencies $R(t)$ of the opening switches were considerably changing (Fig. 7) in comparison with previous experiments. Resistance in these experiments increases to some maximum and then decreases. Perhaps, the value $p \sim 1.5$ GW/cm^2 should be considered critical for this type of EOS.

The possibility of EOS scaling (keeping the current liner density in the foil and parameters of the switching circuit) follows from the comparison of the results of the experiments #1 and #6. EOS sizes (except sizes of RB grooves and ribs, and foil thickness) differed twice, currents in the MCG and in the load also differed approximately twice. At that time characteristics of the current pulse in the load and dependencies $R(t)$ were the same (Fig. 8).

There are results of the experiments #7 and #8 in the Table II, where the possibility of the current pulse sharpening of the helical MCG up to the level of 0.3-0.5 μs is provided. Hereby, the foil length was increased up to (0.65-1) of its width.

The effective work of the opening switch with comparatively thick copper foil (up to 0.3 mm) became possible due to application of the combined barrier with the optimal sizes of ribs and grooves. In the experiments, resistance and its growth rate reach ~100 mΩ and ~7·10^5 Ω/s.

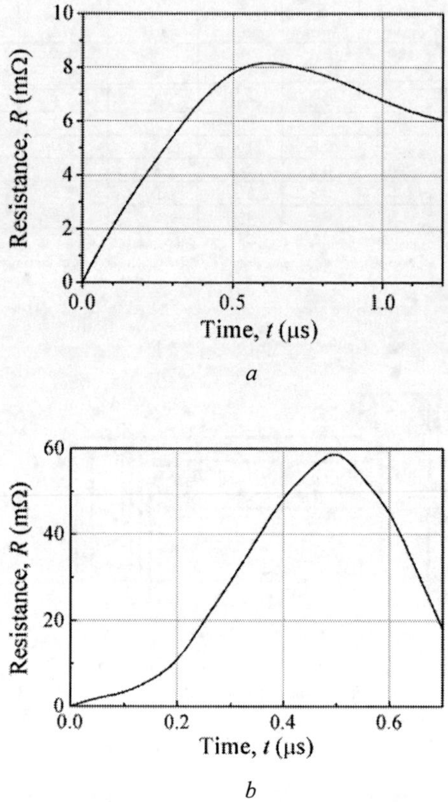

Fig. 7. $R(t)$ dependencies in the experiments #4 (a) and #5 (b)

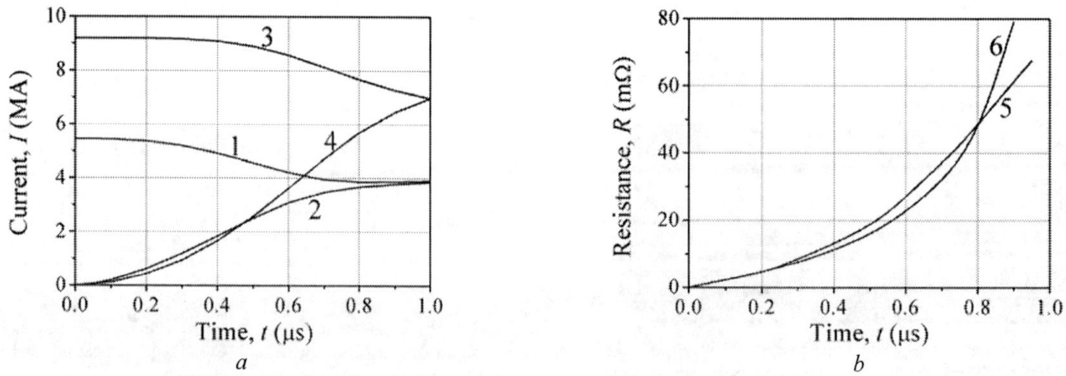

Fig. 8. Time dependencies of the currents (a) and resistance of the opening switches (b) in the experiments #1 and #6. 1, 2, 5 – EMG current, current in the load, and opening switch resistance in the experiment #1; 3, 4, 6 – EMG current, current in the load, and opening switch resistance in the experiment #6

Two kinds of EOS were developed and tested in VNIIEF, aimed to sharpen the current pulses of the disk magneto-cumulative generators (DMCG) and power the liner loads (Z-pinches) by the current of more than 10 MA. They are the one with a foil location on the diameter of 290 mm (EOS290) and the other with the diameter of 560 mm (EOS560). Electro-exploded opening switches [12] are used for the short current pulses formation by the method of DMCG circuit breakage. But in the experiments it is impossible to get the current pulse front less than 1 μs. Besides, electro-exploded opening switches substantially decreases the DMCG energy because of the magnetic flux loss at the heat and vaporization of the exploded conductor.

DMCG scheme with HE-charges of 240 mm diameter (DMCG240) and EOS290 is provided in Fig. 9. The opening switch contains the combined ribbed barrier with an internal diameter of 290 mm and length of 360 mm. The copper foil of 0.26 mm was used in EOS. HE-charge based on HMX had 8 mm thickness. The initiation system of HE-charge is axial.

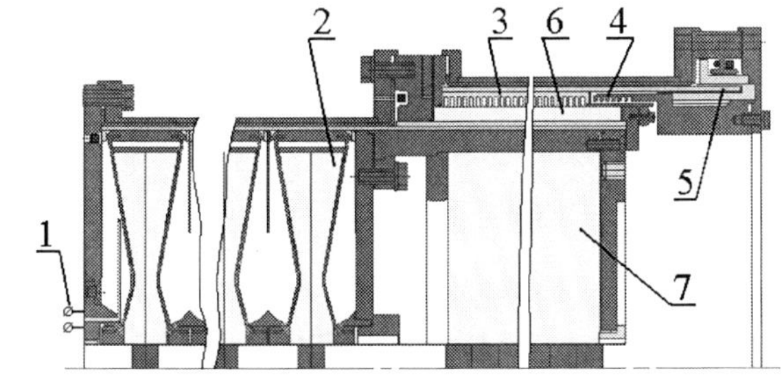

Fig. 9. DMCG240 with EOS290. 1 – electrodes from helical MCG, 2 – ten-elements DMCG, 3 – RB, 4 – closing switch, 5 – load, 6, 7 – HE-charge

By the beginning of the foil breakage, DEMG240-10 provided the current of 18.5 MA in the final circuit with the inductance of ~21 nH at the current powering $I_0 \approx 4.7$ MA from the helical MCG. EOS formed the current pulse of 10 MA, $\tau = 0.55$ μs in the load of 16 nH. Voltage in the opening switch was 320 kV. EOS resistance increases up to 30 mΩ with characteristic time $\tau_R \approx 0.25$ μs. The absorbed energy in the opening switch is 1.9 MJ, flux density of the power is 1.2 GW/cm^2. The peak electric power of EOS is 2 TW.

The opening switch of EOS560 was tested. The goal was to sharp the current of five-element disk generator with HE-charges of 480 mm diameter (DEMG480-5) [13]. The ribbed barrier from Plexiglas with the internal diameter of 560 mm is used in the opening switch, the grooves width is 3 mm, depth is 7 mm, ribs width is 1.5 mm. The ribs are lined by steel rings with a section of 1.5×1.5 mm^2. The breaking conductor is a copper foil of 0.38 mm thickness, 1758 mm width, and 700 mm length. HE-charge thickness is 5.5 mm. The scheme of HE-charge initiation differs from the provided one in Fig. 9. Initiation into the internal surface of HE-charge is performed by the shock wave from the radially scattering aluminum cylinder. HE-charge initiation in the cylinder is axial.

In the experiment, up to the beginning of the foil breakage the disk generator current was 32.5 MA. EOS formed the current pulse of 15 MA in the load 17 nH. The current rise time is $\tau = 0.8$ μs. Voltage in the opening switch reaches 340 kV. Resistance is 18 mΩ ($\tau_R \approx 0.3$ μs). The energy of 5 MJ is dissipated in EOS. The electrical power in the opening switch is 3 TW. Time dependencies of EOS voltage and current in the load are provided in Fig. 10.

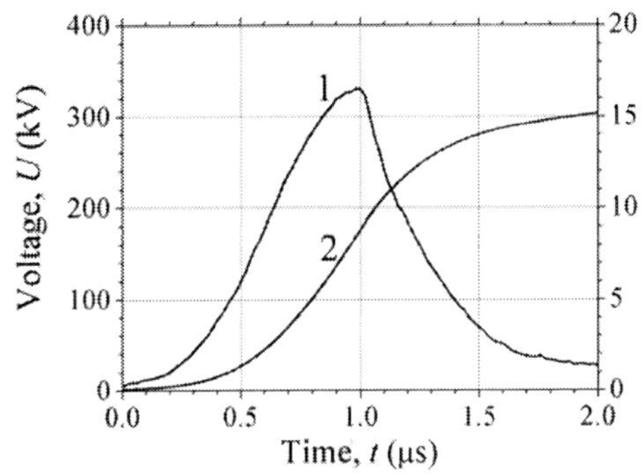

Fig. 10. Time dependencies of the voltage at the opening switch (1) and current in the load (2) in the experiment with DEMG480-5 and EOS560

High-current opening switches with the ribbed barrier got their development in Los-Alamos National Laboratory (USA). In the 80s-90s of the last century, the Procyon system was developed in LANL (J.H. Goforth, et al.). It was intended for Z-pinch loads powering. The system consisted of the helical MCG of Mark-IX type [14], explosive opening switch [15], plasma-flow opening switch (PFOS) [16], and Z-pinch load.

The EOS ribbed barrier is made of fluoroplastic. Its internal diameter is ~300 мм, length is ~760 mm. The groove width is 6.5 mm, depth is 13 mm, and rib width is 1.5 mm. The destructible conductor is the aluminum foil of 0.8 mm thickness. The initiation system of EOS is situated on the device axis. The scheme of EOS powering is so that the magnetic field of the helical generator current presses the foil to the HE-charge unlike the described above constructions that press it to the ribbed barrier.

Reference [17] presents the results of two experiments with the full system of Procyon (without z-pinch). In the first experiment, the generator Mark-IX generated the current of 18.3 MA (magnetic energy is ~12 MJ) into the breaking circuit of 72 nH inductance. EOS formed the current pulse of 13.6 MA with the characteristic rise time $\tau \approx 2.6$ μs. PFOS is its load with the initial inductance of 25 nH. Time dependencies of MCG current and current in the load are provided in Fig. 11. The peak voltage in the opening switch is 130 kV. In the second experiment, an electrical breakdown at the EOS output (in the transmission line to PFOS) took place when MCG current was 20 MA (magnetic energy was 14.4 MJ).

In these experiments, the current linear density in the foil reached 0.2 MA/cm; resistance of the opening switches in both experiments increased up to ~80 mΩ for the time interval of ~4 μs.

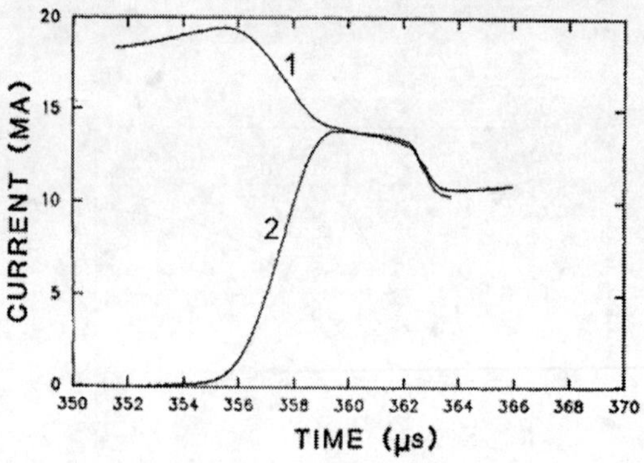

Fig. 11. Time dependencies of MCG current (1) and current in the load (2) in the first experiment with the generator Mark-IX

III. Explosive opening switches with dielectric cumulative jets

In EOS of this type, the effect of cumulative jets implantation into the destructible conductor and jet quencher (made of the dielectric material in the shape of plates or cylinder) is used. The scheme of the opening switch element [18] is presented in Fig. 12. At the axial explosion of the HE-charge 1, the cumulative grooves of the dielectric block 2 are collapsed and the conductor 3 is cut simultaneously in many sections at the device axis by the cumulative jets 4, that are penetrated into the dielectric 5 at the depth z.

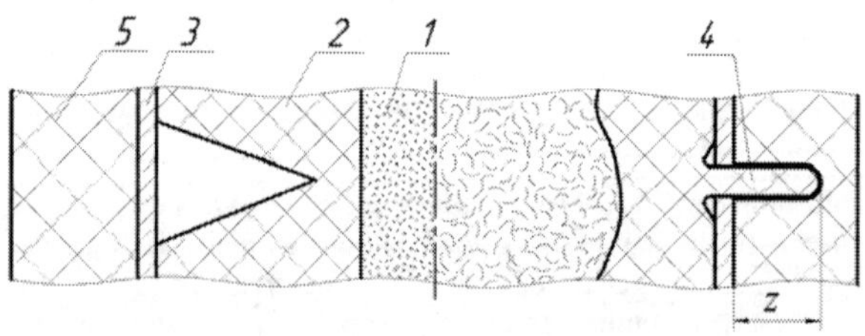

Fig. 12. Scheme of the opening switch element. To the left from the symmetry axis – up to the operation start, to the right – in the process of operation. 1 – HE-charge, 2 – jetformer, 3 – foil, 4 – cumulative dielectric jet, 5 – jetquencher, z – jet length

The investigations revealed that the effective and reliable operation of the opening switch is reached, when the elements 2 and 5 are made of polyethylene, hereby, the angle of the cumulative groove in the jetformer 2 should be 25-30° [19].

X-rays investigations showed that the breakages up to 3.5 mm width are formed in the aluminum plate under the influence of the dielectric jet at the beginning of the current flow in the load. The depth of the jet penetration into the polyethylene reaches 5 mm [8, 19].

Investigations of the current liner density influence in the foil on the EOS resistance with cumulative jet were carried out in the range of i = 120-460 кA/cm [23]. The aluminum destructible conductor of 1 mm thickness was located on the diameter of 200 mm; its length was 600 mm. The EOS resistance dependence on the current liner density in the foil at the moments corresponding to the peak voltage in the opening switch is shown in Fig. 13. This dependence can be approximately expressed as $R = 3.3\ \mathrm{kV{\cdot}cm^{-1}}/i\ \mathrm{кA{\cdot}cm^{-1}}$.

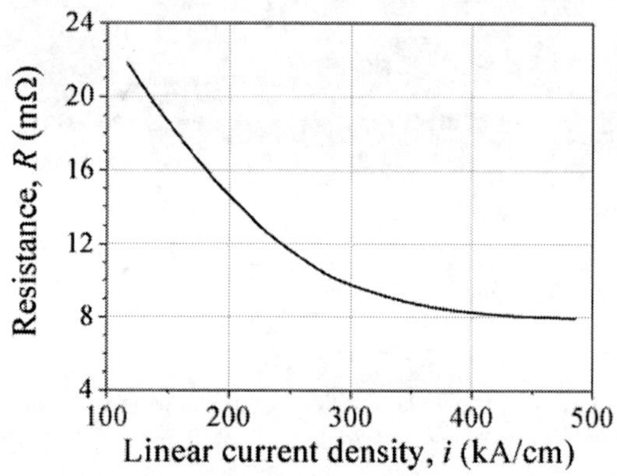

Fig. 13. EOS resistance dependence on the current linear density in the foil at the moments corresponding to the peak voltage

The pulsed power sources based on the helical MCG and EOS with cumulative jet are designed in VNIIEF. They allow generating of the current pulses of 10-15 MA into the load 30-40 nH with the characteristic rise time of 2.5-3.5 μs [19, 21]. MCG of 200 mm and 400 mm diameters are used in these devices that generate the current up to ~28 MA in the breaking circuit of 50 nH, and the opening switch of 200 mm and 300 mm diameters, respectively, with foil of 600 mm and 800 mm length.

IV. Plasma explosive opening switches

In the device, a very thin cylindrical metal shell turns into a plasma shell at the beginning of MCG powering. This shell is located immediately on the cylindrical surface of the HE-charge. The shock wave going out on the plasma shell and presses it at the end of MCG operation. As a result, a fast growth of the opening switch resistance is observed.

The principal device scheme is provided in Fig. 14 [22, 23]. HE-charge of the opening switch is initiated along the axis. The investigations revealed that the resistance change of the plasma channel practically does not depend on the current value in the range up to 10 MA. The experiments [23] with the external conductor breakage of 200 mm diameter and 120 mm length were carried out with the coaxial storage of ~30 nH inductance at the currents from 2 MA to 10 MA. The current with the maximum value of ~5 MA and rise time from 0.1 up to 0.8 of the maximum value of ~0.45 μs was obtained at the current of the storage of 9 MA in the load of 31 nH; the current derivative in the load was $\sim 1 \cdot 10^{13}$ A/s.

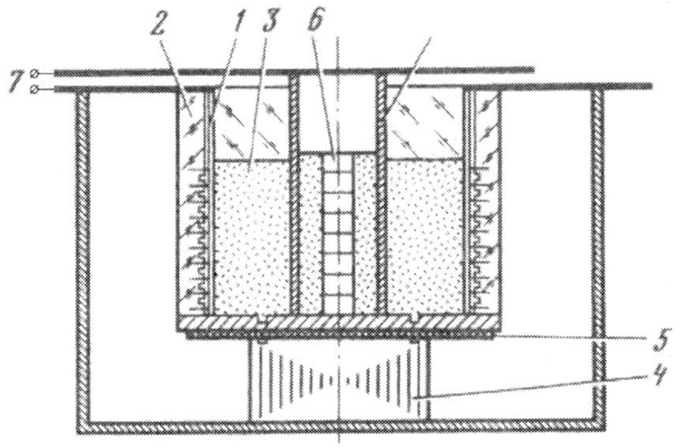

Fig. 14. Scheme of the device that consists of two helical MCG and plasma-explosive opening switch. 1 – cyllindrical plasma channel ignited between the dielectric barrier 2 and HE-charge surface 3. Load 4 is connected in parallel to the obtained channel through the gap-shaper 5, 6 – electrodetonators, 7 – electrodes from MCG

Such opening switches were studied in LANL on the flat models using capacitor banks and MCG [24-26]. The resistance of the opening switch increases approximately 10 times for less than 1 μs in the experiments at $i = 0.16$ MA/cm. The resistance growth decreased fast when i increased up to 0.3 MA/cm and 0.5 MA/cm [24].

A significant disadvantage of such opening switches is that they have high (several mΩ) resistance of the plasma channel at the MCG operation process. It results in severe losses of the magnetic flux and MCG current decrease. Practically, the plasma-explosive opening switches have not been studied and applied for the last 20 years. Their usage is possible for current pulses sharpening of only very fast energy sources.

V. VORTEX OPENING SWITCH

In [27, 28], the opening switch was developed, where the aluminum or copper foil is destroyed at the expense of vortex movement of HE-charge detonation products. The vortex movement was organized near the conductor at the expense of location of circular cross-section elements made of hard metals between the foil and HE-charge (Fig. 15). According to [27], the vortex movement of DP appears at nonstationary flow of elements 4 by the detonation products of HE-charge 2 at both sides from them. The vortex movement of the detonation products near the conductor 1 leads to its stretching, destruction, and intermixing of DP with the conductor materials 1 and base 5; that results in fast growth of the opening switch resistance.

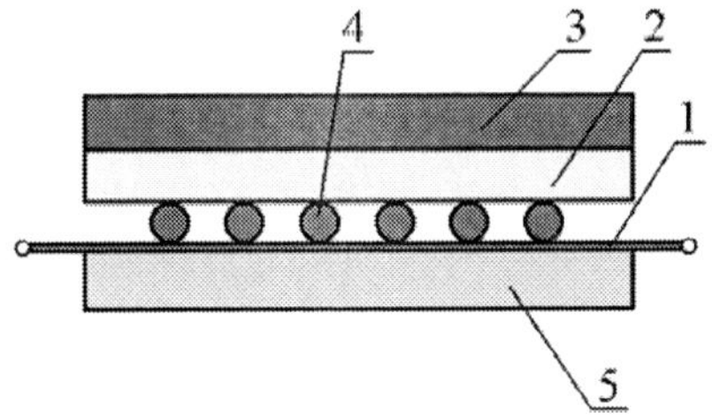

Fig. 15. Flat model of the opening switch. 1 – foil, 2 – HE-charge, 3 – initiation system, 4 – circular cross-section elements, 5 – base

In the experiments on the flat models and with the helical MCG it was shown that the vortex EOS at the current linear density up to 0.2 MA/cm provides the current increase in the load with the time $\tau < 1$ μs. Because of the small thickness of the base (3-4 mm), the vortex opening switch is notable for the low spurious inductance that allows using it for opening of the low inductive energy storage system, i.e. final circuits of the dick MCG with the currents in several tens of megampere.

Reference [29] provides the results of the first experiment of the current pulse sharpening of the disk generator with the help of the vortex EOS. The device scheme is provided in Fig. 16. Two-element disk MCG with HE-charge of 400 mm diameter was used in the experiments. The opening switch was situated above the disk elements. The destructible conductor is a copper foil of 0.2 mm thickness, 1250 mm width, 130 mm length. HE-charge of the opening switch of 3 mm thickness was initiated by shock waves going out on the internal cylindrical surface from HE-charges of the disk MCG. The solenoid located between HE-charges and foil was made with a helix pitch of 3.3 mm of the copper wire of 1.5 mm diameter. The base from foam plastic of 0.15 g/cm^3 density had 3 mm thickness.

The disk generator provided the current of 21 MA in the circuit of 5 nH inductance to the beginning of the foil breakage. The current pulse of ~20 MA with the characteristic rise time of ~0.5 μs was formed by the first microsecond of the EOS operation start time in the load of 1.6 nH. The oscillogram of the current derivative in the load is provided in Fig. 17. The peak voltage in the opening switch is 67 kV, the maximum resistance is 18 mΩ.

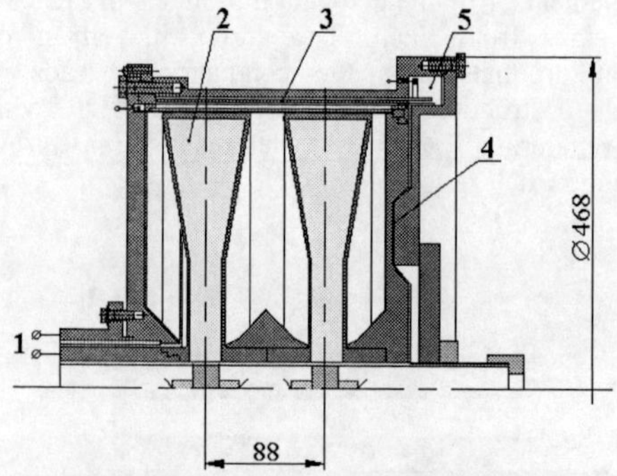

Fig. 16. Disk MCG with low inductive vortex opening switch. 1 – electrodes from helical MCG, 2 – two-element disk MCG, 3 – vortex EOS, 4 – closing switch, 5 – load

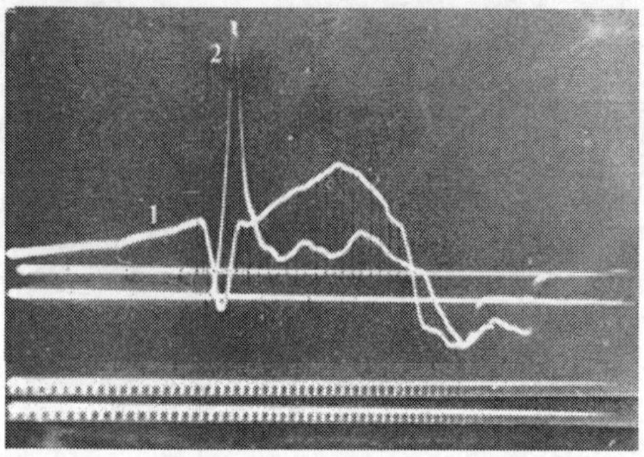

Fig. 17. Oscillogram of DMCG current derivative and current in the load in the experiment with disk MCG and low inductive vortex opening switch. 1 – current derivative of DMCG, $(dI/dt)_{max} = 1.0 \cdot 10^{13}$ A/s, 2 – current derivative in the load, $(dI/dt)_{max} = 4.2 \cdot 10^{13}$ A/s, time marks in 0.2 μs.

VI. Conclusion

It is experimentally shown that EOS with a ribbed barrier at the powering from the helical MCG can form the current pulses with the characteristic rise time of 0.3-0.5 μs in the load of 20-40 nH. The limiting value of the flux density of the power $p = EH$ in EOS with a ribbed barrier is ~1.5 GW/cm^2.

For the first time, the explosive opening switches are applied for the current pulses sharpening of the disk MCG. EOS allows getting of the currents up to 15 MA with the rise time $\tau \leq 0.8$ μs in the real loads according to the inductance value (~20 nH).

References

[1] B.D. Yankovsky, "Pulse voltage generation by explosive opening switches," in Megagauss Field and Pulsed Power Systems, V.M. Titov, G.A. Shvetsov, Eds. N.Y.: Nova Science Publishers, Inc. 1990, pp. 541-548.

[2] P.I. Zubkov, A.M. Kartashov, L.A. Lukyanchikov, et al. "Anomalous behavior of conduction in the detonating wave in trotyl," in Papers of the International conference VII Khariton's topical scientific readings. Sarov: RFNC-VNIIEF, 2005. pp. 58-61.

[3] P.I. Zubkov, L.A. Lukyanchikov, Yu.V. Ryabinin, "Electric strength of flying away detonation products," J. Appl. Mech. Tech. Phys. (JAMTPh), pp. 134-138, No.1, 1976.

[4] A.P. Ershov, P.I. Zubkov, Yu.N. Ilyanovich, et.al. "Current breaking with ark bridging by the conducting detonation products," in Ultra-high magnetic fields. Physics. Technics. Applications, V.M. Titov, G.A. Shvetsov, Eds. Moscow: Science, 1984, pp. 397-401.

[5] V.K. Chernyshev, "Superpower Explosive Magnetic Energy Sources (XXTH Century Resultes. Tasks in the Beginning of XXIST Century)," in Megagauss-10, M. von Ortenberg, Ed. Berlin, 2005, pp. 17-28.

[6] A.G. Ivanov, G.Ya. Karpenko, "About thin plates acceleration by explosion products at different means of HE-charge initiation," Phys. Combust. Explosion, pp. 84-89, No.2, 1980.

[7] Yu. V.Vlasov, V.A. Demidov, V.I. Skokov, "The model of explosive opening switch with ribbed barrier," in Megagauss and megaampere pulse technology and application, V.K. Chernyshov, V.D. Selemir, L.N. Plyashkevich, Eds. Sarov: VNIIEF, 1997, pp. 372-378.

[8] V.K. Chernyshev, L.A. Andreevskikh, G.I. Volkov, et al. "Mechanical conductor disruption character in explosive current commutator," in Megagauss fields and pulsed power systems, V.M. Titov, G.A. Shvetsov, Eds. N.Y.: Nova Science Publishers, Inc. 1990, pp. 533-536.

[9] A.S. Boriskin, Yu.V. Vlasov, V.A. Demidov, S.A. Kazakov, E.V.Shapovalov, "Explosive former of current pulse," RU Patent 2399111, Bulletin of Inventions, No.25, 2010.

[10] T.I. Tucker and B.N. Turman, "Modeling and Scaling Experiments of Explosive Driven Pulse Power Systems," in 3rd IEEE International pulsed power conference, T.H. Martin, A.H. Guenther, Eds. Albuquerque, N.M. 1981, pp. 334-336.

[11] V.K. Chernyshev, V.A. Demidov, V.N. Veselov et al. "Investigation of the Speed Response Dependence of the Explosive Current Opening Switch on Initial Conditions," in Megagauss fields and pulsed power systems, V.M. Titov, G.A. Shvetsov, Eds. N.Y.: Nova Science Publishers, Inc. 1990, pp. 527-531.

[12] V.K. Chernyshev, A.M. Buyko, V.N. Kostyukov, et al. "Investigation of electrically exploded large area foil for current switching," in Megagauss fields and pulsed power systems V.M. Titov, G.A. Shvetsov, Eds. N.Y.: Nova Science Publishers, Inc. 1990, pp. 465-470.

[13] V.D. Selemir, V. A. Demidov, A. S. Boriskin, et al. "Disk Magnetocumulative Generator of 480 mm Diameter for Explosive EMIR Facility," IEEE Transactions on Plasma Science, vol. 38, No.8, pp.1762-1767, August 2010.

[14] C.M. Fowler, R.S. Caird, "The Mark IX generator," in 7th IEEE International pulsed power conference, B.H. Bernstein, J.P. Shannon, Eds. Monterey, California, 1989, pp. 475-478.

[15] J.H. Goforth and S.P. Marsh, "Explosively Formed Fuse Opening Switches for Use in Flux-Compression Generator Circuits," in Megagauss fields and pulsed power systems, V.M. Titov, G.A. Shvetsov, Eds. N.Y.: Nova Science Publishers, Inc. 1990, pp.515-526.

[16] W.L. Baker, J.D. Beason, J.H. Degnan et al. "Plasma Flow Switch Driven Implosions," in Megagauss fields and pulsed power systems, V.M. Titov, G.A. Shvetsov, Eds. N.Y.: Nova Science Publishers, Inc. 1990, pp. 615-622.

[17] J.H. Goforth, H. Oona, J.H. Brownell et al. "Procyon Experiments Utilizing Explosively-Formed Fuss Opening Switches," in 8th IEEE International pulsed power conference, R. White, K. Prestwich, Eds. San Diego, California, 1991, pp. 273-276.

[18] Yu.A. Kotov, A.P. Tsyplenkov, "Investigation of explosive opening switch with cumulative dielectric jets," in Ultrahigh magnetic fields. Physics. Technology. Application, V.M.Titov, G.A.Shvetsov, Eds. Moscow: Science, 1984, pp. 402-405.

[19] G.I. Volkov, P.V. Dudai, V.A. Ivanov, "Main types of explosive opening switches and their switching characteristics," in Megagauss-12, G.A. Shvetsov, Ed. Novosibirsk, 2011, pp.408-419.

[20] V.K. Chernyshev, G.I. Volkov, V.A. Ivanov, et al. "Experimental investigation of explosive opening switch operation," in Megagauss magnetic field generation and pulsed power applications, M. Cowan, R.B. Spilman, Eds. N.Y.: Nova Science Publishers, Inc. 1994, pp. 731-738.

[21] V.K. Chernyshev, V.V. Avdoshin, G.I. Volkov, et al. "Helical explosive-magnetic sources with stored energy 0.5-25 MJ for investigation of gas ponderomotive unit operation with MCG nozzle and acceleration dinamics of solid state liner," in Megagauss and megaampere pulse technology and application, V.K. Chernyshev, V.D. Selemir, L.N. Plyashkevich, Eds. Sarov: VNIIEF, 1997, pp. 279-281.

[22] A.I. Pavlovskiy, V.A. Vasyukov, A.S. Russkov. "Formation of fastgrowing megaampere current pulses from magnetocumulative generators," Letters to JTP, 1977, vol. 3, No. 16, pp. 789-792.

[23] A.I. Pavlovskiy, V.A. Vasyukov, N.F. Popkov, et al. "Investigation of storage and switching of high densities of energy," in Ultrahigh magnetic fields. Physics. Technology. Application, V.M. Titov, G.A. Shvetsov, Eds. Moscow: Science, 1984, pp. 410-415.

[24] J.H. Goforth, R.S. Caird, R.F. Benjamin, et al. "Fast explosive-driven opening switches carrying high linear current densities," in Ultrahigh magnetic fields. Physics. Technology. Application, V.M. Titov, G.A. Shvetsov, Eds. Moscow: Science, 1984, pp. 388-394.

[25] J.H. Goforth, A.E. Greene, "Results from plasma compression opening switch experiments," in Megagauss technology and pulsed power applications, C.M. Fowler, R.S. Caird, D.J. Erickson, Eds. N.Y.-L.: Plenum Press, 1987, pp. 513-520.

[26] J.H. Goforth, R.S. Caird, "Experimental investigation of explosive-driven plasma compression opening switches," in 4th IEEE International pulsed power conference, T.H.Martin, M.F. Rose, Eds. Albuquerque, N.M. 1983, pp. 786-789.

[27] Yu.V. Vlasov, V.A. Demidov, E.E. Meshkov, V.A.Shevtsov, "Explosive opening switch of electric circuit," Author's certificate USSR 1598750 MKI H01 H39/00. Priority 07.06.88. Bulletin of Inventions, No. 23, 1999, p. 332.

[28] Yu.V. Vlasov, V.A. Demidov, T.A. Egorova, et al. "Investigation of a vortex current opening switch having streamlined components between a conductor and an explosive charge," in Megagauss-11, I. Smith, B. Novac, Eds. London. 2007. pp. 192-194.

[29] V.A. Demidov, A.N. Demin, S.A. Kazakov, et al. "Disk Explosive Magnetic Generator with Low Risetime in the Load," in Proceedings of the 12th International Conference on High Power Particle Beams, M. Markovits and J. Shiloh, Ed. Haifa, Israel, 1998, pp. 588-590.

TEST OF DEVICE BASED ON DISK MAGNETO-CUMULATIVE GENERATOR DMCG480 WITH EXPLOSIVE CURRENT OPENING SWITCH

S.N. Golosov, V.A. Demidov, A.S. Boriskin, S.A. Kazakov, O.M. Tatsenko, Yu.V. Vlasov, A.P. Romanov, A.V. Filippov, E.A. Bychkova, A.N. Moiseenko, E.I. Schetnikov, V.A. Yanenko, S.V. Kutumov, N.R. Kazakova, S.I. Volodchenkov, R.M. Garipov, R.R. Yusupov, A.S. Sevastyanov, V.V. Kostin

Russian Federal Nuclear Center – All-Russian Research Institute of Experimental Physics Sarov, Russia

One of the perspective directions in high energy density physics is the creation of powerful electro physical facilities capable to form megampere current pulses with short front. This paper presents test results of the device based on 5-element disk magneto-cumulative generator DMCG480 and explosive current opening switch. The current pulse of 15 MA with characteristic rise time of 0.8 μs was obtained in the equivalent multi-wire array load of 17 nH at circuit breaking with the current of 32 MA.

Key words: disk magneto-cumulative generator, explosive current opening switch, preamplifier, current characteristic rise time

I. INTRODUCTION

The currents of tens megamperes with the rise time at least 0.5 μs are in the liner experiments on the soft X-ray radiation generation [1, 2]. One of the possibilities to generate such pulses is sharpening of the MCG disk generators current (DMCG) using the explosive current opening switches. Earlier, electro-explosive opening switches were used to break the circuit of the DMCG; but in these experiments the current pulse front was higher than 1 μs.

The results of the first experiments on current sharpening of the disk MCG using EOS are presented in [4,5]. The disk generators with the HE-charges of 240 mm in diameter and the opening switches with the ribbed barrier (internal diameter of the barrier was 290 mm) were used in these generators. The current of ~10 MA with the characteristic current rise time of ~0.6 μs (from 0.1 through 0.9 of maximum value) was obtained in the load of 16 nH.

This paper presents experimental results of the device based on 5-elements DMCG480 [6,7] and explosive current opening switch with the foil, located on the diameter of 560 mm (DMCG480-5 – EOS560). Unlike [4,5], linear sizes of the DMCG elements and main sizes of the EOS are increased twice.

II. DEVICE DESCRIPTION

The device includes a preamplifier – high inductive helical MCG-240, DMCG480 with five profiled disk elements, and the EOS. The device load is equivalent in inductance to the multiwire liner.

The device design is presented in Fig. 1. The helical MCG is not presented in the figure.

The preamplifier solenoid internal diameter is 240 mm, its length is 1200 mm. The copper central tube of MCG is cylindrically conical. The HE charge of ~21 kg [8] is fixed inside the central tube.

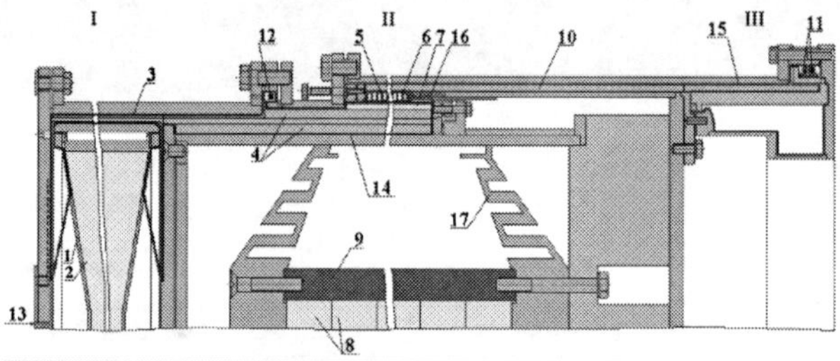

Fig. 1. A device sketch of DMCG 480-5-EOS560. I – DMCG, II – EOS, III – load, 1 – disk shell, 2 – disk HE-charge, 3 – bushing insulator of the transmitting line between DMCG and EOS, 4 – EOS insulator, 5 – EOS foil, 6 – ribbed barrier, 7 – EOS HE-charge, 8 – EOS HE-intermediate charge, 9 – aluminium shell, 10 – transmitting line insulator to the load, 11 – inductive sensors D3 and magneto-optical sensor MO-L in the load circuit, 12 – inductive sensors D2 and magneto-optical sensor MO-D in the DMCG circuit, 13 – preamplifier exit, 14 – DMCG circuit at the moment of a current switching start, 15 –circuit load, 16 – closing switch, 17 – support

The disk MCG consists of 5 sections with HE-charges of 480 mm diameter and ~80 kg weight. HE-charges initiation into the DMCG is provided by the axial initiation device containing 10 electrical detonators (two for each HE-charge) that are electrically connected in series and located along the device axis with 50 mm step. The initial generator inductance is equal to 108 nH.

A ribbed barrier is made of plexiglass with edges covered by steel rings of 1 mm thickness. A ribbed barrier width is 1.5 mm, grooves width is 3 mm, and grooves depth is 7 mm. A destructible conductor is made of the copper foil of 0.38 mm thickness, 1758 mm width and 700 mm length located on the diameter of 560 mm.

Thickness of a plastic high explosive charge is 5.5 mm. A high explosive charge is initiated by a shock wave moving to its inner surface. An intermediate HE-charge is used to create a shock wave. Twenty two flat discs of 240 mm diameter and 40 mm thickness with a central hole of 27 mm diameter are used as an intermediate HE-charge. The intermediate HE-charge initiation is provided by a device of an axial initiation containing 22 electrical detonators that are electrically connected in series and located along the device axis with 40 mm step. Detonation products of the intermediate HE-charge accelerate an aluminum cover of 12.5 mm thickness. An aluminum cover accelerated by the detonation products of HE-charge influences the steel case with a wall thickness of 8.5 mm and copper electric conductor of 1.5 mm thickness. Then a shock wave goes through a multilayer insulator of 7 mm thickness made of a polyethylene terephthalate film and polyethylene insulator of 7 mm thickness. Ribbed supports are used to remove an early damage of the EOS insulator under the influence of the shock wave from the intermediate HE-charge that propagates on the supports ahead of the aluminum cover dispersion.

A load circuit inductance of the EOS is defined by the transmission line inductance from EOS (8 nH) and the inductance equivalent to the multiwire liner (9 nH) and makes up 17 nH. The internal load cavity is filled with transformer oil. The construction of the input load insulator is identical to the input insulator of the liner chamber.

Calculations show that it is necessary to break the circuit with the foil of 0.38 mm in ~25 μs after the start of the DMCG operation to get the impulse of 14-17 MA in the inductive load ~17 nH (with the linear current density in the opening switch ~200 kA/cm). The disk generator should be powered by 6-6.5 MA current, i.e. the initial magnetic flux has to be 2.1-2.2 Wb in the preamplifier and the current in the disk generator ~30 MA. Peak calculated voltage created at the EOS operation is ~450 кV.

III. Testing Results

The measurement of the current derivative in the range of 10^9-10^{13} A/s was provided by the inductive sensors with a different sensitivity. The sensors were evenly circumferentially-spaced: in the energy preamplifier in D1 cross-section, in the disk MCG – D2, in EOS load – D3 (Fig.1). Magneto-optical sensors working on Faraday effect (MO-D and MO-L) were used to measure the current in the disk MCG and in the load. Optical sensors were made of optical fiber. A semiconductor laser with 655 nm wavelength was used as a light resource. Signals were recorded with oscilloscopes.

Signals from all the inductive sensors were registered in the experiment. The units triggering the electric detonators are switched on at setup time values. Averaged time dependences of the current derivatives, EOS current and resistance were created in the result of data analysis; they are presented in Fig. 2.

The energy preamplifier with the initial inductance of 110 μH is powered with the current of 20 kA that corresponds to magnetic flux of ~2.2 Wb. At the start of DMCG operation the peak current in the preamplifier was ~6.6 MA with the characteristic rise time ~34 μs. Energy stored by the preamplifier was ~2.6 MJ.

Magnetic flux compression in the disk MCG started in the calculated time moment, i.e. in 5.6 μs after the current pulse was delivered to electrical detonators. The DMCG operated in accordance with the calculations (Fig.3). A circuit inductance of the DMCG to the beginning of its breaking was ~22 nH, current ~32 MA. Current characteristic rise time was ~9.6 μs. The current derivative peak value of DMCG was ~$3.8 \cdot 10^{12}$ A/s. DMCG operation time from the beginning of disks movement till the current peak value in the generator was 25.5 μs.

EOS foil acceleration in the grooves of the ribbed barrier was ~0.5 μs. The peak linear density of the current through EOS foil was 185 кА/см.

A fast current derivative rise in the load started after the foil fracture. The current pulse of ~15 MA in the load of 17 nH inductance was formed for 1.5 μs. Measurement error by optical and inductive methods is ±0.8 MA. Characteristic current rise time in the load (between the amplitude levels of 0.1 and 0.9) is 0.8 μs. The peak current derivative in the load is $2 \cdot 10^{13}$ A/s that corresponds to ~330 кV on the EOS. Electric power of the EOS is 3TW, the energy of 5 MJ is dissipated in the opening switch.

The authors pay attention to the characteristic current derivative decrease in the load registered by the inductive sensors D3 after ~0.9 μs from the opening start (Fig.2b). At this moment one can see the discrepancy between inductive and optical methods (Fig.4). Perhaps, the inductive sensors being more sensitive than magnetooptical ones recorded the conductance increase in the detonation products along the ribbed barrier; this reflects in the EOS resistance decrease and remainder appearance of ~8 MA between the stationary DMCG current values and current in the load (Fig.5).

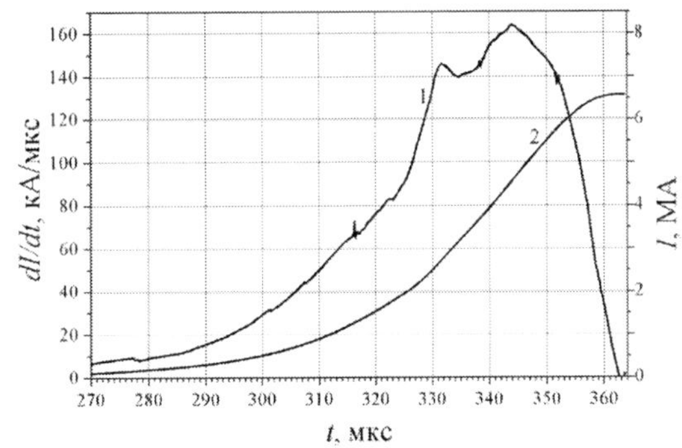

a) current derivative (1) and current (2) in the preamplifier

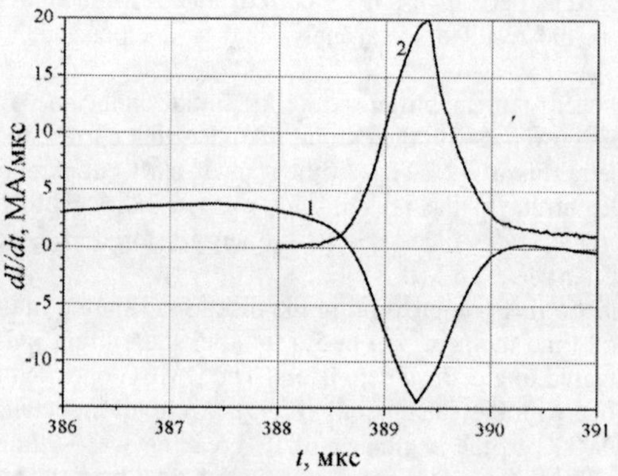

b) current derivative in the DMCG (1) and in the load (2)

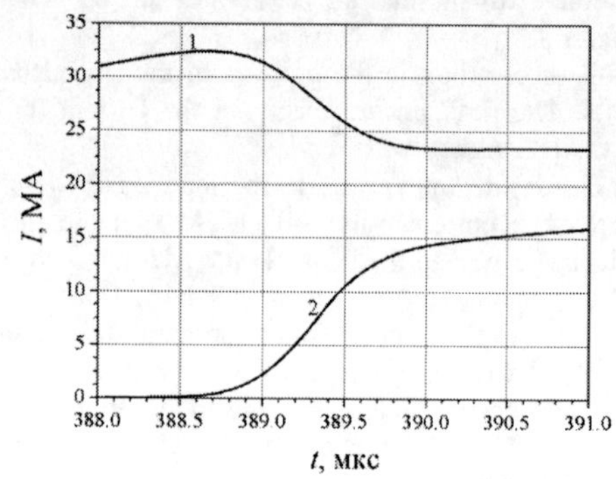

c) average current in the DMCG (1) and in the load (2)

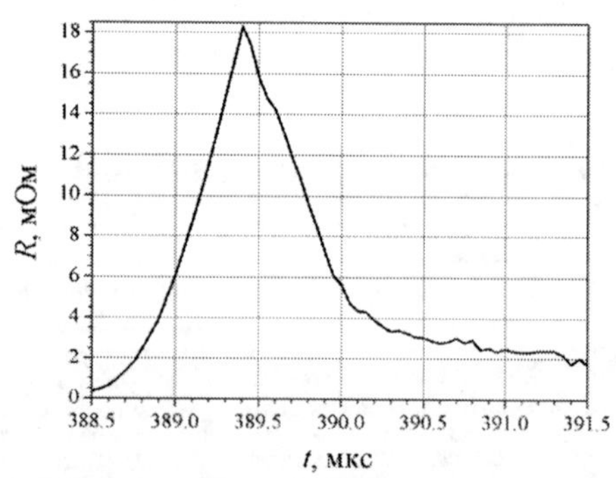

d) EOS resistance

Fig.2 –inductive load testing results of DMCG480-5 with EOS560

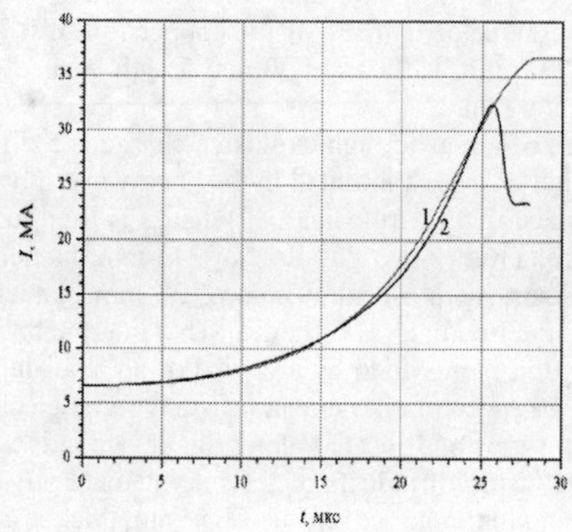

Fig.3. Calculated (at the operation into the load of 18.6 nH inductance) (1) and DMCG experimental current (2)

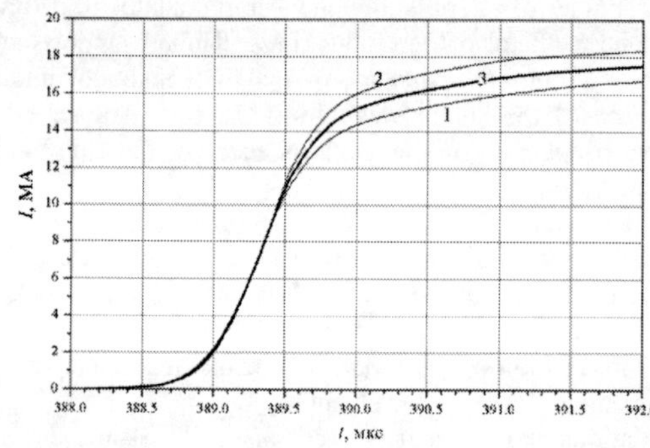

Fig. 4. Current in the load measured with inductive (1) and optical methods (2). 3 – avarage value

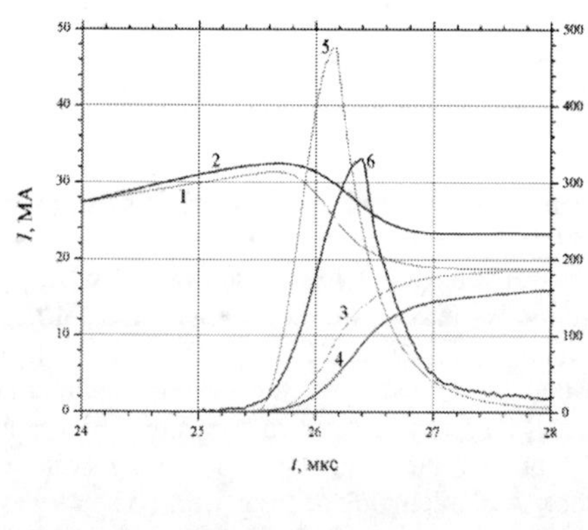

Fig. 5. Comparison with calculation at the conditions realized in the experiment. 1, 2 – DMCG current, 3, 4 – current in the load, 5, 6 – voltage in EOS. Uneven curves – calculation, even curves – experiment

The conductance appearance of the detonation products can be explained if to suppose the following fact. The thermal energy in the destructible conductor is released unevenly because of the time difference of the shock wave exit on the foil surface (such time difference could be realized because of the large square initiation (~1.2 m^2) and the initiation system application with the electrical detonators location step of 40 mm). This leads to appearance of the heat specific capacity higher than the critical value (1.5 GW/cm^2) [9] at different points of the destructive foil, plasma temperature rise at these points and hereby its conductivity increase. Thus, it is necessary to pay special attention to the diversification decrease of the initiation system operation of the explosive charge while developing large-scale EOS.

IV. CONCLUSION

Testing of the explosive current opening switch with a ribbed barrier at its five-element DMCG 480 powering was performed. Measurement results of the device electric characteristics obtained with inductive and magnetooptical sensors match each other.

The current ~32 MA is registered to the beginning of the explosive opening switch operation in the electric circuit of the disk generator with the inductance ~22 nH. Current characteristic rise time in the disk MCG was ~9.6 μs, peak current derivative was $3.8 \cdot 10^{12}$ A/s. Current pulse parameters correspond to the initial powering of the DMCG (I_0 = 6.6 MA).

The explosive opening switch operation at the linear current density through the foil ~200 кA/cm was studied. The current pulse of ~15 MA amplitude with current characteristic rise time between the levels of 0.1 and 0.9 from the peak value of ~0.8 μs is formed in the circuit of ~17 nH inductance by the explosive opening switch. The peak current derivative in the load is $2 \cdot 10^{13}$ A/s, peak voltage is 330kV. Electric power of the EOS is 3 TW, the energy of 5 MJ is dissipated in the opening switch.

The authors showed a possibility of pulse sharpening of the disk MCG using the explosive opening switch with explosive charge initiation by the shock wave. The shock wave is created at the acceleration of a cylindrical liner by the detonation products of the HE-charge. Perhaps, some difference of the experimental current in the load from the calculated one is associated with the conductance increase in the detonation products along the ribbed barrier because of the diversification rise of the shock wave exit on the foil surface. This reflected in the EOS resistance decrease and appearance of ~8 MA difference between the fixed DMCG current value and current in the load. It is noted that it is necessary to pay special attention to diversification decrease of the explosive charge initiation system triggering at large-scale EOS development.

References

[1]. Spielman R. B., Deeney C., Chandler G. A., et al. Tungsten wire-array Z-pinch experiments at 200 TW and 2 MJ // Phys. Plasmas. 1998. Vol.5, N 5. P. 2105–2111.

[2]. Selemir V.D., Demidov V. A., Repin P. B. Explosive Electrophysical Complex EMIR: Current State and Perspectives // IEEE Transactions on Plasma Science. 2010. Vol. 38, N 8. P. 1754-1757.

[3]. V.K. Chernyshev, A.M. Buyko, V.N. Kostyukov, et al. Investigation of electrically exploded large area foil for current switching // Megagauss fields and pulsed power systems // V.M. Titov, G.A. Shvetsov, Eds. N.Y.: Nova Science Publishers, Inc. 1990. P. 465-470.

[4]. Demidov V. A., Boriskin A. S, Kazakov S. A., et al. Disk Magneto-Cumulative generator with explosive current opening switch // IEEE Transactions on Plasma Science. 2010. Vol. 38, N 8. P. 1768-1772.

[5]. Demidov V. A., Boriskin A. S, Kazakov S. A., et al. Disk magneto-cumulative generator with explosive opening switch // PMTP 2015, №1. P. 24-30. (In Russian)

[6]. Demidov V.A., Selemir V.D., Boriskin A.S., et al. Test of disk magneto-cumulative generator of diameter 480 mm at increased powering // Proceedings of the 13-th International Conference on Megagauss Magnetic Field Generation and Related Topics. Suzhou, 6-10 July 2010 / Ed. by Sun Chengwei and Liu Changli. China Academy of Engineering Physics Institute of Fluid Physics, CAEP. 2010. P. 323-330.

[7]. Selemir V. D., Demidov V. A., Boriskin A. S. et al. Disk Magnetocumulative Generator of 480 mm Diameter for Explosive EMIR Facility // IEEE Transactions on Plasma Science. 2010. Vol. 38, N 8. P. 1762-1767.

[8]. Magneto-cumulative generators – pulsed energy resources: Monograph. In 2 volumes./ Eds.: Demidov V.A., Plyashkevich L.N., Selemir V.D. Sarov: RFNC-VNIIEF. 2012. Vol. 1. P.439 (In Russian)

[9]. Patent of Russian Federation RU 2 399 111 C1 MPK HO1H 39/00. Explosive current pulse shaper / Boriskin A.S., Vlasov Y.V., Demidov V.A., Kazakov S.A., E.V. Shapovalov// Discoveries. Inventions. 2010. №25. (in Russian)

[10]. Boriskin A. S., Golosov S. N., Demidov V. A., et al. Pulsed high-current power suppliers based on helical explosive magnetic generators with explosive current opening switches // Megagauss XI. I. R. Smith and B. M. Novac, Eds. London. 2007. P. 199-203.

Impulse breakdown characteristics of propylene carbonate modified by nano-particles

Yanpan Hou, Jiande Zhang, Zicheng Zhang*, Dongqun Chen, Zhen Wang, Hongwei Liu
College of Optoelectronic Science and Engineering
National University of Defense Technology
Changsha, China

***Abstract*—Propylene carbonate (PC) shows good prospects as an energy storage medium in the compact pulsed power sources due to its advantages of a great dielectric constant, high electrical breakdown stress and broad environmental operating ability. In this paper, pulsed power source generating microseconds pulse and the test cell with spherical electrodes were used in the dielectric breakdown experiment. The breakdown properties of PC and PC-based nano-fluids were investigated through changing high-voltage electrode potential increasing rate and how the nano-particles influence the dielectric breakdown of base liquid was analyzed. The results show that for a small high-voltage electrode potential increasing rate, the addition of nano-particles would lead to greatly improved breakdown stability of base liquid; for a large high-voltage electrode potential increasing rate, NFs exhibited much larger mean breakdown voltage. This modification effect was discussed with the energy band theory.**

Keywords—*nano-fluids; dielectric breakdown; propylene carbonate*

I. Introduction

Industrial applications have stimulated intense interest in pulsed power technology towards high average power and compact structure [1-3]. Water is frequently used as energy storage medium in the compact pulsed power source system. Since it characterizes a large dielectric constant and relatively high breakdown stress, the overall size of the pulsed power system can be minimized [4, 5]. Nevertheless, the freezing temperature of 0 ℃ prohibits its use in low temperature environments. In addition, a small fraction of water molecules are always dissociated into H^+ and OH^- and these ions would lead to a residual conductivity of 5.5×10^{-6} S/m, even for highly purified water. Hence, water/ethylene glycol mixtures have been used as storage medium in the compact pulsed power system with the features of higher resistivity and lower freezing point. However, the undesirable viscosity is increasing, so the gases dissolved in the liquid volume or adsorbed on the container surfaces are extremely difficult to be removed. What's worse, the relatively high resistivity of water/ethylene glycol mixtures cannot be maintained for a long time. Therefore, for usage in pulse forming line, that liquid has to be circulated constantly through a deionization filter in order to maintain the required resistivity for microseconds charging.

Propylene carbonate (PC) possesses a large relative permittivity (65), low freezing temperature (-49℃), high resistivity (almost 10^9-10^{10} Ω•m), low viscosity and relatively high chemical stability without chemically reacted with acid gases dissolved in the liquid [6, 7], which can hold the pulse forming line with large resistance for longer periods of time. What is more, Xiao has demonstrated that PC has 1.5 times higher breakdown strength than that of deionized water under 200 ns pulse duration [8].

Recently, various favorable results in electric engineering field have shown that the dielectric properties of insulation oil can largely be enhanced by adding nano-particle suspensions [9-11]. In this study, we have investigated the effect of Al nano-particles on breakdown characteristics of PC and discussed the experimental results with the energy band theory.

II. Experimental Setup

Al nano-particles of average diameter 30 nm were obtained from Deco Island Gold Ltd. and a two-step procedure was used to disperse the nano-particles into PC. Firstly, the nano-particles were vacuum dried at 100 ℃ for 24 h. Then the nano-particles were dispersed into PC with a blender mixing for 1 h and with an ultrasonic machine for 30 minutes' sonication treatment. Finally, the mixtures were placed into an agate jar milling for 48 h and then uniform PC-based nano-fluids (NFs) with 0.1 vol% were obtained.

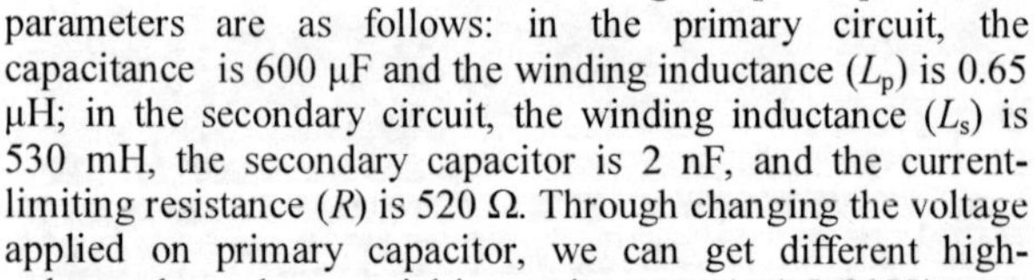

Fig. 1. Dielectric breakdown measurement device.

The circuit diagram of the dielectric breakdown measurement device is shown in Fig. 1 [12-16] and the parameters are as follows: in the primary circuit, the capacitance is 600 μF and the winding inductance (L_p) is 0.65 μH; in the secondary circuit, the winding inductance (L_s) is 530 mH, the secondary capacitor is 2 nF, and the current-limiting resistance (R) is 520 Ω. Through changing the voltage applied on primary capacitor, we can get different high-voltage electrode potential increasing rate, A=1.5-6 kV/μs, on

*E-mail address: zczhang@nudt.edu.cn

the test cell. The gap between the brass spherical electrodes with 26 mm radius was set to be 1 mm during testing.

III. Experimental Results and Discussions

Fig. 2 shows the breakdown voltage plotted as a function of high-voltage electrode potential increasing rate. For each case, twenty times breakdown were conducted and the test samples were well stirred between every breakdown process.

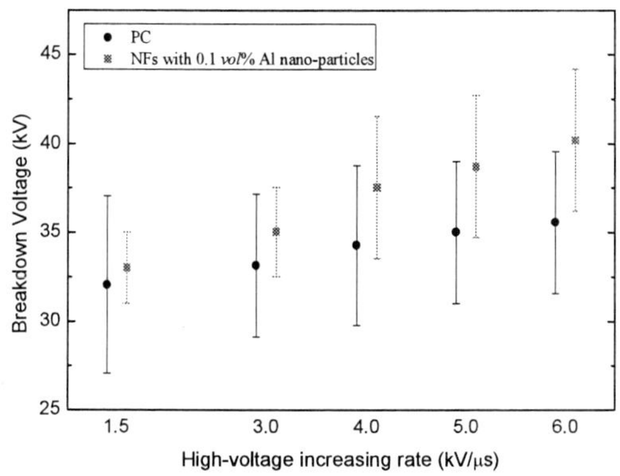

Fig. 2. Breakdown voltage plotted as a function of high-voltage electrode potential increasing rate.

Generally, increasing the value of high-voltage electrode potential increasing rate leads to a marked increase of the dielectric strength, and the breakdown voltage improvements of NFs with larger A are better. It is evident that for NFs, the characteristic breakdown voltages are increased and meanwhile the degree of scatter of breakdown data is improved compared with that of PC for A≤3 kV/μs, which means adding Al nano-particles can increase the breakdown stability of the base liquid. In this case, NFs can be considered as a better candidate than pure PC and the reliability of the pulsed power system can also be enhanced. On the other hand, compared with that of PC, NFs possesses higher breakdown voltage with nearly the same breakdown stability, for A=4 kV/μs and the characteristic breakdown voltages of NFs are largely enhanced with some decrease in breakdown stability for A=6 kV/μs.

We suppose Al nano-particles in PC are arranged in a simple cubic lattice disposition, so interface area of particle and matrix in NFs can be calculated [21]. Under the conditions of particle volume concentration 0.1 vol%, the total interface area in unit volume is 0.2 km^2/m^3. The interaction region with finite mesoscopic thickness would be formed in the interface where the particle and matrix interpose mutually and becomes less ordered from the nano-particle to the matrix and there are large quantities of chemical defects in this area. The energy band theory has been utilized by some researchers to explain electronic processes in liquid [17-20].

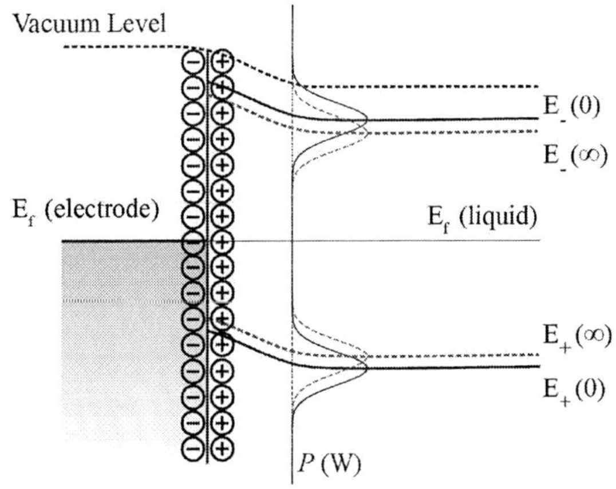

Fig. 3. The electronic equilibrium conditions at the electrode-PC (a) and electrode-NFs interface (b).

Fig. 3 illustrates the electronic energy states for electrode-PC interface. Specifically, the moment when an electron is localized at a PC molecular site, it corresponds to $E_-(0)$. On the other hand, it corresponds to $E_-(\infty)$, the fully localized state, when the surrounding molecular complete reorganization. Similarly, the positive PC ions have delocalized states, $E_+(0)$, and fully localized state, $E_+(\infty)$. Compared with PC, the density of trap sites in NFs is greatly increased due to the large quantities of physical and chemical defects in the interface between nano-particles and the matrix [22, 23]. The chemical defects in the interaction region have inherent dipole moments, which would create trap sites and capture electric charge carriers in the liquid [24, 25]. Due to this capturing effect, the charges trapped and accumulated near the electrodes can weaken the electric field strength in the vicinity of electrodes and reduce field emission currents in NFs effectively. On the other hand, the charge carriers' effective mean free path can be greatly shortened by the scattering effect of nano-particles, which will enhance the dielectric performances of NFs.

IV. Acknowledgments:

This work was supported by the National Natural Science Foundation of China under Grant No. 51307175 and No. 51677190.

V. Conclusions

In this paper, we have dispersed Al nano-particles with diameter of 30 nm into PC, investigated the dielectric breakdown performance of PC-based NFs with high-voltage electrode potential increasing rate, A=1.5-6 kV/μs, and demonstrated that NFs consisting of Al nano-particles in PC exhibited enhanced dielectric properties. Specifically, NFs shows a distinct increase in breakdown voltage when A=4-6 kV/μs and a better breakdown stability when A≤3 kV/μs, compared with pure PC. The main modification effect on dielectric breakdown of NFs may lie in the capture of electric charge carriers by nano-particles and the essential attributes of electron traps have been discussed by electronic energy bands theory.

VI. REFERENCES

[1] Z. C. Zhang, J. D. Zhang, B. L. Qian, C. B. Liu, T. Xun, H. Zhang, and B. Liang, "Compact Rep-Rate GW Pulsed Generator Based on Forming Line With Built-In High-Coupling Transformer," *IEEE Transactions on Plasma Science*, vol. 42, pp. 241-248, 2014.

[2] L. Wang, J. Liu and J. Feng, "A compact 100 kV high voltage glycol capacitor," *Review of Scientific Instruments*, vol. 86, p. 014701, 2015.

[3] W. C. Nunnally, "Critical Component Requirements for Compact Pulse Power System Architectures," *IEEE Transactions on Plasma Science*, vol. 33, pp. 1262-1267, 2005.

[4] Z. C. Zhang, J. D. Zhang and J. H. Yang, "Experimental Study on High Electrical Breakdown of Water Dielectric," *Plasma Science and Technology*, vol. 7, pp. 3161 - 3165, 2005.

[5] B. F. Bolund, M. Berglund and H. Bernhoff, "Dielectric study of water/methanol mixtures for use in pulsed-power water capacitors," *Journal of Applied Physics*, vol. 93, p. 2895, 2003.

[6] R. S. Clark, D. L. Green, M. T. Buttram, R. Lawson, and G. J. Rohwein, "Studies on the use of propylene carbonate as a high-voltage insulator," in *Power Modulator Symposium, 1988. IEEE Conference Record of the 1988 Eighteenth*, Hilton Head, SC, 1988, pp. 381-384.

[7] L. J. Jorat, G. A. Noyel and J. R. Huck, "Dielectric study of propylene carbonate/toluene mixtures and dipole moment of supercooled propylene carbonate," *IEEE Transactions on Electrical Insulation*, vol. 26, pp. 763 - 769, 1991.

[8] X. Shu, J. F. Kolb, M. A. Malik, L. XinPei, M. Laroussi, R. P. Joshi, E. Schamiloglu, and K. H. Schoenbach, "Electrical Breakdown and Dielectric Recovery of Propylene Carbonate," *IEEE Transactions on Plasma Science*, vol. 34, pp. 1653-1661, 2006-01-01 2006.

[9] J. Lee, W. Lee, S. Lee, and S. Lee, "Positive and negative effects of dielectric breakdown in transformer oil based magnetic fluids," *Materials Research Bulletin*, vol. 47, pp. 2984-2987, 2012.

[10] J. Li, Z. Zhang, P. Zou, S. Grzybowski, and M. Zahn, "Preparation of a vegetable oil-based nanofluid and investigation of its breakdown and dielectric properties," *IEEE Electrical Insulation Magazine*, vol. 28, pp. 43 - 50, 2012.

[11] P. Kopčanský, L. Tomčo, K. Marton, M. Koneracká, I. Potočová, and M. Timko, "The experimental study of the DC dielectric breakdown strength in magnetic fluids," *Journal of Magnetism and Magnetic Materials*, vol. 272-276, pp. 2377-2378, 2004.

[12] Y. Hou, Z. Zhang, J. Zhang, Z. Liu, and Z. Song, "Effect of BaTiO3 nano-particles on breakdown performance of propylene carbonate," *Review of Scientific Instruments*, vol. 86, 2015.

[13] Y. Hou, J. Zhang and Z. Zhang, "Significantly improved breakdown performances of propylene carbonate-based nano-fluids," *Micro and Nano letters*, vol. 11, pp. 490 – 493, 2016.

[14] Y. Hou, J. Zhang and Z. Zhang, "Enhanced dielectric breakdown performances of propylene carbonate modified by nano-particles under microsecond pulses," *Journal of Applied Physics*, vol. 119, p. 244307, 2016.

[15] Y. Hou, Z. Zhang, J. Zhang, Z. Wang, and H. Liu, "Experimental investigations into the enhanced dielectric breakdown performances of propylene carbonate modified by TiO_2 nano-particles," *IEEE Transactions on Dielectrics and Electrical Insulation*, 2016.

[16] Y. Hou, Z. Zhang and J. Zhang, "Significantly enhanced impulse breakdown performances of propylene carbonate modified by TiO2 nano-particles," *Chemical Physics Letters*, vol. 662, pp. 192-195, 2016-10-01 2016.

[17] W. F. Schmidt, "Electronic Conduction Processes in Dielectric Liquids," *IEEE Transactions on Electrical Insulation*, 1984.

[18] T. J. Lewis, "Basic electrical processes in dielectric liquids," *IEEE Transactions on Dielectrics and Electrical Insulation*, vol. 1, pp. 630 - 643, 1994.

[19] T. J. Lewis, "The role of electrodes in conduction and breakdown phenomena in solid dielectrics," *IEEE Transactions on Electrical Insulation*, 1984.

[20] T. J. Lewis, "Electronic processes in'dielectric liquids under incipient breakdown stress," *IEEE Transactions on Electrical Insulation*, 1985.

[21] T. Tanaka, M. Kozako, N. Fuse, and Y. Ohki, "Proposal of a multi-core model for polymer nanocomposite dielectrics," *IEEE Transactions on Dielectrics and Electrical Insulation*, vol. 12, pp. 669 - 681, 2005.

[22] T. J. Lewis, "Interfaces are the dominant feature of dielectrics at the nanometric level," *IEEE Transactions on Dielectrics and Electrical Insulation*, vol. 11, pp. 739 - 753, 2004.

[23] T. J. Lewis, "Nanometric Dielectrics," *IEEE Transactions on Dielectrics and Electrical Insulation*, vol. 1, pp. 812-825, 1994.

[24] M. Meunier and N. Quirke, "Molecular modeling of electron trapping in polymer insulators," *The Journal of Chemical Physics*, vol. 113, p. 369, 2000.

[25] M. Meunier, N. Quirke and A. Aslanides, "Molecular modeling of electron traps in polymer insulators: Chemical defects and impurities," *The Journal of Chemical Physics*, vol. 115, p. 2876, 2001.

Investigation on a high power solid-state pulse generator

Jingming Gao*, Hanwu Yang, Song Li, Baoliang Qian, Jun Zhang, Jiande Zhang
College of Opto-Electronic Science and Engineering,
National University of Defense Technology
No.109, Deya Street, Changsha 410073, P.R. China
jingming.gao@ymail.com

***Abstract*—Solid-state pulsed power systems are characterized by high repeatability, high reliability, long lifetime, free maintenance, and easy-to-use. Such systems with moderate peak power are widely applied in high average power region; however, many issues are still to be researched to obtain GW-level peak power. In this paper, a high power solid-state pulse generator delivering GW-level quasi-rectangle pulse is developed to investigate electrical characteristics of the generator which could be applied for High Power Microwave (HPM) generation, X-ray sources and industrial applications such as material surface treatment and Pulse Electric Filed (PEF) sterilization, etc.**

According to the insulation rules, the power compression and pulse forming are accomplished at a relatively low voltage level within a long time scale by a high power low impedance long pulse module, comprised of a AC-link high voltage power supply, a primary unit, a core-type pulse transformer, a two-stage magnetic compression network, a ceramic capacitor based Low Impedance Pulse Forming Network (LIPFN) and magnetic type main switches. Then, the output quasi-rectangle pulse is stepped up by a four-stage Induction Voltage Adder (IVA) in a fairly short time, which is designed with primary windings performing a saturable charging inductor for the Blumlein LIPFN. The above two modules are connected by high voltage coaxial cables to set up the high power solid-state pulse generator.

Repetitive experiments are carried out with on-line DC reset condition for the magnetic switches and the IVA. Preliminary results with repetition rate of 20 Hz are achieved on a dummy load of 50 ohm with output pulse power of 2.1 GW and pulse width of 170 ns showing good repeatability. At present, the system is being upgraded to 5 GW for vacuum diode load.

Keywords—solid-state; magnetic switch; high power low impedance long pulse generator, induction voltage adder

I. INTRODUCTION

Solid-state pulse generators have advantages of high stability, high reliability, high repetition rate and long lifetime, which is also good for maintenance and environmental adaptability. Thus, a lot of solid-state pulse generators are investigated for many kinds of applications[1-6]. However, for high power pulse generators of GW level, it is difficult to accomplish solid-state design due to the limitation on power capability, operation voltage and power compression rate of solid-state switches, as well as on the solid-state pulse forming unit. S-series solid-state generators developed by Y. A. Kotov are excellent for generating GW-level repetitive short pulse (less than 100 ns)[7]; A 12-stage solid-stage Marx generator was set up by F. Hegeler delivering 0.9 GW and 300 ns pulses into a resistive load at a repetition rate of 10 Hz[8]; Another 28-stage solid-state Marx generator was developed by Hongtao Li with power of 0.5 GW and pulse width of 3-10 μs[9]; besides, few researches are reported on GW level solid-state pulse generators. The paper focuses on development of a high power solid-state generator with a relatively long pulse width (larger than 100ns and less than 500ns), delivering a quasi-rectangular pulse to the load, which could be more suitable for HPM generation and also could be applied for X-ray sources, material surface treatment and Pulse Electric Filed (PEF) sterilization, et al.

A high power, low impedance long pulse generator has been investigated previously[10], based on which a four stage IVA is utilized to step up the pulse voltage and transform output impedance for different load requirement. Moreover, the pulse modulation scheme is optimized, where the former charging magnetic switch is replaced by primary windings of the IVA which then performs dual functions. For repetitive operation, the primary unit consisting of a High Voltage Power Supply (HVPS), a capacitor bank and Pseudospark switch is constructed. Under the on-line DC reset condition, repetitive experiments are carried out with results showing good pulse repeatability.

The arrangement of the present paper is as follows. In Sec. II, a full circuit model of the high power solid-state generator is developed via PSpice software. In Sec. III, electromagnetic simulation of the IVA is performed, based on which the IVA is designed. In Sec. IV, repetitive experiments are carried out on a dummy load. Finally, conclusions are presented in Sec. V.

II. SYSTEM CIRCUIT SIMULATION

Based on the circuit model for a high power low impedance long pulse generator, a full circuit model is developed via PSpice software with IVA included. In the circuit schematic shown as Fig.1, the components in the upper section are the same as those before[10] except that the charging magnetic switch MS4 is omitted. In the lower section, FC is the feeding cable, L_c is the connection inductance of feeding port, IVA_C1-4 form a four stage IVA, each of which is a saturable transformer. These four cells are saturated and then unsaturated when the PFN is charging and discharging,

The National High Technology Research and Development Program of China.

respectively, which performs dual functions of charging inductor and pulse voltage adder. The simulation parameters for the lower section is shown in Table I.

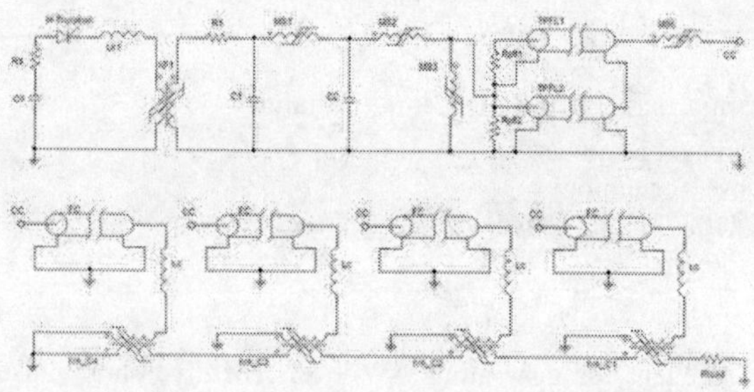

Fig. 1. Schematic of the high power solid-state pulse generator

TABLE I. SIMULATION PARAMETERS OF THE GENERATOR

$T_{\_FC}$	$Z_{\_FC}$	L_C	L_{m_IVA}	L_{s_IVA}	R_{load}
20ns	12Ω	29nH	18μH	100nH	50Ω

A group of typical simulation results are shown in Fig.2. The pulse charging time for PFN is about 900ns, almost the same with the previous process with MS4 as the charging inductor. It could be seen that a quasi-rectangular pulse with stepped voltage is delivered to the load by the IVA, which completes the dual functions well. The pulse power is 2.1 GW, the pulse width is 170ns and the rise time is about 50 ns.

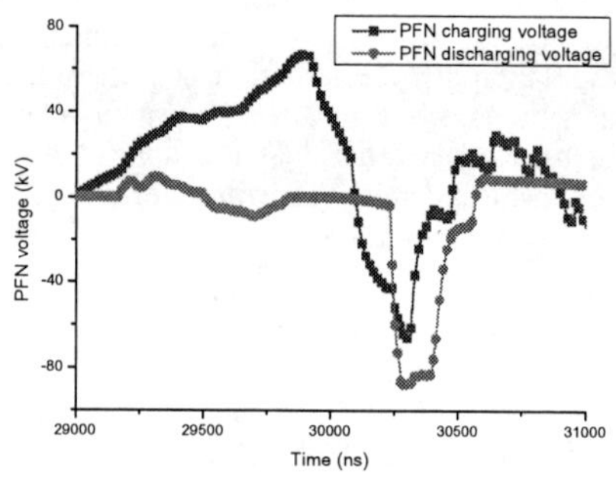

(a) PFN charging and discharging voltage

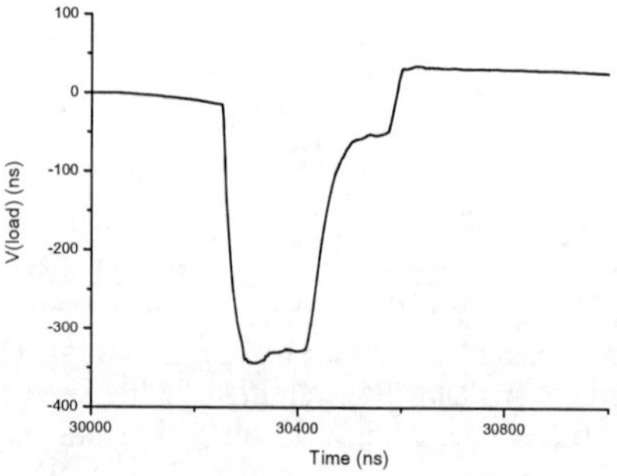

(b) Output pulse voltage on the load

Fig.2. Typical simulation results of the solid-state pulse generator

III. IVA SIMULATION AND DESIGN

Electromagnetic model is setup to investigate pulse response characteristics of the IVA. The magnetic core for IVA is of Fe-based amorphous with parameters shown in Table II. Azimuthal transmission line is utilized to uniformly stimulate the amorphous core. A symmetrical feeding structure of the IVA cell with two azimuthal transmission lines is shown in Fig.3 (a) and four-stage IVA constructed of such four cells is shown in Fig.3 (b).

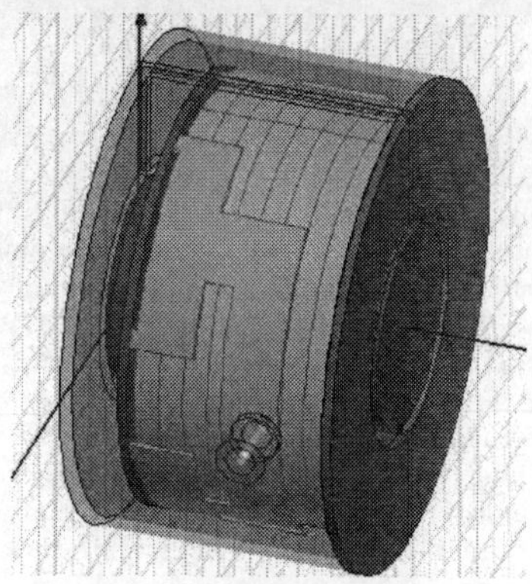

(a) A symmetrical feeding structure of the IVA cell

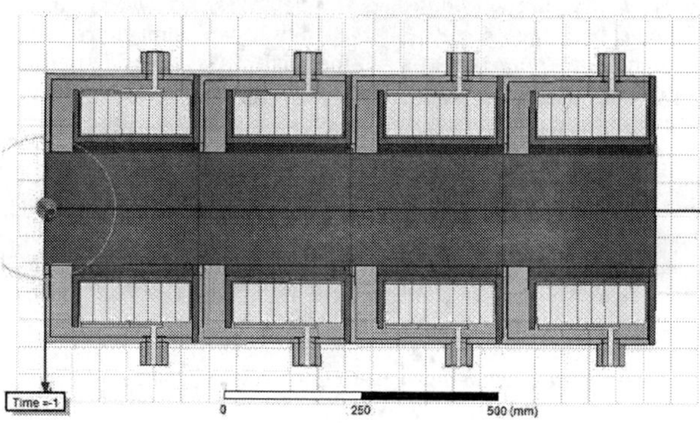

(b) Four-stage IVA

Fig3. Electromagnetic model of the IVA

TABLE II. SPECIFICATIONS OF THE IVA AMORPHOUS CORE

core material	outer diameter	inner diameter	height	thickness
2605SA-1	406 mm	274 mm	20 mm	25μm
width	B_s	B_r	**Insulation**	**stacking factor**
20 mm	~1.56 T	≥1.4 T	120 V/layer	~0.82

Simulation results for several feeding condition are shown in Fig.4. It could be seen that the response characteristics of symmetrical feeding is much better than that of single feeding. Symmetrical feeding stimulates the amorphous core more uniformly which causes a higher coupling coefficient of the IVA resulting in a lower leaking inductance. Based on the above analysis, the four-stage IVA is designed with symmetrical feeding structure.

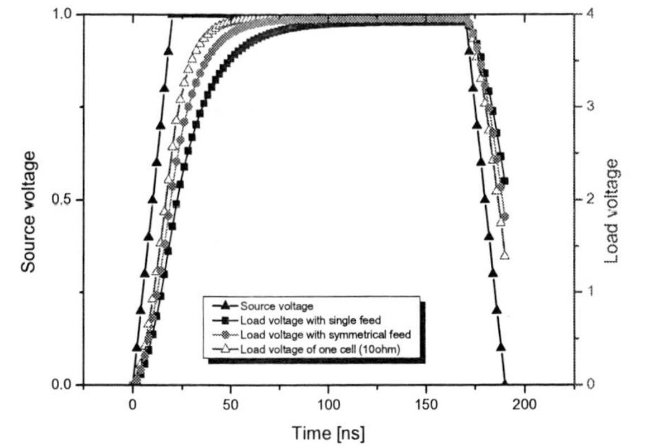

Fig4. Electromagnetic simulation results of the IVA response characteristics

IV. EXPERIMENTAL TEST

The high power solid-state pulse generator is shown in Fig.5, where Pseudospark switch is used to control the discharging of the repetitive primary unit and the low impedance long pulse generator consists of a transformer, two state magnetic compressor, low impedance ceramic capacitors based PFN, magnetic switches for inversion and wave erection, which is connected to the IVA via 8 high voltage coaxial cables with impedance of 24 Ω. The resistance of the dummy load is about 50 Ω. A capacitive divider is installed to measure the pulse voltage at the output of the IVA in front of the load.

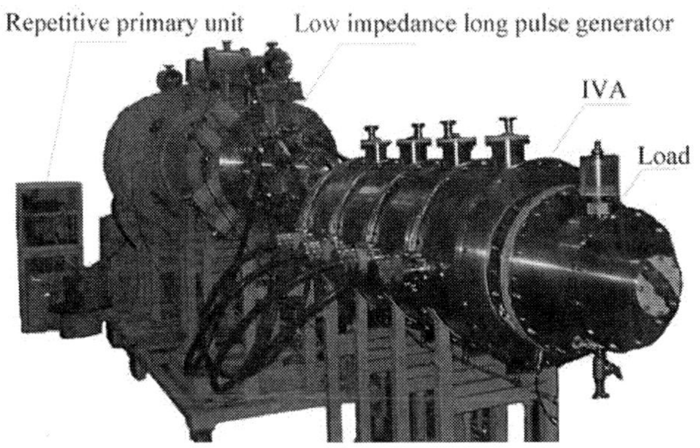

Fig. 5. Photograph of the high power solid-state pulse generator

Under on-line DC reset condition for the magnetic switches and the IVA, preliminary test for repetitive operation is carried out. At a charging voltage of 16 kV for the primary capacitor within 40 ms, 3 pulses of 20 Hz in sequential mode is shown in Fig.6 (a) and those in overlap mode is shown in Fig.6 (b) from which it could be seen the pulse repeatability is fairly good. The output voltage is about 325 kV with pulse power to be 2.1 GW. The pulse width and rise time are 170 ns and 60 ns, respectively.

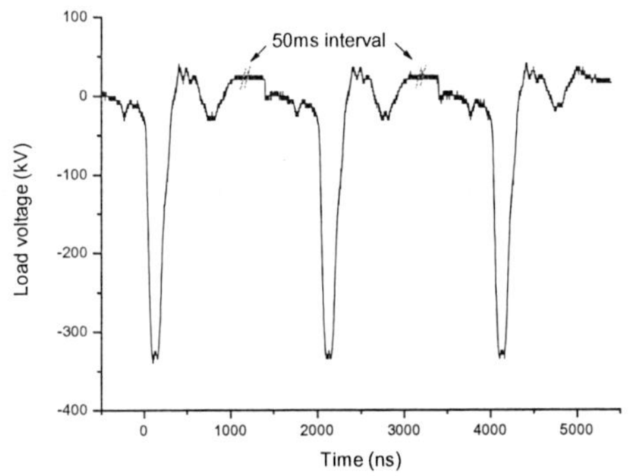

(a) Sequential mode;

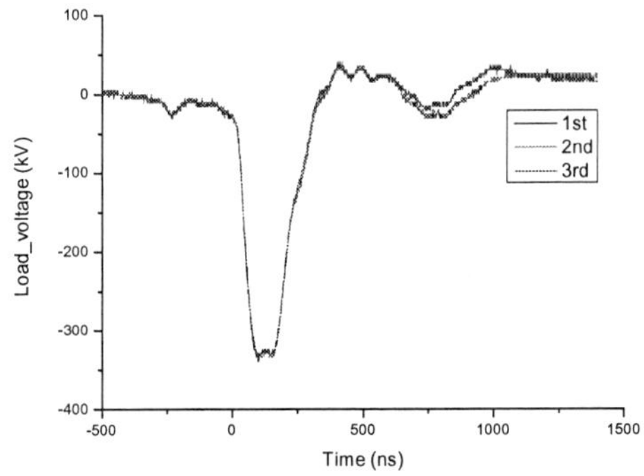

(b) Overlap mode

Fig. 6. Preliminary repetitive results of the high power solid-state pulse generator

V. CONCLUSIONS

A high power solid-state pulse generator-SMART (Solid-state Magnetic switched Accelerator for Repetitive Test) is developed mainly consisting of a high power low impedance long pulse generator and a four-stage IVA. The IVA is designed to be a symmetrical feeding structure, which performs dual functions of charging inductor and pulse voltage adder. Repetitive experiments are carried out with on-line DC reset for the magnetic switches and the IVA. Preliminary results with repetition rate of 20 Hz are achieved on a dummy load of 50 Ω, showing that the output pulse power is 2.1 GW, the pulse width is 170 ns with good pulse repeatability. The Pseudospark switch in the repetitive primary unit will be replaced with a group of serial connected Thyristors to accomplish an all solid-state system in the future.

Acknowledgment

The authors would like to thank Researcher Jianhua Yang for sharing research experience, Professor Huihuang Zhong, Yonggui Liu and Chuanlu Li for their encouragement and valuable suggestions, and Longbo Yan, Jiangnan Kan and Xiwen Tian for their assistance in the assembly.

References

[1] Marcel P.J. Gaudreau, Jeffrey Casey, J. Michael Mulvaney, et al. Solid-state high voltage pulse modulators for high power microwave applications [C]. Proceedings fo EPAC 2000, Vienna, Austria, 2000.

[2] Michael A. Kempkes, Jeffrey A. Casey, Marcel P.J. Gaudreau. Solid-state modulators for commercial pulsed power systems [C]. International Power Modulator Symposium, 2002:689-693.

[3] Ju Won Baek, Dong Wook Yoo, Geun Hie Rim, et al. Solid State Marx Generator Using Series-Connected IGBTs [J]. IEEE Transactions on Plasma Science, 2005, 33(4): 1198-1204.

[4] Y. Zhang, J. L. Liu, "A New Kind of Solid-State Marx Generator Based on Transformer Type Magnetic Switches," Laser and Particle Beams, Vol. 31, No. 2, 2013, pp. 239-248.

[5] Cook E G, Allen F V, Anaya E M, et al. Solid-State Modulator R&D at LLNL [C]. International Workshop on Recent Progress of Induction Accelerators, 2002.

[6] Dongdong Wang, Jian Qiu, and Kefu Liu. All-Solid-State Repetitive Pulsed-Power Generator Using IGBT and Magnetic Compression Switches [J]. IEEE Transactions on Plasma Science, 2010, 38(10): 2633-2638.

[7] Y. A. Kotov, G. B. Mesyats, S. N. Rukin, A. L. Filatov, and S. K. Lyubutin, "A novel nanosecond semiconductor opening switch for megavolt repetitive pulsed power technology: Experiment and applications," in Proc. 9th IEEE Int. Pulsed Power Conf., Mar. 1993, pp. 298–305.

[8] F. Hegeler, M. W. McGeoch, J. D. Sethian, "A durable gigawatt class solid state pulsed power system," IEEE Transactions on Dielectrics and Electrical Insulation, Vol. 18, No. 4, 2011, pp. 1205-1213.

[9] H. T. Li, C. W. Wang, L. Y. Wang, "500 kV all-solid-state Marx generator," High Power Laser and Particle Beams (in Chinese), Vol. 24, No.4, 2012, pp. 921-924.

[10] Jingming Gao, Hanwu Yang, Song Li, Zhaoxi Liu, Baoliang Qian. Investigation on a high power, low impedance, and long pulse generator based on magnetic switches. IEEE Transactions on Plasma Science, 2014, 42(4): 988-992.

Coaxial Pulse Forming Line with High Energy Storage Density Based on Propylene Carbonate

Zicheng Zhang*, Hanwu Yang, Jiande Zhang, Zhen Wang, Yanpan Hou
College of Optoelectronic Science and Engineering
National University of Defense Technology
Changsha 410073, China
*zczhang@nudt.edu.cn

***Abstract*—Applications have stimulated intense interest in pulsed power technology toward high average power and compact structure. Propylene carbonate (PC), due to its great permittivity (65), high dielectric insulating strength and wide range of environmental adaption, has a great potential in improving the compact level of pulse power system as a kind of energy storage liquid dielectric. In this paper, a coaxial pulse forming line with high energy storage density based on propylene carbonate of liquid dielectric is developed. First, the sample PC is purified. The resistivity of PC can be increased to 263 MΩ•cm. Second, the pulsed insulating properties of PC are investigated in microsecond regime. The breakdown strength of PC is higher than de-ionized water. Furthermore, the effect of pressure on the breakdown strength is carried out. Experimental results show that the breakdown strength is proportional to the power of the pressure. Finally, the application of PC in the coaxial pulse forming line (PFL) of capacitive energy storage pulse power system is carried out. The PFL is charged by a pulse transformer and discharged to a load through a high voltage spark gap switch. For the PFL with 40 cm diameter, the maximum operation voltage can reach 1 MV in the 1.5 μs effective time. These efforts set a good foundation for the development of a coaxial PFL based on propylene carbonate and show a promising application in the compact pulse power system for the future.**

***Keywords*—Pulse power science and technology, pulse power system, capacitive energy storage, pulse forming line, propylene carbonate, liquid dielectric, high permittivity.**

I. INTRODUCTION

Applications have stimulated intense interest in pulsed power technology toward high average power and compact structure. Capacitive energy storage type pulsed power system, especially based on liquid pulse forming line (PFL), is a promising one, owing to its advantages including high power operated at low impedance, great permittivity liquid dielectric, rep-rate operation, and simple compact structure[1-4]. HEART series accelerators are developed in National University of Defense Technology, China, where HEART is short brief of High Energy density Accelerator based on Rep-rate pulse Transformer. Of all of the HEART series accelerators, HEART-20 is a typical one, consists of five subsystems, primary energy supply, pulse transformer, pulse forming line, spark gap switch and vacuum diode. In which, pulse forming line occupies approximately 50 % of whole pulse power system in volume or weight. To obtain high power density and compact structure, the accelerator features a coaxial ethanol / water PFL, with a helical central conductor, which would extend the duration and improve the impedance[5-16]. Given the same outer dimensions of 400 mm, this PFL can operate with output voltage 800 kV.

In order to further reduce the freezing point and improve the resistivity and its maintenance time, we explore propylene carbonate as the energy storage medium. Compared with water and water-based mixtures, propylene carbonate (PC) has advantages such as lower freezing temperature with about -50 centigrade degree and higher breakdown strength[17-19]. The most superiority of PC over water and ethylene glycol/water mixtures is relatively low viscosity and stable physical and chemical properties. So it has a great potential in improving the compact level of pulse power system as a kind of capacitive energy storage with liquid dielectric.

In this paper, a coaxial pulse forming line with high energy storage density based on PC liquid dielectric is developed. First, the sample PC is purified. Second, the pulsed insulating properties of PC are investigated in microsecond regime. Finally, the application of PC in the coaxial PFL of capacitive energy storage pulse power system is carried out.

II. SAMPLE PREPARE OF PROPYLENE CARBONATE

A. *Original Propylene Carbonate*

The original sample of propylene carbonate is cell grade. Its purity is more than 99.95%. However, its resistivity is 4 MΩ•cm without special treatment. And the intrinsic time of propylene carbonate, product of the permittivity and the resistivity, is 23 μs. It is not enough to meet the requirement of 10 μs charging time of propylene carbonate-based pulse forming line (PFL). So, it is necessary to improve the resistivity to above 20 MΩ•cm.

B. *Purifying Treatment*

In the case of the low resistivity of sample PC, the PC purification system is designed. It includes water pump, resin, resin trapper, vacuum pump and conductivity gauge. The functions of the purifying system are deionizing, decontamination, degassing, respectively.

C. *Resisitivity Maintaince*

After purification, the resistivity of PC will increase from 4 MΩ•cm to 263MΩ•cm.

This work was supported by the National Natural Science Foundation of China under Grant 51307175 and 51677190.

III. Breakdown Performances of Propylene Carbonate

A. Breakdown Setup

The experimental apparatus mainly consists of a pulsed power source, the test cell, and corresponding diagnostic devices (see Fig. 1). ARC-01 modulator, shown in Fig. 1(a), serves as pulsed power source in our testing system, with the maximum output voltage of 400 kV and the peak charging time about8.4 μs. The circuit diagram of the dielectric breakdown measurement device is shown in Fig. 1(b). At first, the primary capacitor is charged by commercial power supply after the half wave rectification. Then, the test cell, which is in parallel connected to pulse forming line, would be charged through the Tesla transformer after closing the main thyristor controlled by computer. The breakdown of the specimen between the electrodes of the test cell would occur when the voltage of pulse forming line is charged up to the threshold of the dielectric, and the voltage waveforms of whole breakdown process is obtained by the oscilloscope. Through changing the voltage applied on primary capacitor, we can get different high-voltage electrode potential increasing rates. Fig. 1(c) illustrates the test cell, main part of which is a stainless steel box with 200 × 200 × 200 mm^3 holding the test electrodes. The two brass spherical electrodes with 26 mm radius are thus over 100 mm away from any solid dielectric interfaces which might contribute to field anisotropies or cause spurious breakdown paths. The gap between electrodes is visible and spacing is set to be 1 mm during testing.

Fig. 1. Dielectric breakdown experiment setup and the test cell. (a) pulsed power source ARC-01, (b) circuit diagram of the dielectric breakdown experimental setup, and (c) test cell.

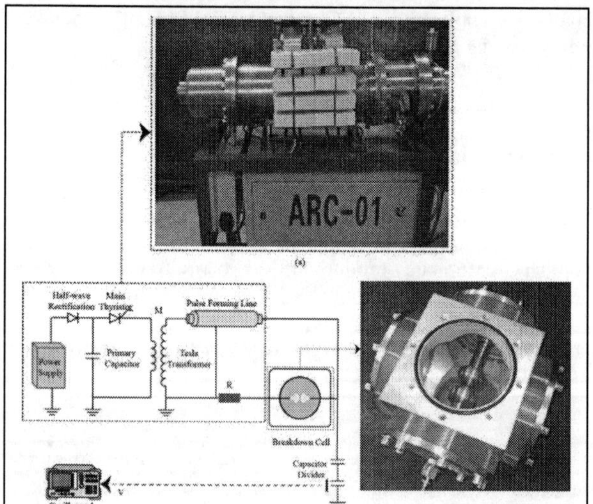

B. Breakdown Experiments of Propylene Carbonate

Weibull statistics have been generally applied to the failure of materials under action of external forces such as mechanical and electrical stresses, where breakdown is assumed to be caused by a weak point of the system and the failure rate increases exponentially with the stress. The breakdown voltage of insulating material can be considered a random variable as it varies from one to another in a random way for apparently identical specimens. In this study, two dimensional Weibull distributions are utilized to estimate the breakdown probability of PC which is written as follows: $P(U)=1-\exp[-(U/\alpha)^{\beta}]$, where U is the voltage applied to the specimen, β is the shape parameter, related to the scatter of the data, and α is the scale parameter, which is the value of U for a dielectric breakdown probability $1-1/e = 0.632$.

The typical breakdown voltage waveform of propylene carbonate in microsecond regime is shown as Fig. 2.

Fig. 2. Breakdown voltage waveform of propylene carbonate in microsecond regime.

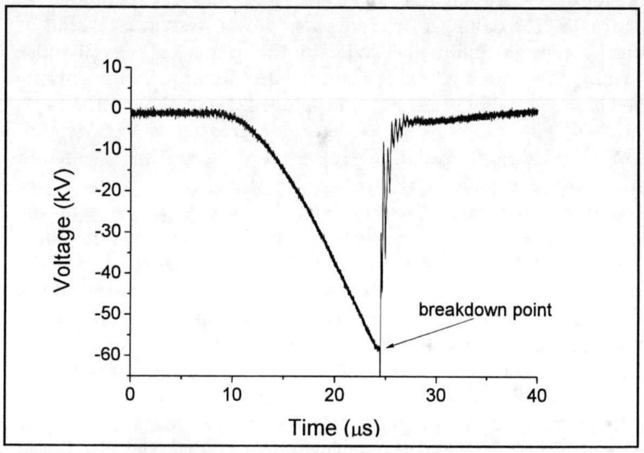

In the effective time of 4.3 μs, the breakdown voltage of PC is 60.7 kV and the normalized breakdown voltage is 77.4 kV. And the breakdown voltage of PC is a little higher than that of deionized water.

C. Pressurization Effect of Propylene Carbonate on the Breakdown Stress

Pressurization effect of propylene carbonate on the breakdown stress is investigated in microsecond regime. The typical breakdown voltage waveforms for different pressure ranged from 1 bar to 11 bar are shown as Fig. 3.

Fig. 3. Breakdown voltage waveform of propylene carbonate in different pressure.

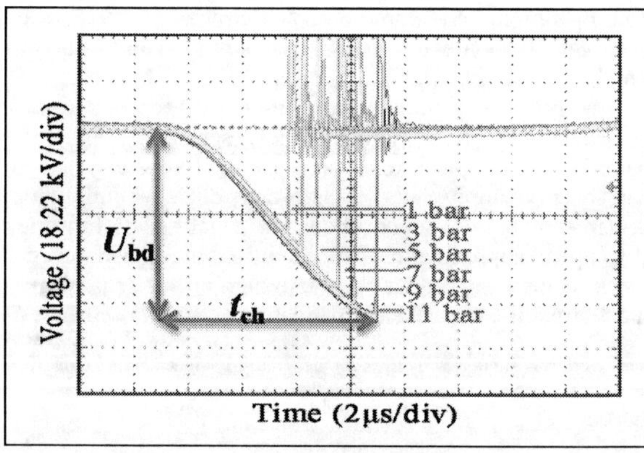

Propylene carbonate's insulating strength can be improved by pressurization. Breakdown strength is proportional to the power of the pressure.

IV. Propylene Carbonate-Based Pulse Forming Line

The application of PC in the coaxial pulse forming line (PFL) of capacitive energy storage pulse power system is carried out. The PFL is charged by a pulse transformer and discharged to a load through a high voltage spark gap switch. The typical charging waveform of PC-based PFL is shown as Fig. 4, where Ch1 is PFL charging voltage waveform and Ch2 charging current waveform. For the PFL with 40 cm diameter, the maximum operation voltage can reach 1 MV in the 1.5 μs effective time.

Fig. 4. Voltage waveforms of PC-based pulse forming line.

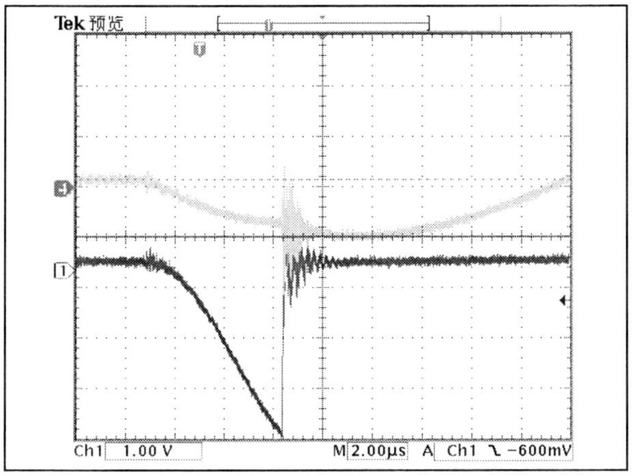

V. Conclusions

First, through purification, the resistivity of PC can be increased to 263 MΩ•cm. Second, Based on pulsed breakdown platform with pulsed power source and test cell, The breakdown strength of PC is higher than de-ionized water. Furthermore, the breakdown strength is proportional to the power of the pressure. Finally, the application of PC in the coaxial pulse forming line (PFL) of capacitive energy storage pulse power system is carried out. The PFL is charged by a pulse transformer and discharged to a load through a high voltage spark gap switch. For the PFL with 40 cm diameter, the maximum operation voltage can reach 1 MV in the 1.5 μs effective time. These efforts set a good foundation for the development of a coaxial PFL based on propylene carbonate and show a promising application in the compact pulse power system for the future.

Acknowledgment

The authors wish to thank Prof. Jun Zhang, Huihuang Zhong, Jianhua Yang and Dr. Zhiqiang Li for their encouragement and valuable suggestions. The assistance of Bo Liang and Xing Hu in the assembly and testing of the pulsed power source was also gratefully acknowledged.

References

[1] Martin J C. Nanosecond Pulse Techniques[M]// J. C. Martin on Pulsed Power. Springer US, 1996:35-74

[2] J.C.Martin. Comparision of Breakdown Voltages for Various Liquids under One Set of Conductions .Dielectric Strength Notes Nov. 1965.

[3] Mesyats G A. Pulsed power[M]. Springer Science+ Business Media, Incorporated, 2005.

[4] Zhang Z, Zhang J, Qian B, et al. Compact Rep-Rate GW Pulsed Generator Based on Forming Line With Built-In High-Coupling Transformer[J]. Plasma Science, IEEE Transactions on, 2014, 42(1): 241-248.

[5] Fenneman D B, Gripshover R J. Experiments on electrical breakdown in water in the microsecond regime[J]. IEEE Transactions on Plasma Science, 1980, 8(3): 209-213.

[6] Espinosa J C. Dielectric breakdown of water[J]. Austin: Doctoral Dissertation of the University of Texas, 2000.

[7] Fenneman D B. Pulsed high-voltage dielectric properties of ethylene glycol/water mixtures[J]. Journal of Applied Physics, 1982, 53(12): 8961-8968.

[8] Bolund B F, Berglund M, and Bernhoff H. Dielectric study of water/methanol mixtures for use in pulsed-power water capacitors[J]. Journal of Applied Physics, 2003, 93(5): 2895-2899.

[9] Miller A R. High energy density, low impedance capacitors using pressuredwater as a dielectric[R]. New York: Inst. of Electrical and Electronics Engineers, Inc., 1974: 471-474.

[10] Mankowski J,Hatfield L, Kristiansen M, et al. High voltage subnanosecond dielectric breakdown[C]//Plasma Science, San Diego: 1997. IEEE Conference Record-Abstracts., 1997 IEEE International Conference on. IEEE, 1997: 317-318.

[11] Lehr J M, Agee F J, Copeland R, et al. Measurement of the electric breakdown strength of transformer oil in the sub-nanosecond regime[J]. IEEE Trans on Dielectrics and Electrical Insulation, 1998, 5 (6) : 857–861.

[12] Tsacoyeanes C W, et al. Capacitor with liquid dielectrics [P]. USA Patent: 3903460, 1975-09-02.

[13] Jones H M, Kunhardt E E. The Influence of Pressure and Conductivity on the Pulsed Breakdown of Water[J]. IEEE Trans. on Dielectrics and Electrical Insulation, 1994, 1(6): 1016-1025.

[14] Jones H M, Kunhardt E E. Development of Pulsed Dielectric Breakdown in Liquid[J]. Journal of Physics, 1995, 28(1): 178-188.

[15] Korobeynikov S M, Sinikh Y N. Bubbles and breakdown of liquid dielectrics[C]//Electrical Insulation, 1998. Conference Record of the 1998 IEEE International Symposium on. IEEE, 1998, 2: 603-606.

[16] Kao K C, Mamath P C. Time-dependent pressure effect in liquid dielectrics[J], IEEE Transactions on Electrical Insulation. 1970, 5(3): 64-68.

[17] Kolb J., Kono S., Xiao S., et al, Water and propylene carbonate as storage and switching media in pulsed power system, 14th IEEE International Pulsed Power Conference, Dallas, Texas, USA, 2003, Vol. 1, 715~718.

[18] Felici N J, Tobazeon R E. Charge carrier elimination and production by electrodialytic polymers in contact with dielectric liquids[J]. Journal of Electrostatics, 1981, 11(2):135–161.

[19] Clark R S, Green D L, Buttram M T, et al. Studies on the use of propylene carbonate as a high-voltage insulator[C] Power Modulator Symposium, 1988. IEEE Conference Record of the 1988 Eighteenth. IEEE, 1988:381 - 384.

An Improved Ku-band MILO with Tapered Choke Cavity and Enlarged First Interaction Cavity

Tao Jiang, Jiande Zhang, Juntao He, Zhiqiang Li and Junpu Ling

***Abstract*—In order to further enhance the power capacity, and speed up the microwave start-up and saturation, an improved Ku-band magnetically insulated transmission line oscillator (MILO) with tapered choke cavity and enlarged first interaction cavity is proposed and investigated in this paper. The simplified plate model of MILO is analyzed to investigate the movement of individual electron. According to the theoretical analysis, the innovative design of Ku-band MILO is presented. In order to make the electrons emitted from the launch point of the cathode closer to the SWS, especially the surface of the first interaction cavity, help to speed up the microwave start-up and saturation, the launch point of the cathode is optimized and designed 1.5 cm in front of the second choke vane. A tapered choke cavity is used to replace traditional uniform choke cavity, which helps further speed up the microwave start-up and saturation. An enlarged first interaction cavity with increased cavity gap is introduced to enhance the power capacity. Typical particle simulation results show that the improved Ku-band MILO can generate a microwave with a power of 1.65 GW and a frequency of 12.3 GHz under the diode voltage of 474 kV and beam current of 42 kA. Compared with the traditional Ku-band MILO, the microwave start-up and saturation of the improved one are 2.5 ns and 5.5 ns faster, respectively. The power capacity of the improved Ku-band MILO is enhanced greatly due to that the radio-frequency (rf) electric field is decreased much from 1.3 MV/cm to 1MV/cm. At last, the experimental investigation of the improved Ku-band MILO is conducted. The device achieves a microwave output with power of 1.2 GW, frequency of 12.36 GHz and pulse width of 35 ns in the experiment. Both the power and pulse width are enhanced significantly.**

***Index Terms*—HPM; Ku-band; MILO; Tapered choke cavity; Enlarged first interaction cavity**

I. INTRODUCTION

The magnetically insulated transmission line oscillator (MILO) is a promising gigawatt-class cross-field device which operates successfully without externally applied magnetic field [1,2]. The magnetic field is generated by its magnetic insulated current. This property makes MILO more compact and portable than other high power microwave (HPM) sources. Because of the above advantages, the MILO has become one of the most attractive HPM sources.

Since the MILO was invented in 1987 [1-3], it has been investigated theoretically and experimentally by researchers all over the world in the past few decades [3-19]. At present, the researches on MILO are primarily focused on L, S, and C-bands and the power level exceeds 1 GW [6-13]. In recent years, the development of the MILO is promoted to higher frequency bands, such as X, and Ku-bands [14-19]. However the output microwave power level is only on the order of 100 MW and the pulse width is less than 20 ns [14,17,19]. Due to the smaller size of high frequency MILOs, the device can support lower power and rf breakdown is more likely to appear in the experiments, which is the main cause for the low power output level. Therefore, the enhancement of the power capacity is the key to achieve high power microwave output in the experiment.

Otherwise, in the aspect of the structure design, the low frequency MILOs except the tapered MILO [5], generally adopt uniform choke cavity and normal SWS with uniform cavity gap [3-19]. Similarly, the reported Ku-band MILOs, also follow the design [14-19]. While, with the shrink of the overall size, especially the period of the SWS, there existing a problem that the electrons emitted from the launch point of the cathode cannot sweep the surface of the first interaction cavity, which will results in the slow-down of the microwave start-up and saturation. So the Ku-band MILO remains to be further investigated in depth.

In order to further enhance the power capacity, and speed up the microwave start-up and saturation, an improved Ku-band MILO with tapered choke cavity and enlarged first interaction cavity is proposed and investigated numerically and experimentally in this paper.

II. MODEL OF THE IMPROVED KU-BAND MILO

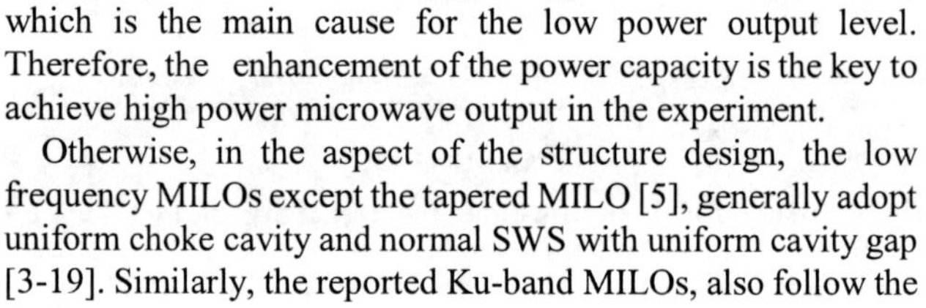

Fig. 1 Model of the improved Ku-band MILO

The basic model of the improved Ku-band MILO is presented in Fig. 1. The device primarily consists of seven parts, which are

Tao Jiang, Jiande Zhang, Juntao He, Zhiqiang Li and Junpu Ling are with the College of Optoelectronic Science and Engineering, National University of Defense Technology, Changsha 410073, China (e-mail: laoda116@163.com; jdzhang12@yahoo.com; hejuntao12@163.com; nudt_lzq@139.com; lingjunpu@163.com).

the SWS, tapered choke cavity, enlarged first interaction cavity, triple-cavities extractor, ladder cathode, limited load, and coaxial output waveguide, respectively. The ladder cathode is verified to be beneficial for the enhancement of the beam-wave interaction efficiency [20]. The triple-cavities extractor is optimized and can improve the extracting efficiency. Besides, the limited load is confirmed to be able to eliminate the electrode erosion of the load by increasing the load anode-cathode gap and reducing the energy deposition density [19]. For the convenience of description, the left side of the cathode is defined as launch point and the right side of the cathode is defined as end point, as shown in Fig. 1.

In order to enhance the power capacity, and speed up the microwave start-up and saturation, 3 innovative designs are introduced to the improved Ku-band MILO. (1) Both the launch point and the end point of the cathode are relatively forward shifted compared with the traditional structures. (2) A tapered choke cavity is used to replace the traditional uniform choke cavity. (3) An enlarged first interaction cavity is adopted to replace the traditional one, in which, the gap of the first interaction cavity is increased appropriately.

III. DESIGN OF THE IMPROVED KU-BAND MILO

A. Theoretical analysis—simplified plate model of MILO

Before express the design of the Ku-band MILO, a simplified plate model of MILO is presented to investigate the movement of individual electron in MILO, which is the theoretical basis of the structure design.

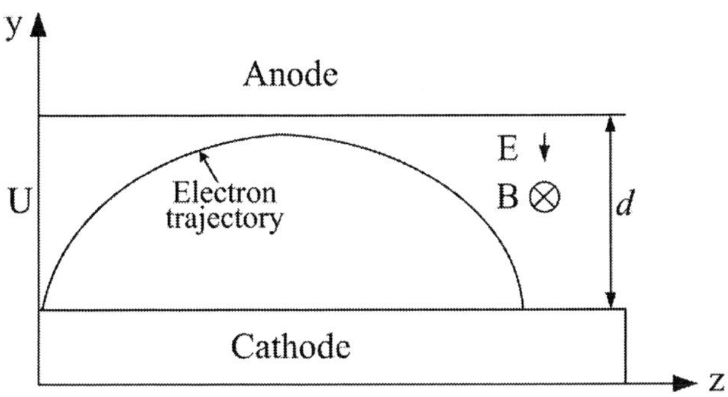

Fig. 2 Simplified plate model of MILO

The simplified plate model is shown in Fig. 2. For simplifying the problem, the anode SWS is degenerated into a plate anode in the model. Under the action of electric field E and magnetic field B (E and B are orthogonal), the individual electron emitted from the cathode will move along a cycloid. The trajectory equation is expressed as [21]:

$$\begin{aligned} y &= R_c(1-\cos\omega_c t) \\ z &= R_c(\omega_c t-\sin\omega_c t) \end{aligned} \quad (1)$$

in which, $R_c = \frac{mE}{eB^2} = \frac{eE}{m\omega_c^2}$, is the cyclotron radius of electron. The electron will roll along the cathode surface with the angular velocity ω_c and the rolling circle radius is R_c. The trajectory curve is also shown in Fig. 2. The height and width of the trajectory curve (cycloid) are $2R_c$ and $2\pi * R_c$, respectively.

In the MILO, when the electrons exactly sweep the surface of the SWS, the device will achieve maximum beam-wave interaction efficiency. In this case, the anode-cathode gap (AK gap) d equals to $2R_c$. So the width of the trajectory can be expressed as:

$$w = \pi \times d \quad (2)$$

And, the axial distance from the launch point to the highest point of the trajectory is $0.5w$.

B. Design of the launch point of cathode

In low frequency band MILOs, the launch point of cathode is always set below the second choke vane, for instance the MILO structure proposed in Ref. 20. Because the SWS period is biggish, the electrons emitted from the launch point of cathode can sweep the surface of the first cavity of the SWS. Taking an example of the L-band MILO proposed in Ref. 20, the period and the AK gap are 2.5cm and 1.8 cm, respectively. Using the above analysis, the width of the individual electron trajectory can be calculated about 4 cm, and the axial distance from the launch point to the highest point of the trajectory is about 2 cm. Therefore, when the launch point is set below the second choke vane, the electrons emitted from the launch point can exactly sweep the first cavity of the SWS, which will make the beam-wave interaction more quickly.

While, in high frequency band MILOs, such as Ku-band MILOs, the AK gap is close to that of the low frequency band MILOs, but the SWS period is much smaller than that of the low frequency band MILOs. Similarly, taking an example of the Ku-band MILO proposed in Ref. 19, the period and the AK gap are 0.65cm and 1.4 cm, respectively. The width of the individual electron trajectory can be calculated about 3 cm, and the axial distance from the launch point to the highest point of the trajectory is about 1.5 cm, which is much larger than the period. Therefore, the launch point must be set in the front of the second choke vane to make the electrons emitted from the launch point close to the surface of the SWS, especially the first interaction cavity of the SWS. In the Ku-band MILO investigated in this paper, the launch point is optimized at the position 1.5cm in front of the second choke vane.

The electrons distributions of the Ku-band MILOs with two different designs of the launch point are shown in Fig.3. Fig. 3 (a) is the situation that the launch point of the cathode is set below the second choke vane (Model 1), which is in accordance with the traditional low frequency MILOs. Accordingly, the electrons emitted from the launch point will be far from the surface of the first two interaction cavities. Consequently, the output microwave power is slightly small about 1.4 GW, and the microwave start-up and saturation of the device are too slow. The start-up and saturation times are 9 ns, and 19 ns, respectively, shown in Fig. 4. Fig. 3 (b) is the situation that the launch point is set 1.5cm in front of the second choke vane (Model 2). The electrons emitted from the launch point are already close to the surface of first two interaction cavities. Accordingly, the output microwave power is increased to 1.65 GW, and the microwave start-up and saturation times are reduced to 5.5 ns, and 13 ns, respectively, shown in Fig. 4.

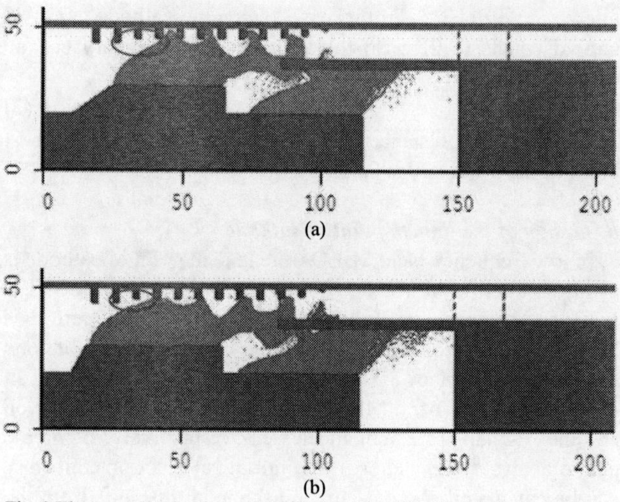

Fig. 3 Electrons distributions of the Ku-band MILO with two designs of the launch point (a) the launch point is set below the second choke vane(Model 1); (b) the launch point is set 1.5cm in front of the second choke vane(Model 2)

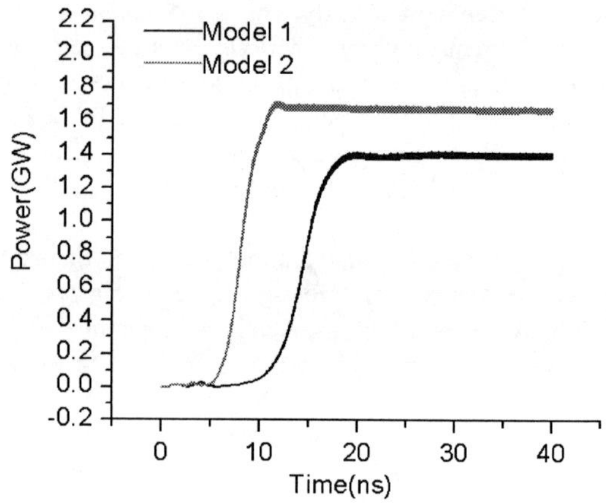

Fig.4 Output microwave power of the Ku-band MILO with two designs of the launch point

Similarly, the end point of the cathode should be set a certain distance in front of the last SWS vane to make the electrons emitted from the end point of the cathode sweep the last several interaction cavities, but not drift to the extracting cavities or the coaxial output structure. Because the electrons drifting to the extracting cavities almost do not participate in the beam-wave interaction and the electrons drifting to the coaxial output structure will absorb and reflect the produced microwave, which is not beneficial for the output of high power microwave and even stop the normal operation of the device. In the improved Ku-band MILO, the end of the cathode is set 2.5 cm in front of the last SWS vane.

C. Tapered choke cavity

In traditional MILO structure, a uniform choke cavity is used to prevent the microwave from propagating into the diode region. When MILO develops to high frequency band (such as Ku band), the launch point of the cathode should be placed in front of the second choke vane according to the above analysis. Because the radii of the two choke vanes are much smaller than those of the SWS, the electrons emitted from the launch point of the cathode are still far away from the surface of the first interaction cavity, after sweeping the choke cavity. Therefore, in order to further pull the electrons emitted from the launch point of the cathode closer to the surface of the first interaction cavity, speed up the beam-wave interaction, a tapered choke cavity is adopted to replace the traditional uniform choke cavity. The tapered choke cavity is made up of two choke vanes with different inner radii, in which, the inner radius of the first choke vane is much smaller and the inner radius of the second choke vane is much larger than that of the uniform choke vane. The effect of the tapered choke cavity can be described as follows.

(1) The first choke vane with smaller inner radius is much closer to the cathode, so the ad electrical field is larger when loading the same voltage on the AK gap. The larger electrical field pulls the electrons emitted from the launch point of the cathode closer to the anode.

(2) The second choke vane with larger inner radius is further from the cathode, which makes the electrons emitted from the launch point of the cathode much easier to sweep the tapered choke cavity and closer to the SWS.

The electrons distribution of the Ku-band MILO with tapered choke cavity is shown in Fig. 5. As can be seen from the figure, the electrons emitted from the launch point of the cathode in the Ku-band MILO with tapered choke cavity are much closer to the SWS, especially the first two interaction cavities, compared with Fig. 3(b). Therefore, this improvement can make the beam-wave interaction faster. As shown in Fig. 6, the start-up and saturation time of the output microwave of the Ku-band MILO with tapered choke cavity are 4.5 ns and 10.5 ns, which are 1 ns and 2.5 ns faster than those of the traditional Ku-band MILO shown in Fig. 3(b), respectively.

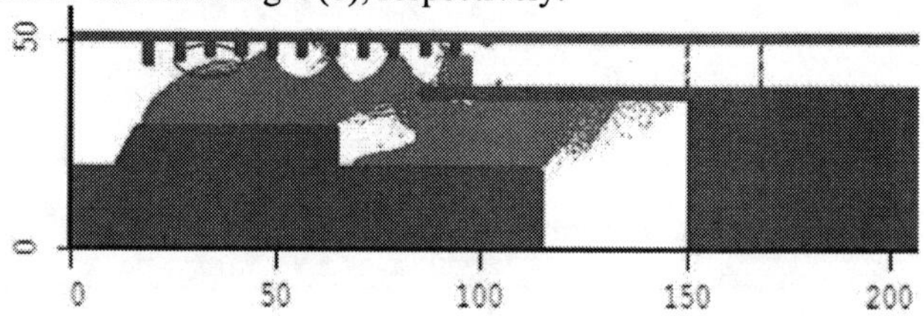

Fig. 5 Electrons distribution of the Ku-band MILO with tapered choke cavity

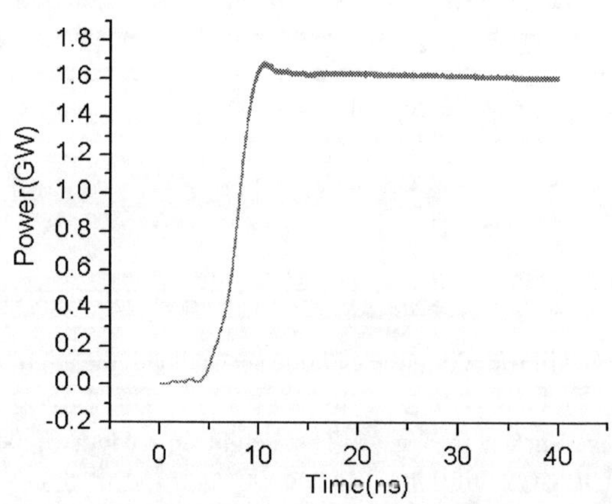

Fig. 6 Output microwave power of the Ku-band MILO with tapered choke cavity

D. Enlarged first interaction cavity

Through the above improvements, the Ku-band MILO has achieved faster start-up and saturation of the output microwave. However, there is still a certain distance between the surface of the first interaction cavity and the electrons emitted from the launch point of the cathode. Take consideration of that the first interaction cavity affects the normal operation of the MILO less due to the small electrical field distribution, but affects the start-up and saturation of the microwave greatly. Therefore, to further speed up the start-up and saturation of microwave, the gap of the first interaction cavity is appropriate increased, which is to make the electrons emitted from the launch point of the cathode much closer to the surface of the first interaction cavity. In addition, the electric field distribution space is increased by adopting enlarged first interaction cavity. So the rf electric field will be decreased and the power capacity will be enhanced accordingly.

The electrons distribution of the improved Ku-band MILO is shown in Fig. 7. The comparison of the output powers of the traditional Ku-band MILO and the improved one (with tapered choke cavity and enlarged first interaction cavity) is presented in Fig. 8.

As can be seen from Fig.7, the electrons emitted from the launch point of the cathode are much closer to the surface of the first interaction cavity. Accordingly, the start-up and saturation of output microwave are further speeded up as shown in Fig. 8. From Fig. 8, one can see that the start-up and saturation time of the output microwave of the improved Ku-band MILO (with tapered choke cavity and enlarged first interaction cavity) are 3.5 ns and 7.5 ns, which are 2 ns and 5.5 ns faster than those of the traditional Ku-band MILO shown in Fig. 3(b), respectively.

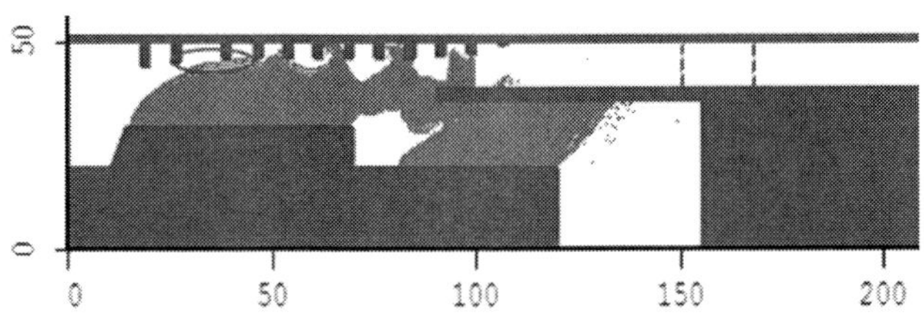

Fig. 7 Electrons distribution of the improved Ku-band MILO

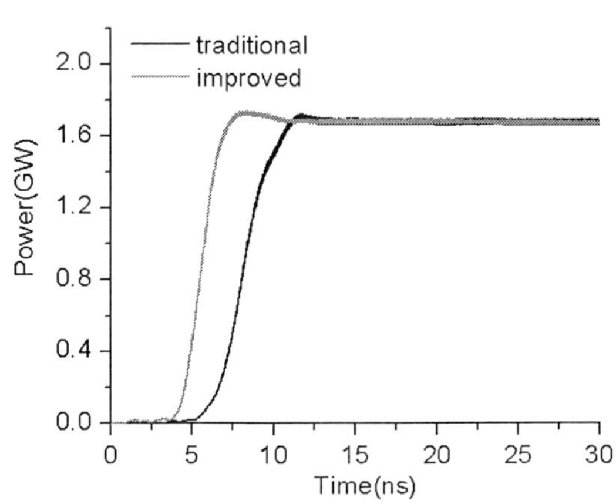

Fig. 8 Output microwave powers of the two MILOs

Additionally, the rf field of the SWS in the improved Ku-band MILO is reduced significantly due to the increase of the beam-wave interaction space. The maximum axial electrical field distributions of the traditional Ku-band MILO and the improved Ku-band MILO are shown in Fig. 9 and Fig. 10, respectively. As can be seen, the maximum axial electrical field of improved Ku-band MILO is 0.3 MV/cm smaller than that of the traditional one. Therefore, the power capacity of the improved Ku-band MILO is enhanced significantly and the experimental output power level is expected to be improved.

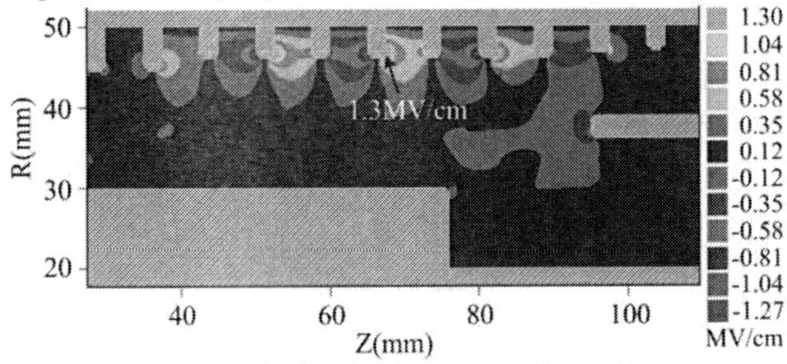

Fig. 9 Maximum axial electrical field distribution of the traditional Ku-band MILO

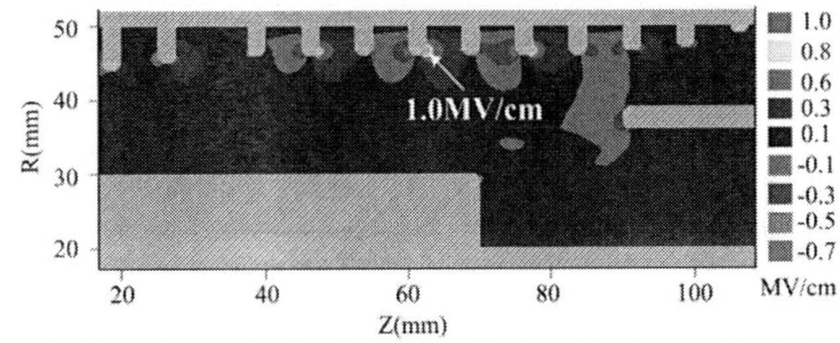

Fig. 10 Maximum axial electrical field distribution of the improved Ku-band MILO

Besides, with the increase of the first interaction cavity gap, the length of the cathode is also enlarged, which will decrease the current density of the cathode and make the electrons emission more uniform, thus increasing the cathode life.

IV. TYPICAL SIMULATION RESULTS

Particle simulation is conducted to investigate the improved Ku-band MILO. The typical simulation results show that under the diode voltage of 474 kV and beam current of 42 kA, the improved Ku-band MILO can generates a microwave with a power of 1.65 GW and a frequency of 12.3 GHz, corresponding to a power conversion efficiency of 8%.

Fig. 11 shows the instantaneous output power and the average output power. The microwave starts to grow at 3.5 ns and achieves saturation at 7.5 ns. The average power is about 1.65GW.

The frequency spectrum of the output microwave is shown in Fig. 12. As can be seen from the figure, the central frequency is 12.3 GHz.

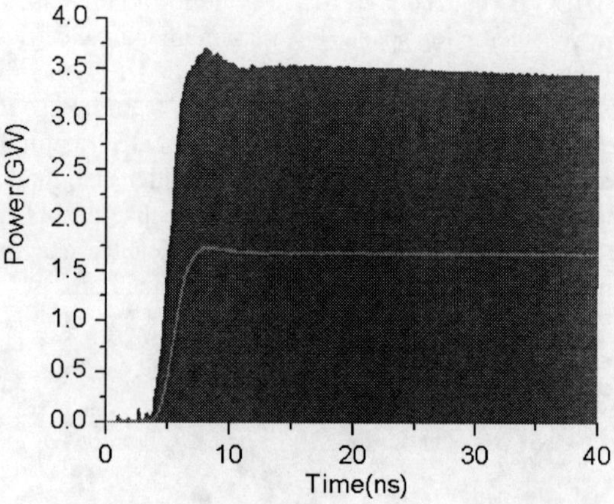

Fig. 11 Instantaneous and average output power of the improved Ku-band MILO

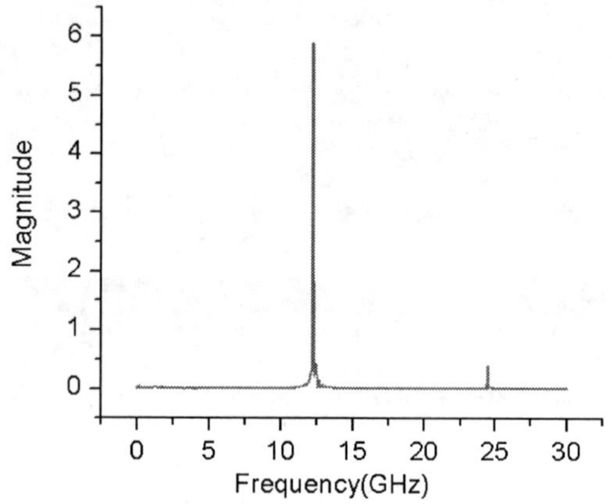

Fig. 12 Frequency spectrum of the output microwave

Tab. 1 shows the performance comparison between the traditional Ku-band MILO and the improved one. Compared with the traditional Ku-band MILO, the improved one proposed in this paper improves performance dramatically. First, the start-up time and saturation time are decreased greatly. Second, the maximum axial electrical field of the SWS is decreased from 1.3 MV/cm to 1.0 MV/cm, which means that the rf breakdown risk is decreased and the power capacity of the device is enhanced by a large margin. Additionally, the current density of the cathode is reduced 10%.

Tab. 1 performance comparison between the traditional Ku-band MILO and the improved one

Ku-band MILO	start-up time(ns)	Saturation time(ns)	Output power (GW)	Cathode current density (kA/cm^2)	Maximum axial field (MV/cm)
Traditional	5.5	13	1.65	0.3	1.3
Improved	3.5	7.5	1.65	0.26	1

V. THE LATEST EXPERIMENTAL RESULTS

The experimental investigation of the improved Ku-band MILO is conducted in a low impedance accelerator. The experimental layout is shown in Fig. 13. The far-field measurement is used to measure the output microwave power and frequency. Two horns (BJ-140 waveguides) are used to receive the HPM generated by the Ku-band MILO device. Two measurement channels are placed 5 m away from the radiation antenna, which satisfy the far-field condition. A digital oscilloscope (Aglient-92504A oscilloscope, the sampling rates is 80 Gs/s.) is used to measure the microwave waveform and the corresponding frequency directly. The detector waveform is measured through a measurement channel by adding a calibrated RF-diode crystal detector.

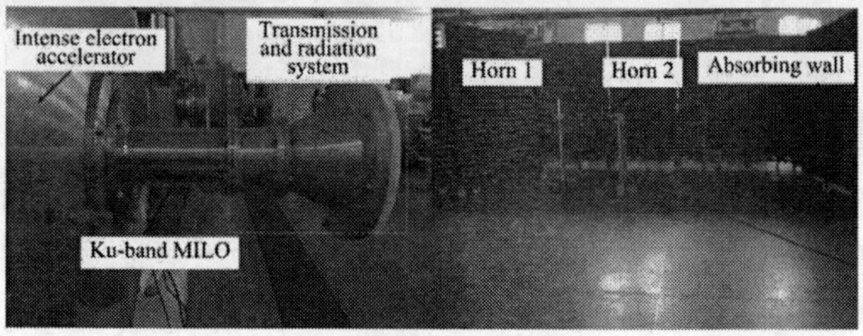

Fig. 13 Experimental layout

In the experiments, when the diode voltage is 515 kV and the diode current is 48 kA, the Ku-band MILO device generates a HPM with power of 1.2 GW, frequency of 12.36 GHz and pulse width of 35 ns.

The experimental results are presented as follows. Figure 14 shows the detector waveform. The generated microwave is about 1.2GW with FWHM pulse duration of 35 ns. The microwave waveform and the corresponding frequency spectrum are shown in Fig. 15. The central frequency is pure 12.36 GHz, which is in good accordance with the particle simulation result.

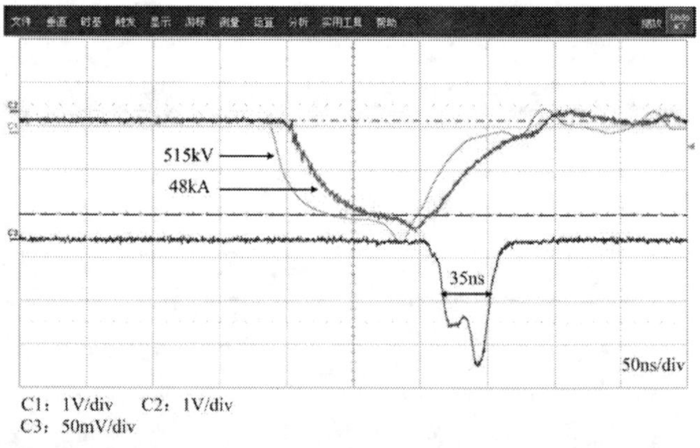

Fig. 14 Detection waveforms with diode voltage and beam current in experiment

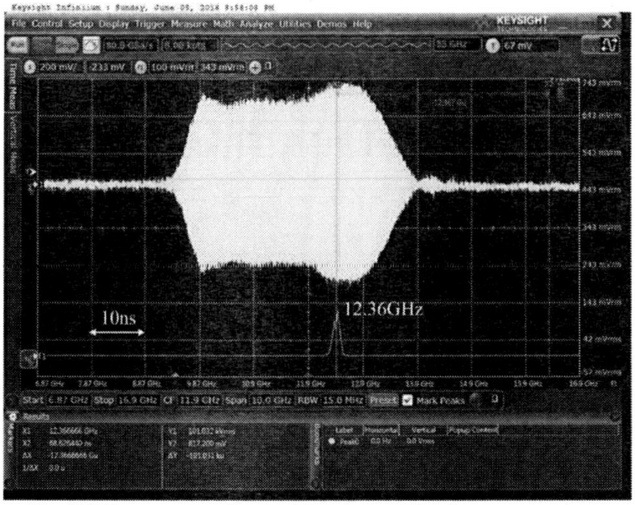

Fig. 15 Microwave waveform and its corresponding frequency spectrum

Compared with the experimental results of Ku-band MILO presented in Ref.17 and Ref.19, the experimental output power and pulse width of the improved Ku-band MILO are enhanced significantly. This verifies that the innovative improvements of tapered choke cavity and the enlarged first interaction cavity really help enhance the power capacity and lengthen the pulse width.

VI. Conclusion

An improved Ku-band magnetically insulated transmission line oscillator with tapered choke cavity and enlarged first interaction cavity is proposed and investigated numerically and experimentally. According to the analysis of the simplified plate model of MILO, the improved design of the Ku-band MILO is presented. A tapered choke cavity and an enlarged first interaction cavity with increased cavity gap are introduced to enhance the power capacity, and speed up the microwave start-up and saturation. Typical particle simulation results show that a microwave with a power of 1.65 GW and a frequency of 12.3 GHz can be radiated, corresponding to a power conversion efficiency of 8% under the diode voltage of 474 kV and beam current of 42 kA. The start-up time is decreased from 5.5 ns to 3.5 ns and the saturation time is reduced from 13 ns to 7.5 ns. In the experimental investigation, the improved Ku-band MILO device generates microwave with power of 1.2 GW, frequency of 12.36 GHz and FWHM pulse duration of 35 ns.

References

[1] R.J.Barker, E. Schanmiloglu, "High Power Microwave Sources and Technologies", New York: IEEE Press, 2001.

[2] James Benford, John Swegle, "High Power Microwaves", Artech House, Norwood, MA, 1991.

[3] Raymond W. Lemke, M. Collins Clark, "Theory and simulation of high-power microwave generation in a magnetically insulated transmission line oscillator," J. Appl. Phys. 62, 3436 (1987).

[4] M. Haworth, K. Allen, G. Baca, J. Benford, T. Englert, K. Hackett, K. Hendricks, D. Henley, R. Lemke, D. Price, D. Ralph, M. Sena, D. Shiffler, and T. Spencer, "Recent progress in the hard-tube MILO experiment", Proc. SPIE, vol. 3158, pp. 28–39, 1997.

[5] J. W. Eastwood, K. C. Hawkins, and M. P. Hook, "The tapered MILO," IEEE Trans. Plasma Sci., vol. 26, no. 3, pp. 698–713, Jun. 1998.

[6] R. W. Lemke, S. E. Calico, and M. C. Clark, "Investigation of a load-limited, magnetically insulated transmission line oscillator (MILO),"IEEE Trans. Plasma Sci., vol. 25, no. 2, pp. 364–374, Apr. 1997.

[7] R. Cousin, J,Larour, and J. Gardelle, "Gigawatt emission from a 2.4-GHz compact magnetically insulated line oscillator (MILO)," IEEE Trans. Plasma Sci.,vol. 35, no. 5, pp. 1467–1474, Oct. 2007.

[8] X. P. Zhang, H. H. Zhong, and C. W. Yuan, "Theoretical design and experimental studies of C-band MILO," High Power Laser Particle Beams, vol. 15, no. 1, pp. 80–84, 2004.

[9] Y. W. Fan, C. W. Yuan, H. H. Zhong, T. Shu, and L. Luo, "Simulation investigation of an improved MILO," IEEE Trans. Plasma Sci., vol. 35, no. 2, pp. 379–383, Apr. 2007.

[10] Z. Q. Li, H. H. Zhong, and T. Shu, "Experimental research of S-band tapered magnetically insulated transmission line oscillator at long-pulse operation," High Power Laser Particle Beams, vol. 20, no. 2, pp. 123–126.

[11] Z. Q. Li, H. H. Zhong, and T. Shu, "Investigation of an S-band tapered magnetically insulated transmission line oscillator," Chin. Phys. Lett., vol. 26, no. 5, p. 055201.

[12] D. B. Chen, Z. K. Fan, and H. J. Zhou, "L-band hard-tube magnetically insulated line oscillator," High Power Laser Particle Beams, vol. 19, no. 8, pp. 1352–1356.

[13] Y. W. Fan, H. H. Zhong, Z. Q. Li, C. W. Yuan, T. Shu, H. W. Yang, Y. Wang, and L. Luo, "Investigation of a 1.2-GHz magnetically insulated transmission line oscillator," IEEE Trans. Plasma Sci., vol. 39, no. 1, pp. 540–544, Jan. 2011.

[14] J. C. Ju, Y. W. Fan, H. H. Zhong, and T. Shu, "An improved X-band magnetically insulated transmission line oscillator," Phys. Plasmas, vol. 16, no. 7, p. 073103, 2009.

[15] Y. W. Fan, H. H. Zhong, and T. Shu, "Investigation of an X-band magnetically insulated transmission line oscillator," Chin. Phys. B, vol. 17, no. 5, pp. 1804–1808.

[16] J. Wen, Y. C. Tian, and Z. K. Fan, "Simulation of Ku-band magnetically insulated transmission line oscillator," High Power Laser Particle Beams, vol. 23, no. 11, pp. 3047–3050.

[17] J. Wen, D. B. Chen, D. Wang, and F. Qin, "Preliminary experimental research on Ku-band MILO," IEEE Trans. Plasma Sci., vol. 41, no. 9, pp. 2501–2505, Sep. 2013.

[18] Tao Jiang, Jiande Zhang, Juntao He, Jinchuan Ju, Zhiqiang Li, and Junpu Ling, "An Improved Ku-Band Magnetically Insulated Transmission Line Oscillator", IEEE Trans. Plasma Sci. 2015, 43 (10).

[19] Tao Jiang, Jiande Zhang, Juntao He, Zhiqiang Li, and Junpu Ling, "Experimental research on Ku-band magnetically insulated transmission line oscillator", Phys Plasmas, 2015, 22, 102112.

[20] Xiaoping Zhang, Fangchao Dang, Jun Zhang, Yuwei Fan and Zhiqiang Li, "Preliminary investigation of an improved metal-dielectric cathode for magnetically insulated transmission line oscillator", Review of Scientific Instruments, 2015, 86, 024705.

[21] Y. H. Guo, "theoretical and experimental investigation of a C-band magnetically insulated transmission line oscillator," Ph.D dissertation (China Academy of Engineering Physics, 2005).

Solid-State Planar Pulse Forming Modules by Utilizing the 2-D Electrode Manufacturing Technology *

Langning Wang, Jinliang Liu [1)], Yongfeng Qiu, Xu Chu
College of Optoelectronic Science and Engineering
National University of Defense Technology
Changsha 410073, People's Republic of China
1) Emai: llle333@163.com

***Abstract*—Recent technological advances have pushed further research to reduce dimensions of pulsed power system to a portable scale. Solid dielectrics have been widely investigated as potential, attractive candidates in compact, portable pulsed power systems. Nowadays, most of the solid-state circuits are fabricated through the application of the 2-D manufacturing technology. In this paper, we present three kinds of solid-state planar pulse forming modules with 2-D electrode manufacturing technology to adjust the output pulse duration and the matching impedance, to produce long pulse duration while less space volume is being occupied. Firstly, on a glass-ceramic (dielectric constant 225) substrate with the dimension 250 mm×95 mm×4 mm, as meander line is a commonly delay-line circuit in printed circuit boards (PCBs), it is adopted to deliver a high voltage pulse with pulse duration longer than 120 ns with simulation analysis and test validation. In addition, the double parallel and the 'V' shaped meander-line structure are also discussed in this section. Secondly, with the same material substrate, one kind LC adjustable pulse forming module is used to generate the high power pulse with good square pulse quality. By changing the L units, this module can be adjusted to deliver pulses with different durations. Thirdly, planar concentric ring pulse forming line is proposed with simulation analysis. In simulation, the planar concentric ring pulse forming line is designed using an attractive material for making PFL - Barium Titanate ($BaTiO_3$) base- ceramics with a dielectric constant about 1000. These designs of solid state pulse forming modules may provide a potential component for the compact pulsed power systems to generate high voltage, long-duration pulses.**

Keywords*—solid dielectric; pulse forming line; pulse forming network; 2-D slow-wave; pulse duration

I. Introduction

It is always a trend to reduce the scale of a pulsed power system and behave the good environmental adaptability with long life.[1, 2] The industrial requirements are also important for both the manufacturing process and the target applications. It means that we not only have to build the pulsed power system to meet the demands of the target applications, but also broaden our views of the mature manufacturing process in the industry to enrich the pulsed power technology. Pulse forming module, including the forms of pulse forming line (PFL) and pulse forming network (PFN), is one of the key components in a capacitive pulsed power system. It acts as an intermediate stage, which is generally charged to several tens or hundreds kilovolts voltage in few microseconds and then discharged to the electrical loads in certain pulse duration time (usually few tens to hundreds of nanoseconds) with suitable pulse quality.

Ceramic attracts strong interest in pulsed power technology due to its essential characteristics such as stable environmental adaptability, high energy storage density since it can behave in a high dielectric constant and high breakdown strength at the same time.[3, 4, 5] Unfortunately, the difficulty of manufacturing large ceramic bulks also influences its extensive use in pulsed power systems.[5] What is worse, the breakdown strength, E_b, defined as breakdown voltage, V_b, per sample thickness, t, of ceramic materials decreases with increasing volume (increasing the size and the thickness).[6, 7, 8] Usually, the volume-effect of the mechanical strength can be explained by the Weibull theory. Volume-effect explained by the Weibull theory has succeeded in the analysis of the mechanical failure of ceramic materials occurs due to micro structural defects. It suggests that the thickness-dependence of the breakdown strength should be explained similar to the mechanical case. Thus, in the case of breakdown strength, the influences of the porosity, grain size, purity and secondary phase on the breakdown strength have been widely investigated, which indicates that the electrical breakdown field is dependent on the purity: the higher the purity, the higher the breakdown field. Some researchers from Hamburg University of Technology have contributed the thickness-dependence on breakdown strength to the space charge injection, not similar to the Weibull concept.[8] Test results show a great difference in the Weibull moduli between breakdown and mechanical strength distribution. In particular, the thickness-dependence of the breakdown strength always leads to a Weibull modulus of two which is not in agreement with the measured Weibull moduli for samples with constant thickness. The thickness-dependence of the extrinsic dielectric breakdown strength scales as $d^{1/2}$ down to the micrometer scale. In a summary, when the pulse forming modules using ceramics are designed to operate under hundreds kilovolts, it cannot be adapted that we only build one ceramic pulse forming modules with large thickness similar to the conventional way used in the liquid pulse forming line. The ceramic pulse forming modules should be manufactured at thinner thickness to hold off tens kilovolts, and then the thin pulse forming modules are stacked to generate the hundreds kilovolts pulse with certain duration time. Currently, PFN-Marx is widely proposed to deliver hundreds kilovolts pulse, which the output voltage of a single stage PFN is often limited by the insulation strength of the ceramic capacitors about tens kilovolts.

PFN, formed by identical inductors and capacitors, is able to adjust electrical parameters of output pulse, since the identical inductors and capacitors can be adjusted.[9, 10, 11] However, the self-inductance of the ceramic capacitor is hard to be reduced, which will affect the fast rise time and the quality of output pulse waveform unpredictably when pulse duration with the magnitude of 100 ns. What is worse, the complex electrode fabrication for deceasing inductance is less possible manufactured inside the solid-state dielectric bulks while the capacitors can hold off high voltage at the same time. In additions, the usage of the large number of ceramic capacitor will waste some volume, since every capacitor has a heavy insulating encapsulation and the arrangement of every *LC*-unit must leave some useless space. The capacitance coupling between capacitors and the inductance coupling do exist, and the distributed coupling electrical parameters will affect the quality of output pulse waveform unpredictably. Therefore, as an intermediate stage, it is suitable to adapt the solid-state pulse forming modules to improve high quality square pulse and stack them to raise the output power.

Typically, planar structure and the coaxial structure are the basic structures. Coaxial structure has the advantages of the less coupling and shorter electrode edges, compared to the planar structure, while it is extremely difficult to make the solid-state pulse forming modules as the coaxial structure and the inner electrode will waste some space. Thus, the planar structure may be available. Moreover, 2-D electrode manufacturing technology is mature to form complex electrode on the planar structure pulse forming module.[12] Based on it, 2-D slow-wave structures are able to adjust the output pulse duration and the matching impedance of the planar pulse forming modules, so that the pulse duration of the pulsed power system can be scaled up, while the size of PFL decreased, similar to the case that the tape helix is applied in pulsed power technology to construct helical pulse-forming line [13, 14]. In addition, narrower lines have a faster response, which converges to an ideal square shape.[15] 2-D slow-wave structures can ensure a narrow line wide in a wide ceramic substrate so that the pulse quality can be improved.

In this paper, we present three typical designs for the planar solid state pulse forming modules and every kind can deliver a pulse with several tens to hundreds nanoseconds pulse duration.

II. TREE TYPES OF PULSE FORMING MODULES

A. *Meander Pulse Forming Line*

The structure of meander pulse forming line (MPFL) is indicated in the Fig. 1. One side of the dielectric substrate is fully covered by a solid layer of a conductor, which may carry out the function of the earth ground electrode. The other electrode, the array of coupled parallel commensurate conductors called a meander line, is formed on the other side of the dielectric substrate with conducting links connecting the ends of adjacent strips alternately at the two sides of that region. This meander line, shown in Fig. 1, is characterized by the height, h, the meander pitch, p(dimension $p=2(w+s)$), line width, w, 5 mm, line separation, s, 5 mm, of the meander electrode and the thickness, h_d, 4 mm, of the dielectric substrate. The relative permittivity of the dielectric substrate is ε_m. The common analysis for meander line is carried out in a quasi-TEM approximation, in which there is simply a fixed phase-change ϕ from one tape to the next. The theoretical calculations of dispersion effects can be obtained with our previous study.

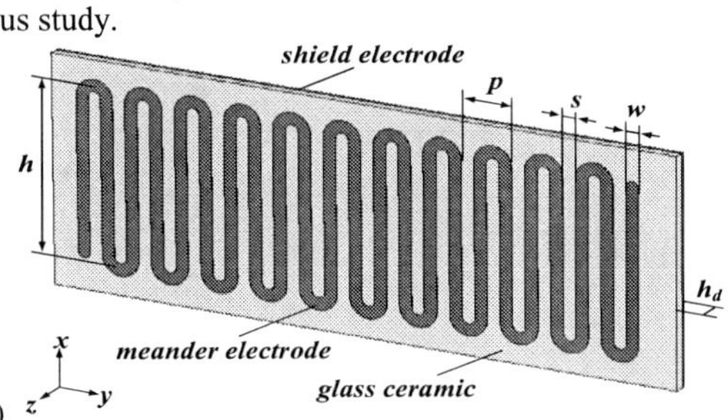

(a)

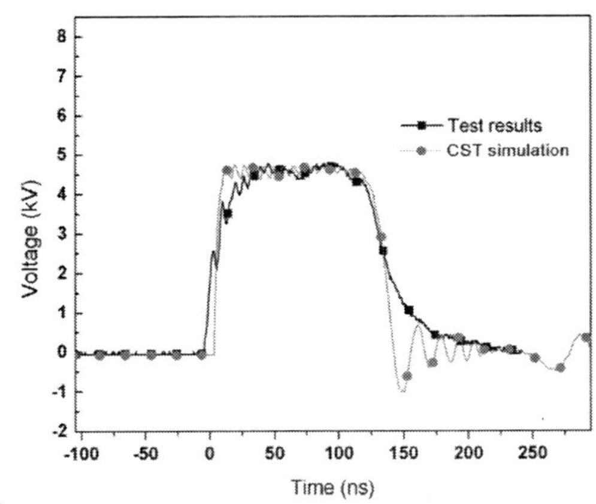

(b)

Fig.1 Geometry for the MPFL (a) and the formed waveforms in the simulation and test (b)

The MPFL is made on a glass-ceramic substrate fabricated by Advanced Electronic Materials Institute, General Research Institute for Nonferrous Metals, of which the dielectric constant is about 225.[16] Previous bulks breakdown tests show high breakdown strength achieved up to the 26 kV/mm level.[17] The candidate glass-ceramic is fabricated differently from the traditional process, named slip casting techniques.

Molten glass at high temperature is used as the base material in which high permittivity nm composites are dispersed in order to achieve the ceramic phase with precipitations size in the range of several tens of nm. Using slip casting techniques, the porosity of glass-ceramic can be significantly reduced by quenching directly from the molten glass and the grain of ceramic phase can be maintained in nm size. The ceramic phase with high dielectric constant is densely enveloped by the glass phase, which makes the ceramic bulk be capable of both high breakdown strength and high dielectric constant. The glass-ceramic has been studied intensively for applying to high voltage capacitors. In our study, MPFL is 250 mm×95 mm in size, fabricated on a substrate of 4 mm thick. Test and simulation waveforms obtained in the 7.5 ohms load are compared in the Fig. 2, and the environmental dielectric is Propylene carbonate (PC) with 69 of dielectric constant. The description of simulation methods can be found in our previous study.[18]

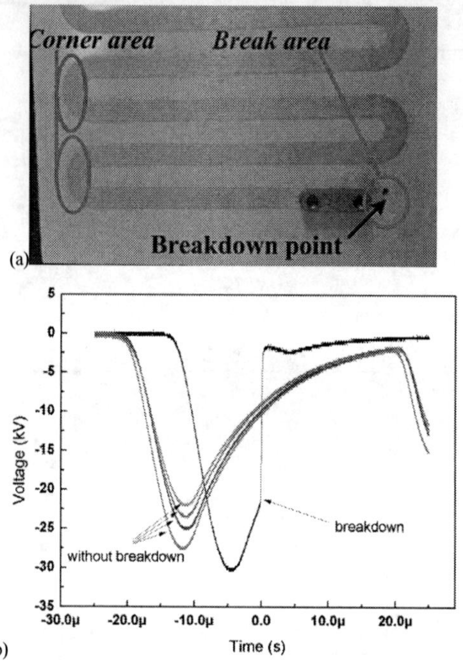

Fig.2 the breakdown point location (a) and the waveforms obtained from breakdown tests (b)

Further work about this MPFL encounter difficulty of the insulting, though previous work shows the same thickness substrate without thin electrode behave a high breakdown strength achieved up to the 26 kV/mm level. In liquid environmental insulting condition, the 4 mm thick MPFL can hold off no more than 23 kV and Fig.2 (a) shows a photo of the breakdown point location. All the breakdown points are located on the corner area or break area. After that, we use epoxy-encapsulated way to promote the withstand voltage and the breakdown test results show an improvement on the high voltage and the waveforms of voltage imposed on the MPFL in the breakdown tests are shown in the Fig.2 (b).

Based on the test results for the locations of break points, the MPFL electrode structure in Fig. 1 should be optimized. [19] Firstly, for the face opposite to the meander line structure, since the thin conductive layer at this face is almost covered the whole face, we also do a surface metal working process to form a corresponding pattern to reduce the field enhancement near the corner. Secondly, as Fig. 2 (a) shows that the 'break' area should be removed. Followed by the ideas mentioned above, a dual parallel MPFL structure has been presented in Fig. 3(a). The red ports indicate the connected locations in one PFL. Additionally, in order to exhibit that the various kinds of the 2-D slow wave structure can be used in the MPFL, the V-shape MPFL is designed and tested with pulse forming experiments. In fact, both the MPFLs shown in Fig. 3 are tested and show a good ability of forming high-quality square pulse.

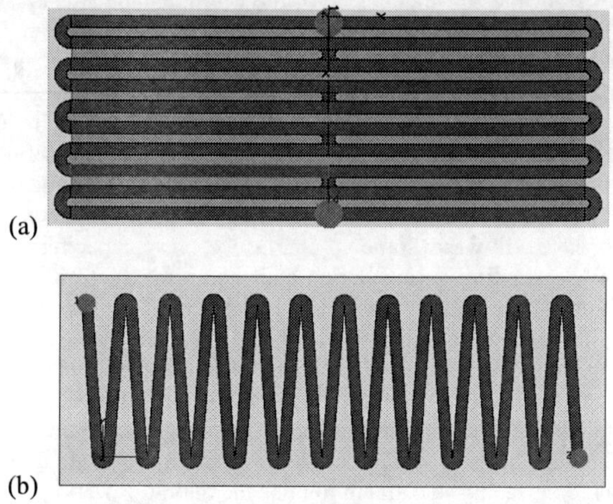

Fig. 3 Dual parallel MPFL structure (a) and V-shape MPFL (b)

B. *LC Pulse Forming Module*

In this section, we use the same glass ceramics substrate. 8 strips are fabricated on the one side of the substrate, shown in Fig. 4. Each strip can be used as capacitance similar to the function of the pulse forming networks and we can add copper connectors between the adjacent capacitors as the L, which the schematic circuits are shown in Fig 4 (b).

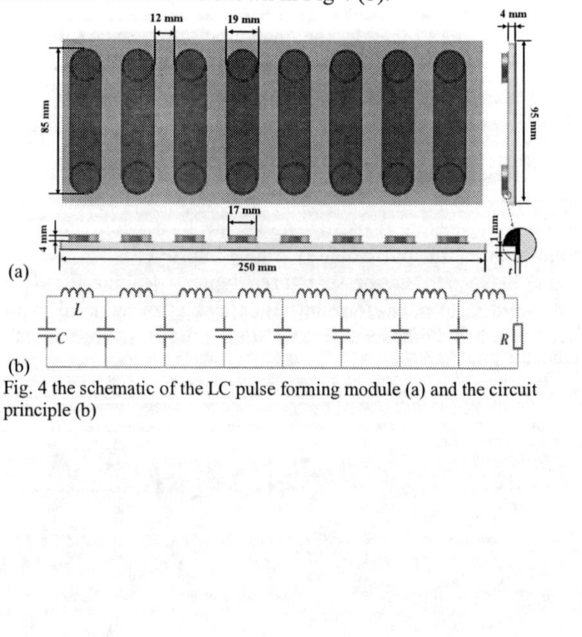

Fig. 4 the schematic of the LC pulse forming module (a) and the circuit principle (b)

In Fig. (5), if the copper connectors are changed that means the inductance L changing, the pulse duration of output pulse is varying. The reason is the same as the PFN principle.

As indicated in the Fig. 5, with the length of the copper connector increasing, the inductance L is increasing. Thereby, the pulse duration of output pulse is larger. Since the storage energy stored in the same 8 strips is basically unvarying, the Fig. 5(b) show a unmatched impedance while Fig. 5(a) show a good match impedance. In detail, with the length of the copper connector increasing, the impedance of the pulse forming modules is also increasing.

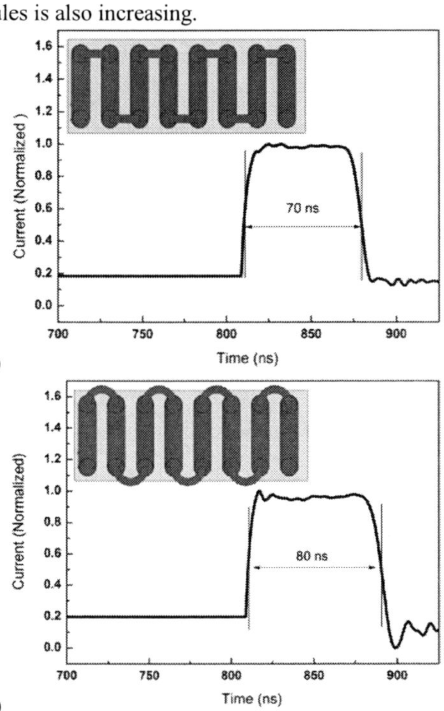

Fig. 5 Waveforms of current on the 3 ohms load by CST simulation with different inductance L

As shown in Fig. 5, the copper connectors can be machined thinner and longer, so the inductance can be enlarged and pulse duration can be added up to more than 100 ns. In particular, if the glass-ceramic substrate is made for this form, the substrate can be machined to form 8 grooves corresponding the 8 strips in the face by slip casting techniques, to embed partial electrode into the ceramic bulk, so that the high E-field enhancement near the thin electrodes and the ceramic bulk can be reduced.

C. Concentric Ring Planar Pulse Forming Line

As improving the design of meander line structure shown in Fig.3 (a), we have supposed a kind of concentric ring planar pulse forming line. This structure will reduce the waste space if we want to increase the radius of the corner, and the red ports indicate the connected locations in one PFL. In fact, the plate structure may be good for the Marx stacked methods. In detail, several plates can be placed across the same axis and then the inner red ports can connected to a switch electrode in the hollow shown in Fig 3(a). Therefore, the numbers of the plates are determined by the voltage requirements and the pulse duration is determined by the electrode length and pattern, which is quite different from the cylinder pulse forming line with liquid dielectric.

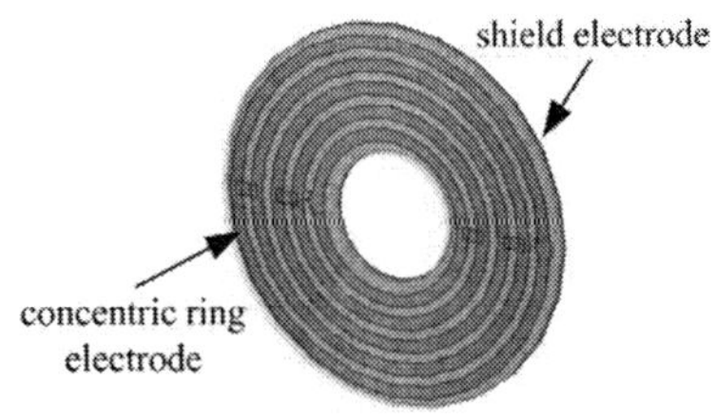

Fig.6. The flow chart how design is varying (a) and the final design (b)

Fig.7 shows a structure of the concentric ring planar pulse forming line and the waveform of the current through the 1.3 ohms load in the CST simulation. In simulation, the dielectric constant of the material is set as 1000, which can be obtained by the Barium Titanate ($BaTiO_3$) base- ceramics.

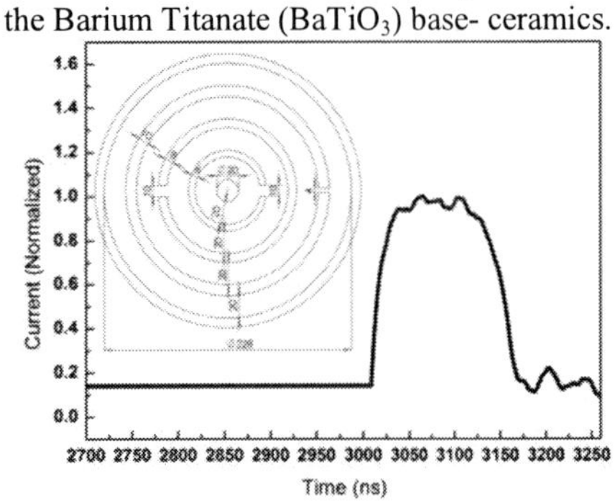

Fig.7 The electrode structure of the concentric ring planar pulse forming line and the output waveform of current

The radial slow-wave structure has carried out on the radial beam wave interaction. This kind of slow-wave structure is very rare, only the logarithmic spiral and log-periodic ring-bar slow-wave structure have been used to design the radial microwave amplifiers tube successfully. The analyses for planar nature of the Cartwheel circuit may contribute to the improving the pulse-forming quality.[20] Further study on it should proceed.

III. Conclusions

Three kinds of solid-state planar pulse forming modules with 2-D electrode manufacturing technology are presented to adjust the output pulse duration and the matching impedance, to produce long pulse duration while less space volume is being occupied. The simple meander pulse forming line has been manufactured on a glass-ceramic (dielectric constant~225) and showed a good pulse-forming quality both in the CST simulation and pulse forming test. In order to promote the working voltage of the meander pulse forming line, some improved MPFL are proposed. Then, on the same glass-ceramic substrate, 8 independent strips are used as the unit capacitance of the PFN. The length of copper connectors between the strips can be adjustable, which means the unit inductance of the PFN can be set as different values to change the pulse-duration of the output pulse. At last, the radial slow-wave structure is introduced and the simulation is carried out to perform a square pulse forming quality. In this paper, these designs show a potential attracts in the ceramic material in the area of pulsed power technology. With the help of the electrode structures, we can design the pulse forming modules to satisfy the requirements of the impedance and the pulse duration based the chosen storage material. As for the high power requirements, how to stack them to provide a good pulse quality need to be researched in the future. At present, we prefer the Marx-forms to increase the output voltage.

References

[1] W. Jiang, K. Yatsui, K. Takayama, M. Akemoto, E. Nakamura, N. Shimizu, et al., "Compact solid-state switched pulsed power and its applications," Proc. IEEE, vol. 92, no. 7, pp. 1180-1196, Jul. 2004.

[2] J. A. Gaudet, R. J. Barker, C. J. Buchenauer, C. Christodoulou, J. Dickens, M. A. Gundersen, R. P. Joshi, H. G. Krompholz, J. F. Kolb, and A. Kuthi, "Research issues in developing compact pulsed power for high peak power applications on mobile platforms," Proc. IEEE, vol. 92, no. 7, pp. 1144-1165, Jul. 2004.

[3] L. Xia, H. Zhang, J. Shi, H. Zhang, J. Deng, H. Liu, and M. Cao, "A compact, portable pulse forming line," Rev. Sci. Instrum., vol. 79, no. 8, pp. 086113, Aug. 2008.

[4] S. K. Sharma, P. Deb, R. Shukla, T. Prabaharan, and A. Shyam, "Compact pulse forming line using barium titanate ceramic material," Rev. Sci. Instrum., vol. 82, no. 11, pp. 115102, Nov. 2011.

[5] J. Su, X. Zhang, R. Li, L. Zhao, X. Sun, L. Wang, et al., "An 8-GW long-pulse generator based on Tesla transformer and pulse forming network," Rev. Sci. Instrum., vol. 85, p. 063303, 2014.

[6] A. Young, G. Hilmas, S. C. Zhang, and R. W. Schwartz, "Effect of liquid-phase sintering on the breakdown strength of barium titanate," J. Am. Ceram. Soc., vol. 90, no. 5, pp. 1504-1510, May. 2007.

[7] D. Malec, V. Bley, F. Talbi, and F. Lalam, "Contribution to the understanding of the relationship between mechanical and dielectric strengths of Alumina," J. Eur. Ceram. Soc., vol. 30, pp. 3117-3123, 2010.

[8] C. Neusel, and G. A. Schneider, "Size-dependence of the dielectric breakdown strength from nano-to millimeter scale," J. Mech. Phys. Solids, vol. 63, pp. 201-213, Feb. 2014.

[9] J. Hammon, S. Lam, and S. Pomeroy, "A transportable 500 kV, high average power modulator with pulse length adjustable from 100 ns to 500 ns," in Proc. IEEE Int. Pulsed Power Conf., 1995, pp. 429-434.

[10] H. Zhang, J. Yang, J. Lin, and X. Yang, "A compact bipolar pulse-forming network-Marx generator based on pulse transformers," Rev. Sci. Instrum., vol. 84, p. 114705, 2013.

[11] L. Wang, J. Liu, and J. Feng, "A compact 100 kV high voltage glycol capacitor," Rev. Sci. Instrum., vol. 86, no. 1, p. 014701, Jan. 2015.

[12] J. H. Booske, "New opportunities in vacuum electronics through the application of microfabrication technologies," in Proc. IEEE Int. Vac. Electron. Conf., Monterey, CA, 2002, pp. 11–12.

[13] L. Wang and J. Liu, "Dispersion analysis of the solid helical pulse forming line," Laser Part. Beams, vol. 33, no. 03, pp. 405-413, Sep. 2015.

[14] L. Wang and J. Liu, "Two Ways to Increase Pulse Duration of the Solid Pulse Forming Line based on Dispersion Analysis," IEEE Trans. Plasma. Sci., Vol. 43, No. 11, pp. 3901–3907, 2015.

[15] M. Joler. Analysis and Optimization of Parallel-Plate Blumlein Line for CompactPulsed-Power Systems [D]. Ph.D. Dissertation, The University of New Mexico, Albuquerque, New Mexico, USA, 2006.

[16] Q. Zhang, L. Wang, J. Luo, Q. Tang and J. Du, "Improved energy storage density in barium strontium titanate by addition of BaO–SiO2 – B2O3 glass", J. Am. Ceram. Soc., Vol. 92, pp. 1871-1883, 2009.

[17] S. Wang, T. Shu, J. Zhang, Z. Zhang, and H. Yang, "Breakdown characteristics of niobate glass-ceramic under pulsed condition," IEEE Trans. Dielectr. Electr. Insul., vol. 20, no. 1, pp. 275-280, Feb. 2013.

[18] L. Wang, J. Liu, X. Chu and Q. Zhang, "Effect of the Different Environmental Dielectric on the Pulse-Forming Characteristics of Solid State Meander Pulse Forming Line," IEEE Trans. Plasma Sci., Vol. 44, No. 5, pp. 821–828, 2016.

[19] L. Wang, J. Liu and Q. Zhang, "Analysis of the Electrode Design to Increase the Withstand Voltage of Solid State Meander Pulse-Forming Line," IEEE Trans. Dielectr. Electr. Insul., to be published.

[20] T. Wessel-Berg, "Basics of radial sheet beam interactions with potential applications in the microwave K and W bands," High Energy Density High Power RF: 7thWorkshop, Kalamata, Greece, 2006, vol. 807, pp. 55–64.

Simulation on Magnetic Field Distribution of the 20-kJ Pulse Forming Unit for Electromagnetic Launcher

Guo Xiaolong, Dai Ling , Lin Fuchang, Li Jian, Huang Yucheng,Liu Wenxun

(Central Southern China Electric Power Design Institute of China Power Engineering Consulting Group, Wuhan 430071 , Hubei Province , China)

***Abstract*—A 20-kJ pulse forming unit (PFU) for electromagnetic launcher has been designed. The PFU is comprised of a high energy density capacitor (3270 μF/3.5 kV), a 4 inches thyristor, a pulse shaping inductor (1.5 μH) and a crowbar diode. Parameters of each part in the PFU have been elaborated to maintain a high efficiency of the power system. The peak value of the output current is 60 kA with a peak time of 245 μs. The PFU will produce strong magnetic field intensity when discharging, which will affect the performance of control system. It is necessary to reduce the magnetic field intensity to an acceptable value. To study the magnetic field distribution of the PFU, a PFU model has been established based on Ansoft Maxwell. The solution type is transient. Eddy current effect and skin effect are taken into consideration in simulation. A proper shielding measure is used to reduce the magnetic field intensity based on the simulation results. After manufacturing the PFU, measurement of magnetic field experiments are carried out to verify the correctness of the simulation results. The results show that finite element simulation can be effectively used in the design of the PFU.**

***Index Terms*—Pulse forming unit; magnetic field distribution; finite element simulation; shielding measure**

I. Introduction

ELECTROMAGNETIC launching is one of the main applications of pulsed power supply. Due to the special application situation, the reliability of the PFN should be guaranteed. However, PFN will produce strong magnetic field intensity when discharging, which will affect the performance of the control system. Therefore, it is particularly important to study the distribution of magnetic field of PFN when designing and suitable shielding measures should be taken to reduce the influence. Finite element is used as the main method to simulate the distribution of magnetic field currently. However, the PFN is complex, it is difficult to establish finite element model. In the meantime, high performance server should be used to solve the model. Therefore, few works has been carried out to analysis the distribution of magnetic field of PFN as we investigated.

A stable and reliable 20kJ-PFN used for electromagnetic launching has been developed in this paper. The peak value of the output current is 60 kA with a peak time of 245 μs. To study the distribution of magnetic field, the finite element model of PFN has been established based on Ansoft Maxwell. Distributions of magnetic field were analyzed in the case of unshielded and shielded. The simulation provides reference to the designing of PFN.

X. Guo, J.Li, Y.Huang, and W.Liu, are all with Central Southern China Electric Power Design Institute of China Power Engineering Consulting Group,Wuhan Hubei, 430071, China.

L. Dai and F.Lin are with State Key Laboratory of Advanced Electromagnetic Engineering and Technology, Huazhong University of Science and Technology,Wuhan Hubei, 430074, China(e-mail: dailing@mail.hust.edu.cn).

II. Pulsed Forming Unit

The electrical circuit of a 20-kJ PFU is shown in Fig. 1.

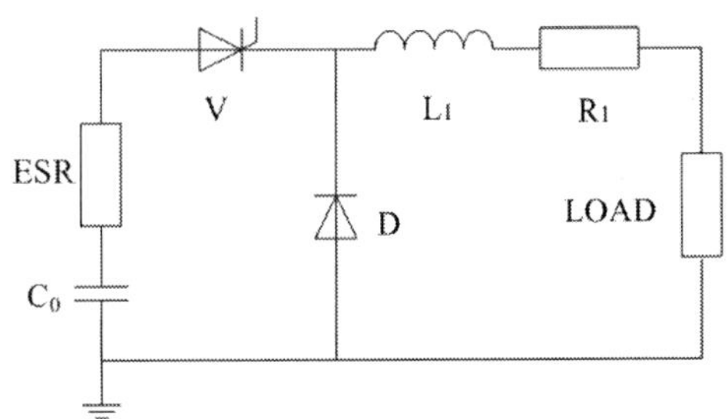

Fig.1. Circuit model of 20 kJ-PFU

The configuration of a 20 kJ PFU is shown in Fig.2, and the PFU is composed of a high energy density pulsed capacitor (3270 μF), a thyristor, a crowbar, an inductor（1.5μH）and a trigger system.

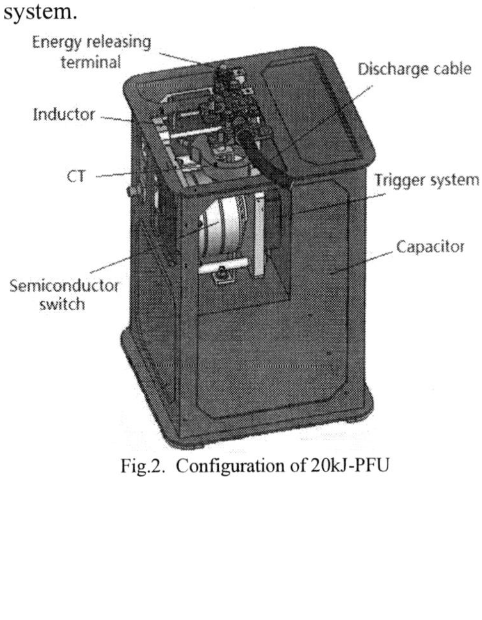

Fig.2. Configuration of 20kJ-PFU

A. Capacitor

Dry metalized capacitor is used as an energy stored element in the PFU. The self-healing property of metalized capacitor guarantees the safety and economy. It's possible to monitor the capacitance of the capacitor, dielectric loss and other electrical parameters to maintain the insulation condition of the capacitor, which partially avoided the sudden breakdown of the capacitor. The rated capacitance of capacitor is 3270 μF and the rated voltage is 3.5 kV, with the energy 20 kJ. To decrease the leakage of voltage after stopping charging, a low energy storage density (1.0 MJ/m^3, 3.5 kV) capacitor is used. For this energy storage density at the voltage of 3.5 kV, the leakage current is 17.63 mA in 60 s (Voltage drop rate curve is shown in Fig.3. Lifetime of this capacitor is larger than 1000 charge/discharge cycles at rated voltage.

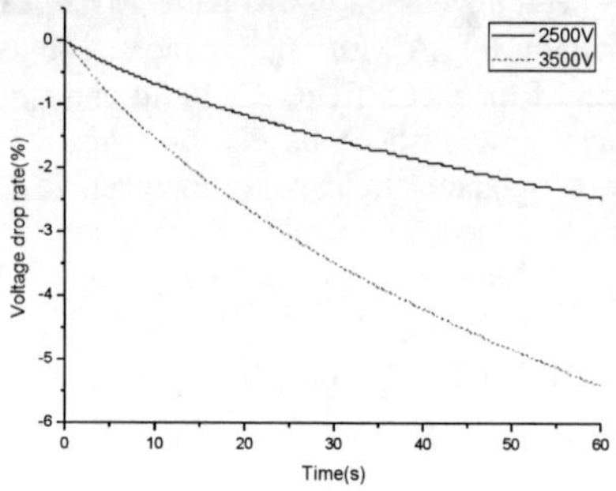

Fig.3 Voltage drop rate curve

B. Thyristor

The forward blocking voltage of the thyristor used as discharge switches in PFU is 5.2 kV. The diameter of the valve is 4 inches. Limit value test show that this thyristor can endure pulse current of 146 kA/5 ms and the maximum di/dt is 1.5 kA/μs. Life test displays that the thyristor were not damaged after 1000 shots. Parameters of thyristor are shown in Tab. I.

TABLE I
PARAMETERS OF THYRISTOR

Type	Limit surge current		Limit di/dt (kA/μs)	Limit di/dt	
	Peak (kA)	Width (μs)		Peak (kA)	Width (μs)
4 inches	146	400	1.5	15	30

C. Pulse Shaping Inductor

Inductor is used to adjust the load waveform and output energy in PFU, therefore, the electrical characteristics and reliability of the inductor will affect the overall performance to PFU directly. The inductor should have a good current-through capability and impact-resist capability. Warren F.T. proposed a disc structure to improve the impact-resist capability [10]-[11]. Inductor models of different structures were built in Ansoft Maxwell, the simulation result of magnetic field distribution shows that the magnetic flux leakage of disc inductor is higher than solenoid structure. What's more, the volume of solenoid structure of low inductance is smaller. Therefore, solenoid inductor is adopted. The diameter of the inductor is 95 mm, the height is 46 mm.

D. Crowbar Diode

The reverse blocking voltage of the diode is 5.2 kV. The diameter of the valve is 4 inches. Crowbar diodes in PFN conveniently perform important functions. They automatically prevent voltage reversal on capacitors and trap energy in the pulse shaping inductor.

III. SIMULATION

As it can be seen from the circuit model of 20 kJ-PFU, PFU is composed of a high energy density pulsed capacitor, a thyristor, a crowbar, an inductor and a trigger system. When the thyristor is triggered, the energy storaged in the capacitor will drain on the load quickly and the inductor is charged simultaneously. The current increases rapidly. At this time, the diode is cut off because of the reverse voltage. When the current reaches at the peak value, the magnetic field intensity reaches to the maximum value. However, the diode is still cut off at this time. Therefore, when establishing the finite element model, the diode is not taken into consideration. In practical situations, coaxial cable will be used to connect PFU with the railgun. In order to analyze the magnetic field distribution of PFU individually. The PFU will short-circuit discharge in simulation. The discharging current is shown in Fig.4.

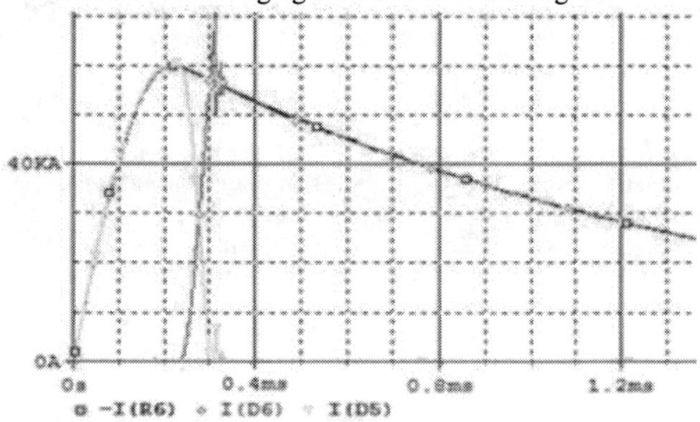

Fig.4 Discharging current

Where the green line shows the current which through the thyristor. The blue line shows the current which through the diode. As it analyzed before, the green line can be used as the exciting current in simulation.

A. Model

Finite element model of PFU has been established based on Ansoft Maxwell. The model are shown in Fig.5(unshielded) and Fig.6(shielded). The thickness of shield is 0.5mm. The material is iron. Base on the navy specifications (RE101), the magnetic field intensity (7cm and 50cm from the PFU) are assessed in the simulation.

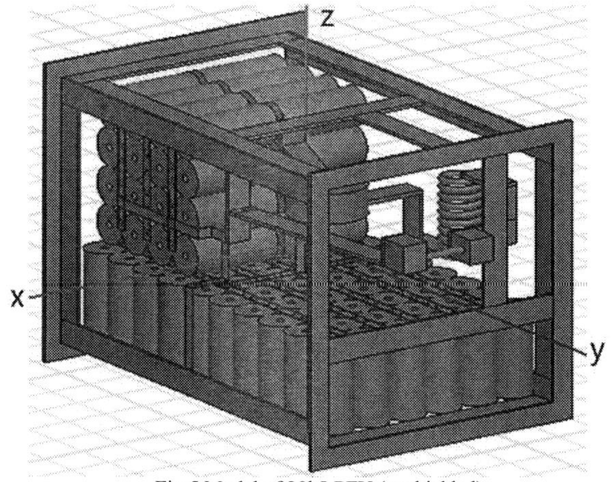

Fig.5 Model of 20kJ-PFU (unshielded)

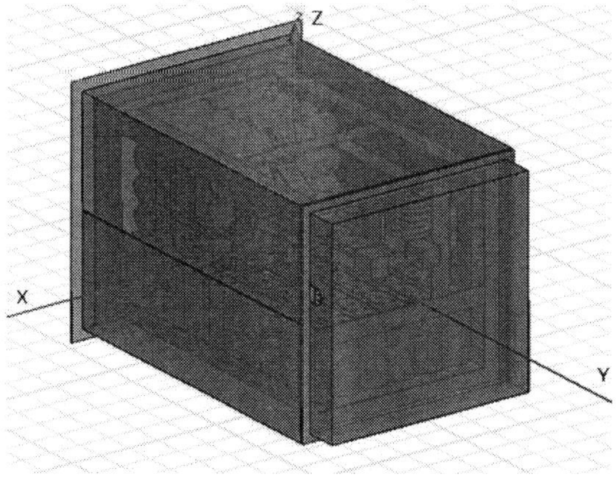

Fig.6 Model of 20kJ-PFU (shielded)

B. Solution conditions

Solution conditions are shown as below,

1) Solver: Transient field solver.

2) Meshing: Curved area use 'surface meshing', the remaining area use 'length meshing'.

3) Eddy current effect is taken into consideration.

4) Excitation:
if(Time>128us,60000*cos(44880*(Time-128us)),60000*sin(16411.4*Time)). (Fig.7)

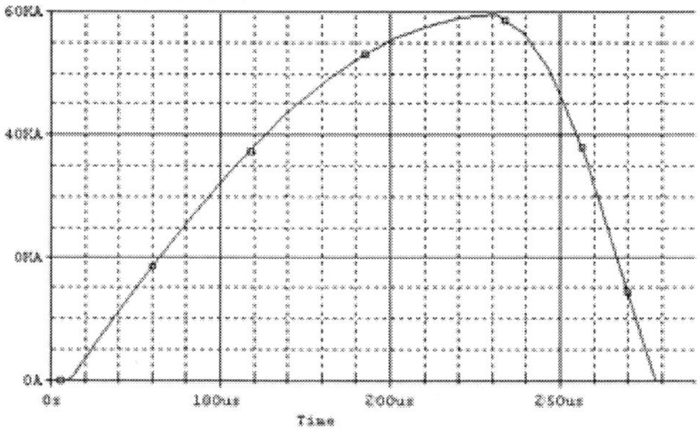

Fig.7 Excitating current

5) Time step: 1.28us.

6) Radiation boundary.

C. Simulation results

When the discharge current reach the maximum value, the distribution of magnetic field is shown as Fig.5.

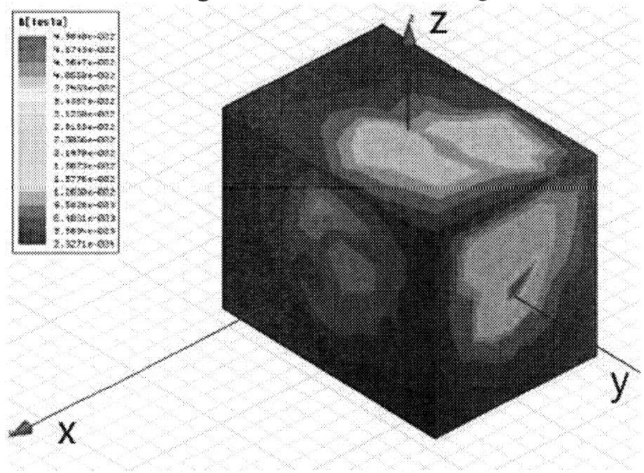

Fig.8 Distribution of magnetic field (unshielded-7cm)

The simulation results are shown in Tab.II.

TABLE II
SIMULATION RESULTS

Distance	Max B when unshielded	Max B when shielded
7 cm	498.40 Gs	20.35 Gs
50 cm	8.20 Gs	0.56 Gs

It can be seen from the simulation results that the shielding measures reduce the magnetic field intensity to an acceptable value effectively.

IV. EXPERIMENT

To verify the correctness of the simulation results. The measurenment of magnetic field experiments are carried out.

The magnetic field intensity have been measured in the case of shielded and unshielded. The measuring points are 7 cm far from each surface centers of PFU. In order to eliminate the influence of the loads, PFU will take a short circuit discharging in this test. The test results will be compared with the simulation results.

Test layout is shown in Fig.9 and the shielded PFU is shown in Fig.10.

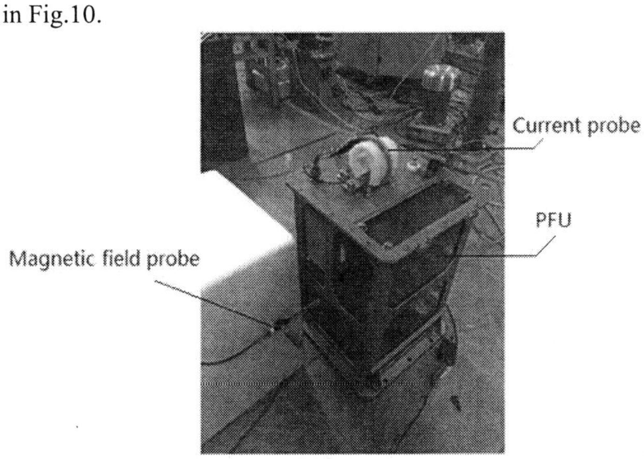

Fig.9 Test layout

Fig.10 The shielded PFU

The typical current waveform and the magnetic field intensity waveform are shown in Fig.11.

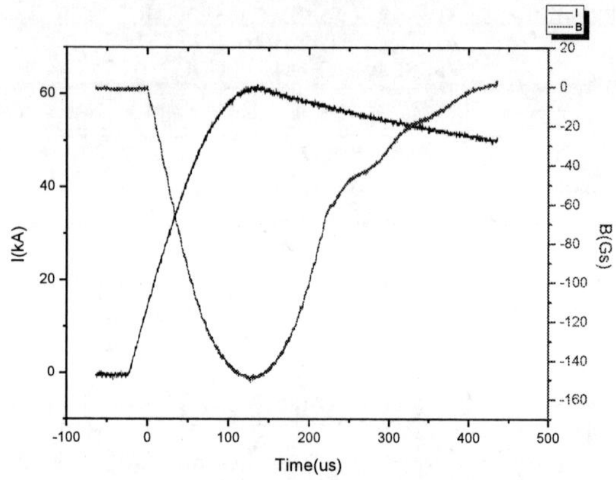

Fig.11 The typical current waveform and the magnetic field intensity waveform

The measuring point is at the positive of z-axis and it is 7 cm far from the PFU. The black line shows the current waveform and the blue line shows the magnetic field intensity in Fig.11.

The magnetic field intensity are calculated in the same conditions based on the finite element model. And the max values of magnetic field intensity in the case of measurement and simulation are shown in Tab.III.

TABLE III
COMPARISON BETWEEN MEASUREMENT AND SIMULATION

position	unshielded		shielded	
	measurement	simulation	measurement	simulation
y-	17.6	17.2	5.4	2.3
z+	149.4	182.3	13.3	7.6
x+	100.2	121.4	5.2	5.1
x-	107.7	136.8	12.1	6.4
z-	16.3	17.1	5.8	2.2
y+	348.2	402.6	7.2	6.2

It can be seen from the above table, the magnetic field intensity will decrease significantly when take the shielding measure, which meet the requirements of the specifications. To some degree, the simulation is accurate enough to provide references to the designing of PFU. The test verifies the correctness of the simulation results.

V. CONCLUSIONS

(1) A new-designed 20kj-PFU is shown in this paper. The PFU is composed of a high energy density pulsed capacitor (C=3270 μF), a thyristor, a crowbar, an inductor（L=1.5μF）and a trigger system. The peak value of the output current is 60 kA with a peak time of 245 μs.

(2) The PFU will produce strong magnetic field intensity when discharging. However, shielding measures can reduce the magnetic field intensity to an acceptable value effectively. The simulation provides reference to the overall optimization of PPS design.

(3) After manufacturing the PFU, measurement of magnetic field experiments are carried out to verify the correctness of the simulation results. The results show that finite element simulation can be effectively used in the design of the PFU.

REFERENCES

[1] E. Spahn, G. Buderer, and W. Wenning, "A compact pulse forming network, based on semiconducting switches, for electric gun applications," Magnetics, IEEE Transactions on, vol.35, no.1, pp.378-382, Jan 1999.

[2] Y.-S. Jin, H.-S. Lee, J.-S. Kim, Y.-B. Kim, and G.-H. Rim, "Compact 200 kJ pulse power system with a simple crowbar circuit," in Dig.14th Pulsed Power Conf. USA, 2003, pp.1239-1242.

[3] M. Shaoliang and L. Baoming, "Setup of pulsed power system for electro-thermal chemical launcher research," Trajectory Trans., vol. 16, p. 86, Mar. 2004, China.

[4] Y.-S. Jin, H.-S. Lee, J.-S. Kim, Y.-B. Kim, and G.-H. Rim, "Novle crowbar circuit for compact 50kJ capacitor bank," Plasma Science, IEEE Transactions on, vol.32, no.2, pp.525-530, April 2004.

[5] E. Spahn, V. Zorngiebel, K. Sterzelmeier, G. Buderer, and V. Brommer, "50 kJ ultra-compact pulsed power supply unit for various applications," in Dig. Eur. Conf. Power Electronics and Applications, Germany, Sept. 2005, p.7.

[6] Weiqun Yuan; Ping Yan, "Simulation of the Eddy Current Effects on the Inductance Gradient of Railgun," Electromagnetic Launch Technology, 2008 14th Symposium on , vol., no., pp.1,3, 10-13 June 2008

[7] Agrawal, P.K.; Das, B.P.; Nakhe, S.V., "Conducted noise studies of high voltage pulsed power supply for Copper vapor laser - A case study," *Electromagnetic Interference & Compatibility, 2008. INCEMIC 2008. 10th International Conference on* , vol., no., pp.161,166, 26-27 Nov. 2008

[8] Cecconi, V.; Matranga, A.; Ragusa, A., "EMI analysis in electrical drives when the grounding system is excited by a lightning strike using a suitable circuital model,"*Power Electronics, Electrical Drives, Automation and Motion, 2006. SPEEDAM 2006. International Symposium on* , vol., no., pp.591,596, 23-26 May 2006

[9] Anderson, R. A.; Clancy, T.J.; Fulkerson, S.; Petersen, D.; Pendelton, D.; Hulsey, S.; Ullery, G.; Tuck, J.; Polk, M.; Kamm, R.; Newton, M.; Moore, W.B.; Arnold, P.; Ollis, C.; Hinz, A.; Robb, C.; Fornes, J.; Watson, J., "Measurements of the radiated fields and conducted current leakage from the pulsed power systems in the national ignition facility at LLNL," *Pulsed Power Conference, 2003. Digest of Technical Papers. PPC-2003. 14th IEEE International*, vol.2, no., pp.793,796 Vol.2, 15-18 June 2003

[10] Murphy, D.P.; Weber, B.V.; Allen, R.J.; Cooperstein, G.; Young, F. C., "Mitigation of background EMI, mechanical and acoustic effects from the Gamble 11 pulsed power generator," *Pulsed Power Conference, 2003. Digest of Technical Papers. PPC-2003. 14th IEEE International*, vol.2, no., pp.891,894 Vol.2, 15-18 June 2003

[11] Hockanson, D.M.; Dreniak, J.L.; Hubing, T.H.; Van Doren, T.P.; Fei Sha; Cheung-Wei Lam, "Quantifying EMI resulting from finite-impedance reference planes,"*Electromagnetic Compatibility, IEEE Transactions on* , vol.39, no.4, pp.286,297, Nov 1997

[12] Feng Ying; Zhang He, "Research on Technique of Pulse Laser Fuze Anti-Electromagnetic Interference," *Photonics and Optoelectronics (SOPO), 2011 Symposium on* , vol., no., pp.1,4, 16-18 May 2011

High-energy proton microscopy for investigation of extreme state of matter

A.V. Kantsyrev, A.A. Golubev, A.V. Bogdanov, V.A. Panyushkin, N.V. Markov, D.S. Kolesnikov, S.M. Savin, Vl.S. Skachkov, A.V. Skoblyakov, A.V. Bakhmutova, I.V.Roudskoy
Institute for Theoretical and Experimental Physics" of NRC "Kurchatov Institute"
Moscow, Russia
e-mail: kantsyrev@itep.ru

D.V. Varentsov
GSI Helmholtzzentrum fur Schwerionenforschung
Darmstadt, Germany
V. B. Mintsev, N.S. Shilkin, A.N. Zubareva
Institute of Problems of Chemical Physics Russian Academia Science
Chernogolovka, Russia

I. Introduction

The study of extreme state of matter generated by intense heavy ion beams or shock wave is one of the most challenging and interesting topic in modern physics [1]. The measurement of density distribution, with high spatial and temporal resolution, is an important task for fundamental understanding of dynamic material properties in extreme states. High-energy proton radiography with magnetic optics exceeds X-Ray diagnostic method in many cases, because it has more transmission ability and high spatial and density resolution. The best spatial resolution was obtained by means of high-energy proton microscopy. The pRad facility [2] at LANL and the PUMA proton microscopy facility [3] at ITEP used a 800MeV proton beams to measure the density distribution. A typical scheme of proton microscope ion optics is a set of quadrupole magnetic lenses (Fig.1). Several lenses of matching section provide the optimal parameters (size and angular characteristics) of the proton beam in the plane of the target. An imaging/magnifying section consists of four quadrupole lenses, which are used for forming and magnifying image of investigation target. In the middle of imaging section a Fourier plane is formed. In the Fourier plane, the protons with different scattering angles are separated by the distance from the beam axis. The collimator installed at this plane can be used to regulate the contrast of the radiographic images. Optimization of matching and imaging sections allows compensating for chromatic aberrations. Magnification provides an increase in spatial resolution.

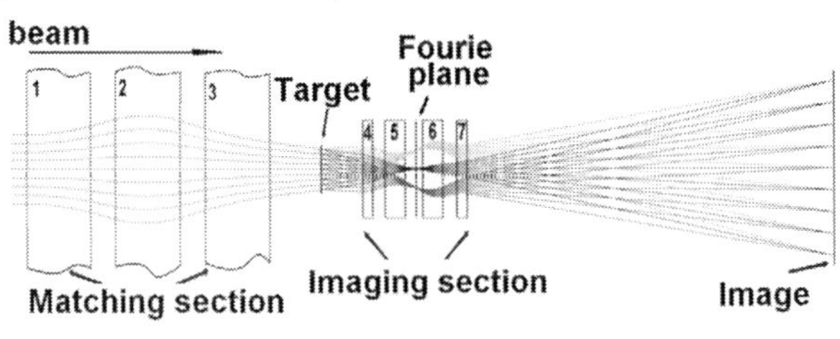

Fig. 1. Typical scheme of ion optics of proton microsope.

II. PRIOR setup for FAIR project

The novel high-energy proton microscope called PRIOR (Proton Microscope for FAIR) [4] will be a key diagnostic tool for the research program of HEDgeHOB collaboration at FAIR (Facility for Antiproton and Ion Research, Darmstadt, Germany). PRIOR prototype with 3.6 GeV proton beam was designed and commissioned on the SIS-18 synchrotron at GSI in 2014 (Fig.2).

Fig. 2. The PRIOR proton microsopy facility at GSI.

The imaging/magnifying section of PRIOR facility consists of four developed quadrupole lenses [5] on permanent NdFeB magnets (PMQ). In the middle of imaging section a tungsten collimator is installed. The field of view for obtained images is 15 mm, magnefication is ~ 4. The registration system consists of a fast plastic or CsI scintillator and two types of PCO DicamPro and Dimax cameras. Tuning of the imaging section is accomplished by longitudinal movement of PMQ lenses. The spatial resolution of 30 μm was obtained in the first experiments [4] at PRIOR facility. As an example, the Fig. 3 shows the proton radiographic image of a static object.

Fig. 3. The photo (at the top) and the proton radiographic image (at the bottom) of a static object (a watch).

Work supported by FAIR-Russia Research Center, Federal Target Program (project no. 14.616.21.002) and the Russian Foundation for Basic Research (project no. 15-08-09030)

The first experiment at PRIOR facility with dynamic target was performed with imaging of underwater electrical wire explosion [4].

III. Permanent Magnetic Quadrupole (PMQ) lenses

The permanent quadrupole lenses (PMQ) were used at all proton microscopes mentioned above [2,3,4]. Their use is due to lower costs in comparison with electromagnetic quadrupoles; no power supply is required, and these lenses are rather compact. Nevertheless, the lifetime of the PMQs under intensified irradiation, which is inherent in proton radiography, is limited. As a result, the PMQs need to be periodically maintained and remagnetized.

In the case of PRIOR facility the axially and radially segmented PMQ lenses [4,5] were used. The modular design of such PMQs (Fig.4) makes it possible to compensate the offset of the magnetic axis from the ion optical axis of the microscope channel and to decrease the variation of the angular position of the magnetic medians (Fig.5). Each module of PMQ has a two-layer structure. It contains trapezoidal magnetic elements that are manufactured of high-coercetive alloys of NdFeB. The effective length of two of PMQ is 144 mm, and the other two lenses have a length of 288 mm, aperture is 30 mm. The magnetic field at the radius of the lens aperture is 1.77 T (gradient 118 T/m), nonlinearity is 0.6%.

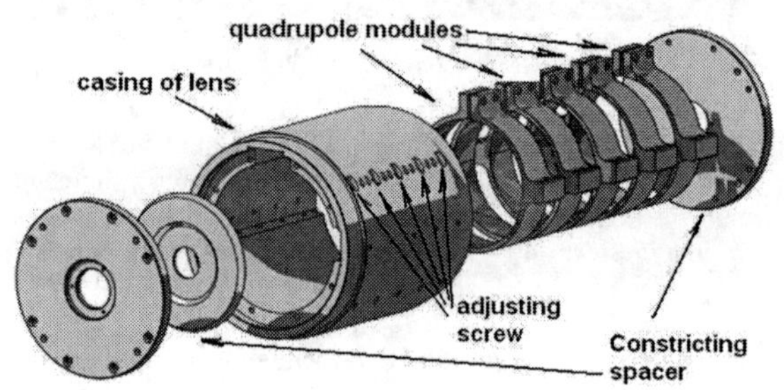

Fig. 4. 3D scheme of Permanent Magnetic Quadrupole (PMQ) lenses of the PRIOR facility

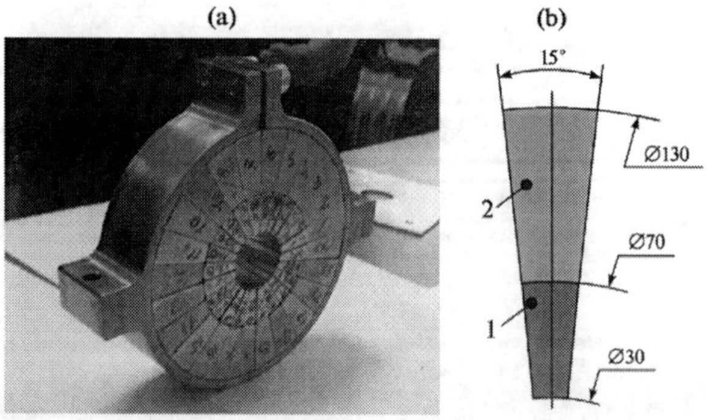

Fig. 5. One module of PMQ of PRIOR. (a)- photo, (b)- the transverse cross sections of the magnetic sectors.

After the first experiments the gradient of the magnetic field is decreased by 10% (Fig. 6) due to radiation damage of magnetic material of PMQ.

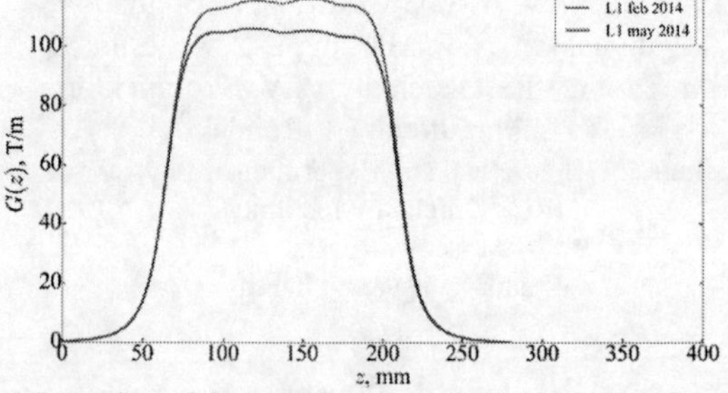

Fig. 6 Comparison of the magnetic field gradient of L1 PMQ before (green line) and after (blue line) the experiments with proton beam.

In order to renewal the performance of PMQ lenses a method and related equipment which allows to remagnetize of the magnetic material of PMQ have been developed. Testing of this method was performed with PMQ lenses of proton microscope PUMA at ITEP [3]. A high voltage generator with the voltage of U=1.8 kV, the pulse duration of 5.5 ms and the maximum current of about 4 kA was developed. The generator can produce a pulse magnetic field of 3.5 T in the cylindrical solenoid (made of copper wire 3.05x1.2 mm) of length 43 mm and inner diameter of 60 mm. The results of remagnetization of PMQ (L1, length 160 mm) are showed in Fig. 7 and Fig. 8. The integral of magnetic field gradient increased from 4.401 to 4.534 T.

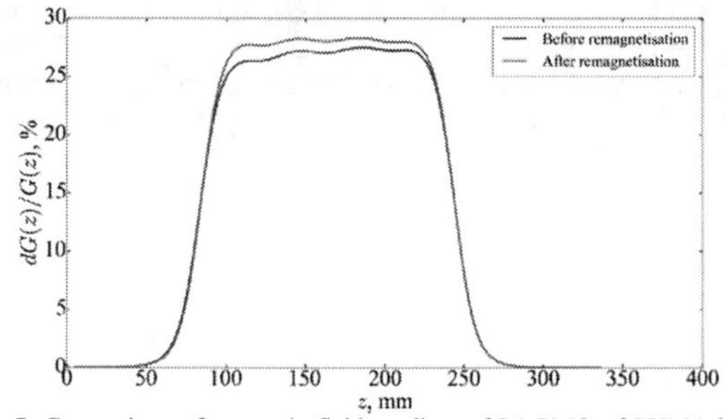

Fig. 7. Comparison of magnetic field gradient of L1 PMQ of PUMA before and after remagnetization.

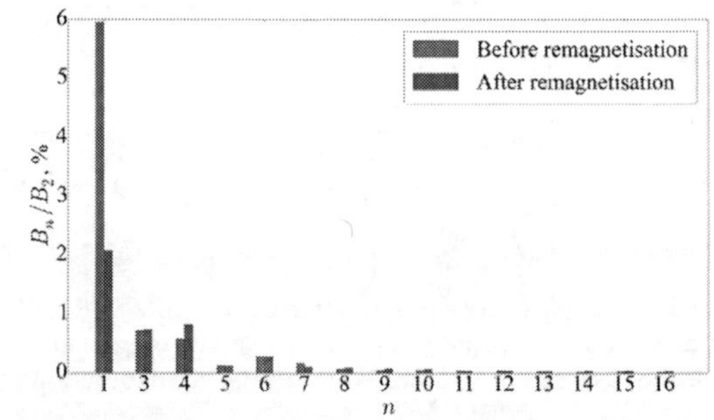

Fig. 8. Comparison of magnetic field harmonics of L1 PMQ of PUMA before and after remagnetization.

IV. ION-OPTICAL DESIGN AND MONTE-CARLO SIMULATION FOR 247 MEV PROTON MICROSCOPE

One of possible application of proton radiography is to study the targets of relatively small areal density (less then 5 g/cm^2). Relevant tasks include the investigation of phenomenon of anomalous compressibility of porous rubber and docosane [7] as well as the compressibility of nonideal plasma of Xe and Ar under shock-wave loading. A new proton microscope with 247 MeV protons will provide such investigations. The proton beam with energy of 247 McV is available at high-current linear accelerator of the Moscow Meson Factory (MMF) [6] in INR Troitsk, Russia. MMF accelerator was designed to operate at frequencies up to 100 Hz, which will also allow to investigate slowly changing dynamic processes such as crystallization and melting. To simulate the ion-optical scheme of the new setup a program for COSY Infinity code [8] was created. This program provides for the optimization of the some setup parameters such as the field of view, the relative position of the magnetic elements, resolution of proton radiographic images etc. The calculated ion-optical scheme includes three electromagnetic quadrupole lenses MQ1-MQ3 of matching section (aperture 150 мм, length 900 мм, field gradient). The imaging/magnifying section consists of PMQ lenses MQ4-MQ7 (aperture 40 mm, length 80 mm &160 mm, gradient 28 T/m) (Fig. 9).

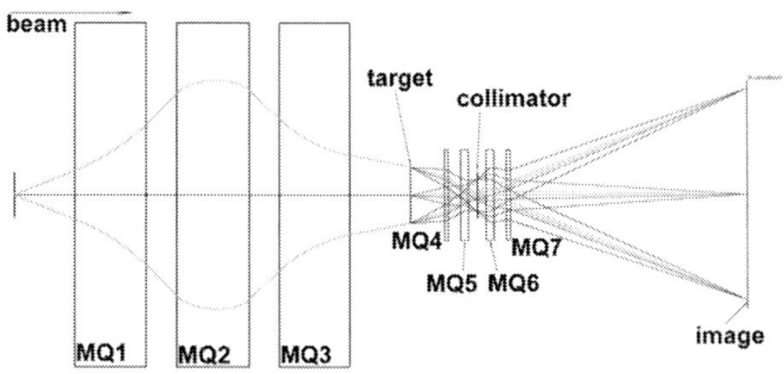

Fig. 9. Ion-optical scheme of 247 MeV proton microscope.

The full-scale Monte-Carlo numerical simulation of experiments was performed with GEANT4 [9] code (statistic of 5*10^8 protons). The 3D model of the setup for GENTN4 is showed in Fig. 10. The image detector is specified as matrix of 2000x2000 virtual pixel. The model of static target is a cupper step wedge with each step of thickness 0.25 mm. The resulting proton-radiography image and the corresponding transmission profile are showed in Fig. 11.

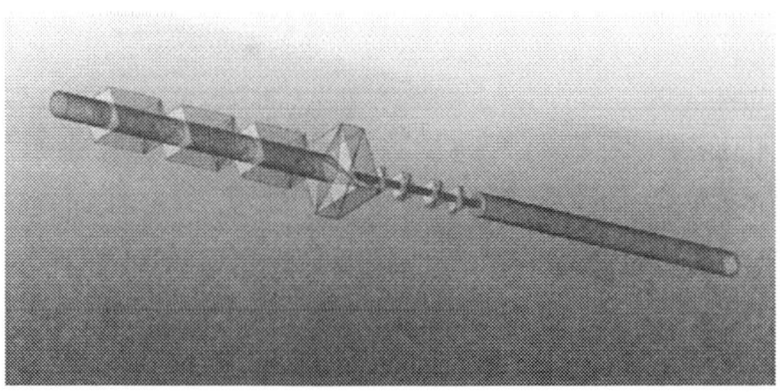

Fig. 10. 3D virtual model of 247 MeV proton microscope at GEANT4

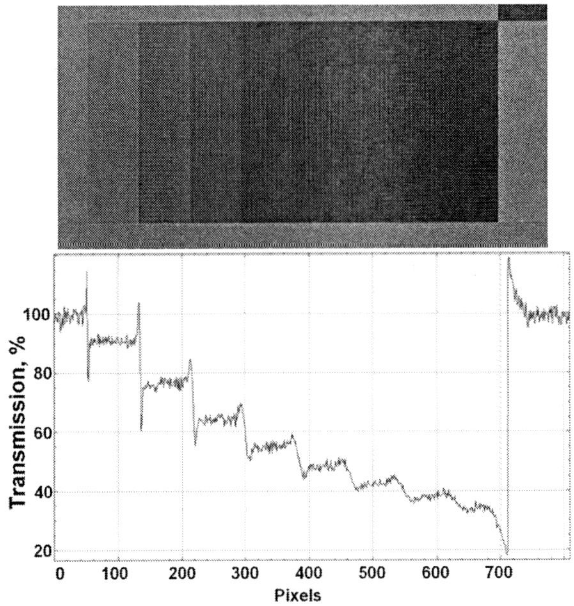

Fig. 11. Proton-radiography image of a cupper steps wedge (one step is 0.25 mm) - top, the corresponding transmission profile - bottom.

The simulation of experiment for investigation of anomalous compressibility of docosane $C_{22}H_{46}$ (ρ_0 = 0.94 г/см^3) [7] under shock-wave loading also was performed (Fig. 12). The density of docosane increases by 5 % under shock-wave loading. The model of dynamic target consisted of two cylinders of equal diameter size of 20 cm and densities, differing by 5 percent. The cylinders tightly adjoined each other. The obtained images showed that the 0.3 % change in transmission signal corresponds to a 1 % change in volume density.

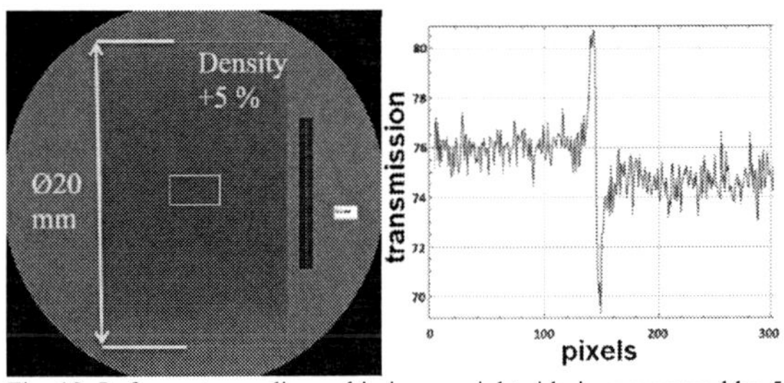

Fig. 12. Left - proton-radiographic image: right side is compressed by 5 % of origin volume density; right - transmission profile at the region bounded by yellow rectangle in left image.

V. CONCLUSION

Proton-microscopy provides a unique opportunity to study the internal structure of matter. The proton microscopy facilities with an image magnification of the objects have been already developed in Russia [3], US [3] and Germany [4]. A new facility, with energy of protons of 247 MeV, to study the low-density objects can be developed on the basis of the MMF linear accelerator at INR (Troitsk, Russia).

References

[1] V.E. Fortov, D. Hoffmann and B.Yu. Sharkov, "Intense ion beams for generating extreme states of matter", UFN, 2008, vol. 178, No. 2, pp.113-138.

[2] Mottershead T., Barlow D., Blind B. et al., "Design and operation of a proton microscope for radiography at 800 MeV", Proc. 2003 Particle Accelerator Conference, Portland (Oregon, USA), May 12-16 2003, pp. 702-704.

[3] Kantsyrev, A.V., Golubev, A.A., Bogdanov et al., "TWAC-ITEP proton microscopy facility", Instrum. Exp. Tech., 2014, no. 1, pp. 1-10.

[4] D. Varentsov et al., "Commissioning of the PRIOR proton microscope", Review of Scientific Instruments, 2016, 87, issue 2, pp. 023303/1–023303/8.

[5] A. V. Kantsyrev, Vl. S. Skachkov, V. A. Panyushkin, A. A. Golubev et al., "Quadrupole Lenses on the Basis of Permanent Magnets for a PRIOR Proton Microscope Prototype", Instruments and Experimental Techniques, 2016, Vol. 59, No. 5, pp. 712–723.

[6] L.V.Kravchuk, "INR Proton Linac operation and applications", NIM A, v.562, issue 2, 2006, p.932-934

[7] A.N. Zubareva et al., Journal of Physics: Conference Series 653 (2015)

[8] K. Makino, M. Berz, "COSY INFINITY Version 9", NIM A558, 2005, pp. 346-350.

[9] Geant4 Collaboration. Geant4 User's Guide for Application Developers, 2014

Fast semiconductor switching modules for transformer coupled LC inversion generators

R. Bischoff, V. Brommer, M. Stoll, and S. Scharnholz
French-German Research Institute of Saint-Louis (ISL)
Saint-Louis, France

Abstract—**The recent availability of fast high-power semiconductor switches makes the transformer coupled LC inversion generator (TCLCG) topology a very promising candidate for the realization of reliable, multi-kHz repetition rate pulse generators with an output voltage in the range of 200 kV and more. Usage of 1:1-transformers instead of closing switches in each generator stage reduces the total number of switches inside the TCLCG to only one, independent of the numbers of stages. Thus, problem of synchronization as in Marx-generators or classical LC inversion generators can be omitted. In principle, capacitors are initially charged to the charging voltage, whereas the odd-numbered of each stage are charged with the opposite polarity to the even-numbered capacitors. Fast voltage inversion of the even-numbered capacitors leads to efficient voltage multiplication provided the semiconductor switching module will be sufficiently fast to minimize switching losses. For initial investigations, a special asymmetrically compensated 3-stage test TCLCG was designed and set up. Three high-voltage semiconductor switching modules were experimentally tested in detail; two 20-kV high voltage switching modules consisting of a series arrangement of 15 Si-based IGBTs or 15 SiC-based MOSFETs, with in each case 1700 V breakdown voltage, and a fast experimental 10-kV thyristor switching module with kilo ampere current carrying capability. The achievable output signal rise time and generator efficiency for different electrical loads (resistive, capacitive or inductive) are discussed.**

Keywords—high voltage, pulsed power; semiconductor switches; IGBT, SiC MOSFET, thyristor, vector inversion generator; TCLCG

I. Introduction

From the historical point, the transformer coupled LC inversion generator (TCLCG) belongs to the group of generator using a LC-vector inversion technology which was for the first time described by Fitch und Howell [1]. This technology has been revived in the 1980's by a modified design which needs only one closing switch because of the coupling via transformers inside each single TCLCG stage. After its inventor Y. Pichugin the transformer-coupled LC inversion generator is also named "Pichugin pulser" in the literature [2]. During the last about twenty years it has been investigated by research groups in United States [3] and France [4]. At ISL the TCLCG was a research topic for the first time from 2006 to 2008, with special focus on fundamental Spice simulation of the generator principle, research on transformer ferrite core material and the development of a repetitive multi-stage TCLCG [5]. First experiments have already been carried out in the past using a wired-up prototype version of the ABB-ISL IGBT high voltage switching module (see. Fig. 3, on the left), however with a slow rise time of about 400 ns. It became clear that turn-on time of the closing semiconductor needs to be smaller if comparing with the rise time of the generator in order to achieve a higher generator efficiency. Now, with the recent availability of fast high voltage semiconductor switching devices the TCLCG comes back into focus as a promising topology for a reliable high-repetitive high voltage generator.

The need of only one closing switch also makes the TCLCG a perfect test environment for the functional tests of the different types of innovative fast semiconductor switching modules that are currently under development. Furthermore, the oscillating output voltage signal of the TCLCG is of special interest for driving electromagnetic effector systems and for voltage reversal capability tests of semiconductor switching modules.

II. Generator Principle

In order to describe the operational principle it is useful to have a look on one single stage of a TCLCG, which consists of two capacitors C_1 and C_2 and one 1:1 pulse transformer in in-phase winding orientation as marked by the two "dots" side by side on the primary and secondary winding. During the charging phase the switch S_1 is open and the two capacitors C_1 and C_2 are getting charged up to the charging voltage V_{ch} in opposite polarity. Then S_1 is getting closed and shortcuts the capacitors. Due to their special winding orientation of the transformer the capacitor C_1 discharges through the total inductance L of the transformer primary winding, whereas the capacitor C_2 discharges through the transformer leakage inductance L_{TL} (see Fig. 1). This is due to flux compensation between the primary and secondary winding. In other words, the magnetic field produced by the current through the primary winding is opposite to the field produced by the current through the secondary winding. Assuming a near perfect coupling of the transformer with a coupling coefficient k to be larger than 0.99 the total Inductance L can be expected to be much larger than L_{TL} and due to $f_1 / f_2 = (L_{TL} / L)^{1/2}$ it will be $f_1 << f_2$.

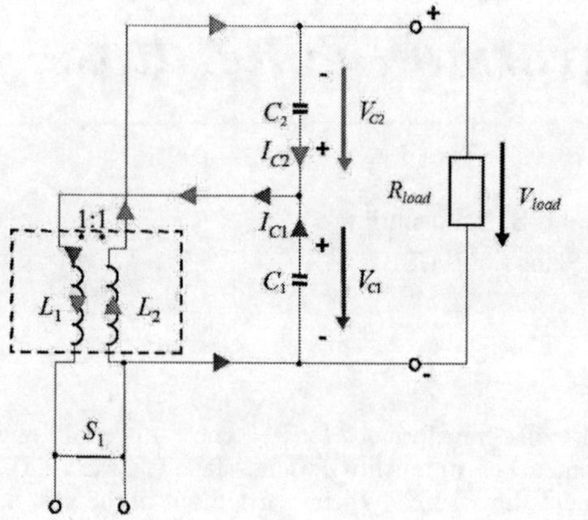

Fig. 1: One-stage TCLCG. Discharging phase.

As a result, the reversal of the voltage polarity of the "even numbered" capacitor C_2 in combination with the virtually not inverted polarity of the "odd numbered" capacitor C_1 leads to voltage amplification in the generator.

III. EXPERIMENTAL SETUP

A. 3-stage TCLCG

In principle, a multi-stage TCLCG can be set up by simply adding more stages. However, with increasing stage number the contributions of even capacitors to the total output voltage is getting more and more out of phase which lowers generator efficiency [4]. One solution is an asymmetrically compensated version of the TCLCG [6, 7]. A version of such an asymmetrically compensated TCLCG with a stage number $N = 3$ has been set up as test environment. The generator is shown in Fig. 2.

Each transformer consists of two stacked EPCOS N87 ferrite toroids with outer diameter $d_a = 102$ mm, inner diameter $d_i = 65.8$ mm and height $h = 15$ mm with 6 bifilar windings of a HV cable of the type 178-8073 by Teledyne Reynolds. The relative permeability of the ferrite material is $\mu_r = 2200$ (± 25%).

Theoretically, the capacitor values for asymmetric compensation are given by the relation $L_{TL1} \cdot C_2 = (L_{TL1} + L_{TL2}) \cdot C_4 = (L_{TL1} + L_{TL2} + L_{TL3}) \cdot C_6$ [7]. Experimentally, capacitor values deviate from the ideal case simply due to commercial availability. In detail, high voltage ceramic capacitors by AVX were used. The odd capacitors C_1, C_3, C_5 and the even capacitor C_2 had a value of about 1830 pF, C_4 was 911 pF and C_6 was 553 pF.

The charging voltage of the each single capacitor V_{ch}, which is also the switch voltage V_{SW}, and the generator output voltage across the load V_{load} were measured with high voltage probes of the type PVM-5 by Northstar with an internal divider ratio of 10000:1. The switch current I_{SW} was measured with a Person current monitor model 110. A fast 3-GHz oscilloscope of the LeCroy Wavemaster 7300A type was used to record the probe signals.

Fig. 2: Experimental setup of the 3-stage, asymmetrically compensated TCLCG.

Fig. 3: Left: IGBT switching module (18 kV/500 A) [8].
Center: SiC MOSFET switching module (18 kV/5 A).
Right: Thyristor switching module prototype (10 kV/1.4 kA).

B. Semiconductor switching modules

The following three high voltage semiconductor switching modules have been tested as closing switch inside of the TCLCG and were compared with the performance of a classical air spark gap at atmospheric pressure in self breakdown mode.

The first one is an 18-kV high voltage ABB-ISL switching module consisting of a series arrangement of 15 Si-based IGBTs of the type ABB 5SMX12M1701 with 1700 V rated blocking voltage (see Fig. 3, on the left). The switching module and the developed trigger generator, which is inductively coupled to the switches of the module for synchronous switching via a ferrite core, are described in detail here [8]. The switching module has an active clamp circuit for protection from overvoltage in case of an uneven voltage distribution between the switches on the module. The diameter of the circular circuit board is 155 mm.

The second HV switching module is a modification of the upper described IGBT switching module, removing the IGBTs and replacing them by SiC-based MOSFETs of the type CREE C2M1000170D with 1700 V rated blocking voltage in TO-247 packaging (see Fig. 3, in the center). The synchronous switching with a ferrite core connection of the trigger unit stayed the same.

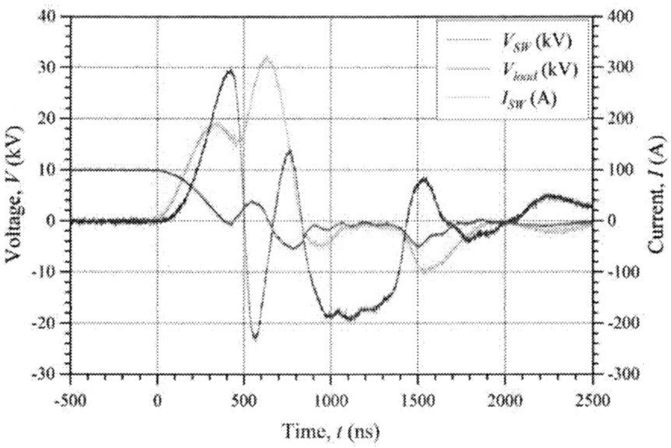

Fig. 4: Output signals of the 3-stage TCLCG with IGBT switching module at V_{ch} = 10.0 kV and R_{load} = 5.16 MΩ.

The third switch is an experimental thyristor switching module. Four versions have been set up with 2, 4, 6 or 10 standard Si-based thyristors by IXYS with 1600 V blocking voltage in series. The 10-kV version with two layers of each 5 thyristors is shown in Fig. 3, on the right. Die diameter is about 90 mm. The thyristors are in each case getting triggered by the overhead ignition of 1000-V breakover diodes, with the anode of the breakover diode being connected to the anode of the thyristor and the cathode of the breakover diode being connected to the gate of the thyristor. This switching module was protected by five IXYS DSEI-20 12A fast recovery epitaxial diodes (FRED) in series against voltage reversal.

IV. Experimental Results

A. *Characteristic generator output voltage and switching behavior*

Fig. 4 shows an example for the typical oscillating output voltage signal of the 3-stage TCLCG. In this case the switch was the 18-kV high voltage ABB-ISL switching module operating on a static high resistive load of R_{load} = 5.16 MΩ. The charging voltage was V_{ch} = 10.0 kV.

In the following, introductory investigations on the influence of different static electrical loads - resistive, capacitive and inductive - and of the different semiconductor switching technologies on the generator efficiency, defined as $\eta = V_{out,peak} / 2 \cdot N \cdot V_{ch}$, are presented.

B. *Generator efficiency as function of load type and value (air spark gap)*

According to [6, 7] TCLCGs have a capacitive output impedance, given by the formula $Z(\omega) = N / j \cdot \omega \cdot C_1$ and typically exhibit a dramatic decrease in generator efficiency if operating on a low-resistive load. For the TCLCG at hand the output impedance is about $Z_{exp} \approx 115\ \Omega$. Interestingly, values for the resistive load need to be significantly higher than Z_{exp} for good generator efficiency, as a rule of thumb about $R_{load} \geq 10 \cdot Z_{exp}$ (see Fig. 5). Here, the maximum achievable generator efficiency was about η = 77% for V_{ch} = 10.0 kV.

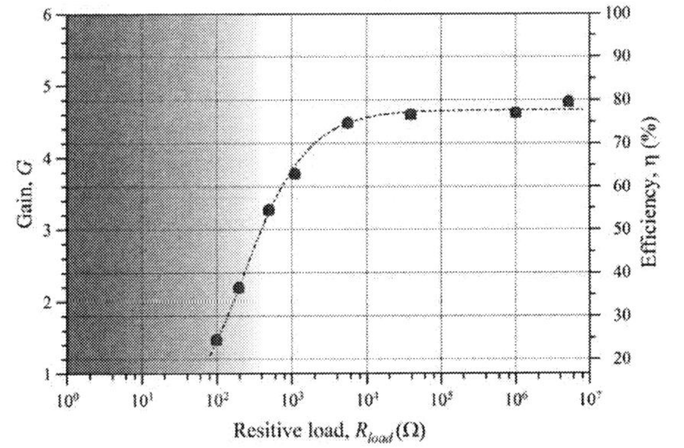

Fig. 5: Generator efficiency η and gain G as a function of the value of a resistive load R_{load}.

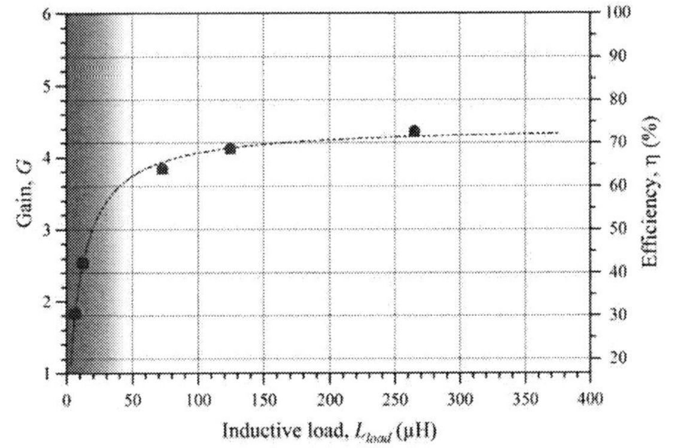

Fig. 6: Generator efficiency η and gain G as a function of the value of an inductive load L_{load}.

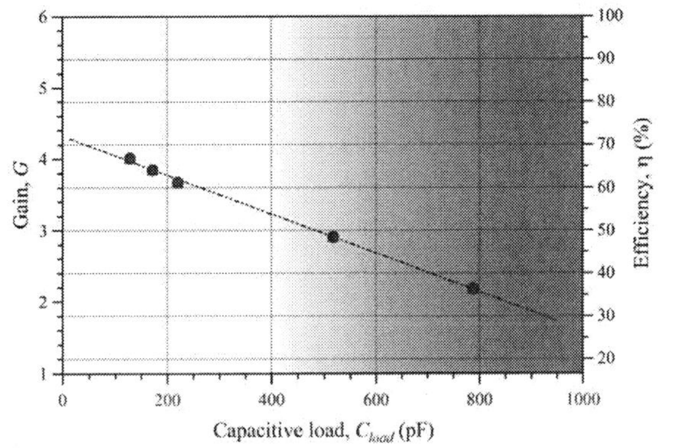

Fig. 7: Generator efficiency η and gain G as a function of the value of a capacitive load C_{load}.

Fig. 6 and Fig. 7 show, for the very first time, experimental data for the efficiency of a TCLCG operating on a capacitive or inductive load. The quantitative values are only valid for the specific 3-stage TCLCG (see Fig. 2). Nevertheless, the tendency of the curve shapes complies with what can be expected since capacitive impedance is $Z_C = 1 / \omega \cdot C$ and inductive impedance is $Z_L = \omega \cdot L$ and it proves the option of a

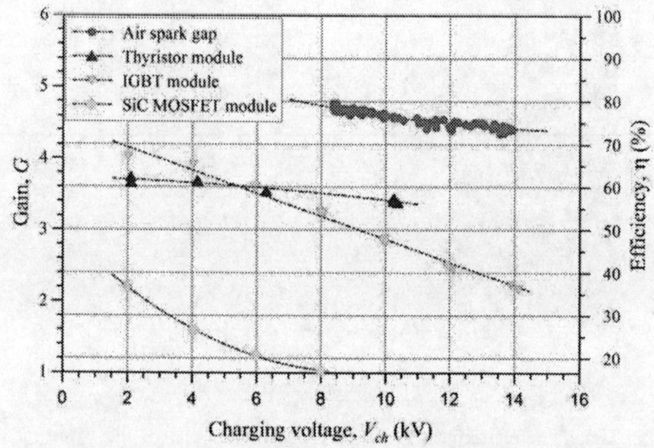

Fig. 8: Gain G and efficiency η [%] of the 3-stage TCLCG as a function of the implemented semiconductor switching technology in comparison to a classical air spark gap. R_{load} = 5.16 MΩ.

successful operation of TCLCGs on low capacitive and high inductive loads.

C. Generator efficiency for different semiconductor switching technologies

The dependency of gain G and efficiency η as a function of the switching technology was investigated. For these experiments a high resistive load of R_{load} = 5.16 MΩ was used. The results are shown in Fig. 8. The 10-90% current turn-on times for each switch $t_{SW,I}$, the switching current I_{SW}, the generator output voltage rise time t_{load}, gain G and efficiency η at a charging voltage of about V_{ch} = 10 kV (unless otherwise stated) are compiled in Table 1.

In general, efficiency η and gain G decrease with increasing charging voltage V_{ch} because of the saturation of the ferrite cores and enhanced for the IGBT modules because of their larger turn-on times with increasing switching voltage.

The SiC MOSFET switching module turned out to be unsuitable for operation as the closing switch in a TCLCG. Please note that the used CREE C2M1000170D SiC MOSFETs only have a rated pulsed drain source current of 5 A and therefore at a peak switching current of 31.5 A (see Tab. 1) are significantly stressed beyond specification. As a result, the turn-on time SiC MOSFET switching module is also much slower as could have been expected from the specification of the manufacturer with the turn-on time of a single switch of 46 ns.

TABLE I. TRANSFORMER COUPLED LC-INVERSION GENERATOR AND SWITCHING MODULES DATA.

Switching technology	V_{ch} (kV)	I_{SW} (A)	$t_{SW,I}$ (ns)	t_{load} (ns)	Gain	η (%)
Air spark gap	10.1	680	16	81	4.61	76.8
Thyristor module	10.2	285	127	129	3.42	57.0
IGBT module	10.0	186	211	205	2.90	48.3
SiC MOSFET Module	8.0	31.5	296	421	1.03	17.2

For the case of the IGBT and Thyristor module, the generator output voltage rise time t_{load}, and as a result gain G and efficiency η, are directly limited by the switching rise time (see Table 1). The rise time t_{load}= 81 ns for the case of the use of an air spark gap with a turn-on time of 16 ns indicates the minimum achievable rise time of the TCLCG at hand. In order to close the existing gap to the classical spark gaps semiconductor switches still need to speed up.

V. CONCLUSION & OUTLOOK

A 3-stage asymmetrically compensated TCLCG as testing environment for fast high power semiconductor switching modules was established. Functional tests of the TCLCG with an IGBT and thyristor switching module were performed. A fast turn-on time and a high current capability turned out to be indispensable for successful operation. Load behavior test showed that TCLCs are capable to operate not only on high resistive but also on low capacitive and high inductive loads.

In depth theoretical studies on the fundamental load behavior of a TCLCG are planned. The development and investigation of innovative new fast high power semiconductor switching modules are already ongoing, making use of the last generation of wide band gap semiconductor switches and gate boosting techniques to reduce turn-on time. The design and realization of a high-repetitive fast TCLCG of the 250-kV class is planned for the future.

References

[1] R.A. Fitch, and V.T.S. Howell, "Novel principle of transient high voltage generation," Proc. IEE, Vol. 111, No. 4, pp. 849-855, 1964.

[2] Yu. P. Pichugin, I. V. Rezyukov, and I. N. Romanenko, "*Pulse voltage generator*," USSR patent number 812725 (1985 translated by Foreign Technology Division, FTD-ID(RS)T-0541-85), 1981.

[3] T. G. Engel, Ch. Kaplicki, and W. C. Nunnally, "High voltage pulse production using transformer-coupled LC vector inversion generators," IEEE Trans. Plasma Sci., Vol. 28, No. 5, pp.1377-1381, 2000.

[4] M. Rivaletto, P. Domens, and P. Pignolet, "Inner wave shape characterization and realization of a Pichugin pulse generator," Meas. Sci. Technol., 9, pp. 824-831, 1998.

[5] R. Bischoff, M. Schael., and S. Pinguet, "The transformer-coupled LC-Inversion generator: A promising concept for high repetitive high-power high voltage pulse generation," NATO SCI-177 Symposium on "High Power Microwaves, Threat to Infrastructure and Military Equipment", Mannheim, DE, 15-17 October 2007.

[6] M. Rivaletto, P. Domens, and P. Pignolet, "A asymmetrical 175 kV, 210 ns Pichugin pulse generator," Meas. Sci. Technol., 9, pp. 17765-1771, 1998.

[7] M. Rivaletto, P. Domens, and P. Pignolet, "Characterization and realization of an asymmetrical-compensated Pichugin pulse generator," Eur. Phys. J. AP, 3, pp. 273-281, 1998.

[8] V. Zorngiebel, M. Hecquard, E. Spahn, A. Welleman, and S. Scharnholz, "Modular 50-kV IGBT switch for pulsed-power applications," IEEE Trans. Plasma Sci., Vol. 39, No. 1, pp. 364-367, 2011.

Development of Trigger Circuit of Triggered Vacuum Switch Used in High Temperature Environment

Zhang Kai-lun, Dai Ling, Xiang Ling-feng, Mao Xin-guo, Zhang Chi

(State Key Laboratory of Advanced Electromagnetic Engineering and Technology

Huazhong University of Science and Technology Wuhan 430074 China)

Abstract—**In some special applications, the pulsed power supply works in high temperature environment more than 120℃. This paper aims at the application of surface-flashover triggered vacuum switch (TVS) in high temperature, and developed a kind of special trigger circuit for TVS. Because most of semiconductor devices cannot work reliably under high temperature, the trigger circuit in this paper doesn't use any semiconductor device. In the trigger circuit, an energy storing capacitor discharged to the trigger electrodes of TVS by a three-electrode spark gap, and the pulsed trigger current will be up to several hundred amperes. The control signal of the spark gap also be obtained from a small energy storing capacitor discharging through a current-limiting resistor and a designed high-voltage relay, which can pass current of several amperes. When the charging voltage of the pulsed power system reaches the set value, a control signal will be put on the high voltage relay, then the spark gap will be triggered and the trigger circuit will generate a pulsed current to the TVS. In order to reduce the volume of the trigger circuit, this paper developed a new kind method of charging energy storage capacitor in the trigger circuit from the main capacitor in the PPS. This method solves high-voltage isolation problems at the same time. On the basis of the simulation, the trigger circuit prototype is developed. Conduction tests for TVS are done with the prototype under the 120℃ high temperature environment. The results shows that the trigger circuit works steadily and reliably, fully meeting the trigger requirements of TVS working in high temperature environment.**

Keywords—**Triggered vacuum switch, Trigger circuit, Three-electrode spark gap, High temperature environment**

I. INTRODUCTION

Pulsed power closing switch technology is an important research content in pulse power techniques [1]. The switch is one of the crucial components of the pulsed power devices and is also one of the bottlenecks in the development of pulsed power technology at present. It plays a part of connecting the power storage unit with the load, used for conducting and isolating the circuit, and it can affect the steep degree of waveform. Its performance directly affects the performance of the whole system [2-3]. The switch used in the technical field of pulse power must have the ability of withstanding high voltage, high current and high power, and have the requirements of little closing time delay, little delay jitter, working stably, being able to work continuously, meeting the repetition frequency requirements, and possessing certain service life [4-5]. Triggered vacuum switch (TVS) is a new type of switching devices, combining vacuum switch technology with three electrode spark gap technology to develop. And its characteristics is using vacuum as the insulation and arc quenching medium among the main contact, and using the special designed trigger electrode to control switch for quick making [6-7]. It has many advantages, such as wide working voltage range, great through current capacity, triggering simply, rapid media recovery, compact structure, short time delay and jitter of closing, high working reliability, long life time, low price and so on. TVS has obtained widespread application of the best in pulsed power technology field, attracting more and more attention in the high voltage and large current switching field [8-10]. In some special applications, such as electrohydraulic forming, mineral exploration, electrohydraulic cleaning, rock fragmentation and soil compaction, the pulsed power devices need to work in a high temperature environment for a long time [11-16]. In particular, for the pulse discharge device applied for oil well plugging removal and stimulation, the maximum operating depth is under the well 3000m and the maximum long-term working temperature is of 120 ℃ [17-20]. In order to realize the pulse discharge under high temperature, the relatively harsh demands on the pulse discharge device has been put forward, in which the main discharge switch is required to work in a wide temperature range. In this paper, the trigger source of TVS is studied aimed at the triggered vacuum switch working in high temperature environment. Because most of semiconductor devices cannot work reliably under high temperature, a TVS trigger source without any semiconductor devices is devised. The trigger source uses a three electrode spark gap instead of the thyristor to control the energy-storage capacitor discharging along the trigger surface of TVS, and the three electrode spark gap is controlled by the trigger signal formed from the discharge of a small capacitor that is controlled by a specially designed high-voltage relay to discharge. In order to further reduce the trigger source volume, according to the applied pulse power supply circuit, a trigger source used in oil well stimulation device is developed. The energy storage capacitors of the trigger source charges from the main capacitor of the pulse source in the voltage dividing way, avoiding to adopt the traditional way that uses high voltage isolation transformer of large volume to charge the energy storage capacitors of the trigger source, which can simplify the circuit and has no potential isolation problem.

II. Design and Study of Trigger Source Used under High Temperature

A. *Trigger circuit design*

The typical trigger circuit of TVS [21-24] is shown in Fig. 1. In the diagram, T_1 is a boosting transformer, D_1 is a high-voltage silicon heap, and R_1 is a charging resistor; C_1 is the primary energy-storage capacitor for the pulse transformer T_2, and the capacitance is 3 μF; silicon SCR serves as the discharge switch of the primary capacitor C_1 and its conduction is controlled by pre-amplifier control circuit using optical fiber; D_2 is the freewheeling diode used to prevent transformer winding to oscillate when C_1 is discharging; T_2 is a pulse transformer that generates a high voltage pulse output functioning along the triggering surface of TVS; secondary capacitor C_2 is mainly used to ensure the stability of the pulse transformer output voltage, and the capacitance is 2nF. When it works, C_1 is charged by the half wave rectifier circuit composed of T_1, D_1 and R_1; after the completion of charging to C_1, the thyristor SCR is controlled to conduct, which leads to the C_1 discharging to the pulse transformer T_2; and then secondary winding of T_2 generates a high voltage pulse output functioning along the triggering surface of TVS. Since C_2 is connected in parallel with both ends of the trigger electrode and the cathode, T_2 will continue charging to C_2 until the voltage on C_2 reaches the breakdown voltage of the trigger surface. When the voltage on C_2 is higher than the breakdown voltage of the trigger surface, insulation breakdown will occur on the surface, C_2 and T_2 discharging to the trigger surface simultaneously.

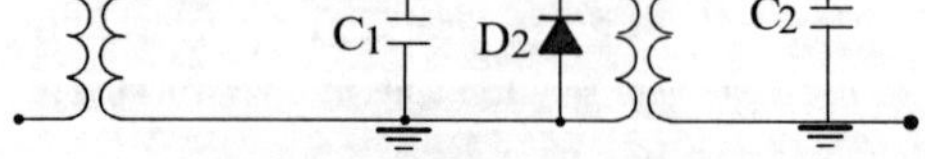

Fig. 1. Typical trigger circuit of TVS

In some special applications, TVS serves as the main discharge switch of the pulsed power devices needs to work in a high temperature environment for a long time(the maximum long-term working temperature is of 120℃).Thus, the corresponding trigger source elements selection must be high temperature resistant components. Because most of semiconductor devices such as thyristor and diode cannot work reliably under high temperature, the commonly typical trigger circuit (As shown in Fig. 1) is not available. So the trigger circuit is studied aimed at the high temperature working environment of the TVS. A trigger circuit without any semiconductor device is designed as shown in Fig. 2. In the diagram, C_1 is trigger capacitor and is charged by the half wave rectifier circuit composed of boosting transformer T_1、 charging resistor R_1 and high-voltage silicon heap D_1; GAP is a three electrode spark gap, controlling the discharge of C_1 to the trigger surface of TVS; the trigger signal used to control the conduction of GAP is given by the discharge of C_2; the relay is used to control C_2 discharging along the trigger surface of GAP; boosting transformer T_2, high-voltage silicon heap D_2 and charging resistor R_2 consist of a half wave rectifier circuit to charge C_2; R_3 is current-limiting protection resistor. When the charging voltage of the main capacitor in the pulsed power source reaches the set value, a control signal formed from the voltage divider and comparator will be put on the high voltage relay to make it conduct; and then the C_2 discharges upon the trigger surface of GAP, which triggers GAP to conduct; after the GAP conducting, C_1 discharges upon the trigger surface of TVS, generating initial plasma to trigger TVS to conduct.

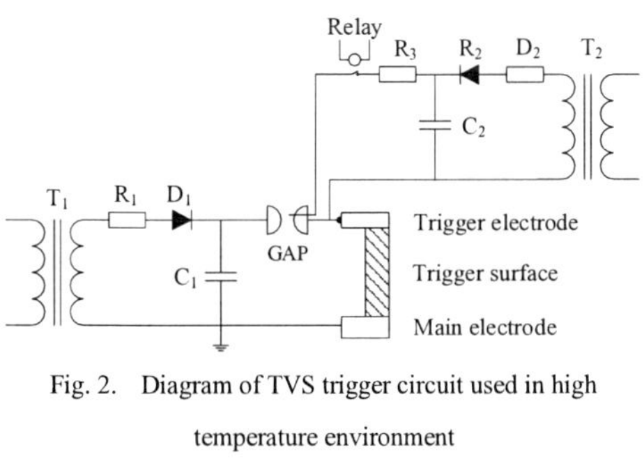

Fig. 2. Diagram of TVS trigger circuit used in high temperature environment

All the components in the trigger circuit need to work under high ambient temperature about 120℃. C_1 and C_2 both choose 2nF high temperature ceramic capacitor; GAP is selected as trigger tube type three electrode spark gap whose trigger gap is 5mm air gap; relay is a kind of high temperature and high voltage relay specially designed; D_1 is of 100kV/1A and D_2 is of 10kV/1A; R_1 and R_2 are both 2 MΩ resistors; the resistance value of R_3 is 5K.

B. Trigger and conduction tests on TVS

Use the designed TVS trigger circuit under high temperature as shown in Fig. 2 to take the trigger and conduction tests on the 25kV triggered vacuum switch. The test monitoring equipment used to monitor the current and voltage contains P6015 high voltage probe, PERSON110 CT current sensor (two kinds, one is of 1000:1 ratio and the other is of 10:1 ratio), PEM current probe (0.05mV/A). And use the oscilloscope whose model is TBS2042 /200M to record the voltage and current waveforms.

1) Trigger source output test

When the trigger source output test was taken, the output ends of the trigger source was connected to the trigger surface of the TVS, and the output high voltage side of the output ends was connected with the TVS trigger electrode. Firstly the test needed to determinate the trigger energy for the GAP conduction. Charge C_1 to 10kV voltage, close the relay, and then give C_2 a step-up voltage until the GAP conduct. After multiple tests, when C_2 charging voltage was up to 2kV-2.5kV, the GAP could be successfully triggered to conduct. The voltage waveform of C_2 was shown in Fig. 3.

Thus, while test being taken, C_1 was charged to 10kV and C_2 was charged to 2.5kV, and then closed the relay, triggering the GAP to conduct. C_1 discharged upon the trigger surface of TVS, generating a pulsed current. Waveforms of TVS trigger voltage and trigger current flowing through TVS trigger surface are shown in Fig. 4. As can be seen from Fig. 4, the trigger current flowing through the trigger surface is a high frequency damped oscillation current with a short period about hundreds ns. It is due to the small capacitance of trigger capacitor C_1. In order to test the trigger reliability, 50 times output tests were carried out on the designed trigger source. The trigger voltage could reliably spark over the trigger surface of TVS, and the amplitude and rising edge of the trigger current were also relatively stable. The results indicate that the trigger source can trigger the TVS reliably.

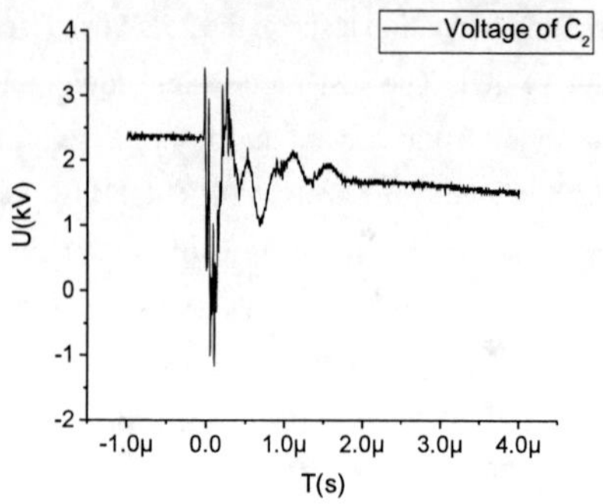

Fig. 3. The voltage waveform of C_2

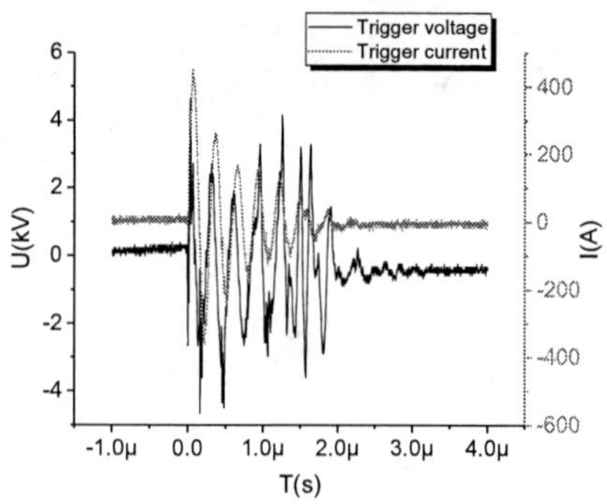

Fig. 4. Waveforms of trigger voltage and trigger current

2) TVS conduction test

a. Experimental circuit

The main discharge circuit for the TVS conduction test is shown in Fig. 5, where the charging capacitors bank C is of 150μF, the inductor L is of 68μH and the silicon heap D is of 100kV/1A. The trigger source used the designed trigger circuit shown in Fig. 2. During the test, put the trigger source into the heat oven of 120℃. The main circuit capacitor C was charged to a certain voltage value; trigger capacitor C_1 was charged to 10kV and C_2 was charged to 2.5kV. After the charging completion, the relay was controlled to be closed so that the GAP was triggered to conduct and then the TVS was triggered. Observe whether the circuit is conducted and record the trigger voltage waveform and the main circuit current waveform. Simultaneously read the TVS trigger and conduction time-delay. The trigger and conduction time-delay is defined as the time difference from the moment when the trigger voltage waveform starts rising to the moment when the main current starts rising smoothly.

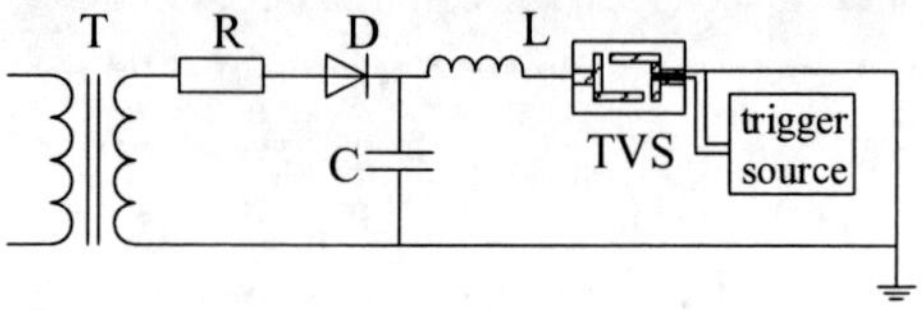

Fig. 5. Experimental main discharge circuit

b. Experimental results

Add the charging voltage of the main circuit charging capacitor C to 12kV, and then control the relay to be closed so that the GAP and TVS were triggered to conduction in turn. Get the typical experimental waveform as shown in Fig. 6 and Fig. 7. Fig. 6 is the main circuit current whole waveform, which shows that the peak value is about 16 kA and the pulse width is about 340μs. It indicates that the TVS conducted successfully and the arc quenching occurs at the time of the current zero crossing. Fig. 7 is the rising edge waveform of the main circuit current and the trigger voltage waveform, which shows that the trigger and conduction time-delay is about 4μs.

In order to test the work reliability and stability of the designed TVS trigger source used in high temperature environment, 50 times trigger and conduction tests on the TVS were carried out making use of the trigger source. The waveforms and data corresponding to the main circuit current, trigger voltage and trigger time-delay were recorded for each test. The TVS could be triggered and conducted every time in the 50 tests.

The amplitude and pulse width of the main circuit current were also relatively stable, and the time-delay joggle was little.

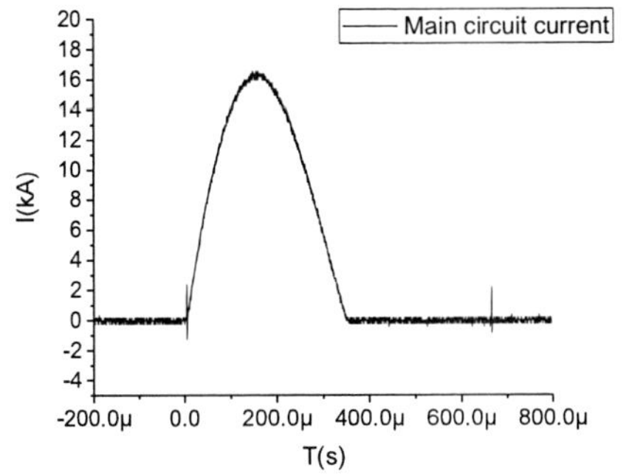

Fig. 6. The main circuit current waveform

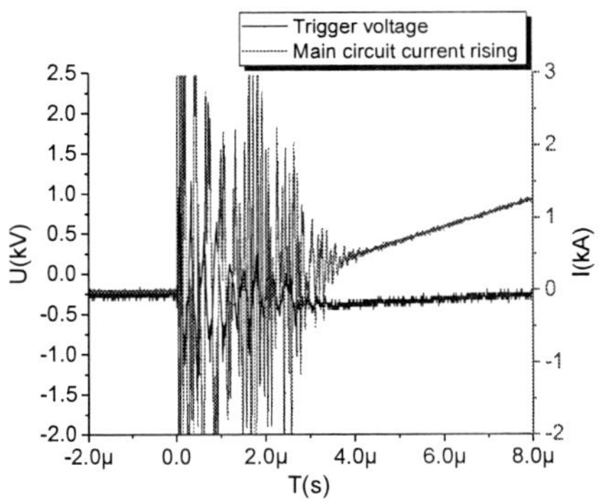

Fig. 7. Main circuit current rising and trigger voltage waveforms

III. DEVELOPMENT OF THE OIL WELL TRIGGER SOURCE

The designed TVS trigger circuit used in high temperature environment as shown in Fig. 2 needs to use high voltage isolation transformer of large volume to charge the energy storage capacitors of the trigger source. However, in some special applications, the volume of the pulsed power device is limited, which puts forward demand on the volume of the TVS and its trigger source. For example, for the pulse discharge device applied for oil well plugging removal and stimulation, the main discharge switch TVS and its trigger source work in the well deep to3000m underground, the maximum long-term working temperature being up to 120 ℃ , and are encapsulated in a tube whose outside diameter is less than 70 mm. In order to further reduce the trigger source volume, according to the applied pulse power supply circuit, a trigger source used in oil well stimulation device is developed on the basis of the trigger circuit under high temperature shown in Fig. 2. The energy storage capacitors of the trigger source charges from the main capacitor of the pulse source in the voltage dividing way, avoiding to adopt the traditional way that uses high voltage isolation transformer of large volume to charge the energy storage capacitors of the trigger source, which can simplify the circuit and has no potential isolation problem.

A. Trigger circuit used in oil well

The designed TVS trigger circuit of series trigger capacitors charging from the main capacitor used in oil well is shown in Fig. 8. In the diagram, C_1 is the main capacitor of the pulse source and is charged by the half wave rectifier circuit composed of boosting transformer T_1、charging resistor R_1 and high-voltage silicon heap D_1; C_2 and C_3 are series trigger capacitors, connected in parallel to both ends of the TVS main electrode through a resistor R_0; R_4 and R_5 are divider resistors, ensuring the voltage distribution of C_2 and C_3 after charging; GAP is three electrode spark gap, controlling the discharge of C_3 to the trigger surface of TVS; the trigger signal used to control the conduction of GAP is given by the discharge of C_4; the relay is used to control C_4 discharging along the trigger surface of GAP; boosting transformer T_2, high-voltage silicon heap D_2 and charging resistor R_2 consist of a half wave rectifier circuit to charge C_4; R_3 is current-limiting protection resistor.

Before the water gap is of breakdown, the gap presents resistance characteristics. So when the charging device charges the main capacitor C_1 (The setting value of charging voltage is 30kV), trigger capacitors C_2 and C_3 are charged by C_1 through R_0 and the water gap at the same time. After the completion of charging to the whole circuit, the voltage on both ends of the series capacitors C_2 and C_3 is up to 30kV and the respective voltage on C_2 and C_3 is distributed in inverse proportion to their own capacitance, the voltage on C_3 being higher than the breakdown voltage of GAP. When the charging voltage of C_1 reaches the setting value, a control signal formed from the voltage divider and comparator will be put on the high voltage relay to make it conduct; and then C_4 discharges upon the trigger surface of GAP, which triggers GAP to conduct; after the GAP conducting, C_3 discharges upon the trigger surface of TVS, generating initial plasma to trigger TVS to conduct. Due to the voltage on capacitor C_2 can't mutate, in the process of C_3 discharging that is the process of TVS triggering, the voltage on both ends of TVS main electrode can be maintained at a high level until the main electrode is of breakdown, and then the TVS is of conduction. Finally, after the conduction of TVS, voltage on the C_1 is applied to the water gap until the gap is of breakdown, and then the entire circuit conduction, resulting in shock pressure wave in the water.

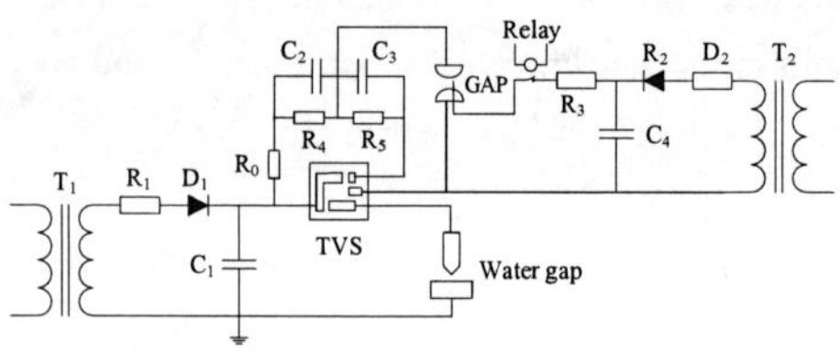

Fig. 8. Diagram of the TVS trigger circuit used in oil well

All the components in the trigger circuit need to work under high ambient temperature about 120℃.The parameters of the components in the trigger circuit:

R_0: 200kΩ, R_4: 10MΩ, R_5: 10MΩ

C_2: 2nF, C_3: 2nF

D_2: 10kV/1A, R_2: 2MΩ

C_4: 2nF, R_3: 2kΩ

B. Trigger circuit simulation

The designed trigger circuit scheme shown in Fig. 8 is simulated in order to verify whether the scheme is feasible and whether the circuit can trigger the TVS and further cause the discharge of the water gap. The simulation model is established as shown in Fig.9. In the model, TVS, GAP and water gap are all replaced by ideal switch, in which the water gap is equivalent to a resistor before the conduction; trigger capacitors C_2 and C_3 are set containing the same capacitance of 2nF; resistors R_2 and R_3 are set the same resistance of 10 MΩ. After charging to the whole capacitors in the circuit by 30kV AC supply through a half wave rectifier circuit is completed, respectively control the GAP, TVS and water gap to conduct, in which the conduction time of TVS is set for 10μs lag behind the GAP conduction and the conduction time of water gap is set for 15μs lag behind the GAP conduction.

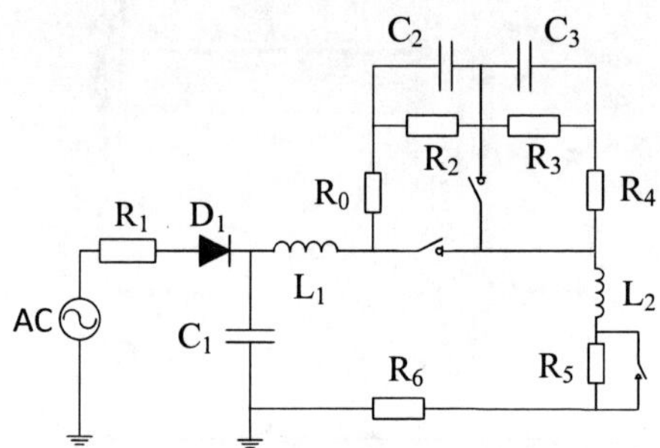

Fig. 9. The simulation model of the TVS trigger circuit used in oil well

The simulation results are shown in Fig. 10, Fig. 11 and Fig. 12. The waveforms of voltage on capacitors C_1, C_2 and C_3 are shown in Fig. 10, in which the Fig. 10 (b) shows the capacitors voltage waveforms at trigger conduction moment marked out by the green imaginary line in Fig. 10 (a); Fig.11 shows the waveform of the main circuit current; The waveforms of voltage on TVS and water gap are shown in Fig.12, in which the Fig.12 (b) shows the voltage waveforms at trigger conduction moment marked out by the green imaginary line in Fig.12 (a). From the simulation results, it can be seen that: when the main capacitor C_1 is charged, trigger capacitors C_2 and C_3 can be charged normally in inverse proportion to their own capacitance; in the process of trigger, main capacitor C_1 generates no overvoltage and trigger capacitors C_2 and C_3 can discharge normally; the designed trigger circuit can trigger TVS and further cause the conducting of the water gap. The simulation results show that the designed TVS trigger circuit of the series trigger capacitors charging from the main capacitor can mainly meet the requirements of the oil well pulse discharge stimulation device.

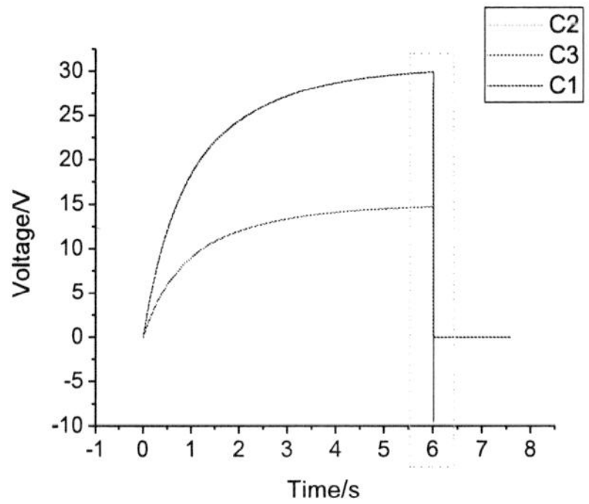

(a) Capacitors voltage waveforms

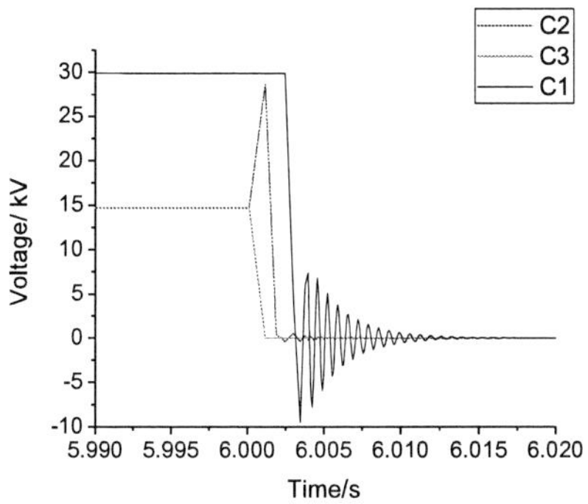

(b) Capacitors voltage waveforms at trigger conduction moment

Fig. 10. Waveforms of voltage on capacitors C_1, C_2 and C_3

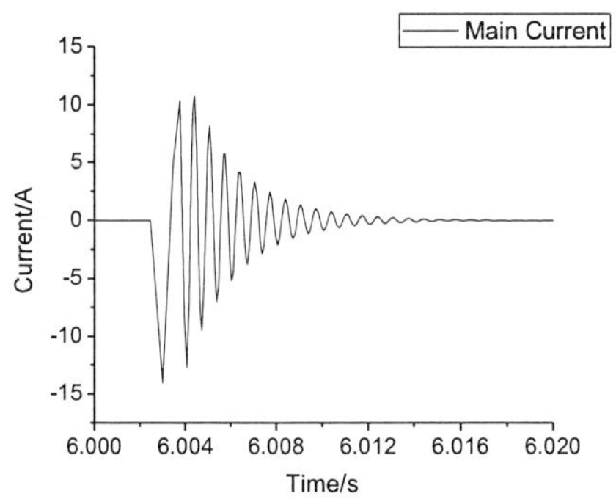

Fig. 11. The main circuit current waveform

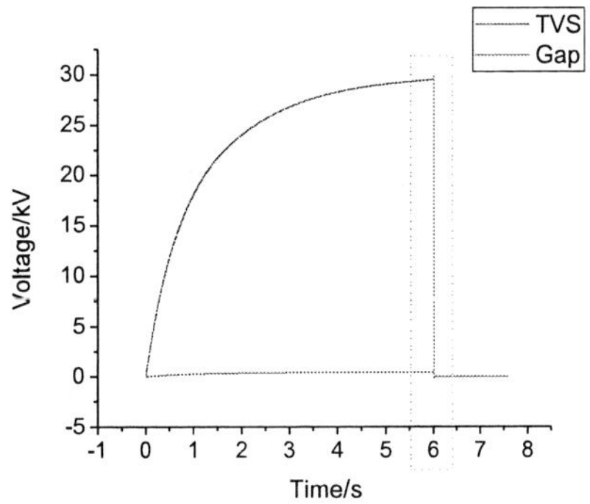

(a) Waveforms of voltage on TVS and water gap

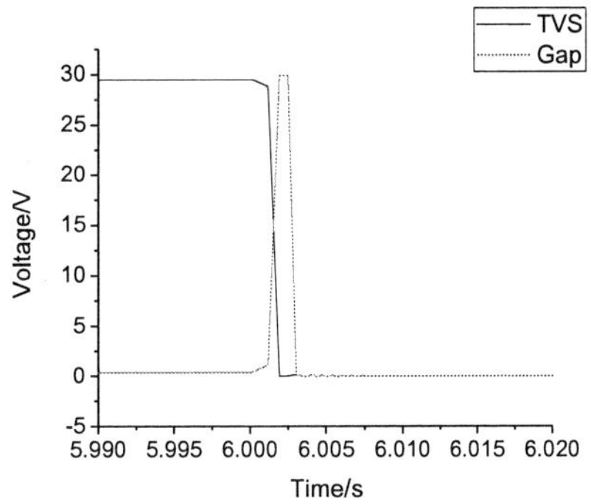

(b) Voltage waveforms at trigger conduction moment

Fig. 12. Waveforms of voltage on TVS and water gap

C. *Experimental test*

The experimental platform is set up to test the trigger scheme according to the designed trigger circuit shown in Fig. 8, where the main charging capacitors C_1 is of 3μF, the charging resistor R_1 is of 2MΩ and the silicon heap D_1 is of 100kV/1A. The test monitoring equipment used to monitor the current and voltage contains P6015 high voltage probe, PERSON110 CT current sensor (two kinds, one is of 1000:1 ratio and the other is of 10:1 ratio), PEM current probe (0.05mV/A). And use the oscilloscope whose model is TBS2042 /200M to record the voltage and current waveforms.

1) Trigger and conduction test on TVS

While using the oil well trigger source shown in Fig. 8 to take the trigger and conduction tests on TVS, replace the water gap with 1kΩ resistor and put the trigger source into the heat oven of 120℃. Add the charging voltage of the main capacitor C_1 to 12kV and charge C_4 to 1.5kV. Then control the relay to be closed so that the GAP was triggered to conduct and then the TVS was triggered. Observe whether the circuit is conducted and record waveforms of trigger voltage, trigger current and the main circuit current. Meanwhile read the TVS trigger and conduction time-delay. Get the typical experimental waveform as shown in Fig. 13, Fig. 14 and Fig. 15. Waveforms of TVS trigger voltage and trigger current are shown in Fig. 13, which show that the trigger current is a high frequency damped oscillation current with a short period about hundreds ns and the trigger voltage could reliably spark over TVS trigger surface. Fig. 14 is the main circuit current whole waveform with a 4 kA peak value and a 10μs pulse width about, which indicates that the TVS conducted successfully and the arc quenching occurs at the time of the current zero crossing. Fig. 15 is the rising edge waveform of the main circuit current and the trigger voltage waveform, which shows that the trigger and conduction time-delay is about 5μs.

In order to test the work reliability and stability of the designed TVS trigger source used in high temperature environment, 50 times trigger and conduction tests on the TVS were carried out making use of the oil well trigger source. The TVS could be triggered and conducted every time in the 50 tests. The amplitude and pulse width of the main circuit current were also relatively stable, and the time-delay joggle was little.

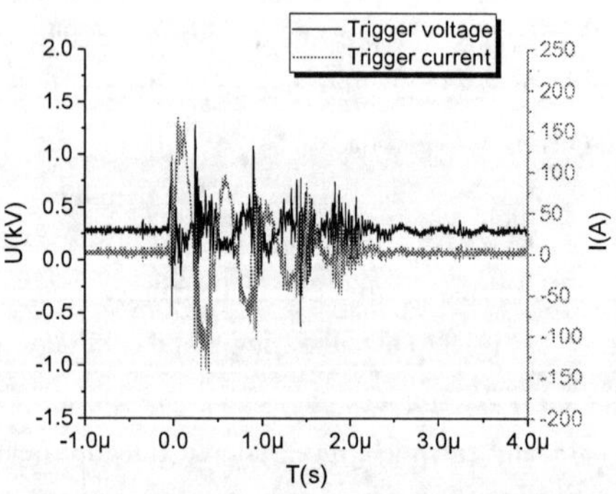

Fig. 13. Waveforms of trigger voltage and trigger current

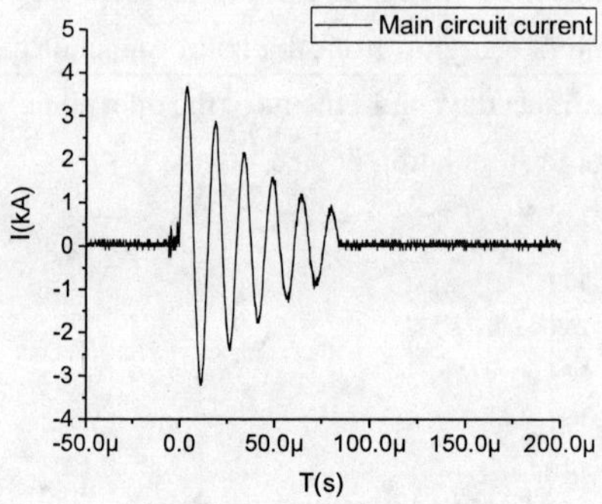

Fig. 14. The main circuit current waveform

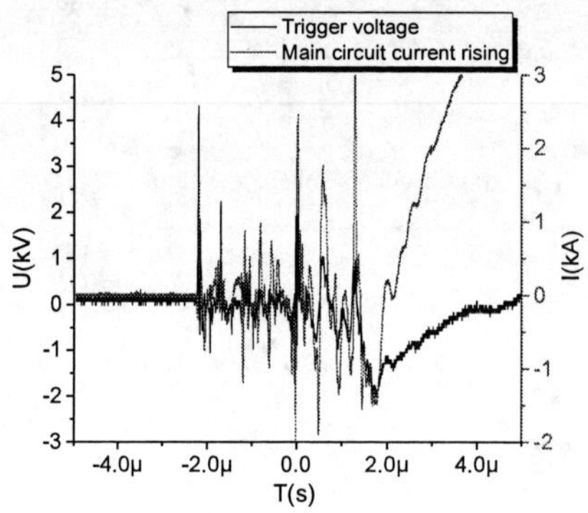

Fig. 15. Main circuit current rising and trigger voltage waveforms

2) Test of pulse discharge in water

The experimental platform was set up to take the test of pulse discharge in water according to the circuit shown in Fig. 8. Add the charging voltage of the main capacitor C_1 to 25kV and charge C_4 to 2kV. Then control the relay to be closed so that the GAP and TVS were triggered to conduction in turn. Finally water gap was of breakdown and then the entire circuit conduction, resulting in shock pressure wave in the water. Waveforms of the shock pressure wave and the main circuit current were recorded by the oscilloscope, and were contrasted with the waveforms obtained by using the typical trigger source (shown in Fig. 1). The waveforms of the test results are shown in Fig. 16, where the Fig. 16 (a) shows waveforms of the shock pressure wave and the main circuit current obtained by using the designed oil well trigger source and Fig. 16(b) shows waveforms obtained by using the typical TVS trigger source.

From the experimental test waveforms, it can be seen that waveforms obtained using two kinds of different trigger modes are not very different from each other and that the designed TVS trigger circuit of the series trigger capacitors charging from the main capacitor can normally trigger TVS and further achieve the discharge of the water gap.

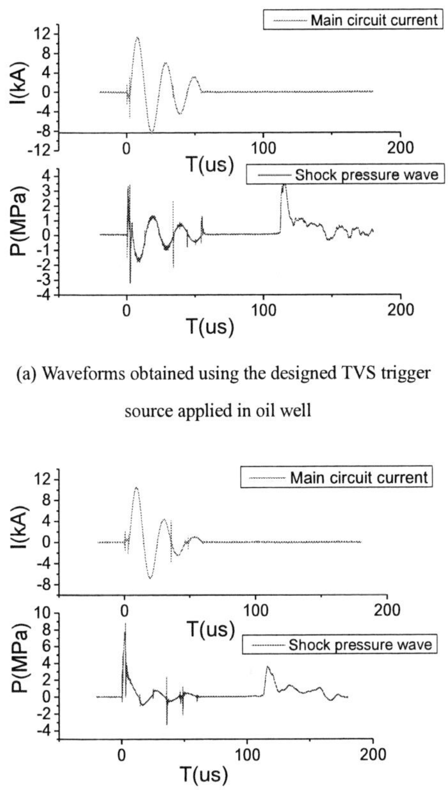

(a) Waveforms obtained using the designed TVS trigger source applied in oil well

(b) Waveforms obtained using the typical TVS trigger source

Fig. 18. Waveforms of the shock pressure wave and the main circuit current

IV. CONCLUSION

In this paper, aimed at the triggered vacuum switch applied in high temperature environment, the TVS trigger source is studied, developing a kind of high temperature trigger circuit without any semiconductor device. In the trigger circuit, the energy-stored capacitor is controlled to discharge upon the trigger surface of TVS by a three-electrode spark gap instead of a thyristor. And the pulsed trigger current is a high frequency damped oscillation current with a short period about hundreds ns and a high peak value up to several hundred amperes. The control signal of the three-electrode spark gap also be obtained from a small energy-stored capacitor discharging through a current-limiting resistor and a specially designed high-voltage relay, which can pass current of several amperes. Use the designed trigger circuit under high temperature to take the trigger and conduction tests on the 25kV TVS. In the 50 tests, the TVS could be triggered and conducted every time. The amplitude and pulse width of the main circuit current were relatively stable, and the joggle of trigger and conduction time-delay (about 4μs) was little. It shows that the work reliability and stability of the designed trigger source are good, fully meeting the trigger requirements of TVS working in high temperature environment.

In order to reduce the volume of the trigger source, this paper developed a new kind method of charging energy storage capacitor in the trigger circuit from the main capacitor in the oil well stimulation pulsed discharge device on the basis of the trigger circuit under high temperature. This method avoids to use high voltage isolation transformer of large volume to charge the energy storage capacitors of the trigger source and solves high-voltage isolation problems at the same time. On the basis of the simulation, the trigger circuit prototype is developed. Conduction tests for TVS and pulse discharge test in water are done with the prototype under the 120 ℃ high temperature environment. The trigger circuit prototype can normally trigger TVS to conduct and further achieve the discharge of the water gap. The results shows that the trigger circuit works steadily and reliably, fully satisfy the trigger requirements of TVS working in the oil well pulse discharge stimulation devicc.

REFERENCES

[1] Han Min, Zou Xiaobing, Zhang Guixin. Pulsed power technology basis. [M] Beijing, China: TSINGHUA UNIVERSITY PRESS, 2010.

[2] Wang Ying. Pulsed power technology summary. [J]. Electrical Engineering. 2009(04).

[3] Akiyama, Hidenori, Sakugawa, et al. Industrial applications of pulsed power technology. IEEE Transactions on Dielectrics and Electrical Insulation . 2007

[4] Liang Lin. Study on Mechanism and Key Techniques of Pulsed Power Switch RSD [D]. Huazhong University of Science & Technology, 2008.

[5] F.J.Zutavem,K.W.Reed,S.F.Glover,A.Mar,M.H.Ruebush,M.L.Horry,M.E.Swalby,J.A.Alexander,T.L.Smith,F.E. White.fiber-optic controlled pcss triggers for high voltage pulsed power switches. The 15th IEEE international pulsed power conference . 2005

[6] WEI Ronghua, WU Hanji. Triggered vacuum switches[J].High Voltage Apparatus，1983，25(6): 50-56.

[7] YAO Xue-ling, CHEN Jing-liang, SUN Wei. Experimental Research on Triggering Characteristics of Surface Flashover Triggered Vacuum Switch [J]. High Voltage Engineering, 2010,36 (2): 340-344.

[8] HE Junjia, ZOU Jiyan, WANG Hai， et al. Review of development in high quality， high power triggered vacuum switches，advanced technology of electrical engineering and energy[J]. Advanced Technology of Electrical Engineering and Energy, 199716 (2): 28-32.

[9] Lee H S, Jin Y S, Kim J S, et al. Evaluation of a RVU-43 switch as the closing switch for a modular 300 kJ pulse power supply for ETC application [J]. Magnetics, IEEE Transactions on, 2001, 37(1): 371-374.

[10] Alferov D F, Matveev N V, Sidorov V A, et al. Application of triggered vacuum switches in a high-voltage high-speed protective device [J]. Instruments and Experimental Techniques, 2004, 47(3): 360-367.

[11] L. A. Yutkin, Electrohydraulic Effect. Beijing, China: Science Press, 1962.

[12] B. M. Kovalchuk, A. V. Kharlov, V. A. Vizir, et al, "High-voltage pulsed generator for dynamic fragmentation of rocks," *Rev. Sci. Inst.*, vol. 81, no. 10, pp. 103506 1 103506 8, Oct. 2010.

[13] Zhang Zhicheng. Rock Fragmentation by Pulsed High Voltage Discharge and Drilling Equipment Development. Zhejiang University,2013

[14] B. Sun, Discharge Plasma in Liquid and Its Application, (in Chinese). Beijing, China: Science Press, 2013.

[15] Developments of pulsed power industrial applications at the Institute of Pulse Research and Engineering (IPRE)[C]// Pulsed Power Plasma Science, 2001. IEEE Conference Record - Abstracts. IEEE, 2001:1066-1069 vol.2.

[16] Han Min, Han Bo, Wang Xinxin, et al,Applications of impulse discharge with high current technique on oil well plug-releasing [J]. Advanced Technology of Electrical Engineering and Energy, 1998, (01): 36-40.

[17] W. Chen, O. Maurel, T. Reess, et al, "Experimental study on an alternative oil stimulation technique for tight gas reservoirs based on dynamic shock waves generated by Pulsed Arc Electrohydraulic Discharges," J. Petrol Sci. Eng., vol. 88-89, pp. 67-74, Jun. 2012.

[18] K. Dubovenko, Y. Kurashko, L.Onyshchenko, S. Poklonov, I. Shvets, and A. Ivanov, "Equipment developed at IPRE for well stimulation in the process of oil and water production," Presented at the Pulsed Power Plasma Conference, Las Vegas, Nevada, vol. 2, pp. 1070-1073, June 17-22, 2001.

[19] Dubovenko K V, Kurashko Y I, Onyshchenko L I, et al. Equipment developed at IPRE for well stimulation in the process of oil and water production[C]// Pulsed Power Plasma Science, 2001. IEEE Conference Record - Abstracts. IEEE, 2001:1070-1073 vol.2.

[20] Qu Dan-an. Research on the DM-2electric pulse equipment [D]. China University of Geosciences (Wuhan), 2002.

[21] Zhou Zhengyang, Liao Minfu, et al. Switching-on Experiment of a Field-breakdown Triggered Vacuum Switch, High Voltage Engineering, 2010, (12): 3059-3064

[22] Wang Y, Dai L, Lin F, et al. Optimization of a Triggered Vacuum Switch With Multirod Electrodes System[J]. Plasma Science IEEE Transactions on, 2014, 42(1):162 - 167.

[23] Dai L, Zhou Z, Nan J, et al. Characteristics of a Surface-Breakdown Triggered Vacuum Switch with Six-Gap Rod Electrode System [J]. Transactions of China Electrotechnical Society, 2012, (10):128-134.

[24] Mao X, Dai L, Zhang C, et al. Development of trigger circuit of multi-rod triggered vacuum switch [J]. High Power Laser and Particle Beams, 2016, 28 (1)

A fast recovery 60kV charging device for the ISOLDE target extraction voltage

T. Gharsa, E. Carlier, A. Fowler, J. Schipper
TE Department
CERN
Geneva, Switzerland
thierry.gharsa@cern.ch

***Abstract*— The ISOLDE facility at CERN provides radioactive ion beams to a number of experimental stations. These ions are produced by a metal target initially charged to a potential up to 60kV, which is impacted by a 1.4GeV high intensity proton beam. The ions are then accelerated by a grounded extraction electrode before transport to the experimental area. During proton beam impact extremely high ionization of the volume around the target gives rise to significant leakage current which results in loss of charge on the effective target capacitance of approximately 6nF. Recharging the target capacitance up to 60kV and to the required stability of +/- 600mV precludes a direct charging system and an alternative method of re-establishing the target voltage is used.**

I. Introduction

The ISOLDE facility at CERN makes use of new types of ion sources (neutron converter targets) which present ever increasing ionisation loads with the result that it is becoming impossible to respect the maximum 10 ms voltage recovery time. Future increase in beam energy to 2 GeV and higher intensity beams will further aggravate the situation.

To mitigate the problem a 60kV charging device with improved transient response has been developed to replace the actual HV modulator. The ISOLDE community are following this new development with great interest as an improved recovery time will give improved possibilities for the detection of exotic isotopes with extremely short half-lives. The Mass spectrometers situated at ISOLDE are used for high-precision measurements of the masses of the radioactive ions and require a very high stability of the extraction voltage to 10^{-5}, i.e. ± 600mV at 60kV.

A complete prototype of the new device has been built. In a first phase it has been validated with a dynamic test load to reproduce the beam induced leakage current during and after the proton beam impact on the target. In a second phase the validation of the performance of the new charging device has been realized in the real operational environment.

For these tests the HRS (High Resolution Separator) facility was used. The test system was installed in the HT room at ISOLDE replacing the existing HT modulator for the HRS. High intensity proton beam up to 3.1E13 proton per pulse (ppp) at 1.4GeV were extracted from the PSB accelerator with a maximum burst duration of 2.7µs each machine cycle. One machine cycle (1.2 s) was programmed in a Super Cycle with a length of 35 cycles.

The target unit used was a uranium carbide target with a neutron convertor. Radioactive nuclei are produced in a two-step process where neutrons produced by impinging the proton beam on the tungsten convertor induce a fission reaction in the adjacent depleted uranium carbide target. The solid tungsten convertor rod has a diameter of 12.4mm and a length of 125mm.

II. Present High voltage modulator at ISOLDE

The actual modulator [1] discharges the target capacitance 35µs prior to beam impact with a resonant circuit which then restores the voltage close to its nominal value within a few milliseconds. The time to re-establish the HT after the modulation to within 10ppm of its nominal value is called the recovery time. The performance of the modulator in terms of stability and transient recovery is to a large extend determined by the dynamic and static loads seen by the modulator. An attempt to characterize the load was made at the inception of the initial ISOLDE project in 1991. Further series of measurements were able to be made late in 2012. These measurements were made with a neutron converter target, which was assumed to represent the worst-case loading for the modulator, with and without proton beam impacting on the target. The results of this test showed that the loading due to the effective resistance in the first few milliseconds after beam impact was substantially greater in 2012 than in 1991.

Neutron converter targets have resulted in increased dynamic electrical load seen by the actual target voltage modulator, with the result that it is becoming impossible to respect the maximum 10ms voltage recovery time to ensure extraction of short life-time isotopes. Moreover the HIE-ISOLDE upgrade plan to make use of 2 to 3 times higher beam intensities with a possible energy increase to 2GeV. The combination of these two factors will globally produce an increase of power on the target of a factor of 4, aggravating the recovery process even further.

III. DESIGN OF A NEW FAST RECOVERY CHARGING DEVICE

The circuit presented in Fig. 1 is a robust charging device capable of re-establishing the target extraction high voltage with improved transient response.

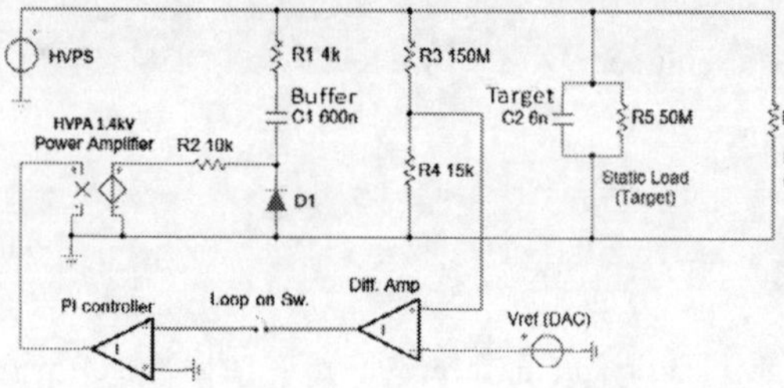

Fig. 1. Simplified circuit of the 60kV charging device

In this circuit, a low power, high precision, high voltage d.c. power supply (HVPS) is used to initially charge the total capacitance (C1+C2, shown in Fig. 1) of the circuit up to the maximum operating voltage of 60 kV. The maximum available output current of the HVPS is 40 mA, and charging time is less than one second provided that the current limit is set to a value higher than 37 mA. When the charge is complete, the HPVS maintains the target voltage constant and it only sources the static leakage current in the resistive part of the circuit, which is approximately 1.6 mA at 60 kV. When capacitors are fully charged, the high voltage circuit is ready for operation in pulsing mode which is described hereafter.

During the beam burst, the target capacitance is rapidly discharged due to the heavy ionized air shunt loading. Previous measurements made with beam impacting the target pre-charged to 60 kV at ISOLDE facility have confirmed a shunt loading with peak leakage current of 30 A. This indicates an average resistance to ground of the ionized air during the beam burst of 2 kΩ. Despite this large leakage, no breakdown voltage occurs and the target voltage is rapidly discharged to ~50 kV.

The fast recovery process can be split into four phases:

Phase 1:

The HVPS is not used during the critical period of the dynamic loading. For this reason prior to the beam impact, the HVPS analogue current reference is set to zero and the HVPS current limitation circuit shall be effective in less than 2 ms, preventing the HVPS to source large current immediately after the beam impact.

Phase 2:

After the beam burst, the 600 nF buffer capacitor, which is partially isolated from the heavy transient shunt loading by virtue of a 4 kΩ series connected resistor, begins to re-establish the voltage on the target capacitance to a voltage close to 99% of its nominal value.

Phase 3:

After a configurable fine delay, the feedback loop controlling the auxiliary High Voltage Power Amplifier (HVPA) is switched on which lifts the floating buffer capacitor potential and the target voltage to the nominal 60 kV in less than 2 ms, achieving the recovery within the required 10 ppm tolerance. The HVPA control loop then remains active and compensates for the dynamic and the static losses thanks to the high precision feedback divider that generates the low level ground referenced voltage feedback signal used to control and regulate the target voltage.

Phase 4:

Finally, after a period of 100 ms the dynamic loading has recovered to a sufficiently high impedance and becomes negligible. The HVPS analogue current reference is then set to a value corresponding to the estimated static leakage current in the circuit (~1.6 mA at 60 kV); moreover an offset of 250 V is applied to the HVPS analogue voltage reference. With these operating conditions, the HVPS regulation will act as a current source whose value is limited to the current reference value. By sourcing the current, the HVPS tries to recharge the capacitors, but at the same time, the HVPA output voltage decreases accordingly and maintains the target voltage constant at the nominal voltage. For this, the current reference value applied to the HVPS is controlled by an external control loop. The dead time between the current reference step changes and the effective current output response is less than 40 ms.

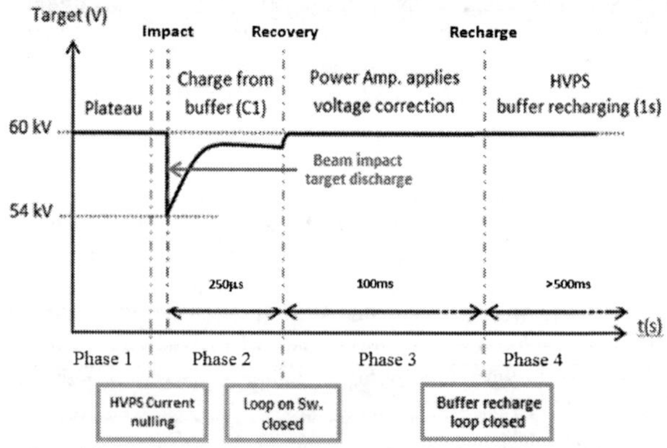

Fig. 2. Simplified diagram of the high voltage recovery process

By sourcing the current demand, the HVPS recharges the buffer capacitor differential voltage up to the initial nominal value in 500 ms. The charging device is then ready for the next proton beam impact on a cycle to cycle basis.

The complete fast recovery process in pulsing mode has been tested in real operational conditions. The HVPS used for this test is a commercial 65 kV, high precision, D.C. capacitor charger power supply with 40 mA current output capability.

A complete and robust control strategy has been defined. It has been successfully implemented into a PXIe controller with FPGA configuration and a real time embedded controller.

The Fig. 2 a scope view can be seen of the main control loop parameters of the recovery process during a machine cycle.

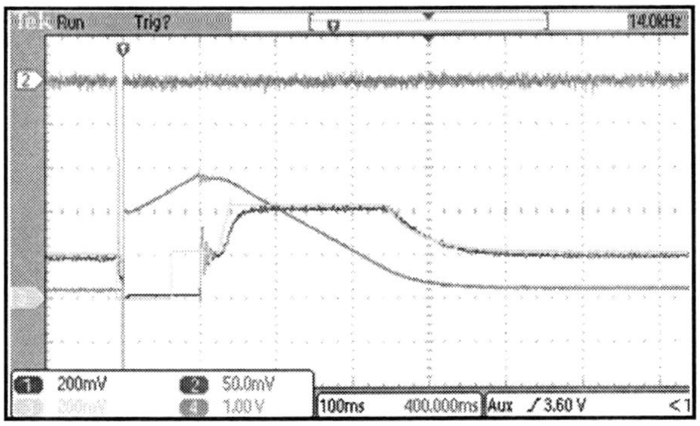

Fig. 3. Control loop main parameters during the recovery process

Scope screenshot legend:

- Red trace is the high voltage recovery error signal
- Magenta trace is HVPA output voltage signal
- Green trace is the HVPS current reference signal
- Blue trace is the HVPS current monitor signal

In the above scope view (Fig. 2) the magenta trace is the HV power amplifier voltage output. The initial jump of the output voltage correspond to the voltage level shifting of the capacitor bank, it compensates for the loss of charge immediately after the beam impact, in the same time fast recovery is achieved by injecting current in the target capacitance. Then, as the HVPS current reference is set to zero immediately prior to beam impact, the power amplifier is the only active current source in the circuit and its voltage output increase proportionally to the integral of the leakage current over the time. One can clearly see that the positive slope of the magenta trace compensate for the buffer capacitor charge loses in the static current leakage.

$$\text{Slope (V/s)} = I_{static\_leakage} / C_{buffer}$$

The green trace and the blue trace represent respectively the HVPS current reference and the HVPS current monitor signal. One can see that the HVPS current output follows the current reference change after a dead time of 30 ms. After this event the buffer capacitor is recharged by the HVPS that is used as a voltage controlled current source. As the HVPA feedback loop maintains a constant voltage on the target, its output voltage decrease proportionally to the integral of the HVPS output current minus the static current over the time and is finally stabilized to the 50 V reference level until next beam impact.

From this measurement it was concluded that the recovery process performed as expected with presence of the proton beam at high pulse intensity and with the neutron converter type target.

IV. Operational Test Results

The HVPA unit used during the test could only deliver half of the expected output voltage, with the consequence that our device was restrained to be operated up to 30kV. Another detrimental effect was that the global recovery loop gain was also reduced by 2, resulting in slower transient response.

Despite this discrepancy, the consistent performance of the device over a wide range of operational conditions has been investigated by varying the nominal target voltage between 30kV and 60kV by steps of 10kV, and at two different proton beam intensities 1.5E13ppp and 3.1E13ppp. The high voltage recovery time measured after beam impact are given in Table I.

TABLE I.

	High voltage recovery time			
Beam	**30 kV**	**40 kV**	**50 kV***	**60 kV***
1.5E13	1.2 ms	1.6 ms	2 ms	n.a.
3.1E13	1.6 ms	2.5 ms	n.a.	n.a.

*not applicable, see text.

Fig. 3 is the target high voltage signal acquired with a dynamic signal analyser. It shows the ionisation loss at the time of the beam impact, and despite the large shunt loading no breakdown voltage occurred.

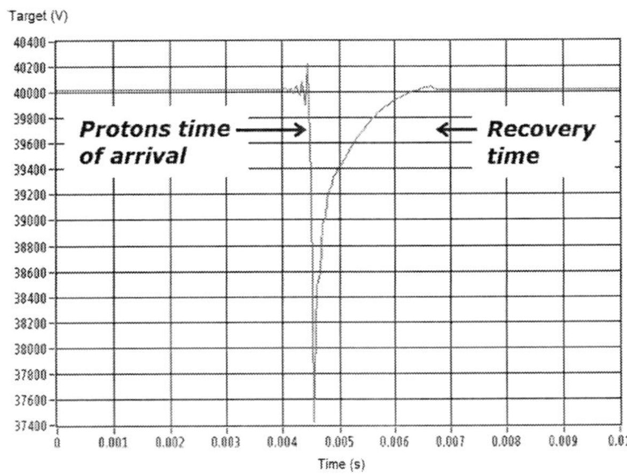

Fig. 4. High voltage recovery signal (HV=40kV, 3.1E13ppp, 1.4GeV)

It was possible to operate the device at a higher voltage than our expectations, the HVPA limited output range finally permitted some tests up to 40kV at 3.1E13ppp and 50kV at 1.5E13ppp. Very small overshoot is observed in the regulation loop, stability margin is large enough and loop gain could be added to reach a faster transient response of the charging device.

The HVPA output current has been monitored during proton pulse. As it is related to the post-impact loading current, this will help for future development of a more realistic dynamic test load. As a first approximation in the first few millisecond after impact, it is proportional to the square root of beam burst intensity $N^{0.5}$ (N being the number of proton per pulse). It is possible to extrapolate the expected loading of future increase of beam burst intensity.

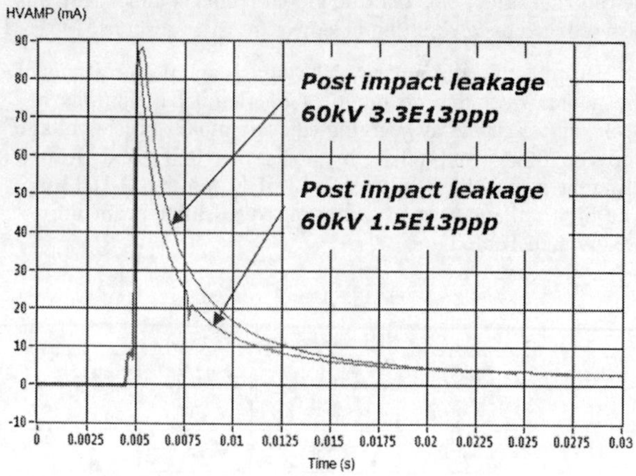

Fig. 5. HVAMP loading current at different beam intensity (HV=60kV)

Based on these results an upgrade of the prototype is envisaged to adapt it to the measured loading. By adding a second HVPA in parallel, the output current will be doubled, and will allow similar performance at 60kV. Fast high voltage recovery along with in-beam detection methods will greatly contribute to the measurements of exotic short-lived isotopes.

V. CONCLUSION

Among the different types of target, use of the neutron converter type has resulted in increased dynamic electrical load seen by the actual extraction voltage modulator, with the consequence that the recovery time of the target voltage to the precision of 10^{-5} is delayed, which in turn reduce the possibilities for the detection of exotic isotopes with extremely short half-lives. This is the primary motivation for the development of a more robust charging device capable of re-establishing the target extraction high voltage with improved transient response. The validation of the performance of the new charging device has been realized in real operational condition at the ISOLDE facility considering the most severe beam induced leakage current at the maximum beam intensity. In this given case, representing the worst-case scenario, the new system can provide a recovery time in the order of 2 ms, which makes a significant improvement in performance by reducing the recovery time by factor of 5 over the actual modulator system.

References

[1] D. Fiander, A. Fowler, "A 60kV modulator for the target voltage of an online isotope separator", Proceedings of the 20th IEEE Power Modulator Symposium; Myrtle Beach South Carolina, USA, June 1992.

[2] R. A. Barlow, B. Bleus, A. Fowler, H. Gaudillet, T. Gharsa, J. Schipper, "A charge-pump 60kV modulator for the ISOLDE target extraction voltage", Proceedings of the EAPPC 2014, 8-12 September 2014, Kumamoto, Japan

[3] T. Gharsa, J. Schipper, "Validation of a new high voltage fast recovery charging device for ISOLDE/HIE-ISOLDE", May 2016, CERN Test Report

Kinetic Modeling of Space Charge Effects in Intense Electron Beams: Distribution Function Method vs. PIC-method

Vasily Yu. Kozhevnikov, Andrey V. Kozyrev, Natalia S. Semeniuk
Laboratory of Low-Temperature Plasma
Tomsk State University
Tomsk, Russia
Vasily.Y.Kozhevnikov@ieee.org

***Abstract*— In this paper, we present the simulation results for the most important one-dimensional problems of current flow in planar vacuum diodes. For the first time, classical problems of vacuum electronics have been solved using the basic principles of nonequilibrium physical kinetics. This approach is based on the accurate numerical solution of the Vlasov-Poisson equation system to find the electron distribution function. We have obtained time dependent numerical solutions that satisfactorily agree with theoretical approximations and with particle-in-cell simulations as well.**

Keywords—vacuum diode; virtual cathode oscillator, Vlasov equation

I. INTRODUCTION

The history of vacuum electronics goes back for more than hundred years [1]. Related researches began much earlier with classical works of Child, Langmuir and Fay. Recently, interest in this field has increased because of the wide range of development prospects in various fields of physics and engineering: microwave radiation generation, accelerator physics, applications in thermonuclear synthesis, diodes with field emission, vacuum microelectronics, processes related to the manufacturing of semiconductors and semiconductor diodes, flat panel display manufacturing technology based on electron emission from carbon nanotubes, etc. From the great variety of theoretical problems in vacuum electronics, virtual cathode oscillators (vircators) have become a high interest research topic [2]-[5]. Apparent advantages of such oscillators lie in their configuration simplicity and wide frequency-tuning ability. Modern understanding of physical processes occurring in vacuum electron devices bases on extensive experiments and simulations.

Until recently, the particle-in-cell (PIC) [6] method was the usual approach used to provide time-dependent solutions for problems arising in vacuum electronics [4]. However, one should take into account that the PIC method has serious disadvantages concerning the use of restricted ensembles of macroparticles. It is extremely difficult to calculate the distribution function for some groups of particles representing a small fraction. Such particles can affect local plasma properties that are needed for correct physical interpretation of processes, even in simple cases (e.g., resonant particles being trapped in a local potential well). Generally, different groups of particles play a major part during the current establishing in vacuum diode at different time points.

In this article, we turn to a different approach to consider the phenomena occurring in vacuum diodes. It is called the "first principle approach." It relies on physical kinetics and the concept of a nonequilibrium electron distribution function (EDF) resulting from it. For collisionless electron transport, the evolution of EDF obeys the kinetic (Vlasov) equation consistent with electromagnetic fields equations (Maxwell-Poisson). We will refer to the results of PIC simulation in order to compare these two approaches. Also, some analytical dependencies and estimations will be provided for the considered problems, e.g. following the [5].

II. NUMERICAL SIMULATION

Following the kinetic approach, our numerical simulations are based on the direct solution of nonrelativistic Vlasov-Poisson equations on a phase space grid. In order to obtain useful dimensionless variables, in this paper, the coordinate is normalized to the diode gap length $x_0 = D$, velocity to the speed of light $v_0 = c$, electrostatic potential to applied voltage (or the voltage corresponding to injected beam velocity) $\varphi_0 = U$, and the electric field value to $E_0 = m_0c^2/qD$, where m_0 and q are electron mass, and elementary charge, respectively. Hence, the Vlasov-Poisson equation system in dimensionless variables reduces to

$$\begin{aligned} &\frac{\partial f}{\partial t}+v\frac{\partial f}{\partial x}-E\frac{\partial f}{\partial v}=0 \\ &\frac{\partial^2 \varphi}{\partial x^2}=-\int_{-\infty}^{\infty} f dv \end{aligned} \qquad (1)$$

where f – is the EDF, φ and E - are the electrostatic potential and field, respectively.

We have to apply proper initial and boundary conditions to both of the equations (1). As for the Poisson's equation in a

The work is supported by Russian Fund of Fundamental Researches project 15-08-03983.

planar one-dimensional case, we use simple Dirichlet boundary conditions to describe the diode without external circuit

$$\begin{aligned} \varphi(x=0) &= 0 \\ \varphi(x=1) &= V_a, \end{aligned} \tag{2}$$

where V_a is a dimensionless applied voltage. For the Vlasov equation, we set emission boundary conditions at cathode and zero at Neumann boundary conditions at anode. Such boundary conditions are apparently artificial, taking into the account the order of the Vlasov equation. Due to the limitations of the computational phase space by velocities, the truncation of EDF at maximum required velocity magnitude is implied also

$$\begin{aligned} & f(x=0) = f_M(v) \\ & \frac{\partial f}{\partial x}(x=1) = 0 \\ & f(|v| = v_{\max}) = 0, \end{aligned} \tag{3}$$

where $f_M(v)$ is staircase cathode emission EDF, and v_{max} is the upper limit of electrons velocity.

Because the problem is formulated in one spatial dimension, the solution of the Poisson equation can be expressed in quadrature by direct integration

$$\begin{aligned} \varphi(x,t) &= x + \int_0^x \int_0^{x'} \int_{-v_{\max}}^{v_{\max}} f(x'',v,t)\, dv dx'' dx' - \\ &- x \int_0^1 \int_0^{x'} \int_{-v_{\max}}^{v_{\max}} f(x'',v,t)\, dv dx'' dx' \\ E(x,t) &= -\frac{qU}{m_0 c^2} \left\{ 1 + \int_0^x \int_{-v_{\max}}^{v_{\max}} f(x',v,t)\, dv dx' - \right. \\ &\left. - \int_0^1 \int_0^{x'} \int_{-v_{\max}}^{v_{\max}} f(x',v,t)\, dv dx'' dx' \right\}. \end{aligned} \tag{4}$$

The numerical solution of the Vlasov equation is performed by backward semi-Lagrangian method Cheng-Knorr method using cubic spline interpolation and Strang operator splitting [7]. Other methods [8] have also been tried, but none proved superior to cubic splines. As a result of the joint numerical solution of the Vlasov equation (1) with the boundary conditions (3), consistently with the electric field (4) , we obtain the EDF $f(x,v,t)$ and compute waveform of diode current density (in dimensionless variables)

$$j(t) = \int_{-v_{\max}}^{v_{\max}} f(x=1,v,t)\, v dv \cdot \tag{5}$$

As for the PIC method, we chose one-dimensional electrostatic code authored by Dr. I. V. Pegel (Institute of High Current Electronics SB RAS), who kindly provided the simulation results. It is intended for transient simulation of the high current electron beam dynamics and their interaction with radio frequency (RF) fields. Our PIC code operates with two types of charged particles (electrons and ions). Particle flow is presented by macroparticles' motion with charges proportional to the value of injection current. The number of particles injecting at each time step is constant. Weighing of the electric field, charge, and current is linear cloud-in-cell (CIC). The electrostatic part is one-dimensional and solved by direct integration of Poisson's equation. Relativistic equations of motion are integrated in three dimensions according to the leapfrog scheme at offset meshes. The total electric field (its longitudinal component only) is defined as the space charge field superposition found from the solution of one-dimensional Poisson's equation with boundary conditions at electrodes. Our code was previously widely applied to the computation of microwave generation in the vircator systems [9].

III. Planar Vacuum Diode

We first apply our approach to a time dependent current establishing in planar nonrelativistic vacuum diode with constant applied voltage. At steady state, the solution of this problem is well-known as the Child-Langmuir relation between current density and applied voltage. In nonrelativistic cases, it is given as

$$j_{C.L.} \approx 2.33 \cdot 10^{-6} \frac{U^{3/2}}{D^2}, \tag{6}$$

where U is applied voltage, and D is interelectrode distance. Our numerical solution should have good agreement with the formula (6) at steady state. In addition, there are some approximate analytical assessments for the diode current in the transition region, for example, when the electron beam reaches the anode (collector), the estimated current amplitude is approximately 2.75 times above stationary Child-Langmuir value [10].

We supposed the emittance of the cathode to be unlimited. In terms of boundary conditions, the emission is defined with a square staircase EDF. The same condition has been applied in PIC simulation. In Fig. 1 the comparison of current densities (5) time profiles obtained with both methods is presented for vacuum diode with parameters $D = 1$ cm and $U = 2$ kV. The emission current was taken nearly fifty times higher than the diode stationary current value providing unlimited emittance, while the thermal spread was fixed at 2 eV. The numerical solution procedure of the Vlasov-Poisson system was performed at 1,000 per 1,001 phase space grid points, which appears to be accurate enough for this case. In PIC simulation, we took 10,000 points of spatial mesh carrying almost 15,000 macroparticles. Both methods run with constant time step of 2 ps.

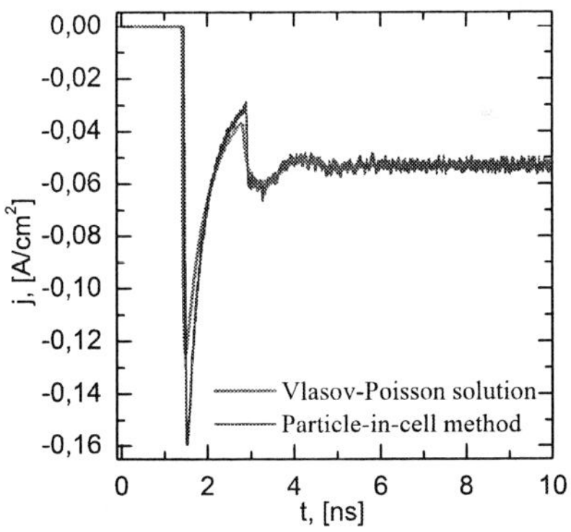

Fig. 1. Anode current density time-profile for planar vacuum diode obtained by both simulation methods.

Both methods gave good agreement as with Child-Langmuir value, as well as with the estimation of current splash in the transition region. Also, they are in good agreement with each other (see Fig. 1). The most important result of the numerical simulation is that during current establishing in diode, the space charge relaxation oscillations always take place. The characteristic frequency of these oscillations is defined in terms of electrons' flight time through the diode gap. It is obvious that frequency of relaxation oscillation is proportional to plasma frequency at the anode.

IV. Electron Beam Injection in Equipotential Diode

Another problem is a propagating of cold electron beam between two equipotential grids. This problem is of great scientific importance due to the direct relation to the physical principles of vircators [1]. It is interesting that some authors point out the existence of multiple modes of electron beam transmission satisfying the same boundary conditions (e.g., [11], [12]). Similar assertions are based on approximate analysis of the dynamic system. Since the transition between current flow modes is a time dependent problem; the correct solution should be based on the formulation in terms of time dependent evolution equations, e.g. (1)-(3).

In our one-dimensional case, this problem is formulated as follows. At the origin ($x = 0$) from the planar electrode (emitter) electron flow comes in the direction of the second electrode (collector) with an energy much greater than possible thermal spread ($qU >> kT$). The collector is located at the distance D ($x = 1$) from the emitter. Electrostatic potentials of emitter and collector are equal (hereinafter they are equal to zero, $V_a = 0$). This problem is also known as the Bursian diode problem.

Since the value of the space charge in a planar diode determines the amount of total current, there is a critical value of the injected current, specifying space charge limiting value depending on the interelectrode distance and the velocity of the injected beam

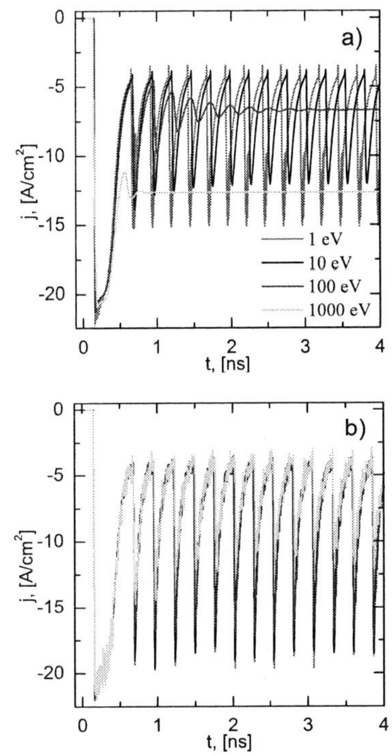

Fig. 2. Time-profile of the collector current density obtained with: a) Vlasov-Poisson system direct solver, and b) with PIC method

$$j_c = 2.71\cdot 10^3 \frac{\left(\int\limits_1^{1+\frac{eU}{m_0c^2}} \frac{d\xi}{\left[\xi^2-1\right]^{\frac{1}{4}}}\right)^2}{D^2} \quad . \qquad (7)$$

If the injection current is of several times more than critical value, sustained virtual cathode oscillations generated [5]. For monoenergetic electron beam, the maximum amplitude of oscillation is achieved when the injection current exceeds the critical value fourfold. In this case, the modulation amplitude of the collector current is 10%. However, the oscillatory state is known to be unstable. So, if thermal spread of electrons in beam exceeds 5% ~ 10% (depending on injecting current value), the self-oscillations fade and the current flow takes on quasi-stationary character [13]. In that case, electron flow separates and is passed through the virtual cathode or reflected from it.

Fig. 2 shows the comparison of collector current time profiles obtained for the electron beam injection corresponding to different thermal spread from 1 eV to 1000 eV. For quasi-monoenergetic electron beam the oscillations at 3.5 GHz frequency coincide with theoretical estimations given in e.g. [14].

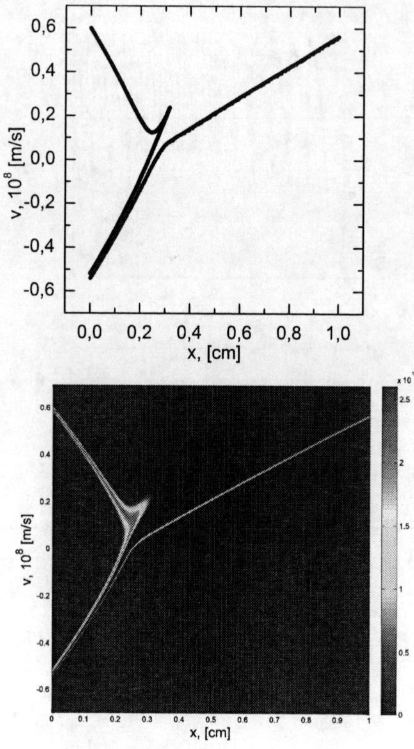

Fig. 3. Virtual cathode oscillation state with 10 eV thermal spread: a) macroparticles distribution in phase-space obtained with PIC method, and b) EDF density plot obtained with Vlasov-Poisson solver.

Further expansion of the thermal spread leads to oscillations damping in the calculations obtained by direct solution of Vlasov-Poisson equations, while in the PIC method results the oscillation damping is observed at a much greater thermal spread equal to 2 keV. The comparison shows the unstable nature of the vircator state, as well as the superior advantage of the EDF approach with respect to PIC simulations. In fact, by increasing the thermal spread the PIC method becomes unacceptably noisy, since its numerical error approaches the damped oscillations amplitude. Following the EDF approach, it turns out that the oscillatory system switches to a quasi-stationary state and the collector current density increases. Oscillation dumping happens because of the increasing number of electrons having wide thermal spread and, as a result, getting an opportunity to overcome the potential barrier between virtual cathode and collector. Phase plots of macroparticles distribution and the EDF are given in Fig. 3. for a 5 ns time point.

The numerical solution of Vlasov-Poisson system was performed at 1,000 per 1,001 points of phase-space grid. The PIC simulation used 4,000 spatial points of mesh carrying about 5,000 macroparticles. Both methods have used 1 ps constant time step.

V. Conclusions

In our paper, we have presented a preliminary comparison of two methods for the simulation of processes in planar vacuum diodes. This was done to show the advantage of the so called first principles computational approach, i.e., based on numerical solution of Vlasov-Poisson equations system. Since the knowledge of the EDF gives comprehensive information about a particular system of charged particles, the full description for some process details can be obtained only in terms of EDF evolution. Comparison of calculation results by examples of two simple fundamental problems shows important limitations of the PIC method, e.g., inaccuracies were found in the transition from an unstable state into a quasi stationary state. The increasing level of internal noise in PIC method significantly reduces the applicability. Despite this, we believe that PIC method is still a convenient instrument to use for the solution of various vacuum electronics and electron beams transportation problems, especially in cases where the EDF and its evolution can be replaced by simpler formulation.

References

[1] A. H. Beck "Physical Electronics," Pergamon Press Ltd., 1968. p. 599

[2] V. L. Granatstein and I. Alexeff, "High-Power Microwave Sources. Norwood, MA: Artech House," 1987

[3] B. V. Alyokhin, A. E. Dubinov, V. D. Selemir, O. A. Shamro, K. V. Shibalko, N. V. Stepanov, and V. E. Vatrunin, "Theoretical and experimental studies of virtual cathode microwave devices," IEEE Transactions on Plasma Science, vol. 22, no. 5, pp. 945–959

[4] R. J. Barker and E. Schamiloglu, "High-Power Microwave Sources and Technologies," 2001.

[5] J. Swegle, E. Schamiloglu, and J. Benford, "High Power Microwaves, Second Edition," Series in Plasma Physics, Feb. 2007

[6] C. K. Birdsall and A. B. Langdon, "Plasma Physics via Computer Simulation," 1991.

[7] C. Z. Cheng and G. Knorr, "The integration of the vlasov equation in configuration space," Journal of Computational Physics, vol. 22, no. 3, pp. 330–351

[8] F. Filbet and E. Sonnendrücker, "Comparison of Eulerian Vlasov solvers," Computer Physics Communications, vol. 150, no. 3, pp. 247–266

[9] S. A. Kitsanov, A. I. Klimov, S. D. Korovin, I. K. Kurkan, I. V. Pegel, and S. D. Polevin, "A vircator with electron beam premodulation based on high-current repetitively pulsed accelerator," IEEE Transactions on Plasma Science, vol. 30, no. 1, pp. 274–285

[10] A. Kadish, W. Peter, and M. E. Jones, "A Generalization of the Child-Langmuir Relation for One-Dimensional Time-Dependent Diodes," IEEE Transactions on Nuclear Science, vol. 32, no. 5, pp. 2576–2578

[11] A. Y. Ender, V. I. Kuznetsov, H. Schamel, and P. V. Akimov, "Switching of nonneutral plasma diodes. I. Analytic theory," Phys. Plasmas, vol. 11, no. 6, p. 3212, 2004.

[12] V. I. Kuznetsov, A. Y. Ender, H. Schamel, and P. V. Akimov, "Switching of nonneutral plasma diodes. II. Numerical results," Phys. Plasmas, vol. 11, no. 6, p. 3224, 2004.

[13] S. D. Korovin, I. V. Pegel, S. D. Polevin, and V. P. Tarakanov, "Numerical simulation of efficient 1.5 GHz vircator," Digest of Technical Papers. 11th IEEE International Pulsed Power Conference (Cat. No.97CH36127)

[14] Ming-Chieh Lin and Pei-Che Chang, "Study on virtual cathode oscillations by using finite difference time domain particle-in-cell codes," 2008 IEEE International Vacuum Electronics Conference, May 2008, pp. 213-214

Semiconductor High Voltage Generators of Subnanosecond Pulses.

A.G. Lyublinsky[1,2], A.F. Kardo-Sysoev[1], I.A. Smirnova[1]
[1]Division of Solid State Electronics
Ioffe Institute
26 Polytechnicheskaya, St. Petersburg, 194021 Russia
[2] Megaimpulse Ltd.
Alexander.Lyublinsky@mail.ioffe.ru

Abstract **— Design and operation principles of subnanosecond and picoseconds rise time high voltage pulse generators are described. These generators are based on two types of semiconductor devices Drift Step Recovery Diodes (DSRD) and Silicon Avalanche Shapers (SAS). DSRD are semiconductor "opening switches". The main advantages of DSRD are very short switching-off time (down to 200 ps), high commutated power density (more than 400 kW per cm² of semiconductor device area) and reliable operation of high voltage stacks consisting of many connected in series DSRD. SAS are very fast semiconductor "closing switches". They have the best in class switching-on time (less than 100 ps) and kV range operation voltage. Typical applications of fast rise rate high voltage pulse generators are Ultra Wide Band radars, communication systems, underground radars, testing and measurement equipment, etc.**

Keywords — Power semiconductor switches, Pulse power systems, High-voltage techniques

I. INTRODUCTION

High voltage subnanosecond and picoseconds ranges pulse generators are used in many application areas where very short high voltage and high rise rate pulses are required including medicine, diagnostics, testing and measurements system, etc. The most promising application where such generators are the key component is Ultra Wide Band (UWB) technology including UWB radars and communication systems.

Drift Step Recovery Diodes (DSRD) and Silicon Avalanche Shaper (SAS) have been discovered in Ioffe Institute in 80th years of the last century. The first experimental results were published in [1, 2]. The detailed description of DSRD and SAS physical principles, applications, circuit design and benefits of these approaches to short high voltage pulse generation are described in [3, 4].

II. DSRD AND SOS OPERATION PRINCIPLES

DSRD is very fast and effective opening switch, which is excellent active component for the output circuit of high voltage subnanosecond pulse generators. DSRD allows direct pulse forming. No additional compression stages are required. DSRD-based circuit includes inductive energy storage L and DSRD stack connected in parallel with the load (please see the inset in Fig.1). When DSRD switches off the energy preliminary stored in the inductance L transfers into the load resulting in high rise rate output pulse generation.

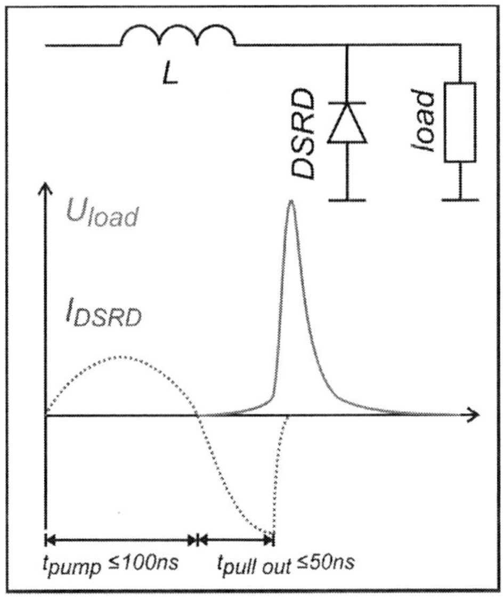

Fig.1. Simplified DSRD-based output circuit of high voltage subnanosecond pulse generator as well as current pulse waveform I_{DSRD} through DSRD during commutation process (red dot line) and pulse voltage on the load U_{load} (blue solid line).

DSRD is a $p^+p'n'n^+$ silicon diode with a deep diffusion *p-n* junction. Special diode structure allows subnanosecond switching. Two stages commutation process is required for the fast current interruption. First, short pulse of the direct current provides the storage of electron-hole plasma in the diode base layers. At the second stage, the reverse pulse current pulls out the stored plasma. During this stage electron and hole plasma fronts moves one towards another in DSRD structure. These fronts should collide exactly at the *p-n* junction plane. In this moment, the space charge region is formed at the *p-n* junction and expands due to the major carrier movement in opposite directions. Major carriers move with the saturated velocity 10^7 cm/s. As a result, the diode recovers sharply and the current switches to the parallel-connected load.

DSRD can be serially connected in a high-voltage assembly. Maximum voltage rise rate (dU/dt) is ~1 kV/ns for single DSRD and this value multiplies proportionally to the number of the diodes in the stack. Forward and reverse current pulses are the same for each DSRD in the assembly.

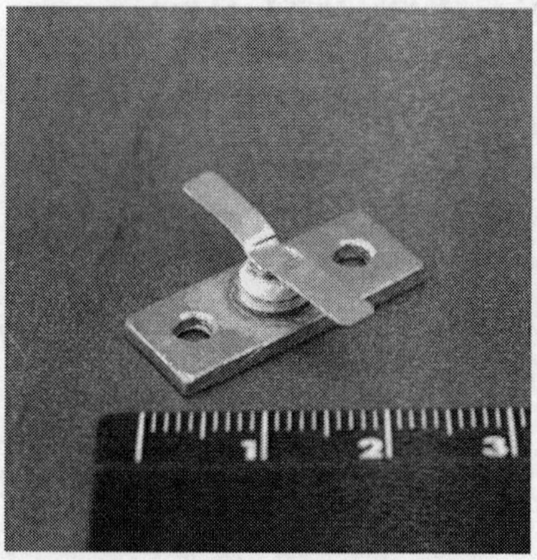

Fig.2. Photo of DSRD assembly with the following main parameters: max pulse voltage 2 kV, peak current 80 A, turn-off time 600 ps, max repetition rate 1 MHz.

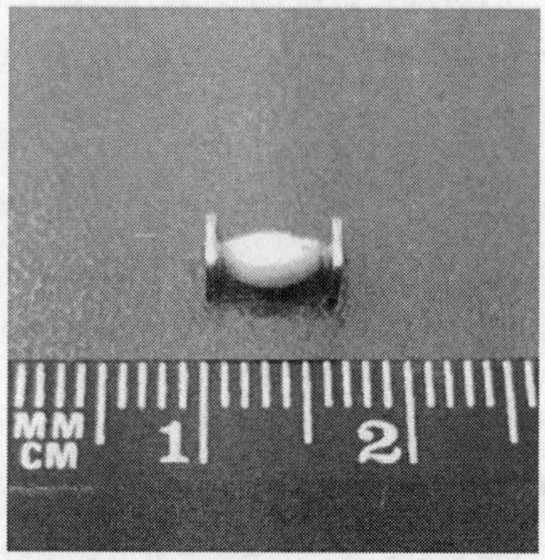

Fig.4. Photo of 5kV SAS diode with the following main parameters: turn-on voltage 5 kV, peak current 100 A, turn-on time 100 ps, max repetition rate 16 kHz.

Therefore, the same charge is injected into all the diodes and pulled out from them resulting in strictly synchronous DSRD triggering.

The photo of 2 kV DSRD assembly is shown in Fig.2. DSRD assemblies are able to switch current pulses up to thousands amperes and generate high voltage pulses with amplitude up to tens of kilovolts.

Another type of semiconductor devices and another operation principle are used in the output circuit of picoseconds range high voltage pulse generators. Picoseconds switching cannot be obtained by the carrier movement in the semiconductor structure. Even moving with the saturated velocity the carriers pass only 1 μm in 10 ps which is few orders less than the width of space charge region required for blocking hundreds or thousands volts. Therefore, another physical phenomenon, which is named delayed impact

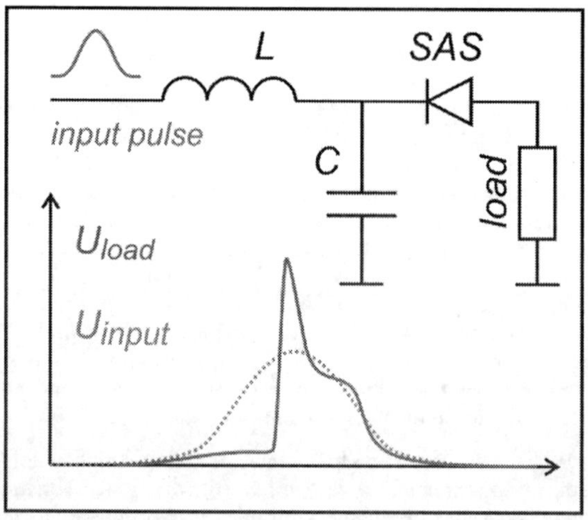

Fig.3. Simplified SAS-based output circuit of high voltage picoseconds rise time pulse generator as well as waveforms of SAS stage input pulse U_{input} (dot blue line) and output sharpened pulse U_{load} (red solid line) on the load.

ionization, is used in SAS diode to get super fast switching from closed to conducting state.

SAS is closing type switch. Simplified SAS-based output circuit is shown in Fig.3 inset. Initially SAS diode is in the reverse biased closed state. Then high voltage pulse with more than 1kV/ns rise rate is applied through the inductance L to the peaking capacitor C and reversely biased SAS diode. Capacitor C is small and input pulse quickly charges it up to near to double voltage. The voltage rise rate of the input pulse applied to SAS-based stage is very high. Therefore, the electric field on the diode *p-n* junction can exceed static breakdown level resulting in the significant SAS overvoltage. This overvoltage is followed by the delayed impact ionization. Ionization front propagates much faster than the carriers saturated velocity, which allows super fast switching of the diode. The discharge current of the peaking capacitor C is essential to support ionization front and effectively fill in the diode base layers by highly conducting plasma. The waveforms of the input pulse and sharpened by SAS diode output pulse are shown in Fig.3.

Currently SAS diodes are the fastest power semiconductor devices. Experimentally registered output pulse voltage rise rate exceeds 50 kV/ns.

III. DSRD- AND SOS-BASED PULSE GENERATORS DESIGN

The design, specifications and output pulse waveforms of several subnanosecond and picoseconds pulse generators based on DSRD and SAS devices are presented.

These generators are commercially available and have been designed for ultra wide band (UWB) radars and phased antenna arrays as well as other applications, which require high voltage subnanosecond and picoseconds rise time pulses. DSRD-based generators produce very stable, clean and ringing free output pulses with up to MHz range repetition rates. SAS-based generators produce high voltage narrow pulses with wider spectrum range, which are required by compact high resolution UWB radars.

DSRD-based subnanosecond pulse generators

The specifications, photos and output pulse waveforms of several DSRD-based pulse generators are shown below.

Specification of PPG-1/500 pulse generator:

Output pulse amplitude	1 kV
Pulse polarity	positive
Pulse rise time	600 ps
Pulse width (FWHM)	2 ns
Max repetition rate	500 kHz
Jitter (RMS)	10 ps
Spectrum range (-6dB)	200 MHz
Mean output power	20 W
Output connector	N type
Triggering	External only
Triggering connector	SMA type
Triggering pulse	+5 V, 10..50 ns width
Power supply	DC +24 V, 2 A
Dimensions	100x144x30 mm^3 (without connectors and mounting flanges)

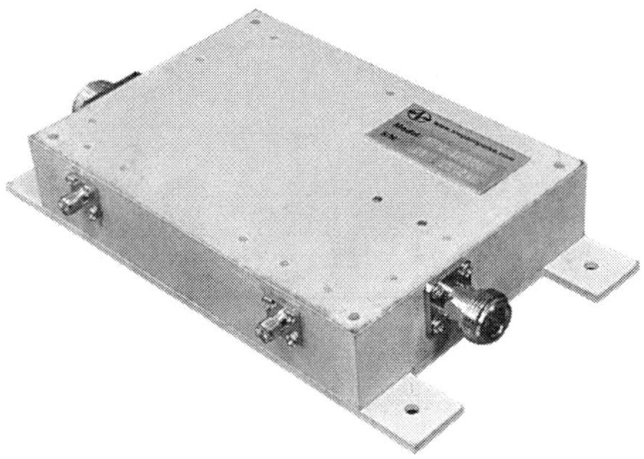

Fig.5. Photo of DSRD-based subnanosecond pulse generator PPG-1/500. The main parameters are: pulse amplitude 1 kV, rise time 600 ps, max repetition rate 500 kHz.

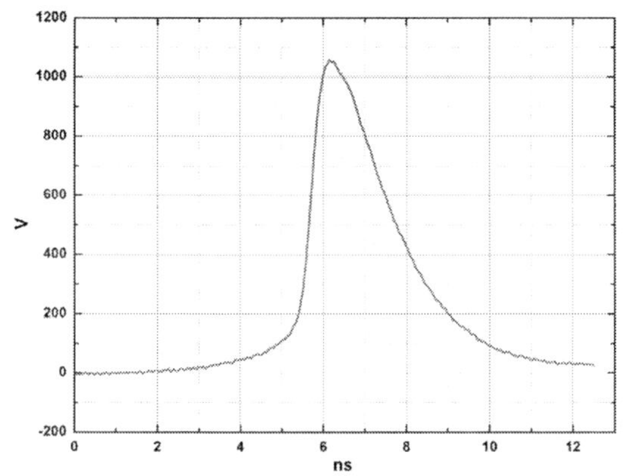

Fig.6. Output pulse waveform of PPG-1/500 subnanosecond pulse generator.

Specification of PPG-2/1000 and PPG-4/250 pulse generators:

Output pulse amplitude	2 kV and 4 kV
Pulse polarity	positive
Pulse rise time	600 ps
Pulse width (FWHM)	1.5 ns
Max repetition rate	1 MHz and 250 kHz
Jitter (RMS)	30 ps
Spectrum range (-6dB)	300 MHz
Mean output power	100 W
Output connector	N type
Triggering	External only, SMA
Triggering pulse	+5 V, 10..50 ns width
Cooling	by forced airflow
Dimensions	250x110x72 mm^3 (without connectors and mounting flanges)

Fig.7. Photo of DSRD-based subnanosecond pulse generators PPG-2/1000 and PPG-4/250. The main parameters are: pulse amplitude 2 kV / 4 kV, rise time 600 ps, max repetition rate 1 MHz / 250 kHz.

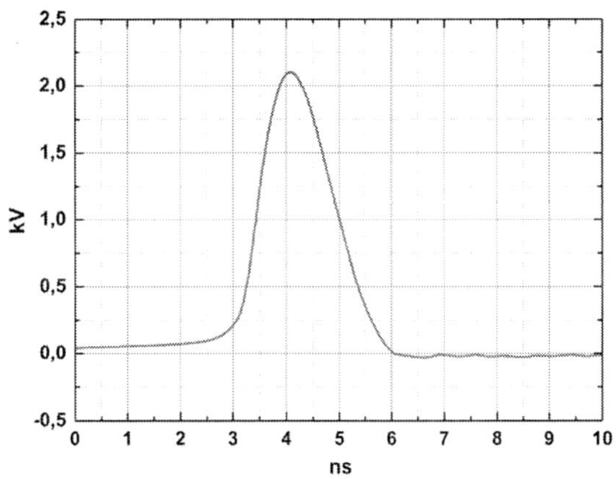

Fig.8. Output pulse waveform of PPG-2/1000 subnanosecond pulse generator.

SAS-based picoseconds pulse generators

The specifications, photo and output pulse waveforms of several SAS-based pulse generators are shown below.

Specification of PPG-5/16 and PPG-7/10 pulse generators:

Output pulse amplitude	5 kV and 7 kV
Pulse polarity	positive
Pulse rise time	100 ps
Pulse width (FWHM)	300 .. 400 ps
Max repetition rate	16 kHz and 10 kHz
Jitter (RMS)	10 ps
Spectrum range (-6dB)	1.5 GHz
Mean output power	10 W
Output connector	N type
Triggering	External only
Triggering connector	SMA type
Triggering pulse	+5 V, 10..50 ns width
Power supply	external AC-DC converter
Dimensions	120x84x49 mm^3 (without output connector)

Fig.9. Photo of SAS-based picoseconds pulse generators PPG-5/16 and PPG-7/10. The main parameters are: pulse amplitude 5 kV / 7 kV, rise time 100 ps, max repetition rate 16 kHz / 10 kHz.

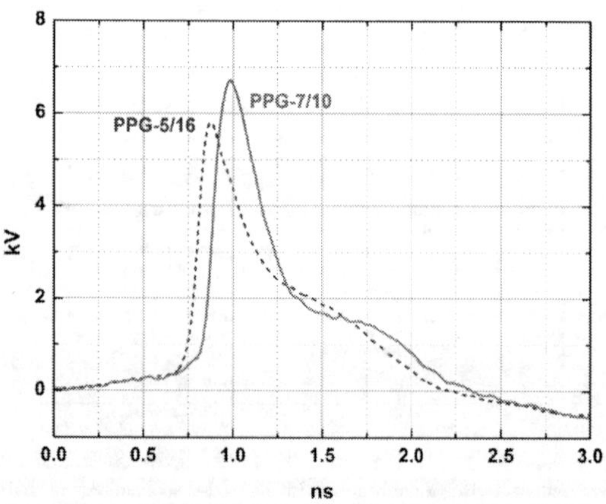

Fig.10. Output pulse waveforms of PPG-5/16 (blue dash line) and PPG-7/10 (red solid line) picoseconds pulse generators.

Specification of PPG-1.5/100 pulse generator:

Output pulse amplitude	1.5..2 kV
Pulse polarity	positive
Pulse rise time	70 ps
Pulse width (FWHM)	300 ps
Max repetition rate	100 kHz
Jitter (RMS)	10 ps
Mean output power	2 W
Output connector	N type
Triggering	External only
Triggering connector	SMA type
Triggering pulse	+5 V, 10..100 ns width
Power supply	AC 110..230V, 50/60Hz
Dimensions	205x204x64 mm^3

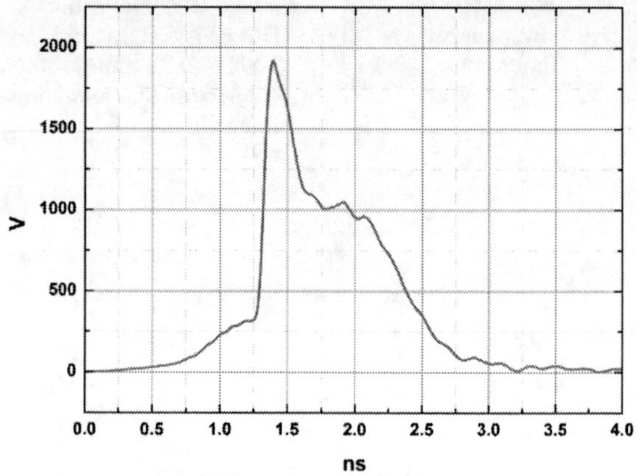

Fig.11. Output pulse waveform of PPG-1.5/100 picoseconds pulse generator.

IV. CONCLUSIONS

Ultra Wide Band (UWB) is a newest promising technology for radars, antenna arrays and communication systems. It has a lot of benefits in comparison with other approaches including better resolution, signal-to-noise ratio, interference immunity, etc. DSRD- and SAS-based subnanosecond and picoseconds pulse generators are perfect transmitters for UWB applications because of supreme pulse parameters, reliability, high efficiency, compact size, high repetition rate as well as low jitter and good pulse stability, which allow synchronization of many generators in the array. A range of DSRD- and SAS-based pulse generators are described in the article.

REFERENCES

[1] I.V. Grekhov, and A.F. Kardo-Sysoev, "Subnanosecond current drops in the delayed breakdown of silicon p-n junction," Sov. Tech. Phys. Let., Vol. 5 (1979), No. 8, p. 395–396.

[2] I.V. Grekhov, V.M. Efanov, A.F. Kardo-Sysoev, and S.V. Shenderey, "Formation of a high nanosecond voltage drop across semiconductor diode," Sov. Tech. Phys. Let. Vol. 9, No. 4, 1983.

[3] I.V. Grekhov, "New principles of high power switching with semiconductor devices," Solid-State Electronics, Vol. 32, No. 11, 1989, pp. 923–230.

[4] Ultra-wideband radar technology / edited by James D. Taylor 2001 by CRC Press LLC.

Pulsed MHD generators for powering rail electromagnetic launchers

A. G Afonin, V.G. Butov, S. V. Sinyaev, V. A. Solonenko
Research Institute of Applied Mathematics and Mechanics of Tomsk State University
Tomsk, 634050, Russia

G. A. Shvetsov
Lavrent'ev Institute of Hydrodynamics, Siberian Branch, Russian Academy of Sciences n
Novosibirsk, 630090, Russia
shvetsov@hydro.nsc.ru

***Abstract*—This paper analyzes the possibility of using Sakhalin type pulsed MHD generators fueled by plasma-generating solid propellants as an energy source for rail electromagnetic launchers. These MHD generators are compact self-contained sources of electrical energy, whose specific energy characteristics are much better than those of capacitors and whose time of start and ramp-up does not exceed a few seconds. The power of pulsed MHD generators can reach hundreds of megawatts. Extensive experience has been gained in the study, development, and production of transportable pulsed MHD generators placed on cars and trains.**

The operation of electromagnetic launchers powered by a pulsed MHD generator alone or in combination with inductive energy storages (IES) is analyzed. The paper presents the results of mathematical simulation of the operation of a rapid-fire rail launcher powered by a combined energy source which features a direct connection of a Sakhalin type MHD generator to inductive energy storage.

The rate of fire of the launcher is 250–400 projectiles per second for 20 sequentially launched projectiles with a total weight of 9.5 kg. The total kinetic energy of the burst is 25 MJ for mean kinetic projectile velocities of more than 2300 m/s, respectively, with stably reproducible dynamic characteristics in each cycle of operation of the device. Separation effects do not occur in the MHD generator channel during operation, including during charging of the IES. Use of the IES as an intermediate storage significantly reduces the weight and size parameters of the energy source.

The results show that high-power pulsed MHD generators are promising pulsed power sources for rapid-fire launchers.

Keywords—MHD generator; plasma-generating propellant; cyclic load; supersonic shock-free flow; inductive energy storage; multirail launcher.

I. Introduction

A number of laboratories are currently engaged in large-scale research projects aimed at developing electromagnetic rail launchers to accelerate single projectiles weighing 10–20 kg to velocities of 2.0–2.5 km/s and groups of projectiles weighing 0.1 to 1.0 kg to velocities over 2.0–2.5 km/s. The implementation of these projects requires a source of electrical energy of tens to hundreds of megajoules. In most of the published studies, capacitor banks are used or intended to be used as an energy source. An apparent advantage of capacitor banks is the relative ease of their operation, but the low weight, size, and energy characteristics of capacitors prevent their use as a basis for designing mobile transportable sources of electromagnetic energy and, hence, for designing mobile transportable electromagnetic launchers of solids.

The potential of pulsed magnetohydrodynamic generators (PMHDGs) as a power supply for rail launchers has been studied in a few papers, e.g., [1–3].

The problem of specialized mobile sources of electrical energy is particularly challenging for the development of rapid-fire electromagnetic rail launchers [4–6].

PMHDGs fueled by plasma-generating solid (powder) propellants (SPP) are compact self-contained sources of electrical energy, whose specific energy characteristics are much better than those of capacitors. Moreover, the time of their start and entry into the nominal mode of electrical energy generation is 0.1 s when using superconducting magnets and a few seconds in the self-excitation mode. The power of pulsed MHD generators can reach hundreds of megawatts [7]. These performance characteristics allow MHD generators to be considered as a promising primary power supply for rapid-fire rail launchers.

This study is aimed at determining the potential of using high-power pulsed Sakhalin type MHD generators [8] as a primary source of energy for rapid-fire multirail electromagnetic launchers [9] (MREL). Calculations were performed for the plasma-generated solid propellant with relatively high consumption used in experiments [10]. The use of advanced propellants [7] provides a severalfold improvement in the weight, size, and consumption characteristics of MHD generators and the electrical energy source as a whole. This has motivated the development of specialized high-energy propellants and pulsed MHD generators of a new generation for the class of launchers considered.

II. Solid propellant fueled pulsed MHD generators

In the USSR, a total of 25 types of large-scale experimental and pilot MHD systems were developed. For

This work was supported by the RF Ministry of Education and Sciences (Government Contract No. 2014/223; Project code. 2382).

self-contained power supply of various loads, over 1000 runs of standard MHD systems were performed in different regions of the USSR from the Kola Peninsula to the Pamirs and East Siberia [11, 12]. The tests have demonstrated the reliable performance of MHD generators under field conditions in various climatic zones and their potential for use in geological exploration, geophysics, etc. Operation with various pulsed electrical loads has shown the possibility of generating electrical pulses of various durations in pulsed, intermittent, and pulsed-periodic modes with an arbitrary duty cycle, which offers unique opportunities for developing electromagnetic launchers of solids. These advantages have formed the basis of the high-energy MHD gun concept [13].

The Sakhalin MHD generator is so far the most powerful in the world (electrical power of the channel up to 600 MW). Using the Sakhalin MHD generator, record-breaking parameters and unique specific characteristic have been obtained, some of which are inaccessible to other types of power plants.

The second step in the development of pulsed MHD generators is associated with the advent of new ecological clean solid propellants with high electrical conductivity. Their use makes it possible to develop MHD systems with a specific energy output of about 1 MJ/kg from SPP unit weight, a ~14% efficiency of conversion of the chemical energy of SPP to electrical energy, and a power density per unit weight of the MHD generator of about 20 MW/ton [12].

III. Formulation of the Problem

Fig. 1 shows a design diagram of a system consisting of an MHD generator and a rail launcher in which the metal-armature multirail launcher is powered directly by two parallel-connected PMHDGs (embodiment 1). In this case, the load is connected to the power supply unit after the self-excitation period (2–3 s for the Sakhalin generator) of the pulsed MHD generator and after the establishment of quasi-steady-state operation with generation of power up to 600 MW.

Equilibrium supersonic plasma flow in the PMHDG is produced by combustion of a solid plasma-generating propellant doped with a fine aluminum powder in the combustion chamber which contains gas and condensed phases. The condensed phase, with a mass concentration $Z = 0.36–0.46$, consists of liquid polydisperse particles of alumina Al_2O_3 with sizes of 0.5–2 μm [10, 11, 14] and melting and boiling points of 2320 K and 3250 K, respectively. The gas phase in the MHD channel is a chemically reactive equilibrium mixture of CO (mole fraction 45%), H_2 (25%), N_2 (14%), H (4%), H_2O (4%), and CO_2 (1%) molecules and K, Cs, Mg, and Fe vapor (≤1%) [11]. The source of free electrons in the gas phase is atoms of alkali metals (Cs, K), with a low ionization potential, which are added as an ionizing agent to the propellant. The remaining components of the gas phase are not substantially ionized [11].

During acceleration of the two-phase flow, there is velocity and temperature nonequilibrium – the particles lag behind the gas and become more heated than the gas. In the conversion region, the trend is reversed – the condensed-phase particles catch up with the gas. Generally, the velocity and temperature nonequilibrium between the gas and condensed phase in the MHD channel is negligible, so that in the calculation, it was neglected and the pseudo-gas model was used [15].

The rail unit consists of several pairs of rails separated by insulation inserts and series-connected by conductive bridges in the inlet section of the launcher. The projectile armature consists of several identical metal brackets insulated from each other and having two peripheral current-collecting parts. Each armature bracket closes its corresponding pair of rail guides of the launcher. Projectiles are fed into the MREL channel sequentially, and in front of the rail unit, they can be pre-accelerated as a single rigid assembly [4].

Before departing from the launcher, the armature of each projectile is virtually de-energized due to the interception of the current by the armature of the next projectile fed into the launcher before the departure of the previous one and by the resistive bridge in the outlet section of the launcher. The bridge diagonally connects the outermost rails with oppositely directed currents. This type of connection minimizes the inductance of the electrical circuit, which speeds up the de-energizing of the next projectile leaving the launcher. Sequential acceleration of projectiles is performed in two alternating modes in which either one or simultaneously two projectiles are accelerated in the MREL channel. Equivalent electrical circuits of the launcher for these two modes are shown Fig. 1b and Fig. 1c, respectively. The first projectile is fed into the launcher channel when the PMHDGs are idling.

A promising development is a combined power supply circuit in which a pulsed MHD generator is the primary source of electrical energy and an inductive energy storage (IES)

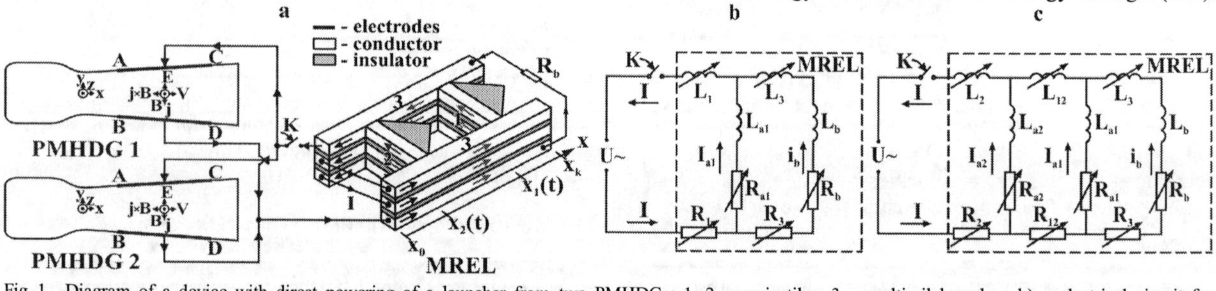

Fig. 1. Diagram of a device with direct powering of a launcher from two PMHDGs: 1, 2 – projectiles; 3 – multirail launcher; b) – electrical circuit for acceleration of one projectile; c) – electrical circuit for simultaneous acceleration of two projectiles.

serves as an intermediate storage.

A schematic diagram of a rail launcher with this combined energy storage (embodiment 2) is shown in Fig. 2. It is assumed that the closing switch K_2 operates instantaneously and the opening switch K_1 starts to operate simultaneously with K_2. The active resistance of K_1 increases by a power law with time.

The unsteady supersonic flow in the pulsed MHD generator was described using the approximation of quasi-two-dimensional flow of single-phase equilibrium ideal pseudo-gas obtained from the three-dimensional equations of magnetohydrodynamics by averaging over the z coordinate directed along the magnetic field $\boldsymbol{B}(0,0,B_z)$. The electromagnetic quantities in the MHD generator channel were calculated in a conventional quasi-steady-state approximation [15]. Within the framework of a semi-empirical approach [11], we used the experimentally measured distribution of the magnetic induction z-component in the MHD channel. This distribution normalized by the maximum induction value (2.2 T) is described by curve 1 in Fig. 3. In the same figure, curves 2 and 3 show the profiles of the electrode and insulator walls of the channel normalized by half the distance between the electrode walls at the outlet of the computational domain (0.9 m). The inlet boundary of the computational domain is the section with coordinate $x = 1.12$ m in the supersonic part of the nozzle. Parameter values at the inlet boundary are determined in the approximation of quasi-one-dimensional isentropic flow in the supersonic nozzle. The pressure and temperature values in the combustion chamber necessary for their determination were 4.76 MPa and 3750 K, respectively. Plasma parameters in the combustion chamber were specified taking into account equilibrium chemical reactions in the boost section of the gas-dynamic channel of the PMHDG. The weight content of the condensed phase in the flow is 0.35. The effective values of the adiabatic index and the gas constant of the pseudo-gas are 1.16 and 238 J/(kg·K), respectively.

Since the thermal properties of the combustion products in the supersonic flow region change only slightly, their values

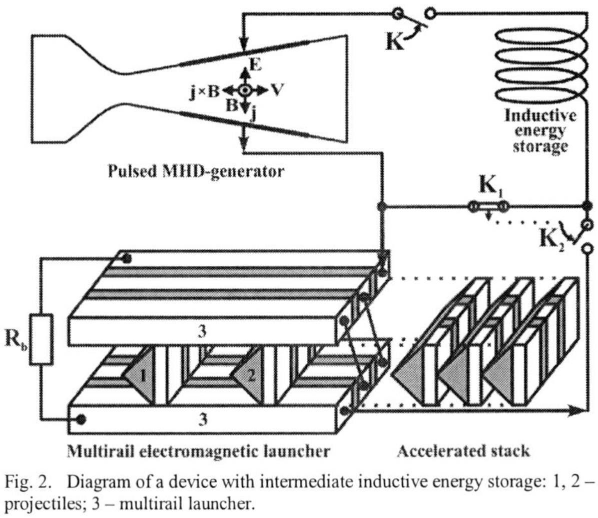

Fig. 2. Diagram of a device with intermediate inductive energy storage: 1, 2 – projectiles; 3 – multirail launcher.

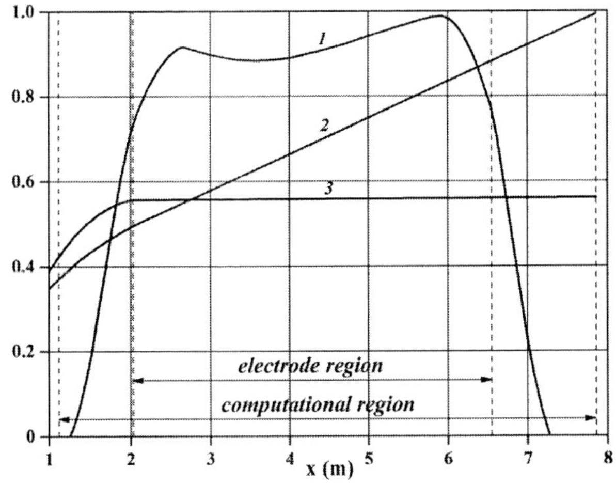

Fig. 3. Channel wall profiles in the Sakhalin PMHDG and magnetic induction: 1 – magnetic induction; 2 – electrode wall; 3 – insulator wall.

are assumed to be constant. Thus, at the inlet of the computational domain, the pressure, temperature, and velocity are found to have values of 0.682 MPa, 2917 K, and 1734 m/s, respectively.

In the rapid-fire launch of projectiles (burst), the rail launcher is a resistive-inductive load with cyclically varying electrical parameters: inductance L and active resistance R. For numerical studies of the operation of the devices considered, a mathematical model has been developed which describes the unsteady processes in a supersonic MHD generator and multirail launcher during their simultaneous operation.

The system of equations describing the simultaneous operation of the MHD generator and rail launcher, the initial and boundary conditions, the necessary relations, the assumptions used, and a description of the calculation procedure are given in [3, 16].

IV. Results of the Numericl Study

Numerical studies were performed to investigate the simultaneous operation of two embodiments of launching devices which use a power supply unit consisting of two Sakhalin pulsed MHD generators with a total power of 1200 MW and one such generator with intermediate inductive energy storage.

The rail launcher has the following characteristics: rail length 5.5 m, number of pairs of rails 5, cross-sectional area of each rail 8 cm^2, rails made of chromium copper, initial bridge resistance 21 mΩ, bridge inductance 2 μH, initial temperature of the rails and bridges 20 C, number of projectiles in a burst 20, projectile weights m_1 = 550 g, $m_2,\ldots,m_{19}$ = 470 g, m_{20} = 525 g, and initial projectile velocity 50 m/s. In the first embodiment of the device, the launcher is powered by two parallel-connected Sakhalin PMHDGs. The idling voltage of the Sakhalin PMHDG is 5 kV.

Fig. 4 shows the changes in the main characteristics of the first embodiment of the PMHDG-MREL launching device.

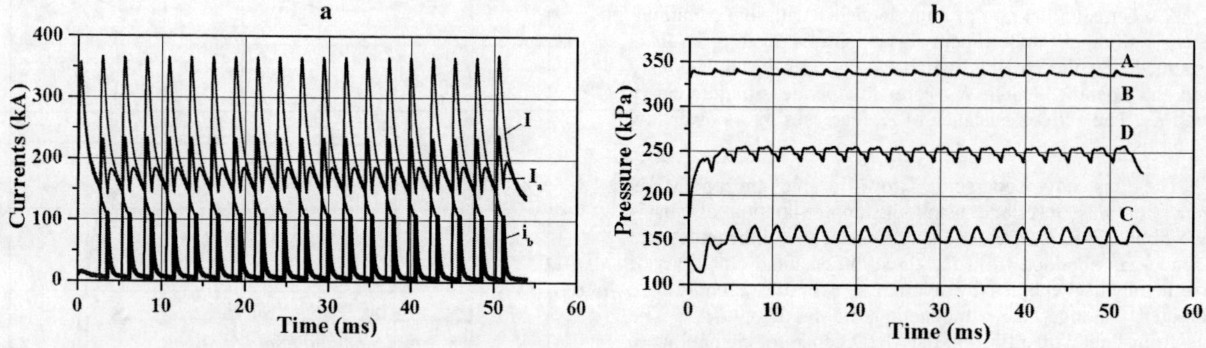

Fig. 4. Calculated parameters of the sequential acceleration of twenty projectiles in embodiment 1 of the device.

Fig. 4a shows the changes in the total discharge current I in the PMHDG and the currents through the armatures I_a and the bridge i_b. It follows from the curves that each next armature acts as a dynamic crowbar by almost de-energizing the previous armature to a current of approximately 100 kA by the time it leaves the launcher. Complete de-energizing of the armatures at the muzzle end of the launcher is provided by the resistive bridge at the exit from the rail unit. Acceleration of each projectile in the burst proceeds in a similar manner to a velocity of about 2.3 km/s and at a rate of 360 Hz. Stabilization of the acceleration modes of the first and last projectiles in the burst is achieved due to an increase in their weight. Fig. 4b shows the corresponding pressure changes at four characteristic points of the MHD channel: points A and B at the entrance and points C and D at the exit of the anode and cathode, respectively (see Fig. 1a). The pressure in the entrance section of the electrode zone is higher than the pressure in its exit section throughout the process, which indicates shock-free flow in the MHD channel [10].

Fig. 5 shows the calculated pressures in the MHD channel at the times of maximum (Fig. 5a) and minimum (Fig. 5b) currents through the armature of the fifth projectile. This pressure distribution in the MHD channel is characteristic of the entire acceleration process, which is further evidence of the shock-free mode of the supersonic flow and hence the stability of its output characteristics in operation with the cyclic load considered. Periodical increase in the (L, R)-parameters of the load reduces the electromagnetic retardation of the flow, which, in turn, prevents the formation of compression and shock waves in the channel of the MHD generator.

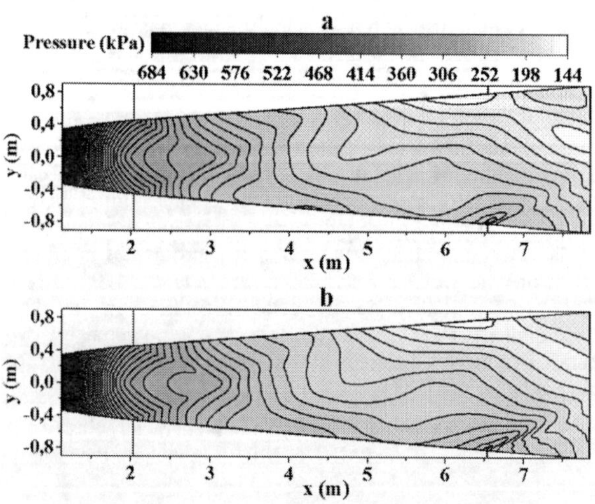

Fig. 5. Pressure distribution in the MHD channel at the times of maximum (a) and minimum (b) current through the armature of the fifth projectile.

Fig. 6 shows the volt-ampere characteristic (VAC) of the PMHDG. There is a slight hysteresis of the VAC due to cyclic changes in the electrical parameters of the launcher. The greatest hysteresis deviations of the VAC from the steady-state characteristic [3] correspond to transients in the operation of the launcher, namely, the acceleration of the first projectile while the MHD system is transitioned from idling to load operation.

The characteristics of the launcher in the second embodiment of the launching device are the same as in the first embodiment. The launcher is powered by one Sakhalin PMHDG and an intermediate IES with an inductance of 1 mH. Before connection of the rail launcher, the IES is charged to a current of 205 kA in a time of 64 ms. Fig. 7–9 show the calculated performance of the second embodiment of the device.

Fig. 7a shows the cyclic changes in the currents at the entrance of the rail unit I, through the armature of each projectile I_a, and through the resistive bridge i_b. The inductive storage significantly stabilizes the current at the launcher entrance and through each armature. Furthermore, the "fullness" of their oscillograms is significantly higher than that in the first embodiment of the device (see Fig. 4), and the

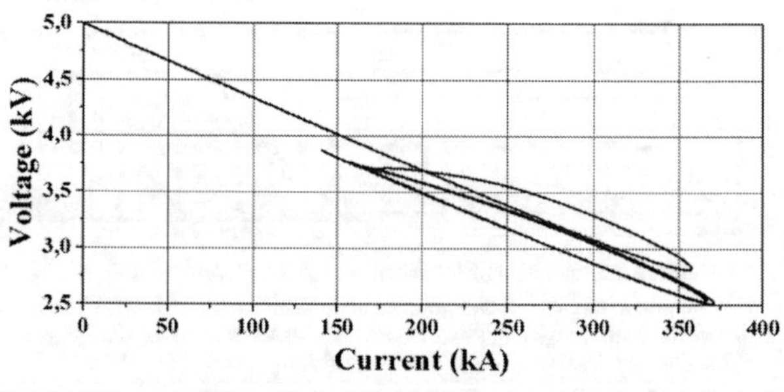

Fig. 6. Volt-ampere characteristic of the PMHDG.

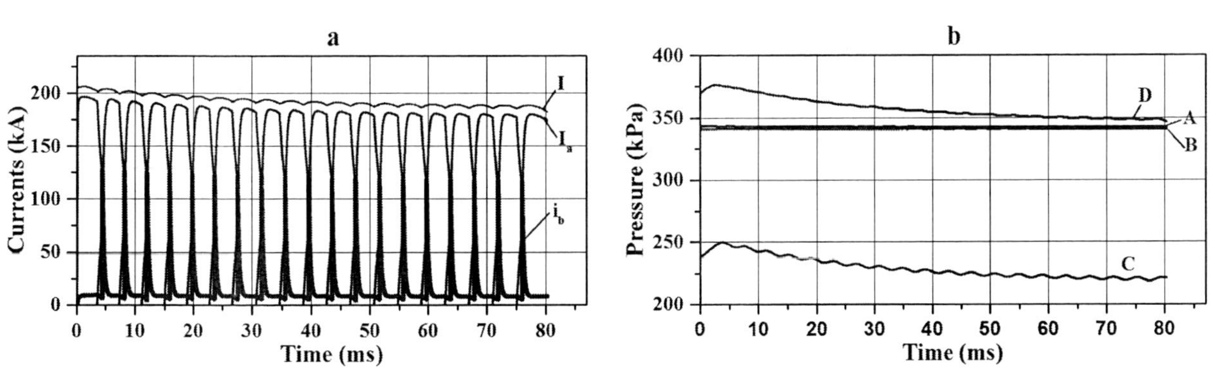

Fig. 7. Calculated parameters of the sequential acceleration of twenty projectiles in launcher embodiment 2.

maximum current values at the launcher entrance and through the armatures of the projectiles are significantly lower than the corresponding values in the first embodiment. At the same time, the flow in the MHD generator channel is more intense (Fig. 7b) because during charging of the IES, there is a constant increase in the electromagnetic force decelerating the MHD flow in the channel. Thus, the pressure in the vicinity of the end of the cathode (curve D) is somewhat greater than the pressure in the vicinity of its beginning (curve B). However, this local pressure gradient in the vicinity of the cathode does not block the flow in the MHD generator channel (see Fig. 8a).

The schlieren pictures shown in Fig. 8b were obtained using the procedure described in [17, 18]. The results of Fig. 8 show that separation effects in the MHD channel do not occur; flow separation is possible beyond the MHD generator channel. The cyclic mode of acceleration of projectiles in the MREL has little effect on the plasma flow pattern in the MHD generator.

Fig. 9 shows a curve of the voltage on the electrodes of the MHD generator versus the generated current – the volt-ampere characteristic of the generator. The VAC curve is smooth at the stage of power supply to the IES and has a negligible hysteresis feature during operation of the launcher.

Fig. 10 shows the muzzle velocities of each projectile for both embodiments of the launching device. In the first case, the acceleration of all 20 projectiles occurs within 55 ms at a rate of 360 Hz, and in the embodiment with IES, within 80 ms at a rate of 250 Hz. From a comparison of these results, it follows that the power supply with IES is preferred in terms of projectile velocity, current load, and weight and size parameters of the energy source.

V. Conclusion

The study led to the following conclusions.

- It has been shown that powerful PMHDGs can be used to power cyclically varying resistive-inductive loads at a rate of several hundred hertz alone, and in combination with an IES.
- The power supply unit based on two Sakhalin PMHDGs and a single generator with an intermediate IES can be effectively used as power supplies for rapid-fire rail launchers.
- Cyclic simultaneous operation with an arcless multirail launcher does not lead to the development of separated flows in the MHD channel and a qualitative change in

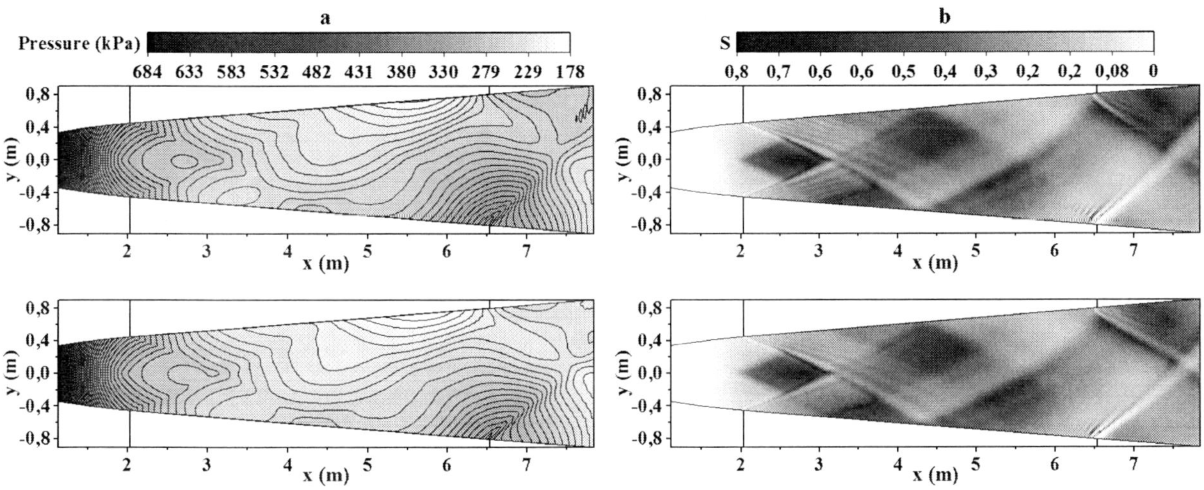

Fig. 8. Pressure distribution (a) and schlieren pictures (b) in the MHD channel (t = 5 ms) at the time of maximum pressure at the cathode end (top) and at the end (t = 80 ms) of the acceleration process (bottom).

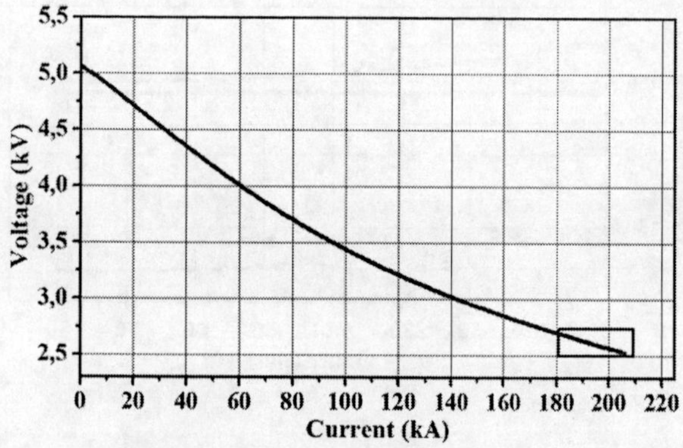

Fig. 9. Volt-ampere characteristic of the generator.

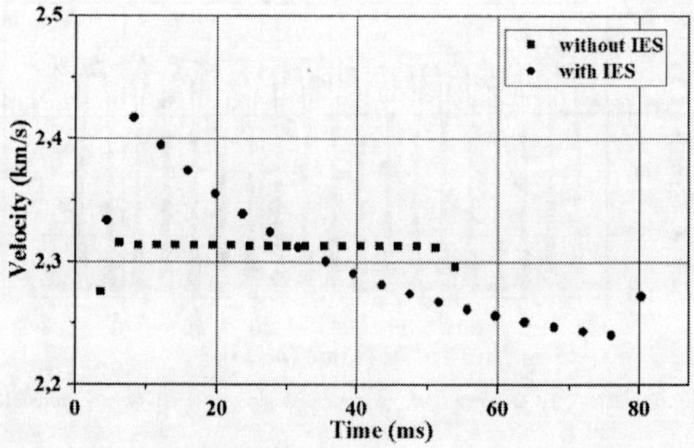

Fig. 10. Comparison of projectile velocities for the two embodiments of power supply circuits of launchers.

the electrical characteristics of the MHD generator.

- The device embodiments considered here allow launching 20 projectiles with a velocity of more than 2300 m/s in a rapid-fire burst mode at a rate of 280–360 shots per second. The total weight of the projectiles is 9.5 kg, and their kinetic energy is 25 MJ.

The results confirm that high-power pulsed MHD generators are promising power supply sources for the type the load considered. In this connection, it seems appropriate to develop specialized pulsed MHD generator using modern solid (powder) plasma-generating propellants and advanced developments and technologies, which will provide a significant (severalfold) improvement in the specific energy, weight, size, and resource characteristics of the power supply unit.

Acknowledgment

We express gratitude to V. P. Panchenko and A. A. Yakushev for useful discussions and valuable comments.

References

[1] G. A. Shvetsov, V. M. Titov, Yu. L. Bashkatov, et al., "Operation of a railgun accelerator for solid projectiles powered from an explosive MHD generator," Ultrahigh magnetic fields. Physics. Techniques. Applications. Proc. of the Third Intern. Conf. on Megagauss Magnetic Field Generation and Related Topics. Nauka, Moscow, 1984. pp. 177-182.

[2] Nob. Harada, "Space application of non-equilibrium MHD generator," 2nd Int. Workshop on Magnetoplasma Aerodynamics in Aerospace Applications. Moscow, Russia, 2000. pp. 325-329.

[3] A. G. Afonin, V. G. Butov, V. P. Panchenko, S. V. Sinyaev, et al., "Multirail electromagnetic launcher powered from a pulsed magnetohydrodynamic generator," J. Appl. Mech. Tech Phys., 2015. Vol. 56, No. 5, pp. 813-822.

[4] S. V. Sinyaev, B. V. Burkin, E. Yu. Pimonov, "Toward the development of a rapid-fire electrodynamic accelerator: Results of experimental and theoretical studies," in: Modern methods of design and development of missile and artillery weapon systems. Proc. Conf. of the Volga regional center of RARAN. Sarov, VNIIEF, 2000. pp. 507-512.

[5] M. Schneider, M. Wötzel and W. Wenning, "The ISL Rapid Fire Railgun Project RAFIRA-Part II: First Results," IEEE Trans. Magn., 2009. Vol. 45, No. 1, pp. 448-452.

[6] Ch. Schupper, F. Alonahabi, M. Schneider, "Electromechanical Aspects of Reliable Loading Procedures for Multishot Railguns," IEEE Trans. on Plasma Science, 2013. V. 41, No. 5, pp. 1387-1391.

[7] E. P. Velikhov, R. V. Dogodaev, V. P. Panchenko, A. A. Yakushev, et al., "Investigation of the properties of combustion products of a solid plasma-generating propellant for pulsed MHD generators," Dokl. Ross. Akad. Nauk, Vol. 434, No. 2, 2010, pp. 1-6.

[8] E. P. Velikhov, O. G. Matveenko, V. P. Panchenko, et al., "Solid propellant fueled Sakhalin pulsed MHD system with 500 MW electric power output," Dokl. Akad. Nauk. 2000. Vol. 370, No. 5, pp.617–622.

[9] A. E. Poltanov, A. K. Kondratenko, et al., "Multi-Turn Railguns: The Concept Analysis and Experiments," IEEE Trans. Magn., 2001, V. 37, No. 1, pp. 457–461.

[10] Gomozov V. A., Matvienko Z. I., Panchenko V. P. Experimental study of the characteristics of the condensed phase in the combustion products of pulsed MHD generators, Teplofiz. Vysok. Temp. Vol. 30, 1992, pp. 1006–1011.

[11] V. V. Breev, A. V. Gubarev, V. P. Panchenko, Supersonic MHD generators, Ed. by A. V. Gubarev. Energoatomizdat, Moscow, 198.

[12] The Way of the Scientist: E. P. Velikhov (collected papers), Ed. by V. P. Smirnov. Kurchatov Institute, Moscow, 2007.

[13] R. V. Dogadaev, V. P. Panchenko, E. P. Polulyakh, A. A. Yakushev, V. V. Topel'berg, "Compact MHD system for generating powerful electric pulses," Izvestiya RAN. Energetika. 2007. N 4. pp. 130–139.

[14] Pulsed MHD converters of chemical energy into electrical energy, Ed. by A. E. and V. E. Sheindlin and V. E. Fortov. Energoatomisdat, Moscow, 1997.

[15] V. G. Butov, V. M. Galkin, V. M. Golovizin, et al., Numerical modeling of spatial two-phase flows in supersonic MHD generators, Preprint IAE-5276/16. Kurchatov Institute of Atomic Energy, Moscow, 1990.

[16] A. G. Afonin, V. G. Butov, S. V. Sinyaev, "Flow in the channel of a pulsed MHD generator in operation with various electromagnetic loads," Izv. Vuzov. Fizika. 2013. V. 56. No. 9/3. pp. 5-8.

[17] L. A. Yates, "Images constructed from computed flow fields," AIAA journal, 31(10): 1877–1884, 1993.

[18] J. J. Quirk, "A contribution to the great Riemann solver debate," Int. J. Numer. Methods in Fluids, Vol. 18 (6), pp. 555–574, 1994.

Design of a Flat-Channel Launcher

Markus J. Löffler
High Voltage and Pulsed Power Technology
Loeffler Consulting
Coesfeld, Germany

Emil Spahn, Mirko Wötzel
Pulsolution GmbH
Othmarsingen, Switzerland

***Abstract*—A study is carried out evaluating the possibility for a flat-channel coil launcher to pre-accelerate nano-satellites for transfer to low-orbit.**

1992-1998 "Flat Channel Accelerators" (FCA) and their follower "Toroidal Coilgun" were investigated in Germany with respect to ordnance applications. This paper reconsiders the idea of an FCA with respect to its applicability as a launcher for e.g. nano-cube satellites. Compared to ordnance applications with peak accelerations of 100 kgee the peak acceleration of FCAs considered here is 10 kgee at average acceleration of 5 kgee.

The principal function of FCAs is reconsidered. Information about optimum coil geometries, current frequencies and amplitudes is given. Results of numerical calculations are shown for a conceptual 10-stage 120-cm accelerator with two stator coil rows accelerating a mass of 1 kg (payload 0.3 kg) to 350 m/s at a kinetic efficiency of 40%. Assuming single-turn twisted-wire coils stator currents of 200 kA will be required generating translator currents of 120 kA at stator voltages up to 1.2 kV at the final stage. By this an average acceleration of 5 kgee per stage is guaranteed at peak accelerations of 12 kgee.

With respect to a 2.5-km/s launcher calculations show the critical role of ohmic losses in the translator.

Keywords—coilgun; flat-channel; launcher; nano-satellite

I. INTRODUCTION

Based on 100 years old patents [1,2] intensive R&D on electromagnetic accelerators now lasts since ~40 years [3]. Many approaches (railguns, coilguns) focused to ordnance applications. Research was also motivated by applications for nuclear fusion [4], space transportation [5], and space debris simulation [6]. The majority of efforts dealt with electromagnetic accelerators where payloads between g and some kg were accelerated up to 9 km/s [5] on distances of acceleration <10m requiring average acceleration rates of ~100 kgee.

Nano-satellites require acceleration rates ~10 kgee. One way to achieve this goal is the use of railguns [5, 7], working with ~MA currents. "Compared to conventional and/or liquid rockets, a railgun offers several advantages in contrast to conventional chemical rockets. The most convincing benefit, however, is the reusability of the entire propulsion part and at such low recurring costs, resulting in launch prices which are expected to be very competitive in the corresponding market segments" [7]. However, at 10 kgee coilguns are another option to accelerate masses in the range of kg. Compared to railguns coilguns don't need sliding contacts. Due to better magnetic coupling between stator coils and translator coil they promise higher efficiencies at lower current levels (<0.1 MA).

II. BASIC CONSIDERATIONS

A. Principle

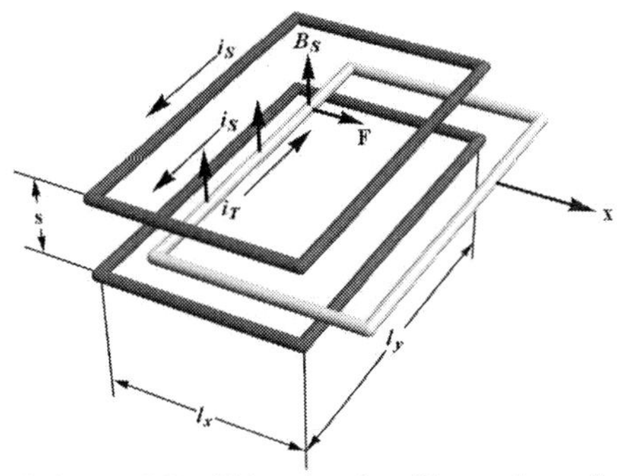

Fig. 1: Single-stage FCA with lateral motion of the translator coil. Structural support and feeding of stator coils not shown.

The FCA principle [8, 9] is shown in fig. 1.

A translator coil of size $l_x \times l_y$ is placed between two stator coils of same size. The stator coils with distance s are switched in series. Stator current i_S induces translator current i_T. The magnetic flux density B_S generated by i_S interacts with i_T generating Lorentz force F propelling the translator coil including payload and structural mass. In the following it shall be presumed that all coils are single-turn coils consisting of very thin twisted wires to avoid skin effect.

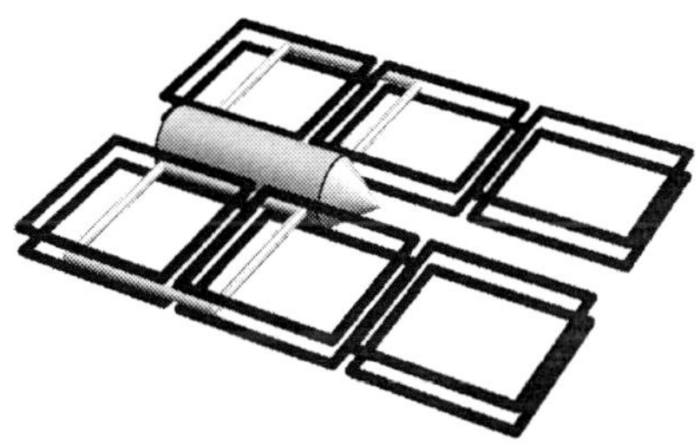

Fig. 2: Flat Channel Accelerator (multiple stages in two rows/wings).

This principle can be multiplied. Fig. 2 shows an accelerator with a payload connected to n=2 translator coils accelerated by two stator rows. The stator coils are switched on consecutively each time when the rear end of the translator coil enters the respective stage.

For best utilization of the mechanical strength of all stages the average force F_{av} on the translator has to be the same per stage providing a constant average acceleration a_{av} per stage.

B. *Mutual inductivity and force*

With mutual inductivity M between stator and translator coil force F on the translator coil can be calculated with [10]

$$F = -\frac{\partial M}{\partial x} i_T i_S = -M' i_T i_S \tag{1}$$

M and derivative M' depend on coil size and on the position of the translator coil relative to the stator coil. M can be

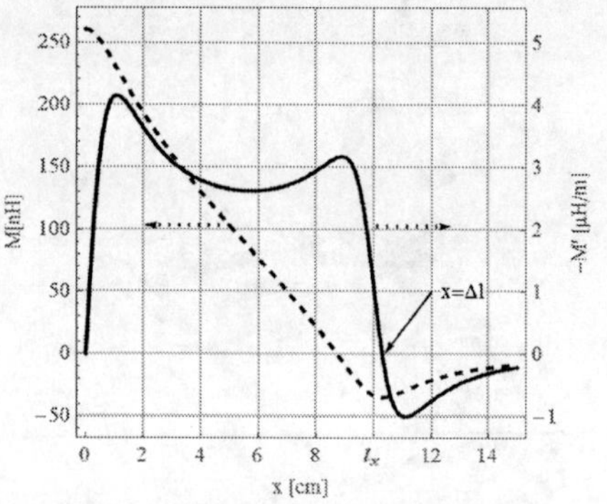

Fig. 3: Mutual inductivity and its negative gradient (l_x=l_y=10 cm, s=2 cm).

calculated using formulas in [11]. Fig. 3 shows an example where stator coils and translator coil have the same edge length l_x=l_y=10 cm at a distance of stator coils s=2 cm.

For stacked coils, x=0 cm, M has its maximum M=M_0. With increasing x M shows a nearly linearly dropping course whereas M' roughly can be considered as a constant. This also holds for arrangements with l_y>l_x. The meaning of M' becomes clearer assuming constant i_S and i_T. In this case F is directly proportional to M', see (1): For x=0 m no lateral force acts on the translator coil. If it is shifted a small portion in direction of x the force appears and rapidly increases to its first maximum. Then it drops down slightly and increases again to a second maximum being smaller than the first maximum. When x=$\Delta l \gtrsim l_x$ F becomes negative. For the operation of the accelerator this means that at this point i_S should become Zero.

C. *Optimum coil geometry*

An optimum coil geometry is given if ohmic losses W_Ω in the coils are minimized and if the per-stage increase $\Delta W_{kin,P}$ of kinetic energy of the payload is maximized:

$$\frac{\Delta W_{kin,P}}{n\, W_\Omega} \to max! \tag{2}$$

With Δl as the point where M' respectively where F become Zero (comp. fig. 3) the kinetic energy of the accelerated overall mass (mass m_P of payload; n times mass m_{coil} of one translator coil including its supporting structure) is given by

$$\Delta W_{kin} = \Delta W_{kin,P} + n\,\Delta W_{kin,coil} = n\int_0^{\Delta l} F\,dx \tag{3}$$

Assuming constant currents i_S=I_S and i_T=I_T and applying (1) to (3) gives the following estimation

$$\Delta W_{kin,P}\left(1 + \frac{n\,\Delta W_{kin,coil}}{\Delta W_{kin,P}}\right) = n\int_0^{\Delta l} -M' I_T I_S\,dx \tag{4}$$

respectively

$$\Delta W_{kin,P} \approx \frac{n\,M_0}{1+\frac{m_{coil}}{m_P/n}} I_T I_S \tag{5}$$

Translator coil and stator coils are identical. The translator coil has an ohmic resistivity R_T. The stator coils are switched in series. Hence their overall resistivity R_S doubles the resistivity of the translator coil.

$$R_S = 2\,R_T \tag{6}$$

At constant currents ohmic losses are

$$W_\Omega = n\,(R_T I_T^2 + R_S I_S^2)\,t = n\,R_T(I_T^2 + 2\,I_S^2)t \tag{7}$$

Applying (5) and (7) to (2) gives

$$\frac{\Delta W_{kin,P}}{W_\Omega} \approx \frac{M_0}{R_T\left(1+\frac{m_{coil}}{m_P/n}\right)} \frac{1}{\left(\frac{I_T}{I_S}+2\frac{I_S}{I_T}\right)t} \tag{8}$$

By this a quality characteristic can be defined:

$$\frac{\Delta W_{kin,P}}{W_\Omega} \sim \frac{M_0}{R_T\left(1+\frac{m_{coil}}{m_P/n}\right)} \to max! \tag{9}$$

As higher M_0, as lower R_T, as smaller m_{coil}, and as smaller n become, as more kinetic energy is delivered to the payload itself compared to the energy delivered to the ohmic resistivities. M_0 depends on l_x and l_y in a nonlinear way. R_T is directly proportional to l_x+l_y. m_{coil} is directly proportional to l_x+l_y. In addition the mass of the structure supporting the translator coil has to be considered being proportional to the area spanned by l_x×l_y. With density ρ_w and cross section A_w of the translator winding, density ρ_s and average thickness d_s of the structure supporting the translator coil this gives:

$$\frac{\Delta W_{kin,P}}{W_\Omega} \sim \frac{M_0(l_x,l_y)}{(l_x+l_y)\left(1+\frac{\rho_w A_w(l_x+l_y)+\rho_s d_s l_x l_y}{m_P/n}\right)} = g \to max! \tag{10}$$

Example:
For n=2, m_p=1 kg, $\rho_w{\cdot}A_w$=1.4 kg/m, and $\rho_s{\cdot}d_s$=13.5 kg/m^2 fig. 4

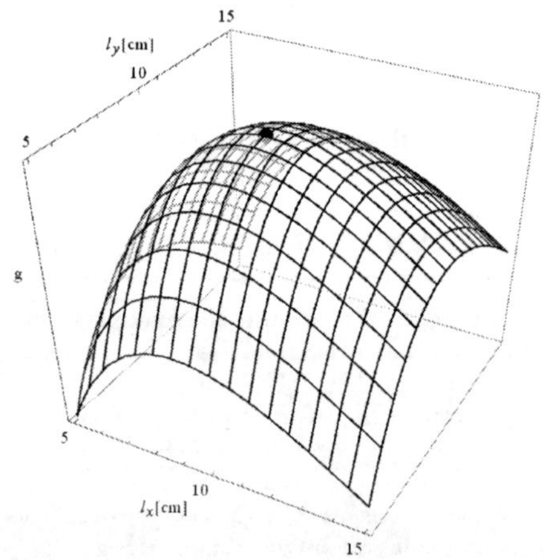

Fig. 4: Coil optimization. Explanation see text.

shows a 3D-plot of function g when varying l_x and l_y.

Maximum of g is marked with a black spot: l_x=l_y≈10 cm. Further interpretation of (10) shows that due to exchangeability of l_x and l_y the optimum is always at l_x=l_y.

D. Stator current frequencies

Sinusoidal current pulses feeding the accelerator are assumed.

For each stage frequency of stator currents has to be adapted to increasing velocity. Stator current of stage j has amplitude $\hat{\imath}_{S,j}$ and frequency f_j. With $\omega_j=2\pi \cdot f_j$:

$$i_{S,j}(t) = \hat{\imath}_{S,j} \begin{cases} \sin \omega_j(t - t_{0,j}) & 0 \le \omega_j(t - t_{0,j}) \le \pi \\ 0 & \pi < \omega_j(t - t_{0,j}) \end{cases} \quad (11)$$

$t_{0,j}$ is the starting time in the jth stage. Duration of the current pulse is half its period T_j:

$$\Delta t_j = \frac{T_j}{2} = \frac{1}{2 f_j} \quad (12)$$

At same average acceleration a_{av} per stage way of acceleration Δl, entrance and exit velocity V_{j-1} respectively V_j are related by classical mechanical equations

$$a_{av}\, \Delta t_j = V_j - V_{j-1} \quad (13)$$

$$\frac{a_{av}}{2 f_j} = \sqrt{2\, a_{av}\, j\, \Delta l} - \sqrt{2\, a_{av}\, (j-1)\, \Delta l} \quad (14)$$

Hence the frequency of the current of the jth stage becomes

$$f_j = \frac{1}{\sqrt{j} - \sqrt{j-1}} \sqrt{\frac{a_{av}}{8\, \Delta l}} \quad (15)$$

E. Stator current amplitudes

Overall magnetic flux across the translator keeps ≈0 Wb:

$$0\, Wb = L_T i_T + M\, i_S \quad (16)$$

with inductivity L_T of the translator coil. It follows

$$i_T = -\frac{M}{L_T}\, i_S \quad (17)$$

With (1), (3), and (11) this yields to

$$\Delta W_{kin} = n\, \frac{\hat{\imath}_S^2}{L_T} \int_0^{\Delta l} M \frac{\partial M}{\partial x} \sin^2 \omega_j(t - t_{0,j})\, dx \quad (18)$$

Replacements $dx=v\, dt$, $\tau=\omega_j\, (t-t_{0,j}$ lead to

$$\Delta W_{kin} = n\, \frac{K}{L_T} \hat{\imath}_S^2 \quad (19)$$

with constant

$$K = \int_0^{\pi} M\, \frac{dM}{d\tau} \sin^2 \tau\, d\tau \quad (20)$$

Finally this yields to

$$\hat{\imath}_S = \sqrt{\frac{L_T}{K}} \sqrt{\left(\frac{m_P}{n} + m_{coil}\right) a_{av}\, \Delta l} \quad (21)$$

Constant L_T/K can be determined experimentally. $\hat{\imath}_S$ is the same for each stator coil. It has to be chosen at higher values when increasing Δl, L_T, m_P, m_{coil}, and a_{av}. It can be chosen smaller with increasing n. More detailed calculations show that this rule is well applicable from stage 3 and higher. Moreover the amplitude has to be adapted to the ohmic resistivity of translator and stator coils being neglected so far. However, detailed calculations exemplarily discussed below confirm that (21) gives a good estimate for $\hat{\imath}_S$.

III. Theoretical Model

An FCA, n=2 (fig. 2), is considered. Although all FCA stages are coupled magnetically each stage is considered independently as they are energized only as long as the rear end of the translator coil resides in an active stage. When leaving this stage current via this stage is switched off avoiding magnetic interaction with successive stages. Mutual inductivity between simultaneously active wing stator packages is neglected due to their distance.

Due to ohmic heating R_T and specific heat c_T of the translator coil depend of conductor temperature T. Respective interpolation formulas are taken from [12]. R_S is assumed to keep constant. At stage j current $i_{T,j}$ starts with the value of current $i_{T,j-1}$ of the preceding stage. Stator coil packages of inductivity L_S are fed with voltages $u_{S,j}$.

For stage j the following equations have to be considered:

Electric circuit equations (single wing):
(L_S ... stator inductivity, $u_{S,j}$ ... stator feeding voltages)

$$0\, V = L_T \frac{d}{dt} i_{T,j} + \frac{d}{dt}\left(M\, i_{S,j}\right) + R_T(T) i_{S,j} \quad (23a)$$

$$i_{T,j}(t_{0,j}) = i_{T,j-1}\left(\frac{\pi}{\omega_j}\right);\, i_{T,1}(0) = 0\, A \quad (23b)$$

$$0\, V = -u_{S,j} + L_S \frac{d}{dt} i_{S,j} + \frac{d}{dt}\left(M\, i_{T,j}\right) + R_S(T) i_{S,j} \quad (24a)$$

$$i_{S,j}(t) = \hat{\imath}_{S,j} \begin{cases} \sin \omega_j(t - t_{0,j}) & 0 \le \omega_j(t - t_{0,j}) \le \pi \\ 0 & \pi < \omega_j(t - t_{0,j}) \end{cases} \quad (24b)$$

Kinetic equations:

$$\left(\frac{m_P}{n} + m_{coil}\right) \frac{dv_j}{dt} = \frac{\partial M}{\partial x} i_{T,j} i_{S,j} \quad (25a)$$

$$v_j(t_j) = V_{j-1};\, v_1(0) = 0\, m/s \quad (25b)$$

$$\frac{dx}{dt} = v_j \quad (26a)$$

$$x(0\, s) = 0\, m \quad (26b)$$

Heat equations:
(single wing; m_T ... mass of translator winding):

$$m_T \frac{d}{dt}\left(c_T(T_j)(T_j - T_0)\right) = R_T(T_j) i_{T,j}^2 \quad (27a)$$

$$T_j(t_j) = T_{j-1}\left(\frac{\pi}{\omega_{j-1}}\right);\, T_1(0) = T_0 \quad (27b)$$

Frequencies/amplitudes of $i_{S,j}$ are defined by (15)/(21). At given a_{av} $\hat{\imath}_{S,j}$ are corrected iteratively comparing the actual a_{av} with the given a_{av}. For calculation Mathematica™ is used.

IV. Results

A 10-stage/2-wing FCA is considered with a_{av}=5 kgee. m_P=0.3 kg at overall mass m_0=1 kg. Coils are made of Cu. l_x=l_y=11.15 cm. L_S=700 nH; L_T=250 nH; R_S=340 µΩ; R_T=170 µΩ at T_0=291 K (18°C); s=19 mm.

From stage 3 $\hat{\imath}_S$ keep constant ensuring constant a_{av} (fig. 5).

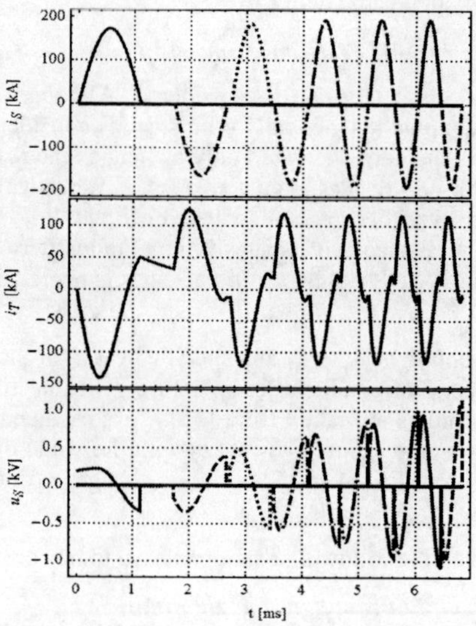

Fig. 5: Waveforms of stator and translator currents and of stator voltages.

i_T starts with negative amplitude compared to i_S. In each stage i_T changes sign whereas i_S does not: Phase shift caused by effective resistivity of ohmic losses and kinetic energy. $\hat{\imath}_T$=120 kA. When entering the next stage i_T keeps at a residual value. To use this effect i_S of the next stage is ignited with negative sign of i_T. The process repeats resulting in alternating stator/translator current profiles. $u_{S,j}$ show an alternating profile with increasing amplitudes. Maximum stage 10 is 1.2 kV.

1st-stage acceleration peaks to 20 kgee (fig. 6). Succeeding it normalizes to 12 kgee. a_{av} keeps at 5 kgee (dashed line). a changes sign for a short time. v increases stepwise. Final velocity V_{10}=350 m/s. Way of acceleration x increases ~t^2. T increases stepwise from 18°C to 120°C. Other calculations where V_{23}=500 m/s show an exponential increase from 18°C to 180°C, and to -137°C when starting at -214°C; at V=2500 m/s T increases by > 500 K.

Overall energy W_Σ increases. Pulsations are due to charging/discharging of coils (W_{mag}). W_{kin} reaches 60 kJ, W_Ω 85 kJ. Overall efficiency is 40%, payload efficiency 12%.

V. Conclusion

- FCAs can be made of rows switched in parallel/serial.
- Optimum shape of coils is square shape.
- Low R_T is mandatory to keep Joule heating low.
- Frequency of stator currents is predictable (see (15)).
- Stator currents have to be kept constant for constant a_{av}.
- Feeding voltages increase linearly.
- Peak acceleration is double the average acceleration.
- High velocities require cooling of translator coil.

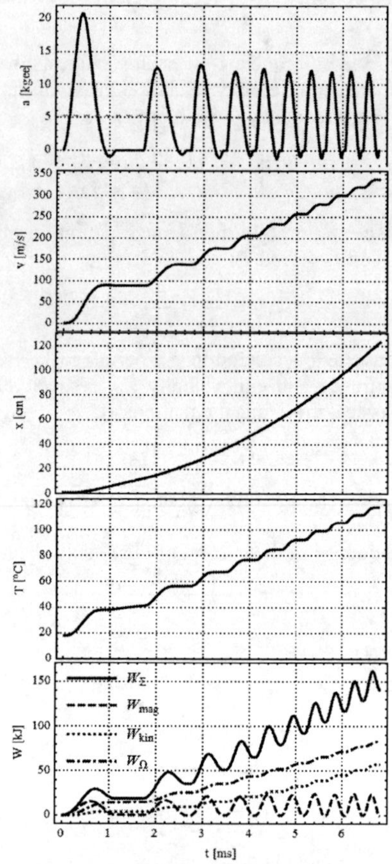

Fig. 6: Resulting kinetic and energetic variables.

References

[1] K. Birkeland: "Verfahren zum Schleudern von Geschossen mittels elektromagnetischer Kraft," Patent No. DE000000139768A, 25.09.1901

[2] A. Fauchon-Villeplée: "Elektrische Kanone," Patent No. DE000000376391A, 20.06.1920

[3] H. Kolm, K. Fine, F. Williams, P. Mongeau "Electromagnetic Guns, Launchers, and Reaction Engines," Proc. of 1st IEEE Pulsed Power Conf. , pp.42 -48, 1979

[4] R.S. Hawke: "Fusion fuel pellet injection with a railgun," J. Vac. Sci. Technol., A1(2)1983, pp. 969-973

[5] I. R. McNab: "Launch to Space With an Electromagnetic Railgun," IEEE Trans. on Magn., Vol. 39, No. 1, January 2003, pp. 295-304

[6] Igenbergs, E.et al., "On the application of plasma pulses generated by electromagnetic and electrothermal launchers for surface treatment", IEEE Trans. On Magn., vol.31, no.1, pp.735-739, Jan. 1995

[7] P. Lehmann, B. Reck, M.D. Vo, J. Behrens: "Acceleration of a Suborbital Payload Using an Electromagnetic Railgun." IEEE Trans. on Magnetics, Vol. 43, No. 1, January 2007, pp 480 ff.

[8] H. Weh, F. Moisson, M. Loeffler, J. Jacquelin: "Coilgun. Concepts." Proc. 4th European Symposium on Electromagnetic Launch Technology, 02.-06.05.1993, Celle/Germany, Paper No. 701

[9] R. Mailfert, F. Moisson, B. Hoffmann, M. Loeffler: "Coilgun. Energy coupling and current-feeding." Proc. 4th European Symposium on Electromagnetic Launch Technology, 02.-06.05.1993, Cell, Paper No. 702

[10] I. Wolff: "Grundlagen und Anwendungen der Maxwellschen Theorie II," Mannheim/Germany, 1970, ISBN 3-411-00731-1

[11] F. W. Grover: "Inductance Calculations. Working Formulas and Tables," New York, N.Y., 1973, ISBN 0-87664-557-0

T. Lindenblatt, "Simulation and stress distribution in coaxial lines loaded with pulsed currents," Westfaelische Hochschule, Laboratory on Pulsed Power and High Voltage Technology, Bachelor Thesis, Nov. 2013

Influence of Target Plasma on Focused Electron Beam in Multiple-pulse Mode of LIA

D.I. Skovorodin, Yu.A. Trunev, V.V. Danilov, V.V. Kurkuchekov, A.E. Kuzmenko
Budker Institute of Nuclear Physics
Novosibirsk, Russia
dskovorodin@gmail.com

S.D. Khrenkov, E.S. Li, P.A. Kolesnikov, A.R. Akhmetov, A.V. Kaplunov
Zababakhin All-Russian Scientific Research Institute of Technical Physics
Snezhinsk, Russia

***Abstract*— Performance of flash radiography system requires high spatial resolution. In present work we consider the defocusing of the e-beam in multiple-pulse mode by target plasma. It is shown that the second beam pulse is strongly affected by expanding plasma. The plasma-beam interaction depends on the delay between pulses and becomes crucial if delay is about 10 microseconds.**

Keywords—radiography; LIA; target; beam spot size.

I. Introduction

Radiographic complex is well known techniques of diagnostic for dynamics test. Strongly focused relativistic electron beam incident on target from large Z material with small spot size as possible is used to produce high brightness source of hard X-ray. Most significant efforts are directed to obtain several snapshots of the exploring object. Linear induction accelerators [1] easy allow to obtain multi-pulse mode, but problem of beam defocusing on the convertor arises [2]. Back streaming ions and plasma expansion form the target were suggested as key problems, in series of shot. Multi-foils target [3] was proposed as workhorse to obtain a pulse train during several microseconds. There are two well known complicated ways to achieve for large break between pulses i.e. for hundred microseconds lag, first one is using of two (or more) accelerators with orthogonal beam pass and second one is dividing of the beam into beamlets with several different directions by fast magnetic kickers. Third way is understanding of target physics, and makes design of target assembly with plasma blow-off suppression ability. This approach requires large experimental data from radiographic facility, it stands to reason to start on thoroughly observation of situation in common mode of LIA operation. This work presents some initial theoretical suggestion of plasma expansion form the Ta target and very first experimental results of X-ray spot size measurements at LIA machine [4].

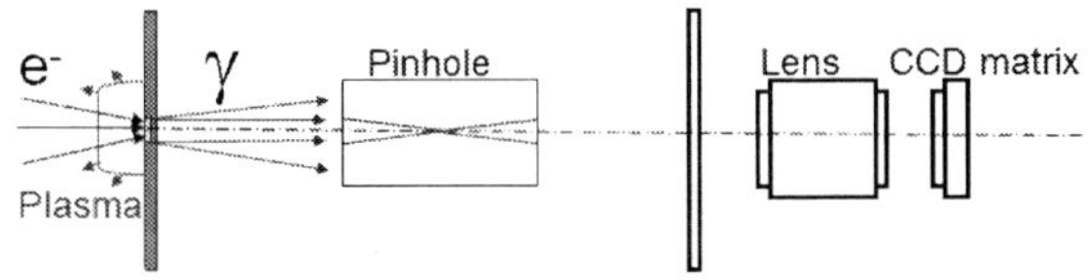

Fig. 1. Experimental set-up for beam spot size measurement

II. Measurment of spot size

A preliminary experiment was done on the Russian radiographic machine LIA described in details elsewhere [4, 5]. Main scenario of LIA operation consisted of two serial shot with break of 3 to 30 microseconds between pulses of 200 ns with standard current of 2 kA and acceleration voltage of 2 MV. Scheme of X-ray spot size measurements is shown on Fig 1. Electron beam impacts on Ta target of circa 0.4 mm thickness and beam spot size is detected by Pinhole with spatial resolution close to suggested at work [6]. Two serial snapshots glowed at the BGO scintillator and focused through objective lens on the CCD matrix. Obtained experimental images were processed by ImageJ software. A Gaussian curve satisfactory fits the intensity profile of x-ray spot. In next shot electron beam interacts with plasma blow-off from convertor therefore beam spot is spread. As could be seen on the figure 2, most dramatic effect is manifested in 5 microseconds run, but in large break, the ratio of spot sizes of two pulses remains to the value of 2.

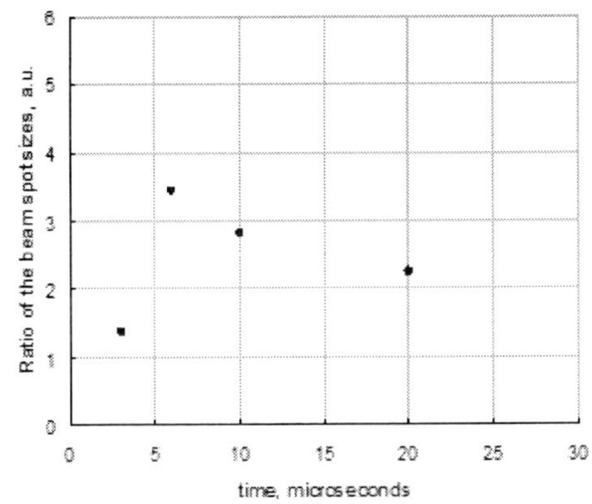

Fig. 2. The ratio of BEAM spot sizes of first and second shots versus time between pulses

III. DISSCUSSION

Experiments confirmed the “standard” model of beam and target plasma interaction during the pulse train [3]. Blow-off plasma produced by the first pulse at Ta target must and has strongly affect the subsequent pulses. Let us estimate properties of a target plasma. A calculation by GEANT4 toolbox shown an average energy deposition of the beam has value of 2.9 MeV/mm into Ta for shot parameters of LIA. Even one does not consider a real beam profile on the target, this value of energy deposition Q has order of 100 eV per atom for tantalum. Thus, so high energy deposition results in transition of material to the state of dense plasma [7]. Almost all internal energy of plasma has been turned to kinetic one. This allows calculating average velocity V of plasma expansion as follows

$$V \sim (Q/m_a)^{1/2} \sim 10^4 \text{ m/s}, \quad (1)$$

,where m_a is atom mass of Ta. Considering expansion as free and inertial, one could calculate the dimension l of plasma cloud after 10 microseconds as:

$$l \sim V \cdot t \sim 0.1 \text{ m} \quad (2)$$

So typical plasma dimension is about of 0.1m which is comparable with spacing between the target and final focus magnet of LIA machine. A plasma expansion could be estimated by self-similar solution for spherical expansion:

$$n \sim n_0 (r_0/V \cdot t)^3 \quad (3)$$

,where n_0 is density of solid tantalum, r0 is characteristic dimension of heated area of the target. Solid line on the Fig.2 illustrates the dependence of cloud density after beam impact versus expansion time. Plasma cools down adiabatically during expansion. Thus the plasma should recombine rapidly according to Saha equation. However the recombination rate decreases because of plasma expansion and the degree of ionization could be “frozen”. The electron density in the target plasma after tens microseconds delay could stay at higher level then density of the e-beam (shown as straight line on the Fig. 2). To estimate this effect we should utilize the model of Zeldovich and Raizer [7]. The model is based on the next kinetic equation:

$$da/dt = -bna^2 \quad (4)$$

,where a is degree of ionization, n is total density of atoms and ions and b is recombination constant. The constant b depends on the plasma temperature as $T^{-9/2}$. Asymptotic value of *a* depends on the plasma cooling regime. If the cooling is slow enough *a* does not tend to zero in the infinity limit and plasma does not totally recombine (so called regime of "frozen of plasma"). But in case of plasma temperature is cooled faster than $\sim t^{-10/9}$, degree of ionization tends to zero and plasma completely recombines. The considered model takes into account plasma heating by recombination itself thus it becomes nonlinear one. We use finite-difference method to solve equations of the model. A dashed curve on the figure 3 shows the density of plasma electrons for Q = 100 eV/atom. In this case the density of plasma is still higher than the density of electron beam even after 100 microseconds of the first pulse. Subsequent pulses injected have to be disrupted in plasma cloud because of neutralization and instabilities. But we should note that density of expanded plasma strongly depends on initial conditions. There are threshold value of plasma as Q = 30 eV/atom to come into “frozen” state.

In near future, we suppose to upgrade pinhole diagnostic for exploring a dynamic of x-ray spot in multi-pulse mode and extend theoretical model of plasma behavior during tens microseconds break between pulses.

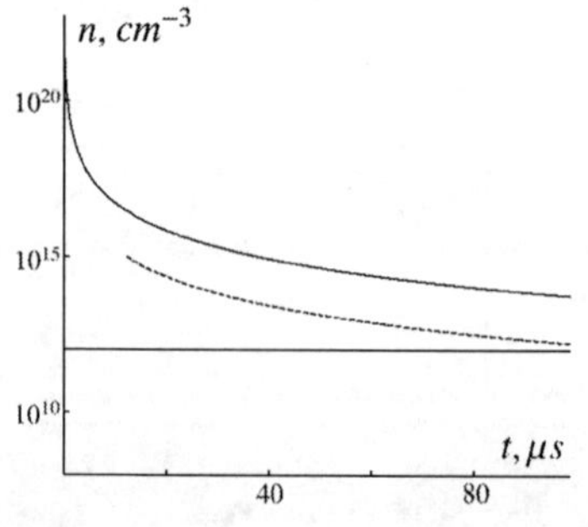

Fig. 3. Density of the cloud vs. time after beam impact. Solid line represents total density of atoms and ions, dashed line corresponds to electron density, gray line illustrates characteristic density of electron beam between the target and final focus magnet.

Acknowledgement

We are indebted to P.V. Logachev and A.V. Burdakov for support this work.

References

[1] C. Ekdahl, "Modem electron accelerators for radiography," IEEE Transactions on Plasma Science, vol. 30, no. 1, pp. 254-261, Feb 2002.

[2] Welch, D. R., & Hughes, T. P.. “Effect of target-emitted ions on the focal spot of an intense electron beam”, Laser and Particle Beams, 16(02), 285-294, 1998

[3] Pincosy, P. A., Back, N., Bergstrom, P. M., Chen, Y. J., & Poulsen, P. “Multiple pulse electron beam converter design for high power radiography”, Review of Scientific Instruments, 72(6), 2599-2604, 2001

[4] P. V. Logachev, et al. "LIU linear induction accelerator." Instruments and Experimental Techniques, vol. 56.6, pp. 672-679, 2013.

[5] D. A. Starostenko, et al. "Results of operating LIA in radiograph mode." Physics of Particles and Nuclei Letters, vol. 11.5, pp. 660-664, 2014.

[6] Crist, C. E., Sampayan, S., Westenskow, G., Caporaso, G., Houck, T., Weir, J., & M. Krogh, “Time resolved, 2-D hard X-ray imaging of relativistic electron-beam target interactions on ETA-II”, Sandia National Labs., №. SAND--98-2054C, 1998

Y. B. Zeldovich and Y. P. Raizer, Physics of Shock Waves and High-Temperature Hydrodynamic Phenomena, Volume 2, New York: Academic Press, 1996

Results from compression of magnetized plasma using imploding solid liner

J. H. Degnan, C. Grabowski, M. T. Domonkos, E. L. Ruden, and D. J. Amdahl
Air Force Research Laboratory, Directed Energy Directorate
Kirtland AFB, NM USA

M. H. Frese and S. D. Frese
NumerEx LLC
Albuquerque, NM USA

G. A. Wurden and T. E. Weber
Los Alamos National Laboratory
Los Alamos, NM USA

***Abstract*— As part of a joint effort to advance magneto-inertial fusion (MIF) technology, the Air Force Research Laboratory (AFRL), Los Alamos National Laboratory (LANL), and their collaborators have developed high trapped flux field reversed configuration (FRC) plasmas and conducted an integrated compression test of an FRC. MIF uses embedded magnetic field to reduce thermal conduction relative to inertial confinement fusion (ICF), thereby alleviating requirements on implosion speed and compression ratio to ranges accessible by existing drivers. The AFRL Shiva Star capacitor bank used at 5 MJ stored energy, ~10 MA current, and 10 μs current rise time, has been used to drive metal shell (solid liner) implosions for compression of axial magnetic fields to multi-megagauss levels, a key capability for MIF. By profiling the thickness of the liner, large electrode apertures ensure current delivery to the liner while enabling the injection of a FRC magnetized plasma. The length of the trapping magnetic field region within the liner was extended, making the FRC trapped flux lifetime was made comparable to the liner implosion time. In order to assess the performance of the integrated MIF test, arrays of magnetic field, spectral (visible and x-ray), interferometric, imaging, and neutron diagnostics were employed. Imaging revealed a luminous core plasma that was compressed cylindrically by more than a factor of ten. Neglecting loss mechanisms, the compression would have increased the density by more than 100x, to over10^{18} cm^{-3} and several keV, but spectroscopically observed temperatures were only in the range of 300-400 eV. Although compression to megabar pressures was inferred by the observed time and rate of liner rebound, we learned that the heating rate was outpaced by the FRC decay mechanisms until the last 3 μs of the compression. Measures that could double the trapped flux lifetime and increase the pre-compression temperature of the FRC are discussed.**

Keywords—magnetized plasma compression, field reversed configuration, FRC, imploding solid liner

I. Introduction

Magneto-Inertial Fusion (MIF) uses embedded magnetic field to reduce thermal conduction relative to inertial confinement fusion (ICF), thereby alleviating requirements on implosion speed and compression ratio to ranges accessible by existing drivers. We have previously reported [1, 2, 3, 4] the development of the MIF approach employing the Field Reversed Configuration (FRC) type of magnetized plasma. The FRC is formed by a reversed field theta pinch discharge, and injected into a magnetic mirror trap within a metal shell or liner for subsequent compression. The liner is imploded using magnetic pressure from a high current axial z-pinch discharge, compressing the magnetized plasma trapped within. The previous work [1,3] included reliably forming, translating, and capturing FRCs with density ~4 x 10^{16} cm^{-3}, ion temperature ~100 eV, magnetic field ~1 Tesla, field exclusion radius ~2.5 cm, and length ~20 cm. Degnan, et al. [2] demonstrated implosion of 30 cm long, 10 cm initial diameter, 1.1 mm initial thickness Al liners to at least 16x radial compression without instability growth, using a 11 to 12 MA, 10 microsecond rise time pulse. The convergence was observed using flash radiography. To facilitate FRC injection into one end of the liner, its axial cross-section was profiled with increased thickness near the ends; the increased mass delayed the initiation of radial implosion relative to the central region of the liner, permitting current drive to continue through the implosion while also providing an open geometry for FRC injection. An experiment reported in [2] demonstrated magnetic field injection and compression. Magnetic field coils configured to capture an FRC were used to inject a 1.36 T axial field within an aluminum liner, representative of the requirements for a MIF demonstration. Using flash radiography to validate 2D-MHD simulations, calculations indicate the initial field was compressed to approximately 540 T [2]. With a sufficiently dense O(10^{17} cm^{-3}) and warm O(300 eV) FRC, the demonstrated liner implosion performance [2] is expected to drive adiabatic compression to fusion conditions for a batch burn in under 1 μs. Much of the intervening work between [2] and this report has concentrated on experiments and numerical simulations to improve FRC formation, capture, and lifetime. FRC rotation is known to be a key performance limiter, and research has been conducted to characterize and mitigate FRC rotation [3, 5]. This report concentrates on the results of an integrated FRC formation-translation-compression experiment using an imploding solid liner.

II. EXPERIMENT DESCRIPTION

The FRC formation, translation, trapping, and liner implosion hardware are shown in a photograph in Figure 1. As shown, the assembly is installed under the center section of the Shiva Star capacitor bank which supplies the discharge for the liner implosion. Single turn theta coils, shown on the lower part of the image, have an increasing inner-bore moving upward in the apparatus and drive the FRC formation and translation. The formation and compression regions are physically separated by 30 cm. Guide magnetic field coils are used to conduct the FRC on its way to the compression region. Within the compression region, the applied magnetic field approximates an unbalanced magnetic mirror, weaker at one end to facilitate initial translation of the FRC into the region and stronger at the other end to prevent its escape. The liner is 30 cm long and 10 cm in diameter and is connected by a co-axial transmission line to the Shiva Star parallel plate transmission line. As described in detail in [1-3, 6], the magnetic mirror field is established in the solid liner interior by means of a 4.7 millisecond rise time discharge through coils surrounding the regions near the upper and lower ends of the liner; due to the diffusion time scale, the mirror coils are energized first in the experiment sequence. The mirror field strength ranges from 0.6 to 1.36 T for FRC capture experiments, and slightly less for FRC compression experiments. To facilitate field diffusion into the liner, the outer coaxial conductor is slotted. 3D magnetic field diffusion calculations using Comsol Multiphysics indicated that field perturbations due to the slotted outer conductor were limited to less than one percent and predominantly at large radius [6]. 3D-MHD simulations using Mach3 indicated that the slots in the return conductor would not significantly affect the solid liner implosion.

To initiate FRC formation, deuterium gas is injected into the apparatus about 2 ms after the mirror field discharge start.

Fig. 1. FRC formation, injection, trapping, and compression load mounted under center of Shiva Star.

The pressure builds to 50 to 100 milliTorr inside a 10 cm diameter quartz tube which is surrounded by a segmented theta coil. The coil is segmented for diagnostic access. The theta coil is first energized with a slow (~130 μs rise) bias field, initiated 3-4 ms after the mirror discharge and producing approximately 0.4 T. Cusp fields to enhance magnetic field reconnection are then applied with a 40 μs rise. A 1 μs rise time ringing pulse in the theta coils pre-ionizes the gas. Ideally, the plasma formed at this point is fully diffused in the axial bias field. Finally, a 3.2-3.3 μs rise, ~1.5 T reverse bias axial field is applied to complete FRC formation. The conical variation in bore diameter of the theta coils creates an axial force that propels the FRC into the guide and mirror regions of the apparatus. For a compression experiment, Shiva Star is triggered several μs prior to the FRC formation and delivers ~12 MA with ~10 μs rise time to the liner. As the liner is imploding the FRC enters through the lower aperture and is captured within the magnetic mirror. The implosion reaches peak compression with an interior mirror field and injected FRC in 24.8 μs.

Diagnostics include pulsed power current and voltage, local (B-dot) and loop magnetic field, fast multi-frame vacuum ultraviolet imaging of a phosphor – pinhole camera [5], optical photodiodes, vacuum ultraviolet and x-ray diodes, optical spectroscopy, neutron activation, and scintillator photomultiplier tube detectors. Auxiliary experiments reported in [1-3] also used flash radiography, fast multi frame optical imaging, and inserted large area magnetic probes. The vacuum ultraviolet imaging system provides μs-resolution through a pinhole observing the plasma from an axial viewpoint.

III. EXPERIMENT RESULTS

The density and temperature of the FRCs used in this investigation lead to accelerated loss mechanisms compared to lower density and lower temperature FRCs [7-8], and the lifetime, as determined by observation of an excluded flux radius, is on the order of 10 μs. The excluded radius is obtained from the magnetic probe data using, for example, the Tuszewski method [9]. By increasing the axial separation of the peak magnetic mirror fields, the FRC lifetime was extended from 10 to 14 μs, as illustrated in Figure 2. Further increases in the separation distance are expected to produce greater improvements in lifetime, however, the existing 30 cm long liner technology constrained the exploration of this parameter. The maximum observed FRC lifetime for 24 cm separation was 11 μs, while 30 cm separation facilitated up to 17 μs lifetime.

As a prelude to the integrated FRC formation-translation-compression experiment, the FRC was captured in a static magnetic mirror with the same strength it will have during the dynamic liner implosion. All of the applicable diagnostics indicated that the FRC would enter the compression region and that the entrance mirror field would be sufficient to retain the rebounding FRC. These preliminary experiments were similar to those reported in [3, 4], indicating an FRC density of 4×10^{16} cm^{-3} and temperature of ~100 eV.

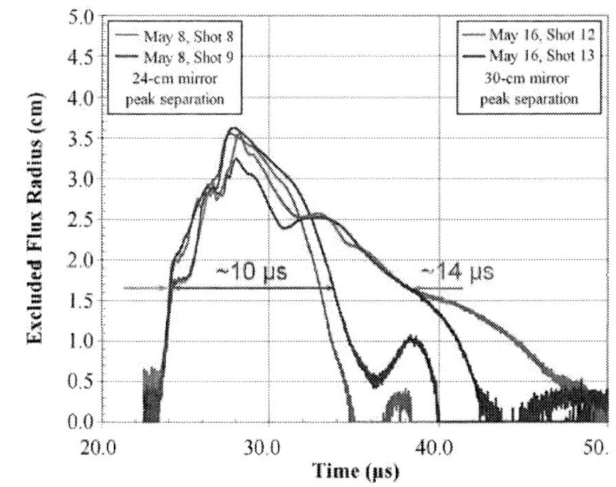

Fig. 3. Increase in FRC closed field lifetime with increased magnetic mirror peak separation.

Shiva Star current driving the implosion exhibited similar characteristics to the prior field injection experiment [3], indicating both implosion and rebound of the flux-excluding liner. The VUV diode, optical photodiode, VUV phosphor pinhole camera fast imaging, and implosion discharge current waveform all indicated a peak compression time of 24.8 μs; 2D-MHD simulations match the experimental observations. The VUV imaging frames, presented in Figure 3, indicated that the FRC was cooling until late in the compression. Due to the viewing angle and field of view, we consider the center of the image to be direct plasma lighting with the ring structures in the images the result of reflections. Initially, frames 1 and 2, the FRC is hollow, and one can estimate the excluded flux radius from the image. Compression is observable in frames 3 to 9 (11.8 to 23.8 μs after start of implosion current discharge). Frames 9 and 10 bracket the time of expected peak compression. Rebound (expansion) is observed in frames 10 to 12 (25.8 to 29.8 μs), with a central rebounding object that becomes hollow, frame 11. While the visual evidence is enticing, measurement of a region of excluded flux would be necessary to conclude that a closed field-line plasma existed after the implosion.

The VUV and optical diode signals were weaker than expected, though the signals increased abruptly preceding the time of peak compression. Stark broadening of the D beta line, 100 nm FWHM, indicated a density of 6×10^{17} cm^{-3} density 5 μs before peak compression, in the cooler regions of the plasma.

Observation of the neutron production provided another measure of performance. Based on static measurements of FRC density, temperature, and lifetime, 10^9 DD neutrons were expected. Only the most sensitive neutron detector, a Los Alamos arsenic activation detector, recorded neutron activation above the background. The arsenic detector is a 25 cm diameter, 8.4 kg epoxy embedded arsenic powder system. It was located 90 cm from the center of the liner. It recorded 38 counts. Background counts were 5. The neutron yield was approximately 3×10^6, indicative of a cooler and possibly less dense plasma than expected.

The observations consistently argue that cooling occurred until late in the implosion. Assuming the neutrons were thermonuclear, the estimated peak temperature was 300 to 400 eV.

IV. DISCUSSION

We hypothesize that 3D and non-MHD loss mechanisms were responsible for shortening the FRC closed field lifetime. Additional diagnostics such as polarimetry employed in both the static-capture and dynamic-implosion configurations may shed light on the loss mechanisms. Measures that could

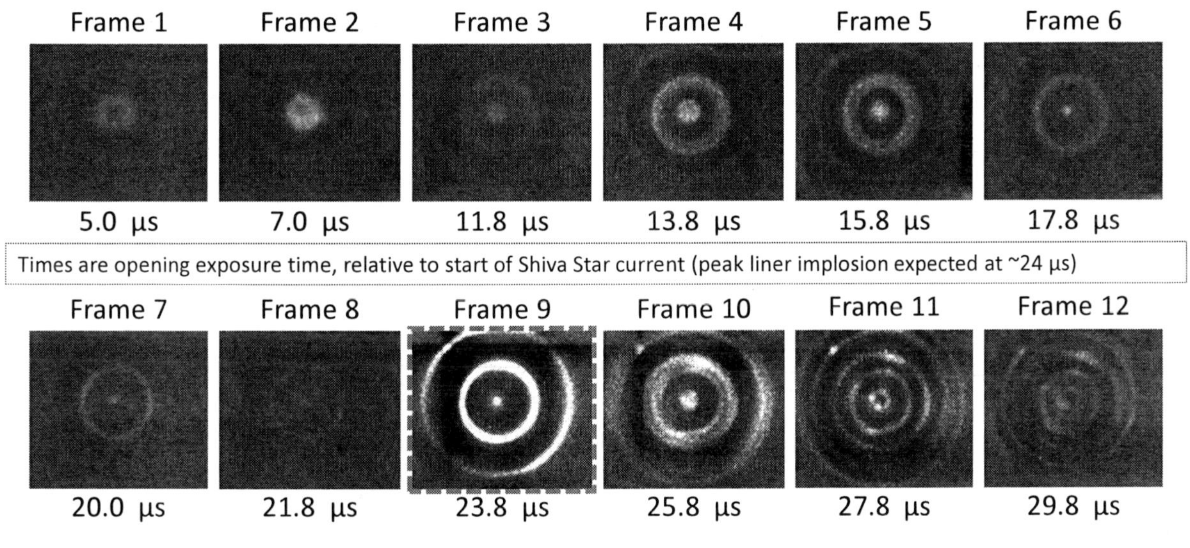

Fig. 2. Vacuum Ultra Violet time resolved pinhole camera images of FRC implosion compression experiment. 2 μs frames.

improve the performance of the compressional heating of an FRC include a longer liner and FRC capture region, a higher initial FRC temperature and axial magnetic field. Injecting two FRCs from opposite directions and merging them is one avenue to achieving better FRC performance for MIF. Such merging of counter propagating FRCs has been proposed before [7, 8], though at lower density and not in combination with imploding solid liner compression.

Acknowledgment

This work was supported by The Department of Energy, Office of Fusion Energy Science, Grant IA-DE-AI02-04ER54764.

References

[1] G. A. Wurden, T. P. Intrator et al, LA-UR-08-0796, "FRCHX Magnetized Target Fusion HEDLP Experiments," IC/P4-13, IAEA 2008 Fusion Energy Conference, Geneva, Switzerland, Oct. 13-18, 2008.

[2] J. H. Degnan et al, "Experimental and Computational Progress on Liner Implosions for Compression of FRCs," Transactions on Plasma Science, 36, Part 1, p. 80, Feb 2008, doi: 10.1109/TPS.2007.913814.

[3] J. H. Degnan et al, "Recent magneto-inertial fusion experiments on the field reversed configuration heating experiment", Nuclear Fusion 53, 093003 (2013); doi: 10.1088/0029-5515/53/9/093003.

[4] C Grabowski, J. H. Degnan, D.J. Amdahl, et al., "Addressing Short Trapped-Flux Lifetime in High-Density Field-Reversed Configuration Plasmas in FRCHX," IEEE Transactions on Plasma Science, 42, pp. 1179-1188, 2014, doi: 10.1109/TPS.2014.2305402.

[5] G.A. Wurden and S.K. Coffey., "A multi-frame soft x-ray pinhole imaging diagnostic for single-shot applications," Review of Scientific Instruments, 83, 10E516, 2012; doi: 10.1063/1.4733536.

[6] M.D. Domonkos, et al., "Applied magnetic field design for the field reversed configuration compression heating experiment," Review of Scientific Instruments, 84, 043507 (2013); doi: 10.1063/1.4801952.

[7] G. Votroubek, J. Slough, S. Anderson, et. al., "Formation of a Stable Field Reversed Configuration through Merging," Journal of Fusion Energy 27, pp. 123-127, 2008; doi:

[8] H.Y. Guo, M.W. Binderbauer, D. Barnes, et. al., "Formation of a long-lived hot field reversed configuration by dynamically merging two colliding high-β compact toroids," Physics of Plasmas 18, 056110, 2011; doi: 10.1063/1.3574380.

[9] M. Tuszewski, "Excluded Flux Analysis of a Field-Reversed Plasma," Physics of Fluids 24, 11, p. 2126-7, 1981; doi: 10.1063/1.863283.

Analysis of a Gyromagnetic Nonlinear Transmission Line Based on Experimental Results

Fernanda Sayuri Yamasaki
José Osvaldo Rossi
Plasma Associated Laboratory
National Institute for Space Research
São José dos Campos, Brazil
fernandayamasaki@hotmail.com

Joaquim José Barroso
Technological Institute of Aeronautics
São José dos Campos, Brazil

Edl Schamiloglu
Electrical and Computer Engineering Department
University of New Mexico
Albuquerque, USA

Abstract— **Advantages of the continuous non-dispersive lines (also known as gyromagnetic lines) in relation to the discrete nonlinear transmission lines have motivated a deeper study on the first one. Some of these advantages are stronger pulse oscillations at higher frequencies and higher radiofrequency conversion efficiency. Different models have been proposed along the years by several authors with different approaches in order to better understand the gyromagnetic line behavior, using simulation to provide the precession movement. The models already available from other works were studied (including the parameters) by comparison between simulation and numerical analysis techniques. The goal of this paper is to perform the experimental analysis of the gyromagnetic lines previously studied in order to validate the proposed method of analysis. Such a technique could be useful for communications in aerospace systems and/or mobile defense platforms.**

Keywords*— coaxial line; ferrite beads; gyromagnetic line; magnetic field; NLTL.*

I. Introduction

The first investigation on the use of ferrite-loaded lines to obtain pulse sharpening effect was made by Katayev in 1966 [1] using a basic mathematical description. In 1981, Weiner [2] based on Katayev explanations proposed a simplified theory with resistances predicting the output pulse rise time based only on the nonlinear magnetization effect. On the other hand, Pouladian et al. [3] and Pouladian et al. in 1991 [4] showed a complete solution including the ferrite magnetic precession with azimuthal bias, which characterizes the gyromagnetic lines. For this, they used two approaches, where RLC resonant branches were placed in parallel with the capacitors of the model circuit. In the first approach, fixed values were used for the RLC components while in the second approach the values varied along the line gradually. In 2000 Dolan et al. [5] examined a method with instantaneous voltage sources in series with linear inductors to investigate these effects using a numerical program including the presence of an external axial bias to boost the magnetic precession as it induces oscillation along the line which can be used to irradiate RF from an antenna installed at the line output. In summary, several studies with different configurations were proposed in the last decades to investigate the gyromagnetic line, where the more interesting up to date is the one proposed by Dolan et al.[5] as external bias so far is more used to produce RF oscillations at higher efficiencies. In fact, the same results have been obtained by using SPICE simulation (LT) or numerical computation (Mathematica) as given in [6] with good agreement, validating both methods. One interesting aspect of simulations in [6] is that they do not require the use of the voltages sources (nonlinear element) in series with the linear inductor as in Dolan´s model, since the flux equation considers the ferrite magnetization effects. In short, the goal of this paper is to validate the simulated model described in [6] by means of experimental results obtained with a small gyromagnetic line prototype of 20 cm in length.

II. Line Design And Experimental Set-up

The coaxial gyromagnetic line is built by inserting ferrite beads though an internal conductor, where a tape sleeve is used to insulate the ferrite outer surface from the braiding. The first step for the line design is to calculate the saturated inductance in H/m given approximately as [7]:

$$L_i = \frac{\mu_0}{2\pi}\left[ln\left(\frac{do}{di}\right)\right] \quad (1)$$

where $\mu_0 = 4\,\pi \times 10^{-7}$H/m is the magnetic permeability of free space, do is the outer diameter of the ferrite and di is the inner diameter of the ferrite. Also, assuming ε_r=1 for the ferrite, the linear capacitance in pF/m was calculated as [7]

$$C_0 = \frac{2\pi\,\varepsilon_0}{\ln\left(\frac{do}{di}\right)} \quad (2)$$

where $\varepsilon_0 = 8.854 \times 10^{-12}$ F/m is the permittivity of free space. The values used in simulations were based on the

* Work supported by AFOSR-US Air Force under Grants nos. FA9550-14-1-0133 (INPE) and FA9550-15-1-0094 (UNM)

parameters and dimensions of the ferrite beads, where the line model has to be equivalent to the length of the line (20 cm in this case). For the simulation, the line has to be represented by a large number of LC sections (40). For the line construction, we used Amidon ferrite beads (FB43201) of 3.8 mm in length and with inner and outer diameters of 1.93 and 1.09 mm, respectively. To extend up to 20 cm across the length of the internal copper wire, 53 ferrite beads were placed side by side (see Fig, 1 (a)) and their outer surface isolated from the braiding by two layers of insulating tape, forming a plastic sleeve (Fig. 1 (b)). The braiding was built using a layer of flat metallic solder wire (CT-BRAND) made of copper of 2.5 mm wide (Fig. 1(c)). This setting was kept firmly by welding a thin layer of tin on the cooper braiding (Fig. 1 (d)). To complete the coaxial line construction, two layers of insulating tape were also wound on the outer screening braiding. As line input and output work at high voltages, inner terminals were isolated and kept apart from outer terminals (braiding). The set-up of the experimental gyromagnetic line was completed by placing small pieces of magnets over the extension of the coaxial to produce the magnetic bias axial field.

Fig. 2 gives the scheme employed to test the gyromagnetic line, showing the 50 Ω HV fast pulse generator, diodes and load. The general view of the experimental set-up is shown in Fig. 3. Both signals at input and output were observed and extracted through a 4 kV/250 MHz Agilent probe linked to a 1.0 GHz digital scope from Agilent Technologies. The line was fed by the high voltage pulse generator FPS 5-1 NM, from FID Technology, capable of producing triangular pulses with amplitudes varying in the range of 1.5 kV - 4 kV and with 2.5 ns of rise and fall times. As a testing load, carbon resistors were used because of their small stray inductance as they have solid bulk for conducting the current instead of spiral wires as in the case of metallic resistors with higher inductance. Five power carbon resistors of 10 ohms/3 W were connected in series to provide a 50 Ω load. In order to protect the pulse generator output against reflections a BYW56 diode of 1 kV breakdown reverse voltage was used in series with the gyromagnetic line input. Other five diodes of the same model connected in series at the line output and in parallel with the 50 Ω load carbon resistors were also used as free-wheeling diodes for generator protection.

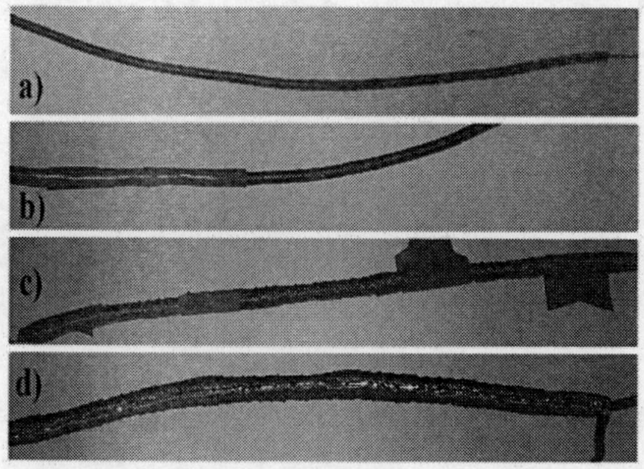

Fig. 1. Construction steps of the coaxial line. a) copper wire inside the ferrite beads, b) insulating tape layer, c) solder wire layer, d) welded tin line.

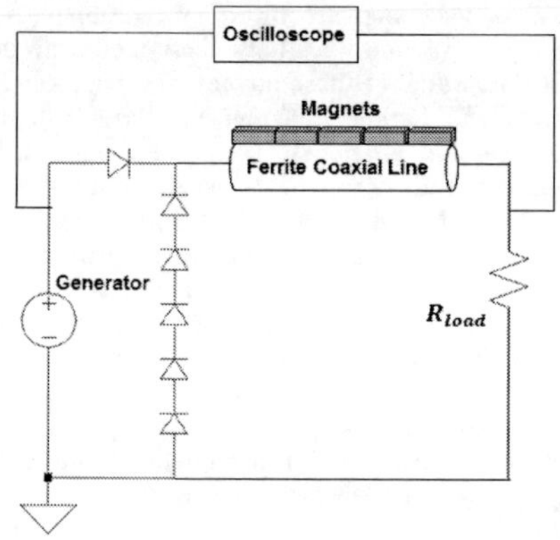

Fig. 2. Squeme employed to test the gyromagnetic line.

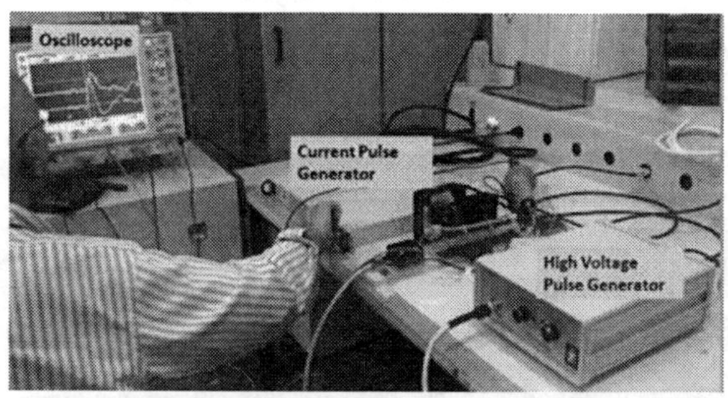

Fig. 3. Set-up used to test the gyromagnetic line.

III. EXPERIMENTAL RESULTS AND SIMULATIONS

Herein the experimental results of the setup-up without and with magnetic axial bias are presented. For magnetic bias, permanent magnets are used and SPICE validation model is tested.

A. Line Operation Without Magnetic Bias

First, Fig. 4 displays the output and input voltages without any magnetic bias, for comparison to the biased cases. In this figure, one can see that there is a line delay of approximately 2.0 ns between pulses as expected and considerable drop of peak amplitude of about 1.6 kV for the output pulse due to the line losses and reflection as the line is not matched to the load because of the nonlinearity.

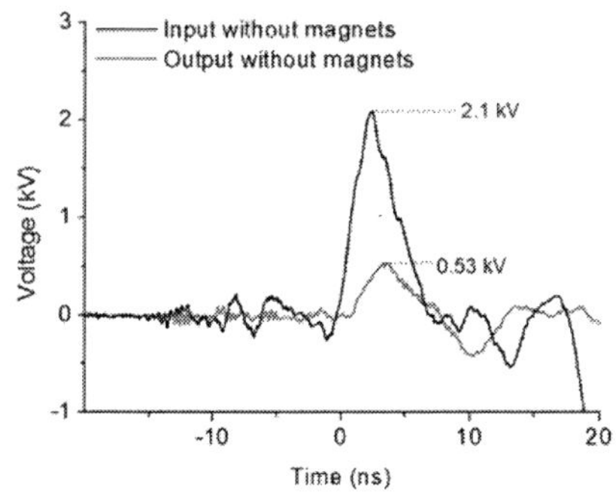

Fig. 4. Input and output of the coaxial line without magnetic bias.

B. Line Operation With Permanent Magnets

For line operation with axial magnetic bias without any DC current source, five permanent magnets with transversal square section of 1.5 ×1.5 cm^2 and length of 3.5 cm were placed side by side along the line (on top) as shown in Fig. 5.

Fig. 5. Several magnets placed along the coaxial line (on top).

Fig. 6 shows the results (input and output pulses) obtained with the set-up given in Fig. 5. It is possible to observe an improvement on the output pulse compared to operation without bias, which demonstrates the considerable magnetization influence on line performance as in this case the output rise time is shortened and the peak amplitude relative to the input is increased.

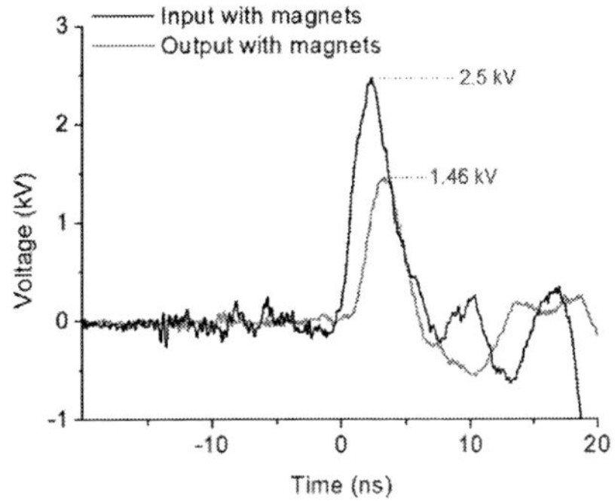

Fig. 6. Input and output pulse using permanent magnets as external bias.

In order to validate the SPICE gyromagnetic simulation described in [6], the circuit model with the same parameters of the coaxial line built and biased with magnets was reproduced (see corresponding scheme in Fig. 7).

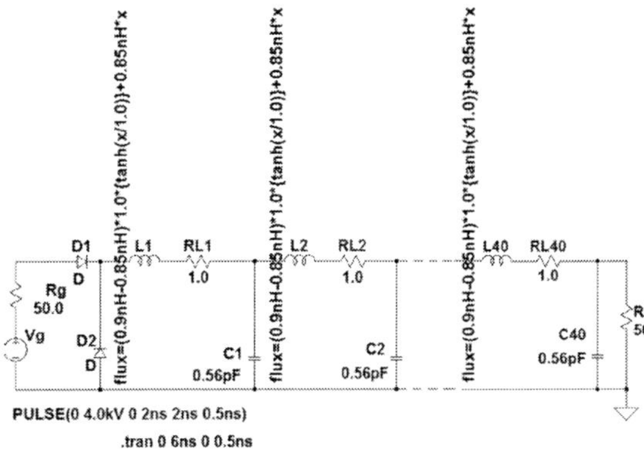

Fig. 7. SPICE scheme used to simulate the gyromagnetic line with magnets.

In the simulation, the generator output impedance and load are fixed at 50 ohms according to the experimental parameters. The input pulse simulated has the same specification according to the experimental result in Fig. 6. The values of the saturated inductance (0.85 nH) and the capacitance (0.56 pF) were calculated according to (1) and (2), based on the ferrite and line dimensions already described. The initial inductance in the flux equation displayed on the schematics was adjusted so that the simulations are fitted to the experimental results since the ferrite magnetic relative permeability (μ_r=160) drops when the permanent magnets are placed along the line. Also, in the scheme of Fig. 7, the resistance of 1.0 Ω in series with the line inductance represents the ferrite losses. A comparison between the experimental results of Fig. 6 and the corresponding simulation scheme in Fig. 7 is given in Fig. 8, showing a good agreement and thus proving the viability of using the SPICE circuit model proposed.

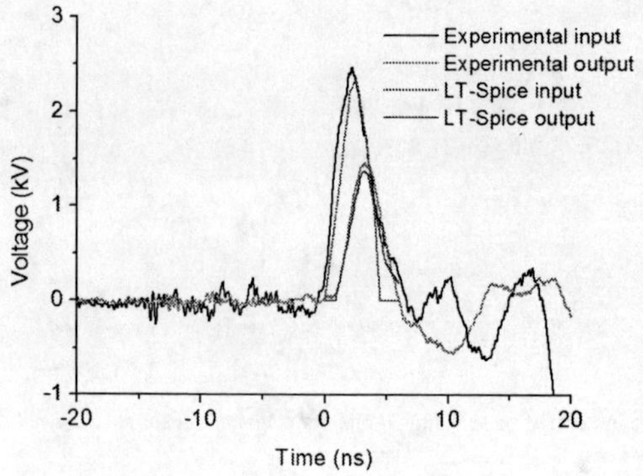

Fig. 8. Comparison between simulations (using squeme of Fig. 7) and experimental results of Fig. 6.

IV. CONCLUSIONS

Despite the lack of RF oscillations at the output, the circuit model for gyromagnetic lines was successfully tested, demonstrating the feasibility of the SPICE simulations to assess device performance. Absence of induced oscillations is explained by the use of permanent magnets of low magnetic induction. In future work, this will be addressed by inserting the line into a solenoid coil fed by a DC current source to provide a stronger axial field. Higher input voltages will be also needed to provide a higher azimuthal field due to the pulse current induced and to excite the gyromagnetic effect. SPICE simulation could be used to check the line performance on these conditions to test again model validity for RF generation.

REFERENCES

[1] Katayev, I.G., Electromagnetic shock waves, London, U.K.: Iliffe Books Ltd., 1966.

[2] Weiner, M. Pulse Sharpening in Ferrite Transmission Line, in Proc. of the 1989 IEEE Int. Pulsed Power Conf., pp. 91-95.

[3] Pouladian-Kari R.; Benson, T.M.; Shapland, A.J.; Parkes, D.M., The electrical simulation of pulse sharpening by dynamic lines, in 1989 Proc. IEEE International Pulsed Power Conference, pp.178-181, 1989.

[4] Pouladian-Kari R.; Shapland, A.J. and Benson, T.M. Development of ferrite line pulse sharpeners for repetitive power applications, in IEEE Proceedings-H, vol 138, no. 6, Dec. 1991.

[5] Dolan, J.E. and Bolton, H.R. Shock front development in ferrite-loaded coaxial lines with axial bias, in IEEE Proc.-Sci. Meas. Technol., vol. 147, no. 5, Sept. 2000.

[6] F. S. Yamasaki; E. Schamiloglu; J.O. Rossi ; J.J. Barroso, "Simulation Studies of Distributed Nonlinear Gyromagnetic Lines Based on LC Lumped Model", in: IEEE Transactions on Plasma Science, DOI:10.1109/TPS.2016.2556320, to be published.

[7] A. Vaselaar, "Experimentation and modeling of pulse sharpening and gyromagnetic precession within a nonlinear transmission line", M. S. thesis, Electrical Eng. Depart., Texas Tech University, Lubbock, TX, Aug. 2011.

Influence of MOCVD technological conditions on magnetoresistance of nanostructured La-Ca-Mn-O films used for magnetic field sensors

N. Zurauskiene, D.Pavilonis, J. Klimantavicius, V. Stankevic, S. Kersulis, S. Balevicius, M. Skapas, R. Juskenas and V. Plausinaitiene
Semiconductor Physics Institute, Center for Physical Sciences and Technology
A. Gostauto Str. 11, LT-01108 Vilnius, Lithuania
e-mail: nerija.zurauskiene@ftmc.lt

N. Zurauskiene, S. Balevicius and V. Stankevic
Vilnius Gediminas Technical University
Naugarduko 41, LT-03227 Vilnius, Lithuania

V. Plausinaitiene and A. Abrutis
Vilnius University
Vilnius LT- LT-03225, Lithuania

***Abstract*—The results of structure and magnetoresistance (*MR*) of nanostructured $La_{0.78}Ca_{0.22}MnO_3$ films, grown at different gas pressure (from 3 to 7 Torr) by metal–organic chemical vapor deposition (MOCVD) technique, are presented. The *MR* was investigated in pulsed magnetic fields up to 60 T in the temperature range 1.5–294 K. The results were analyzed from the perspective of using these films for magnetic field sensors operated at low temperature. It was demonstrated that with the increase of the $Ar+O_2$ gasses pressure, the surface morphology of the films becomes rougher and grain size increases. Meanwhile at low temperature the magnetoresistance of the films grown at low pressure has a higher value. It was demonstrated that in the range of lower magnetic fields (up to 10 T) the films grown at low gas pressure have higher sensitivity. However, for measurement of higher magnetic fields, the properties of films grown at higher gas pressure are favored. The obtained results allow evaluating the influence of MOCVD technological conditions on magnetoresistance of La-Ca-Mn-O manganites used for pulsed magnetic field sensors.**

Keywords—Colossal magnetoresistance, magnetic field sensors, manganites, thin films, MOCVD technology.

I. Introduction

It has been recently demonstrated that nanostructured (polycrystalline with nano-size grains) $La_{1-x}Sr_xMnO_3$ (LSMO) and $La_{1-x}Ca_xMnO_3$ (LCMO) films, which exhibit the colossal magnetoresistance (CMR) effect, can be successfully used for the development of CMR-B-scalar sensors [1], [2]. Such sensors allow measuring the absolute magnitude of the magnetic flux density during high magnetic field pulses. They were used to measure the magnetic field distribution in railguns [3], [4] and in non-destructive pulsed-field magnets [5]. However, for applications in other fields, such as in plasma science or condensed matter physics, it is of great interest to have sensors operating at cryogenic temperatures and wide range of magnetic field amplitudes (0.1–100) T. It was shown, that for low temperature applications the polycrystalline LCMO films with lower critical temperature (T_C) are more preferable than the LSMO films [6]. It was demonstrated, that the LCMO films have higher sensitivity to magnetic field, lower memory effects and faster resistance relaxation during the measurements of pulsed magnetic field. Moreover, the electric and magnetic properties of these films depend on the doping element Ca concentration (x). The optimal concentration for the application of LCMO films for fast magnetic field sensors operating at cryogenic temperatures was found to be x=0.22 [7]. However, the properties of manganite films grown by using the Metal-organic chemical vapor deposition (MOCVD) technique depend not only on the Ca concentration, but also on a number of processes occurring during film deposition. The MOCVD processes are extremely complex and involve a series of gas-phase and surface reactions. The film growth rate, chemical composition and structure are determined by several parameters, primarily temperature of the substrate, the operating pressure (transported gas pressure) of the reactor as well as the composition and chemistry of the gas-phase. A crucial factor that determines the relative importance of film properties is the pressure in the MOCVD reactor [8]-[10]. It was shown in [8] that kinetics, gas phase reactions and mass transport depend on the gas pressure and can play a significant role in the deposition. Therefore, for development of fast magnetic field sensors operating at cryogenic temperatures, it is very important to determine the optimal deposition conditions of these films.

In this paper, the results of the investigation of the influence of gas pressure of Pulse Injection (PI) MOCVD on morphology, microstructure, electrical and magnetic properties of $La_{1-x}Ca_xMnO_3$ films are presented.

This work was partly supported by the Research Council of Lithuania (grant No. MIP-095/2015).

II. EXPERIMENTAL

The nanostructured manganite $La_{0.78}Ca_{0.22}MnO_3$ (LCMO) thin films were grown on lucalox (99.9% Al_2O_3 + 0.01% MgO) substrates by PI MOCVD technique [11]. The injector was used to inject microdoses (~3 mg) of an organic solution containing a dissolved mixture of organometallic precursors. La(thd)3, Ca(thd)2, and Mn(thd)3 (thd is 2,2,6,6-tetramethyl-3,5-heptandionate) dissolved in monoglyme were used as precursors. After flash evaporation (at ~270 °C) of the microdoses, the resulting vapor mixture was transported by an Ar + O_2 (5:1) gas flow (~95 l/h) towards the heated substrate. During the growth, the temperature of substrate was kept at 775 °C and injection frequency was 2 Hz. A series of films were deposited at three different total pressures of Ar + O_2 gases: 3, 5 and 7 Torr. It has to be pointed out that the deposition conditions were at first optimized in order to obtain a pure perovskite phase of the film by using monocrystalline substrates of $LaAlO_3$ and only then the films on lucalox substrate were grown. Total gas flow rate, injection frequency and precursor concentration were changed for this goal. The thickness of the deposited films was 280 – 340 nm.

The morphology of the film surface was studied using the Atomic Force Microscopy (AFM) (VEECO Dimension 3100) and Scanning Electron Microscope with attached Energy Dispersive X-Ray Spectrometer (SEM/EDS) (Hitachi SU70). EDS analysis and ICP mass spectrometry were used for the evaluation of the film composition. The crystal structure of the samples was investigated using X-ray diffraction (SmartLab diffractometer) and RHEED diffraction (EMR-100 electron-diffraction camera). The detailed microstructure and composition analysis of the samples was characterized in cross section geometry by using Transmission electron microscope (TEM) (Tecnai G2 F20 X-TWIN) with energy dispersive X-ray spectrometer EDX.

For electrical transport and magnetoresistance measured, two electrodes were prepared by the thermal deposition of silver (Ag) using a Cr sublayer. After Ag deposition the samples were annealed in argon atmosphere at 420 °C for 1 h.

The electrical and magnetic characterization of the samples was performed by measuring the resistivity (ρ) and magnetoresistance (MR) in a temperature range of $T = 5 – 315$ K and magnetic flux density range $B = 0 – 0.8$ T using a closed cycle helium gas cryocooler and electromagnet. The magnetoresistance measurements in the pulsed magnetic fields up to 60 T were performed using a 65 T/ 20 ms/ 20 mm non-destructive pulsed magnet fabricated and installed at the Dresden High Magnetic Field laboratory (HLD) [5].

III. RESULTS AND DISCUSSION

For all the investigated films, X-ray diffraction spectra are characteristic for the orthorhombic Pnma phase. X-ray and RHEED spectra show that the films are not textured, indicating the polycrystalline structure. By indexing the LCMO peaks the lattice parameters were obtained to be a = 0.548, b = 0.774, c = 0 .549 nm (in pseudocubic a = 3.88 Å).

Fig. 1 shows the SEM images of the LCMO films grown at the total pressure of Ar + O_2 gases 3 Torr (a) and 7 Torr (b), respectively. The grains are densely packed, however, some pores are found at the film surfaces. Furthermore, the grain size is more homogeneous for the LCMO films grown at pressure of 3 Torr. The LCMO grain sizes for these films were approximately in the range of 40 to 100 nm (see Fig.1a). With the increase of the pressure of Ar + O_2 gasses, the surface morphology of the films becomes more rough, grain width increases and the polygonal shape of the crystallites turns to triangular shape (see Fig.1b). The grain size of these films range from approximately 40 to 170 nm.

The low magnification cross-sectional TEM images of the films demonstrate a columnar structure with typical column width of 40 – 150 nm (Fig. 2a, b). These columns are spread throughout the whole film with their long axis perpendicular to the substrate. The contrast of TEM images within the columns reveals that columns are not single crystalline and consist from several single crystal slabs (most probably twins). The high-resolution TEM (HRTEM) showed 5 – 7 nm grain boundaries between the columns. The faceted tips of the columns result in the surface roughness of the films. For the films grown at pressure of 3 Torr, the roughness is approximately 20 nm (Peak-peak), while the roughness of films grown at pressure 7 Torr is as high as 50 nm. The inspection of the electron diffraction patterns taken from the columns showed that for films grown at 7 Torr the column tips are truncated with (111c) and (110c) facets (in pseudo-cubic indexes). In the case of films grown at pressure of 3 Torr, no pronounced planes of column faceting were observed. Such dependence of film surface roughness on the gas pressure can be caused by specifics of PI MOCVD. At the transport gas pressure of about 10 Torr the gas phase reactions of precursors cannot be neglected and, in addition, a significant boundary gas layer over the film surface is present [8]. These effects can cause limited surface diffusion of adatoms. In this case, the column tips become faceted due to the kinetic roughening, which in combination with atomic shadowing results in deep cusps between columns and open column boundaries. The

a)

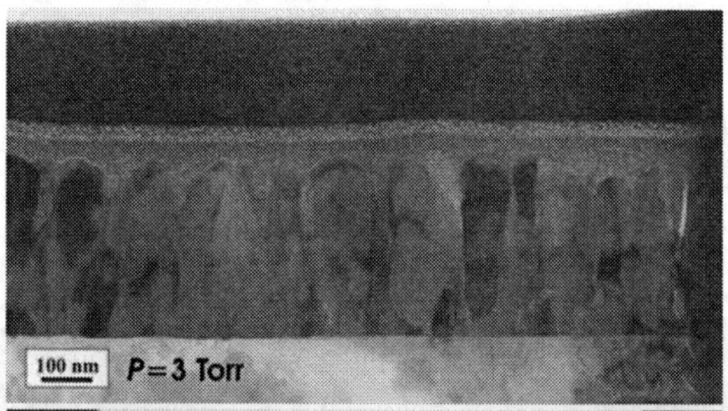

b)

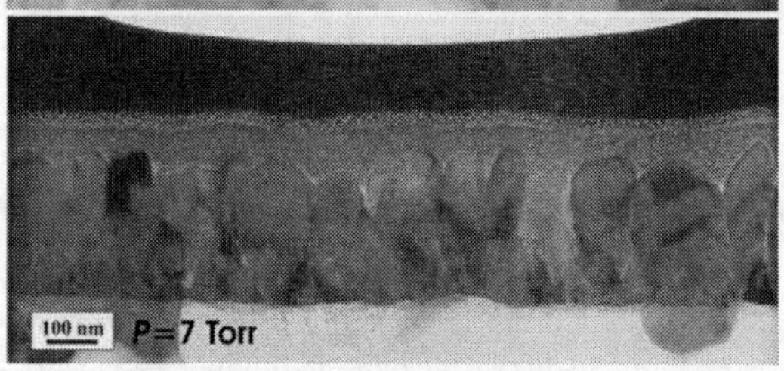

Fig.2. Low magnification cross-sectional TEM images of the films grown at Ar+O_2 gasses pressure of 3 Torr (a) and 7 Torr (b).

individual columns, however, are dense indicating sufficient adatom surface mobility to sustain local crystal growth. However, a grain boundary (GB) is a region of structural or even chemical disorder. As the pressure falls, gas phase reactions tend to become less significant, and the film growth is often controlled by the surface reactions. This can influence the surface morphology. Moreover, the low gases pressure of Ar+O_2 can causes decomposition of the precursor in the reactor volume [8], which can result in the changes of chemical composition or appearance of the oxygen vacancies.

The resistivity (ρ) dependences on temperature of films grown at different gas pressure are shown in Fig. 3. It is observed that the resistivity of films grown at 5 and 7 Torr has almost the same value of maximal resistivity (ρ_m=2.8 Ωcm) and it is achieved at nearly the same temperature T_m~225 K. However, electrical transport properties of films grown at 3 Torr are significantly different. The films grown at 3 Torr have much higher resistivity (ρ_m=19 Ωcm) and the ρ_m is shifted to the lower temperature (T_m~175 K). It is known that the resistivity of polycrystalline films is mainly determined by the grain boundary resistivity and quantity thereof. However, as it follows from the structural studies, all films, independently from the growth pressure, have almost the same width of the columns and GB. Therefore, the high resistivity and low T_m can be a result of possible deviation from stoichiometry of the films or appearance of the oxygen vacancies due to low gas pressure [10]. Moreover, taking into account film growth time and thickness, it was obtained that average growth rate of the films was also different keeping different pressure: 22 nm/min, 20 nm/min and 17 nm/min at 3 Torr, 5 Torr and 7 Torr, respectively. All these conditions can result in more disordered grain boundaries and as a result, higher resistivity values of films grown at gas pressure of 3 Torr.

The dependencies of *MR* on magnetic flux density *B* measured at temperatures of 100 K and 1.3 K applying 60-Tesla magnetic field pulses are shown in Fig. 4 a,b for LCMO films grown at 3, 5 and 7 Torr. The *MR* is defined as: $MR = 100\% \times [\rho(B) - \rho(B = 0)]/\rho(B=0)$, where $\rho(B)$ and $\rho(B = 0)$ correspond to electrical resistivity in magnetic field and without it, respectively. One can see that at these temperatures

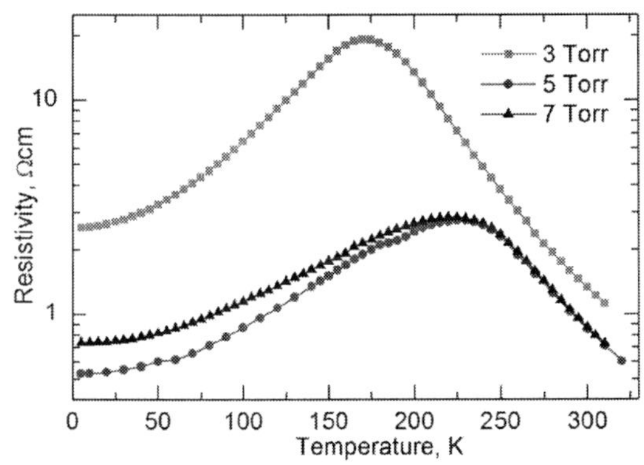

Fig. 3. Resistivity *vs.* temperature dependences for LCMO films grown at different gas pressures.

the *MR* of the films depends nonlinearly on the magnetic field and does not fully saturate up to 60 T. The largest magnetoresistance slope is at low magnetic fields (in the range up to 10 T. At a higher magnetic field (more than 20 T) the *MR* change with the field becomes lower and tends to saturation. In addition, as can be seen from Fig. 4b, at very low temperature (T = 1.3 K), hysteresis of *MR* is observed in magnetic fields below 20 T. This phenomenon shows the memory effect at these temperatures and is related with short-range interaction of the magnetic moments in disordered grain boundaries having spin-glass properties [6]. It is evident that at cryogenic temperatures the manganite films grown at pressure of 3 Torr have the higher *MR* in comparison with the films grown at 5-7 Torr.

For the development of magnetic field sensors operating at low temperatures in wide magnetic field range, it is important to obtain the sufficiently high sensitivity of the sensors over the large range of magnetic fields. The highest sensor's sensitivity corresponds to the largest derivative of the magnetoresistance with respect to magnetic field (see Fig. 5). One can see that in magnetic field range up to 10 T, the higher sensitivity show the films grown at pressure of 3 Torr. However, for higher than 10 T magnetic fields, the films grown at 7 Torr have the higher sensitivity. Therefore, during the growth of the films used for magnetic field sensors, PI MOCVD conditions have to be chosen depending on the range of the magnetic field which has to be measured. For

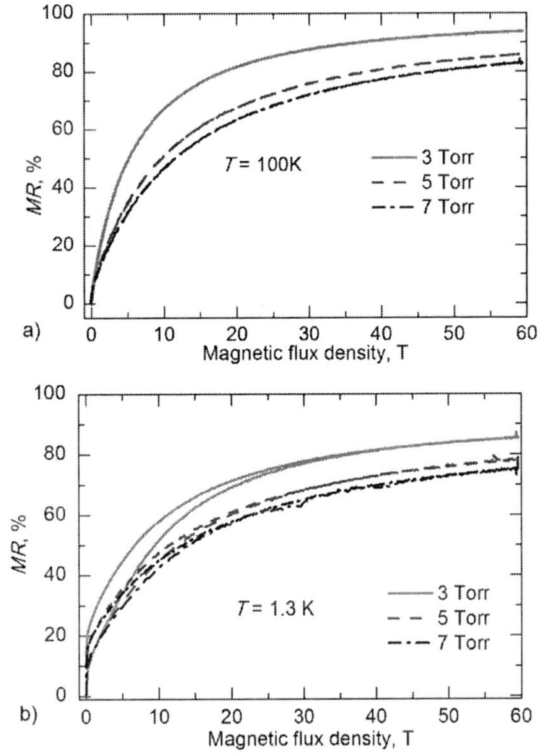

Fig. 4. *MR vs.* magnetic flux density dependences of LCMO films grown at different gas pressures measured at temperatures: a) 100 K, b) at 1.3 K.

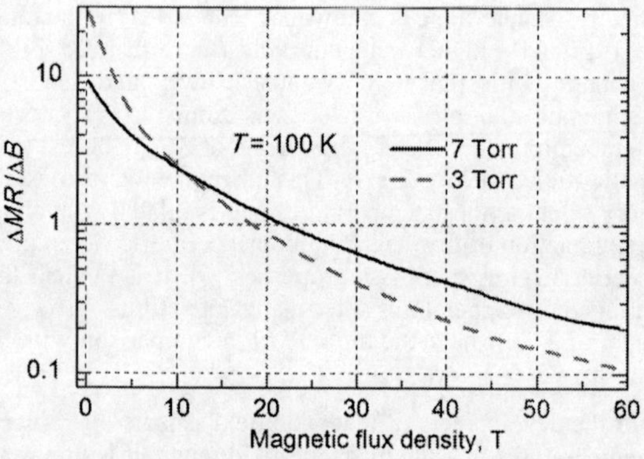

Fig. 5. Change of magtenoresistance *vs*. magnetic flux density for LCMO films grown at gases pressure 3 and 7 Torr.

measurements of the magnetic flux density up to 10 T, films grown at transport gas pressure of 3 Torr are favored. For higher magnetic fields, the higher gas pressure (7 Torr) leads to higher sensitivity.

IV. Conclusions

In conclusion, we have demonstrated that surface morphology, electrical transport and magnetic properties of nanostructured $La_{1-x}Ca_xMnO_3$ films to a large extent depend on transport gas pressure during the growth of the films using PI MOCVD technique. It was obtained, that for the applications of the LCMO films as magnetic field sensors operating at cryogenic temperatures, the technological growth conditions have to be chosen depending on the required magnetic field range. For measurements of magnetic flux density up to 10 T, films grown at transport gas pressure of 3 Torr are preferred. For a higher magnetic field measurement, especially approaching megagauss, the films grown at higher gas pressure (7 Torr) have to be used.

Acknowledgment

The authors gratefully acknowledge the staff of Dresden High Magnetic Field Laboratory, Helmholtz-Zentrum Dresden-Rossendorf, for the possibility to use high-field magnet facilities.

References

[1] S. Balevičius, N. Žurauskienė, V. Stankevič, S. Keršulis, V. Plaušinaitienė, A. Abrutis, S. Zherlitsyn, T. Herrmannsdörfer, J. Wosnitza and F.Wolff-Fabris, "Nanostructured thin manganite films in megagauss magnetic field," Appl. Phys. Lett., vol. 101, pp. 092407-1–4, August 2012.

[2] T. Stankevič, L. Medišauskas, V. Stankevič, S. Balevičius, N. Žurauskienė, O. Liebfried, M. Schneider, "Pulsed magnetic field measurement system based on colossal magnetoresistance-B-scalar sensors for railgun investigation," Rev. Sci. Instrum., vol. 85, pp. 044704-1–5, April 2014.

[3] M. Schneider, O. Liebfried, V. Stankevic, S. Balevicius and N. Zurauskiene, "Magnetic Diffusion in Railguns: Measurements Using CMR-Based Sensors," IEEE Trans. Magn., vol. 45, iss. 1, pp. 430-435, Januar 2009.

[4] T. L. Haran, R. B. Hoffman, S. E. Lane, "Diagnostic capabilities for electromagnetic railguns," IEEE Trans. Plasma Science, vol. 41, iss. 5, pp. 1526-1532, May 2013.

[5] S. Balevicius, N. Zurauskiene, V. Stankevic, T. Herrmannsdörfer, S. Zherlitsyn, Y. Skourski, F. Wolff-Fabris, J. Wosnitza, "CMR-B-Scalar sensor application for high magnetic field measurement in non-destructive pulsed magnets," IEEE Trans. Magn., vol. 49, iss. 11, pp. 5480–5484, November 2013.

[6] N. Zurauskiene, S. Balevicius, D. Pavilonis, V. Stankevic, V. Plausinaitiene, S. Zherlitsyn, T. Herrmannsdörfer, J. M. Law, J. Wosnitza, "Magnetoresistance and resistance relaxation of nanostructured La-Ca-MnO films in pulsed magnetic fields," IEEE Trans. Magn., vol. 50, iss. 11, pp. 6100804, November 2014.

[7] N. Zurauskiene, D. Pavilonis, S. Balevicius, V. Stankevic, A. Maneikis, V. Plausinaitiene, J. Novickij, "Fast resistance relaxation in nanostructured La–Ca–Mn–O films in pulsed magnetic fields," IEEE Trans. Plasma Science, vol. 43, iss. 10, pp. 3445-3450, October 2015.

[8] Chemical Vapour Deposition: Precursors, Processes and Applications Edited by Anthony C. Jones and Michael L. Hitchman, Royal Society of Chemistry, 2009, Published by the Royal Society of Chemistry, www.rsc.org.

[9] P. Lu, H. Li, Y.-M. Wang, S. Sun, B. Tuttle, "Novel single-solid-source metalorganic chemical vapor processing of PT and PZT thin films," J. Cryst. Growth, vol. 171, p. 453, 1997.

[10] A. Bartasyte, R. Bouregba, E. Dogheche, M. Boudard, G. Poullain, O. Chaix-Pluchery, C. Jimenez, V. Plausinaitiene, D. Remiens, A. Abrutis, Z. Saltyte, F. Weiss, "Ferroelectric PbTiO3 films grown by pulsed liquid injection MOCVD," Surf. Coat. Technol., vol. 201, pp. 9340–9344, 2007.

[11] A. Abrutis, V. Plausinaitiene, V. Kubilius, A. Teiserskis, Z. Saltyte, R. Butkute, and J. P. Senateur, "Magnetoresistant $La_{1-x}Sr_xMnO_3$ films by pulsed injection metal organic chemical vapor deposition: Effect of deposition conditions, substrate material and film thickness," Thin Solid Films, vol. 413, no. 1–2, pp. 32–40, June 2002.

Liner Performance of Ranchero-S Experiments*

J. H. Goforth, E. V. Baca, P. Dickson, T. A. Gianakon, B. B. Glover, J. A. Gunderson, E. H. Haroz, D. H. Herrera, R. K. Meyer, H. Oona, P. J. Rae, C. L. Rousculp, L. J. Tabaka, R. G. Watt
Los Alamos National Laboratory, box 1663,
Los Alamos, NM, USA

***Abstract*—A new high energy imploding liner effort is beginning at Los Alamos, and the "Ranchero-S" flux compression generator (FCG) has been chosen to perform these tests. The first liner experiment has not been conducted at the time of this conference, but performance characteristics have been verified using the Los Alamos Roxane MHD code and preparations are being finalized. The test uses the Los Alamos Point 88 capacitor bank to supply 3.5 MA of seed current to the 87 nH FCG. Peak current is predicted to be 42 MA into the imploding liner load which has an initial inductance of 2.5 nH and increases to 3.5nH by current peak. The 7.85 cm outer radius, 0.35 cm thick aluminum liner is expected to exceed 1 cm/μs implosion velocity. The liner is 4 cm tall initially (182 g), 2 cm tall at full implosion, and at 1.5 cm/μs will have 20.5 MJ kinetic energy. This paper describes the expected performance of the FCG and liner. In addition, scaling Ranchero FCGs for pursuing high energy liners into the 100s of Megajoule implosion regime is discussed. Separate papers in the conference provide further descriptions of subsystem tests and the suite of diagnostics being used to characterize both the generator and liner performance.**

Keywords—Flux Compression Generator

I. INTRODUCTION

Los Alamos has initiated a renewed effort to perform high energy density physics experiments using solid liners imploded at high velocity in a z-pinch geometry. Liner experiments in the late 1990s were conducted using the Pegasus and Atlas capacitor banks [1], and the Ranchero FCG [2]. A brief effort funded by Laboratory Directed Research and Development (LDRD) funds and using Ranchero generators was also successfully completed in 2011 [3]. The renewed program employs high explosive pulsed power (HEPP) techniques, and the first liner test will be powered by an updated Ranchero FCG configured for currents at or near the practical current limit for FCG operations. This paper presents the design details for the first experiment, which attempts to drive a 182 g aluminum liner to a velocity of 1.5 cm/μs at impact on a one cm radius central measuring unit (CMU) diagnostics housing. In addition, projections for Ranchero FCG designs to operate at higher currents are presented in anticipation of improved liner design with increased performance. Papers in this conference by Haroz [4] and Gunderson [5] present details of subsystem tests that have been conducted and improved diagnostics techniques that are being pursued.

II. RANCHERO-S GENERATOR

The Ranchero module for the upcoming liner test is designated Ranchero 43-S. Fig. 1 shows an original 43 cm long Ranchero module, the new "Swoop" (S) configuration, and two frames from a hydrocode calculation that illustrate how the explosive is initiated and how the armature zippers along the stator in the final microseconds of operation. The purpose of the swoop at the output is to eliminate a flux pocket observed in simulations of very high current Ranchero performance [6]. A sub-system test of the armature is presented by Haroz, in this conference [4].

III. LINER DESIGN AND PERFORMANCE

The liner to be tested is fabricated from a near pure Aluminum stock [7] and will have a 3.5 mm thick wall with an O. D. of 15.7 cm. The height at the initial glide plane contacts is 4 cm, and will be 2 cm on contact with the 2 cm diameter diagnostics housing. Brass glide planes following the philosophy of Atchison have been used successfully on previous tests [8], and have been scaled to suitable dimensions for this application. A detail of the liner configuration is given in fig. 2, and the results of an end-to-end calculation of system performance using the 2-D MHD code Roxane are given

---

*This work sponsored by the U. S. DOE

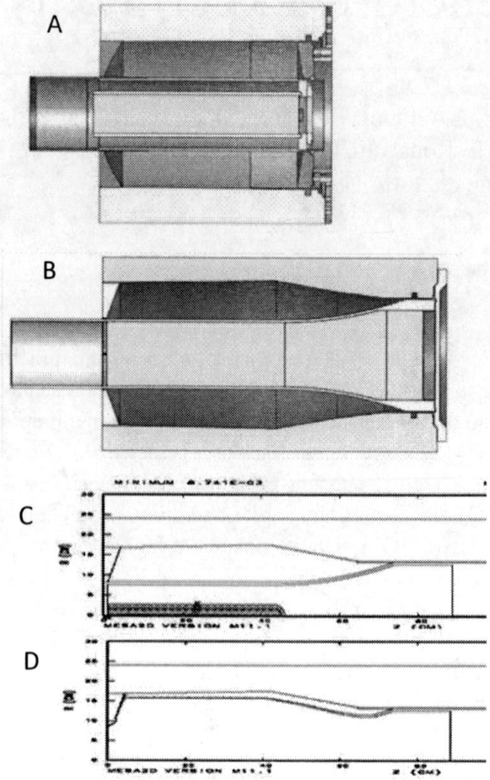

Figure 1. Standard 43 cm Ranchero module (A) and 43-S module (B). Both modules use the same 43 cm long line detonator. Frame C shows the initial setup for a MESA 2-D hydrodynamic code calculation and D shows the position of the armature as it approaches the stator.

in fig. 3. The current in the system is shown in fig. 4, and liner velocity versus radius is given in fig. 5. The end-to-end calculation starts when the FCG begins flux compression, and shows the armature position at 82 μs, just before peak current at 83.5. The liner has moved by ~1 cm at peak current, and is just reaching the position of the CMU at the last frame, 90.8 μs. At this time, the v vs r plot shows that the liner is predicted to be moving at 1.5 cm/μs, at which condition it will have 20.5 MJ of implosion kinetic energy.

IV. Liner Diagnostics

The CMU houses 12 PDV probes to characterize the liner implosion. Four probes will look radially outward, and two of these will be operated in the downshifted mode to allow speeds over 1 cm/μs to be recorded. Four of the remaining probes will look up at a ten degree angle and the other four will look down at the same angle. These probes allow features seen in the Roxane simulations to be verified experimentally. In particular, a step caused by glide plane interactions propagates toward the liner center as it implodes, and an observable inflection should be seen on the angled PDV probes. The diagnostics will provide achieved velocity and symmetry, and good code verification.

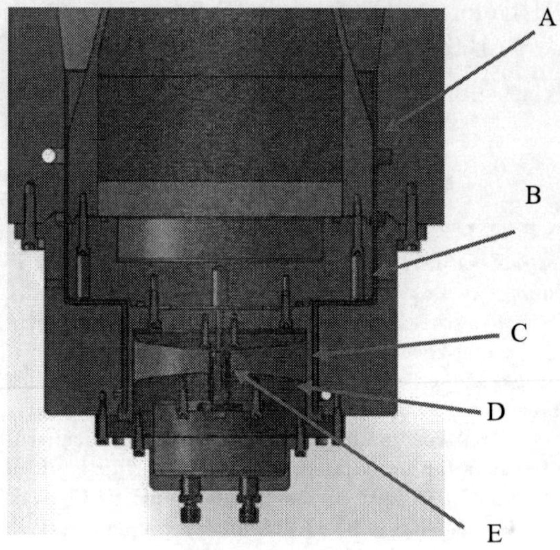

Figure 2. Liner configuration for first test. High current joints connect the FCG output to the brass liner base (B) and an Al outer transmission line. The output is insulated with a machined Delrin insulator (A). The liner assembly (C) attaches to the liner base and the upper glide plane is placed inside the assembly. The lower glide plane (D) and CMU (E) are inserted into the bottom of the liner assembly. A vacuum pump attaches at the bottom.

V. Ranchero Scaling

The original Ranchero configuration has been demonstrated to work at 76 MA, but simulations indicate that an aneurism may form at the output glide plane at higher currents. To avoid this problem, the "swooped" Ranchero was designed, but a necessary feature was the reduced diameter output seen in fig. 1. A long standing rule of thumb for Los Alamos generators is that good performance should not be expected beyond 1 MA/cm of conductor width. The original module had a circumference of 96 cm, although the circumference was less at the glide plane and there was an enhanced field region at the glide plane contact point in addition. The S model Ranchero has a circumference of 83 cm, and no enhanced field region. The expectation is that this configuration should work up to about 83 MA. 2-D MHD simulations have been conducted to explore what actual performance limits would be [9], [10]. The result is that for external loads of 1 nH, the 43-S generator will produce 88 MA given an initial current of ~6 MA. However, larger initial currents result in reduced output because early magnetic backpressure has extracted kinetic energy from the armature and there is not enough KE left to close the tapered section. Similarly, with a 5

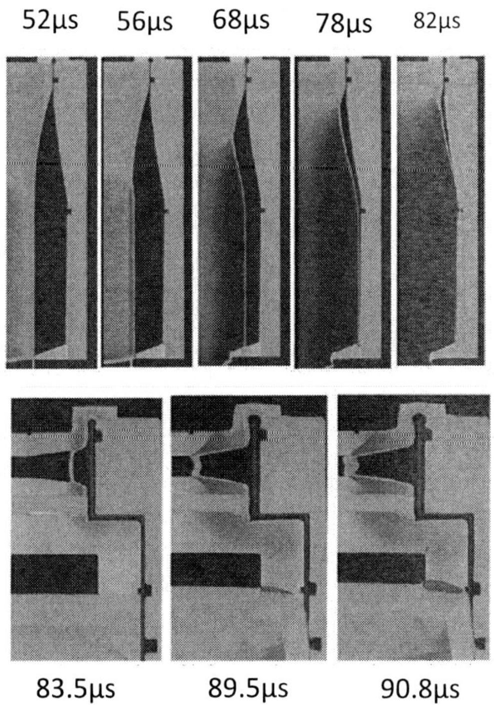

Figure 3. Material plots from an end-to-end Roxane simulation. Armature fist motion is at 52 µs, and the armature contour is followed to near closure at 82 µs. The liner motion is captured at peak current (83.5 µs) and followed to approximate time of impact with the CMU (90.8 µs).

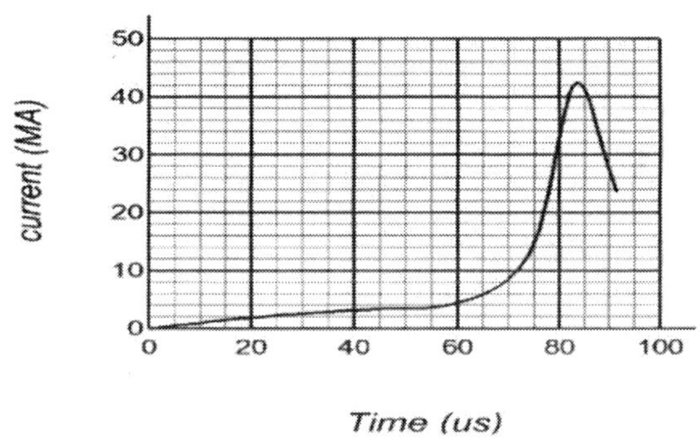

Figure 4. System current. Peak seed current is 3.5 MA at 52 µs and peak FCG current is ~42 MA at 83.5 µs.

nH external load the maximum current expectation is ~83 MA with 8 MA initial current, and with a 10 nH load only ~73 MA is generated from 10-12 MA seed. Even though the new design brings the detonation front near the high current density region, the pressure in the detonation products falls off very quickly, and the magnetic field still stops the armature before the gap is closed. Longer FCGs (detonators exist for 100-S or 144-S modules) have increased KE in the armatures and higher gain reduces

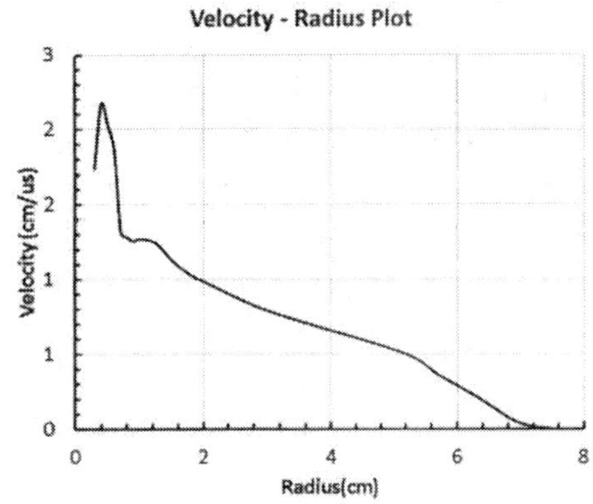

Figure 5. Liner velocity versus radius. The impact on the CMU should occur in the relatively flat region at 1.5 cm/ µs

early time backpressure, but the upper limit remains around the 1 MA/cm level. Thus, to produce higher currents with a Ranchero-S generator, the overall diameter must be increased. There are many ways to scale Ranchero to higher current. One can increase the diameter and keep the gain the same by preserving the ratio of the initial diameters of the armature and the stator, but this increases the function time of the generator. To keep the function time the same, the armature-stator separation can be held fixed as radius is increased, but the gain is reduced. In this case the length can be increased to recover gain. Many simulations have been run, and a practical compromise is shown in figure 6. In this simulation, the length of the simultaneous section was fixed at 1.44 m, which is the length of the longest detonation system on hand. Likewise, the initial current was fixed at 12 MA and the load was 10 nH for all runs. The armature-stator separation is fixed at its original value, and the radius is increased in 0.5 cm steps. The peak current shown is 118 MA, which represents 1.1 MA/cm in that module. Because these calculations were run at reduced resolution, it is believed that actual values would be ~10% higher, and no studies were made to explore the limit. At 120 MA in 10 nH, the stored magnetic energy is 100 MJ. A philosophy for optimizing such a generator is given by Watt [10], and practical considerations, such as length of available detonators and real losses, are considered.

VI. Conclusions

A renewed high energy imploding liner program is beginning at Los Alamos. The first test will utilize a new configuration of the Ranchero generator with armature and stator contoured to zipper armature closure with no output glide plane. The contour added to the generator

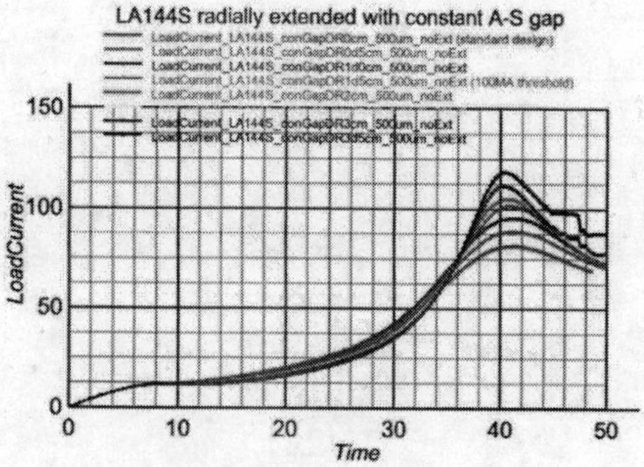

Figure 6. Predicted currents for a 144-S generator with stator-armature gap held constant while the radius is increased in 0.5 cm increments. The initial current is 12 MA in all cases and the load is a static 10 nH cavity. Largest peak current is 118 MA, for a 3.5 cm increase in radius above the initial 43-S design. All predicted currents are believed to be low by ~10% due to reduced resolution in the computation.

adds ~30 nH to the inductance of the simultaneous section for any length coax, but reduces the maximum current expectation to around 83 MA. On the first test, the 2.4 MJ capacitor bank at the firing point 88 complex will be used to seed the Ranchero 43-S generator with 3.5 MA. The load will be initially 2.5 nH, and will expand to ~3.5 nH at the 42 MA peak current. The liner is predicted to reach 1.5 cm/μs at a radius of 1 cm where it will impact the CMU, and at that velocity the 182 g liner will have 20.5 MJ kinetic energy. Ranchero generators will be used for higher current experiments in the future, and a booster FCG will be required to increase the seed current to a level above that which the capacitor bank will deliver. Detonation systems for cylindrical systems of 100 and 144 cm have been developed, and provide extra gain, but do not provide currents much in excess of 83 MA for useful size loads. To achieve substantially larger currents, Ranchero can be scaled up in diameter and calculations show that scaling follows straightforward expectations for increased current capacity.

References

[1] R. E. Reinovsky, W. L. Atchison, R. J. Faehl, and I. R. Lindemuth, "Advanced Liner Performance for Extended Experiments in Hydrodynamics and Material Properties," Proceedings of the14th IEEE International Pulsed Power Conference, 2003, Dallas Texas, Giesselmann and Neuber eds., pp 107.

[2] J. H. Goforth, W. L. Atchison, C. M. Fowler, R. K. Keinigs, H. Oona, D. G. Tasker, D. B. Reisman, P. T. Springer, and R. C. Cauble, "Design of High Explosive Pulsed Power Systems for 20 MB Isentropic Compression Experiments," Proceedings of the Ninth International Conference on Megagauss Magnetic Field Generation and Related Topics, 2002, Moscow-St Petersburg, Selemir and Plyashkevich eds., pp, 137.

[3] J. H. Goforth, W.L. Atchison, S. A. Colgate, J. R. Griego, J. Guzik, D. H. Herrera, D. B. Holtkamp, G. Idzorek A. Kaul, R. C. Kirkpatrick, R. Menikoff, R. K. Meyer, H. Oona, P. T. Reardon, C. R, E. Reinovsky, L. Rousculp, A. G. Sgro, L.J. Tabaka, T. E. Tierney, D. T. Torres, R. G. Watt, "A New 40 MA Ranchero Explosive Pulsed Power System," Los Alamos Unlimited Distribution Report, LA-UR-09-04121. (Also in the digest of papers for the 17th IEEE International pulsed Power Conference, 2009, Washington D.C.)

[4] E. H. Hároz, P. J. Rae, C. L. Armstrong, E. V. Baca, P. Dickson, T. A. Gianakon, B. B. Glover, J. A. Gunderson, D. H. Herrera, R. K. Meyer, C. Rousculp R. G. Watt & J. H. Goforth, "Ranchero 43-S Flux Compression Generator: Full-Scale Armature Test," this conference.

[5] J. A. Gunderson, B. B. Glover, and J. H. Goforth, "Measuring Flux Pocketing and Current Symmetry on Ranchero-S Experiments," this conference.

[6] J. H. Goforth, T.A. Gianakon, B. B. Glover, D. H. Herrera, R.L. Holmes, J.L. Johnson, E. C. Martinez, R. K. Meyer, H. Oona, P. J. Rae, C. L. Rousculp,, and R. G. Watt, "Improvements in Ranchero Magnetic Flux Compression Generators," Los Alamos Unlimited Distribution Report, LA-UR-15-20313.

[7] R. G. Watt, "Preshot Predictions for LA-43S-1 using the Eulerian AMR Code Roxane," Los Alamos Unlimited Distribution Report, LA-UR-15-29079.

[8] W. L. Atchison et al., "The Quest For The Wholly Stable Liner," Proceedings of the 2006 International Conference on Megagauss Magnetic Field Generation and related topics, 2006, Santa Fe, NM, Kiuttu, Reinovsky, and Turchi Eds., pp. 57.

[9] R. G. Watt, "Predictions for the drive capabilities of the RancheroS Flux Compression Generator into various load inductances using the Eulerian AMR Code Roxane," Los Alamos Unlimited Distribution Report, LA-UR-16-23924.

[10] R. G. Watt, "The search for a 100MA RancheroS magnetic flux compression generator," Los Alamos Unlimited Distribution Report, LA-UR-16-26685.

Multi-step accelerated ageing of magnetic field sensors based on nanostructured La-Sr-Mn-O thin films

J. Klimantavicius, V. Rudokas, N. Zurauskiene, S. Balevicius, C. Simkevicius, S. Kersulis, V. Stankevic, R. Vasiliauskas and V. Plausinaitiene
Center for Physical Sciences and Technology
Sauletekio Ave. 3, LT-10257 Vilnius, Lithuania
e-mail: nerija.zurauskiene@ftmc.lt

N. Zurauskiene, S. Balevicius, V. Stankevic, C. Simkevicius, S. Kersulis and S. Tolvaisiene
Vilnius Gediminas Technical University
Naugarduko 41, LT-03227 Vilnius, Lithuania

V. Plausinaitiene
Vilnius University
Naugarduko str. 24, LT- LT-03227 Vilnius, Lithuania

***Abstract*—We present results of investigation of influence of accelerated ageing on electrical resistivity and magnetoresistance of nanostructured La-Sr-Mn-O films used for magnetic field sensors. The 400 nm thick manganite films were grown by MOCVD technique. To investigate the effectiveness of accelerated ageing, the two methods were tested: 1) annealing the samples at 100 °C temperature in Ar atmosphere when film surface was uncoated, and 2) annealing of the samples coated prior to treatment by polymer. The samples were studied without and in pulsed magnetic fields up to 20 T. To describe the underlying processes, the kinetics of resistivity change during accelerated ageing was analysed using stretched exponent relaxation function. It was found that the change of resistivity of coated samples saturates after significantly shorter accelerated ageing time than uncoated samples. Both ageing methods resulted in minor influence (<1%) on magnetoresistance of the films in magnetic fields up to 20 T.**

***Keywords*—Colossal magnetoresistance; ageing effects; thin films; magnetic field sensors; pulsed magnetic fields.**

I. Introduction

Thin (up to few hundred nm thick) nanostructured $La_{1-x}Sr_xMnO_3$ (LSMO) manganite films have demonstrated unique properties of colossal magnetoresistance (CMR) in a wide range of temperatures and magnetic fields [1]. This feature makes them perspective materials for the high pulsed magnetic field sensors, for example, for the investigation and testing the operation of electromagnetic railguns [2], [3], high power induction coils, etc. CMR-B-scalar sensors [3] measure magnetic field amplitude independently on the field direction; the phenomenon is based on the film resistivity change as reaction to magnitude of applied magnetic field. However, both the initial resistance and the sensitivity to magnetic field depend on ambient temperature. This imposes a challenge in applications, because temperature caused resistance drift has to be taken into account for accurate measurement of the magnetic field.

In case of pulsed magnetic fields, the temperature can be calculated from measured initial resistance of the sensor at zero magnetic field [3], [4]. In such case zero-field resistance of the film is a measure of ambient temperature while the resistance change during the pulse corresponds to magnitude of the applied magnetic field. For this reason, it is very important to know exact values of the initial resistance of the sensor at particular ambient temperature values.

However, it was observed that over time the resistivity of structurally imperfect nanostructured manganite films undergoes dramatic changes which can limit technological applicability of the material. Ageing reduces life time of the sensor and requires its frequent recalibration. It was demonstrated that annealing of the manganite films in inert atmosphere at temperatures higher than the operation temperature accelerates the ageing processes and stabilizes main parameters of the sensors [4], [5]. During the accelerated ageing the maximal resistance changes depend on the annealing temperature; sample treatment at 100 °C temperature for about 24 hours was found to be optimal [5]. The process of stabilization of parameters was explained by the oxygen release, displacement or redistribution, which, most likely, takes place in the grain boundaries of polycrystalline manganite films [4], [5].

It was obtained that after accelerated ageing procedure the samples achieved a good stability and the magnetoresistance (*MR*) of the samples in 0-2.3 T magnetic field range have changed insignificantly [4], [5]. On the other hand, for sensors used to measure high pulsed magnetic fields, the investigations of the magnetoresistance stability in high magnetic fields still should have been done.

In this work we have investigated influence of accelerated ageing on electrical resistivity and magnetoresistance of nanostructured LSMO films used for magnetic field sensors. For these purposes two thermal treatment procedures were applied, and the effects induced by the annealing on main sensors' parameters are investigated and discussed.

II. Experimental

The samples were made from 400 nm thick LSMO film ($La_{x-1}Sr_xMnO_3$, where x=0.17), deposited by a pulsed injection metal organic chemical vapor deposition (PI MOCVD) technique [6] onto a special 400 µm thick polycrystalline substrate (lucalox). The PI MOCVD injector injected micro doses (~3 mg) of an organic solution containing a dissolved mixture of organometallic precursors into the reactor. During deposition the temperature of the substrate was kept 750 °C, the total pressure of gases was ~5 Torr.

Due to the polycrystalline structure of the substrate nanostructured films were produced. Morphology of the film surface was studied by VEECO Dimension 3100 Atomic Force Microscopy (AFM) and Hitachi SU70 Scanning Electron Microscope with attached Energy Dispersive X-Ray Spectrometer (SEM/EDS). EDS analysis was used for the evaluation of film composition.

Planar photolithography was used for formation of the samples used for the investigation. Electrical contacts were prepared by silver thermal deposition on Cr sublayer in vacuum at pressure of ~10^{-5} Pa. All resistance measurements were performed with two contacts method. Our earlier results [4] showed that contact resistance of this design is about 1000 times lower than the resistance of the active area of the manganite film between the contacts, so its impact on the total resistance in negligible.

The prepared samples were characterized by measuring temperature dependences of resistivity (ρ) and magnetoresistance (MR) which is defined as MR = 100% × [$\rho(B)/\rho(0)$-1], where $\rho(B)$ is the sample resistance at magnetic flux density B and $\rho(0)$ is the sample resistance at "zero" magnetic field. The resistivity and magnetoresistance was measured in the temperature range of 5-300 K using closed cycle helium gas cryo-cooler and a DC electromagnet up to 0.8 T magnetic field. The MR was also measured in permanent field up to 2.3 T at temperatures of 263–313 K. The investigations of the MR in high pulsed magnetic fields were performed in the temperature range of 78 – 300 K up to magnetic flux density of 20 T by using pulsed magnetic field generator based on capacitor bank discharge through a coil (for more details of the setup see [7]).

The samples prepared from the LSMO films were treated by annealing at temperature T_a = 100 °C which was found optimal for accelerated ageing of manganite films [4], [5]. For the investigation of the effectiveness of accelerated ageing, the two methods of sample treatment were tested. The samples were split into two groups: I and II. To test the first method, the LSMO samples of group I with uncovered film surface were annealed for total duration of 24 h in Ar atmosphere at temperature T_a = 100 °C. To test the second method, prior to annealing the film's surface of the samples from the group II was coated with a special polymer – hot melt adhesive. Afterwards coated samples were annealed for total duration of 9 h in Ar atmosphere at temperature T_a.

In order to investigate resistivity (ρ) and magnetoresistance (MR) changes during accelerated ageing procedures, the samples at certain time intervals were slowly (during ~30 min) cooled down to the room temperature and measurements of ρ and MR were performed.

III. Rezults and Discussion

A. The accelerated ageing influence on the resistivity

Fig. 1 presents the resistivity *vs.* temperature dependences of LSMO samples of group I (uncovered) measured after different time intervals of accelerated ageing. One can see that the resistivity change is the greatest during the first 8 hours of annealing and then it is slowed down. The stability of the resistance was achieved after annealing for 24 h. For the LSMO samples of group II (covered), the smaller changes of resistivity during annealing were obtained after shorter annealing times.

The relative resistivity changes $(\rho_t - \rho_0)/\rho_0$ during the annealing of both sample groups measured a 293 K are presented in Fig. 2. Here ρ_0 is the initial value of the resistivity measured before accelerated ageing, ρ_t – resistivity value measured after a certain different time t. The kinetics of the resistivity changes was analyzed by using normalized relaxation function expressed as 'stretched' exponent [8], [9]:

$$(\rho_t - \rho_0) / \rho_0 = [(\rho_\infty - \rho_0) / \rho_0] \times \{1\text{-}\exp[-(t/\tau)^k]\}, \quad (1)$$

where τ is characteristic time, k – stretched exponent ratio and ρ_∞ – ultimate resistivity after the accelerated ageing (when the $t \to \infty$). The fitting parameters for group I were found as follows: $(\rho_\infty - \rho_0)/\rho_0$ = 0.296, τ = 6.9 h and k = 0.85; for group II: $(\rho_\infty - \rho_0)/\rho_0$ = 0.182, τ = 2.4 h and k = 0.85. As it was shown in [4]-[5], [8]-[9], the process described by this kind of

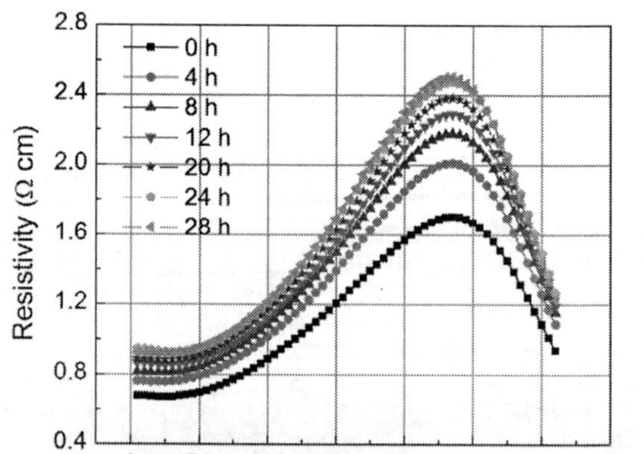

Fig. 1. The resistivity vs. temperature dependences after different annealing duration of nanostructured LSMO films of group I during accelerated ageing process.

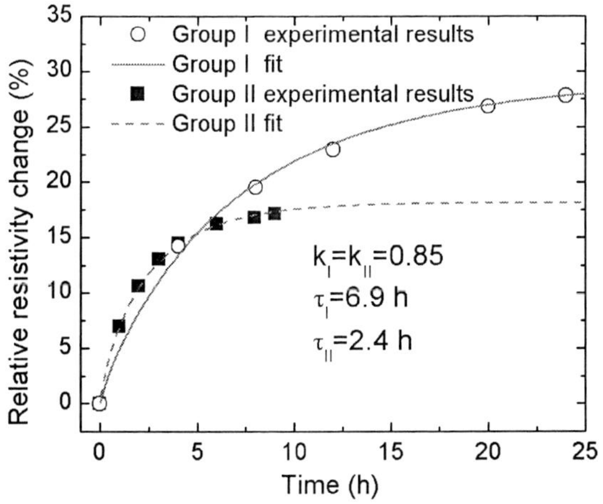

Fig. 2. The kinetics of relative change of resistivity measured at temperature 293 K during accelerated ageing process of LSMO samples of groups I and II. Points—experimental results, curves—fitting to Eq.(1).

function, is typical for degradation in topologically-disordered systems. The investigated nanostructured LSMO films consist of crystallites (grains) having almost perfect structure and intergranular areas (grain boundaries, GBs), with high number of various defects, created during film deposition process. It was shown that the mechanism of relaxation depends on the k parameter [8]-[9]. For $k \geq 0.6$, most likely, that the hierarchically limited relaxation mechanism takes place [8]. Within this model, each individual relaxation act in disordered solid is possible after previous relaxation events.

The obtained fitting results show that for both groups of the samples the relaxation of resistance during the ageing takes place according to the same hierarchically limited mechanism (k = 0.85). Oxygen plays the main role in the resistivity increase during accelerated ageing process, since it takes part in a double exchange mechanism between manganese ions: Mn^{+3}-O-Mn^{+4}. During annealing of the samples at 100 °C only oxygen bonded with low binding energies in the disordered grain boundaries and having higher diffusion coefficient [4] can be released, what courses the resistivity increase in nanostructured manganites.

The oxygen release velocity depends not only on the diffusion coefficient, but also on the conditions on the surface of the film. As can be seen from Fig. 2, the samples covered by special polymer (group II) exhibit saturation of resistivity change after significantly shorter annealing time than samples with uncovered surface (group I). Characteristic times τ obtained by fitting to Eq. (1) are as follows: τ_I=6.9 h for group I, τ_{II} =2.4 h for group II. The apparent τ difference between groups indicates that polymer layer acts as a buffer, preventing the oxygen from leaving the film, and causing reach of steady-state within shorter annealing time. On the other hand, since polymer contains polyethylene as a base and some additive copolymer which contains carbonyl or hydroxyl groups, the oxygen can be released from the polymer and introduced into the GBs of the film, causing decrease of resistivity during long-term ageing [4].

B. *The accelerated ageing influence on the magnetoresistance*

The results of the investigation of accelerated ageing influence on the magnetoresistance measured up to magnetic flux density of 20 T are presented in Fig. 3 (group I) and its inset (group II). One can see that, over the whole range of magnetic field, the *MR* change for both groups of the samples is insignificant (<1 %), in the range of measurement error, during all accelerating ageing process.

The obtained result can be explained with double exchange mechanism of manganites electrical transport. The manganites exhibit spin-polarized electron transport; this suggests that an electron transfer from Mn^{3+} to Mn^{4+} ion is realized via an intermediate oxygen ion (double exchange model). If the spins of the Mn ions are not parallel, the electron transfer becomes more difficult and the mobility decreases. The magnetic field aligns the core spins and, therefore, decreases the resistivity.

The annealing of the films at relatively low temperatures (100 °C) activates only weekly bonded oxygen depletion and its diffusion in the grain boundaries. Such processes reduce number of Mn^{3+}-O-Mn^{4+} bonds taking part in the double exchange transport mechanism of manganite films. As a result, the resistivity increases, but the ratio $\rho(B)/\rho(0)$ remains the same and the magnetoresistance changes insignificantly.

The obtained stability of magnetoresistance in all magnetic field range up to 20 T, when resistance changes during accelerated ageing process, is promising result for the development of high magnetic field sensors.

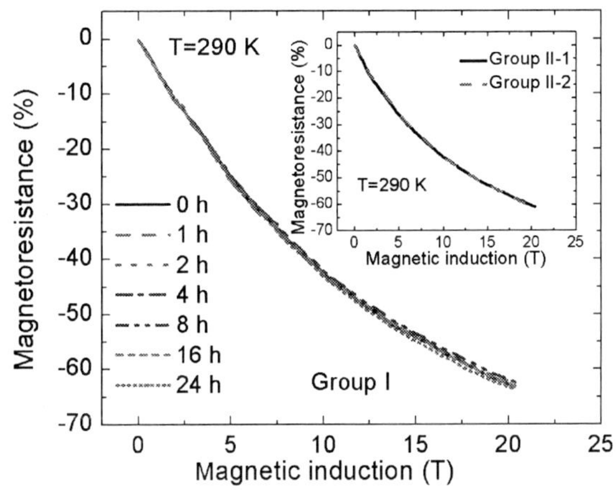

Fig. 3. Magnetoresistance change of samples of groups I and II in magnetic field 0 – 20 T measured at 290 K temperature during accelerated ageing.
Group II-1 – after coating with polymer, Group II-2 - after accelerated ageing during 9 h.

IV. Conclusions

Annealing in inert atmosphere of the LSMO samples used for production of CMR-B-scalar magnetic field sensors, at temperatures higher than the operating temperature accelerate the ageing process of nanostructured manganite films. The stabilization of resistivity when annealing the covered samples (group II) is faster in comparison with thermal treatment of uncovered samples (group I). Despite the resistivity increase, the magnetoresistance remains stable up to high magnetic fields (20 T) during the both accelerating ageing procedures. Therefore, the CMR-B-scalar magnetic field sensors can be produced with sufficiently lower resistance drift (< 1% of value per year) and stable sensitivity (magnetoresistance).

Acknowledgment

This work was partly supported by the Research Council of Lithuania (grant No. MIP-095/2015).

References

[1] S. Balevičius, N. Žurauskienė, V. Stankevič, S. Keršulis, V. Plaušinaitienė, A. Abrutis, S. Zherlitsyn, T. Herrmannsdorfer, J. Wosnitza, and F. Wolff-Fabris, „Nanostructured Thin Manganite Films in Megagauss Magnetic Field," *Appl. Phys. Lett.*, vol. 101, pp. 092407 (4), Aug. 2012.

[2] T. L.Haran, R.B. Hoffman, and S.E. Lane, "Diagnostic capabilities for Electromagnetic Railguns," *IEEE Trans. Plasma Sci.* , vol. 41, no. 5, pp. 1526-1532, May 2013.

[3] T. Stankevič, L. Medišauskas, V. Stankevič, S. Balevičius, N. Žurauskienė, O. Liebfried, and M. Schneider, "Pulsed magnetic field measurement system based on colossal magnetoresistance-B-scalar sensors for railgun investigation," *Rev. Sci. Instrum.*, vol 85, pp. 044704, Apr. 2014.

[4] V. Stankevič, Č. Šimkevičius, S. Keršulis, S. Balevičius, N. Žurauskienė, D. Pavilonis, and S. Tolvaišienė, "Improvement in the long-term stability of parameters of encapsulated magnetic field sensors based on La-Sr-Mn-O thin films," *Sensors Actuators*, vol. A228, pp.112-117, March 2015.

[5] Č. Šimkevičius, S. Keršulis, V. Stankevič, D. Pavilonis, N. Žurauskienė, S. Balevičius, and S. Tolvaišienė, "Accelerated ageing effects in nanostructured $La_{0.83}Sr_{0.17}MnO_3$ films," *Thin Solid Films*, vol. 589, pp. 331-337, May 2015.

[6] A. Abrutis, V. Plausinaitiene, V. Kubilius, A. Teiserskis, Z. Saltyte, R. Butkute, and J.P. Senateur, "Magnetoresistant $La_{1-x}Sr_xMnO_3$ films by pulsed injection metal organic chemical vapor deposition: effect of deposition conditions, substrate material and film thickness," *Thin Solid Films*, vol. 413, pp. 32-40, June 2002.

[7] J. Novickij, S. Balevičius, N. Žurauskienė, R. Kačianauskas, V. Stankevič, Č. Šimkevičius, S. Keršulis, and S. Bartkevičius, "Vilnius High Magnetic Field Centre facilities," *J. Low Temp. Phys.*, vol. 159, no. 1-2 1, p. 406-409, Jan. 2010.

[8] J.C. Phillips, "Stretched exponential relaxation in molecular and electronic glasses," *Rep. Prog. Phys.*, vol.59, pp. 1133-1207, Sep.1996.

[9] M.M. Vakiv, O.I.Shpotyuk , V.O.Balitska. B.Butkiewicz, L.I.Shpotyuk, Ageing behavior of electrical resistance in manganite NTC ceramics, J. Eur. Ceram. Soc. **24** (2004) 1243-1246.

Warm X-ray Source Development on the Saturn Accelerator

B. V. Oliver, V. Harper-Slaboszewicz, K. Struve, B. Ulmen
Sandia National Laboratories
Albuquerque, NM

***Abstract*— There is a continuing long-term need to develop high fluence, short pulse (~40 ns) warm x-ray sources with photon energies in the ~100 keV range for electronics and materials testing. The Saturn high current accelerator at Sandia National Laboratories has been used extensively for driving high current (8 MA, 0.3-1.7 MeV) electron beam bremsstrahlung diodes for such testing. The primary diode fielded on Saturn is comprised of three concentric cylindrical cathode rings and a thin anode disk. The anode is separated from the ring cathodes by a small (0.5 cm) vacuum gap and acts both as the return current path for the emitted electrons and as a bremsstrahlung x-ray converter. In order to both increase x-ray fluence while simultaneously lowering the spectral energy of the x-rays we have been conducting research on new diode geometries. We will present advances in reflex triode development and the fielding of a multi-rod-pinch diode configuration and discuss the potential advantages of these sources over our standard 3-ring diode.**

Keywords—Saturn accelerator; pulsed power; reflex triode; multi rod-pinch diode

I. INTRODUCTION

The Saturn accelerator is a low impedance, very high current pulsed power driver [1]. It has been a workhorse for x-ray environment testing and experimentation since it began service in 1987. The current and voltage at peak are nominally 10 MA and 1.6 MV with a power pulse lasting 40 ns. The machine is axially symmetric with 36 parallel lines converging on a central vacuum stack, with each line providing about 50 kJ of forward-going power to the stack.

II. SATURN PULSED POWER

Each line on Saturn (Fig. 1) begins with a Marx bank storing 150 kJ at 2.7 MV. Upon triggering, the capacitor bank discharges into a 19 nF intermediate store water capacitor until a gas switch is electrically triggered. The energy flows into two parallel 2 Ω transmission lines until the voltage reaches approximately 2.3 MV at which point 20 parallel water switches self-break, transfering the power into the output 2 Ω section of the pulse forming line (PFL). This discharges through a prepulse suppression rail gap into the vertical triplate transmission line. This 3 m long, 2 Ω line tapers to 4 Ω and terminates in two 8 Ω rods. Each rod connects to an 0.5 Ω radial disk feed formed by 36 adjacent "water bottles" attached to each electrode ring of the central vacuum stack.

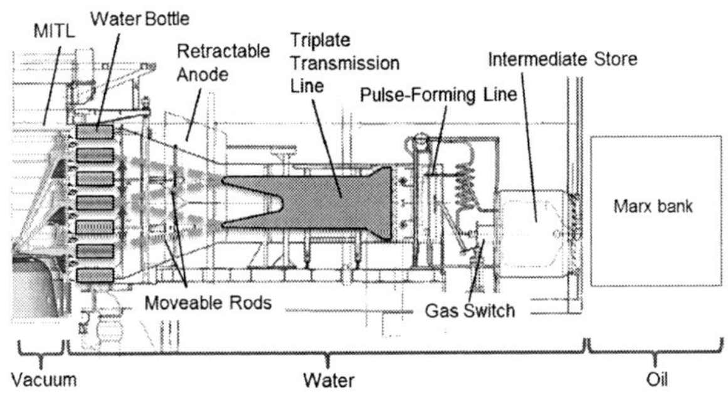

Fig. 1. Schematic of a single power flow line showing the energy storage, pulse compression and power flow sections of the machine. Each of the rods (dotted lines) can be attached to one of two cathode rings.

The disc feed is 18.6 ns long, so the vacuum section is transit time isolated from remainder of the water section. Inside the vacuum stack (Fig. 2), seven nested conical anodes and cathodes form six magnetically insulated transmission lines (MITLs) that converge to the load. The design impedance of the top two MITLs is 6 Ω, the middle two MITLs is 3 Ω, and the bottom two MITLs is 2 Ω. These parallel MITLs provide a reduced impedance to allow high currents at low voltage.

Saturn is a very flexible current driver because rods in the water section can be moved to distribute the current to a desired ratio between the upper, middle and lower cathodes. Up to 36 of the total 72 rods can be connected to each cathode ring. This has allowed a large variety of loads to be fielded, most recently x-ray sources based on a reflex triode and a multi rod-pinch diode.

III. BASELINE BREMSSTRAHLUNG DIODE

The standard x-ray source utilized on Saturn is a bremsstrahlung diode consisting of three nested triaxial diodes. The baseline pulsed power configuration for this source is to drive the bottom cathode with 36 rods, the middle cathode with 24 rods, and the top cathode with 12 rods. The intent is to achieve an even distribution of linear current density on the cathode rings, which have a ratio of radii of 3:2:1 from outer to inner. The gaps on the MITLs are set to achieve balanced impedances to support this current distribution.

Sandia National Laboratories is a multi-mission laboratory managed and operated by Sandia Corporation, a wholly owned subsidiary of Lockheed Martin Corporation, for the U.S. Department of Energy's National Nuclear Security Administration under contract DE-AC04-94AL85000. SAND NO. 2016-8949 C

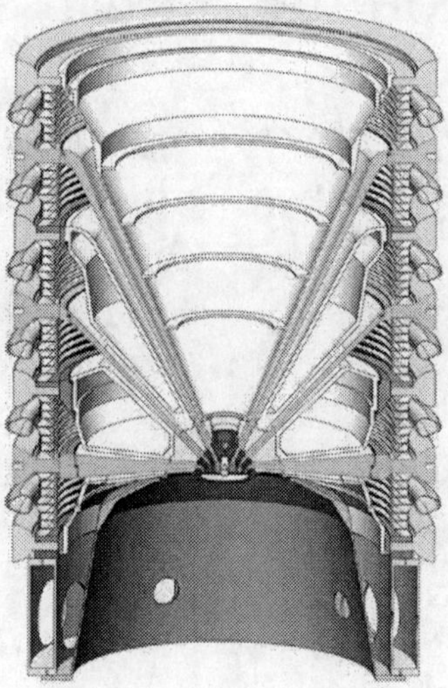

Fig. 2. Saturn vacuum stack showing the MITL configuration of the baseline bremsstrahlung diode. The MITLs have several sections indicated by the white, yellow and red colored hardware.

The MITLs are joined from three bolted sections with compressed current contacts. This makes servicing the MITLs easier as well as allowing for changes to load power flow with only minimal hardware changes. Fig. 3 shows the hardware configuration of the 3-ring bremsstrahlung diode. The only hardware that requires replacement after each shot are the anode and cathode tips shown in white, the tantalum converter, and the aluminum beam stop.

The compact geometry and high radiation dose rate make fielding current and voltage diagnostics within the diode impossible, so evaluation of diode performance must rely on other diagnostics. One diagnostic particularly useful for characterizing the performance of x-ray sources on Saturn is a fast framing camera which indirectly images the x-ray source region using a fast scintillator and an 8-frame segmented, gated image intensifier read out using a single large format CCD camera [3]. The maximum frame rate of the camera is 500 MHz with zero interframe time. This allows evaluation of the relative timing and intensity of the radiation produced by each of the three rings, which in turn can be used to evaluate the relative current and voltage.

Fig. 3. A sectioned view of Saturn's baseline 3-ring bremsstrahlung diode x-ray source. The cathodes are shown in green. The tip hardware replaced for each shot is shown in white.

Fig. 4. Saturn fast framing camera showing the entrance pinhole, the radiation shielding of the optical path, and the screen box containing the imaging system.

A. Operating parameters

The operating parameters of this load are deduced from the currents measured in the vacuum section on the lower anode and upper anode, and the voltages measured in the radial disc feed in the water section on the corresponding lines. The voltage is time-shifted and inductively corrected using the current signal to give the voltage and current at the current monitor, which is close to the load. Based on this approach, near peak power, the operating impedance of the outer diode is about 0.25 Ω and the inner diode is about 0.6 Ω, significantly undermatched to the MITLs.

IV. Reflex Triode Development

The bremsstrahlung spectrum produced by the baseline 3-ring diode has the majority of its energy content at high energies. Warm x-ray sources are desirable for many applications and this has spurred the development of the reflex triode, which produces a bremsstrahlung spectrum of much lower endpoint energy with relatively higher warm x-ray spectral content.

The reflex triode consists of a thin radiator foil anode between a pair of cathodes shown conceptually in Fig. 5. Electrons emitting from the cathodes make multiple passes through the foil, converting their energy into bremsstrahlung x-rays. The design allows for a larger fraction of low energy photons to escape the converter.

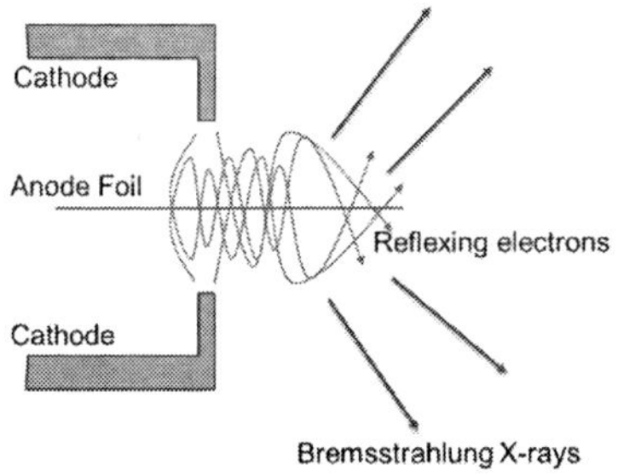

Fig. 5. Conceptual diagram of a reflex triode.

A. Positive polarity power flow

In order to operate a reflex triode on Saturn, the machine must operate in positive polarity. Normally, the anodes are grounded through the water section and the cathodes transmit the negative polarity power pulse to the diode. For a reflex triode however, the anodes must be driven positive relative to the grounded cathodes. This is not straightforward on Saturn; the Marx bank cannot be charged positive without significant reconfiguration.

The solution is to use the large dimensions of the machine, the relatively long time scales of the forward going power pulse to transit time isolate the anodes from the water section ground. The upper and lower anodes are removed, and the upper and lower cathodes are connected directly to machine ground creating a large ballast inductor (~250 nH). The power pulse sees this large inductance and that drives the voltage on the anodes positive during the time that power pulse takes to go out to the water section and reflect. The positive polarity on the two anodes in turn drives the center cathode to ground.

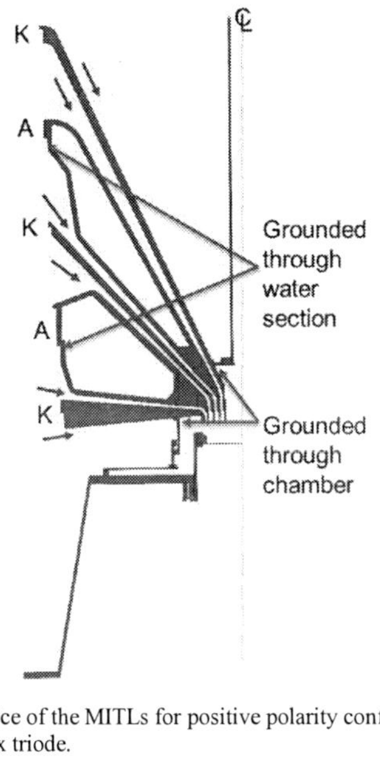

Fig. 6. Section slice of the MITLs for positive polarity configuration on Saturn used for the reflex triode.

This configuration was conceptually validated by models before being implemented on the machine. For short pulses, this mode does result in reduction in available current to the load by about one third since the power on two of the six transmission lines reflects back to the water section. This power is considered lost since the radiation pulse has ended by the time it comes back to the load.

B. LANTERN source

The Large Area Nested Triode with Electron Reflexing eNhancement source is an implementation of the reflex triode on Saturn consisting of two nested cylindrical reflex triodes. The anode foils are wrapped around the anode tips and hang down through a gap in the cathodes. This is shown in detail in Fig. 7.

There are several advantages to this configuration on Saturn. First, the hardware is not significantly different than the standard 3-ring bremsstrahlung diode. Only minor modifications were made to the innermost section of the MITL meaning that power flow to the tips will be similar. The 13 μm tantalum foil is conveniently available in strips which are easy wrapped around the anode tips. Additionally, the cylindrical geometry has a low susceptibility to premature shorting of the AK gaps. The framing camera image of this source (Fig. 9) shows that even late in the pulse both foils continue to radiate symmetrically.

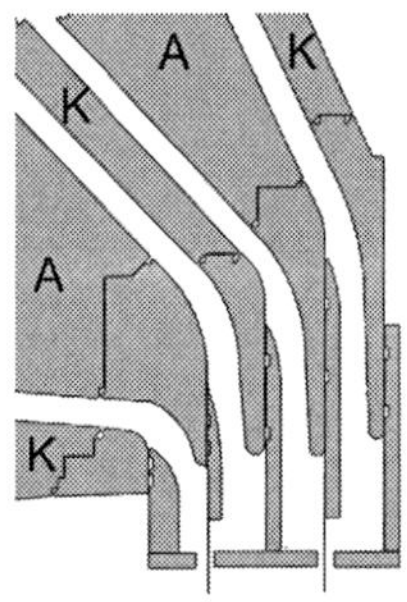

Fig. 7. Diagram of LANTERN showing the configuration of anodes and cathodes and the reflexing foils. This is a zoomed in portion of Fig. 6.

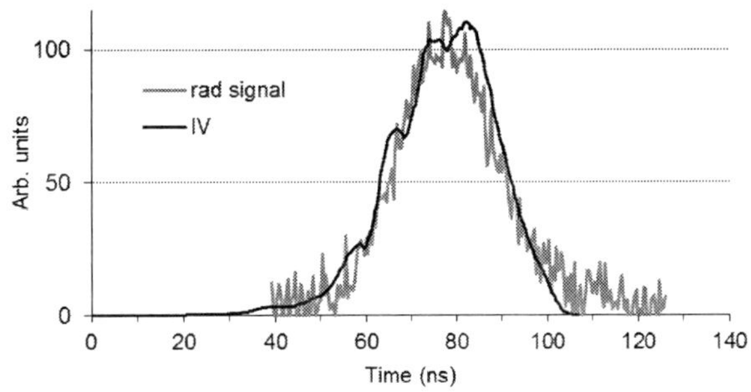

Fig. 8. The LANTERN source produces a radiation pulse that closely tracks the power source with a FWHM of approximately 28 ns.

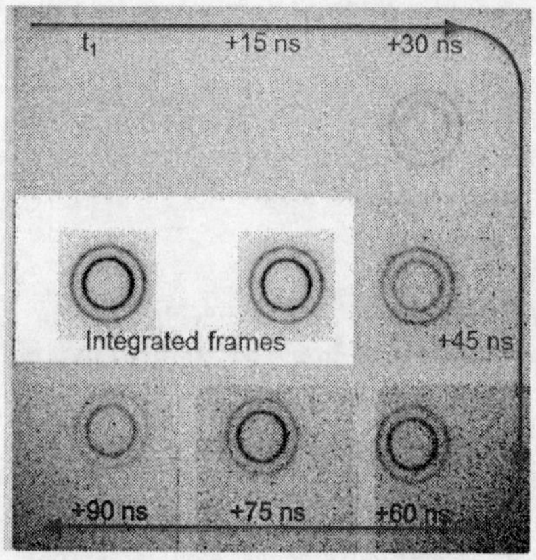

Fig. 9. Framing camera image of nested cylindrical reflex triode radiation source.

V. MULTI ROD-PINCH DIODE DEVELOPMENT

While the reflex triode performs well, there are several drivers that lead toward the development of another type of source, a massively parallel rod pinch array. The two big drivers for the new source are enhancing the warm x-ray yield and increasing the areal density. On the operational side, it is highly desirable to reduce the mechanical complexity and reduce refurbishment requirements of the hardware. This allows for reduced labor and faster machine turn-around times.

Building a higher flux warm x-ray source presents an immediate physics challenge, because the efficiency of bremsstrahlung sources goes down as electron energy goes down. Because the bremsstrahlung flux scales with the current, the key to increasing x-ray output is increasing the areal current density of the source. However, the current density of cylindrical intense electron beams across a vacuum gap is limited by pinching of the electrons in the self-magnetic field of the beam,

$$I_c = \alpha \left[\frac{2\pi m_e c}{e\mu_0}\right] \frac{R_c}{d} (\gamma_0^2 - 1)^{1/2}$$

where d is the gap, R_c is the cathode radius, and α is an empirical correction factor (~1.4) [4]. Experience shows that there are practical limits to the minimum gap, and as a result, for a given electron energy there is a maximum current per unit length that can be transported across the gap. In order to obtain a higher bremsstrahlung flux at a given electron energy, it is necessary to put more pinch length in the field of view of the test region. Saturn's baseline bremsstrahlung diode uses three concentric rings to increase this pinch length. However, increasing the number of rings introduces mechanical complexity, increasing the machine turn-around time. An alternate approach that we explored is a closely packed array of small aspect ratio rod pinches.

A. Initial testing on URSA Minor

The initial testing of the multi rod-pinch was performed using components of the Ursa Minor accelerator. The experiments used five LTD cavities and the central stalk to examine the pinch behavior of small aspect ratio rod pinches using solid tungsten rods of varying radius: 0.635 cm, 0.95 cm, and 1.27 cm [5]. It was noted that the rods sustained very little damage, and could be reused.

The next set of experiments constructed 4 rod and 10 rod, close-packed arrays of anode radius of 0.635 cm and cathode radius 1.27 cm [6]. The voltage and current were approximately 300 kV, 45 kA/rod on the 4 rod array and 150 kV, 15 kA/rod on the 10 rod array. No problems were found with current penetration to the center of the array and the pinch behavior between the two arrays was qualitatively similar. The pinch traveled out to the end of the rods in all cases. Pinhole imaging showed that the initial pinch is asymmetric in the A-K gap. The pinch symmetrizes as it moves toward the tip. These experiments gave confidence that the current would penetrate uniformly into the larger array planned for Saturn and that the Saturn pulsed power parameters would be sufficient to see pinching at the tips.

B. Diode simulation

Early simulation of the flow in a rod pinch array showed that the coupling of the magnetic fields produced by the multiple diodes would result in some current asymmetry around each rod. The result would be increased current density on the side of the rod toward the center of the array and reduced current density toward the edge of the array. In addition, the simulations showed the formation of magnetic nulls between the rods. Based on these simulation results, the cathode geometry was modified to confine electron emission to the desired area and to increase isolation between the rods. This modified geometry was implemented in all of the multi-rod experiments.

The small-scale multi-rod experiments using Ursa Minor showed the expected asymmetry around each rod, with increased current density on the side toward the array axis. However, they also showed that the pinch would symmetrize as it moved toward the tip of the rod.

An early concern in the development of the multi-rod array was that the current would not penetrate to the center of the array, however the simulations showed that the current on the center rods was virtually identical to that on the edge rods. Additionally, the small-scale experiments showed that the currents on all the rods were the same, to within the accuracy of the measurements.

C. Experiments on Saturn

Obtaining the required pulsed power drive on Saturn was a concern. The multi-rod-pinch array is most naturally driven by a single anode and cathode line, operated in positive polarity. However, the post-hole convolute developed to drive z-pinch sources on Saturn is at too small a radius to drive the multi-rod-pinch array. In addition, it operated in negative polarity. To obtain the required drive on Saturn, the pulsed power is configured for positive polarity operation with the lower cathode grounded through the chamber and the lower middle anode driving the anode plate. To avoid damaging the accelerator, these experiments were conducted with only some of the Marx generators charged, so the unpowered lines could absorb the reflected power from the mismatched load.

The multi-rod-pinch array deployed on Saturn consists of 51, 0.635 cm radius tungsten anode rods bolted to a stainless steel plate (Fig. 10). The stainless steel cathode plate has 1.27 cm radius apertures giving a nominal radial A-K gap of 0.32 cm. The accuracy of the gaps after installation in the machine was found to be 0.32 ± 0.005 cm. The voltage and current design values for 12 lines are 200 kV and 33 kA/rod; for 18 lines they are 400 kV and 50 kA/rod.

D. Saturn experimental results

Several observations from the Ursa Minor experiments were confirmed on the Saturn experiments. The anode and cathode hardware was reusable, with only slight damage to the anode rods and debris deposition on the cathode plate. Based on the appearance of similar damage to all anode rods and the uniformity of the framing camera image (shown in Fig. 11), the current penetration to the center of the array was generally uniform. The rod damage was greater on the side facing the axis of the array than on the side facing away from the axis.

Unlike the Ursa Minor experiments, the current did not pinch to the end of the rods, and appeared to be confined mostly to the region under the cathodes. This was a surprise, since the rod currents and voltages were designed to be similar to the Ursa Minor configuration. The reason for this failure to pinch is not known at this time. Limited data were collected on the Saturn experiments. The current monitors on the cathode feed plate failed, so the current reaching the diode area was not measured. Although there was no indication of a localized loss of current in the feed, it is possible the current reaching the array was substantially less than the design value. In addition, the Ursa Minor experiments used a longer current pulse of nominally 150 ns, while the Saturn current pulse was only 30 ns. It is possible that the pulse was not long enough to allow pinching to the end of the rod.

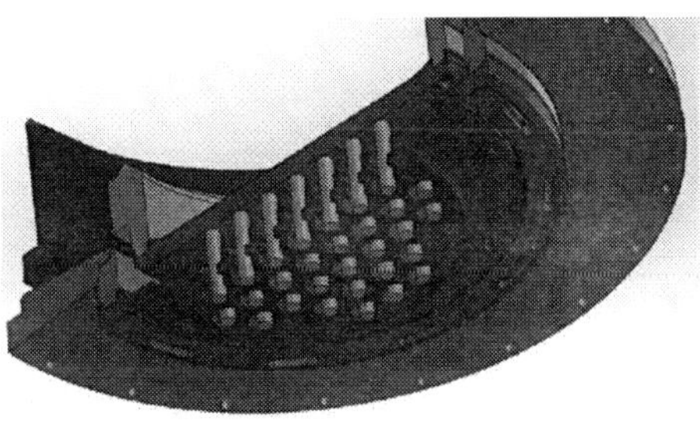

Fig. 10. Sectioned 3-D model of the Saturn multi rod-pinch experiment hardware.

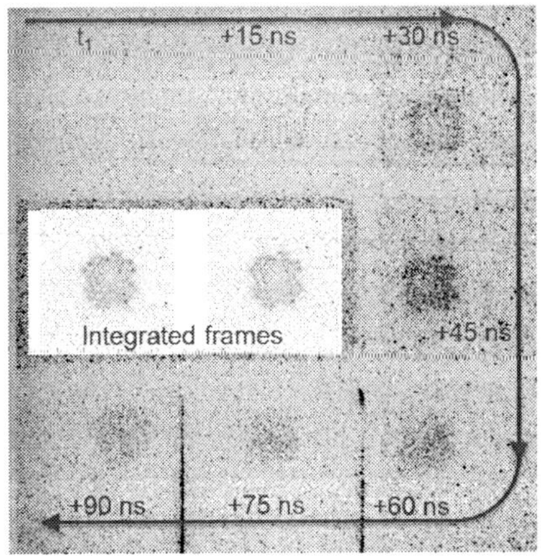

Fig. 11. Framing camera image of Saturn multi-rod-pinch radiation source showing good current penetration to the center of the array.

VI. CONCLUSION

Saturn is a very flexible pulsed power driver for short pulse bremsstrahlung radiation sources. It can be configured to drive up to half of the power to any of three cathode levels. By not charging some of the Marx generators, it can be operated at reduced power without having to adjust any switches. With some loss of power, it can be configured to operate in positive polarity without the use of a posthole convolute. It can be operated in short pulse (30 ns) or long pulse (110 ns) modes. With the implementation of "combo rods", two rods connected to a single water bottle, it will be possible to drive all of the power to a single cathode without the use of a vacuum convolute.

Saturn's flexibility has been exploited to drive a variety of bremsstrahlung radiation sources requiring different pulsed power configurations, including nested ring diodes, nested cylindrical reflex triodes, and a large multi-rod-pinch array. It is anticipated that Saturn will continue to be adapted to drive other radiation loads in the future.

Acknowledgment

We acknowledge input from B. Weber and others at the Naval Research Laboratory, Washington D.C. on the reflex triode implementation on Saturn. We also acknowledge the excellent LSP simulation work performed by Voss Scientific, LLC, Albuquerque, NM.

References

[1] D.D. Bloomquist, R. W. Stinnett, D. H. McDaniel, J. R. Lee, A. W. Sharpe, J. A. Halbleib, L. G. Schlitt, P. W. Spence, and P. Corcoran. "Saturn, a large area X-ray simulation accelerator." In Proc. 6th IEEE Pulsed Power Conference, Arlington, VA, p. 310. 1987.

[2] J.D. Boyes, J.S. Cap, G.M. Douglas, T.L. Franklin, J.M. Hart, H.C. Ives, S.J. Robischon, B.L. Smith, W.O. Stoppkotte, and D.M. VanDeValde. "Engineering Design of the Saturn Accelerator" .In Proc. 6th IEEE Pulsed Power Conference, Arlington, VA, 1987.

[3] G. W. Smith, D. S. George, D. Harrison, S. Hill, R. J. Hohlfelder, V. Harper-Slaboszewicz, R. R. Gallegos, M. B. Ingle and P. Simpson, "Ultrafast imaging of nanosecond pulse x-ray simulators", Proc. SPIE 7126, 28th International Congress on High-Speed Imaging and Photonics, 712603 (February 10, 2009).

[4] Swanekamp, S. B., Commisso, R. J., Cooperstein, G., Ottinger, P. F., & Schumer, J. W. (2000). Particle-in-cell simulations of high-power cylindrical electron beam diodes. *Physics of Plasmas (1994-present)*, *7*(12), 5214-5222.

[5] V. J. Harper-Slaboszewicz, J. Leckbee, P. W. Lake, and A. L. McCourt. 2014. "Effect of Rod Material on the Impedance Behavior of Small Aspect Ratio Rod Pinches." IEEE Transactions on Plasma Science, 42(9), 2207-2212, (2014).

[6] V. J. Harper-Slaboszewicz, J. Leckbee, N. Bennett, E.A. Madrid, D. V. Rose, C. Thoma, D. R. Welch, P.W. Lake, and A.L. McCourt, ."Parallel Operation of Multiple Closely Spaced Small Aspect Ratio Rod Pinches". IEEE Transactions on Plasma Science, 43(1), 422-432 (2015).

A DEVELOPMENT OF A BIPOLAR PULSE GENERATOR USING MOS-FETS BASED LTD

Keisuke Kise (*Author*), Akira Tokuchi
Kusatsu Factory
Pulsed Power Japan Laboratory Ltd.
Kusatsu 525-0058, Japan
Email:kise@myppj.com

Weihua Jiang
Extreme Energy-Density Research Institute,
Nagaoka University of Technology,
Niigata 940-2188, Japan

***Abstract*— We are developing a bipolar pulse generator using MOS-FETs based LTD. A out put pulse with +/- 10 kVp, 300 ns each, 1 kHz was generated. Maximum output voltage of each LTD board is 1 kV. We used 15 LTD boards for positive pulse and 15 LTD boards for negative pulse.**

Keywords— Pulsed Power; LTD; Bipolar Pulse Generator;

I. INTRODUCTION

The pulsed power supply done based on LTD at present has been developed. The LTD substrate of the maximum output voltage is 1 kV, the positive pulse generator 15- stages, using a 15-stage negative pulse generator, the output voltage is±10 kVp, pulse width 300 ns, bipolar repetition was realized 1 kHz, the which is the content related to the development of the pulse generator.

II. ABOUT LTD POWER SOURCE

LTD is a combination of the induced voltage superimposing circuit and semiconductor switch, high voltage, a large current, the pulse power source capable of controlling waveform in the order of nanoseconds.

Unprecedented capable of generating a high voltage pulse of an arbitrary waveform in real time by combining a personal computer is a revolutionary power source system.

It amplifies the current by a substrate placed in parallel with the semiconductor switching element such as a lot of the FET and the IGBT, a power source for amplifying and outputting a voltage by stacking substrates in series.

Each substrate and the output terminal is a non-contact, the pulse power core mounted on the board, to form a pulse transformer. Not required to isolate the charging circuit is a feature of the LTD power source is difference from the MARX circuit. (A schematic circuit and features of the LTD power source shown in Figure 1.)

When you use the LTD power source, it can generate any of the high-voltage ultra-short pulse adjusted by 5ns unit. Thereby, it is possible to long-term stability to generate the plasma. (Figure 2).

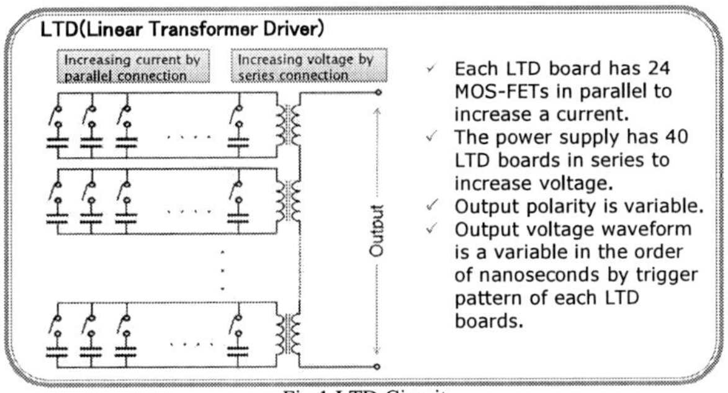

Fig.1 LTD Circuit

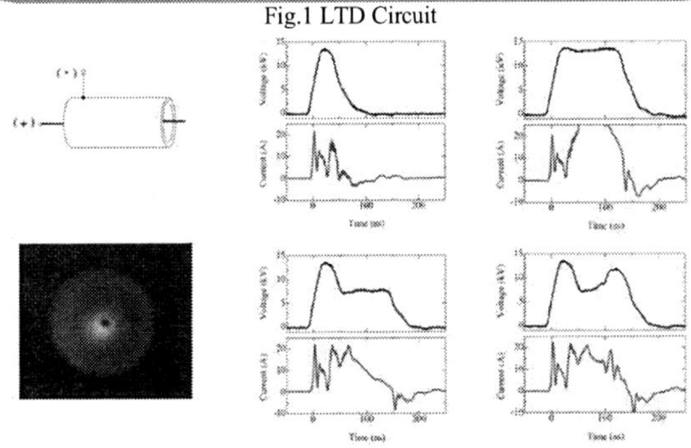

Fig3 Plasma generation test corrugation

We have developed LTD substrate of various specifications so far. The performance of the LTD board is shown in Table 1.

TABLE.1 LTD BOARDS PERFORMANCE LIST

Device Name	PPJ-FET-LTD-2	PPJ-FET-LTD-3	PPJ-FET-LTD-4	PPJ-IGBT-LTD-1
Vdss	1000 V	1000 V	1000 V	900 V
Idc	6 A	15 A	22 A	165 A
Ip	25 A	45 A	144 A	360 A
Turn on time	8.6 ns	10 ns	5 ns	103 ns
Devices	24	24	25	36
Board current	576 A	1,100 A	3,600 A	13,000 A
Board size	ϕ 300 mm	ϕ 300 mm	ϕ 360 mm	ϕ 360 mm

A LTD power supply can expect to be utilized for combustion efficiency improvement of an engine, exhaust gas treatment and an environment application study of effluent treatment.

III. BIPOLAR PULSED POWER SUPPLY

(1) Target specification

The output voltage ± 10 kV is output by pulse width 300 ns and repeat 1 kHz more than an output terminal. Rise time is done below 100 ns. A load is a capacitive, 100 pF. Target specification is indicated in Table 2.

TABLE.2 TARGET SPECIFICATION

Output Voltage	±10 kV
Pulse Width	300 ns
Repetition	1 kHz
Rise Time	100 ns

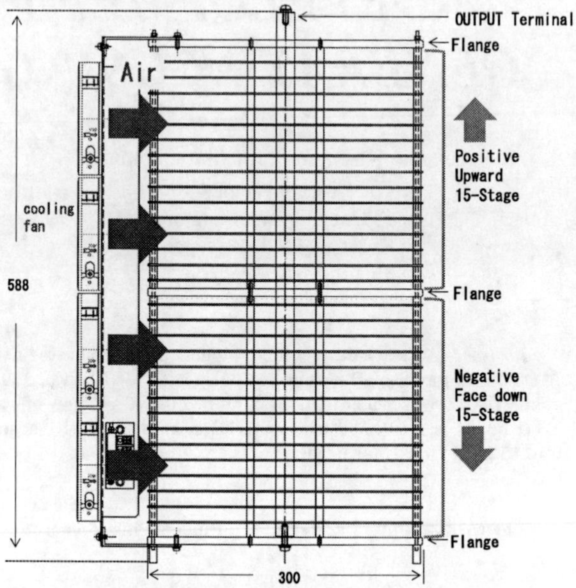

Fig.4 Bipolar pulsed power supply assembly drawing

Fig.5 photographs of a Bipolar pulsed power supply

A LTD substrate uses 30 of PPJ-FET-LTD-3 for which 24 of FET(1kV,15A) was used.

A pulsed power core is using thickness 13.5mm because a pulse width is long when it's compared with 300ns and the past's items.

A charger used 2 power supplies of DC800V (1kW) output, and made sure that charging voltage can be controlled every respective polarity.

I charge 15-stages of LTD substrate by 720V, and 10kV is output more than an output terminal.

15-stages of negative electrode is arranged in the bottom more than a middle flange and 15-stage positive poles are being arranged in the top more than a middle flange. A high-pressure output terminal is a point of a central stick.

An assembly drawing of the power supply body, an outward appearance picture is indicated in Figure4 in Figure 5.

A diode is mounted on a LTD substrate, and when it tries to output fellow's polarity, an electric current can't be stopped and output any more a pulse by a diode of a substrate of reverse polarity.

27.5 Ω of resistance is mounted in parallel with a diode, it was done. An electric current process of reverse polarity is made, bipolarity, a pulsed output was achieved.

A circuit diagram is indicated in Fig6.

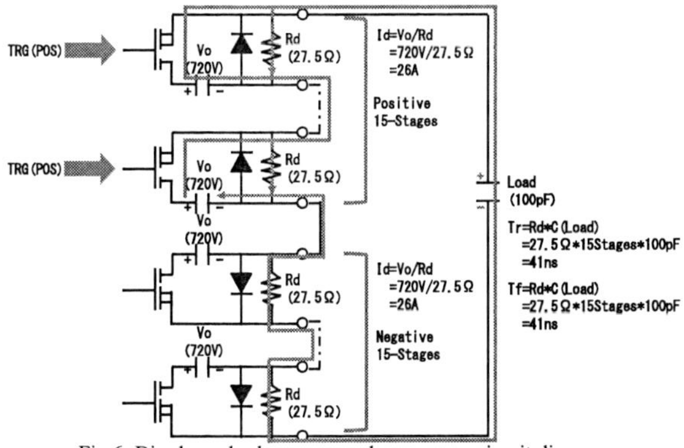

Fig.6 Bipolar pulsed power supply summary circuit diagram

IV. Experimental Result

Test outcome by an imitation load of 100pF is indicated below. Pulse width 300ns and repeat 1kHz were input more than charging voltage 720 V and a function generator, and it was confirmed that it's the output voltage ± 10 kV.

Measurement of output voltage waveform of an output terminal was put into effect by a high-voltage probe.

Output voltage waveform is indicated in Figure7 and test outcome is indicated in Tabele3.

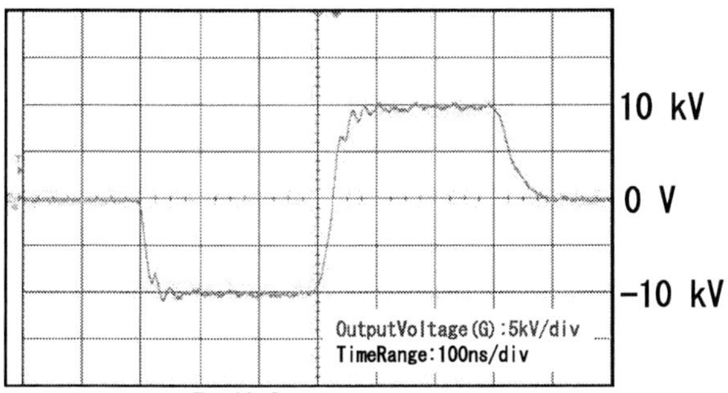

Fig.12 Output voltage waveform

Table.3 Test result

Output Voltage	+10 kV -10 kV
Pulse Width	300 ns
Repetition	1 kHz
Rise Time	25 ns

V. Conclusions

Industry, this available power supply, in real load, it's connected, and, I'm estimating.

It was possible to be to use a LTD substrate of our original development this time and achieve compact and high-performance high-voltage bipolar pulsed power supply.

It's expected to put an experimental production and a test of further high efficiency and high repetition into effect during this fiscal year.

I'd like to make the most suitable power supply with the equipment which are environment application studies by performance improvement and make an effort toward the spread of a LTD power supply..

References

[1] Weihua Jiang "Repetitive Linear Transformer Driver Using Power MOSFETs"; IEEE Transactions on Plasma Science (Volume:40 , Issue: 10)

Development of SOS type High-speed High-voltage Pulse Generator

Yo Sawamura (*Author*), Akira Tokuchi
Kusatu Factory
Pulsed Power Japan Laboratory Ltd.
Kusatsu, Japan

Weihua Jiang, Taichi Sugai
Extreme Energy-Density Research Institute
Nagaoka University of Technology
Nagaoka, Japan

***Abstract*—Effective radical generation by streamer discharge are required pulse power supply of a high voltage rises as steep as possible. In order to improve the decomposition speed is required to generate a high voltage pulse at a high repetition. By combining a plurality of semiconductor switches, improve efficiency, and developed an optimum high-voltage pulse power source to streamer discharge. Further, we developed a power supply with variable output voltage control in the high repetition burst mode.**

Keywords—Pulsed power; Semiconductor opening switch (SOS); High-voltage pulse generator; Streamer discharge

I. Introduction

Exhaust gas, persistent environmental pollutants of the waste water is discharged from factories, power plants and automobiles, and streamer discharge method capable of degrading it effectively has been attracting attention recently. This presentation is a development of compact and high repetition ultra-high voltage pulsed power supply which is suitable and innovative for streamer discharge method.

- Improvement of combustion efficiency by plasma ignition
- Removal of gaseous environmental pollutants

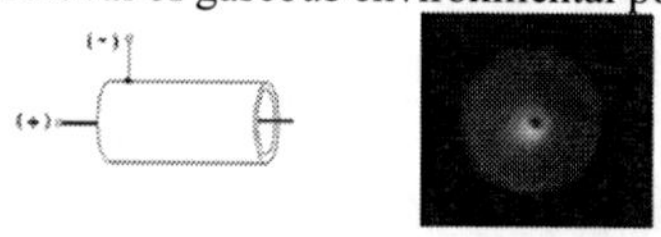

Fig. 1. Gass Discharge for Removal of gaseous environmental pollutants.

- Water treatment [2]

Sample water;
Indigo carmine solution
Concentration 20 mg/L
Volume 3 L

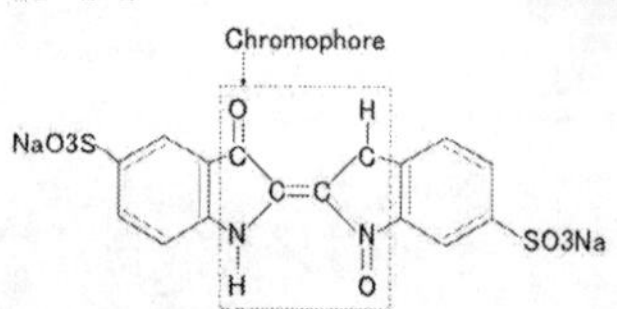

Fig. 2. Indigo carmine solution

Chromophore is decomposed by ozone. Benzene rings are decomposed by OH radical.

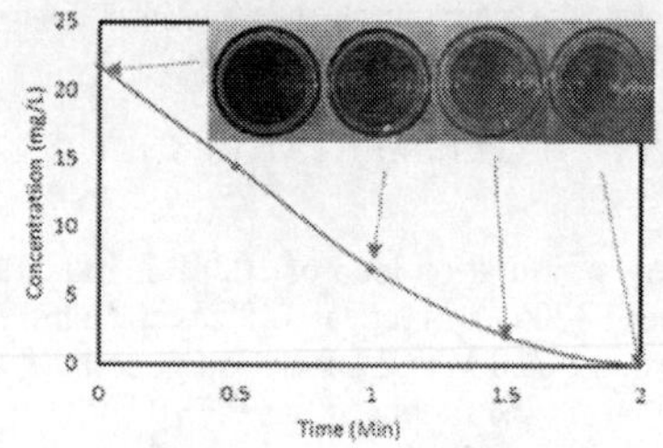

Fig. 3. Decreaseing curbe of Indigo carmin concentration

II. SOS type pulse generator

A. Operating principle of SOS-Diode

Followings are described operating principle of SOS-diode as shown with Fig. 4.

A) $MS^{+}ON \Rightarrow$ Cp charge with I^{+} SOS

B) Cp charged $\Rightarrow MS^{-}ON \Rightarrow I^{-}$ SOS

C) I^{-} SOS is several times larger than I^{+} SOS

D) I^{-} is interrupted by SOS, energy of L^{-} is transferred to load

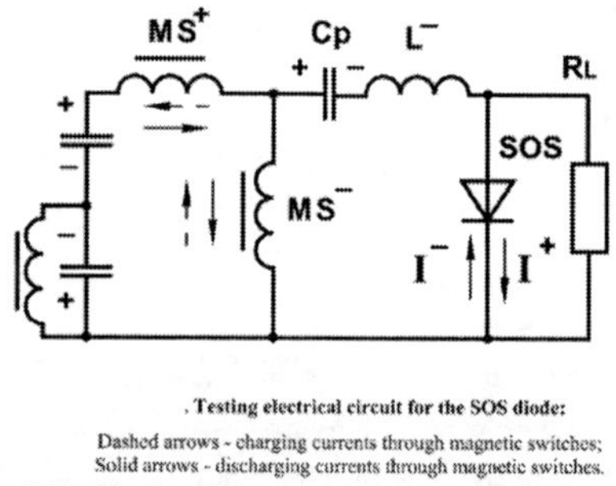

Fig. 4. General circuit for high-speed high-voltage pulse generator using SOS-Dioce.

B. Characteristic of SOS-Diode

Table I is Excerpt from test report of SOS-40-4 and SOS-60-4 by Electrophysics Institute Pulsed Power Lab. Those data shows peak current is 1.38kA in SOS-40-4, 1.36kA in SOS-60-4, and peak voltage is 45kV in SOS-40-4, 72kV in SOS-60-4.

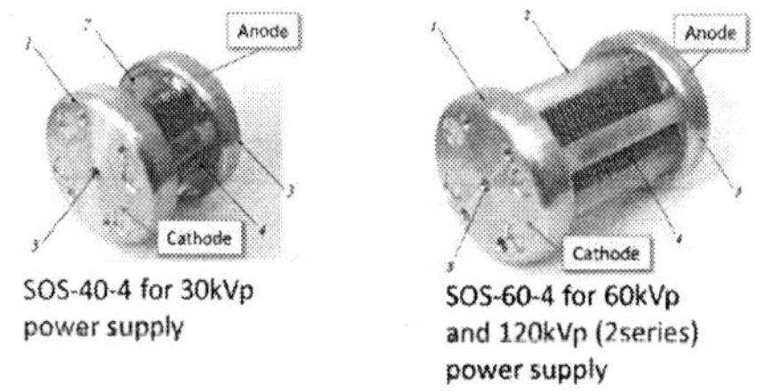

Fig. 5. SOS-Diode which are developed in Institute of Electro physics, Russian Academy of Science.[1].

By generating a high voltage pulse in the exhaust gas and wastewater, research to decompose these non-biodegradable substance by streamer discharge to generate a radical, has been promoted by some of the universities, research institutions and companies. PPJ has been using two types of SOS diode which were designed by Russian Institute, shown Fig. 5 and Table I in development of SOS power supply.

TABLE I. SPECIFICATION OF SOS-DIODES

Specification of SOS-diodes		
Parameter	***SOS-40-4 value***	***SOS-60-4 value***
Maximum peak reverse voltage	40kV	60kV
Pulse duration (FWHM)	10 - 60 ns	10 - 60 ns
Maximum forward pumping current	1 kA	1 kA
Maximum reverse (cutoff) current	4 kA	4 kA
Forward pumping time	150 - 400 ns	150 - 400 ns
Reverse pumping time	40 - 100 ns	40 - 100 ns
Current cutoff time	5 - 15 ns	5 - 15 ns
Maximum PRF / Cutoff current	Single pls. / 4 kA 100Hz / 4 kA 300Hz / 2 kA 1kHz / 1 kA 5kHz / 200 A	Single pls. / 4 kA 100Hz / 4 kA 300Hz / 2 kA 1kHz / 1 kA 5kHz / 200 A
Maximum dissipation power in transformer oil (10%burst mode)	70W (700 W)	100W (1 kW)
Dimensions	54mmΦ Length 47mm	54mmΦ Length 59mm
Stack mass	147 g	170 g

III. DEVELOPMENT OF VARIABLE OUTPUT VOLTAGE FUNCTION ON BURST MODE

In order to provide an optimum high-voltage pulse power source to streamer discharge, we developed a power supply with variable output voltage control in the high repetition burst mode. Based on an ordinary SOS power supply, new function was taken with adding a few devices. These are described in this section.

A. Ordinary SOS power supply diagram

Followings are described operating principle of ordinary SOS power supply as shown with Fig. 6.

A) SI-Thy ON ⇒ C_2 charge with I^+ SOS ⟶

B) C_2 charged ⇒ MS^-ON -> I^- SOS - - ->

C) I^- SOS is several times larger than I+ SOS

D) I^- is interrupted by SOS, energy of L^- is transferred to load ⟶

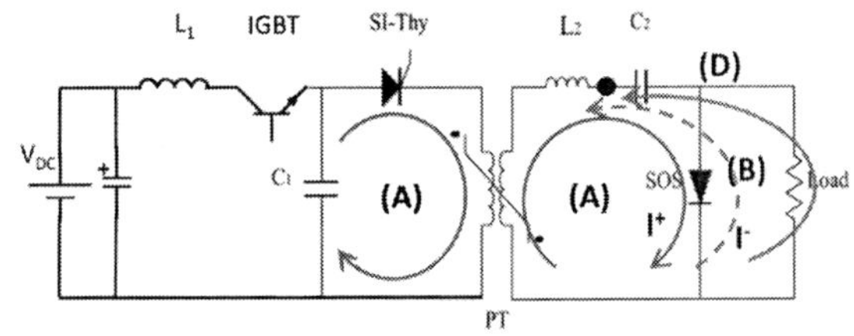

Fig. 6. Ordinary SOS power supply diagram.

B. SOS power supply with variable output voltage function diagram

Fig. 7 shows SOS power supply with variable output voltage function diagram.

1) Addition of regenerative diode

2) Exchanging IGBT and L_1

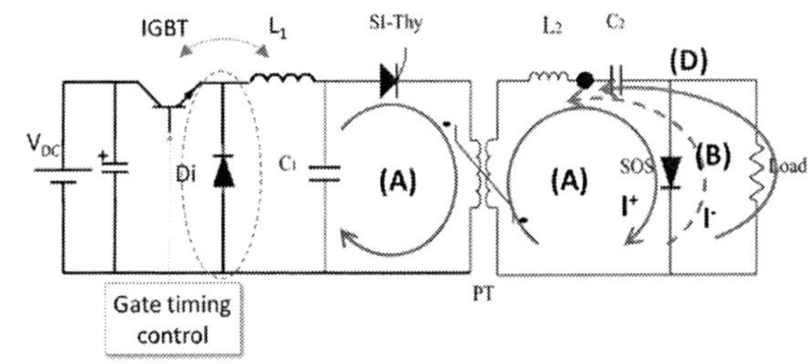

Fig. 7. SOS power supply with variable output voltage function diagram.

To support variable output voltage, change the C1 charged voltage by IGBT gate timing control.

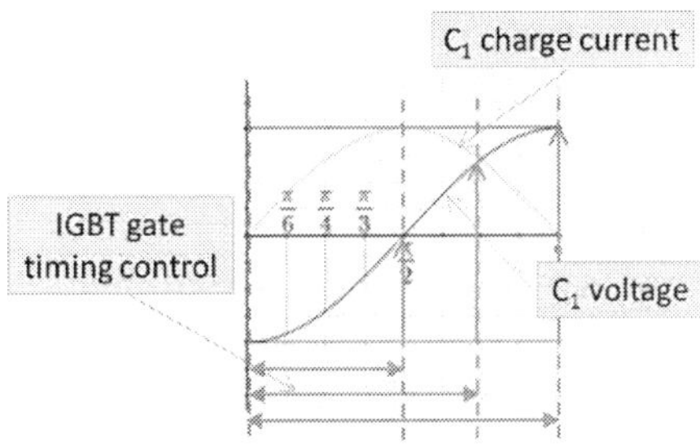

Fig. 8. IGBT gete timing control diagram.

C. IGBT gate is generated by FPGA and CPU on control box and Pulse Pattern Maker (PC application)

To control IGBT gate timing precisely, we developed control box and tool application running on Windows PC shown in Fig. 9.

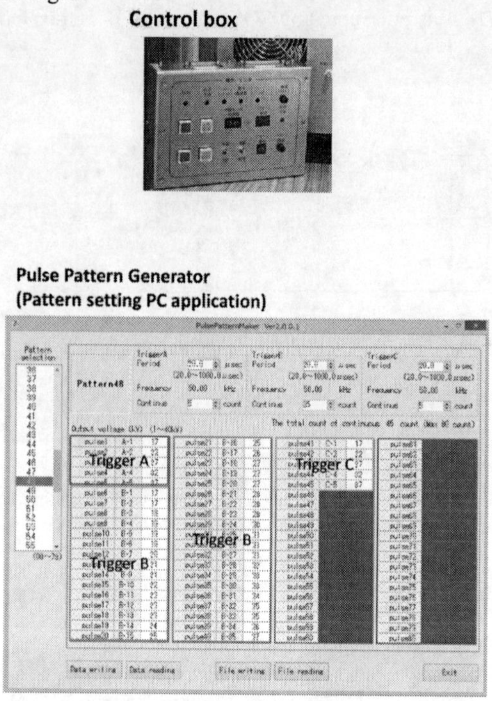

Fig. 9. IGBT gete timing control box and tool application.

- Maximum 80 burst pulses are set by PC application
- Trigger can be set from A, B, C trigger input

D. Output waveform of variable output voltage

Fig. 10 shows variable output voltage waveform of developed SOS type High-speed High-voltage Pulse Generator equipped with voltage controlled 50 kHz high rate burst mode. Output voltage is changing according with programed sequence on Windows PC application in advance.

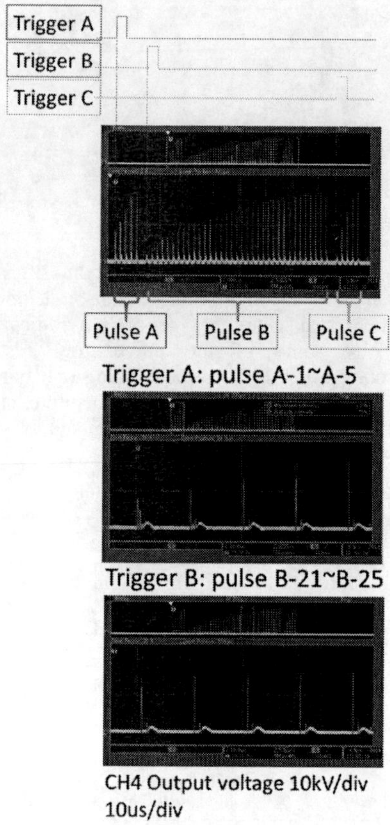

Fig. 10. Outut waveform of variable output voltage.

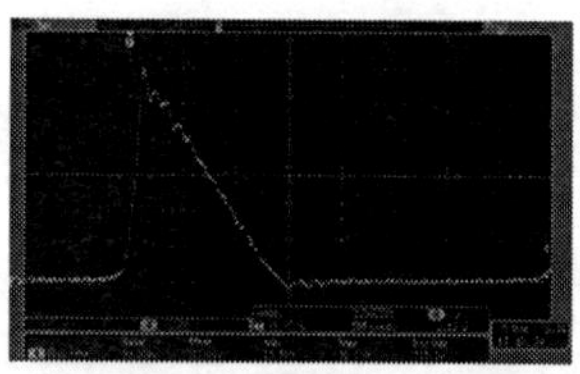

CH4 Output voltage 10kV/div

100ns/div

Fig. 11. Outut waveform of 60kV output pulse.

IV. PRODUCT REOADMAP OF SOS TYPE POWER SUPPLY

PPJ has been developing High-speed High-voltage Pulse Generators using SOS-Diode in over the last few years which are shown in Fig. 12 and Fig. 13.

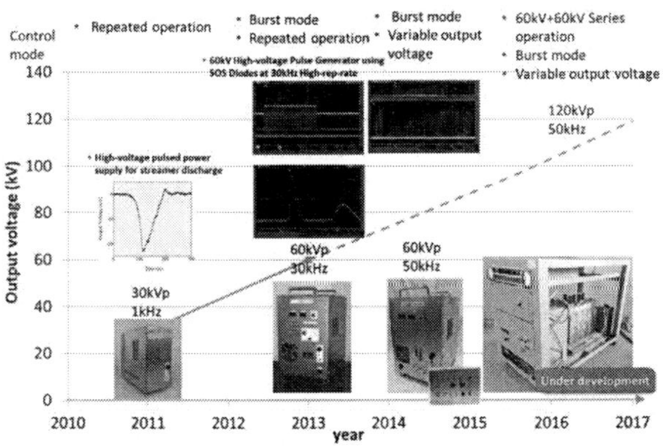

Fig. 12. PPJ's developmet roadmap for high-speed high-voltage pulse generator using SOS-diode.

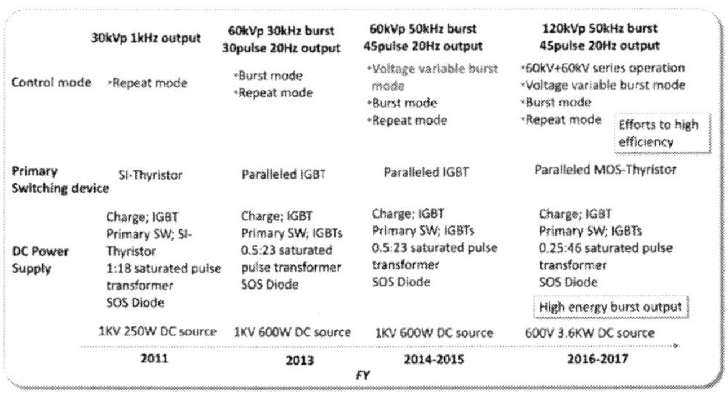

Fig. 13. PPJ's development for SOS pulse power products.

Fig. 14 shows PPJ's latest developing product. We are targeting to obtain 120kV 50kHz output in this SOS power supply. Main feature is as below.

- 60kV+60kV series operation
- Voltage variable burst mode
- Burst mode
- Repeat mode

Acknowledgment

SOS type High-speed High-voltage Pulse Generator equipped with voltage controlled 50 kHz high rate burst mode has been developed with Extreme Energy-Density Research Institute of Nagaoka University of Technology. Thanks them for developing.

Furthermore we will keep developing SOS type power supply targeting with higher voltage and functions along with requirement of market. Fig.13 shows PPJ's development for SOS pulse power products.

Fig. 14. PPJ's latest developing product (undder development).

References

[1] Institute of Electro physics, Russian Academy of Science http://eng.iep.uran.ru/

50 kV Solid-state Switch using Static Induction Thyristors for The Klystron Modulator

Kaoru Tenjin (*Author*), Akira Tokuchi
Kusatsu Factory
Pulsed Power Japan Laboratory Ltd.
Kusatsu 525-0058, Japan

***Abstract*—We are developing a solid-state switch with static induction thyristors(SI-Thy) for the klystron modulator of X-ray free electron laser SACLA in Japan. The switch is designed to meet the specifications that the maximum voltage is 50 kV and the maximum current is 6 kA at a repetition rate of 120 pps for forced oil cooling, so that 24 thyristors are connected in series, with eight such series connected in parallel; the solid-state switch uses 192 SI thyristors with measured characteristics of 2.5 kV and 1 kA at the maximum. The detail of the design and the tested results of the solid-state switch will be reported.**

Keywords—Solid-state switch, Static induction thyristor, Klystron modulator

I. Introduction

The klystron modulator often equipped with a thyratron as a high-voltage switch. Which is comprised gas-enclosed switching discharge tube, and have been weakened due to aged deterioration like a following problems. (1)Low stability of conduction (2) Large electrical noise (3)Periodical maintenance (4)Circumference circuit. In the X-ray free electron laser facility SACLA which needs a low-emittance and short-pulse accelerator, Long term and high precision stability of conduction is the important issue. A solid-state high voltage switch using a static-induction thyristor (SI-Thy) has been developed for klystron modulators, instead of thyratrons. In order to satisfy the specification, such as conditions of an applied voltage of 50 kV, a pulse current of 6 kA, 4.5 μs, and the maximum pulse repetition of 120 pps, the switch employs the 192 SI-thyristors (8 parallel, 24 series) with gate trigger circuits and snubber circuits. We experimentally tested a prototype SI-Thy switch with the rated voltage of 50 kV and current of 6 kA.

II. Design Configuration

A. Requrements of the switch module

A required specifications of the solid-state switch for the Klystron Modulator in SACLA(SPring-8 Angstrom Compact free electron LAser、RIKEN、Japan) is defined in TABLE 1.

TABLE 1. Requirements of the switching module for the klystron modulator of SACLA

Parameter	*Requirement value*
Blocking voltage	50 kV at the minimum
Peak current	6 kA at the minimum
Pulse width	4.5 μs(FWHM)
Pulse repetition freq.	120 pps at the maximum
Switching Time *1	500 ns at the maximum
Swiching Jitter	3 ns at the maximum
Cooling	Insulation oil
Istallation capacity (W×D×H)	700 × 260 × 800 mm at the maximum

B. Simiconductor device Static Induction Thyristor (SI-Thy)

We chose SI-Thy(Manufacturer/ Sindengen Co.Ltd,)as the semiconductor device replaced for thyratron. SI-Thy is a semiconductor device with which low loss, high Blocking-voltage and high-speed operation. And that structure is a type of PIN diode that is equipped with the gate [1]. TABLE 2 shows the main specifications of SI-Thy device.

TABLE 2. Specifications of the SI-Thy device

Parameter	*Requirement value*
Operating Temperature [℃]	-40 ~ +125
Blocking voltage [kV]	3.2
Average Current [A]	50
ON-Resistance [Ω]	0.1

C. Design of the 50 kV Solid-state Switch

There is a case at Osaka University as a past example of a high voltage switch equipped with SI-thy [2]. TABLE 3 shows comparison of the solid-state switches between a case at Osaka University and a case at SACLA.

TABLE 3. COMPARISON OF THE SOLID-STATE SWITCHS BETWEEN THE OSAKA UNIVERSITY CASE AND THE SACLA CASE

Parameter	*Osaka University L-band SW*	*SACLA modulator*
Blocking Voltage	25 kV	50 kV
Peak Current	6 kA	6 kA
Pulse Width	8 μs	4.5 μs
Repetition Freq.	10 pps	120 pps
Number of thyristors	60	192
Cooling	Force air cooling	Insulation oil circulation
Total height	540 mm	800 mm

The important points of this design specifications is as follows.

- 1. High Blocking Voltage:

  We chose 24 as a number of a series connection of the SI-Thys in order to make 2.08kV of a applied voltage on each Si-Thy that is lower than 2.5kV of an experimentally aimed safety maximum voltage for a single SI-Thy.

- 2. Cooling design which corresponds to high Repetition Freq.

  We chose 8 as a number of a parallel connection of the SI-Thys in order to make 750A of a anode current in each Si-Thy that is lower than 1kA of an experimentally aimed safety maximum current for a single SI-Thy under high-repetitive operation..

  Fig. 1 shows circuit diagram of SI Thyristor stack.

  The general loss of the SI-Thys stack is as follows.
  Anode-Cathode Voltage / 1-device : 2.08 kV
  Anode-Cathode Current / 1-device : 750 A
  On Resistance / 1-device : 0.1 Ω
  Pulse Width / 4.5 μs
  Repetition Freq. / 120 pps
  Total loss $24 \times 8 \times 30 = 5.8$ kW

  The total loss mentioned above does not include a switching loss of the Si-Thys.

  In case of SACLA, an insulation oil is enclosed with the klystron modulator in which this equipment is installed. Therefore, the devices(SI-Thys) on the heatsinks are cooled by a circulation of the insulation oil in the heatsinks.

- 3.Downsizing by the restrictions in installation space

  There is another subject concerned in a size of the SI-Thys stack as shown in TABLE 3. Restriction of the height direction is severe above all. So we designed this switch the structure as shown in Fig. 2. Two SI-Thys are connected in series and 4 for each standing in parallel is arranged in both sides of the each heatsink.

So we defined a connected block as "1-Stage" in this 2 series 8 parallels. Thy. Consequently, we were able to make half with the number of the stacking of the heatsinks with causing increase in SI-Thy series connections, and meet it in the requirement of the height direction.

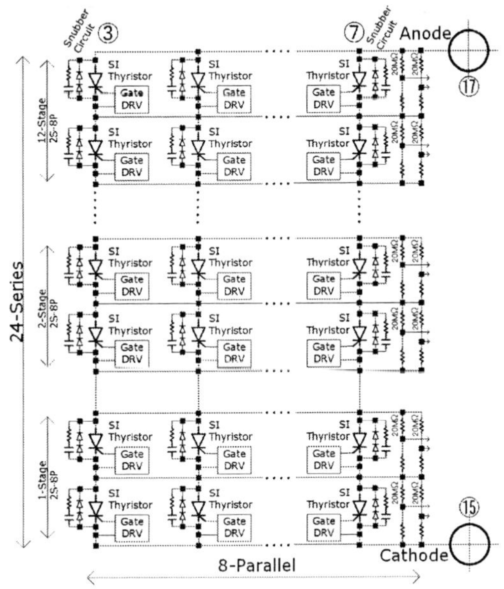

Fig. 1. Circuit Diagram of SI Thyristor Location 24S 8P

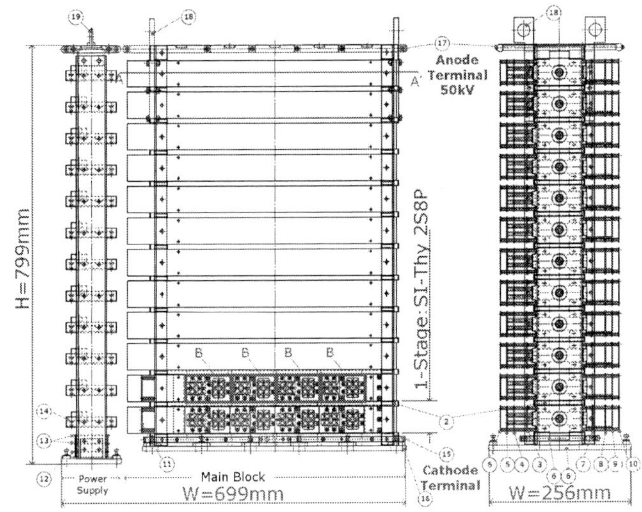

Fig. 2. Overview of the 50 kV solid-state switch

Fig. 3 shows a block diagram of the trigger and interlock circuit for the individual SI-Thy switch. Figure 4 shows Appearance of the 50kV solid-state switch.

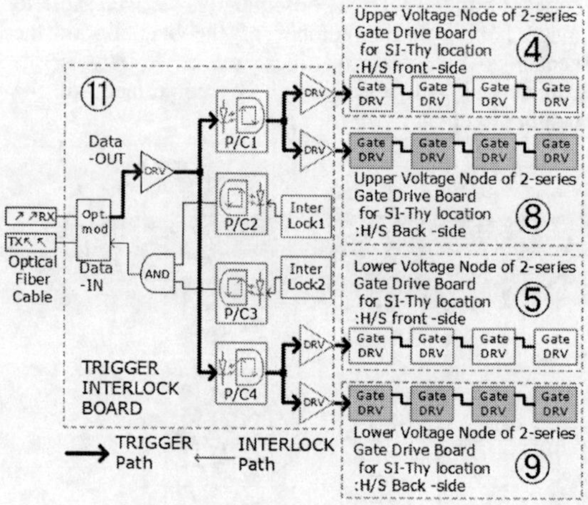

Fig. 3. Block Diagram of the trigger and interlock circuit for the individual SI-Thy switch

Fig. 4. Appearance of the 50kV solid-state switch

III. Test Resuls (Factory Test Run)

A. Test Circuit

We prepared a test circuit shown in Fig. 7. The high voltage SI-thy switch is connected to PFN (Pulse Forming Network) and A resistive load were used for a dummy load instead of the klystron.

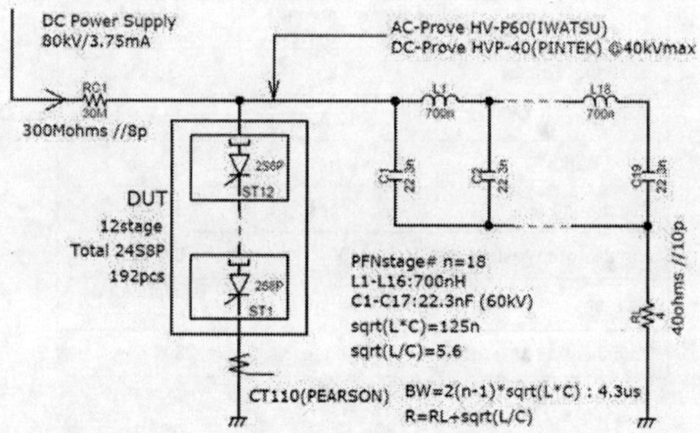

Fig. 5. Test circuit for the high-voltage SI-Thy switch

B. Test Results 1

Fig. 6 shows waveforms of the anode voltage(purple) and the pulse current(blue) at a load resistance(RL) value of 4 Ω and a charging voltage of 50 kV.

A peak current of 5 kA was obtained in this test.

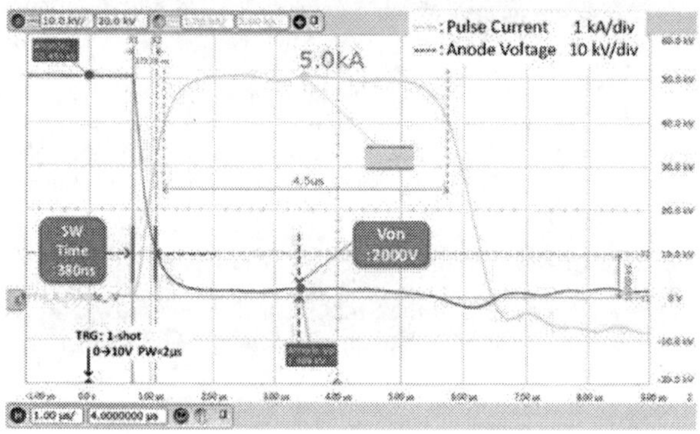

Fig. 6. Waveforms of the anode voltage and the pulse current

TABLE 4. Test conditions summary1

Parameter	*Requirement Specifications*	*Results of Test Run*
Anode Voltage	50 kV	51 kV
Peak Current	6 kA	5.0 kA
Pulse Width	4.5 μs	3.2 μs [a]
Swiching Time	500 ns	380 ns

[a] Due to the PFN circuit

C. Test Results 2

To confirm a peak current of 6kA, we replaced the load resistance in Fig. 5 to 2.1 Ω from 4 Ω. TABLE 5 shows test conditions summary of the SI-Thy switch at the various anode voltages and RL=2.1 Ω. Fig. 7 shows waveforms of anode voltage. And fig. 8 shows waveforms of the peak current.

A peak current of 6.1kA was obtained in this test under the anode voltage of 50kV.

TABLE 5. Test conditions summary 2

Anode Voltage [kV]	10	20	30	40	50
Peak Current [kA]	1.2	2.3	3.6	4.8	6.1
Swiching Time [ns]	870	560	470	460	410
Von [V]	820	1100	1200	1500	2000
ON-Resistance [Ω]	0.23	0.15	0.11	0.10	0.11

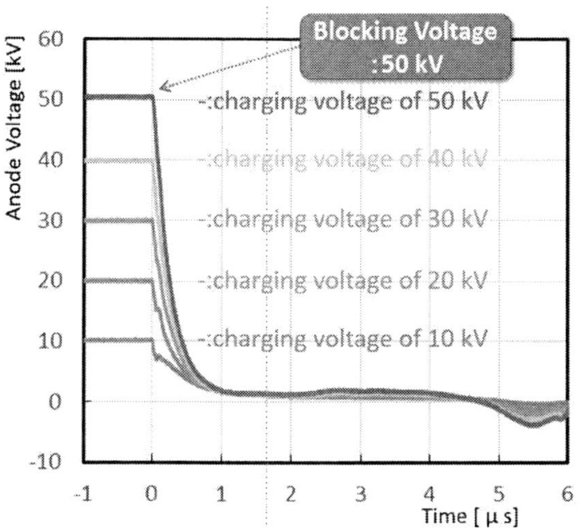

Fig. 7. Waveforms at the various anode voltages

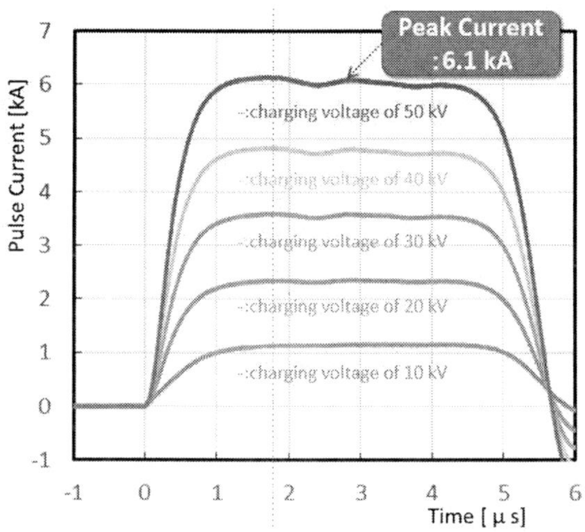

Fig. 8. Waveforms of the pulse currents at the various anode voltages

IV. CONCLUSIONS

We have developed the high-speed and the high-voltage semi-conductor switch which is aimed to replace the thyratron of the klystron modulator in SACLA. This switch employed the 192 Static Induction Thyristors (8 parallel,24 series).

We grappled with the circuit and mechanical structure design in consideration of the blocking voltage and cooling performance.

This switch was manufactured so that it might fit into the klystron modulator for SACLA .And we confirmed the target performance, an anode voltage of 50kV, an anode current of 6.1kA, a pulse width of 3.2 μ s and a switching time of 410ns in the factory test. In this factory test, a pulse repetition rate was 1 pps or less that was lower than required value of 120pps.

As the next examination, we are going to confirm the other characteristics in an actual operation using klystron load with a continuous running with higher repetition rate in the insulation oil .

References

[1] J.Nishizawa and K.Nakajima. Static induction thyristor. Revue de Physique Appliquee. Vol13,No.12PP.725-728,1978.

[2] Akira Tokuchi, et al. "Development of high-power solid-state switch using induction thyristor for klystron modulator" (Nuclear Instruments and Methods in Physics Research A 769 (2015) 72-78)

Development of SiC MOS FET MARX type Klystron Modulator for International Linear Collider

Yo Sawamura(*Author*), Akira Tokuchi
Kusatsu Factory
Pulsed Power Japan Laboratory Ltd.
Kusatsu, Japan

Mitsuo Akemoto, Hiromitsu Nakajima, Masato Kawamura
Accelerator Laboratory
High Energy Accelerator Research Organization (KEK)
Tsukuba, Japan

Weihua Jiang, Ryutaro Suzuki, Takumi Hayashi
Extreme Energy-Density Research Institute
Nagaoka University of Technology
Nagaoka, Japan

***Abstract*—I ILC (International Linear Collider) plan, the main linear accelerator of the ILC plan, 378 units 1.3GHz 10MW multi-beam klystron system is mounted. RF power for generating the acceleration field on the superconducting accelerating cavity is composed of multi-beam klystron and klystron power source for driving it. We propose a low-cost pulsed power supply in a compact that combines the structure and the chopper circuit of the MARX-type power supply, aim to reduce the ripple by droop compensation and phase control by PWM control. The power supply is made up of 20 units a chopper control method MARX circuit of four stages. Aiming at miniaturization of the size, adopting SiC devices to reduce the PWM switching losses, and also adopted the charging voltage digital feedback control using an A/D converter.**

Keywords—ILC; Klystron; Marx; SiC; Pulsd power

I. INTRODUCTION

ILC (International Linear Collider) plan, in the linear accelerator of the total length of about 30km, plan to perform the collision experiments of electrons and positrons in the highest energy that can be currently achieved.

Fig. 1. Underground 100m and narrow space.

Multi beam Klystron
- ✓ Main accelerator: 380 units
- ✓ Others: 270 units

High-frequency power supply for ILC

Output voltage [kV]	-120
Current [A]	140
Flat top part [ms]	1.65
Max. average power [kW]	139
Repetition rate [pps]	5
Ripple ratio [%]	< ±0.5
Rise time [us]	< 100

TABLE I. SPECIFICATION OF PULSED POWER MODULE.

Since a number of plus power supplies are operated for a long time, high efficiency, high reliability, small size, light weight and low cost are required.

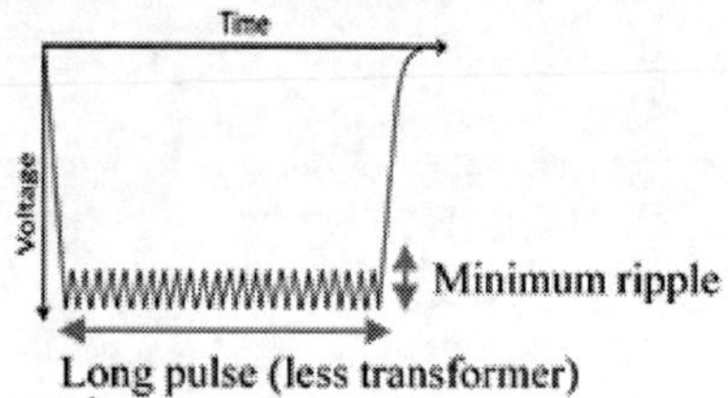

Fig. 2. Output pulse.

Klystron power supply is called the Marx modulator to generate a pulse voltage of 120kV 140A 1.65ms, supplied to the cathode of the multi-beam Klystron.

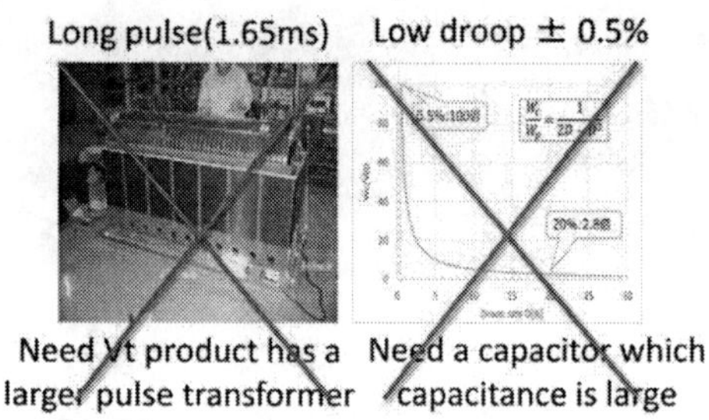

Fig. 3. Requirements of Klystron modulator.

In the power supply to be mounted, downsizing, cost reduction and high reliability has been strongly desired.

And with respect to the power output to be the pulse of, which in a very long pulse width with a flat-top extends to 1.65ms, the voltage variation rate is high precision output of less than 1% is required.

Chopper Marx power supply has the following features.

- Circuit is blocked, it enables onboard.
  - ✓ Possible mass production (low-cost)
- Withstand voltage of each element is equal to the charging voltage
  - ✓ Low breakdown voltage element (low cost)

- It can be switched ON / OFF □ for each stage
  - ✓ Low ripple (small size, light weight)

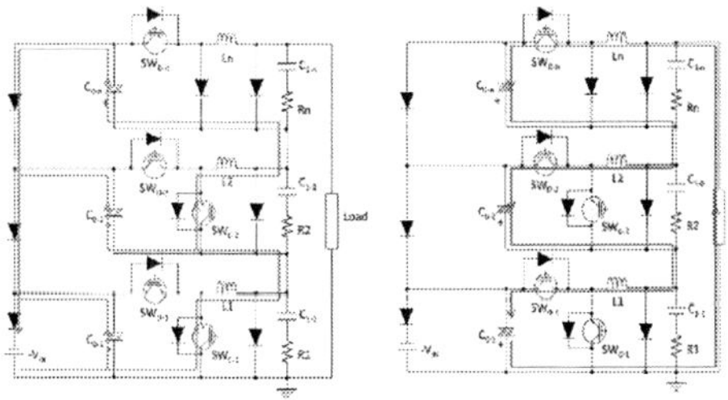

Fig. 4. Chopper type MARX circuit in Charge and Discharge mode.

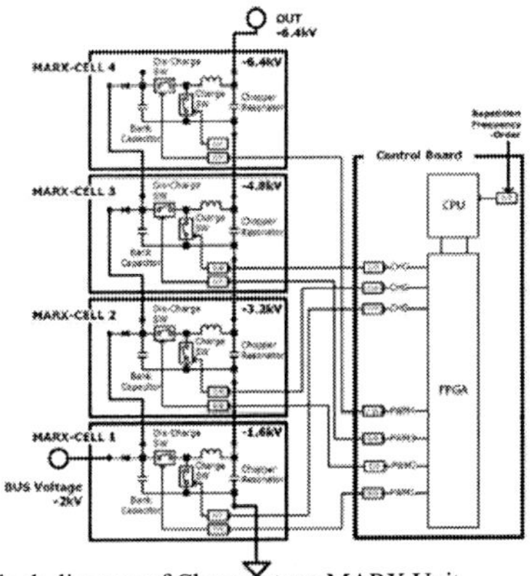

Fig. 7. Block diagram of Chopper type MARX Unit.

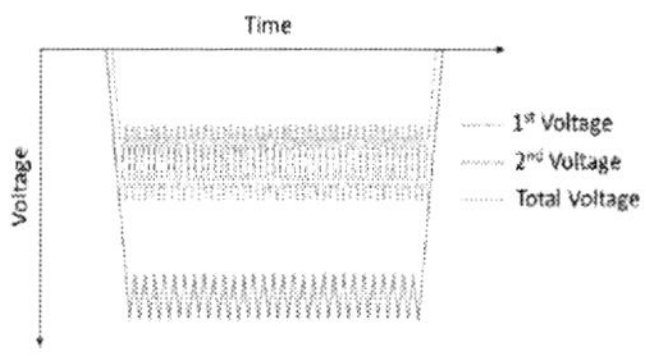

Fig. 5. Explanation drawing of Chopper type MARX circuit in two stacked operation mode..

II. Configuration Study of the Chopper-Type Marx Power Supply

A. Configuration of the chopper Marx prototype board and the unit

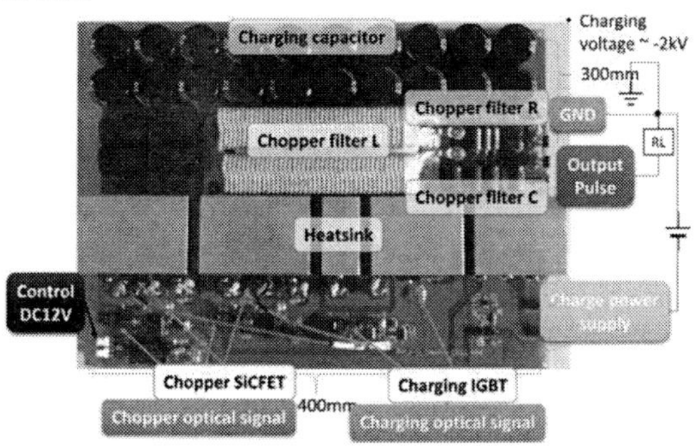

Fig. 6. Explanation drawing of Chopper type MARX board.

Chopper Marx power supply by using four stages MARX board of 300mm × 400mm shown in Fig. 6, MARX unit. And Klystron Modulator power supply is consisted from 20 units of MARX unit.

TABLE II. Specification of Chopper type MARX Cell and Unit.

CELL	Bus Voltage	-2.0kV
CELL	Nominal Output Voltage	-1.6kV
CELL	Bank Capacitor Drop	>20%
UNIT	Number of CELL stage	4
UNIT	Output Voltage	6.4kV
UNIT	Size	$445W \times 650D \times 382H$

Mass 47.5kg

B. Device and loss measurement in the chopper type Marx prototype board

SiC, the IGBT, the result of comparison Si of each loss, was adopted SiC devices to reduce the PWM switching losses , and aims to further miniaturize.

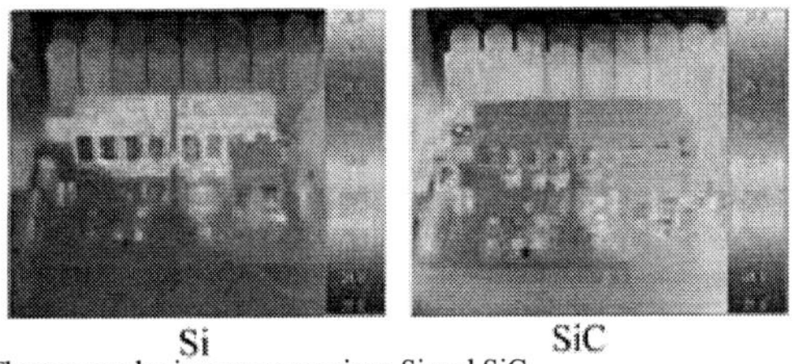

Fig. 8. Thermography image comparison Si and SiC.

III. Development of MARX Unit

In this study, at the suggestion of cheaper and compact pulse power supply that combines MARX structure and chopper circuit of Marx power, do a droop compensation by PWM control, and a ripple reduction of by the phase control.

A. MARX unit configuration

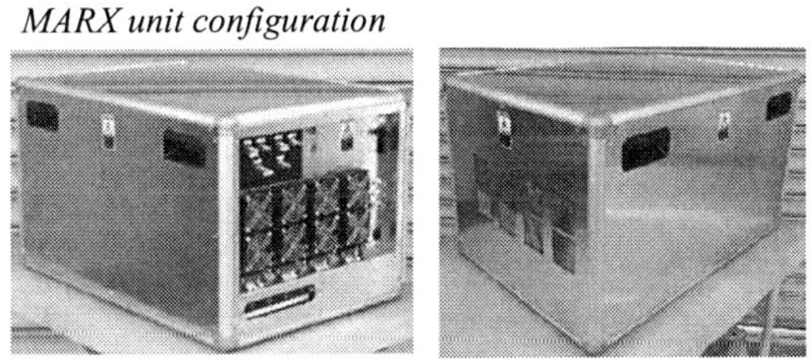

Fig. 9. Appearance of MARX Unit.

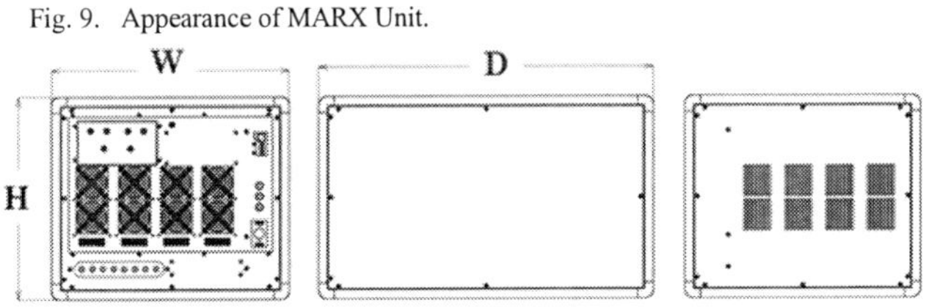

Fig. 10. Outline drawing of MARX Unit.

MARX unit uses four stages MARX board of 300mm × 400mm. Generating an output voltage of -6.4kV in each MARX unit.

B. MARX unit interface

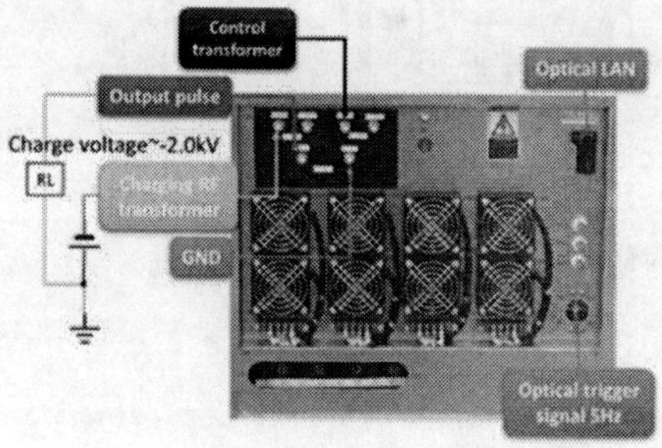

Fig. 11. Interface of MARX Unit.

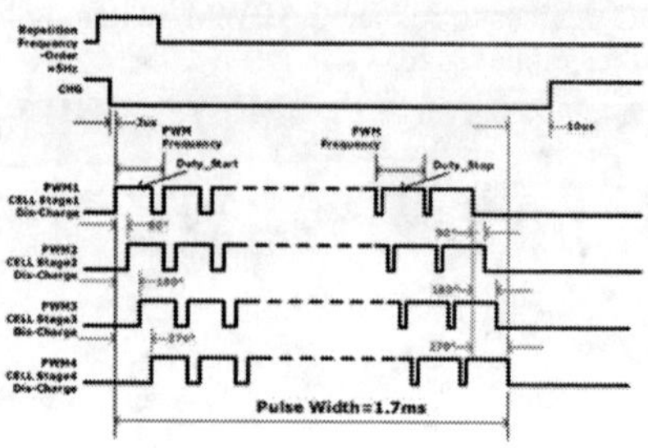

Fig. 12. Timing chart of Control signal.

By isolation transformer and optical cable, interface of MARX unit has made it possible to keep suppling power and control, even though the MARX unit is in a high voltage state of -120kV applying.

C. MARX unit operation check

1) MARX unit 2kV charging operation waveform

a) MARX unit; MARX board 4-stack

b) Charging voltage 2.0V Dummy load 43Ω

c) Chopper frequency 50kHz, PWM Duty 77-97% variable, Output pulse number 85pulse (1.7ms)

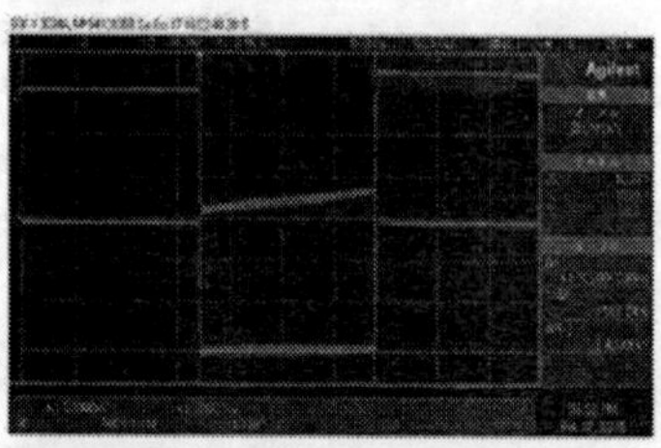

CH4; Output voltage6.4kV2.0kV/div
CH3; Outputcurrent140A50A/div 500us/div
Output voltage ripple: ±3.7%

Fig. 13. MARX Unit -2kV charging operation.

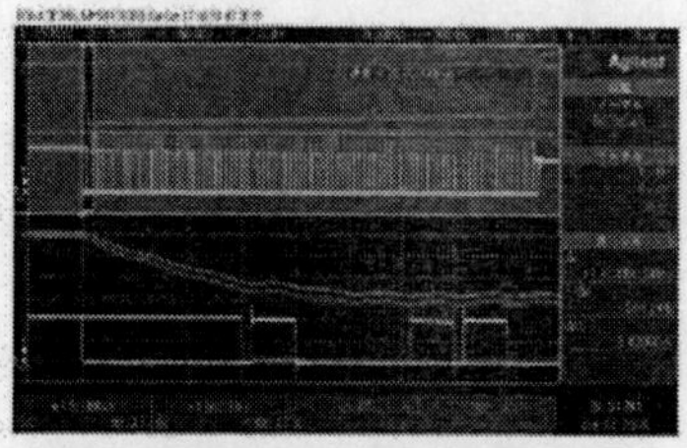

Fig. 14. A MARX Unit -2kV charging operation. (Expanded)

CH1; MARX1 SiC-FET S-D voltage 500V/div
CH2; MARX2 SiC-FET S-D voltage 500V/div 5us/div
Output voltage rise time (10% -90%): 19us

Drop of the output current is due to the characteristics of the CT. The operation of the output voltage of -6.4kV and the operation of the output current 140A could be confirmed.

2) MARX unit 5Hz repeat operati

In heat run test of the rated output, the temperature rise of the discharge SiC-FETs was confirmed to be about several degrees.

3) Load short test of MARX unit using a gap SW

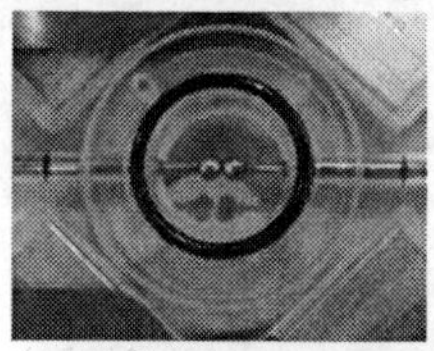

Fig. 15. Gap switch for load short test.

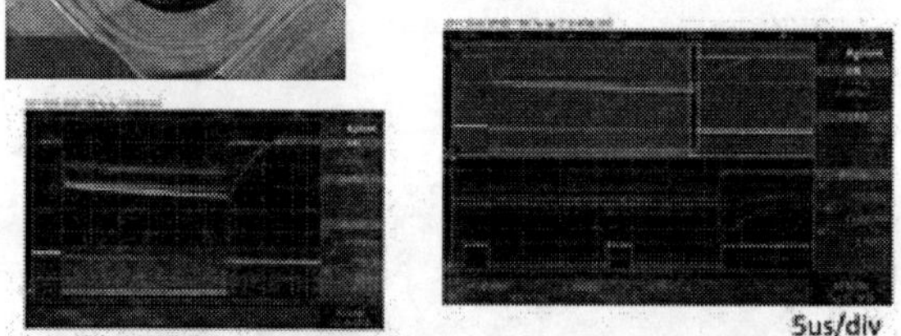

CH1; SiC-FET S-D voltage 500V/div
CH2; Output voltage 2kV/div
CH3; Output current 50A/div 200us/div

Fig. 16. Load short test wave form.

The gap SW of Fig. 15 connected to the load was carried out load short-circuit (overcurrent detection) test of MARX unit. Delay time between the overcurrent detection 150A (3.75V) and the gate turn-off time was within 600ns.

IV. DEVELOPMENT OF THE POWER SUPPLY FOR KLYSTRON MODULATOR

A. Unit-grounded enclosure withstand voltage test DC120kV

1) MARX unit DC withstand voltage test

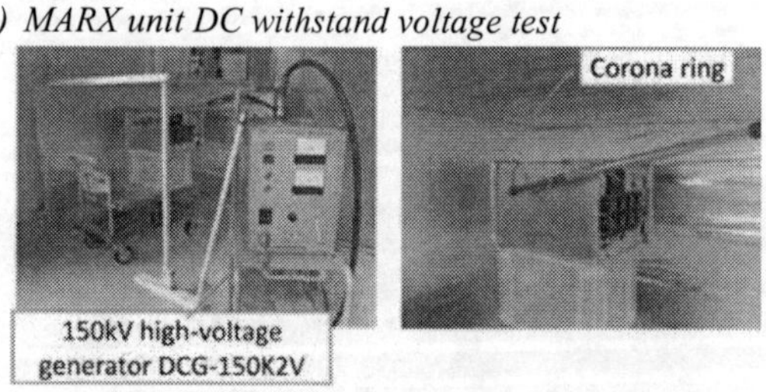

Fig. 17. Unit-grounded enclosure withstand voltage test DC120kV.

By using simulated unit-grounded enclosure, MARX unit withstand voltage was tested. Applying DC120kV high voltage, withstand voltage was confirmed.

2) Unit-grounded enclosure high voltage applied, unit operation test

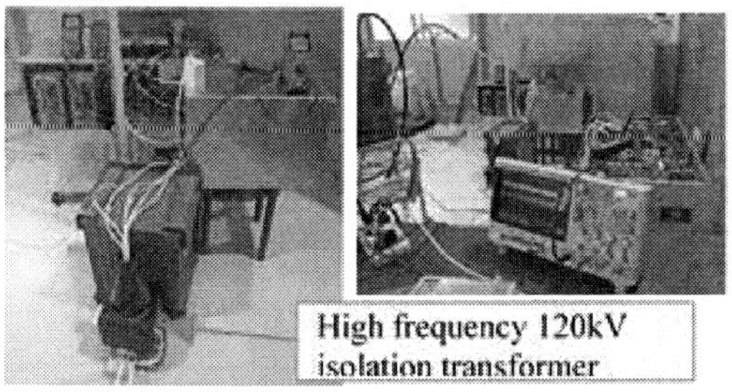

Fig. 18. Unit-grounded enclosure DC120kV applied, unit operation test.

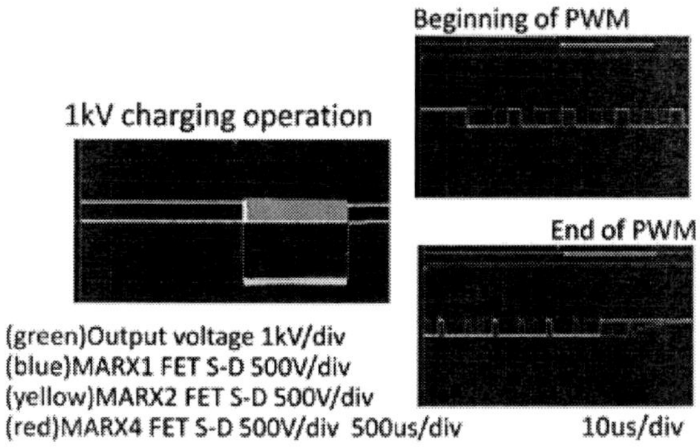

Fig. 19. Unit-grounded enclosure DC120kV applied, unit operation test waveform.

50kV inverter -> TV150 (120kV isolation transformer) -> supply the charging voltage and the control voltage from the rectifier circuit. Conducted an operation confirmation test of the unit in the state in which floated a MARX unit to a high voltage of DC120kV, was confirmed MARX unit operation at 1kV charge.

B. Configration of Power supply for Klystron Modulator

Generate the output voltage of -6.4kV in each MARX unit, a prototype system to obtain an output voltage of -120kV as Klystron Modulator power supply by constructing 20 units of MARX unit.

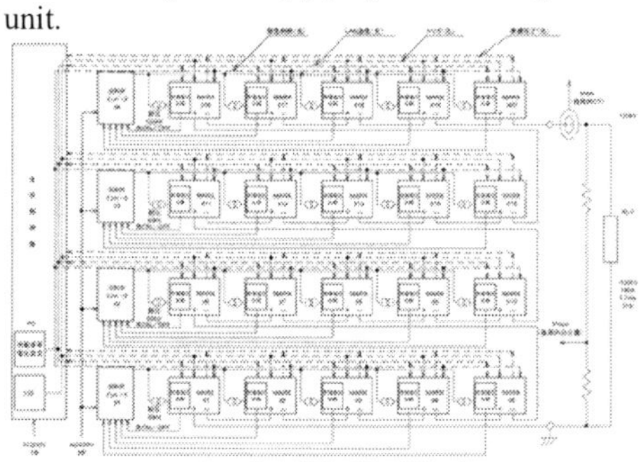

Fig. 20. Pulsed power supply block diagram as the Klystron Modulator.

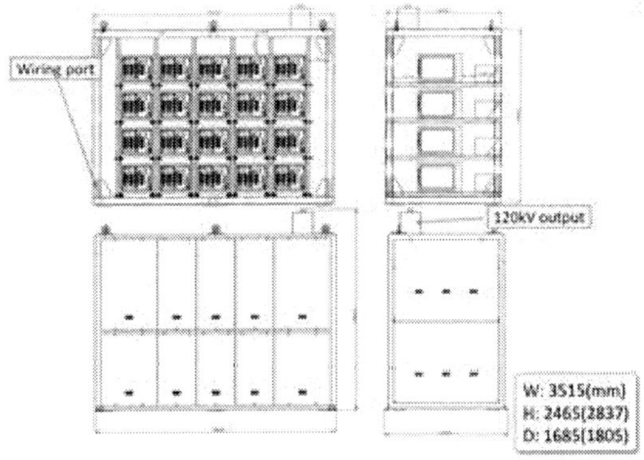

Fig. 21. Pulsed power supply rack configuration as the Klystron Modulator.

Fig. 22. Pulsed power supply as the Klystron Modulator.

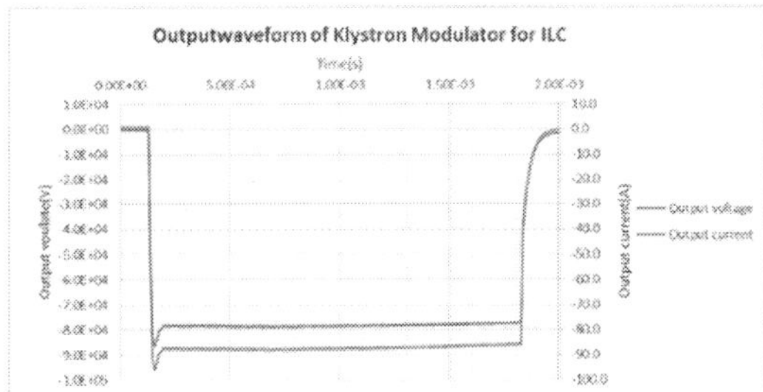

Fig. 23. Output waveform of Klystron Modulator for ILC.

At the moment, in a single operation, an output voltage of 80kV in simulated load, to confirm an output current of 88A.

Output voltage ripple was ±0.23%.

Acknowledgment

Prototype of the power supply for Klystron Modulator which is in joint research and development with High Energy Accelerator Research Organization (KEK) and Nagaoka University of Technology, is installed in KEK Tsukuba campus, and having a performance test.

Aims to make a target of rated operation 120kV140A output 5Hz continuous operation during this year.

2-section Pulse Current Generator for Concrete and Rocks Destruction by Splitting off*

Artem S. Yudin, Nikita V. Voitenko, Natalia S. Kuznetsova
Department of high voltage electrophysics and high current electronics
Tomsk polytechnic university
Tomsk, Russia
tevn@hvd.tpu.ru

***Abstract*—The relevance of electro-discharge technology usage for construction and demolition work in the dense urban areas is discussed. The technical possibilities and the equipment based on 2-section pulse current generator for the multi-borehole electro-discharge splitting off and destruction of rocks and concrete have been described. The ways of efficiency increase of electro-discharge solid materials destruction method have been proposed. To increase the effectiveness of the crack initiation the high-current pulse should have the high rate of rising edge (practically well-founded 3–7 kA/μs). For crack propagation the pulse duration should be as long as possible (good value is more than 500 μs). Achievement of such controversial requirements in one generator is not a trivial task. To solve this contradiction and to increase the discharge circuit energy the separation of the storage energy into two capacitor batteries as well as twice increase of the capacitance has been done. The throttle of 28 μH was installed in series with one battery section, which led to the enlargement of the channel energy release duration up to 400 μs. To increase the pulse duration even greater the system of pulse delay between two batteries is implemented. It is shown that in modes with triggering pause between the batteries the tensile stresses are formed earlier and covered almost all distance from the borehole to the free border and acting on this space longer period of time.**

Keywords*—*discharge; rocks splitting off; high current generator.

I. INTRODUCTION

In building and construction sector as well as mining industry the conventional blasting causes environmental hazard, like dust, gases and heightened noise, as well as ground vibration and debris, that during the explosion can damage surrounding objects, especially in dense urban areas. Moreover storing and handling explosives is expansive because strong safety regulations have to be considered. Furthermore it require to ensure security zone and evacuation of staff while carrying out of blasting. So it motivates researchers to find alternative methods for rocks and concrete destruction, that will be free from this shortcomings. The promising one for splitting off and destruction of solid materials is using pulse electric-discharge blast. The method is based on employing of the electro-burst wave dynamics principles [1-3].

This work was supported by Government, Federal program project № 14.575.21.0059 (RFMEFI57514X0059).

Physical base of the method is employing the destructible force of electrical discharges in solids. The principle of destruction by the tensile stresses gives the opportunity of achieving the lower power-intensity of destruction in comparison with traditional methods (in the same conditions). The energy efficiency of destruction have a weak relation with mechanical strength of the material. Due to this fact the best technological effect could be obtained on hard and extremely hard rocks and materials. It should be noted however that despite this fact the real productivity of the technology limited by attendant operations like boreholes boring and electrodes replacement if needed.

To store the energy, a capacitor storage is commonly used in such technologies. During the discharge the energy stored in the capacitors feeds the plasma channel, that generates shock-waves, which in turn destroy the material. To destruct the materials effectively one should meet some requirements according to energy release parameters. First of all the pressure pulse should have enough amplitude to form tensile stresses for cracks initiation. Then pulse duration should be as long as possible for providing cracks growing [4]. According to our experiments practically well-founded pulse duration is more than 500 μs.

In this paper, 2-section pulse current generator with separate ignition and with the inductor for pulse width duration increasing will be presented. This generator is appropriate for pulsed power applications especially for rock blasting and splitting off from monolith. The generator could operate both on two separate loads (electrode systems) and on common load.

II. EXPERIMENTAL EQUIPMENT

The elementary circuit diagram of the pulse current generator is presented in Fig. 1.

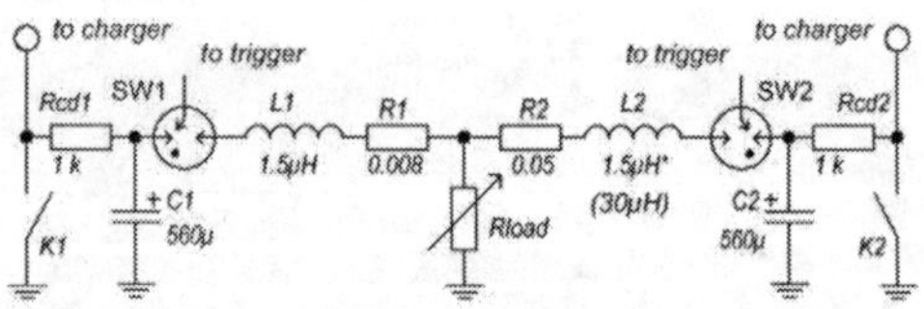

Fig. 1. The elementary circuit diagram of the 2 section generator.

In the Fig. 1 the situation when the generator is connected to the common load is shown. The *L1, L2* and *R1, R2* represents the inductances and resistances of the *R-L-C* circuits respectively. R_{load} is the resistance of the discharge channel that actually not constant with the time. The capacitor battery of each section *C1* and *C2* consists of 40 nonpolar capacitors of K75-101-15kV-14uF type, developed by ELCOD CSC (St Petersburg, Russia). This capacitors can work with deep ringing mode of up to 80 % of reverse voltage from initial value. The maximum operating voltage is 15 kV. To prolong the lifespan of the capacitors the charging voltage under working conditions is reduced to 13 kV. The switches SW1 and SW2 is RVU-43-1 type is the vacuum switch developed by VEI-AVIS JSC (Moscow, Russia). It is known, that vacuum switches have a diode effect – brakes the circuit when current reaches zero value and therefore in ringing modes can pass only half cycle of the current pulse. To overcome this limitation one can either apply to the switch a trigger pulse when current crossing each zero value or provide the ringing mode when the current change speed *di/dt* while crossing zero value is greater some critical meaning. The RVU-43-1 vacuum switches have such critical meaning of about 100 A/μsec. In our operation modes with current amplitude in average 120-150 kA and oscillation period *T* of 230 μsec the *di/dt* at time point of *T/2* was about 300 A/μsec, so the switch passed from 2 to 3 half cycles of the current pulse. The typical voltage and current waveforms in operation mode are shown in Fig. 2.

The modular arrangement of the capacitors was implemented in case of protection requirements and for ability to vary the capacitance by a number of connected modules in easy way. Each module included 4 capacitors protected by a common fuse. The total quantity of modules in one battery was 10. The total battery capacitance was 560 μF. An external view of the capacitor batteries of the 2-section pulse current generator is presented in Fig. 3.

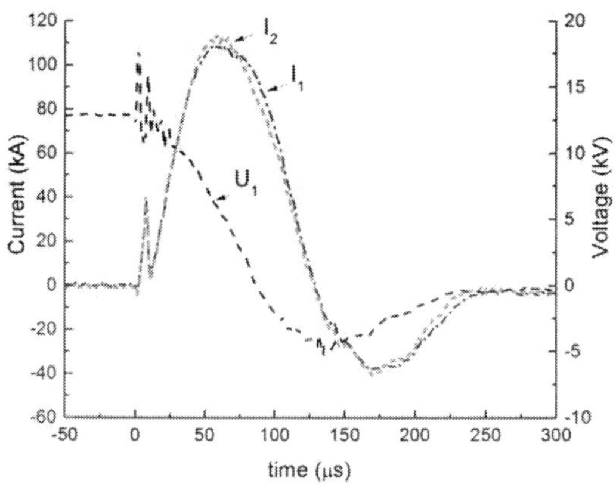

Fig. 2. Typical voltage and current waveforms in operation mode of 2-section pulse current generator: I_1 I_2 – current waveforms of the 1st and 2nd battrey respectively; U_1 – voltage waveform on the 1st battery.

Fig. 3. An external view of the capacitor batteries of the 2-section current generator: 1) - fuse on capacitor module (4 capacitors); 2) - vacuum switch RVU-43-1; 3) fluid charging-dumping resistor.

A two channel constant current charging power supply has been also developed to charge the batteries simultaneously. The maximum current provided by the supply was 400 mA per channel, thereby the maximum output power under 15 kV was 12 kW, and minimum charge time of two 560 μF batteries became of about 20 sec. The charger was equipped with a digital controller, which enables both local and remote modes of control. The fluid charging-dumping resistors (*Rcd1, Rcd2* in Fig 1.) were made to protect the charger from reverse current and for dumping the battery either residual charge or full charge in case of misfire.

The pulse shaping inductor *L2* was made from 4 copper bars with 150 mm width and 0,6 mm of thickness coiled spirally. The value of inductance measured on 5 kHz was about 30 μH, the Q-factor was 5.8.

III. Simulation of Operating Modes

Four operating modes were implemented and compared with each other: 1) – discharge of one battery section with inductance *L1* of 1,5 μH; 2) – discharge of two sections in parallel with inductances *L1, L2* of 1,5 μH and 30 μH respectively; 3) – discharge of two sections in parallel with inductances *L1, L2* of 1,5 μH in both sections; 4) – discharge of two sections with inductances *L1, L2* of 1,5 μH in both sections and with 50 μs delay of 2nd discharge. The results of simulation are sown in Fig. 4.

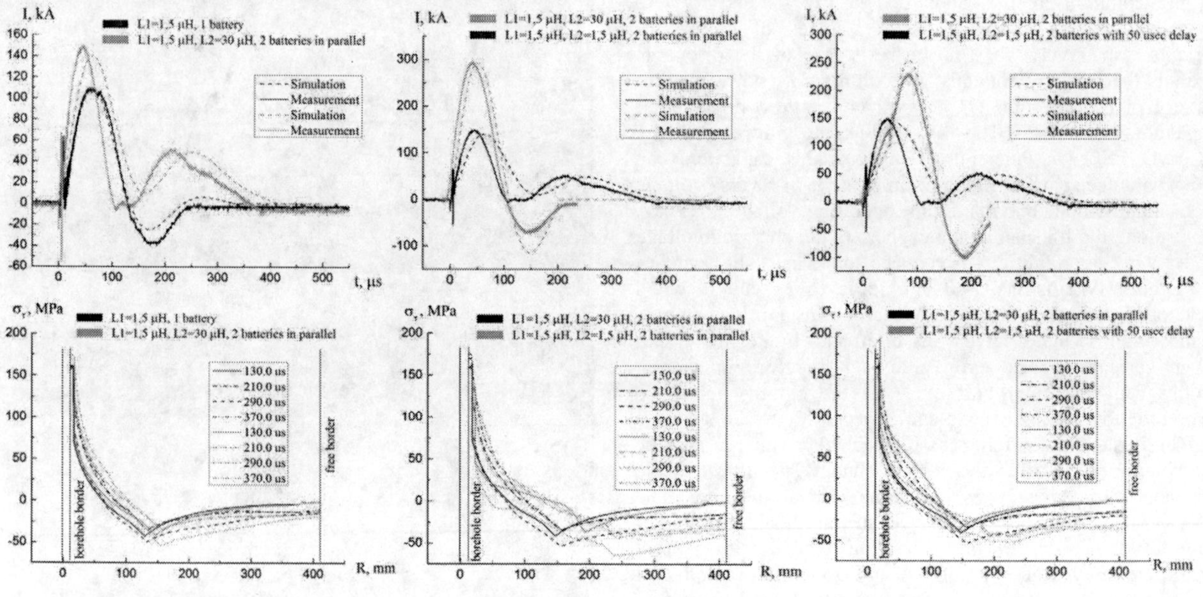

(a) (b) (c)

Fig. 4. The simulation results of current waveforms and distribution diagrams of tangential stresses in the pressure wave in concrete (*(a) – the comparision of modes 1) and 2); (b) – the comparision of modes 2) and 3); (c) – the comparision of modes 2) and 4)*)

The computer simulations was made under the parameters of concrete listed in Table I. This parameters were used in the state equation of solid in form of barotropic dependence [5]. The complete description of the electro burst model used in this study is given in [3].

In comparison of the first two modes (Fig. 4(a)) one can see that before the shock-wave comes to the free border there are no significant difference in tangential tensile stresses (130 μs). But when the wave comes back to the borehole border (290 μs) the tensile stresses in span from 150 mm to the free border on around of 10 MPa greater and reach of about 50 MPa in peak and of 19 MPa on the free border. The comparison of modes 2) and 3) (Fig.4(b)) shows that at time of 130 μs the region covered with tensile stresses is significantly smaller, but at the time of 290 μs the almost all space from the borehole to the free border covered with tensile stresses with the peak value of 64 MPa and on the border of about 39 MPa. If one compare modes 3) and 4) (Fig.4(c)) he can see that in mode 4) tensile stresses are formed earlier covered almost all space from borehole to the free border and acting on this space longer period of time. The tensile stress peak value is about 50 MPa and on the border of about 30 MPa.

So in modes with triggering pause between the batteries the tensile stresses are formed earlier and covered almost all span from the borehole to the free border and acting on this space longer period of time

TABLE I. SIMULATION PARAMETERS OF THE CONCRETE

ρ, g/cm³	*μ, GPa*	*Y, MPa*	*c, m/sec*	*n*
2200	17	77	2330	1.51

IV. CONCLUSION

A 2-section pulse current generator was successfully designed, fabricated, and tested. The generator is consisted of two 63-kJ capacitor batteries. The usage of two sections of the generator with ability to set the pause between discharges have the best perspectives for concrete and rock destruction by splitting off. It leads to decrease of the generator operating voltage which allows to increase its maintainability and efficiency of lumps destruction and splitting off operations. The usage of two sections of the generator in parallel with throttle on one of them can prolong the pressure acting time and at the same time reduce the amplitude and steepness of the pressure wave and therefore can decrease induced cracks in partial dismantling operations.

References

[1] H. Bluhm, W. Frey, H. Giese, P. Hoppé, C. Schultheiß and R. Sträßner, "Application of pulsed HV discharges to material fragmentation and recycling," IEEE Trans. on Diel. and El. Insul. vol **7**, N5, pp. 625-636, October 2000.

[2] C.M.M Silva, A Stellin, W.T. Hennies and E.G. Costa, "Electrohydraulic rock blasting: an alternative for mining in urban areas," Int. J. Min. Reclamat. Environ. vol 16, No 4, pp. 261-269, 2002.

[3] V.V. Burkin, N.S. Kuznetsova and V.V. Lopatin, "Dynamics of electro burst in solids: I. Power characteristics of electro burst," J. Phys. D: Appl. Phys., vol 43, pp. 185-204, 2009.

[4] B. V. Sjomkin, A. F. Usov and V. I. Kurets, "The Principles of Electric Impulse Destruction of Materials," Russian Academy of Sciences, Kola Sci. Center, 1995, pp. 215-219 (in Russian).

[5] Born M and Geppert-Meyer M "Theory of Solid," Moscow: ONTI, 1938 (in Russian).

Four channel gyromagnetic nonlinear transmission line high power RF source with beam steering

M. R. Ulmaskulov, K. A. Sharypov, S. A. Shunailov, V. G. Shpak, M. I. Yalandin, M. S. Pedos, and S. N. Rukin
Institute of Electrophysics, UB RAS
106 Amundsen St., Ekaterinburg 620016, Russia

I. V. Romanchenko, V. V. Rostov, and V. Yu. Konev
Institute of High Current Electronics, SB RAS
2/3 Akademichesky Ave., Tomsk 634055, Russia

***Abstract*—Synchronized operation of four gyromagnetic nonlinear transmission lines (NLTLs) was tested with pulse repetition frequency up to 1 kHz during 1 second bursts. High voltage pulses with duration of ~5 ns from the solid state driver S-500 were split into four 50 Ohm channels reaching about -200kV in each channel with ~10% jitter. Maximum peak voltage at the NLTL output was within 220-235 kV with maximum modulation depth of decaying oscillations up to 70% at center frequency near 2.1 GHz. The relative delay between channels reached the half period of center frequency of oscillations. The associated beam steering by four element array of conical helical antennas was demonstrated and detailed together with corresponding radiation patterns.**

Keywords — nonlinear transmission lines; high power microwaves; beam steering; helix antenna

I. Introduction

Nonlinear transmission lines (NLTLs) were shown to be a promising candidate for high power microwave (HPM) and RF generation as alternative to traditional relativistic tubes [1-2]. Whereas the operation of a single gyromagnetic NLTL is thoroughly studied [3-5] the next step of this technology includes development of NLTLs arranged in a phased array [6-8]. It was found, that identical NLTLs produce RF signals with an uncontrollable shift in phase because of unpredictable shock wave velocity inside the NLTL [6,8]. In order to avoid this shift dispersion in an NLTL array additional NLTL sections for controllable time delay usually are installed. This sections can be separate [6] or combined with the main NLTL [7,8]. Another function of these sections is bias-controlled inter-NLTL phase shifting providing the beam steering of the irradiated RF pulse [6,8]. In this paper we present experimental results on synchronized operation of 2x2 array of gyromagnetic NLTLs with beam steering with 1000 Hz pulse repetition rate and subgigawatt power level.

II. Experimental Setup

The schematic diagram of the experimental setup is shown in Fig. 1. The S-500 high voltage driver with double pulse-forming line in pumping circuit and semiconductor opening switch (SOS) was used [7]. After sharpening by unsaturated ferrite line the high voltage pulse was split into four 50 Ohm channels including NLTL delay section combined with generating NLTL section, band-pass filter and irradiating helix antenna. The amplitude of the high voltage pulse in each channel was set to -200 kV with duration of 5 ns. The bias currents in both the delay and RF generating NLTL sections of each channel were independently supplied and controlled by computer control system through fiber optic connection. The NLTLs and band-pass filters were filled with oil insulation under additional nitrogen pressure (5-10 bar) to avoid electric breakdown. The high voltage pulses were registered by wideband capacitive dividers placed in transmission lines (5 ns double electric length) before and after NLTLs.

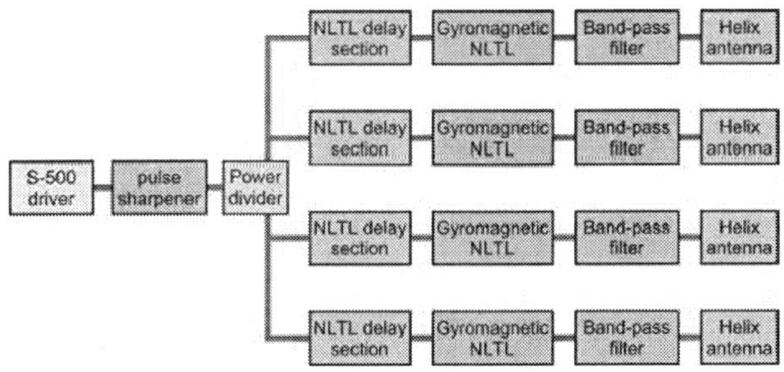

Fig. 1. Shematic diagram of the experimental setup.

III. Half-period Phase Shift Between Channels

At first, the operation of NTLTs was tested when they were loaded on a resistive load. After finding bias current sets for synchronous operation the vertical pairs were shifted by a half period of oscillations at frequency of 2.1 GHz, Fig. 2.

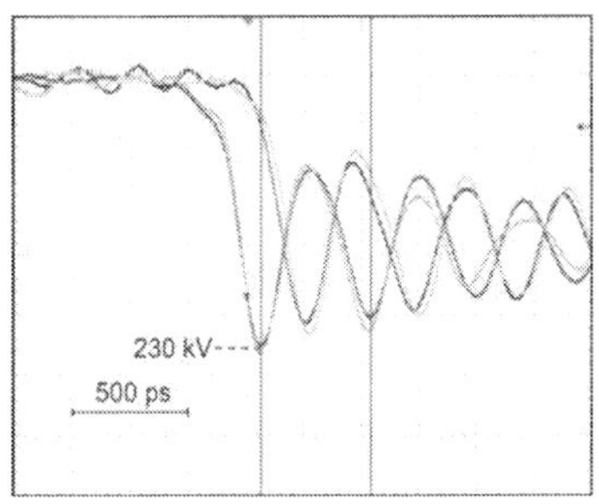

Fig. 2. Half-period shift of waveforms from vertical NLTL pairs.

This work was supported by RFBR Grant No. 16-08-00058 and IHCE SB RAS Project No. SP-3354.2015.1.

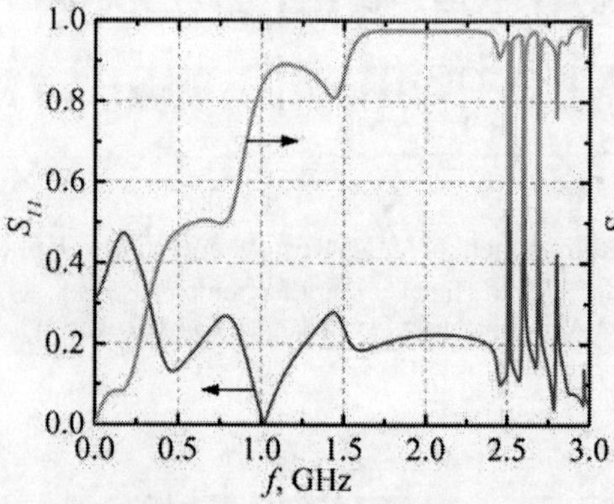

Fig. 3. Calculated S-parameters of the band-pass filters.

After this preliminary experiment the whole assembling as shown in Fig. 2 was set up and the waveforms were recorded by sensors placed after band-pass filters. The band-pass filters were calculated by CST Microwave Studio and designed to transmit RF pulse with center frequency near 2 GHz to the irradiating antenna and absorb the rest video-pulse in the resistive load of the filter (HVR). The calculated S-parameters of the filter are shown in Fig. 3.

Again the currents in the solenoids were set to synchronous operation and half-period shift between vertical pairs of NLTLs. The corresponding waveforms are presented in Fig. 4. The maximum peak voltage reaches 100 kV and higher, implying about 200 MW RF peak power per channel.

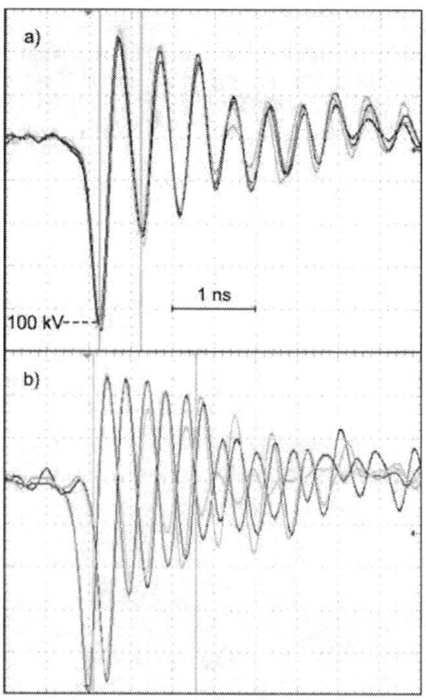

Fig. 4. a) Synchronous waveforms after band-pss filters and b) half-period shift between vertical pairs.

IV. IRRADIATION OF RF PULSES BY HELIX ANTENNAS

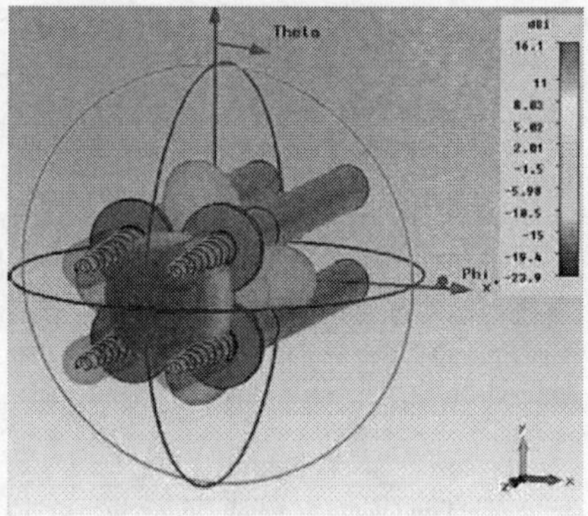

Fig. 5. Helix antenna array with calculated radiation pattern.

Conical helix antennas with oil insulation, Fig. 5, were chosen for irradiation of RF pulses produced by NLTL which usually have about 0.4 GHz bandwidth. The calculated reflection from helix antenna input does not exceed -15 dB in the frequency range from 1.5 to 2.5 GHz. The measurements of irradiated pulses were carried out by a pair of two orthogonal tiny strip-line antennas in order to have signals simultaneously in two polarizations. The distance between irradiating and receiving antennas were varied from 1 to 4 meters. The sources of reflections in the room were placed at distances not disturbing the measurement of a true signal. The measured peak power density distribution corresponding to synchronous NLTL operation shown in Fig, 4 a) together with peak electric field dependence upon distance are presented in Fig. 6.

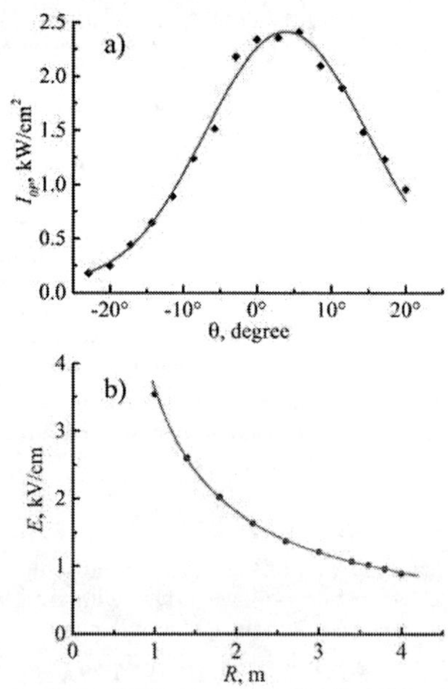

Fig. 6. a) Peak power density distribution at 4 m distance with Gauss approximation. b) Peak electric field at the axis upon distance to antnenna.

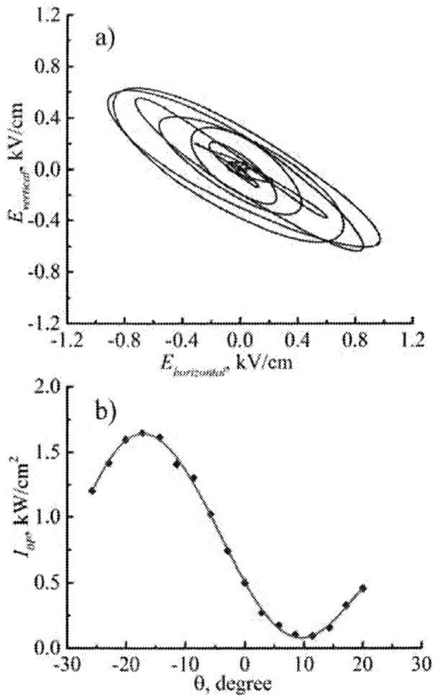

Fig. 7. a) Hodograph at raition axis at 4 m distance and b) -17° steered peak power density ditriburion for half-period shift between vertical NLTL pairs.

The hodograph of the RF pulse measured at the radiation axis at 4 m distance for synchronous operation of NLTLs, Fig. 7 a). The measured polarization ellipticity of 0.3 can originates from not ideal azimuthal symmetry of helix antennas. When the current sets for half-period shift were applied, as in Fig. 4 b), the maximum of the power density distribution was steered to -17°. The high side lobe level of the steered beam is due to low number of antennas in the array.

The irradiation from one channel when the other three were not biased was compared with synchronous four-channel operation at the radiation axis at 1 m distance from the focal point, Fig. 8. One can see 4-fold increase in the measured electric field, which corresponds to 16-fold increase in the power density.

The tests of the system at 1000 Hz pulse repetition rate during 1 s bursts exhibited stable operation with no breakdown in any component.

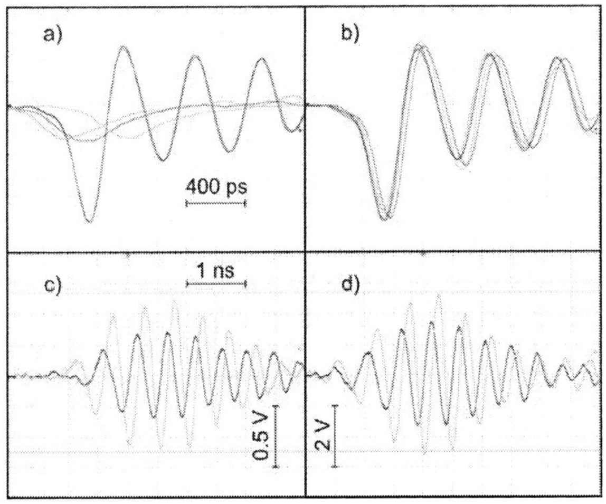

Fig. 8. a) One channel operation with three others unbiased; b) syncronous 4-channel operation; c) waveform of the RF pulse in vertical and horizontal polarization for 1-channel operation and d) for 4-channel operation.

References

[1] N. Seddon, C. Spikings, and J. Dolan, "RF pulse formation in nonlinear transmission lines," in Proceedings of the IEEE Pulsed Power Conference, Albuquerque, NM, USA, pp. 678–681, June 2007.

[2] I.V. Romanchenko, V.V. Rostov, V.P. Gubanov, A.S. Stepchenko, A.V. Gunin, and I.K. Kurkan, "Repetitive sub-gigawatt rf source based on gyromagnetic nonlinear transmission line," Rev. of Sci. Instrum. 83, 074705, 2012.

[3] V.P. Gubanov, A.V. Gunin, O.B. Koval'chuk, V.O. Kutenkov, I.V. Romanchenko, and V.V. Rostov, "Effective transformation of the energy of high-voltage pulses into high-frequency oscillations using a saturated-ferrite-loaded transmission line," Tech. Phys. Lett., vol. 35, p.626, 2009.

[4] J.-W. Bragg, J. Dickens, and A. Neuber, "Material selection considerations for coaxial, ferrimagnetic-based nonlinear transmission lines," J. Appl. Phys. 113, 064904, 2013.

[5] S.J.F. Chadwick, N. Seddon, and S. Rukin, "A novel solid-state HPM source based on a gyromagnetic NLTL and SOS-based pulse generator," in Proceedings of the IEEE Pulsed Power Conference, Chicago, IL, USA, pp. 178–181, June 2011.

[6] I.V. Romanchenko, V.V. Rostov, A.V. Gunin, and V.Yu. Konev, "High power microwave beam steering based on gyromagnetic nonlinear transmission lines," J. Appl. Phys. 117, 214907, 2015.

[7] M.R. Ulmaskulov, M.S. Pedos, S.N. Rukin, K.A. Sharypov, V.G. Shpak, S.A. Shunailov, M.I. Yalandin, I.V. Romanchenko, and V.V. Rostov, "High repetition rate multi-channel source of high-power rf-modulated pulses," Rev. Sci. Instrum. 86, 074702, 2015.

[8] J.M. Johnson, D.V. Reale, J.T. Krile, R.S. Garcia, W.H. Cravey, A.A. Neuber, J.C. Dickens, J.J. Mankowski, "Characteristics of a four element gyromagnetic nonlinear transmission line array high power microwave source," Rev. Sci. Instrum. 87, 054704, 2016.

Relaxation of ferromagnetic and paramagnetic state of thin La-Sr-MnO films exposed by high power picosecond duration optical pulses

S. Balevicius, S. Kersulis, M. Malisauskas, J. Klimantavicius, V. Stankevic, N. Zurauskiene, Z. Balevicius, V. Plausinaitiene, K. Pozela, S. Tolvaišienė,
Center for Physical Sciences and Technology
Sauletekio Ave. 3, LT-10257 Vilnius, Lithuania
e-mail: saulius.balevicius@ftmc.lt

S. Balevicius, S. Kersulis V. Stankevic,
N. Zurauskiene, Z. Balevicius and S. Tolvaisiene
Vilnius Gediminas Technical University
Naugarduko str. 41, LT-03227 Vilnius, Lithuania

V. Plausinaitiene
Vilnius University
Naugarduko str. 24, LT- LT-03227 Vilnius, Lithuania

***Abstract*—We have investigated resistivity rise and relaxation of thin epitaxial (EP) La-Sr-MnO films after thermal shock. Samples of 400 nm thickness La-Sr-MnO were grown on $LaAlO_3$ substrate. The thermal shocks were induced by Nd:YAG laser generating 30 ps pulses at 1064 nm wavelength. The pulse energy at the sample was about 1 mJ. Samples were treated by single laser pulse at different ambient temperatures when the film was in ferromagnetic state. Immediately after the laser pulse resistance of the sample has increased. At temperatures (T_0) lower than phase transition temperature (T_m) time of resistance response increases with rising of T_0 in the range from few hundred ps to several ns. Meanwhile the time of total relaxation of resistance back to initial level was in the range from 10 μs to 35 μs. The experimental results showed longer relaxation time than theoretical modelling of the film cooling process. This led us to conclude that, besides vibrational heat dissipation mechanism, magnetic ordering process in La-Sr-MnO films has to be taken into account. The obtained results help to evaluate the influence of fast temperature changes on high pulsed magnetic field measurements using La-Sr-MnO film sensors.**

***Keywords*—thin films; manganites; optical pulse; magnetic field sensor; thermal shock.**

I. Introduction

Thin manganite films had demonstrated good characteristics to be applied for measurement of wide amplitude range high pulsed magnetic fields and wide temperature range [1]-[6]. The main advantage of sensors based on colossal magnetoresistance (CMR) phenomenon [7], is possibility to measure magnitude of magnetic pulse independently on magnetic field orientation relative to the sensor's film plane. Moreover, changing the composition and preparation conditions of the film it is possible to design sensors operating up to megagauss fields at temperatures ranging from cryogenic to room temperature. These, so called CMR-B-scalar, sensors can be used for magnetic field measurement during operation of non-destructive high magnetic pulse coils, railgun, electromagnetic metal forming or contactless high power electrical current pulses.

However, in all these cases the generation of high magnetic field pulse is always accompanied with thermal shock. Temperature increase time and level depends on type of the device: destructive and non-destructive coils [8], [9], railgun [10], plasma gun [11] or magnetic pulse welding instruments [12]. The peak temperature level could be from several hundred to thousands of Kelvins degrees [13]-[15]. The rise time of thermal shock could be from several tens of seconds [13] to milliseconds (typical time of railgun rails surface heating due to high speed projectile movement [16]). In case of railgun operation the plasma cloud and hot gases speeding along with plasma as well as metallic wear debris are able to generate fast increase of temperature in the vicinity of rails [17]. For example, it was found that the air temperature pulse rise time is less than 1 ms [18]. The response of CMR-B-scalar sensor also depends on ambient temperature; to account for this the sensor response to magnetic field is calibrated at different temperatures.

If time of magnetic pulse and resistance response to thermal shock overlap, the resultant signal is a sum of CMR response and thermal response. Usually the CMR-B-scalar sensor is placed far enough from heat source and the temperature pulse is significantly delayed in time relative to magnetic field pulse; this helps to avoid aforementioned errors. However, in case of repetitive pulses (magnetic welding or multi-shoot railgun) or if magnetic measurement are performed in the vicinity of rapidly heated parts of the device thermal resistance response has to be taken into consideration. Evaluation could be done by investigating dynamics of manganite film resistivity response to thermal shock and cooling characteristic of the CMR-B-scalar sensor.

Our approach to study temperature shock impact on resistivity of manganite films was to heat the film with short high power laser pulse as described in [19]-[21]. It was noted in [22] that manganite films resistivity response to IR light pulse has to be mainly associated with heating, because wavelength of the light is smaller than the band gap (~1 eV) of these films. In this paper we present the investigation of La-Sr-MnO epitaxial films resistivity response to fast heating by picosecond duration 1064 nm wavelength high power laser pulses.

II. Experimental

The epitaxial (EP) 400 nm thick $La_{0.83}Sr_{0.17}MnO_3$ films were used for investigations. The films were deposited by MOCVD technique [23] on 400 µm thickness $LaAlO_3$ substrate. Samples used for investigation were prepared as two terminal devices with 1×1 mm^2 square thin film silver electrodes deposited in the same plane and separated from each other by 50 µm. Resistance (R) *vs.* temperature (T) dependence (see Fig.1) measured by low DC electric field method demonstrated that films exhibited typical for ferromagnetic to paramagnetic phase transition at critical temperature $T_m \approx$ 320 K.

Schematic diagram of equipment used to measure thermal shock impact on electrical conductivity of the films is presented in Fig. 2. The set-up consists of close cycle helium cryocooler (Janis with LakeShore controller) with sample mounted inside, I1-15 type electric pulse generator, which generated 0.1-10 µs duration rectangular waveform pulses with amplitude up to 10 V, Ekspla Nd:YAG laser generating 1064 nm wavelength 30 ps duration 100 mJ energy single pulses and four channels real time oscilloscope Tektronix 16 GHz, 100 GS/s. The equipment operates in the following way: after stabilizing the $La_{0.83}Sr_{0.17}MnO_3$ sample's ambient temperature, the sample was biased by 10 µs duration 1 V amplitude electrical pulse. This pulse was synchronized with 30 ps duration 1 mJ energy laser pulse, which was by 1 µs delayed relative to the front of the electrical pulse. The electrical response to laser pulse was recordered by real time oscilloscope. For measurement of sample's resistance both incident and transmitted electrical pulses were measured.

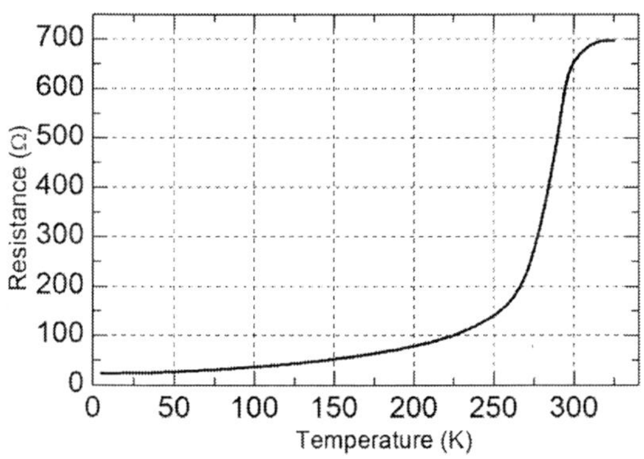

Fig. 2. Resistance *vs.* temperature dependence for $La_{0.83}Sr_{0.17}MnO_3$ epitaxial film.

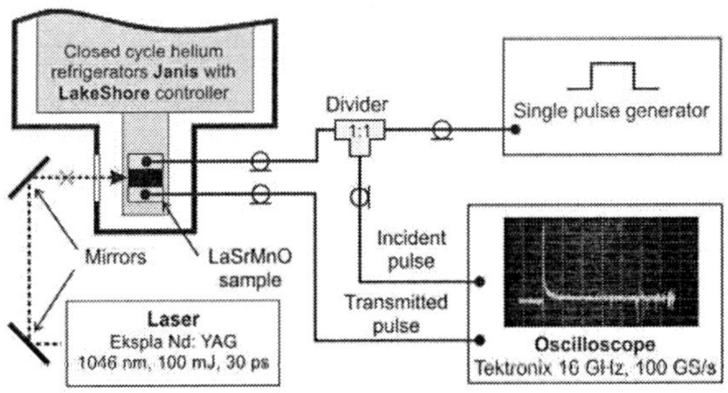

Fig. 1. Schematic diagram of the experimental equipment.

The equation for resistance calculation is: $R = Z \times (V_{in}+V_{tr})/(V_{in}-V_{tr})$, here V_{in} and V_{tr} is incident and transmitted signal voltage, respectively, and Z=50 Ω is the impedance of the transmission line.

III. Results

The investigation of the resistance response of epitaxial (EP) film to illumination by 30 ps laser pulse was performed in two temperature ranges corresponding to ferromagnetic (FM) state (from 50 K to 230 K) and transition from ferromagnetic to paramagnetic (TFP) state (from 250 K to 300 K). It was obtained that the rise time of the response pulse (see Fig.3a, b) depends on initial temperature (T_0) of the film and increase with the increase of this temperature. In the FM region it was less than 1 ns meanwhile in the TFP region it changes from 1 ns to 2.5 ns. Typical waveforms of this

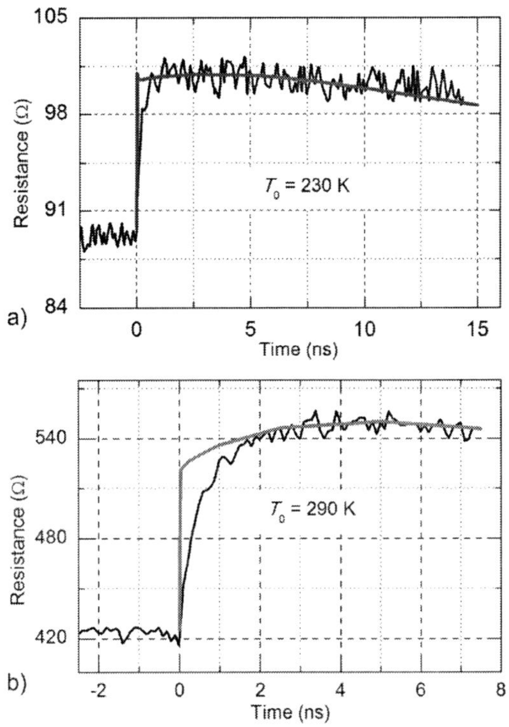

Fig. 3. The waveforms of resistance change response to optical pulse at different T_0: a) 230 K, b) 290 K.

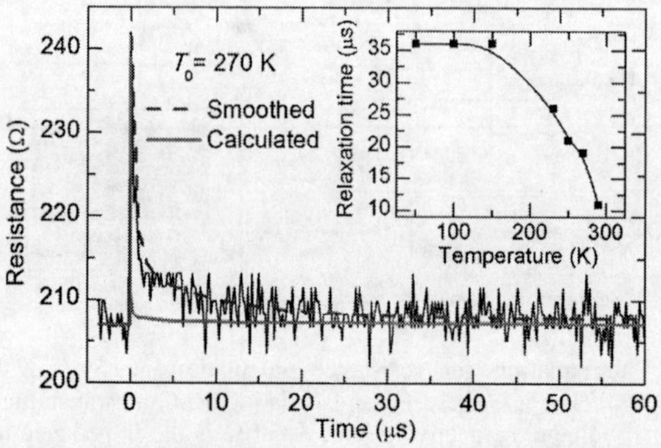

Fig. 4 Waveform of resistance relaxation process after laser pulse impact for the $La_{0.83}Sr_{0.17}MnO_3$ films at T_0=270 K. The inset presents relaxation time dependence on T_0.

response immediately after laser pulse are presented in Fig.3 a, b.

The resistance relaxation back to the lower initial (prior to laser pulse) state process is significantly longer than resistance response to laser pulse duration. Typical waveform of this relaxation at T_0 = 270 K is presented in Fig.4. It can be seen, that after a certain time period (relaxation time) the resistance of the film completely returns to the initial state. The relaxation time is a very weak function of T_0 in the FM state, but decreases significantly with increases of T_0 in TFP state (see Fig.5).

IV. Discussion

The experimental investigations showed that illumination by τ=30 ps duration laser pulse of manganite film, kept at temperature lower than T_m, always causes the increase of film resistance. The duration of this raised state is significantly longer than the duration of laser pulse and at temperatures close to T_m exceeds several nanoseconds. This leads to conclude, that heating of the film and, particularly, heat propagation deep into the film is mainly responsible for resistance rise. Fig.5 shows theoretical temperature increase (ΔT) profile at different time instants during laser pulse, calculated using equation (1) presented in [24]; the equation takes into account decrease in propagated light intensity inside the film sample:

$$T_{1,2}(z,t) = \frac{\alpha I_0(1-R)}{k}\left\{2(Dt^*)^{0.5} ierfc\left[\frac{z}{2(Dt^*)^{0.5}}\right] - \frac{1}{\alpha}e^{-\alpha z} + \frac{1}{2\alpha}e^{\alpha^2(Dt^*)}\left(e^{-\alpha z} erfc\left[\alpha(Dt^*)^{0.5} - \frac{z}{(Dt^*)^{0.5}}\right] + e^{\alpha z} erfc\left[\alpha(Dt^*)^{0.5} + \frac{z}{(Dt^*)^{0.5}}\right]\right)\right\} \quad (1)$$

Here α is light absorption coefficient, I_0 is light intensity, D is heat diffusion coefficient, k is thermal conductivity, R is reflectivity coefficient, z is coordinate along axis perpendicular to film plane and t^* is time. The temperature distribution during laser pulse is calculated at certain time instants t^*. $\Delta T_{1,2}$ indicates temperature increase ΔT at different time instants $t^*=t$ and $t^*= t-\tau$. For resistance relaxation process the temperature distribution is calculated by the following equation:

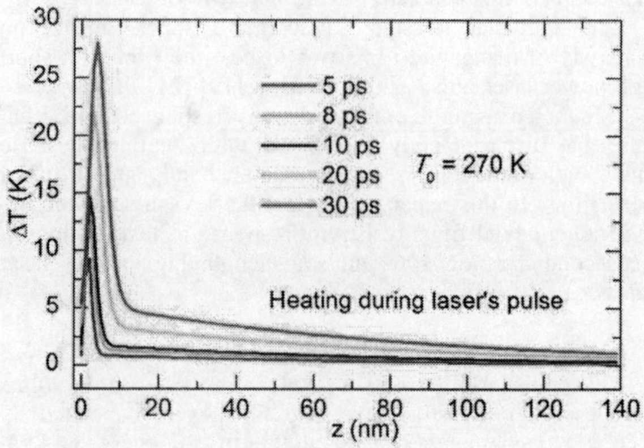

Fig. 5 Temperature change distribution inside the sample of the film along axis perpendicular to film surface during 30 ps laser pulse impact (time in ps is counted starting from the laser pulse front).

$$T(z,t) = T_1(z,t) - T_2\,[z,(t-\tau)] \quad (2)$$

For calculations the following parameters of epitaxial $La_{0.83}Sr_{0.17}MnO_3$ films were used: k=2.2 W/(mK); heat capacity C_p=10^3 J/(kgK); density ρ=6.5×10^3 kg/m^3 and laser beam intensity I_0=1.18×10^4 W/m^2.

The temperature distribution inside the sample film immediately after the laser pulse is strongly non-homogeneous (see Fig. 5). This means that further propagation of heat into volume of the substrate is responsible for resistance rise and relaxation time values. This propagation is demonstrated in Fig. 6 showing that after 5-10 ns the temperature distribution in the film is very close to uniform.

The sample resistance change (ΔR) due to ΔT redistribution during heat propagation deep into the film was calculated using $La_{0.83}Sr_{0.17}MnO_3$ film $R(T)$ relationship presented in Fig.1. It has been shown, that the rise time of the response shall be much longer than optical pulse duration (see solid line in Fig 3a, b). The mismatch between experimental and calculated curves in the beginning of the rise time can be attributed to bandwidth limitation of the transmission line of measurement system, which was not accounted in calculation method. An increase of this rise time with increase of T_0 is a

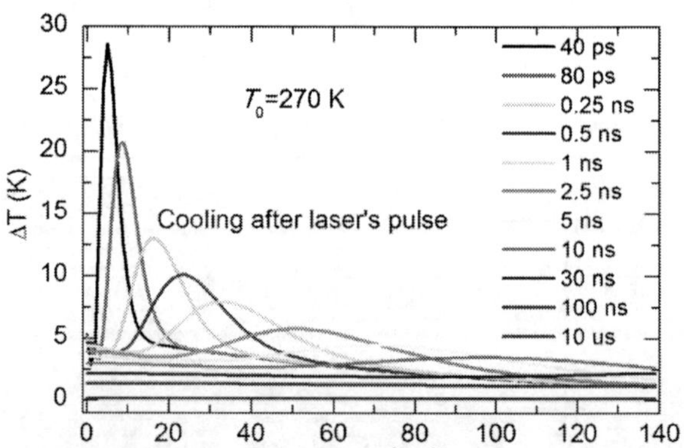

Fig. 6. Temperature change distribution inside the sample of the film along axis perpendicular to film surface at different time instances after laser pulse impact (time in ps is counted from the laser pulse front).

result of heat flow rate decrease due to increase of film heat capacity and subsequent decrease of maximal ΔT.

The comparison of experimental and calculated risen resistance relaxation back to the initial state (level prior to laser pulse) behavior shows substantial disagreement as calculated relaxation is significantly faster than obtained experimentally (see „smoothed" and „calculated" lines in Fig.4). We explain the difference in the following way: in the calculations the cooling down process is the only cause of resistance relaxation, and magnetization relaxation is not accounted for. So the difference can be explained by magnetization relaxation to initial state being slower than film cooling down process as it was noted in [25]. This assumption also is in agreement with the decrease of resistance relaxation time with increase of T_0, demonstrating that steady state magnetic ordering is achieved faster at higher temperatures of the film.

In conclusion, we found that characteristic time of resistance relaxation of submicron thickness epitaxial $La_{0.83}Sr_{0.17}MnO_3$ films prepared by MOCVD technique on $LaAlO_3$ substrate after thermal shock induced by picosecond duration laser pulse is of the order of several tens of microseconds. Relaxation process cannot be explained only by film cooling down process, and magnetic system ordering has to be taken into account. As a result, if measured magnetic pulses are accompanied by substantial thermal shocks, highest repetition rate of magnetic pulses and operating frequency of high magnetic field sensors shall be limited to 50 kHz.

Acknowledgments

This work was partly supported by the Research Council of Lithuania (grant No. MIP-095/2015).

References

[1] M. Schneider, R. Schneider, V. Stankevič, S. Balevičius, N. Žurauskienė, "Highly local measurements of strong transient magnetic fields during railgun experiments using CMR-based sensors," *IEEE Trans. Magn.*, vol. 43, pp. 370-375, 2007.

[2] O. Liebfried, M. Löffler, M. Schneider, S. Balevičius, V. Stankevič, N. Žurauskienė, A. Abrutis, V. Plaušinaitienė, "B-scalar measurements by CMR-based sensors of highly inhomogeneous transient magnetic fields", *IEEE Trans. Magn.*, vol. 45, iss. 1, p. 5301-5306, 2009.

[3] N. Žurauskienė, S. Balevičius, V. Stankevič, S. Keršulis, M. Schneider, O. Liebfried, V. Plaušinaitienė, A. Abrutis, "B-Scalar Sensor Using CMR Effect in Thin Polycrystalline Manganite Films," *IEEE Trans. Plasma Sci.*, vol. 39, iss. 1, pp. 411-416, 2011.

[4] S. Balevicius, N. Zurauskiene, V. Stankevic, S. Kersulis, V. Plausinaitiene, A. Abrutis, S. Zherlitsyn, T. Herrmannsdorfer, J. Wosnitza, F. Wolff-Fabris, "Nanostructured thin manganite films in megagauss magnetic field," *Appl. Phys. Lett.*, vol. 101, 092407, 2012.

[5] S. Balevičius, N. Žurauskienė, V. Stankevič, T. Stankevič, J. Novickij, M. Schneider, "High frequency CMR-B-Scalar sensor for pulsed magnetic field measurement", *IEEE Trans. Plasma Sci.*, vol. 41, iss. 10, pp. 2885–2889, 2013.

[6] T.L. Haran, R.B.Hoffman, S.E.Lane, "Diognostic capabilities for electromagnetic railgun," *IEEE Trans. Plasma Sci.*, vol. 41, iss. 5, pp. 1526-1532, 2013.

[7] A-M Haghiri-Gosnet[1] and J-P Renard, "CMR manganites: physics, thin films and devices," *Journal of Physics D: Appl. Phys.*, vol. 36, iss. 8, pp. R127-R150, 2003.

[8] CH. Mielke, and BM. Novac, "Experimental and numerical studies of megagauss magnetic-field generation at LANL-NHMFL," *IEEE Trans. Plasma Sci.*, vol. 38 (8 PART 1), pp. 1739-1749, 2010.

[9] S. Zherlitsyn; T. Herrmannsdorfer; B. Wustmann; J. Wosnitza, "Design and Performance of Non-Destructive Pulsed Magnets at the Dresden High Magnetic Field Laboratory," *IEEE Trans. Appl. Supercond.*, vol. 20, iss. 3, pp. 672 – 675, 2010.

[10] O. Liebfried, M. Schneider, S. Balevičius, "Current Distribution and Contact Mechanisms in Static Railgun Experiments With Brush Armatures," *IEEE Trans. Plasma Sci.*, vol.39, iss.1, pp.431-436, 2011.

[11] D.A.Wetz, Jr.F.Stefani, I.R.McNab, "Experimental Results on a 7-m-Long Plasma-Driven Electromagnetic Launcher," *IEEE Trans. Plasma Sci.*, vol.39, iss. 1, pp.180-185, 2011.

[12] R. M. Miranda, B. Tomás, T. G. Santos, N. Fernandes, "Magnetic pulse welding on the cutting edge of industrial applications," *Soldag. Insp. Sao Paulo*, vol. 19, iss. 01, pp. 069-081, 2014.

[13] S. Gies, A. Langolf, C. Weddeling, A.E. Tekkaya, "Measurement of working coil temperature in electromagnetic forming processes by means of optical frequency domain reflectometry," *Case Studies in Nondestructive Testing and Evaluation*, vol. 3, pp. 15–20, 2015.

[14] R. Kratz, Y. M. Eyssa, L. LI, H. J. Schneider-Muntau, M. R. Vaghar, S. W. Van Sciver, "Optimal use of magnetic energy in magnet, Megagauss magnetic field generation, its application to science and ultra-high pulsed-power technology," *Proceedings of the VIIIth International Conference on Megagauss Magnetic Field Generation and Related Topics*, Tallahassee, Florida, USA 18-23 October, World Scientific, Edited by Hans J. Schneider-Muntau, pp. 112-115, 1998.

[15] J. J. Weimer, I. L. Singer, "Temperatures From Spectroscopic Studies of Hot Gas and Flame Fronts in a Railgun", *IEEE Trans. Plasma Sci.*, vol.39, iss. 1, pp. 174-179, 2011.

[16] M. D. Bryant, "A Bond Graph Model of an Electromagnetic Launcher-Part 2: Simulation Results," *IEEE Trans. Plasma Sci.*, vol. 39, iss.1, pp. 40-47, 2011.

[17] Ch. Schuppler, F. Alouahabi, M. Schneider, "Novel Sensor for Projectile Detection in a Multishoot Railgun," *IEEE Trans. Plasma Sci.*, vol.41, iss. 5, pp. 1382-1386, 2013.

[18] Ryan B. Hoffman, Terence L. Haran, J. Christopher James, R. Brandon Vaughan, Cody W. Lamb, Nathan Meraz, "In Situ Measurement of Strain and Temperature for Railgun Launcher Diagnostics," *IEEE Trans. Plasma Sci.*, vol. 43 iss. 5, pp. 1302 – 1309, 2015.

[19] L. Méchin, J-M. Routoure, B. Guillet, F. Yang, S. Flament, D. Robbes, R. Chakalov, "Uncooled bolometer response of a low noise $La_{2/3}Sr_{1/3}MnO_3$ thin film," *Appl. Phys. Lett.*, vol. 87, p. 204103, 2005.

[20] F. Yang, L. Méchin, J.-M. Routoure, B. Guillet, and R. A. Chakalov, "Low-noise $La_{0.7}Sr_{0.3}MnO_3$ thermometers for uncooled bolometric applications," *J. Appl. Phys.*, vol. 99, iss. 2, p. 024903, 2006.

[21] A. Aryan, J-M. Routoure, B. Guillet, P. Langlois, C. Fur, J. Gasnier, C. Adamo, D. G. Schlom, L. Méchin, "Optical Characterisation of $La_{0.7}Sr_{0.3}MnO_3$ Thin Film Based Uncooled Bolometers," *SENSORDEVICES 2012: The Third International Conference on Sensor Device Technologies and Applications*, pp. 61-65, 2012.

[22] Kun Zhao, Kui-Juan Jin, Yan-Hong Huang, Hui-Bin Lu, Meng He, Zheng-Hao Chen, Yue-Liang Zhou, Guo-Zhen Yang, "Laser-induced ultrafast photovoltaic effect in $La_{0.67}Ca_{0.33}MnO_3$ films at room temperature," *Physica B* vol. 373, pp. 72–75, 2006.

[23] N. Žurauskienė, S. Balevičius, P. Cimmperman, V. Stankevič, S. Keršulis, J. Novickij, A. Abrutis, V. Plaušinaitienė, "Colossal magnetoresistance properties of $La_{0.83}Sr_{0.17}MnO_3$ thin films grown by MOCVD on lucalox substrate," *Journal of low temperature physics*, vol. 159, iss. 1-2, pp. 64-67, 2010.

[24] Afaf M. Abd El-Hameed, "Analytical study of pulsed laser irradiation on some materials used for photovoltaic cells on satellites," *NRIAG Journal of Astronomy and Geophysics*, vol. 4, pp. 256–265, 2015.

[25] N. Kazantseva, U. Nowak, R. W. Chantrell, J. Hohlfeld, A. Rebei, Slow recovery of the magnetisation after a sub-picosecond heat pulse, *EPL (Europhysics Letters)*, vol. 81, iss. 2, pp. 27004, 2007.

Pulsed Power Source Voltage Forming Methods for Six Phase Two Level Inverter

A. Baskys, T. Lipinskis, V. Bleizgys, S. Balevicius, S. Tolvaisiene
Faculty of Electronics, Vilnius Gediminas Technical University
Vilnius, Lithuania
algirdas.baskys@vgtu.lt

A. Baskys, V. Bleizgys, S. Balevicius,
Dept. of Material Science and Electrical Engineering
Center for Physical Sciences and Technology
Vilnius, Lithuania
saulius.balevicius@ftmc.lt

T. Lipinskis
Military Technology Centre
Gen. Jonas Zemaitis Military Academy
Vilnius, Lithuania
tadas.lipinskis@mil.lt

***Abstract*—Two voltage forming methods for pulsed power sources used for realization of inverters for speed and torque control of six phase AC induction motors have been proposed. They are based on space vector modulation principle. The space vector diagram of a six phase inverter loaded with a symmetrical single neutral load is obtained. The first voltage forming method is based on large space vectors, the second one – on medium space vectors. The developed methods were analyzed using simulation and experimentally.**

***Keywords*— pulsed power, inverter, six phase, AC inductin motor.**

I. Introduction

The pulsed power sources used for realization of inverters for speed and torque control of AC induction motors allow us to generate not only generally used three phase voltage but multiphase (more than three phase) voltage as well. Therefore, the multiphase AC induction motors can be introduced. Multiphase AC induction motors have following advantages as compared to commonly used three phase motors: higher reliability due to ability to operate with one or more faulty phases; lower acoustic noise; lower torque pulsations; smoother start-up and higher torque at low speed. These features make the multiphase AC induction motors promising for use in electric vehicle drives. Six phase AC induction motors are perspective candidates for commercial applications. The reason is that stator of generally used three phase motor can be applied for production of six phase motor because windings slots of stator are spaced geometrically correct for six phase application. One more advantage of six phase AC induction motor drives is that high power inverter transistor modules are manufactured for three phases, so a six phase inverter could use two of these modules without any unused transistor pairs.

Despite advantages, six phase AC induction motor drives are not widely researched. There are only few publications on the topic of six phase voltage forming methods using space vector modulation principle [1-4]. Therefore, the development of voltage forming methods for a six phase inverter is relevant.

The both six phase voltage forming methods proposed in this work are based on space vector modulation principle, which is used usually for three phase voltage forming [5, 6]. It is necessary to stress that the complexity of the space vector modulation principle realization for the forming of multiphase voltage increases heavily with the increase of phase number.

II. Six Phase Voltage Forming Methods Based on Space Vector Modulation Principle

A. Space vector diagram for six phase inverter

The space vector diagram of an inverter is obtained by applying Clarcke's transformation for all phase voltages and every inverter switch combination, for a given load topology [7]. Six phase two level inverter includes six switch arms, because of this has 2^6 possible inverter switch combinations, which represent 64 different space vectors in the space vector diagram. The developed space vector diagram of a six phase inverter loaded with a symmetrical single neutral load is presented in Fig. 1. The space vectors of six phase inverter are mapped on two planes: main plane d_1-q_1 (Fig. 1a) and secondary plane d_2-q_2 (Fig. 1b). Vectors of the same magnitude and angle in the main plane have completely different magnitude and angle in the secondary plane, the only exception are zero vectors (vectors number 1 and 64), and pseudo-zero vectors (vectors number 22 and 43), which have zero magnitude in both planes. Space vectors with nonzero amplitude are named as active vectors.

Vectors in the main plane can be divided into three groups according their magnitude: there are large vectors with the magnitude of $0.667U_{DC}$ (vectors, which compose the largest hexagon in Fig. 1a), medium vectors with the magnitude of $0.577U_{DC}$ and small vectors with the magnitude of $0.33U_{DC}$,

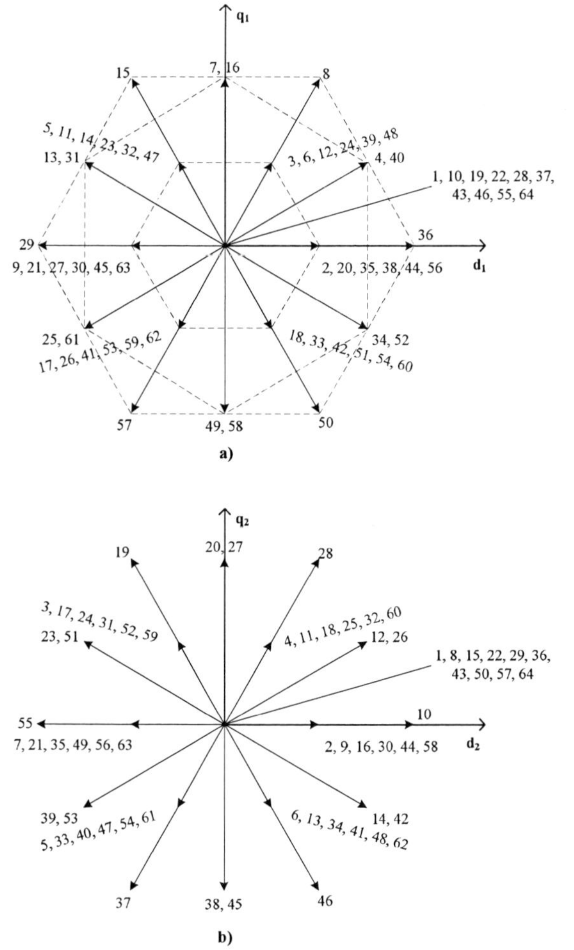

Fig.1. Space vector diagram of a six phase inverter loaded by a symmetrical single neutral load: a) main d_1-q_1 plane; b) secondary d_2-q_2 plane

where U_{DC} is DC supply voltage of inverter. Vectors of the secondary plane can be divided by their magnitude into the same three groups as well, only the numbers of vectors differ.

The angular rotation speed ω of generated reference vector is determined by inverter output voltage frequency f_{AC} acording to the realation $\omega = 2\pi f_{AC}$. The frequency f_{AC} is called usually as phase frequency. Reference vector length determines the amplitude of the output voltage. The reference vector has to draw a perfect circle when rotating. Therefore, the maximum value of inverter output voltage amplitude is limited by the maximum radius of the circle that can be drawn inside the largest hexagon (Fig 1a). Harmonic distortions of the output voltage will occur if the reference vector would be longer than radius of maximum circle.

B. *Six phase voltage forming method using large space vectors*

The voltage forming method employed in voltage source inverter should provide the sinus waveform of motor windings current. This can be achieved if the inverter output voltage does not include the low frequency harmonics. The high frequency harmonics caused by switching of inverter switches are eliminated by inductances of motor windings, which operate as low pass filter. The voltage generated by inverter should have appropriate amplitude and frequency to provide the desired speed and torque of motor. The large vectors 8, 15, 29, 57, 50 and 36 (Fig. 1a) are used for generation of six phase voltage in the first proposed voltage forming method. The space vector diagram includes only 6 space vectors, which form six sectors, with the angle π/60 each, in such a case.

The reference vector in every space vector diagram sector is formed by two adjacent active space vectors, which form the sector. The reference space vector position inside of sector is determined by activity duration of space vectors, which create the sector. The reference vector angle is changed discreet with period T_s, which is reverse proportional to the switching frequency of inverter switches f_s. During the period $T_s = 1/f_s$ not only two active space vectors but two zero vectors are used as well. As an example (Fig. 2), in the first sector active space vectors are 36 and 8, therefore, when the reference space vector is in the sector 1, it is generated using active vectors 36, 8 and two zero vectors 1 and 64. Zero vectors are used to fill the gaps of time, which is left in the switching period T_s, after active space vectors action is over. The length of reference space vector determines the amplitude of generated six phase voltage. It is controlled by variation of ratio between the active and zero vectors action duration. The parameter - modulation index m, which varies in range from 0 to 1, is used to set this ratio.

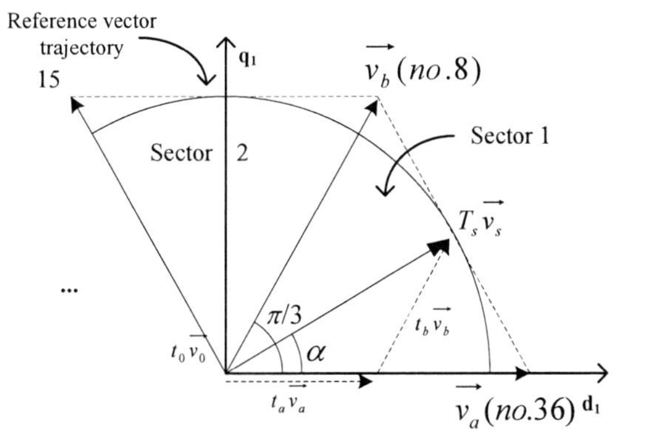

Fig.2. Generation of reference space vector in the first sector using large space vectors

Maximum amplitude of the output voltage is the radius of circle that can be inscribed into the hexagon. This circle represents the reference vector trajectory when the output voltage amplitude is set to maximum (m=1). As it can be seen from the Fig. 2, the circle touches the hexagon at an angle α = 30°. Therefore, the maximum voltage amplitude, which coincides with maximum magnitude of reference space vector, is

$$\left|\vec{v}_{\mathrm{s\,max}}\right| = 0,667U_{\mathrm{DC}}\cos(\pi/6) = 0,577U_{\mathrm{DC}}. \quad (1)$$

Duration of active space vectors action is estimated by applying the volt-second rule on the d and q axes. The equations for calculation of active space vectors and zero vectors action duration for the case when large space vectors are used for forming of six-phase voltage are as follows:

$$t_a = T_s m \frac{\left|\vec{v}_{s\max}\right|}{\left|\vec{v}_a\right|} \frac{\sin(k\pi/3-\alpha)}{\sin(\pi/3)}, \quad (2)$$

$$t_b = T_s m \frac{\left|\vec{v}_{s\max}\right|}{\left|\vec{v}_b\right|} \frac{\sin\left[\alpha-(k-1)\pi/3\right]}{\sin(\pi/3)}, \quad (3)$$

$$t_0 = T_s - t_a - t_b, \quad (4)$$

where T_s is the switching period of inverter switches, m is modulation index, $\vec{v}_{s\max}$ is maximum magnitude of reference space vector, $\vec{v}_a$ and $\vec{v}_b$ are active space vectors, which form the reference vector in given sector, k is number of sector, α is angle of reference vector.

C. Six phase voltage forming method using medium space vectors

The developed six phase voltage forming method using medium space vectors offers lower utilization of inverter DC supply voltage since the space vectors used in the generation of reference space vector are shorter, i.e. $0.577U_{DC}$ (Fig. 1). Therefore, the maximum magnitude of the reference space vector is

$$\left|\vec{v}_{s\max}\right| = 0,577U_{DC}\cos(\pi/6) \approx 0,5U_{DC}. \quad (5)$$

Medium space vectors, which are used for six phase voltage forming overlap in the main plane d_1-q_1 (Fig. 1a), i. e. there are two vectors that have the same magnitude and angle in the space vector diagram, e.g. space vector 4 and 40, 7 and 16 (Fig. 1a), which form the sectors similar as in the method based on large vectors. The inverter output voltage can be formed using one vector from every couple of overlapping vectors. On the other hand, it can be formed using both overlapping vectors as well, under condition that activity duration of every vector is shortened twice.

The problem with the application of medium space vectors for six phase voltage forming is that they appear as small magnitude vectors in the secondary plane (Fig 1b). Therefore, during the generation of reference space vector in the main plane, this vector inevitably will be generated in the secondary plane also and will lead to introduction of unwanted low order harmonics. The solution of this problem can be found analyzing space vector diagram of six phase inverter (Fig. 1). It is seen that medium vectors overlapping in the main plane d_1-q_1, e.g. 4 and 40, in the secondary plane d_2-q_2 are presented by vectors with the same magnitudes but with the angles shifted by 180°. In the case when the output voltage is generated using both overlapping space vectors, which lie on the sector's border, e.g. 4 and 40, 7 and 16, this couple of vectors eliminates each other in the secondary plane, i.e. the resultant vector magnitude in secondary plane is 0. Since the angles of these vectors in the main plane coincide, the resultant vector is a sum of these two vectors, i.e. it would be twice longer as it is necessary. This problem can be solved by decreasing of vectors action time, i.e. one half of calculated action time should act one vector and another half - another vector according following equations:

$$t_a\left|\vec{v}_{a12}\right| = 0,5t_a\left|\vec{v}_{a1}\right| + 0,5t_a\left|\vec{v}_{a2}\right|, \quad (6)$$

$$t_b\left|\vec{v}_{b12}\right| = 0,5t_b\left|\vec{v}_{b1}\right| + 0,5t_b\left|\vec{v}_{b2}\right|, \quad (7)$$

where $\vec{v}_{a1}$, $\vec{v}_{a2}$ and $\vec{v}_{b1}$, $\vec{v}_{b2}$ are overlapping vectors, which create the sector.

The action time of active vectors t_a, t_b and zero vectors t_0 in the proposed six phase voltage forming method based on medium space vectors can be calculated using the same equations (2) – (4) as in case of method based on large vectors.

III. Investigation of Six Phase Voltage Forming Methods

Simulation and experimental investigation was conducted to analyze the proposed six phase voltage forming methods. Simulation was performed using software Simulink. A model of a six phase voltage source inverter, which realizes developed voltage forming methods, was created for simulation. To analyze the low frequency harmonics in inverter output voltage, the low pass RC filter was connected in parallel to load of inverter for filtering of high frequency harmonics caused by switching of inverter switches. Simulation of inverter was performed for 30V inverter supply DC voltage and 5 kHz switching frequency.

The waveforms of inverter phase voltage in respect to neutral voltage obtained using simulation are presented in Fig. 3. It can be seen that the shape of inverter output voltage is not pure sinus in the case when six phase voltage forming method based on large space vectors is used. The distortions of voltage are inherent to that caused by third harmonic. It was confirmed after analyzing the data using FFT that output voltage contains the main 30 Hz harmonic, as well as the third 90 Hz harmonic, with amplitude, which is about 15% of main harmonic amplitude. The impact of harmonics on output voltage shape caused by switching of inverter switches is eliminated by RC filters connected to inverter load.

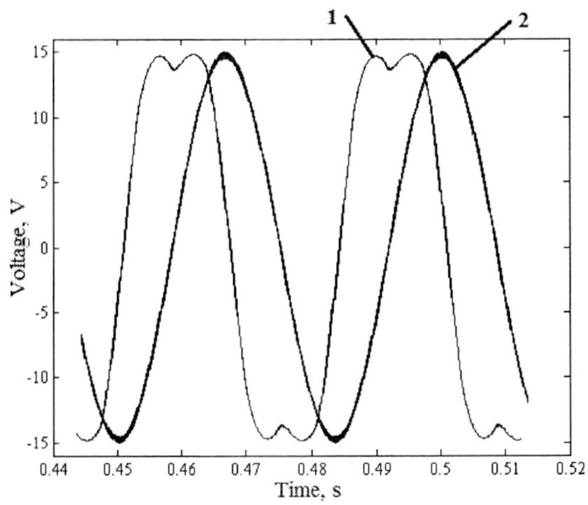

Fig.3. Waveforms of inverter phase ouput voltage in respect to neutral voltage filtered by low pass filter, generated using six phase voltage forming methods: 1- using method based on large space vectors; 2 - using method based on medium space vectors. Results are obtained using simulation.

The waveform is close to pure sinus without tangible distortions caused by low frequency harmonics when six phase voltage forming method based on medium space vectors is used (Fig. 3). The FFT analysis shows that the voltage contains the main harmonic and low amplitude harmonics with the frequency, which is close to switching frequency of inverter switches.

The experimental investigation of six phase inverter prototype has been performed. The obtained experimental waveforms support the results of simulation of a six phase inverter. Consequently, it can be stated that the shape of inverter output voltage is distorted in the case when six phase voltage forming method based on large space vectors is used. However, the waveform has sinus shape without visible distortions when the six phase voltage forming method based on medium space vectors is applied.

IV. Conclusions

1. Space vector diagram of six phase inverter loaded with a symmetrical single neutral load includes 64 different space vectors, which are mapped on two planes: main plane and secondary plane.

2. There are three groups of vectors in the main plane of space vector diagram: large vectors with the magnitude $0.667U_{DC}$; medium vectors with the magnitude $0.577U_{DC}$; small vectors with the magnitude $0.33U_{DC}$.

3. The six phase voltage forming method based on large vectors allows us to achieve high inverter output voltage amplitude, however, the voltage includes third harmonic.

4. The six phase voltage formed using medium space vectors does not contains low frequency harmonics, therefore, is more applicable for employment in six phase inverter.

References

[1] D. Dujic, A. Igbal, E. Levi and L. J. Moores, "A space vector PWM technique for symetrical six-phase voltage source inverters", EPE Journal, vol.17, no.1, 2007, pp. 24-32.

[2] M.V. Aware, "Six-phase inverter operation with space vector pulse-width modulation for a symmetrical single neutral load", Electric Power Components and Systems, vol.41, no.16, 2013, pp.1635-1653.

[3] M. B. R. Correa, C. B. Jacobina, C. R. da Silva, A. M. N. Lima and E. R. C. da Silva, "Vector and scalar modulation for six-phase voltage source inverters" IEEE 34th Power Electronics Specialist Conference, PESC '03, vol.2, 2003, pp. 562 – 567.

[4] M. A. Khan, "Pulse width modulation scheme for quasi six-phase voltage source inverter", International Conference on Innovative Applications of Computational Intelligence on Power, Energy and Controls with their Impact on Humanity (CIPECH14), 2014, pp.296-303.

[5] A. Baskys, V. Bleizgys and V. Gobis, "The impact of output voltage modulation strategies on power losses in inverter", Elektronika ir Elektrotechnika, vol. 94, no. 6, 2009, p. 47-50.

[6] K. Zhou and D. Wang, "Relationship between space-vector modulation and three-phase carrier-based PWM: a comprehensive analysis", IEEE Transactions on Industrial Electronics, vol. 49, no. 1, 2002, pp. 186-196.

[7] E. Levi, R. Bojoi, F. Profumo, H. A. Toliyat and S. Williamson, "Multiphase induction motor drives – a technology status review", IET Electr. Power Appl., vol. 1, no. 4, 2007, pp. 489–516.

Gyromagnetic RF source for interdisciplinary applications and its approbation on isolated liver mitochondria

I. V. Romanchenko, V. Yu. Konev, M. A. Bolshakov, L. P. Zharkova, A. V. Gunin and V. V. Rostov
Institute of High Current Electronics, SB, RAS, 2/3 Akademichesky Ave.,
Tomsk 634055, Russia

***Abstract* — RF pulses from a coaxial gyromagnetic nonlinear transmission line (NLTL) were converted to TE_{10} mode of the rectangular air-filled waveguide loaded onto RF spirit-filled load. The maximum amplitude of the electrical field strength in the waveguide was varied from 0.1 to 40 kV/cm with central frequency of RF pulses from 0.6 to 1.0 GHz. The uniform section of the waveguide was created as a chamber for testing of biological objects. The investigation of the effect of the applied RF pulses on the functionality of mouse liver mitochondria exhibited the change in the mitochondria respiration level depending on the field strength and number of pulses. A significant decay of mitochondria oxygen consumption after RF pulses exposure was detected.**

Keywords* — *nonlinear transmission lines; RF pulses; mitochondrial respiration; electroporation

I. Introduction

The development of compact sources of short high power RF pulses is stimulated by growing interest in interdisciplinary research of different biological objects: food treatment, enhancement of cancer chemotherapy, cell electroporation research [1,2]. Though nanosecond pulsed electric fields (nsPEF) are more common means in these fields [3], high power RF pulses have some advantages for they can be used in the electrodeless systems capable of treating large volumes of the target. One of the promising solution to generate high power RF pulses are nonlinear transmission lines (NLTLs) as they do not use an electron beam and are free of accompanying X-ray radiation [4-6]. The most powerful NLTLs were shown to be gyromagnetic ones [7-9]. Gyromagnetic NLTLs produce nanosecond RF pulses with hundreds of megawatts of power with the possibility of gradual decreasing in power (up to several times) by keeping the frequency in the 0.5 to 2 GHz range [10], when the radiation penetration depth into biological objects lies in several centimeter range (2 to 6 cm for muscle-like tissue) [11]. Controlling the level of power can be of great value in the treatment of biological objects where the studied effects usually come into play at certain threshold values of the electrical field strength. Therefore, RF sources based on gyromagnetic NLTLs could serve as a possible instrument to discover and study different threshold effects on the cell level. RF pulses can be produced in a free space or in a guided structure. In this paper we present the source of RF pulses traveling through the rectangular waveguide, which affect the respiration of isolated mice mitochondria.

II. Experimental Setup

The appearance and the general schematic of the RF source are presented in Fig. 1. The source of high voltage nanosecond pulses was SINUS-200 high voltage driver [8,12], which includes a pulse forming line with Tesla transformer inside switched by self-breakdown of a N_2-filled gas gap. Gyromagnetic NLTL is a coaxial line partially filled with ferrite saturated by axial bias magnetic field [7-9]. The optimum length for 1 GHz center frequency was found to be close to 1 meter.

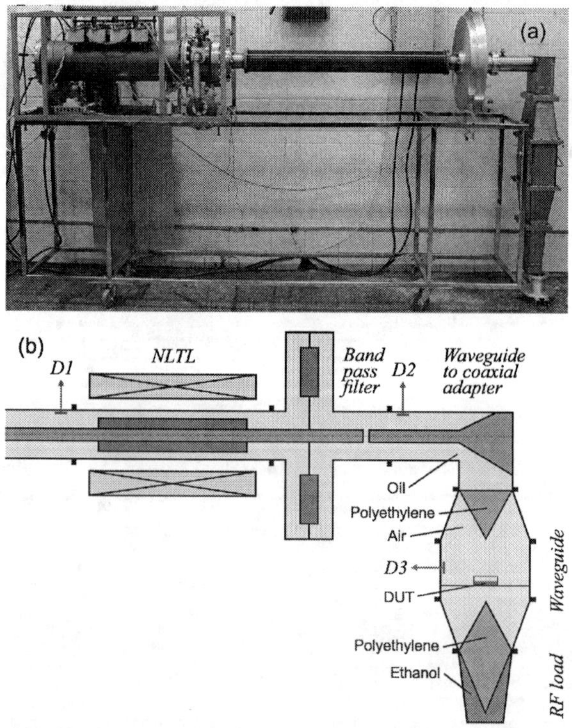

Fig. 1. Shematic diagram of the experimental setup.

This work was supported by RFBR Grant No. 16-38-60040.

Since no commercial waveguide to coaxial adapters are available for such a high power an original design of an oil-filled adapter has been simulated and manufactured. The transition from oil-filled to air-filled waveguide was made gradually with polyethylene triangular prism in order to exclude reflections. The biological object (device under test, DUT) is placed on 2 mm thick organic glass plate in the center of the rectangular waveguide, where the TE10 mode has maximum value.

III. RF Pulse Generation

The RF pulses were measured in the center of waveguide by D-dot D3 where the electrical field strength of the TE_{10} mode has the maximum value. There are two ways to change the amplitude of RF pulses produced by gyromagnetic NLTLs [10]: changing the bias magnetic field and changing the amplitude of the incident high voltage pulse. Usually the changing of bias magnetic field allows to change the amplitude of RF pulses by tens percent whereas the high voltage amplitude change can lead to RF pulse amplitude change for tens times. Therefore, we focused on changing the incident pulse amplitude by control of the pressure in the gas gap switch, while the bias magnetic field takes its optimal value lying in the 20-40 kA/m range. The results for RF pulse measurements are presented in Fig. 2. The measured RF pulses are falling into two groups. The first group with higher amplitudes, (Fig. 2 a), corresponds to the transmission band of the system and the absorption band of the ethanol load. The frequency of these pulses is gradually decaying starting from 1 GHz with decreasing of amplitude of the incident pulse and correspondingly the amplitude of the RF pulse, which is usual for RF oscillations produced by gyromagnetic NLTL [10]. The duration of the RF pulses in this this group amounts to several nanoseconds which corresponds to the duration of RF oscillations at the output of the NLTL.

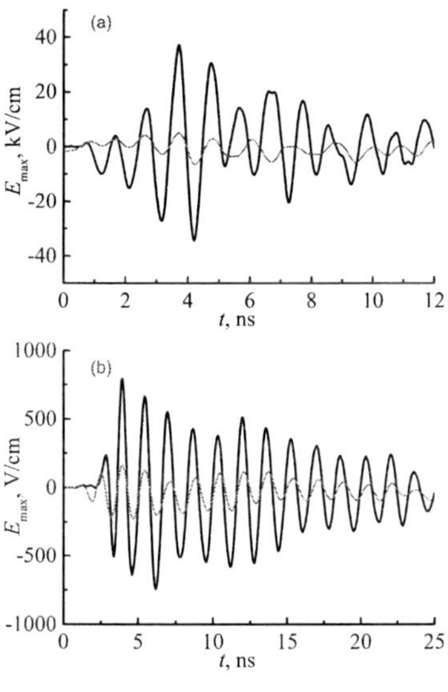

Fig. 2. (a) RF pulses with no reflections from the load (b) elongated RF pulses with decaying standing wave.

The RF pulses in the second group, (Fig. 2 b), have almost the same frequency of 0.6 GHz, which is little above the cut-off frequency of the waveguide. The pulse duration in this group is longer than 20 ns that could be caused by standing wave excitation due to lower absorption in ethanol at this frequency and not ideal transmission of the waveguide to coaxial adapter near cut-off frequency. Another possible reason for RF pulse elongation is dispersion of the group velocity near the cut-off frequency of the assembly.

IV. Mitochondrial Respiration Inhibition

Isolated by the differential centrifugation [13] mitochondria of outbred white mice liver were used for the experiments. Eppendorf test tube with the suspension was placed in the center of the waveguide, where the amplitude of the electric field has a maximum value. Mitochondria respiration was measured by an oxygen meter AKPM-02 (Russia). Respiratory control (the ratio of oxygen consumption, RC) was calculated. Irradiated and sham mitochondrial suspensions were used. Sham suspensions were subjected to the same manipulations as irradiated one, but without activating the radiation source. Six replicates were performed for each exposure experiment and sham.

The respiratory control by Chance [14] is the most informative indicator for the analysis of a mitochondria functioning. RC is equal to the ratio of mitochondrial oxygen absorption rate in the presence of ADP to the oxygen absorption rate after the exhaustion of ADP in suspension. RC equal to 5 shows good and effective conjugation of oxidation and phosphorylation [13] for our isolated mitochondria (Fig. 3, sham).

It was found that nanosecond RF pulses affect the rate of mitochondrial respiration. The effect was shown as a decrease of oxygen consumption by mitochondria in the presence of ADP. As a result, the value of the RC was decreased. The effect depends on the electric field. RC decreased monotonically with increasing field strength to a value of 0.15 kV/cm or more (Fig. 3). This indicates the inhibitory effect of UWB pulses on mitochondria. The obtained effect of the inhibition of respiration could be caused by either electric field or total absorbed RF radiation energy. Clarifying experiments were conducted in order to determine the contribution of the energy impact on the effect. In these experiments the effect was measured after exposure to pulses of different intensity and different amounts, provided the same energy in the bunch (RF power * pulse number = const).

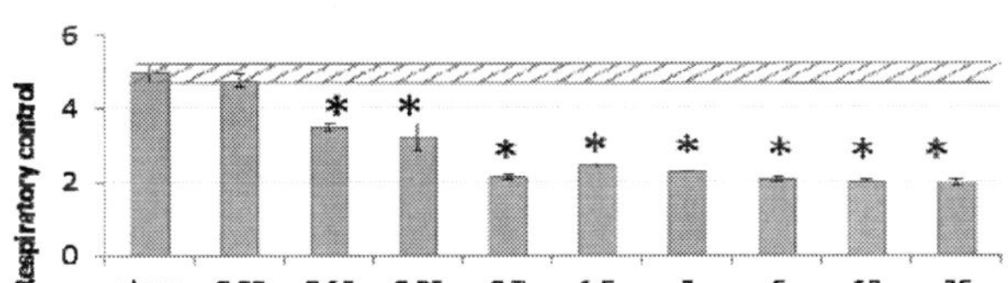

Fig. 3. Changes of the RC value after exposure to 1000 nanosecond UWB pulses with field of 0.09–36 kV/cm per pulse. Shaded space is ± standard error of the mean of RC for sham mitochondria. * – differences of indicators are significant as compared with sham ($p \leq 0.05$).

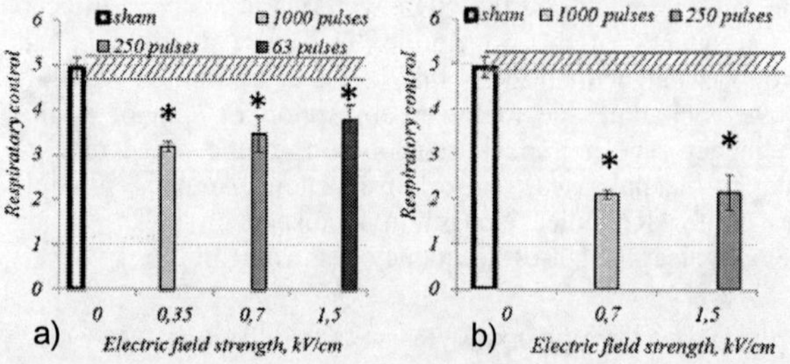

Fig. 4. RC changes after exposure to 1000 nanosecond UWB pulses with total energy of a) 2,5 mJ and b) 10 mJ. Field intensity from 0.35 to 1.5 kV/cm per pulse, the number of pulses from 1000 to 63, center frequency 0.6 GHz, pulse duration 25 ns. * – differences of indicators are significant as compared with sham ($p \leq 0.05$).

It was found that the effect of the electromagnetic pulse with the same total energy provided formation of the equal biological effect. In particular, there was a 30% reduction of the RC after exposure to RF pulses with total energy of 2.5 mJ per session (Fig. 4 a) and 61% reduction after exposure to RF pulses with larger total energy of 10 mJ per session (Fig. 4 b). Thus, increasing the energy leads to an increase in biological effects.

The drop of the RC indicates decreased conjugation of an oxidation and phosphorylation, as well as a broken integrity of the mitochondrial membrane. Under such conditions it should be disturbed ATP production by irradiated mitochondria. A confirmation of this assumption validity could be the data of [15], in which the decrease in the ATP level after exposure to microsecond pulses was described. The authors of the work linked this effect with electroporation of cell membranes and intracellular structures.

References

[1] D.M. French, M.D. Uhler, R.M. Gilgenbach, and Y. Y.Lau, "Conductive versus capacitive coupling for cell electroporation with nanosecond pulses," J. Appl. Phys. 106, 074701, 2009.

[2] L. Rems, and D. Miklavčič, "Tutorial: Electroporation of cells in complex materials and tissue," J. Appl. Phys. 119, 201101, 2016.

[3] S.J. Beebe, P.F. Blackmore, J. White, R.P. Joshi, and K.H. Schoenbach, "Nanosecond pulsed electric fields modulate cell function through intracellular signal transduction mechanisms," Physiol. Meas. 25(4), 1077, 2004.

[4] N. Seddon, C. Spikings, and J. Dolan, "RF pulse formation in nonlinear transmission lines," in Proceedings of the IEEE Pulsed Power Conference, Albuquerque, NM, USA, pp. 678–681, June 2007.

[5] J.M. Johnson, D.V. Reale, W.H. Cravey, R.S. Garcia, D.H. Barnett, A.A. Neuber, J.C. Dickens, and J.J. Mankowski, "Material selection of a ferrimagnetic loaded coaxial delay line for phasing gyromagnetic nonlinear transmission lines," Rev. Sci. Instrum. 86, 084702 (2015).

[6] S.F. Yamasaki, L.P. Silva Neto, J.O. Rossi and J.J. Barroso, "Soliton Generation Using Nonlinear Transmission Lines," IEEE Trans. Plasma Sci. 42, 3471, 2014.

[7] S.J.F. Chadwick, N. Seddon, and S. Rukin, "A novel solid-state HPM source based on a gyromagnetic NLTL and SOS-based pulse generator," in Proceedings of the IEEE Pulsed Power Conference, Chicago, IL, USA, pp. 178–181, June 2011.

[8] I.V. Romanchenko, V.V. Rostov, V.P. Gubanov, A.S. Stepchenko, A.V. Gunin, and I.K. Kurkan, "Repetitive sub-gigawatt rf source based on gyromagnetic nonlinear transmission line," Rev. of Sci. Instrum. 83, 074705, 2012.

[9] J.-W. Bragg, W. Sullivan III, D. Mauch, A. Neuber, and J. Dickens, "All solid-state high power microwave source with high repetition frequency," Rev. Sci. Instrum. 84, 054703, 2013.

[10] I.V. Romanchenko, V.V. Rostov, A.V. Gunin, and V.Yu. Konev, "High power microwave beam steering based on gyromagnetic nonlinear transmission lines," J. Appl. Phys. 117, 214907, 2015.

[11] C. Gabriel, "Compilation of the dielectric properties of body tissues at RF and microwave frequencies", Armstrong Laboratory, Brooks Air Force Base, TX, Brooks Air Force Technical Rep. AL/OE-TR-1996-0037.

[12] G. A. Mesyats, S. D. Korovin, A. V. Gunin, V. P. Gubanov, A. S. Stepchenko, D. M. Grishin, V. F. Landl, and P. I. Alekseenko, "Repetitively pulsed high-current accelerators with transformer charging of forming lines," Laser Part. Beams 21, 197, 2003.

[13] I.R. Knjazeva, V.V. Ivanov, L.P. Zharkova, O.P. Kutenkov, V.V. Rostov and M.A. Bolshakov, "The effect of the repetitive pulsed microwaves on functional activity of isolated mitochondria," *Tomsk state university journal of biology*, vol. 4, pp. 113–123, 2011.

[14] B.Chance, "Electron transfer: Pathways, mechanisms, and controls," Annu. Rev. Biochem., vol. 46, pp. 967–980, 1977.

[15] S.K. Frandsen, L. Gibot, M. Madi, G. Gehl and M.P. Rols, "Calcium electroporation: evidence for differential effects in normal and malignant cell lines, evaluated in a 3D spheroid model," Plos ONE 10 (12): e0144028, 2015.

Power Dynistors with Subnanosecond Switching on Time and Based on Them Pulsed Power Devices

S.V. Korotkov, A.G. Lyublinsky, Yu.V. Aristov, D.A. Korotkov
Division of Solid State Electronics
Ioffe Institute
St. Petersburg, Russia
Alexander.Lyublinsky@mail.ioffe.ru

Abstract— **The results of investigation of thyristor-like silicon switches with subnanosecond switching-on time, so-called Deep Level Dynistors (DLDs), which were discovered in Ioffe Institute, are presented. So fast switching on is obtained due to generation in the semiconductor structure impact ionization wave initiated by short overvoltage pulse with the amplitude much higher than the static breakdown voltage. Several high voltage DLD-based switches and generators of many kiloamperes current pulses of nanoseconds duration are described.**

Keywords— High-voltage techniques, Power semiconductor switches, Pulse power generators.

I. Introduction

It was shown in [1-3] that high current pulses of nanoseconds duration could be formed by using Deep Level Dynistors (DLD) □ semiconductor devices with subnanosecond switching-on time. So fast switching on is possible due to generation in the semiconductor structure of impact ionization wave initiated by short overvoltage pulse with the amplitude much higher than the static breakdown voltage [4, 5].

Like a power thyristor DLD is closing type semiconductor switch. DLD operation principle is the following. Initially DLD is in off state and static high voltage is applied to DLD semiconductor structure. A space charge region (SCR) is formed near the collector *p-n* junction. SCR is free from the carriers, has maximum electric field on the collector junction and occupies most of DLD base layers. Two narrow neutral regions remain in the base layers near to both emitters. These regions prevent a static breakdown of the semiconductor structure. The switching-on process is realized by applying to DLD anode nanoseconds duration high voltage pulse. Under these conditions, the electric field on the collector junction abruptly increases above the critical level, but the semiconductor structure breakdown is delayed because SCR could not expand immediately up to the emitters due to saturated velocity of the carriers and very high triggering pulse voltage rise rate.

When the electric field exceeds static breakdown level in 1.5-2 times, the field emission of the electrons from the deep levels in the semiconductor band gap occurs. These electrons initiate fast impact ionization that provides prompt filling up of DLD base layers with electron-hole plasma. As a result, DLD switches to a high conductivity state in less than 1 ns. The triggering pulse current amplitude should be several hundred amperes to obtain high voltage rise rate on DLD anode contact. So high triggering pulse current is required to charge quickly the intrinsic capacitance of the reversely biased DLD structure. Then the carriers are injected intensively by the emitters and modulate conductivity of the base layers. Very short DLD switching-on time provides low commutation losses and high allowable current rise rate in the power circuit (up to 250A/ns).

II. Research Objectives

In spite of lot of research works and intensive studying of DLD and DLD-based pulse power devices, the following issues have not been investigated in details yet:

- the uniformity of DLD switching-on process over the device area;

- the optimal parameters of the triggering current pulses for the effective DLD switching-on;

- design of optimal triggering circuit for power DLD-based switches.

These research directions are the main objectives of the current work. The design and experimental testing results of several power DLD-based switches and generators are also presented.

III. Experimental Results

The specific of DLD switching-on process requires special triggering circuit which should provide reliable operation of DLD in power circuits with high rise rate current pulses as well as ensures minimum commutation losses.

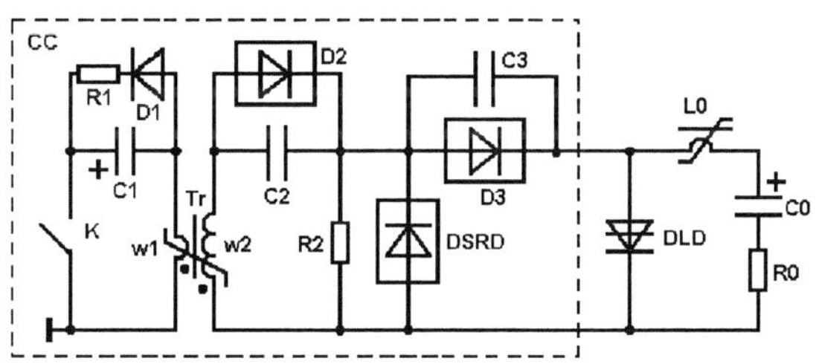

Fig.1. The electric circuit of the DLD-based generator of high rise rate current pulses.

The simplified electric circuit of the designed DLD-based generator of power high rise rate current pulses is shown in Fig. 1. Nanosecond duration high voltage triggering pulses, which are required for DLD switching-on, are formed by the control circuit (CC). The main components of CC are high voltage Drift Step Recovery Diodes (DSRD) assembly [6], low voltage switch K, which is made of several IGBT transistors connected in parallel, and saturable core transformer Tr.

In the initial state, capacitor C1 is charged, K is switched off. Power capacitor C0 is charged up to the operation voltage and this voltage is blocked by both DLD and diode assembly D3. Capacitor C2 is discharged. When K switches on, a short pulse current charges capacitor C2, flows in forward direction through DSRD and pumps the charge into DSRD structures. After that, Tr core saturates, capacitor C2 discharges during 100 ns through DSRD and winding w2. C2 discharge current is reverse current for DSRD which pull-out the charge stored in DSRD structures. At the moment when the reverse current reaches its maximum, the charge is pulled-out completely and capacitor C2 discharges almost down to zero voltage. As a result, all the stored electron-hole plasma is pulled-out from DSRD base layers and DSRD switches off within a few nanoseconds. The energy stored in w2 winding goes into D3-DLD circuit and forms DLD triggering current pulse which quickly charges the intrinsic capacitance of DLD structure and provides a high rise rate voltage pulse on DLD anode.

L0 choke is in unsaturated state during DLD triggering process and has high inductance which prevents triggering current flowing through the power circuit C0-R0. DLD switches on when the voltage reaches the triggering level. The triggering current continues to flow through DLD even after its triggering, while C0 discharge current in the power circuit is limited by the high inductance of L0. Then L0 core saturates, the inductance of the choke abruptly reduces and C0 capacitor discharges through switched-on DLD into the load.

The switching-on process of DLD with 24 mm diameter of semiconductor structure and up to 2.5 kV operation voltage has been thoroughly investigated. The oscillograms of the current through the load (I1, I2 and I3) and the voltage drop on DLD (U1, U2 and U3) at different triggering pulse amplitudes (350 A, 250 A and 150 A correspondingly) are shown in Fig. 2.

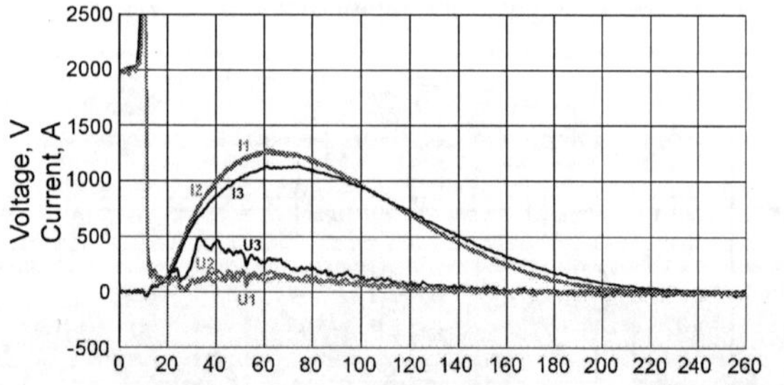

Fig.2. Oscillograms of the commutated current (I1, I2 and I3) and the voltage drop on DLD (U1, U2 and U3) at different triggering pulse amplitudes 350A, 250A and 150A correspondingly.

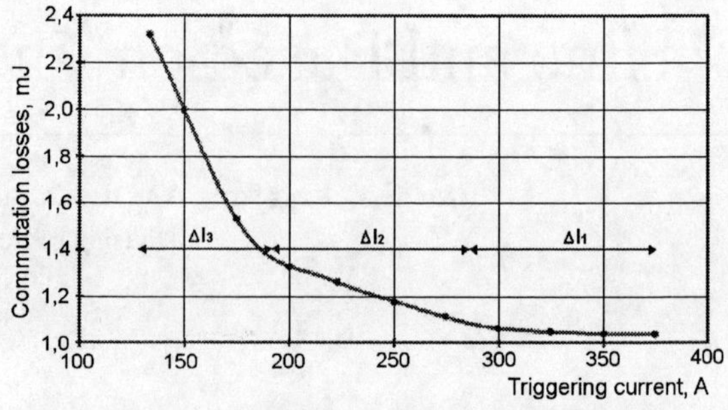

Fig.3. DLD commuatation losses vs triggering pulse current amplitude.

The dependence of triggering pulse current amplitude on commutation losses is shown in Fig. 3. This dependence has three characteristic ranges: ΔI1 corresponds to minimum energy losses, ΔI2 is the intermediate interval with slightly higher losses and ΔI3 is the interval with significant increase of the losses. The optimum triggering current amplitude corresponds to the left part of ΔI1 interval.

Fig. 4 shows oscillograms of DLD voltage drop at different triggering pulse current amplitudes: 350 A (U1), 250 A (U2) and 150 A (U3), which correspond to the ranges ΔI1, ΔI2 and ΔI3 in Fig. 3. One can see that lower triggering pulse amplitude with lower voltage rise rate results in DLD switching-on at lower overvoltage pulse level and leads to higher commutation losses. It can be explained by lower electric field near the collector junction at the switching-on moment and less effective and non-uniform impact ionization which produces lower density plasma. If the triggering pulse amplitude is high enough and we operate in ΔI1 region, then slight reducing of triggering pulse energy results in minor increase in commutation losses. If the triggering pulse amplitude is too low, then only small part of the total device area is involved into the impact ionization switching-on process and the energy losses increase significantly.

Very fast switching into the high conductivity state provides high commutation capability of DLD. Our experiments demonstrate that in case of optimal triggering DLD commutation capability is directly proportional to the DLD device area. This fact indicates that the switching-on process and power current are distributed uniformly over the

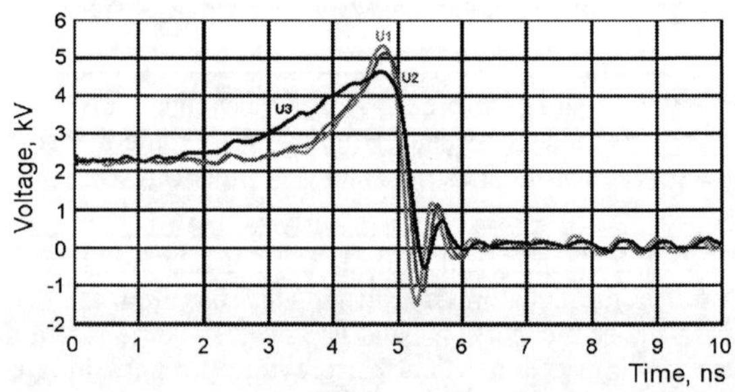

Fig.4. Oscillograms of the voltage drop on DLD during switching-on process at different triggering pulse currents 350 A (U1), 250 A (U2) and 150 A (U3).

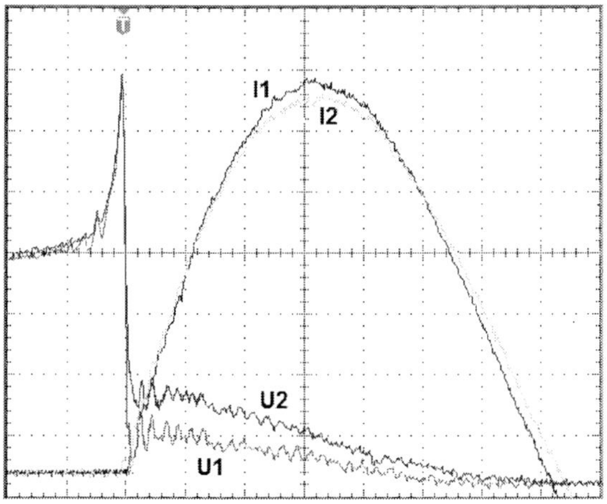

Fig.5. Comparative oscillograms of the commutated current pulses and voltage drop on two different types of DLD with 24 mm (I1 and U1) and 16 mm (I2 and U2) diameters of the semiconductor structure. The scales are 500 A/div., 600 V/div. and 10 ns/div.

device area, which is very important for the reliable operation of the device and low energy losses. Two series of experiments have been made to confirm uniform switching-on process distribution, including comparison of different diameters DLD operation and DLD operation with long square-like pulse current.

Comparative oscillograms of the commutated current pulses and voltage drops on two different types of DLD with 24 mm (I1 and U1) and 16 mm (I2 and U2) diameters of the semiconductor structures are shown in Fig. 5. These oscillograms were obtained with the optimal triggering pulse current amplitudes which provide minimum energy losses in DLD. One can see that voltage drop just after DLD switching-on is lower for 24 mm diameter device and the amplitude is about inverse proportional to the device area. This fact demonstrates good average distribution of the commutated current over the device area, but in principle, the current may flow through many uniformly distributed small areas, i.e. many "hot spots", and only this experimental result is insufficient to confirm quite uniformity of the switching-on process over the device area.

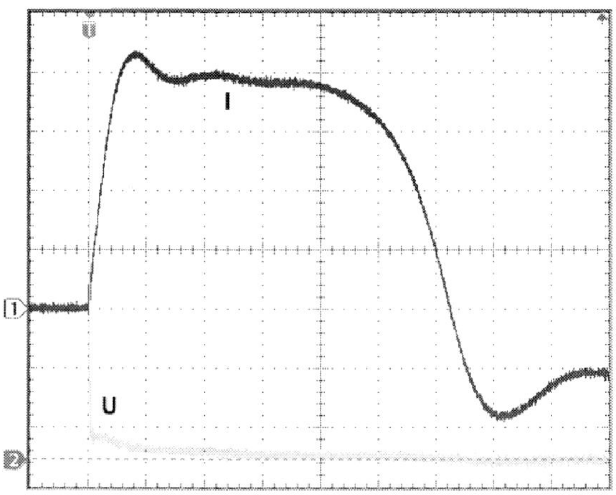

Fig.6. Oscillograms of the long square-like current pulse I and the voltage drop U on 16mm diameter DLD. The scales are 1 kA/div., 250 V/div. and 2 μs/div.

Therefore, we have additionally tested DLD operation with long (microseconds duration) square-like pulse current. The oscillogram of the voltage drop on 16 mm diameter DLD during the commutation of 4kA square-like pulse current of 10 μs duration is shown in Fig. 6. Multi-stage pulse forming line has been used instead of capacitor C0 (see the circuit in Fig. 1) to provide this pulse current waveform.

After the triggering the voltage on DLD drops in two stages. The first stage is abrupt fall in several times within less than 1 ns followed by the second stage of slower reducing down to steady state voltage drop. The first stage is governed by the impact ionization process, while the second stage duration is determined by the carrier injection by the emitters and complete filling up of DLD base layers by the electron-hole plasma. In case of uniform over the device area impact ionization process, the characteristic time constant of the second stage is determined by the base layers conductivity modulation by the carrier injection only. Otherwise, in case of non-uniform impact ionization, the duration of the second stage is determined also by expansion of switched-on "hot spot" region(s) over the all device area. This expansion rate is no more than 0.5 mm per microsecond. About 1 μs time constant proves the uniform impact ionization process or even "hot spots" exist the distance between them is less than 1 mm.

Low commutation losses allow designing of compact high pulse current DLD-based generators. As an example, the photo of 2.5 kV pulse generator based on 16 mm diameter DLD is shown in Fig.7. It forms single current pulses with 4.5 kA amplitude and 300 ns pulse duration.

DLD operation principles ensure reliable switching of

Fig.7. Compact DLD-based generator. The operation voltage is 2.5 kV, current pulse amplitude is 4.5 kA, nominal current pulse duration is 300 ns.

Fg.8. A high voltage DLD switch: the operating voltage is 24 kV, the current pulse amplitude is 2 kA, the pulse duration is 200 ns, the operating repetition rate is 1 kHz.

DLD stack consisting of many devices connected in series and switched on by common triggering circuit. As an example, DLD-based switch with 12 kV operation voltage consisting of six DLD with 24 mm diameter of the semiconductor structure connected in series has been designed and successfully tested. It is able to commutate single current pulses with 20 kA amplitude and 500 ns pulse width.

To eliminate DLD overheating in case of high frequency, high average output power operation the additional heat sink for each DLD in the assembly is required. Unfortunately, the increased stray inductance of DLD assembly with the heat sinks prevents effective using of common triggering circuit. Therefore, several synchronous triggering circuits should be used, i.e. one triggering circuit individually for each DLD or for a group of DLDs. Saturable core choke L0 makes a delay of the current in the power circuit and solve the possible problem of non-strictly synchronous switching of the triggering circuits.

A photo of DLD-based switch with 24 kV operation voltage is shown in Fig. 8. It consists of three serially connected DLD assemblies with four 24 mm diameter devices in each. This switch is able to commutate 2 kA current pulses with 200 ns pulse width and 1 kHz repetition rate.

IV. Conclusion

The following results have been obtained:

- the effective triggering circuits of DLD and high voltage DLD stacks have been designed;

- the details of switching-on processes of DLD with 16 mm and 24 mm diameters of semiconductor structure have been investigated;

- the optimal parameters of the triggering pulse which provides uniform switching-on process over the device area, have been determined;

A range of DLD-based pulse generators and switches has been designed and tested, including:

- compact 2.5 kV single pulse operation mode generator of 4.5 kA current pulses with 300 ns pulse width;

- power 12 kV single pulse operation mode switch of 20 kA current pulses with 500 ns pulse width;

- 24 kV operation voltage switch of 2 kA amplitude, 200 ns duration current pulses with up to 1 kHz repetition rate.

References

[1] Yu.V. Aristov, V.B. Voronkov, I.V. Grekhov, A.K. Kozlov, S.V. Korotkov, and A.G. Lyublinskii, "A High-Power Semiconductor Switch of High-Voltage Pulses with a Rise Time of Nanosecond Duration," Instrum. Exp. Tech., vol. 50, no. 2, pp. 224-227, 2007.

[2] I.V. Grekhov, S.V. Korotkov, and P.B. Rodin, "Closing Switches Based on Propagation of Fast Ionization Fronts in Semiconductors," IEEE Trans. Plasma Sci., vol. 36, no. 2, part 1, pp. 378-382, 2008.

[3] S.V. Korotkov, Yu.V. Aristov, V.B. Voronkov, A.L. Zhmodikov, D.A. Korotkov, and A.G. Lyublinskii, "Dynistors with Nanosecond Response Times," Instrum. Exp. Tech., vol. 52, no. 5, pp. 695-698, 2009.

[4] V. Tuchkevitch, and I. Grekhov, "New principles of high power switching with semiconductor devices", Bulletin of Academy of Sciences of USSR, no. 4, pp.18-23, 1987 (in Russian).

[5] V. Efanov, A. Kardo-Sysoev, I. Tchashnicov, P. Yarin, "New superfast power closing switches – dynistors on delayed ionization," Proc. of 22nd Int. Power Modulator Symp., Boca Raton, Fl, p. 247, 1996.

[6] I.Grekhov, V.Efanov, A.Kardo-Sysoev, and S.Shenderey, "Power drift step recovery diode," Solid State El., vol. 28, pp. 597-599, 1985.

Simulations of the Interaction of High-Velocity Condensed-Matter Liners with Current-Carrying Walls

A.M.Buyko and S.F.Garanin

FSUE RFNC-VNIIEF, Sarov, Russia.

Abstract — The paper presents an overview of publications on VNIIEF numerical simulations of one of the major 2D effects observed in magnetically driven implosion of cylindrical condensed-matter liners: their interaction with glide planes. We consider liners accelerated to 8-20 km/s by currents of 30-100 MA delivered by disk explosive magnetic flux compression generators with azimuthal magnetic fields of 1-6 MG (magnetic pressures of 0.04-1.4 Mbar): experimental liners HEL-1, ALT-1,2 and liners HEL-2, ALT-3 proposed for high energy density physics research.

Key words—azimuthal magnetic field; liner implosion, liner/wall interaction; two—dimensional numerical simulations.

I. Introduction

Implosion of cylindrical condensed-matter liners driven by azimuthal magnetic fields is used for multiple purposes, for example, studies of interaction of liners with glide planes and liner instabilities [1-9], see Fig. 1; generation of 5 to 30 Mbar pressures to measure Hugoniots of materials, as in the ALT-1,2 and HEL-1 experiments [5, 6] and in the HEL-2 [2, 4] and ALT-3 [8, 9] projects; materials strength measurements based on the growth of perturbations [3].

Fig. 1. LT-1 EMG experiment [4]. Radiograms (a) and 2D HD implosion simulations (b) of two liners: with wall slopes $\theta = 2°$ and $8°$

Magnetically driven liner implosion has a number of advantages over HE-driven implosion, such as the nearly instantaneous (with the speed of light) equilibration of magnetic pressure on the liner, smooth (without shock waves) liner acceleration, higher pressures and better control of their magnitude and rise time. During the liner implosion of interest – in the z-pinch geometry – magnetic field $B(z,t)$ and magnetic pressure $P_B(z,t)$ as a rule grow steadily with time up to 1-6 MG and 0.04-1.4 Mbar, which can be estimated conveniently using the relations [cm, MA, MG, Mbar]:

$$P_B(z,t)=B^2(z,t)/8\pi, \quad B(z,t)=I(t)/5R(z,t)\,, \qquad (1)$$

where $I(t)$ is the current flowing through the liner, $R(z,t)$ is the outer liner radius. One- and two-dimensional (1D and 2D) implosion simulations can be done in the hydrodynamic (HD) approximation – ignoring the magnetic diffusion and assuming that the pressure (1) acts on the surface $R(z,t)$ if the liner has a thickness $\Delta_l(z,t)$ much greater than the skin layer thickness $\Delta_B(z,t)$. Otherwise, one should use magnetic hydrodynamics (MHD): the magnetic field $B(z,t)$ acts on the boundary $R(z,t)$, and the pressure $P_B(z,t)$ is the result of bulk forces proportional to $\boldsymbol{j}\times\boldsymbol{B}$, where $\boldsymbol{j}$ is the current density.

The highest-current explosive magnetic flux compression generators (EMGs) at VNIIEF are disk EMGs, which are unique liner drivers, see Table I. In the **HEL-1** experiment, the current delivered to the liner by such an EMG having a diameter of 1 m was ~105 MA with a rise time of τ_{ef}~40 µs (B~1 MG, see Fig. 2a), and the liner having a mass of ~1 kg (kinetic energy E_{kin} >20 MJ!) was driven to a velocity of ~8 km/s. In the **ALT-1,2** experiments with a 0.4 m diameter disk EMG and an opening switch the currents in the Al liner were as high as 31-32 MA (τ_{ef}~2 µs, B ~2 MG, see Fig. 2b), and the liner velocity was 12 km/s.

TABLE I. Implosion Parameters of Experimental Al Liners

Liner	R_l, cm	Δ_l, cm	R_{imp}, cm	H_l, cm	S_l/H_l	m_l, g/cm
LT-1	3.0	0.1	0.6	2.4	1	5.0
HEL-1	24	0.4	5.5	10	1.8	160
ALT-1,2	4.0	0.2	1.0	4.0	0.7	13

TABLE I. (continued)

Liner	I_{max}, MA	τ_{ef}, µs	B_{max}, MG	v_{imp}, km/s	θ(Z), degrees
LT-1	5	40	1	5	2-8(Cu)
HEL-1	105	40	1	8	6 (Fe)
ALT-1,2	31	2	2	12	8(Cu)

Designations in the Table I: R_l and R_{imp}(v_{imp}) are the outer and the impact liner radius (impact velocity); H_l and Δ_l are the liner height and thickness; $S_l = R_l - \Delta_l - R_{imp}$; m_l is the liner mass per unit length; I_{max} and τ_{ef} are the peak current and the rise time from 0.1 to 0.9 of I_{max}; B_{max} is the maximum magnetic field on the liner surface; θ (Z) is the wall slope (material), see Fig. 1.

Basic liner parameters (see Table I) can be used to identify the conditions to minimize the wall effects on the implosion, for example, condition (2) is violated only for the LT-1 and HEL-1 liners, which are entirely exposed to the wall effects during implosion, see Fig. 1 and section II.

$$S_l = (R_l \text{-} \Delta_l \text{-} R_{imp}) < H_l \qquad (2)$$

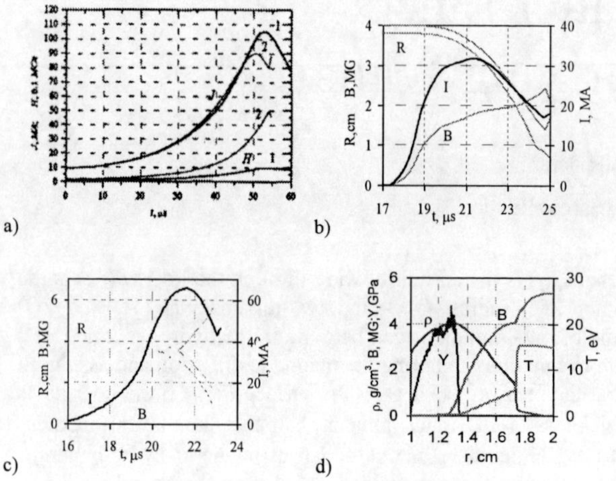

Fig. 2. 1D MHD simulations of Al liner implosion. a) current and magnetic field on the **HEL-1** and HEL-2p liners (curves 1 and 2) [2, 6]; b,c) current and magnetic field on the **ALT-1,2** and ALT-3 liners, (R-t) diagrams of the liner boundaries [5, 7]; d) profiles of density ρ, magnetic field B, temperature T and yield strength Y in the ALT-3 liner before impact (R_{imp} =1 cm, v_{imp} ~20 km/s)

In the ALT-3 project, similar to ALT-1,2, one can deliver currents up to 60-70 MA and magnetic fields up to 6 MG, which can drive a 3 mm thick Al liner to ~20 km/s, see Fig. 2c. The temperature at the outer liner surface can reach 15 eV by the end of implosion, but the inner part ~40% of the liner mass remains solid and dynamically hardened (Fig. 2d): here, the temperature is $T<T_{melt}$~0.34 eV and the yield strength is Y<4 GPa (under normal conditions, the melting point is T_{melt}~0.08 eV, Y_0=0.27 GPa). If this liner part is replaced with a higher-density material of similar mass, the liner will preserve its velocity, and potential shock pressures will be as high as 10-30 Mbar (Fe-Fe, Mo-Mo, W-W,…) [5, 7].

However, the desired 1D picture of liner implosion is distorted by 2D effects, such as the development of a Rayleigh-Taylor-like instability (simulated by initial perturbations) and the liner/wall interaction, which should preferably be simulated without initial perturbations in order to focus on the wall effects [1-9].

Preliminary (simplified) 2D liner implosion simulations were done in the HD and MHD approximations mostle using the ALE code MIMOZA [10, 3] and the Lagrangian code DRAGON [11] employing the Mie-Gruneisen equations of state, linear dependence of resistivity on internal energy, and elastoplastic models of shear strength [12]. **Revised** 2D MHD liner simulations were done using the MIMOZA code with a wide-range EOS-conductivity complex for aluminum [13] consistent with the strength model [12]. The initial stage of such simulations was done on a Lagrangian mesh with an Eulerian mesh used afterwards; simulations on Eulerian meshes were done as well [7].

This paper reports and discusses the 2D liner implosion simulations that focus on the interaction of liners with glide planes (without initial perturbations). It considers experimental and proposed "High-Energy Liners" HEL-1 and HEL-2 driven by currents generated by 1 and 0.4 meter diameter disk EMGs (τ_{ef}~40 and ~12 μs) and the ALT-1,2,3 ("Advanced Liner Technology") liners driven by a 0.4 m diameter EMG with an opening switch (τ_{ef}~2μs).

II. HEL-1 AND HEL-2 LINER IMPLOSION

In the HEL-1 experiment [6], implosion of the Al liner was driven by a current of ~105 MA (B~1 MG, see Fig. 2a). Parameters of this liner (made of the AMg-6 alloy) and the walls (see Table I) were chosen considering the 2D HD implosion simulations - with the magnetic pressure (1) at current amplitude of 135 MA. Results of the simulations – using different codes, with different materials and wall geometries – are qualitatively close to Fig. 3a. The simulations give a close wall/liner contact during implosion, but the wall effects extend across the whole liner and at later implosion times result in considerable outrunning of the liner's near-wall parts due to the growth of near-wall perturbations and the development of an instability (condition (2) is not satisfied here, see Table I). Results of the 2D MHD simulation of liner implosion with a current waveform similar to the experiment (Fig. 2a, curve 1) are shown in Fig. 3b. In the revised simulation, the longitudinal asynchronism of the liner/target impact is τ_{imp}~2 μs (the longitudinal asymmetry is ΔR_{imp} ~$\tau_{imp} v_{imp}$~0.3 R_{imp}), which is close to the experiment.

The HEL-2 device [4] was designed to accelerate the liner to ~14 km/s and to produce a pressure of ~10 Mbar at molybdenum/molybdenum impacts. The HEL-1 experiment, however, left the following question open: is it possible to produce a synchronous liner/target impact directly from a 1 m diameter EMG? The 2D HD simulations by the DRAGON code (with inclusion of the liner Joule heating from the 1D MHD simulations) gave a positive answer to this question, for example, for the HEL-2p Al liner having the following parameters: R_l=10 cm, Δ_l=10 mm, m_l=160 g/cm, H_l= 8cm, R_{imp}=1 cm; θ=0°(Cu). Driven by a current of 90 MA (B~ 5MG, see Fig. 2a, curves 2), this liner hits the target virtually synchronously at a velocity of ~14 km/s (condition (2) is satisfied, S_l/H_l =1.0), as distinct from the same liner having a height of H_l=4 cm, cf Figs. 4a, 4b. The required shock pressure of ~10 Mbar can be produced by replacing the HEL-2p liner with a two-layer Al+Mo liner having a similar mass and by re-optimizing the wall geometry based on the 2D MHD implosion simulations.

For the HEL-2 experiment – with a 0.4 m diameter disk EMG delivering current ~80 MA with a rise time of τ_{ef}~12 μs (B~4 MG) – the following parameters of the two-layer Al+Mo liner and the walls were chosen: R_l=7 cm, R_{imp}=1 cm; Δ_l=3.5 mm(Al)+0.7 mm(Mo), H_{Al}=60 mm, H_{Mo}=58 mm; θ=2°(Fe). The 2D MHD implosion simulations of this liner by the MIMOZA code (Fig. 4c) give a good liner/wall contact during the implosion and a small longitudinal impact asynchronism at a velocity of ~14 km/s (E_{kin}=24 MJ). The walls effects on the implosion in this case are minor – condition (2) is satisfied: S_l/H_l=0.9.

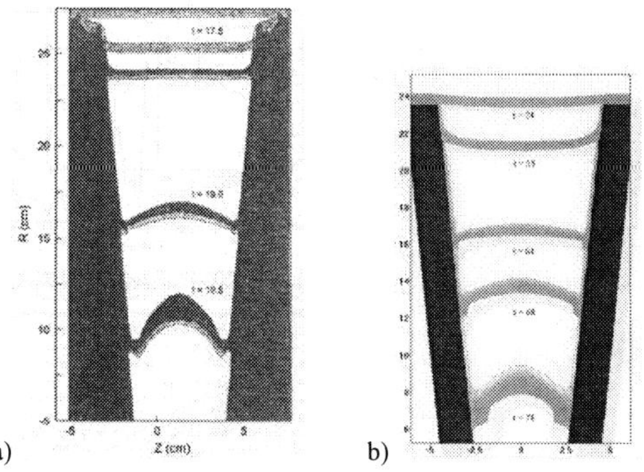

Fig. 3. 2D implosion simulations of the HEL-1 liner: HD [6] (a) and MHD (b) simulations by the MIMOZA code

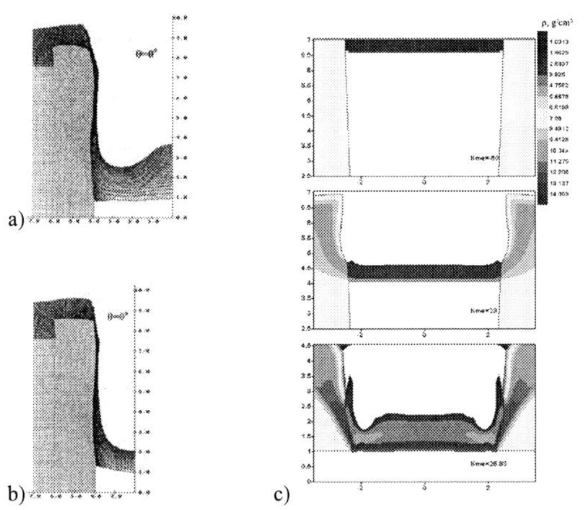

Fig. 4. a,b) 2D HD simulations of the HEL-2p liner and a similar liner having a height of H_l=4 cm [2]; c) 2D MHD simulation of the HEL-2 liner [4]: distributions of density during implosion (top-down)

III. Implosion of the ALT-1,2 and ALT-3 Liners

The ALT-1,2 experiments [5] met the key goals: they delivered currents of 31-32MA (Fig. 2b) and enabled a virtually synchronous impact of the Al liner on the target at a velocity of ~12 km/s. Liner and wall specifications are listed in Table I (the walls have a 4 mm×4 mm “step”, see Fig. 5a).

The revised 2D MHD simulations of the ALT-1,2 liner with a current waveform similar to the experiment (Fig. 2b, 5a-b) confirmed the basic results of the preliminary simulations: the liner/wall contact during implosion is good; the late-time implosion velocity is ~12 km/s, and the impact is virtually synchronous, which is close to the experiment (condition (2) is satisfied); ~40% of the liner mass remains solid. These simulations, however, revealed a small delay in the contact of the whole liner with the wall (because of a void forming between them) and perturbations of the outer liner surface (resulting from the liner/wall interaction and simulation errors). The perturbations grow with time due to the instability the stronger the closer is the wall and the smaller is the grid resolution h, see Fig. 5b. They are weak along the main body of the liner, but strong at the ends: here, they can intensify the liner/wall interaction effects, including the outrunning of the near-wall inner part of the liner (the results do not converge with respect to h, which is indicative of simulation errors).

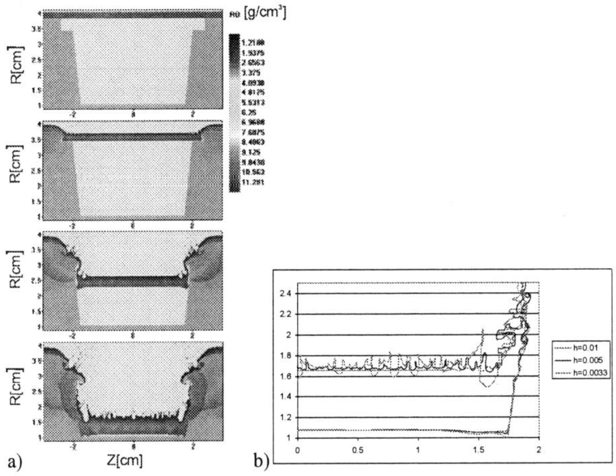

Fig. 5. 2D MHD simulations of the ALT-1,2 liner on Eulerian meshes, MIMOZA code. a) liner implosion process (h=33μm); b) liner shapes by the end of implosion (h=100, 50 and 33μm)

The ALT-3 device has been developed to test the efficiency of magnetic liner implosion at currents of 60-70 MA (B~6 MG, Fig. 2c) and the possibility of precision Hugoniot measurements at a velocity of ~20 km/s and pressures up to 10 Mbar. The liner differs from the ALT-1,2 liner in that its thickness is increased to 3 mm. Wall effects on implosion are minor: condition (2) is satisfied, S_l/H_l= 0.7. The pre-chosen geometry of the Cu walls with a slope of θ=0° provides an acceptable shape of the inner liner surface without destructive jets by the end of implosion (Fig. 6a), as distinct from the walls without “steps” at different wall slopes [8].

The walls with a 4mm×4mm “step” produce undesirable effects during implosion (Fig. 6a). The liner is “cut” after the high-velocity impact (~6 km/s) with the lower step producing shock waves and rarefaction waves in the liner and in the walls. As a result of their interaction, a void forms between the liner and the wall, which delays the sliding contact between the whole liner and the wall. Compression waves propagate along the liner from the walls towards its symmetry plane and introduce longitudinal perturbations that grow with time due to the instability development and result in the asynchronism of the liner/target impact. The revised MHD simulations give the value of liner asynchronism at the effective height of H_{imp}≈24 mm (impactor) that equals τ_{imp}≈ 8 ns.

The walls with a "tooth" (see Fig. 6b,c) are predicted to "cut" the liner on the "tooth" at low velocities, which significantly mitigates the above undesirable effects and improves the performance of the impactor. Simulations give good sliding contact of the whole liner with the wall during implosion and substantially smaller longitudinal impact asynchronism $\tau_{imp} \approx 1$ ns ($H_{imp} \approx 24$ mm). There is, however, a small jet at the wall that lags behind the impactor by ~ 10 ns. One cannot exclude that in reality it could outrun the impactor and impede the output of measurement data. To prevent this risk, a jet catcher, which was shown to be quite efficient, see Fig. 6c, Z>0, was introduced into the chosen wall geometry with the "tooth".

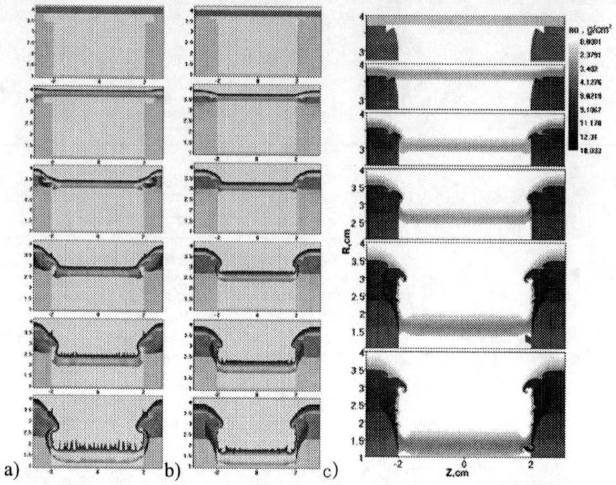

Fig. 6. 2D MHD simulations of the ALT-3 liner with different wall geometries, MIMOZA code. a,b) the walls with a "step"or a "tooth" [8], h = 33μm; c) the chosen wall geometry – with a jet catcher (Z>0) and without it (Z<0), h=20 μm [9]

IV. Conclusion

This paper reports and discusses 2D cylindrical liner implosion simulations modeling the effects of liner interaction with walls (without initial perturbations). The experiments and the simulations demonstrated that wall effects on implosion are minor, when the height of the liner H_l is greater than the distance it travels S_l.

In the experiments, the walls had a slope of θ=6° and 8° to provide good liner/wall contact during implosion. The **HEL-1** liner (S_l=1.8H_l) demonstrated great longitudinal liner/target impact asynchronism, τ_{imp}~2 μs (longitudinal asymmetry of $\Delta R_{imp} \sim \tau_{imp} v_{imp} \sim 0.3 R_{imp}$), as the liner sides flew faster. The **ALT-1,2** liner (S_l=0.7H_l) approached the target nearly synchronously, except for small side regions running a little ahead of the main liner body. These results agree with the simulations.

The MHD implosion simulations demonstrated that the chosen parameters of the proposed HEL-2 (S_l=0.9H_l, θ=2°) and ALT-3 (S_l=0.7H_l, θ=0°) liners can provide good liner/wall contact during implosion and virtually synchronous liner/target impact.

Acknowledgment

We thank V.V. Zmushko and V.B. Yakubov for discussions of the content of this paper.

References

[1] V.A. Arinin, V.N. Buzin, A.M. Buyko, et al. "Near-wall effects influence on longitudinal symmetry of a solid liner compressed by slowly increasing EMG current". Digest of Technical Papers 12th IEEE Int. Pulsed Power Conf., Eds. C. Stalligs and H. Kirbie, pp. 880-883, 1999.

[2] A.M. Buyko, V.V. Zmushko, P.N. Nizovtsev, et al. "On feasibility to achieve high longitudinal symmetry of cylindrical metal liners compressed by currents from most powerful disk EMG". Ibid. pp. 1145-1148, 1999.

[3] A.M. Buyko, S.F. Garanin, V.V. Zmushko, et al. "2D computations for perturbation growth of magnetically driven cylindrical aluminum and aluminum alloy liners". Proc. 7th Intern. Workshop on The Physics of Compressible Turbulent Mixing, St.Petersburg, 1999, Eds. E.Meshkov, Yu.Yanilkin & V.Zhmailo, Sarov, Russia, pp. 237-243, 2001.

[4] A.M. Buyko, O.M. Burenkov, V.V Zmushko, et al. "On the feasibility to achieve high pressures with disk EMG driven impacting liners". Ibid. pp. 516-519, 2001.

[5] A.M. Buyko, Yu.N. Gorbachev, V.V. Zmushko, et al., B.G. Anderson, W.L. Atchison, D.A. Clark et al., "Simulation of Atlas parameters in explosive magnetic experiments ALT-1,2". Proc. IX Int. Conf. Megagauss Magnetic Field Generation and Related Topics, M.–St. Petersburg, 2002. Eds V.D.Selemir and L.N.Plyashkevich, VNIIEF, Sarov, pp. 747-751, 2004.

[6] R.J. Faehl, B.G. Anderson, D.A. Clark, et al., V.K. Chernyshev, V.N. Mokhov, V.N. Buzin et al., "Results of a 100-megaampere liner implosion experiment". IEEE Trans. Plasma Sci., vol. 32, No. 5, pp. 1972-1985, 2004.

[7] A.M. Buyko, S.F. Garanin, V.V. Zmushko, et al. "Simulations of condensed-matter liner implosion for experiments with disc explosive magnetic flux compression generators". Proc. XII Int. Conf. on Meg. Mag. Field Gen. and Related Topics. Novosibirsk, 2008. Ed. G.A.Shvetsov, pp. 465-474, 2010.

[8] A.M. Buyko, S.F. Garanin, and V.V. Zmushko, "Magnetic liner implosion simulations for the ALT-3 experiment". Megagauss-XIV, Maui, HI, USA, 2012. Eds. G.F.Kiuttu, K.W.Struve, J.H.Degnan, IEEE, Inc., pp. 172-177, 2012.

[9] A.M. Buyko, S.F. Garanin, A.M. Glybin, et al. "Revision of the parameters of a system for liner driving to 20 km/s" PMTF, v. 56, No.1, pp. 116-124, 2015.

[10] I.D. Sofronov, O.A. Vinokurov, V.V. Zmushko, et al. "MIMOZA code." Multidimensional hydrodynamic simulations. Transactions of Mathematical Modeling Issues in Computational Mathematics and Informatics. Moscow - Arzamas-16, RFNC-VNIIEF, pp. 94-96, 1994.

[11] A.I. Abakumov, P.N. Nizovtsev, A.V. Pevnitsky, and V.P. Soloviev. "The code DRAGON to simulate elastoplastic flow under shock loading in the two-dimensional and three-dimensional approacimation". Proc. Int. Conf. IV Zababakhin Talks. Snezhinsk, pp. 227-228, 1995.

[12] V.A. Rayevsky, A.I. Lebedev, P.N. Nizovtsev, et al. "Studies of Rayleigh-Taylor instability in copper and aluminum at pressures up to 45 GPa and strain rates of $10^5 - 10^8$ s^{-1}». Book of Abstracts, Int.Conf. V Khariton Scientific Talks. RFNC-VNIIEF, Sarov, pp. 180-181, 2003.

[13] A.M. Buyko, S.F. Garanin, V.A. Demidov, et al. "Investigation of the dynamics of a cylindrical exploding liner accelerated by a magnetic field in the megagauss range". Megagauss-V, Novosibirsk, 1989. Eds. V. M. Titov and G. A. Shvetsov, New York: Nova Science Publishers, pp. 743-748, 1990.

Magnetoresistance relaxation anisotropy of nanostructured La-Sr(Ca)-Mn-O films induced by high pulsed magnetic fields

N. Zurauskiene, D. Pavilonis, J. Klimantavicius,
S. Balevicius, V. Stankevic, V. Plausinaitiene,
R. Vasiliauskas, M. Skapas, R. Juskenas
Center for Physical Sciences and Technology,
Vilnius, Lithuania
nerija.zurauskiene@ftmc.lt

N. Zurauskiene, S. Balevicius, V. Stankevic
Faculty of Electronics,
Vilnius Gediminas Technical University,
Vilnius, Lithuania

Abstract—**The results of investigation of colossal magnetoresistance relaxation in nanostructured La-Sr(Ca)-Mn-O films upon removal of magnetic field pulse are presented. The 280 nm thick films grown by PI MOCVD technique were studied in pulsed magnetic fields having duration of 0.9 ms and amplitudes up to 10 T. It was obtained, that 'fast' relaxation process occurring in hundred microseconds time scale exhibits anisotropy: the magnitude of remnant resistivity is 3 times smaller and the process is faster when magnetic field is applied perpendicular to the film plane in comparison with in-plane direction. The dynamics of this relaxation process was analyzed using Kolmogorov–Avrami–Fatuzzo model, taking into account reorientation of magnetic domains into equilibrium state. The 'fast' remnant relaxation was not observed after the application of longer pulses (20 ms) having amplitude of 60 T.**

Keywords—Colossal magnetoresistance; resistance relaxation; thin films; magnetic field sensors; pulsed magnetic fields.

I. Introduction

Recently, it has been demonstrated that nanostructured lanthanum manganite films which exhibit the colossal magnetoresistance effect (CMR) can be successfully used for the development of CMR-B-scalar sensors [1], [2]. Such sensors have been used at room temperatures to measure the magnetic diffusion processes in electromagnetic launchers [3] and the distribution of transient magnetic fields in non-destructive pulsed-field magnets [4]. However, for the applications in plasma science or condensed matter physics, sensors operating at cryogenic temperatures and measuring high pulsed magnetic fields up to megagauss [1] are required. Therefore, it is important to avoid magnetic memory effects which limit the accuracy and speed of such sensors. The memory effects in manganite films are related to magnetization relaxation of these films when external magnetic field is switched-off. Due to close relation between transport properties (conductivity) and magnetization of manganite materials, the investigation of magnetoresistance relaxation provides indirect method for the study of the magnetic relaxation processes in these films [5]. There have been performed numerous investigations proposing different relaxation mechanisms in thin magnetic films and, therefore, different mathematical expressions were used to characterize the time-dependences of the magnetization / resistance relaxation: logarithmic, power-law-like, exponential (Debye), 'stretched' or 'compressed' exponential [5]-[12]. These all depend on the mechanisms involved: the nucleation and ferromagnetic domain wall propagation [6]-[9], spin-glass-like behavior, local disorder and others [6], [10]-[12]. It has to be pointed out that the magnetization reversal was investigated in a number of out-of-plane and in-plane magnetized ultrathin epitaxial magnetic films and multilayers [7]-[9], however, there is a lack of information about investigations of magnetic relaxation anisotropy of thin nanostructured manganite films.

In this paper, we present the results of an investigation of colossal magnetoresistance relaxation anisotropy in thin nanostructured La-Sr(Ca)-Mn-O films upon removal of pulsed magnetic field. As will be demonstrated, magnetic relaxation dynamics can reveal important information about possible relation between the film morphology and magnetic properties.

II. Experimental

The $La_{0.8}Ca_{0.2}MnO_3$ (La-Ca-MnO) and $La_{0.83}Sr_{0.17}MnO_3$ (La-Sr-MnO) films having thicknesses of 280 nm were deposited using the Pulsed Injection Metal Organic Chemical Vapour Deposition technique onto a lucalox (99.9 % Al_2O_3 + 0.1 % MgO) substrate at deposition temperatures of 775 °C and 750 °C, respectively. Due to the polycrystalline structure of the substrate, nanostructured films were produced. The morphology and microstructure of the films were investigated using Scanning Electron Microscopy (SEM) and Transmission Electron Microscopy (TEM). The TEM images (see Fig.1) demonstrated the complex structure of the films: column-shaped crystallites having width of 40–60 nm were separated by a 5-10 nm thickness grain boundaries. The lower part of the image in Fig.1 shows the substrate while the upper part is Pt layer covered the film for the TEM-sample preparation. The structure of La-Sr-Mn-O and La-Ca-Mn-O films was similar.

This work was partly supported by the Research Council of Lithuania (grant No. MIP-095/2015).

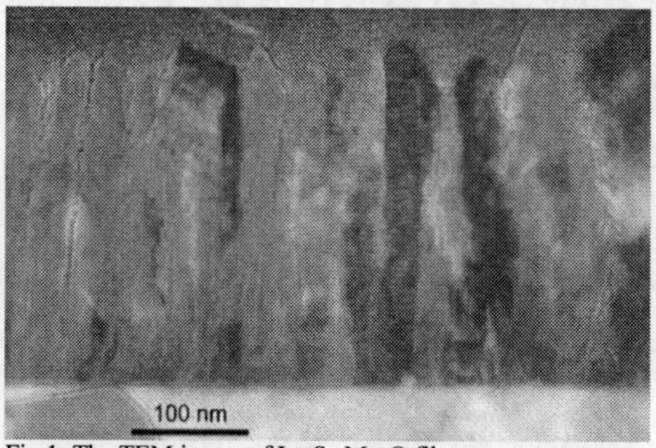

Fig.1. The TEM image of La-Sr-Mn-O film.

The samples were fabricated in co-planar shapes onto top of which two 0.5 mm×0.5 mm square Ag electrodes were deposited by thermal evaporation. These were separated by a distance of 50 μm. The resistivity ρ dependence on temperature was investigated by using a low dc electric field in a closed cycle helium gas cryo-cooler in the temperature range 5–300 K. The dependence revealed transition which is typical for manganites, i.e. from metal-like to an insulator-like state at temperature T_m. The T_m and corresponding resistivity maximum ρ_m were the following: 1) 237 K, 1.16 Ω·cm for the La-Sr-Mn-O film, 2) 225 K, 2.8 Ω·cm for the La-Ca-Mn-O film. The magnetoresistance (*MR*) and resistance relaxation measurements were performed by using a pulsed magnetic field generator based on capacitor bank discharge through a special multi-shot magnetic field coil [11] which ensures generation of a half-sine waveform magnetic field pulses without any "tail" after the current is switched-off. The experiments were performed at temperature of 100 K when magnetic field was applied in three different directions to the film plane and bias current *J*: 1) parallel to film plane and *J* direction, 2) parallel to film plane, but perpendicular to *J* direction, and 3) perpendicular to the film plane. For measurements at high pulsed magnetic fields, a 65 T/ 20 ms/ 20 mm non-destructive field coil fabricated and installed at the Dresden High Magnetic Field laboratory (HLD) was used [4]. In this case the magnetoresistance was measured in the temperature range of 1.5-290 K using 60 T amplitude magnetic field pulses. The magnetic field was applied parallel to the film plane, but perpendicular to the current direction.

III. Results and Discussion

The typical resistivity dynamics of the investigated manganite films after the application of a 0.9 ms duration magnetic field pulse with an amplitude of 10 T is presented in Fig. 2. The measurement is performed at temperature of 100 K when magnetic field is applied along three different directions in respect to the film plane and current direction. One can distinguish the relaxation processes occurring in different time scales: 'ultrafast' (< 1 μs), which follows the magnetic field pulse; 'fast' (~ 100 μs) and slow (> 1 ms) which are observed when magnetic field is switched-off. It is evident that the slowest component of resistance relaxation is independent on magnetic field direction. This relaxation process is usually attributed to the short-range interactions of disordered grain boundaries having spin-glass properties [5]-[6]. The kinetics of 'slow' relaxation in manganite films was studied and reported by us [11], as well as by the other authors, and described by logarithmic [5] or stretched exponential [10]-[11] expressions. Therefore, in this study we will mainly discuss the results obtained during investigation of the "fast" relaxation process. It was obtained that the 'fast' resistance relaxation processes are similar for both in-plane magnetic field directions (see Fig.2). However, when magnetic field is applied perpendicular to the film plane, the remnant resistivity is much smaller. This result is important not only from fundamental point of view, but also for the sensor application in order to avoid memory effects. For the development of magnetic field sensors operating at cryogenic temperatures, it is important to determine the conditions under which the remnant resistivity and characteristic time of the relaxation processes can be minimized to ensure sufficient measurement accuracy. The results of resistance (conductance) relaxation, namely, the dependences of remnant resistivity magnitudes and time constants of these processes on magnetic field direction in respect to the film plane as well as influence of film structure and morphology on the mechanisms of relaxation, are presented and discussed below.

A. Anisotropy of resistance relaxation

The dynamics of 'fast' relaxation was analyzed using the well-known Kolmogorov–Avrami–Fatuzzo (KAF) model [13], taking into account reorientation of magnetic domains into equilibrium state. This model adapted by Labrune to study ferromagnetic relaxation [14] considers both nucleation and subsequent domain wall propagation processes. The results of remnant conductivity kinetics (see inset in Fig.2) were fitted to a 'compressed' exponential decay [6]:

$$\sigma(t) = \sigma_{0fast} + \sigma_{fast}\exp[-(t/\tau_{fast})^{\beta}], \ 1<\beta<3 \quad (1)$$

where τ_{fast} is the time constant of the process, σ_{0fast} is the conductivity, when the 'fast' relaxation process is considered to be finished, and σ_{fast} is the remnant conductivity magnitude. The corresponding resistivity ($\rho_{fast} = 1/\sigma_{fast}$) is presented in Fig. 2 for clarity. The inset presents kinetics of remnant

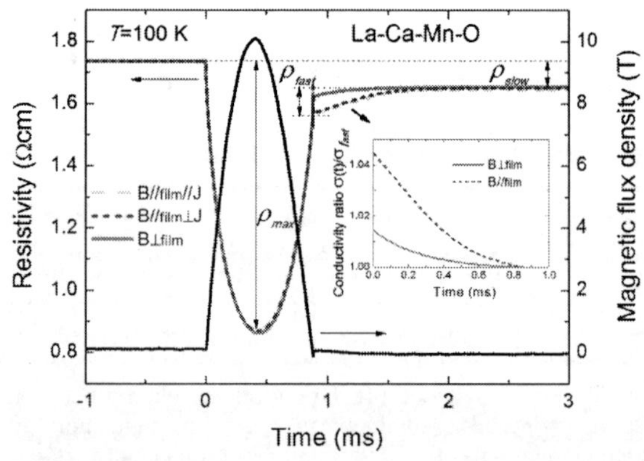

Fig.2 Magnetic field pulse (right scale) and resistivity response (left scale) for La-Ca-Mn-O film, when magnetic field is applied in three different directions in respect to film plane and current *J*. The inset presents corresponding conductivity dynamics (t=0 when B is switched-off) for two B directions.

conductivity ratio $\sigma(t)/\sigma_{fast}$. The time instant t=0 was chosen as the moment when the magnetic field pulse was switched-off. This time instant corresponds to 0.9 ms for the resistivity change pulse presented in Fig. 2. It is clearly seen that remnant conductivity amplitude is higher when magnetic field is applied in-plane. By fitting relaxation curves, it is possible to determine the parameter β and hence the predominant relaxation mechanism (magnetic domain propagation or nucleation). Also, one can extract the time $t_{1/2}$ needed to relax half of the film remnant conductivity magnitude. In order to compare times $t_{1/2}$ of relaxation processes for different films and magnetic field directions, we normalized the remnant conductivity $\Delta\sigma_{fast}=\sigma(t)-\sigma_{0fast}$ to its initial value σ_{fast}. Fig. 3 presents the results for La-Sr-Mn-O and La-Ca-Mn-O films measured at 100 K temperature (ferromagnetic state).

One can see that kinetics of conductivity $\Delta\sigma_{fast}$ when magnetic field is applied in-plane has more pronounced S-shape and is longer ($t_{1/2}$=350 μs) for La-Sr-Mn-O films in comparison with the La-Ca-Mn-O ($t_{1/2}$=280 μs). The characteristic time τ_{fast} and exponent β obtained by fitting the experimental results using Eq. (1) are 430 μs and 1.87, respectively for the La-Sr-Mn-O and 400 μs and 1.36, respectively for the La-Ca-Mn-O films. For the case when magnetic field is applied perpendicular to the film plane, the process is faster if compare with the in-plane cases, however, in this case it is faster for the La-Sr-Mn-O film.

The obtained results can be explained as follows: In the ferromagnetic state, the relaxation process is determined mostly by the reorientation of the magnetic domains into their equilibrium state. It is known that for polycrystalline or nanostructured films having disordered grain boundaries, as in our case, various defects behave like pinning sites and at first, some reversed spins or small domains appear in the films because of them [6]-[9] and act as nuclei. Due to their short-range interaction, the adjacent spins reverse more often than others, and magnetic domains grow from these reversed spins. As it was mentioned above, such magnetization relaxation was explained by the KAF model. Usually two limiting cases are considered [6]: 1) when nuclei do not grow and relaxation

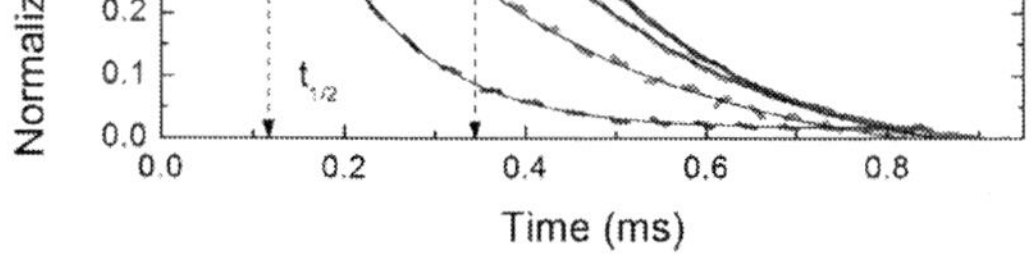

Fig.3 Normalized conductivity dynamics for different films when magnetic field is applied parallel or perpendicular to film plane. The inset presents half-magnitude times $t_{1/2}$, characteristic times τ_{fast} and exponent β values obtained by fitting experimental results to Eq. (1).

follows the simple Debye exponent (β=1); and 2) when domains grow fast and relaxation takes place as an 'compressed' exponent with β=3, i.e. $\sim\exp[-(t/\tau)^3]$. For the intermediate cases, the experimental results are "fitted" by means of Eq. (1) when exponent 1< β <3 [6]. The β values obtained by fitting experimental curves of our investigated films are presented in the inset of Fig.3. The higher β values for La-Sr-Mn-O films (1.87) in comparison with La-Ca-Mn-O films (1.36) for the in-plane magnetic field case (B//) indicate that nucleation process is faster in the latter case. Moreover, the crystallites of the nanostructured films are small enough and thus the domain wall propagation is limited. For the case when the film is magnetized perpendicular to the plane (B⊥), the faster decrease of magnetization and, as a result, the faster decrease of conductivity upon the removal of external magnetic field can be related to the rotation of magnetization vector towards the in-plane direction which coincides with easy-magnetization axis for our investigated films. It is known that the shape anisotropy of thin non-strained manganite films favors the magnetization to lie in-plane, while for strained epitaxial films it is very dependent on the level of tensile or compressive strain, which originates from the mismatch with the substrate [9]. Therefore, for our investigated nanostructured films the relaxation process is faster for out-of-plane direction: half-magnitude characteristic times $t_{1/2}$ are 120 μs and 180 μs for La-Sr-Mn-O and La-Ca-Mn-O films, respectively. Moreover, the demagnetizing field for out-of-plane magnetized films acts to reduce the domain nucleation field in respect to the in-plane case [8]. As a result, domain nucleation is easier for out-of-plane case and this makes out-of-plane relaxation nucleation dominated ($\beta\approx1$), while in the process of in-plane relaxation domain wall propagation acts together with the nucleation. The faster out-of-plane relaxation process results in the smaller remnant conductivity magnitude in comparison with the in-plane-case (see inset in Fig.2).

B. Ultrahigh magnetic fields

The resistance change upon application of high magnetic field was measured at Dresden High Magnetic Field laboratory (HLD). The magnetic field direction was chosen in-plane in order to observe remnant resistivity with highest possible magnitude.

Fig. 4 presents magnetic field pulse (right scale) and response signal (recalculated to resistance values) of the La-Sr-Mn-O sample (left scale) measured at temperature of 100 K. Due to negative colossal magnetoresistance effect the resistance decreases with increase of magnetic field. The small (up to 4 T) magnetic field pulse of opposite direction also induces resistance decrease, therefore, we observe signal from the manganite film of the same polarity. One can see that 'fast' resistance relaxation process is finished until the end of the positive magnetic field pulse (20 ms). This result is confirmed by the figure presented in the inset: The resistance dependence on magnetic flux density for both rising and falling edges of the pulse coincide well. The remnant resistivity also was not observable at higher temperatures in the range of 100-290 K. Therefore, we can conclude that for longer pulses (≥20 ms) there is no observable remnant resistance at T≥100 K. The similar results were found also for

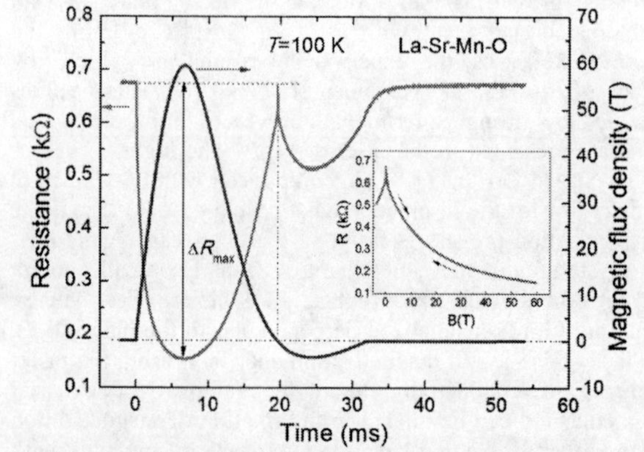

Fig.4. Magnetic field pulse (right scale) and response signal presented as resistance change (left scale) of La-Sr-Mn-O film. Inset shows resistance change *vs.* magnetic flux density during rising and falling edges of magnetic field pulse.

La-Ca-Mn-O films. For the lower temperatures ($1.5 \le T \le 80$ K) the observed remnant resistivity was assigned to 'slow' process (see ρ_{slow} in Fig.2). Fig. 5 presents the resistance change upon application of 60 T magnetic field at temperature T=4.4 K. The remnant resistivity magnitude ΔR_{slow} is more clearly seen in the inset. It exceeds 4 % of the maximum resistivity change (ΔR_{max}) upon application of 60 T magnetic field. At 1.5 K it exceeded 6 % for both La-Sr-Mn-O and La-Ca-Mn-O films.

This result is very important for the development of magnetic field sensors based on nanostructured manganite films and operating at cryogenic temperatures. For the measurement of high magnetic fields up to megagauss, the rising edge of the sensor's response pulse can be used. However, for the measurement of entire pulse shape, the accuracy at lower fields (<10 T) at the falling edge of the pulse is limited to 6 %.

In conclusion, it was found that dominated mechanism in 'fast' resistance relaxation process of nanostructured manganite films after removal of magnetic field pulse is magnetic domain nucleation and reorientation to equilibrium state. The observed anisotropy of relaxation process when magnetic field is applied perpendicular or parallel to film plane is mostly determined by "shape" anisotropy of the films.

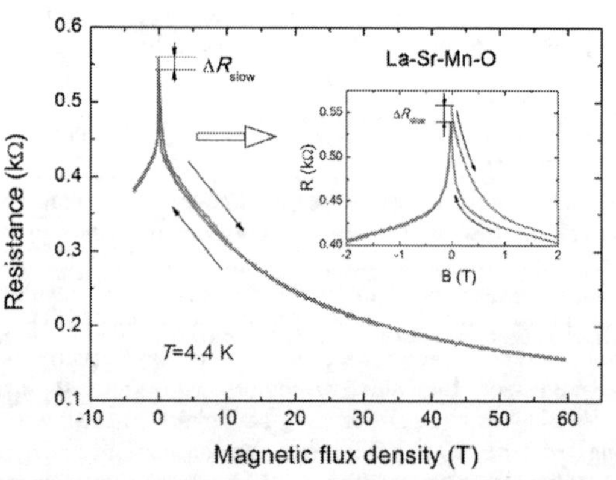

Fig.5. Resistance change of La-Sr-Mn-O film *vs.* magnetic flux density during rising and falling edges of magnetic field pulse having 60 T amplitude.

Acknowledgment

The authors gratefully acknowledge the staff of Dresden High Magnetic Field Laboratory, Helmholtz-Zentrum Dresden-Rossendorf, for the possibility to use high-field magnet facilities.

References

[1] S. Balevičius, N. Žurauskienė, V. Stankevič, S. Keršulis, V. Plaušinaitienė, A. Abrutis, S. Zherlitsyn, T. Herrmannsdörfer, J. Wosnitza and F.Wolff-Fabris, "Nanostructured thin manganite films in megagauss magnetic field," Appl. Phys. Lett., vol. 101, pp. 092407-1–4, Aug. 2012.

[2] T. Stankevič, L. Medišauskas, V. Stankevič, S. Balevičius, N. Žurauskienė, O. Liebfried, M. Schneider, "Pulsed magnetic field measurement system based on colossal magnetoresistance-B-scalar sensors for railgun investigation," Rev. Sci. Instrum., vol. 85, pp. 044704-1–5, Apr. 2014.

[3] M. Schneider, O. Liebfried, V. Stankevic, S. Balevicius and N. Zurauskiene, "Magnetic Diffusion in Railguns: Measurements Using CMR-Based Sensors," IEEE Trans. Magn., vol. 45, pp. 430-435, Jan. 2009.

[4] S. Balevicius, N. Zurauskiene, V. Stankevic, T. Herrmannsdörfer, S. Zherlitsyn, Y. Skourski, F. Wolff-Fabris, J. Wosnitza, "CMR-B-Scalar sensor application for high magnetic field measurement in non-destructive pulsed magnets," IEEE Trans. Magn., vol. 49, iss. 11, pp. 5480–5484, Nov. 2013.

[5] M. Sirena, L. B. Steren and J. Guimpel, "Magnetic relaxation in bulk and film manganite compounds," Phys. Rev. B, vol. 64, pp. 104409-1–6, Aug. 2001.

[6] Haiwen Xi, Kai-Zhong Gao, Jun Ouyang, Yiming Shi and Yizhang Yang, "Slow magnetization relaxation and reversal in magnetic thin films", J. Phys.: Condens. Matter., vol. 20, pp. 295220-1–8, 2008.

[7] M. Czapkiewicz, T. Stobiecki, and S. van Dijken, Thermally activated magnetization reversal in exchange-biased [Pt/Co]$_3$/Pt/IrMn multilayers, Phys. Rev. B, vol. 77, pp. 024416-1–8, Jan. 2008.

[8] R. P. Cowburn, J. Ferre´, S. J. Gray and J. A. C. Bland, "Domain-wall dynamics, pinning, and nucleation in ultrathin epitaxial Fe films", Phys. Rev. B, vol. 58, pp. 11507–11513, Nov. 1998.

[9] S. R. Bakaul, B. F. Miao, W. Lin, W. Hu, A. David, H. F. Ding, T. Wu Domain-related origin of magnetic relaxation in compressively strained manganite thin films, Appl. Phys. Lett., vol. 101, pp. 012408-1–4, July 2012.

[10] J. Baszyński, „"Slow Relaxation of Resistance in Manganite Perovskite Nanoconstrictions," Acta Phys. Polonica A, vol. 115, pp.339-341, 2009.

[11] N. Zurauskiene, S. Balevicius, D. Pavilonis, V. Stankevic, S. Kersulis, J. Novickij, "Magnetoresistance Relaxation in Thin La-Sr-Mn-O Films Exposed to High-Pulsed Magnetic Fields", IEEE Trans. Plasma Sci., vol. 41, pp. 2830-2835, Oct. 2013.

[12] N. Žurauskienė, S. Balevičius, D. Pavilonis, V. Stankevič, V. Plaušinaitienė, S. Zherlitsyn, T. Herrmannsdörfer, J. M. Law, J. Wosnitza, Magnetoresistance and Resistance Relaxation of Nanostructured La-Ca-MnO Films in Pulsed Magnetic Fields, IEEE Trans. Magn., vol. 50, pp. 6100804-1–6100804-4, Nov. 2014.

[13] E. Fatuzzo, "Theoretical considerations on the switching transient in ferroelectrics," Phys. Rev., vol. 127, pp. 1999–2005, Sept. 1962.

[14] M. Labrune, S. Andrieu, F. Rio, P. Bernstein, "Time dependence of the magnetization process of RE-TM alloys," J. Magn. Magn. Mater. vol. 80, pp. 211-218, Aug. 1989.

EXPLODING WIRE COMPRESSION FOR EVALUATING ELECTRICAL CONDUCTIVITY IN WARM-DENSE DIAMOND-LIKE-CARBON

Toru Sasaki, Takumi Ohuchi, Kazumasa Takahashi, Nob. Harada
Department of Electric Engineering
Nagaoka University of Technology
Nagaoka, Japan
sasakit@vos.nagaokaut.ac.jp

Mayuko Koga
Department of Electrical Engineering and Computer Science
University of Hyogo
Himeji, Hyogo

Takashi Kikuchi, Arata Watabe, Satoshi Sugimoto
Department of Nuclear Energy System Safety Engineering
Nagaoka University of Technology
Nagaoka, Japan

Shinsuke Fujioka
Institute of Laser Engineering
Osaka University
Suita, Osaka

***Abstract*—To measure the electrical conductivity in warm dense diamond-like-carbon (DLC), we propose an exploding wire compression method. The warm dense DLC was generated by a shock compression driven by an exploding wire discharge with confined by a rigid capillary. To determine the generation condition for warm dense DLC, we have carried out a one-dimensional magnetohydrodynamic simulation. From the numerical estimation, we found the condition of measuring electrical conductivity for DLC plasma. The exploding wire compression method was demonstrated. From the comparison of time-evolution of voltage and current for gold and gold+DLC samples, the voltage-current evolutions were different after the ablation. The DLC temperature estimated from the absorption spectroscopy was estimated to be 8000 - 9000 K. From the experiment associating the numerical simulation, the electrical conductivity of DLC plasma was measured. The results show that the electrical conductivity of DLC plasma was estimated to be 106 S/m, which is relatively high for that of conventional carbon plasma. On the other hand, the comparison of the theoretical result for diamond at solid density and the experimental results shows similar.**

Keywords*—*Warm dense matter, Exproding wire compression, Diamond-like carbon, Electrical conductivity

I. Introduction

Fast ignition of inertial confinement fusion [1] is the transport of the laser-generated fast electrons and their associated heating of compressed DT fuel. The coupling efficiency from laser energy to fast electron energy deposited in the fuel should need to be improved the cone materials. The reduction of unstoppable, un-ejected, and diverging fast electrons are key issue to improve the energy coupling efficiency. From the numerical results, the low-Z cone to generate the fast electrons is expected to increase the coupling efficiency, because the fast electrons are scattered by the large Coulomb potential of highly charged ions in high-Z cone wall [2]. The transport of fast electrons in the cone also depends on the electrical conductivity in warm dense matter (WDM) state [3].

From above predictions, a diamond-like-carbon (DLC)[4] cone, which is one of the low-Z cone, is promised to increase the coupling efficiency due to the redaction of stopping power in the cone. The properties of DLC depend on not only the fabrication process but also the composition ratio. Generally, the properties of DLC respect the diamond and the graphite, however, materials in WDM state also are in a complex area. Thus, the properties of vapor DLC is difficult to obtain the conventional method.

To measure the electrical conductivity in warm dense DLC, we propose an exploding wire compression method. The warm dense DLC was generated by a shock compression driven by an exploding wire discharge with confined by a rigid capillary [5]. To determine the generation condition for warm dense DLC, we have carried out a one-dimensional magnetohydrodynamic simulation. The qualitatively evaluation of the electrical conductivity for the WDM DLC requires a small electrical conductivity of the exploding wire. From the estimation, we have selected a sample and a method.

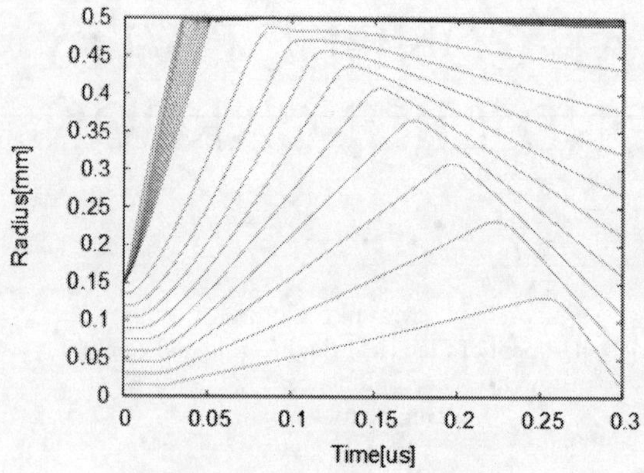

(a) Hydrodynamic behavior up to 0.3 μs

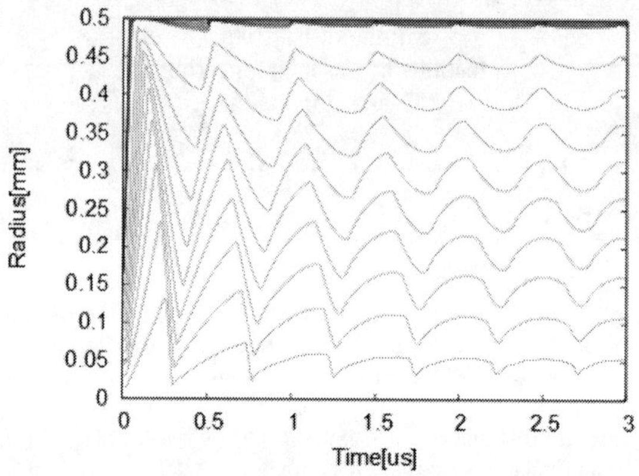

(b) Hydrodynamic behavior up to 3 μs

Fig. 1. Hydrodynamic behavior after the exploding wire discharge, (a) is the hydrodynamic behavior up to 0.3 μs and (b) is the hydrodynamic behavior up to 0.3 μs. Red lines indicate the stream lines of gold, and blue lines reveal the stream line of DLC.

II. Experimental Method

To generate the warm dense DLC, the exploding wire compression method was considered because DLC wire is difficult to make a fine wire. In order to observe the electrical conductivity of warm dense DLC, a resistance of the exploding wire should be higher than that of DLC membrane. From the previous researches, the gold was small electrical conductivity in warm dense state [5-9]. Therefore, a gold wire was selected as an exploding wire. To confirm the hydrodynamic behavior and the difference of resistance between warm dense gold and ablated DLC, one-dimensional hydrodynamic simulation was carried out.

Figure 1 shows a typical hydrodynamic behavior after the exploding wire discharge. The initial temperatures of gold and DLC membrane were 5000 K and 300 K, respectively. The thickness of DLC membrane was set to be 5 μm. The boundary conditions were reflective boundaries assuming rigid wall at 0.5 mm in radius and axisymmetric condition at the center. The result shows that the gold plasma rapidly expanded to the rigit wall with the DLC membrane. The gold plasma is compressed the DLC membrane at 0.1μs from the beginning of discharge. After the compression, the DLC membrane/plasma seemed a small perturbation around the wall. The radius of DLC is almost 10 μm after a few μs from beginning of discharge.

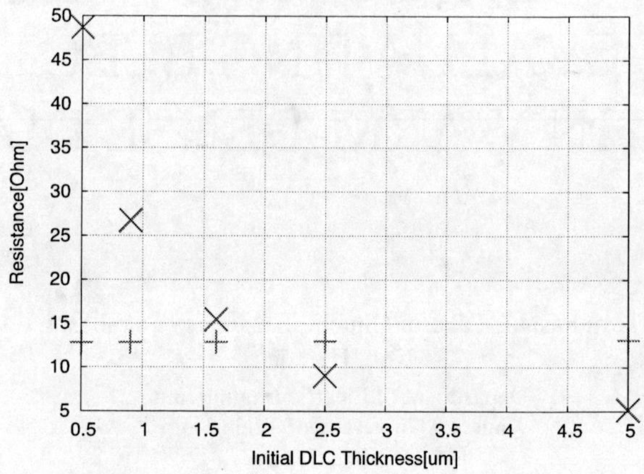

Fig. 2. Initial DLC thickness as a function of the resistances of gold and DLC. The red cross shows the resistance of gold, and the blue cross means the resistance of DLC plasma.

Figure 2 shows an initial DLC thickness as a function of the resistances of gold and DLC. The electrical conductivites of warm dense gold and warm dense DLC were assumed 10^3 S/m [5] and 10^5S/m [3], respectively. The results show that the resistance of warm dense gold is almost same as 12 Ω. On the

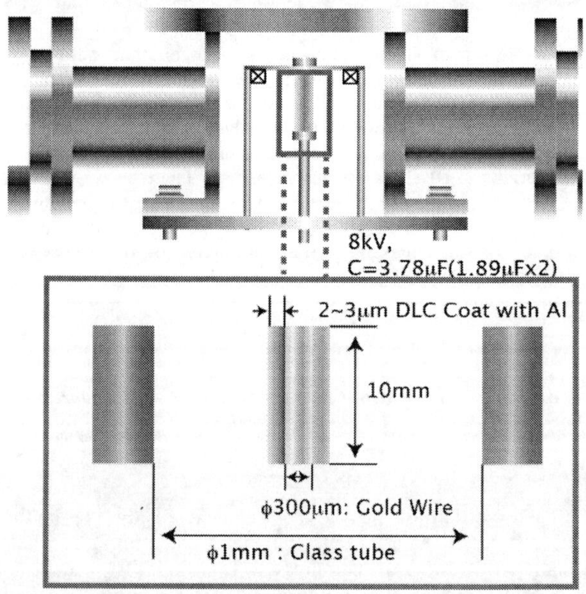

Fig. 3. Experimental setup of exploding wire compression for evaluating electrical conductivity for warm dense DLC

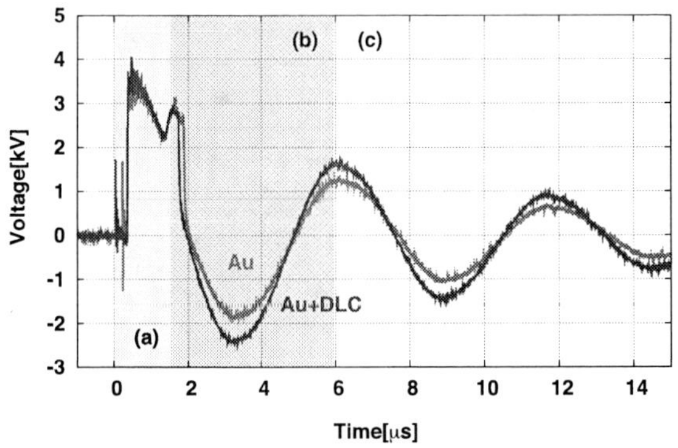

(1)Voltage waveforms for gold and gold+DLC samples.

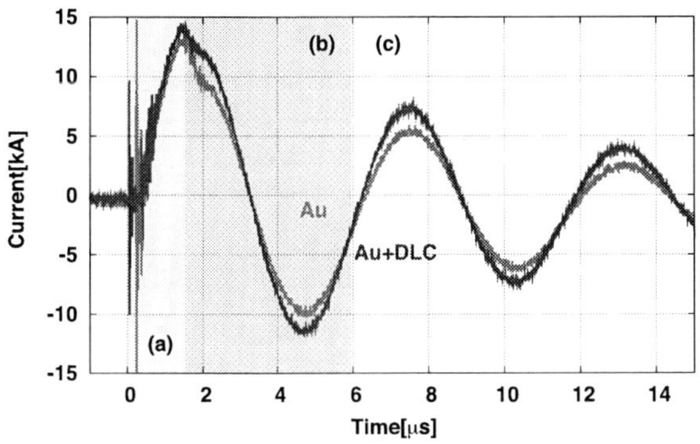

(2)Current waveforms for gold and gold+DLC samples.

Fig. 4. Time-evolutions of voltage and current for gold and gold+DLC samples.

other hand, the resistance of warm dense DLC decreases with increase in the thickness of the DLC membrane. To observe the resistance of warm dense DLC, the initial DLC thickness should be over 2 μm.

Figure 3 shows an experimental setup of exploding wire compression for evaluating electrical conductivity for warm dense DLC [5]. Low inductance capacitors (2 x 1.89 μF) were charged up to 8 kV. The charged energy is ensured to the ablation of gold wire. The time evolutions of voltage and current were observed with a high voltage probe and a Rogowskii coil, respectively. The stray inductance of this system was 212 nH. To observe the temperature of ablated DLC membrane, a spectrometer mounted on a streak camera was used.

A DLC membrane was made by plasma based ion implantation and deposition method [10]. Because the electrical conductivity of gold in 0.1 ρ_s with warm dense state is relatively insulating condition compared to the theoretical prediction of that warm dense DLC, a gold wire, which was 300 mm in diameter having 15 mm in length, was coated with

This work was partly supported by the NIFS Collaboration Research programs (NIFS13KUGK073).

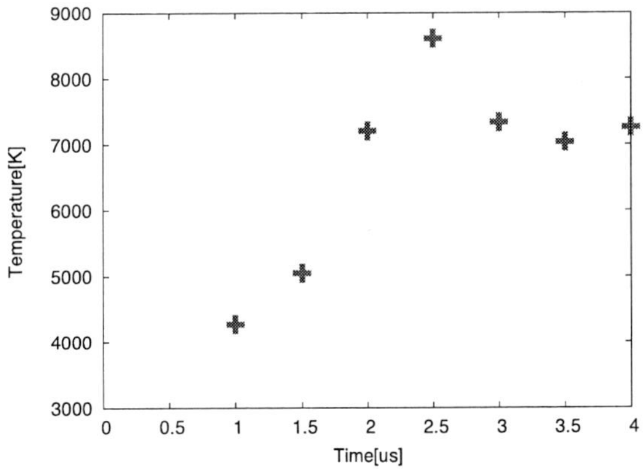

Fig. 5. Time-evolution of temperature for DLC sample estimated with the absorption spectroscopic measurement

the DLC membrane. The exploding wire has a huge ablation pressure approximately a few GPa. Thus, the pressure of exploding wire drives the shock heating for the insulator as the DLC membrane which is coated on the wire. The heated DLC membrane state is observed by an absorption spectrum for the temperature.

III. Experimental Results

Figure 4 shows a time-evolution of voltage and current for gold and gold+DLC samples. As shown in Fig. 4 (1), observed voltage and current waveforms are same in the region (a). The region (a) in Fig. 4 means that the wire is heated up to vaporization. After the region (a), we can see that the voltage and current waveforms between gold wire and gold wire + DLC as shown the region (b) in Fig. 4 change due to the difference of the resistance of the sample. After the region (b), the capillary was broken by the expansion pressure and the wire/plasma behaves free expansion from the streak image.

Figure 5 shows a time-evolution of temperature for DLC sample estimated with the absorption spectroscopic measurement. The optical radiator was assumed the exploding wire of gold. From the spectroscopic measurement, the spectrum of the exploding wire seems to be a black body radiation. The observed spectrum for gold wire + DLC membrane was found in the spectra, which were the transitions of singly ionized carbon as 426 nm (4f→3d) and 657 nm (3p→3s). As shown in Fig. 5, the DLC temperature is up to 9000 K at 2.5 μs from the beginning of discharge. Its temperature behavior almost corresponds to the current waveform. It means that the warm dense DLC is generated by using the proposed method.

To estimate the resistance, the voltage and the current waveforms are fitted with over damping waveform of LCR circuit. The estimation of resitance for DLC plasma was 0.1 ohm. From the simulation result, the inner and outer DLC radii are 4.99 mm and 5 mm, respectively. From the above estimation, the electrical conductivity of DLC plasma is estimated to be 3.2 x 10^6 S/m. From the comparison of the experimental estimation and the previous results for conventional carbon[11], the electrical conductivity for the obtained experimental results is higher than that for conventional carbon results. On the other hand, the comparison of the theroretical result for diamond at solid density [3] and the experimental results shows similar. To understand the electrical conductivity dependence of the temperature, we will measure the exploding wire compression method.

IV. CONCLUSION

To measure the electrical conductivity in warm dense DLC, we propose an exploding wire compression method. The warm dense DLC was generated by a shock compression driven by an exploding wire discharge with confined by a rigid capillary. To determine the generation condition for warm dense DLC, we have carried out a one-dimensional magnetohydrodynamic simulation. From the numerical estimation, we found the condition of measuring electrical conductivity for DLC plasma.

The exploding wire compression method was demonstrated. From the comparison of time-evolution of voltage and current for gold and gold+DLC samples, the voltage-current evolutions were different after the ablation. The DLC temperature estimated from the absorption spectroscopy was estimated to be 8000 - 9000 K. From the experiment associating the numerical simulation, the electrical conductivity of DLC plasma was measured. The results show that the electrical conductivity of DLC plasma was estimated to be 10^6 S/m, which is relatively high for that of conventional carbon plasma. On the other hand, the comparison of theroretical result for diamond at solid density and the experimental results shows similar.

References

[1] R. Kodama, P. A. Norreys, K. Mima, A. E. Dangor, R. G. Evans, H. Fujita, Y. Kitagawa, K. Krushelnick, T. Miyakoshi, N. Miyanaga, T. Norimatsu, S. J. Rose, T. Shozaki, K. Shigemori, A. Sunahara, M. Tampo, K. A. Tanaka, Y. Toyama, T. Yamanaka and M. Zepf, “Fast heating of ultrahigh-density plasma as a step towards laser fusion ignition” , Nature, **412**, p. 798, 2001.

[2] T. Johzaki, H. Nagatomo, A. Sunahara, H.-B. Cai, H. Sakagami, Y. Nakao and K. Mima, “Pre-plasma effects on core heating and enhancing heating efficiency by extended double cone for FIREX”, Nuclear Fusion, **51**, p. 073022, 2011

[3] P. McKenna, A. P. L. Robinson, D. Neely, M. P. Desjarlais, D. C. Carroll, M. N. Quinn, X. H. Yuan, C. M. Brenner, M. Burza, M. Coury, P. Gallegos, R. J. Gray, K. L. Lancaster, Y. T. Li, X. X. Lin, O. Tresca, and C.-G. Wahlström, “Effect of Lattice Structure on Energetic Electron Transport in Solids Irradiated by Ultraintense Laser Pulses”, Physical Review Letters, **106**, p.185004, 2011

[4] S. Fujioka, Z. Zhang, N. Yamamoto, S. Ohira, Y. Fujii, K. Ishihara, T. Johzaki, A. Sunahara, Y. Arikawa, K. Shigemori, Y. Hironaka, Y. Sakawa, Y. Nakata, J. Kawanaka, H. Nagatomo, H. Shiraga, N. Miyanaga, T. Norimatsu, H. Nishimura, and H. Azechi, "High-energy-density plasmas generation on GEKKO-LFEX laser facility for fast-ignition laser fusion studies and laboratory astrophysics", Plasma Physics and Controlled Fusion, **54**, 124042, 2012

[5] T. Sasaki, K. Takahashi, T. Kudo, T. Kikuchi, T. Aso, Nob. Harada, S. Fujioka, and K. Horioka, "Study on Exploding Wire Compression for Evaluating Electrical Conductivity in Warm-Dense Diamond-Like-Carbon", Journal of Physics: Conference Series, **688**, p. 012102, 2016

[6] T. Sasaki, Y. Yano, M. Nakajima, T. Kawamura, K. Horioka, "Warm-dense-matter studies using pulse-powered wire discharges in water", Laser and Particle Beams, **24**, p. 371, 2006

[7] T. Sasaki, M. Nakajima, T. Kawamura, K. Horikoka, “Electrical conductivities of aluminum, copper, and tungsten observed by an underwater explosion”. Physics of Plasmas, **17**, p. 084501, 2010

[8] Y. Amano, Y. Miki, T. Takahashi, T. Sasaki, T. Kikuchi, and Nob. Harada,"Isochoric heating of foamed metal using pulsed power discharge as a making technique of warm dense matter", Review of Scientific Instruments, **83**, p. 085107, 2012

[9] Y. Miki, , H. Saito, T. Takahashi, T. Sasaki, T. Kikuchi, Nob. Harada, "Evaluation of transport properties of warm dense matter using isochoric pulsed-power discharges", Nuclear Instruments and Methods in Physics Research Section A: Accelerators, Spectrometers, Detectors and Associated Equipment, **733**, p. 8, 2014

[10] M. Koga, T. Ono, T. Tokunaga, H. Kadota, T. Hashimoto, K. Kanda, T. Norimatsu, "Development of DLC cone for fast ignition experiment", APS Gaseous Electronics Conference, abstract #GT1.153, 2015

[11] J. Haun, H.-J. Kunze, S. Kosse, M. Schlanges, and R. Redmer, “Electrical conductivity of nonideal carbon and zinc plasmas: Experimental and theoretical results”, Physical Review E, **65**, p. 046407, 2002

Wide band resistive sensors for double ridged waveguides

Paulius Ragulis, Rimantas Simniškis, Mindaugas Dagys, Žilvinas Kancleris
Department of Physical Technologies
Center for Physical Sciences and Technology
Vilnius, Lithuania
paulius.ragulis@ftmc.lt

***Abstract*—Resistive sensors with flat frequency response for WRD840 and WRD250 double ridged waveguides were developed. The flat frequency response of the sensors was achieved by changing the parameters of the sensing element. Manufactured sensors can measure high power microwave pulses covering frequency ranges 0.84 – 2.0 GHz and 2.6 – 7.8 GHz.**

Keywords—resistive sensors, double ridged waveguide, high power microwaves, electric field measurements

I. Introduction

A resistive sensor (RS) is a device based on electron heating effect in semiconductors. It found applications for high power microwave (HPM) pulse measurements [1]. The RSs measure HPM pulses directly, demonstrate long term stability and are overload resistant. The frequency band were these sensors can be used is limited only by the waveguide in which they are mounted. In order to measure HPM pulses in a wide frequency range a few different RS mounted in rectangular waveguide should be used. At least five different RSs are required to measure microwave power in the range 1 to 18 GHz. Implementation of the RSs into a double ridged waveguide allows us to increase frequency range of the single RS more than two times comparing it with a standard rectangular waveguide [2].

In this study we investigated resistive sensors mounted in double ridged waveguides WRD840 and WRD250. Pass bands of these waveguides are from 0.84 GHz to 2.0 GHz and from 2.6 GHz to 7.8 GHz. Sensing element of these RSs is made of two n-Si samples. This type of configuration allows us to achieve smoother frequency response.

II. Sensitivity and Frequency Response

When the HPM pulse is traveling inside the waveguide its electric field component heats electrons in the sensing element of the RS. Heated electrons establish a new state with mean energy higher than equilibrium and for this reason the resistance of the sensor increases. By measuring this resistance change, we can determine power of the HPM pulse. Finite-difference time-domain (FDTD) method was used to calculate electric field distribution inside the segment of the double ridged waveguide with the sensing element. The sensitivity of the RS,

This work was partly funded by EU project No 284802 "Protection of Critical Infrastructures against High Power Microwave Threats" (HIPOW).

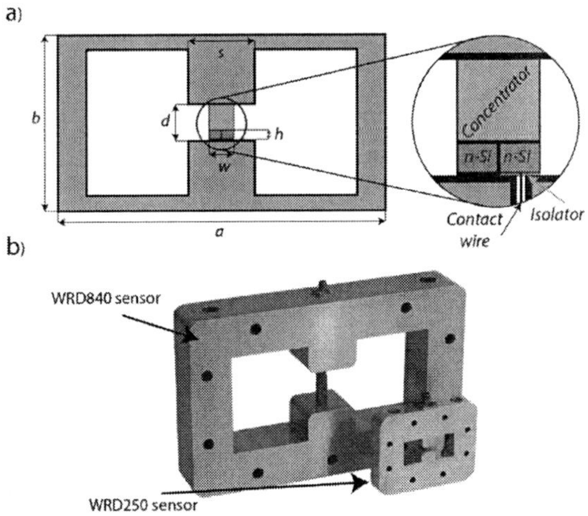

Fig. 1. Investigated resistive sensors: a) schematic view of RS with concentrator, b) manufactured sensors for WRD840 and WRD250 waveguides.

in the linear region where output signal linearly depends on power, can be expressed using formula [2]:

$$\zeta = \frac{3\beta E_0^2}{2P_\infty} \times \frac{1}{\sqrt{1-(\lambda/\lambda_c)^2}} \times \frac{\langle E\rangle^2}{E_0^2} \tag{1}$$

where β is a warm electron coefficient, P_∞ is power flow at infinite frequency, λ is a wavelength of electromagnetic waves, λ_c is the critical wavelength in waveguide, $\langle E\rangle$ is an average of the electric field amplitude inside the sensing element, E_0 is an amplitude of electric field in the center of an empty waveguide. The second multiplier accounts for the wave dispersion in the waveguide. Due to this phenomena, the electric field strength in the waveguide decreases when frequency increases, even if the microwave power is maintained at the same level. In order to design the RS which sensitivity is independent of frequency one should compensate this electric field decrease by adjusting physical parameters and configuration of the sensing element.

In our previous works we investigated cross-waveguide type RS [2-4] where the sensing element covers the entire gap between the ridges of the waveguide. Unfortunately, for this type of sensor we were unable to achieve less than 12% sensitivity variation within WRD250 waveguide's frequency

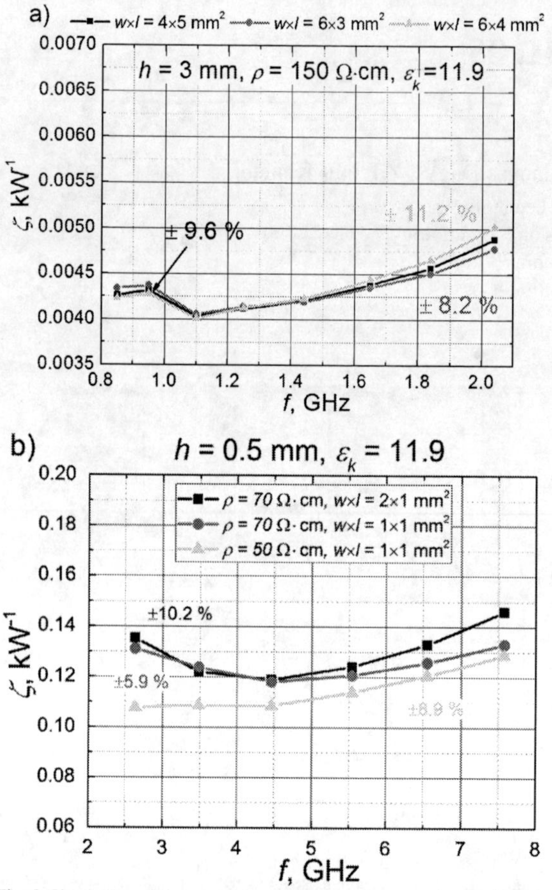

Fig. 2 Simulation of the sensitivity of the resistive sensor using FDTD method: a) sensitivity of the RS for WRD840 waveguide b) sensitivity of the RS for WRD250 waveguide.

band [3]. Performed additional simulations demonstrated better results for the shorter sensing element with concentrator. It should be noted that the thermal regime of the shorter sensor was also improved in comparison with the longer cross-waveguide type sensor.

The layout of proposed sensor is shown in Fig. 1a. It is seen that the sensing element is made of two separated n-Si samples. The upper contacts of both samples are connected with metal foil. One of the bottom contacts is grounded, whereas the other one is isolated. Isolated contact is connected to the measurement circuit and DC source which is used for sensor feeding. In order to increase the electric field strength inside the sensor, we used the concentrator. It is made in the form of the rectangular piece of high resistivity silicon placed on a top of the sensing elements, which together with sensors covers the entire gap between the ridges.

Simulation results of the sensitivity of the RS for WRD840 and WRD250 waveguides are shown in Fig. 2. In the figure the results demonstrating the lowest sensitivity variation on frequency are presented. Sensing element of the RS for WRD840 waveguide was chosen from n-Si the specific

TABLE I. PARAMETERS OF RESISTIVE SENSORS

Waveguide type	Dimensions, mm			Specific resistivity, Ω·cm	Predicted sensitivity variation, %
	height h	*width w*	*length, l*		
WRD840	3.0	4.0	5.0	150	±9.6%
WRD250	0.5	2.0	1.0	70	±10.2%

resistivity of which is $\rho = 150$ Ω·cm and the height $h = 3$ mm. Concentrator of this sensor was made from high resistivity $\rho = 10$ kΩ·cm Si the dielectric permittivity of which is $\varepsilon_k = 11.9$. From Fig. 2a one can see that the smallest sensitivity variation (8.2%), in the frequency range from 0.84 GHz to 2.1 GHz, demonstrates the sensor, the sensing element of which has cross-sectional dimensions $w \times l = 6 \times 3$ mm^2. Voltage standing wave ratio (VSWR) calculated for such a sensor is less than 1.2 in an entire frequency range. In the case of waveguide WRD250 (ref to Fig. 2b), the sensor, with dimensions $h \times w \times l = 0.5 \times 1.0 \times 1.0$ mm^3 and $\rho = 70$ Ω·cm demonstrates the lowest sensitivity variation (5.9%) within frequency range from 2.6 GHz to 7.8 GHz. The VSWR of such a sensor is less than 1.14 within considered frequency range.

For practical realization, we chose the sensing elements with larger cross sectional area. In this way, we increase possible sensitivity variation but decrease the resistance of the RS, which is very important parameter when measuring short HPM pulses. The parameters of the manufactured sensors are collected in TABLE I. Their external view is shown in Fig. 1b.

A few sensors for each frequency band were manufactured and investigated experimentally by measuring their output signal dependences on pulse power propagating in a waveguide, frequency response, and reflection coefficient.

III. Measurement Setup

Measurements of the dependence of the RS resistance change on pulse power propagating in the waveguide were performed using magnetron generators in S (2.75 GHz) and C (5.7 GHz) frequency bands. The measurement setup is shown in Fig. 3. Magnetron generator was used to generate high power microwave pulses. The RS under test was connected directly to the magnetron generator via transition from the rectangular to the double ridged waveguide. Pulse duration was 4 μs and the repetition rate was 25 Hz. The reference resistive sensor was connected to the main port via a directional coupler.. It controls pulse power in the main waveguide. By changing pulse power in the main waveguide with the help of a precise attenuator, the dependence of the output signal on the pulse power was measured using the oscilloscope TDS520A. The RS was fed by a current source keeping $U_0 = 10$ V voltage drop on it. The measurements have been performed using an additional amplifier, the amplification of which was $K = 10$ and $K = 100$. Output signal dependence on propagating power was measured for WRD250 RS only, since we do not have suitable HPM sources below 2.75 GHz.

Since the RS operation is based on a resistance change in the electric field of the microwave pulse, the sensitivity of the

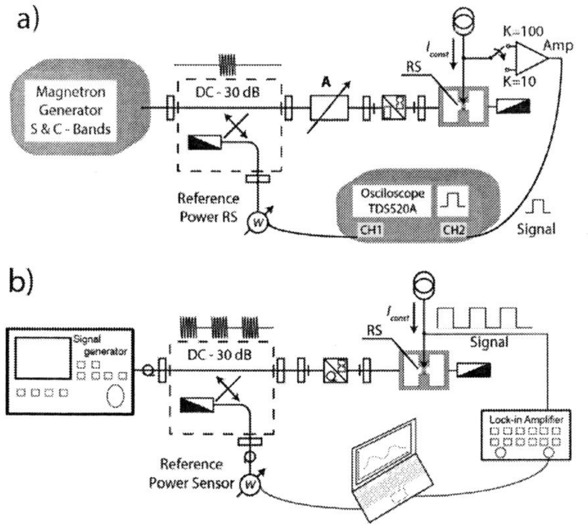

Fig. 4 Measurement setups: a) measurement of the dependence of the RS output signal on pulse power propagating in the waveguide, b) measurement of the frequency response of the RS.

RS in the linear region can be related to the experimentally measured quantities in the following way:

$$\zeta = \left(\frac{\Delta R}{R}\right)/P = \frac{U_s}{U_0 P}, \tag{2}$$

where $\Delta R/R$ is a relative resistance change of the RS in the microwave electric field, P represents power of microwave pulse in the waveguide, U_s denotes the signal from the RS measured with the oscilloscope, and U_0 is a DC voltage drop on the RS.

To measure the frequency response of the RS the measurement setup shown in Fig. 3b was employed. We used a tunable microwave generator producing roughly 200 mW average power. Employing meander modulated microwave signal, the significant increase of the measurement sensitivity was achieved. The modulation frequency of a microwave signal was chosen 10 kHz. It let us avoid additional resistance modulation due to lattice heating, which depends only on the specific resistance of n-Si used for sensors manufacturing and decreases when modulation frequency grows. The tunable microwave source was connected to the coaxial directional coupler. The RS under test was connected to the main port of the directional coupler with the help of an adaptor from the coaxial line to the double ridged waveguide. The microwave power supplied to the RS was measured using an average power sensor from Rhode & Schwarz. The output signal from the RS was measured using a lock-in amplifier. By measuring the output signal from the lock-in amplifier and the power transmitted to the RS, the dependence of the sensitivity on frequency was determined. The sensitivity of the RS can be related with an output signal U_m measured with the lock-in amplifier in the following way:

$$\zeta = \frac{\pi U_m}{\sqrt{2} U_0 P}, \tag{3}$$

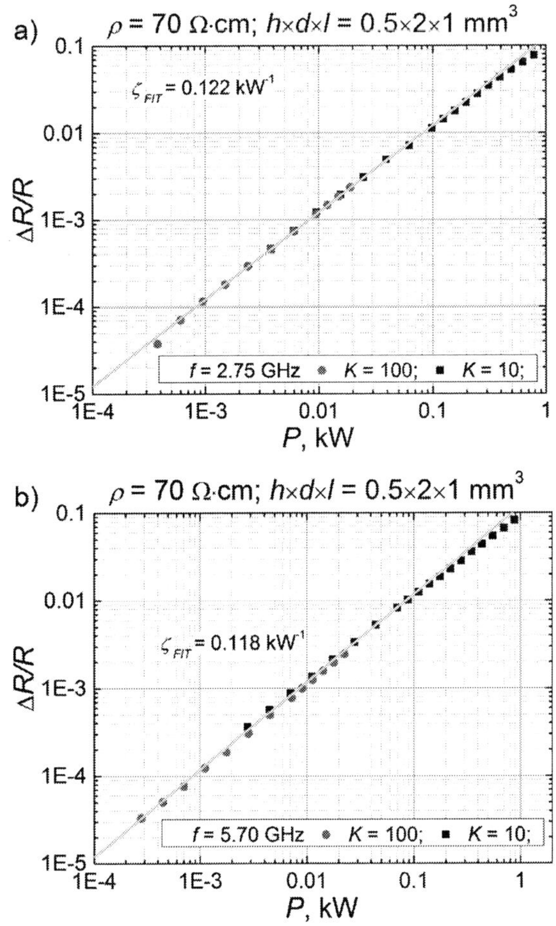

Fig. 3. Dependences of the RS resistance change on pulse power propagating in the waveguide for WRD250 sensor: a) measurement at 2.75 GHz and b) measurement at 5.7 GHz. Points show measurement results, lines correspond to linear approximation.

IV. MEASUREMENT RESULTS

The reflection coefficient of manufactured sensors was measured using vector network analyzer. The VSWR for the WRD840 sensor was less than 1.12 in a whole frequency range of the waveguide that is in good agreement with the simulation results. The VSWR for WRD250 sensor was below 1.35. The measured value was higher than it was expected from the modelling results but, in general, such value of VSRW is acceptable for the measurements in a waveguide.

The output signal dependence on pulse power transmitted through the waveguide was measured for WRD250 sensor using microwave pulse power sources available in the lab at 2.75 GHz and 5.7 GHz. Calculated from the measurement results the dependences of the relative resistance change on pulse power is shown in Fig. 4. As it was mentioned in the previous section we used the additional amplifier, the amplification of which was $K = 10$ and $K = 100$. As it is seen from the figure good coincidence between results measured

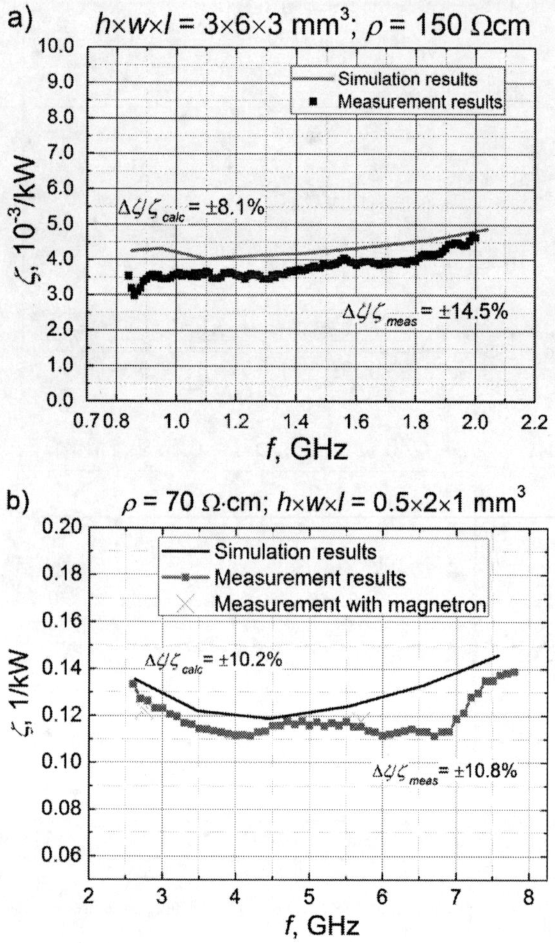

Fig. 5 Dependences of the RS sensitivity on frequency: a) WRD840 resistive sensor, b) WRD250 resistive sensor. Lines show simulation results, squares demonstrate measurement results using low power source, and crosses present measurement results with magnetron generators.

with different K was achieved. It is seen that using high amplification amplifier we can start our measurements when the relative resistance change of the RS in a microwave electric field is less than 0.01%. In this particular case the output characteristic was measured up to 10% increase of the resistance of the RS under test. Nevertheless, at a higher pulse power level some observable deviation from the linear dependence is seen.

From the measurement results using (2), the sensitivity of the RS can be determined. We got that $\zeta = 0.122$ kW^{-1} at 2.75 GHz and $\zeta = 0.118$ kW^{-1} at 5.7 GHz. It is seen that sensitivity values are almost equal at both frequencies.

The dependences of the sensitivity on frequency for the manufactured RS are shown in Fig. 5a and b for sensors mounted in waveguides WRD840 and WRD250, respectively. Together with measurement results, the calculated data using FDTD method are presented in the figure. In the case of WRD250 sensor, the sensitivities determined from experiments with magnetron generators are also shown in Fig. 5b by crosses. It is seen that they fairly well correspond to the results obtained using the meander modulated technique. As follows from Fig. 5, in both cases, for sensor WRD840 and for sensor WRD250, a good agreement between simulation using FDTD method and experimental results was obtained. In general, the measured sensitivity variation is larger than predicted from the modeling; it is roughly ±14% for WRD840 sensor and ±11% for WRD250 sensor. A few samples for each RS were fabricated and tested experimentally. They demonstrate similar sensitivity variation with the data for the sensors shown in Fig. 5. Although the measured variation of the sensitivity of the RS implemented in the double ridged waveguide is larger than expected, their good reproducibility and increased frequency range are promising features for the application of the proposed sensors in HPM experiments.

Sensors proposed and investigated in the present paper can find application for the pulse microwave electric field measurement in free space. In this case the RS should be connected to a horn antenna. It seems prospective accounting for dependence of the antenna gain on frequency, in the design of the RS, and in this way constructing the system for electric field measurement in free space with desirable frequency response.

V. Conclusions

In the present paper the resistive sensors with concentrator for double ridge waveguides WRD840 and WRD250 were proposed, manufactured and investigated. It was found that the RS for WRD250 waveguide is able to measure pulse power starting from a few watts. The frequency response of the manufactured RSs was measured. The results were compared with calculated ones, using FDTD method, and a good coincidence between them was obtained. Proposed sensors can find their application for measuring microwave pulse power propagating in the waveguides. It seems that connecting them to the corresponding horn antenna, they can be also used for the measurement of pulse power density or electric field strength in free space.

References

[1] M. Dagys, Z. Kancleris, R. Simniskis, E. Schamiloglu, and F. J. Agee, "The resistive sensor: a device for high-power microwave pulsed measurements," *Antennas and Propagation Magazine, IEEE,* vol. 43, pp. 64-79, 2001.

[2] Ž. Kancleris, P. Ragulis, R. Simniškis, and M. Dagys, "Resistive Sensor for High-Power Microwave Pulse Measurement in Double-Ridged Waveguide," in *Ultra-Wideband, Short-Pulse Electromagnetics 10*, ed: Springer, 2014, pp. 369-377.

[3] Ž. Kancleris and P. Ragulis, "Interaction of semiconductor sample with TE10 mode in double ridged waveguide," *Lithuanian Journal of Physics,* vol. 52, 2012.

[4] Ž. Kancleris, P. Ragulis, R. Simniškis, and M. Dagys, "Wide band waveguide sensor for microwave pulse measurement," *Lithuanian Journal of Physics,* vol. 53, 2013.

Numerical Simulation of Hybrid Vircator System with Photonic Crystal Grid in CST Particle Studio

Nikita S. Frolov, Alexander E. Hramov,
Semen A. Kurkin
Department of Automation, control, mechatronics
Saratov State Technical University
Saratov, Russia
phrolovns@gmail.com

Alexey A. Koronovskii
Faculty of Nonlinear Processes
Saratov State University
Saratov Russia

***Abstract*—: In this work we provide the numerical simulation of virtual cathode oscillator (VCO) with photonic crystal as a microwave output system. It is shown that this approach determines significant output power gain. We consider the classical scheme of axial vircator with cylindrical waveguide where the periodically disposed foil grids form the photonic crystal structure in the drift tube. The results of numerical PIC-simulations in the framework of CST Particle Studio are shown.**

Keywords—virtual cathode; vircator; relativistic electron beam; photonic crystal; efficiency gain; CST Particle Studio simulation.

I. INTRODUCTION

According to [1,2], high-power microwave (HPM) electronics is the promising scientific direction in the area of modern microwave physics. HPM sources are applied in the wide spectrum of fundamental science tasks such as particle acceleration in linear colliders, fusion heating, plasma confinement, etc [1,2]. These sources are also of the strong interest for applied techniques (radar and communications, power beaming, spectroscopy, etc.). One of the most attractive classes of HPM devices is the virtual cathode oscillator (VCO) or vircator – the bremsstrahlung microwave device which operation mechanism is based on the formation and oscillations of virtual cathode in the intensive beam with overcritical current [1-6]. This microwave source has a number of advantages besides high output power level. Vircators are characterized by the simple construction and low demands on the beam quality, the possibility of operation in the absence of external magnetic field, frequency and oscillations regime tuning by controlling the beam parameters or geometrical parameters of waveguide or resonator [3-6]. In spite of the denoted advantages, vircators also have the weak points. The most significant are relatively low operation frequency (about 1-20 GHz) and low efficiency of microwaves generation (about 1- 5%). The latter is the consequence of the virtual cathode nature that radiates electromagnetic energy in all directions while the existing output structures (different types of slow-wave structures or waveguides) do not cope with output of a significant part of the virtual cathode oscillations energy.

Earlier the hybrid modification of VCO with photonic crystal has been proposed and considered theoretically in the group of Prof. V.G. Baryshevsky [7]. Different effects of relativistic beam and photonic crystal interaction [8] has been studied in the framework of theoretical analysis.

In this Report we provide the numerical 3D particle-in-cell (PIC) simulation of the hybrid HPM system that represents the axial vircator with the photonic crystal foil grid structure for microwave oscillations power output via CST Particle Studio. In the framework of the numerical simulation, we have shown high efficiency of the observed microwave oscillator with photonic crystal.

II. SYSNEM UNDER STUDY

We have carried out the numerical PIC simulation of the hybrid "axial vircator - photonic crystal" system in the framework of CST Particle Studio. In Fig. 1a the scheme of the simulated hybrid VCO in the presence of the relativistic electron beam with overcritical current is shown. We have carried out the calculations neglecting processes in the electron gun. So, in the framework of the observed model the solid relativistic electron beam was injected into the cylindrical waveguide with the photonic crystal from the cylindrical emitter area. In this case we have used the DC electron beam emission model with beam current I0=1.5 kA and particle energy V0=150 keV. The injected relativistic electron beam is confined by the magnetic field B0=1.0 T for the better excitation of the photonic crystal structure. The output photonic crystal structure consists of periodically located in the y and z directions brass foils. In our numerical treatment we have used the model of photonic crystal described in [9]. Notable, that there is a free space between emission area and photonic crystal for the virtual cathode formation. We analyze the output microwave signal by means of the output waveguide port simulation located after the photonic crystal and by means of power flow integration over the surface that represents z-axis cross-section of the waveguide near the waveguide port. In Fig. 1b the phase portrait of the electron beam in the hybrid system under study is presented. One can see that the formation of the virtual cathode occurs approximately at 15 – 20 mm from the emission surface. The interaction of the electron beam with electromagnetic wave in the photonic crystal leads to the deep velocity modulation of the electron flow (the sagging of the particles velocity distribution at 200

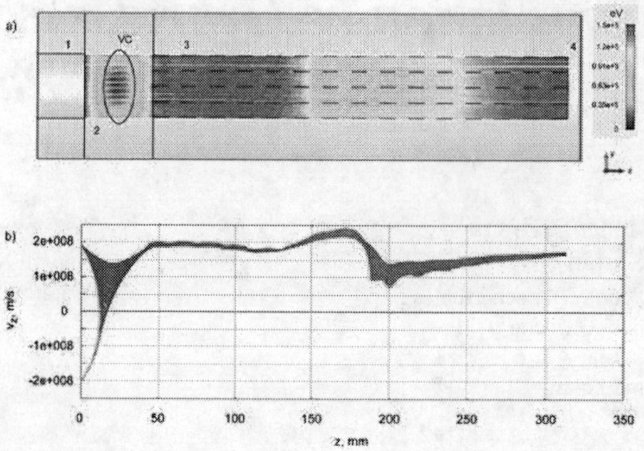

Fig 1. (a) The model of the hybrid microwave system "axial vircator with photonic crystal output structure" in the presence of the electron beam realized in CST Particle Studio: 1 – DC emission particle source; 2 – relativistic electron beam with virtual cathode (marked as "VC"); 3 – photonic crystal foil grid structure; 4 – output cylindrical waveguide port; t=1.6 ns. (b) phase portrait of the electron beam at t=1.6 ns in (v_z; z) plane.

velocity distribution at 200 mm). It is important to note, that such modulation of the electron beam increases the efficiency of vircator generation due to the involvement of particles

III. Main Results

The numerical simulations of the vircator system with the photonic crystal foil grid structure in the framework of CST Particle Studio show high efficiency of interaction between the virtual cathode electromagnetic radiation and the electromagnetic waves in the photonic crystal. Besides, the self-consistent process of interaction between the particles passing through the photonic crystal and the photonic crystal electromagnetic waves has been shown: at first passing beam excites travelling waves in the periodic crystal structure and then these waves modulate the electron beam and cause it to give up energy to the electromagnetic field. Also the passing beam excites backward waves providing distributed feedback to the virtual cathode oscillations. The results of output signal Fourier analysis are presented in Fig. 2. One can see that the output spectrum is rather narrow and peak power about 45 MW at ~1.6 GHz corresponds to the virtual cathode oscillations frequency. Calculation of microwave generation efficiency gives the value about 15-20%.

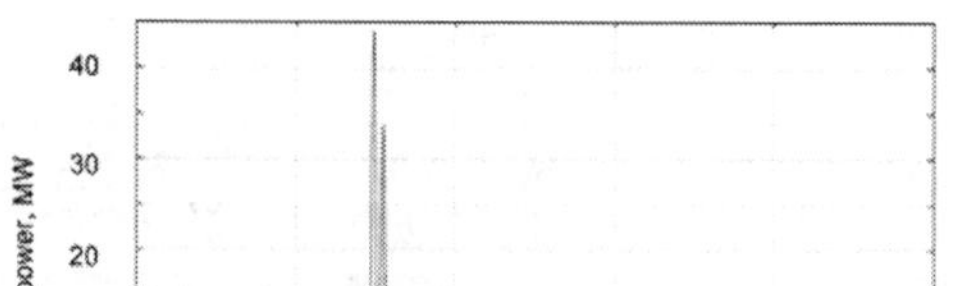

Fig 2. Power spectrum of output microwave oscillations of the considered hybrid vircator with the photonic crystal system.

Acknowledgment

We thank Prof. V.G. Baryshevsky, Dr. A.A. Gurinovich and Dr. P.V. Molchanov for the fruitful discussion. This work has been supported by Russian Science Foundation (grant No. 14-12-00222). NSF acknowledges individual financial support from the Russian Foundation for Basic Research (Grant No. 16-32- 60107).

References

[1] Benford, J., J. A. Swegle, and E. Schamiloglu, "High Power Microwaves," 3rd Edition, CRC Press, Taylor and Francis Group, 2016.

[2] Granatstein V.L. , Alexeef I. High Power Microwave Sources, Artech House Microwave Library, 1987.

[3] Dubinov, A. E., V. D. Selemir, Journal of Communications Technology and Electronics, vol. 47, № 6, p. 575, 2002.

[4] Hoeberg R.F., Fazio M.V. Advances in virtual cathode microwave sources, IEEE Trans. Electromagnetic Compatibility 34 (3) (1992) P 252-258.

[5] Hramov A.E., Koronovskii A.A., Kurkin S.A. Numerical study of chaotic oscillations in the electron beam with virtual cathode in the external non-uniform magnetic fields. Phys. Lett. A 374 (2010) 3057-3066.

[6] Phrolov N.S., Koronovskii A.A., Kalinin Ju.A., Kurkin S.A., Starodubov A.V., Hramov A.E. The effect of an external signal on output microwave power of a low-voltage vircator. Phys. Lett. A. 378, 32-33 (2014) 2423–2428.

[7] Baryshevsky V. G., Gurinovich A. A. Hybrid systems with virtual cathode for high power microwaves generation //arXiv preprint arXiv:0903.0300. – 2009.

[8] Baryshevsky V. G. Spontaneous and induced radiation by electrons/positrons in natural and photonic crystals. Volume free electron lasers (VFELs): From microwave and optical to X-ray range //Nuclear Instruments and Methods in Physics Research Section B: Beam Interactions with Materials and Atoms. – 2015. – T. 355. – C. 17-23.

[9] Baryshevsky V. G. et al. Experimental studies of volume FELs with a photonic crystal made of foils //Proc. 32 Int. Conf. FEL2010. – 2010. – 23-27.

Investigations of high current discharge in high density gas

Mikhail E. Pinchuk[1,2], Alexander A. Bogomaz[1], Alexander V. Budin[1]
[1]Institute for Electrophysics and Electric Power of Russian Academy of Sciences, IEE RAS
St.-Petersburg, Russia,
pinchme@mail.ru

[2]Peter the Great Saint-Petersburg Polytechnic University
Saint-Petersburg, Russia

***Abstract*— Pulsed powerful discharge with megaampere current range in high pressure gas is very interesting and challenging for investigations. The discharge can be used as a laboratory model of astrophysical objects for modeling radiative transport in star photosphere, equations of state for matter, shock wave propagation in plasma, etc. Investigation results for discharge in hydrogen with current up to 1.6MA at pressure up to 160MPa are presented. Original high-speed diagnostic methods in the optical and x-ray regions have been developed. Also it has been designed contact methods for measure magnetic and electric fields in the discharge channel. The discharge has very complex channel structure. Narrow axial zone with high temperature generate X-ray radiation flux. Transient photoionized plasma zone between current channel and surrounding gas produced by the flux. Acoustic oscillations in discharge volume and channel radius oscillations produce X-ray radiation modulation by convergence of acoustic compression wave on the discharge axis.**

Keywords—high pressure discharge; high current channel; plasma diagnostics

I. Introduction

Megaampere current range discharge in high pressure gas is very interesting and challenging object for investigations [1,2]. It is used as a laboratory model of astrophysical objects [3], for modeling radiative transport in star photosphere, equations of state for matter, shock wave propagation in plasma, etc. Discharge in hydrogen and helium at high initial density of gas (initial pressures of ~1–100 MPa) is the most promising for the development of hypervelocity launcher of projectiles [4].

Some selection results of investigations for discharge in hydrogen with current up to 1.6 MA at pressure up to 160 MPa are presented in this paper.

II. Experimental set-up and diagnostics

Experiments were performed on two installations, theirs detailed descriptions are presented in [4,5]. The electrode system have axisymmetric geometry, the distance between electrodes could vary from 0.5 cm up to 5.0 cm. The central electrode insulated from the chamber's housing serves as the cathode, and the chamber casing plays the role of the anode. Steel, copper and tungsten have been used as the electrode material. Arc ignition is performed by wire explosion. At the initial instant in time the wire joins the cathode with the anode. The time length of the first half-period of the discharge current is 70-150 μs and current rise rate is 10^{10}-10^{11} A/s. The energy input in discharge achieved 0.5 MJ. Energy source is the capacitive battery [6].

Phenomena arising from the interaction of intense energy fluxes with matter under extreme conditions at high pressures and temperatures are very challenging diagnostic tasks [1-3]. Basic difficulties of the diagnostics [7] are due to the high density of plasma. Optical methods give information only about peripheral areas of the discharge channel because of the strong absorption of radiation from the central area. It is necessary to use X-ray methods of diagnostics for direct measurements in the central area of the discharge channel.

Original high-speed diagnostic methods in the optical and X-ray regions have been developed [8-11]. Also it has been designed contact methods for measure magnetic [12-14] and electric fields [15] in the discharge channel.

III. Some experimental results

Typical experimental pictures are presented in Fig. 1 and 2. The experiments in the figures were performed at an initial hydrogen pressure of 5 and 32 MPa, correspondingly. In the first case, the tungsten electrodes were used, and the current amplitude was 500 kA. In the second case, the steel electrodes were used, and current was 1300 kA. Wire explosion stage of the discharge are shown by marker 1. After secondary breakdown the discharge channel starts to expand (between marker 1 and 2). After this the discharge channel contractions are observed on the photostreaks. Some

The work is partially supported by Russian Foundation for Basic Research (grant 15-08-04219-a, 16-08-00767-a).

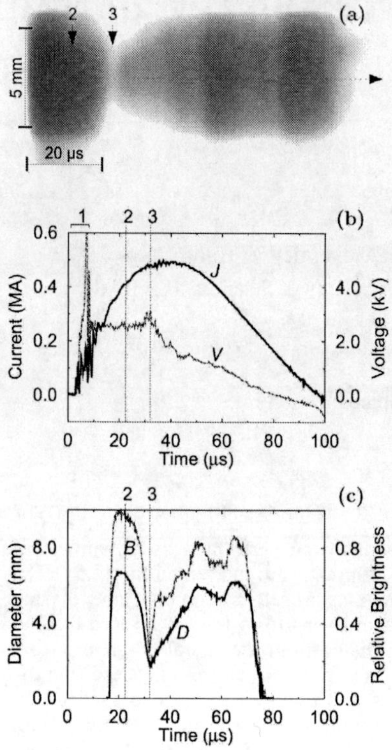

Fig. 1. Discharge in hydrogen at initial pressure of 5 MPa: (a) – photostreak, (b) – current J and voltage V, (c) - relative brightness B, discharge channel light diameter D_{xx} for relative brigtness level xx, and brightness temperature T_{br}.

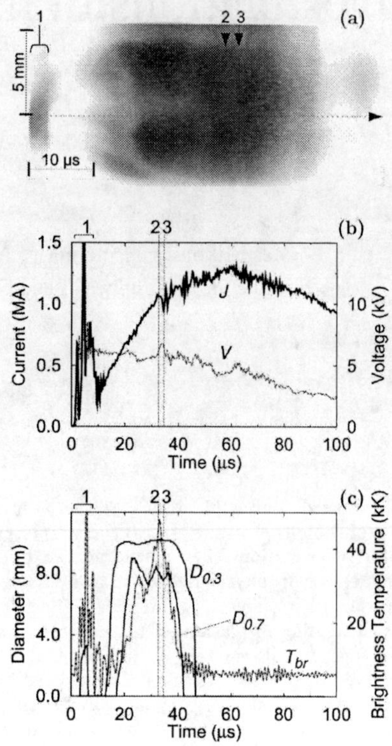

Fig. 2. Discharge in hydrogen at initial pressure of 32 MPa: (a) – photostreak, (b) – current J and voltage V, (c) - relative brightness B, discharge channel light diameter D_{xx} for relative brigtness level xx, and brightness temperature T_{br}.

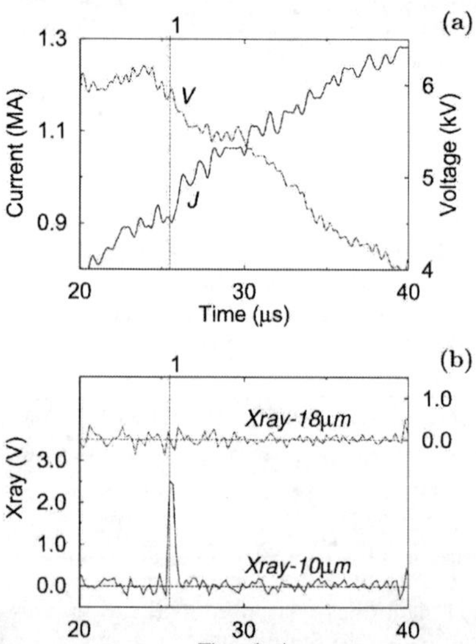

Fig. 3. Discharge in hydrogen at initial pressure of 7 MPa. (a) – current J and voltage V; (b) - X-ray signals: Xray-10μm and Xray-18μm – signals with 10μm- and 18μm-thickness Al-foil filters on sensors respectively; 1 – X-ray flash instant.

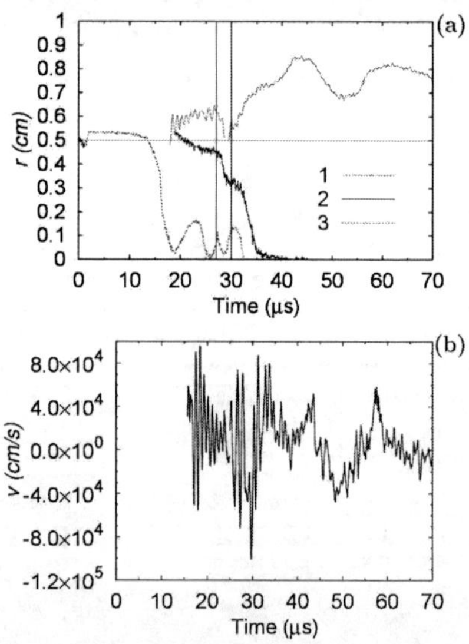

Fig. 4. Channel radius (a) and speed of current channel border by magnetic probe (b) for discharge in hydrogen with current amplitude of 1 MA at initial pressure of 5 MPa: 1 - by magnetic probe; 2,3 - by photostreak at relative brightness of 0.5 and 0.6, correspondingly; two vertical lines mark a time interval one of contraction; the horizontal line shows magnetic probe position.

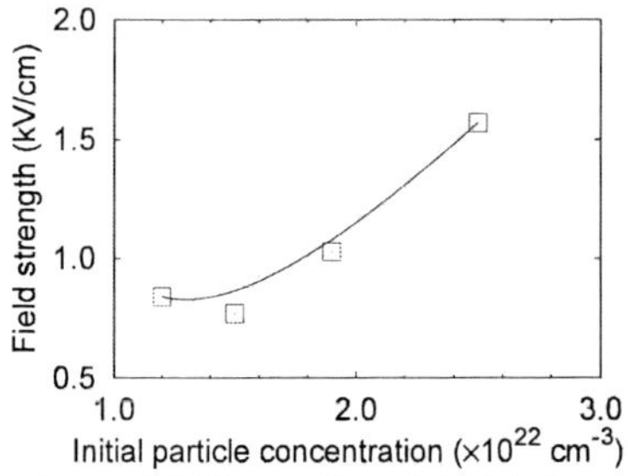

Fig. 5. Dependence of field strength vs initial hydrogen density for discharge with current amplitude of 500 kA at current maximum.

contractions and expansions follow one by one. Each contraction on photostreak was accompanied by the fall of the channel brightness and by peculiarities on the curves of voltage and current [2].

The discharge channel image [2] on the photostreaks is disappeared during the contraction (marker 3 in Fig. 1) and there is a strong brightness decreasing over time at high energy input into discharge channel (Fig. 2). It is concerned with formation of transmitting layer between channel and surrounding gas. Channel radiation had been heating this layer for several time interval before the layer absorption was sufficient for full disappearance of image. The central channel area consists of electrode material and igniting wire vapor plasma [9,10,11].

Soft X-ray radiation (SXR) was registered form the discharge channel [8]. Typical "peculiarities" on the current and voltage curves durin the observed contraction (marker 3 in Fig. 1 and 2) correspond to soft X-ray radiation burst from the channel (marker 1 in Fig. 3). Absorption analysis in the hydrogen (7 MPa, 25 cm) and aluminum foil (10 μm) shows that the sensor with the filter of thickness 10 μm can register the X–ray quanta with energy no less than 900 eV. The value of oscillation amplitude of the current radius is in agreement with an estimation of SXR intensity from channel axial zone [2].

Magnetic probe measurements (Fig. 4) supplement the optic and X-ray data. The discharge channel radius and the amplitude of its oscillations decrease with growth of initial pressure and grow with increase in discharge current amplitude [12,13,14].

A magnitudes of electric field strength in the discharge channel and near electrode voltage drops increase with initial particle density increases and with increase of current rise rate too (Fig. 5). It is connected with decreasing of a radiation loss level due to absorption into the transmitting layer between channel and surrounding gas [2]. Magnitude of near-electrode voltage drops has been determined by extrapolation of interelectrode gap length on zero.

IV. Conclusion

The discharge has very complex channel structure. Narrow metal plasma axial zone with high temperature generate X-ray radiation flux. Transient photoionized plasma zone between current channel and surrounding gas is produced by the flux. Surrounding gas decrease energy losses from discharge channel, assisting channel plasma overheating.

References

[1] Ph. Rutberg, Physics and Technology of High-Current Discharges in Dense Gas Media and Flows. New York: Nova Science Publishers, Inc., 2009.

[2] Ph. G. Rutberg, A. A. Bogomaz, M. E. Pinchuk, A. V. Budin, A. G. Leks and A. A. Pozubenkov, "High-current discharge channel contraction in high density gas," Physics of Plasmas, vol. 18, n. 12, p. 122702, December 2011, DOI 10.1063/1.3662053.

[3] B.A. Remington, "High energy density laboratory astrophysics," Plasma Phys. Control. Fusion, vol. 47, pp. A191-A203, 2005, DOI 10.1088/0741-3335/47/5A/014.

[4] A.V. Budin, A.F. Savvateev, and P.G. Rutberg, "A Two-Stage Launcher–Accelerator Working on Hydrogen," Instrum. Exp.Techn., vol. 47, n 4, pp. 534-538, 2004, [Prib. Tekhn. Eksp., 47(2004) n4 125-129], DOI 10.1023/B:INET.0000038404.52440.ce.

[5] A.V. Budin, S.Y. Losev, M.E. Pinchuk, Ph.G. Rutberg, and A.F. Savvateev, "An Experimental Stand for Studying a High-Current Discharge in a Dense Gas," Instrum. Exp. Tech., vol. 49, n 4, pp. 549-552, 2006, [Prib. Tekhn. Eksp., 49(2006) n4 106-109], DOI 10.1134/S0020441206040178.

[6] 5. P.Yu. Emelin, B.E. Fridman, and Ph.G. Rutberg, Instrum. Exp. Tech., 36(1993) n5 730-733 [Prib. Tekhn. Eksp., 36(1993) n5 109-115].

[7] Diagnostics of Dense Plasma [Diagnostika plotnoi plazmy], N.G. Basov, Ed., Moscow: Nauka, (1989) (in Russian).

[8] A.A. Bogomaz, A.V. Budin, V.V. Zabrodskii, I.V. Kuznetsova, S.Yu. Losev, M.V. Petrenko, M.E. Pinchuk, and Ph.G. Rutberg, "Detection of X Rays from a High-Current Discharge in a Dense Gaseous Medium," Instrum. Exp. Tech., vol. 51, n 5, pp. 744–747, 2008, [Prib. Tekhn. Eksp., 51(2008) n5114-117], DOI 10.1134/S0020441208050163.

[9] M.E. Pinchuk, A.A. Bogomaz, A.V. Budin, L.A. Shirochin, M.A. Polyakov, A.G. Leks, S.Yu. Losev, Ph.G. Rutberg. "X-Ray Flash Radoigraphy System for a High-Current Discharge in a Dense Gas," Instrum. Exp. Tech., vol. 53, n. 5, pp. 723–727, 2010, [Prib. Tekhn. Eksp., 53(2010) n5 p109-114], DOI 10.1134/S0020441210050192.

[10] Ph. G. Rutberg, A. V. Budin, M. E. Pinchuk, A. A. Bogomaz, L. A. Shirochin, M. A. Polyakov, A. G. Leks, S. Y. Losev, "X-ray Flash Radiography System for High Pressure Arcs Diagnostics," IEEE Trans. Plasma Sci., vol. 39, n1, pp. 394-398, January 2011, DOI 10.1109/TPS.2010.2084109.

[11] M.E. Pinchuk, A.A. Bogomaz, A.V. Budin, P.G. Rutberg. "Electrode Plasma Jets in Powerful Pulsed Discharge in High-Pressure Gas," IEEE Trans. Plasma Sci., vol. 42, n 10, pp. 2434-2435, 2014, DOI 10.1109/TPS.2014.2316064.

[12] M.E. Pinchuk, A.V. Budin, V.V. Leontev, A.G. Leks, A.A. Bogomaz, Ph.G. Rutberg and A.A. Pozubenkov, "Magnetic probe diagnostics of pulsed powerful discharge in high pressure hydrogen [Magnitozondovaja diagnostika moshhnogo impul'snogo razrjada v vodorode vysokogo davlenija]," Izvestia Vuzov, Physica, vol. 57, n 12/2, pp. 240–244, 2014, (in Russian).

[13] M.E. Pinchuk, A.A. Bogomaz, A.V. Budin, V.V. Leontev, A.G. Leks, A.A. Pozubenkov and Ph G Rutberg, "High-current channel characteristics in high-pressure gas," Journal of Physics: Conference Series, vol. 653, n 1, p. 012149, 2015, DOI 10.1088/1742-6596/653/1/012149.

[14] A.A. Bogomaz, M.E. Pinchuk, A.V. Budin, A.G. Leks, V.V. Leontev, A.A. Pozubenkov and N.K. Kurakina, "Measuring of high current channel parameters in high pressure gas by combined using of magnetic probe and high speed streak photography," Journal of Physics: Conference Series, 2016, (in press).

[15] M.E. Pinchuk, A.V. Budin, A.A. Bogomaz, V.V. Leont'ev, A.G. Leks, A.A. Pozubenkov, "Measurements in high-current channel at high pressure gas by contact diagnostics," at Proceeding of this conference, 2016.

Measurements in high-current channel at high pressure gas by contact diagnostics

Mikhail E. Pinchuk[1,2], Alexander V. Budin[1], Alexander A. Bogomaz[1],
Vladimir V. Leont'ev1, Anatoly G. Leks[1], Alexander A. Pozubenkov[1]
[1]Institute for Electrophysics and Electric Power of Russian Academy of Sciences, IEE RAS
St.-Petersburg, Russia,
pinchme@mail.ru

[2]Peter the Great Saint-Petersburg Polytechnic University
Saint-Petersburg, Russia

***Abstract*— It was investigated discharge initiated by copper wire explosion at high pressure hydrogen. Discharge chamber was designed with axisymmetric geometry. Energy input was up to 300kJ. Current amplitude was up to 1 MA with current rise rate 10^8 - 10^{10} A/s and at initial pressure up to 30 MPa. Original electric and magnetic probes, working in condition of high-current high pressure discharge with extreme heat and shock load, was designed and created. Measurements of current density and electric field in channel distribution had been carried out.**

Keywords—high pressure discharge; high current channel; plasma diagnostics; magnetic probe; electric probe

I. Introduction

The diagnostics of a high-current discharge in a high pressure gas is a very difficult task [1]. The basic difficulties of the diagnostics are connected to the high density of plasma.

Earlier a series of discharge channel contractions at current rise stage had been observed in our experiments at initial working gas pressure of 5-30 MPa and current amplitude of 0.5-1.6 MA [2,3]. The contraction was accompanied by soft X-ray radiation [4].

Each contraction was accompanied by the fall othe channel brightness. And the channel image in photostreak disappeared over time at high current amplitudes because of radiation absorption in heated transmitting zone between the channel and the surrounding gas. It causes some problems in the analysis of the channel behavior.

Therefore for advance in researches original magnetic and electric probe diagnostic techniques has been developed.

II. Experiments

Scheme of magnetic probe measurements is presented in Fig. 1 and scheme of electric probe measurements - in Fig. 2.

The discharge was initiated by copper wire. Discharge chamber was designed with axisymmetric geometry.

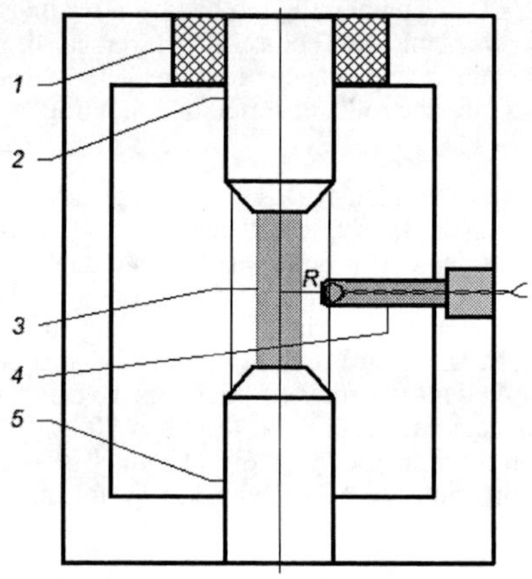

Fig. 1. Scheme of magnetic probe measurements: 1 - insulation, 2 - cathode, 3 - discharge channel, 4 - magnetic probe, 5 - anode, R - distance from discharge chamber axis to probe coil.

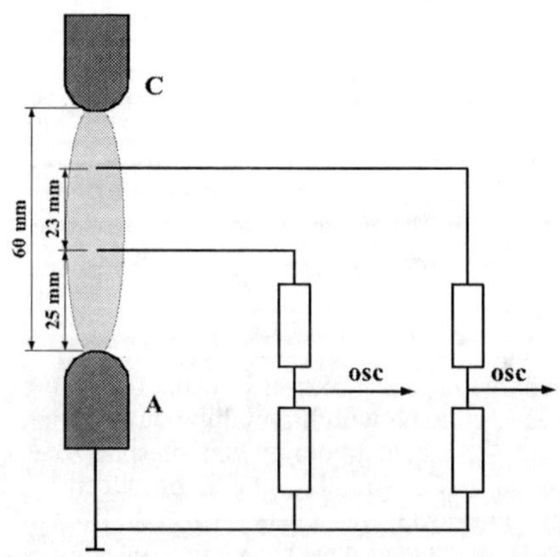

Fig. 2. Scheme of electric probe measurements.

The work is partially supported by Russian Foundation for Basic Research (grant 15-08-04219-a, 16-08-00767-a).

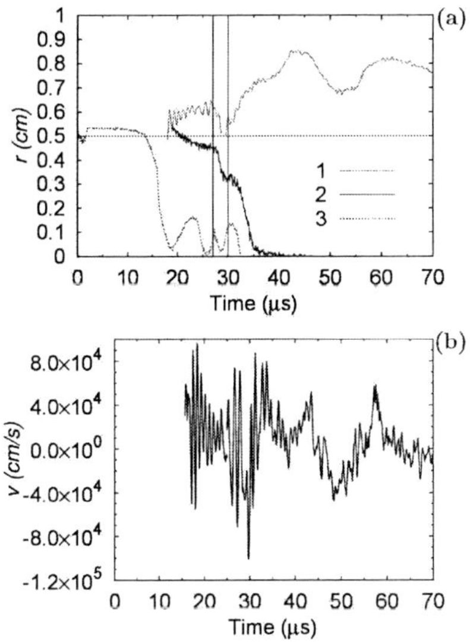

Fig. 3. Channel radius (a) and speed of current channel border by magnetic probe (b) for discharge in hydrogen with current amplitude of 1 MA at initial pressure of 5 MPa: 1 - by magnetic probe; 2,3 - by photostreak at relative brightness of 0.5 and 0.6, correspondingly; two vertical lines mark a time interval one of channel contraction; the horizontal line shows magnetic probe position.

Interelectrode gap was changed from 2 cm to 6 cm. Hemispherical steel electrodes with radius 1 cm was used.

Capacity of energy storage was changed. It was 1.2, 2.4, 4.8 and 40 mF. The current pulse duration was in the range from 70 μs to 2 ms. Charging voltage was from 1 to 15 kV. Energy input was up to 500 kJ.

Channel radius and speed of current channel border had been measured by magnetic probe diagnostics up to current amplitude of 1 MA and initial gas pressure up to 30 MPa. An experimental plot are presented in Fig. 3 for discharge with current amplitude of 0.95 MA and initial hydrogen pressure of 5 MPa.

Electric probe signals are shown in Fig. 4 for discharge with current amplitude of 80 kA and initial hydrogen pressure of 1 MPa. A magnitudes of electric field strength in the discharge channel and near electrode voltage drops increase with initial particle density increases and with increase of current rise rate too.

III. Conclusion

Original electric and magnetic probes, working in condition of high-current high pressure discharge with

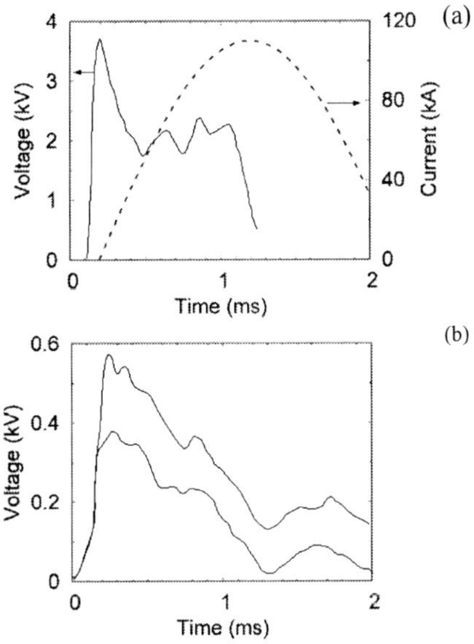

Fig. 4. Discharge in hydrogen at initial pressure of 1 MPa: (a) – current and voltage across the discharge gap; (b) - electric probe signals.

extreme heat and shock load, was designed and created. Measurements of current density distribution and electric field in channel had been carried out.

References

[1] Diagnostics of Dense Plasma [Diagnostika plotnoi plazmy], N.G. Basov, Ed., Moscow: Nauka, (1989) (in Russian).

[2] A. A. Bogomaz, A. V. Budin, S. Yu. Losev, M. E. Pinchuk A. A. Pozubenkov, F. G. Rutberg and A. F. Savvateev, "Attainment of the Pease-Braginskii Current in an Ultra-High Pressure Discharge," Plasma Physics Reports, vol. 34, n 5, pp. 366-375, 2008, [Fiz.Plasmy, 34(2008), n5, pp.404-413] DOI 10.1134/S1063780X08050036

[3] Ph. G. Rutberg, A. A. Bogomaz, M. E. Pinchuk, A. V. Budin, A. G. Leks and A. A. Pozubenkov, "High-current discharge channel contraction in high density gas," Physics of Plasmas, vol. 18, n. 12, p. 122702, December 2011, DOI 10.1063/1.3662053.

[4] A.A. Bogomaz, A.V. Budin, V.V. Zabrodskii, I.V. Kuznetsova, S.Yu. Losev, M.V. Petrenko, M.E. Pinchuk, and Ph.G. Rutberg, "Detection of X Rays from a High-Current Discharge in a Dense Gaseous Medium," Instrum. Exp. Tech., vol. 51, n 5, pp. 744–747, 2008, [Prib. Tekhn. Eksp., 51(2008) n5114-117], DOI 10.1134/S0020441208050163.

Power Nanosecond Range Semiconductor Opening Switch Based on p-type Silicon (p-SOS).

A.G. Lyublinsky, I.V. Grekhov, E.I. Belyakova
Division of Solid State Electronics
Ioffe Institute
26 Polytechnicheskaya, St. Petersburg, 194021 Russia
Alexander.Lyublinsky@mail.ioffe.ru

***Abstract*— Effect of very fast breaking of the reverse current with a density of 10^3–10^4 A/cm^2 in SOS-diodes (SOS – Silicon Opening Switch) upon switching from direct to reverse applied voltage is widely used in nanosecond duration gigawatt power pulsed devices. SOS-diode is a $p^+p'Nn^+$ structure formed by deep diffusion of boron and aluminum from one side and phosphorus from another side into *N*-type silicon wafer. The detailed experimental and theoretical study of the SOS-effect has been previously published by Institute of Electrophysics, Ural Branch, Russian Academy of Sciences.**

In this work, we present the first experimental investigation of the alternative SOS-diode based on *P*-type silicon (p-SOS). p^+pn^+ diode structure is formed by the diffusion of boron and phosphorus into *P*-type silicon wafer. The operation cycle of p-SOS diode is similar to n-SOS diode (SOS-diode based on *N*-type silicon) and consists of the short forward pumping current pulse followed by high rise rate reverse current pulse which forms plasma front moving through the highly doped p^+ region to p^+P boundary. The holes density in the flux is much higher than the base *P*-layer doping level. Therefore, the space charge region is formed immediately just after plasma front reaches p^+P boundary, the voltage drop on p-SOS diode increases and the current is commutated into the parallel-connected load.

It was shown that switching-off time and energy losses are lower in p-SOS diodes in comparison with n-SOS diodes in the same operation regimes. The additional benefit is much simpler manufacturing technology of p-SOS diodes.

Keywords — Power semiconductor switches, Pulse power systems, High-voltage techniques

I. INTRODUCTION

The possibility of fast nanosecond range interruption of high pulse current with a density up to few tens kiloamperes per square centimeter by the semiconductor switches was first demonstrated in [1]. This effect, called SOS-effect (SOS - Semiconductor Opening Switch), was observed in the silicon $p^+p'Nn^+$ diode with a deep (180–200 μm) *p-N* junction fabricated by simultaneous diffusion of aluminum and boron from the one side and by phosphorus from the other side into *N*-type silicon wafer. The typical thicknesses of diffusion layers in such SOS-diode are shown in Fig. 1. Standard SOS manufacturing technology includes long-time diffusion from doping solutions into grinded silicon wafer. The grinded surface allows diffusion of aluminum even in an oxidizing atmosphere [2]. In this process, the resulting surface concentration of aluminum dopant is about 7×10^{16} cm^{-3}, whereas the surface concentrations of boron and phosphorus can be varied in the wide range depending on the respective doping solution concentration. Usually, the surface concentrations of boron and phosphorus are up to 10^{20} cm^{-3}. Thus, p-layer of the SOS-diode consists of two parts: high concentration boron-doped p^+-part and aluminum-doped deep p'-part. The thickness of p'-layer is much higher than that of p^+-layer because aluminum diffusion coefficient is significantly higher than that of boron. n^+-layer is formed on the opposite side of the silicon wafer by phosphorus diffusion. The thicknesses of n^+ and p^+ layers are about the same. Therefore, the characteristic feature of SOS-diode based on *N*-type silicon is a very deep (180–200 μm) p^+p'-layer which requires very long-time diffusion at up to 1320ºC.

Two stage operation cycle is required to realize SOS-effect in the diode. First, forward current pulse passes through the $p^+p'Nn^+$ structure and the electron–hole plasma (Fig. 1, curve 1) is stored into the diode. Then, high rise rate reverse current pulse pulls out injected carriers and two plasma fronts (Fig. 1, curve 2 and 3) move one toward another. While left front (Fig. 1, curve 2) moves through the p^+-layer, the resistance of plasma-free part is very low because of the high doping level

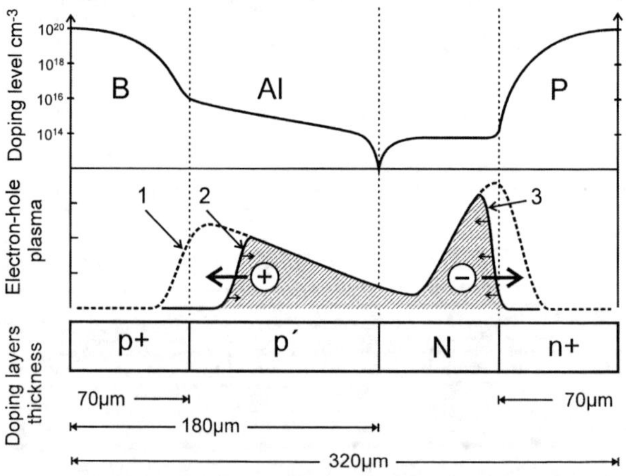

Fig. 1. SOS-diode semiconductor structure and moving electron-hole plasma fronts. On the figure:

1 – injected high-density electron-hole plasma during diode pumping by the forward current;

2, 3 – plasma fronts moving towards each other during pull-out process by the reverse current; remaining in the diode plasma is shaded; holes move to the left, plasma front 2 moves to the right; electrons move to the right, plasma front 3 moves to the left.

of p^+-layer. Thus, SOS diode remains in highly conducting state. However the situation changes drastically, when the plasma front goes into the p'-layer. The holes density in the flux increases continuously in time due to continuous increasing of pull-out current. In contrary, the dopant concentration decreases with the depth. When the hole flux density exceeds local dopant concentration, the space charge region (SCR) with a high electric field is formed, the diode's voltage drop increases and the current switches into the load connected in parallel to the diode. Similarly, SCR can be formed in the Nn^+-layers near plasma front 3, but usually it occurs much later, because electron flux density is about three times lower due to their three times higher mobility. As a rule, *p-N* junction is still filled with the electron–hole plasma at the moment of SOS-diode switching-off and current interruption. Typical current density interrupted by SOS-diode ranges from few kiloamperes up to few tens of kiloamperes per square centimeter.

The results of the SOS-process computer simulation are presented in [3, 4]. The photo of one of the SOS-based pulse power generator made in Institute of Electrophysics, Russian Academy of Sciences is shown in Fig.2.

II. EXPERIMENTAL SETUP

Specialized experimental setup has been designed and assembled to provide the detailed investigation of the SOS-effect in single diode structure with up to 5 kA/cm^2 pulse current density and about 50 ns duration of operation cycle. Simplified principal electric circuit of the setup is shown in Fig. 3. It is based on two parallel-connected RF switching power MOSFETs DE475-501N44A and has two power supplies U_1 and U_2, which allow independent and smooth adjustment of forward and reverse current pulses through the SOS-diode.

Fig. 2. S-5N SOS-based nanosecond range pulse power generator.

Output voltage	up to 1000 kV
Peak current	up to 3 kA
Pulse width (FWHM)	30 ÷ 50 ns
Pulse repetition rate	up to 300 Hz (continuously)
Average power	18 kW
Dimensions	3.5 × 1.4 × 1.1 m^3
Weight	2800 kg

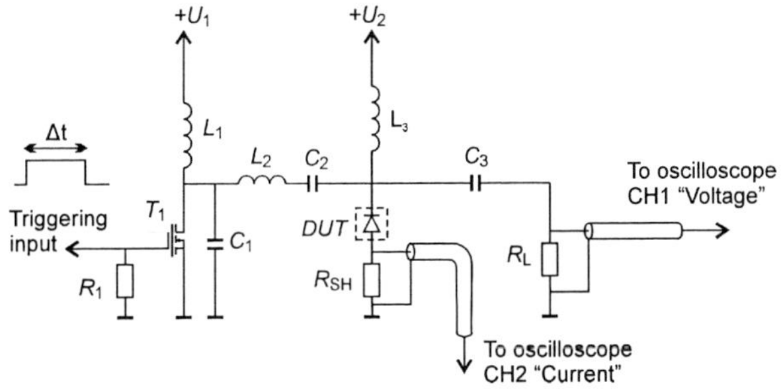

Fig. 3. Simplified principal electric circuit of the setup for the invesigation of the SOS-effect in a single diode structure at high current density and short time operation cycle.

Initially, MOSFET transistor T_1 is in off-state and capacitor C_2 is charged to a voltage equal to the difference between U_1 and U_2. When positive triggering pulse is applied to the triggering input, transistor T_1 switches on and diode *DUT* (Device Under the Test) is pumped by the forward current through the circuit T_1–L_2–C_2–DUT–R_{SH}. The duration of forward pumping current is determined by the duration of input triggering pulse Δt and the amplitude I_F is equal to:

$$I_F = \frac{U_1 - U_2}{R_{T_1,ON} + R_{DUT,F} + R_{SH} + \sqrt{\dfrac{L_2}{C_2}}}, \quad (1)$$

where:

I_F is *DUT* forward current amplitude

$R_{T_1,ON}$ is T_1 drain-to-source resistance in on-state;

$R_{DUT,F}$ is equivalent resistance of forward biased SOS-diode.

Therefore, *DUT* forward current pulse amplitude can be easily controlled by C_2 charging voltage. Simultaneously, current flows through the L_1–T_1 circuit, and energy is stored in the inductance L_1. When transistor T_1 switches off, the energy stored in the inductance L_1 causes high rise rate current flowing through L_1–L_2–C_2–DUT–R_{SH} circuit. This is reverse pull-out current for the SOS-diode. The reverse pulse current duration is determined by T_1 switching-off time and L_1C_1 time constant; amplitude I_R is equal to:

$$I_R = \frac{U_1}{L_1} \cdot \Delta t, \quad (2)$$

where:

I_R is *DUT* reverse current amplitude

Δt is triggering pulse duration.

Therefore, *DUT* reverse current pulse amplitude can be precisely controlled by U_1 voltage. Then, *DUT* goes into the high impedance state due to the SOS-effect and the current is switched to the parallel-connected load R_L. R_{SH} is a low inductance shunt used for the direct *DUT* current waveform measurement. The triggering pulse is fed from the external square pulse generator via a coaxial cable; resistor R_1 = 50 Ω is used for triggering input impedance matching. Capacitor C_3

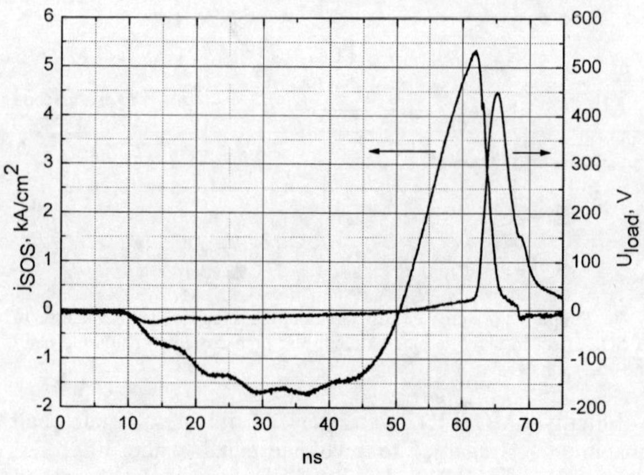

Fig. 4. Typical forward and reverse current pulse waveform through the SOS-diode and pulse voltage on the parallel-connected load.

blocks direct current flow from the source U_2 though the load R_L.

External oscilloscope with 18GHz bandwidth attenuators has been used for *DUT* voltage and current pulse registration. The maximum pulse amplitude and minimum pulse duration are limited by the MOSFET transistor T_1. Two parallel-connected DE475-501N44A transistors allow obtaining high rise rate, high amplitude pulse current required for the SOS operation mode of *DUT*. Typical forward and reverse current waveform is shown in Fig. 4. The forward pulse duration is 40 ns; reverse pulse duration is 12 ns. The experimental setup is described in more details in [5].

III. N-SOS AND P-SOS DIODES INVESTIGATION

1.2 mm diameter n-SOS diodes were cut from large size $p^+p'Nn^+$ diode structures made by simultaneous diffusion of boron and aluminum from one side and phosphorus from the opposite side into *N*-type silicon wafers with $\rho = 40\ \Omega\cdot\text{cm}$ resistivity. *p'*-*N* junction depth is 180 μm; p^+-layer is 70 μm and n^+-layer is 60 μm. The total thickness of the wafer is 320 μm and width of *N*-base region is about 80 μm.

In the experiments, plasma is pumped by 40 ns width forward current pulse with an amplitude about 1.8 kA/cm^2 and pulled out by high rise rate reverse current pulse with 12 ns duration. Typical oscillograms of the current interrupted by n-SOS diode and pulse voltage on the parallel-connected load are shown in Fig.5. Measured n-SOS diode switch-off time is 3 ns at 5.2 kA/cm^2 reverse current density. Pulse voltage amplitude on the load is 450 V and so called "pedestal" (pre-pulse, i.e. voltage amplitude just before current interruption) is 100 V. It is important to note that "pedestal" amplitude should be as low as possible, especially for high repetition rate applications, because it is associated with the "losses" inside the SOS-diode. Therefore, "pedestal" amplitude and switching-off time characterize SOS-diode commutation capability and commutation losses. The top line in Fig.5 is calculated energy losses during the switching-off process. In the experiment, the total commutation losses are equal to 25.6 μJ.

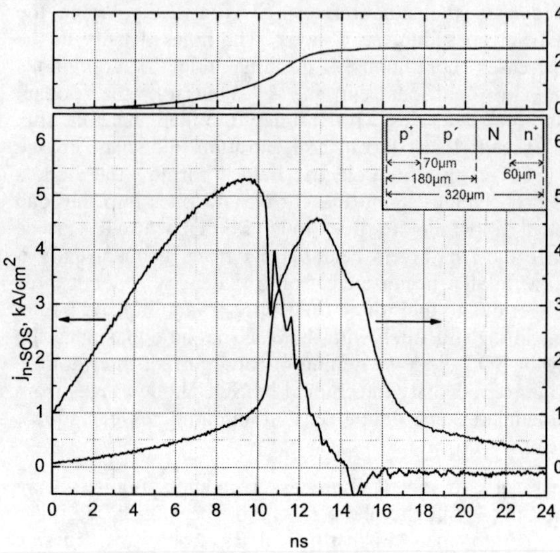

Fig. 5. Oscillograms of the current interrupted by n-SOS diode, pulse voltage on the parallel-connected load and calculated energy losses during switching-off process (on the top). n-SOS diode semiconductor structure is shown in the inset.

The "pedestal" amplitude is determined mainly by the resistance of a plasma-free part of *p'*-layer just before the current interruption. This resistance may be high enough in case of very thick *p'*-layer with low gradient of dopant concentration. SOS-diode switching-off time is determined by SCR formation, which depends on the durations of forward and reverse current pulses, reverse current rise rate as well as doping profile of the SOS-diode. Presumably, the minimum obtainable switching-off time is also limited by low gradient of dopant concentration.

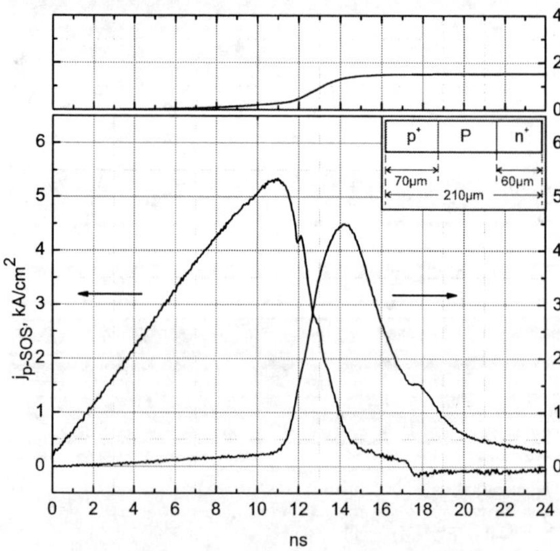

Fig. 6. Oscillograms of the current interrupted by p-SOS diode, pulse voltage on the parallel-connected load and calculated energy losses during switching-off process (on the top). p-SOS diode semiconductor structure is shown in the inset.

Step gradient concentration of p-layer can improve the diode switching parameters. Therefore, such p-SOS diode structures based on p-type silicon have been made and investigated. p-SOS is a p^+Pn^+ diode with 70 µm p^+-layer made by boron diffusion, 60 µm n^+-layer made by phosphorus diffusion and 210 µm total thickness. The main difference with n-SOS diodes is the absence of deep p'-layer which results in more simple manufacturing technology as an additional advantage.

The operation cycle of the p-SOS diode is similar to n-SOS diode and consists of the short forward pumping current pulse followed by the high rise rate reverse current pulse which forms plasma front moving through highly doped p^+ region to p^+P boundary. The hole density in the flux is much higher than the base P-layer doping level. Therefore, SCR is formed just after the plasma front reaches p^+P boundary. In this moment, the voltage drop on p-SOS diode increases and the current is commutated into the parallel-connected load. The oscillograms of the current interrupted by p-SOS diode and pulse voltage on the load are shown in Fig.6. The switching-off time is about 2.5 ns at 5.2 kA/cm^2 reverse current density. One can see that "pedestal" amplitude of p-SOS diode is much lower than that of n-SOS diode and is equal to 20V only. Total commutation losses in p-SOS diode are about 15.3 µJ, which is 40% less than the losses in n-SOS diode in the same operation regime. Pulse voltage amplitude on the load is 450 V.

SOS-diode switches off due to the SCR formation in p-layer, but reverse current flowing through the diode leads to P-n^+ junction recovery and this process has to be taken into account. Fast recovery at high reverse current density is dangerous for the semiconductor structure and can damage the diode. Fortunately, P-n^+ junction usually is filled with the electron-hole plasma at the moment of p-SOS diode switching-off and current interruption, because n^+-layer plasma front moves three times slowly due to three times higher electron mobility. Therefore, P-n^+ junction recovers significantly later after the main current interruption by SOS-effect. One can see in Fig.6 the nanoseconds range decay of the current followed by the abrupt fall to zero at about 200 A/cm^2 current density. Obviously, this current break corresponds to the P-n^+ junction recovery, which occurs at relatively low current density and does not cause overvoltage and dangerous operation regime for semiconductor structure.

IV. CONCLUSIONS

This work presents the first experimental evidence of the SOS-effect in p^+Pn^+ diode structures. SOS-diode based on P-type silicon (p-SOS) has faster switching-off time and lower commutation losses in comparison with the conventional $p^+p'Nn^+$ SOS-diode based on N-type silicon. Investigated p^+Pn^+ semiconductor structure is formed by simultaneous 60-70 µm depth diffusions of boron and phosphorus from the opposite sides of P-type silicon wafer. This manufacturing technology is much simpler than conventional SOS-diode technology and does not require long-time deep diffusion for fabrication of thick p'-layer. Better commutation capability of p-SOS diodes is associated with step doping profile p-layer in comparison with the low gradient profile p-layer in n-SOS diodes.

Additional study is required to investigate the current interruption process by p-SOS diodes in more details and optimize the p-SOS diodes structure.

ACKNOWLEDGMENTS

The authors thank to S.N. Rukin and Institute of Electrophysics for providing n-SOS diode structures.

This work was supported by the Russian Science Foundation, grant no. 14-29-00094.

REFERENCES

[1] Yu. A. Kotov, G. A. Mesyats, S. N. Rukin and A. A. Filatov, Doklady Physics, Vol.38, No.3, 1993, p.315-317.

[2] I. V. Grekhov, I. A. Liniichuk, L. V. Lebedeva, V. P. Tuchkevich, V. E. Chelnokov, V. B. Shuman, and N. I. Yakivchik, "Method of formation of a diffusion source of aluminum into silicon," USSR Patent No. 176989, 1964.

[3] S. N. Rukin and S. N. Tsyranov, "Effect of Space Charge on Subnanosecond Current Cutoff in Powerful Semiconductor Diodes", Technical Physics, 2009, Vol. 54, No. 11, pp. 1591–1596.

[4] S. N. Rukin and S. N. Tsyranov, "Simulation of the Subnanosecond Cutoff of Current in High Power Semiconductor Diodes", Semiconductors, 2009, Vol. 43, No. 7, pp. 957–962.

[5] I. V. Grekhov, A. G. Lyublinskii and I. A. Smirnova, "Analysis of Nanosecond Breaking of a High-Density Current in SOS Diodes", Technical Physics, 2015, Vol. 60, No. 11, pp. 1677–1681.

RF Soliton Generation in NLTLs using Negative Input Pulse of Fast Rise Times Provided by the Discharge of a Storage Capacitor*

Lauro P. Silva Neto
José Osvaldo Rossi
Plasma Associated Laboratory
National Institute for Space Research
São José dos Campos, Brazil
silvaneto007@yahoo.com.br

Joaquim J. Barroso
Technological Institute of Aeronautics
São José dos Campos, Brazil

Edl Schamiloglu
Electrical and Computer Engineering Department
University of New Mexico
Albuquerque, USA

***Abstract*— High power soliton generation has been studied for applications in communications systems of small satellite (CubeSats), pulse radars, and military communications, etc. Nonlinear transmission line is a periodic network of capacitors C(V) and inductors L(I), where at least one of the components has a nonlinear behavior. In this paper, a nonlinear capacitive line (NLCL) is implemented and studied, in which the capacitance decreases with the applied voltage. The NLCL was built using commercial ceramic-capacitor based on barium titanate (BT) with 10 nF nominal capacitance and inductors of fixed values at 3.3 μH. A high voltage switch system was projected and built to feed the line to produce a negative input pulse of the order of 1.0 kV with a rise time of less than 100 ns. This is much faster than the pulse rise time reduction factor (ΔT) of a calculated value of 3.27 μs for the line, which results in soliton generation on the line output. The frequency of generated soliton oscillations reached 4.0 MHz with 300 V of voltage modulation depth (VMD). The obtained experimental results with the conventional system of pulse generation with positive polarity are also given for comparison with those from the negative pulse generation system. It is shown that both systems are able to generate RF oscillations if the pulse rise time satisfies the soliton generation condition, being less than the pulse rise time reduction factor. The performed tests made on single shot condition provided a FFT output spectrum with an output oscillation frequency measured in the range of 3-4 MHz for both systems.**

Keywords—Nonlinear transmission line; soliton; nonlinear capacitor; high-power RF generation

*Work supported by the Southern Office of Aerospace Research and Development (SOARD – USAF) under contract number FA9550-14-1-0133.

I. Introduction

Nowadays radiofrequency (RF) is used mostly in communications systems to provide link between different devices such as in mobile phones, television, satellites, battlefield communication disruption and radars. One way to produce this type of signals could be made by the use of nonlinear transmission lines (NLTLs) that are lumped lines built with nonlinear capacitors C(V) in most cases, as shown in Fig. 1. In this work, a nonlinear capacitive nonlinear line (NLCL) was built using 10 nF commercial ceramic capacitors and 3.3 μH linear inductors. The capacitance nonlinearity of the line produces an output pulse with shorter rise time as shown in Fig. 2(a) and in some cases an output pulse superposed by a burst of oscillations called soliton waves (see Fig. 2 (b)) if the input rise time is less than the pulse rise time reduction factor ΔT calculated as [1]

$$\Delta T = n(\sqrt{LC_0}) - \sqrt{LC(V_{\max})}) \qquad (1)$$

where n is the number of sections of the line, C_0 the unbiased capacitance, and C(Vmax) the capacitance at the maximum amplitude of the propagating pulse.

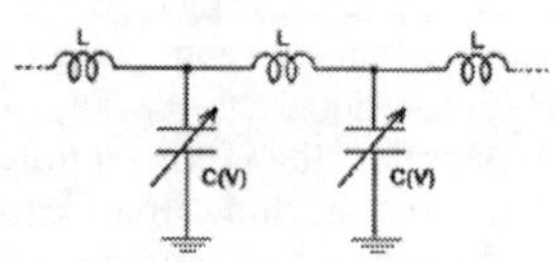

Fig. 1. Nonlinear capacitive line (NLCL).

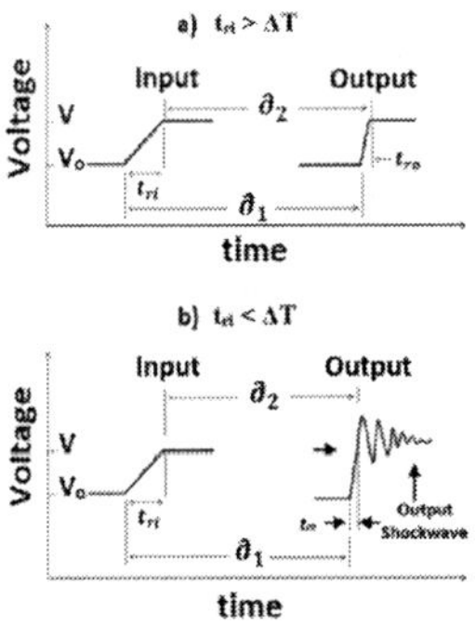

Fig. 2. Waveforms at the output of NLTL (a) pulse compression and (b) soliton generation.

II. METHODOLOGY

The commercial-off-the-shelf ceramic capacitors (COTS) with strong nonlinearity were selected for the line construction, which are similar to those used in [2]. In particular, based on US EIA standards, 10 nF capacitors with Y5V dielectric (+22 % to -82 %) over a temperature range between -30 °C and +85 °C were used. The NLCL consists of 30 sections, where each one is composed by a linear inductor L=3.3μH and a nonlinear capacitor C= 10 nF as shown in Fig. 3. In this work, two switching systems were constructed, one to generate output pulses of positive polarity and the other of negative polarity.

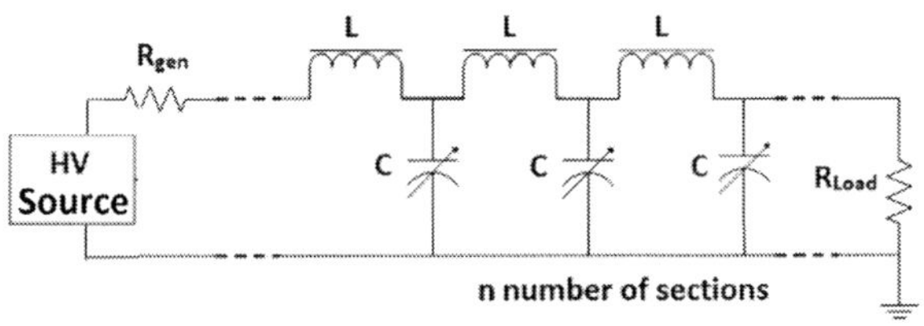

Fig. 3. NLCL using ceramic capacitors and linear inductors.

The electric circuit used to produce high-voltage positive pulse is similar to that used in [3] shown in Fig. 4. In this scheme, a high voltage DC source (Gamma HV- Model RR10-6R) represented by V1 charges a capacitor bank of two oil capacitors (1.5 μF/ 1.5 kV each) connected in series, where the discharge capacitor bank followed by a solid state IGBT switch (model IRGP30B120KD-E) produces an input pulse into the line. The Pulse Generator (TTI- model TGP 110), represented by V2, is used to trigger the IGBT switch, and the diodes D5-D7, D8-D10 (model DSEI-60), D4 and D11 (model 1N4007) are employed to protect the switch against reverse voltage. Similarly, the diodes D1-D3 (model 1N4007) are used for protection of the 12 V low voltage pulse generator against reverse voltage spikes, while the 32 Ω input resistor is used to limit the current in the IGBT switch.

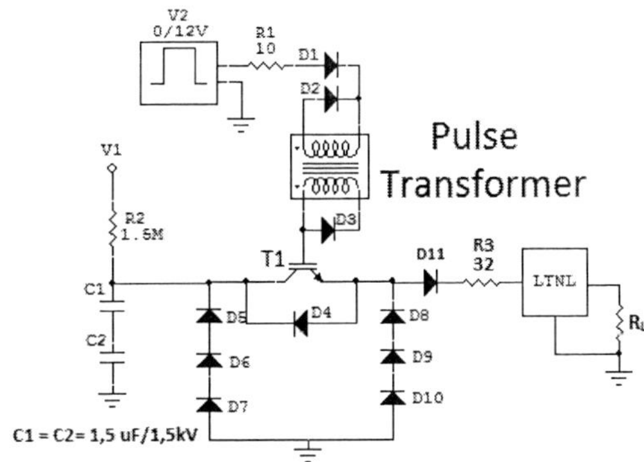

Fig. 4. High voltage circuit used to produce positive pulse on the NLCL.

Decreasing the input pulse rise-time facilitates the soliton formation on the line output. For this, an additional circuit driver was used to feed the IGBT gate in order to provide higher IGBT gate current in such a way that switching rise time is improved, leading to a shorter rise time. To avoid the use of pulse transformer that would otherwise affect switching rise time, a new scheme is used for this configuration, as shown in Fig. 5. The IGBT emitter is grounded and the discharge capacitor bank operates at the HV voltage floating potential, where a pulse with negative polarity is produced to feed the NLCL. As described elsewhere [4] the circuit gate driver UCC37321 from Texas Instruments was used.

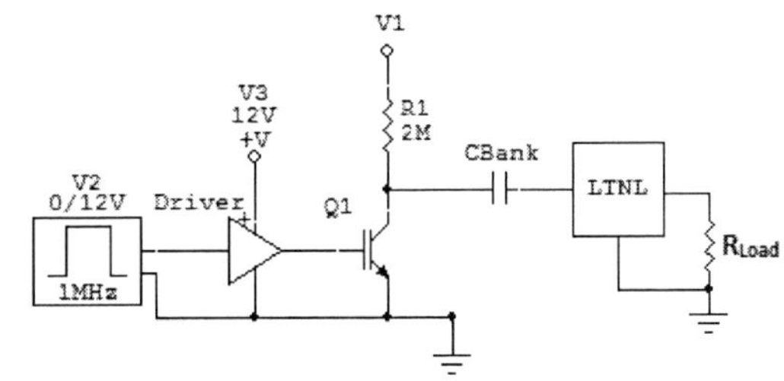

Fig. 5. High voltage circuit used to produce negative pulse on the NLCL.

III. RESULTS

Before presenting the experimental results with both schematic pulser circuits, the curve C versus V measured for the nonlinear capacitor with 10 nF nominal capacitance used in the NLCL is shown in Fig. 6. The initial capacitance is near to 7.5 nF and close to the maximum nominal working voltage of 1.5 kV the capacitance decreases approximately to 0.51 nF, giving a nonlinearity factor of 93 %, which is suitable for nonlinear applications. More details about the schematics used for the C versus V measurement can be found in [5].

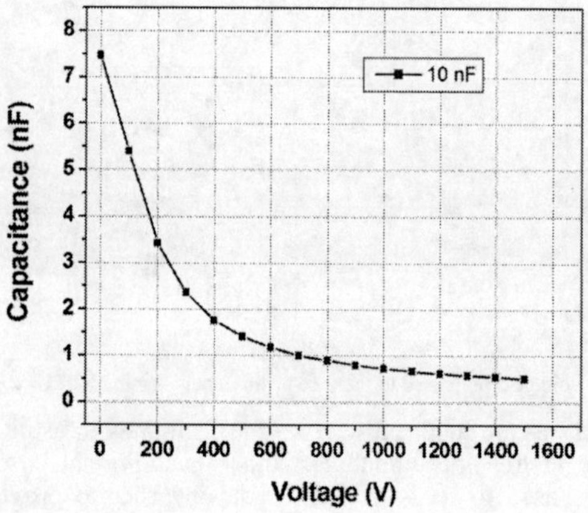

Fig. 6. C-V curve measured for commercial ceramic capacitor of 10 nF.

The calculated characteristic impedance of the stressed line tested (see Fig. 3) is of the order of 60 Ω. For the NLCL tests, a resistive load of 82 Ω close to the stressed line impedance was used for NLCL operation. Using the first system in Fig. 4 to produce a positive pump pulse applied to the input of the line, one obtains a pulse rise time of the order of 1.5 μs with a peak of 0.42 kV as shown by the blue line sown in Fig. 7. As the input pulse rise time t_{ri}= 1.5 μs is shorter than the pulse rise time reduction factor calculated ΔT as 3.5 μs, soliton generation is produced along the line and at its output as shown by the oscillations superposed on the output pulse amplitude (red line) in Fig. 7. The soliton oscillations generated has a voltage modulation depth (VMD) of the order of 210 V, approximately with five oscillation peaks at a frequency of about 3.0 MHz with decreasing amplitudes due to the line attenuation.

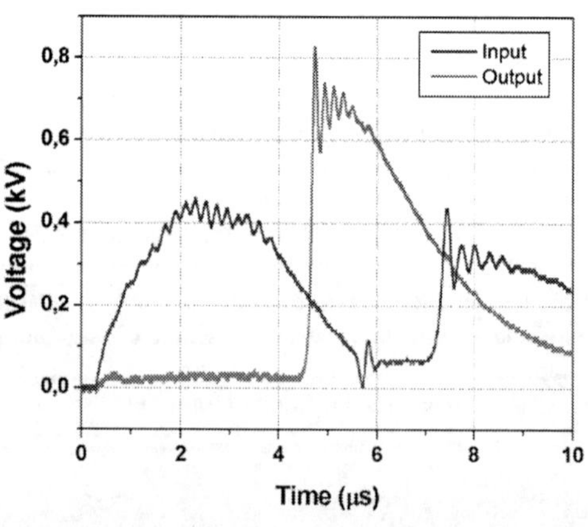

Fig. 7. Experimental input and output waveforms on the NLTL with positive input pulse peak of 0.4 kV, 5 μs width and a load of 82 Ω.

The RF generated in the middle of the line has a greater number of oscillations and better VMD than in the load. However, at this section the peak amplitudes of oscillations are lower as shown by the green line in Fig. 8 compared with the red line in Fig. 7. At the load, part of the reflected voltage pulse is added to the main incoming pulse, which gives higher peak amplitudes. In Fig. 8, the reflection effect is also observed by the subsequent peaks after 6.0 μs, which is better detailed in [6].

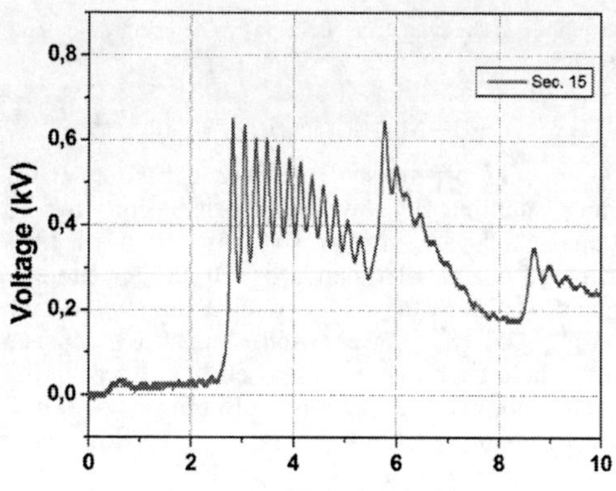

Fig. 8. RF soliton oscillations generated in the middle of the line.

In Fig. 9, the corresponding faster Fourier transform (FFT) spectrum of the output pulse confirms the frequency generated of the order of 3.3 MHz at a point of 3 dB approximately below the FFT curve knee. As in this case the load is close to the matched value, the generated frequencies on the load and on section 15 are approximately of the same order of 3.3 MHz.

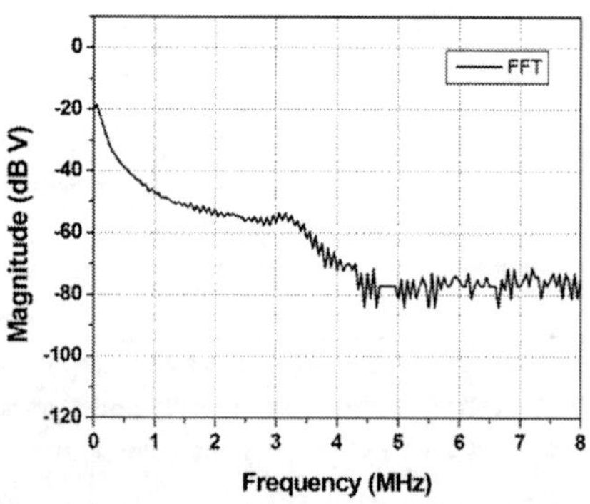

Fig. 9. Measured frequency spectrum for the output signal on the NLTL using a resistive load of 82 Ω.

By using the second pulser circuit system with negative polarity presented in Fig. 5, one can see that the rise time of the input pulse is now significantly much shorter (< 100 ns) than the pulse rise time reduction ΔT. This is explained by the use of the IGBT gate driver, which provides higher gate current during switching, producing pulses with faster rise time, when compared to the scheme of Fig. 4 without the driver. Fig. 10 shows the rise time of the negative input pulse (blue line) of less than 100 ns, with again $t_{ri}<\Delta T$ such that soliton oscillations of approximately 4 MHz are generated in the middle of the line (section 15) as shown by the red line in Fig. 10. The corresponding FFT spectrum is shown in Fig. 11, which confirms the measured oscillation frequency of about 4 MHz, a slightly higher value than the previous case because of the bigger unmatched load of 1kΩ used in this case.

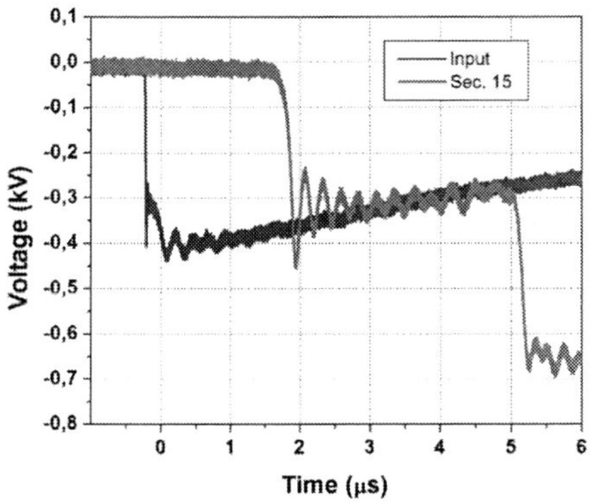

Fig. 10. Experimental input and waveforms obtained in the middle of the NLTL with negative input pulse amplitude of 0.4 kV and a load of 1.0 kΩ.

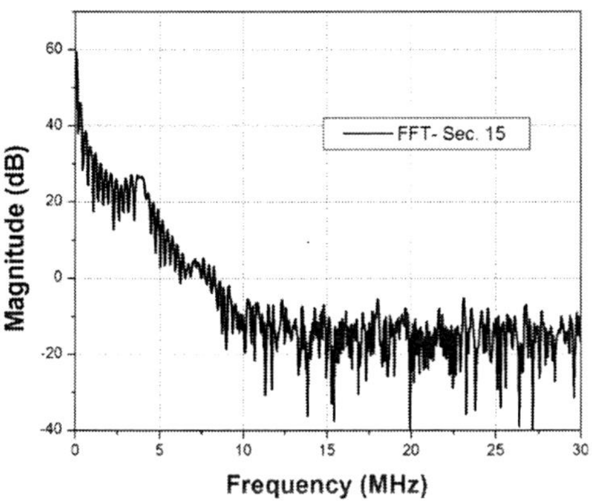

Fig. 11. Measured frequency spectrum on section 15 for the NLTL using resistive load of 1.0 kΩ.

IV. Conclusion

Commercial ceramic capacitors of high nominal capacitance (10 nF) were used to build the NLTL. It was shown that commercial ceramic capacitors of strong nonlinearity are needed for soliton generation. The characterization of the 10 nF COTS ceramic capacitor showed a nonlinear decrease in capacitance of the order of 93 % near its nominal working voltage, which is ideal for NLTL applications. Two systems were used to switch the high voltage source to the NLTL input: one using pulse of positive polarity and the other of negative polarity. It was demonstrated that RF soliton generation occurred in both cases. In particular, it was seen that with the use of a faster pulse of negative polarity, soliton generation remained practically the same in the middle of the line as generated frequency, number of oscillations and VMD did not vary significantly. This is explained by the fact that pulse rise time does not influence much soliton generation once the condition $t_r << \Delta T$ has been satisfied. Finally, the results showed that by using COTS ceramic capacitors in the NLTL, RF soliton generation with frequency of about 3-4 MHz and VMD near to 0.3 kV can be produced in the middle of the line with positive or negative pulses since $tr < \Delta T$.

References

[1] F. S. Yamasaki, L. P. Silva Neto, J. O. Rossi, J. J. Barroso, "Soliton generation using nonlinear transmission lines," *IEEE Trans. on Plasma Science*, vol. 42, no. 11, pp. 3471-3477, Nov. 2014.

[2] N. S. Kuek, A. C. Liew, E. Schamiloglu, "Experimental demonstration of nonlinear lumped element transmission lines using COTS components." in Proc. *of the IEEE International Pulsed Power Conference*, 2011. p. 1-4.

[3] N. S. Kuek, A. C. Liew, E. Schamiloglu, J. O. Rossi. "Oscillating pulse generator based on a nonlinear inductive line," *IEEE Transaction on Plasma Science*, vol. 41, n. 10, pp. 2619-2624, 2013.

[4] Q. R. Marksteiner, B. Carlsten, S. Russell, "Efficient generation of RF using a biased soliton generating nonlinear transmission line with a bipolar input." *Microwave and Optical technology letters*, v. 52, n. 6 Jun. p. 1411-1413, 2010.

[5] L. P. Silva Neto, J. O. Rossi, J. J. Barroso. High power RF generation in nonlinear lumped transmission lines using commercial ceramic capacitors as nonlinear elements. In Proc. *of the IEEE International Pulsed Power Conference*, 2015, pp. 1-5.

[6] P. W. Smith, "Pulsed, high Power, RF generation from nonlinear dielectric ladder networks- performance limits," in Proc. *of the 18th International Pulsed Power Conference*, 2011, pp. 167-172.

Study on the Package Design of High-Power Semiconductor Switches

Youngseok Bae, Seong-Ho Kim, Byungha Lee, Sanghyuk Ahn, Young Hyun Lee, and Kyung-Seung Yang
Institute of Defense Advanced Technology Research, ADD, Daejeon, Korea

***Abstract*—Switches using the discharge characteristics of gas, vacuum, or pseudospark have been widely used for the purpose of high power switching. Recently they are replaced with high power semiconductors due to reliable operations and conveniences. Semiconductors used for the high power switching need disk-type packaging in common. The components inside the package are usually separated, but they should ensure good electrical conduction, and address structural and thermal problems through the high contact pressure. In addition, the electrodes should have a high hold off voltage. The voltage depends mainly on the electrode geometry and the gas inside the package. The properties can be studied with Paschen's law. In this paper, we have studied on the structural, thermal, and electrical properties of the components inside the package of the high power semiconductors considering the given hold off voltage and the high contact pressure on them.**

Keywords*— *power electronics; semiconductor switch; packaging

I. INTRODUCTION

Accelerators or medical equipment requiring high power usually utilize gas, vacuum, or pseudospark switches yet. However, these have some problems in the power loss and a reliable triggering. To overcome these problems high power semiconductors showing good reliabilities are being developed, and replacing the conventional switches [1].

Though the semiconductor technologies are changing diversely, the method of packaging is almost common. In switch packaging, the characteristics of the semiconductor wafer should be considered first. The wafer should have a high hold-off voltage in the reverse bias state. When it is triggered electrically or by light, it should accommodate high current which accompanies a considerable amount of heat. The components of the package exposed to the heat can be damaged due to high temperature or differences in thermal expansion rates. Besides the wafer, insulation of the components near the wafer should be considered. The voltage hold-off property can be investigated using the Paschen's law which describes the breakdown voltages in the gas environment. To keep good hold-off property in the given electrode geometry, a good sealing should be established by using the ceramic housing. To transfer the heat generated in the wafer region to outside rapidly, a large pressure should be applied enhancing the contact of the components. To avoid the cracking of the thin wafer, prudent structural analysis on the components is required. In this study, the electrical, thermal, structural analyses were conducted with the help of finite element method [2, 3]. The results will be used in the design of the switch packaging.

II. ELECTRICAL ANALYSIS

The voltage hold-off property of the semiconductor can be estimated by using Paschen's law. It describes the breakdown voltage for a given electrode geometry by using a function of the product of the gas pressure and the spacing between the electrodes as shown in Fig. 1.

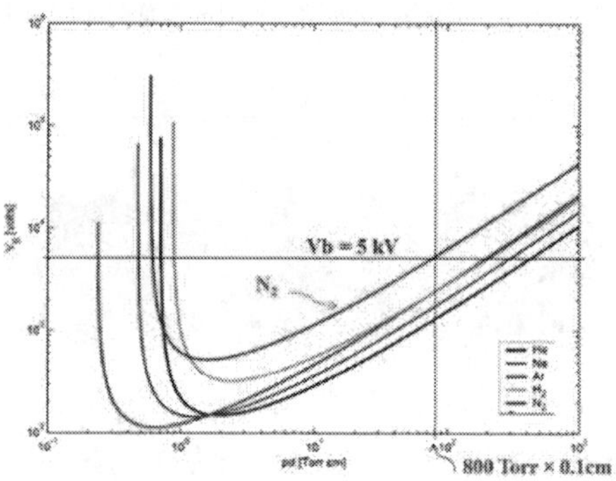

Fig. 1. Breakdown voltage by Paschen's law

High power semiconductor switches are packaged by the ceramic housing and the disk-type electrodes. Inside the package the nitrogen gas of about 800 torr in pressure is filled to protect arc discharges. In Fig. 1, the breakdown voltage for the spacing of 1 mm is about 5 kV. To achieve a hold-off voltage of 7 kV at such a spacing due to the thickness of the wafer, the distance of about 3 mm is needed to avoid the surface breakdown between the electrodes keeping in contact with the wafer. The increased distance can be achieved by designing the diameters of the electrodes differently. Fig. 2 shows the structure of the electrodes with different diameters.

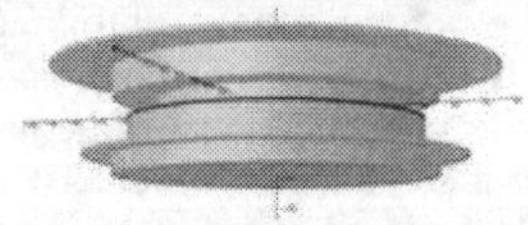

Fig. 2. Geometry of the electrodes with different diameters

III. Thermal Analysis

To see the thermal property, the analysis of thermal conduction has been done. Fig. 3 shows a typical structure of electrodes in the semiconductor package where sheets of silver, brazing material, and molybdenum are included to enhance electrical and mechanical properties.

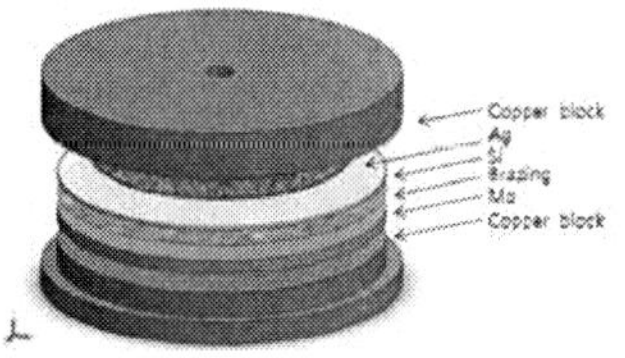

Fig. 3. Structure of Electrode Components

For an abrupt step-like temperature rise in the Ag-Si region, the transient conduction of heat at the outer surfaces of the copper electrodes was studied. The source of heat was assumed to have constant temperatures from 70 to 150 °C. The outer surfaces of the electrodes were assumed to be in the condition of convection boundary condition with the air of 45 °C, or the water of 25 °C. The initial temperature of all the components was assumed to be 20 °C. Fig. 4 shows the temperature change of the outer surfaces of the copper electrodes after the temperature of the source region was abruptly increased.

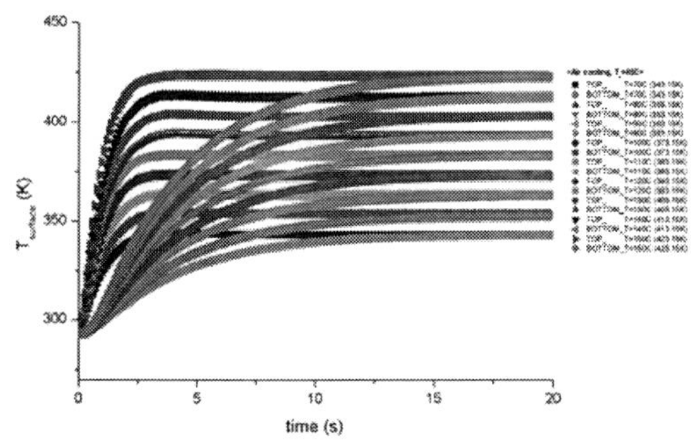

Fig. 4. Temperature rise of the outer surfaces in an air-cooled condition

The outer surfaces reached an equilibrium with the source region soon in several seconds, which means the cooling can also be done in the same way. For pulsed operations like several shots in a minute, cooling is not so critical. More accurate estimation can be done when the heat generation in the source region is described based on the condition of the switching operation. The thick copper disk plays an important role for the rapid heat extraction inside the package.

IV. Mechanical Analysis

The wafers of high power semiconductors are about 1 mm in thickness. To enhance the capability of the electrical and heat conduction an application of large pressure on the electrodes is required. However, the wafer will be damaged if the stress larger than the allowed one is applied or other supporting components are deformed severely.

To prevent the internal arching, the distance between the electrodes are increased in two ways. The first type is the electrodes with the same diameters which are smaller than the diameter of the wafer. The second type is one electrode of a smaller diameter and another electrode of the same diameter with the wafer. Fig. 5 shows the electrode structures of such types.

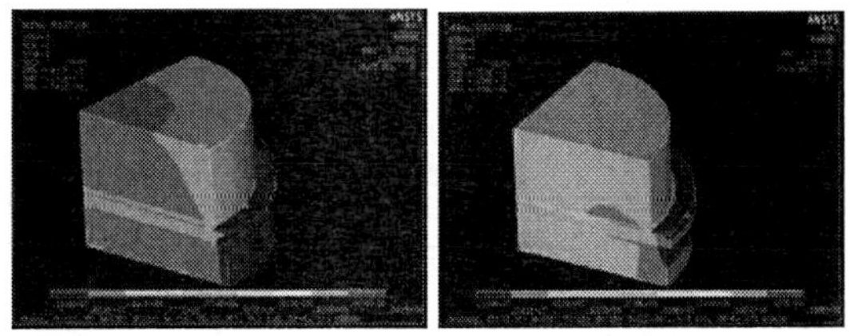

(a) Type 1 (b) Type 2

Fig. 4. Electrode geometries pressing the wafer

Considering the safety factor and the maximum allowed stress, the allowed force was calculated to be about 89 kN for the type 1 geometry, and 66 kN for the type 2 geometry. In general, the required force on the wafers of 4 inch commercial products is known to be about 70 kN.

V. Summary

To start the studies on the package structure, basic investigations to find and understand the reasonable electrical and mechanical parameters in the high power semiconductors have been done.

Acknowledgment

This study was supported by Institute of Civil Military Technology Cooperation in Korea (No. of contract : UM14103RD1).

References

[1] J. Przybilla, J. Dorn, R. Barthelmess, U. Kellner-Werdehausen, H.-J. Schulze, F.-J. Niedernostheid, "Diodes and Thyristor - past, presence and future," 13th European Power Electronics and Applications, 2009.

[2] N. S. Nobeen, K. Ahmad, D. C. Whalley, D. A. Hutt, and Haworth, "Polymeric Packaging of High Power Semiconductor Devices: Material Selection & Reliability Assessment," IEEE 15th Electronics Packaging Technology Conference, 2013.

[3] N. S. Nobeen, D. C. Whalley, D. A. Hutt, and Haworth, "Computational Modeling of Electrical Field Intensity for High Voltage Semiconductor Package Design," International Symposium on Advanced Packaging Materials: Microtech, 2010.

Development of Bipolar Pulse Accelerator with Gas Puff Plasma Gun for High-Purity Intense heavy Ion Beam

Tarou Honoki, Keito Okajima, Yoshiki Imai and Hiroaki Ito*
Dep. of Electric and Electronic Engineering
Univ. of Toyama, 3190 Gofuku, Toyama 930-8555, Japan
hiroaki@eng.u-toyama.ac.jp

***Abstract*— We have developed a new type of a pulsed ion beam accelerator named "bipolar pulse accelerator" for improvement of the purity of the intense pulsed ion beam. The system utilizes a magnetically insulated acceleration gap and was operated with the bipolar pulse. A coaxial gas puff plasma gun was used as an ion source, which was placed inside of the grounded anode. Source plasma (nitrogen) of current density of about 20 A/cm^2 and pulse duration of about 2.0 μs was injected into the acceleration gap. When the bipolar pulse with voltage of about −110 kV and +90 kV and pulse duration of about 70 ns each was applied to the drift tube, the ions were successfully accelerated from the grounded anode to the drift tube in the 1st gap by the negative pulse of the bipolar pulse. The pulsed ion beam with current density of 70 A/cm^2 and pulse duration of about 50 ns was obtained at 48 mm downstream from the anode surface. The ion energy was in reasonable good agreement with the acceleration voltage, i.e., 1st pulse (negative pulse) voltage of the bipolar pulse.**

Keywords—bipolar pulse accelerator; intense pulsed heavy ion beam; Magnetically insulated ion diode; plasna gun

I. Introduction

Intensity pulsed heavy ion beam (PHIB) technology has been developed over the last two decades primarily for nuclear fusion and high energy density physics research [1,2]. Compared with the traditional ion implantation method, IPIB enables the accumulation of energy in very short time into the near surface region while it maintains a low substrate temperature. Recently, PHIB with an ion energy of several 100 keV, a high ion current density of several 100 A/cm^2, and a short pulse duration of < 1 μs has been widely used as a tool for material processes including the surface modification, thin film deposition and ion implantation [3-5]. Especially, PHIB has received extensive attention as a new ion implantation technology named "pulsed ion beam implantation" for next generation semiconductor materials including silicon carbide (SiC) and diamond, since the ion implantation and the surface annealing can be completed in the same time. In the case of SiC, a conventional high temperature annealing at 1700-1900 K causes several disadvantages such as the redistribution of implanted dopants and increasing of surface roughness. The use of IPIB enables the accumulation of energy in very short time into the near surface region while it maintains a low substrate temperature. The pulsed ion beam implantation is expected to be novel annealing techniques.

A number of PHIB sources have been developed so far for the requirements of research and industrial application. The producible ion species, however, is limited to the material of electrode (anode), since the anode plasma is produced by a high-voltage flashover and an electron bombardment to the anode surface. In addition, the purity of the PIB is usually deteriorated by absorbed matter on the anode surface and residual gas molecules in the diode chamber. For example, the pulsed heavy ion beam produced in a point pinch ion diode contains many kinds of ions including protons, multiply ionized carbons, and organic ions [6]. Therefore, the conventional PHIB is not suitable for various applications to material processes.

It is very important for the application of IPIBs to the semiconductor implantation to develop the IPIB technology to generate high-purity ion beams with useful ion species for donor and acceptor in SiC such as nitrogen, phosphorous, boron and aluminum. We have developed a new type of the magnetically insulated ion diode (MID) using the ion source of gas puff plasma gun and vacuum arc plasma gun each and were successful in generating the high-purity pulsed nitrogen and aluminum ion beam with purity of more than 90 % [7,8]. In order to improve the purity of the intense pulsed ion beam, we have proposed a new type of pulsed ion beam accelerator named "bipolar pulse accelerator (BPA)" [9] and developed a prototype of the accelerator and a bipolar pulse generator as the power supply of the BPA [10-12]. The bipolar pulse generator consists of a Marx generator and a pulse forming line (PFL) with a rail gap switch on its end. The BPA utilizes a magnetically insulated ion diode with the coaxial gas puff plasma gun. In this paper, the experimental results of the evaluation of the pulsed ion beam accelerated by the bipolar pulse are presented.

II. Principle of Bipolar Pulse Accelerator

Fig. 1 shows the concept diagram of the bipolar pulse accelerator. A conventional PIB diode is also shown for comparison. As shown in Fig. 1(a), proposed ion accelerator consists of a grounded ion source, a drift tube and a grounded cathode and is an electrostatic two-stage accelerator. In the system, a bipolar pulse of voltage $\pm V_0$, duration τ_p each is applied to the drift tube. At first the negative voltage pulse of duration τ_p is applied to the drift tube and ions on the

This work was supported in part by JSPS KAKENHI Grant Number 24540534.

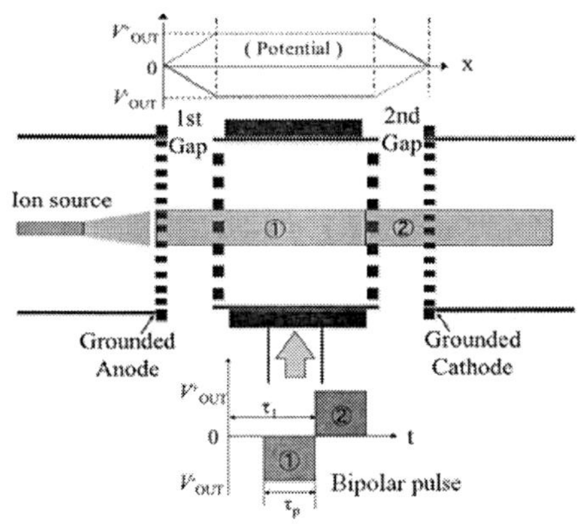

Fig. 1 Conceptual drawing of bipolar pulse accelerator

grounded ion source are accelerated toward the drift tube. If τ_p is adjusted to the time of flight delay of the ions to pass the drift tube, the pulse is reversed and the positive voltage of duration τ_p is applied to the drift tube when top of the ion beam reaches the 2nd gap. As a result, the ions are again accelerated in the 2nd gap toward the grounded cathode. As seen in Fig. 1(b), in the conventional PIB diode, the ion source is placed on the anode where high voltage pulse is applied, while in the proposed ion diode, the ion source is on the grounded anode, which extremely enhances the accessibility to the anode.

Fig. 2 illustrates the principle of the improvement of the purity of the ion beam. Let us now consider the acceleration of ions in the case that the ion beam produced in the ion source consists of N^+ ion and impurity of H^+ ion. Each ion of N^+ and H^+ is accelerated in the 1st gap toward the drift tube when the negative voltage is applied, where N^+ and H^+ ion beams are schematically described in Fig. 2. As seen in Fig. 2, the length of H+ beam is much longer than that of N^+ beam due to the difference of the velocity. We assume that the length of the drift tube is designed to be same as the beam length of N^+ beam with a beam pulse duration τ_p at an acceleration voltage V_0. It is, for example, calculated to be 11.6 cm when $V_0 = 200$ kV and $\tau_p = 70$ ns. On the other hand, the length of H+ beam at $V_0 = 200$ kV and $\tau_p = 70$ ns is 43.3 cm. When the voltage is reversed and the positive voltage is applied to the drift tube ($t = t_1$), N^+ beam with the length of 11.6 cm in the drift tube is

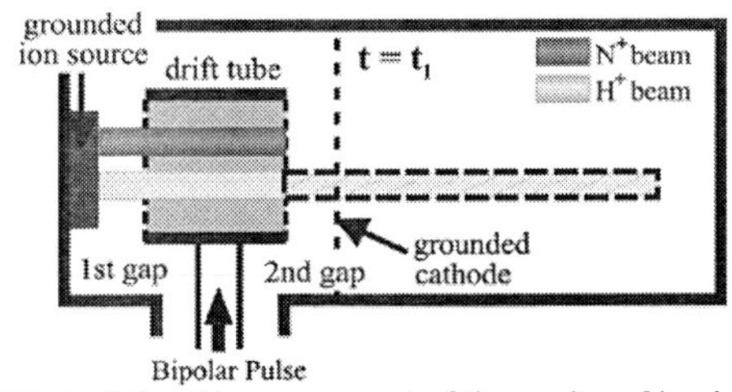

Fig.2 Principle of improvement of the purity of ion beam

accelerated in the 2nd gap. In contrast, 73 % of the beam is out of the drift tube and decelerated in the 2nd gap by the first pulse (negative voltage pulse). Hence 73 % of H^+ beam is not accelerated in the 2nd gap by the positive voltage pulse of the bipolar pulse and is removed in the bipolar pulse accelerator. As a result, the purity of the ion beam is improved.

III. Experimental Apparatus

Fig. 3(a) shows the schematic configuration of the BPA in the present experiment. The system consists of a bipolar pulse generator and an accelerator. The bipolar pulse generator consists of a Marx generator and a pulse forming line (PFL). The designed output of the bipolar pulse generator is the negative and positive pulses of voltage ±200 kV with pulse duration of 70 ns each. In the present system, the double coaxial type is employed as the PFL for the formation of the bipolar pulse. The line consists of three coaxial cylinders with a rail gap switch on the end of the line, which is connected between the intermediate and outer conductors. The characteristic impedance of the line between the inner and intermediate conductors and one between the intermediate and outer conductors are 6.7 Ω and 7.6 Ω, respectively. The PFL is filled with the deionized water as a dielectric and charged positively by the low inductance Marx generator with maximum output voltage of 300 kV through the intermediate conductor. The rail gap switch is filled with pure SF_6 gas and the pressure can be adjusted to control the optimum trigger timing for each experimental condition. The bipolar pulse voltage (V_0) and charging voltage of the PFL (V_{PFL}) are measured by the resistive voltage divider and capacitive voltage divider placed near the rail gap switch, respectively.

Fig 3(b) shows in detail the acceleration gap design and experimental setup of the BPA, which consists of a grounded

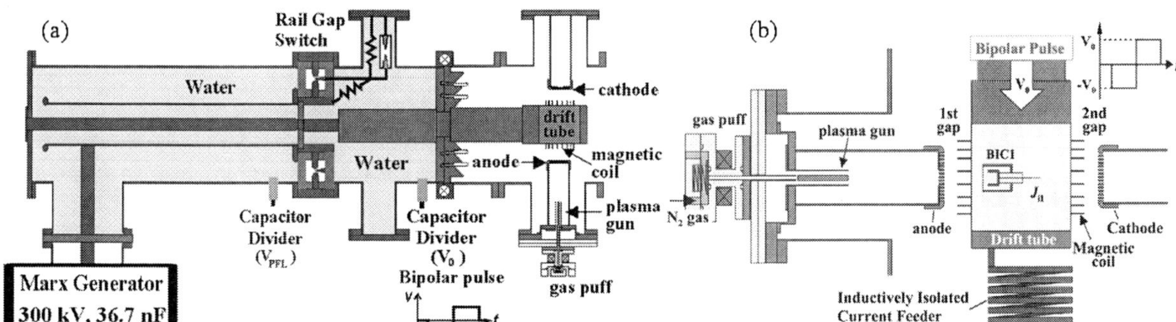

Fig. 3 (a) Schematic configuration and (b) acceleration gap design and experimental setup of bipolar pulse accelerator

anode, a drift tube, a grounded cathode and a magnetically insulated acceleration gap (MIG). The drift tube is connected to a terminal of a high voltage pulsed power generator. The anode and the cathode are the copper electrodes of diameter 78 mm, thickness 5 mm. The electrodes are uniformly drilled with apertures of diameter 4 mm, giving beam transmission efficiency of 58 %. In order to produce insulating magnetic fields in both acceleration gaps for suppression of the electron flow, a magnetic field coil of grating structure is used, which produces a uniform magnetic field in direction transverse to the acceleration gap. The magnetic coil of the MIG is installed on the rectangular drift tube. The uniform magnetic field with strength of 0.3-0.4 T is produced in the acceleration gap of gap length d_{AK}=10 mm by a capacitor bank with capacitance of 500 μF and charging voltage of 5 kV. At peak of the magnetic field, the bipolar pulse voltage supplied from bipolar pulse generator is applied to the drift tube. To obtain higher transmission efficiency of the ion beam, right and left sides of the coil consist of 8 blades each and have a grating structure. Each of the blades (10 mm^W x 118 mm^L x 1 mm^T) is connected in series and works as an 8-turn coil. Since high voltage pulse is applied to the drift tube, the pulsed current produced by the capacitor bank is applied to the coil through an inductively isolated current feeder (IC). A gas puff plasma gun was used as the ion source and installed inside the anode.

The plasma gun is composed of a coaxial plasma gun and a high-speed gas puff valve as shown in Fig. 3(b). The plasma gun has a pair of coaxial electrodes, i.e. an inner electrode of 80 mm length by 6 mm outer diameter and an outer electrode of 18 mm inner diameter. The inner electrode has six gas nozzles of 1 mm diameter. The gas puff valve consists of a nylon vessel, an aluminum valve and a drive coil and the vessel is pre-filled with N_2 gas. By applying a pulse current to the driver coil, a pulsed strong magnetic field is produced and the magnetic pressure pushes the aluminum valve to open the valve. As a result, the gas expands with a supersonic velocity and is injected into the plasma gun via the nozzles on the inner electrode of the plasma gun. The ion source plasma is produced by discharging the capacitor bank of the plasma gun with a delay time of τ_{PG} around 260-320 μs, since it takes about a hundred μs to open the valve and several tens μs for N_2 gas to reach the gas nozzle on the inner electrode of the plasma gun. To apply pulsed current to the gas puff coil and the plasma gun, capacitor banks of 5 μF and 1.5 μF were used, respectively. Each capacitor bank is usually charged up to 6 kV and 15 kV, respectively.

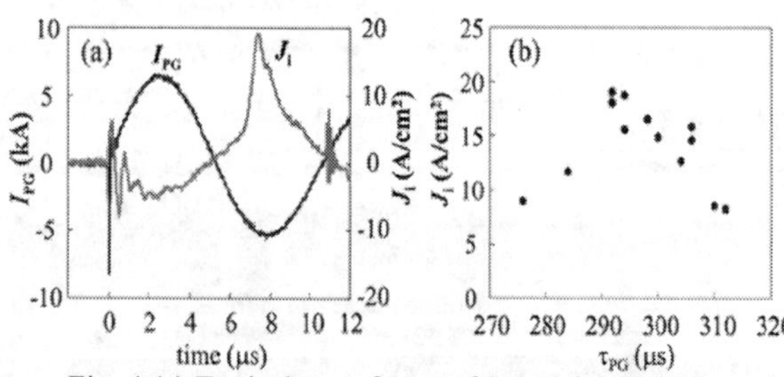

Fig. 4 (a) Typical waveforms of I_{PG} and J_i and (b) dependence of J_i on τ_{PG}

The ion current density of the source plasma produced by the plasma gun was evaluated by a biased ion collector (BIC) placed at 130 mm downstream from the top of the plasma gun where the anode is placed in the acceleration experiment. The collector of BIC was biased at a voltage of −200 V to remove accompanying electrons. Fig. 4(a) shows the waveforms of the discharge current of the plasma gun (I_{PG}) and the ion current density (J_i) of the plasma gun at the condition of τ_{PG}=292 μs. Here the filling gas pressure of the vessel is 2.0 atm. As seen in Fig. 4(a), the discharge current I_{PG} has a sinusoidal waveform of peak current 6.5 kA and quarter cycle 2.6 μs. The ion beam with a peak current density of J_i=20 A/cm² and a pulse duration of about 2 μs is observed at about τ_d=7.5 μs after the rise of I_{PG}. The result suggests that it takes 7.5 μs after the rise of the discharge current of the plasma gun (I_{PG}) for the ion beam produced in the plasma gun to reach the acceleration gap. Fig. 4(b) shows the dependence of J_i on the delay time of the discharge current rise from the rise of the gas puff current (τ_{PG}) and the filling N_2 gas pressure. As seen in Fig. 7, J_i rises at $\tau_{PG} \approx$ 270 μs and has a peak around 290 μs, and after that decreased. The result suggests that it takes 270 μs after the rise of the gas puff coil for the gas to reach the nozzles.

IV. EXPERIMENTAL RESULTS

In order to confirm the acceleration of ions in the 1st gap by the negative pulse of the bipolar pulse, BIC was installed inside the drift tube to measure the ion current density (J_i), as seen in Fig. 3(b). Since the high voltage pulse is applied to the drift tube, the inductively isolated coaxial cable of same structure as IC was used to transport the BIC signal. Insulating magnetic field with strength of 0.3-0.4 T was applied to the acceleration gap of gap length d_{AK}=10 mm. The plasma gun was operated at the condition of $\tau_{PG}\approx$290 μs and the Marx generator was fired at τ_{PM}=8-20 μs after the rise of the discharge current (I_{PG}) of plasma gun. The bipolar pulse generator was operated at 70 % of the full charge condition of the PFL.

Fig. 5 shows the typical waveforms of the charging voltage of the PFL (V_{PFL}), the output voltage (V_0) and the ion current density (J_i) accelerated in the 1st gap. They were obtained at the condition of τ_{PM} =15 μs. J_i was measured at 48 mm downstream from the anode surface. As seen in the Fig. 5, the

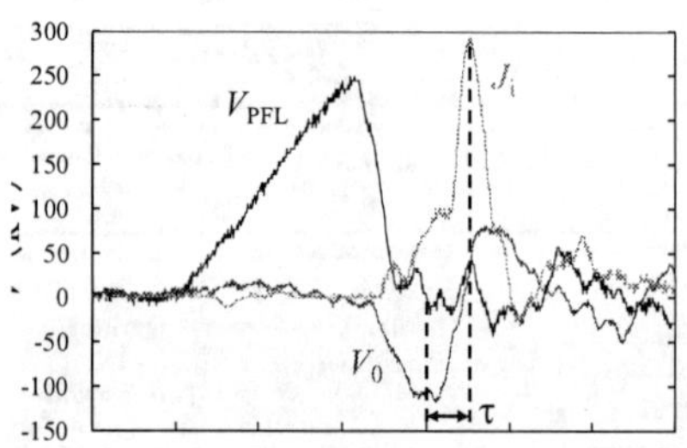

Fig. 5 Typical waveforms of charging voltage of PFL (V_{PFL}), output voltage (V_0) and ion current density (J_i)

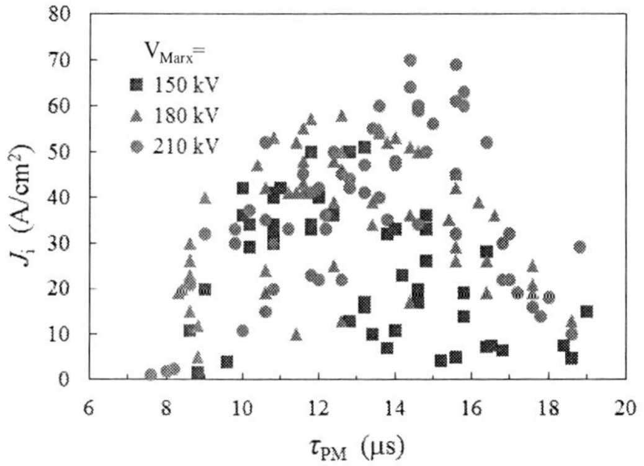

Fig. 6 Dependence of the peak values of ion current density (J_i) on the delay time (τ_{PM})

first pulse of V_0 rises at t=210 ns and the bipolar pulse, which consists of the first pulse with a voltage of about −110 kV and a pulse duration of about 70 ns (FWHM) and the second pulse with a voltage of about +80 kV and a pulse duration of about 60 ns is applied to the drift tube. The ion beam with a current density of J_i=60 A/cm^2 and a pulse duration of 50 ns(FWHM) was obtained at around 50 ns after the peak of the negative voltage pulse.

Fig. 6 shows the dependence of the peak value of J_i on the delay time (τ_{PM}) between the plasma gun and Marx generator for each output voltage (V_{Marx}) of Marx generator, where the other conditions of τ_{PG} and insulating magnetic field were fixed at 292 μu and 0.3 T, respectively . As seen in the Fig. 6, J_i increases monotonically with increasing τ_{PM} and reaches a peak and after that decreased. J_i has a peak at appropriate delay time (τ_{PM}) for each V_{Marx}. This seems to be due to that quantity of the plasma on the anode or in the 1st gap increases with increasing τ_{PM}. The maximum current density of 70 A/cm^2 was obtained with V_{Marx}=210 kV and τ_{PM} =14.4 μs.

The peak current density corresponds to nitrogen ions (N^+, N^{2+}) on the result reported in [7,10]. Considering the time of flight delay, the ion beam corresponding the peak of J_i seems to be accelerated around the peak of the negative pulse. In the case of Fig. 5, we can estimate from the time of flight delay

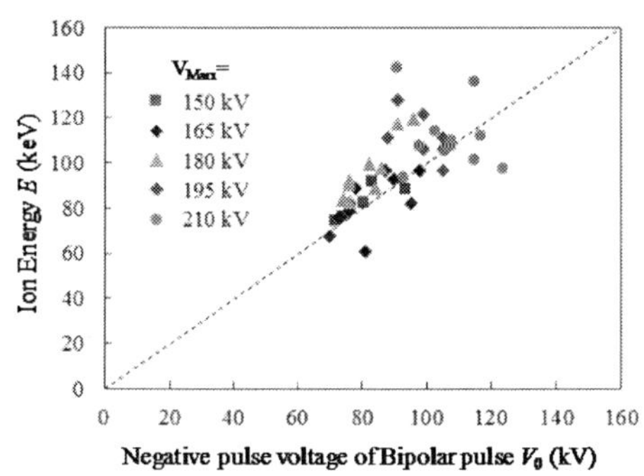

Fig. 7 Relationship between estimated ion energy and applied negative voltage (V_0) of bipolar pulse

that the energy of accelerated nitrogen ion beam is about 108 keV/Z, where Z is the charge state of the ions. The estimated ion energy seems to be in reasonable good agreement with the accelerating voltage (negative pulse voltage of bipolar pulse) and the ion energy evaluated by a magnetic energy spectrometer [12]. Fig. 7 shows the relationship between estimated ion energy and applied negative pulse voltage of bipolar pulse. Even though the output voltage (V_0) of the bipolar pulse was varied, the accelerated ion energy and V_0 is on proportional line with the coefficient of 1. Therefore, the singly ionized nitrogen ion was the dominant component of accelerated ion beam.

V. SUMMARY

We have developed a prototype of the bipolar accelerator to perform proof of principle experiments. When the bipolar pulse of −114kV, 70 ns (1st pulse) and +85 kV, 62 ns (2nd pulse) was applied to the drift tube, the ions were successfully accelerated in 1st acceleration gap by the negative pulse voltage of the bipolar pulse. To confirm the principle of the BPA, we are planning to evaluate the ion species and the energy spectrum of the accelerated ion beam in detail.

References

[1] S. Humphries, Jr, "Intense pulsed ion beams for fusion applications", Nucl. Fusion, Vol. 20, pp.1549-1612 (1980).

[2] F.M. Bieniosek, J. J. Barnard, A. Friedman, E. Henestroza, J. Y. Jung, M. A. Leitner et al., "Ion-beam-driven warm dense matter experiments", Journal of Physics: Conference Series Vol.244 032028 (2010).

[3] Y. Nakagawa, T. Ariyoshi, H. Hanjo, S. Tsutsumi, Y. Fujii, M. Itami, A. Okamoto, S. Ogawa, T. Hamada, and F. Fukumaru, "Modification of solid surface by intense pulsed light-ion and metal-ion beams", Nucl. Instrum. Method B, Vol.39, pp.603-606 (1989).

[4] K. Yatsui, X.D. Kang, T. Sonegawa, T. Matsuoka, K. Masugata, Y. Shimotori, T. Satoh et a., "Applications of intense pulsed ion beam to materials science", Phy. Plasma Vol.1, 1730 (1994).

[5] R. M. Bayazitov, L. Kh. Zakirzyanova, I. B. Khaibullin, and G. E. Remnev, "Formation of heavily doped semiconductor layers by pulsed ion beam treatment", Nucl. Instrum. & Methods in Phys. Res. B, Vol.122, pp.35-38 (1997).

[6] Y. Hashimoto, M. Sato, M. Yatsuzuka, and S. Nobuhara, "Generation and focusing of intense ion beam with an inverse pinch ion diode", Jpn. J. Appl. Phys., Vol.31, pp.1922-1927, (1992).

[7] H. Ito, H. Miyake, and K. Masugata, "Diagnosis of high-intensity pulsed heavy ion beam generated by a novel magnetically insulated diode with gas puff plasma gun", Rev. Sci. Instrum., Vol.79, 103502, (2008).

[8] H. Ito, K. Fujikawa, H. Miyake, and K. Masugata, "Characteristic observation of intense pulsed aluminum ion beam in magnetically insulated ion diode with vacuum arc ion source", IEEE Trans. Plasma Sci., Vol.37, No.10, pp. 1879-1884, (2009)

[9] K. Masugata, "A high current pulsed ion beam accelerator using bi-directional pulses", Nucl. Instrum. & Methods in Phys. Res. A, Vol.411, pp.205-209, (1998).

[10] K. Masugata, Y. Shimizu, Y. Fujioka, I. Kitamura, H. Tanoue, and K. Arai, " Development of bipolar-pulse accelerator for intense pulsed ion beam acceleration", Nucl. Instrum. & Methods in Phys. Res. A, Vol.535, pp.614-621, (2004).

[11] H. Ito, K. Igawa, I. Kitamura, and K. Masugata, "Bipolar pulse generator for intense pulsed ion beam accelerator", Rev. Sci. Instrum., Vol.78, 013502, (2007).

[12] H. Ito and K. Kitajima, "Bipolar pulse generator for intense pulsed ion beam accelerator", IEEJ Transactions on Fundamentals and Materials, Vol.135, pp.136-141, (2015) (in Japanese).

Study of Fast Rising Voltage Circuit Using Fast Recovery Diode

Takashi Sakugawa, Hideniri Akiyama
Institute of Pulsed Power Science
Kumamoto University
Kumamoto, Japan
sakugawa@cs.kumamoto-u.ac.jp

Koichro Yamashita, Kengo Tasaki
Graduate School of Science and Technology
Kumamoto University
Kumamoto, Japan

***Abstract*— There are various methods of pulsed power generation using semiconductor switch. Semiconductor switching device is indispensable to realize a high repetitive operation of the pulsed power modulator for industrial applications. High repetitive operation of the modulator have contributed to industrial application such as excimer laser for photo lithography. However, the semiconductor switching device is impossible to generate fast rising high voltage pulse directly. In this study, we have employed a fast recovery diode (FRD) for fast pulse rising circuit of the compact pulsed power modulator. Primary switching devices were use Silicon-Carbide based Metal Oxide Semiconductor Field Effect Transistor (SiC-MOSFET), the pulse compression circuit was using a saturable transformer (ST) and FRD. We have investigated the effective saturation timing of the ST for applied reverse current to the FRD. We have realized generating the fast rising high voltage pulse using fast recovery characteristics of the FRD. The developed modulator can be obtained several ten kV output voltage with several ns rise time.**

Keywords—fast recovery diode; Siilicon-Carbide; saturable transformer;

I. INTRODUCTION

Recently, the development of all-solid-state pulsed power generators are in progress. In particular, there are remarkable developments of a wide-band gap semiconductor switch such as SiC-MOSFET. However a plasma generation under atmospheric pressure and an excitation of gas laser cannot be in the semiconductor switch itself. Therefore, an assistance of magnetic switch and a step-up transformer is required in all-solid-state pulsed power generator using semiconductor switch [1-3]. Their circuit are capacitive energy stores, such as low inductance and high permittivity capacitors. On the other hand, there is an inductive energy store pulsed power generator such as accumulated in the magnetic field of the inductive current-carrying circuit and delivered to a load with the help of semiconductor opening switches (SOSs) [4-7]. However, these almost generators has a complicated circuit, or using specialized SOS. We have developed short-pulse compression circuit using a saturable transformer (ST) and a fast recovery diode (FRD). The short-pulse compression circuit consists of a semiconductor switch circuit, a ST and a FRD. The semiconductor switch of the primary circuit is a SiC-MOSFET. The switching characteristics of the SiC-MOSFET were improved such as capable of high current and decreasing switching loss. The ST made of a ferromagnetic core has two functions, those of step-up transformer and magnetic switch. We have used two kinds of ferromagnetic core as Co-based amorphous and Fe-based nanocrystalline for ST. These core material has a high squareness ratio (Br/Bs). This circuit is used as the opening switch the FRD to the output stage. Reverse recovery time of the FRDs used in this circuit is several ns. However, high change rate of current during recovery time from peak to zero is essential in addition to the reverse recovery time is extremely short in order to obtain a high voltage. Therefore, increasing the reverse current of the FRD is important. In this study, we present the results of research on a fast rising output voltage pulsed power generator for pulsed power generator of simple configuration using ST and FRD. We describe the method of optimal combining the ST and the FRD, and present experimental results on the performance of a short-pulse compression circuit.

II. FAST RISING HIGH VOLTAGE CIRCUIT

A. Principles of Generation of Fast Rising High Voltage

We are explain the basic principles of short-pulse compression in a circuit using a FRD as shown in Fig. 1. Fig. 1 (a) shows a simple circuit diagram of FRD as SOS. The principle of operation of a FRD is as follows: At first, charged to capacitor C. The triggering switch S, whose diagram is given in Fig. 1 (a). Generally, such a FRD is a set of series connected p-n junctions produced by diffusion of donors and acceptors into n-type low doped silicon. The triggering circuit is designed so that the current passing through the FRD is oscillatory in character: during the positive and negative trigger half-wave, the FRD conducts in the forward and in the reverse direction, respectively (Fig. 1 (b)) [4]. Where trr is reverse recovery time of the FRD, V is output voltage to Load.

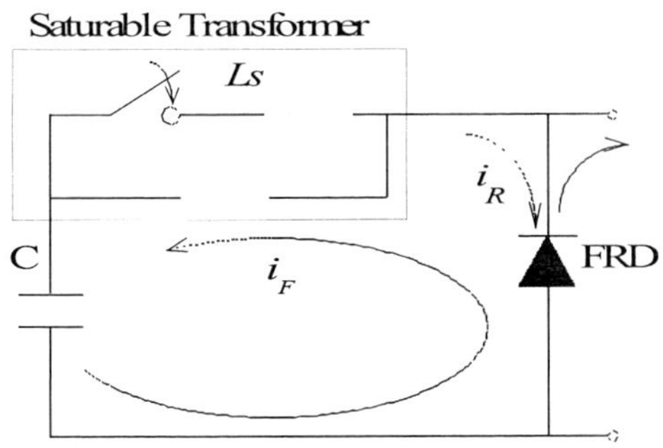

(a) Simple circuit model using fast recovery diode.

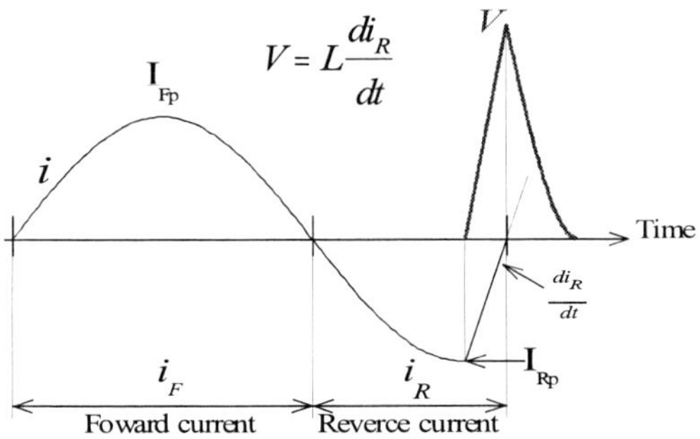

(b) Voltage and current waveforms of fast recovery diode.

Fig. 1. Basic principles of fast rising voltage circuit.

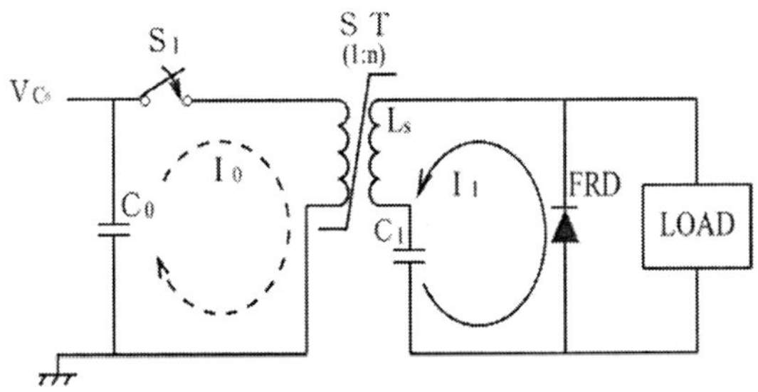

Fig. 2. Short-pulse compression circuit using ST and FRD.

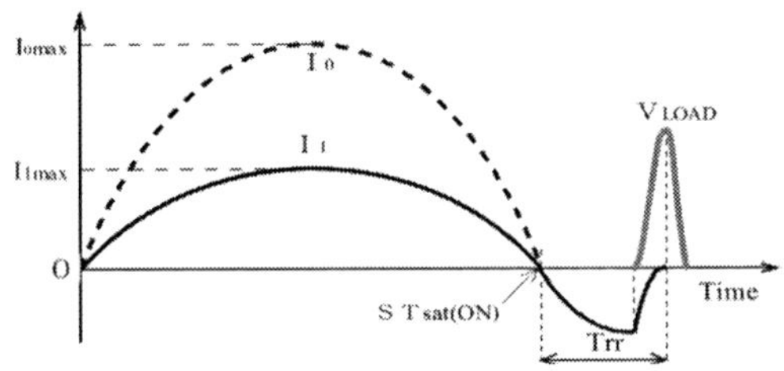

(a) Current and voltage waveforms of FRD by normally circuit.

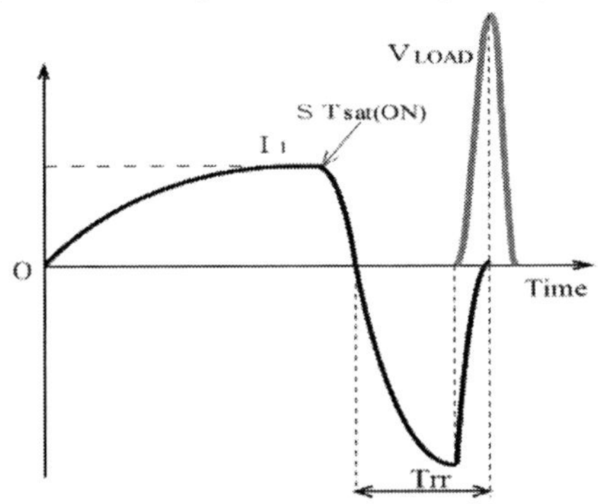

(b) Current and voltage waveforms of FRD by improved circuit.

Fig. 3. Current and voltage waveforms of pulse compression circuit using ST and FRD.

B. Fast Rising Voltage Circuit Using a ST and FRDs

We are describe an optimum design method of short-pulse compression in a circuit using a ST and a FRD. Basic circuit is shown in Fig. 2. Fig. 3 (a) shows normally current and voltage waveforms of the short-pulse compression circuit. For a ST with a primary : secondary winding ratio of 1 : n, the capacitance of capacitors are related as C0=n2C1. By inserting the FRD in parallel to the LOAD, a charging loop of C1 is provided. After capacitor C0 is charged to VC0, first turn on switch S1 so that the ST remains unsaturated until the electric charge is completely transferred from C0 to C1. The saturating ST, whose reverse current of the FRD and output voltage V_{LOAD} are as shown in Fig. 3 (a). Generally, such a design of ST is remains unsaturated until the electric charge is completely transferred from C0 to C1. We performed a design improvement to advance the saturation time of the ST to increase the reverse current of the FRD. Result of saturation of the ST was early by the design improvement, the output voltage and the reverse current of the FRD are increased as shown in Fig. 3 (b).

III. Experimental Results

A. Early Saturation of ST

We were developed all-solid-state short-pulse generator is shown in Fig. 4. The all-solid-state generator consists of a Charger, SiC-MOSFETs, SI0, capacitors, ST and FRD. Specifications of the FRD is shown in Table I . Primary pulse generate by SiC-MOSFETs switching circuit.

We are used two kinds of core for the ST such as Fe-based nano-crystalline (Fig. 5 (a)) and Co-based amorphous (Fig. 5 (b)). Because the saturation flux density is small, Co-based amorphous can be saturated earlier ST compared to Fe-based nano-crystelline. In Fig. 5(b), the value of the peak voltage and rise time attained 30 kV and 3 ns respectively.

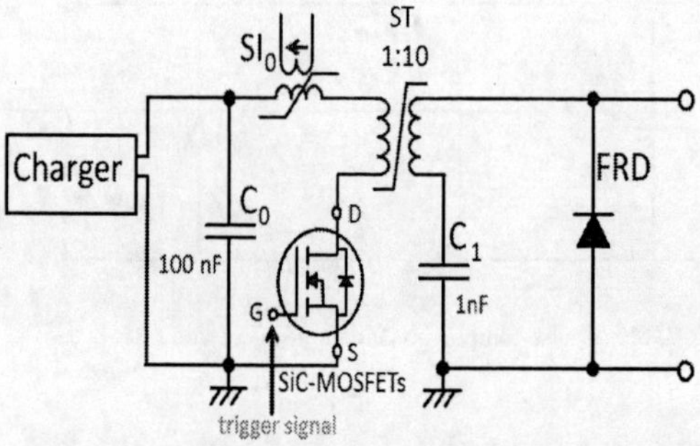

Fig. 4. All solid-state short-pulse generator using ST and FRD.

TABLE I. SPECIFICATIONS OF FAST RECOVERY DIODE

Working Reverse Voltage	10000 (V)
Average Rectified Current	180 (mA)
Reverse Recovery Time	30 (ns)

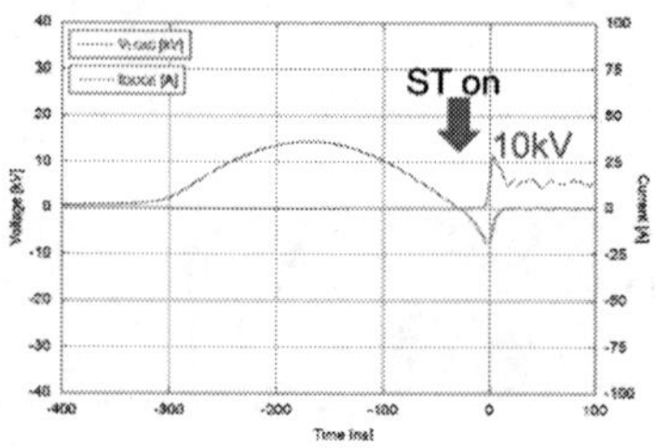

(a) Normal saturation of ST (Fe-based nano-crystalline).

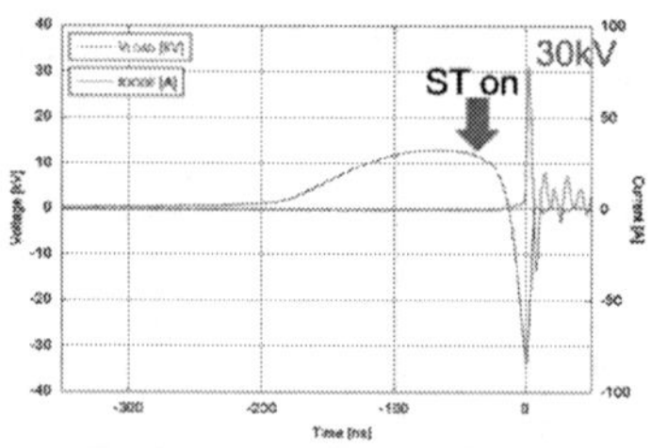

(b) Early saturation of ST (Co-based amorphous).

Fig. 5. Current and voltage waveforms of FRD.

B. Parallel Connection of FRD

In this section, we describe the experimental results of the parallel connection of FRDs. Fig. 6 shows the waveforms of the voltage and current of the FRD in the case of increasing the number of parallel FRDs. Reverse current of the FRDs is increased by increasing the number of parallel FRD, the maximum value of the output voltage rises. When four parallel-connected FRDs, the reverse current of FRDs is 108 A, the output voltage is 45 kV with 2.6 ns the voltage rise time.

Parallel connection of FRD is effective for improving the output voltage of short-pulse compression circuit.

The output voltage and four parallel FRDs current waveforms at high repetition operation are shown in Figure 7. Repetition rate is 1000 pulses per second (pps). The waveforms are overwriting number of 499 pulses. In this study, also it has a stable output voltage is obtained in high repetitive operation.

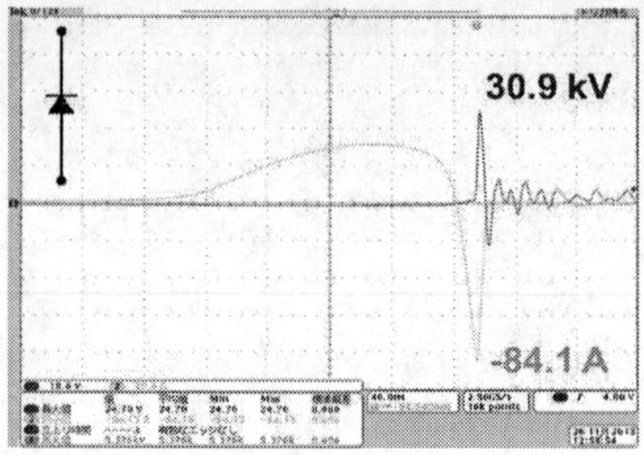

(a) Single FRD.

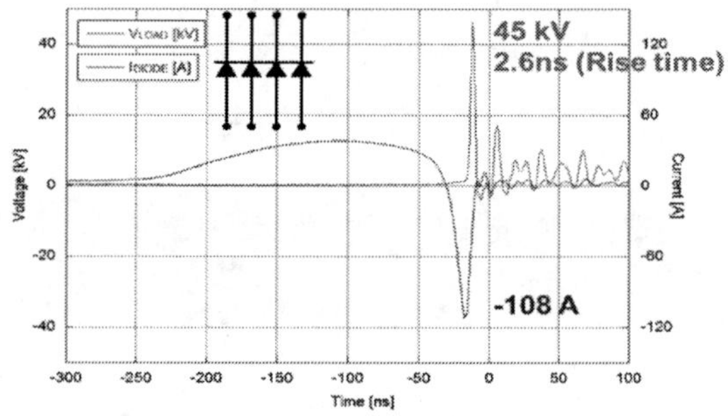

(b) Four parallel FRDs.

Fig. 6. Effect of parallel connection of FRDs.

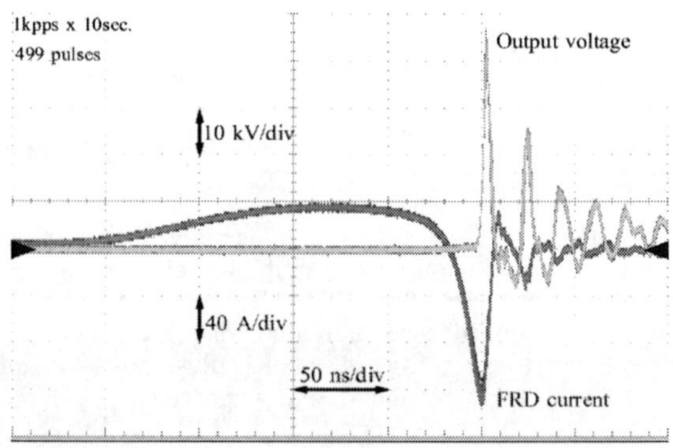

Fig. 7. Output voltage and FRDs current waveforms at 1000 pps.

C. Output Characteristics

Fig. 8 shows load voltage and current waveforms of the short-pulse generator using ST and FRD. The load is used 270 Ω resister. It was confirmed that a current sharp cut-off characteristic by the FRD, output voltage is increased to high speed. And it is gradually decreased by the time constant of the load resistance is after the peak value. Connect a single FRD, were measured load voltage, load current, and peak power respectively, when it was changed load resistance from 27 Ω to 5 kΩ. These results of the measurements are shown in Fig. 9. It was confirmed that the resistance value increases from the experimental results, the peak voltage of the output voltage rises. On the other hand, current is reduced along rising resistance of load. Considering the application of the short-pulse generator, the optimal load impedance is around 270 Ω. Fig. 10 shows generation of atmospheric discharge plasma using the short pulse generator. Discharge electrode configuration is the point-to-plane electrode gap of 15mm. It has applied a high voltage nanosecond pulse to the point electrode. On the other hand, plane electrode is grounded. It was observed that the glow-like discharge plasma to have been generated between the electrodes.

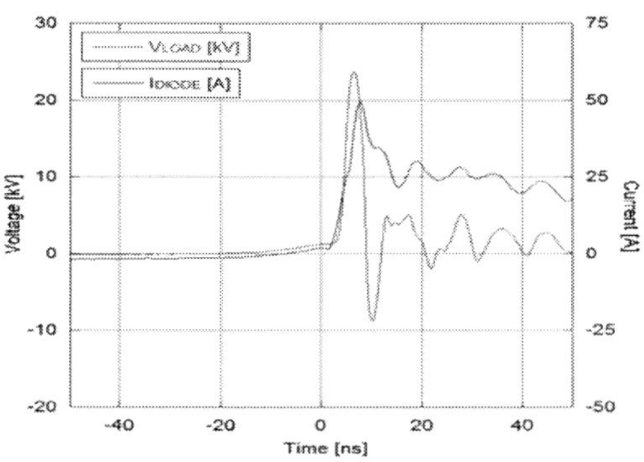

Fig. 8. Voltage and current waveforms of load resistor [270Ω].

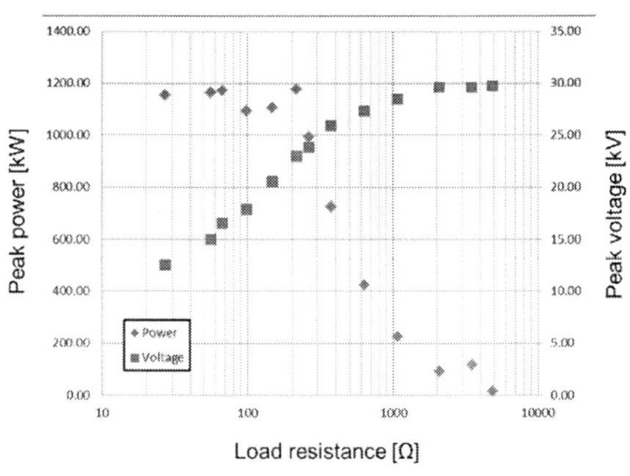

Fig. 9. Output characteristics of short-pulse compression circuit.

Fig. 10. Generation of atmospheric discharge using shot-pulse generator.

IV. SUMMARY

We have developed the short-pulse compression circuit using a FRD. The results of this study are summarized as follows,

a) We have realized generating the fast rising high voltage pulse using fast recovery characteristics of the FRD.

b) We have been generate a fast rising high voltage pulse using the new developed pulsed power modulator. And we have obtained output voltage of several 10kV with fast rise time by the modulator.

The developed modulator can be obtained several ten kV output voltage with several ns rise time. We expect to short-pulse compression circuit can contribute for many applications.

References

[1] W. Partro, R. Sandstrom, I. Fomenkov, "A low cost of ownership KrF excimer laser using a novel pulse power and chamber configuration", Proc. SPIE No. 2440, 95, 1995.

[2] H. Mizoguchi, O. Wakabayashi, T. Aruga, T. Sakugawa, T. Koganezawa, "High power KrF excimer laser with a solid state switch for microlithograpy", Proc. SPIE No. 2726, 831, 1996.

[3] T. Sakugawa, H. Akiyama, "An All-Solid-State Pulsed Power Generator Using a High-Speed Gate-Turn-Off Thyristor and a Saturable Transformer", IEEJ, Vol. 140, No.4, pp. 17-26, 2002.

[4] Gennady A. Mesyats, "PULSED POWER", Springer, 2005.

[5] Igor V. Grekhov, Gennady A. Mesyats, "Physical Basis for High-Power Semiconductor Nanosecond Opening Switches", IEEE Transactions on Plasma Science, Vol. 28, No. 5, pp. 1540-1544, 2000.

[6] S. K. Lyubutin, S. N. Rukin, B. G. Slovikovskii, "High-Frequency Pulse Generators Based on SOS Diodes with Subnanosecond Current Cutoff Time", Instruments and experimental techniques, Vol. 43, No. 3, pp. 331-338, 2000.

[7] K. Takaki, I. Yagi, T. Fujiwara, T. Go,"Influence of Circuit Parameter on Ozone Synthsis Using Inductive Energy Storage System Pulsed Power Generator", Transactions on Dielectrics and Electrical Insulation, Vol. 18. No. 5, pp. 1732-1758, 2011.

A Variable Trigger Pulse Generator for the Triggered Vacuum Switch*

Yoon Kyoo Son, Moo Sang Kim, Wung Hoa Park, Byung-Joon Lee
Pohang Accelerator Laboratory
POSTECH
Pohang, Korea
bjlee707@postech.ac.kr

***Abstract*—The prototype of a variable trigger pulse generator (TPG) is suggested to research various trigger conditions for a triggered vacuum switch (TVS). The TVS is significantly used in a high power system, it discharges current of more than 100 kA within several microseconds after a trigger signal is provided in the trigger unit (TU) composed of a trigger pin, an insulator and a holder. To understand the trigger system of the TVS, the relationship between the TPG and the TU should be figured out. In this paper, to research it, we have adopted the hard tube type of a pulse generator. This type has an advantage of unfixed load impedance. The required specification of the TPG is that the voltage amplified up to 18 kV, the width variable from 1 µs to 20 µs, and a voltage rising time fixed at below 5 kV/µs. The fabricated prototype of the variable TPG is applied to research the characteristics of the TU, see the paper [1]. And the result is used to find optimized parameters of the TPG.**

Keywords— *triggered vacuum switch; hard tube type;IGBT*

I. Introduction

A TVS is one of the important switch in the field of high pulsed power systems. General requirements of the high power switches are low-cost, high-coulomb transfer, long lifetime, low jitter, and flexible environment. Furthermore, the sealed-off TVS is enable to be operated in any places and easily replaced [1]. It holds a high voltage and current by vacuum and discharged by an plasma arc. The generation of plasma arc begins in the trigger unit (TU) of TVS after a trigger signal is supplied by a trigger pulse generator (TPG). In this paper, we have designed the TPG system, and researched for triggered conditions of the TU with the trigger pulse width and amplitude of the TPG signal. Although the components of trigger circuit are simulated by the LTspice IV program, especially, we have focused on the trigger device which is a critical component to control the trigger voltage, and the trigger circuit topology is chosen to hard type pulser.

II. TVS Structure and Trigger Experiment

A. Spcification and fabrication of the TPG

Though the previous research, the TPG is selected to a hard tube type because the plasma impedance is not fixed. And some parameters are confirmed, the amplitude is set up to 18 kV, the variable range of a pulse width is 1~20 µs, a rising time is 2 µs, and other parameters are shown in the table 1.

Table 1. Specification of the TPG

Parameter	Specification
Supply Voltage	220 V, 60 Hz
Output Voltage	18.0 kV
Output Current	500 A
Isolation(housing ↔ ground)	20 kV
Switching frequency	0.5 Hz
Rising time	2 µs
Control signal	Optical
Pulse width	1~20 µs
Switching element	IGBT

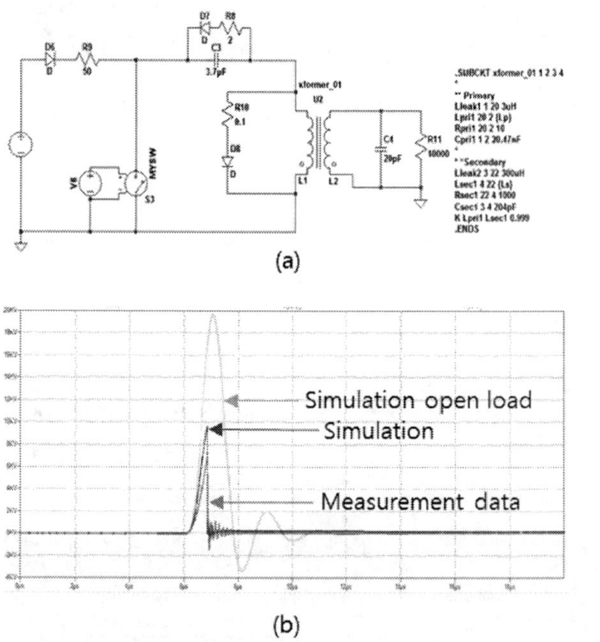

Fig. 1. (a) Simulation circuit (b) output signal

This project was supported by the Agency for Defense Development (ADD).

The basic concept of TPG is introduced in a various kinds of papers [2][3]. The simulation circuit diagram of the TPG is proposed through research papers, which is shown in Fig. 1 (a). The input current is charged in the storage capacitor bank assembled 6 capacitor arrays and the discharge is accomplished after the gate is switched on which is controlled by a TTL signal generator. The output voltage is amplified about 10 times by the output transformer whose turn ratio is 10. The output load is assumed a high impedance to match with no load. The output signal is 18.5 kV. And the rising time 2 μs and the pulse width is 30 μs at 90 % of the max peak voltage. The signal is showed in Fig. 1 (b) which is simulated with the LTspice IV program for 6 capacitors.

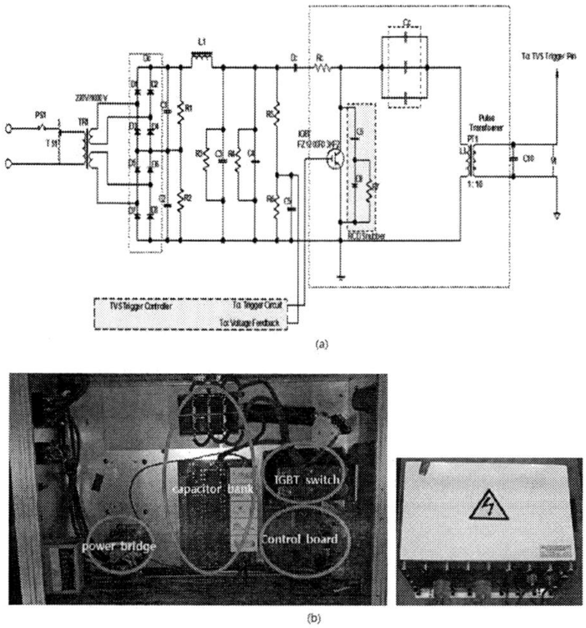

Fig. 2. (a) Circuit diagram for fabrication (b) fabricated TPG

The designed TPG is separated with a control circuit and an output transformer, each components are like that PS1 is a charge control switch, TS1 is a voltage regulator, TR1 is a Step-up transformer, D_c is a rectification diode, R_c is a charging protection resistance, C_c is a storage capacitor, R_5 and R_6 are divider resistors for measuring charging voltage. L_1 is a output transformer, and V_t is the trigger pulse signal which supplies to the TU. Fig. 2 (a) shows them.

The fabricated control circuit is divided 4 blocks which are a power bridge, a control board, an IGBT switch, and a capacitor bank. The power bridge is a connector between a 220 VAC power and the control board. The control board makes a gate signal to control the IGBT switch. The generated gate signal controls the discharge of the capacitor bank. The capacitor bank is a charge storage which is connected with the output transformer.

The output transformer whose leakage inductance of the primary winding is about 3 μH and stored energy capacitance is 3 μF, respectively. It is detached from the main circuit and the winding coil is stored in a tank filled an oil. This design is proposed to endure up to 18 kV and reduce leakage inductance.

B. Measured signal of the TPG

The output signal is measured with changing from 1 to 4 capacitors, Fig. 3 (a) shows one capacitor and (b) is four capacitors. The pulse width is adjusted by the combination of this numbers of capacitor. However, in the RC circuit, the pulse width and the amplitude are simultaneously affected by the capacitor. The amplitude is peak at the one capacitor, on the contrary, the pulse width is maximum at the four capacitors. The rising time is 2 μs at 90 % of the max peak voltage. These results are not enough to satisfy the original goal, although the simulation result is satisfied. The amplitude and the pulse width should be optimized, while the amplitude keeps maximum voltage.

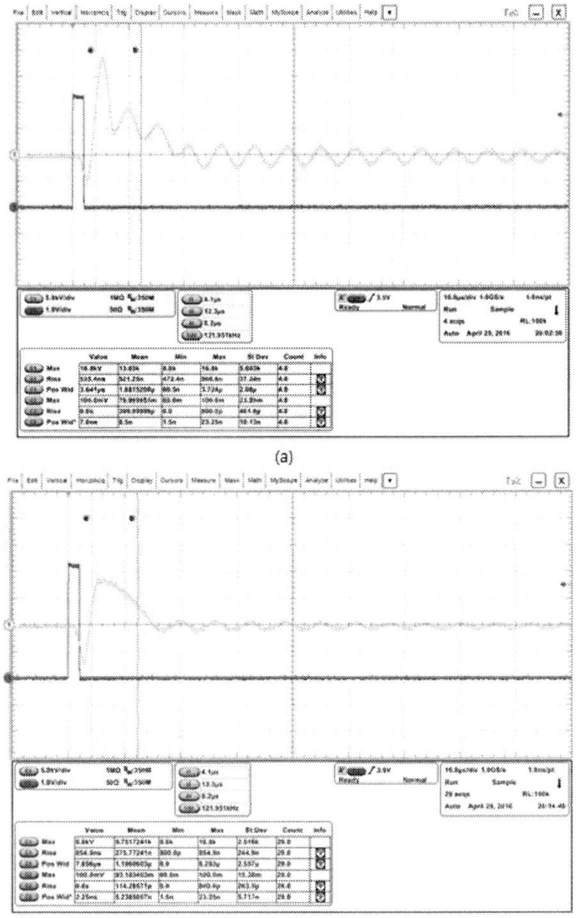

Fig. 3. Measured signal for the fabricated TPG

C. Characteristics of TU

The main electrodes conducts high current, each of them made up of 3 pairs of rods, interfacing each other and forming

6 vacuum gaps [4]. The TU assembled with the holder, the insulator, and the trigger pin is set in the center of cathode and connected to an external TPG. It ignites an initial plasma, after the trigger signal is supplied from the external TPG. The generation of the initial plasma is begun by the electron emission at the triple point in which a vacuum, a ceramic insulator, and the holder are contacted [5]. Fig. 4 shows the structure of the TVS and the TU.

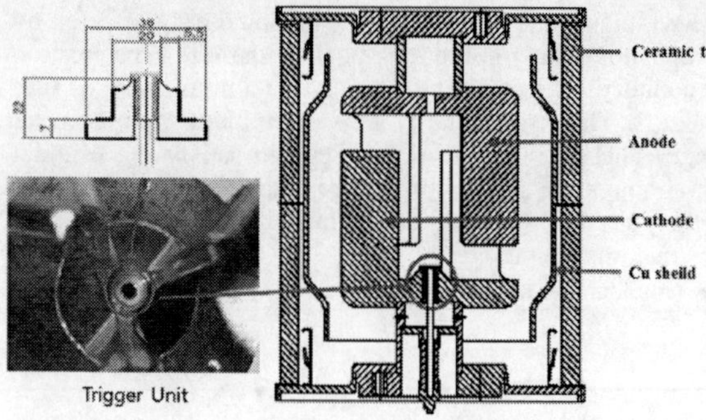

Fig. 4. Structure of the TVS and the TU

The condition of the electron emission is decided by the electric filed strength which is enable to enhance through the TPG and TU. Actually, the TPG voltage is depended on the structure and material of the TU. To investigate the relation, the holder shape is proposed up and down types, whereas, other TU parameters are fixed

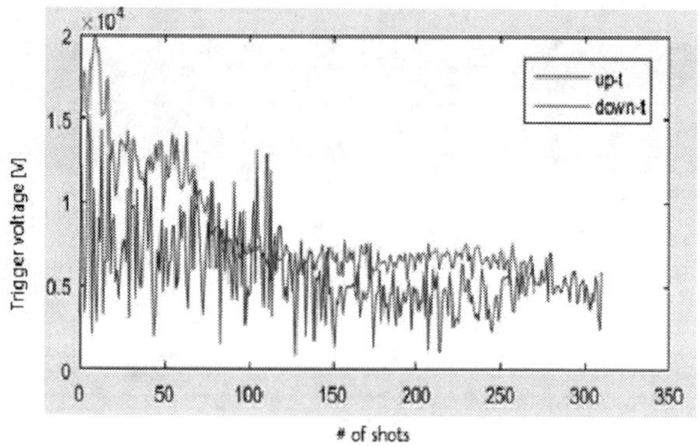

Fig. 5. Trigger voltage for holders

The initially produced prototype TPG is fixed at 42 kV output voltage without a load, 4 us rising time and, 300 A max current. Unfortunately, these values are not determined by the optimized TU, because the virgin condition of the trigger signal was unknown, so it just produced to understand trigger conditions. Fig. 5 shows the trigger voltage for shots and trigger signals are measured with the Cu holder, whereas, the insulator is fixed with ceramic and trigger pin is tungsten. The result shows the dependency of the between the trigger voltage and holders, however, still it is not enough to understand the characteristics of the trigger signal.

III. SUMMARY

The triggered condition of the TU is strongly depended on the external TPG. To research the relation, the holder shape which one of TU components is chosen up and down type, and proposed the variable TPG. The required variable parameters are the amplitude and the pulse width, others are fixed. Through the simulation, the circuit has been checked the specification and the TPG is fabricated. To endure high voltage and reduce the leakage inductance, the output transformer is separated from the control circuit. The output signal is satisfied the required condition of the amplitude and the pulse width, individually. However, it is not enough to fulfill the specification, because the amplitude and the pulse width have to be simultaneously satisfied the requirement. Therefore, the TPG should be more optimized for amplitude and the pulse width with maintaining the maximum voltage, and it will be done in the next work.

Acknowledgment

This project was supported by the Agency for Defense Development (ADD) under grant number 14-BR-EN-22.

References

[1] Soung Soo Park, Sang Hoon Nam, Yeong Jin Han, "Characterization of a 30kV, 75C triggered vacuum switch," IEEE Trans. on Plasma Science, vol. 33, no. 1, pp. 192-196, 2005.

[2] Yao, Xueling, Jingliang Chen, and Wei Sun. "A high-voltage and high-current triggered vacuum switch." *IEEE Transactions on Plasma Science* 38.10 (2010): 2901-2905.

[3] Jiyan, Zou, Duan Xiongying, and Qin Shihong. "Experimental investigations on operation capacity of triggered vacuum switches." *Discharges and Electrical Insulation in Vacuum, 2000. Proceedings. ISDEIV. XIXth International Symposium on*. Vol. 2. IEEE, 2000.

[4] Zhou, Zhengyang, et al. "The lifetime of a high-current triggered vacuum switch with multi-gap." *Instruments and Experimental Techniques* 57.3 (2014): 296-299.

[5] Zhou, Zhengyang, et al. "Time delay of a field-breakdown triggered vacuum switch with flat electrodes." *Instruments and Experimental Techniques* 54.6 (2011): 803-807.

Investigation of the effect of material in miniature lightning strike model using 3D printing topography model

F. Tanaka
Department of Electrical Engineering
National Institute of Technology, Ishikawa College
Ishikawa, Japan
f_tanaka@gm.ishikawa-nct.ac.jp

***Abstract*—Recently, efficient natural energy such as power generation and solar power generation have attracted attention. Therefore, it is important to grasp the tendency of lightning strike in the area for installing the power generation. In addition, generally in Japan, the thunder is thought to the natural disaster, and the lightning strike measures for residential is thought to useless. It is important to advance measures against lightning, and the lightning simulation system is required to understand about thunder and to change people's mind about thunder. From these reasons, it is performed a low cost lightning strike simulation system research. In this study, using a 3D Printer in order to develop a low cost simulation system However, an inexpensive 3D printers using insulation materials like abs resin, it is predicted that the surface potential of the model produced is different from the ground level. Therefore, to investigate the position of the lightning strike, it was investigated techniques for showing a simulation system equivalent tendencies produced in metal. As a result, by covering the model produced in ABS resin with a thin aluminum foil, it was possible to obtain results similar to the model produced in aluminum.**

Keywords—Lightning strike, 3D printer , ABS resin

I. Introduction

In recent years, introduction of renewable energy has been promoted in Japan. Among them, wind power and solar power generation increase is noticeable. One of the reasons of solar power generation increase, it is conceivable that it is installed on the roof when a house is rebuild.

One of the major causes of these power failure is a lightning strike[1]. However generally in Japan, the thunder is thought to the natural disaster , and the lightning strike measures for residential is thought to useless. Therefore, tell the basic knowledge easy to understand the general public about the lightning strike, there is a necessary to teach the need for lightning protection.

In addition, inexpensive simulation model for advancing the simulation models and lightning studies to demonstrate the efficacy of Lightning stably is required.

For these reasons, to create an inexpensive simulation model using 3D printer is performed. In this study, investigation of the effect of material in miniature lightning strike model was done.

II. Experimental Method and Setup

A. Preparation of 3D model

In this study, the red square in fig.1 shows the area output by 3D printer, and The bottom of Fig.2 is the south side. Using this 3D data, the Three different 3D models was printed. The output data is shown fig. 2. (a) was output using a 3D printer of the uPrint SE Plus. (a) what was covered with aluminum is the (b). (c) made of only aluminum is what was printed by the outsourcing. The width and length of all of the model is approximately 180 mm. And the scale of the output map is one of 17,640 minutes (10 mm on map = 176.4 m).The area created by the 3D model is an Interesting area which a wind power generator and a steel tower to prevent lightning are adjacent.

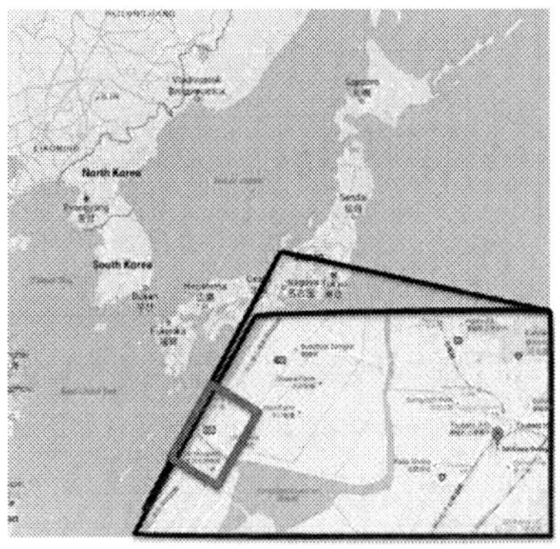

Fig. 1 The output area of 3D printer

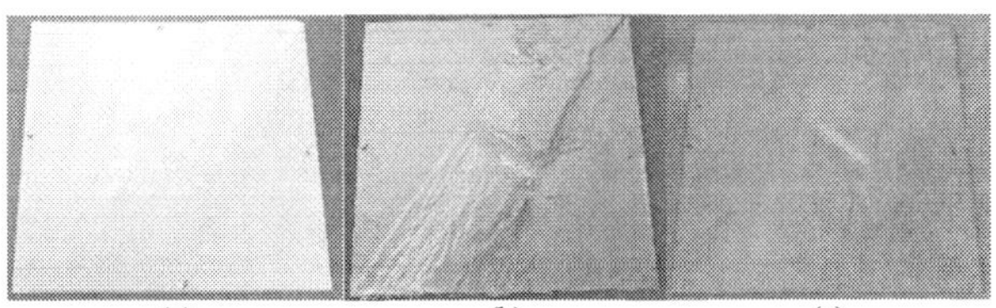

(a) (b) (c)

Fig. 2 Outputted 3D model

(a)ABS, (b)ABS covered with aluminum foil, (c)All aluminum model

B. Experimental setup

To simulate lightning, the six stage Marx generator which own National Institute of Technology, Ishikawa College has been use.[2] Typical output waveform from the six stage Marx generator when occurred arc discharge between electrode and 3D model is shown fig.3. The maximum voltage is about140 kV, the maximum current is about 400 A.

The distance from the electrode to the surface of the 3D model was 100 mm. this value is considered the scale of the 3D model and a height of thundercloud.

C. Experimental method

Using these models two experiments about lightning strike were performed. One is to investigate the influence of the only topography using the 3D model and the other is to investigate the 3D model that is in addition of power generation and lightning protection tower to approximate the actual environment.

Experiment simulating the real environment is performed using a model of Fig. 4. The distance of lightning protection tower and wind power generation is 2.6 mm, the height is 6 mm and this distance is calculated from the scale of the map. The modeling of the wind power generation and lightning protection tower is using a nail.

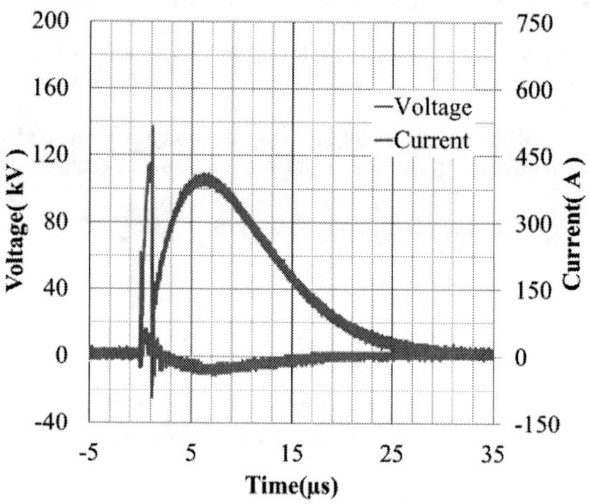

Fig. 3 Typical output waveform from the Marx generator

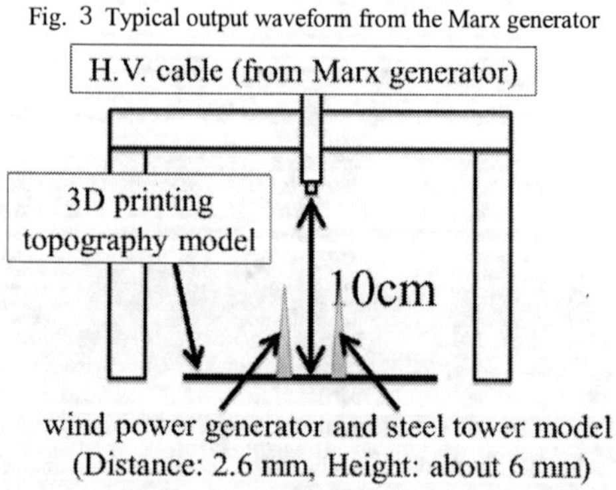

Fig. 4 Setup of experiment simulating the real environment

III. Experimental Results

A. Experimental results of using 3d models only

The experimental results are shown in Fig. 5. Red circles show the position of the lightning strike and blue circles show a position of the wind power generator and the lightning protection tower. Grids show a 10mm on 3D each models. By using this system, the voltage was applied 25 times .

Results of FIG. 5, when comparing (a) and (b), and shows that scattering of A slightly larger. Also, three data when compared, show that scattering of (c) is small.

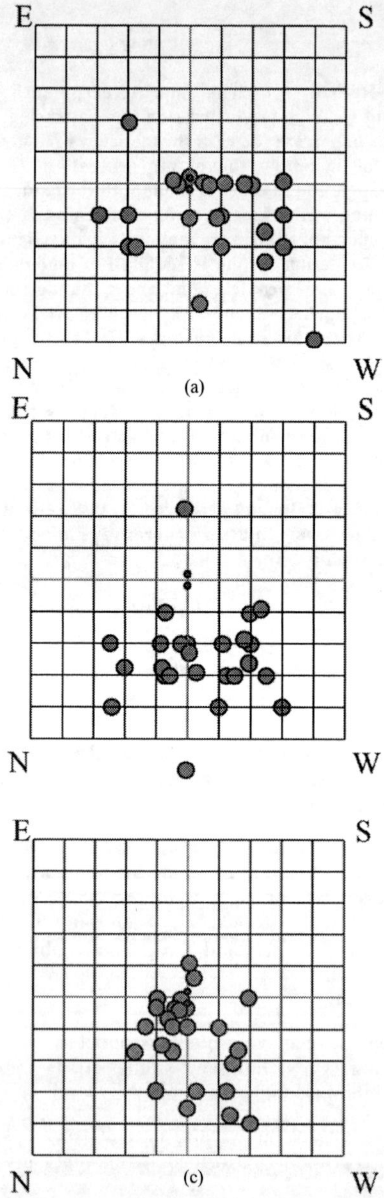

Fig. 5 Experimental results of using 3d models only
(a)ABS, (b)ABS covered with aluminum foil, (c)All aluminum model

Become of such a result, it is considered that the rough state of the surface of the terrain model. Causes of the roughness of the surface is due to the performance of the 3D printer,

B. *Experimental results of simulating the real environment*

The experimental results are shown in Fig. 6, Table 1, Table 2, Table 3. In this experiment, as the center between the towers and towers, at the position of Fig. 6 (a), it was applied 25 times voltage, respectively. Table 1, Table 2, Table 3 are a respectively, counted the number of lightning strikes the ground. Fig.6 (b) and table 1 are the result of a model using ABS resin model, Fig.6 (c) and table 2 are the result of a model using ABS resin model covered with aluminum foil, and Fig6 (d) and table 3 are All aluminum model.

From the results of FIG. 6 (b), when the simulation of discharge by using only ABS resin model , it can be seen that come to lightning well nail. As a result, when compared with other conditions, it is seen that little discharge to ground. Also, variations in discharge even when compared with other results, it can be seen larger.

Fig.6 (c) is a result of the ABS resin model was discharged in those covered with aluminum foil. However, the probability of a lightning strike to the ground, has a similar result as in Table 3.

Fig.6 (d) is aluminum model. Compared with other experimental results, it has resulted dispersion of lightning strike position is small.

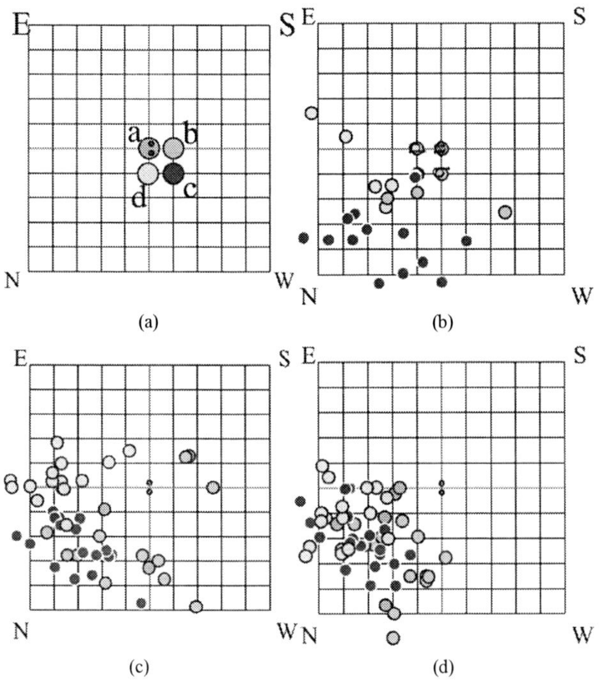

Fig. 6 lightning strike point of experiment simulating the real environment
(a) Position to apply a voltage (b)ABS,
(c)ABS covered with aluminum foil, (d)All aluminum model

IV. Consideration

The results of experiment A, the case of using ABS resin, it was shown that the variation is large. This trend is the same trend even experiment B. However, in Experiment B, and comparing the results of lightning strike number to the ground, when covering the ABS resin with aluminum foil, it showed the model and close the tendency of aluminum.

From these things, for the variation of the lightning strike position on the ground, smoothness of the surface of the terrain model is considered to be important. Therefore, when performing the discharge experiment in consideration of the influence of terrain, the resolution in the vertical direction of the 3D printer is considered to be important.

Furthermore, experiments B, from Table 2, the results of Table 3, model covering the ABS resin with aluminum foil, showed the similar tendency as aluminum model. In addition, with respect to aluminum foil and ABS model, things can be confirmed there is no significant damage.

For that reason, that there is likely to be of aluminum model of the alternative has been shown. Further, even when comparing costs, ABS resin model was found to be superior those covered with aluminum foil.

Table 1 ABS model

Electrode position	The number of lightning strike to nail	The number of lightning strike to Other point
a	25 (100%)	0
b	22 (88%)	3
c	12 (48%)	13
d	20 (80%)	5

Table 2 ABS covered with aluminum foil

Electrode position	The number of lightning strike to nail	The number of lightning strike to Other point
a	20 (80%)	5
b	12 (48%)	13
c	3 (12%)	22
d	8 (32%)	17

Table 3 Aluminum model

Electrode position	The number of lightning strike to nail	The number of lightning strike to Other point
a	22 (88%)	3
b	13 (42%)	12
c	7 (28%)	18
d	11 (44%)	14

In addition, as the future works, the ABS resin model is considered to influence of distribution of surface charge. Therefore, it is necessary to experimental results when remove the surface charge.

References

[1] New Energy and Industrial Technology Development Organization, "Wind Power Guideline for Japan" 2008, In Japanese

[2] Kotaro Omote, F. Tanaka " Inspection of the Effectiveness of the Thunderbolt Prevention System Using a 3D Printing Topography Model" 5th Euro - Asian Pulsed Power Conference, 2014, P2-40

Development of compact pulsed power modulator for biological application

Y. Kusaba, R. Fukumura
Graduate School of Science and Technology,
Kumamoto University
2-39-1, Kurokami, Chuo-ku,Kumamoto, 860-8555, Japan

H. Akiyama, K. Yano, and T. Sakugawa
Institute of Pulsed Power Science,
Kumamoto University
2-39-1, Kurokami, Chuo-ku, Kumamoto, 860-8555, Japan

***Abstract*— Pulsed electric fields are known to induce apoptosis when applied to cancer cells, a phenomenon which has been actively studied. This paper shows the development of a pulse power modulator with variable pulse width, frequency and voltage; it also shows output results. Two MOSFETs were used on the switch unit to develop an all solid-state circuit, one for rising pulse and the other to form a square wave. Output voltage was 600V, and the pulsed width was varied up to 100ns. The main circuit board area was 110mm x 170mm. A microscope system incorporated on the circuit board enabled microscopic observation during pulse power application.**

Keywords—compact modulator; variable pulsed width; nano second;

I. Introduction

Recently, studies of nanosecond pulsed power have led to applications in such areas as medical, plasma and recycling. Specifically, cell applications such as drug administration using electro pores and cancer treatment by apoptosis are highly anticipated [1-4]. Observation of cells when an electric field is applied will elucidate cellular mechanisms and effects. Therefore, the cells are observed by microscope during electric field application. It is known that influence to cells changes by differing parameters; e.g., electropore size varies by pulse width, and cell mortality varies by electric field strength. Thus, control of the applied voltage and pulse width are important in the development of pulsed power modulators. Especially, the ability to control pulse width will enable adjustment of cell hole size in accordance with the size of the molecule of the drug. Thus, we chose MOSFET to control pulse width because the switching speed is faster than switches such as IGBT, bipolar transistors and thyristors. The goal of this study was to narrow pulse width to 100ns. Fast rise and fall times are necessary to maintain a square wave with a short pulse width, and our previous study, which used a single MOSFET, was unable to form a square wave due to a slow fall time. In this study, we used two MOSFET in order to achieve both; in addition, we took steps to downsize the modulator.

II. Experimental Setup

In this section, we describe our pulsed power modulator using MOSFET. Figure 1 is a schematic diagram of the modulator. The principle of operation, C1 is charged from a charger via the R1when S1 is OFF. When sending TTL signal of a square wave of the requested pulsed width to gate control circuit from multifunction generator, S1 is turned ON and a current flows to Rout from C1 via coaxial cable. At that time, Rout is at high voltage when the TTL signal is high. When the TTL signal falls, S1 is turned OFF and S2 is turned ON. Then, the current which had flowed to Rout flows to S2, and Vout is low. Rated voltage of the modulator is 1000V. In this experiment, we used a charger able to output up to 650V. Figure 2 shows a circuit of the modulator, and Figure 3 shows our experiment setup.

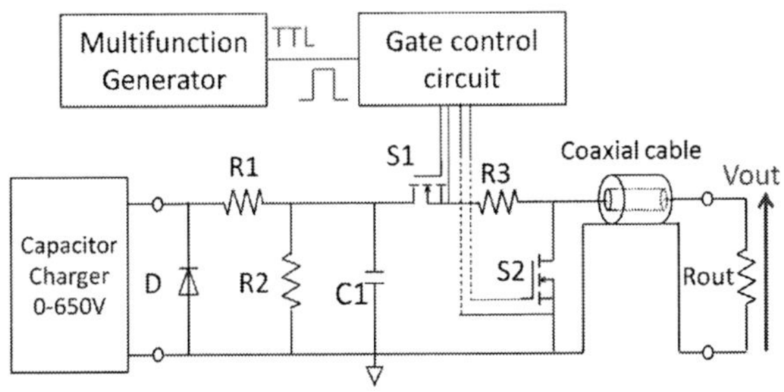

Fig. 1. Schematic diagram of the modulator

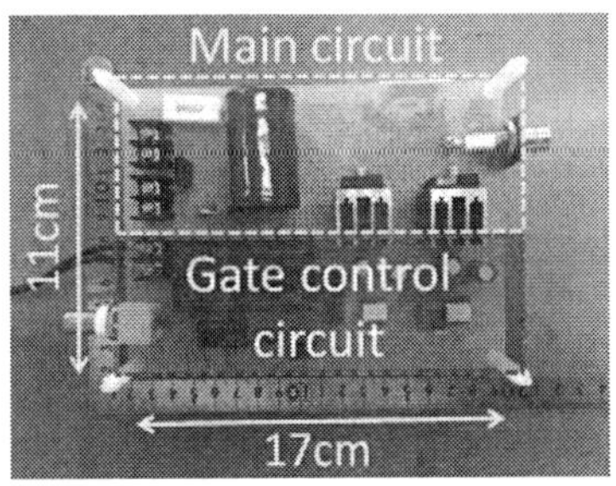

Fig. 2. Circuit of the modulator

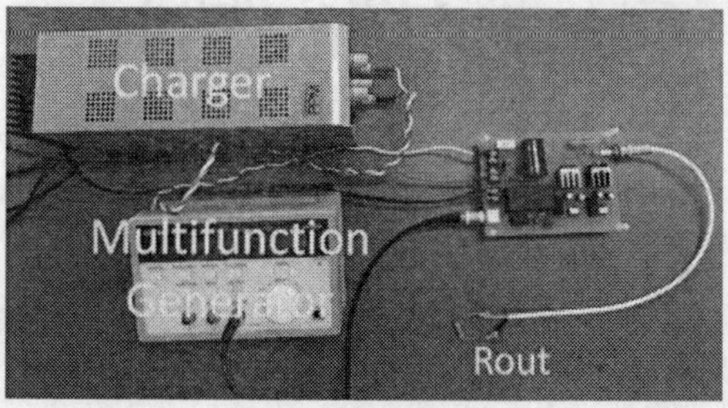

Fig. 3. Experiment setup

III. Experiment Result

Figure 4 shows output voltage when TTL signal is 1μs and charger is 600V. Delay time was 260ns to both rising and falling pulses. Output pulse width was about 980ns against the TTL signal 1μs. Rise time was about 23ns, and fall time about 4ns. Figure 5 shows output voltage when TTL signal is 500ns and charger is 600V. The pulse width of output was about 480ns against the TTL signal 500ns. Thus, we succeeded in controlling the pulse width. This is because the S1 turn OFF and S2 turn ON were on the same timing. Figure 6 shows voltage of MOSFETs (S1, S2) between gate source when TTL signal is 500ns. When TTL signal was 500ns, S1 ON time was also 500ns. Then, when S1 turned OFF, S2 turned ON. S2 is turned ON only 300ns at the moment of the fall of TTL signal.

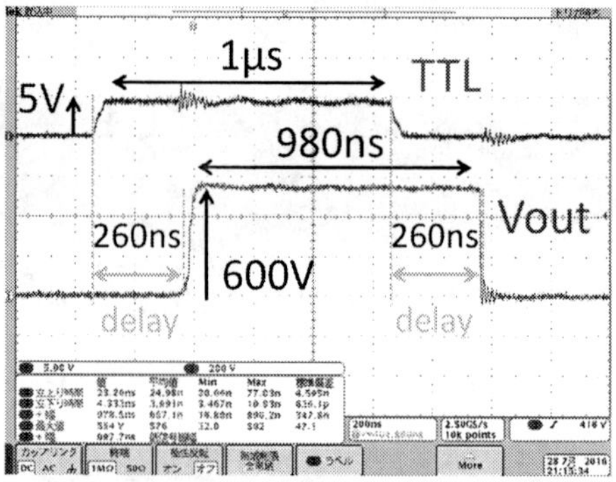

Fig. 4. Output voltage [TTL:1μs, Charger:600V]

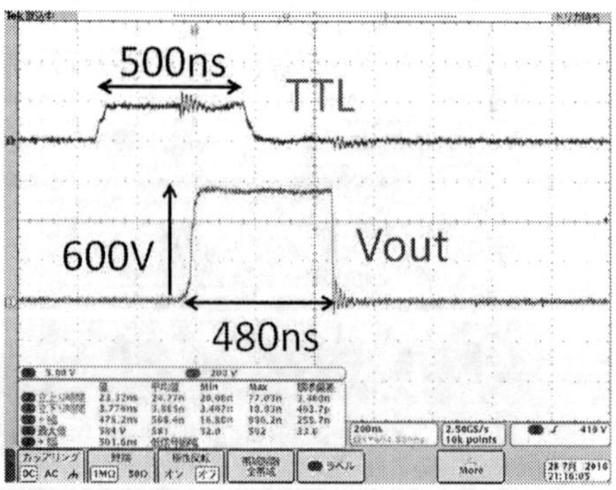

Fig. 5. Output voltage [TTL:500ns, Charger:600V]

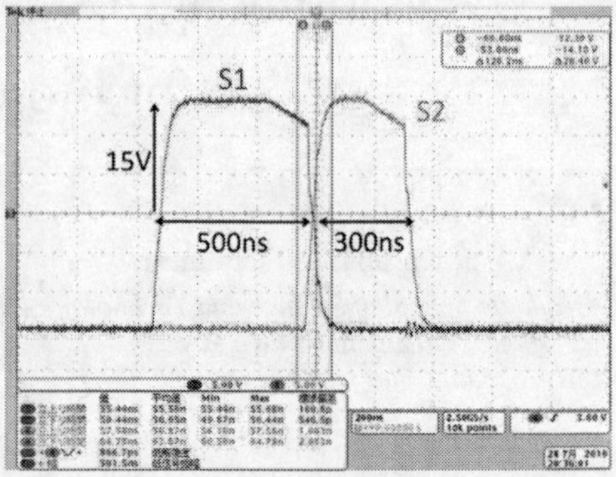

Fig. 6. Voltage of MOSFETs (S1, S2) between Gate-Source [TTL: 500ns, Input:±15V]

IV. Discussion

In this section, we discuss pulse formation and cell experiments.

A. Pulse forming

S2 is very important for the modulator. Here, we explain circuit operation both without and with S2 using each output voltage.

1) Without S2

The left side of Figure 7 shows output voltage and circuit operation without S2. C1 current flows to coaxial cable when S1 turns ON. In this case, the load becomes a high voltage (①). When the supply of the TTL signal expires, S1 is Turn-OFF. At this time, current also flows slightly. Therefore, the load is a high voltage (②). Then, when the current becomes smaller, S1 turns OFF completely. As a result, the load becomes low voltage (③). The pulse width of output is approximately 1μs against the TTL signal 500ns. The rise time is about 24ns, and fall time is about 530ns. This configuration is unable to control pulsed width, and it is difficult to say that the output waveform is a square wave.

2) With S2

The right side of Figure 7 shows output voltage and circuit operation with S2. The current of C1 flows to coaxial cable when S1 turns ON. In this case, the load becomes a high voltage because S2 is OFF (①). When the supply of the TTL signal expires, S1 is Turn-OFF and S2 turns ON. At this time, the current flows toward S2 through S1. Therefore, the load is a low voltage because the current doesn't flow to coaxial cable (②). Then, when the current becomes smaller, S1 turns OFF completely, and S2 also subsequently turns OFF. Thus, the load is low voltage (③). The pulse width of output is approximately 500ns against the TTL signal 500ns. The rise time is about 24ns, and fall time is about 4ns. Unlike the result of without S2, the output is a square wave which is the same pulsed width as TTL signal.

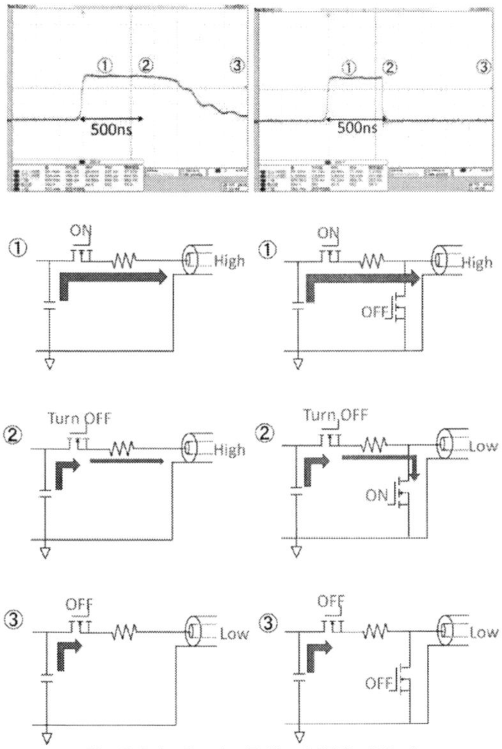

Fig. 7. Pulse forming(Tail cut) [TTL: 500ns]
Left: Without S2, Right: With S2

B. *For cell experiments*

The modulator, which was developed for cell experiments, can output a square wave of 600V and 100 ns as shown in Figure 8. Cells were observed using a microscope. Figure 9 shows the microscope for cell experiments. Its electrode distance is 0.2 mm. The electric field intensity generated between the electrodes is shown as follows.

$$E = \frac{V[\mathrm{V}]}{d[\mathrm{mm}]} = \frac{600}{0.2} = 30[\mathrm{kV/cm}] \quad (1)$$

The modulator is able to produce an electric field strength equivalent to that used in previous studies. Therefore, it is possible to freely vary the pulse width (such as 500, 450, 300.... and 100ns) while satisfying electric field strength necessary in cell experiments.

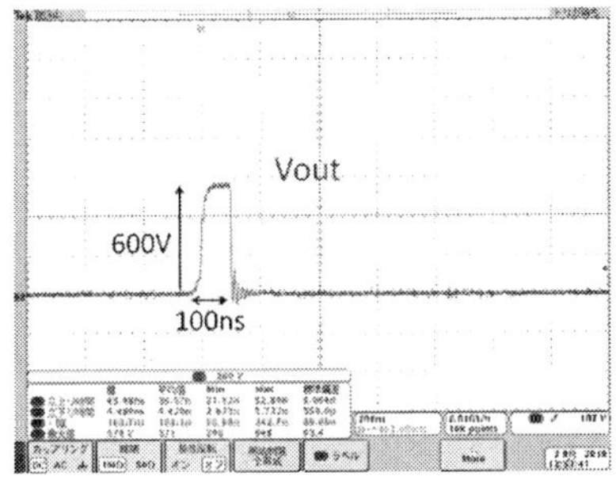

Fig. 8. Output voltage (Pulse width: 100ns)

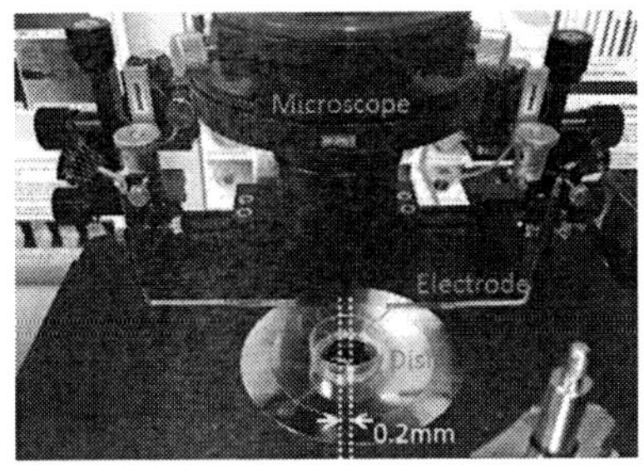

Fig. 9. Microscope for cell experiments

V. SUMMARY

We have developed a compact pulsed power modulator using MOSFET. The results of this study are summarized as follows:

- The modulator can output during incoming TTL signal, so it can control pulse width.
- In this experiment setup, it was able to output a square wave up to 600V with variable pulse width.
- The pulse width can be narrowed up to 100ns, the rise time is about 24ns, and the fall time is about 4ns.

References

[1] Keiko Morotomi-Yano, Hidenori Akiyama, Ken-ichi Yano, "Nanosecond pulsed electric fields activate MAPK pathways in human cells", Archives of Biochemistry and Biophysics 515 (2011) 99–106

[2] Keiko Morotomi-Yano, Hidenori Akiyama, Ken-ichi Yano, "Nanosecond pulsed electric fields induce poly(ADP-ribose) formation and non-apoptotic cell death in HeLa S3 cells", Biochemical and Biophysical Research Communications 438 (2013) 557–562

[3] Bennett L. Ibey, Jody C. Ullery, Olga N. Pakhomova, Caleb C. Roth, Iurii Semenov, Hope T. Beier, Melissa Tarango, Shu Xiao, Karl H. Schoenbach, Andrei G. Pakhomov, "Bipolar nanosecond electric pulses are less efficient at electropermeabilization and killing cells than monopolar pulses", Biochemical and Biophysical Research Communications 443 (2014) 568–573

[4] Andrei G. Pakhomov, Elena Gianulis, P. Thomas Vernier, Iurii Semenov, ShuXiao, OlgaN.Pakhomova, "Multiple nanosecond electric pulses increase the number but not the size of long-lived nanopores in the cell membrane", Biochimica et Biophysica Acta 1848 (2015) 958–966

Study on Acceleration Mechanism of Particles from Fast Plasma Flow Generated by Tapered Cone Plasma Focus Device

Taichi Takezaki, Yutaro Hatakeyama, Kazumasa Takahashi, Toru Sasaki, Takashi Kikuchi, and Nob. Harada
Nagaoka University of Technology
NUT
Kamitomioka 1603-1, Nagaoka Niigata 940-2188, Japan
ttakezaki@stn.nagaokaut.ac.jp

***Abstract*—Behavior of a one-dimensional fast plasma flow generated by a tapered cone plasma focus device flowing in a perpendicular magnetic field has been investigated. A velocity of the plasma flow is about 30 km/s using a helium gas discharge with a pressure of 0.5 Pa. The magnetic Reynolds number is estimated to be 4.2, and the plasma pressure is higher than the magnetic pressure. The plasma flow has decelerated by the magnetic pressure, and the magnetic field is frozen-in at the front of the plasma flow. The plasma flow affects the magnetic field, and induced electromagnetic fields occur the acceleration of ions from the plasma flow.**

Keywords—collisionless shock; plasma focus; fast plasma flow;

I. Introduction

To generate high energy particles in outer space, collisionless shocks as heliospheric or supernova shocks have been considered to play an important role [1-3]. The particles are accelerated by three-dimensional non-linear interactions between fast plasma flow and electromagnetic fields. In the relativistic region, the particles obtain its energy reflected by upstream and downstream magnetohydrodynamic (MHD) turbulences at the shock surface [4-5]. This mechanism, which is called as Fermi acceleration or diffusive shock acceleration, has explained the energy distribution function of the high-energy particles with the power law spectrum. In the non-relativistic region, full particle simulations have been carried out for understanding the wave-particle and the wave-wave interactions [6-7].

To understand astrophysical phenomena, laboratory-scale experiments with well-defined experimental conditions are helpful because numerical modelings to understand its phenomena are simplified [8-11]. For generating a collisionless shock in a laboratory-scale experiment, the required conditions as a function of a fast plasma flow and a magnetic field with suitable strength were estimated [9].

To generate a fast plasma flow in the laboratory-scale experiment, methods using laser ablation or pulsed-power discharges have been demonstrated [12-17]. The plasma beta is required to be higher than unity to produce a perturbation for a precursor of MHD turbulences. To generate a fast plasma flow in a laboratory-scale experiment, a tapered cone plasma focus device using a guiding acrylic tube drives a quasi-one-dimensional plasma flow with a velocity of 40 km/s using a xenon gas discharge. The device provides a one-dimensional fast plasma flow for understanding astrophysical phenomena and for a simple comparison with numerical simulations [18,19].

In this study, to understand an interaction between fast plasma flow and perpendicular magnetic field, we have investigated a one-dimensional plasma flow generated by the tapered cone plasma focus device flowing through a perpendicular magnetic field. The magnetic Reynolds number is estimated to evaluate the relation between the plasma flow and the magnetic field, and the pressure balance are discussed. A numerical simulation based on an electromagnetic hybrid particle-in-cell method has been carried out for the comparison with the experimental result, and for the analysis of the relation between the propagation of plasma flow and the evolution of perpendicular magnetic field.

II. Experimental Setup

To obtain a fast plasma flow in a laboratory-scale experiment, the tapered cone plasma focus device was developed [18,19]. Figure 1 shows a schematic image of the tapered cone plasma focus device. The tapered cone plasma focus device consists of an outer tapered electrode, an inner cone electrode, and an acrylic tube with the length of 20 mm. This device generates a thermalized plasma by using a plasma focus to improve the energy density for a gas discharge. To provide the plasma with a directional motion and to converge the energy in the acceleration process of the current sheet, the electrodes have a cone shape. The accelerated current sheet is pinched and compressed at the cone tip, and its kinetic energy changes to thermal energy. The plasma flow is driven by the pressure gradient between the thermalized plasma and an ambient gas. The driven plasma flow propagates with quasi-one-dimensional behavior in the acrylic tube. The electrodes and the acrylic tube are coaxially arranged in a vacuum chamber. The vacuum chamber is filled with helium gas, and its pressure is controlled in a range from 0.1 Pa to 0.5 Pa by using turbo-molecular and rotary pumps. A capacitor bank is

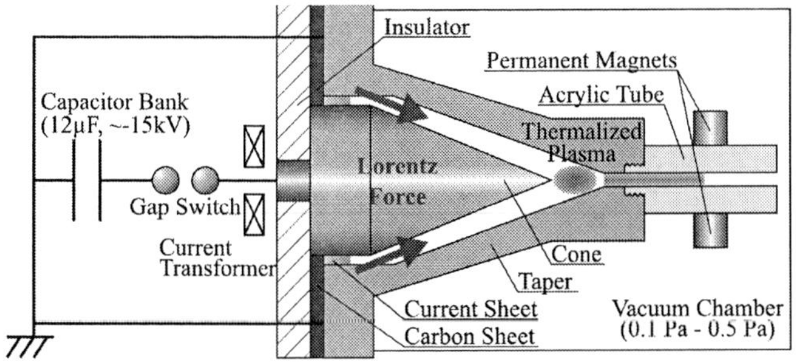

Fig. 1. Schematic image of the tapered cone plasma focus device.

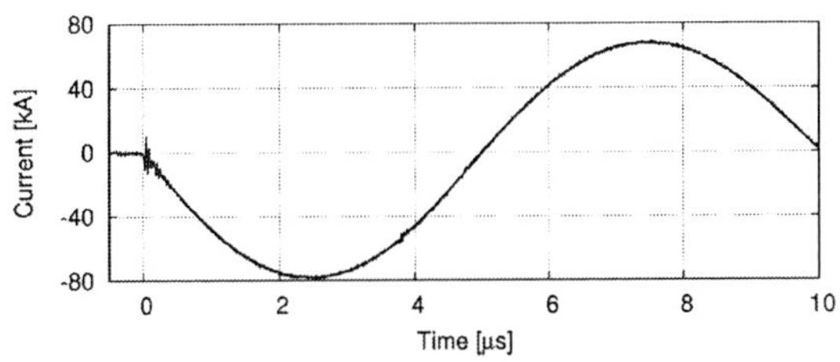

Fig. 2. Typical discharge current waveform.

charged up to -15 kV for driving the plasma focus discharge. Figure 2 shows a typical discharge current waveform measured by a current transformer. The velocity of the plasma flow depends on the discharge current due to the pinched process. The behavior of the plasma flow in the acrylic tube is observed by a streak camera.

In order to apply a magnetic field B_z perpendicular to the plasma flow direction x, permanent magnets (diameter of 5 mm and height of 2.5 mm) are set on the center of the acrylic tube. The peak of the magnetic flux density in the acrylic tube is about 25 mT at 10 mm from the connection point of the acrylic tube and the tapered electrode. The magnetic pressure P_B is defined as follows:

$$P_B = \frac{B^2}{2\mu_0} \tag{1}$$

where μ_0 is the permeability in vacuum. The magnetic pressure is estimated to be P_B = 250 Pa with B = 25 mT. The perpendicular magnetic field affects the plasma flow with balancing each pressure.

III. Results and Discussions

Figure 3 shows a streak image of the plasma flow with the perpendicular magnetic field in the acrylic tube. The plasma flow has decelerated from 28 km/s to 10 km/s at about 15 mm, after passing through the peak of the magnetic field. In addition, accelerated ions flowing ahead of the main plasma flow after through the perpendicular magnetic field have been detected by using an ion collector measurement [19].

To evaluate the relation between the plasma flow and the magnetic field, the magnetic Reynolds number R_M is estimated as follows:

$$R_M = \frac{\mu_0 UL}{\eta} \tag{2}$$

where U (= 28 km/s) is the velocity of the flow, L is the system scale length (= 20 mm), and η is the electrical resistivity. The electrical resistivity η is estimated using the Spitzer resistivity [20] as follows:

$$\eta = \frac{e^2 \sqrt{m_e} \ln \Lambda}{16\pi \varepsilon_0^2 (k_B T_e)^{3/2}} \tag{3}$$

where e is the elementary charge, m_e is the electron mass, ln Λ is the Coulomb logarithm, e_0 is the permittivity in vacuum, k_B is the Boltzmann constant, and T_e is the electron temperature.

The electron temperature T_e has been estimated to be 3 eV from the optical emission spectrum of the plasma flow [19]. The electrical resistivity η and the magnetic Reynolds number R_M are estimated to be $\eta \sim 1.7 \times 10^{-4}$ Ω·m and $R_M \sim 4.2$. These estimates indicate that the plasma flow has decelerated by the magnetic pressure, and the magnetic field has been frozen-in the plasma.

Considering the pressure balance between the plasma flow and the magnetic field, we define the pressure ratio of the plasma pressure to the magnetic pressure β_{eff} as follows:

$$\beta_{\text{eff}} = \frac{P_{\text{flow}}}{P_B} \tag{4}$$

where the plasma pressure P_{flow} is defined as a total pressure as follows:

$$P_{\text{flow}} = \frac{1}{2}\rho_i u^2 + n_i k_B T_i \tag{5}$$

where $\rho_i = m_i n_i$ is the mass density, m_i is the ion mass of helium, n_i is the ion number density, and T_i is the ion temperature. We assume that the ions are fully single-ionized and thermodynamic equilibriium due to the plasma focus. Therefore, the plasma pressure P_{flow} is estimated to be P_{flow} = 310 Pa with u =28 km/s, $n_i = n_0 = 10^{20}$ m^{-3}, and $T_i = T_e$ = 3 eV. The estimated plasma pressure P_{flow} is higher than the magnetic pressure P_B estimated from Eq. (1).

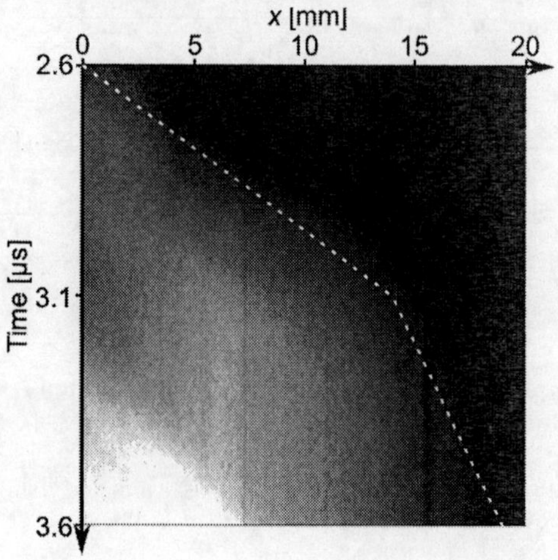

Fig. 3. Streak image of the plasma flow with the magnetic field in the acrylic tube.

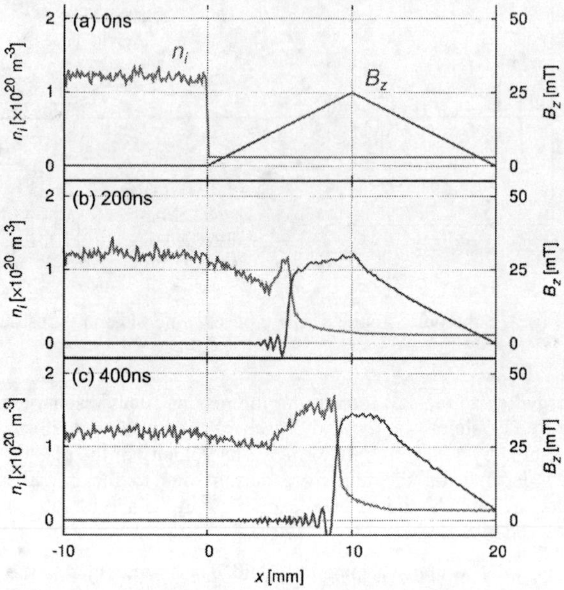

Fig. 4. Time evolution of the ion number density n_i and the perpendicular magnetic field B_z simulated by the hybrid PIC method. Fig. 4(a) is at 0 ns, 4(b) is at 200 ns, and 4(c) is at 400 ns.

IV. Numerical Simulation for Fast Plasma Flow in Perpendicular Magnetic Field

To understand the interaction between the fast plasma flow and the perpendicular magnetic field, an electromagnetic hybrid particle-in-cell (PIC) method is employed [21]. In the hybrid PIC method, ions are treated as particles and electrons are treated as a fluid, which immediately satisfies charge neutrality. Therefore, the hybrid PIC method is able to describe a non-equilibrium distribution of ions.

The numerical simulation was carried out in four-dimensional phase-space (x, v_x, v_y, v_z). In the initial condition of the numerical simulation, the plasma flow generated by the tapered cone plasma focus device is from -10 mm to 0 mm. The thermodynamic equilibrium is assumed due to the plasma focus with the temperature $T_i = T_e = 3$ eV. We consider fully-single ionized helium ions with the density $n_i = n_e = 10^{20}$ m^{-3}. The ions have an average drift velocity of 25 km/s, which means the velocity of the plasma flow. The collision between the ions and ambient neutral particles is neglected. The region from 0 mm to 20 mm represents the acrylic tube. We applied a triangle-shaped magnetic field amplitude, which has the peak of 25 mT at 10 mm, to simulate the experiment. Figure 4 shows the time evolution of the ion number density n_i and the perpendicular magnetic field B_z. The magnetic field is compressed by the plasma flow, and the shape of the magnetic field changes until 400 ns. These results indicate that the magnetic field is frozen-in at the front of the flow. As shown in Fig. 4, the ion density is increased with the compressed magnetic field. We can see that the ion density in front of shock front has a precursor region. In the region, some ions have been accelerated by the induced electric field, and the numerical results has agreed with the experimental results [19].

The plasma flow shown in Fig. 3 has decelerated after through the peak of the magnetic field. It appears that the magnetic pressure due to the compressed and increased magnetic field affected the deceleration of the plasma flow. The time variation of the magnetic field with the propagation of the plasma flow causes the perturbation and the induced electric field. As a result, the acceleration of the ions occured.

V. Conclusion

To understand the interaction between fast plasma and perpendicular magnetic field, the the behavior of the fast plasma flow generated by the tapered cone plasma focus device flowing in the perpendicular magnetic field was experimentally evaluated. The plasma flow was decelerated from 28 km/s to 10 km/s after through the peak of the magnetic field. The magnetic Reynolds number was estimated to be 4.2, and the plasma beta was higher than 1. These estimates indicate that the plasma flow has decelerated by the magnetic pressure, and the plasma flow has strongly affected the magnetic field.

The numerical simulation result indicated the modulation of the perpendicular magnetic field with the propagation of the plasma flow. The magnetic field has been frozen-in the front of the plasma flow, and the perturbation of electromagnetic fields are induced. Some ions in the plasma flow have been accelerated by these induced electromagnetic fields modulated with the plasma flow.

For the experimental evaluation of the modulated magnetic field, we will measure a fluctuation of the magnetic field and its plasma parameters.

Acknowledgment

This work was partly supported by the NIFS Collaboration Research Program (NIFS14KLEH042).

References

[1] W. I. Axford, E. Leer, and J. F. McKenzie, "The structure of cosmic ray shocks", Astron. Astrophys. **111**, 317, 1982.

[2] L. A. Fisk, "On the acceleration of energetic particles in the interplanetary medium", J. Geophys. Res. **81**, 4641, 1976.

[3] M. A. Lee, "The association of energetic particles and shocks in the heliosphere", Rev. Geophys. Space Phys. **21**, 324, 1983.

[4] A. R. Bell, "The acceleration of cosmic rays in shock fronts - I", Mon. Not. R. Astron. Soc. **182**, 147, 1978.

[5] R. D. Blandford, and J. P. Ostriker, "Particle acceleration by astrophysical shocks", Astrophys. J. **221**, L29, 1978.

[6] T. N. Kato, and H. Takabe, "Nonrelativistic collisionless shocks in unmagnetized electron-ion plasmas", Astrophys. J. Lett. **681**, L93, 2008.

[7] R. A. Treumann, "Fundamentals of collisionless shocks for astrophysical application, 1. Non-relativstic shocks", Astron Astrophys. Rev. **17**, 409, 2009.

[8] D. Ryutov, R. P. Drake, J. Kane, E. Liang, B. A. Remington, and W. M. Wood-Vasey, "Similarity criteria for the laboratory simulation of supernova hydrodynamics", The Astro-Phys. J. **518**, 821 1999.

[9] R. P. Drake, "The design of laboratory experiments to produce collisionless shocks of cosmic relevance", Phys. Plasmas **7**, 4690, 2000.

[10] A. Stockem, F. Fiuza, A. Bret, R. Fonseca, and L. O. Silva, "Exploring the nature of collisionless shocks under laboratory conditions", Sci. Rep. **4**, 3934, 2014.

[11] S. Bulanov, T. Z. Esirkepov, M. Kando, J. Koga, K. Kondo, and G. Korn, "On the problems of relativistic laboratory astrophysics and fundamental physics with super powerful lasers", Plasma PHys. Rep. **41**, 1, 2015.

[12] A. Bell, P. Choi, A. Dangor, O. Willi, d. Bassett, aand C. Hooker, "Collisionless shock in a laser-produced ablating plasma", Phys. Rev. A **38**, 1363, 1988.

[13] C. Courtois, R. Grundy, A. Ash, D. Chambers, N. Woolsey, R. Dendy, and K. McClements, "Experiment on collisionless plasma interaction with applications to supernova remnant physics", Phys. Plasmas **11**, 3386, 2004.

[14] K. Kondo, M. Nakajima, T. Kawamura, and K. Horioka, "Compact pulse power device for generation of one-dimensional strong shock waves", Rev. Sci. Instrum. **77**, 036104, 2006.

[15] T. Morita, Y. Sakawa, Y. Kuramitsu, S. Dono, H. Aoki, H. Tanji, T. Kato, Y. Li, Y. Zhang, X. Liu, *et.al.*, "Collisionless shock generation by a high-power laser", J. Phys.: Conf. Ser. **244**, 042010, 2010.

[16] Y. Kuramitsu, Y. Sakawa, T. Morita, C. Gregory, J. Waugh, S. Dono, H. Aoki, H. Tanji, M. Koenig, N. Woolsey, *et.al.*, "Time evolution of collisionless shock in counterstreaming laser-produced plasmas", Phys. Rev. Lett. **106**, 175002, 2011.

[17] K. Adachi, M. Nakajima, T. Kawamura, J. Hasegawa, and K. Horioka, "A Table-top Pulsed Power Device for Counter-streaming Plasma Experiments", IEEJ Trans. Fundam. Mater. **135**, 366, 2015.

[18] T. Sasaki, H. Kinase, T. Takezaki, K. Takahashi, T. Kikuchi, T. Aso, and N. Harada, "Laboratory Scale Experiments for Collisionless Shock Generated by Taper-Cone-Shaped Plasma Focus Device", J. Phys. Soc. Proc, **1**, 015096, 2014.

[19] T. Takezaki, K. Takahashi, T. Sasaki, T. Kikuchi, and N. Harada, "Accelerated ions from pulsed-power-driven fast plasma flow in perpendicular magnetic field", Phys. Plasmas **23**, 062904, 2016.

[20] W. Baumjohann, and R. A. Treumann, *Basic Space Plasma Physics* (World Scientific, 1996), Vol. 57.

[21] D. Winske, L. Yin, N. Omidi, H. Karimabadi, and K. Quest, *Space Plasma Simulation* (Springer, 2003), pp. 136-165.

Influence of Cavity Structure on Microwave Pulse Emitted from Axial Virtual Cathode Oscillator

Tsukasa Nakamura, Shougo Shinokawa, Motohiro Teramae, Fumiya Niwa, and Hiroaki Ito*
Dep. of Electric and Electronic Engineering
Univ. of Toyama, 3190 Gofuku, Toyama 930-8555, Japan
hiroaki@eng.u-toyama.ac.jp

***Abstract*—The vircator is considered to be very attractive due to its conceptual simplicity, output power capability, and frequency tunability. However, the efficiency of converting an electron beam to microwave does not reach sufficient levels. In our study, the high power microwaves generated in an axially extracted vircator have been studied experimentally. The vircator is driven by a Marx generator and pulse forming line. To improve the efficiency, we installed a cavity in the vircator since the narrowing of the output microwave frequency leads to the improvement of the efficiency. The electron beam diode was an axially extracted vircator with the solid type aluminum cathode of 60 mm in diameter and the stainless steel mesh anode. The disc resonator with an extraction window was placed at the distance away from an anode with anode-cathode gap of 8 mm. The electronic field of output microwaves were measured varying the length and the aperture of cavity to analyze the cavity effect. We evaluate the characteristics of output microwave with the resonator by analysis of the time-dependent frequency analysis and using a particle-in-cell simulation.**

Keywords*—High-power microwave, virtual cathode oscillator, vircator, resonance cavity

I. INTRODUCTION

Radiation sources of the high-power microwave are being developed for applications in plasma heating, particle acceleration, high-power radar, and many other industrial and military fields [1,2]. A number of high-power microwave sources have been developed, such as Virtual Cathode Oscillator, Magnetron, Klystron, Gyrotron and so on. The vircator has remarkably attracted attention over other kinds of high-power microwave sources. The vircator has the attribute of high-power capability, wide frequency tuning ability, and conceptual simplicity [3]. It is possible to generate high-power microwaves ranging from a few tens of megawatts up to a few gigawatts at frequency from a few hundreds of MHz to a few tens of GHz. However, the vircator has difficult problems in low efficiency of beam-to-microwave power conversion and wide frequency bandwidth [3]. The many researchers have been making research efforts and experimental studies to improve the microwave conversion efficiency and oscillation frequency control [4-8]. Despite these efforts, the best conversion efficiency is still not higher than 10% and an order of magnitude less than that of other high-power generators.

It has been found that the conversion efficiency depends on the electromagnetic field strengths in the interaction region around the virtual cathode. This field plays important roles on both modulating the electron beam and extracting energy from the electron beam. In order to improve the conversion efficiency and narrow the output frequency, the vircator with a resonance cavity has been proposed and proven successful by several research groups [9, 10]. When the resonant mode has an electric field component along the direction of the virtual cathode oscillation, the beam-field interaction can be enhanced by feeding back the microwave field to the virtual cathode and the field can extract higher energy from the virtual cathode oscillation at the resonant frequency. Thus, the choice of the mode is an important consideration in the design of the drift space. The experiments were performed using an axial vircator with the resonance cavity, which was installed by inserting a movable disc plate with an extraction window in the drift space. This results in an axial electron beam and a virtual cathode oscillating along the vircator axis. The axial position of this plate was varied during the experimental series. This paper reports the experimental results for the effect of length and aperture size of the resonance cavity on the field strength and frequency of the output microwave. In addition, the obtained output microwave was evaluated by the time-dependent frequency analysis (TFA) [11, 12] and 2.5-dimensional particle-in-cell (PIC) simulation code " MAGIC".

II. PRINCIPLE OF VIRCATOR

The electron beam diode consists of the cathode and the anode. When a high pulsed negative voltage is applied to a cathode, electrons are emitted from the surface of the cathode. The electron beam passes through the anode, which is usually a thin foil or a mesh, and is injected into drift space on the other side of the anode. When the injected electron beam current exceeds the space charge limited current in the region behind the anode, a virtual cathode, which is an unstable region of negative potential, is formed at the same distance as AK gap. Most of the electrons in the beam lose their kinetic energy and are then reflected back toward the anode. The mechanism for microwave generation in the vircator can generally be explained by two dynamical mechanisms [2, 13]. The first is electron reflection in the potential well formed between the real and virtual cathodes. The radiation frequency due to electron reflexing depends on the AK gap and the beam energy. The frequency is approximately given by

$$f_e = \frac{v_e}{4d_{AK}} \approx \frac{c\sqrt{\gamma^2 - 1}}{4d_{AK}\gamma}$$

where v_e is the velocity of the electrons, d_{AK} is the distance between the anode and the cathode, and c is the velocity of light. The relativistic factor γ is given by

$$f_e = \frac{v_e}{4d_{AK}} \approx \frac{c\sqrt{\gamma^2 - 1}}{4d_{AK}\gamma}$$

The second is an oscillation of the virtual cathode in time and space. The oscillation frequency is given by

$$f_{vc} = \frac{5}{4\pi}\omega_p = \frac{5}{4\pi}\sqrt{\frac{e^2 n_b}{\varepsilon_0 m_e \gamma}}^{-1/2}$$

$$= \frac{5}{4\pi}\sqrt{\frac{e^2}{\gamma\varepsilon_0 m_e}}\sqrt{\frac{1}{e\beta c}}\left(\frac{4\varepsilon_0}{9}\left(\frac{2e}{m_e}\right)^{1/2}\frac{V_0^{3/2}}{d_{AK}^2}\right)^{1/2}$$

where ω_p is the electron plasma frequency, n_b is electron beam density derived by the space charge limited current, *e* and me are the electron charge and mass, respectively, ε_0 is the vacuum permittivity, V_0 is the applied voltage across the AK gap, and β=v/c. These two oscillation frequencies are fundamentally determined by the geometric structure, the applied voltage and the electron beam parameter.

III. Experimental Arrangement

The pulse power system for generation of the high-power microwave consists of a high-voltage generator, a single pulse forming line (PFL), a gas SF_6 spark gap and a pulse transmission line. The high-voltage generator is an eight-stage bipolar Marx generator with the maximum output of 800 kV and the stored energy of 5 kJ. The PFL is filled with the deionized water as a dielectric. The designed parameters of the PFL are characteristic impedance of 3 Ω, and electrical length of 50 ns. In this experiment, the pulsed power system was operated in negative-output mode and the charging voltage of Marx generator was set to 24 kV and to operate in a single-shot mode.

Fig. 1 shows the schematic structure of the axial vircator and the microwave measurement setup. The high-voltage pulse generated from the pulsed power system is applied to the cathode of the electron beam diode. A vacuum explosive electron emission diode was used to generate an intense relativistic electron beam. The diode consists of an aluminum planar disk cathode of 60 mm in diameter and a stainless-steel meshed anode with a transparency of ~65%. As shown in Fig. 1, both anode and cathode are plane surfaces that are at right angles to the vircator symmetry axis. To improve the microwave efficiency and control the output frequency, the resonance cavity was installed by placing a movable disc plate with a center hole at the distance (*L*) away from an anode. In this resonator configuration, it is expected that the dominant mode might be the transverse magnetic (TM) mode, which have an axial electric field that strongly couples to the oscillating virtual cathode's electric field. The output microwave was measured varying both the distance (*L*) and the diameter (*a*) of extraction window to analyze the resonator effect. From previous experiments where the AK gap was varied from 4 mm to 12 mm in 2 mm steps, the AK gap d_{AK} was set to be 8 mm regarding small shot-to-shot variations and good durability of mesh anode. The diode region and output waveguide are maintained at a base pressure of 5.5×10^{-3} Pa.

A resistive $CuSO_4$ voltage divider was used to measure the output voltage (V_{PFL}) of the pulsed power system and placed at the edge of PFL. The diode voltage (V_d) applied to the cathode is calculated by the inductive correction, since the inductance of the transmission line from the resistive divider to the cathode is ~100 nH. The diode current (I_d) and the beam current (I_b) passed through the anode were monitored by Rogowski coil at the upstream and downstream of the anode. The output microwave from the vircator traveled along a circular waveguide of 100 mm in diameter and 375 mm in length and is emitted into free space through an acrylic window. All microwave measurements were carried out with the receiving antenna placed at a distance of 0.7 m away from the vircator window. The temporal waveform of the output microwave was obtained by using a crystal diode and recorded by a digital oscilloscope with a bandwidth of 200 MHz and a sampling rate of 2 GS/s, together with the beam parameters (voltage and current). In order to carry out Fourier transforms and TFA, the signal from the antenna was also recorded by a high-speed digital oscilloscope with a sampling rate of 40 GS/s through a coaxial RF cable.

IV. Experimental Results

Fig. 2 shows the typical waveforms of diode voltage V_d, diode current I_d and electric field of microwave with no resonance cavity. The diode voltage increases up to 200 kV within 100 ns and the diode current rises with the diode voltage and has a peak of 14 kA. It can be seen from the waveform of the microwave field that the microwave is emitted at the rise time of V_d. Fig. 3 shows the Fast Fourier transform of the microwave waveform shown in Fig. 2. As shown in Fig. 3, the bandwidth of microwave frequency is

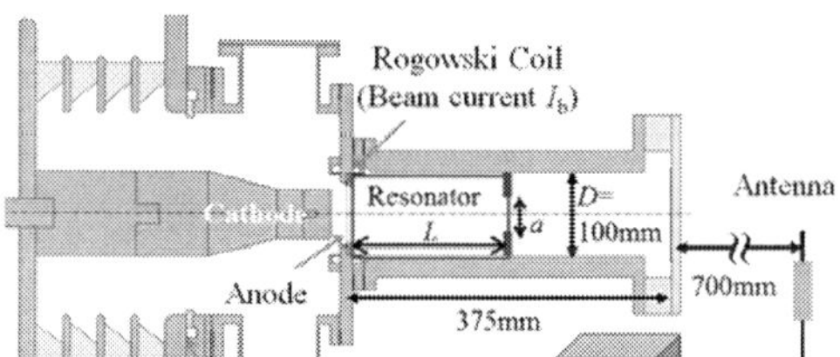

Fig. 1 Schematic structure of axial vircator and microwave measurement setup

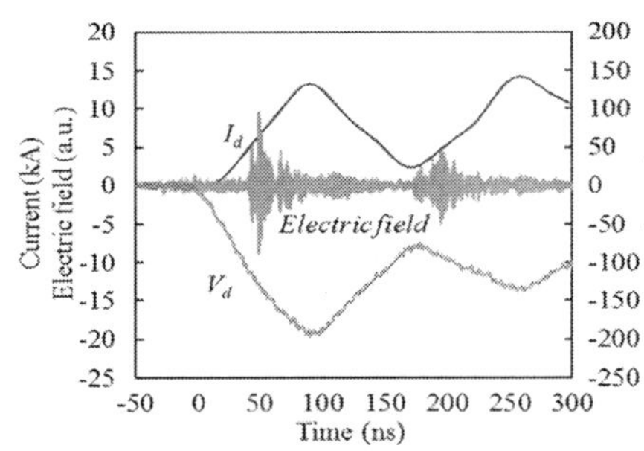

Fig. 2 Typical waveforms of diode voltage V_d, diode current I_d and electric field of microwave with no cavity

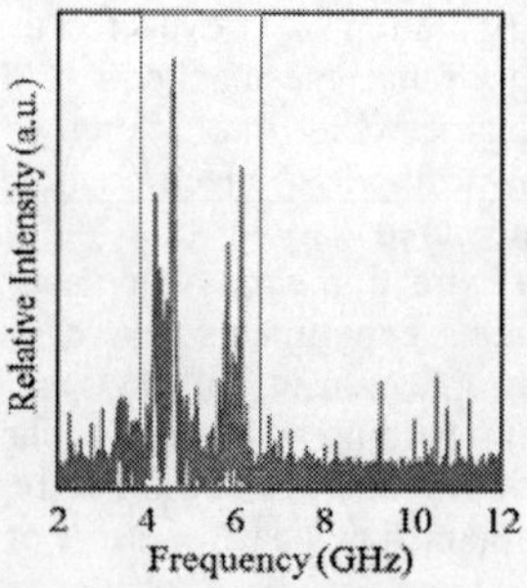

Fig. 3 FFT of microwave field shown in Fig.2

wide and the main frequency components can be identified as ~4.4 and ~5.9 GHz. This result indicates that two mechanisms of microwave radiation exist together, namely, the electron reflex between the real and virtual cathodes as well as the virtual cathode oscillation. From above-mentioned equations, the low frequency is related to the virtual cathode oscillation, while the high frequency corresponds to the electron reflex process.

Fig. 4 shows the microwave output signal recorded by the high-speed digital oscilloscope for three diameters (a=4, 6, 8 cm) of the cavity extract window at various cavity lengths. The waveforms of the diode voltage applied to the cathode and the microwave field without the cavity are shown as the reference. It can be clearly seen from Fig. 4 that the amplitude of the microwave field with the cavity is higher than that without the cavity. The experimental results show that the emission time of microwave is more stable at the extract window diameter a=6 cm than at other two diameters. Fig. 5 shows the Fast Fourier transform of each microwave E-field shown in Fig. 4. It turned out from Fig.6 that the frequency spectrum has narrow bandwidth and sharp peak by installing the resonance cavity. The mode structure of the cavity can be evaluated by varying the length, since the cavity diameter is fixed to D=10 cm. Considering that the microwave frequency is in range of 4-5 GHz and $(D/L)^2$ is varied from 0.1 to 1, it is found that the frequency corresponding to TM_{01p} or TM_{11p} modes is tuned to resonate with the broadband virtual cathode oscillation, while nonresonant frequencies are suppressed and the bandwidth of the emitted microwave decreases.

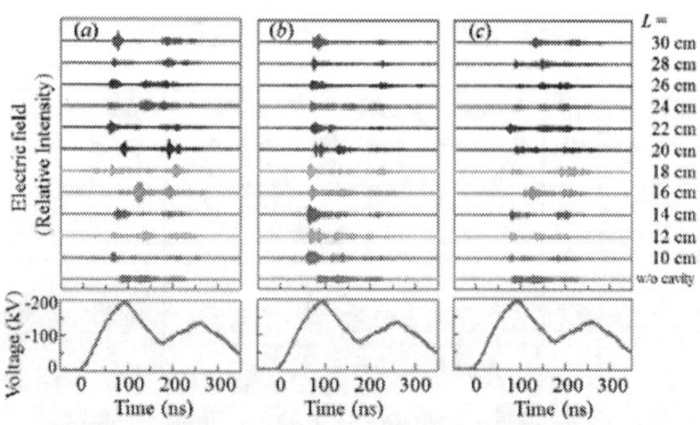

Fig. 4 Waveform of microwave electric field as a function of cavity length at extract window diameters (a) a=4 cm, (b) a=6 cm and (c) a=8 cm

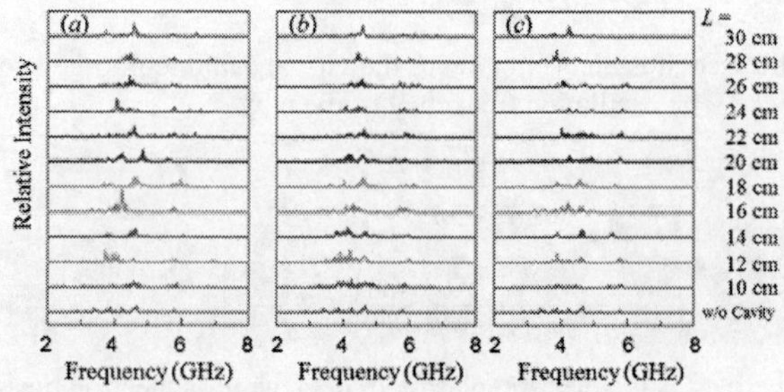

Fig. 5 FFT of the microwave E-field of Fig.5 as a function of cavity length at extract window diameters (a) a=4 cm, (b) a=6 cm and (c) a=8 cm

To investigate the effect of the cavity, the evaluation for microwave power is carried out. However, the absolute microwave power has not been obtained yet, since the calibration of our microwave power measurement system is under way. We evaluated the power spectral density (PSD) obtained by using FFT results. The PSD is calculated by the following equation,

$$P(f) = \frac{1}{N}\sum_{k=0}^{N-1}\frac{|X_k(f)|^2}{2\pi}\Delta f$$

where $X_k(f)$ is FFT value, N is the number of data, and Δf is the sampling frequency. Fig. 6 shows the PSD as function of the cavity length for three diameters (a=4, 6, 8 cm) of the cavity extract window. Each data point and error bar in Fig. 6 are an average of 4 shots and shot-to-shot variations, respectively. In addition, the PSD without the cavity is 0.63 and is shown in Fig. 6 as the reference. Although the PSD obtained at each cavity length changes shot by shot, it is evident that the PSD at a=4 and a=6 cm is larger than that without the cavity. At the extract window diameter of a=6 cm and the cavity length of 12 cm, The power spectral density is 4.7 times larger than that without the cavity. On the other hand, the disc plate with a larger extract window does not work as the resonance cavity.

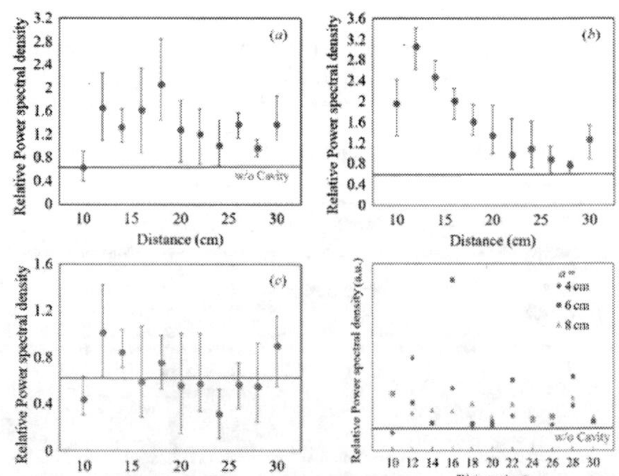

Fig. 6 PSD as a function of cavity length at extract window diameters (a) a=4 cm, (b) a=6 cm and (c) a=8 cm and (d) simulation result obtained by MAGIC code

Fig. 6(d) shows the simulation result obtained by using MAGIC code, which is in almost good agreement with the experimental results.

The oscillation frequency in the vircator depends on electron-beam parameters such as voltage, current density, and the geometric structure of the diode. As a result, the virtual cathode oscillations vary with time during the output pulse because of the diode gap closure due to cathode plasma expansion. According to previous experimental results, at the AK gap distance of d_{AK}=4 mm, the microwave frequency increases from 6 to 12 GHz during the period of 0-50 ns and the emission is broadband. The time frequency analysis was carried out for evaluation of the frequency stabilization of the virtual cathode oscillation. Fig. 7 shows the time evolution of the microwave frequency spectrum with no cavity and with the cavity, where the AK gap is d_{AK}=12 mm and the cavity is set to the length of L=10 cm and the extract window of a=6cm. In Fig. 7, a time-window width of 12.8 ns (512 data points) was shifted by the step of 0.25 ns and the horizontal and vertical axes represent the center of the time window and the spectrum intensity obtained within each window by the blackness variation, respectively. It can be seen from Fig. 7 that in the case of no cavity, the output frequency increases from 4.1 to 4.9 GHz with time and the bandwidth of the emission is broad, while in the case of the vircator with the cavity, the constant frequency of ~4.3 GHz was observed over the microwave emission and the spectrum became sharp. This result shows that the frequency stabilization of the virtual cathode oscillator is achieved by installing the resonant cavity.

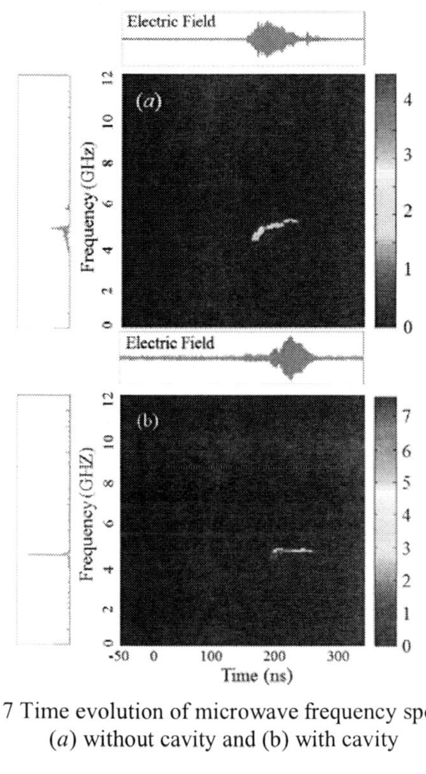

Fig. 7 Time evolution of microwave frequency spectrum (a) without cavity and (b) with cavity

V. SUMMARY

We have investigated experimentally and numerically the influence of cavity structure on output characteristics of microwave emitted from axial virtual cathode oscillator. By installing the cavity, the microwaves of a certain frequency in the resonance condition are enhanced whereas nonresonant frequencies are suppressed and the bandwidth of the emitted microwave becomes narrow. The power spectral density is 4.7 times larger than that without the cavity, which is in good agreement with results obtained by the simulation using MAGIC code. The dominant emission mode from the virtual cathode oscillator is found to be mainly TM mode by the resonance condition of cavity and simulation results. We plan to measure the microwave output power and the conversion efficiency and to find a practical way of improving its frequency characteristics and microwave efficiency.

References

[1] R. J. Barker and E. Schamiloglu, Eds., "High- Power Microwave Sources and Technologies", New York: Wiley-IEEE Press (2001)

[2] J. Benford, J. Swegle, and E. Schamiloglu, "High Power Microwaves" 2nd ed., New York: Taylor & Francis (2007)

[3] D. J. Sullivan, J. E. Walsh, and E. A. Coutsias, "Virtual cathode oscillator (vircator) theory", in High Power Microwave Sources, V. Granastein and I. Alexeff, Eds. Norwood, MA: Artech House, p. 441 (1987)

[4] B. V. Alyokhin, A. E. Dubinov, V. D. Selemir, O. A. Shamro, K. V. Shibalko, N. V. Stepanov, and V. E. Vatrunin, "Theoretical and experimental studies of virtual cathode microwave devices", IEEE Trans. Plasma Sci., vol.22, pp.945–959 (1994)

[5] E. H. Choi, M. C. Choi, Y. Jung, M. W. Chong, J. J. Ko, Y. Seo, G. Cho, and H. S. Uhm, "High-power microwave generation from an axially extracted virtual cathode oscillator", IEEE Trans. Plasma Sci., vol.28, pp.2128–2134 (2000)

[6] W. Jeon, J. E. Lim, M. W. Moon, K. B. Jung, W. B. Park, H. M. Shin, Y. Seo, and E. H. Choi, "Output characteristics of the high-power microwave generated from a coaxial vircator with a bar reflector in a drift region", IEEE Trans. Plasma Sci., vol.34, pp.937-944 (2006)

[7] D. Biswas and R. Kumar, "Efficiency Enhancement of the Axial VIRCATOR", IEEE Trans. Plasma Sci., vol.35, pp.369-378 (2007)

[8] R. Menon, A. Roy, S. K. Singh, S. Mitra, V. Sharma, S. Kumar, A. Sharma, K. V. Nagesh, K. C. Mittal, and D. P. Chakravarthy, "High power microwave generation from coaxial virtual cathode oscillator using graphite and velvet cathodes", J. Appl. Phys. vol.107, p.093301 (2010)

[9] J. Benford, D. Price, H. Sze, and D. Bromley, "Interaction of a vircator microwave generator with an enclosing resonant cavity," J. Appl. Phys., vol. 61, pp. 2098–2100 (1987)

[10] W. Jiang, N. Shimada, S. Prasad, and K. Yatsui, "Experimental and simulation studies of new configuration of virtual cathode oscillator," IEEE Trans. Plasma Sci., vol. 32, pp. 54–59 (2004)

[11] C. W. Peters, R. L. Jaynes, Y. Y. Lau, R. M. Gilgenbach, W. J. Williams, J. M. Hochman, W. E. Cohen, J. I. Rintamaki, D. E. Vollers, and T. A. Spencer, "Time-frequency analysis of modulation of high-power microwaves by electron-beam voltage fluctuations", Phys. Rev. E, vol.58, pp.6880-6883 (1998)

[12] W. Jiang, "Time–Frequency Analysis of Virtual-Cathode Oscillator", IEEE Trans. Plasma Sci., vol.38, pp.1325-1328 (2010).

[13] H. Sze, J. Benford, W. Woo, and B. Harteneck, "Dynamics of a virtual cathode oscillator driven by a pinched diode", Phys. Fluids, vol.29, pp.3873-3880 (1986)

A high-power high-efficiency Ku-band transit-time oscillator with low guiding magnetic field

Junpu Ling, Jiande Zhang,Juntao He, Chen Geng
College of Optoelectronic Science and Engineering
National University of Defense Technology
Changsha, China
E-mail address: zhangjiande@nudt.edu.cn

***Abstract*—A high-power high-efficiency Ku-band transit-time oscillator (TTO) with low guiding magnetic field is proposed. The physical mechanism to increase the conversion efficiency and power capacity is investigated by theory and simulation. It is shown that by introducing three-cavity extractor instead of the dual-cavity one, the novel device has higher power capacity. Meanwhile, the three-cavity extractor is more favorable of increasing the beam-wave conversion efficiency of the device. In addition, analysis indicates that the distance-tunable reflector employed in the device plays important roles in enhancement of conversion efficiency. Particle-in-cell simulations show that with the three-cavity extractor and distance-tunable reflector, this proposed device generates a 12.4GHz, 2.2 GW microwave, while the diode voltage is 570 kV, beam current is 10.5 kA and the guiding magnetic field is 0.7 T. The corresponding power efficiency reaches 37%.**

Keywords—Ku-band; transit-time oscillator; high power capacity; high conversion efficiency; low guiding magnetic field

I. INTRODUCTION

In our previous research, we proposed a coaxial Ku-band foilless TTO with the non-uniform buncher. Owing to the coaxial inner conductor, relatively few interaction-cavities, non-uniform buncher and a quasi body wave adopted as the operating mode, the device has the virtues of high power capacity, large space-charge limiting current, monochromatic operation mode and low guiding magnetic field [1-4]. In the experiments, with the 500 kV diode voltage, 10 kA diode current, and 0.7 T guiding magnetic field, a microwave pulse with power of 1GW, frequency of 14.3 GHz and efficiency of 20% was generated [5].

Generation of the HPM with higher power can further expand the application prospects of the device and has always been the research hotspot in the HPM domain [6]. Hence, increasing the output power of the Ku-band coaxial TTO with low guiding magnetic field to multi-gigawatt level is greatly meaningful. Besides, under the condition that outputting the same multi-gigawatt power, the device with high conversion efficiency needs less input energy supplied by the intense electron accelerator than that with lower efficiency [7]. Thus, the investigation on the high-power high-efficiency Ku-band TTO with low guiding magnetic field becomes more attractive to the researchers. In this paper, with a three-cavity extractor and a distance-tunable reflector, a novel high-efficiency Ku-band TTO needing low guiding magnetic field is proposed to generate multi-gigawatt power microwave.

II. DESCRIPTION OF THE DEVICE

The schematic structure of the proposed Ku-band TTO is shown in Fig.1. Compared with the previous Ku-band TTO, there are some improvements in this novel one. Above all, a three-cavity extractor instead of the dual-cavity one is employed into the novel Ku-band TTO. Besides, a distance-tunable coaxial reflector is introduced into the device. Details of the description of each improved part of this proposed device will be given later.

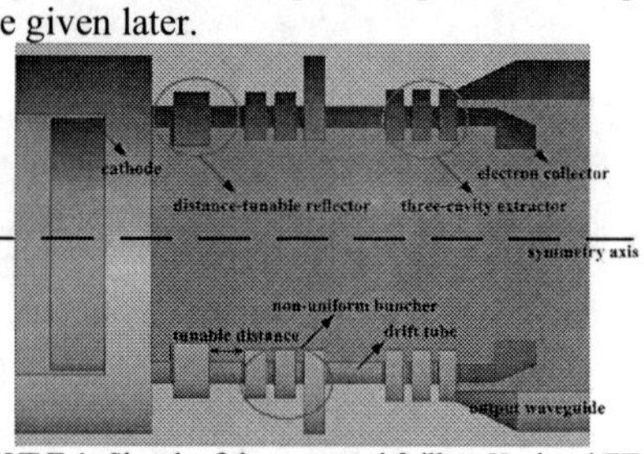

FIGURE 1. Sketch of the proposed foilless Ku-band TTO.

A. Peculiarity of the three-cavity extractor

For our previous device, we found that the dual-cavity extractor could not sustain such high power capacity, when the output power was increased to higher level that above 2 GW. Hence, to obtain Ku-band multi-gigawatt power, the extractor needs to be improved. Due to this, we employ a three-cavity extractor in this paper, whose operation mode is the π mode of the TM_{01} mode instead of the dual-cavity one with same operation mode. Compared with the dual-cavity one, this proposed extractor with three resonant cavities has a larger volume of accommodating electromagnetic field. With the same storage, the surface electric field in the three-cavity extractor should be lower. To verify this conjecture, we conduct a comparative investigation by the 2.5D relativistic fully electromagnetic code CHIPIC [8], which is not only used for electromagnetic field analysis software but also for PIC simulation. The computation results are show in Fig.2. As shown in the figure, under the same energy storage, operation mode and resonant frequency, the maximum surface electric

This work was supported by the National Natural Science Foundation of China under Grant No. 61171021.

field in the three-cavity structure is remarkably lower than that in the dual-cavity structure.

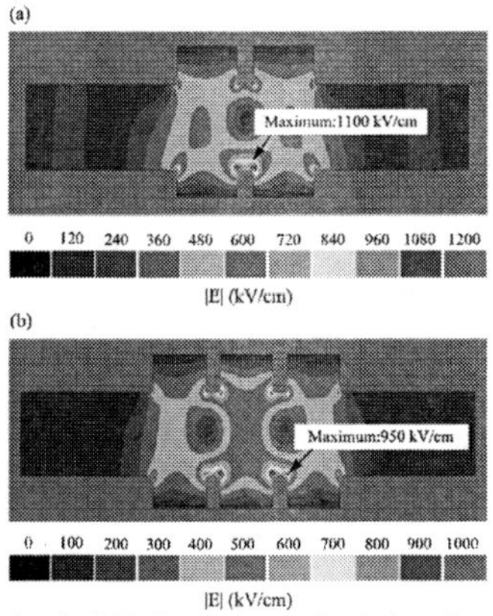

FIGURE 2. Electric field distribution in the dual-cavity structure (a) and the three-cavity structure (b).

Besides of its virtue of high power capacity, this three-cavity extractor is more favorable of increasing the beam-wave conversion efficiency of the device. To verify this function, firstly, the theory analysis on the physical mechanism is carried out.

Based on the small-signal theory and equivalent circuit method [9], we can obtain the factors affecting the output microwave power in the extractor. The relationship between them can be expressed as

$$P_{out} = \frac{1}{2} M^2 \hat{i}_1(0)^2 \frac{(R/Q)Q_{ext}}{Y_0} \frac{1}{\left[1+\left(\frac{2Q_{ext}}{Y_0 f_0}\right)^2 \Delta f^2\right]}, \quad (1)$$

Where M is the beam coupling coefficient, $\hat{i}_1(0)$ is the amplitude of the fundamental harmonic current in the initial point of the electron beam traveling into the extractor, R/Q is the characteristic impedance of the extractor, Q_{ext} is the externally loaded quality factor, Y_0 is the characteristic admittance of the coaxial output waveguide, f_0 is the resonant frequency of the extractor and $\Delta f = f - f_0$ is the difference between the frequency of the harmonic current and f_0.

From the Eq. (1), we can find that the smaller Δf is , the higher power we get. In the TTO, the frequency of the fundamental harmonic current usually is determined by the resonant frequency of the buncher. Thus, to achieve high conversion efficiency, we should adjust the resonant frequency of extractor as close as possible to that of the buncher during the designing process. While Δf is small enough, Δf^2 will approach to zero. As an approximation, Eq. (1) can be given by

$$P_{out} = \frac{1}{2} M^2 \hat{i}_1(0)^2 \frac{(R/Q)Q_{ext}}{Y_0}, \quad (2)$$

Equation (2) indicates that for the same fundamental harmonic current and coaxial output waveguide, the extractor with the large value of M^2 , R/Q and Q_{ext} will have high conversion efficiency.

Based on the calculation equation of the R/Q [10], we obtain the values of the characteristic impedance for both the three-cavity extractor and the dual-cavity one, which are 5.76Ω and 3.95Ω, respectively. In addition, their M^2 and Q_{ext} are also calculated by the numerical methods, respectively. From Fig. 3, we can find that the three-cavity extractor has higher M^2 in the operation region. As shown in Fig.4, the Q_{ext} of the three-cavity extractor is also larger than that of the dual-cavity one. To sum up, our proposed three-cavity extractor has larger value of M^2, R/Q and Q_{ext}, so it is more advantageous for improving the conversion efficiency of the device.

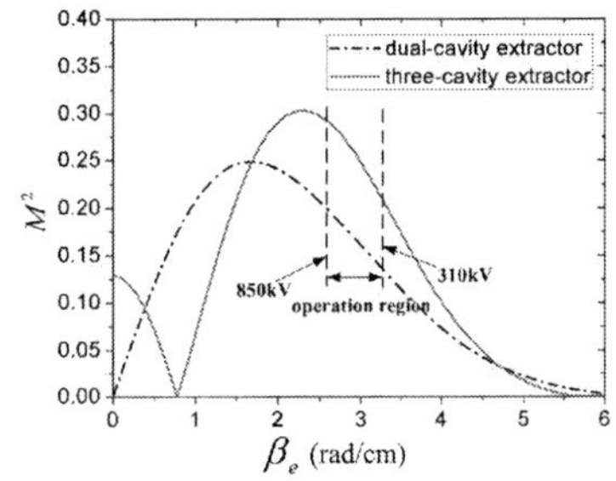

FIGURE 3. M^2 versus the electron phase constant

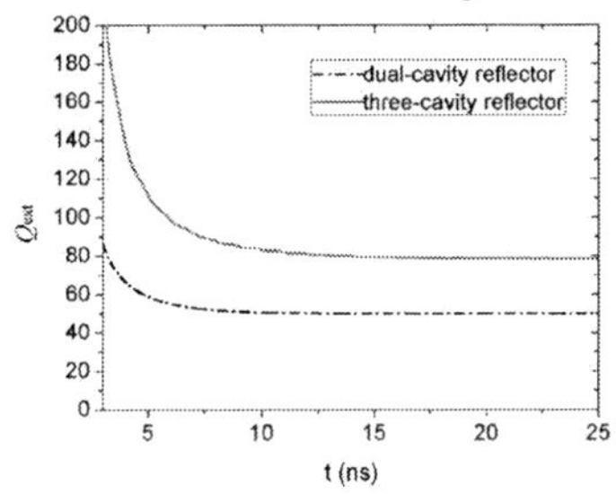

FIGURE 4 Q_{ext} versus time for the dual-cavity extractor and three-cavity extractot

B. Function of the distance-tunable reflector

Considering an annular electron beam with the initial velocity modulation by the voltage and density modulation traveling in the coaxial drift tube, we assume the beam voltage and current as [11]

$$U = U_0\left[1-\varepsilon_1 \cos(2\pi ft)\right], \quad I = I_0\left[1-\varepsilon_2 \cos(2\pi ft+\varphi)\right] \quad (3)$$

where U_0 is the dc beam voltage, I_0 is the dc beam current,

ε_1 and ε_2 are the initial velocity modulation and density modulation coefficients, f is the modulation frequency and φ is the initial phase difference between the velocity modulation and density modulation. As the beam radius, inner tube radius and outer tube radius are 5.6 cm, 5.2cm and 6.0cm, and the parameters U_0, I_0 and f are 570 kV, 10.5 kA, and 12.4 GHz, respectively, the harmonic current distributions for different cases are shown in Fig. 5. As shown in this figure, for the beam with both of the above initial modulations, the amplitude and position of the harmonic current peak varies with the initial phase difference. Its amplitude of the harmonic current peak is larger that that with only initial velocity modulation or density modulation, by optimizing the initial phase difference between them. For our proposed TTO, the main velocity modulation of the beam occurs in the buncher. While the electron beam pre-modulated by the reflector travels into the buncher [1], it will have some density modulation. Consequently, when the beam enters into the coaxial drift tube, it simultaneously has the two initial modulations. Since, the initial phase of the density modulation is mainly determined by the distance between the reflector and the buncher, by tuning the distance between them, we can get the optimal harmonic current distribution for the beam-wave interaction. According to Eq.2, improving the amplitude of the harmonic current peak in the extractor is helpful for enhancing the output microwave power. In brief, by introducing the distance-tunable reflector and tuning its distance between it and the buncher, the conversion efficiency of the device can be optimized to the maximum.

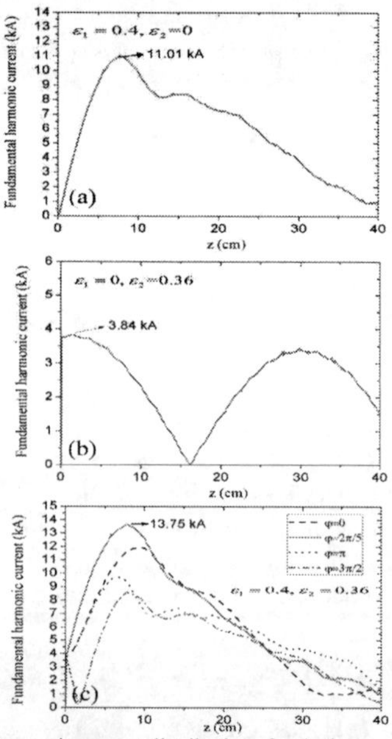

FIGURE 5 Harmonic current distributions for the beam with only velocity modulation (a), for the beam with only density modulation (b) and for the beam with both the velocity modulation and density modulation (c).

III. Simulation Results

The novel device is simulated with the PIC code CHIPIC. Under the diode voltage 570 kV, current 10.5 kA and external guiding magnetic field 0.7 T, the main simulation results are shown as follows.

As shown in Fig.6, the average output power is 2.2GW and the microwave frequency equals to 12.4 GHz, which is close to the cold resonant frequency of the buncher (12.37 GHz). The power conversion efficiency reaches 37%.

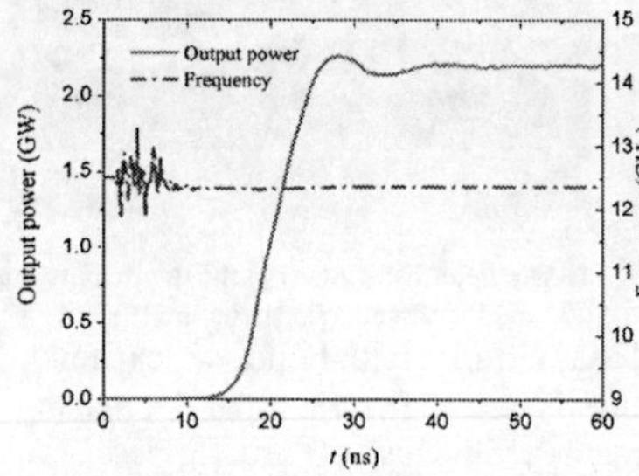

FIGURE 6 Average output power and the microwave frequency versus time

Figure 7 shows the fundamental harmonic current distributions and the conversion efficiencies of the device for three different distances. As shown in the figure, the amplitude and the position of the fundamental harmonic current peak vary with the different distance. The device has the highest efficiency, while the distance equals 1.9 cm. By observing its harmonic current distribution, we find the amplitude of the harmonic current peak has maximum, and its harmonic current peak appears just right at the beginning of the extractor. The peak of this harmonic current is 13.77 kA, and the corresponding modulation coefficient is 131%. Such high fundamental harmonic current is helpful for high-efficiency beam-wave interaction. Yet, for the device with other distance, the amplitude of the harmonic current peak is lower, and the current peak position dose not appear at the beginning of the extractor. The above simulation results illustrate by introducing this reflector and optimizing the distance between it and the buncher, the device can obtain a favorable harmonic current distribution to meet the requirement of high-efficiency beam-wave interaction.

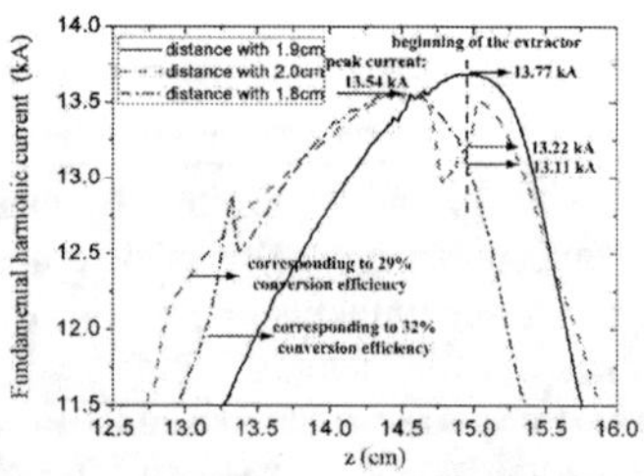

FIGURE 7 Fundamental harmonic current and the conversion efficiency versus the distance between the buncher and the reflector

Employing the three-cavity extractor is the main feature of our novel device. The electron power distributions for the device with three-cavity extractor and the one with the dual-cavity one is shown in Fig.8. It can be seen that the dominating energy loss of the beam electrons happens in the region of the extractor. Compared with the device with dual-cavity extractor, the IREB loses larger energy in the one with three-cavity extractor. This verifies the three-cavity extractor has higher conversion efficiency. Figure 9 displays the electric field distribution in this improved TTO with the three-cavity extractor, the highest surface electric field in the extractor is 980 kV/cm. This value is lower than the experiential breakdown threshold (1 MV/cm) in vacuum. Nevertheless, for the device with the dual-cavity extractor, when it output 2.2 GW power, the highest surface electric field in it is over 1.15 MV/cm. Hence, by comparing the above simulation results, we can get the conclusion that the three-cavity extractor has higher power capacity.

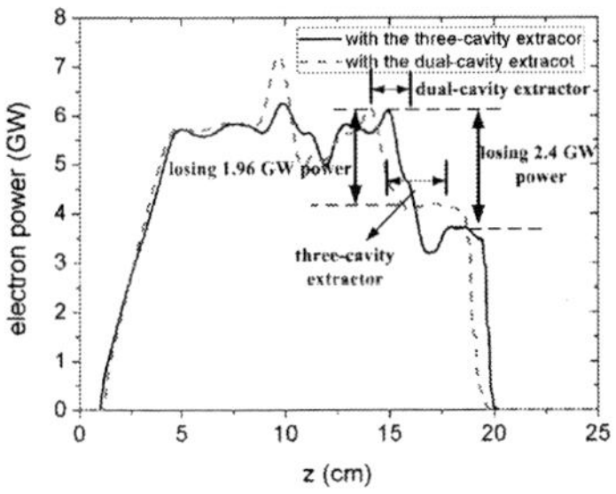

FIGURE 8 Electron power distributions for the two kinds extractors

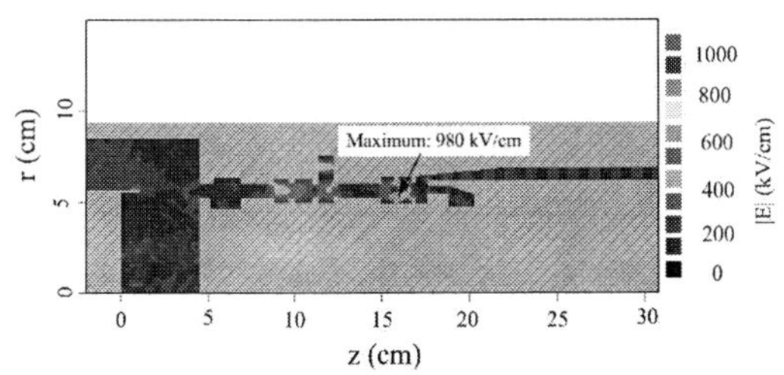

FIGURE 9 Electric field distribution in the device

IV. CONCLUSION

In this paper, a novel foilless Ku-band TTO with the three-cavity extractor and distance-tunable reflector is proposed to generate multi-gigawatt level HPM. The device is featured by its high output power, high conversion efficiency, and low guiding magnetic field. In order to increase the power capacity and conversion efficiency, a three-cavity extractor is adopted instead of the previous dual-cavity one. Owing to the property of the distance-tunable reflector and the transmission electron beam with the modulation, by tuning the distance between the reflector and buncher, we can make the device have the highest efficiency. PIC simulations show that the device can produce 2.2 GW HPM at frequency of 12.4 GHz when the diode voltage is 570 kV, beam current 10.5 kA and guiding magnetic field 0.7 T corresponding to an efficiency of 37%.

Reference

[1] J. P. Ling, J. T. He, J. D. Zhang, T. Jiang, and L. L. Song, "A Ku-band coaxial relativistic transit-time oscillator with low guiding magnetic field," *Laser Part. Beams.*, vol. 40, pp. 295-303, March 2014.

[2] J. P. Ling, J. D. Zhang, J. T. He, L. Wang, and B. F. Deng, "Focusing electrode and coaxial reflector used for reducing the guiding magnetic field of the Ku-band foilless transit-time oscillator," *Rev. Sci. Instrum.*, vol. 85, pp. 08470201-08470207, August 2014.

[3] J. P. Ling, J. T. He, J. D. Zhang, T. Jiang, and Y. Hu, "Improved foilless Ku-band transit-time oscillator for generating gigawatt level microwave with low guiding magnetic filed," *Phys. Plasmas.*, vol. 21, pp. 0931071-0931079, September 2014.

[4] J. P. Ling, J. T. He, J. D. Zhang, T. Jiang, and L. Wang, "Suppression of the asymmetric competition mode in the relativistic Ku-band coaxial transit-time oscillator," *Phys. Plasmas.*, vol. 21, pp. 1031081-1031086, October 2014.

[5] J. P. Ling, J. D. Zhang, J. T. He, T. Jiang, and L.L. Song, "Gigawatt-class microwave generation from a novel Ku-band coaxial transit-time oscillator with low guiding magnetic field," *Phys. Plasmas.*, vol. 23, pp. 1031031-1031037, October 2016.

[6] Y. B. Cao, J. T. He, J. D. Zhang, J. Zhang, and Z. X. Jin, "An oversized X-band transit radiation oscillator," *Appl. Phys. Lett.*, vol. 101, pp. 1735041-1735043, October 2012.

[7] X. J. Ge, H. H. Zhong, J. Zhang, and B. L. Qian, "A high efficient relativistic backward wave oscillator with coaxial nonuniform slow-wave structure and depth-tunable extractor," *Phys. Plasmas.*, vol. 20, pp. 0231051-0231054, February 2013.

[8] J. Zhou, D. G. Liu, C. Liao, and Z. H. Li, "CHIPIC: An efficient code for electromagnetic PIC modeling and simulation," *IEEE Trans. Plasma Sci.*, vol. 37, pp. 2002-2011, October 2009.

[9] Z. H. Zhang, J. Zhang, Z. M. Qi, and J. Zhu, "Matching conditions of the RKA input cavity based on the cavity absorbing property under intense beam loading," *IEEE Trans. Plasma Sci.*, vol. 40, pp. 3121-3126, August 2012.

[10] Z. T. Zhang, The measurement for high frequency system of microwave. Beijing, China: Defence Industry Press., 1982, pp. 135-136.

[11] R. Z. Xiao, C. H. Chen, and X. W. Zhang, "Improved fundamental harmonic current distribution in a klystron-like relativistic backward wave oscillator by two pre-modulation cavities," *Appl. Phys. Lett.*, vol. 102, pp. 1335041-1335044, April 2013.

Progress on the Design of a Stabilized Liner Compressor for Low-Cost Controlled Fusion at Megagauss Field-Levels

P.J. Turchi, S.D. Frese, M.H. Frese
NumerEx
Santa Fe and Corrales, New Mexico, USA

***Abstract* - The notion of employing very high magnetic fields for fusion has been extended to so-called Magnetized-Target Fusion (MTF), which may comprise both magnetic and inertial confinement fusion schemes, and Magneto-Inertial Fusion (MIF) in which the inertia of the liner is explicitly recognized for compressing and holding fusion plasma at relatively high density. Recently, the US Department of Energy through ARPA-E has initiated the ALPHA program for technologies that will enable development of low-cost controlled fusion by MIF. While it is certainly possible to continue the past history of single-shot implosions of liners onto plasma targets, it has become clear that some means for performing laboratory experiments at multi-megajoule levels is needed for reasonable progress. To develop the necessary plasma targets for liner compression requires hundreds of shots, so technology for low-cost, repetitive experiments must be created and demonstrated. Furthermore, to satisfy overall program goals, these techniques must extend to break-even experiments and economical fusion power reactors. The *Stabilized Liner Compressor* (SLC) seeks to accomplish these goals by means of pneumatically-driven, annular free-pistons imploding rotationally-stabilized liquid metal liners. We report progress in confronting the many engineering challenges for developing SLC.**

Keywords: imploding liner, controlled fusion, megagauss

I. Introduction

In the 1970's, there was significant interest in both the US and USSR in the use of imploding liner, magnetic flux compression to achieve controlled thermonuclear fusion at megagauss magnetic field levels [1]. The basic notion was originally applied to the open-ended theta-pinch concept for which megagauss fields would provide higher densities and lower times for fusion breakeven, thereby reducing the length and energy of the system dramatically. Indeed, the term *liner*, now used in many concepts, derives from the proposal at IAE Kurchatov for "theta-pinch with liner" indicating the use of an imploded, conducting sleeve inside a theta-pinch coil.

Recognition that there may be a lower cost approach to controlled fusion has led to a revival of interest in concepts based on pulsed high magnetic fields. This lower cost approach (Fig.1) from [1] is based on the decrease in capital cost with higher magnetic field and an increase of system cost with higher power density at much higher (equivalent) fields. The possible minimum in cost appears to occur at megagauss magnetic field levels [1]. Such fields are difficult to achieve in (quasi-) steady-state, but are readily attained in useful volumes by magnetic flux compression using the implosion of electrically-conducting shells (aka, liners) surrounding trapped magnetic flux. The basic problem for application of such techniques to controlled fusion, however, is the growth of perturbations in the liner surface due to instabilities during implosion [2]. This problem was solved at the Naval Research Laboratory, c. 1979 [3, 5] where repetitive implosions of liquid metal liners were achieved by means of a free-piston drive and liner rotation (Fig. 2). This approach avoids Rayleigh-Taylor instabilities during acceleration and re-capture of the liner, and near peak compression [1, 3] where energy can be exchanged reversibly between liner kinetic energy and payload internal energy (Fig. 3). In the fusion reactor concept (Fig. 4), the liner serves as the first-wall and blanket to protect the reactor from high-energy, neutrons [3, 6]. It also is the principal medium for heat transfer and chemical processing for tritium extraction.

System cost: $\$ = C_w W_p + C_p P = C_w W_p + C_p'/W_p^{3/2}$

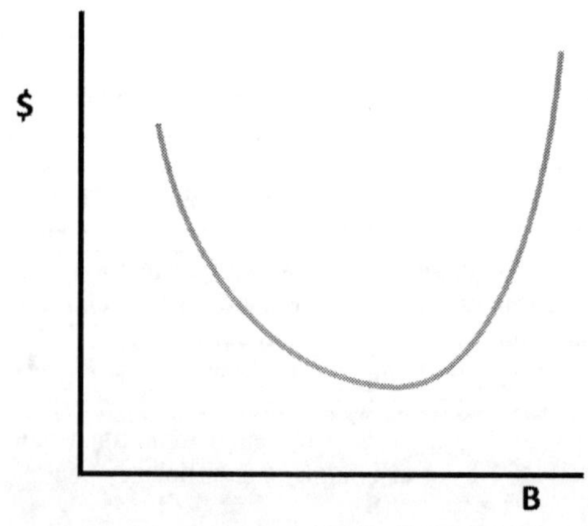

Fig. 1: The cost of a fusion system decreases with higher magnetic field due to reduced capital costs, $C_W W_p$ but increases at higher (equivalent) magnetic fields if higher power densities, P, are needed, from [1].

Sponsored by ARPA-E ALPHA Program

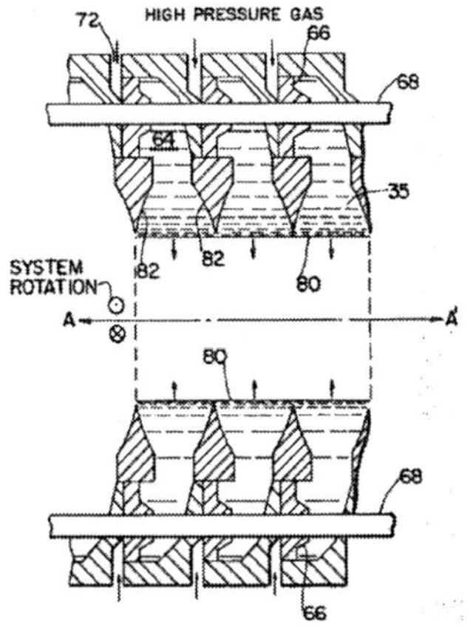

Fig. 2: Annular free-pistons, displaced axially by high-pressure gas to drive implosion of rotating liquid liner, from 1979 patent [4]. The use of "ganged" pistons depicted here allowed for plasma targets of arbitrary length.

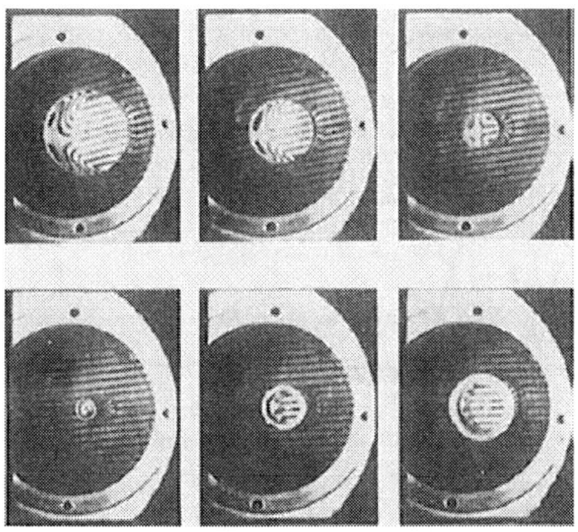

Fig. 3: Experimental movie from NRL "Water Model" of rotationally-stabilized liner implosion, driven by an axially-displaced, annular free-piston; liner of water compressing trapped air. Photo is at slight angle to axis, displaying reflection of line pattern by inner surface of liner [1, 3, 5].

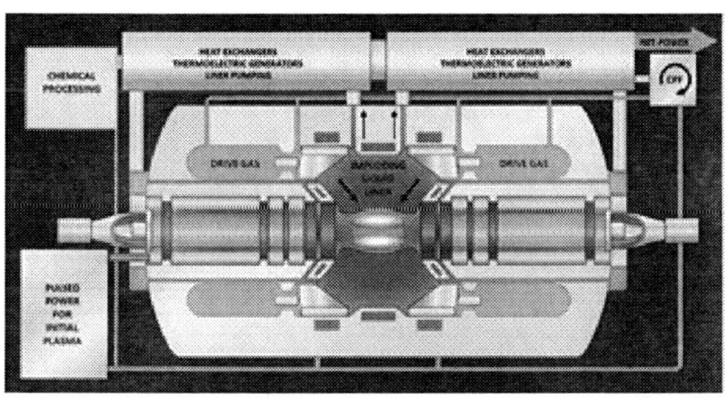

Figure 4: Present conceptual design of a fusion power reactor based on stabilized liner implosion technology follows earlier design [3, 6].

The design of an SLC, however, has many significant challenges, in part because progress, over the last forty years since the NRL experiments, in the development of plasma targets, e.g., the Field-Reversed Configuration (FRC) has not yet resulted in lifetimes greater than tens of microseconds for plasma diameters of tens of centimeters appropriate for near term liner compression experiments [7, 8]. Thus, SLC will need to use a free-piston drive-pressure upwards of 25 kpsi (1.7 x10^8 Pa) and liquid Na (or sodium-potassium alloy) or Li in order to achieve liner surface speeds of 2 km/s and implosion times for final compression of tens of microseconds.

II. MHD Design

In raising the driving pressure and implosion speed to values much higher than attempted in the NRL experiments, we encounter the effects of liner compressibility and the associated finite transit-times through the liner from its inner surface at very high pressure to the pistons (at only moderately high pressure). These effects were first noticed in the Helius experiments and modeled with a specialized 1-D hydrodynamics code [9]. Other numerical simulations [10] at the time pointed to the importance of compressibility in terms of a parameter $\zeta = p_f/\rho c^2$, where p_f is the peak pressure attained by the payload, and ρ and c are the mass density and sound speed of the liner material, respectively. For the envisioned power reactor, the use of Pb-Li (at 90:10, by atom) reduces the effects of compressibility at a desired peak pressure. For plasma target development and near term fusion breakeven experiments, however, shorter implosion times and higher implosion speeds require low mass density liners, e.g., Li and Na, for which compressibility is a major concern.

To design for such conditions, we employ the 2-1/2 dimensional code MACH2 [11], both to analyze the liner dynamics and to provide predictions of pulsed pressure loadings for use in a dynamic-structures code. In addition to the basic design work, we also need to confront a variety of operating issues, including safety protocols for nominal and off-design behavior. Fig. 5 depicts an arrangement for SLC in which pairs of ganged pistons allow for separate control of the axial and radial motions of the liner. An example of the results from a MACH2 simulation of the operation of SLC is provided Fig. 6 of Na compressing trapped magnetic flux. From a starting field of 0.2 T, peak fields over 50 T are obtained, including the effects of magnetic diffusion into the liner surface.

Earlier modeling based on incompressible flow [6, 9] estimated that the axial extension of the liner after peak compression could be overcome by angling the channel to provide an initial, axially contracting component to the flow.

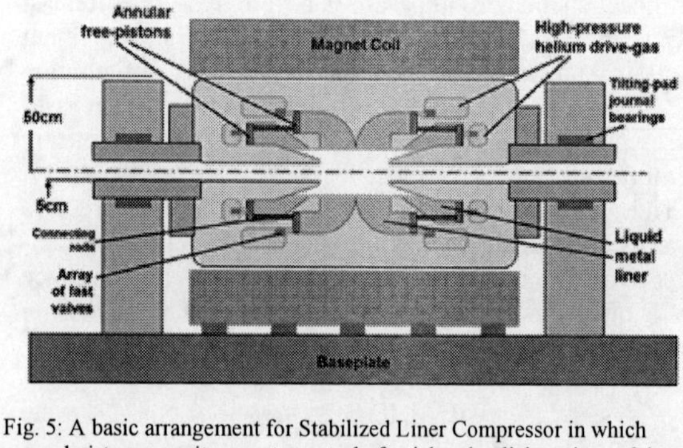

Fig. 5: A basic arrangement for Stabilized Liner Compressor in which ganged pistons permit separate control of axial and radial motions of the liner. As with the Linus-0 system [3] at NRL, the implosion chamber is rotated with a horizontal axis and clear bore for diagnostic and plasma access.

For compressible liners, however, a "water-hammer" is launched at peak compression. This allows more time at high pressure for axial expansion than the incompressible case. For the near term experiments, safe re-capture of the liner material is still possible and enables the desired repetitive operation for plasma target development.

III. Plasma Considerations

A principal difficulty for many decades now has been the creation of a plasma target appropriate for liner compression. This has led to two categories of endeavor: faster liner implosions to match shorter-lived plasma targets, and plasma targets of longer life that will tolerate slower implosions. The former category becomes linked to the need for economical replacement of the liner implosion apparatus because of the very high energy-densities involved. The latter accepts the lower energy-densities for repetitive implosions, e.g., SLC, but may demand more than plasma physics will allow for the target. We follow this latter path in the present project and offer that repetitive implosions in the laboratory may permit the necessary plasma target development. In particular, essentially all the physics of importance for liner compression of plasma occurs within a final diameter or two of the implosion. Albeit with a greater burden on the initial plasma target, we could inject it in the later stages of implosion and increase the injection radius as we learn more, (reducing the initial plasma energy).

Typically, in assessing the possibility of useful compression of a plasma target, comparison is made between the lifetime of a plasma configuration in a plasma generation experiment and the expected implosion time of the liner. Such comparison, however, does not account for changes during implosion in the several processes determining the plasma life. For example, if the lifetime is based on Bohm-like diffusion, the characteristic times for loss of energy, particles and magnetic flux can all scale as r^2B/T, where r is a plasma dimension for transport, and B and T are the characteristic magnetic field and temperature, respectively. The magnetic flux scales as Br^2 and the temperature as E/N, where E is the plasma energy and N the associated number of particles. With similar loss rates for

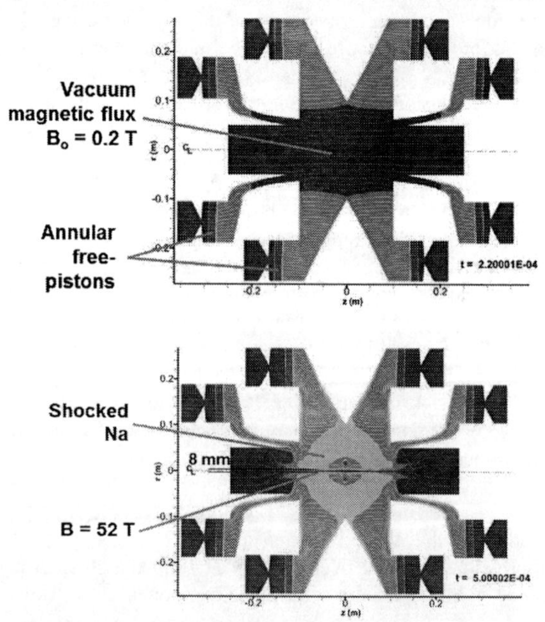

Fig. 6: MACH2 calculation of the compression of magnetic flux by SLC. Upper left has inner surface of liner starting inward from r = 10 cm onto initial field of 0.2T. Lower left shows peak compression to 52 T. Lighter blue region is quasi-spherical shell of Na at 20% above normal density.

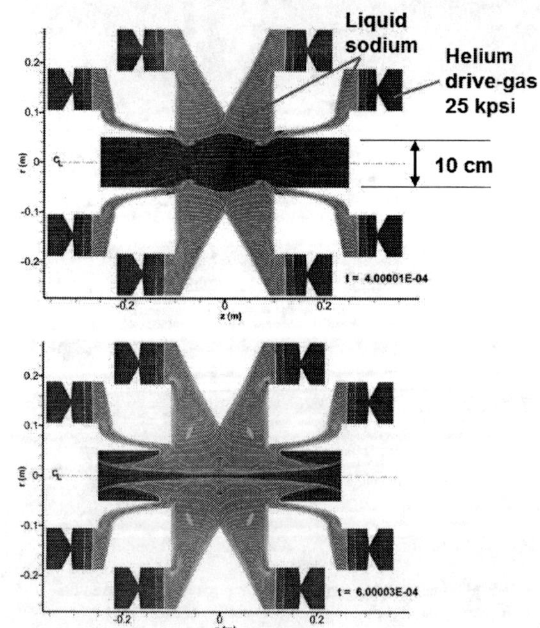

Fig. 6: MACH2 calculation of the compression of magnetic flux by SLC. Upper right shows combination of conical end-flows and central implosion, providing yoke around payload region. Lower right indicates the axial expansion, including radial return, of liner edges.

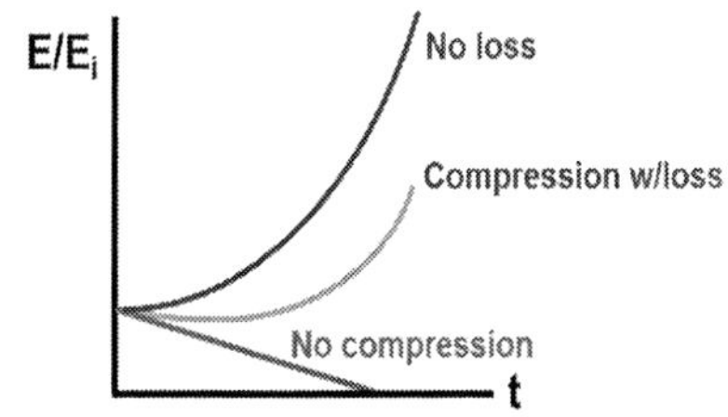

Fig. 7: Compression of an FRC plasma target to increase temperature and energy must work against losses of energy, particle number and magnetic flux to diffusion processes. Based on the loss time from static experiments, the target may first lose energy until the later stages of compression and then gain significantly.

energy, particles and flux, and assuming a cylindrical implosion $u(t)r(t) = u_c r_c$ = constant, we may solve for the plasma energy compared to its initial energy E_i. This is sketched in Fig. 7. Without loss, the energy increases adiabatically; while, without compression, the plasma energy decreases linearly with time because of the change in transport values. For the intermediate situation, the plasma may initially lose energy, but then gain it during the later, faster stages of compression.

IV. Operational Issues

Apart from the challenges of high speed, compressible fluid flow (> 1 km/sec), high values of mechanical stress (with drive pressures of 25 kpsi), and details of a unique rotating machine, we have to allow for the possibility of failure due to latent defects. Such defects, which by definition have eluded inspection procedures, could result in catastrophic failure of an apparatus involving tens of megajoules and alkali metals (Li or Na). We have therefore planned to operate SLC within a structure (Fig. 8) of stainless steel beams and plates capable of containing the mechanical and pneumatic energies and preventing release of alkali metal to the surrounding laboratory site.

References

[1] P.J. Turchi, "Imploding Liner Compression of Plasma: Concepts and Issues," *IEEE Trans. on Plasma Science*, 36, 1, 52 (2008).

[2] P.J. Turchi,"Liner Stability Problems for Megagauss Fusion," *IEEE Trans. on Plasma Science*, 43, 1, Part II, 369 (2015).

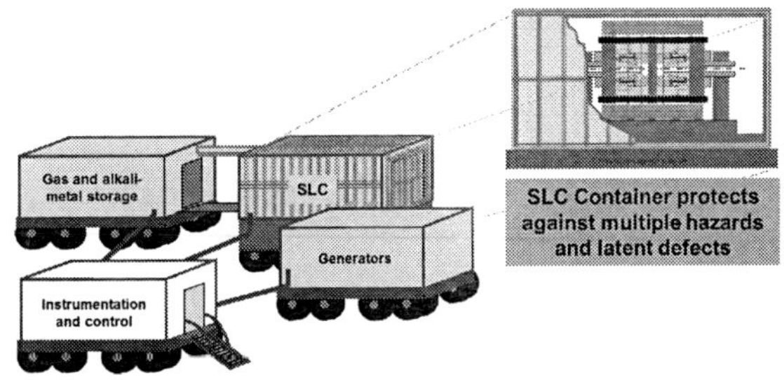

Fig. 8: In order to visit locations for plasma target generation, the SLC system will need to be largely self-sufficient, including trailers for gas and alkali-metal handling, electric power and instrumentation/control.

[3] P.J. Turchi, et al, "Review of the NRL Liner Implosion Program", in Megagauss Physics and Technology, P.J. Turchi, ed., Plenum, NY (1980). P. 375.

[4] P.J. Turchi and D.J. Jenkins, "Stabilized Liner Implosion System," US Patent 4,140,057 (1979).

[5] P.J. Turchi, A.L. Cooper, R.D. Ford, D.J. Jenkins, W.L. Warnick, "Stabilized Liner Implosions Driven by Axially-Moving Free-Pistons," Memo Report 3511 (June 1977), Naval Research Laboratory, Washington, DC.

[6] P.J. Turchi, "A Compact-Toroid Fusion Reactor Design at 0.5 Megagauss Based on Stabilized Liner Implosion Techniques," in Ultrahigh Magnetic Fields, V.M. Titov and G.A. Shvetsov, eds., Nauka, Moscow (1984). P. 184.

[7] Slough, John, George Votroubek, and Chris Pihl, "Creation of a high-temperature plasma through merging and compression of supersonic field reversed configuration plasmoids." Nuclear Fusion 51.5 (2011): 053008.

[8] J.H. Degnan, et al, "Results from Compression of Magnetized Plasma using Imploding Solid Liner," These proceedings. Paper 151.

[9] A.L. Cooper, J.M. Pierre, P.J. Turchi, J.P. Boris, R.L. Burton, "Modeling of Linus-Type Stabilized Liner Implosions," in Megagauss Physics and Technology, P.J. Turchi, ed., Plenum, NY (1980). P. 447.

[10] D.L. Book and P.J. Turchi, "Dynamics of Rotationally Stabilized Implosions of Compressible Liquid Shells," *Phys. Fluids* 22, 68 (1979).

[11] R.E. Peterkin, M.H. Frese, C.R. Sovinec, "Transport of magnetic flux in an arbitrary coordinates ALE code," *J. Comp. Phys.*, 140, 148 (1998).

Some History of Alternatives for Stabilized Liner Implosions

P.J. Turchi
NumerEx
Santa Fe, New Mexico, USA

Abstract **- Recently, ARPA-E, US Department of Energy, has revived support for the use of imploding liner magnetic flux compression as an enabling technology for low-cost controlled fusion. This has led to a return to concepts in the Linus program at the Naval Research Laboratory during the 1970's. In particular, the Stabilized Liner Compressor (SLC) has been proposed to provide repetitive implosions for laboratory development of appropriate plasma targets for compression to megagauss field levels. While the SLC is based on successful concepts at NRL, it is useful to review alternative schemes for liner implosion technology that were explored at NRL. Some of these schemes were attempted experimentally before the successful approach was adopted and demonstrated of an annular free-piston, accelerated axially by high-pressure gas, driving a rotating liquid liner. Other schemes were set aside for various reasons, including lack of stability, difficulties with seals, and uncertainties for mechanical design. Concepts that were considered include an azimuthal array of pistons moving radially (which was actually operated), mechanical devices, and pistons moved azimuthally by gas pressure or mechanically. This history will be reviewed with the aim of both displaying possibilities and offering cautions.**

Keywords: imploding liner, controlled fusion, megagauss

I. INTRODUCTION

Recognition that there may be a lower cost approach to controlled fusion has led to a revival of interest in concepts based on pulsed high magnetic fields. This lower cost approach has been depicted in Figure 7 [1] in terms of the decrease in capital cost with higher magnetic field and an increase of system cost with higher power density at much higher (equivalent) fields. The possible minimum in cost appears to occur at megagauss magnetic field levels. Such fields are difficult to achieve in (quasi-) steady-state, but are readily attained in useful volumes by magnetic flux compression using the implosion of electrically-conducting shells (aka, liners) surrounding trapped magnetic flux. The basic problem for application of such techniques to controlled fusion, however, is the growth of perturbations in the liner surface due to instabilities during implosion. Penetration of the plasma by high-Z material of the liner can prevent the attainment of necessary fusion temperatures. For liners that present a liquid surface to the enclosed plasma, Rayleigh-Taylor instability in the final 1-2 diameters of plasma compression sets severe requirements on the initial smoothness of the surface [2].

Paper sponsored by ARPA-E ALPHA Program

A related problem is the launch and re-capture of the liner during which liquid liners are subject to the disruption of Rayleigh-Taylor instability. Liners initially at solid-density exhibit elastic-plastic buckling instability [3] leading to unacceptable perturbations on the inner surface of the liner (feeding instability growth there, if the liner has liquefied) and the return of fragments/droplets after peak compression. With energies on the order of 100 MJ (~ 25 kg of high-explosive) per shot required to develop a fusion break-even experiment, failure to control the liner material would preclude a practical experimental program. Recognition of this difficulty led the Linus program at the Naval Research Laboratory to develop and demonstrate the concept of a stabilized liner implosion system [3].

II. INITIAL STABILIZED LINER IMPLOSION SYSTEM

Inspired by the Soviet program (c. 1968) on imploding liners for fusion at megagauss field-levels, the initial experiments at NRL used theta-pinch style, electromagnetic implosion of thin-walled, aluminum cylinders at 500 kJ to achieve megagauss fields over cm-diameter bores. It was clear, however, that the subsequent explosion of liner material could not be tolerated for future, higher energy experiments. In regard to the use of liquid metal liners, R.A. Shanny (NRL) had asked E.P. Velikhov (IAE Kurchatov) how the liquid liner would be stabilized. Velikhov replied in the context of simply forming the initial liner with a horizontal axis that they would spin it. Shanny was actually asking about Rayleigh-Taylor instability. This misunderstanding led to calculations at NRL of rotational stabilization of Rayleigh-Taylor modes [4]. The subsequent experiments [5], using rotating liners of eutectic sodium-potassium (NaK) driven electromagnetically (Fig. 1), provided the first demonstration of such stabilization of the inner surface of a decelerating liquid liner.

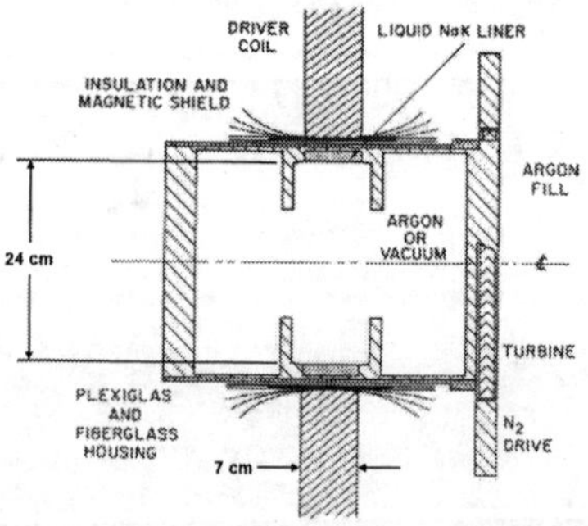

Fig. 1: First demonstration of rotational stabilization of Rayleigh-Taylor modes on the inner surface of a decelerating liquid liner was driven by a theta-pinch style coil at 3 MA [5].

The outer surface of the liner, however, was still subject to Rayleigh-Taylor instability during both launch and re-capture of material. This problems was solved by eliminating the free outer surface of the liner, driving the liquid instead with solid pistons [6,7]. The use of such free-pistons (Fig. 2) offered the opportunity to perform the necessary work by means of pneumatic vs electromagnetic energy. High–pressure helium, for example, could store and deliver the necessary multi-megajoule energies far more compactly and at much less cost than large capacitor banks. Furthermore, there was the potential to recover the liner kinetic energy, including the work done by fusion alpha-particles trapped in the magnetized-plasma, and thereby substantially reduce the size, energy and cost of a fusion power reactor [3].

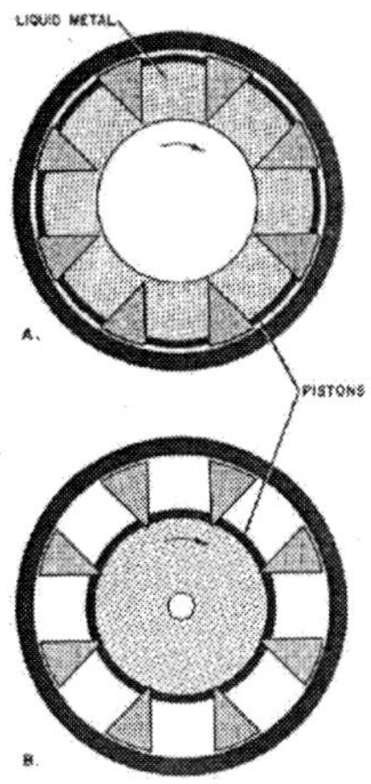

Fig. 2: Early representation of pneumatically-driven, free-pistons displaced radially to implode rotating liquid liner [6].

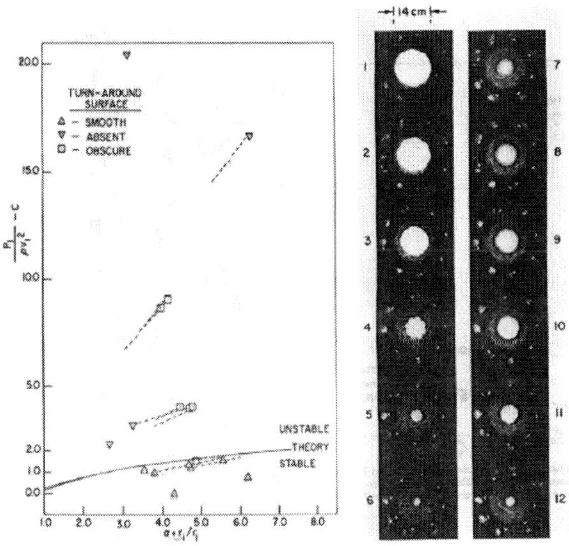

Fig. 3: Experimental demonstration and analysis of rotationally-stabilized liner driven by radially-displaced free-pistons in NRL "Water Model" [7] was in accord with theory.

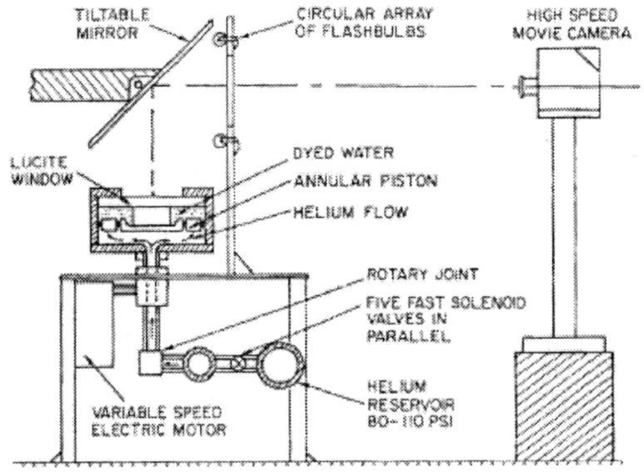

Fig. 4: "Water Model" experimental apparatus at NRL used for first demonstration of axially-displaced, annular free-piston for stabilized, rotating liner implosion of water onto trapped air [3].

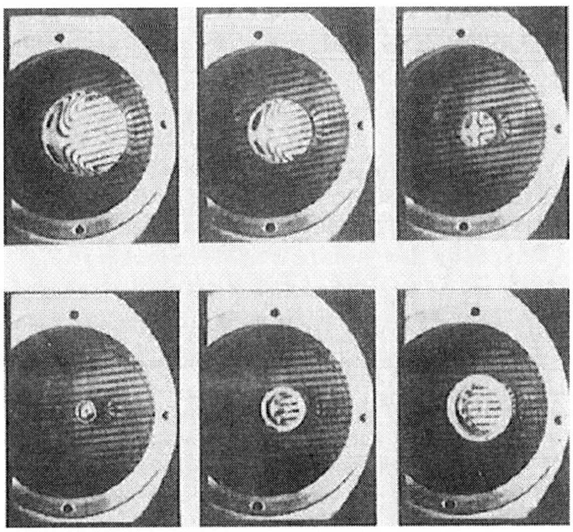

Fig. 5: Experimental movie from NRL "Water Model" of rotationally-stabilized liner implosion, driven by an axially-displaced, annular free-piston; liner of water compressing trapped air. Photo at slight angle to axis, displaying reflection of line pattern by inner surface of liner [3].

III. Axially-Driven Annular Pistons

While the initial demonstration of the stabilized liner implosion system was successfully demonstrated using pistons that moved radially (Fig. 3), there was still the opportunity for the array of several pistons to lose axisymmetry. This corresponded in part to a low-mode number form of Rayleigh-Taylor instability. For the case certainly of rotating the entire implosion chamber (as in the experiments), the imbalance due to such asymmetry would be intolerable. Instead, the approach of driving the liquid by axially-displaced annular free-pistons (Fig. 4) was adopted and successfully pursued [3, 8], first with liners again of water (Fig.5), then at higher pressures (Fig. 6), with liners of

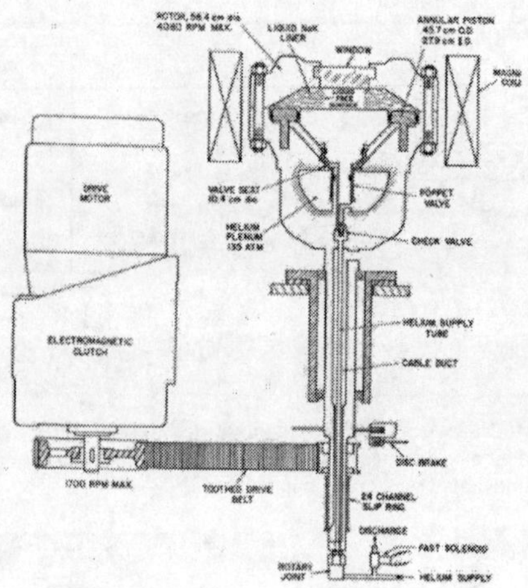

Fig. 6: Helius apparatus at NRL used to implode liners of NaK for magnetic flux compression [3].

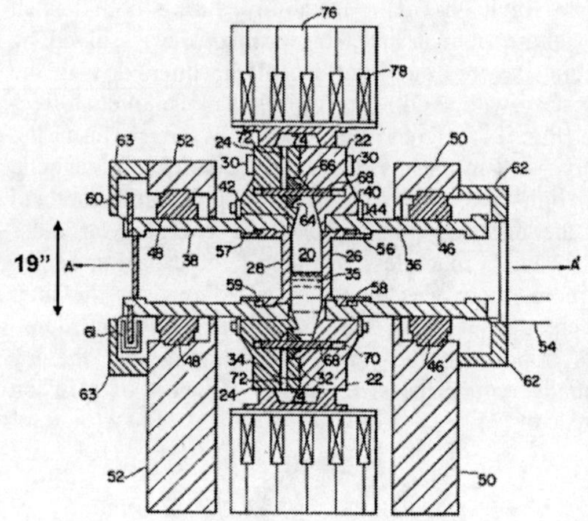

Fig. 7: Linus-0 stabilized liner implosion system at NRL used single annular free-piston, displaced axially by high=pressure gas generated using combustion of high explosive in non-rotating volumes surrounding the rotor/implosion chamber [3].

water and (separately) eutectic NaK, and up to a large system called Linus-0 (Fig. 7). This approach continues in present work as the Stabilized Liner Compressor (SLC) within the revived program for low-cost fusion [9].

IV. Alternatives to SLC

Other notions for stabilized liner compression systems were also entertained. For example, suppose the radial piston motions could be synchronized by some sort of mechanism (Fig. 8). This was proposed by A.E. Robson [10] and pursued with a small model (that apparently jammed [11], perhaps due to stress-concentrations and complexity). In this system, the energy was still supplied by high-pressure gas.

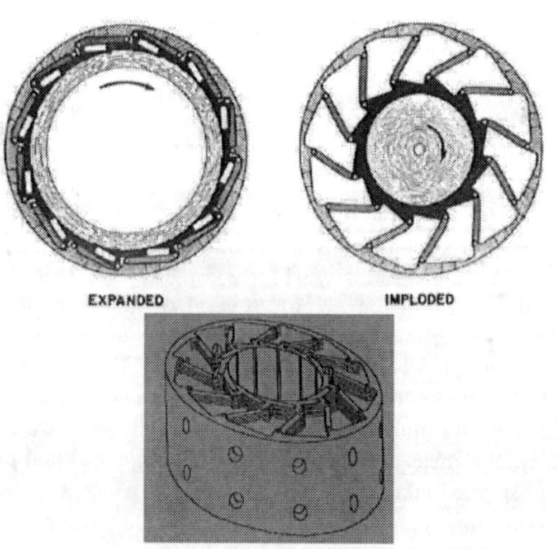

Fig. 8: Proposed mechanism to couple gas-driven, radially-displaced pistons imploding liquid liner [10].

The use of direct mechanical action between solid parts that would result in the implosion of the liquid liner was also considered, but suffered from issues of stress-concentrations between these parts. So-called "Hertzian stresses" occur when a solid bears locally on another solid. Simple modeling offers estimates of such stress levels, which appear to involve values that are 50-150 times the basic driving pressure required for the desired liner speeds. Calculated values are well above ultimate strengths of materials, which makes the simple modeling insufficient as a guide for design. Numerical techniques can, of course, substitute for analytical estimation, but depend on the exact conditions of roughness and adhesion at the contact points. An alternative concept that is more amenable to prediction for design would use azimuthal motion of pistons (Fig. 9) to force liquid out of an axisymmetric array of channels feeding the radial motion of the liner material.

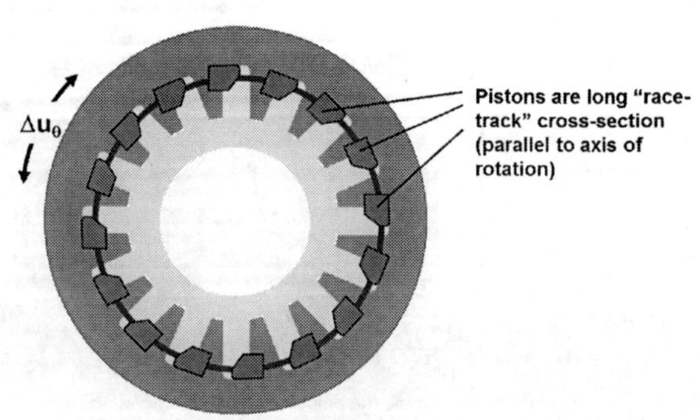

Fig. 9: An array of linked pistons displaced azimuthally by relative speed Δu_θ between concentric shells forces liquid liner inward without requiring metal-to-metal contact and stress concentrations.

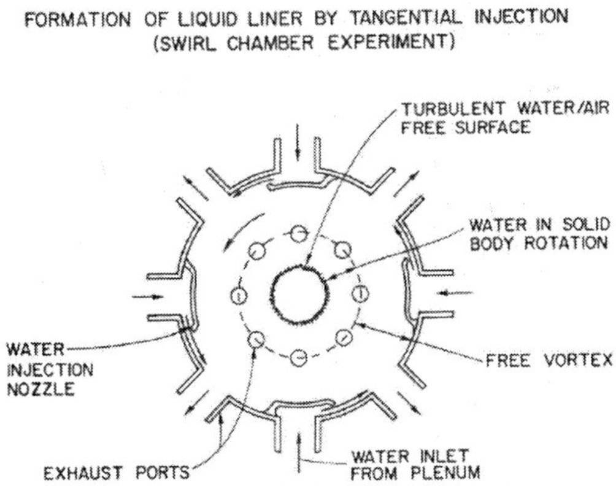

Fig. 10: Experimental demonstration of tangential injection to form a rotating liquid liner [13].

V. Future Developments

In the near future, the axially-driven annular-piston approach will be developed to extend the success at NRL to much higher pressures and speeds [9]. The requirement for inner surface speeds upwards of 2 km/s is driven by the anticipated lifetimes of plasma targets (< 50 μsec for initial diameters ~ 10 cm). For the eventual fusion power reactor [3, 12], it becomes more appropriate to create the rotating liquid liner by tangential injection (Fig. 10), explored briefly at NRL [13], as an alternative to spinning the implosion chamber. This approach is consistent with the use of the liner material as the reactor blanket and heat transfer material (Fig. 11).

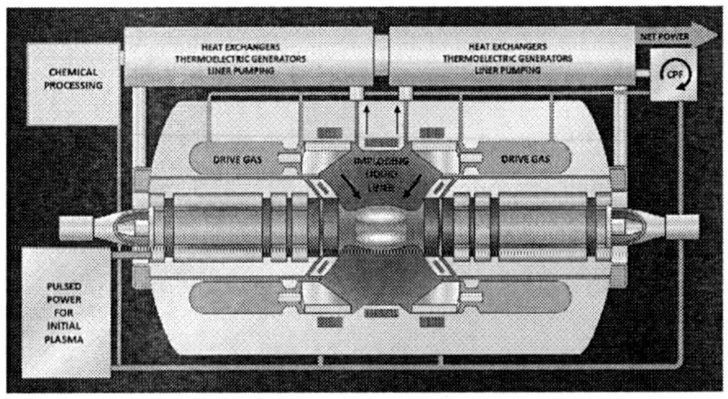

Fig. 11: Conceptual design of a controlled fusion power reactor based on the stabilized liner compressor [8, 12].

References

[1] P.J. Turchi, "Imploding Liner Compression of Plasma: Concepts and Issues," *IEEE Trans. on Plasma Science*, 36, 1, 52 (2008).

[2] P.J. Turchi, "Liner Stability Problems for Megagauss Fusion," *IEEE Trans. on Plasma Science*, 43, 1, Part II, 369 (2015)

[3] P.J. Turchi, A.L. Cooper, R.D. Ford, D.J. Jenkins, R.L. Burton , "Review of the NRL Liner Implosion Program", in Megagauss Physics and Technology, P.J. Turchi, ed., Plenum, NY (1980). P. 375.

[4] A. Barcilon, D.L. Book, and A.L. Cooper, "Hydrodynamic Stability of a Rotating Liner," *Phys. Fluids*, 17, 1707 (1974).

[5] P.J. Turchi, et al, "Rotational Stabilization of an Imploding Liquid Cylinder," *Phys. Rev. Letters* 36:1546 (1976).

[6] A.E. Robson and P.J. Turchi, "The NRL Linus Program", in Pulsed High Beta Plasmas, D.E. Evans, ed., Pergamon (1976). P. 477.

[7] P.J. Turchi, D.J. Jenkins, R.D. Ford, A.L. Cooper, "Rotational Stabilization of the Inner Surface of a Piston-Driven Imploding Liner," Memo Report 3536 (June 1977), Naval Research Laboratory, Washington, DC.

[8] P.J. Turchi, A.L. Cooper, R.D. Ford, D.J. Jenkins, W.L. Warnick, "Stabilized Liner Implosions Driven by Axially-Moving Free-Pistons," Memo Report 3511 (June 1977), Naval Research Laboratory, Washington, DC.

[9] P.J. Turchi, S.D. Frese, M.H. Frese, "Progress on the Design of a Stabilized Liner Compressor for Low-Cost Controlled Fusion at Megagauss Field-Levels", Paper 225, this conference.

[10] A.E. Robson, "A Conceptual Design for an Imploding-Liner Fusion Reactor," in Megagauss Physics and Technology, P.J. Turchi, ed., Plenum, NY (1980). P. 425.

[11] G. Moskowitz, personal communication (1979).

[12] P.J. Turchi, "A Compact-Toroid Fusion Reactor Design at 0.5 Megagauss Based on Stabilized Liner Implosion Techniques," in Ultrahigh Magnetic Fields, V.M. Titov and G.A. Shvetsov, eds., Nauka, Moscow (1984). P. 184.

[13] R.L. Burton, et al, "Generation of a Rotating Liquid Liner by Tangential Injection," Memo Report 4025 (June 1979), Naval Research Laboratory, Washington, DC.

The Effect of Holder in the Triggered Vacuum Switch*

WungHou Park , Moo-Sang Kim, Yoon-Kyoo Son and Byung-Joon Lee
Pohang Accelerator Laboratory
POSTECH
Pohang, Korea
bjlee707@postech.ac.kr

Abstract- A triggered vacuum switch (TVS) is a switching device widely used in a high power system. It is switched-on by a plasma arc generated in an internal trigger unit (TU) and main electrode. The main electrode is composed of an anode and a cathode. The TU is assembled a trigger pin, an insulator and a holder. The initial plasma is generated in the TU and propagated into the gap between the anode and the cathode. The beginning process of plasma generation in the TU is delicate, because it is affected by all TU components. However, the initial generating process of the main electrode is simple, because it is induced by the plasma generated in the TU. In this research, we have investigated the TU, especially have focused on the structure and material of the holder. In general, the life time and stable operation in TVS are decided by the metal layer coated on the ceramic insulator which is originated from main electrode and holder. Before researching the metal layer, we have searched for the relations of between operation conditions and two holder types. Operation conditions are investigated on a trigger voltage, current and resistance for fabricating a concave and a convex type holder with Al and Mo.

Keywords— *triggered vacuum switch; trigger unit; holder; plasma arc*

I. Introduction

A TVS insulated by the vacuum is enable to hold a high voltage and is switched on by a plasma arc. The switching operation is sequentially processed by a TU and a main electrode. At first stage, the TU ignites an initial plasma and diffuses plasma into the gap between the anode and cathode of the main electrode. And the next stage, the filled plasma induces the discharge of the main electrode [1]. At this moment, the operating condition is unfixed, because the TU igniting condition is changed in shot by shot. Therefore, to stably operate TVS for long time, the plasma generation in the TU should be understood.

The ignition of the plasma in the TU is begun by an electron emission at the triple point which is the junction of insulator, holder and vacuum [2]. The emitted electron is amplified by an interaction with a monolayer gas covered an insulator surface, and in consequence of the interaction, proton clouds are produced [3]. In the next sequence, proton clouds are accelerated to the cathode by an electromagnetic field and collided on the cathode surface, which makes cathode spots [4]. The initial cathode spot is randomly taken place close to the insulator, and developed all round of the TVS within few microseconds. The cathode spot ejects a lot of fragments included charged particles, molecules, metal vapor, and so on. In particular, the ceramic surface covered by the metal vapor causes the transition of the operating condition [5].

In this research, to stably operate for long time, we have focused on the proper structure and material of the TU. To find the proper conditions of the TU, we have suggested a UP and a DOWN type holder with an Al and a Mo material. The selection of materials is decided to distinguish from the main electrode material Cu. On the contrary, the insulator is fixed with a ceramic. A vacuum pressure is kept in the order of 10^{-7} mbar at room temperature.

II. Main Electrod and Holders

The assembled TVS system is designed to research various TVS parts. Especially, the hole in the middle of the cathode is devised to adapt alternative TU components. The fabricated holder is inserted in this hole. Fig. 1 shows assembled TVS and holders made of Al and Mo. Each Al and Mo holders have been fabricated with a UP and a DOWN type, respectively.

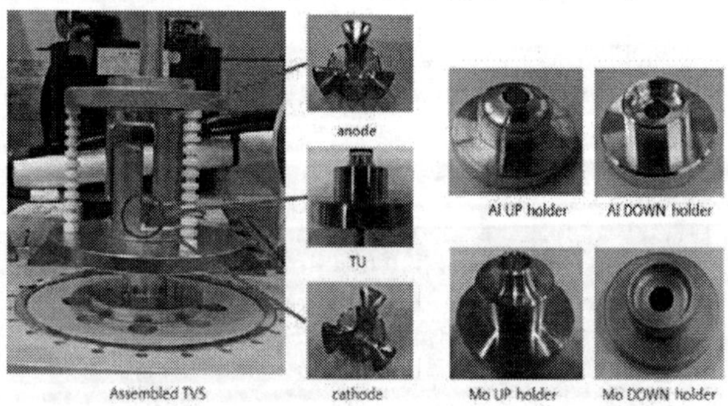

Fig. 1. Structure of the TVS and holders

The holder type is discriminated by the upper shape. A convex shape is the UP holder and concave one is the DOWN holder. In the Fig. 1, the assembled TU is used the UP holder, but it can be easily replaced by the DOWN holder. The research of the structure is begun with CST program to understand the distribution of electric field strength in the TU.

This project was supported by the Agency for Defense Development (ADD).

III. EXPERIMENTAL SETUP

A TU test system has been constructed to investigate TU components, which is supplied with 20 kV voltage and maximum current is reached at 88.

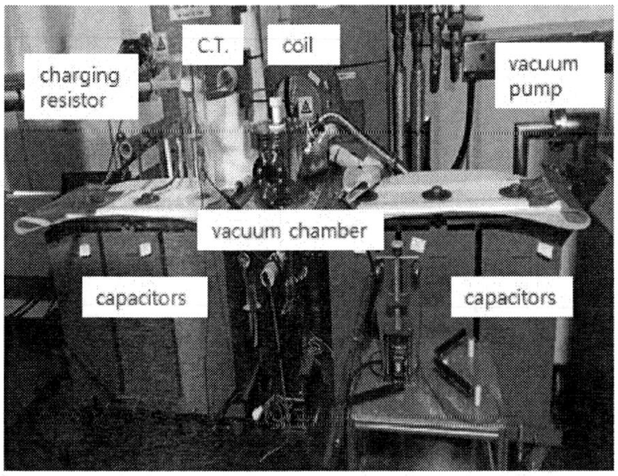

(a) TVS trigger unit test system

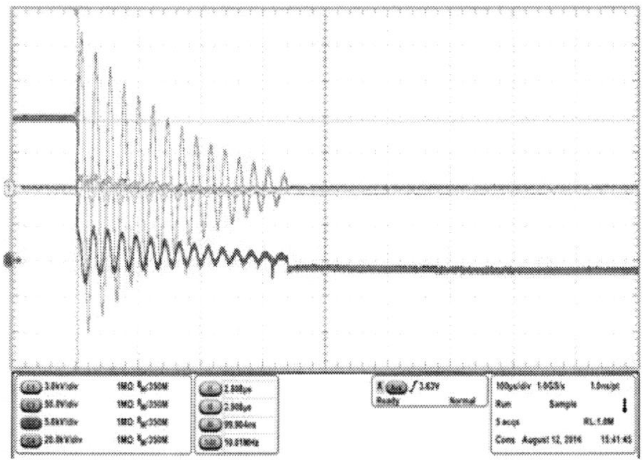

(b) Measured oscilloscope signal

Fig. 2. Trigger voltage for holders

The constructed TU test system is shown in Fig. 2 (a), it is combined with a charge storage bank assembled with 6 capacitor cells, and a vacuum chamber included TVS. An external vacuum pump is connected to the vacuum chamber to maintain the vacuum pressure. In addition, two high voltage probes and two current transformers are installed to measure the main voltage and current and the trigger voltage and current. The measured signals are shown in the Fig. 2 (b), where yellow is the trigger voltage, blue is the trigger current, red is main voltage, and green is main current. The trigger voltage is 5.6 kV and trigger current is 96 A at the maximum. However, these values are not fixed, those are varied for every shots. On the contrary, the main voltage is fixed at the 20 kV and the maximum main current is 88 kA.

IV. MEASURED RESULTS

To understand the effects of trigger condition for Al and Mo materials, we have investigated the trigger voltage, the current and the resistance according to the number of shots. Measured signal values are involved in the coated layer of the ceramic surface. After shot finishing, the ceramic and holder surfaces are observed with a microscope.

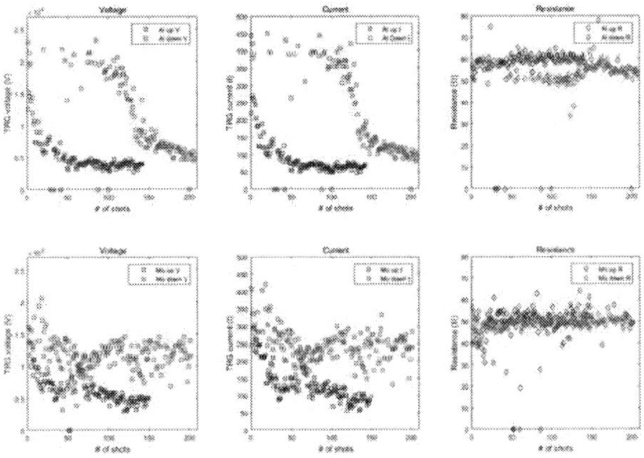

Fig. 3. measurements of the Al and Mo holders

At first, the characteristics of holder shapes are researched with Al and Mo, it is shown in Fig. 3. The trigger voltages of two holders are decreased and converged at 5 kV, as the number of shots is increased. Before converging, the trigger voltage of the Al DOWN holder is higher than the Al UP. And after converging, the voltage of the Al DOWN holder is kept stable compared to the Al UP. The current trend of the Al holder is the same of the voltage. The resistance of the Al UP holder is kept higher than the Al DOWN holder before 150 shots. The trigger voltage and current of the Mo DOWN holder are also higher than the Mo UP holder for all shots. However, to compare the stable region of the Mo holder, we need more shots, because the Mo DOWN holder is still not converged in the stable voltage region. The resistances of Mo holders are almost same, but the Mo UP holder is more fluctuated. These results are caused by the ceramic surface of the DOWN holder is slowly coated, it is expected that the DOWN holder is enable to keep thin metal layer for a long time operation. It means that the DOWN holder is proper to the long time operation.

The next result is the comparison of between the Al and Mo material under same holder type, it is shown in the Fig. 4. The melting points of Al and Mo are different, the Al is 900 ˚C and the Mo is 2900 ˚C. The trigger voltage and current of the Mo UP holder is higher than the Al. However, the DOWN holder is difficult to compare for the voltage and current, because the Mo DOWN holder should be more shots to converge. However, our test system is difficult to prolong the number of shots, the system is happened a self-breakdown at more than 250 shots for the main of 20 kV. Therefore, the Mo DOWN result is enable to be seen after removing the breakdown.

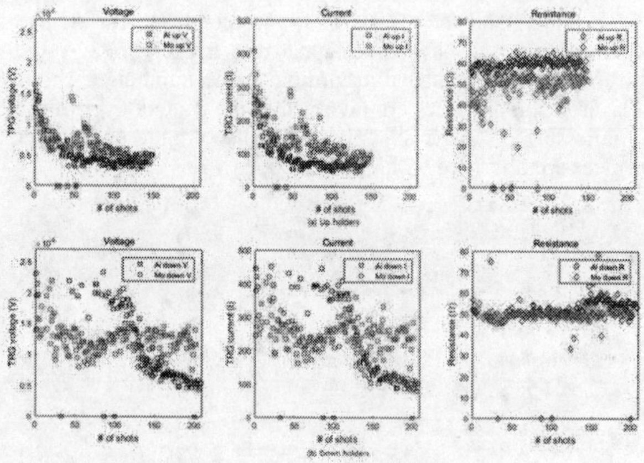

Fig. 4. Compared results for the Al and Mo material

The ceramic insulator is inserted between the trigger pin and the holder in the TU. The initial insulation resistance of the ceramic is more than GΩ, and it goes down until few hundreds MΩ after hundreds shots. Fig. 5 shows microscope pictures for the ceramic surfaces and the upper surface of holders after finishing experiments.

Fig. 5. Surface of the ceramic insulators

The color images of Al holders are brighter than the Mo holders. However, it is an inappropriate information to understand the characteristics of holders. It can be changed by a microscopic setup, a sample position, and so on. To intensively understand the characteristics of ceramic surface, it is needed to research for the ingredient ratio, depth and roughness with an X-ray spectrometer and a roughness tester. However, the analysis of the X-ray spectrometer is restricted, the Al material is improper to discriminate with the ceramic material, because the ceramic includes an Al material. To complement the result of the Al material, we have scheduled to measure with a Fe holder.

V. SUMMARY

In this research, we have focused on the trigger holder condition for a long time and stable operation. This condition is dominantly decided by the coated metal layer of the ceramic surface. To stably operate TVS for long time, the coated layer should be maintained thin and uniform. To keep these conditions, we have fabricated the UP and the DOWN holder with Al and Mo, and measured for the trigger voltage, current and resistance for different holders. The result of holder shape is shows that the DOWN structure is proper to keep thin and uniform for a long time. The Mo UP holder is better than the Al UP holder in maintaining the thin and uniform layer, however, the comparison of the DOWN type is postponed. The research of the ceramic surface is still ongoing. In the next experiments, the test system should be upgrade to increase the number of shots and the experiment of the Fe holder is scheduled.

Acknowledgment

This project was supported by the Agency for Defense Development (ADD) under grant number 14-BR-EN-22.

References

[1] B M Coaker, C Bell, R J Seddon & JS Bower, "Miniature triggered vacuum switches for precise initiation of insensitive loads in demanding environments" 39th IEEE International Conference on Plasma Science, DOI: 10.1109/PLASMA.2012.6383436. 2012.

[2] Nicholas M. Jordan, Y. T. Lau, David M. French, R. M. Gilgenbach, and P. Pengvanich, "Electric field and electron orbits near a triple point", J. Appl. Phys., vol. 102, 2007.

[3] R. V. Latham, *High Voltage Vacuum Insulation*, Academic Press, 1995.

[4] J. M. Lafferty, *Vacuum Arcs-Theory and Application,* New York: Wiley, 1980.

[5] J. M. Lafferty, "Triggered vacuum gaps," Proc. IEEE **54**, 23. 1966.

Solid State Bipolar Marx Modulators: Overvoltage Tolerance Capability

Hiren Canacsinh[1,2], Luís M. Redondo[2]
[1]INESC-ID, Lisboa
[2]Instituto Superior de Engenharia de Lisboa, ISEL, GIAAPP
Lisboa, Portugal

J. Fernando Silva[1,3]
[1]INESC-ID, Lisboa
[3]Instituto Superior Técnico, Universidade de Lisboa
Lisboa, Portugal

Abstract—**Two solid-state bipolar Marx modulators are compared in terms of semiconductors overvoltage tolerance capability in abnormal operating conditions such as switching synchronization failure. Simulation results are presented for 4 stages of two Marx circuits using 500 V per stage with 1 kHz bipolar pulse repetition rate, with 5 μs pulse width and 10 μs relaxation time into resistive load.**

I. Introduction

Through the years, power semiconductors increased their importance in pulsed power topologies named the Marx generator, conferring flexibility, higher pulse frequency and duty cycle, [1-5]. The increase of interest, for instance, from food and water purification industry applications for compact modulators and for bipolar pulses led to development of solid-state bipolar Marx modulators, [6-10]. However, in one hand topologies with additional switches for fault capability require complex triggering circuit on another hand optimized topologies, with reduced number of switches in each cell, uses some switches for more than one operating mode, contributing to the lack of current modularity of Marx type circuits, [11]. Aggravating this feature, the parallel charging technique increases the current in charging switches that are near the power supply [12]. Thus, when designing a solid-state Marx modulator, power semiconductors selection is also to be considered, not only by current rating but also by overvoltage across the switches in abnormal switching operation among operating modes.

The purpose of this paper is to present an analysis of semiconductors voltages in two solid-state bipolar Marx modulators presented in Fig. 1, under failure condition such as switch synchronism failure and open fault, considering that the semiconductors remain open-circuited when defective.

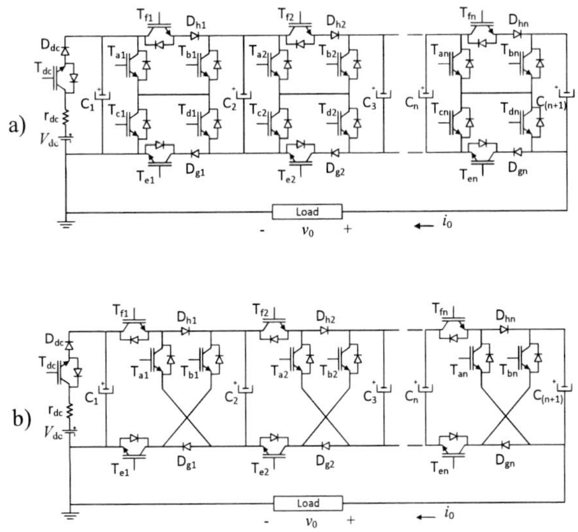

Fig. 1. Bipolar solid-state Marx modulators topologies, a) generalized with 6 on-off semiconductors and b) optimized, with 4 semiconductors on-off, per stage.

II. Semiconduct's Voltage

Fig. 1 presents two solid-state bipolar Marx modulators, having each with *n* stages, with IGBTs as on-off switches as described in detail elsewhere [8-9]. The main three operating modes of circuits of Fig. 1 are summarized in Table I. For clearness the IGBT anti-parallel diodes are not labelled in the circuits of Fig. 1, being named after the IGBT device number (e.g., D_{a1} corresponds to the anti-parallel diode of IGBT T_{a1}).

TABLE I

SEMICONDUCTORS'TURNED ON OF THE CIRCUITS OF FIG. 1 FOR CHARGING AND PULSE OPERATING MODES

Circuit	Operating mode		
	Charging	Positive Pulse	Negative Pulse
Fig. 1a)	T_{fi} & T_{ei} or T_{ai} & D_{bi} & T_{ei} or T_{fi} & D_{di} & T_{ci}	T_{ai} & T_{di}	T_{bi} & T_{ci}
Fig. 1b)	T_{fi} & T_{ei}	T_{fi} & T_{ai}	T_{bi} & T_{ei}

The solid-state bipolar Marx modulators presented in Fig. 1 and the switch path for charging and pulse operating modes shown in Table I are considered. Also, the collector to emitter switch ON voltage and anode-cathode voltage of ON diodes are considered to be approximately zero. The following sections will present an analysis of the switch voltages in the circuits of Fig. 1 under faulty condition such as switch synchronization failure (leading to brief open faults).

1) Charging the capacitors

Considering the circuit of Fig. 1a) and in case of synchronization failure (delay) of switch T_{fi} or T_{ei} of any independent (i) stage, the hold-off voltage across the delayed switch is the difference between voltages of capacitors C_i and C_{i+1}, $V_{C_i} - V_{C_{i+1}}$. In case of failure of both T_{fi} and T_{ei} switches, these switches share the above voltage difference. In any of these individual or simultaneous switch failures, the set of switches T_{ai} and T_{ci} or switches T_{bi} and T_{di} share the voltage of V_{C_i} or $V_{C_{i+1}}$, respectively, as listed in Table II. Considering now the circuit of Fig. 1b), similar situation happens in case of individual or simultaneous failure of switches T_{fi} and T_{ei}, as shown in Table II.

2) Positive pulse

Considering circuit of Fig. 1a) in positive pulse operating mode, and considering delays in switches T_{ai} or/and T_{di}, these switches hold-off the voltage of V_{dc} in case of individual or simultaneous failure as listed in Table II. However, this is different for circuit of Fig. 1b), because the failure of switch T_{fi} implies that this delayed switch has to sustain a voltage nearly equal two times the power supply voltage, i.e. $2V_{dc}$. Considering yet the circuit of Fig. 1b) and for situations of individual failure of switch T_{ai} or simultaneous failure of switches T_{fi} and T_{ai}, the delayed switch(es) hold-off the V_{dc} voltage, as shown in Table II.

TABLE II

FAULT CLEARANCE PATH AND HOLD-OFF VOLTAGE OF SEMICONDUCTORS FOR CHARGING AND PULSE OPERATING MODES OF THE CIRCUITS OF FIG. 1 AND THE OUTPUT VOLTAGE UNDER FAULT CONDITION

Operat. mode	Circuit	Switch in fault condition	Hold-off voltage of faulty switches	Hold-off voltage of switches in OFF-state driven
Charg.	Fig. 1a)	T_{fi}	$V_{CE_{T_{fi}}} = V_{C_i} - V_{C_{i+1}}$	$V_{CE_{T_{ai}}} + V_{CE_{T_{ci}}} = V_{C_i}$ and $V_{CE_{T_{bi}}} + V_{CE_{T_{di}}} = V_{C_{i+1}}$
		T_{ei}	$V_{CE_{T_{ei}}} = V_{C_i} - V_{C_{i+1}}$	
		T_{fi} & T_{ei}	$V_{CE_{T_{fi}}} + V_{CE_{T_{ei}}} = V_{C_i} - V_{C_{i+1}}$	
	Fig. 1b)	T_{fi}	$V_{CE_{T_{fi}}} = V_{C_i} - V_{C_{i+1}}$	$V_{CE_{T_{ai}}} = V_{C_{i+1}} + V_{AK_{D_{hi}}}$ and $V_{CE_{T_{bi}}} = V_{C_{i+1}} + V_{AK_{D_{gi}}}$
		T_{ei}	$V_{CE_{T_{ei}}} = V_{C_i} - V_{C_{i+1}}$	
		T_{fi} & T_{ei}	$V_{CE_{T_{fi}}} + V_{CE_{T_{ei}}} = V_{C_i} - V_{C_{i+1}}$	
Pos. pulse	Fig. 1a)	T_{ai}	$V_{CE_{T_{ai}}} = V_{dc}$	$V_{CE_{T_{bi}}} \approx V_{dc}$ $V_{CE_{T_{ci}}} = V_{CE_{T_{di}}} = V_{CE_{T_{ei}}} = V_{CE_{T_{fi}}} \approx 0$ $V_{AK_{D_{gi}}} = V_{AK_{D_{hi}}} \approx 0$
		T_{di}	$V_{CE_{T_{di}}} = V_{dc}$	$V_{CE_{T_{ci}}} \approx V_{dc}$ $V_{CE_{T_{ai}}} = V_{CE_{T_{bi}}} = V_{CE_{T_{ei}}} = V_{CE_{T_{fi}}} \approx 0$ $V_{AK_{D_{gi}}} = V_{AK_{D_{hi}}} \approx 0$
		T_{ai} & T_{di}	$V_{CE_{T_{ai}}} = V_{CE_{T_{di}}} = V_{dc}$	$V_{CE_{T_{ei}}} \approx 0$ $V_{CE_{T_{fi}}} \approx V_{dc}$ $V_{AK_{D_{gi}}} \approx -V_{dc}$ $V_{AK_{D_{hi}}} \approx 0$
	Fig. 1b)	T_{fi}	$V_{CE_{T_{fi}}} = 2V_{dc}$	$V_{CE_{T_{ai}}} = V_{CE_{T_{bi}}} = V_{CE_{T_{ei}}} \approx 0$ $V_{AK_{D_{gi}}} \approx -V_{dc}$ $V_{AK_{D_{hi}}} \approx -V_{dc}$
		T_{ai}	$V_{CE_{T_{ai}}} = V_{dc}$	$V_{CE_{T_{bi}}} + V_{CE_{T_{ei}}} \approx V_{dc}$ $V_{CE_{T_{fi}}} \approx 0$ $V_{AK_{D_{gi}}} = V_{CE_{T_{bi}}} - V_{dc}$ $V_{AK_{D_{hi}}} \approx 0$
		T_{fi} & T_{ai}	$V_{CE_{T_{fi}}} + V_{CE_{T_{ai}}} = 2V_{dc}$	$V_{CE_{T_{bi}}} = V_{CE_{T_{ei}}} \approx 0$ $V_{AK_{D_{gi}}} \approx -V_{dc}$ $V_{AK_{D_{hi}}} = V_{CE_{T_{ai}}} - V_{dc}$
Neg. pulse	Fig. 1a)	T_{bi}	$V_{CE_{T_{bi}}} = V_{dc}$	$V_{CE_{T_{ai}}} \approx V_{dc}$ $V_{CE_{T_{ci}}} = V_{CE_{T_{di}}} = V_{CE_{T_{ei}}} = V_{CE_{T_{fi}}} \approx 0$ $V_{AK_{D_{gi}}} = V_{AK_{D_{hi}}} \approx 0$
		T_{ci}	$V_{CE_{T_{ci}}} = V_{dc}$	$V_{CE_{T_{di}}} \approx V_{dc}$ $V_{CE_{T_{ai}}} = V_{CE_{T_{bi}}} = V_{CE_{T_{ei}}} = V_{CE_{T_{fi}}} \approx 0$ $V_{AK_{D_{gi}}} = V_{AK_{D_{hi}}} \approx 0$
		T_{bi} & T_{ci}	$V_{CE_{T_{bi}}} = V_{CE_{T_{ci}}} = V_{dc}$	$V_{CE_{T_{ei}}} \approx V_{dc}$ $V_{CE_{T_{fi}}} \approx 0$ $V_{AK_{D_{gi}}} \approx 0$ $V_{AK_{D_{hi}}} \approx -V_{dc}$
	Fig. 1b)	T_{bi}	$V_{CE_{T_{bi}}} = V_{dc}$	$V_{CE_{T_{ai}}} + V_{CE_{T_{fi}}} \approx V_{dc}$ $V_{CE_{T_{ei}}} \approx 0$ $V_{AK_{D_{gi}}} \approx 0$ $V_{AK_{D_{hi}}} = V_{CE_{T_{ai}}} - V_{dc}$
		T_{ei}	$V_{CE_{T_{ei}}} = 2V_{dc}$	$V_{CE_{T_{ai}}} = V_{CE_{T_{bi}}} = V_{CE_{T_{fi}}} \approx 0$ $V_{AK_{D_{gi}}} \approx -V_{dc}$ $V_{AK_{D_{hi}}} \approx -V_{dc}$
		T_{bi} & T_{ei}	$V_{CE_{T_{bi}}} + V_{CE_{T_{ei}}} = 2V_{dc}$	$V_{CE_{T_{ai}}} = V_{CE_{T_{fi}}} \approx 0$ $V_{AK_{D_{gi}}} = V_{dc} - V_{CE_{T_{bi}}}$ $V_{AK_{D_{hi}}} \approx -V_{dc}$

3) Negative pulse

To apply negative pulse in circuit of Fig. 1a), switches T_{bi} and T_{ci} are turned ON. However, if any of these switches in each independent stage individually or simultaneously fails to turn ON the hold-off of voltage of faulty switch or switches is V_{dc}, as shown in Table II. This voltage condition is similar to the circuit of Fig. 1b) in case of failure switch T_{bi} or simultaneous switches T_{bi} and T_{ei}, but it is not verified in case of individual failure of switch T_{ei}, as this switch has to hold-off the V_{dc} voltage, as illustrated in Table II.

III. SIMULATION RESULTS

The circuit of Fig. 1, with four stages was simulated in SIMULINK, MATLAB software. The simulated circuit was operated with V_{dc} ~500 V, 1 kHz, and 5 µs pulses width. The switch synchronism failure was simulated by delaying 1 µs to turn ON switches, T_a, T_b, T_c, and T_d of the first stage in regard of other stages switches.

In Fig. 6 are shown the simulated results for switches: a) T_{a1}, T_{b1}, T_{c1} and T_{d1}, and b) T_{e1} and T_{f1}, voltages, during normal charging and abnormal pulse (positive and negative) operating modes.

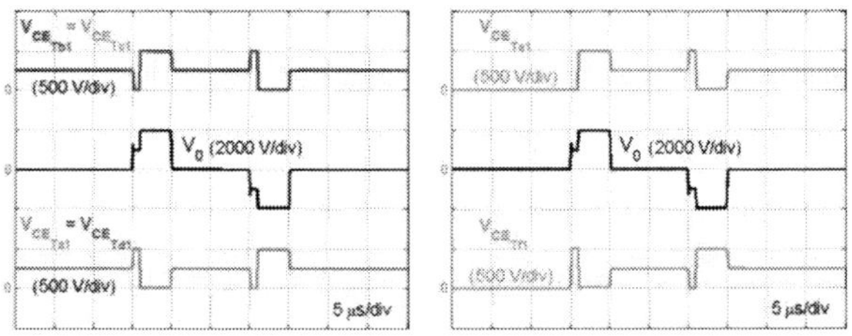

(a) (b)

Fig. 6 – Simulation results for the hold-off voltage of switches of the circuit of Fig. 1a) with lack of synchronism of simultaneous switches T_{a1} and T_{d1} for positive pulse and T_{b1} and T_{c1} for negative pulse: a) switches T_{a1}, T_{b1}, T_{c1} and T_{d1} hold-off voltage and the output voltage; b) switches T_{e1} and T_{f1} hold-off voltage and the output voltage. The scales are 5 µs/div (horizontal) and 500 V/div (vertical).

Considering Fig. 6a) and during charging operating mode, the set of switches T_{a1} and T_{d1} or T_{b1} and T_{c1} shares the voltage of V_{dc}. Considering pulse operating mode, and during the lapse of synchronism of switches T_{a1} and T_{d1} or T_{b1} and T_{c1} for positive and negative pulse, respectively, these switches hold-off the V_{dc} voltage accordingly to section II. Regarding switches T_{e1} and T_{f1} (Fig. 6b)), during negative delayed condition, switch T_{e1} holds-off the V_{dc} voltage, but the voltage across switch T_{f1} is approximately zero, because the anti-parallel diode D_{f1} is forward biased. This is similar for switches T_{a1} and T_{d1} for abnormal positive pulse condition, as these switches hold-off a voltage of V_{dc} and switch T_{f1}, but

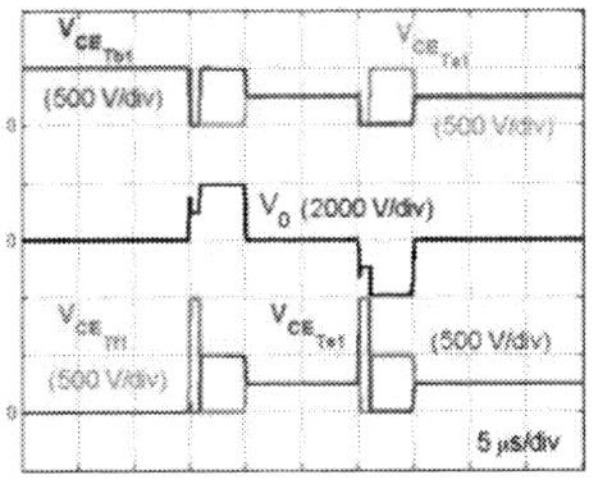

Fig. 7 – Simulation results for the hold-off voltage of switches T_{a1}, T_{b1}, T_{e1} and T_{f1} of the circuit of Fig. 1b) with lack of synchronism of switches T_{f1} and T_{e1} for positive pulse and negative pulse, respectively. The scales are 5 µs/div (horizontal) and 500 V/div (vertical).

switch T_{e1} voltage is zero, due to the anti-parallel diode D_{e1} becomes forward biased.

Fig. 7 shows the simulated waveforms for hold-off voltage of switches T_{a1}, T_{b1}, T_{e1} and T_{f1} during different operating modes and the output voltage of circuit of Fig. 1b). Considering charging period, switches in OFF-state (T_{a1} and T_{b1}) hold-off a voltage of V_{dc}, as described in section II. In negative pulse operating mode and during synchronization failure of switch T_{e1}, this switch has to hold off the double ($2V_{dc}$) of the V_{dc} voltage while the switches T_{a1} and T_{f1} voltage is approximately zero because the anti-parallel diodes D_{a1} and D_{f1} are forward biased. This situation is analogous for switch T_{f1} with respect of lack of synchronism during positive pulse operating mode, where the switch T_{f1} has to hold-off the voltage of $2V_{dc}$ and the voltage across the OFF-state driven switches T_{b1} and T_{e1} is approximately zero, as can be seen in Fig. 7.

IV. CONCLUSIONS

A comparative analysis of two solid-state Marx modulators was presented in terms of hold-voltage in abnormal switching operating condition such as switching synchronization failure.

Simulation results shows that in circuit of Fig. 1a) the maximum hold-off voltage of any switch within fault condition is V_{dc} against the hold-voltage of $2V_{dc}$ for faulty switches T_{f1} and T_{e1} of circuit of Fig. 1b) for positive and negative operating mode, respectively, presenting a hold-off voltage asymmetry between semiconductors at the same stage.

V. ACKNOWLEDGEMENTS

This work was supported by national funds through Fundação para a Ciência e a Tecnologia (FCT) with reference UID/CEC/50021/2013.

References

[1] Goebel, D. M.: "Pulse Technology", Capítulo 8 de "Handbook of Plasma Immersion Ion Implantation & Deposition", Editor Anders, André, 1ª edição, John Wiley & Sons, New York, 2000, p. 760, ISBN 0-471-24698-0.

[2] Jong-Hyun Kim; Myung-Hyo Ryu; Byung-Duk Min; Sergey V. Shenderey; Jong-Soo Kim; Geun-Hie Rim: "High Voltage Pulse Power Supply Using Marx Generator & Solid-State Switches", IEEE Transactions on Power Electronics, 2005, pp. 1244-1247, ISBN 0-7083-9252-3.

[3] Cassel, R. L.: "An All Solid State Pulsed Marx Type Modulator for Magnetrons and Klystrons", IEEE Pulsed Power Conference, pp. 836-838, 2005.

[4] L. M. Redondo, J. Fernando Silva, P. Tavares and E. Margato, "All Silicon Marx-bank topology for highvoltage, high-frequency rectangular pulses," in Proceedings of the 2005 IEEE 36th Annual Power Electronics Specialists Conference, 12-16 June, 2005, Recife, Brasil, pp. 1170-1174.

[5] H. Canacsinh; L. M. Redondo; J. Fernando Silva: "Isolated Autonomous Capacitive Power Supplies for Trigger Floating Semiconductors in a Marx Generator" in 2007 IEEE International Symposium on Industrial Electronics, Vigo, Spain, June 2007.

[6] Barbosa-Cánovas, V. Gustavo; Tapia, S. María; Cano, Pilar M.: "Novel Food Processing", CRC Press, 2005, Capítulo 1, ISBN 0-8247-5333-X.

[7] Shin, Wanho; Choi, Jaeho; Kim, Taewoong: "Biderectional Pulse Plasma Power Supply for Treatment of Air Pollution", in 37th IEEE Power Electronic Specialists Conference, 18-22 June, Jeju, South Korea, 2006, pp: 1-6.

[8] Zhang, Ruo-bing; Wu, Yan; Li, Guo-feng; Wang, Ninghui; LI, Jie: "Plasma induced degradation of Indigo Carmine by bipolar pulsed dielectric barrier discharge (DBD) in the water-air mixture", in Journal of Environment Sciences, Vol. 16, No. 5, pp. 808-812, 2004.

[9] H. Canacsinh; L. M. Redondo; J. Fernando Silva: "New Solid-State Marx Topology for Bipolar Repetitive High-Voltage Pulses", in 39th IEEE Power Electronics Specialists Conference, Greece, June 2008.

[10] Sakamoto, T.; Nami, A.; Akiyama, M.; Akiyama, H.: "A Repetitive Solid State Marx-Type Pulsed Power Generator Using Multistage Switch-Capacitor Cells", IEEE Transactions on Plasma Science, Vol. 40, No. 10, pp. 2316 – 2321, October 2012.

[11] H. Canacsinh; L. M. Redondo; J. Fernando Silva: "Marx type solid-state bipolar modulator topologies: performance comparison", in IEEE Transaction on Plasma Science, Vol. 40, Issue: 10, October 2012, pp. 2603-2610.

[12] H. Canacsinh; L. M. Redondo; J. Fernando Silva; E. Schamiloglu: "Solid-State Bipolar Marx Modulator Modeling", in IEEE Transaction on Plasma Science, Vol. 42, Issue: 10, August 2014, pp. 3048-3056.

*LLNL High Explosive Pulsed Power Capabilities and Experiments at the Ancho Canyon Test Site**

A.J. Young, A.D. White, D.P. Milhous, R.A. Anderson, R.M. Kuklo, J.J. Trueblood, A.J. Ferriera, R.D. Speer, E.V. Baluyot, G.R. Mease, A.CS. Ray, R.K. Hicks, D.B. Norton, A.D. Wiltse, C.K. Black, R.A. Richardson, J.P. Lewis, M.W. Danforth, D.N. Hulsey, D.J. Gire, K.M. Hood, R.A. Bonner
Lawrence Livermore National Laboratory
Livermore, California, USA

P.J. Rae, J.H. Goforth, D.H. Herrera, E.V. Baca, P. Dickson, J.A. Gunderson, B.B. Glover
Los Alamos National Laboratory
Los Alamos, New Mexico, USA

E.P. Daykin, C. Perez
National Security Technologies
Las Vegas, Nevada, USA

***Abstract*—High explosive pulsed power (HEPP) is a unique branch of pulsed power technology that requires specialized facilities that: (1) can safely fire high explosive (HE) devices, (2) have the subsystems necessary to initiate HE, (3) have subsystems which can inject electrical energy into an experiment, (4) have a control, timing and firing system to sequence a number of events with high precision, (5) have a number of auxiliary systems needed for experiments and diagnostics (e.g. gas systems, lasers, etc.), and (6) have the infrastructure needed to capture data from a wide range of diagnostics. Since 2014, Lawrence Livermore National Laboratory (LLNL) and Los Alamos National Laboratory (LANL) have undertaken a collaborative effort to refurbish the Ancho Canyon Point 88 firing site located near Los Alamos, New Mexico. This effort led to the establishment of a state-of-the-art facility capable of fielding high energy HEPP experiments. This paper will provide an overview the Ancho Canyon facility, and highlight the HEPP subsystems, including custom initiation systems (firesets), high energy capacitor banks, and Faraday rotation diagnostics. Also, a discussion of the grounding and shielding philosophy LLNL holds for HEPP experiments, and its impact on the facility refurbishment, will be given. To date, LLNL has fielded three experiments at the refurbished Ancho Canyon Point 88 facility. Data and waveforms will be shown from a recent experiment, where a staged magnetic flux compression generator system delivered 88 mega-amperes of current to a load.**

Keywords—High Explosive Pulsed Power; Magnetic Flux Compression Generator

I. Introduction

High explosive pulsed power (HEPP) is a unique subset of the pulsed power science and technology field which utilizes explosives to create fast pulses of electrical energy. Magnetic flux compression generators (MFCGs) are a specific class of HEPP which utilize high explosives (HE) to compress magnetic flux, thereby amplifying current and magnetic fields. The applications for MFCGs are numerous, and include creating multi-Megagauss magnetic fields for the purpose of studying matter in extreme states. HEPP can be used to drive equation-of-state (EOS) experiments, where shock or isentropic compression is used to study the properties of materials at high density and pressure. Lawrence Livermore National Laboratory (LLNL) has developed a suite of MFCGs to provide researchers with platforms in which to perform isentropic compression EOS experiments [1-4], namely the Full Function Test (FFT), Mini-G and Flat-Plate Generator (FPG) systems.

Facilities where HEPP experiments are conducted are unique in the fact that they must contain a broad range of capabilities and infrastructure. This includes the capabilities that one finds in a conventional pulsed power laboratory, e.g. capacitor banks, electromagnetic diagnostics, faraday cages, etc. In addition, an HEPP experimental facility must also provide the infrastructure required for high explosive tests, which aside from an area suitable for detonating HE, includes areas shielded from HE products to house experimental subsystems (e.g. a bunker), HE safety and initiation systems (firesets). Moreover, ancillary equipment for capturing other data of interest from the experimental campaign should be available, as the application requires. For LLNL's EOS work, this includes velocimetry diagnostics such as photonic Doppler velocimetry (PDV) systems.

In 2010, LLNL began fielding experiments at the Bunker 851 facility at Site 300 [5]. These experiments were focused on the FPG and Mini-G platforms discussed in [3, 4]. In 2014, LLNL focus shifted back to the FFT, a large-scale, 1000 lb HE weight experiment reported in [2]. For various reasons, the FFT experiments could not be fielded at the Bunker 851 facility. The solution, in the form of a collaborative effort between LLNL and Los Alamos National Laboratory (LANL), was to transition experimental efforts to the Ancho Canyon Point 88 test site, a facility which has a rich history of fielding HEPP experiments [6]. In late 2014, a refurbishment effort began at Ancho, with the aim of establishing a world-class facility where HEPP experiments, including large-scale high HE weight experiments, could be fielded. The result of the refurbishment, seen in Fig. 1, is a state-of-the-art facility with enhanced capabilities for HEPP experiments. As of the time of this report, LLNL has fielded three experiments at Ancho Canyon, including two FFT experiments. The Ancho Canyon facility has performed to expectations, enabling large, complex experiments to be executed with precision, and returning high quality data.

* This work was performed under the auspices of the U.S. Department of Energy by Lawrence Livermore National Laboratory under contract DE-AC52-07NA27344.

Fig. 1. The Ancho Canyon Point 88 test site being prepared for an LLNL HEPP experiment.

II. LLNL Experimental Subsystems

A. Capacitor "Seed" Bank

In order to provide the initial "seed" magnetic flux into the MFCG system for compression, a modular 450 kJ capacitor bank is used. From a facility perspective, this capacitor bank supplements the capabilities of the 2.4 MJ capacitor bank already in existence at Ancho [7]. As seen in Fig. 2, the bank is comprised of four, parallel 333 μF capacitors which can be charged to 26 kV. The interconnecting buss-work is designed such that any number of the four capacitors may be used, so that the bank may be tailored to the requirements of a given experiment. The capacitors are charged via an 8 kW high voltage power supply. To protect the capacitors in case of a fault in the circuit, a fault-current limiting inductor is connected in series with each capacitor. These inductors are of the same model as those that are used as ballast inductors in the National Ignition Facility's power conditioning systems [8]. The system is housed within a Faraday cage, which is maintained from the bank's location within the bunker up to the experiment area. The Faraday cage is maintained this way in order to manage the risk that electromagnetic interference (EMI) produced during discharge of the bank will upset other experimental subsystems. One important feature of the seed bank is that, during experiment execution, the ground of the bank is electrically isolated from facility ground. The margin of isolation is greater than 50 kV DC. The isolation requirement stems from the desire to eliminate ground-paths (or loops) within the facility during experiment execution. The margin of isolation is derived from actual measurements of the potential between facility ground and seed bank ground during experiments.

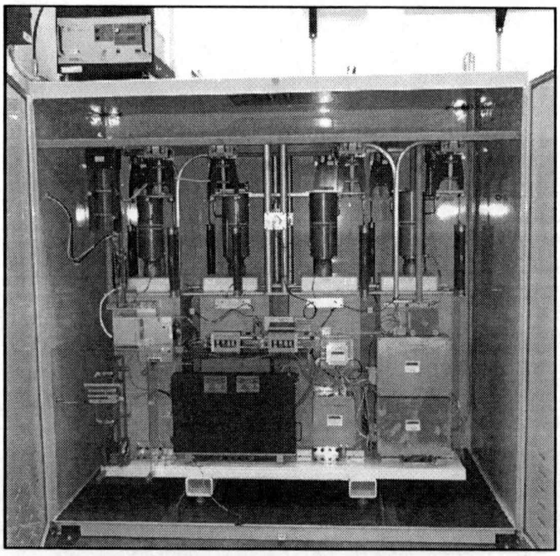

Fig. 2. Photograph of the 450 kJ capacitor bank installed at Ancho Canyon, with 8 kW power supply external to the Faraday cage in the upper left of the picture.

The energy stored in the capacitor bank is switched into the MFCG system by a solid dielectric puncture switch. The dielectric used in the switch is Mylar with a ~15 mil thickness. Actuation of the switch occurs through the use of exploding bridge-foils (EBFs). Two striplines, which each contain two 3.175 mm x 3.175 mm x 0.005 mm copper EBFs, are embedded into the switch assembly. A 6 kV, 100 J capacitor discharge unit (CDU) is used to drive the EBFs. The electrical energy deposited into the EBFs by the CDU causes them to "fuse", i.e. the copper material is driven to a plasma state through ohmic heating. The mechanical pressure of the fusing process is applied to the switch assembly, ultimately causing the dielectric failure and the switch to close. With the present design of the CDU and EBF circuit, the switch function time (from EBF fusing to switch closure) is a few microseconds, with a few hundred nanoseconds of jitter, which is suitable for the timescales of seed bank discharge (many tens to hundreds of microseconds). Since the switch and switch CDU are all electrically connected to the seed bank during an experiment, they are also electrically isolated from bunker ground with same degree of isolation as the capacitor assembly.

B. Initiation Systems (Firesets)

The MFCG platforms used by LLNL for EOS research are all staged systems, that is multiple MFCGs are connected serially to achieve the final desired output current pulse. Thus, at least two initiation systems are required for each experiment. The firesets installed at Ancho for LLNL experiments are identical to those described in [5, 9], which can be referenced for a more detailed discussion of their design and function. For single point initiation of HE, used primarily in LLNL helical MFCGs, a 1 μF, 3.5 kV CDU is used, which is used to drive an RP-1 type detonator through ~27.5 m of Teledyne Reynolds Type C cable. For multi-point initiation of HE, which is required for LLNL coaxial MFCG designs, a 14 μF, 10 kV CDU is used. This fireset has up to fifty-four outputs. For the LLNL FFT experiments, this system is used to drive 54 RP-1 type detonators. As with the seed bank, these initiation systems are isolated from ground during an experiment, with the same margin of electrical isolation.

C. Control, Timing and Trigger Systems

Any given LLNL experiment is ultimately composed of many pulsed power (e.g. seed bank), diagnostic (e.g. PDV) and data acquisition (DAQ) systems. In order for an experiment to be successful, all of these systems must be controlled and sequenced with high reliability and temporal precision. Where possible, fiber optic or pneumatic systems are used for control, feedback and trigger distributions. These efforts are made to reduce EMI insult to subsystems during an experiment. The trigger system installed at Ancho has sub-nanosecond precision, and to-date has seen no failures after tens of thousands of triggers distributed amongst experimental subsystems.

D. Pulsed Power Diagnostics

Since HEPP experiments are inherently single-shot events, extra emphasis is placed on the implementation of diagnostics into the MFCG system and facility. In addition, the quality of the data which are captured from the diagnostics is of the upmost importance, given that the function of the experiment (intended and otherwise) is inferred almost exclusively from recorded data, post-execution. Detailed discussions of the functionality and implementation of MFCG system and facility diagnostics has been published elsewhere [5, 9], and only a few examples, relevant to the discussions in this report, will be described briefly here.

For MFCG systems, two primary diagnostics fielded are: B-dot probes and Faraday rotation sensors. The B-dot probes are used for deriving the current derivative from a measurement of the time varying magnetic field in the experiment. In general, 8-14 B-dot probes are integrated into experiments, and are typically installed in the MFCG systems. Faraday rotation sensors are used to derive the current in the system by measuring the change in the state of polarization of light due to the interaction of the light and the magnetic field generated in the experiment. There is a four sensor system installed at Ancho, which enables these measurements to be taken at multiple locations within the MFCG system.

For other experimental subsystems, a variety of diagnostics are used to measure the performance of the experiment and bunker systems. In the seed bank, for example, the individual capacitor currents are measured via current transformers, while the total current is derived from two redundant Rogowski coil measurement at the seed bank switch. A resistive voltage probe provides a measurement of seed bank voltage near the capacitors, and a capacitive voltage divider is used to measure the potential formed between the seed bank ground and bunker ground, which are electrically isolated from one another during an experiment.

To record the data from all diagnostics, 120 channels of digitizers are available at Ancho. The DAQ is composed of Teradyne digitizers [10], with either 2 GS/s (8-bit) or 400 MS/s (14-bit) sample rates. The DAQ system is housed within a Faraday cage in the bunker, which is maintained all the way to the MFCG system. A few select diagnostics (e.g. B-dots) are hard-wired to the digitizers, while a majority of diagnostic signals are transmitted into the Faraday cage optically. Once in the Faraday cage, optical signals are converted back to electrical signals and recorded on the digitizers.

E. Additional Diagnostic Systems

The primary system for diagnosing the EOS load is PDV in LLNL experiments. At Ancho, there exists the capability to field hundreds of channels of PDV on a given experiment, of both the conventional [11] and multiplexed [12] technologies. Livermore experiments executed thus far have only required tens of PDV probes.

Another diagnostic, which was just recently fielded on LLNL experiments, is an HE shock probe. This sensor is used to measure the arrival of the detonation front at a desired location in the MFCGs. The sensor is composed of a conventional multi-mode (62.5/125 μm) fiber optic cable, with one end having a standard ST/UPC connecter, and the other end having only a ferrule-optical fiber termination. The tip of the latter is coated with 150 nm of aluminum. As discussed in [13], these probes provide a fast-rise (nanosecond range) optical signal upon arrival of the HE detonation front at the probe. For experiments at Ancho, these probes are used to provide information on HE burn timing and symmetry in the MFCG systems.

III. LLNL Experiments at Ancho Canyon

To-date, LLNL has conducted three experiments at the refurbished Ancho Canyon Point 88 test site, including two FFT experiments. The facility has performed to expectations, with high quality data captured during each experiment.

Figure 3 shows two sets of waveforms captured from the seed bank from an LLNL FFT experiment. The upper plot shows an overlay of the integrated Rogowski coil signals with the sum of the individual capacitor currents. The data are within 1-2% of one another during the MFCG seeding phase. The lower plot shows the measured potential difference between the seed bank ground and facility ground during the experiment. Note that the peak voltage seen is roughly -30 kV, which illustrates the need for high voltage isolation of the system.

Figures 4 and 5 shows the current and current derivative waveforms taken from the staged MFCG system during an LLNL FFT experiment. In Fig. 4, an overlay of current derivative waveforms, taken from B-dot probes and Faraday rotation sensors, is presented. One aspect of these waveforms that speaks to the fidelity of the captured data is the quality of the current derivative taken from the processed Faraday rotation data. Faraday rotation is a direct measurement of

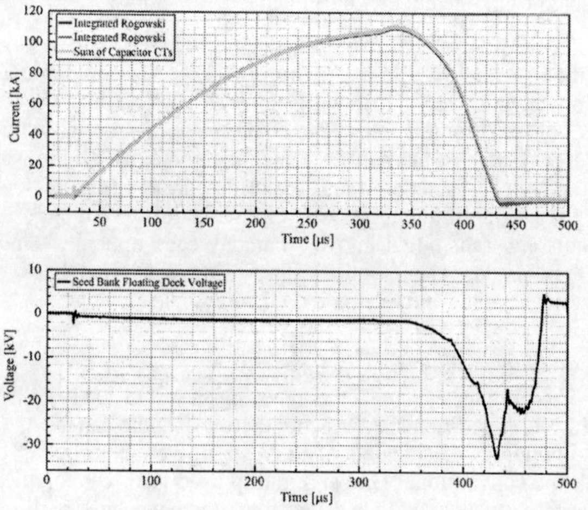

Fig. 3. Upper: current waveforms captured from seed bank discharge (traces superimposed). Lower: voltage waveform of the potential difference between seed bank ground and bunker ground.

current, thus the waveform presented in the figure was obtained by taking the derivative of the processed data. Typically, taking the derivative of measured data results in extremely "noisy" waveform, due to the amplification of noise that is present in the original data. Here, one finds that the current derivative of the Faraday rotation data has a high signal-to-noise ratio, and is in excellent agreement with B-dot probe data. The waveforms presented in Fig. 5 show an overlay of current data as taken from the B-dot and Faraday rotation sensors. The waveforms are in excellent agreement, and showed less than a 1% difference at peak current, which was ~88 MA in this particular experiment.

IV. Conclusion

In conclusion, LLNL and LANL have collaborated to refurbish the Ancho Canyon Point 88 test site located near Los Alamos, New Mexico. The facility is operational, and has conducted three LLNL experiments thus far, with more

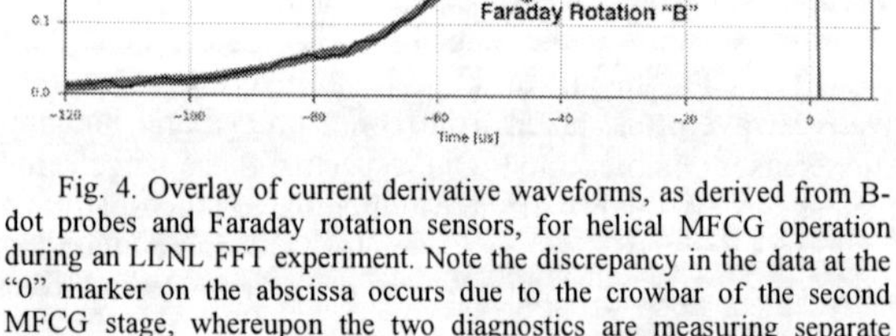

Fig. 4. Overlay of current derivative waveforms, as derived from B-dot probes and Faraday rotation sensors, for helical MFCG operation during an LLNL FFT experiment. Note the discrepancy in the data at the "0" marker on the abscissa occurs due to the crowbar of the second MFCG stage, whereupon the two diagnostics are measuring separate circuits.

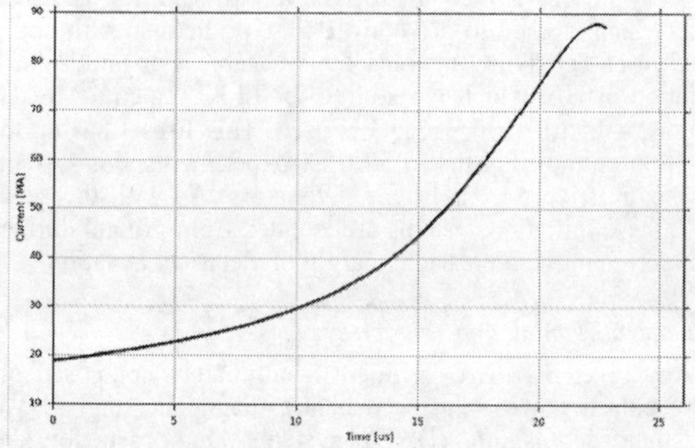

Fig. 5. Overlay of current waveforms, as derived from B-dot and Faraday rotation sensors, for the coaxial MFCG stage of an LLNL FFT experiment.

planned in the near future. Additional infrastructure has modernized the facility, enabling complex, valuable experiments to be conducted reliably and expeditiously, with high quality data return.

References

[1] D. B. Reisman, et al., "The advanced helical generator", Rev. Sci. Instr., 81 (2010), p. 034701.

[2] D. B. Reisman, et al., "Note: The full function test explosive generator", Rev. Sci. Instr., 81 (2010), p. 036109.

[3] D. B. Reisman, et al., "Explosive Flux Compression Generators at LLNL," Proc. of the 14th Conference on Megagauss Magnetic Field Generation and Related Topics, 2012.

[4] D. A. Goerz, et al., "Flat plate FCG experimental system for material studies", Proc. of the 14th Conference on Megagauss Magnetic Field Generation and Related Topics, 2012.

[5] A. D. White, et al, "Explosive pulsed power experimental capability at LLNL," Proc. of the 14th Conference on Megagauss Magnetic Field Generation and Related Topics, 2012.

[6] D.J. Erickson, et al., "Firing Complex for Explosive Pulsed Power," Proc. of the 4th Conference on Megagauss Magnetic Field Generation and Related Topics, 1986.

[7] A. Ramrus, W. Caton, R. Robinson, B. Freeman, M. Fowler, "A 2.4 MJ Bank with Fiber Optic and Pneumatic Control Links," Proc. of the 4th Conference on Megagauss Magnetic Field Generation and Related Topics, 1986.

[8] P.A. Arnold, et al., "An update on the status of the NIF power conditioning system, " IEEE Transactions on Plasma Science, Vol. 36, No. 2, 2008.

[9] D.A. Goerz, et al., "Implementation of Pulsed Power Diagnostics on Explosive Flux Compression Generators at LLNL," Proc. of the 13th Conference on Megagauss Magnetic Field Generation and Related Topics, 2010.

[10] http://www.teradyne.com/products/defense-aerospace/zt-series-oscilloscopes-and-digitizers

[11] O. T. Strand, D. R. Goosman, C. Martinez, T. L. Whitworth, W. W. Kuhlow, "Compact system for high-speed velocimetry heterodyne techniques," Rev. Sci. Instr., 77, 2005.

[12] E. P. Daykin, A. Diaz, C. Gallegos, C. Perez, A Rutkowski, "A multiplexed many-point PDV – techniques and technologies," 5th Annual PDV Workshop, Ohio, 2010.

[13] F. Roeske, "A Fast Rise-Time, Fiber Optic Pin," Proc. of the 11th International Detonation Symposium, 1998.

Generation of Series and Series Trains of Nanosecond Pulses Terawatt Level on Inductive Storages and Combination Opening Switches: Semiconductor Opening Switch and Opening Switch Based on Skin – Effect Phenomenon

Egorov O.G.
Atomic energy department
TRINITI
Troitsk, Moscow, Russian Federation

***Abstract*—The present paper considers a pulse generator design based on an IS of transformer type, i.e. a combination of a semiconductor opening switch and skin-effect as a phenomenon of opening switch resulted in providing current break for generating both single pulses and a series of pulses. As a fact, the given design proves to be simpler than other paper was offered to consideration.**

Keywords—skin-effect; opening switch; series pulses; pulsed power generator.

I. Introduction

A large number of energy storage devices with different parameters (different in physical principle of energy storage, energy density, the unit cost of stored energy and so on) have been developed by this time. Each of them has its own application area. All the additional elements used for energy storage operation are adapted for the particular application area and compose the pulsed generator being the single technical unit. For example, the application area for the inductive storages (IS) and transformer inductive storages (TIS) is associated with the properties of the opening switch. Opening switch is selected by the load pulse parameters (power, duration, etc.) and can be a single switch or a combination of several switches. One of the main switch parameters is the ratio $\sim t_{store}/t_{out}$, where t_{store} is the time of energy storage and t_{out} is the time of power switching to the load. For a single opening switch, this ratio does not exceed $\sim 10 \div 20$ [1-4].

The task to generate series of multiterrawatt power pulses and each having nanosecond duration cannot be solved on the basis of on TIN or IN, since there no opening switches providing $t_{stor}/t_{out} \sim 10^7 \div 10^8$ [5]. And, as result, there is no efficient electrical generator circuit which could: -generate a pulse and switch the energy to the load with the maximal efficiency; -provide damp of the energy reflected from the load.

The present work is the continuation of the efforts for finding technical solutions to generate a nanosecond pulse of multiterrawatt power based on TIS. The main switch parameter in this case is $t_{store}/t_{out} \sim 10^8$ and even higher for swiched power at multiterrawatt level. In this paper it is offered to consider the contactless opening switch on the basis of the skin-effect. Current interruption in the primary TIN circuit could be realized for the time interval equal to pulse duration. The skin–effect in a highly resistive layer can be used as an opening switch along with other opening switches such as plasma opening switch (POS) or semiconductor opening switch (SOS). One of possible variants enabling to apply such approach is offered in this paper.

II. The Theory of the Skin Effect and Its Use as an Opening Switch

A possible technical solution enabling to use skin-effect was shown in [5]. The outside surface of primary winding and inner surface of secondary winding of TIN were covered with a highly resistive film. Let us analyze the efficiency of this approach.

The current action on the highly resistive coating can be considered in three steps. The first step is associated with the energy accumulation; the second one corresponds to current transformation from the primary to the secondary winding and the third one is for with the energy switching to the load. To estimate energy losses at each step of the current transformation, consider the time –depended current density distribution at different depth of the high -resistance layer. Consider a one-dimensional solution of the problem on the penetration of a plain wave of the pulse electromagnetic field into a bimetallic plate. In this case we is assume, that the thickness of the skin-layer is significantly smaller than the size of the conductor; - uniform plane wave incident on the surface is perpendicular; - electrical parameters of the conductors are constant in time. Consequently, the leaner current density **j=[H•n]** [6].

Fig.1 shows a section of the bipolar plate where d_1 is the thickness of the high resistance layer, d_3 is the thickness of the low resistance part of the bimetallic conductor.

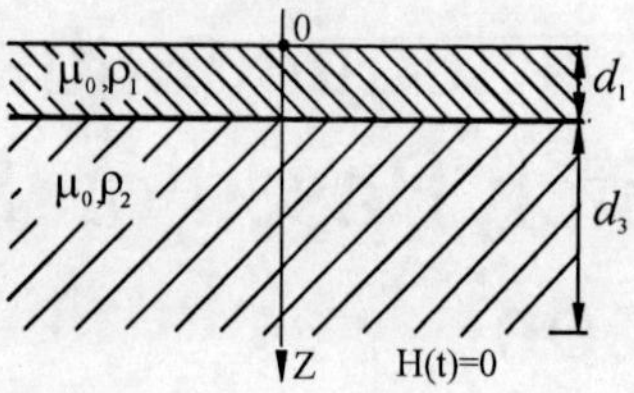

Fig. 1. Design of the bimetallic conductor for the analysis of the pulsed magnetic field penetration.

The penetration of the pulse magnetic field into the bipolar plate is described by the following equations:

$$\frac{\partial^2 H_k}{\partial z^2} - \mu_0 \Big/ \rho_k \frac{\partial H_k}{\partial t} = 0 \ , \qquad (1)$$

$$E_k = -\rho_k \frac{\partial H_k}{\partial z}, \qquad (2)$$

where $H(z,t)$, and $E(z,t)$ is the intensity of the magnetic and electric fields, respectively; μ_0 and ρ_k are the magnetic permeability and resistivity, respectively, k=1 at $0<z<d_1$, k=2 at $d_1<z<d_3$.

Initial conditions $H_k(z,0)=0$. Boundary conditions

$H_1(0,t)=H_1(t)$; $H_1(d_1,t)=H_2(d_1,t)$; $H_2(\infty,t)=0$;$E_1(d_1,t)$-$E_2(d_1,t)=0$

This system of equations may be solved by both analytical and numerical methods. In this case, the task is solved by numerical methods [7, 8]. Equation (1) is transformed into a differential form:

$$\rho_{i-1/2}H_{i-1}^{j+1} - \left(\frac{\mu_0 h^2}{\tau} + \rho_{i-1/2} + \rho_{i+1/2}\right)H_i^{j+1} + \rho_{i+1/2}H_{i+1}^{j+1} = -\frac{\mu_0 h^2}{\tau}H_i^j \qquad (3),$$

where $\rho_{i\pm1/2}=0.5(\rho_i+\rho_{i\pm1/2})$.

This expression has an error of approximation $O(h^2+\tau)$ and is solved by the sweep method in a "pass-through" account without explicit highlighting break points $\rho(z)$. The sweep algorithm is well- known.

In Fig.2 is the time-dependent of magnetic field having expression $H(t)=A\cdot\sin(\omega t)\cdot\exp(-\alpha t)$ at $z=0$ -(1) and $z=d_1$ –(2). The incident wave is formed when outputting energy from TIN to the load (third step). In Fig.2 b) is time-dependent of magnetic field having expression $H(t)=A\cdot\sin(\omega_2 t)$ at z=0 –(1) and z=d_1 –(2) that equivalence the conversion of the current from the primary winding to the secondary (the second step). The trace analysis shows that ΔH is corresponds of current loss in high resistance layer. Obviously, the less duration of pulse (Fig.2a) and ticker high resistance layer d_1, the grater loss in it. The other hand, similar energy losses increase for the second step (Fig.2b) and for the first step. Among the aspects the problems of generation a series of high power pulses is cooling the high-resistance layer. This is especially important in the generation of a series pulses and pulse train.

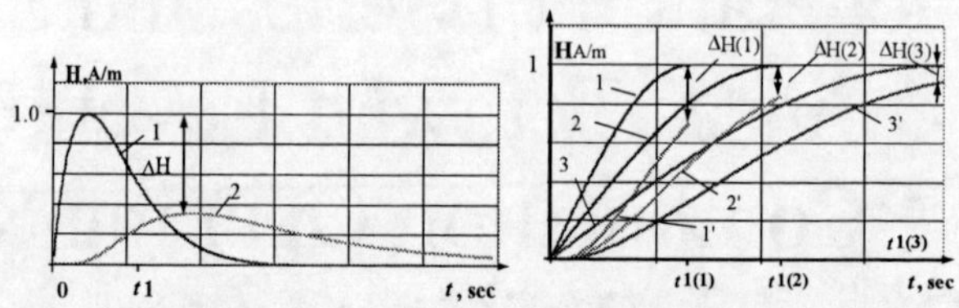

Fig. 2. The pulses of the external magnetic field both on the surface of the high resistance layer - (1) at z=0, and in the conducting layer – (2) $z=d_1$, at various time duration of t and form of the pulsed magnetic field.

The primary goal offered for consideration in these paper to decrease energy loses first and second steps and to make easy high resistance layer cooling. So, it is proposed to shift the energy loss in the first and second steps to the area of high frequencies where their amplitude is insignificant. With this purpose the high-resistance layer is separated from the primary winding surface, that makes its cooling more convenient, and each coil of the layer is separated from each other by a gap. Thus, the gaps between the high-resistance layers are perpendicular to the current induced by magnetic field. In the third step the divided parts of the high- resistance layer are short circuited thereby, during the pulse the layer is converted to inductance with a time-dependent resistance. In other words, the time interval $[0\div t_S]$ the high resistance layers is open inductance where t_S is the sum of the energy storage time in the primary winding and the time of current conversion from primary to secondary winding. Obviously, the pulse generator is effective (in the third stage, namely when outputting power to a load), provided that $R_l << R_S(0<t<t_p)$, where R_l is load resistance, $R_S<t<t_p$ is the resistance of high resistance layer, and t_p is the pulse duration.

The technical solution is illustrated by Fig.3. Fig. 3 displays the section of the axisymmetric cable where ***a-b*** is the secondary winding, ***a'-b'*** the high resistance winding and ***a''-b''*** the primary winding. For clarity, a diode is located in the gap of the high- resistance layer ***a'-b'***. Such sizes as thickness of high resistance layer (d_1), high resistance layer and first turn (d_2), primary winding (d_3) are select from pulse parameters.

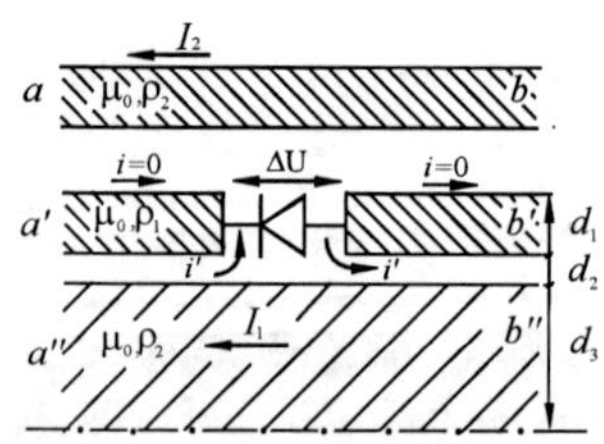

Fig. 3. Design of technical solution.

It's known, that the screens are effective with gaps, when gaps are arranged parallel to currents induced [10]. In the presented version the gaps are arranged perpendicular to the currents **j**=[**H·n**]. Thus, the loss of current at the first and second steps are possible at frequencies comparable with the transverse dimensions of the storage coil ~$c/\pi D$. With a diameter of coil in the storage D~1m, losses are possible at frequencies of ~10^8 s^{-1} and more. The time of the first stage is

1-2 seconds, while during the second step it does not exceed 250-300 ns, i.e. less than ~10^6 s^{-1} (for SOS) and 1-1.5mks i.e. less than ~10^5 s^{-1} (POS). Thus, the loss at both steps are substantially reduced and may be neglected in the first approximation.

Solution to reduce emerging currents has been proposed in [10]. At other frequencies arising currents will charge the capacity at the joint and between the high -resistance layer and the primary and secondary windings. At the third step, the nature of magnetic field penetration will be determined by her diffusion of the magnetic field through high –resistance layer and the penetration of the magnetic field through the inductance of diode *D*.

The coupling coefficient *k* between the primary and secondary windings is dependent on value of d_2. Since the voltage between the high resistance layer and the primary winding at the first and second steps is insignificant, one can meet the condition $d_2 << d_1$ which leads to a slight decrease in *k*. In the first approximation one can also neglect the impact of gap d_2 magnetic field penetration to in metal [9].

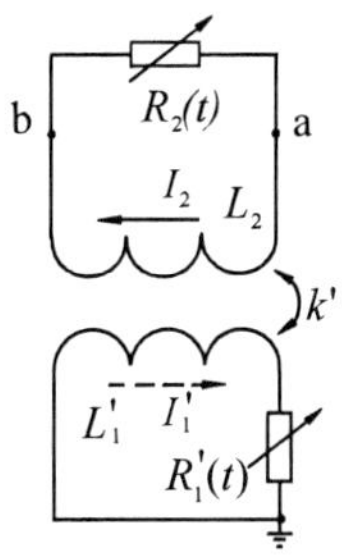

Fig. 4. Circuitry for modelling the transition process.

Modelling the form of voltage pulse on the resistive load has been performed in the circuit diagram (see Fig.4). The load resistance $R_1(t)$ changes from 0 to R_{load}. To assess the nature of the current changes in primarily and secondary windings, it was assumed that the instantaneous value of skin layer equal

$$d(t) = \frac{1}{2}\sqrt{\frac{\rho\pi t}{\mu_0}} \quad (4),$$

where ρ is the resistivity of the material.

In the given case it is assumed that the current is uniformly distributed with the a density

$$j(0,t) = 2\sqrt{\frac{\mu_0}{\pi t\rho}}H(0,t) \quad (5)$$

where $H(0,t)$ is the magnetic field on the material surface.

The transition process (Fig.4) is described by a system of differential equations

$$L_1\frac{dI_1}{dt} + M\frac{dI_2}{dt} + I_1R_1(t) = 0 \quad (6),$$

$$L_2\frac{dI_2}{dt} + M\frac{dI_1}{dt} + I_2R_2(t) = 0 \quad (7),$$

Initial conditions of currents in the circuits are as follows: $I_1(t=0)=0$ and $I_2(t=0)=I_0$, respectively. The resistance $R_2(t)$ is a combination of the constant load and time -varying resistance of the switch SOS. The value of $R_2(t)$ is specified numerically and changes linearly from 0 to R_{load}=const during the break Δt. The given system of the equations was solved by numerical methods, i.e. by replacing the differential expressions on the implicit difference scheme with a uniform time grid. In general, the error that occurs when replacing, depends on the distribution of the grid nodes and the function smoothness. Given that all the functions of current and resistance are smooth, the resulting errors are quite acceptable for evaluation of the nature of current changes [8]. In calculations it was take into account that the conductor of the primary winding is not sold, but it is a set of wires thickness equal one skin –layer of pulse.

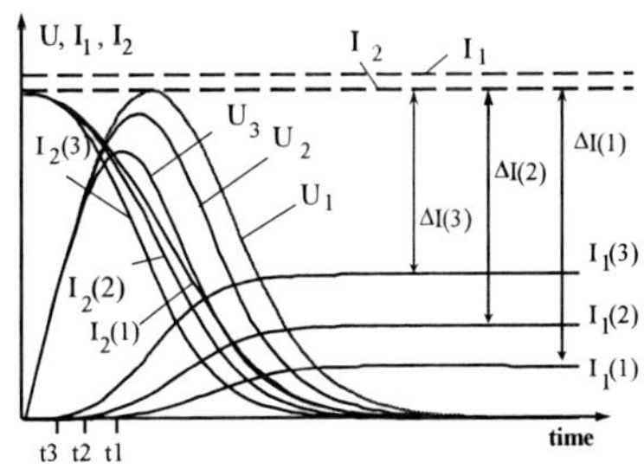

Fig. 5. Time- dependence changes of currents I_1 and I_2 at different electrical conductivity of the high -resistance layer $\rho_1 > \rho_2 > \rho_3$.

The calculation results obtained are shown on Fig.5, where U_1, U_2, U_3 are the voltage pulses on resistive load, $I_1(1)$, $I_1(2)$, $I_1(3)$ –current in primary winding, and $I_2(1)$, $I_2(2)$, $I_2(3)$ –the current flowing through R_2(t). The value t_1, t_2, t_3 correspond to a phase shift arising with the magnetic field of penetration through high -resistance layer. The value of current and voltage pulses correspond to a change in the electrical conductivity of high -resistance layer values $\rho_1 > \rho_2 > \rho_3$.

As seen from curves, the higher the value of the layer resistance, the lower value of current back to primary winding. The current difference $\Delta I(1)$, $\Delta I(2)$, $\Delta I(3)$ correspond to a sum of energy transfer to the load and energy lost in both high -resistance layer and the opening switch SOS (or POS).

III. TECHNICAL APPLICATION - PULSE GENERATOR MODULE

The electrical circuit of generator is given on Fig.6. The device operates as follows: when the energy is being accumulated in the primary winding L_1, (see Fig. 6), in direction from **b**" to **a**" current I_1 flows the increasing in time to value of I_0. At the same time on high -resistance layer of **a'** - **b'** and secondary winding **a-b** a voltage of opposite is induced (see Fig.3). The first case the diode D prevents from current flowing, in the second –the opening switch SOS or POS (see Fig.6). The potential difference ΔU *is* equal to the potential difference of the charger. At the second step, the current I_1 is converted from the primary winding to the secondary one. In Fig.6 the switch S_1 is closed, the capacitor C_1 is discharged through the plasma- saturated opening switch SOS or POS. Without regard to resistive loss its value is equal $I_2=k{\cdot}I_1{\cdot}(L_1/L_2)^{1/2}$. The current excited in the high -resistance layer of ***a'-b'*** is zero. The diode D prevents the current flow because its direction is opposite to the direction of I_2. The potential ΔU is equal the potential of the charge device or voltage capacitor bank C_2. As result, the transformer with the primary and secondary windings L_1 and L_2' respectively, and the coupling coefficient k' has following conditions: the current of secondary winding is equal to I_2, while the current I_1 =0 and I_1'=0.

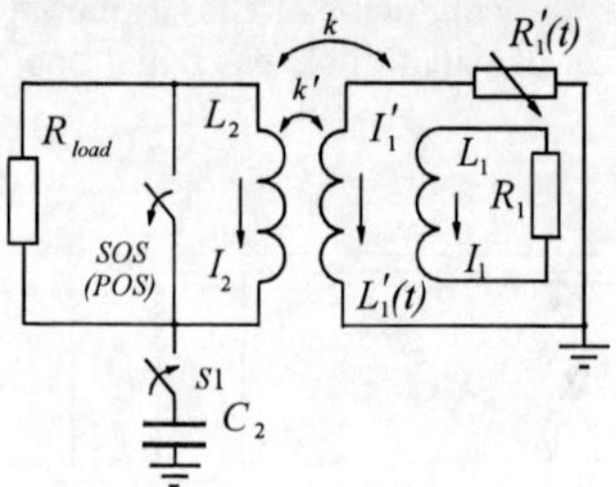

Fig. 6. Scematic diagram of the device in steps of operation.

The third step- energy output to the load. The current is excited in the high -resistance winding ***a'-b'*** and flows in the same direction with the current I_1 which is shorted with the diode D. Penetration of the magnetic field in the primary winding is affected by two factors, namely, the diffusion magnetic field through high -resistance layer and the diode design, its inductance, structure of contacts, etc. The lower inductance and the contact resistance, the lower the value of the magnetic field that penetrates through the diode.

Fig. 6 shows a circuit diagram of the third step: when shorting the switch SOS (POS), the current of the secondary winding is converted to the current inductance L_1' (in general case, it is time -dependent) with a high -resistance layer R_1' with coupling coefficient equal k' between primary winding and inductance L'. Inside of high resistance layer is the inductance L_1 with its current to zero and own resistance equal to R_1 (see Fig.3). Obviously, the condition is satisfied $L'_1(t)/R'_1(t) \ll L_1/R_1$, in this case the energy is outputted to the load with maximum efficiency.

A general view and elements of the module is ***a***) and the unit of open circuit and energy output to the load is ***b***) are shown in Fig.7. Based on the considered method of the current gap the attention is given to the pulse generator module with a single transformer ring.

Figs. 7 ***a***) and b) show: a transformer ring -1 with two units of current conversion and energy output to the load is view I, a wear ring -1, which is intended to compensate the excess inductance of the energy storage -2. In Fig.7 ***b***) is view I - units of current conversion and energy output to the load. This figure also shows: primary winding -3, high -resistance winding -4, secondary winding -5, cable of energy output to the load -7, and circuit break cable connected with SOS-6.

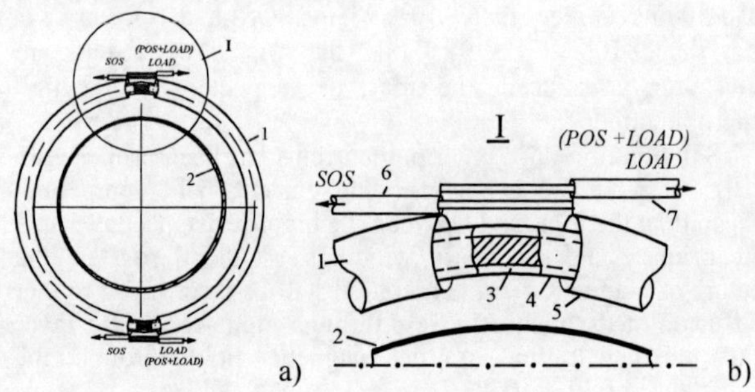

Fig. 7. General views of the module- a) and the unit of open circuit and energy output to the load -b)

IV. CONCLUSION

The technical solution in which the skin-affect is considered as a phenomenon that can used as a non-contact high-speed opening switch for breaking primary of TIS is proposed in this paper. The combination SOS or POS and skin -effect as a phenomenon of noncontact opening switch allows for the ratio to t_{store}/t_{out} ~$10^7 \div 10^8$ and more. The use of this phenomenon the vacuum, explosive and other high-speed opening switches, which are commonly used for this purpose. Such solutions can be applied to generation a series of pulses and series of pulses train terawatt level and nanosecond duration. Peculiarity of this solution is capability realize thickness of high resistance layer equal 1.5-2 skin layer, this is increase efficiency of pulse generation. Such solution of combination opening switches has self synchronization of parallel working SOS (POS). At energy output magnetic coupling connection between the turns of TIS is an additional load to the total load. The asynchronous behavior in working SOS (especially POS) causes an electromagnetic wave in TIS, which, consequently, gives rise to an additional current pulse in parallel opening switches and, as a result, leads to acceleration of their operation.

References

[1] D.But D.A at al. "Storages of energy", M: Energoatomizdat, 1991, 400p. (rus)

[2] J. "Physics and technology of high powers" // Ed. E.P. Velihov, M:. Energoatomizdat, 1987, p.352.

[3] "Physics and technology of electromagnetic energy of high density" //Ed. V.D.Selemir, L.N.Pljashkevich, M:. Atomizdat, 2003, p.555.

[4] "Opening switches" //Ed. A.Guenther, M.Kristiansen, and T.Martin, PP, NY, 1987, 313p.

[5] O.G. Egorov // Proc. Euro-Asian Pulsed Power Conference, Kumamoto, Japan, September 8-12, 2014.

[6] H.Buchholz "Electrical and magnetic field" Berlin, 1957, 712p.

[7] V.M.Mikhaylov "Pulses magnetic fields", Kharkiv, 1979, 140p.

[8] A.A.Samarsky, A.V.Gulin "Numerical methods", Nauka, 1989, 432p.

[9] V.M. Mikhaylov, Ed.I. Pismenny // Proc of the Russian Academy of Sciences, "Energy and transport" , 1974, No.6, pp 147-151.

[10] N.Kaden "Electromagnetic screens". 1950, p.111.

Solid-State Pulsed Power System for Dielectric Wall Accelerator *

Yi Liu, Liansheng Xia, Yi Chen, Wei Wang, Mao Ye, Huang Zhang, Chao Yang, Jinshui Shi, Linwen Zhang
Institute of Fluid Physics
China Academy of Engineering Physics
Mianyang, China

Abstract—**Benefiting from solid-state pulsed power system and high gradient insulator (HGI), dielectric wall accelerator (DWA) as a new kind of high acceleration gradient accelerator has a compact construction that makes it quite suitable for some certain applications, such as proton tumor treatment and industrial inspection. The solid-state pulsed power system is made up of several groups of pulsed power units. Each group contains 2 to 4 units, and each unit is set up as a 2-stage Blumlein line. It is constituted by GaAs photoconductive semiconductor switches and plate pulse forming lines, which are made of high-performance dielectric ceramic. The plate pulse forming line can withstand maximum 55 kV. The GaAs PCSS triggered by a high power laser diode driver works in nonlinear mode and its bias voltage and conduction current can be up to 20 kV and 800 A respectively. An ECR ion source is used for providing proton beams. In the experiments, the 2-BL unit can provide a 34 kV, 10 ns pulse. The performance of the pulsed power group containing 2 to 4 units is well tested and optimized. Several groups have been used to load HGIs at different intervals so as to accelerate the proton beams which are just passing through the HGIs. Until now, the solid-state pulsed power system has performed well and a system with 6 groups (14 units in total) can accelerate proton from 40 keV (after the kicker) to 380 keV.**

Keywords—dielectric wall accelerator; solid-state pulsed power system; 2-BL unit; proton beam;

I. Introduction

The application background of dielectric wall accelerator (DWA) is compact proton accelerator for radiation therapy[1]. Proton radiation therapy is proved better than X-ray, because of Bragg peak, a peak dosage at the end of proton range in human body. So proton beams do less damage to healthy organic tissue but more damage to target tumor. However, traditional proton accelerator is quite huge, complex and expensive, that makes it can't be widely used. So some compact and affordable ones are needed. The idea of DWA came from LLNL[2], that beams can be accelerated during the whole transmission tube. The most crucial advantage of DWA is its high accelerating gradient. It was initially designed as 100 MV/m, so the proton accelerator for radiation therapy could be shorten to 2 m. In this way, the target tumor could be aimed at by moving accelerator rather than rotating human body, leading to more accurate and comfortable treatment. The assumption of DWA means less cost for installation and operation, more cure opportunity for common patients.

After redesigning many details of the initial design, our team started a prototype research. An ECR ion source is used for producing proton beams. Proton beams will be injected to accelerating tube after LEBT and kicker. The accelerating tube is constituted by High Gradient Insulators. The HGIs are loaded by solid-state pulsed power system. The length from ion source to the end of accelerating tube is about only 80 cm.

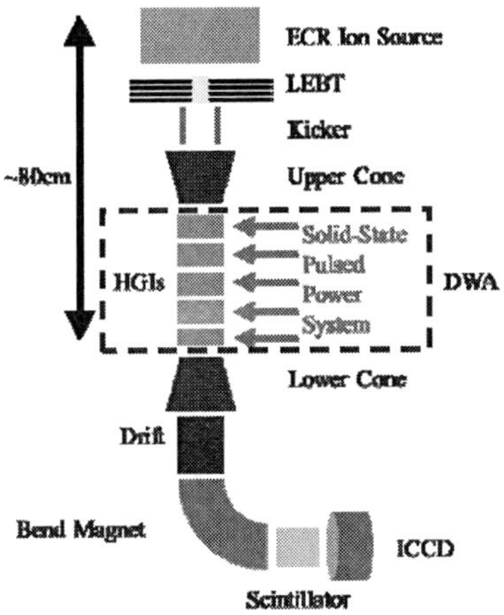

Fig. 1 Layout of the DWA prototype.

II. Key Technologies of the Pulsed Power System

A. High Gradient Insulator

The whole accelerating tube consists of a few high gradient insulators (HGI)[3]. Every HGI is an accelerating cavity. HGI is composed of periodically stacked thin alternate layers of floating copper electrodes and dielectric insulators. The emitted electrons will be repelled away from the insulator by the electric fields near the floating electrodes. The flashover process is inhibited, because secondary electrons are avoided. The principle indicates that HGI has a higher flashover threshold than traditional insulator.

Our HGI is an annular cylinder. Its capacity is a few picofarads related to different height from 5 to 15 mm. Any little burrs of floating electrodes can bring about spark,

Supported by National Nature Science Foundation of China (No. 51507162, 11035004 and 51407169), Science and Technology Foundation of CAEP (No. 2013A0402018), Key Laboratory of Inorganic Function Material and Device, CAS (No. KLIFMD-2014-01).

flashover or even breakdown. After a lot of work to improve the manufacturing technology, now the flashover voltage of our HGIs is 25 MV/m for 10 ns pulse, or 18 MV/m for 120 ns pulse.

Fig. 2 (A) Photographs of HGIs with different ratios of insulator and metal. (B) Enlarged view of the surface of HGI. The interface between insulator and metal layer is smooth and no obvious burr is found. (C) A few HGIs are stacked to form the accelerating tube. The copper electrodes between them are used for connecting the solid-state pulsed power system.

B. Dielectric Ceramic Plate Pulse Forming Lines

We cooperated with Institute of ceramics, CSA, to develop dielectric ceramic plate pulse forming line (PFL)[4][5]. The material is composite ceramic. The size is $300 \times 15 \times 1$ (or 3) mm^3. The electrodes on both sides are made of silver, with 280 mm in length and 4 or 6 mm in width. The relative permittivity of the ceramic is 23 or 100 and it remains constant from 1 to 1000 MHz. The characteristic impedance is about 12.5 Ω, and the transit time is about 4.5 ns or 9 ns. 1 mm thick PFLs can withstand over 20 kV DC. 3 mm ones can withstand 50 kV DC. Good manufacturing technologies ensure high surface flashover and bulk breakdown thresholds.

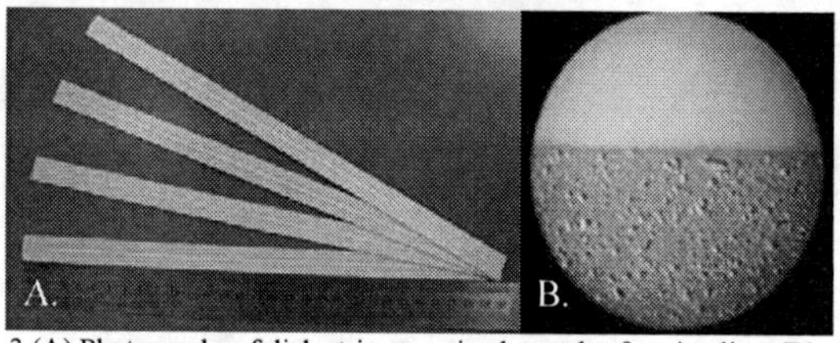

Fig. 3 (A) Photographs of dielectric ceramic plate pulse forming line. (B) Enlarged view of the silver electrode shows the interface between the electrode and the ceramic is smooth.

C. GaAs PhotoConductive Semiconductor Switches

Photoconductive semiconductor switches are used in the solid-state pulsed power system for achieving high-voltage nanosecond pulses[6][7]. PCSS has a faster response than traditional high-power switches and a higher conducting power than IGBT and MOSFET. Small volume, simple structure and optical controlling are advantages too. The size of our GaAs-PCSS is $12 \times 8 \times 0.6$ mm^3. Dark resistivity and carrier mobility are over 10^8 Ω·cm and 5000 $cm^2/(V \cdot s)$ respectively. Two Ge-Ni-Au alloy electrodes are set on the opposite faces of the substrate. The gap between electrodes is 5 mm.

Dozens of PCSSs are needed for the prototype DWA. It would be quite complex and expensive to using traditional laser as trigger. Fortunately, GaAs-PCSS can operate in lock-on mode that much less laser energy is needed. So we developed trigger device based on laser diode (LD). Our LD trigger device can generate 20 ns infrared laser. The highest laser energy is 360 μJ. The rise time is about 3 nm. The intensity distribution of the spot is almost uniform and its length is over than 7 mm which is enough for the gap of the PCSS.

The GaAs-PCSS can operate at 10 to 20 kV bias voltage. Its on-state resistance is no more than 10 Ω. The highest on-state current is 800 A. The conducting jitter is about 200 ps. The rise time is no more than 1 ns[8]. The longevity is 1000 to 100,000, related to current, laser energy and bias voltage[9].

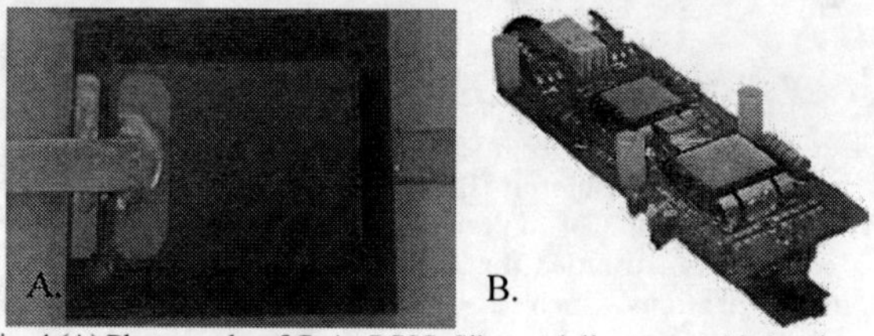

Fig. 4 (A) Photographs of GaAs-PCSS. Silver striplines are welded to the electrodes. (B) Photographs of a LD trigger device.

III. Solid-State Pulsed Power System

Two stacked PFLs and a GaAs-PCSS compose a 25 Ω Blumlein line (BL). Two BLs are stacked together at the load end which is a coaxial cable with magnet rings, to form a 50 Ω 2-BL pulsed power unit. The size of a 2-BL unit, including poly substrate, is about $50 \times 15 \times 5$ cm^3. Coupling loss among BLs can reduce output efficiency. That`s why only two BLs are stacked. A long coaxial cable is used to keep the pulse forming process from reflected wave and magnet rings are used to reduce the coupling loss between units. A 2-BL unit charged 15.3 kV can output a 10 (or 20) ns, 34 kV, 480 A pulse to a 70 Ω load.

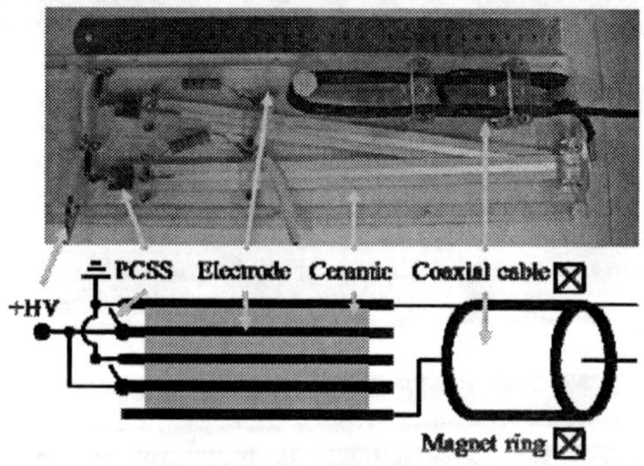

Fig. 5 Photographs of a 2-BL unit

2 to 4 2-BL units are connected in series at the load end, becoming into a pulsed power source. In experiments, we use the power sources with different numbers of 2-BL units to load HGIs with different capacity values. The output voltage increase with cascaded units. A power source with 4 units can generate 100 kV voltage and it doesn`t reduce much for HGIs with higher capacity. However, the efficiency of each unit decreases along with more units cascaded, because the coupling loss between units unavoidably get more.

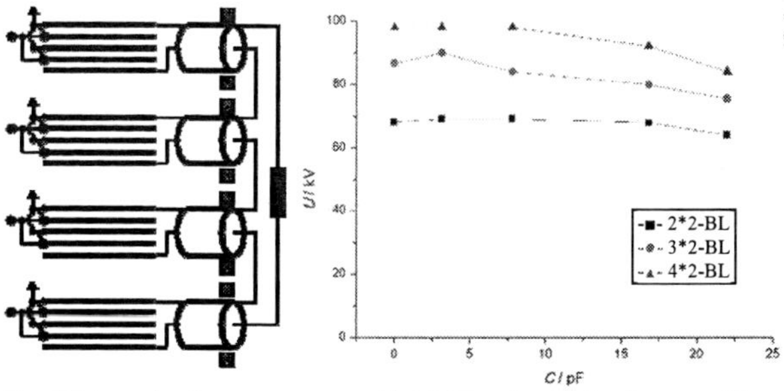

Fig. 6 Schematic of a power source with 4 2-BL units and the diagram of the experiment results. With more units, the power source can generate higher output voltage but the efficiency decreases. For a power source with 2 2-BL units, each unit outputs 34 kV equally. But for a power source with 4 units, each unit output only 25 kV equally.

IV. Results of Accelerated Proton Beams

In initial experiments, we used 3 power sources with 2, 4 and 4 units respectively to load 3 HGIs in series. 10 units were used in total and all of them were 16.4 kV charged. 300 keV proton beams were expected but the highest energy of the detected protons was only 240 keV. After some extended experiments of measuring the output of every power source, it is noticed that power sources interfere with each other. As shown in Fig. 7, the accelerating tube consists of 8 HGIs in series. The output voltage of each individual group of power source and its HGI is measured at first. Then groups are added one by one. When one groups are added, the output of every group decrease. The more groups are used, the output of each decreases more, especially for the middle ones. We calculated the efficiency of each group. The efficiency of every group decrease when more groups are used, especially for the middle ones. The reason is that the coupling loss between two connected power sources is unavoidable and very group is also a load for other groups.

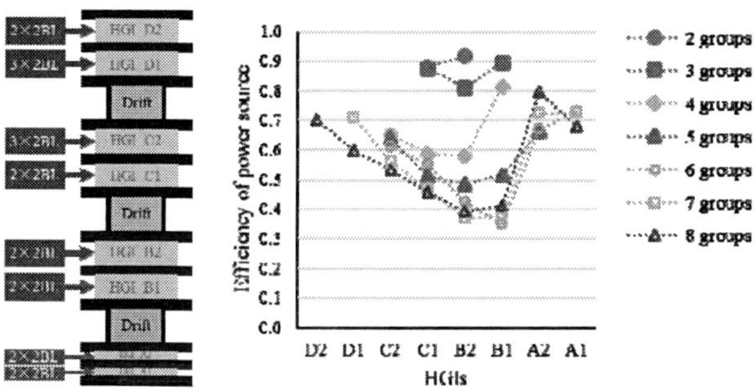

Fig. 7 Schematic on the left shows that 8 power sources with 18 2-BL units in total are used to load 8 HIGs. Diagram on the right shows the efficiency of the output voltage of each power source. When 2 or 3 power sources are used, the efficiency is around 0.8 to 0.9. But when more groups are operating, the efficiency of every group decreases, especially for the groups at the middle of the series.

As shown in Fig. 8, we tried to accelerate proton beams by the power system charged at 11.5 kV and 14.1 kV, with different numbers of power sources. It`s obviously the energy will be higher when the power system is charged at a high voltage. With increasing the number of power sources, the energy of accelerated proton increased too. As predicted, the efficiency of the whole pulsed power system decreases. The highest energy of accelerated protons is 380 keV when the pulsed power system has 6 groups (14 units) and is charged at 14.1 kV.

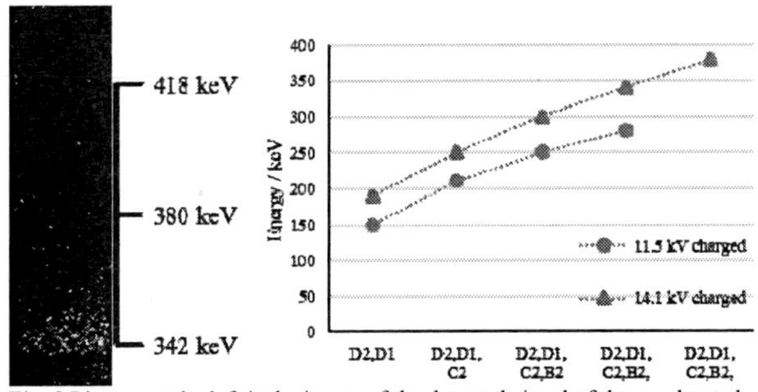

Fig. 8 Picture on the left is the image of the detected signal of the accelerated proton beams from scintillator and ICCD. The highest energy is calculated to about 380 keV. Diagram on the right shows the energy of the proton beams accelerated by different numbers of power sources charged at 11.5 kV and 14.1 kV. With increasing the number of power sources, the energy of accelerated proton increased, but the efficiency of the whole pulsed power system decreases. The highest energy of accelerated protons is 380 keV when the pulsed power system has 6 groups (14 units) and is charged at 14.1 kV.

V. Conclusion and Discussion

For three key technologies of the solid-state pulsed power system for dielectric wall accelerator, the HIG can withstand pulses of 25 MV/m. The dielectric ceramic plate pulse forming line with 1 or 3 mm can withstand 20 or 50 kV DC voltage, respectively. Triggered by our LD device, the GaAs PCSS can conduct 5 to 16 megawatts power with rise time less than 1 ns, but the longevity is still the disadvantage. By integrating these key technologies, we make out the 2-BL unit that can output 34 kV, 10 or 20 ns pulse. The power source consisting of different numbers of 2-BL units is discussed. The coupling loss between units reduce the output efficiency. Loaded by the solid-state pulsed power system, the prototype DWA can accelerate the proton beams from 40 keV to 380 keV.

For the future work, We are focusing on the longevity and stability of PCSS and HGI, to improve the solid-state pulsed power system. And we are still looking for some other methods for reducing coupling loss, such as better way to insulate the power source or faster and narrower pulse.

Acknowledgment

This work is supported by National Nature Science Foundation of China (No. 51507162, 11035004 and 51407169), Science and Technology Foundation of CAEP (No. 2013A0402018) and Key Laboratory of Inorganic Function Material and Device, CAS (No. KLIFMD-2014-01).

References

[1] Sampayan S, Caporaso G, Chen Y J, et al. Development of a compact radiography accelerator using dielectric wall accelerator technology[C]//Proc of Pulsed Power Conference, 2005 IEEE: 50-53.

[2] Sampayan S, Caporaso G, Chen Y J, et al. Development of Compact Pulsed Power for the Dielectric Wall Accelerator[C]//Power Modulator Symposium, 2006: 411-414.

[3] Sampayan S, Caporaso G, Carder B, et al. High gradient insulator technology for the dielectric wall accelerator[C]// Particle Accelerator Conference, 1995., Proceedings of the 1995. IEEE, 1995:1269 - 1271.

[4] Yi L., Liansheng X., Yi S.. et al. "Transmission character of glass-ceramic Blumlein lines", HPLPB, 26(4), 045018 (2014).

[5] Yi S., Wei W., Yi L., et al. A compact 300 kV solid-state high-voltage nanosecond generator for dielectric wall accelerator.[J]. Review of Scientific Instruments, 2015, 86.

[6] Loubriel G. M., Zutavern F. J., Baca A. G. et al. "Photoconductive semiconductor switches", IEEE Trans. Plasma Sci., 25(2), 124-130(1997).

[7] Nunally W. C., "Critical component requirements for compact pulse power system architectures", IEEE Trans. Plasma Sci., 33(4), 1262-1267(2005).

[8] Wei W., Liansheng X., Yi S., et al. Research on synchronization of 15 parallel high gain photoconductive semiconductor switches triggered by high power pulse laser diodes[J]. Applied Physics Letters, 2015, 106(2).

[9] Yi L., Wei W., Yi S., et al. Lifetime of high-power GaAs photoconductive semiconductor switch triggered by laser of different power density[C]// The 20th International Symposium on High Power Laser Systems and Applications. International Society for Optics and Photonics, 2015.

Optical Diagnosis on Breakdown Characteristics of Propylene Carbonate

Hongwei Liu, Yanpan Hou, Zicheng Zhang*, Jiande Zhang, Zhen Wang
College of Opto-electronic Science and Engineering, National University of Defense Technology
Changsha 410073, China
*E-mail address: zczhang@nudt.edu.cn

***Abstract*—Liquid dielectric, due to its high energy storage density and strong self-healing ability, has been widely used in high power pulse conditioning system as a kind of energy storage medium. In recent years, studies have shown that propylene carbonate(PC), compared with the commonly used liquid dielectric such as water, has larger breakdown field strength, higher resistivity and better temperature adaptability. In this paper, the experimental research on the breakdown characteristics of PC was carried out by a compact Tesla Transformer typed pulsed power source charging on the needle-plate electrode test cell under microsecond pulse. Clear image of the streamer during the breakdown process has been obtained by ultra-high-speed camera and the formation of breakdown voltage has been recorded by the oscilloscope, measured by a standard high pressure transducer. The streamer formation, propagation and the polarity effects of PC were studied. It is found that the positive breakdown voltage is 51% higher than the negative one and the growth rate of the positive streamer is much faster. In terms of morphology, the positive streamer of PC is filamentary and dense, but the negative streamer is less branched and sparse. The experimental results further explore the breakdown mechanism of PC, which is of great practical significance.**

Keywords*—propylene carbonate; breakdown characteristics; streamer; optical diagnosis;

I. INTRODUCTION

The development of high power microwave technology in the process of practical application requires high power pulse modulator with compact structure and high output power [1-4]. However, due to the limitation of energy storage density, the volume of the device is so big in order to meet the requirements of that. We know that pulse forming line in the storage process is equivalent to a capacitor, so the maximum stored energy density is proportional to the relative permittivity of the energy storage medium. Therefore, Liquid dielectric has become the first choice of high energy storage medium, because of the high permittivity and strong self-healing ability.

Propylene carbonate(PC) with high permittivity, high breakdown strength and high resistivity, becomes a hotspot in liquid media, its physicochemical properties such as TABLE I. In the 21st century, the Old Dominion University in the United States used small gap breakdown test cell for the experimental study of PC and discovered its breakdown strength up to a maximum of 2.2MV/cm, but actual resistivity only 240kΩ·cm [5]. However, N.J.Felici et al demonstrated the resistivity of PC by electrochemical purification treatment can be reached 20~30MΩ·cm [6], which made the application of PC in pulse forming line possible.

This work was supported by the National Natural Science Foundation of China under Grant No. 51307175 and No. 51677190.

TABLE I. The physical and chemical properties of PC

Parameter	value
Freezing point	-48.8℃
The boiling point	242℃
The relative density	1.2
The viscosity	2.5mPa·s
The permittivity	65(25℃)

It is worthy to carry out optical diagnosis on breakdown characteristics of the PC and streamer development analysis. From the point of insulation, we should prevent the development of streamer so that the liquid dielectric is not broken down to improve the energy storage density effectively. Although the breakdown of the liquid has certain research at home and abroad, a very perfect explanation for the formation of discharge streamer channel and its development mechanism has not yet at present [7]. Previous studies showed that the emergence and development of streamer discharge in liquid was a very complex process. The streamer development is related to the electric field of streamer head, which is a thermal or electrical process. Complex events such as collision, ionization, recombination, drift and adsorption, occur in the process of discharge channel development. It is difficult to make a detailed quantitative analysis and simulation of these events in the present study [8-11].

As an accurate type of non-contact diagnostic technology, optical diagnosis is widely applied to the test of transient physical quantity. The outstanding advantages are that the test quantity is not disturbed, and the transient process of the experiment is recorded by the ultra-high-speed camera. It provides a feasible way for the deeper analysis of the experimental phenomena under microsecond pulses. Using the optical diagnosis to record streamer images in this article, we can analysis experimental rules and then discuss its physical mechanism.

II. EXPERIMENTAL METHOD

The use of optical diagnosis for data acquisition is the highlight of this paper, shown in Fig.1. Combining with the working mode and performance of the camera, we adjust the sensitivity and exposure of the ultra-high-speed camera to achieve the best photography effect.

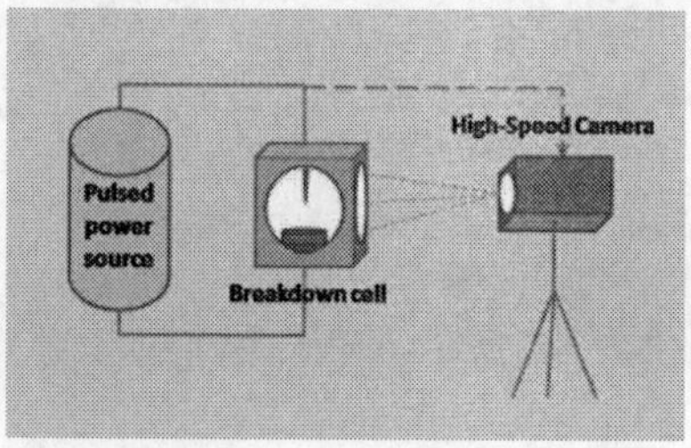

Fig.1 Schematic diagram of optical diagnosis on breakdown

In the experiments, a needle-plate electrode system was used for the generation of streamer discharge, shown in Fig.2 [3]. The breakdown test cell system consists of a needle-plate electrode made of brass, stainless steel body with 20cm length and PC as liquid dielectric filled in the box. The gap is 5mm. ARC-01 modulator serves as pulsed power source in our testing system, with the maximum output voltage of 400 kV and the peak charging time 8.4μs [1].

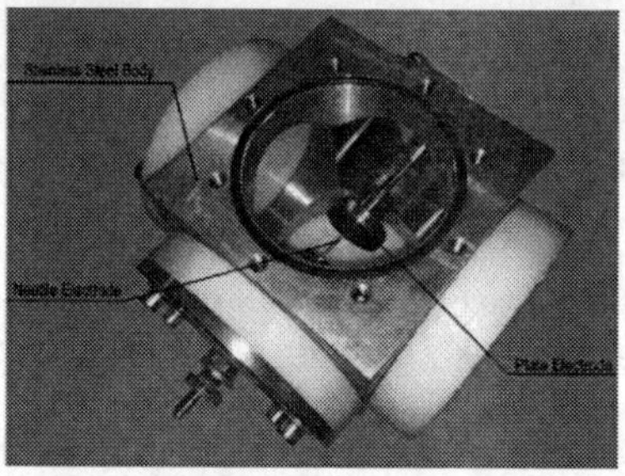

Fig.2 The breakdown test cell

III. EXPERIMENTAL RESULTS

According to Fig.1, the experiment platform was set up and the experimental results were recorded by optical diagnostic technique. The voltage waveform is recorded by oscilloscope in the whole process.

A. The voltage waveform of PC in different polarity

When we charge the needle electrode with negative voltage and positive voltage, the breakdown voltages are recorded. Each condition is tested 20 times. The experimental results after calculating the average value are listed in TABLE II. It is obvious that the positive breakdown voltage is 51% higher than the negative one in this article, and the effective time, defined as the time suffering over 63% of the maximum voltage, is longer. The voltage waveforms are recorded by the oscilloscope, presented in Fig.3. From the image, it can be known that the voltage has a relatively slow rising process. When the voltage reaches a certain critical value, the voltage suddenly drops to zero. Obviously, the test cell is charged until the breakdown occurs in this process.

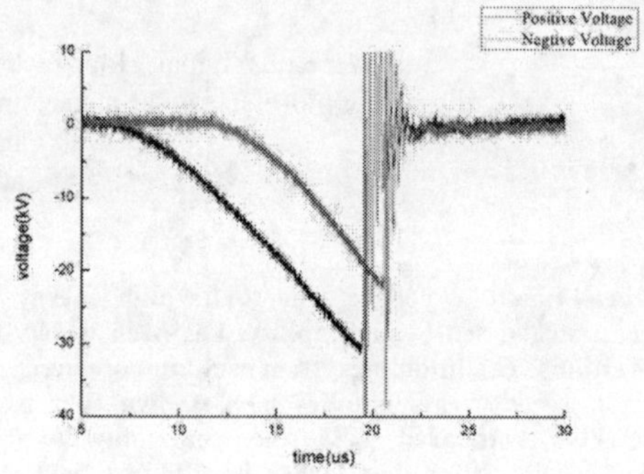

Fig.3 The voltage waveforms in different polarity

B. Integral streamer images of PC in different polarity

The ultra-high-speed camera recorded integral images of the streamer in different polarity under 2μs exposure, shown in Fig.4. It can be obtained that streamer is starting from the needle electrode. Positive streamer(a) of PC is silky and dense. Carefully observe the positive streamer image, it has many discharge channels to reach plate electrode. And the more close to the plate electrode, the more discharge channel. Negative streamer(b) of PC is less branch and sparse. It has a very obvious main discharge channel. Moreover, the more close to the plate electrode, the less the branch.

Comparing the two pictures, energy release for negative streamer is more concentrated, but it is more dispersed for the positive streamer. In the experiment, it found that the higher the breakdown voltage, the more dispersed the energy release, whether positive or negative streamer discharge.

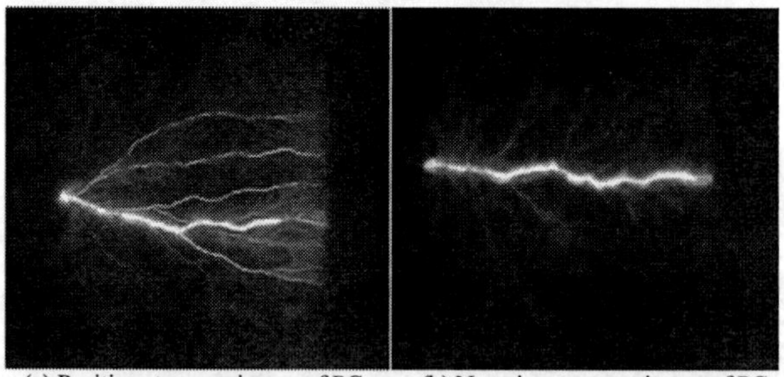

(a) Positive streamer image of PC (b) Negative streamer image of PC
Fig. 4 Streamer images in different polarity under 2μs exposure

TABLE II. Breakdown test results of PC in different polarity

Needle polarity	Breakdown voltage (kV)	Standard deviation (kV)	Effective time (μs)	Breakdown strength Ebd/(kV/cm)
Needle positive	32.18	4.89	5.51	64.36
Needle negative	21.30	1.73	2.63	42.60

C. Streamer development images of PC in different polarity

The ultra-high-speed camera recorded the partial image of streamer development process, shown in Fig.5. Breakdown moment is seen as a zero time in the experimental images. It is known that streamer is starting from the needle electrode to the end of the plate electrode in the whole development process. At first, the needle appears a small bright spot, and the brightness is very strong. With the passage of time, the volume of the bright spot continues to expand outward, and streamer begins to appear irregular bifurcation. Gradually, the length and width are increasing, and there is a new generation of small branches. When a branch reaches the plate electrode, the two electrode is switched on.

In the experiment, the shapes of positive streamer are more diverse, and there are often many discharge channels, that all reach the plate electrode. All this shows that the development of streamer in positive polarity is very casual, and less constrained. However, there is generally only one main discharge channel for negative streamer. Even if many branches are generated in the process of streamer discharge, but in the end there will be only one. And it's obvious that the growth rate of the positive streamer(a) is much faster than the negative streamer(b). For negative streamer, its speed has been relatively slow. Until it is about to reach the plate electrode, the speed suddenly become very fast.

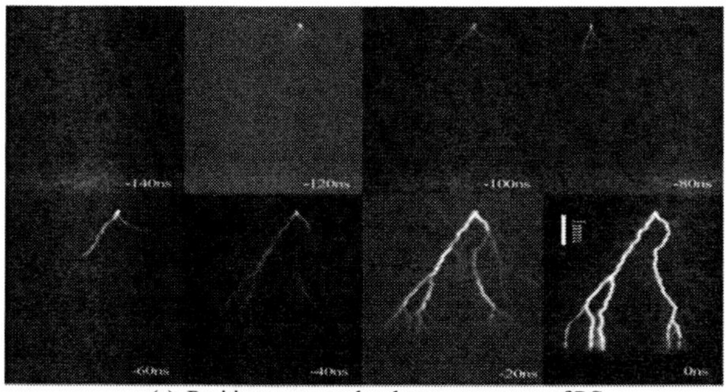

(a) Positive streamer development process of PC

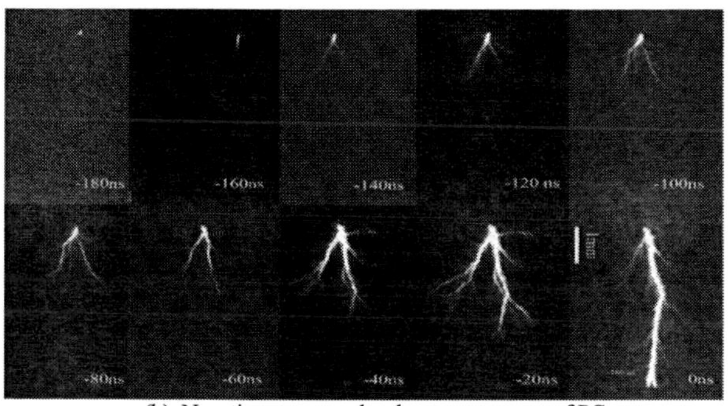

(b) Negative streamer development process of PC

Fig.5 Streamer development process of PC under different polarity

IV. DISCUSSION

The streamer theory originated from the gas discharge, and then applied to the liquid breakdown. The core of the streamer theory is collision ionization, considering the light phenomenon during the process of discharge[12]. Needle-plate model is the most common model for gap discharge. In the case of the needle electrode as anode, the electric field near it is particularly strong. So nearby molecules first start ionization and the electron moves toward the needle electrode. When the electron reaches the needle electrode, it is absorbed. Therefore, the positive ion is gathered near the needle electrode, which weak the nearby electric field. This makes it difficult to generate initial streamer. However, the field strength produced by these space charges is consistent with the applied field strength and make the external electric field of the discharge area be strengthened. This distribution of electric field causes the high breakdown voltage and the rapid development. The high field strength after enhancement between the electrodes allows the release of energy from the streamer to be more random, which makes the branch more. Conversely, when the needle electrode is cathode, positive ions gather near the needle electrode, which enhances the electric field in the vicinity. The generation of the initial streamer becomes very easy. But these charges weaken the space electric field outside the discharge area, that make it so hard to cause ionization again. That is why the speed of the negative streamer is slow and branch less.

The streamer velocity is an important quantity describing the characteristics of streamer propagation. In the initial stage, the distortion of space charge generated is very small, the streamer development depends mainly on the applied electric field to maintain, so the streamer speed is slow[13]. With the ongoing photo-ionization, streamer head field has been greatly enhanced, so the velocity of streamer development has greatly increased.

V. CONCLUSION

in this paper, the experimental research on the breakdown characteristics and the streamer of PC were carried out. According to the experimental data and streamer images, the polarity effect of PC was analyzed. It is found that the positive streamer of PC is filamentary and dense, but the negative streamer is less branched and sparse. The positive breakdown voltage is 51% higher than the negative one and the growth rate of the positive streamer is much faster. Not only the development process of the streamer has a more in-depth understanding, and its mechanism is also discussed. It is found that electronic ionization strength of streamer head has great influence on its morphology. And the distribution of charged particles caused different breakdown voltage in different polarity. In the experiment, we also see the advantages of the optical diagnosis for a subtle analysis of the field characteristics. The experiment provides some theoretical basis for the application of PC to the energy storage device.

References

[1] Z. C. Zhang, J. D. Zhang, B. L. Qian, C. B. Liu, T. Xun, H. Zhang,and B. Liang, "Compact Rep-Rate GW Pulsed Generator Based on Forming Line With Built-In High-Coupling Transformer," IEEE Transactions on Plasma Science, vol. 42, pp. 241-248, 2014.

[2] V. P. Gubanov, S. D. Korovin, I. V. Pegel, A. M. Roitman, V. V.Rostov, and A. S. Stepchenko, "Compact 1000 pps high-voltage nanosecond pulse generator," IEEE Transactions on Plasma Science,vol. 25, pp. 258 - 265, 1997.

[3] Z.Wang, Z. C. Zhang, and J. D. Zhang, "Polarity Effects of Propylene Carbonate on Breakdown Strength in Microsecond Range, "IEEE Chinese journal of chemical physics,vol.28,pp.557-660,2015.

[4] S. D. Korovin, V. P. Gubanov, A. V. Gunin, I. V. Pegel, and A. S.Stepchenko, "Repetitive nanosecond high-voltage generator based on spiral forming line," in Pulsed Power Plasma Science, 2001. PPPS-2001. Digest of Technical Papers, Las Vegas, NV, USA, 2001, pp.1249-1251 vol.2.

[5] J. Kolb, S. Kono, S. Xiao, et a1, "Water and propylene carbonate as storage and switching media in pulsed power system," IEEE International Pulsed Power Conference Proc 14th,pp.715- 718, 2003.

[6] F. W. Bian, M. Zhou, L. Lei," Formations of Active Species and By-Products in Water by Pulsed High-Voltage Discharge," Plasma Chemistry & Plasma Processing, vol.27,pp.337-348, 2007

[7] M. D. Cevallos, M. Butcher, J. Dickens, A. Neuber, and H. Krompholz, "Imaging of Negative Polarity dc Breakdown Streamer Expansion in Transformer Oil Due to Variations in Background Pressure," IEEE Transactions on Plasma Science, vol. 33, pp. 494-495, 2005.

[8] I. V. Lisitsyn, H. Nomiyama, S. Katsuki, and H. Akiyama, "Thermal processes in a streamer discharge in water", IEEE Trans. Dielectr. Electr. Insul., vol. 6, pp. 351-356, 1999.

[9] T. J. Lewis, "Breakdown initiating mechanisms at electrode interfaces in liquids," IEEE Trans. Dielectr. Electr. Insul., vol. 10, pp. 948-955, 2003.

[10] O. Lesaint, "Propagation of positive discharges in long liquid gaps," IEEE 12th Intern. Conf. Conduction and Breakdown in Dielectric Liquids, Roma, Italy, 1996, pp. 161-166.

[11] O. Lesaint and G. Massala, "Positive streamer propagation in large oil gaps: Experimental characterization of propagation modes," IEEE Trans. Dielectr. Electr. Insul., vol. 5, pp. 360-370, 1998.

[12] Y. P. Raizer, Gas discharge physics. New York, USA: Springer-Verlag, 1991.

[13] S. K. Dhali, P. F. Williams, "Numerical simulation of streamer propagation in nitrogen at atmospheric pressure," Physical Review A, vol. 3l, pp. 1219-1221, 1985

High Magnetic Field-Driven Liner-Impactors for Shock Wave and Warm Dense Matter Applications

R.E. Reinovsky, C.R. Rousculp, S. Kreher, J. Jacobs
Los Alamos National Laboratory,
Los Alamos, NM, USA

Abstract – Electromagnetically driven cylindrical liners provide a useful, and advantageous, technique for shock-wave physics experiments and for the production of warm dense matter. The applications of electromagnetically driven liners as an experimental technique in hydrodynamics and material properties experiments can be limited by the development of magneto-Rayleigh-Taylor (MRT) instabilities on the outer (field/liner). Those instabilities, if present, can, under some condition can feed through the thickness of the liner producing severe density non-uniformities and distorting the inner (impacting) surface. In this paper we explore electromagnetic drive techniques that may reduce the growth of the MRT, extending the application of magnetically driven liners.

I. INTRODUCTION

Shock wave physics experiments generally require the high precision (uniform) impact of a solid driver at well known and controllable velocity upon a sample. For example, Tera-Paschal pressures requires achieving 10 km/sec (1mm/40ns) shock speed in a tungsten sample or 25 km/sec shock speed in an aluminum sample. Such shock speeds requires 20 km/sec impact for an Al liner on a Tungsten target or 15 km/sec for a Cu liner on a Tungsten target. The 1% accuracy demanded by many shock experiments translates to 400 pS timing uncertainties or 10 μm dimensional uncertainties

Similarly, one approach to producing warm dense matter conditions in a sample may be a combination of an initial shock followed by extended compression driven by a very high energy imploding liner. Warm Dense matter is a state characterized by densities from 5% of normal to about twice normal density and temperatures below about 50 eV. The plasma parameter (Γ) is typically less than 0.10 and the inter-particle potential energy is significant compared to thermal kinetic energy. Warm dense matter occurs in the core of giant planets and in the evolution of plasma from materials that is initially solid such as the ohmic explosion of a fine wire.

Shock wave experiments are frequently driven by powder or gas gun accelerated impactors or by the direct application of chemical high explosive -- and these technology is well established. Electromagnetically (pulse power) driven impactors can offer significant advantages compared to conventional techniques for some direct shock or shock/compression situations. Notably, pulse power drive provides highly flexible, predictable and controllable drive (because the magnitude of the driving magnetic field can be precisely controlled), and is usually continuously adjustable. Driving magnetic fields are transparent to visible IR, UV and X-rays -- making some configurations easier to diagnose. Magnetic drive delivers energy at the speed of light -- making higher impact velocities possible under some circumstances. Further, in configurations requiring confinement of the experiment either because of the experimental environment, or because of the materials used, magnetic drive delivers energy without associated mass -- reducing collateral damage or late time pressure delivered to the confinement vessel.

The principle obstacle to achieving high precision (accurately cylindrical) impact by a liner on the sample/target is the development of hydrodynamic instabilities. The magneto-Rayleigh Taylor-like (MRT) instability at the (outer) magnetic field / liner interface can grow, feeding through to the inner surface, or the surface-conserving, crenulation, instability can form at the inner surface. Hydro-instability growth increases with acceleration and is inhibited by material strength and by refining the surface precision, (to reduce the magnitude of initial perturbations). High precision implosions have been reported for cases in which the driving field was sufficiently low that the material at the liner field interface is not melted, maintaining material strength through the thickness of the liner and through the duration of the experiment.

Where large amounts of kinetic energy, high efficiency of conversion of electrical to kinetic energy, and/or high final implosion velocities are required, the traditional approach is to match the liner implosion time to the energy delivery time of the pulsed power driver, but this approach favors MRT growth. An alternate approach is to introduce the driving impulse, and hence momentum to the liner early in the implosion ("kick") when radii are large, circumferential current density and acceleration is relatively low, but allowing the nearly incompressible liner to "coast" during the later stages of the implosion, accelerating the inner surface by the effect of incompressible convergence while maintaining overall liner momentum and kinetic energy unchanged in these later stages.

II. THE MAGNETO-RAYLEIGH-TAYLOR INSTABILITY

The magneto-Rayleigh-Taylor instability at the field-liner interface imposes a , perhaps severe, limitation on the integrity of the liner and upon the precision of the liner's impact with the target when impact at high velocities is required. The MRT is a fluid instability arising when an interface between fluids of different densities (ρ_1, ρ_2) such as the metal liner (ρ_1) and vacuum field (ρ_2=0) experiences acceleration. Perturbations at the interface grow when the lighter fluid is accelerating the heavy fluid such as when the liner is accelerated toward the axis by the azimuthal magnetic field generated by the axial current flowing in its outer surface. The current that drives the acceleration also ohmically heats the outer surface of the liner, and if that heating is sufficient, can melt the outer surface leading to a fluid layer of time dependent thickness ($\delta_{(t)}$) adjacent to the region containing the vacuum field. Depth of melting on the back surface is important because material strength (present in the inner un-melted liner material, but absent in the melted layer) reduces the MRT growth rate leading to the description of the MRT as "growing in the melted layer"

The dispersion relation for the MRT instability connects growth rate (γ) to acceleration (g) and scale length (λ or $k=2\pi/\lambda$) by:

$$A_{(t)} = A_0\, e^{\gamma t} \qquad \gamma = (Nkg)^{1/2} \qquad (1)$$

Where $A_{(t)}$ is the perturbation amplitude, A_0 is the initial perturbation amplitude; and N is the Atwood number $N=(\rho_1-\rho_2)/\rho_1+\rho_2)$. For the vacuum magnetic field/metal liner interface, ρ_2 = 0 and N=1; while for an (imaginary) interface where $\rho_1=\rho_2$, N=γ=0. Straightforwardly, for constant acceleration the number of e-foldings of the initial perturbation amplitude scales just as the total distance traveled. Short wavelength (large k) grow most rapidly but are limited by the depth of the melted layer. Longer wavelength grow more slowly, but are the most disruptive and hence the most important

The linear region of MRT growth leads to the traditional approach for accelerating liners, directly matching the implosion time to the driver's energy delivery time. In this approach the liner initial radius is minimum, resulting in minimum implosion time (minimum time for the instability to grow) and minimum distance traveled. However this traditional approach also results in highest average acceleration on the liner, but also highest linear current densities (A/cm, circumferentially) and hence highest ohmic heating rates in the liner and deepest melting of the back surface of the liner.

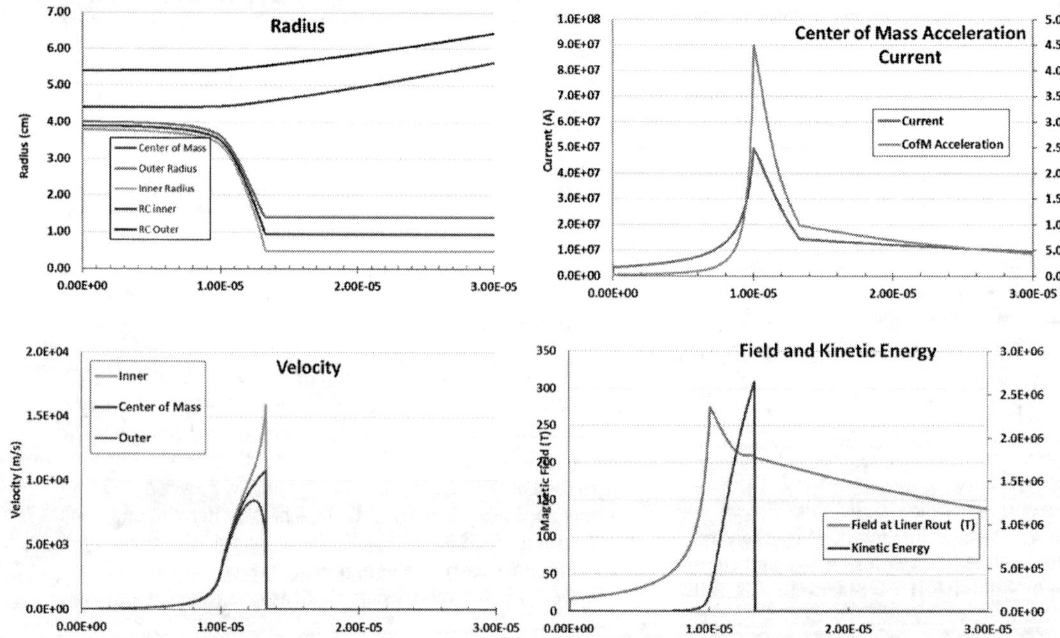

Fig. 1 Performance of Traditional liner driven by representative flux compression generator

If all the liner's kinetic energy could be delivered to the material in a brief period, as Lindl and Mead [1] pointed out in an ICF context "...the smallest growth (of the Rayleigh Taylor) is obtained in a shell if it receive the entire acceleration in a single shock (kick)". Conceptually, MRT instability growth might be mitigated by an alternate approach to matching liner dynamics to driver energy delivery characteristics if most of the acceleration can be delivered early in the implosion (kick), relying on "quasi-incompressible" convergence during the later, "coasting" phase, to accelerate the impactor inner surface to the needed velocity with little or no acceleration is applied to the liner center of mass.

In this alternate approach initial liner radius might be substantially larger than in the traditional approach resulting in a thinner, but potentially more massive liner, and certainly longer implosion times. Magnetic energy is converted into kinetic energy early in the implosion. Later, when the drive (current) is much lower – or can even be interrupted-- the liner coasts under low (or zero) acceleration, decreasing MRT growth rates. During the acceleration phase, the larger initial radius gives rise to lower current densities, lower magnetic pressures and lower average accelerations. Lower current densities also result in lower ohmic heating rates – perhaps never melting the outer surface – and possibly maintaining strength stabilization in the un-melted material (Nearly) incompressible behavior of the liner during convergence, after the main acceleration continues to accelerate the inner (impactor) surface to the required velocities even as the center of mass velocity remains almost constant.

III. COMPARISON OF TWO APPROACHES

Using a simplified model of a flux compression generator, performance of the traditional and alternate approaches to matching liner dynamics to driver characteristics can be compared for an implosion designed to produce a 16 km/sec impact on a target of radius 5 mm. The flux compression driver is described by:

$L_0 = 50$ nH $\quad$ L_f=~1nH $\quad$ Δt=10μs

$I_0 = 3.5$ MA $\quad$ E_0= 315 kJ

Where L_0 and L_f are the initial and final driver inductances, I_0 and E0 are the initial current, and inductive energy and Δt is the run time of the generator. The aluminum liner geometry following the traditional approach has an initial radius of 40 mm an axial height of 30 mm, a mass of 15.16 grams and an initial thickness of 2.2 mm which increases to about 9 mm upon arrival at the target.

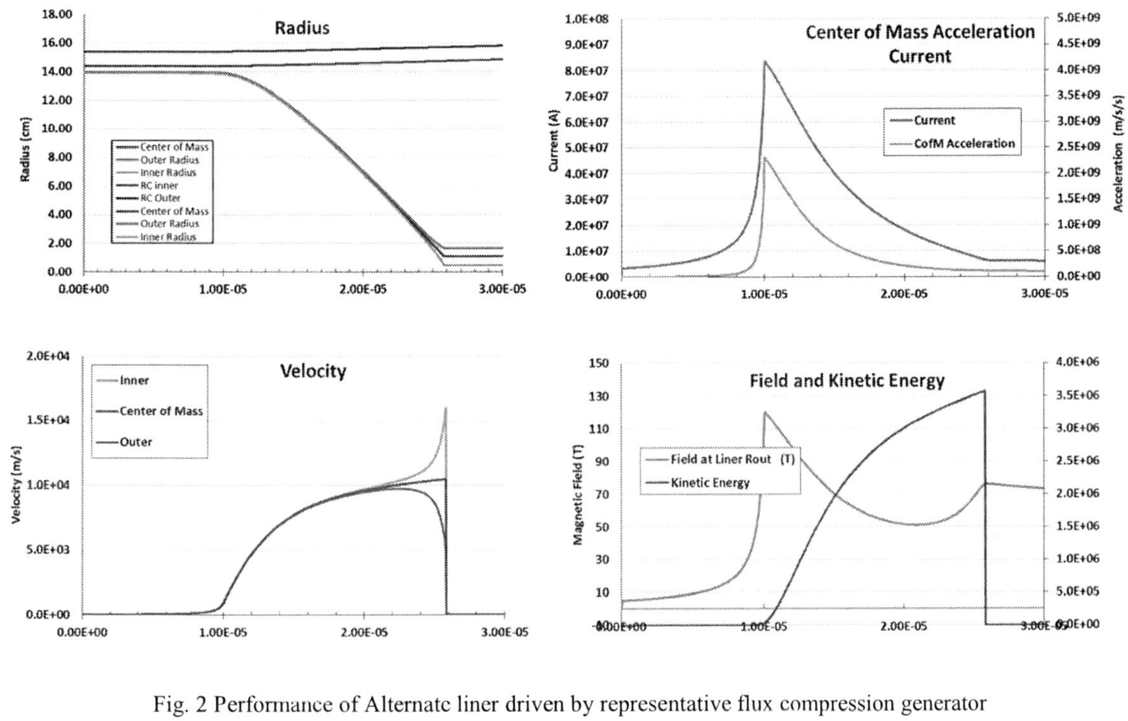

Fig. 2 Performance of Alternate liner driven by representative flux compression generator

Figure 1 illustrates the performance for the traditional approach. The liner arrives at the target at a velocity of 16 km/sec with 2.5 MJ of kinetic energy. The peak generator current is 50 MA and the implosion time is 13 μs and the peak acceleration is about $4.5x10^9$ m/sec^2. The peak magnetic field at the interface is greater than 275 Tesla and the time averaged field is almost 200 T. Using a very simple, action based, melt model the melted layer has penetrated about 6 mm into the 9mm thick liner

The aluminum liner geometry following the alternate approach has a much larger initial radius of 140 mm an axial height of 30 mm, a mass of 21.74 grams and an initial thickness of 0.9 mm which increase to almost 12 mm upon arrival at the target. Figure 2 illustrates the performance for the alternate approach. The liner arrives at the target, by design, at about the same velocity of 16 km/sec but with 3.5 MJ of kinetic energy because of its larger mass. The peak generator current is 85 MA, the implosion time is 26 μs and the peak acceleration is about $2.3x10^9$ m/sec^2. Because the initial liner radius is larger, the peak magnetic field at the interface is about 120 T, less than half that of the traditional approach, and the time-averaged field is about 75T. Using a very simple, action based, melt model the melted layer has penetrated about 2.5 mm into the almost 12 mm thick liner, resulting in a much smaller fraction of the liner being melted at the time of impact.

Two dimensional resistive MHD modeling gives an estimate of the growth of the MRT instability on the back surface of the liner. The MRT grows from initial perturbation characteristic of the surface finish of a high precision liner (a few micro-inches RMS). For the traditional approach the amplitude of the MRT is 3-4 mm and the dominant wavelength is 2-4 mm. For the alternate approach the amplitude, growing from the same initial perturbations is only about 1mm and the dominant wavelength is shorter, about 1-2 mm

IV. SUMMARY

The "Alternate Approach" attempts to apply lower accelerations, over somewhat longer times to larger radii, initially thinner, but potentially more massive liners to reduce MRT growth. For the alternate approach, the (time) average magnetic field at the driving surface is about 80T compared to 200 T for the traditional approach – and the peak acceleration is about half. Larger fraction of the liner mass is un-melted at impact (r=5 mm, V=16km/sec) for the alternate approach although that may not be entirely true mid-way through the implosion. Full 2D R-MHD calculations indicate a more apparent advantage for the alternate approach than indicated by simple modeling

V. REFERENCES

[1] J.D. Lindl, W.C, Mead, "Two—Dimensional Simulation of Fluid Instability in Laser-Fusion Pellets", Phys Rev Letters, Vol 34, No20, pp 1273-1276, May 1975.

Research of Electron Beam Pulse on Dielectric Discharge Cold Cathode

Ryosuke Horie, Akira Sugawara, Takeru Otsuki, Kiyoyuki Yambe and Kazuo Ogura
Graduate School of Science and Technology, Niigata University
8050 Ikarashi-2, Niigata, 950-2181, Japan

***Abstract*—Electron beam generation in weakly relativistic region less than 100kV utilizing a new dielectric discharge cold cathode (DDCC) which is made of a ceramic cylinder and a cupper disk is researched. From burn patterns with guide magnetic field of 0.189T, electron beams are generating from triple junction and have a uniform shape. Electron beam characteristics are following the Child-Langmuir law. A DDCC is applied to a backward wave oscillator (BWO) with W-band corrugate. Microwave is observed in especially low voltage region of around 20kV. And by using the DDCC, microwave can be emitted at low beam voltage under 30kV stably. The BWO with the DDCC generates around 100GHz microwave with the almost same value as the BWO using a velvet cathode.**

Keywords—Electron beam, cold cathode, dielectric discharge, microwave, weakly relativistic region, backward wave oscillator (BWO)

I. INTRODUCTION

Recently, microwave sources which have high-power and variable frequency for plasma heating on nuclear fusion, radar, communication systems, measurement systems and medical equipment are desired.

Various microwave sources are researched such as backward wave oscillators (BWOs) [1], [2]. For practicability and miniaturization, BWOs are driven by a weakly relativistic electron beam. It is essential to use the reliable cathode which is able to produce a uniform and a high current electron beam, and BWO generates microwave under a weakly relativistic region stably. One way to increasing electron beam is installing velvet to the circumference of the electron emission edge of the cathode [3]. On W-band (75-110GHz), Min Thu San et al. [4] obtained microwave radiation power of about 20kW employing the oversized surface wave oscillator driven by a weakly relativistic electron beam less than 100kV using the disk cold cathode with velvet.

In this study, for increasing electron beam density with a uniform and annular electron beam, a new cold cathode with a dielectric cylinder guide tube (dielectric discharge cold cathode: DDCC) is used. The dielectric cylinder is set outside of a copper disk. The purpose of this study is to apply the DDCC to a broadband pulse surface wave oscillator, so that the electron emission characteristics of the beam diode are measured. An electron beam is emitted from plasma which is made from a metal part of around triple junction. The triple junction consists of intersection by vacuum, metal and dielectric tube. The burn patterns by the electron beam on a thermal paper at the beam collector are observed. And then, the DDCC is applied to a BWO with a magnetic guide, as density of magnetic flux is 0.189T. And W-band microwaves generated by the BWO with the DDCC are measured.

II. EXPERIMENTAL SET-UP

A. Dielectric Discharge Cold Cathode (DDCC)

The structure of the DDCC is shown in Fig. 1. The DDCC is made of a copper disk and a ceramic cylinder. The ceramic cylinder is length of 40mm, outside diameter of 35mm, and inside diameter of 28mm. The copper disk is thickness of 5mm and diameter of 28mm. A copper column is used to install to the experimental equipment. The surface of copper disk is set to 5mm inside of ceramic edge. The boundary between the ceramic and copper is formed a triple junction. When voltage is applied to the DDCC, electric field concentrates on the triple junction and electron emission surface is formed. Plasma is formed by flashover at the triple junction and then electrons are emitted from plasma.

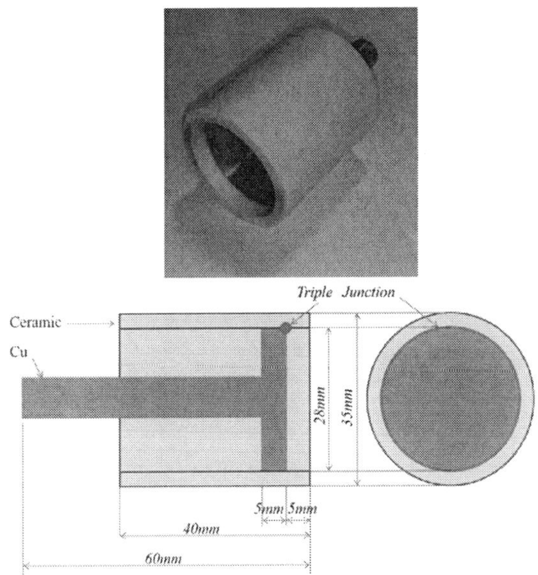

Fig.1. The dielectric discharge cold cathode

B. Beam Generation System

The schematically experimental set-up is shown Fig. 2. An electron beam diode and a beam collector are installed in a vacuum vessel made of stainless steel. The vacuum system consists of a turbo molecular pump and a rotary pump. The pressure is kept under 4×10^{-3}Pa during experiment. The electron beam diode consists of the DDCC and a beam limiter. The beam limiter is used to trim electron beams for applying into the BWO and also combines with an anode. The beam limiter is 29.7mm in diameter and made of stainless steel. A Marx generator or a MPC (Magnetic Pulse Compression Circuit) is used as a high-voltage power source. The Marx generator is used for applying voltage approximately over 30kV, the MPC is for approximately under 30kV. In this experiment, the cathode voltage range is about -20kV to -75kV. Solenoid magnetic field coils of 10 pieces are used to provide axial magnetic field for electron beam transmitting. For optimal electron beams transmitting to beam collector, the magnetic field is adjusted 0.819T uniformly. The electron beam shape is observed as a burn pattern on a thermal paper. Distance from the edge of the DDCC to the beam limiter is 10mm. Electron beam currents into the beam collector are measured by a current monitor.

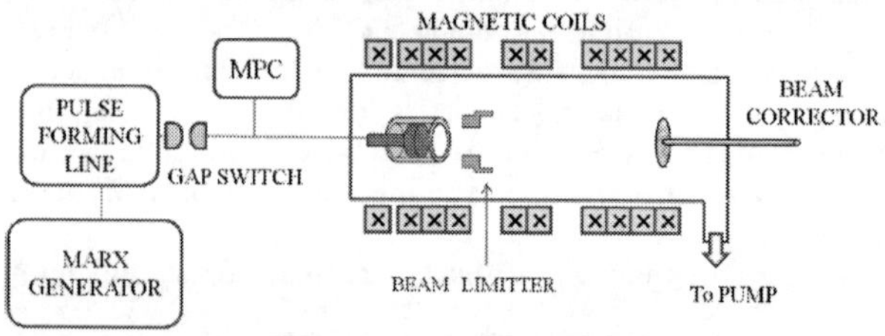

Fig.2. Schematic diagram of the experimental set-up for beam generation

C. W-band slow-wave structure

A corrugated hollow waveguide is used in this experiment. The slow-wave structure is shown in Fig. 3. The corrugation has dimensions: R_0=15.1mm, h = 0.3mm, d = 0.3mm, z_0 = 0.5mm, and the length $L = 80\ z_0$.

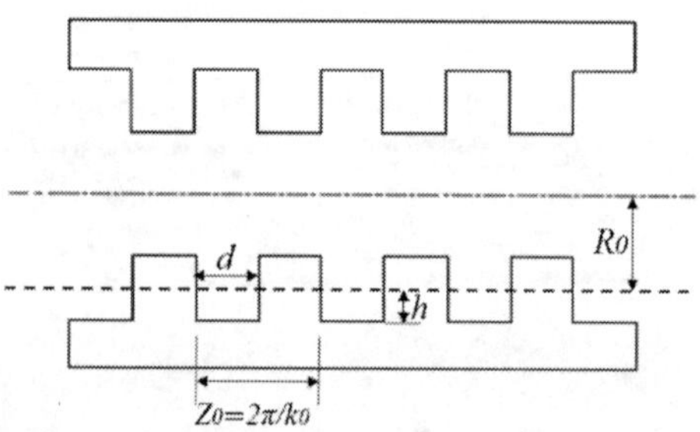

Fig. 3. Slow-wave structure

Fig. 4 shows the dispersion curve of the slow-wave structure. Though the corrugated hollow waveguide is made of aluminum, the dispersion relation is calculated with the resistivity of zero by the same calculation code of references [4] – [8]. The upper cutoff frequency is about 100GHz for theTM_{01} surface wave. The beam lines are depicted as annular electron beams with energies of 20, 40, 60keV, respectively.

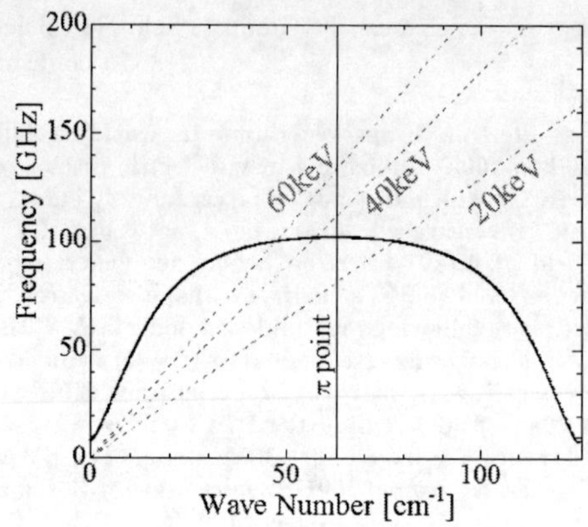

Fig. 4. Dispersion curve of the slow-wave structure and beam lines with energies of 20, 40, 60 keV

D. Microwave Genaration System

The DDCC is applied to the BWO. The schematically experimental set-up is shown Fig.5. The BWO consists of the vacuum system, a high voltage source, the DDCC, a beam limiter, and a corrugated hollow waveguide. A W-band (75-110GHz) corrugated hollow waveguide with length $L = 80\ z_0$ is used. The W-band corrugated hollow waveguide was already studied [4]. The same equipment as the beam generation experiment, i.e. the high voltage pulse source, the vacuumed vessel, the vacuum system, the solenoid coils, and the DDCC, is used. The pressure in the vacuum chamber is also kept under 4×10^{-3}Pa. The radiated microwaves are detected by rectangular horn antennas connected to waveguides of G-band (140-220GHz) and F-band (90-140GHz), which are EIA standard rectangular waveguides. The cutoff frequencies of the F-band and G-band waveguides used as filters are 74GHz and 116GHz, respectively.

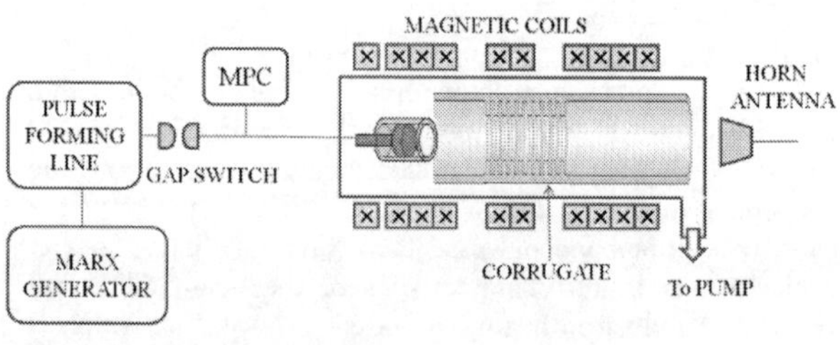

Fig.5. Schematic diagram of the experimental set-up for microwave generation

III. EXPERIMENTAL RESULT

A. Beam Generation

Fig. 6 shows an example of the applied voltage and the beam current waveforms. The peak values of voltage and beam current are about 66kV and 208A, respectively. Fig. 7 shows characteristics of applied voltage versus beam current using the DDCC. Curve lines in Fig. 7 is the Child- Langmuir law. The curve is defined as

$$\text{Beam Current} = \alpha V^{\frac{3}{2}} \tag{1}$$

and drawn as the coefficient **α** of 0.5, 0.55, 0.6. The plots fit the curve of the Child-Langmuir law well. From Fig. 7, the characteristic of the beam diode is determined by the Child-Langmuir law, so electron beam current is limited by space charge.

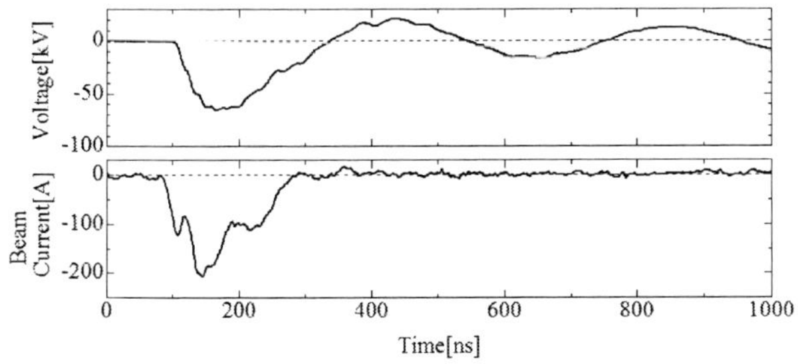

Fig.6. An example of waveforms on beam generation experiment, (upper) beam voltage and (lower) beam current

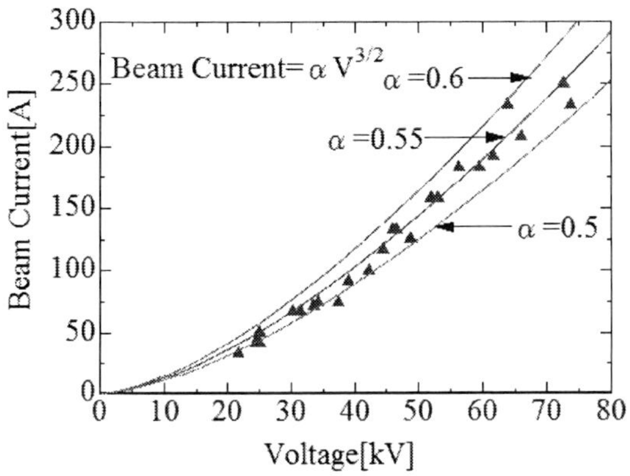

Fig.7. Characteristics of applied voltage versus beam current and curves of Child-Langmuir law with various coefficients

Fig. 8 shows burn patterns of the electron beams observed by thermal papers. Fig. 8 (a) is the 5 shots burn pattern using the MPC voltage source of about 25kV. Fig. 8 (b) is the 1 shot using the Marx generator of about 69.3kV. In Fig. 8 (a), the maximum outside diameter of the burn pattern is 28mm. It seems that the value of the outside diameter is determined by inside diameter of the DDCC. Therefore, electrons are emitted from the triple junction effectively and progress along the inside ceramic surface of the DDCC. In Fig. 8 (b), the maximum outside diameter is 29.8mm. It seems that the beam is trimmed by the beam limiter of 29.7mm in inside diameter. From this, electron beams spread with increasing beam current. And the DDCC generates electron beam uniformly.

	Max ↔ (mm)	min ↔ (mm)		Max ↔ (mm)	min ↔ (mm)
Inside Diameter	26.6	25.0	Inside Diameter	24.5	24.0
Outside Diameter	28.0	26.8	Outside Diameter	29.8	29.5

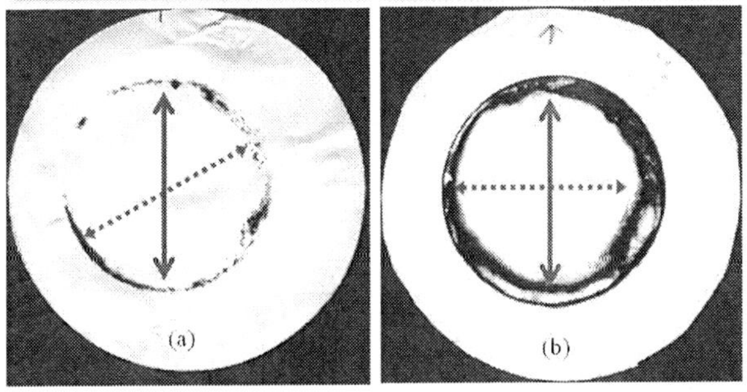

Fig.8. Burn patterns of electron beams by using the DDCC, (a) Electron beams of 5 shots using the MPC voltage system to apply about 25kV, (b) A electron beam of 1 shot using the Marx generator to apply about 69.3kV

B. Microwave Genetration

Fig. 9 shows typical waveforms of microwave signals, beam current and applied voltage using the corrugated hollow waveguide. The microwave signals from the output window are detected by F-band (90-140GHz) rectangular horn antenna. An attenuator of 30-dB is used for comparatively high-voltage over 30kV and 20-dB is used for comparatively low-voltage under 30kV in this experiment. The peak value of the beam voltage and current are about 54.3kV and 240A, respectively.

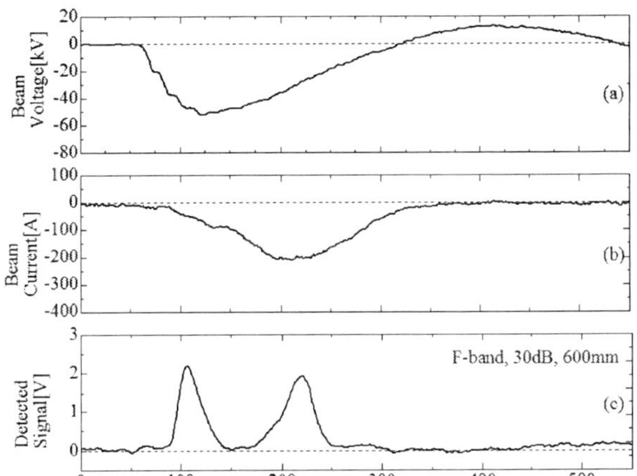

Fig.9. Typical waveforms of microwave generation experiment measured on F-band, (a) beam voltage, (b) beam current and (c) microwave signal

Fig. 10 shows the detected microwave power by an F-band (90-140GHz) rectangular horn antenna and beam current versus beam voltage. From Fig. 9, the values of voltage and beam current are changing with time. The values of the plots are measured at the time of the microwave peak in Fig.9. The previous waveform of microwave signal is 1st Phase, later is 2nd Phase. When the cathode voltage is applied under 30 kV with using the MPC, only single signal is observed. In Fig.10, the maximum detected microwave power is 5880mW when applied voltage is 60.7kV and beam current is 82.8A, and the minimum detected power is 91.1mW when applying voltage is 15.2kV and beam current is 13A. We find that the BWO with using the DDCC generates W-band microwave in especially low voltage region of under 20kV. And by using the DDCC, electron beams can be emitted at low beam voltage under 30kV stably. The DDCC can generate almost same detected microwave power, i.e. estimated power of 20kW to the velvet cathode in [4]. The actual radiation power by the DDCC will be measured by integrating the detected power over the broad pattern using the same method in [4].

After experiments, the DDCC did not take conspicuous damage of burning seen on the velvet cathodes. Therefore, the DDCC has more durability than the velvet cathodes.

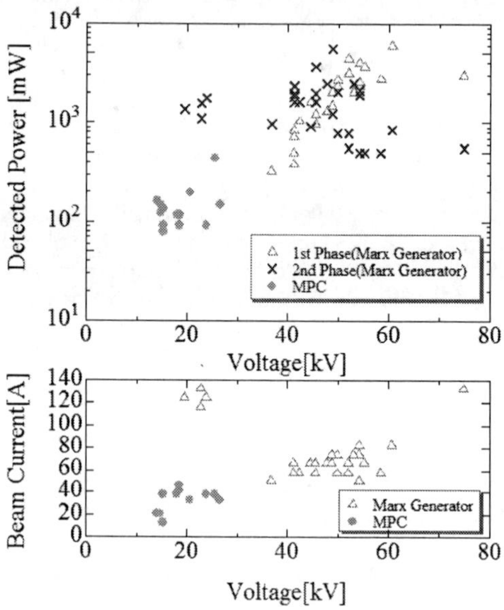

Fig.10. Detected microwave power and beam current versus beam voltage

IV. CONCLUSION

To develop THz microwave sources, a BWO utilizing a new DDCC which was made of a ceramic cylinder and a cupper disk in weakly relativistic region of less than 100kV was researched.

Experiments of electron beam generation and microwave generation using the DDCC were conducted. For microwave generation, the W-band (75-110GHz) corrugated hollow waveguide was used. As a result of electron beam emission, the electron beam was generated at higher applied voltage of about 15 kV, and characteristics of voltage versus beam current were following to the Child-Langmuir law.

From the burn patterns of electron beams using the DDCC with applied voltages of about 25kV and 69.3 kV, the electron beam was uniform and clear even low applied voltage of 25 kV. The emission surface was constructed from the triple junction, and the DDCC utilizing the triple junction was efficient to generate electron beams. Merits of DDCC are as follows; Burning damage on the DDCC is not seen like velvet cathodes, the DDCC has longer lifetime than previous cathodes, the DDCC works at low beam voltage under 30kV stably.

From the microwave generation experiment, for applying the DDCC to the BWO with W-band corrugated hollow waveguide, W-band microwave was observed. At the detection angle of 0 degrees, the BWO with the DDCC generated around 100GHz microwave detected power of 5880mW which was the almost same value by the BWO with velvet cathode estimated microwave power of 20kW in [4]. Especially, the BWO with the DDCC generated microwave in low voltage region less than 20kV stably.

From these experimental results, the DDCC suits for BWOs working in a weakly relativistic electron beam especially.

References

[1] H. Yoshimura, K. Ogura, K. Bansho, H. Iiduka, M. Takahashi and A. Sugawara, "Experimental Study of Oversized Backward Wave Oscillator with Coaxial Slow-Wave Structure", Journal of Plasma and Fusion Research, Vol. 5, pp. S2093-1-S2093-4, 2010.

[2] M. Takahashi, K. Ogura, H. Yoshimura, H. Iiduka, A. Sugawara, K. Yambe and W. S. Kim, "Experimental Study on Weakly Relativistic Oversized Backward Wave Oscillator with Coaxial Rectangular Corrugations", Journal of the Korean Physical Society, Vol.59, pp.3573-3577, 2011.

[3] R. B. Miller, "Mechanism of explosive electron emission for dielectric fiber (velvet) cathodes", J Appl. Phys., Vol. 84, pp.3880-3889,Jul. 1998.

[4] Min Thu SAN, Kazuo OGURA, Kiyoyuki YAMBE, Yuta ANNAKA, Shaoyan GONG, Jun KAWAMURA, Takuro MIURA, Shin KUBO, Takahashi SHIMOZUMA, Sakiji KOBAYASHI and Kohji OKADA, "Experimental Study on W-band(75-110GHz) Oversized Surface Wave Oscillator Driven by Weakly Relativistic Electron Beams", Plasma and Fusion Reserch:Regular Articles, Vol.11, 2406085, 2016.

[5] H. Oe, K. Ogura, Y. Kazahari, K. Bansho, H. Iizuka, A. Sugawara, and W. S. Kim, "Experimental Study on Disk Type Cold Cathode in Weakly Relativistic Energy Region", Journal of Plasma and Fusion Research SERIES, Vol.8, pp.1477-1482, 2009.

[6] K. Yambe, K. Ogura, S. Hasegawa, T. Shinada, T. Iwasaki, and T. Furuichi, "Experimental Study on Generation of Electron Beam Utilizing Cold Cathode in the Weakly Relativistic Energy Region", IEEE Transactions on Plasma Science, Vol.41, pp.2781-2785, 2013.

[7] Shaoyan Gong, Kazuo ogura, Kiyoyuki Yambe, ShintaroNomizu, Akihiro Shirai, Kosuke Yamazaki, Jun Kawamura, Takuro Miura, Sho Takahashi, and Min Thu San, "Experimental Study of Intense Radiation in Terahertz Region Based on Cylindrical Surface Wave Resonator", Journal of Applied Physics, Vol. 118, 123101, 2015.

[8] Shanoyan Gong, Kazuo Ogura, Shintaro Nomizu, Akihiro Shirai, Kosuke Yamazaki, Kiyoyuki Yambe. Shin Kubo, Takashi Shimozuma , Sakuji Kobayashi, and Kohji Okada, "Oscillation-Starting Conditions for Oversized G-band (140-220 GHz) Backward Wave Oscillator Driven by Weakly Relativistic Electron Beam", IEEE Transactions on Plasma Science, Vol.43, No.10, Oct. 2015.

Corrugated Waveguide Mode Purifier for TEM Output in a Dual-mode Operation Overmoded Coaxial Millimeter-wave Generator

Zhen Bai , Jun Zhang , Huihuang Zhong , Dongqun Chen, and Dian Zhang
College of Optoelectronic Science and Engineering
National University of Defense Technology
Changsha, China

***Abstract*—A coaxial corrugated waveguide mode purifier is designed for a dual-mode operation overmoded coaxial millimeter-wave generator. With the purifier, the mixed TEM and TM_{01} modes output are purified into a pure TEM mode. Particle-in-cell (PIC) simulation shows that the purifier would not decrease the total output power of the generator, and play an independent role to the upstream structure. Effects of mode composition ratio and phase difference on the purification ability of the purifier are also researched by both electromagnetism simulation and PIC simulation, which show that the purifier has a certain tolerance for both mode composition ratio and phase difference.**

Keywords—Millimeter wave; corrugated waveguide; mode purification

I. INTRODUCTION

Overmoded coaxial high power microwave (HPM) generators are able to achieve high power handling capacity in high frequency band because of their large transverse dimension [1]-[5]. Compared with fundamental mode operation, dual-mode operation mechanism could efficiently improve the beam-wave interaction efficiency [5][6]. Recently, we presented an overmoded coaxial millimeter-wave generator with dual-mode operation [5], in which the electron beam synchronously interacts with $7\pi/8$ mode of quasi-TEM, at the meanwhile exchanges energy with $3\pi/8$ mode of TM_{01}. Inevitably, dual-mode operation results in impure mode composition of the output power. However, mixed modes output can influence the microwave mode conversion and radiation. In fact, it has been found that a resonant cavity converter can purify the mixed modes output into pure TEM mode.

However, resonant cavity converter, no matter single cavity [5][7], dual-cavity [8], or cascaded cavities[9], always brings forth power reflection more or less, which would severely affect the working state of the interaction region upstream, especially at high frequency. Of cause, the effect may be positive, but also can be negative. Therefore, the design of resonant cavity converter is not an independent process. Sometime a few changes have to be made in the upstream structure in order to obtain a positive effect.

As a kind of non-reflection mode converter, tapered waveguide [10], nested coaxial waveguide [11], and corrugated waveguide [12], could be independently designed. Tapered waveguide is usually used to purify two mixed mode and one of them must play an absolute dominant role. Nested coaxial waveguide could purify more than two mixed modes, but it is comparatively complicated to design. Corrugation waveguide could be adequate to purify two mixed modes with arbitrary power ratio. Owing to that there are only two modes in the output power of the generator to be purified, and the mode composition ratio is close to one to one [5], the corrugation waveguide is the best choice to achieve mode purification.

In this paper, a corrugated waveguide mode purifier for TEM output in a dual-mode operation overmoded coaxial millimeter-wave generator is proposed, and the paper is organized as follows. In Sec. II, method of designing a corrugated waveguide mode purifier is presented. In Sec. III, the mixed modes output of a dual-mode operation overmoded coaxial millimeter-wave generator is purified into a pure TEM mode. In Sec. IV, effects of mode ratio and phase difference on the purification ability of the purifier are studied by electromagnetism simulation soft, and dependence of mode purification effect on the variation of diode voltage and guiding magnetic field is also investigated by Particle-in-cell (PIC) simulation. Finally, conclusions are given in Sec. V.

II. DESIGNING METHOD OF A CORRUGATED WAVEGUIDE MODE PURIFIER

Corrugation waveguide used in purification is first presented in [12], and the figure and designing method of the corrugated waveguide mode purifier originate from those of the TM_{02}-TM_{01} mode converter [13]-[17]. Different from that in [12], the output waveguide in the generator to be purified [5] in this paper is coaxial waveguide. Simultaneously considering the convenience of manufacture, a coaxial inner corrugation waveguide mode purifier is proposed. As the mode composition ratio is close to one to one, corrugation waveguide with one period always brings a non-negligible power reflection when the good purification effect is achieved. Then a two period coaxial corrugation waveguide is designed (shown in Fig. 1).

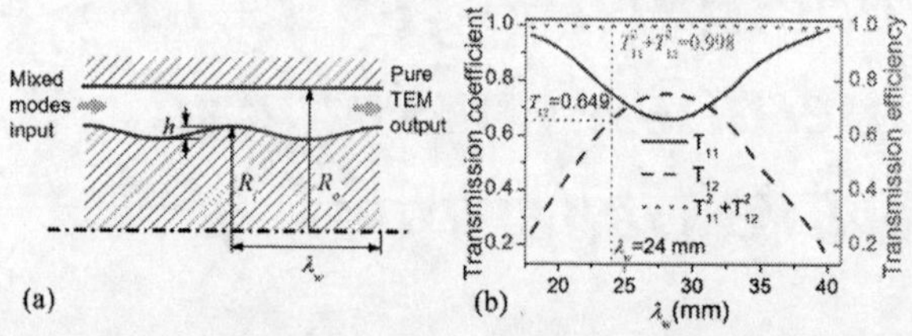

Fig. 1. Properties of the proposed purifier. (a) Sketch of the two period coaxial inner corrugation waveguide mode purifier. (b) Variations of transmission coefficient T_{11}, T_{12} and transmission efficiency versus the period length λ_w when h = 1.8 mm.

According to the theories in [12], the TM_{01}-TEM mode transmission coefficient should satisfy the following equation:

$$T_{12} = \sqrt{1/(1+P_1/P_2)},\qquad T_{11}^2 = 1-T_{12}^2. \tag{1}$$

Where T_{12} denotes the magnitude of the TM_{01}-TEM mode transmission coefficient, T_{11} denotes the magnitude of the TEM-TEM mode transmission coefficient, P_1 and P_2 are the powers of the TEM and TM_{01} modes output, separately. Powers of TEM and TM_{01} mode of the generator in [5] are 346 MW and 252 MW. Based on (1), the TM_{01}-TEM mode transmission coefficient of the coaxial inner corrugation waveguide should be T_{12} = 0.649.

The corrugation waveguide with desired T_{12} is optimized by a study version of the Ansoft's HFSS software, which is based on finite-element method (FEM) [18]. The outer radius R_o and inner radius R_i are equal to those of the output waveguide of the generator to be purified. In order to achieve a negligible power reflection, the corrugation depth h should be small. Accordingly, the corrugation depth h is set at 1.8 mm. Then the variations of T_{11} and T_{12} versus period length λ_w are shown in Fig. 2, and the transmission efficiency ($=T_{11}^2+T_{12}^2$) is also laid out. It indicates that when λ_w = 24 mm, the value of T_{12} satisfies the desired value, and the transmission efficiency is almost equal to 1, which means the power reflection can be negligible.

III. Purification Effect In Pic Simulation

The structure and typical results of the generator to be purified are specified in [5]. When the diode voltage is 400 kV, and guiding magnetic field is 1 T, the output power is 598 MW with mixed TEM and TM_{01} modes at the frequency 32.26 GHz.

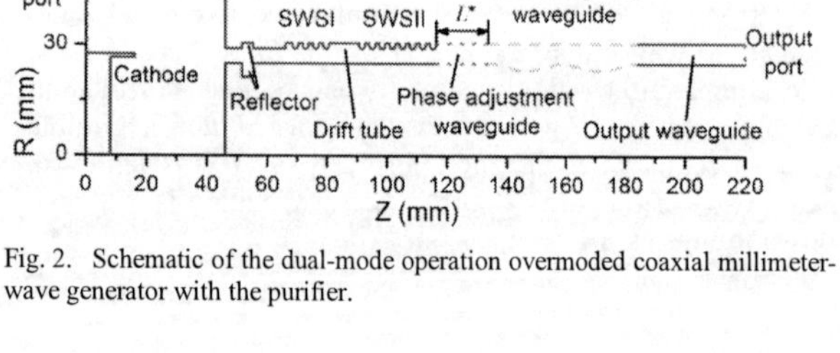

Fig. 2. Schematic of the dual-mode operation overmoded coaxial millimeter-wave generator with the purifier.

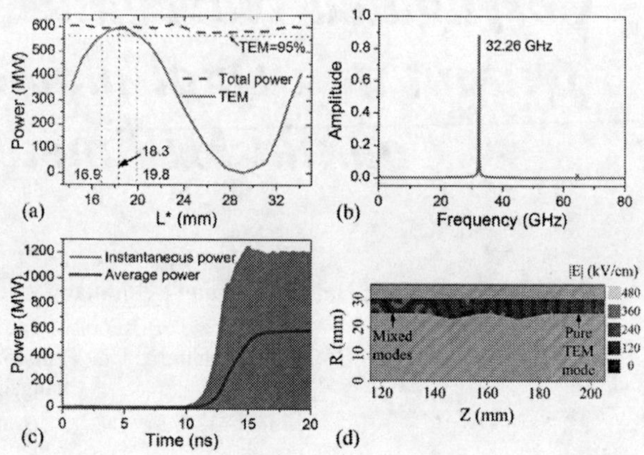

Fig. 3. PIC simulation results of the generator with the purifier. (a) Variations of the power of TEM mode and total power versus L^*. (b) FFT spectrum of output electric field. (c) Time evolution of output power. (d) Contour of the electric field distribution in the purifier and output waveguide.

The coaxial inner corrugation waveguide mode purifier designed above is applied in the generator, and a phase adjustment waveguide is inserted between the generator and the purifier. The whole structure is shown in Fig. 3 and is simulated and investigated by a 2.5 dimensional PIC code CHIPIC [19]. The diode voltage is still 400 kV and guiding magnetic field is 1 T. As the length of the phase adjustment waveguide L^* increases from zero to λ_B (≈ 20 mm, the beat wavelength of TEM and TM_{01} modes), the variation of power of TEM mode is illustrated as the red solid line in Fig. 3(a), and the total output power is denoted as the blue dashed line. When L^* = 18.3 mm, the power carried by TEM mode reaches a maximum, and the ratio of TEM is 99.3%. Besides, the total output power remains about 598 MW as L^* varies. This is certainly since the power reflection of the corrugated waveguide designed above is negligible. We set L^* to 18.3 mm to achieve single TEM mode output. The FFT spectrum and time evolution of the output power are shown in Fig. 3(b) and Fig. 3(c), respectively. They demonstrate that the purifier does not affect the work state of the generator at all. The contour of the electric field distribution in the purifier and output waveguide is illustrated in Fig. 3(d). Obviously, the mixed TEM and TM_{01} modes are purified into a pure TEM mode after propagating though the corrugation waveguide.

IV. Tolerance Ability Of The Purifier

In practice experiment, it is possible to happen that the mode composition ratio and phase difference would be a little different from the simulation results. Therefore, the purifier should have certain tolerance ability.

A. Tolerance on mode composition ratio and phase difference

First, we study the influence of mode composition ratio and phase difference on the purification effect by HFSS. A model of two period coaxial inner corrugation waveguide with straight coaxial waveguide added to both ends is established. The mixed TEM and TM_{01} mode is input from one port, and the electric field distribution is monitored at the other port.

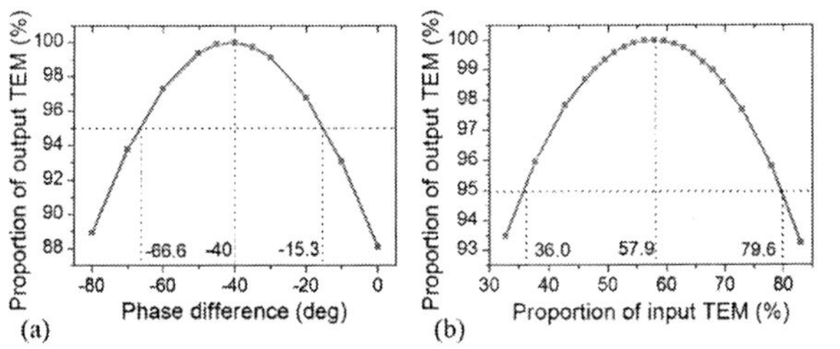

Fig. 4. HFSS simulation results of the purification effect. Variation of the proportion of output TEM mode versus (a) phase difference and (b) proportion of input TEM mode.

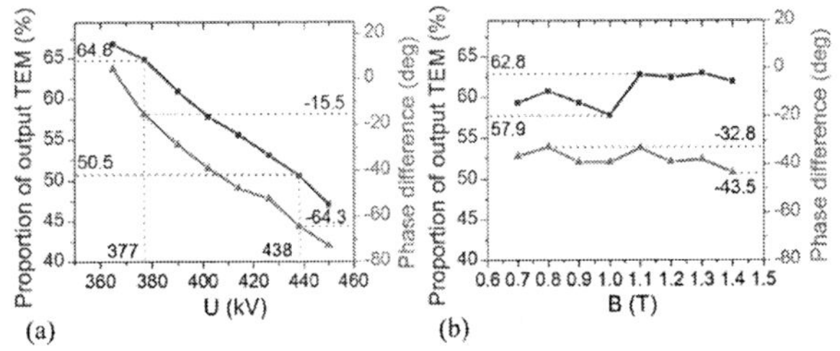

Fig. 5. PIC simulation results of the generator without the purifier. Variation of the proportion of output TEM mode and phase difference versus (a) diode voltage and (b) guiding magnetic field.

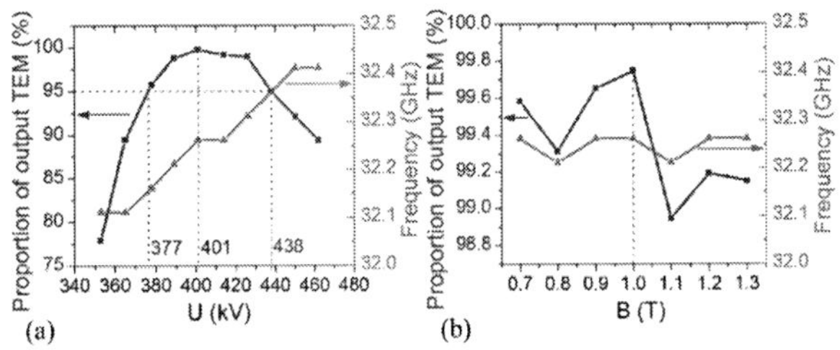

Fig. 6. PIC simulation results of the purification effect. Variation of the proportion of output TEM mode and microwave frequency versus (a) diode voltage and (b) guiding magnetic field

The mode composition ratio and phase difference of input modes can be modulated in HFSS, accordingly the purification effect changes and the mode composition can be analyzed by a mode composition analysis code based on the electric field distribution at the output port. The proportion of output TEM mode versus the phase difference and proportion of input TEM mode are shown in Fig. 4. Fig. 4(a) indicates that if the phase difference is within the range of (-40° ± 25°), the percentage of TEM mode can keep above 95%. Similarly, Fig. 4(b) denotes that when the proportion of input TEM mode is between 36% and 79.6%, a good purification effect can be obtained.

B. Dependence of mode composition ratio and phase difference on diode voltage and guiding magnetic field

The variations of diode voltage and guiding magnetic field would certainly affect the mode composition ratio and phase difference of generated microwave. The extent of their influence determines the tolerance ability of the purifier. Aiming at the generator without purifier [5], the dependences of mode composition ratio and phase difference on diode voltage and guiding magnetic field are investigated by PIC simulation. As the diode voltage and guiding magnetic field varies, the proportion of TEM mode and the phase difference between TEM and TM_{01} modes in the output power are illustrated in Fig. 5.

Fig. 5(a) demonstrates that the influence of diode voltage on phase difference is greater than that on mode composition ratio. In the range of 377 kV to 438 kV, the phase difference varies from -15.5° to -64.3°, which almost reaches the maximum range with pure TEM output (-15° to -65°), while the proportion of TEM mode only changes from 64.8% to 50.5%, far away the maximum range (36% to 79.6%). Although the diode voltage makes a greater impact on phase difference, it is convenient to compensate the phase difference by modulating the adjustment waveguide.

Fig. 5(b) shows that guiding magnetic field has little impact on both mode composition ratio and phase difference, which is coincident with our desire.

C. Tolerance on diode voltage and guiding magnetic field

After the coaxial inner corrugation waveguide is added to the generator, the purification effect versus diode voltage and guiding magnetic field is also studied. PIC simulation shows that when the diode voltage is between 377 kV and 438 kV, the proportion of TEM mode in the output power is above 95% (shown in Fig. 6(a)). That means the acceptable diode voltage is about (407 ± 30) kV. And when guiding magnetic field varies, the percentage of TEM mode in output power changes little (shown in Fig. 6(b)), which implies that the purifier has a remarkable tolerance on guiding magnetic field.

V. Conclusions

In conclusion, a coaxial inner corrugation waveguide mode purifier for a fundamental mode output in a dual-mode operation overmoded coaxial millimeter-wave generator is presented. The purifier can be designed independently because the reflection is negligible. With the purifier, the mixed TEM and TM_{01} modes output are purified into a pure TEM mode without decreasing the total output power. Tolerance ability of the purifier on mode composition ratio and phase difference is also studied. HFSS simulation shows that the phase difference tolerance is ±25°, and the mode composition tolerance is ±22%. PIC simulation demonstrates that diode voltage makes a greater impact on phase difference than mode composition, while guiding magnetic field has little impact on both mode composition ratio and phase difference. Finally, if the modes purification standard is that the proportion of TEM mode in the output power is above 95%, the acceptable diode voltage is about (407 ± 30) kV.

References

[1] T. Jiang, J. Zhang, J. He, Z. Li, and J. Ling, "Analysis and suppression of the higher order competition modes in Ku-band magnetically

insulated transmission line oscillator," *IEEE Trans. Plasma Sci.*, vol. 44, no. 5, pp. 755–760, May 2016.

[2] L. Song, J. He, and J. Ling, "A novel Ka-band coaxial transit-time oscillator with a four-gap buncher," *Phys. Plasmas*, vol. 22, no. 5, pp. 053107-1–053107-5, May 2015.

[3] Z. Qi, Jun Zhang, Qiang Zhang, Huihuang Zhong, Liurong Xu, and Lei Yang, "Design and experimental demonstration of a long-pulse, X-band triaxial klystron amplifier with an asymmetric input cavity", *IEEE Electr. Device L.*, vol. 37, no. 6, pp. 782-784, June 2016.

[4] F. Dang, X. Zhang, H. Zhong, Y. Li, and Z. Qi, "Simulation investigation of a Ku-band radial line oscillator operating at low guiding magnetic field," *Phys. Plasmas*, vol. 21, no. 6, pp. 063307-1–063307-7, June 2014.

[5] Z. Bai, J. Zhang, and H. Zhong, "A dual-mode operation overmoded coaxial millimeter-wave generator with high power capacity and pure transverse electric and magnetic mode output," *Phys. Plasmas*, vol. 23, no. 4, pp. 043109-1–043109-7, Apr. 2016.

[6] R. Xiao, J. Li, X. Bai, X. Zhang, Z. Song, Y. Teng, H. Ye, X. Li, J. Sun, and C. Chen, "An overmoded relativistic backward wave oscillator with efficient dual-mode operation," *Appl. Phys. Lett.*, vol. 104, no.9, pp. 093505-1–093505-4, March 2014.

[7] H. Ye, Y. Teng, C. Chen, H. Ning, Z. Song, Y. Cao, and P. Wu, "A millimeter wave relativistic backward wave oscillator operating in TM_{03} mode with low guiding magnetic field," *Phys. Plasmas*, vol. 22, no. 6, pp. 063104-1–063104-9, June 2015.

[8] J. Li, R. Xiao, X. Bai, Y. Zhang, X. Zhang, Q. Zhu, H. Shao, C. Chen, and W. Huang, "Dual-cavity mode converter for a fundamental mode output in an over-moded relativistic backward-wave oscillator," *Appl. Phys. Lett.*, vol. 106, no. 11, pp. 113505-1–113505-4, March 2015.

[9] Y. Teng, Y. Cao, Z. Song, H. Ye, Y. Shi, C. Chen, and J. Sun, "A Ka-band TM_{02} mode relativistic backward wave oscillator with cascaded resonators," *Phys. Plasmas*, vol. 21, no. 12, pp. 123108-1–123108-8, Dec. 2014.

[10] D. Zhang, J. Zhang, H. Zhong, and Z. Jin, "Analysis of the mode composition of an X-band overmoded O-type Cerenkov high-power microwave oscillator," *Phys. Plasmas*, vol. 19, no. 10, pp. 103102-1–103102-8, Oct. 2012.

[11] S. Peng, C. Yuan, T. Shu, X. Zhao and Q. Zhang, "A novel TM_{0n}-TEM mixed-mode converter," *IEEE Trans. Microw. Theory Tech.*, vol. 64, no. 4, pp. 1163–1169, Apr. 2016.

[12] D. Zhang, J. Zhang, H. Zhong, Z. Jin, and Y. Yuan, "Purification of the output modes of overmoded relativistic backward wave oscillators," *Phys. Plasmas*, vol. 21, no.2, 023115-1–023115-6, Feb. 2014.

[13] M. Thumm, "High-power millimetre-wave mode converters in overmoded circular waveguides using periodic wall perturbations," *Int. J. Electron.*, vol. 57, no. 6, pp. 1225–1246, Dec. 1984.

[14] H. F. Li and M. Thumm, "Mode coupling in corrugated waveguides with varying wall impedance and diameter change," *Int. J. Electron.*, vol. 71, no. 5, pp. 827–844, Nov. 1991.

[15] M. J. Buckley, D. A. Stein, and R. J. Vernon, "A single-period TE_{02} - TE_{01} mode converter in a highly overmoded circular waveguide," *IEEE Trans. Microw. Theory Techn.*, vol. 39, no. 8, pp. 1301–1306, Aug. 1991.

[16] M. K. Thumm and W. Kasparek, "Passive high-power microwave components," *IEEE Trans. Plasma Sci.*, vol. 30, no. 3, pp. 755–786, June 2002.

[17] T. U. Haq, K. J. Webb, and N. C. Gallagher, "Compact circular waveguide mode converters," *Microwave Opt. Tech. Lett.*, vol. 13, no. 5, pp. 251–255, Dec. 1996.

[18] Y. Zhu and A. C. Cangellaris, "Macro-elements for efficient FEM simulation of small geometric features in waveguide components," *IEEE Trans. Microw. Theory Techn.*, vol. 48, no. 12, pp. 2254–2260, Dec. 2000.

[19] J. Zhou, D. Liu, C. Liao, and Z. Li, "CHIPIC: An efficient code for electromagnetic PIC modeling and simulation," *IEEE Trans. Plasma Sci.*, vol. 37, no. 10, pp. 2002–2011, Oct. 2009.

EVIDENCE OF ANODE INITIATED BREAKDOWN OF VACUUM GAPS WITH A 50-NS, 1-MV PULSE GENERATOR*

R.J. Allen, D.D. Hinshelwood, J.W. Schumer,
I.M. Rittersdorf and S.L. Jackson
Plasma Physics Division
Naval Research Laboratory
Washington, DC 20375 USA
allen@nrl.navy.mil

P.F. Ottinger
Independent consultant to NRL through Syntek/Icarus.
Syntek Technologies
Arlington, VA 22203

***Abstract*— Breakdown measurements between metal electrodes in high vacuum with a 50-nanosecond, one-megavolt pulse generator based test stand have provided evidence that the anode initiates breakdown with non-field-enhanced electrodes. The widely held belief is that breakdown in this situation is initiated by explosive electron emission at some field threshold on the cathode and that there is no anode involvement. However, in recent experiments, large changes in cathode roughness and cathode material did not have a noticeable effect on the breakdown field threshold. Also, the breakdown threshold did not change as expected with varying AK gap. Finally, coating the anode with graphite particles caused a very significant reduction in the breakdown threshold. These observations provide strong evidence that particles from the anode are actually responsible for initiating breakdown of the vacuum gap. Although anodized aluminum cathodes did have a higher breakdown threshold, it is possible that the coating acts to shield the cathode from anode particles rather than to suppress explosive emission.**

Keywords—explosive emission; total voltage effect; ion field emission; vacuum breakdown

I. INTRODUCTION

A test stand capable of applying 50-ns pulses of up to one megavolt was constructed to investigate turn on of emission from metal anodes and cathodes in vacuum [1]. Emissions from anode and cathode surfaces were generally thought to begin with explosive electron emission of the cathode in pulsed power systems. However, our first measurements of the field threshold for explosive emission showed emission appearing to originate from the sharp edge of an anode [2]. This forced us to change the electrode geometry to one with a curved anode [3].

Explosive emission is a process by which plasma forms on a cold cathode surface when very high electric fields are applied. The usual idea is that there are tiny whiskers or imperfections on the cathode surface that cause the local electric field to be increased to the point where electrons can tunnel into vacuum from the tips. When this field emission current becomes very large, the whisker becomes very hot and eventually explodes. The explosion creates a plasma spot on the cathode surface that expands sideways and outwards. If there are many whiskers, the cathode surface becomes covered with plasma and is then a space-charge-limited source of electron current. Explosive electron emission from bare metal surfaces is known to be spotty and often some field enhancing treatment is applied to create many enhanced sites on the cathode. Common treatments include graphite spray and micro machining of the surface to produce sharp edges.

Our recent measurements described herein provide further evidence that for the case of a parallel-plate-like geometry and bare, non-field-enhanced metal cathodes it is actually ion emission from the anode that initiates cathode turn on. Previous work by Sandia also found that breakdown was initiated by anode emission, although they used much longer pulse widths [4]. They found results consistent with a process termed the Total Voltage Effect from DC breakdown studies, where it is believed that macro particles from the anode initiate breakdown [5]. This same effect with our short pulses puts a limitation on the size of the particle that is responsible, most likely atomic or small molecule ions due to the consideration of transit time across the AK gap. This emission mechanism is believed to be anode field desorption.

II. ELECTRODE GEOMETRY

The electrode geometry is an important consideration when comparing results of breakdown studies. Most works in the literature use nearly flat anodes and cathodes with rounded edges. Our geometry uses a flat cathode so that we get a clear view of the smaller anode from our x-ray pinhole camera, which looks through the cathode. When our first tests with a flat disk anode with a sharp outer edge showed unexpected signs of emission, we modified our anode geometry to have a uniform field Chang profile [6]. The field profiles for our original button anode geometry and the new Chang profile anode geometry were calculated with a 2D electrostatic field solver and are shown in Fig. 1 for 8.8 mm AK gap. Note that the cathodes are the same flat disks in both of these cases. The x-axis is the distance, from the center of the electrode,

Work supported by the Naval Research Laboratory Base Program moving outward along the surface. One can define a global field enhancement factor, β, for an electrode geometry as the maximum field compared that of parallel plates, or voltage divided by AK gap. Our original button geometry gave a β of nearly one for both anode and cathode on axis, or very near to parallel plates. With the Chang anode, we have β=0.91 for the cathode and β=1.16 for the anode. So, we are still very close to a parallel plate geometry, but have eliminated the high field at the edge of the anode. In the results that will be presented below, we see that enhancing β by applying a graphite spray to either the cathode or anode can reduce the breakdown threshold and turn-on time although the reasons for this are different.

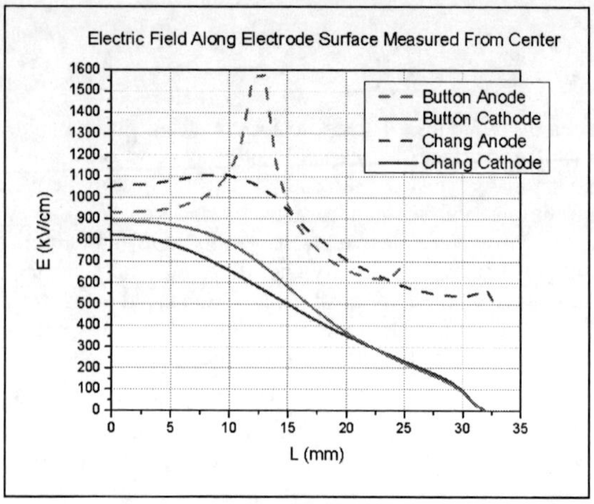

Fig. 1. Eletrostatic fields on 8.8 mm AK gap electrodes, plotted as distance outward from center, for both the original button anode and the current Chang profile anode configurations showing elimination of the highly enhanced anode edge (near L=13mm) while maintaining nearly parallel plate field enhancements.

III. Breakdown Threshold Measurements

Several breakdown threshold measurements were made with both stainless steel and aluminum bare metal electrodes and combinations of the two. We had the cathodes fabricated with either a #32 or a #64 finish. A surface roughness tester revealed that the actual roughness of the two groups were separated by more than this factor of two. The general test procedure was to repeatedly pulse the electrodes with increasing charge voltage on the generator until breakdown was observed. Tests were performed with 13.5, 8.8 and 5.6 mm AK gaps. However, the 13.5 mm AK gap tests did not show any signs of emission, even at full charge voltage.

Two different threshold were observed in this testing. The first threshold we saw was for very low level emission from a small number of point emission sites on the cathode. This emission was visible with our gated optical imager (GOI), the x-ray pinhole camera and the x-ray PIN diode. The current probe also showed this emission, although the level was often on the same order as the displacement current during the main pulse and often only becoming obvious at the very tail of the pulse. The initial start of this current appears to be just after peak voltage with a current that is believed to rise during the pulse due to plasma expansion of the point emitters. Although the next shots would often show the same behavior, just at a slightly higher current level, eventually a second threshold was reached where there was a jump to a large level of current with many emitters on the cathode surface. Looking closely at this second threshold revealed that it was preceded by visible damage on the cathode. When we stopped the test at the first sign of cathode damage and took the samples to a scanning electron microscopy (SEM), it became clear that what looked like cathode damage was actually the result of molten anode material being sprayed onto the cathode.

For comparison with data from other groups, we will equate this second threshold to much higher current to what others call the breakdown field. Our results showed no significant dependence on either electrode material or surface finish to either the first threshold for the spotty emission or for the breakdown threshold. The first turn on of spotty emission is now believed to be due to the impact of particles from the anode, likely field emitted atoms or molecules. The second threshold, associated with breakdown, is believed to be the point where emission from a single point on the cathode can cause the anode surface to melt and spray molten metal onto the cathode where it creates many highly field enhanced sites. Monte Carlo simulations of electron impact on the anode using the ITS code (Integrated Tiger Series) have shown that the melting point is consistent with our measured current and voltage history and the deposition area observed with the x-ray pinhole camera. Also, this calculated threshold is nearly the same for both stainless and aluminum.

For DC voltages on vacuum gaps greater than 1 mm, it well known that the breakdown field is not constant as predicted by an explosive emission model, but decreases with increasing gap. Our breakdown threshold measurements provide scaling similar to what is referred to as the Total Voltage Effect (TVE) due to anode interactions. The simple theory for the TVE is that macro-particles are pulled from the anode with a positive charge, Q, proportional to the applied field, E, and then impact the cathode, leading to plasma formation followed by breakdown at some threshold energy, which equals Q times the applied voltage, V. Therefore, the breakdown energy scales as EV. Since the breakdown electric field, E_B, is just the breakdown voltage, V_B, divided by AK gap, d, we have $V_B \sim d^{0.5}$ or $E_B \sim d^{-0.5}$.

The recent results from Sandia [4] also show evidence of this TVE, although their measurements were with a 160-ns pulse. As mentioned in the introduction, our 50-ns pulses further limits the charge to mass ratio of the anode particles involved. The average of our breakdown measurements for 5.6 and 8.8 mm AK gaps (red squares) with a TVE-like fit (olive) is shown in Fig. 2 along with a TVE-like fit from Sandia to their data and also the old idea of a constant breakdown field, independent of AK gap (brown). The difference in our bare metal 5.6 and 8.8 mm AK gap data provides further evidence of anode interactions because it can be fit by $E_B=665*d^{-0.5}$ as shown in Fig. 2.

Also shown in Fig. 2 is the average of measurements at both 5.6 and 8.8 mm AK gaps when the anode was coated with graphite spray (blue circles). Graphite spray is commonly used on the cathode to enhance turn-on, but should have had no effect on the anode if only the cathode were involved in turn-on. This provides further evidence that anode interactions are involved in this breakdown measurement.

We also made two measurements with an anodized aluminum cathode and 5.6 mm AK gap. Anodized aluminum is commonly used where cathode emission is not desired as it is known to have a higher explosive emission turn-on field than bare metal. The resulting breakdown field (green triangle) is only for one of these as the other one would not break down, even at full charge voltage.

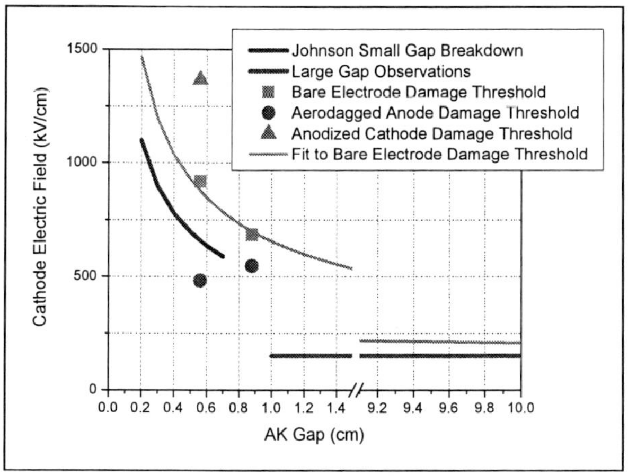

Fig. 2. Breakdown field measurements at 5.6 and 8.8 mm AK gaps (red squares) with TVE-like fit (olive) along with TVE-like fit from Sandia (dark blue) and measurements for anodes treated with graphite spray (blue circles) and that for an anodized cathode (green triangle).

IV. Full Charge Measurements

We also made several series of tests where, instead of slowly raising the charge voltage of the generator after every shot, we simply fired the first shot at full charge voltage. This is useful data for application to cases in the field where the electrodes are replaced on every shot. The primary purpose of these tests were to see if graphite spray on the anode would affect turn on. In one series with 8.8 mm AK gap, with results shown in Figs. 3 to 6, we also compared results with graphite spray on the cathode and graphite spray on both the anode and cathode. The results show that graphite spray on the anode clearly increases the total current. Additionally, x-ray pinhole and GOI diagnostics show that the emission with graphite spray on either anode or cathode or both is from a large number of emission sites with uniform x-ray emission from the anode whereas with bare metal electrodes the emission was from only a few or in some cases just one. Although the case of graphite on only the cathode resulted in faster and higher turn on than the case of graphite on only the anode, the case of graphite on both electrodes resulted in the fastest turn on and most loading of the generator voltage.

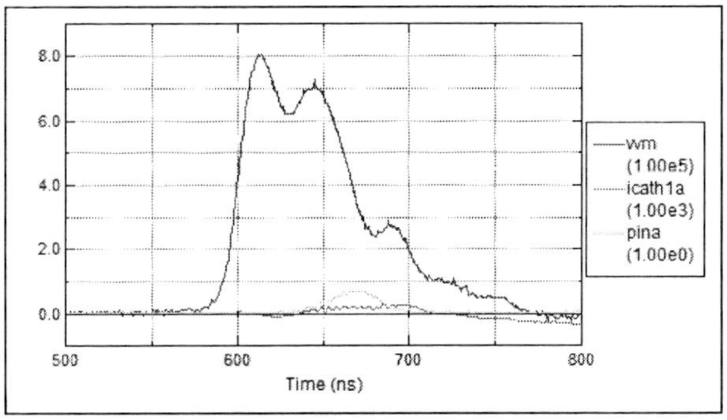

Fig. 3. Measured voltage (blue, V), current (red, A) and x-ray PIN signal (green, arb.) for full charge voltage shot with 8.8 mm AK and bare metal electrode.

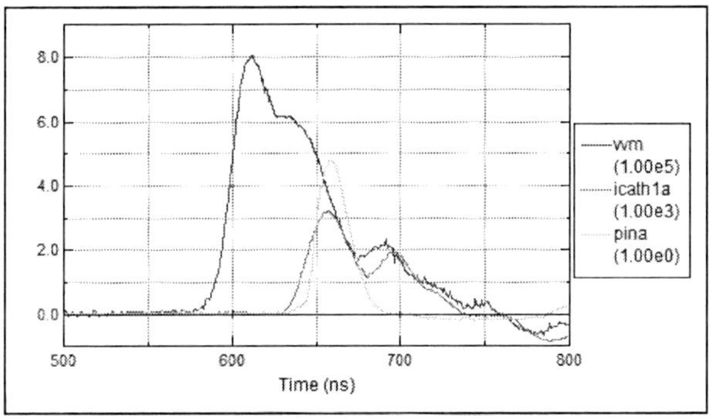

Fig. 4. Measured voltage (blue, V), current (red, A) and x-ray PIN signal (green, arb.) for full charge voltage shot with 8.8 mm AK and graphite spray applied to anode only.

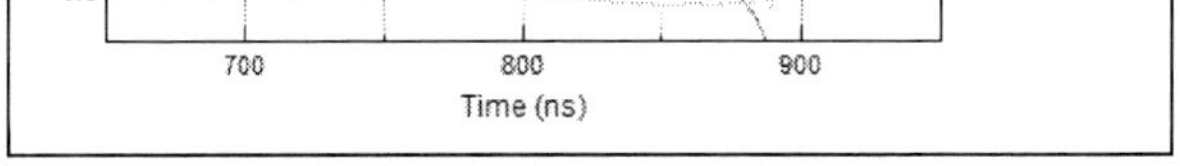

Fig. 5. Measured voltage (blue, V), current (red, A) and x-ray PIN signal (green, arb.) for full charge voltage shot with 8.8 mm AK graphite spray applied to cathode only.

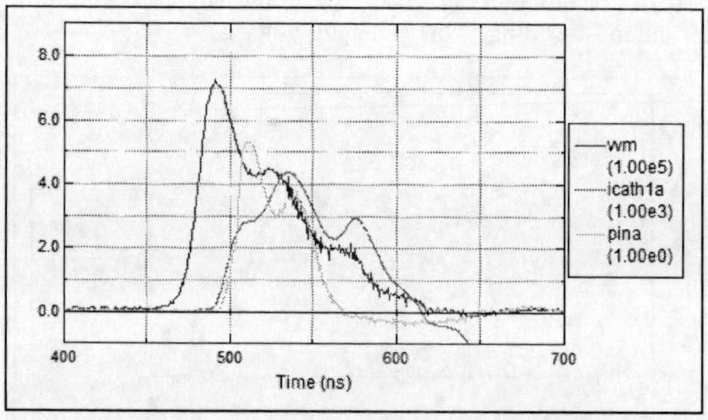

Fig. 6. Measured voltage (blue, V), current (red, A) and x-ray PIN signal (green, arb.) for full charge voltage shot with 8.8 mm AK graphite spray applied to both anode and cathode.

It should be noted that the test setup includes a vacuum voltage meter (VVM) that, besides directly measuring the applied voltage to the AK gap, acts as a dummy load due to its liquid resistor. The resistivity of the liquid used in the resistor determines the value of the load resistance and also sets the maximum applied voltage. Although the maximum voltage can be up to 1 MV, the resistance used here limited the maximum voltage to 800 kV. In cases where the test AK gap turns on, the additional current loads down the generator and reduces the applied voltage.

V. CONCLUSIONS AND DISCUSSION

We have now compiled many experimental results that point to anode particles initiating cathode turn on for the case of bare metal electrodes in a parallel plate geometry. Evidence includes x-ray emission at low applied fields for an anode with a highly enhanced edge, no change in turn on or breakdown fields when cathode roughness changed by more than a factor of two, no dependence on electrode material, a change in breakdown field with AK gap that is consistent with the scaling of the TVE model and a dramatic lowering of the turn on field when the anode is coated with graphite spray. Our previous understanding that explosive emission at the cathode was the only process involved in cathode turn on appears to be only true for highly enhanced cathodes and unenhanced anodes.

We have observed two distinct field thresholds in our breakdown field measurements. The first is when spotty emission first begins and a second for when the cathode becomes damaged by molten anode material. It is this second threshold that seems to scale like the TVE of DC breakdown. It may be that the TVE scaling is actually due to the field required for point emitters on the cathode to melt the anode. The prerequisite for that is the formation of these point emitters on the cathode, which may also play a role in the TVE scaling, especially with smaller AK gaps. It appears that the actual field threshold for explosive emission of bare metal is much higher than previously thought, perhaps 1000 kV/cm or more. Also, the role of anodized cathodes in raising the breakdown field may be different than previously thought. The standard idea is that this dielectric coating reduces the electric field on the cathode. But, it could be that the coating is actually acting as a shield to impinging anode particles, preventing plasma formation.

VI. FUTURE WORK

We intend to identify the species of anode particle responsible for initiating turn on of bare metal electrodes. One diagnostic that can do this is a Thompson parabola. The Thompson parabola contains permanent magnets and high voltage electrodes allowing parallel electric and magnetic fields to be applied over a drift region. We will need to arrange for emitted anode particles to enter the small entrance aperture of the Thompson parabola. At the exit of the Thompson parabola, we place CR-39 film to detect the impact of ionized particles. Different species of particles will trace out different parabolic paths on the CR-39 film, allowing us to determine both species and energy range of the particles. We will also investigate other anode and cathode treatments for either raising or lowering the turn-on thresholds. We are constructing a deposition system for applying different coatings to our test electrodes. We have also acquired a UV electrode cleaning device that may affect turn on. Finally, we now have spectrometers connected to the test chamber and we will see what information they can give us.

References

[1] R.J. Allen, D.D. Hinshelwood, J.W. Schumer and I.M. Rittersdorf, "A Megavolt Test Stand for Measuring Cathode and Anode Emissions with Nanosecond Pulses," Proceedings of the International Conference on Plasma Science and High Power Particle Beams 2014.

[2] R.J. Allen, D.D. Hinshelwood, J.W. Schumer and I.M. Rittersdorf, "Initial Results from a Megavolt Test Stand for Measuring Pulsed Electrode Emissions in Vacuum," Proceedings of the 2014 Euro-Asian Pulsed Power Conference.

[3] R.J. Allen, D.D. Hinshelwood, J.W. Schumer, I.M. Rittersdorf, S.L. Jackson and P.F. Ottinger, "Pulsed Vacuum Breakdown Measurements with a 1-MV Generator," Proceedings of the 2015 Pulsed Power Conference, to be published.

[4] Johnson, D.J.; Savage, M.E.; Sharpe, R.A; Batrakov, AV.; Proskurovsky, D. I, "Pulsed HV vacuum breakdown of polished, powder coated, and e-beam treated large area stainless steel electrodes with 0.5 to 7 mm gaps", IEEE Transactions on Dielectrics and Electrical Insulation, vol.13, no.1, pp.52,64, Feb. 2006.

[5] A.Maitland, J.Appl.Phys. 32, pp.2399-2407 (1961)

[6] T.Y. Chang, "Improved Uniform-Field Electrode Profiles for TEA Laser and High-Voltage Applications", Review of Scientific Instruments", Vol. 44, No. 4, pp. 405-407, April, 1973.

Spectroscopic analysis and sterilization of DBD plasma jet enveloped by water flow

K. Minami, A. Yamamoto, M. Mori
Graduate School of Science and Technology,
Kumamoto University
2-39-1, Kurokami, Chuo-ku,Kumamoto, 860-8555, Japan

H. Akiyama, T. Sakugawa
Institute of Pulsed Power Science,
Kumamoto University
2-39-1, Kurokami, Chuo-ku, Kumamoto, 860-8555, Japan

***Abstract*— Atmospheric pressure non-thermal plasmas have been widely studied for heat-sensitive material processing and sterilization due to attractive characteristics such as minimal thermal damage to the substrate and simplicity of operation, e.g., no need for vacuum. The plasmas produce many active species types which contribute to sterilization. Particularly, the OH radical is a strong oxidant, and water plays an important role for its generation. In this study, gas was flowed through a cylindrical acrylic tube and emitted from the nozzle center. The outer surface of the tube was wrapped by a copper ring electrode, with the inner wire tungsten electrode inserted to the tube center. A pulsed high voltage was applied to the inner electrode with the outer electrode grounded, after which dielectric barrier discharge occurred. A plasma jet then emerged from nozzle into ambient air. The plasma jet was enveloped by water flow in order to increase water intramembrane humidity and prevent plasma diffusion. As the target of our work was hand sterilization, we investigated characteristics using optical emission spectroscopy both prior to and following plasma jet envelopment, which was conducted with Argon gas. In addition, sterilization of Escherichia coli was carried out with and without discharge. Results show that OH radical generation is promoted by wrapping plasma jets with running water, thus contributing to sterilization.**

Keywords—plasma jet; OH radical; spectroscopy; sterilization;

I. INTRODUCTION

Industrial applications of plasma technology have expanded the technology to a diversity of fields including mechanical engineering applications such as rockets, material device process applications such as semiconductor device manufacturing, environment and chemical engineering applications such as waste disposal, and medical care applications such as surgical scalpels. For plasma use in the processing of the materials which are thermally weak, non-thermal plasma is necessary. Because realization of glow discharge at atmospheric pressure was previously considered difficult, such thermally weak materials have to date been processed by glow discharge under low pressure using a vacuum pump in the container. Recently, though, generating non-equilibrium plasma at atmospheric pressure has become possible.

This study was aimed at generating plasma jets at atmospheric pressure by forming a dielectric barrier discharge between the needle electrodes and the ring electrode covering the dielectric tube. Here, we investigated the characteristics prior to and following envelopment by plasma jets which were generating using Argon gas and were targeted against E. coli using optical emission spectroscopy and sterilization.

II. EXPERIMENTAL SETUP

This section introduces our plasma jet experimental apparatus and explains the method of spectroscopic measurement.

A. Apparatus of Plasma jet

A schematic diagram of our Dielectric Barrier Discharge (DBD) plasma jet driven Magnetic Pulse Compressor (MPC) is shown in Fig. 1. In this experiment, plasma jets were generated using argon as the working gas. Flow rates of gas and water were set at 5L/min. As shown in Fig. 1, voltages and currents between the electrodes were measured by oscilloscope.

B. Spectroscopic Measurement

A schematic diagram of plasma jet spectroscopic measurements is shown on the left of Fig. 2. We investigated characteristics of prior to and following envelopment by the plasma jet by optical emission spectroscopy. The spectrometer receiver was placed perpendicular to the jets, vertically 1cm and horizontally 3cm away from the center of the nozzle. Thus, the active species to be generated were analyzed.

C. Sterilization

A schematic diagram of sterilization effects of the plasma jet is shown on the right of Fig. 2. E. coli was used as the sterilization target. E.coli on petri dishes were placed inside the water curtain, 2cm away from the nozzle. E. coli was evenly spread on the entire medium in the dish and evaluated by comparing the size of the sterilized zone. Sterilizing effect changes by increasing the processing time (or the number of shots) were also investigated.

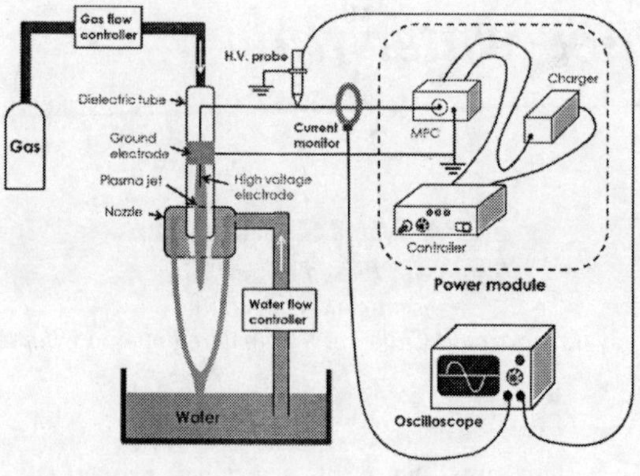

Fig. 1. Schematic diagram of the plasma jet

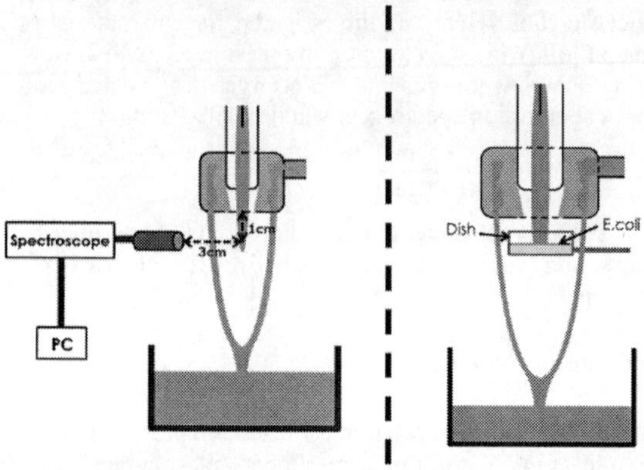

Fig. 2. Left figure — Schematic diagram of spectroscopy
Right figure —Schematic diagram of sterilization

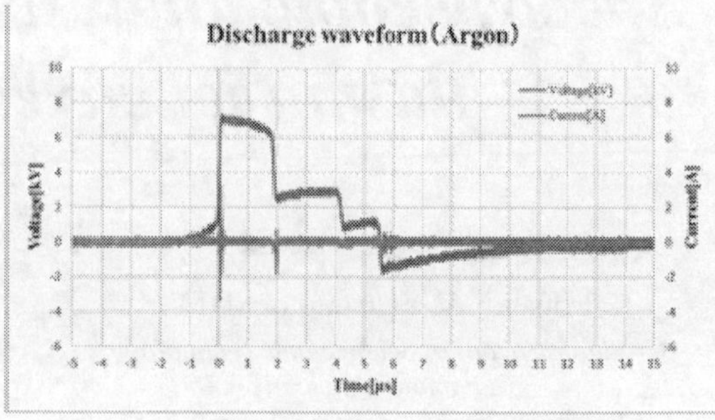

Fig. 3. Discharge waveform without water

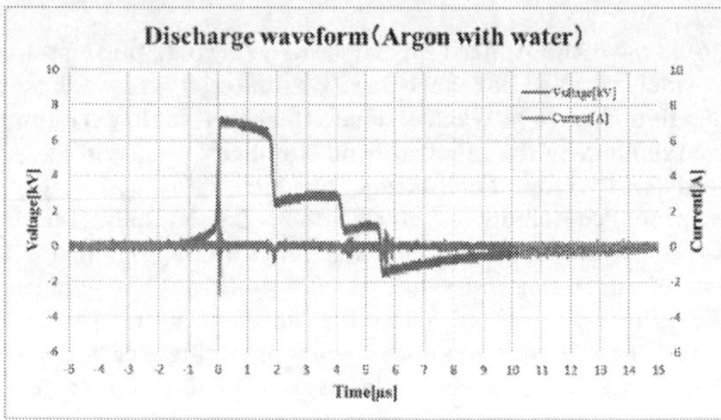

Fig. 4. Discharge waveform with water

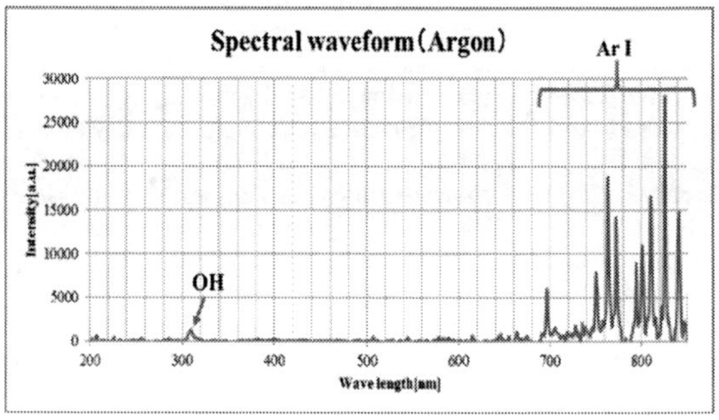

Fig. 5. Spectral waveform without water

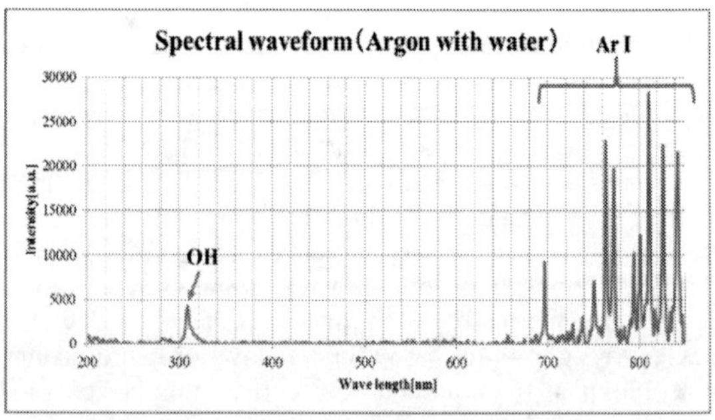

Fig. 6. Spectral waveform with water

III. RESULT AND DISCUSSION

A. Voltage —Current Characteristics

Voltage–current characteristics without water are shown in Fig. 3, and those with water are shown in Fig. 4. Without water, the maximum voltage is 7.32kV and the maximum current is 5.64A. With water, the maximum voltage is 7.32kV and the maximum current is 5.88A. The changes between two waveforms aren't almost visible.

B. Spectroscopic Measurement

The spectral waveform without water is shown in Fig. 5, and the waveform with water is shown in Fig. 6. Emission of wavelengths exceeding about 690 nm is strong; this is considered a spectrum of Ar. Importantly, a spectrum of 309nm can be observed in both cases, which is considered to be OH radicals; however, the spectrum after wrapping is significantly stronger than before wrapping by water.

C. Sterilization

Photographs of sterilization before wrapping with water are shown in the upper photograph of Fig. 7, and those after wrapping are shown below. As the processing time (or the number of shots) increases, the sterilization zone spreads. Sterilization zones achieved under the process with water are a little wider than those achieved without water. According to the results of the spectroscopic measurement, the increase of OH radicals can be considered as a cause of these discrepancies. The processing time used in this experiment, 200 seconds, was insufficient to achieve complete sterilization.

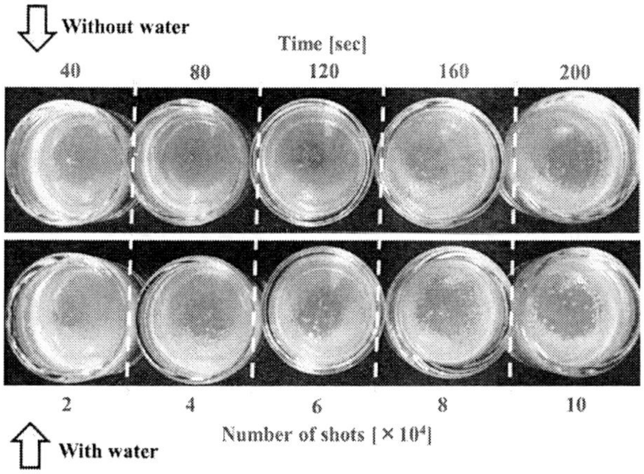

Fig. 7. Upper photograph — Sterilization without water
Lower photograph — Sterilization with water

IV. CONCLUSION

Argon plasma jet generates OH radicals which contribute to sterilization. Wrapping the argon plasma jet with water suggests the possibility of an increase in OH radicals which would enhance effects of sterilization. For complete sterilization, some adjustments such as increasing processing time are necessary.

References

[1] A. Schutze, J. Y. Jeong, S. E. Babayan, J. Park, G. S. Swlwyn and R. F. Hicks, "The Atmospheric-Pressure Plasma Jet", A Review and Comparison to Other Plasma Sources. IEEE Trans. Plasma Sci., 26 [6] (1998) 1685-1694

[2] C. Tendero, C. Tixier, P. Tristant, J. Desmaison and P. Leprince, "Atmospheric pressure plasmas", A review. Spectrochimica Acta Part B, 61 [1] (2006) 2-30

[3] J. H. Choi, I. Han, H. K. Baik, M. H. Lee, D. Han, J. Park, I. Lee, K. M. Song and Y. S. Lim, "Analysis of sterilization effect by pulsed dielectric barrier discharge", Journal of Electrostatics, 64 (2006) 17-22

[4] E. Stoffels, I. E. Kieft, R. E. J. Sladek, L. J. M. van den Bedem, E. P. van der Laan and M. Steinbuch, "Plasma needle for in vivo medical treatment: recent developments and perspectives", Plasma Sources Sci. Technol., 15 [4] (2006)

[5] F. Iza, G. J. Kim, S. M. Lee, J. K. Lee, J. L. Walsh, Y. T. Zhang and M. G. Kong, "Microplasmas: sources, particle kinetics, and biomedical applications", Plasma Process. Polym., 5 ［4］ （2008） 322-344

[6] G. Fridman, G. Friedman, A. Gutsol, A. B. Shekhter, V. N. Vasilets and A. Fridman, "Applied plasma medicine", Plasma Process. Polym., 5 [6] (2008) 503-533

[7] M. Laroussi, "Low-temperature plasmas for medicine", IEEE Trans. Plasma Sci., 37 [6] (2009) 714-725

[8] L. Ḃardos and H. Bařankova̕, "Cold atmospheric plasma: sources, processes, and applications", Thin Solid Films, 518 [23] (2010) 6705-6713

Matching a Nanosecond Pulse Source to a CO_2 Plasma Load

B. J. Hooimeijer, T. Huiskamp and A. J. M. Pemen
Electrical Engineering Department
Eindhoven University of Technology
Eindhoven, The Netherlands

***Abstract*—This paper presents the investigation of the efficiency of energy transfer, or matching, between a nanosecond pulse source and a corona plasma reactor flushed with CO_2. In order to optimize this matching, the effects of multiple parameters are investigated, such as pulse duration and amplitude, and reactor length, as well as various inner electrodes. We conclude that, within the range of parameters investigated, matching increases with pulse length and duration, as well as with reactor length. Furthermore, the use of multiple inner electrodes resulted in matching of over 90 percent.**

Key words*—*Non-thermal plasma, CO_2, pulsed power.

I. Introduction

As concerns about climate change increase, more sustainable sources are used for energy production. Combustion of fossil fuels is replaced by e.g. wind turbines and solar cells, in order to reduce emission of greenhouse gasses. However, supply may fall short of demand in case of ill weather conditions or peak hours. This could be solved by proper energy storage [1], the lack of which is one of the major challenges in the sustainable energy transition. Many techniques are being explored to arrange such storage, one of which is the conversion of CO_2 to synthetic gas (syngas) when energy supply surpasses demand [2]. Syngas can then be used to generate energy when supply is short of demand. Hence, syngas production essentially reverts the fossil fuel combustion process, and thus has the added benefit of reducing CO_2 levels.

Currently, syngas is often formed either through coal gasification or natural gas reforming [3], which both depend on fossil fuels. Thus, Lebouvier *et al.* suggest the use of non-thermal plasmas to dissociate CO_2 into CO, from which syngas can be made using the water gas shift reaction [2], [3]. Many thermal and catalytic processes exist for CO_2 dissociation [4]. However, usage of non-thermal plasmas offers much more flexibility, as such setups can often be switched on and off fast and easily. This allows the setup to utilize the varying surplus of energy in the network [3].

Multiple techniques to use plasmas for CO_2 dissociation exist, such as radiofrequency discharges or microwave discharges [3]. Dielectric barrier discharges plasmas have been used extensively for CO_2 dissociation as well [5]. Different techniques often go hand in hand with different pressures, temperatures, throughput and efficiency. For instance, for air purification purposes and ozone production, nanosecond pulsed plasmas, with subnanosecond rise times, have proven to be an efficient

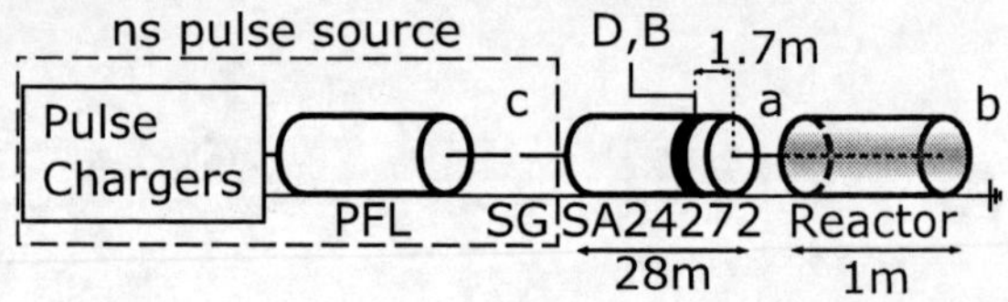

Fig. 1: A schematic representation of the setup. A nanosecond pulse source is connected to a cylindrical reactor using a SA24272 coaxial cable. The D-dot and B-dot sensors are mounted upon the cable to measure the pulses. The nanosecond pulse source consists of two microsecond pulse chargers charging a pulse forming line (PFL), which is connected to the cable with a spark gap (SG). If the PFL is sufficiently charged the spark gap breaks down, resulting in a pulse.

alternative [6], [7]. Nevertheless, much is unknown about its application for CO_2 dissociation. In this (and future) studies we will investigate the effect of nanosecond pulses on CO_2 dissociation, where we will start with the energy transfer, or matching, of the pulses to the plasma.

With ns-pulse durations, pulses act as waves and can reflect upon interfaces between components. Such reflections reduce the overall efficiency of the setup, as less energy is transferred into the plasma reactor. In order to prevent such reflections, the impedance of components have to be matched. However, the impedance of the plasma is transient, making it a highly dynamic load. This makes overall matching a challenge. We define matching as optimal, when most of the incident energy is transferred into the reactor.

Our previous study showed that very high matching of a nanosecond pulse source to a corona plasma can be achieved for air-plasmas [8]. In the current paper we investigate whether this can also be achieved for a CO_2 corona plasma load. We investigate how parameters such as pulse amplitude and duration influence matching. Other parameters that are investigated are reactor length, pulse polarity, wire diameter and the amount of parallel wires in the reactor as well as their thickness.

II. Experimental Setup & Methods

A. Experimental Setup

In our setup, CO_2 is flushed through a 50 mm-cylindrical corona plasma reactor with a wire-electrode inside [7, Fig. 6.1]. The reactor is connected to a nanosecond pulse source using a SA24272 coaxial cable. This pulse source outputs variable amplitude (up to 60 kV), variable duration

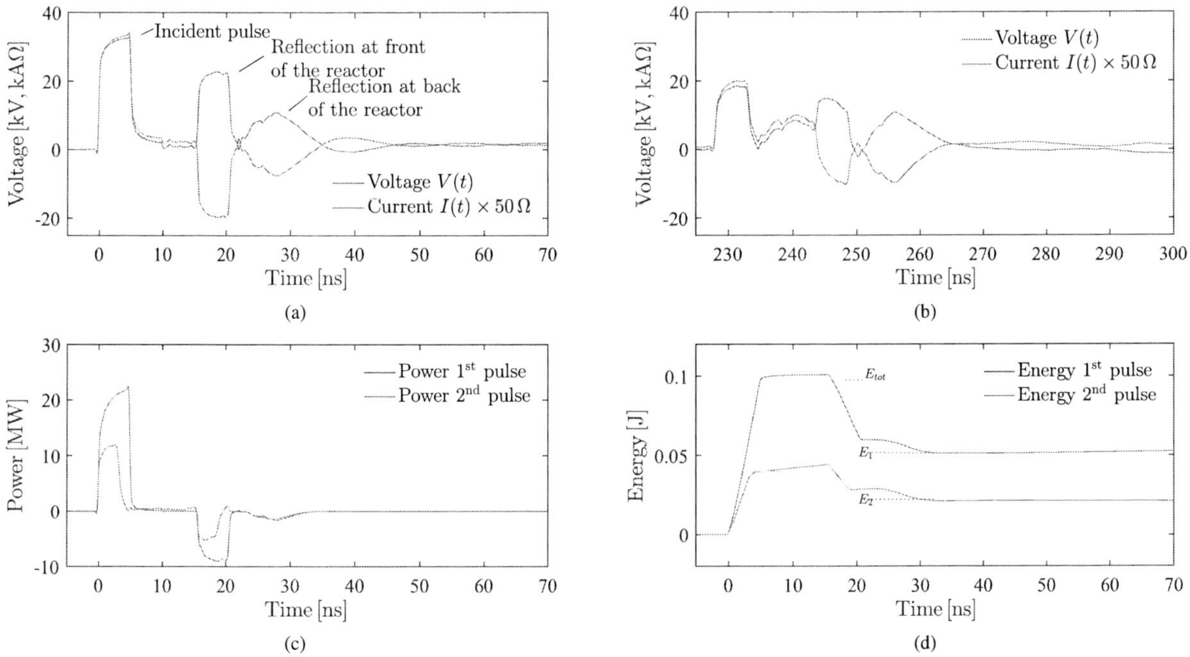

Fig. 2: Pulse waveforms as measured on the SA24272 cable. (a) The voltage and current waveforms, measured using the D-dot and B-dot sensors. After the incident pulse, two reflections can be distinguished. The first occurs at the interface between the SA24272 cable and the reactor. The second reflection occurs at the back of the reactor. Both reflections travel up and down the cable, resulting in a second pulse incident on the reactor, shown in (b). (c) The power waveforms calculated from (a) and (b). (d) The energy calculated from (c). E_{tot} is the energy in the incident pulse. E_1 and E_2 represent the energies dissipated in the first (a) and second (b) pulse respectively.

(0.5–10 ns) pulses. The pulse source consists of two microsecond pulse chargers in series, that charge a pulse forming line (PFL) [9], [10]. The PFL is connected to the SA24272 cable with a spark gap which breaks down if the PFL is sufficiently charged. The spark gap is flushed with oil to achieve sufficiently fast switching. The complete setup is shown schematically in Fig. 1.

Sensors are mounted on the SA24272 cable, measuring both the magnetic field (B-dot), inductively, and the electric field (D-dot), capacitively. The design is described in [11]. The voltage, $V(t)$, and current, $I(t)$, waveforms in the cable can be reconstructed from these fields. By multiplying these waveforms and subsequent integration we find the power $P(t)$ and energy $E(t)$ waveforms respectively.

B. *Reflections*

A pulse incident on the reactor will reflect both on the front and on the back side of the reactor. In order to distinguish these reflections from the incident pulse, the D-dot and B-dot sensors are placed at 1.7 m from the reactor. These reflections, shown in Fig. 2a, travel down the cable towards the pulse source, where they will reflect upon the spark gap, leading to a second pulse incident on the reactor. This pulse, shown in Fig. 2b, has a lower amplitude and carries less energy, but can still ignite a plasma and thus contribute to the matching. Further reflections of this second pulse have even lower energies and are unlikely to result in a plasma. Hence, only the first and second incident pulses are taken into account when investigating the matching.

The power and energy of the two incident pulses of Figs. 2a and 2b are shown in Figs. 2c and 2d respectively. The incident pulse results in a positive incident power pulse in Fig. 2c which represents the total energy incident on the reactor, shown in Fig. 2d and denoted E_{tot}. The reflected pulses are described by negative pulses in Fig. 2c, and thus represent energy traveling away from the reactor. This energy is not dissipated in the plasma and thus does not contribute to the energy transfer into the plasma. The remaining energy, E_1 and E_2, for the first and second pulse respectively, was incident on the reactor but is not reflected. Thus we conclude that the sum of these energies is dissipated by the plasma.

C. *Matching*

In order to quantify matching, we define it as the amount of energy dissipated by the plasma with respect to the total incident energy. Previously, we found the dissipated energy to be the sum of E_1 and E_2, hence the total matching, η_{tot}, can be determined using

$$\eta_{tot} = \frac{E_1 + E_2}{E_{tot}}. \tag{1}$$

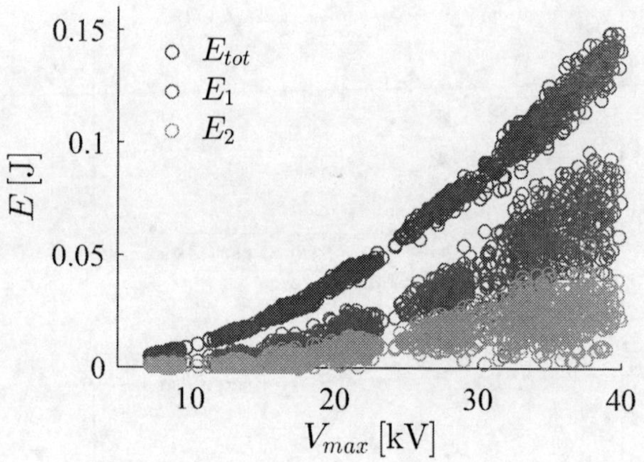

Fig. 3: Energies E_{tot}, E_1 and E_2 as determined from e.g. Fig. 2d, for 5 ns positive pulses.

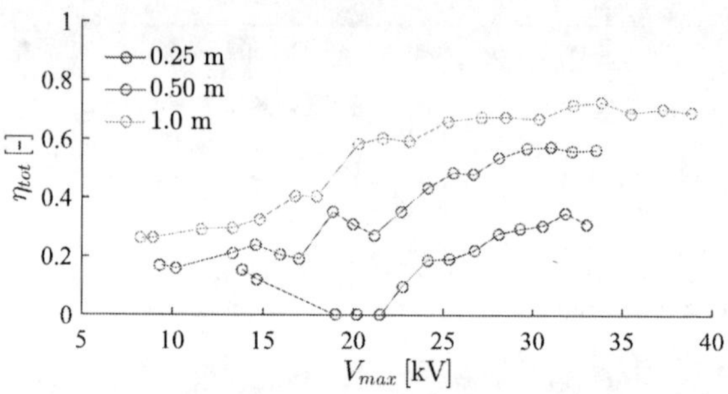

Fig. 4: Matching as function of reactor length using positive pulses. $\Delta t = 5$ ns, $F = 2.0$ L·min^{-1}, $f_{rr} = 10$ Hz, $d_i = 0.5$ mm.

D. Measurements

All measurements are conducted with pulse amplitudes, V_{max}, ranging from 5–40 kV, which are achieved by varying the spark gap distance. For each such distance 500 pulses are measured using a LeCroy WaveRunner 620Zi oscilloscope. The results in Fig. 3 give an impression of the spread due to stochastic variations in the plasma development, as well as noise. As the breakdown of the spark gap is also a stochastic process, pulse amplitudes vary naturally, which we use to our benefit. The effect upon matching is investigated for the following parameters: CO_2 flow, F; pulse repetition rate, f_{rr}; reactor length, L; pulse duration, Δt; wire thickness of the inner electrode, d_i; and pulse polarity.

III. Results and Discussion

A. Pulse Amplitude and Reactor Length

Fig. 4 shows the matching for reactor lengths of 0.25 m, 0.50 m and 1.0 m, using positive 5-ns pulses. We conclude that matching increases as the pulse amplitude increases, although gains are lower for higher amplitudes. The former is in accordance with results in [8].

We can account for this increase in matching by considering that larger pulse amplitudes result in a stronger electric field

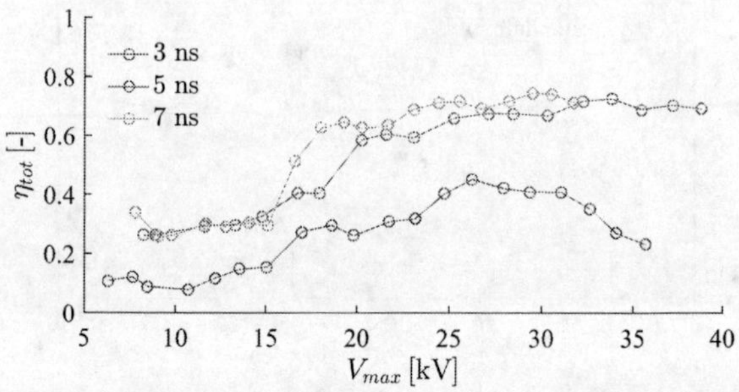

Fig. 5: Matching as function of pulse duration using positive pulses. $L = 1$ m, $F = 2.0$ L·min^{-1}, $f_{rr} = 10$ Hz, $d_i = 0.5$ mm.

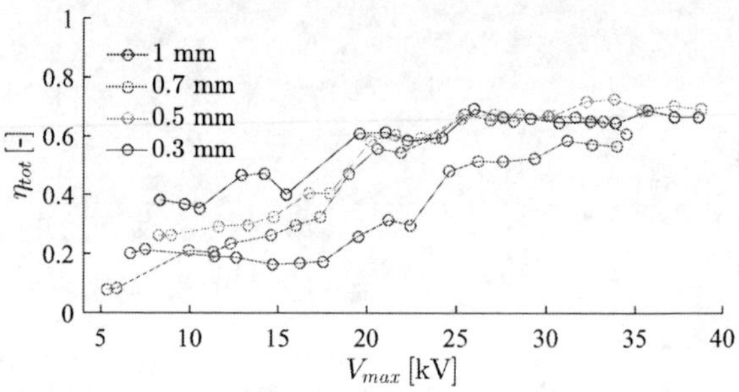

Fig. 6: Matching for different inner electrodes, using positive 5 ns pulses. $L = 1$ m, $F = 2.0$ L·min^{-1}, $f_{rr} = 10$ Hz.

inside the reactor. Consequently, more streamers are formed which propagate further [7, Fig. 7.24]. This leads to higher energy dissipation, ΔE_{str}, and thus matching.

Another conclusion which we can draw from Fig. 4, is that matching decreases as the length of the reactor becomes shorter. For shorter reactor lengths the time needed for the pulse to travel up and down the reactor becomes shorter and hence less energy is dissipated. Consequently, matching is lower.

B. Pulse Duration

In Fig. 5 the matching is shown for pulse durations of 3, 5 and 7 ns. First, we note that for 5 and 7 ns pulses, the matching increases again with pulse amplitude. For 3 ns this does not seem the case, as the matching is maximum at 26 kV. Secondly, we note that matching is worse at 3 ns, compared to 5 and 7 ns.

C. Diameter of the Inner Conductor

Finally, we use different diameters for the inner electrode of the reactor. Both positive and negative pulses are used, of which the results are shown in Fig. 6 and Fig. 7 respectively. For negative pulses, the matching at high amplitudes slightly increases as the pulse diameter increases. This is as expected, as a larger diameter leads to a lower vacuum impedance.

For positive pulses the matching appears to be very similar for different diameters, especially at high voltages. Only the 1-mm electrode performs significantly worse. This is in contrast with what we would expect, if we would only consider the lower impedance at 1 mm. However, for larger diameters the

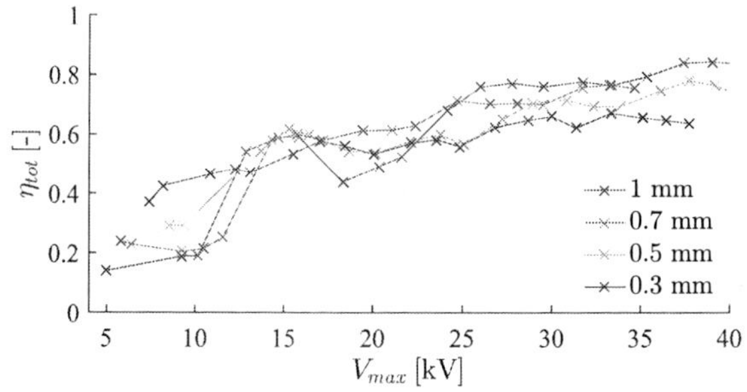

Fig. 7: Matching for different inner electrodes, using negative 5 ns pulses. $L = 1\,\text{m}$, $F = 2.0\,\text{L·min}^{-1}$, $f_{rr} = 10\,\text{Hz}$.

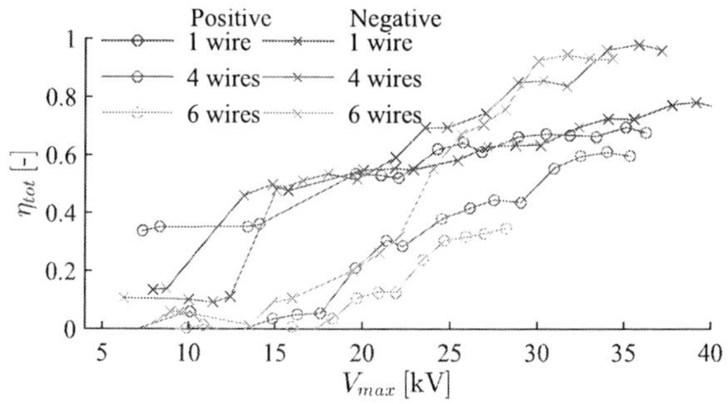

Fig. 8: Matching for different amounts of inner electrodes, using 5 ns pulses. $L = 1\,\text{m}$, $F = 2.0\,\text{L·min}^{-1}$, $f_{rr} = 10\,\text{Hz}$.

magnitude of the electric field at the inner electrode, E_{elect} decreases: $E_{elect} = \frac{2V_0}{d_i \ln \frac{d_o}{d_i}}$. A weaker electric field leads to a lower matching, according to what we have seen in section III-A. Hence, two competing effects influence matching.

D. Multiple Parallel Inner Electrodes

Ideally, the wire diameter has to be as small as possible to maximize the electric field. However, we also concluded that a lower reactor impedance improves matching. Fortunately, the impedance can also be lowered by using multiple parallel inner electrodes, while maintaining small wire diameters [8]. The results are shown in Fig. 8. We have tightened the wires in a circular arrangement with a diameter of 15 mm [8, Fig. 19].

Remarkably, for negative pulses matching has increased significantly and reaches values over 90 percent. However, for positive pulses the matching becomes significantly worse, contrary to the results in [8]. This result might be explained by the differences between air and CO_2 and will be investigated in future work.

IV. Conclusions and Research Prospects

In this study we have taken a first step in assessing the viability of using a nanosecond pulsed plasma for dissociation of CO_2 and thereby the production of syngas by investigating the efficiency of energy transfer, or matching, between a nanosecond pulse source and a plasma made in CO_2 for various parameters. From the results we can conclude that two effects improve matching: a higher energy transfer into the reactor, and a higher energy dissipation in the reactor.

First, the energy transfer into the reactor can be improved by lowering the impedance, as this decreases the reflected energy. The energy transfer has been increased most by using multiple parallel inner conductors. Second, the energy dissipation in the reactor improves as the electric field at the inner conductor increases. We have achieved this by using a higher pulse amplitudes. We have also concluded that the reactor, as well as the pulse, have to be sufficiently long to be able to dissipate the energy. Finally, we have been able to reach a matching of over 90 percent by taking both effects into account. We have managed to do so by using multiple wires with a sufficiently small diameter and negative pulses with a high pulse amplitude, in combination with a sufficiently high reactor length and pulse duration.

Further research might be able to confirm many of the findings presented here by using imaging techniques. For example, the contribution of the reflection through the reactor upon the matching can be determined. This can then be used to find optimal reactor lengths. In order to fully asses the viability of pulsed plasmas for CO_2 dissociation, further research into the chemical efficiencies is required. Only then the overall system efficiency can be determined and compared with other techniques.

References

[1] M. Beaudin, H. Zareipour, A. Schellenberglabe, and W. Rosehart, "Energy storage for mitigating the variability of renewable electricity sources: An updated review," *Energy Sustain. Dev.*, vol. 14, no. 4, pp. 302–314, dec 2010.

[2] G. Centi, E. A. Quadrelli, and S. Perathoner, "Catalysis for CO2 conversion: a key technology for rapid introduction of renewable energy in the value chain of chemical industries," *Energy Environ. Sci.*, vol. 6, no. 6, pp. 1711–1731, jun 2013.

[3] A. Lebouvier, S. A. Iwarere, P. D'Argenlieu, D. Ramjugernath, and L. Fulcheri, "Assessment of Carbon Dioxide Dissociation as a New Route for Syngas Production: A Comparative Review and Potential of Plasma-Based Technologies," *Energy & Fuels*, vol. 27, no. 5, pp. 2712–2722, may 2013.

[4] S. Rayne, "Thermal Carbon Dioxide Splitting: A Summary of the Peer-Reviewed Scientific Literature," *Nat. Preced.*, apr 2008.

[5] S. Paulussen, B. Verheyde, X. Tu, C. De Bie, T. Martens, D. Petrovic, A. Bogaerts, and B. Sels, "Conversion of carbon dioxide to value-added chemicals in atmospheric pressure dielectric barrier discharges," *Plasma Sources Sci. Technol.*, vol. 19, no. 3, p. 034015, jun 2010.

[6] T. Matsumoto, D. Wang, T. Namihira, and H. Akiyama, "Energy efficiency improvement of nitric oxide treatment using nanosecond pulsed discharge," *IEEE Trans. Plasma Sci.*, vol. 38, no. 10 PART 1, 2010.

[7] T. Huiskamp, "Nanosecond Pulsed Power Technology For Transient Plasma Generation," Ph.D. dissertation, 2015.

[8] T. Huiskamp, F. J. C. M. Beckers, W. F. L. M. Hoeben, E. J. M. van Heesch, and A. J. M. Pemen, "Matching a (sub)nanosecond pulse source to a corona plasma reactor," *Plasma Sources Science and Technology*, vol. 25, no. 5, p. 054006, 2016.

[9] T. Huiskamp, E. J. M. Van Heesch, and A. J. M. Pemen, "Final implementation of a subnanosecond rise time, variable pulse duration, variable amplitude, repetitive, high-voltage pulse source," *IEEE Trans. Plasma Sci.*, vol. 43, no. 1, pp. 444–451, jan 2015.

[10] T. Huiskamp, F. J. C. M. Beckers, E. J. M. van Heesch, and A. J. M. Pemen, "A Solid-State 0120 kV Microsecond Pulse Charger for a Nanosecond Pulse Source," *IEEE Trans. Plasma Sci.*, vol. 41, no. 12, pp. 3666–3674, dec 2013.

[11] T. Huiskamp, F. J. C. M. Beckers, E. J. M. van Heesch, and A. J. M. Pemen, "B-Dot and D-Dot Sensors for (Sub)Nanosecond High-Voltage and High-Current Pulse Measurements," *IEEE Sens. J.*, vol. 16, no. 10, pp. 3792–3801, may 2016.

GW X band TEM mode phase shifter

Xuelong Zhao, Chengwei Yuan, Lie Liu, Jingming Gao, Shengren Peng
College of Optoelectronic Science and Engineering
National University of Defense Technology
Changsha, China
xlong_z@163.com

***Abstract*—An X band TEM mode phase shifter is investigated which can adjust the output TEM mode High Power Microwave (HPM) phase in a range of 0-360°.The phase shifter is composed of two identical coaxial TEM-circular polarized TE_{11}(Cir.TE_{11}) mode converter. The electrical behaviors are analyzed and simulated by CST software, the results show that the phase shifter at frequency 9.375GHz can adjust the microwave phase in a range of 0-360°, the transmission in this range of phase shifting process is over 99%, and the power handling capacity is as high as 2.2GW which would be the power handling capacity limitation.**

Keywords—HPM; power combination; phase shifter; power handling capacity

I. Introduction

HPM as a new technology generating new applications and offering innovative ways to existing applications has emerged in recent years. Such as, directed-energy weapons, plasma heating, atmospheric monitoring and space applications [1]-[4]. Lots of work in improving the output microwave power [5]-[7] has been done by researchers. However, in some applications, the power is still unable to meet the needs. Power combination technology especially the spatial power combination is a promising way to pursue higher level power of microwave [8]. In the power combination technology, phase shifter which is a device used to control the output phase of the microwave plays a key role. Many kinds of phase shifters, such as ferrite phase shifter [9] and p-i-n diode phase shifter [10] are commonly used. However, these phase shifters are designed for radar or communication and cannot be applied directly to HPM applications since their low power-handling capacity. Therefore, in HPM domain, the phase shifter with high power handling capacity is necessary. In authors' previous work [11], a GW-class TEM mode phase shifter with high power handling capacity and high transmitting efficiency worked in L-band was proposed which could adjust the output phase of TEM mode from 0 to 360 degree linearly. However, in higher frequency band such as X, Ku bands, owing to the power handling capacity of this kind of phase shifter is not sufficient as TE_{31} and other higher modes are prevented by controlling the waveguide radius in order to work efficiently. In this paper, a TEM mode phase shifter is proposed with GW-class power handling capacity and high transmitting efficiency in X-band which provides a good choice for high power microwave combine technology especially in high frequency bands.

II. Analysis And Design

The brief schematic of phase shifter is shown in Fig.1, which is assembled by two identical TEM-circular polarized TE_{11} mode converters (TCTC). The TCTC with n (n=3, 4,…) 90°-90°-bend rectangular waveguides (Arms) can convert TEM mode to coaxial circularly polarized TE_{11} (Cir.TE_{11}) mode in over-mode state in input and output waveguides. The input waveguide is a coaxial waveguide with $d1$ and $d2$ diameters inner conductor and outer conductor. The output waveguide is also a coaxial waveguide composed of inner conductor and outer conductor with the diameters of $d3$ and $d4$. The cross section dimension of the Arm is a, b and n Arms with lengths Ri (i=1, 2, 3,…n) are distributed angularly. Two TCTCs are identical and reversely connect each other along the TEM mode transmission direction. The total longitudinal size of the phase shifter is about 3~4 free space wavelength.

When a TEM mode microwave is injected into the input port of the phase shifter, the microwave would propagate in the phase shifter undergoes the following mode conversions: Firstly, the TEM mode microwave would be converted into a Cir.TE_{11} mode microwave by the first TCTC. Then, after the Cir.TE_{11} mode microwave propagating through the second reversely connected TCTC, a TEM mode would be formed in the output port. By rotating the second TCTC η degree azimuthally, the output phase of the TEM mode microwave would shift η degree.

To generate a Cir.TE11 mode by one TCTC, the parameter Ri should meet the requirements below which would make a

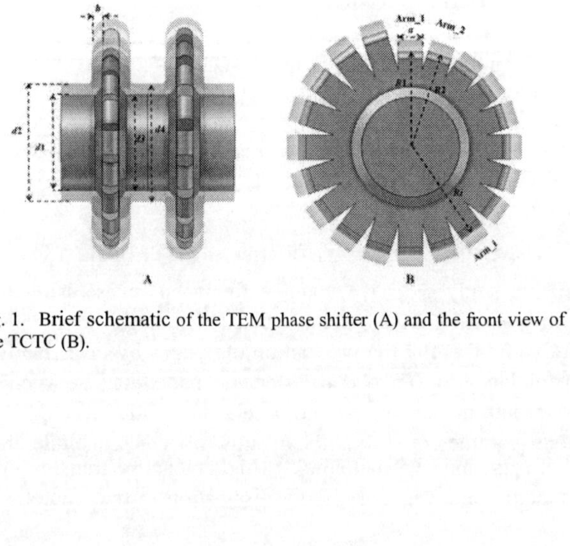

Fig. 1. Brief schematic of the TEM phase shifter (A) and the front view of one TCTC (B).

phase shift of $\frac{2\pi}{n}$ between adjacent Arms output microwave

$$\Delta R = R_i - R_{i-1} = \frac{\pi}{n\beta} \tag{1}$$

Where

$$\beta = \frac{2\pi}{\lambda_0}\sqrt{1-(\frac{\lambda_0}{2a})^2} \tag{2}$$

and λ_0 is free space wavelength at the central frequency. In order to prevent undesired high-order modes, such as TE_{30}, TE_{11}, or TM_{11} mode generated in the Arms, the cross section dimensions of the Arm should meet the following requirements:

$$\lambda_0 / 2 < a < 3\lambda_0 / 2 \tag{3}$$

$$b < \lambda_0 / 2 \tag{4}$$

Based on the equivalent ports network and mode matching method, a n-arms TCTC with a phase shift of $\frac{2\pi}{n}$ between adjacent Arms output microwave would simultaneously excite circular polarized TE_{11} and circular polarized $TE_{(n-1)1}$ in the output port of the TCTC. In order to ensure the purity of the output microwave mode, the $TE_{(n-1)1}$ mode should be prevented by controlling the waveguide radius like TE_{31} mode is prevented in the waveguide of the n=4 converter [11]-[12]. In order to prevent the coaxial $TE_{(n-1)1}$ mode and other higher order modes like TM_{01} generating and propagating in the Input and Output waveguides, $d1$, $d2$, $d3$, $d4$ should meet:

$$\lambda_0 > \lambda_{cTM01} \tag{5}$$

$$\lambda_{cTE(n-1)1} < \lambda_0 < \lambda_{cTE11} \tag{6}$$

where λ_{cTE11}, $\lambda_{cTE(n-1)1}$, λ_{cTM01} are the cutoff wavelengths of coaxial TE_{11}, $TE_{(n-1)1}$, TM_{01} mode which related to $d1$, $d2$, $d3$, $d4$.

Aim at a HPM source in the Lab which output TEM mode microwave at 9.375GHz, a n=20 TCTC is proposed based on the parameters selected principles above, and the $TE_{(10)1}$ mode would not be prevented in the input and output waveguides.

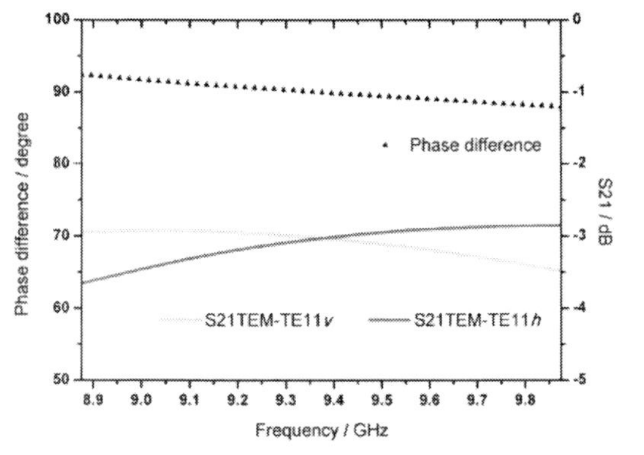

Fig. 2. The S_{21} parameter magnitudes from TEM to two degenerate TE_{11} modes and the phase difference of two TE_{11} modes

The electronic behaviors of the TCTC are obtained by the CST-MS software [13].

The injected mode is TEM mode. The S_{21} parameter magnitudes from TEM to two degenerate TE_{11} modes ($TE_{11(v)}$ and $TE_{11(h)}$) and the phase difference between two TE_{11} modes are shown in Fig.2. It is shown that the output modes are two degenerate TE_{11} modes, the amplitudes of the two components are approximately equal -3dB and the phase difference between them is about 90 degree at the central frequency 9.375GHz. In a word, the CirTE_{11} mode is formed by the TCTC.

From the simulation above, a Cir.TE_{11} mode is formed by the first TCTC, based on the theory in authors' previous paper, after the Cir.TE_{11} mode propagating via the second reversely TCTC with η degree rotating azimuthally, the output phase of TEM mode would shift η degree, where $0 \le \eta \le 360$ degree.

The phase shifter electronic behaviors are simulated by the CST software. The injected mode is TEM mode and the second TCTC is reversely located along the TEM mode propagating direction relative to the first TCTC. The electric field distribution of the phase shifter after rotating the second TCTC azimuthally η degree is shown in Fig.4.

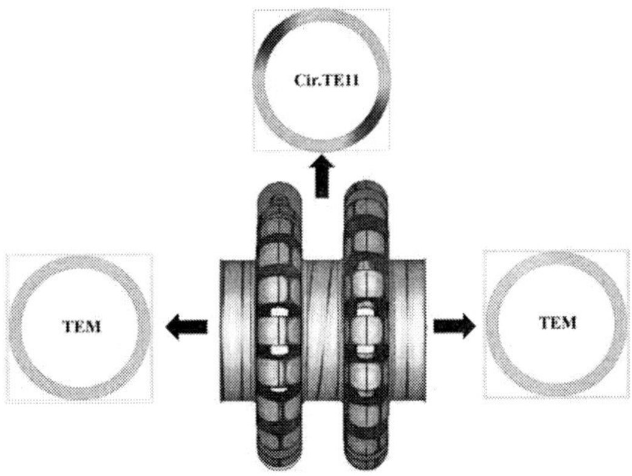

Fig. 3. The electric field distribution of the phase shifter after rotating the second CTCP azimuthally η degree

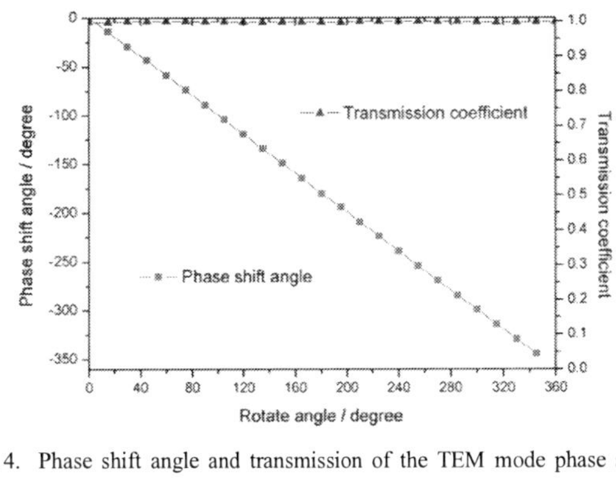

Fig. 4. Phase shift angle and transmission of the TEM mode phase shifter when the rotating angle η varies from 0 to 360°

From Fig.3, one can see that the TEM mode is converted into Cir.TE_{11} mode (rotating electric field) after propagating through the first TCTC. After the generated Cir.TE_{11} mode propagating through the second TCTC, the electric field is converted back into TEM mode, but the output phase of TEM mode shifts η degree contrast to no rotating case, where η varies from 0 to 360 degree, is shown in Fig.4. The phase shifted angle is linear dependent on the rotating angle η.

The transmission of the phase shifter at the central frequency 9.375GHz is also shown in Fig.4 which shows that the transmission coefficient is higher than 99% in the η range of 0-360 degree. Moreover, from the electric field distribution of the phase shifter after rotating the second TCTC azimuthally η degree (Fig.3), there is no obvious electric field enhancement, and the maximum field is no more than 750V/m when the injected power is 0.5W. Generally, the phase shifter would be pumped into vacuum level together with the HPM system, as the RF electric field breakdown strength is about 1MV/cm class [2], [14]. For insurance, assuming the breakdown strength is 0.5MV/cm, the power handling capacity of the phase shifter is about $[(0.5MV/cm)/(750V/m)]^2 \times 0.5$ W=2.2GW.

III. Conclusions

Power combination especially the space power combination is a promising way to increase the power of the microwave. High power handling capacity phase shifter plays a key role in the power combination technology. In this paper, a high power handling capacity TEM phase shifter worked in X-band is put forward, which can adjust the output phase of TEM mode from 0 to 360 degree linearly. The transmission in this range is more than 99% and the power handling capacity is about 2.2 GW. The analysis and simulation are matched well, and both results verified its feasibility which provides a good choice for high power microwave combine technology especially in high frequency bands.

References

[1] J. Benford, J. A. Swegle, and E. Schamiloglu, "High Power Microwaves," Taylor and Francis Press, 2007.

[2] R. J. Barker and E. Schamiloglu, "High-Power Microwave Source and Technologies," IEEE Press, New York, 2001.

[3] N. Kumar, U. Singh, T. P. Singh, A. K. Sinha, "A Review on the Applications of High Power, High Frequency Microwave Source: Gyrotron," *J Fusion Energ.*, vol. 30, no, 4, pp. 257-276, Jan. 2011.

[4] J. Benford, "Space Applications of High-Power Microwaves," IEEE Trans. Plasma Sci., vol. 36, no. 3, pp. 569–581, Jun. 2008.

[5] J. Zhang, H. H. Zhong, Z. Jin, T. Shu, S. Cao, and S. Zhou, "Studies on efficient operation of an X-band oversized slow-wave HPM generator in low magnetic field," *IEEE Trans. Plasma Sci.*, vol. 37, no. 8, pp. 1552–1557, Aug. 2009.

[6] Y. W. Fan et al., "Repetition rate operation of an improved magnetically insulated transmission line oscillator," *Phys. Plasmas*, vol. 15, pp. 083102, Aug. 2008.

[7] X. J. Ge et al., "An L-band coaxial relativistic backward wave oscillator with mechanical frequency tub ability," *Appl Phys Lett.*, vol. 97, no. 10, pp. 101503, Sep. 2010.

[8] T. A. Treado, R. A. Bolton, T. A. Hansen, P. D. Brown, and J. D. Barry, "High-power, high efficiency, injection-locked secondary-emission," *IEEE Trans. Plasma Sci.*, vol. 20, no. 3, pp. 351-359, June. 1992.

[9] E.G.Erker et al., "Monolithic Ka-band phase shifter using voltage tunable BaSrTiO3 parallel plate capacitors," *IEEE Microw. Guided Wave Lett.*, vol. 10, no. 1, pp. 10–12, Jan. 2000.

[10] A. Malczewski, S. Eshelman, B. Pillans, J. Ehmke, and C. L. Goldsmith, "X-band RF MEMS phase shifters for phased array applications," *IEEE Microw. Guided Wave Lett.*, vol. 9, no. 12, pp. 517–519, Dec. 1999.

[11] X. L. Zhao et al., "GW TEM-Mode Phase Shifter for High-Power Microwave Applications," *IEEE Trans. Plasma Sci.*, vol. 44, no. 3, pp. 268–272, Mar. 2016.

[12] S. R. Peng, C. W. Yuan, H. H. Zhong and Y. W. Fan, "Design and experiment of a cross-shaped mode converter for high-power microwave applications," *Rev Sci Instrum.*, vol. 84, pp. 124703, Dec. 2013.

[13] CST Microwave Studio. [Online]. Available: https://www.cst.com/, accessed 2012.

[14] R. Z. Xiao, C. H. Chen, J. Sun, X. W. Zhang, and L. J. Zhang, "A high-power high-efficiency klystronlike relativistic backward wave oscillator with a dual-cavity extractor," *Appl. Phys. Lett*, vol. 98, no. 10, pp. 101502, Mar. 2011.

The Electric Vector Potential for Calculating Inductive Electric Fields

Revisited

Gerald F Kiuttu
VariTech Services
Albuquerque, USA
gerald.kiuttu@varitech-services.com

***Abstract*— The electric vector potential for calculating inductive electric fields in coaxial geometries was presented at the 2001 IEEE International Pulsed Power Conference. That paper introduced the mathematical tool that allowed relatively simple calculation of inductive electric fields, and the technique was subsequently applied to the design of coaxial magnetic flux compression generators. The technique used an approximation for boundary conditions that allowed solution of the fields in 2-D geometry using partial differential equation solvers. In this paper, we explain how to eliminate the boundary condition simplifications by solving the full diffusion equations for electric vector potential and magnetic flux density in 2-D coaxial geometry. The results are fully self-consistent.**

Keywords*— pulsed power, computational techniques, inductive fields, electric vector potential

I. Introduction

In a paper at the 2001 IEEE International Pulsed Power Conference [1], we presented a simple, but powerful method for calculating inductive electric fields in axisymmetric systems. The technique involves using an electric vector potential (EVP). It is appropriate for determining electric fields when, for example, a capacitor is discharged into an inductive-resistive load characteristic of coaxial accelerators, Z-pinches, or coaxial magnetic flux compression generators (FCGs).

Because the highest field stresses occur in such systems when the current – and hence, magnetic flashover inhibition – is zero, it is important to know the electric field distributions, directions, and values in order to optimize the design of the power feed. The EVP technique has been used to design the power feed for a coaxial magnetic flux compression generator [2], and it is quite general. Furthermore, because in axisymmetric systems the EVP is purely azimuthal, it can be easily calculated using relatively inexpensive partial differential equation solvers.

The Electric Vector Potential equation follows from Maxwell's Equations and identity relations from vector calculus. The full derivation is given in [1], but the summary is reproduced here for background.

$$\vec{\nabla} \bullet \vec{D} = \rho = 0 \tag{1}$$

$$\Rightarrow \vec{D} = -\vec{\nabla} \times \vec{F} \tag{2}$$

$$\vec{D} = \varepsilon \vec{E} \tag{3}$$

$$\vec{\nabla} \times \vec{E} = -\frac{\partial \vec{B}}{\partial t} \tag{4}$$

$$\vec{\nabla} \times \frac{\vec{\nabla} \times \vec{F}}{\varepsilon} = \frac{\partial \vec{B}}{\partial t} \tag{5}$$

Equation (1) is Gauss's Law in the absence of space charge, which is normally the case, and **D** is the electric displacement. If the divergence of a vector is identically zero, then it can be expressed as the curl of another vector quantity – in this case the negative Electric Vector Potential, **-F**, (2). Equation (3) is the constitutive relation between displacement and electric field, **E**, where ε is the permittivity. Equation (4) is Faraday's Law, where **B** is the magnetic induction, or magnetic flux density. Substituting for **E** from the previous equations, we arrive at the defining equation for **F**, (5). Since we are free to choose a gauge condition for (5), we choose the Lorenz Gauge, $\vec{\nabla} \bullet \vec{F} = 0$, for convenience. In this case, and using a vector calculus identity for the double curl, (5) reduces to Poisson's equation with a source term given by the negative time derivative of the induction.

II. Calculation of the EVP Using Boundary Conditions from the Magnetic Diffusion Equation

A. Incorporating Magnetic Diffusion

In [1], we solved (5) for the simple coaxial problem shown in Fig. 1. Solution is straightforward using a finite-element generalized partial differential equation solver, such as

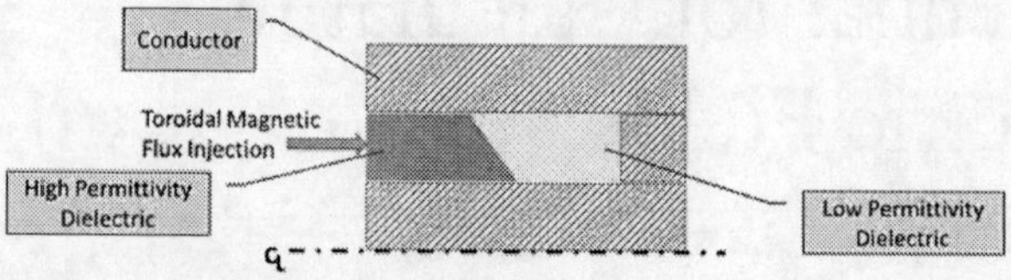

Fig. 1. Axisymmetric problem domain.

FlexPDE [3]. To complete the solution, boundary conditions (BCs) are required on the problem domain boundary. Earlier, we used approximations for these BCs, and we remarked that generally a more accurate solution for the inductive electric field could be obtained using the electric field given by Ohm's Law at the conductor surfaces; i.e., the tangential electric field at the conductor surface must be given by the product of the surface value of the current density and the resistivity,

$$E_{tangential} = \eta \, j_{surface}, \quad (6)$$

where η is the resistivity.

In this paper, we present the results for the Electric Vector Potential when the magnetic diffusion equation is simultaneously solved. From the curl of the induction, we easily obtain the current density in the conductor,

$$\bar{j} = \frac{\bar{\nabla} \times \bar{B}}{\mu_0}, \quad (7)$$

where μ_0 is the permeability of free space.

The magnetic induction is given by the solution of the diffusion equation (or Laplace's equation in nonconducting media), derived from Maxwell's equations and Ohm's Law,

$$\bar{\nabla} \times \left(\bar{\nabla} \times \bar{B}\right) = -\nabla^2 \bar{B} = -\mu_0 \sigma \frac{\partial \bar{B}}{\partial t}. \quad (8)$$

Here, σ is the electrical conductivity.

B. Sample Problem Calculation Results

We used a simple quarter-sine-wave current waveform rising to a maximum in 1 μs, as shown in Fig. 2, to inject magnetic flux into the coaxial cavity of Fig. 1, and then solved the induction diffusion equation. We note that (8) can easily be generalized to include nonlinear diffusion due to Joule heating, but such effect was negligible in the test problem.

The current density from the induction solution was then used to specify the boundary condition for the tangential electric field on the inner surface of the cavity (insulator boundary) for solution of the inductive electric field within the cavity. The induction (Dirichlet, or Value BC for **B**) and Electric Vector Potential (Dirichlet, or Value BC for **F**) were approximated as zero on the outer boundary of the thick

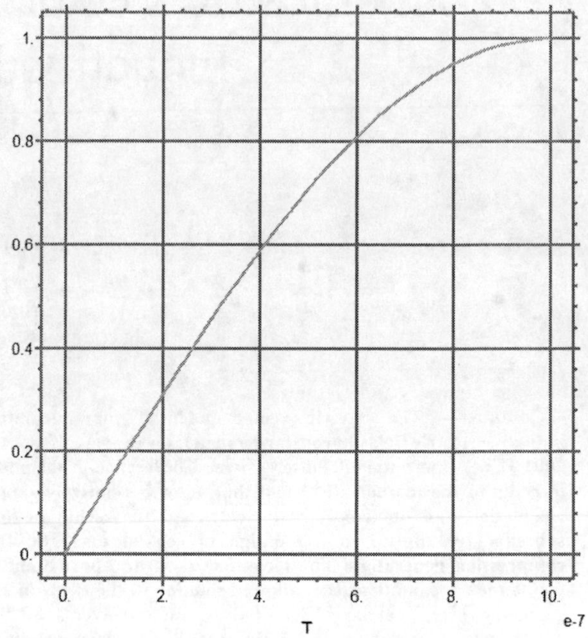

Fig. 2. Quarter-cycle sine wave current waveform used in the calculations.

conductor over the relatively short diffusion times, except on the left face (upstream end) of the problem, where a mirror BC was used to approximate a long system.

Additionally, since only the derivative (field) for vector potential was specified for the insulator boundary, the input face boundary value of the potential was specified to be zero. This forces the normal component of the field (given by the tangential derivative of the potential) to be zero.

The induction at the surfaces of the dielectric within the cavity was related to the current via the integral form of Ampere's Law on the conductor-insulator interface,

$$B_{\phi} = \frac{\mu_0 I}{2\pi r}. \quad (9)$$

Figure 3 shows the purely azimuthal induction at the end of the simulation. (Note that the material properties of the conductor and insulators were adjusted to better illustrate features of the calculation results.) A radial scan of the induction profile along the dashed line in Fig. 3 is shown in Fig. 4, and one can clearly see the development of the skin effect by the end of the calculation, as well as the $1/r$ dependence of the induction inside the cavity. The associated poloidal current density amplitude at the same time is shown in Fig. 5.

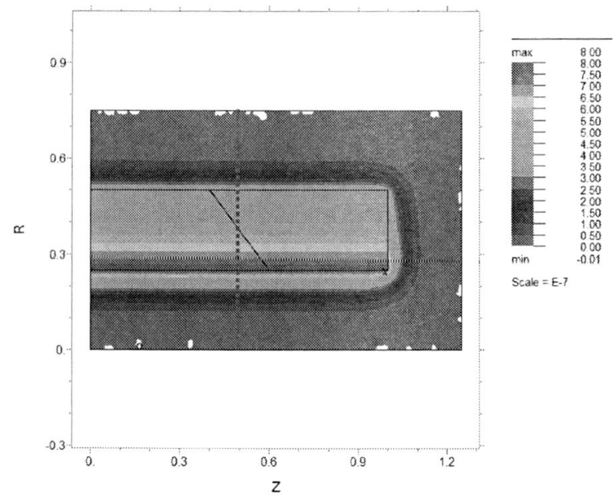

Fig. 3. Magnetic induction at the end of the simulation.

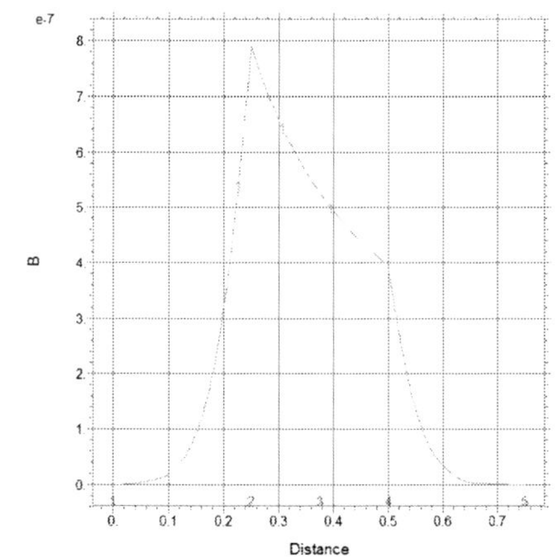

Fig. 4. Radial scan of the induction at end of simulation.

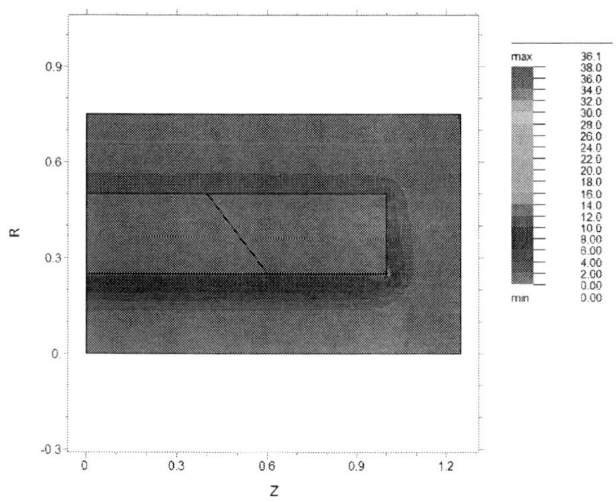

Fig. 5. Current density at the end of the simulation.

Figure 6 shows the solution for the EVP at 1/10 of the time through the simulation, 100 ns, when the inductive fields are relatively high, and Fig. 7 shows the resulting inductive electric field intensity in the cavity. Finally, Fig. 8 shows the electric field streamlines at 100 ns. One can see that the inductive electric field at the input to the cavity is purely radial because of the boundary condition there. As expected, the field drops to zero by the end of the cavity, and this can be thought of as the L dI/dt inductive voltage drop decreasing toward the end of the cavity where the downstream inductance goes to zero.

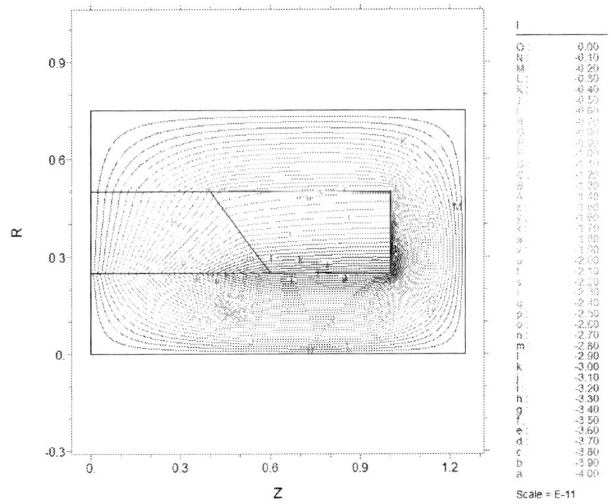

Fig. 6. EVP at 100 ns.

Other features of the calculation can be seen in Figs. 7 and 8. For example, the expected discontinuity in the electric field magnitude and direction across the dielectric-dielectric interface due to the difference in permittivity is readily apparent (relative permittivities of 6 and 1 were used in the calculation). Additionally, one can see the relative concentration of electric field at the vacuum side of the outer (maximum radius) triple point, as well as at the inner surface (minimum radius) of the dielectric at the input end of the cavity, where the voltage is highest.

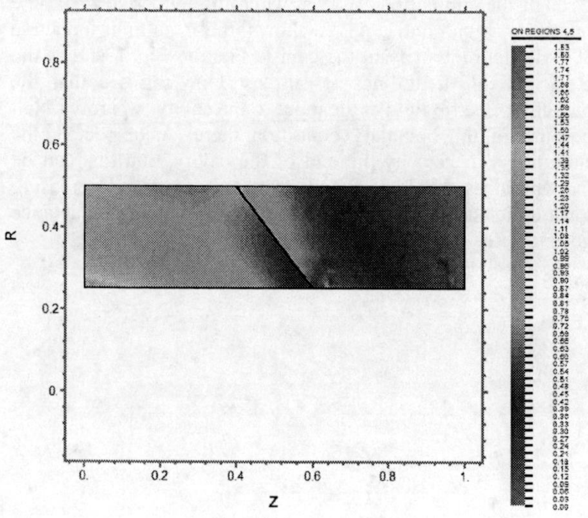

Fig. 7. Inductive electric field intensity at 100 ns.

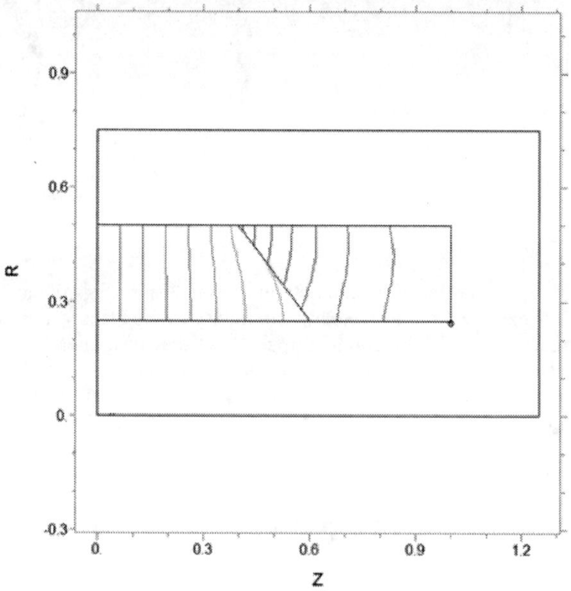

Fig. 8. Inductive electric field streamlines at 100 ns.

III. CONCLUSION

In this paper we have shown that it is relatively straightforward to include magnetic induction diffusion to more accurately set boundary conditions at dielectric boundaries in calculations of the inductive electric fields using the Electric Vector Potential. The inclusion of the magnetic diffusion improves the accuracy of the technique, and the calculations can be more self-consistent.

The technique lends itself to calculating electric fields in the power feed regions of axisymmetric pulsed power systems, facilitating better designs.

References

[1] 1. G.F. Kiuttu, "Calculation of inductive electric fields in pulsed coaxial devices using electric vector potentials," Pulsed Power Plasma Science, 2001. PPPS-2001. Digest of Technical Papers, Vol. 2 (2001), pp. 1506-1508.

[2] 2. J.B. Javedani, T.L. Houck, B.R. Poole, and A.D. White, "The application of Kiuttu's formulation to study coaxial flux compression generators," IEEE Trans. Plasma Sci. 43 (2015), pp. 3339 – 3343.

[3] 3. Available from PDE Solutions Inc, 9408 E. Holman Rd, Spokane Valley, WA 99206 USA, Tel: +1-509-891-0160.

Test of a Pulsed Corona Driven Process for Industrial Emission Control

Processing behavior, model and verification

E.J.M van Heesch, F.J.C.M. Beckers, W.F.L.M. Hoeben, T. Huiskamp, A.J.M. Pemen
Electrical Engineering Dept.
Eindhoven University of Technology
Eindhoven, Netherlands
e.j.m.v.heesch@tue.nl

A.G.A. Bonkestoter
Solybright BV
Heesch, Netherlands
info@solybright.nl

***Abstract*—We describe results of a robust pulsed plasma system for on-site application. It concerns VOC (volatile organic compound) decomposition at an industrial site. We show that 90% of the VOC content has been removed at an acceptable energy requirement. To explain the test data a convenient solution of global reaction kinetics was found. It concerns the Lambert function and a simple approximation of it that is a generalization of two existing models. The latter shows that the amount of conversion, in good approximation, is a function of a single variable. This variable is electric plasma power divided by gas flow divided by input concentration.**

Keywords*— pulsed corona, pulsed plasma, pulsed power, VOC abatement, pollution control, plasma chemistry, plasma processing, Lambert function, pulsed processing

I. Introduction

Pollution control is an ever-increasing concern for industry. Regulations meet up with new technologies on two sides: source and control. A control technology that has been emerging for a long time is pulsed plasma driven pollution abatement [1,2,3,4]. The plasma is a streamer-corona type transient gas discharge that is energized by repetitive pulses of high voltage. Unfortunately lack of adequate pulsed power technologies has hampered progress.

Technological breakthroughs are needed to make the technology useable. Important targets here are to raise the levels of performance in the areas of High-Voltage (HV), peak pulse power and repetition rate, Electromagnetic compatibility (EMC), overall reliability, and energy efficiency.

The present paper reports about our results which show that important technology goals have been met. A reliable device has been developed that can operate continuously at 1000 pulses per second and 150 MW peak pulse power. It operates reliably and achieves air cleaning down to satisfactorily levels. A challenging task for future work is scaling up.

The applied plasma type is pulsed corona. It is a non-thermal plasma that can be produced by short duration high-voltage pulses. Non-thermal means that the plasma electrons gain relatively high energies (are hot) as they are effectively accelerated during the very short high-voltage pulse. The molecules, ions and their excited states and fragments can remain cold (low thermal energy) since they are insensitive to the short duration pulse. The energetic electrons are the actors that initiate the chemical reactions. This is, because they have energies in the range of the bonding energies of the molecules. This is the mechanism responsible for the conversion or abatement of chemical compounds in a pulsed corona reactor.

Pulsed corona discharges for chemical processing were reported since the late 1980s [1,2,3,4]. Very high power during the discharge was first reported in 1989 [3]. By investigating the effect of high-voltage pulsation of a corona discharge it was found that during the high-voltage pulse, the plasma electric current showed transient peak values of 50 Amp and up [3]. The power level during this pulse was uncommon. This meant that a powerful source of chemical energy was found.

II. Method

The applied pulsed corona technology relies basically on the combination of a pulsed power source and a plasma reactor. The pulsed power source delivers a flow of electrical energy to the reactor. This flow of energy occurs as a continuing stream of short duration high-power pulses. Each pulse is 100 ns wide and has a power level that peaks at 150 MW. These pulses are repeated at a rate of 10 to 1000 pulses per second.

The pulsed power source takes its energy from the grid and delivers it to the plasma in the reactor at an efficiency of around 80%. The plasma delivers this energy to the electrons with an efficiency between ca. 1-50 % (see the work in [5]).

The electrons initiate the chemical process. This process is a chemical conversion from reactants into products. The final result is conversion. This conversion can be expressed as the concentration of reactants converted divided by the concentration of reactants on input.

This conversion will be compared with the electrical energy needed to perform this conversion. We will use the energy per volume element as the unit for applied energy. Concentration is the applied unit for reactants and products. The energy density is the crucial parameter for pulsed corona processing. The energy density is defined as electrical energy divided by treated volume which is equal to electrical power into the reactor divided by volumetric gas flow into the reactor.

Pulsed power source and plasma reactor are described in detail in [6]. The device has been housed in a freight container for easy transport. It has been equipped with fully automated control. The control covers: reactor flow, energy flow, safety, electrical measurements, flow measurements, and full pulsed-power control. The entire system is remotely controlled via a protected Internet connection.

III. Tests

We describe here one of our tests at an industrial site. The emission problem concerns a flow that is contaminated with volatile organic compounds (VOC), being mainly triglycerides. We only treated a partial flow of the system: 180 – 400 st.m^3 per hour. The applied power from the pulsed power source into the reactor is 0 - 4 kW. The input VOC concentration is 6 – 296 ppm (parts per million). We monitored the VOC concentrations simultaneously on input to the reactor and on output. The measuring devices were two similar FID (flame ionization detector) analyzers. The composition of the waste gas flow is very complex. We apply integral FID detection and apply the propane equivalent to obtain ppm (parts per million). In special cases, where knowledge about the actual composition is required, a differential FID technique is needed. In that case the detection is preceded by chromatographic separation, an example is given in [7].

Simultaneously, we also measure the electrical energy density using the detection system that is an integral part of the pulsed power source.

The container with the pulsed corona system was brought to the test site. Test sessions covered one week, in which three days of 8 measuring hours each.

IV. Results

For a range of input parameters, the total conversion was measured and documented. The results have been summarized in fig. 1. The input concentration will be denoted by *C0*, the output concentration by *CN*. The remaining fraction will be denoted by *R* and by definition *R* equals *CN/C0*. The energy density will be denoted by *E* and equals, by definition *P/F*, where *P* is the power into the reactor and *F* the volumetric flow into the reactor.

The unit we use for the energy density is Joules per Liter. For concentrations, we use ppm, parts per million. We collected 82 data points, for *C0* and for the accompanying *CN*. The behavior of the conversion process can be approximated by (1). The equation was derived in [8]. Equation 1 is valid in the approximation of $a.N > 1$, where a is the plasma volume of one pulse divided by the reactor volume and N is the number of pulses. The approach is based on:

1. Global chemical kinetics developed in [9].
2. Multiple pulsed plasma volumes in which reactive species are generated by a high-voltage pulse.
3. Reactive species induced oxidative degradation of target compounds.
4. Termination of reactive species.
5. Remixing of the plasma volumes into the reactor after each pulse.
6. Dimensionless parameters.

The resulting behavior is given as:

$$(C0 - CN) + \gamma.LN(C0/CN) = k.E \tag{1}$$

where γ and k are characteristic numbers for a specific process. γ is a characteristic value of concentration. k is the radical production per unit of energy density, as given in the already mentioned publication [8]. The unit of k is ppm.L/J. Both parameters are constants for a given process and can be evaluated for this specific process in two ways: from the theoretical expression and from a data fit. However, to evaluate the theoretical expressions for these parameters (given in [8]), detailed knowledge of rate constants would be needed. In the present case, we performed a least squares fit on the data.

Equation 1 is implicit in the unknown *CN*. The solution involves the application of the Lambert function $W(z)$. This function is defined as the inverse of $f(z) = z.exp(z)$. We only use the positive part of the principal branch of W, denoted by (W_0), and apply real arguments. We used Mathematica from Wolfram [10] to obtain the explicit solution of (1):

$$R = CN/C0 = \gamma.W_0(Y)/C0 \tag{2}$$

where

$$Y = (C0/\gamma).exp(C0/\gamma - k.E/\gamma) \tag{3}$$

Fig. 1 shows the behavior of the value of *R*, the remaining fraction, as a function of the parameter *E/C0*. Three data sets have been plotted. The measured data, the behavior according to the Lambert function and a proposed simplified model that is given by:

$$R = exp\{-k.E/C0\} \tag{4}$$

This parameter k actually equals the global radical production rate. The expression (2) was fitted (least squares, linear) to the measured data to obtain $\gamma = 19$ ppm and $k = 1.19$ ppm.L/J. The coefficient of determination of the fit r^2 is 0.68. The resulting linear fit of (4) and the test data yields $r^2 = 0.70$.

We conclude that the test data fit the model according to the Lambert function with some scatter. The scatter in the data points according to the Lambert function stems from the fact that it is a function of independently *E* and *C0* and not exactly a function of the single variable *E/C0*. However, the new function (4), is a very simple function of the single variable *E/C0*, and quite well follows the more complicated Lambert function. This new function (4), now is proposed as the replacement of the two previous models (see e.g. [8, 9, 11]):

Model 1 $$R = exp(-E/SIE) \tag{5}$$

Model 2 $$R = 1 - k.E/C0 \tag{6}$$

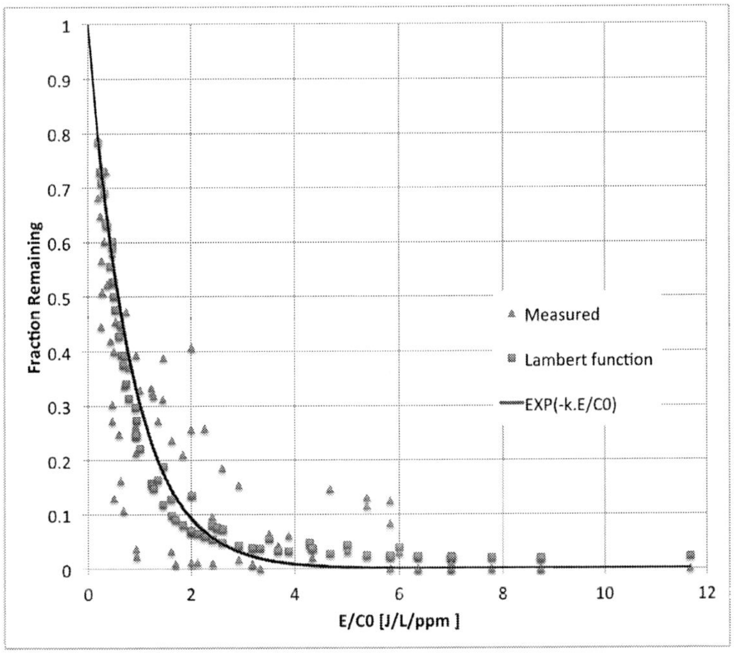

Fig. 1. The remaining fraction R=CN/C0 as a function of the parameter E/C0. The triangles are the 82 measured data points from 3 days of on-site testing. The squares are the values according to the solution (Lambert function) of Eq. 1. The line is the approximation according to Eq. 4.

where *SIE* (specific energy density, sometimes denoted as beta in literature) is an engineering constant for a specific process. *SIE* is the specific input energy, defined as the energy density needed to obtain a 1/e reduction of the pollutant concentration. The *k* in Eq. 6 is the same *k* as used in (1-4), actually a reaction rate for a specific process of pollutant removal. We notice the following: Equation 4 has a power series expansion that for small arguments ε, $\varepsilon = (k.E/C0)$, simplifies to (6).

Furthermore, it can be noticed that (5) is the analogue of (4) for $SIE = C0/k$. This latter observation shows that the model based on an *SIE* value only can be used if *C0* is constant during the test, and that the actual *SIE* value depends on the value of *C0* of the specific test.

Moreover, these findings show that the model according to (4) combines both former models (5) and (6). However, the elaborate model based on the Lambert function, (1), should be applied firstly in a particular application, as a check of the validity of (4).

V. Conclusion

The developed apparatus for VOC abatement performed very well. Tests conducted on VOC removal at an industrial site in The Netherlands proved 90% reduction at an energy cost that is acceptable. The operation principle of the applied apparatus is based on a pulsed corona technique. It involves a non-thermal plasma that is created using a pulsed high-voltage energy source. The processing behavior has been analyzed theoretically and experimentally. It turns out that, in both these aspects, the pulsed plasma process behaves according to the principal branch of the Lambert function. The argument of the Lambert function in our case, is a combination of input pollutant density C0 and energy density E. A good approximation of this function is an exponential function of the argument E/C0. That means, the process behavior of pulsed corona chemical conversion is governed by a parameter that is the ratio of energy density and pollutant concentration. This latter observation proofs to be a generalization of two previously known models.

Acknowledgment

The authors gratefully acknowledge the support received during the development phase of the devices. The supporting parties were: STW, IOP-EMVT and Antea.

References

[1] A. Mizuno, J.S. Clements, and R.H. Davis, A method for the removal of sulfur dioxide from exhaust gas utilizing pulsed streamer corona for electron energization, IEEE Transactions on Industry Applications, pp. 516-522, Vol. IA-22, No. 3, 1986

[2] S. Masuda, Pulse Corona Induced Plasma Chemical Process: a Horizon of New Plasma Chemical Technologies, Pure and Appl. Chem., Vol. 60, No. 5, pp. 727-731, 1988

[3] E.J.M. van Heesch, A.J.M. Pemen and P.C.T. van der Laan, Pulsed Corona Existence Up To 850 C, paper 14.07, Sixth International Symposium On High Voltage Engineering, New Orleans, LA, USA August 28 - September 1, 1989

[4] J. Chang, P.A. Lawless, and T. Yamamoto, Corona Discharge Processes, IEEE transactions on Plasma Science, Vol 19, No. 6, December 1991

[5] E.J.M. van Heesch, G.J.J. Winands and A.J.M. Pemen, Evaluation of pulsed streamer corona experiments to determine the O-radical yield, J. Phys. D: Appl. Phys. 41, 2008

[6] F. J. C. M. Beckers, W. F. L. M. Hoeben, T. Huiskamp, A. J. M. Pemen, E.J.M. van Heesch, Pulsed Corona Demonstrator for Semi-Industrial Scale Air Purification, IEEE Transactions on Plasma Science, Vol. 41, No.10, pp. 2920-2925, 2013

[7] W.F.L.M. Hoeben, F.J.C.M. Beckers, A.J.M. Pemen, E.J.M. van Heesch, W.L. Kling, Oxidative degradation of toluene and limonene in air by pulsed corona technology, J. Phys. D: Appl. Phys. 45 055202 2012, pp. 1–14

[8] E.J.M. van Heesch, K. Yan, A.J.M. Pemen, G.J.J. Winands, F.J.C.M. Beckers, W.F.L.M. Hoeben, From Pulsed Power to Processing: Plasma Initiated Chemical Process Intensification, International Journal of Plasma Environmental Science & Technology, Vol.6, No.2, September 2012

[9] K. Yan, E.J.M. van Heesch, A.J.M. Pemen and P.A.H.J. Huijbrechts, From Chemical Kinetics To Streamer Corona Reactor And Voltage Pulse Generator, Plasma chemistry and plasma processing, 21(1), 107-137, 2001

[10] E. W. Weisstein, Lambert W-function, mathworld.wolfram.com, 2002

[11] L. A. Rosocha, Nonthermal Plasma Applications to the Environment: Gaseous Electronics and Power Conditioning. IEEE Transactions on Plasma Science, Vol. 33, No.1, pp. 129-137, 2000.

Different operation modes of cylindrical GESA IV facility with converging electron beam

Renate Fetzer, Wladimir An, Alfons Weisenburger, Georg Mueller
Institute for Pulsed Power and Microwave Technology
Karlsruhe Institute of Technology
76021 Karlsruhe, Germany

***Abstract*—The cylindrical triode-type electron accelerator GESA IV was specially designed for the treatment of metallic rods, typically about 10 mm in diameter. Due to the large difference between cathode and grid diameter on the one hand and the comparatively small target (anode) diameter on the other hand, unstable operation and inhomogeneous target treatment are often observed. This may be caused by the formation of a virtual cathode between grid and anode or by the large distortion of electron trajectories due to the self-induced magnetic field. Another crucial phenomenon is the generation of ions at the target. In the present study, we systematically analyze the different operation modes of the converging radial electron beam generated in GESA IV by means of a numerical study using the PIC code software package MAGIC. In case ion generation at the target and ion flow towards the cathode are negligible, i.e., for unipolar flow, two operation modes are found. In the regime of low cathode-grid voltage, laminar flow and homogeneous power density at the target are obtained. For cathode-grid voltages exceeding a critical value, a virtual cathode forms between grid and anode, which leads to unstable currents and an inhomogeneous power density at the target. In order to obtain melting of the target surface layer, pulse durations in the microsecond timescale are usually chosen. In this operation regime, ion flow cannot be neglected and bipolar flow needs to be studied. Again, in a regime of low cathode-grid voltage, laminar flow and a homogeneous power density at the target are obtained. For increasing cathode-grid voltage, however, magnetic insulation becomes significant and finally dominates the operation. In this presentation, the different operation modes of the converging radial electron beam are introduced and discussed.**

Keywords—pulsed electron beam; cylindrical triode; unipolar flow; bipolar flow; PIC code simulations.

I. Introduction

Intense pulsed electron beams can be used to improve the material properties of metallic targets at the surface. Electron beam irradiation leads to local heating and melting of the surface layer followed by rapid cooling and re-solidification. As a result of this treatment surface hardness, wear, and corrosion resistance can be improved. For surface treatment of cylindrical tubes such as cladding tubes of GEN IV fission reactors, the cylindrical pulsed electron beam facility GESA IV was specifically designed [1]. The cylindrical target with diameter ~10 mm serves as anode and is surrounded by a cylindrical cathode and controlling grid. A high voltage pulse is generated and applied to the explosive emission cathode, where plasma is formed and serves as emission source. The grid potential controls the beam current density. The electron beam is then accelerated towards the target for the specified surface treatment. No external magnetic field is applied.

Operation of the cylindrical GESA IV turned out to be unstable and inhomogeneous in some cases [2]. In order to improve the beam performance and to extend the region of treatment, a new design with a one-meter long cathode has been developed recently [3]. In this presentation, a systematic numerical study of the operation performance of the updated version of GESA IV is presented. Stable and homogeneous operation regimes are identified. Regimes of inhomogeneous beam power density at the target are studied in detail.

II. Methods

For this study, the PIC code simulation package MAGIC2D by Orbital ATK, USA, is used [4]. Azimuthal symmetry of the cylindrical system is assumed as well as mirror symmetry at z=0. Figure 1 shows the system geometry. The target anode is 2.2 m long with radius 5 mm. The cathode has a length of 1 m and an inner radius of 15 cm. The controlling grid has a length of 1.2 m and radius 10 cm. It is composed of individual conductor rings, connected via line resistors. In the shown case, each ring is 2 mm wide (the chosen cell size in axial direction) and the periodicity is 26 mm. The line resistors fix the electric potential between the rings, which minimizes the electric field distortion around the grid rings. A high voltage pulse is applied to the cathode via the input port. The potentials of cathode and grid against the grounded anode are controlled via the resistors R_{cg} and R_{ga}. Beam current reaching the target flows in axial direction along the target to the outer boundary. Primary beam electrons are generated at the cathode via the explosive emission routine. The beam electrons then move in accordance with the relativistic Lorentz equation in the free space of the simulation area until they hit a conductor. There is no background gas or ionization/recombination processes included in the simulation. Thus, electrons are deleted if and only if they hit a conductor. Backscattering of electrons at the target is included, i.e., when electrons hit the surface, secondary electrons (backscattered electrons) are emitted according to the target material specifications (in this case aluminum). The third type of charged particles considered in the simulation are protons. If included, they are emitted from the target surface obeying the space charge limit. Once emitted, ions move according to the Lorentz equation and are deleted from the simulation if and only if they hit a conductor.

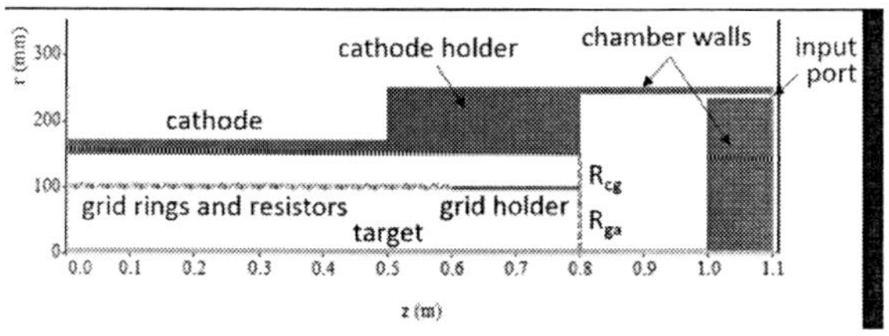

FIG. 1. System geometry. The accelerator is composed of conductors and line resistors. Conductors are the outer chamber walls, cathode and cathode holder, grid rings and grid holder, and the target anode. Line resistors R_{cg} and R_{ga} connect cathode holder, grid holder, and target. Additional line resistors connect the individual grid rings. The system is connected to a generator via a transmission line at the input port.

III. Results and Discussion

A. Unipolar Flow

First, unipolar electron flow is considered, i.e., protons are not included in the simulations. The performance of unipolar flow is of interest for the initial stage of pulsed electron beam application when ion generation at the target and ion motion towards the cathode are negligible. The cathode voltage is fixed at U_{ca} = 120 kV in the simulations, while the ratio of the controlling resistors R_{cg} and R_{ga} is varied. In figure 2, results of the simulation with U_{cg} = 18 kV are shown.

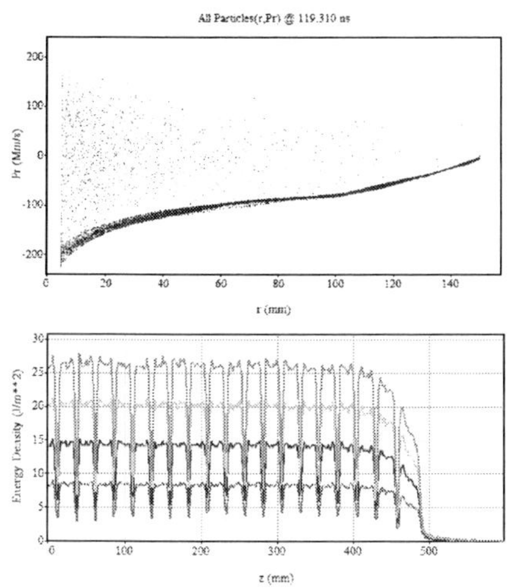

FIG. 2. Simulation results for unipolar flow with U_{cg} = 18 kV: (a) radial momentum versus radial position; blue dots represent primary beam electrons, red dots indicate electrons backscattered at the target; (b) accumulated beam energy density at target surface versus axial position at four times between 47 ns and 119 ns.

The beam electrons (blue dots) are continuously accelerated from the cathode at r = 150 mm via the grid at r = 100 mm to the anode at r = 5 mm. The radial momentum of all beam electrons is negative, i.e., no primary electrons return towards the cathode. Backscattered electrons (red dots) are found mainly in the vicinity of the target. They show both positive and negative radial momentum with a wide spread. Starting at the target, they move towards the cathode until they stop and finally return to the target. As shown in fig. 2b, the beam energy density measured at the target is homogeneous (except the positions shaded by the grid rings) up to z~400 mm and is continuously decreasing to zero in the axial range ~400 – 500 mm. The beam power density at the target is 0.74 MW/cm². Note that the quantitative numbers of the energy density given in fig. 2b require some re-normalization to the size of the simulation cells. To summarize the situation shown in figure 2, laminar flow of beam electrons is obtained.

Figure 3 shows various currents of the full accelerator (not only half of the source as shown in fig. 1). Two different regimes are found, with the transition occurring at a cathode-grid voltage of about 32 kV. The above described situation of laminar flow is obtained below this threshold. The currents of the laminar flow regime are analyzed more closely in the following.

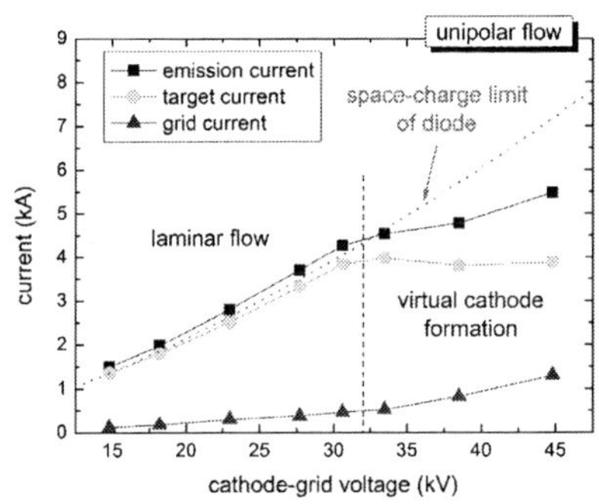

FIG. 3. Emission, target, and grid currents of the full length accelerator as a function of cathode-grid voltage for constant cathode-anode voltage 120 kV, Al target, and unipolar flow.

Due to the grid transparency of 92%, about 8% of the current emitted at the cathode hits the grid rings and is conducted along the grid and resistor R_{ga} to ground. The rest of the beam current reaches the target directly, where it flows along the target to ground. A very small amount of electrons (mainly backscattered electrons) escapes to the sides before reaching any conductor. The respective current of escaped electrons is below 40 A and is not shown in the figure. All three currents of the GESA IV triode shown in figure 3 increase with increasing cathode-grid voltage. The simulation results are compared with the theoretical expectation of space-charge limited flow in infinite-length cylindrical diodes with cathode radius 15 cm and anode radius 10 cm (the controlling grid in our system). Apart from a very small offset, the simulation results for the GESA IV triode in the laminar flow regime agree very well with the predicted space-charge limit of cylindrical diodes. Thus, processes between grid and target do not influence the beam characteristics in the cathode-grid gap and the space charge limited emission current of the triode can be predicted by the corresponding diode configuration.

For cathode-grid voltages above the laminar flow threshold of ~32 kV, the emission current stays below the space charge limit, see figure 3. This regime will be analyzed more closely in the following. As the cathode-grid voltage is increased and the emission current becomes larger, a potential well is formed between grid and target due to space charge. By further increasing the cathode-grid voltage and thus the emission current, the potential well reaches the cathode potential and a virtual cathode is formed. The transition from the laminar flow regime to a regime where a virtual cathode is formed is shown in figure 4.

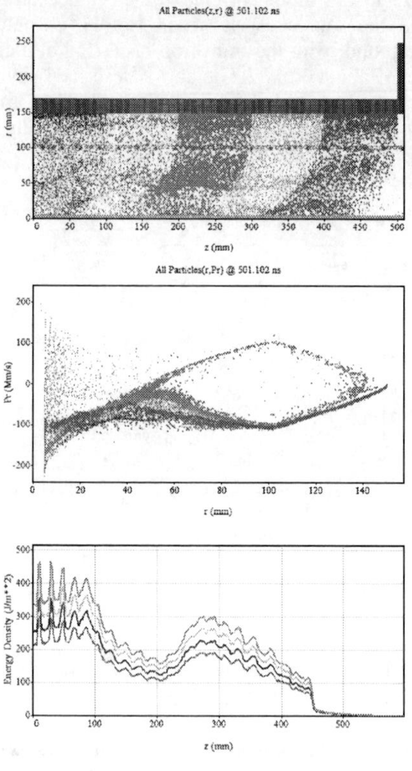

FIG. 4. Virtual cathode formation, observed for unipolar flow with U_{cg} = 33.5 kV: (a) particle distribution in (z,r) space, (b) radial momentum distribution versus radial position; beam electrons are represented by, respectively, blue, green, pink, orange, and grey dots; electrons backscattered at the target are represented by red dots; (c) accumulated beam energy density distribution at the target surface versus axial position at four times between 329 ns and 501 ns.

Here, the beam electrons are represented in different colors (blue, green, pink, orange, and grey) depending on the position along the cathode where they were emitted, see fig. 4a. From the radial momentum distribution (fig. 4b) it becomes clear that mainly "pink" electrons see a virtual cathode between grid and target. Thus, a virtual cathode is formed only locally in the z-range between ~150 and ~270 mm. The beam energy density at the target (fig. 4c) shows a depression at z~100-250 mm due to the local virtual cathode.

By further increasing the cathode-grid voltage and beam current, the local virtual cathode axially expands until almost the full length of the beam is subject to a virtual cathode. As soon as a virtual cathode is formed (even locally), the emission current stays below the space charge limit predicted for the corresponding diode, see figure 3. This is easily explained by the increased space charge in the cathode-grid gap due to electrons returning towards the cathode. Another observation is the increase of the grid current well beyond the geometrical ratio of 8%, which is caused by the returning electrons that oscillate around the grid.

B. Bipolar Flow

In order to achieve melting during target treatment, pulse durations in the microsecond range are chosen. On such time scales, ion flow from the target cannot be neglected [5]. Therefore, space charge limited ion emission from the target is introduced in the simulations and the situation of bipolar flow is studied. Figure 5 shows the emission, target, and grid electron currents of bipolar flow for the full range of cathode-grid voltages investigated. The current lost to the side stays below 100 A and is not shown. Again a regime of laminar flow is found for low cathode-grid voltages. In this regimes, the presence of counter-streaming ions results in an increase of the electron emission current by roughly 50% against unipolar flow (fig. 3).

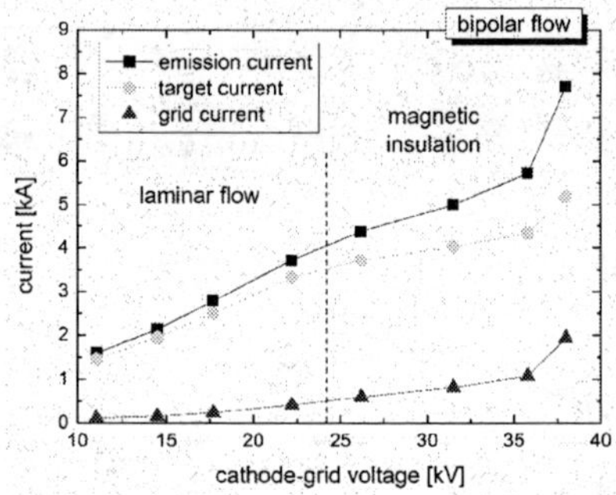

FIG. 5. Emission, target, and grid currents as a function of cathode-grid voltage for constant cathode-anode voltage 120 kV, Al target, and bipolar flow.

Beyond a cathode-grid voltage of ~24 kV the electron emission current of bipolar flow seems to be reduced, although virtual cathode formation is not observed due to space-charge compensation by the ions. As will be shown in the following, the reason is magnetic insulation. The further away from the central axial position, the higher is the accumulated target current. The higher induced azimuthal magnetic field results in stronger bending of the electron trajectories on their way towards the target. Beyond a certain magnetic field strength, electrons are unable to reach the target and return towards the cathode (magnetic insulation). In the particular case shown in figure 6, all electrons emitted at axial positions larger ~350 mm are unable to reach the target directly (i.e., all 'grey' electrons and some of the 'orange' electrons), see figs. 6a and 6b.

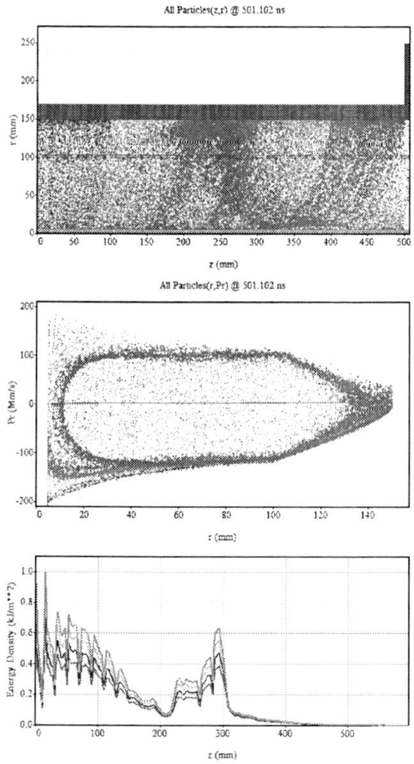

FIG. 6. Magnetic insulation, observed for bipolar flow with U_{cg} = 35.8 kV. Presentation details see figure 4, with light blue dots added for ions.

The increased electron space charge in the cathode region at axial positions around 210-250 mm due to returning electrons from the beam edge leads to a locally reduced emission current, which is also observed in the local minimum of the beam energy density at the target at z~200 mm (fig. 6c). Returning electrons also result in an increase of the grid current beyond the geometrical 8% of the emission current, which reduces the percentage of emission current reaching the target. For increasing cathode-grid voltage, the z-range of the described magnetic insulation extends until the electron beam reaches the target only in a focused spot at the target center (pinch). In the presented system, electron beam pinching is observed for cathode-grid voltages beyond about 37 kV.

IV. CONCLUSION

Simulations of the beam performance of the cylindrical pulsed electron beam accelerator GESA IV at given cathode voltage 120 kV showed different operation regimes depending on the cathode-grid voltage applied. For low cathode-grid voltages and low beam current, laminar flow conditions with homogeneous energy density at the target are obtained. The laminar flow regime exists for both unipolar electron beams and bipolar beams with counter-streaming ions generated at the target. It is the only feasible operation regime for homogeneous treatment of metal surfaces. For the considered case of 120 keV at the target, 20-40 J/cm² are required to obtain melting and alloying of the target surface [6,7]. With a maximum power density at the target of 1.2 MW/cm² found for the laminar flow regime, a pulse duration of ~35 µs is enough to achieve this goal.

For unipolar flow and cathode-grid voltages above ~32 kV, a virtual cathode is formed between grid and target, which results in an inhomogeneous beam energy density at the target. For bipolar flow, the laminar flow regimes already ends at cathode-grid voltages of ~24 kV. Due to the larger target current compared to unipolar flow, magnetic insulation at the beam edge is found. By further increasing the cathode-grid voltage, the magnetically insulated region extends until beam pinching occurs.

References

[1] V. Engelko, V. Kuznetsov, G. Mueller, "Electron source of triode type with radial converging electron flow for irradiation of cylindrical targets", J. Appl. Phys., vol. 105, pp. 023305, 2009.

[2] W. An, V. Engelko, G. Mueller, A. Weisenburger, "Optical investigation of plasma formation process by interaction of intense electron beams with metallic targets", Acta Physica Polonica A, vol. 115, pp. 1183-1185, 2009.

[3] V. Engelko, "Source of radial converging electron beam for modification of long-length cylindrical targets", 2014 10th International Vacuum Electron Sources Conference, IVESC 2014 and 2nd International Conference on Emission Electronics, ICEE 2014 – Proceedings.

[4] B. Goplen, L. Ludeking, D. Smithe, G. Warren, "User-configurable MAGIC for electromagnetic PIC calculations", Computer Physics Communications, vol. 87, pp. 54-86, 1995.

[5] V. Engelko, G. Mueller, "Formation of plasma and ion flux on a target, irradiated by an intense electron beam", J. Appl. Phys., vol. 98, pp. 013303, 2005.

[6] A. Weisenburger, A. Jianu, W. An, R. Fetzer, M. Del Giacco, A. Heinzel, G. Mueller, V. G. Markov, and A. D. Kasthanov, "Creep, creep-rupture tests of Al-surface-alloyed T91 steel in liquid lead bismuth at 500 and 550 °C", J. Nucl. Mater., vol. 431, pp. 77-84, 2008.

[7] R. Fetzer, W. An, A. Weisenburger, G. Mueller, „Surface layer dynamics during e-beam treatment", IEEE Trans. Plasma Sci., vol. 41, pp. 2858-2862, 2013.

Magnetic Pulse Welding: Solutions for Process Monitoring within Pulsed Magnetic Fields

Jörg Bellmann, Eckhard Beyer
Institute of Manufacturing Technology, TU Dresden and
Fraunhofer Institute for Material and Beam Technology
Dresden, Germany
Joerg.bellmann@tu-dresden.de

Jörn Lueg-Althoff, Soeren Gies, A. Erman Tekkaya
Institute of Forming Technology and
Lightweight Construction, TU Dortmund
Dortmund, Germany

Sebastian Schulze
Fraunhofer Institute for Material and Beam Technology
Dresden, Germany

***Abstract*—For Magnetic Pulse Welding (MPW), Lorentz forces are utilized for the rapid acceleration and controlled high velocity collision of two metallic parts. Due to the absence of external heat, solid state weld seams can be produced without the formation of critical intermetallic phases. In order to adjust the process parameters for a sound weld, optical measuring techniques are preferred, since they are not influenced by the strong magnetic fields of the working coils. Unfortunately, existing technologies are either expensive, difficult to evaluate, or limited in their accessibility to the interesting areas. To overcome these drawbacks, a new method was applied taking advantage of the hypervelocity impact (HVI) flash that is characteristic for the process. Experiments were performed using phototransistors for the detection of the HVI flash and the established photonic Doppler velocimetry in parallel. The flash characteristic was found to be a reliable measure for the evaluation of the local impact time as well as the weld seam length. It has a great potential for both process development and monitoring. Furthermore, disturbances, which are unfavorable for the weld seam formation, e.g. oily contaminants on the surfaces of the parts to be joined, can be detected.**

Keywords*— Magnetic Pulse Welding, Process Monitoring, Collision Conditions, Dissimilar Metal Joining, Materials Testing

I. Introduction

Magnetic pulse welding (MPW) is a solid state joining technology based on the controlled high speed collision of two metallic joining partners. The deformation and acceleration of the material is realized by the magnetic pressure of a tool coil when a pulsed current runs through it. MPW is a clean and safe [1] process for similar and dissimilar metal joining without the formation of critical brittle intermetallic phases (Fig. 1). It is reliable at the same time if all relevant geometrical and tool related values are adjusted properly by simulations and experiments. Unfortunately, certain geometrical and physical part properties or wear of the tool coil have a strong influence on the welding result. For large scale production it is therefore mandatory to ensure that at least the collision velocity v_i, one of the most important parameters, is in an appropriate range.

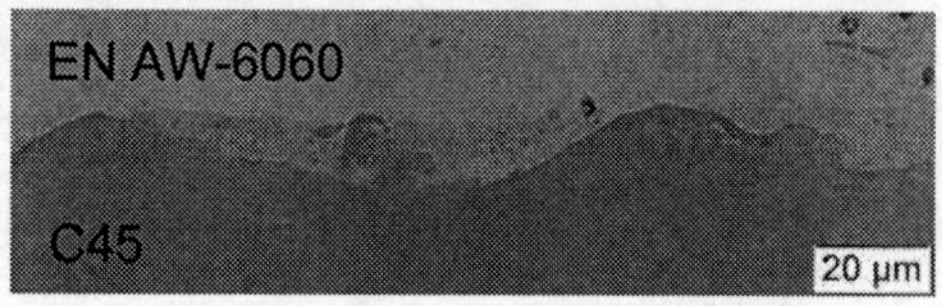

Fig. 1. Magnified polished cross section of a magnetic pulse welded area between aluminum (top) and steel (bottom) with a typical wavy interface

Since electrical signals are influenced by the strong magnetic field [2], optical solutions are preferred. To record surface velocities up to several kilometers per second within a few microseconds, photonic Doppler velocimetry (PDV) is an established method [3]. One of its main challenges is the integration into the working coil that can lead to an unwanted increase of the current densities due to boreholes in the concentration zone for the installation of the probes [4]. Furthermore, the required laser is expensive and the sensor is not applicable at certain positions, for example the coil's slot. The equipment for high speed imaging is even more expensive and requires lighting lasers or flashes as well as optical band pass filters to fade out the process glare that naturally occurs in the joining region [5]. Interestingly, this flash phenomenon is described as an obstacle for the process monitoring in MPW, while it is explicitly utilized in other fields of research. Eichhorn found that during high speed impact experiments the primary impact velocity and the projectile mass are in correlation with the maximum light intensity and the time integrated light intensity of the hypervelocity impact (HVI) flash [6]. He also performed measurements with spectrometers to evaluate the temperature of the radiating material. This work was supplemented by Jean who explored two phases of the HVI flash [7]. The first event is a pulse in the submicrosecond range, which is associated with the radiation of high density, strongly ionized gas. The second event is a slowly rising, long enduring light pulse that is due to the radially expanding jet of debris. Auluck took advantage of this light pulse to monitor the jetting effect at different locations [8]. The present paper will demonstrate the applicability of the HVI flash measurement as a useful tool for quality assurance and process adjustment during MPW.

This work was sponsored by the German Research Foundation (DFG).

II. Experimental Setup

Joining experiments were performed in order to correlate the characteristic of the HVI flash with the weld seam formation during MPW. Therefore, aluminum tubes were used as flyer parts and accelerated inwards to steel cylinders as shown in Fig. 2. The proper welding parameters collision angle β and impact velocity v_i for that configuration were investigated before [9] and serve as a basis here. The chemical compositions of both alloys are given in Table I. Both flyer and parent material were used in uncoated state and were cleaned in ethanol before the joining experiments to remove debris from the workpieces' surfaces.

The magnetic pressure is generated by a single turn working coil made of a CuCrZr alloy (2.1293) with an integrated PDV collimator placed in the concentration zone and the geometric dimensions shown in Fig. 2. Thus, the radial flyer velocity can be monitored. The same setup and evaluation used in the present study is described by Lorenz et al. [12]. The pulsed current is supplied by a pulse generator with its characteristic values listed in Table II.

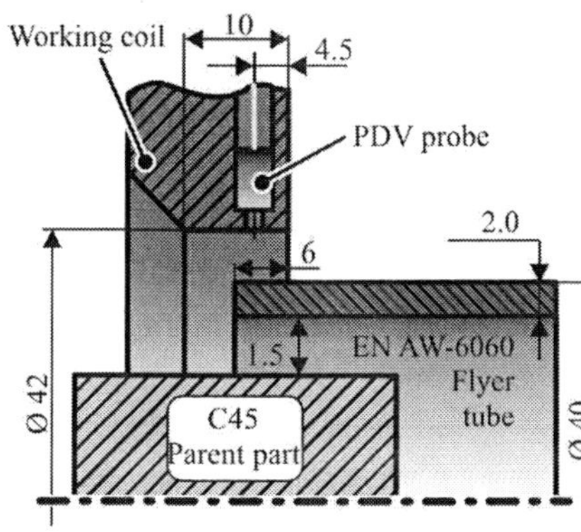

Fig. 2. Geometrical definitions of the welding setup and location of the PDV probe for measurement of the radial flyer velocity (all values in mm)

TABLE I. ALUMINUM EN AW-6060 ALLOY COMPOSITION [10] AND STEEL C45 (1.0503) ALLOY COMPOSITION [11]

Flyer part EN AW-6060, T4 [a], static yield strength approx. 60 MPa [b]		**Parent part C45 (1.0503), normalized, surface polished (Ra = 1)**	
Element	***Weight %***	***Element***	***Weight %***
Mg	0.35-0.6	C	0.42-0.5
Mn	≤0.1	Mn	0.5-0.8
Fe	0.1-0.3	P	<0.045
Si	0.3-0.6	S	<0.045
Cu	≤0.1	Si	<0.4
Zn	≤0.15	Ni	<0.4
Cr	≤0.05	Cr	<0.4
Ti	≤0.1	Mo	<0.1

a. T66, heat treated one hour at 500°C

b. determined by tube tensile test

TABLE II. CHARACTERISTIC OF PULSE GENERATOR BMAX MPW 50/25

Setup	Value	Unit
Capacity	160	µF
Maximum charging energy	32	kJ
Applied charging energy – E	2.4 ... 11.5	kJ
Corresponding maximum tool coil current I (E)	≈ 35·E(kJ) + 218	kA
Discharge frequency with coil and workpieces - $f_{discharge}$	20	kHz

Current measurements were conducted for each trial using a Rogowski current probe CWT 3000 B from Power Electronic Measurements Ltd.. The rising current signal triggered the recording of the PDV signal as well as the flash detection during the joining experiments. The recording of the light emission with two simple silicon phototransistors (SDP8405, Honeywell) is used to evaluate the temporal points of impact (Fig. 3). Therefore, the optical signals are collimated and transmitted with two polymeric optical fibers (POF, diameter 1 mm, length 4 m, by Telegaertner Geraetebau GmbH, Germany) to the phototransistors. Due to their photosensitivity they act as optical/electrical converters and thus generate signals that can be timed to the current signal from the tool coil on the oscilloscope, see Fig. 4. The voltage is measured at a 220 Ω resistor that is series connected with the phototransistor and a 9 V battery and thus reaches maximum values of 9 V. Three characteristic values are defined for each flash measurement: the start time, the flash duration above an intensity that is equivalent to 4 V, and the maximum intensity.

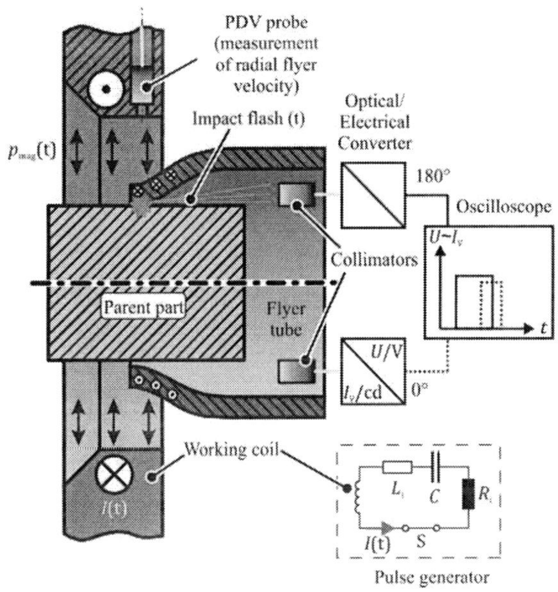

Fig. 3. Experimental setup with PDV measurement and two positions for flash detection via phototransistors

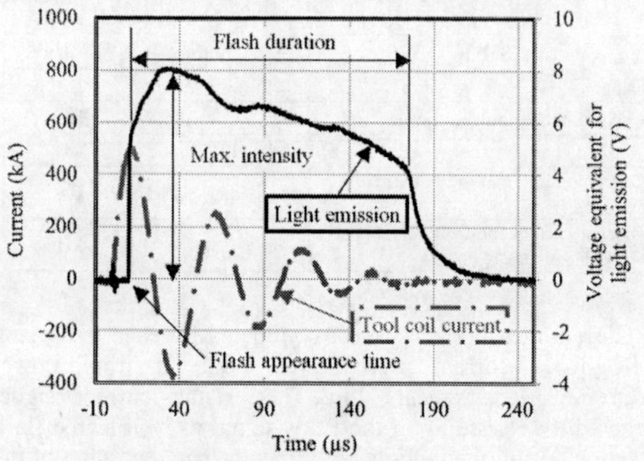

Fig. 4. Tool coil current and light intensity vs. time

After the magnetic pulse joining, small kerfs are cut in axial direction at the 180° position into the flyer tube to generate a metal strip on each sample with a width of 5 mm. They are peeled off perpendicularly to the parent surface as described in [9]. With this procedure, non-welded samples as defined in DIN EN ISO 6520-2 [13] can be identified easily, since no residues of aluminum are visible on the parent's surface. For successfully welded samples failure occurs in the aluminum base material and residues are apparent. A shear test like described in [9] is applied to remove residues of the flyer material on the parent's surface and evaluate the weld seam length according to DVS 2902-3 [14].

III. Results and Discussion

Fig. 5 shows the radial flyer velocity measured with the PDV system and the flash intensities with respect to the time course of the tool coil current. The first pulse of the HVI flash as described by Jean [7] is too short to be monitored by the applied phototransistors. But the second flash occurs approx. 1…2 µs after the impact and the drop of the flyer velocity detected by the PDV-system at 180°. Another 2 µs later, the light intensity at the coil's slot (0°) rises, too. Since the acceleration gap of 1.5 mm was equal at both positions, this delay is the consequence of the lower magnetic pressure and the impact velocity, respectively.

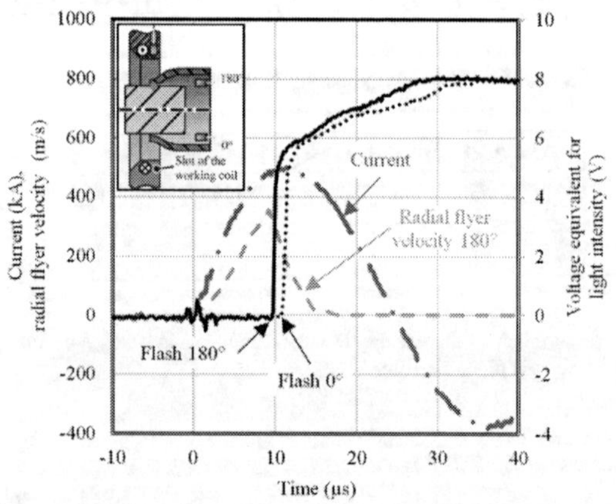

Fig. 5. Time courses of tool coil current, flyer velocity and light intensities

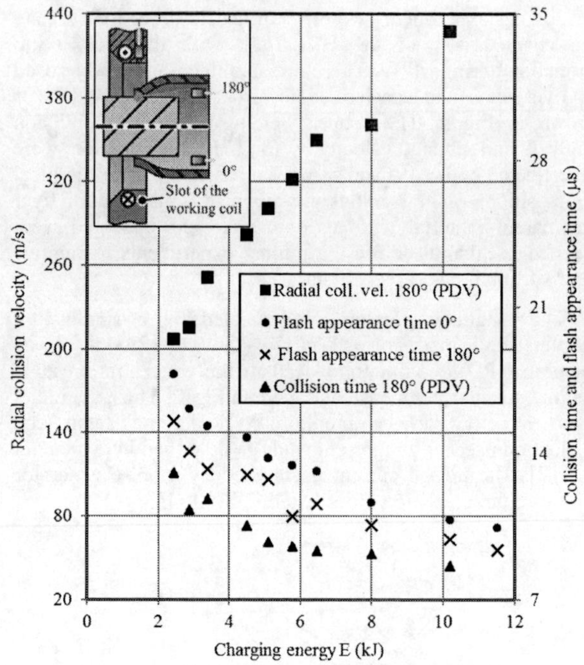

Fig. 6. Collision time, collision velocity and flash appearance time for different charging energies

The observation was validated for different charging energies as shown in Fig. 6. It is apparent that the start time of the flash varies, depending on the location and charging energy. The collision time and collision velocity gained from the PDV measurements are used as reference values here. The radial collision velocity increases with increasing charging energy, while the collision time decreases for a higher acceleration, as expected. The flash at 180° shows an average delay of 2.1±0.6 µs compared to the PDV signal, on the opposite direction 3.7±0.9 µs due to the lower magnetic pressure at the coil's slot as explained before. The delay of the flash for both positions gets smaller for increasing energies due to the increasing magnetic pressure. This assimilation is probably an indication of a good weld quality around the circumference. The results show that the flash measurement is a reliable measuring tool for positions that are hard to access with PDV probes like the slot of the working coil (0°). For the chosen charging energies, the correlation between the flash characteristics and the weld seam length at the opposite direction of the coil's slot (180°) are illustrated in Fig. 7. Three categories can be defined:

a) Shortage of kinetic flyer energy for charging energies $E < 5$ kJ: The flashes have medium intensities but their durations are far below 100 µs and welding does not take place.

b) Beginn of weldability 5 kJ < E < 7 KJ: The durations of the flashes increase up to 125 µs while the level of intensity is almost equal to category a. The peel and shear tests revealed welded zones with a length > 1 mm.

c) Stable welding for E > 7 kJ: The weld seams become longer and the flash durations and intensities increase significantly.

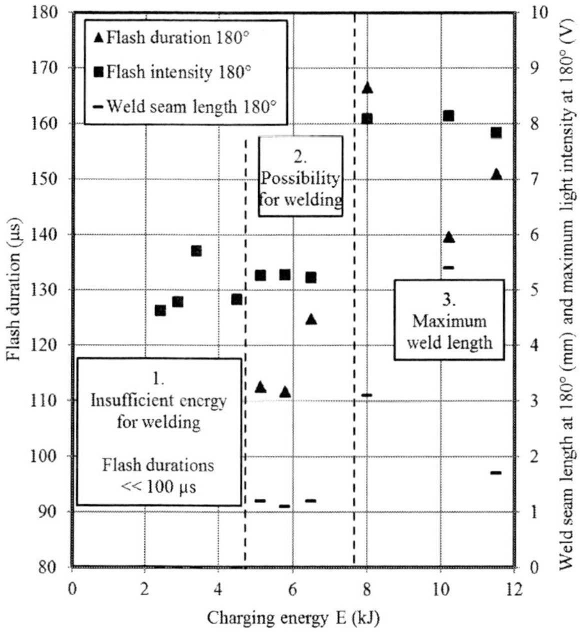

Fig. 7. Flash characteristics and welding results depending on the charging energy

The results are in good correlation with Eichhorn [6], who defined the total light energy as the time integral of the light intensity. If the impact mass is constant, which is guaranteed by a constant flyer thickness for all tests, the total light energy increases with increasing collision velocity at the initial impact zone. Transferring the explanation of Jean [7] to MPW would mean that the total light energy is a measure for the jet emission. Since jetting is a mandatory effect for the welding process, the monitoring of the total light energy with phototransistors is a suitable tool for quality assurance during MPW.

IV. Conclusion and Outlook

Aluminum tubes were magnetic pulse welded on steel cylinders with successively increased discharge energy. PDV measurements proved that the impact velocity increased as well. The hypervelocity impact flash was recorded during all exeriments. It was found that the appearance time of the flash is in good correlation with the impact time. This can help to adjust and monitor welding processes. For example, the acceleration distance between flyer and parent part can be optimized easily with respect to the current course, even at locations with bad accessibility like the coil's slot. Magnetic field inhomogeneities as well as incorrect part alignments can be detected when using at least two measuring devices. Additionally, the time integral of the light intensity is a good measure for the correct weld seam establishment. First experiments showed significant effects on the flash characteristic, if surface related disturbances that are unfavorable for welding, were apparent, e.g. an oil film or an increased roughness.

Acknowledgment

This work is based on the results of subproject A1 of the priority program 1640 ("joining by plastic deformation"); the authors would like to thank the German Research Foundation (DFG) for its financial support. Furthermore, the authors would like to thank "Telegaertner Geraetebau GmbH" (Hoeckendorf, Germany) for providing the fiberoptic components.

References

[1] A. Kapil and A. Sharma, "Magnetic Pulse Welding: An efficient and environmentally friendly multi-material joining technique," J Cleaner Prod, vol. 100, pp. 35–58, 2015.

[2] O. Gafri, A. Izhar, Y. Livshitz, and V. Shribman, "Magnetic Pulse Acceleration," in M. Kleiner (Ed). High Speed Forming 2006: Proc of the 2nd Int Conf. Dortmund, Germany, pp. 33–40, 2006.

[3] O. Strand, D. Goosman, C. Martinez, T. Whitworth, and W. Kuhlow, "Compact system for high-speed velocimetry using heterodyne techniques," Review of Scientific Instruments, 2006.

[4] A. Jaeger and A. E. Tekkaya, "Online Measurement of the Radial Workpiece Displacement in Electromagnetic Forming Subsequent to Hot Aluminum Extrusion," in A.E. Tekkaya, G.S. Daehn, M. Kleiner (Eds.), High Speed Forming 2012: Proc of the 5th Int Conf. Dortmund, Germany, pp. 13–22, 2012.

[5] P. Groche, M.-X. Wagner, C. Pabst, and S. Sharafiev, "Development of a novel test rig to investigate the fundamentals of impact welding," J of Mater Proc Technol, vol. 214, no. 10, pp. 2009–2017, 2014.

[6] G. Eichhorn, "Analysis of the hypervelocity impact process from impact flash measurements," Planetary and Space Science, vol. 24, no. 8, pp. 771–781, 1976.

[7] B. Jean and T. L. Rollins, "Radiation from hypervelocity impact generated plasma," AIAA Journal, vol. 8, no. 10, pp. 1742–1748, 1970.

[8] S. K. H. Auluck, T. C. Kaushik, L. V. Kulkarni, and A. S. Rav, "Conical Electric Gun: A New Hypervelocity Macroparticle Launcher Based on the Munroe Effect," IEEE Trans on Plasma Science, vol. 31, no. 4, pp. 725–728, 2003.

[9] J. Bellmann and J. Lueg-Althoff, "Influence of Selected Coatings on the Welding Result During Magnetic Pulse Welding (MPW)," in International Laser Symposium & International Symposium "Tailored Joining" 2016, Dresden, 2016.

[10] Seeberger, Datasheet AlMgSi (EN AW-6060). Available: http://www.seeberger.net/_assets/pdf/werkstoffe/aluminium/de/AlMgSi.pdf. Accessed on: Sep. 08 2016.

[11] Guenther + Schramm, Datasheet C45 (1.0503). Available: http://www.guenther-schramm-stahl.de/files/datasheets/C45(1.0503).pdf. Accessed on: Sep. 08 2016.

[12] A. Lorenz et al., "Workpiece Positioning during Magnetic Pulse Welding of Aluminum-Steel Joints," Welding journal, vol. 95, no. 3, pp. 101–109, 2016.

[13] "Welding and allied processes - Classification of geometric imperfections in metallic materials -part 2: Welding with pressure," DIN EN ISO 6520-2, 2013.

[14] DVS Media GmbH, „Resistance Spot welding of steel up to 3 mm thickness – design and calculation," DVS 2902-3 (in German), 2016.

Magnetic Pulse Welding of Tubes: Ensuring the Stability of the Inner Diameter

Jörn Lueg-Althoff, Soeren Gies, A. Erman Tekkaya
Institute of Forming Technology and Lightweight Construction, TU Dortmund
Dortmund, Germany
Joern.Lueg-Althoff@iul.tu-dortmund.de

Jörg Bellmann, Eckhard Beyer
Institute of Manufacturing Technology, TU Dresden and Fraunhofer Institute for Material and Beam Technology
Dresden, Germany

Sebastian Schulze
Fraunhofer Institute for Material and Beam Technology
Dresden, Germany

***Abstract*—Magnetic Pulse Welding (MPW) is a solid-state joining process using pulsed magnetic fields. The process is especially attractive for the welding of dissimilar metals because no external heat, which might lead to brittle intermetallic phases or other defects, is transferred into the joining zone. Multi-material joints have an increasing importance for the implementation of lightweight concepts in automotive and aerospace industry, for instance. Specifically tubular joining partners are widely used in these sectors.**

The manuscript focuses on MPW of two cylindrical parts made of aluminum and steel. Potentials are shown, how unwanted deformations of the inner part can be significantly reduced by an optimized process design. Removable and reusable polyurethane inserts are applied for backing the parent. Different pulse generators with distinctively different pulse characteristics are compared. Moreover, the manuscript shows how the monitoring of the deformation of the outer part by photonic Doppler velocimetry can support the process analysis.

Keywords—magnetic pulse welding; joining; process design

I. Introduction

Economic and efficient resource management requires the use of the best material for every specific application in technical structures. As a result, multi-material joining operations gain in importance. In the case of materials with dissimilar thermo-physical properties, conventional fusion welding processes reach their technological limits with respect to quality, accuracy, and reliability [1]. Solid-state impact welding processes like magnetic pulse welding (MPW) do not feature these limitations and are of increasing importance in multi-material design for the automotive or aerospace industry [2]. MPW was patented by Lysenko et al. in 1970 [3]. In this process, one of the joining partners (the "flyer") is accelerated by Lorentz forces resulting from the pulsed discharge of a capacitor bank via a tool coil (Fig. 1b). The flyer is deformed plastically and, after overcoming an initial standoff *a* of few millimeters, collides with the other joining partner, the "parent" part. Under correct conditions with respect to radial impact velocities $v_{i,r}$, collision angles β, and impact pressures p_i (Fig. 1a), a jet is formed blowing out particles of both surfaces and leaving them chemically pure. A prerequisite of the cracking of oxide layers or other contaminants is a certain amount of plastic deformation leading to a magnification of the surfaces [4]. This leads to the characteristic wavy interface (Fig. 1c and d). Under the prevailing high pressure, a metallic bond is formed without excessive heating of the parts.

The process physics are closely related to other high-energy joining processes like explosion welding [5]. Current amplitudes in the range of several hundred kiloamperes and magnetic flux densities of few ten Tesla are commonly used [6]. Flat sheets as well as profiles and tubes can be pro-

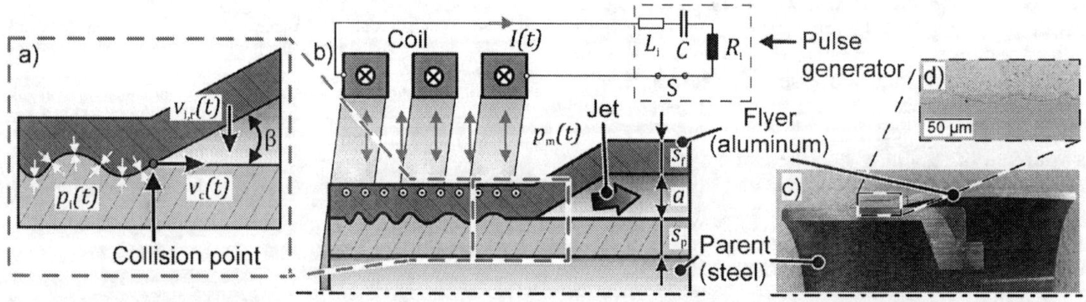

Fig. 1. Principle of MPW for tubular parts: a) Key parameters at the collision point, b) geometrical setup, c) joined structure, d) interface waves

This work was sponsored by the German Research Foundation (DFG).

cessed by MPW. In MPW of sheets or plates, unwanted deformations or fracture of the parent part can be reduced or prevented by backing it [7]. For tubular parts supporting structures (i.e. mandrels) might get jammed inside the parent. However, for the fabrication of tubular joints, e.g. for transportation of fluids or for structural parts in automotive engineering, the wall thickness s_p of the stationary parent part is a crucial parameter. During dimensioning of the joint, a tradeoff between minimum material usage (and thus a reduced wall thickness s_p) and a sufficient mechanical stability is required. Ben-Artzy et al. investigated the influence of hollow parent parts with varying wall thickness on the formation of interfacial waves [8]. Cui et al. (2016) calculated a critical thickness of the parent part based on Lamé's equations for stresses in tubes due to external pressure [9]. They considered any plastic deformation of the parent part as critical, but Lueg-Althoff et al. (2016) [10] showed that a permanent deformation of the parent part is not always associated with a failure of the weld. However, in cases where the integrity of the welded part's cross-section is of importance, the deformation should be limited to a minimum. This paper presents an approach how this can be achieved by the help of supporting elements made of plastic materials.

II. Experimental Methodology

MPW experiments with flyer tubes (40 mm outer diameter) made of aluminum EN AW-6060 and parent tubes made of steel C45 were conducted. The aluminum parts were heat-treated for one hour at 500 °C in order to reduce the strength (final yield strength approx. 60 MPa, determined by tube tensile testing). The flyer wall thickness s_f as well as the initial standoff a were varied between 1.5 and 2 mm, the parent wall thickness s_p was 2 mm. Two different pulse generators with distinctly different characteristics were used, see Table I. The authors already proved a strong influence of the discharge frequency on the feasibility of tube-to-tube MPW using the example of comparably high and low discharge frequencies [11].

The charging energies E were chosen based on successful welding experiments with solid parent parts. An eight-turn compression coil was used in combination with a fieldshaper with an inner diameter of 41 mm and a concentration zone of 10 mm width. The working length l_w (overlap of the coil edge, see Fig. 2) was set to 6 mm. During the experiments the discharge current was measured using a Rogowski current probe CWT 3000 B from PEM Ltd. The velocity of the flyer during the radial compression (v_f) was recorded inline by a photonic Doppler velocimetry system, as described in [10]. The welding result was analyzed in instrumented mechanical tests applying a 90° peel force on a Zwick Z250 universal testing machine. A failure in the aluminum base material was regarded as proof of a successful weld. A 3D digitizer ATOS of GOM mbH was used for the digitization of the welded specimens in order to analyze the degree of permanent deformation of the parent part. For the purpose of minimizing the deformation of the parent part, supporting inserts made of polyurethane (PU) were applied, see Fig. 2. Two different hardnesses (80 Shore A and 92 Shore A) were tested; the parts were prestressed applying a torque of approx. 50 Nm via a washer and a nut, see Fig. 2. After the MPW process, the inserts could be removed easily by the help of compressed air and were used again.

TABLE I. Characteristics of the Deployed Pulse Generators

	Maxwell Magneform 7000 (Bank 2+3)	Poynting SMU 0612 FS
E_{max}	12 kJ	6 kJ
U_{max}	8.16 kV	12.25 kV
C_i	362 µF	80 µF
L_i	78 nH	73 nH
R_i	5.4 mΩ	6.7 mΩ
$f_{discharge}$	Approx. 7 kHz	Approx. 17 kHz

E_{max} = maximum charging energy, U_{max} = maximum discharge voltage, C_i = inner capacitance, L_i = inner inductance, R_i = inner resistance, $f_{discharge}$ = frequency of the discharge current

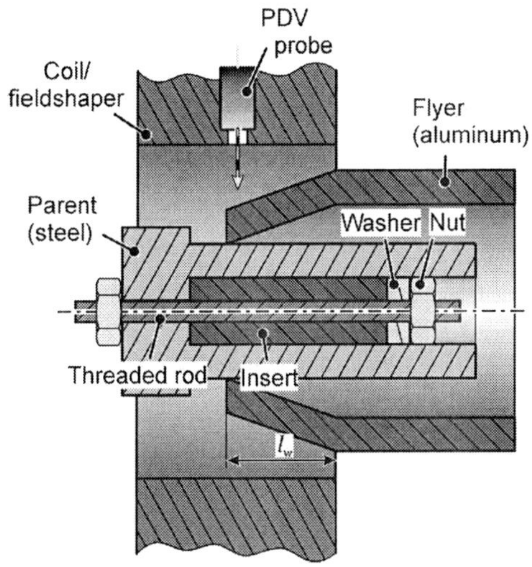

Fig. 2. Joining setup with PDV measuring device for recording of the radial flyer deformation velocity

III. Results and Discussion

Fig. 3 shows a comparison of the permanent deformation of the parent parts without and with supporting inserts. The experiments were performed at the Maxwell machine with the same discharge energies, leading to radial impact velocities $v_{i,r}$ of approx. 245 m/s. The parts were scanned from the inside after the MPW process. It can be seen that all parent parts experienced a permanent deformation. This deformation is not distributed uniformly around the circumference, which is due to the slot of the fieldshaper. The unsupported parent part is compressed at the area of impact and expands at the free end (Fig. 3a). The amount of deformation is significantly reduced by the PU inserts, see Fig. 3b and c. The harder PU leads to a slightly lower deformation at the same prestress level.

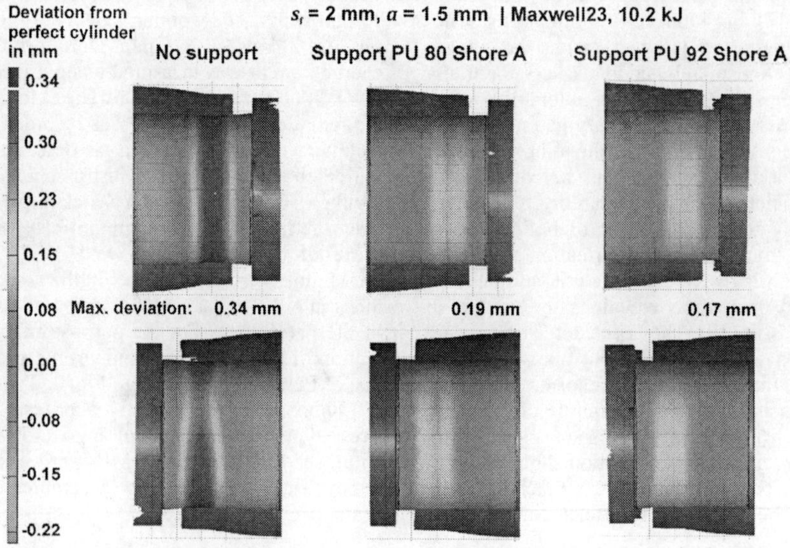

Fig. 3. 3D scans of the welded specimens (views from left and right), the color scale shows the residual radial deviation from a perfectly cylindrical parent, a) without supporting insert, b) with insert made of PU, hardness 80 Shore A, c) with insert made of PU, hardness 92 Shore A

The reduction of the deformation can also be seen from the PDV curves, see Fig. 4. The diagram shows the recorded flyer velocity for the three experiments. The flyers are accelerated to the peak velocity of approx. 245 m/s within 13 µs. Then, the impact takes place, followed by a rapid deceleration. At this stage, the necessary pressure for welding is formed. Before the end of the deceleration phase the three curves are almost identical. However, towards the end of the movement a deviation between the curves becomes visible. The deceleration phase of the unsupported experiment lasts longer than the other ones. As shown in [11], the deformation of the parent part takes place at this stage, the deviation between the velocity curves reflects the stronger plastic deformation of the unsupported parent visualized in Fig. 3.

The evaluation of the PDV curves can, thus, be used for answering the question if the parent has plastically deformed or not. This has advantages for cases in which the inner parent surface is not directly accessible, e.g. small tubes. In order to evaluate the degree of plastic deformation of the parent part, the PDV curves shown in Fig. 4 were further analyzed. Integrating the velocity curves yields the displacement. At the time of impact the flyer displacement is equal with the initial standoff a, which was identical for the experiments. Fig. 5 shows the calculated amount of displacement starting from the time of impact for the three velocity curves. This displacement is the sum of the radial parent deformation and the axial deformation of the flyer during impact. It can be seen that the total deformation is reduced if a supporting insert is used. Using a PU insert with the hardness 92 Shore A reduces the deformation by more than 40 %, which is in good agreement with the 3D scans (Fig. 3). Subtracting the measured deformations of the parent parts obtained by 3D scanning from the displacements shown in Fig. 5 leads to approx. the same value of about 0.12 mm in each case. This is an estimation of the thinning of the flyer during impact, which is supposed to be identical in the experiments, since the impact conditions are the same. With the PDV, both elastic and plastic deformations are recorded. This also contributes to the deviations between

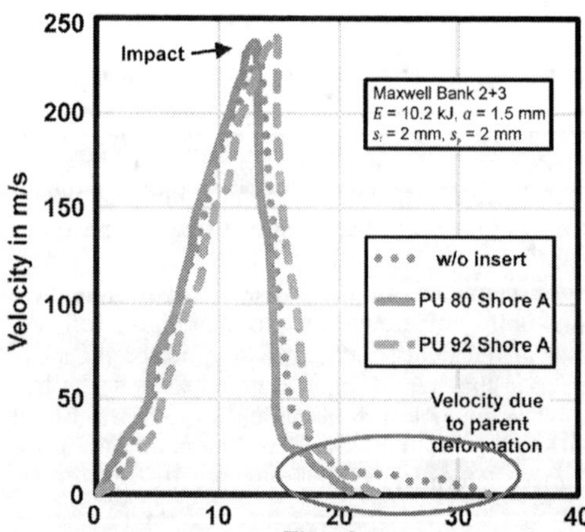

Fig. 4. Radial flyer velocities for experiments with and without supported thin-walled parent parts

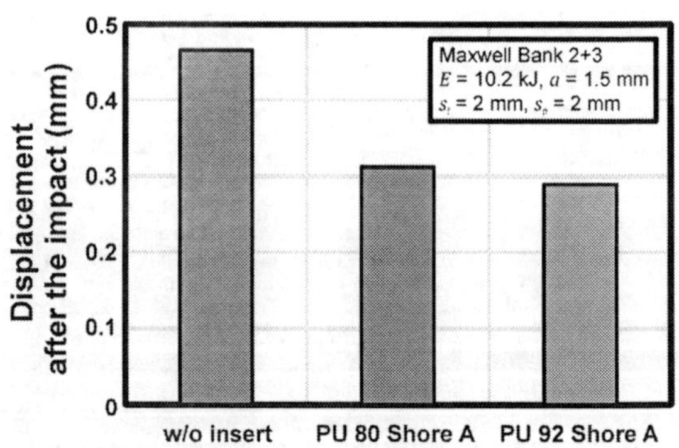

Fig. 5. Maximum (elastic and plastic) radial flyer displacement after impact, extracted from the PDV curves presented in Fig. 4

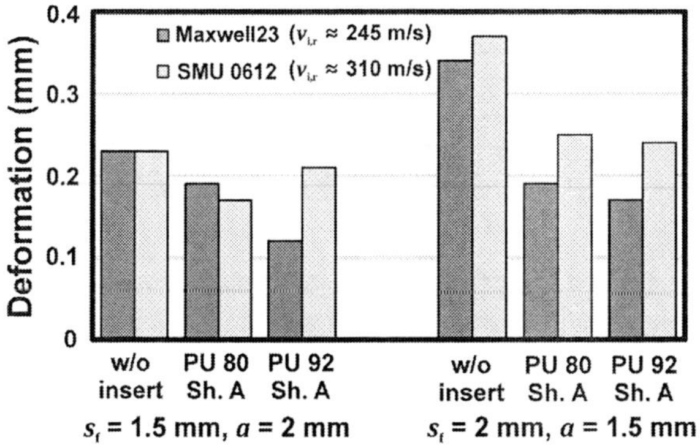

Fig. 6. Comparison of maximum residual radial parent deformations measured by 3D scanning for different pulse generators and geometrical parameters

the PDV measurements and the 3D scans (Fig. 3). The latter are solely depticting the permanent deformation.

The tendencies shown in Fig. 5 are the same for the Maxwell and the SMU machine as well as for different geometrical configurations, see Fig. 6. At the Maxwell pulse generator, lower radial impact velocities $v_{i,r}$ are sufficient to create sound welds. These lower velocities (approx. 245 m/s compared to 310 m/s at the SMU machine) and corresponding lower kinetic energies result in a slightly decreased parent deformation in the unsupported case. An increase of the flyer wall thickness leads to an increase of the amount of parent deformation. The kinetic energy of the flyer is larger since the required impact velocity for welding stays approximately the same.

Fig. 7 shows a comparison of the recorded coil current during the experiments with unsupported parent part. As shown in Table I, the frequencies of the currents differ significantly. The peak current at the Maxwell machine is higher, but this peak is reached after the impact has occurred. At the SMU machine, despite a lower peak current, within the first 20 µs a higher current is flowing and thus a higher magnetic pressure is acting on the flyer. This leads to higher impact velocities and higher deformations (Fig. 6).

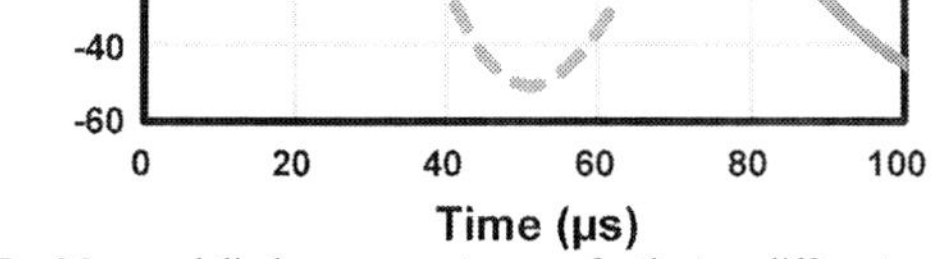

Fig. 7: Measured discharge current curves for the two different pulse generators. Only the first 100 µs are depicted.

IV. Conclusion

Based on the presented experimental investigations, the following conclusions can be drawn:

- PU inserts can be used for a significant reduction of the deformation of thin-walled parents. In contrast to metallic inserts, these parts can be removed easily and used repeatedly after removing.
- The parent deformation during the impact can be assessed indirectly by a PDV probe recording the movement of the outer flyer surface.
- Different pulse characteristics were compared regarding their influence on the parent deformation. A lower necessary impact velocity for welding is also benefitial for a reduced parent deformation.

Acknowledgments

The presented results originate from research carried out within the subproject A1 of the Priority Programme 1640 ("Joining by plastic deformation") funded by the German Research Foundation (DFG). The financial support is greatly acknowledged. The PU with hardness 92 Shore A was kindly provided by P+S Polyurethan-Elastomere GmbH & Co. KG, Diepholz, Germany.

References

[1] K.-i. Mori, N. Bay, L. Fratini, F. Micari, and A. E. Tekkaya, "Joining by plastic deformation," CIRP Ann – Manuf Technol, vol. 62 (2), pp. 673–694, 2013.

[2] A. Kapil and A. Sharma, "Magnetic Pulse Welding: An efficient and environmentally friendly multi-material joining technique," J Cleaner Prod, vol. 100, pp. 35–58, 2015.

[3] D. N. Lysenko, V. V. Ermolaev, and A. A. Dudin, "Method of pressure welding" (US Patent No. 3,520,049), 1970

[4] M. P. Bondar and V. F. Nesterenko, "Contact deformation and bonding criteria under impulsive loading", Comb, Expl, Shock Waves, vol. 27 (3), pp. 364–376, 1991.

[5] B. Crossland, "Explosive welding of metals and its application", Clarendon Press, Oxford (UK), 1982.

[6] T. Aizawa, M. Kashani, and K. Okagawa, "Application of magnetic pulse welding for aluminum alloys and SPCC steel sheet joints", Weld J, vol. 86, pp. 119–124 (2007).

[7] M. Watanabe, S. Kumai, G. Hagimoto, Q. Zhang, and K. Nakayama, "Interfacial Microstructure of Aluminum/Metallic Glass Lap Joints Fabricated by Magnetic Pulse Welding", Mat Trans, vol. 50 (6), pp. 1279–1285, 2009.

[8] A. Ben-Artzy, A. Stern, N. Frage, V. Shribman, and O. Sadot, "Wave formation mechanism in magnetic pulse welding," Int J Imp Engin, vol. 37 (4), pp. 397–404 (2010).

[9] J. Cui et al., "A study on the critical wall thickness of the inner tube for magnetic pulse welding of tubular Al–Fe parts," J Mater Proc Technol, vol. 227, pp. 138–146, 2016.

[10] J. Lueg-Althoff et al., "Influence of the wall thicknesses on the joint quality during magnetic pulse welding in tube-to-tube configuration," in A. E. Tekkaya, M. Kleiner (Eds.). High Speed Forming 2016: Proc of the 7th Int Conf. Dortmund, Germany, pp. 259–268, 2016.

[11] J. Lueg-Althoff et al., "Magnetic pulse welding of dissimilar metals in tube-to-tube configuration", 10th Int Conf on Trends in Welding Res. Tokyo, Japan, 2016. In press.

Computational Justification of the Common Vacuum Diode Load for GAMMA-4 Experimental Facility

Nikolay V. Zavyalov, Vyacheslav S. Gordeev, Alexander V. Grishin, Alexander L. Mozgovoy, Sergey L. Glushkov, Sergey Yu. Puchagin, Kirill V. Strabykin, Sergey T. Nazarenko, Vladimir S. Pavlov, Vyacheslav A.Demanov
Federal State Unitary Enterprise «Russian Federal Nuclear Center-VNIIEF»
Sarov, Russian Federation
otd4@expd.vniief.ru

***Abstract*— The variant of the design of the common vacuum diode load for the electro-physical GAMMA-4 facility is presented. On the base of telegraph equations the one-dimensional computational model of the facility has been developed. The calculations of output electrical parameters of the facility in the mode of simultaneous operation of all modules into common vacuum diode load have been performed with taking into account electron losses in vacuum transmission lines. For the presented design of the current adder the estimations of the diode voltage and current, electron loss currents and energy, as well as an evolution of the process of magnetic insulation in vacuum part of the facility have been obtained.**

Keywords*— *vacuum diode load; vacuum current adder; vacuum transmission line; electron losses

I. INTRODUCTION

Four-module electro-physical facility GAMMA-4 is being created at RFNC-VNIIEF. The facility is primarily designed to produce bremsstrahlung pulses with duration of ~50 ns [1]. At first step all modules will operate autonomously with separate vacuum diode loads. In this mode the x-ray source consists of four discrete disc emitters that located at the corners of a square with a side of 250 mm. In this case the energy density of the radiation produced by 4 modules in near field is not increased with comparison to radiation produced by single module. Only irradiated area is increased. The mode with autonomous modules is optimized for irradiating of large area objects (~10^3-10^4 cm^2). To increase energy density at small areas (~10-10^2 cm^2) the mode with common vacuum diode load should be implemented.

In this paper the variant of design of the vacuum current adder and common vacuum diode load for GAMMA-4 facility is presented.

II. VACUUM CURRENT ADDER AND COMMON VACUUM DIODE LOAD

The proposed design of the vacuum current adder and common vacuum diode load is shown in Fig.1. The outputs of 4 coaxial magnetically-insulated transmission lines (MITLs) are joined together with the aid of current adder unit. The electrodes (grounded and high-voltage) of the current adder form disc MITL to transport the EM pulse to the load.

Fig.1. The design of current adder and common vacuum diode load

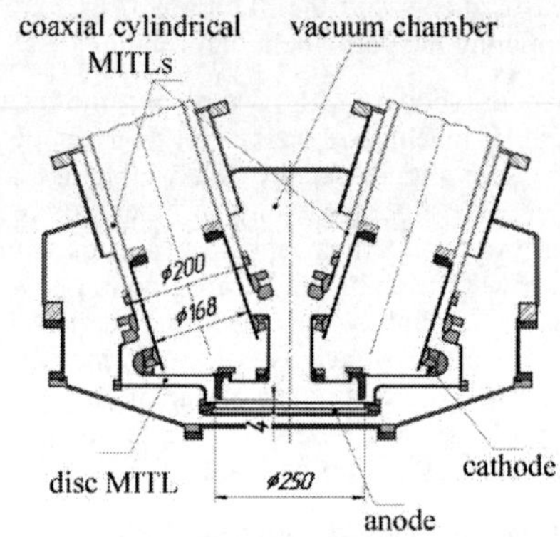

Dependence of the diode current on the voltage in steady state is well described by parapotential flow model [2, 3]. According to this model, in order to provide the diode impedance matched to the generator at the level of 0.7 Ohm, the cathode diameter should be of 250 mm, while the gap is 4 mm.

III. THE MODEL DESCRIPTION

In Fig.2. the geometry of the central part of the facility used in calculations is shown. The vacuum insulators and coaxial cylindrical MITLs are also included into the model but not shown. All components have axial symmetry and therefore can be easily converted from 3D to 1D representation with continuous parameters distributed along the path of wave travelling: $L_x(x)$ – inductance per unit length, $C_x(x)$ – capacitance per unit length, $Z(x)$ – line impedance, $d(x)$ – anode-cathode gap. Inductances and capacitances are the coefficients in telegraph equations and the gap is used for calculations of electron loss current density. The telegraph equations in such conditions are fully equivalent to Maxwell's equations [4].

Fig.2. The geometry of the central part of the facility

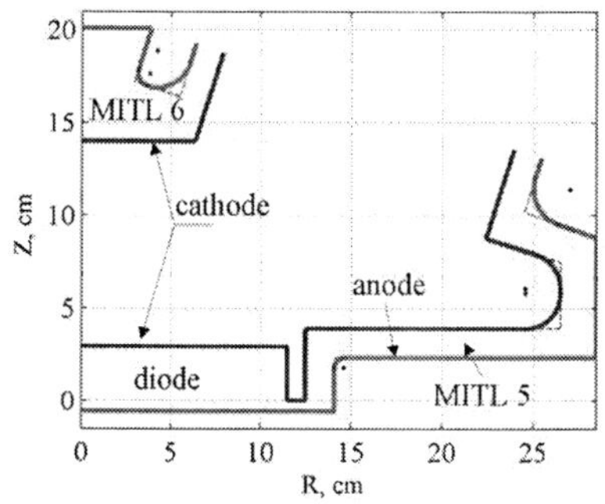

The graph of the model is shown in Fig.3. The telegraph equations are solved at the edges of this graph with proper boundary conditions at the nodes. System of PDEs is solved with finite difference technique on the non-regular spatial grid using two-layer acoustic scheme [5]. The scheme is stable in case it satisfies the Courant condition, and it has the second order of convergence for both spatial and temporal coordinates. Also components of the model, their locations on the axis, and connections between them are shown in this figure.

Fig.3. The graph of the model

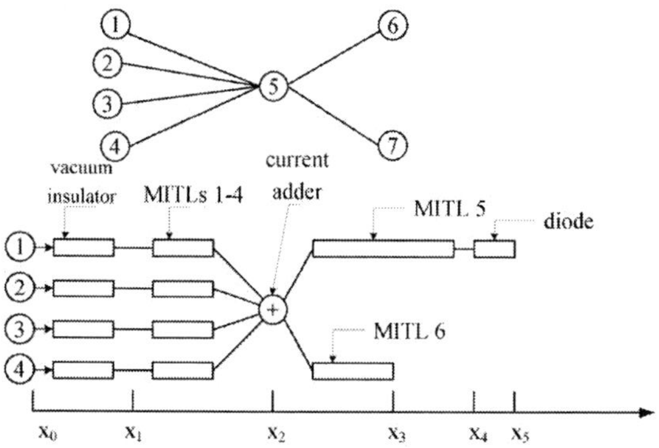

The non-regular spatial grid with given global and local parameters is created automatically using the coordinates of nodes. Parameters of electron losses are calculated by method proposed in [6] for ANGARA-5-1 facility with only one exception that we used relativistic form of Child-Langmuir law for current density. The conversion of 3D components into 1D representation was performed by means of following procedure: for vacuum insulators – on the base of electrostatic and magnetostatic calculations, for MITLs 1-4 (coaxial) – analytically, for MITLs 5,6 (disc) – semi-analytically (numerical calculations of the $d(x)$), for diode – on the base of electrostatic calculations. The distributions of the model parameters $L_x(x)$, $C_x(x)$, $Z(x)$, $d(x)$ along the spatial coordinate are shown in Fig.4.

The expected impedance of the vacuum diode with hollow cathode predicted by parapotential model should be $Z_d \approx 1$ Ohm.

Fig.4. Model parameters distributions

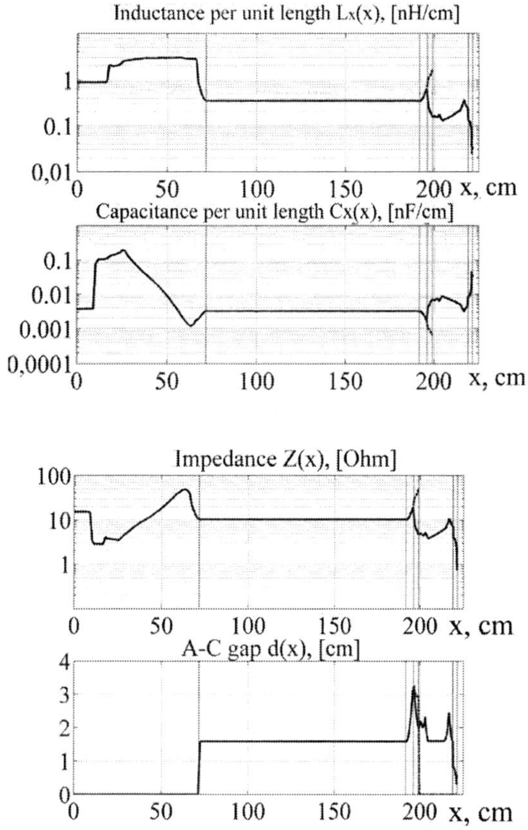

III. THE RESULTS

There were only 2 "bottlenecks" in calculations: the conjunction of the initial and boundary conditions at the node #5 – the current adder, and short length of MITL #6 with open end (see Fig.3). The short distance between open end of MITL#6 (X_3) and the adder (X_2) increase the stiffness of the system of equations that cause rapidly growing instabilities on this edge of the graph.

In the calculations 4 input coaxial ports with the impedance of 2.9 ohm are fed simultaneously with the equivalent voltage sources. The incident voltage curve V_{inc} is shown in Fig.5a together with diode current I_d, diode voltage U_d and electron loss current I_{loss}. In Fig.5b electron loss current in MITLs #1-6 are shown.

Fig.5. Voltage and current curves in MITLs and diode

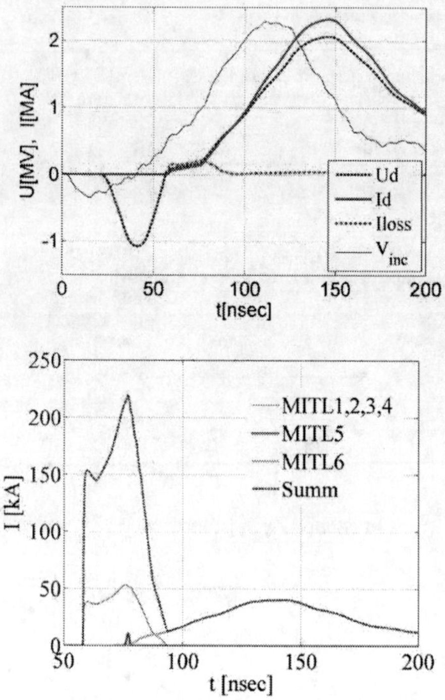

The calculated output electric parameters of the facility that characterize the efficiency of energy delivering from modules to the load are shown in Table 1.

TABLE 1. OUTPUT ELECTRIC PARAMETERS

Parameter name	Value
Input energy, kJ	328,7
Energy delivered to the diode, kJ	289,5
Energy of electron losses, kJ	6,5
Peak input power, TW	6,2
Peak diode power, TW	4,8
Diode current, MA	2,3
Diode voltage, MV	2,1
Electron loss current, kA	220
Electron loss current in MITLs #1,2,3,4, kA	53
Electron loss current in MITL #5, kA	10
Electron loss current in MITL #6, kA	40,5

According to the calculations, at the diode voltage 2.1 MV the diode current is of 2.3 MA, diode impedance is of 0.9 Ohm. The total electron loss current is about 220 kA, the loss energy is 6.5 kJ or 2% of input energy. The energy delivered to the diode is of 290 kJ or 88% of input energy.

IV. CONCLUSION

On the base of telegraph equations 1D model of the central part of the facility GAMMA-4 that includes current adder and common vacuum diode load is developed. Using this model the parameters of pulse delivered to the load are calculated with taking into account electron losses in the vacuum transmission lines. According to the results the efficiency of energy delivering from modules to the common load with the impedance of 0.9 Ohm is about 88%. The energy of electron losses doesn't exceed 2%.

The obtained results make it possible to consider the presented design of the vacuum current adder and common vacuum diode load as a perspective variant for advanced research.

References

[1] N.V. Zavyalov, V.S. Gordeev, V.T. Punin et al. "Design of electro-physical facility GAMMA-4", in Proc. of 2014 IEEE 41st International Conference on Plasma Science and 2014 IEEE 20th High Power Particle Beams (ICOPS/BEAMS 2014), pp.674-678.

[2] J.M. Creedon. Relativistic Brillouin flow in high ν/γ diode. Journ. Appl. Phys. 1975. V.46, №7, pp.2946-2955.

[3] J.M. Creedon. Magnetic cutoff in high-current diodes. Journ. Appl. Phys. 1977. V.48, №3, pp.1070-1077.

[4] L.A.Vainstein. *Electromagnitniye volny* [Electromagnetic waves]. Moscow, Nauka, 1988. 440 p.

[5] N.N.Kalitkin. *Chislenniye metody* [Numerical methods]. Moscow, Nauka, 1978. 512 p.

[6] V.V. Aleksandrov, E.V. Grabovski, A.N. Gribov, G.M. Oleinik, A.A. Samokhin, "Transmission Efficiency of an Electromagnetic Pulse in the Vacuum Concentrator of Angara-5-1 Facility", in Proc. of the 18th International Conference on High-Power Particle Beams, pp. 436-445, 2010.

Nuclear microprobe applied to the study of corroded metallic surfaces

J. Cruz
Laboratório de Instrumentação, Engenharia Biomédica e Física da Radiação (LIBPhys-UNL)
Departamento de Física, Faculdade de Ciências e Tecnologia, FCT
Universidade NOVA de Lisboa
Caparica, Portugal

V. Corregidor
IPFN, IST-UL, Campus Tecnológico e Nuclear
Sacavém, Portugal

J. Lopes
Instituto Superior de Engenharia de Lisboa
GIAAPP/ISEL
Lisboa, Portugal

Abstract - **This work reports on the use of a nuclear microprobe to the study of corroded metallic surfaces. The mapping and resolution microbeam capabilities combined with the simultaneous and self-consistent PIXE and EBS spectra analysis allowed a fast and non-destructive procedure to determine the corrosion layer elemental composition and concentration depth profiles.**

Keywords—nuclear microprobe; corrosion; IBA techniques

I. Introduction

Corroded surfaces composition and thickness are highly dependent on the alloy, on the exposure time and on the environmental conditions, such as humidity, temperature or contact with other metals. The study of corrosion products gives important information about degradation mechanisms that occurred or are still occurring, their extension, and which conservation/restauration procedures can be effective to stabilize the metallic surface. Very often, due to the intrinsic value of the artifact being studied, a non-destructive analysis is required. PIXE (Particle Induced X-ray Emission) is a IBA (Ion Beam Analytical) technique commonly used in these studies [1], but since it does not give information about elemental depth distribution and about elements lighter than sodium, it is sometimes combined with EBS (Elastic Backscattering Spectrometry), which gives information about the depth distribution of the elements detected by PIXE, and can also detect light elements such as carbon and oxygen. In order to get a self-consistent solution from the PIXE and EBS spectra that can describe the concentration profile, a simultaneous analysis of these spectra is required.

Due to the heterogeneity always present in a corroded surface, the proton beam used in these techniques must have dimensions at the micrometer level, thus requiring a nuclear microprobe.

The results from this microprobe for the spatially resolved analysis of a highly inhomogeneous coin corroded surface is presented here. 2D-PIXE maps were acquired, followed by point analyses in selected regions where µ-PIXE and µ-EBS spectra were taken simultaneously.

II. Materials and Methods

A. Studied Coin

Examples of corroded metallic surfaces are the ancient coins, which in most of the cases require the use of non-destructive techniques to study them due to their high historical value. The metallic sample taken under study is a 2 Reais copper coin (m = 3.26 g; ∅ = 25 mm) minted during D. António regency (1580-1583) – Fig. 1. The coin was in reasonable state of preservation, presenting a corrosion layer of unknown thickness and composition. The storage/preservation conditions are unknown.

B. Experimental Procedure

The nuclear microprobe (Oxford Microbeams type) used was commissioned in 2000 on one of the three beam lines of the 2.5 MV single-ended Van de Graaff Accelerator (Fig. 2) located at the Laboratory of Accelerators and Radiation Technologies at CTN-IST (Sacavém - Portugal) [2].

For a 2 MeV proton beam, the microprobe lenses (high excitation magnetic quadrupole triplet), scanning coils (magnetic dipoles) and two sets of micrometer driven slits, allows beam scans at the target surface up to 2640×2640 μm^2. A spatial resolution of 3×4 μm^2 was achieved for analytical techniques as PIXE and EBS under vacuum conditions ($P \approx 5\times10^{-6}$ mbar). The brightness is ~ 1 pA/(μm^2×$mrad^2$×MeV).

Fig. 1. Photo of studied coin. Obverse (left) and reverse (right) of the 2 Reais copper coin minted during D. António regency (1580-1583). The square indicate the area covered by the 2D PIXE maps (scan 2640×2640 μm²); the dots indicate the locations of PIXE+EBS point analysis (3×4 μm²).

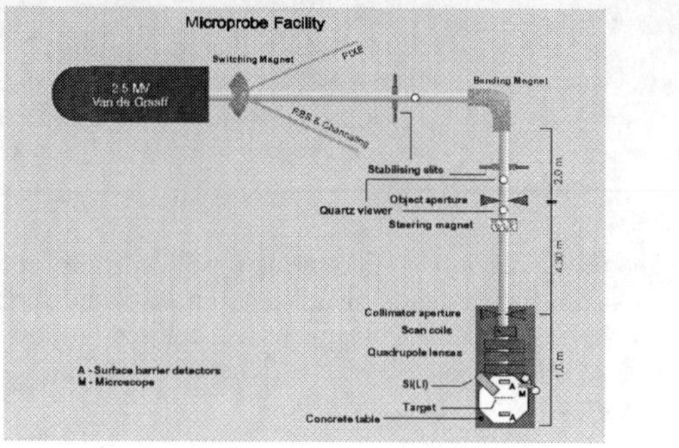

Fig. 2. Location of the CTN microprobe and its associated components on the 2.5 MV single ended Van de Graaff accelerator [2].

The coin was cleaned with alcohol in an ultrasonic cleaning bath before the measurements and its surface optically inspected with an Olympus SZX9 stereo microscope for area selection to be probed by the ion microbeam and then obtain the local X-ray surface elemental distribution maps. These maps allowed to identify regions with different corrosion patterns.

The PIXE spectra were collected using an 8 μm thick Be windowed Si(Li) detector with 145 eV resolution placed at 135° to the beam direction. The EBS spectra were collected with a 200 mm² PIPS detector with 30 keV resolution in Cornell geometry placed at 140° to the beam direction.

Microprobe operation and basic data manipulation, including X-ray surface elemental distribution mapping, was achieved through the OMDAQ software code [3]. Evaluation and quantification of PIXE spectra was done with the GUPIXWIN [4] code, assuming a bulk sample. From these results, the peak area of each X-ray line is used as input in the DataFurnace code [5] to fit the EBS spectra, thus providing a self-consistent elemental depth profiles for each selected point of interest. This procedure was implemented assuming local search fitting, and some constraints were imposed depending on the studied coin (see details in next section). The beam straggling was calculated assuming Chu correction with Tschälar effect.

In terms of sensitivity, for PIXE the lowest limit of detection (LOD) can be as low as a few μg/g. The accuracy of the measurements was considered to be only dependent on the counting statistics and fit error. So, in general, major elements concentrations are obtained with accuracy below 2%. For the trace elements, this accuracy degrades, being about 5% for absolute concentrations around 0.1% and about 20% for elements at a lower concentration level (< 0.01%). For EBS, the LOD is around 3% and the depth resolution of concentration profiles is estimated to be around 100 nm. For both techniques, the accessible depth is a few micrometers.

III. Results and Discussion

A 2640×2640 μm² surface scan was taken in the area highlighted by the open square in Fig.1. The detected X-rays indicated the presence of Cu (coin's major element) and also traces of Si, S, Cl, Ag, Ca, Ni, Cr, Fe, As and Pb with heterogeneous surface distributions. Fig. 3 shows the X-ray surface elemental distribution maps (relative yield) for the K_{α} lines Cu (8.04 keV), Fe (6.40 keV), Ca (3.69 keV) and S (2.31 keV). A heterogeneous surface elemental distribution is clearly observed for these four elements. μ-PIXE and μ-EBS spectra were simultaneously taken for each of the three points (3×4 μm²) marked with dots in Fig.1 and Fig. 3. Fig.4 shows the μ-PIXE fitted spectrum (identified as having a clear X-ray Fe signal), and table I gives the elemental composition obtained from this fit. Due to the altered surface, a gradient concentration is expected and these results represent only average values over the proton range in the analyzed material. The EBS spectrum shown in Fig. 5 was fitted considering the information from the PIXE fit and included carbon and oxygen (not detectable by PIXE). The self-consistent elemental depth profile thus obtained is shown in Fig. 6 and table II. The DataFurnace code used a total 8 layers for describing the concentration profiles in the surface before reaching the coin bulk composition (fitted with 1 layer). It corresponds to an iron incrustation that extends as deep as ≈ 36×10^{18} at/cm² (≈ 4.6 μm), showing a maximum concentration at a depth of ≈ 20×10^{18} at/cm² (≈ 2.6 μm). This depth also defines the extension of the corroded layer, due to oxidation from the surface. Elements present in very small concentrations (S, Cl, Ca, Cr, Ni and As) have little influence on the EBS spectra fit which is largely insensitive to their presence.

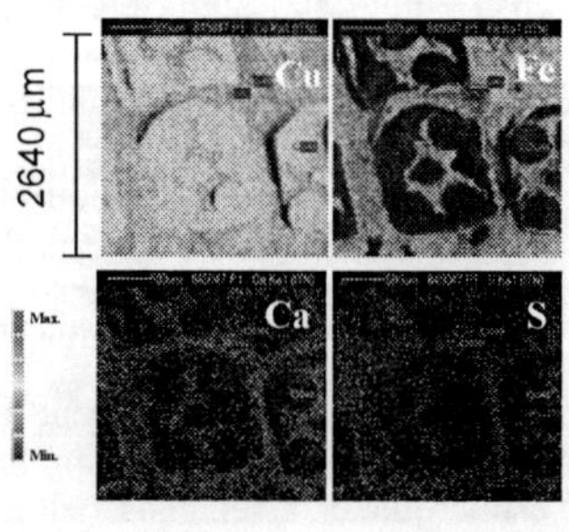

Fig. 3. PIXE elemental surface maps for Cu, Fe, Ca and S obtained with a 2 MeV proton beam (scan 2640×2640 μm²).

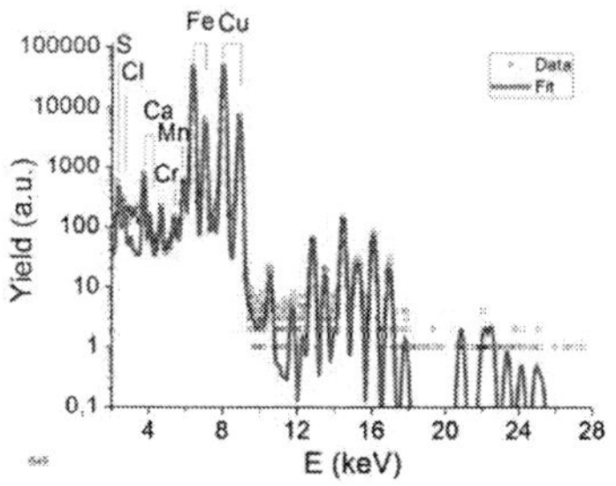

Fig. 4. μ-PIXE spectrum aquired in a location (3×4 μm^2) where Fe was clearly identified. Spectrum fitted with GUPIXWIN. Beam energy = 2.0 MeV.

TABLE I. ELEMENTAL CONCENTRATIONS OBTAINED FROM THE μ-PIXE SPECTRUM FIT

Element	Conc. (wt.%)
S	1.14±0.03
Cl	0.25±0.02
Ca	0.339±0.006
Cr	0.034±0.002
Mn	0.248±0.006
Fe	24.31±0.08
Ni	0.14±0.02
Cu	73.3±0.2
As	0.11±0.01

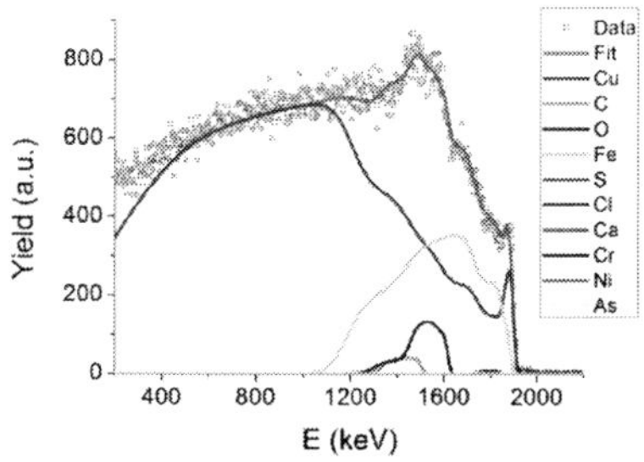

Fig. 5. μ-EBS spectrum taken in a location (3×4 μm^2) where Fe was clearly identified. Spectrum fitted with DataFurnace. Beam energy = 2.0 MeV.

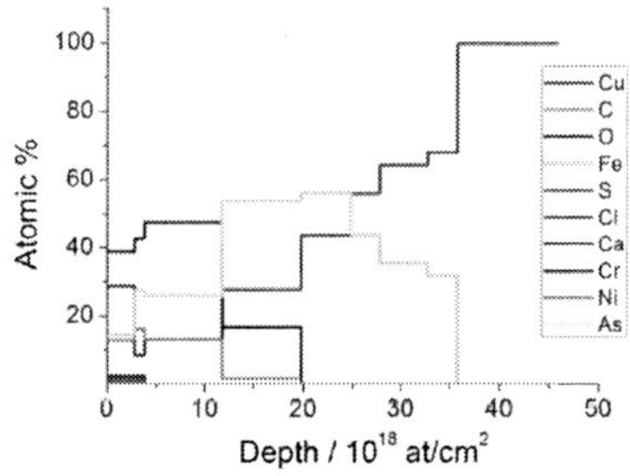

Fig. 6. Elemental depth profile obtained from the μ-EBS spectrum fit of Fig.5.

TABLE II. ELEMENTAL DEPTH DISTRIBUTION OBTAINED FROM THE μ-EBS SPECTRUM FIT.

Layer	t (10^{15} at.cm^{-2})	Cu	C	O	Fe	S	Cl	Ca	Cr	Ni	As
1	2800	28.8	12.7	38.8	14.4	2.6	0.71	1.79	0.03	0.11	0.05
2	1000	8.5	16.1	42.7	27.3	2.6	0.69	1.82	0.03	0.11	0.06
3	8000	13.2	13.2	47.5	25.9	0	0	0	0.04	0.13	0.07
4	8000	27.7	1.66	16.6	53.8	0	0	0	0.03	0.10	0.05
5	5000	43.7	0	0	56.1	0	0	0	0.04	0.13	0.07
6	3000	55.8	0	0	43.9	0	0	0	0.04	0.15	0.07
7	5000	64.2	0	0	35.5	0	0	0	0.04	0.14	0.07
8	3000	67.9	0	0	31.8	0	0	0	0.04	0.12	0.06
9	Bulk	99.8	0	0	0	0	0	0	0	0.12	0.06

IV. CONCLUSIONS

This work showed the potential of using a nuclear microprobe for corrosion studies.

Simultaneous and self-consistent PIXE and EBS analysis was applied to study the superficial corrosion layer of a XVI century Portuguese coin. The results revealed a complex multielemental depth concentration structure that would have been very hard to be obtained if we had not followed a self-consistent approach. The combination of PIXE and EBS at the experiment level and at the data analysis level showed that it is possible to characterize the superficial corroded layers in metals in terms of elemental composition and depth distribution in a fast and non-destructive way. However, some care should be taken when considering the depth distribution of elements whose concentration is below 1-2%, since small variations on these contents usually do not contribute significantly to the variation of global EBS spectra. For these elements, we obtain only a tentative approach to the real distribution.

The use of a micrometer sized proton beam for this study was also very important because it allowed probing regions with different degrees of corrosion.

Acknowledgment

J. Cruz gratefully acknowledges the FCT support through the UID/FIS/04559/2013. V. Corregidor acknowledges FCT for the Ciência program and project UID/FIS/50010/2013.

References

[1] C. Flament and P. Marchetti, "Analysis of ancient silver coins" in Nucl. Instr. and Meth. in Phys. Res. B 226, 2004, pp. 179–184.

[2] L.C. Alves, M.B.H. Breese, E. Alves, A. Paúl, M.R. da Silva, M.F. da Silva, and J.C. Soares, "Micron-scale analysis of SiC/SiCf composites using the new Lisbon nuclear microprobe" in Nucl. Instr. and Meth. B 161-163, 2000, pp. 334-338.

[3] G.W. Grime, and M. Dawson, "Recent developments in data acquisition and processing on the Oxford scanning proton microprobe" in Nucl. Instr. and Meth. B 104, 1995, pp. 107-113.

[4] J.A. Maxwell, W.J. Teesdale, and J.L. Campbell, "The Guelph PIXE software package II " in Nucl. Instr. and Meth. B 95, 1995, 407-421.

[5] C. Pascual-Izarra, M. Reis, and N. Barradas, "Towards truly simultaneous PIXE and RBS analysis of layered objects in cultural heritage" in Nucl. Instr. and Meth. B 249, 2006, pp. 780-783.

Study of physical processes at high energy densities with the use of explosive magnetic generators

A.V. Ivanovskiy
Russian Federal Nuclear Center – VNIIEF,
37 Mira ave., 607188 Sarov, Nizhniy Novgorod region, Russia
*Corresponding author: ivanovsky@elph.vniief.ru

Abstract- **One of the methods to produce high energy density during modeling of physical processes and studies of material properties is to convert the kinetic energy of a high-speed shell or a liner. Traditionally, the liners are driven by:**

- an explosive charge (HE) on gas-dynamic complexes;

- a current pulse on electrophysical facilities;

- radiation on laser facilities.

The paper will present these methods, will evaluate the maximum attainable velocities and pressures, will compare the possibilities of providing symmetry and effect controlling, as well as recovery of the explored samples for subsequent analysis.

The electrophysical methods of research:

- provide better monitoring of parameters and obtained data in comparison with gas-dynamic complexes;

- allow studying the big-size samples (scale factor) in comparison with laser facilities.

The already realized and the planned experiments with the use of explosive magnetic generators (EMG) will be overviewed.

In the area of relatively low specific energies (to ~10 kJ/g) of interest are the studies of the rheological properties of matter, i.e. spall and shear strength, ejecta processes at shock wave release to the surface, etc. The use of EMG ensures high accuracy of control and reproducibility of loading parameters, and in some cases the recovery of the explored samples for further metallographic analysis.

In the intermediate region of specific energy (up to hundred kilojoules per gram) it would be interesting to investigate the equations-of-state of materials (EOS) and turbulent mixing. The paper will present the experimental test bench on the basis of a small-class disk EMG, making it possible to conduct high-precision measurements of the EOS (2-3%) at shock and isentropic compression by 2-3 Mbar pressure and to add significantly the existing experimental data (to 10^5 g_0) for turbulent mixing of substances at acceleration to 10^8-10^9 g_0.

Lasers are the most appropriate tool to conduct studies in the region of high (more than 1 MJ/g) energy densities. The laser facilities allow reaching the radiation temperatures 250-300 eV. The experiments on Z-machine (SNL, USA) demonstrated that the electrophysical facilities, able to produce the currents of tens of MA for the time of ~100 ns, can be an alternative to lasers. The achieved level of energy generated by Z-pinch in the X-radiation compares well with lasers. However, in view of worse organization of energy for realization of radiation temperature of 250-300 eV it is necessary to construct the electrophysical facilities one order more powerful that Z-machine. To realize the above-mentioned temperatures, we need the energy of tens of megajoules in the X-radiation. At this the electrophysical facilities may add the lasers in scale effects modeling.

To achieve the laser energy densities using the EMG, it is necessary to resolve the problem of current pulse sharpening to 100 ns. The paper will discuss the ways to solve this problem. In case the current pulse sharpening technology is realized, the middle-class disk EMGs can serve the basis for a creation of an explosive analogue of Z facility, and this opens the way to realization of one-shot explosive experiments with the use of a super-power EMG and generation of Z-pinch X-radiation of tens of MJ.

Keywords—high energy density, explosive magnetic generator, current pulse peaking

I. Introduction

One of the methods to produce high energy density during modeling of physical processes and studies of material properties is to convert the kinetic energy of a high-speed shell or a liner.

Traditionally, the liners are driven by an explosive charge (HE) on gas-dynamic complexes; by a current pulse on electrophysical facilities; by radiation on laser facilities.

On the gas-dynamic complexes the possibilities of achieving high-velocities are limited by the detonation wave velocity, which is ≤ 9 km/s. In practice it is possible to realize the velocities of the flat plates not higher than 6 km/s (the specific energy ~10 kJ/g).

The laser facilities allow achieving easily the radiation temperatures of 250-300 eV. For this reason the lasers are the most suitable for the studies in the areas with high (≥1 MJ/g) energy densities. However, due to a relatively low total energy of laser radiation (≤1 MJ) the spatial scales of the studied objects are small. This raises the question about the effect of the scale factors on the obtained research data.

The electrophysical facilities on the basis of explosive magnetic generators (EMG) with foil current opening switches (FOS) shape the current pulses of tens of MA for the time of ~1 µs. This allows covering the area of research on gas-dynamic complexes and move forward in specific energy density to several hundred kJ per gram.

The experiments on Z-machine (USA) demonstrated that the electrophysical facilities, able to produce the currents of tens of MA for the time of ~100 ns, can be an alternative to lasers. The achieved level of energy generated by Z-pinch in the X-radiation compares well with lasers. However, in view of worse organization of energy for realization of radiation temperature of 250-300 eV it is necessary to construct the facilities one order more powerful that Z-machine. In case such facilities are constructed, they may add the lasers in scale effects modeling.

Below presented is the evaluation of the specific energy (pressure) parameters achieved on the gas-dynamic complexes, lasers and electrophysical facilities. The advantages and the disadvantages of these methods will be discussed. The realized and the planned research with the use of the EMG will be presented. The ways have been directed to development of the electrophysical facilities on the basis of EMG in order to achieve the laser energy densities.

II. OVERVIEW OF THE RESEARCH METHODS

Gas-dynamic complexes. Figure 1 shows the elementary schematic of studies of the matter properties with the use of HE. The metal shell driven by HE impacts the sample from the explored material. At this the shock wave velocity, the mass velocity, the instabilities development, the ejecta formation on the surface and other parameters are diagnosed.

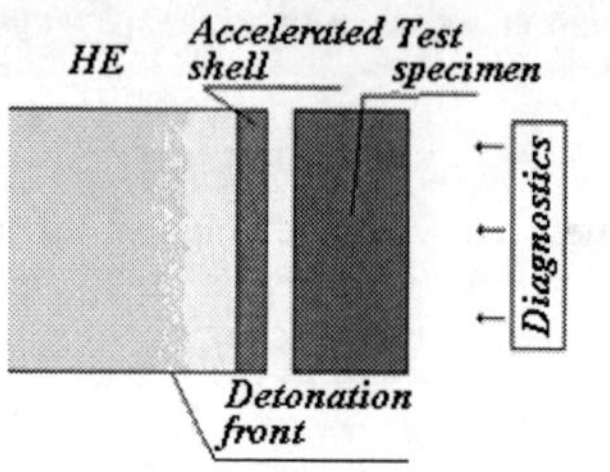

Fig. 1 Scheme of experiments with the use of HE

In accordance with the solution proposed by E.I. Zababakhin [1], the maximum velocity of the incompressible shell Wk depends on the detonation rate WD as

$$\frac{W_k}{W_D}=\left[1+\frac{27}{16\alpha}\left(1-\sqrt{1+\frac{16\alpha}{27}}\right)\right], \qquad (1)$$

where α is the ratio between HE mass and shell mass.

Dependence of Wk/WD on α is presented in Fig. 2. To α~10 the shell velocity grows fast with the increase of HE mass. Further increase of the HE mass is not effective. For the sake of evaluation we take α=12.5, that corresponds to Wk/WD =0.6. At the detonation rate WD=9 km/s we get shell velocity Wk≈5.4 km/s.

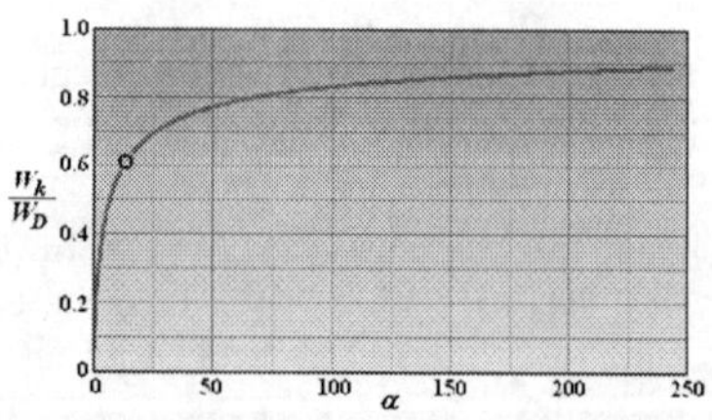

Fig. 2 Shell velocity Wk as a function of detonation rate WD depending on α

If the shell and the explored specimen are made of the same material, the mass velocity of substance in the shock wave will make v1=Wk/2. In case of an intensive shock wave for pressure p1 and specific energy ε1 we get [2]

$$\begin{cases} v_1 = W_k/2, \\ p_1 = \frac{1}{2}\rho_0 W_k D, \\ \varepsilon_1 = W_k^2/8, \end{cases} \qquad (2)$$

where D is shock wave front velocity.

Velocity v1, pressure p1 and specific energy ε1 are summarized in Table 1. Pressure is estimated in approximation D≈v1 for Al (ρ0=2.7 g/cm3) and Fe (ρ0=7.8 g/cm3).

TABLE 1. PARAMETERS REALIZED ON GAS-DYNAMIC COMPLEXES

v_1, km/s	p_1, Mbar (*Al-Fe*)	ε_1, kJ/g
2.75	0.2-0.6	4

The advantage of the scheme is its simplicity, but it has got some disadvantages:

- the perturbations introduced by the initiation system and by the instabilities on the HE detonation front may significantly affect the accuracy of the results;

- a big amount of HE placed next to the explored specimens makes it difficult to recover them for the post-shot visual and metallographic analysis.

Laser facilities. The typical experimental scheme is presented in Fig.3. During absorption of laser radiation by the target the plasma is formed, that moves towards radiation. The evaporation front moves inside the target with velocity D. At the same time, due to the law of conservation of momentum, the shock wave velocity D1 propagates through the explored specimen.

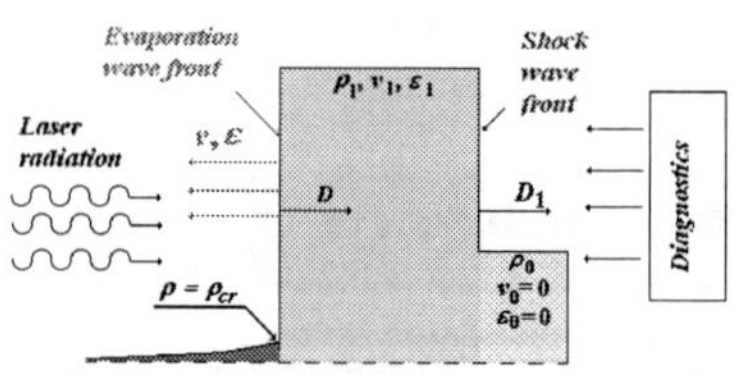

Fig. 3 Scheme of the experiments on laser facilities

In case of flows providing complete ionization the absorption of energy happens in the vicinity of the point with critical density

$$\rho_{cr}=\frac{A}{Z}\cdot\frac{\pi c^2 m_p m_e}{\lambda^2 e^2} \qquad (3)$$

where mp, me are masses of proton and electron, Z, A are the charge and the atomic mass of the target material, λ is the radiation wave length. Let's write down the laws of conservation of mass, momentum and energy on evaporation front [2]

$$\begin{cases} (v-D)\cdot\rho=(v_1-D)\cdot\rho_1, \\ p+\rho\cdot(v-D)^2=p_1+\rho_1\cdot(v_1-D)^2, \\ \varepsilon+\frac{p}{\rho}+\frac{(v-D)^2}{2}=\varepsilon_1+\frac{p_1}{\rho_1}+\frac{(v_1-D)^2}{2}+\frac{S}{\rho_1\cdot(v_1-D)}, \end{cases} \quad (4)$$

where ρ, v, p, ε are density, mass velocity, pressure and specific energy; index «1» refers to the substance compressed by the shock wave.

The rate of evaporating plasma outflow v-D is determined by the sound speed c0= v-D. In adiabatic case $c_0=\sqrt{\gamma p/\rho_{cr}}$ and for ideal - p=(γ-1) ρcrε evaporating plasma we get the outflow conditions (γ is adiabatic index)

$$\begin{cases} v-D=c_0, \\ p=\frac{1}{\gamma}\cdot\rho_{cr}c_0^2, \\ \varepsilon=\frac{c_0^2}{\gamma(\gamma-1)}=\frac{v_1^2}{2}. \end{cases} \quad (5)$$

In a strong shock wave the specific energy is determined from (δ is compression)

$$\varepsilon_1=\frac{1}{2}\frac{p_1}{\rho_0}\frac{\delta-1}{\delta} \quad (6)$$

From the conditions on the evaporation wave front (4) with regard for the outflow condition (5) for a strong shock wave (6) we have

$$\begin{cases} p_1=\rho_{cr}\cdot c_0^2\left(\frac{\gamma+1}{\gamma}-\frac{\rho_{cr}}{\delta\cdot\rho_0}\right), \\ \varepsilon_1=\frac{1}{2}\frac{\rho_{cr}}{\rho_0}\cdot\frac{\delta-1}{\delta}c_0^2\left(\frac{\gamma+1}{\gamma}-\frac{\rho_{cr}}{\delta\cdot\rho_0}\right), \\ S=\frac{1}{2}\rho_{cr}c_0^3\left(\frac{\gamma+1}{\gamma-1}-\frac{\rho_{cr}}{\rho_0}\cdot\frac{\delta+1}{\delta}\frac{\gamma+1}{\gamma}+\frac{1}{\delta}\left(\frac{\rho_{cr}}{\rho_0}\right)^2\right). \end{cases} \quad (7)$$

For the sake of evaluation we take $\rho_{cr}/\rho_0<<1, (\delta-1)/\delta\approx 1$. In this case

$$p_1=\frac{\gamma+1}{\gamma}\rho_{cr}\cdot c_0^2,\quad \varepsilon_1=\frac{1}{2}\frac{\gamma+1}{\gamma}\frac{\rho_{cr}}{\rho_0}\cdot c_0^2,\quad S=\frac{1}{2}\frac{\gamma+1}{\gamma-1}\rho_{cr}c_0^3. \quad (8)$$

Taking that A/Z=2, for the wave length λ=1 μm from formula (3) we have ρcr≈4·10-3 g/cm3. At flows S≥S0≈1015 watt/cm2 a big portion of radiant energy is spent for a production of fast electrons. The sound speed c0, pressure p1, specific energy ε1 and mass velocity v1, obtained by (8) at λ=1 μm, S=S0, γ=5/3, are summarized in Table 2.

TABLE 2 PARAMETERS REALIZED ON LASERS AT λ=1 μM, S=S_0, γ=5/3

c_0, cm/s	p_1, Mbar	ε_1, MJ/g (ρ_0=2.7·g/cm^3)	v_1, km/s
1.26·10^8 cm/s	100	1.9	62

The laser facilities allow achieving pressures up to several hundred megabars and specific energies of megajoule level. The laser radiation flow S is time-regulated. In addition, there are some problems related to:

- provision and control of spatial uniformity of laser radiation flow, required for the accuracy of the obtained results;

- adequacy of reproduction of large-scale (≥1 cm) effects at small dimensions (0.1-1 mm);

- recovery of the explored specimens for post-shot metallographic analysis.

Electrophysical facilities. Typical scheme of experiments to study the properties of matters is presented in Fig.4.

The Lorentz force, produced at the interaction of the current flowing in the liner I with the generated magnetic field Hφ, accelerates the liner. The liner hits the target surface. In the explored specimen-target the shock wave is formed. The diagnostic sensors are located on the system axis.

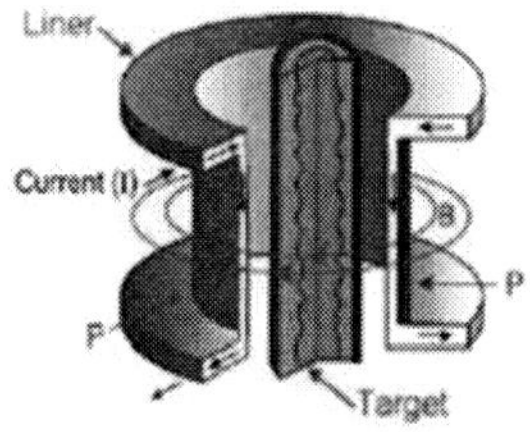

Fig. 4 Scheme of experiments on electrophysical facilities

To describe the liner acceleration, it is necessary to solve the following system of equations:

$$\begin{cases} m\frac{dv}{dt}=-\frac{\mu_0 I^2}{4\pi r}, \\ v=\frac{dv}{dt}. \end{cases} \quad (9)$$

where t – time, r – liner coordinate, v – liner velocity, m =2πρ0r0h – linear mass (ρ0, h and r0 – matter density, thickness and initial position of the liner).

In dimensionless form $\tilde{r}=r/r_0, \tilde{t}=t/\tau, \tilde{v}=v/v_0=v\cdot\tau/r_0$ (τ - time of maximum current achievement Im) the system of equations can be rewritten as

$$\begin{cases} \frac{d\tilde{v}}{d\tilde{t}}=-\Pi\frac{\tilde{I}^2}{\tilde{r}}, \\ \tilde{v}=\frac{d\tilde{v}}{d\tilde{t}}, \end{cases} \quad (10)$$

where $\Pi=\frac{\mu_0}{8\pi^2}\cdot\frac{\tilde{I}_m^2\tau^2}{\rho_0 h r_0^3}$ is the parameter determining the scaling of liners.

Solution of equation (10) with current $\widetilde{I} = I/I_m = \sin(\pi/2\cdot\widetilde{t})$ and parameter Π=3 is presented in Fig.5. At the current maximum the liner travels half o the initial radius and has the velocity $\widetilde{v} = 2$. It means that the liner is stable during acceleration on the base not more than N=5 of initial thicknesses, i.e. h=r0/(2N), and we get the velocity at the moment of impact on the target

$$W_k = \left(\frac{4\mu_0 N}{\pi^2 \Pi}\right)^{1/4}\frac{1}{\rho_0^{1/4}}\cdot\sqrt{\frac{I_m}{\tau}} \Rightarrow W_k[\text{км/с}] = \frac{5.4}{\rho_0^{1/4}[\text{г/см}^3]}\cdot\sqrt{\frac{I_m[\text{MA}]}{\tau[\text{мкс}]}}. \quad (11)$$

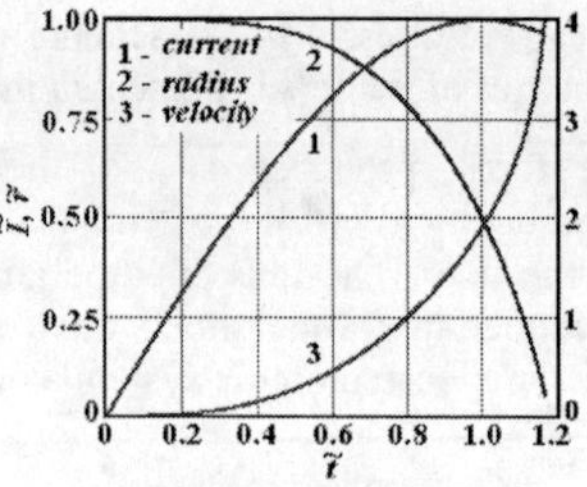

Fig. 5 Liner current, radius and velocity as a function of time

If the target and the liner are made from the same material, then the mass velocity of the shock wave in the target is equal to v1=Wk/2.

The current sources on the basis of EMG with the current opening switches (explosive and electrically exploded) of microsecond range produce the voltage U0 to 500 kV in the load L0~10 nH [3]. Making estimates of ∂I/∂t~I/t, from U0=L0·∂I/∂t we get I/t~ U0/ L0~50 MA/μs. For aluminum ρ0=2.7 g/cm3 we have Wk ≈30 km/s. Solution of the problem of pulse front reduction to 100 ns will make it possible to increase the velocity of the shell to ~90 km/s.

TABLE 3 THE PARAMETERS REALIZED WITH THE USE OF EMG (AL, ρ0=2.7 G/CM3)

	Wk, km/s	v1, km/s	ε1, MJ/g	p1, Mbar
EMG with current front ~1 μs	30	15	0.11	6.1
EMG with current front ~100 ns	90	45	1.0	55

Table 3 shows the specific energies ε1, pressure p1, obtained in EMG. The same table gives the values during realization of current pulses with the front ~100 ns.

Fig. 7 Experimental facility: 1- HEMG; 2– units of current sharpening and interruption; 3- wave line; 4– explored specimens shielding; 5– load

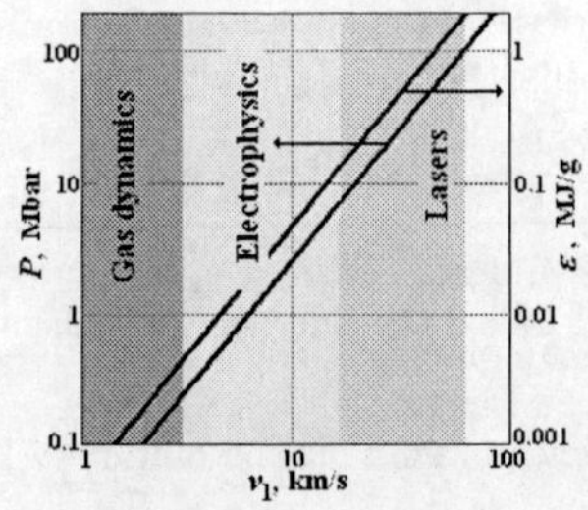

Fig. 6 Research areas

In Fig.6 one can see the areas available to conduct research on gas-dynamic complexes, EMG-based facilities and lasers. The lasers are the most universal tool for the studies – the attainable pressures up to 100 Mbar.

Solution of the problem of EMG current pulse sharpening to ~100 ns will make it possible to add and verify the laser facilities data for almost all possible pressure ranges accessible on lasers. It will also provide a possibility to test the effect of the scale factors on the obtained results.

The electrophysical methods of research:

- provide higher accuracy of control of the loading parameters, as compared with laser and gas-dynamic complexes;

- allow controlling the impact, as compared with gas-dynamic complexes;

- make it possible to study the specimens of bigger size (scale factor), as compared with lasers.

III. EXPERIMENTS WITH THE USE OF EMG

Low specific energies (to 10 kJ/g). Of interest are the studies of spall and shear strength of materials, the processes of ejecta at shock wave release to the surface of materials.

The view of the experimental facility is presented in Fig.7. The pulsed power source on the basis of HEMG 1 and units of current peaking and interruption 2 shapes a trapezoidal current pulse with the amplitude to 10 MA and controlled duration to ~30 μs. The time of current rise and drop is ~2 μs. The current pulse transferred by the wave line 3 drives liner 1 (see Fig.8). The liner produces the shock wave in targets 2 made of the explored materials. To preserve the targets from destruction by the shock wave, load 5 is protected by an armored plate 4. At the target velocities of ≤ 1 km/s the test specimens are recovered for metallographic analysis. The dynamic picture of impact can be reconstructed due to velocities of inner surfaces of the targets and the liner recorded by the PDV technique.

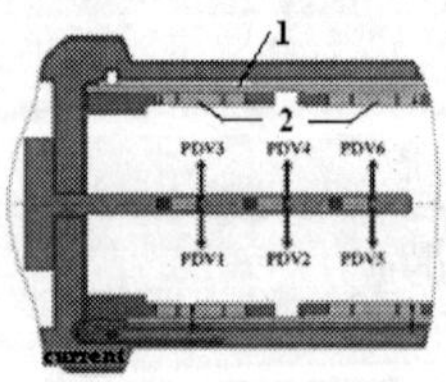

Fig. 8 Load scheme:
1. –driven liner; 2 – cylindrical targets from explored material

Such experiments for the first time provided representative information on initiation, growth, full and partial recollection of a spall in metal [4]. The results allowed verifying the numerical models of spall damage and recollection of damaged medium [5].

We have started the experiments studying the shear strength of Be. The specimens 3, 4, 5 mm thick were enveloped into aluminum shell 1 mm thick to collect the debris in case of fragmentation. In order to avoid the spall damage of Be, the targets were placed on the aluminum "sacrificial" substrate 3 mm thick. The external view of the target unit before the experiment is presented in Fig. 9a. The current was flowing in the 2 mm-thick aluminum liner driven to velocity of ~ 1 km/s. That provided pressure of ~10 GPa during impact on the target.

a b

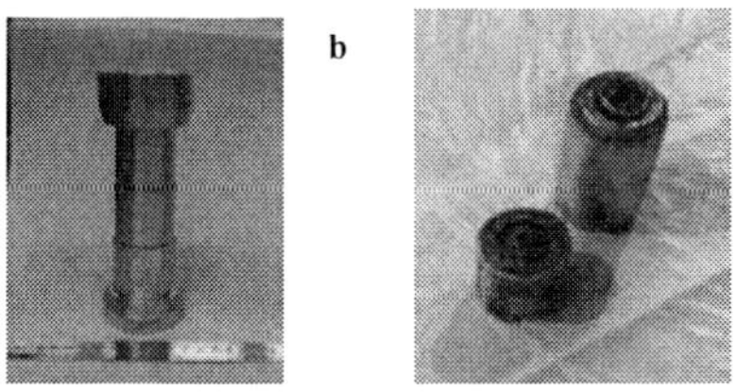

Fig. 9. Target unit with Be: (a) before experiment;(b) after experiment

The first experiment gave the expected information on the velocities of the liner and the targets. We managed to avoid the targets implosion to the axis, i.e. to recover the test specimens (Fig. 9b) for metallographic analysis. The information was obtained on the behavior of Be at strain rates 103÷104s-1 and strain intensity ε~0.3÷0.4.

A series of tests has been planned to study the ejecta of lead surface at pressures of 20-40 GPa. 5-6 specimens are loaded in every experiment; they differ in the inner surface finish. The specimens are loaded by a 1 mm-thick copper liner driven by the current of 10 MA with duration of ~30 μs. the experiments will explore the ejecta process per se, i.e. without different dynamic impacts caused by the HE initiation system and the detonation front instabilities. It is planned to measure current, velocity of the liner and ejecta, mass of ejecta due to the pulse transferred to lavsan film.

IV. INTERMEDIATE SPECIFIC ENERGIES (TO 100 KJ/G)

The study of turbulent mixing (TM) with solid bodies on gas-dynamic complexes involves difficulties, as at the velocity of ~5 km/s the elastic-plastic effects impede the development of TM. Velocities ≥10 km/s can be achieved in the three-cascade systems [6] or on light gas guns [7]. However, the thicknesses of the shells (fractions of mm) are not enough for radiographic imaging. For this reason the models are verified using data obtained at acceleration to 105 g0 in the experiments with gasses.

Disk EMG allow driving condensed cylindrical liners ~1.5 mm thick to velocities of 10-20 km/s. This will broaden the range of studies to 109g0. Fig.10 shows the experimental test bench being constructed on the basis of a small-class disk EMG [8] to explore the processes of TM.

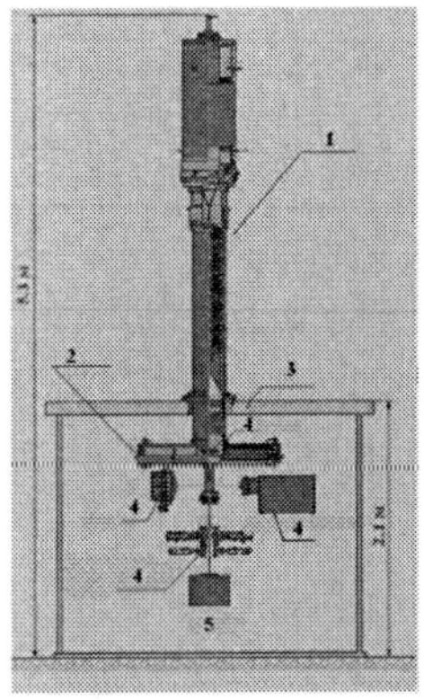

Fig. 10 Scheme of experimental test bench to study TM and EOS of matters

V. HIGH SPECIFIC ENERGIES (≥ 1 MJ/G)

Fig. 11 presents the project of the experimental test bench to conduct the studies. The energy basis of the test bench is formed by a current source on the basis of a helical EMG -1 and a disk EMG -2 of middle class (0.4 m in diameter), equipped with a unit of current peaking -4 to ~ 100 ns located in the protective construction -3. In order to ensure vacuum in the unit of current peaking and to realize measurements, the test bench is equipped with the vacuum system consisting of a flexible vacuum section -5, gate valves in the armored plate -7,protective constructions -6, vacuum station -9 located inside the bunker -8.

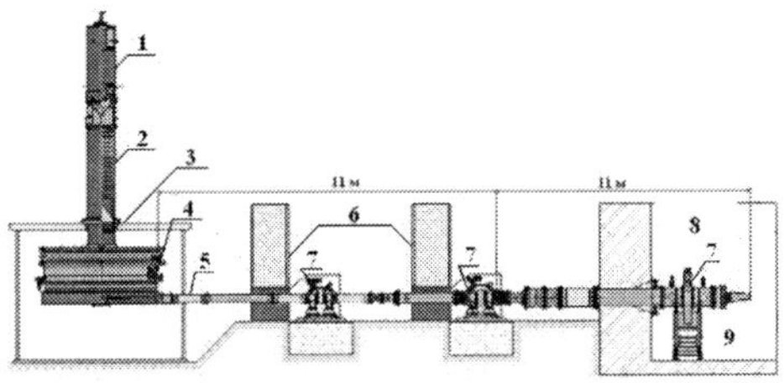

Fig. 11 Layout of experimental test bench on the basis of a middle-class disk EMG

It is expected that only the current source on the basis of EMG will be destroyed during the operation.

The estimates demonstrate that the test bench can be used to conduct experiments on Z-pinch implosion by the current ~ 25 MA for the time of ~120 ns. It will make it possible to achieve pressures to 40 Mbar and to generate X-rays with the energy to ~3 MJ for the time ~10 ns. Further increase of pressure and X-radiation energy is possible in single experiments with the use of super-power disk EMG 1 m in diameter.

The problematic point of realization is current peaking to ~ 100 ns. It is planned that a corrugated electrically exploded [9]-[11] or a magneto-dynamic current opening switch will be used for this purpose [12].

References

[1] E.I. Zababakhin. Some questions of explosion gas-dynamics. Snezhinsk, 1997.

[2] Ya.B. Zeldovich, Yu.P. Raiser. Physics of shock waves and high-temperature hydrodynamic phenomena, Moscow, 1963.

[3] V.K. Chernyshev, M.S. Protasov, V.A. Shevtsov, P.N. Piskarev, E.I. Zharinov, G.I. Volkov, V.V. Vakhrushev, V.A. Ivanov, V.A. Demidov, S.V. Pak, N.P. Bidylo, B.E. Grinevich, V.P. Pogorelov, A.A. Petrukhin, A.I. Kuzayev, V.B. Yakubov, V.I. Shpagin. Explosive magnetic generators of "Potok" family//Questions of atomic science and technology (VANT). Ser. Mathematical modeling of physical processes. 1992, №4, p. 33-41.

[4] V.A. Vasyukov, A.M. Glybin,, P.V. Duday et al. Study of rheological properties of aluminum with the use of explosive magnetic generators//DAN, 2013, v.448, №3, p.285-288.

[5] A.M. Kaul , A.V. Ivanovsky , W.L. Atchison et all. Damage growth and recollection in aluminum under axisymmetric convergence using a helical flux compression generator//J. Appl. Phys. 115.023516(2014).

[6] B.L. Glushak, A.P. Zharkov, M.V. Zhernokletov et al. Experimental study of thermodynamics of dense plasma of metals at high energy concentrations // ZhETF. 1989.V.96.Iss.4. P.1301-1318.

[7] L.C. Chhabildas, L.N. Kmetyk, W.D. Reinhart, C.A. Hall. Launch capabilities to 16km/s// Shock Compression of Condensed Matter – 1995. New York, 1996, AIP.P.1197-1200.

[8] A.V. Ivanovsky, V.K. Chernyshev, V.V. Vakhrushev, et al. Pulsed power source with current rise rate to 10^{14} A/s for acceleration of light liners to high velocities on the basis of disk EMG of second generation //VIII Khariton's Topical Scientific Readings on High Energy Density Physics, Sarov, P. 563-566 (2006).

[9] A.V. Ivanovsky Explosive pw-power current source – myth or reality?//Proc. of the 13th Inter. conf. on Megagauss Magnetic Field Generation and Related Topics, Suzhou (China), 6-19 July 2010. Changsha: Press of Nat. Univ. of Defense Technol., 2010. P. 32-43.

[10] S.G. Garanin, A.V. Ivanovsky and L.S. Mkhitariyan An ICF system based on Z-pinch radiation produced by an explosive magnetic generator//Nuclear Fusion., 2011, V.51, N10 (15pp).

[11] S.G. Garanin, A.V. Ivanovsky Explosive magnetic pulsed power system to achieve thermonuclear ignition by Z-pinch radiation //PMTF, 2015, V.56, №1, p. 7-16.

[12] A.A. Bazanov, A.V. Ivanovsky, V.Sh. Shaidullin. Magnetodynamic current opening switch with submicrosecond switching time // ZhTF, 2010.

Plasma Focus Neutron Source Powered by Explosive Magnetic Generator

V.E. Ablesimov, A.V. Andrianov, A.A. Bazanov, Yu.N. Dolin, P.V. Duday, A.Yu. Fevralev, A.M. Glybin A.M., A.V. Ivanovskiy, A.E. Kalinychev, G.V. Karpov, A.I. Krayev, S.S. Lomtev, V.N. Nudikov, N.I. Pozdov, S.M. Polyushko, A.F. Rybakov, A.N. Turov, A.A. Zimenkov
Russian Federal Nuclear Center
All-Russia Research Institute of Experimental Physics
607190 Sarov, Russia
E-mail: karpov@elph.vniief.ru

Abstract -- **The paper will describe the results of an explosive experiment with a plasma-focus neutron source powered by an explosive magnetic generator. During the preparation for the explosive experiment the regime of stable operation of the plasma-focus discharge chamber with deuterium filling is tested under the laboratory conditions. The inductance of the plasma-focus chamber as a function of time has been measured. The obtained dependence was used to adjust the explosive magnetic generator able to provide in the explosive experiment the same rate of energy input to the chamber as on the capacitor bank under laboratory conditions. The paper will present the results of the preparatory laboratory experiments with the Mather plasma-focus chamber with DD-neutrons generation and the results of the explosive experiment with the same chamber at deuterium-tritium gas filling and discharge current amplitude of 1,3...1,4 MA. In the laboratory experiments the yield of neutrons with the energy 2,5 MeV achieved 10^{11} neutrons per pulse; in the explosive experiment the integral yield of DT-neutrons with the energy 14 MeV was about $2 \cdot 10^{12}$ neutrons per pulse.**

Key words -- plasma focus neutron source, neutron yield, explosive magnetic generator

INTRODUCTION

When the explosive magnetic generators (EMG) are used to power the plasma focus neutron sources with currents of more than one megaampere, there is a problem of providing a guaranteed value of neutron yield in the explosive experiment. It often happens that the plasma chamber working normally under laboratory conditions gives zero or almost zero neutron yield in the explosive experiment at the same amplitude of discharge current. This can be associated with the fact that in the explosive and in the laboratory experiments one and the same discharge current is achieved at considerably different maximum voltages at the chamber input. For example, the current maximum in the chamber in the explosive experiment can be achieved during a microsecond, while under laboratory conditions the same current can be generated during several microseconds at, respectively, lower voltage at the chamber input. Comparatively high voltage in the explosive experiments can be the reason for the secondary breakdowns inside the chamber that results in the neutrons generation failure. In this connection the EMG with the regulated amplitude and duration of the output voltage pulse was developed for the plasma focus experiments. The amplitude and duration of the voltage pulse can be selected so that at a decrease of the voltage amplitude the maximum current delivered to the load could remain unchanged due to a corresponding increase of duration. In the EMG comprising the helical generator and the sectionalized explosive current opening switch (EOS), the profile of the voltage pulse can be changed due to the time diversity of the EOS sections operation. The plasma focus discharge chamber represents a load with variable inductance *L(t)*, and the law of time variation of inductance depends on current in the chamber. Selection of the optimal diversity of EOS sections operation represents a self-consistent problem. This problem is planned to be solved experimentally in several iterations thus approaching gradually the optimal variant. The data obtained during plasma focus chamber powering by a capacitor bank will be used as the initial data. These data allow determining the dependence *L(t)* that is later used in the calculations to select the diversity of the EOS sections operation. The amplitude of the capacitor bank current is provided to be approximately equal to the expected maximum current in the chamber in the upcoming explosive experiment. The paper presents the basic results obtained under laboratory conditions and the results of the explosive experiment with the plasma focus discharge chamber powering by the EMG.

LABORATORY EXPERIMENTAL PROCEDURE

The «Mather» chamber [1] with long ($l > d$) cylindrical coaxial electrodes was used. Potential electrode is anode 70 mm in diameter. Grounded electrode is cathode in the form of an armature with punched holes all along the length. The inner diameter of cathode is 120 mm, the length is 280 mm. The ends of the anode and cathode are located in one plane at a distance of 280 mm from the input grounded flange. The anode and the input flange are separated by a ceramic insulator. The diameter of the insulator is 70 mm; the height is 70 mm. The electrode system of the chamber is placed into a sealed case. The photo of the electrode system is shown in Figure 1.

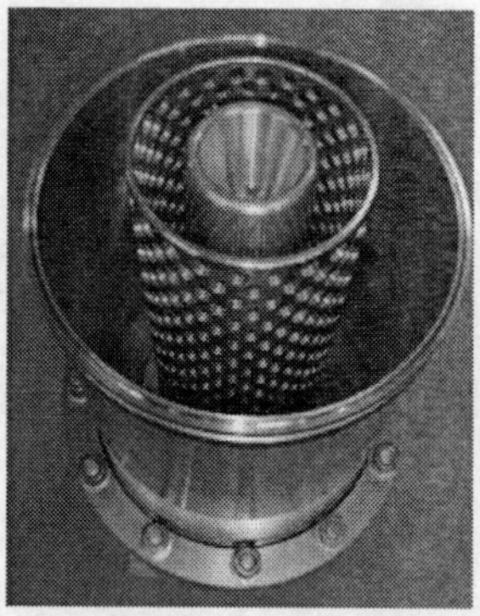

Fig.1. Photo of the chamber' electrode system.

The inorganic vacuum gaskets are used; the pump-down is realized by the oil-free turbo-molecular pump. The experiments were realized on CASCADE facility representing a powerful capacitor bank. The facility consists of 16 similar energy sections with capacity of 41 μF each. The maximum charge voltage of the bank is 50 kV. The number of the used energy sections can be varied. The sections are connected to the current collector by means of a low-inductive cable line; the cables are attached to the collector so that the switching of sections does not affect the azimuthal distribution of current supplied to the load. The maximum energy capacity of the facility is 820 kJ. Each section is switched by its solid discharger. The switching is realized through an electric breakdown of the replaceable insulation by a high-power electric discharge produced in each section by its own firing unit synchronized with the firing units of other sections. A set of several layers of polyethylene film is used as a replaceable insulation. The total thickness of insulation varies depending on the charge voltage of the bank. For the charge voltage of 46…50 kV the thickness of replaceable insulation is 750…900 μm.

RESULTS OF LABORATORY EXPERIMENTS

The objective of experiments on the capacitor bank, in addition to heating and degassing of the working surfaces of the electrode system by electric discharges, was to reveal the mode of stable operation of the discharge chamber with rather high neutron yield and to determine the law of *L(t)* variation in this mode. Deuterium was used as a working gas. When passing over to deuterium-tritium gas mixture, the neutron yield at the same discharge current usually increases by two orders. As in the explosive experiment, the induction probes are used to measure the chamber current in the experiments on the capacitor bank. The voltage at the chamber input was determined from the readings of the voltage probe. The parameters of the generated radiation were measured with the use of scintillators [2]; the integral neutron yield was determined with the use of activation techniques. The techniques of silver and indium activation were used. The results of the second activation technique seem to be more reliable, as the data provided by this technique are less exposed to the effect of scattered neutrons. Figure 2 presents the current curve $I(t)$ and the oscillogram of signal from the scintillation detector obtained in one of the electric discharges.

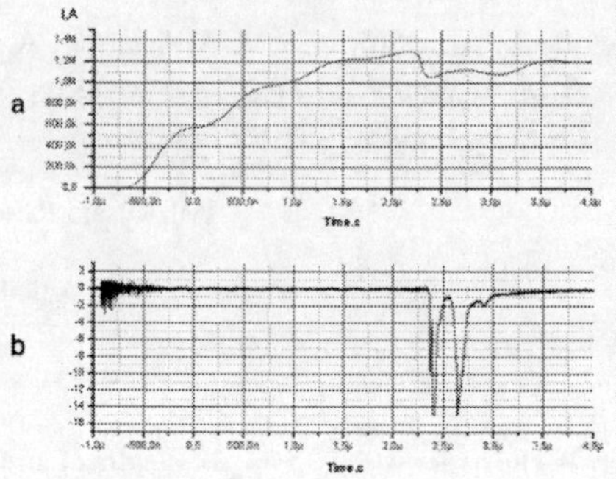

Fig. 2. Current curve (a) and oscillogram of signal from scintillation detector (b).

The scintillation detector was placed at a distance of 5,5 meters from the estimated place of neutrons generation. Current curve $I(t)$ was obtained by an integration of the signal from one of the induction probes. The initial pressure of deuterium in the discharge chamber was 10 Tor. Five sections of the capacitor bank were used; they were charged to the voltage of 44 kV. The discharge current reached the maximum value of 1,3 MA about three microseconds after the operation of the dischargers. One can see a feature on the current curve: a characteristic bend in the form of an abrupt drop of current. At the moment of this current feature the neutron radiation and the X-radiation were observed; they were recorded by the scintillation detector. Due to a flight base of 5,5 meters the neutron and the x-ray pulses on the oscillogram were shifted relative to each other by 236 ns. The neutron yield data obtained from the technique of indium activation and from the neutron pulse of the scintillation detector were almost in agreement. The neutron yield corresponding to the presented oscillograms was $1{,}1\cdot10^{11}$ n/pls. Uncertainty of measurements of the neutron yield did not exceed 20%. Stability of the neutron yield, estimated by the activation technique, under this regime of the device operation was no worse than 10%. It did not seem to be expedient to strive for better stability, so a decision was made to prepare the explosive experiment based on the data obtained under this regime of operation of the plasma focus chamber. Figure 3 shows the curve of voltage $U(t)$ at the chamber input and the curve of the chamber inductance *L(t)* calculated by formula $L(t) = I^{-1} \cdot \int U dt$.

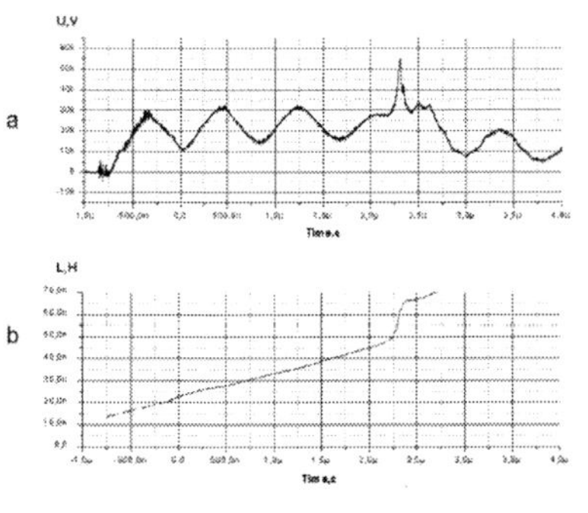

Fig. 3. Voltage at the chamber input (a) and chamber inductance (b)

Wavelike form of the input voltage is explained by occurrence of current oscillations in the transmission cable line; the capacity of this line is rather big as the unused cables are not disconnected from the current collector at the switching of the sections. The maximum voltage at the input to the plasma focus chamber is 30 kV and this value is deciding during determination of the EOS parameters for the explosive experiment. In addition, the law of variation of the plasma focus chamber' inductance in time is essential. Figure 3, curve (b), shows that in the process of calculation of the EOS parameters the law of variation of $L(t)$ can be approximated by linear time dependence with rather good accuracy.

DEVELOPMENT OF AN EXPLOSIVE CURRENT GENERATOR

In the process of development of an explosive current generator the task was to provide the same discharge current amplitude as in the experiments on the capacitor bank. At this the maximum voltage at the chamber input should not exceed the corresponding maximum achievable under laboratory conditions. The ideal variant of meeting the requirements could be to generate a rectangular-shape voltage pulse at the chamber input. It is desirable that the voltage could stay at the same level up to the moment of plasma shell coming to the anode end. In accordance with the data from laboratory experiments the duration of the process till the moment of shell implosion is about 3 μs. Such times of current generation in the load are common to the sources on the basis of helical explosive magnetic generators (HEMG) with explosive current opening switches (EOS), however, the shape of the voltage pulses they generate is far from rectangular. In this connection we have developed the so-called sectionalized current opening switch on the basis of the existing EOS. In fact it represents a set of series-connected opening switches or sections with programmed moments of operation. Due to variation of the sections' operation moments we can change the shape of voltage pulse on the load within the time interval of pulse durations from 1 to 5 μs. the equivalent electric circuit of sectionalized opening switch is presented in Figure 4. Structural layout of the sectionalized EOS is given in Figure 5.

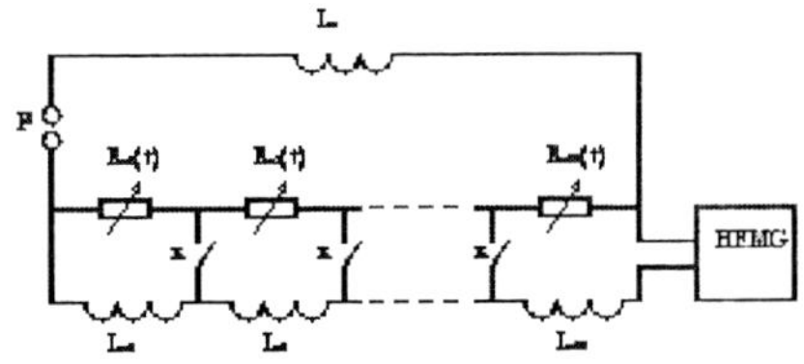

Fig. 4. Electric diagram of sectionalized EOS.

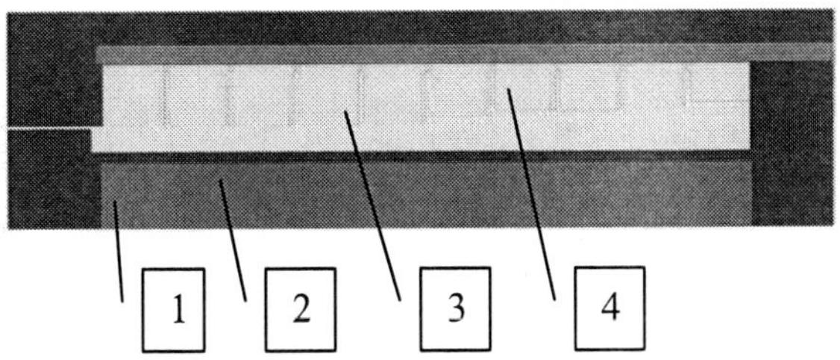

Fig. 5. Sketch of sectionalized EOS.
1 – central current conductor, 2 – HE charge, 3 – jet-former, 4 – breakable foil.

The EOS sections' operation moments can be varied due to a time diversity of impact of central conductor 1, expanding under the effect of HE charge 2, on jet-former 3; the inner diameter of the jet-former is different for different sections. In the calculations of the sectionalized EOS for the experiment with the plasma focus chamber we used the time dependence of its inductance obtained on the capacitor bank (see Figure 3 (b)). Prior to explosive plasma experiment several preliminary explosive experiments with the fixed load were carried out. The main objective of preliminary experiments was to check the calculations under conditions close to those realized in the explosive experiment with the plasma focus discharge chamber: maximum current ~ 1,4 MA achieved during 2 ÷ 3 μs, maximum load voltage does not exceed 30 kV.

RESULTS OF EXPLOSIVE EXPERIMENT

In the explosive experiment the equal-component deuterium-tritium mixture was used as the working gas in the plasma focus discharge chamber at the total pressure of about 8 torr. The helical EMG 200 mm in diameter with sectionalized EOS was used as the current generator. The total number of the EOS sections was 10. The chamber anode was connected to the EOS potential electrode at the moment of its operation through an uncontrolled solid discharger made in the form of a lavsan film 50 μm thick. All electrical characteristics necessary to get a comprehensive idea of operation of the helical generator and of sectionalized EOS were recorded in the experiment. To measure the current in the plasma focus chamber, the induction probes distributed uniformly across the azimuth at the chamber input were used. The measurements of the penetrating radiations were done with the use of three scintillation detectors SSDI-8 [3]; the integral neutron yield was also recorded by a set of activation detectors from aluminum, copper, molybdenum, indium, Teflon and zinc. The activation detectors made in the form of disks 20 mm in diameter and 1 mm thick were placed inside the cadmium cover and attached to the end part on the plasma focus chamber case. After the experiment the activation detectors were removed from the cadmium cover, and the activation integrals were determined for every working section. The calculation of neutron yield from the source was based on the calculation of values of activation integral for each working reaction with normalization per 1 neutron of the source. The calculation was performed with the use of Monte-Carlo method; the calculation model took into account the effect of the construction elements of the explosive magnetic generator, of plasma focus chamber, of activations detectors in cadmium cover and of earth ground under the chamber. The source of neutrons in the calculations was taken to be point and isotropic. The neutrons energy spectrum in the calculation was specified in accordance with the data from Ref. [4], [5]. Scintillation detectors SSDI-8 were installed at the angle of 20° to the axis of experimental assembly at a distance of 58,33 m from the source. The detectors had three different thresholds of sensitivity for the measurements to 10^{11}, to 10^{12} and to 10^{13} n/pls. Figure 6 shows the oscillograms of the scintillation detectors D2 and D3 having sensitivity to $10^{12}$and to 10^{13} n/pls, respectively.

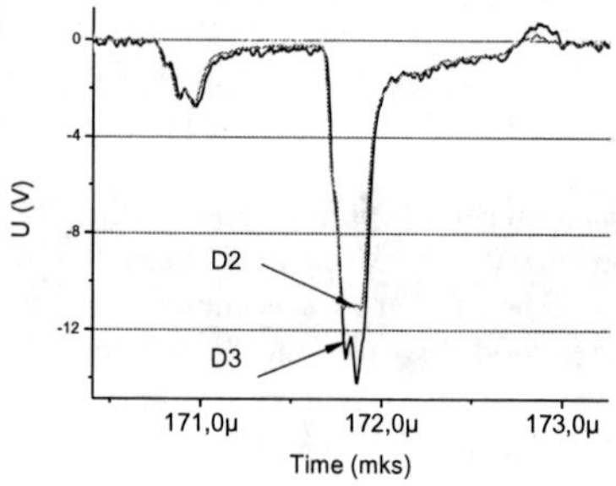

Fig. 6. Oscillograms of scintillation detectors.

In accordance with the recorded signals the time-of-flight interval between the X-ray and neutron pulses was 939±6 ns that corresponds to the neutrons energy of 14,16±0,15 MeV. Duration of the neutron pulse at half-height is ~ 185 ns. For comparison, the duration of pulse of DD-neutrons in the laboratory experiments was about 110÷130 ns. The integral neutron yield calculated on the basis of signals from scintillation sensors was $1{,}42\cdot10^{12}$ n/pls. The inaccuracy of measurements of the neutron yield on the basis of time-of-flight technique is about 13%. The data from the activation measurements give a somewhat higher neutron yield and evaluate it as $1{,}79\cdot10^{12}$ n/pls. The root-mean-square deviation based on the selection of activation detectors is 2,8%. In accordance with the same measurements the fluence of neutrons with energies within the range 0,3÷15 MeV in the zone of sensors location is $3{,}74\cdot10^{8}$n/cm^2. The average energy of neutrons within the specified range of spectrum is equal to 9,7 MeV. The neutron yield obtained in the explosive experiment is about one order smaller than the expected one. The most probable reason for this could be shunting breakdowns in the plasma focus chamber. This could happen because the voltage at the chamber input in the explosive experiment turned out to be somewhat higher than in the laboratory experiments on the capacitor bank. In Figure 7 the current curves obtained in the explosive and laboratory experiments testify to the fact of exceeded voltage.

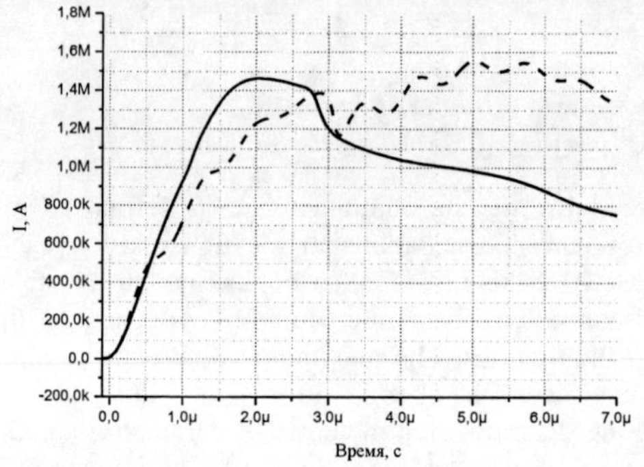

Fig. 7 Current curves from explosive (solid line) and laboratory (dotter line) experiments.

The rates of currents rise agree only at the initial parts of the current curves. Further, the current curve corresponding to the laboratory experiment has a much smaller slope than the current curve from the explosive experiment. The current in the explosive experiment reached the maximum of 1,46 MA 2,04 µs after the beginning of discharge. The oscillogram shows that the implosion of the shell at the anode end starts 2,74 µs after the beginning of discharge. By this moment the current drops to 1,4 MA. Under similar conditions in the laboratory experiments with deuterium filling the current reached the maximum of 1,38 MA by the moment of 2,88 µs after the beginning of discharge.

The obtained results testify to a necessity of making corrections to the conditions of the next explosive experiment planned to be conducted with similar plasma focus chamber. Firstly, it is necessary to change the sectionalized EOS to reduce the voltage at the plasma focus chamber' input. Secondly, the corrections should be introduced into the design of the chamber in order to increase the electric strength of its inner areas. To do this, it is planned to try several configurations of cathodes of "squirrel cage" type. It is possible that due to this it would be possible to somewhat reduce the effect of shunting breakdowns.

CONCLUSION

The technique for preparation and realization of experiments on plasma focus chamber powering by the explosive magnetic generator has been perfected. The explosive experiment with deuterium-tritium filling of the plasma focus chamber has been conducted. The neutron yield exceeding 10^{12} neutrons has been achieved. The experimental results have been used to determine the directions of further work on bringing the neutron yield to 10^{13} neutrons.

References

[1] J.W. Mather, P.J. Bottoms, Characteristics of the Dense Plasma Focus Discharge., in Phys. Fluids, Vol. 11, № 3, 1968, pp.611-618.

[2] A.I. Veretennikov, K.N. Danilenko, Tools for diagnostics of a single pulse of radiation, Moscow: Publ. House AT, 1999, p. 45.

[3] V.K. Chernyshev, G.I. Volkov, V.A. Ivanov et al., Explosives Opening Switches for Fast-Operating Helical EMG in Proc. Ninth IEEE Pulsed Power Conference, Albuquerque, New Mexico, USA, 1993.

[4] V.V. Bragin, V.V. Glushikhin, V.I. Golubev, V.M. Gorbachev et al., Study of anisotropy and time dependence of spectrum of DT-neutrons in plasma focus, VANT, series "Physics of nuclear reactors" TIYaS-XI, Special issue, 1997, pp.153-157.

[5] V.I. Golubev, N.G. Makeev, E.S. Pashchenko et al., Study of spectral distribution of DT-neutrons in plasma focus, in Letters to ZhETF, 1983, 37, #1, pp.18-21.

Study of rheological properties of materials with the use of explosive magnetic generators

Baranov V.K., Duday P.V., Dudin V.I., Glybin A.M., Nadezhin S.S., Tagirov R.R., Tyupanova O.A., Zimenkov A.A.
Russian Federal Nuclear Center
Sarov, Nizhniy Novgorod region, Russia
e-mail: duday@elph.vniief.ru

Abstract- **The availability of a big class of pulsed power sources on the basis of helical explosive magnetic generators (HEMG) developed over a period of the last decades at RFNC-VNIIEF provides considerable opportunity to study the rheological properties of materials under the effect of shock-wave and quasi-isentropic loading over a wide range of variations of the amplitude-time characteristics of compression and tension pulses. The use of the magnetically driven cylindrical liners-impactors gives a unique chance to explore the dynamic and spall strength of materials and the processes of recollection and ejecta under axisymmetric loading.**

The paper will give an overview of studies of the rheological properties of various materials carried out with the use of the pulsed power sources on the basis of helical EMG. The results of the experimental series «*R-Damage-0-9*», realized jointly by LANL and VNIIEF in 2004-2011, will be presented [2]. The goal of the experiments was to study the processes of initiation and evolution of a spall damage and subsequent recollection of aluminum by a converging liner. The paper will present the setup and the results of the first experiment exploring the shear strength of beryllium in cylindrical geometry under shock-wave loading. The configuration of a pilot series of experiments studying the process of lead ejecta under shock loading of cylindrical samples will be demonstrated. It is planned that the experiments will investigate the effect of the inner surface finish on the velocity and mass of ejecta cloud in case of the shock wave amplitude variation.

Keywords—rheological properties, explosive magnetic generator, spall strength, recollection, ejecta.

I. INTRODUCTION

In different high-energy processes the solid bodies are exposed to the action of intensive pulsed loadings. The reaction of the structural materials (metals and their alloys) to such kind of action is an individual characteristics of each of them and, as a rule, it differs drastically from the reaction to static and quasi-static loadings in qualitative and quantitative sense. In this relation the study of the rheological properties of materials, such as dynamic and spall strength, and of the processes of recollection and ejecta with the use of different experimental techniques and modern systems of registration is at the moment a promising and topical problem.

Nevertheless, in the world research laboratories the method of impactor drive by magnetic field in a classical Z-pinch scheme is being developed.

The method of impactor drive by magnetic field pressure, as applied to the studies of rheological properties of materials, possesses a number of important advantages:

- a possibility to set the current pulse of required shape (amplitude, rise and droop time, duration) means the availability of a broad range of characteristics of the impactor drive;
- high symmetry of loading conditioned by the method of magnetic field generation and the absence of additional factors;
- "transparency" of magnetic field allows recording the velocity not only of the inner but also of the outer surface of the impactor;
- recovery of the explored targets for metallographic analysis of their post-shot condition is possible without any additional methods.

The experiments with the liner systems driven by magnetic field add to and verify the research results obtained with the use of HE, gas guns and laser facilities.

The availability of a big class of various pulsed power sources on the basis of helical or disk explosive magnetic generators [1,2], developed in RFNC-VNIIEF in the course of several decades, offers considerable possibilities for the studies of the behavior of materials under the effect of shock-wave and quasi-isentropic axisymmetric loading in a wide range of variations of amplitude-time characteristics of the compression and tension pulses.

II. EXPERIMENTAL TEST BENCH TO STUDY OF PROCESSES OF SPALL DAMAGE AND RECOLLECTION OF ALUMINUM

The first experience of application of the current source on the basis of helical EMG to study the rheological properties of materials was the experimental series «R-Damage-0-9», conducted jointly by the teams from LANL and VNIIEF in 2004-2011 [3]. The objective of those experiments was to study the features of nucleation and development of spall damage and of recollection of the damaged medium under conditions of axisymmetric loading of the targets from aluminum ALCAN by a converging liner.

In the described series R-Damage two hollow cylindrical targets were loaded by an impact of cylindrical liners driven by the electromagnetic field produced by the explosive magnetic pulsed power source (EPPS).

EPPS comprised the helical explosive magnetic generator (HEMG), the explosive current opening switch (OS), explosive current interrupter (CI) and the coaxial transmission line (TL) along which the current from the explosive magnetic device is delivered to the liner load (Figure 1).

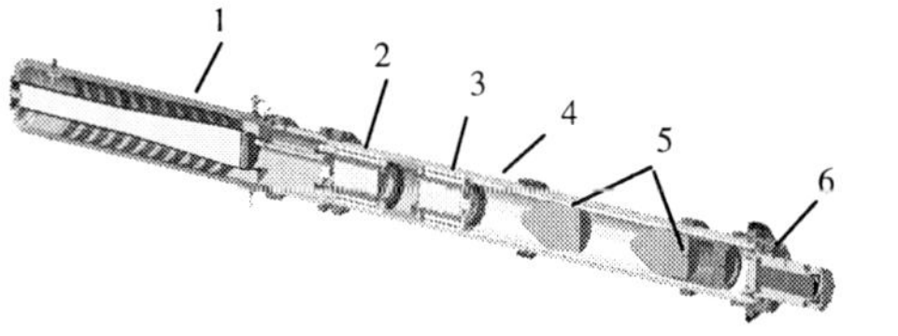

Fig. 1 Explosive magnetic pulsed power source

1 – helical explosive magnetic generator (HEMG), 2 – explosive current opening switch (OS), 3 – explosive current interrupter (CI), 4 – coaxial transmission line (TL), 5 – LPU protection system elements, 6 – liner load

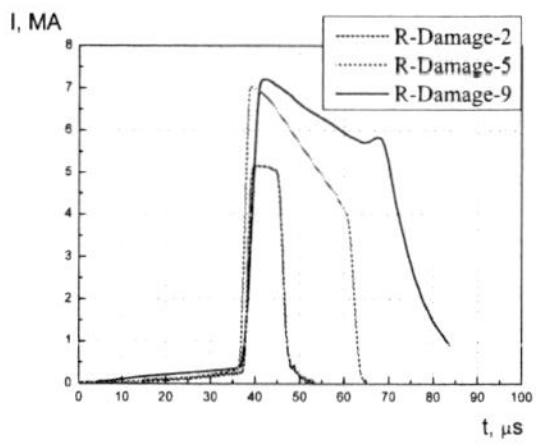

Fig.5 Characteristic current pulses in experiments R-Damage

With the use of the explosive opening switch we shaped the leading front of current pulse in the liner load, while the explosive interrupter, in its turn, provided current interruption in the liner load and shaping of the current drop profile. At this by setting the specific moments of their operation we can shape a quasi-trapezoidal pulse of current with required duration in the liner [4].

To protect the liner load from debris and shock action of the helical EMG and explosive current opening switches, the system of protection [5] that allows preserving the liner unit in the explosive experiments has been developed and tested. The use of the load protection system made it possible to recover the targets and to realize metallographic analysis of their cross sections in order to evaluate the nature and the degree of damage.

The main source of information on the processes occurring during liner-target the interaction were the time dependences of the velocity of motion W(t) of their inner surface recorded by the VISAR [6] and PDV [7] systems.

In the experimental series «R-Damage» LANL/VNIIEF team obtained for the first time a representative amount of experimental information on the processes of initiation, growth, full and partial recollection of the main spall in the studied material [8]. The data provided a possibility to verify the numerical models of spall damage and to offer the recollection model [9].

The conducted experimental series R-Damage demonstrated that the pulsed power sources on the basis of explosive magnetic generators are a promising tool for a creation of dynamic loads with controlled amplitude and duration in the constructional materials. Successful realization of the experimental series «R-Damage» gave a new stimulus to the studies of the rheological properties of constructional materials with the use of EMG.

III. Study of the Shear Strength of Beryllium

In recent years the world scientific community has displayed an increased interest in such material as beryllium, showing a number of unique properties and characteristics under conditions of dynamic and pulse loading [10]. Few studies have revealed high specific shear strength of beryllium, an anomalously low Poisson's ratio, high sound speed, and a complex dependence of strain on strain rate.

In order to explore the shear strength, we have used the experimental setup, in which loading of hollow cylindrical targets from beryllium is realized by deceleration of a hollow cylindrical liner from aluminum, driven by the explosive magnetic pulsed power source' pressure, on them (Figure 10). Beryllium specimens were placed inside the aluminum shells (envelops) localizing the products of beryllium destruction. In order to exclude the spall damage of the enveloped specimens, the aluminum sacrificial substrate is adjacent to them. The laser interferometric technique of "PDV" type was used in the experiment to record the velocity of motion of the outer surface of the liner compressing the entire experimental assembly and the velocity of the sacrificial substrate' inner surface.

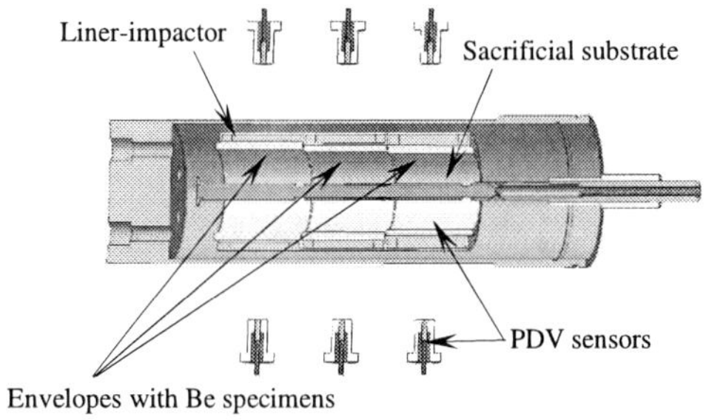

Fig. 10 Scheme of experiments to study shear strength of beryllium

The motion velocity and the radius of the liner stopping in this setup are to a large extent determined by the shear strength of beryllium at fixed impact velocity and known parameters of the liner and shell materials. In addition, recovery of the explored specimens makes it possible to estimate the maximum value of plastic strain the specimen experienced in the process of loading.

In the first experiment we obtained the dependences of velocity of the liner' outer surface and the sacrificial substrate' inner surface (Figure 12), recorded by PDV technique. Together with the measurements of the stopping radii preserved in the experiment with Be specimens (Figure 13) the presented experimental velocity curves give full information on the liner and specimens dynamics before and after the impact. This helps to determine the dynamic shear strength of beryllium in this loading range and to verify different mathematical models of the rheological properties of beryllium (Figure 14).

To date there is a number of models of the dynamic strain of beryllium. To perform numerical simulations, we used the PTW model [11] with two sets of parameters differing in the strength characteristics. Each of them (PTW 1, PTW 2) describes well its experimental data domain. Comparison between computational data and experimental results testify to the fact that the preference should be given to the data set PTW1, having lower strength parameters in the domain of strain rates $\dot{\varepsilon}\sim$ $10^3 \div 10^4$s-1 and intensity of plastic strain $\varepsilon \sim 0.3 \div 0.4$.

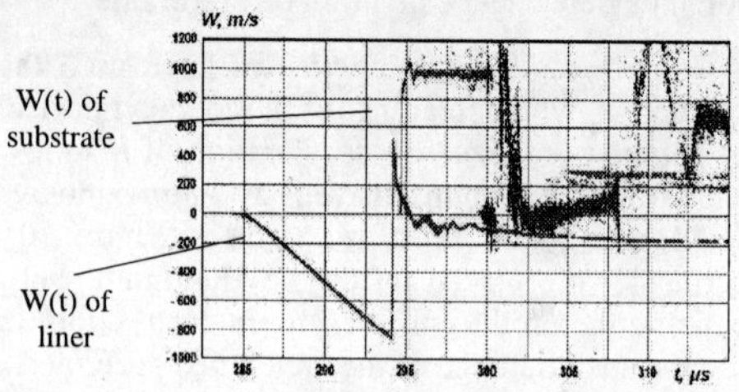

Fig. 12 Typical experimental dependences of velocity of the liner surface and sacrificial substrate (PDV technique)

Fig. 13 Be-specimen after the experiment

TABLE I. RADII OF THE ASSEMBLY STOPPING (LINER+SPECIMEN)

Specimen thickness, mm	Radius of liner inner surface, mm		
	Calculation		Experiment
	Set of parameters PTW 1	Set of parameters PTW 2	
3	25.6	27.3	25
4	28.6	29.2	26.3
5	28.8	29.4	27.7

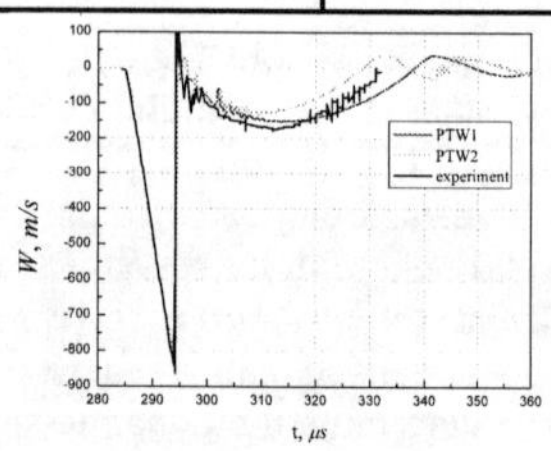

Fig. 14 Computational and experimental dependence of velocity of motion of the liner inner surface

Figure 15 presents the experimental results (strain degree from experiment, yield strength from calculation) in comparison with experimental and computational yield strength of beryllium as a function of accumulated plastic strain from Ref. [11, 12].

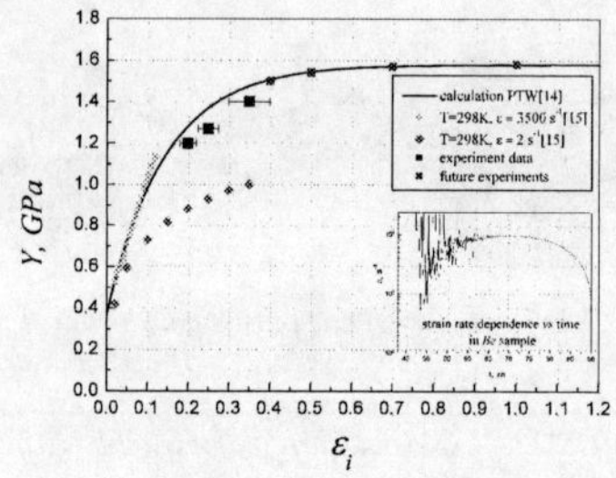

Fig. 15 Yield strength vs. plastic strain

As is seen from the presented data on the strength of beryllium at high-rate strain ($\dot{\varepsilon}\sim$ $0,5 \div 1 \cdot 10^4$ s-1), the method of studies of the rheological properties of structural metals with the use of EMG allows obtaining the strength data in the range of strong plastic strains. In the experiment the specimens of different thickness gave final strains much exceeding the limiting value obtained by standard methods of strength analysis (Hopkinson split bar method).

In the next experiment we plan to get higher levels of plastic strain $\varepsilon \sim 0.4 \div 1.0$ (Figure 15 – blue dots) while preserving the strain rate within the range $\dot{\varepsilon}\sim$ $10^3 \div 10^4$ c-1.

IV. STUDY OF EJECTA PROCESSES

Ejection of particles from the metal surface under the effect of shock wave (SW) is a complex process influenced by the combination of many factors – profile of the free surface (FS) of metal and inner defects of material, dynamic diversity and time diversity of SW, profile of SW, phase state of material and corresponding rheological properties [13,14]. The influence of each factor is a subject of a separate research. Of maximum attractiveness are the setups of experiments that allow reducing to a minimum the effect of various structural features on the experimental results.

At the present time RFNC-VNIIEF has a plan to conduct a series of experiments to study the process of matter "ejecta" from the free surface of lead under conditions of axisymmetric shock loading produced with the use of EMG. The objective of the experiments is to determine general mechanisms and quantitative characteristics of the "ejecta" process, such as velocity distribution and disperse composition of the ejected particles, depending on the parameters of surface finish, amplitude and profile of shock wave, phase state of the shock-loaded material.

The experimental configuration is given in Figure 16. It is planned to install 6 cylindrical lead specimens with different finish of the inner surface ($0 \leq A \leq 80$ мкм, $0 \leq \lambda \leq 400$ мкм) in every experiment. Precise measuring of the lead specimens' inner surface profile will be carried out before the experiments with the help of a laser surface finish gage. Shock loading will be realized with a cylindrical copper liner driven by magnetic field of the pulsed power source on the basis of a helical EMG. The parameters of shock loading (W(t) of liner) and the velocity distribution of particles in the dust cloud will be recorded by the PDV technique. In some experiments it is planned to have gas medium in the volume of the loaded specimen in order to study the dynamics of deceleration of disperse particles ejected from the surface.

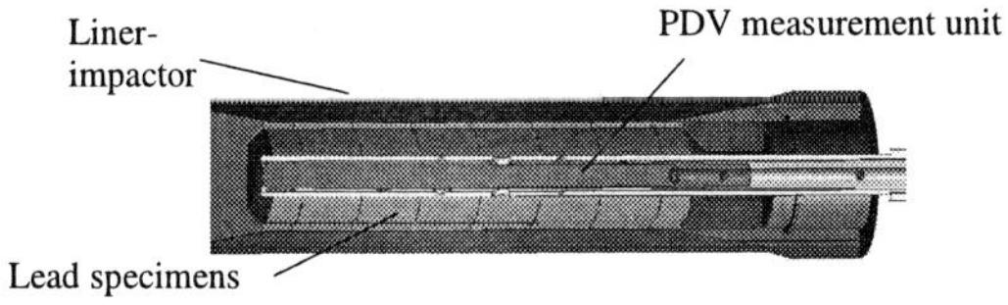

Fig. 16 Experimental configuration to study the "ejecta" process in lead

The variation of the inner surface radius in specimens from experiment to experiment makes it possible to change the velocity of the liner at the impact moment and, respectively, to shock wave pressure in the specimens within the range from 20 to 40 GPa. As it follows from the phase diagram in Figure 17, within this range the lead is able both to stay in solid phase and to go over to a liquid phase in the unloading wave. The planned series of experiments (green dots in Figure 17) will allow evaluating the effect of the phase condition of lead on the parameters of the ejecta process.

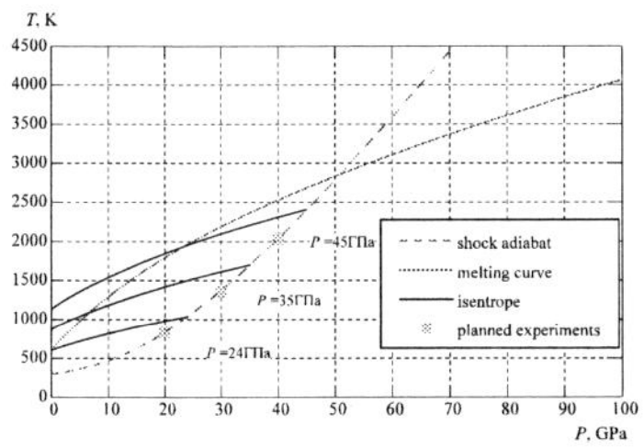

Fig. 17 Shock adiabat, lead melting line and unloading isentropes [15]. Future experiments are marked with green dots

V. CONCLUSION

The joint VNIIEF/LANL experimental series «R-Damage» has demonstrated considerable promise of application of the pulsed power sourced on the basis of helical EMG for a creation of pulsed loads with controlled amplitude and duration in the constructional materials. The representative scope of experimental data on the processes of initiation, growth, full and partial recollection of the main spall in aluminum has been obtained.

The studies of the shear strength of beryllium under conditions of shock-wave loading with the use of EMG have been successfully started. The first experiment has provided the data that have allowed verifying the model of high-rate strain in the range of strain rates $10^3 \div 10^4$ s-1.

The configuration of the pilot series of experiments studying the process of lead "ejecta" under conditions of axisymmetric shock loading produced with the help of a pulsed power source on the basis of helical EMG has been presented. The objective of the experiments is to determine velocity distribution and disperse composition of the ejected particles, depending on the parameters of surface finish, amplitude of shock wave and phase state of the material.

REFERENCES

[1] V.K. Chernyshev. Super-power explosive magnetic sources for thermonuclear and physical studies // Magagauss and megampere pulsed technology and application (edit. by V.K. Chernyshev, V.D. Selemir, L.N. Plyashkevich). Sarov, VNIIEF, 1997, p.41-58.

[2] A.V. Ivanovskiy. Complex of explosive magnetic hydrodynamics EMHD. VIII Khariton's Topical Scientific Readings. Digest of technical papers. Sarov, pp. 547-562.

[3] V.A. Vasyukov, A.M. Glybin, P.V. Duday, et al. Study of rheological properties of aluminum with the use of explosive magnetic generators // DAN (Russ: Doklady Akademii Nauk), 2013, V. 448. №3. P. 285-288.

[4] P.V. Duday, A.M. Glybin, V.A. Ivanov et al. Quasi-trapezoidal shape current pulse former. Proceedings of the 2006 International Conference on Megagauss Magnetic Field Generation and Related Topics, November 5-10, 2006, Santa Fe, New Mexico, USA.

[5] P.V. Duday, V.A. Ivanov, A. V. Ivanovsky, A.I. Kraev, et al. Experimental test bench on the basis of a helical EMG to study spall mechanism in cylindrical geometry//Digest of technical papers "VIII Khariton's Topical Scientific Readings on the problems of high energy density physics", Sarov, 21-24 March 2006, p.503-509.

[6] L.M. Barker, R. E. Hollenbach, "Laser interferometer for measuring high velocities of any reflecting surface" J.Appl.Phys. 43 [11] 4669-4678 (1972).

[7] D.B. Holtkamp. // Proceedings of 2006 International Conference on Megagauss Magnetic Field Generation and Related Topics. Santa Fe: Omnipress, 2006, P. 119-128.

[8] A.M. Kaul, A.V. Ivanovsky, W.L. Atchison, A.A. Petrukhin, P.V. Duday et al. Damage growth and recollection in aluminum under axisymmetric convergence using a helical flux compression generator // JOURNAL OF APPLIED PHYSICS 115, 023516 (2014).

[9] S.S. Nadezhin, O.A. Tyupanova, P.V. Duday, R.E. Reinovsky, A.M. Kaul, D. Holtkamp et al. "Numerical analysis of nucleation, growth and recollection of damage in aluminum under axis-symmetric convergence," Intern. Conf. "Shock Waves in Condensed Matter", Kiev, Ukraine, 2012, Pp. 298-301.

[10] L.M. Hull, J.R. Faulkner, M.E. Briggs, E.N. Ferm. Explosively Driven Damage in Beryllium // The 2009 APS Shock Compression of Condensed Matter..

[11] Dean L. Preston, Davis L. Tonks, and Duane C. Wallace. Model of plastic deformation for extreme loading conditions. J. Appl. Phys., 93:211-220, 2003.

[12] M. Fugate, B. Williams et. al. Hierarchical Bayesian analysis and the Preston-Tonks-Wallace model. Los Alamos Technical Report, LA-UR-05-3935.

[13] V.A. Ogorodnikov, A.G. Ivanov, A.L. Mikhailov et al. Particles ejection from metal free surface during shock wave release to it and the diagnostics methods. Physics of combustion and explosion, 1998, v.34, №6, p.103-107.

[14] J.R.Asay, L.P.Mix, F.C.Perry. Ejection of material from shocked surfaces. Appl. Phys. Lett. V.29, №5, p.284-287, 1976.

[15] D.G. Gordeev, L.F. Gudarenko, A.A. Kayakin, V.G. Kudelkin. Semi-empirical model of equations-of-state of metals with effective recording of ionization. // Questions of atomic science and technology. Ser. Theoretical and applied physics. 2010. Iss. 3. p.19-25.

Experimental Test Bench on the Basis of a Small-Class Disk Explosive Magnetic Generator to Study Processes of Turbulent Mixing and Material Properties.

E.I. Bochkov, B.E. Grinevich, P.V. Duday, B.T. Egorychev, A.V. Ivanovskiy,

K.N. Klimushkin, A.I. Krayev, V.B. Kudelkin*, V.I. Mamyshev, Yu.I. Mattsev, S.M. Polyushko, E.V. Shapovalov,

A.A. Zimenkov

Russian Federal Nuclear Center –All-Russia Research Institute of Experimental Physics

607190 Sarov, Russia.

Abstract **– The paper will present the circuit diagram of the test bench. The results of calculations of operation of a disk EMG with an electrically exploded current opening switch will be described. The calculation results will be compared with the experimental results. The test bench possibilities to get high-precision equations-of-state and the rheological properties of materials will be demonstrate. The parameters of X-ray radiography will be determined.**

Key words - explosive magnetic generator; solid liner; current opening switch

INTRODUCTION

Many astrophysical phenomena and the operation of some engineering equipment are characterized by intensive turbulent mixing (TM) of compressed media under the effect of high gasdynamic acceleration on the order of $10^6 \div 10^{15}$ g_0. The computational-theoretical description of corresponding flows is possible only when using different semiempirical models. To calibrate and verify these models, the experimental data obtained at low acceleration on the order of $10^0 \div 10^2$ g_0 should be used.

In this connection of interest are the experiments studying radiographically the TM zone developed from deceleration of a thin copper plate with initial velocity of 10 km/s by a massive layer of light material (polyethylene or water). The plate is driven by the EMG current [1]. In this case it is possible to provide the initial velocity of the plate up to 20 km/s and higher.

The paper presents a circuit diagram of the test bench on the basis of a small-size DEMG with flat disk modules [2-4], equipped with a low-inductive electrically exploded current opening switch [5,6] and meant to explore the processes of turbulent mixing under conditions of loading by cylindrical liners.

1. TEST BENCH CIRCUIT DIAGRAM

The proposed scheme of the test bench designed to form the current pulses with the amplitude to 10 MA and rise front ~ 2 µs is shown in Fig. 1.

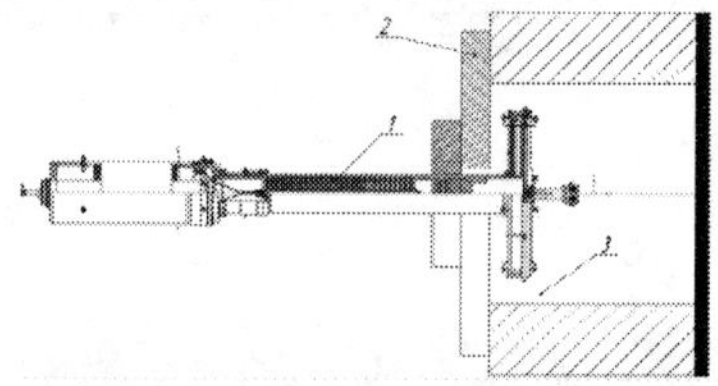

Fig. 1. Test bench schematic: 1 – pulsed power source on the basis of 30-element disk EMG Ø250 mm with low-inductive explosive current opening switch (LECOS) and liner PU; 2 – protection system; 3 – place for X-ray devices and film cassettes.

The schematic of the pulsed power source is shown in Figure 2.

The initial current in the DEMG circuit is produced by the helical EMG. All HE charges of the disk elements in the DEMG are initiated along the axis at the moment of HEMG current maximum. At the same time HEMG is disconnected from DEMG by means of an explosive disconnection unit DU. The neighbor disks implode under the effect of the explosion products, thus delivering the magnetic flux to the wave line. Current increase in DEMG circuit results in an electrical explosion of the foil in LECOS and in operation of the explosive closing switch. At this the magnetic energy goes to the ponderomotive unit PU.

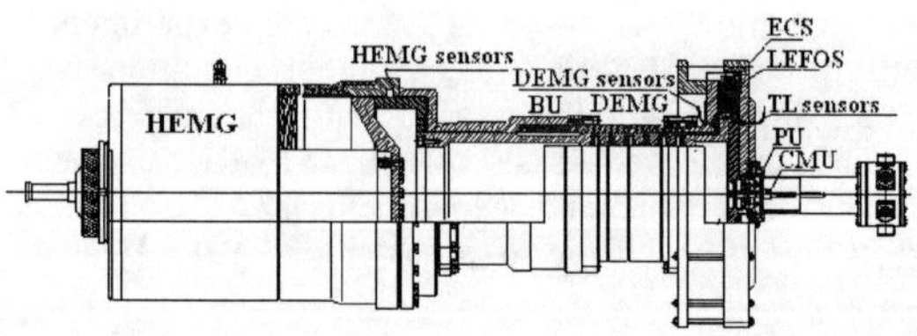

Fig. 2. Pulsed power source schematic:
HEMG – helical EMG; DU – disconnection unit; DEMG – 30-element disk EMG; ECS – explosive closing switch; LECCS –low-inductive electrically exploded current closing switch; PU – ponderomotive unit.

2. Testing of Pulsed Power Source Elements

Prior to conducting full-scale experiments on the test bench it was necessary to check the operability of elements of the explosive magnetic device. The explosive experiments were realized with the current source models comprising 10 disk elements ∅250 mm.

A. Testing of 10-module DEMG ∅250

The scheme of the experiment to check 10-module DEMG ∅ 250 mm and the photo of the source before the experiment are shown in Fig. 3.

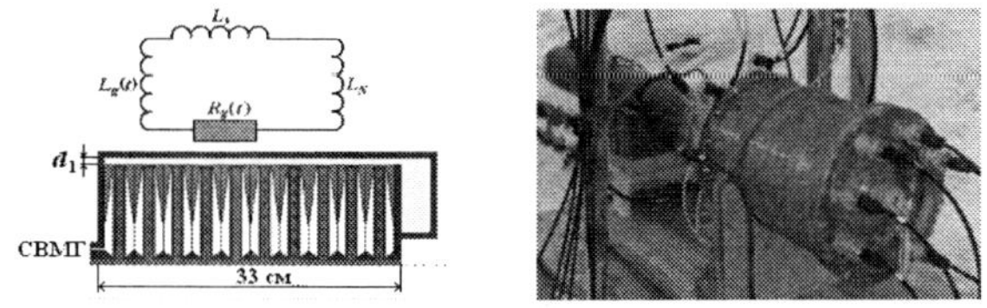

Fig. 3. The scheme of the experiment and the photo of the source before the experiment

The DEMG feeding current in the experiment was 4,0 MA. The DEMG current in the inductive load (L_L=1,87 nH) exceeded 40,0 MA [7]. The calculated and the experimental load current waveforms as a function of time are presented in Figure 4. The reconstructed dependence of resistance in the 10-element DEMG as a function of time taking into account the calculated dependence of the generator inductance as a function of time is given in Fig. 5.

Fig. 5 shows the conventionalized dependence of resistance I the 10-element DEMG on time $R_{int}(t)$, that will be used to perform the calculations.

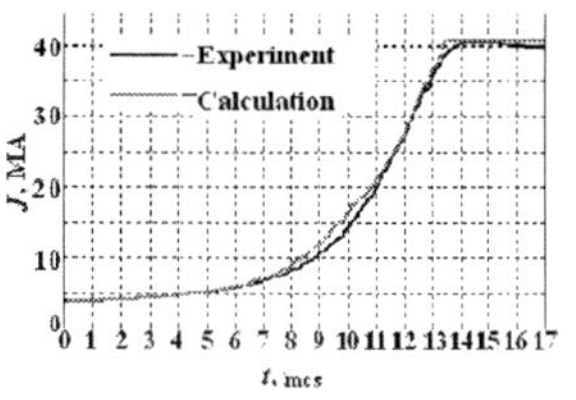

Fig. 4. Calculated and experimental generator current as a function of time

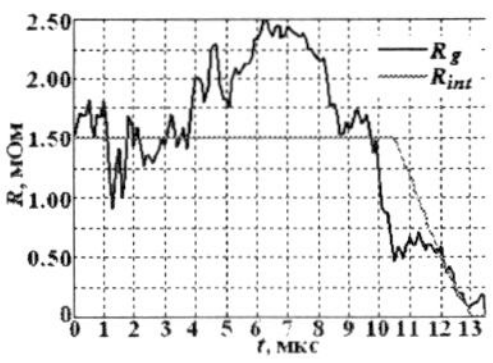

Fig. 5. DEMG resistance as a function of time and its interpolation R_{int}.

B. Testing of 10-module DEMG ∅250 mm with LECOS

To check a possibility to form the current pulse with duration ~ 2 μs, the experiment was conducted with 10-element DEMG and LECOS. The scheme of the experimental device to test the operation of the pulsed power source comprising DEMG and LECOS, and the photo of the source before the experiment are presented in Fig.6.

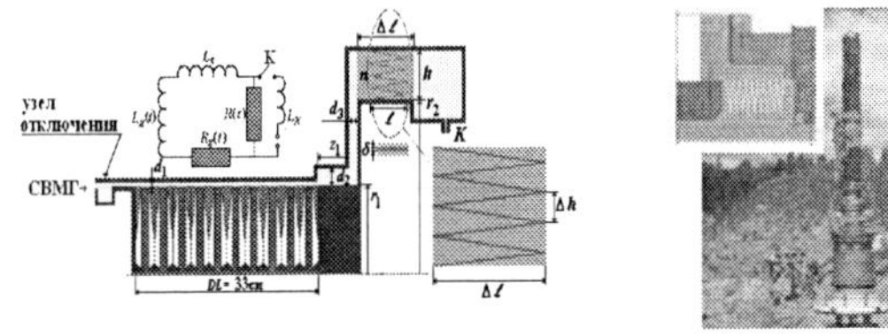

Fig. 6. Scheme of experiment and photo of the source comprising DEMG with LECOS before the experiment

The DEMG initial current was J_0=4.6 MA. The load inductance was L_N=4.68 nH. In the experiment the LECOS with n = 4 "teeth" was used (see Fig.6): foil length ℓ_f=19.4 cm, total foil width h_f=66 cm, foil thickness δ=0.015 cm.

In the experiment the explosive closing switch (K) connecting the load was used. The switching time in the experiment was t_0=9 μs. The DEMG current was ~ 20 MA, the current in the load was ~ 10 MA with characteristic rise time 2,2 μs [5]. The experimental and calculated currents as a function of time are given in Fig. 7.

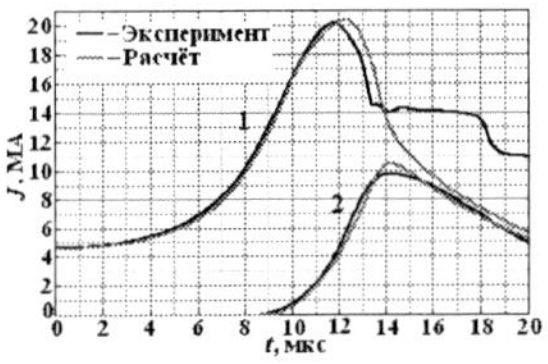

Fig. 7. Experimental and calculated currents as a function of time: 1-currents in the DEMG circuit; 2 – currents in the load

As is seen, the calculation model describes satisfactorily the experimental data. Divergence of currents in the DEMG circuit at the times t > 14 μ, can, apparently, be related to the damage of the induction sensors.

3. Optimization of Source Parameters

Copper liner with initial thickness δ_ℓ=0.4 cm was taken to perform the calculations. The length ℓ_f (number of teeth n) and length δ of the foil were optimized. The optimization was based on liner velocity at the moment of its coming to the radius r_f=2 cm. At this the effective thickness of the liner δ_{eff}, determined from the ratio $(r_3+\delta_\ell)^2-r_3^2=(r_\ell+\delta_{eff})^2-r_\ell^2$, makes δ_{eff}=0.23 cm.

Fig. 8 shows the calculated velocity of the liner as a function of the foil thickness at its length ℓ_f=30 cm (n=6). It is clear that the maximum velocity v=9.14 km/s on the radius $r=r_f$=2 cm is realized at δ=60 μm.

The calculation demonstrated that this foil thickness is optimal when ℓ_f is varied within the range 15 cm≤ℓ_f ≤50 cm. Velocity as a function of foil length is presented in Fig. 9. The maximum is observed at ℓ_f=30 cm.

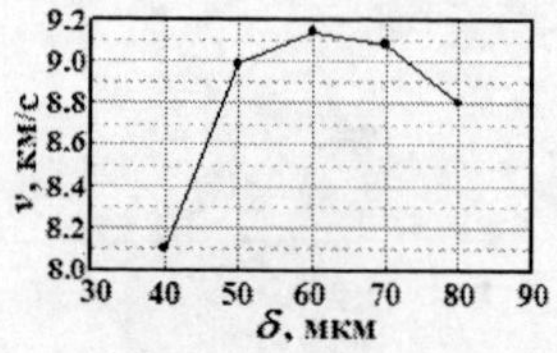

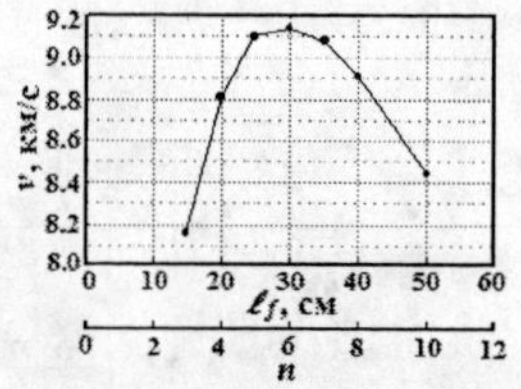

Fig. 8. Liner velocity as a function of foil thickness at its length of 30 cm.

Fig. 9. Liner velocity as a function of foil length at its thickness of 60 μm.

So, the following foil parameters are the optimal:

– length 30 cm (number of teeth n=6);

– thickness 60 μm..

The maximum voltage on the foil is V_m=128 kV. At this voltage the electric field intensity in the wave lines does not exceed 85 kV/mm. The maximum current in the liner system is J_m=45 MA. The time of current rise to the level 0.1-0.9 is $\tau_{0.1\text{-}0.9}$=2.2 μs.

In the result of the electric explosion the foil expands by a factor of 15. The beginning of electric explosion correlates with an abrupt increase of voltage on the foil.

The liner acceleration is effective to the radius r ~2 cm.

At the initial liner thickness δ_l=0.3 cm and DEMG feeding current J_0=6.5 MA, the maximum current was 42,9 MA, liner velocity at r=2 cm was v=10 km/s.

4. Two-dimensional calculations of the system protecting recording equipment

Two-dimensional calculation was performed to determine the deformation of the plug protecting the equipment from shock-fragmentation effect during actuation of HE in the 30-element DEMG Ø 250 mm. The calculation model of the device is presented in Fig. 10.

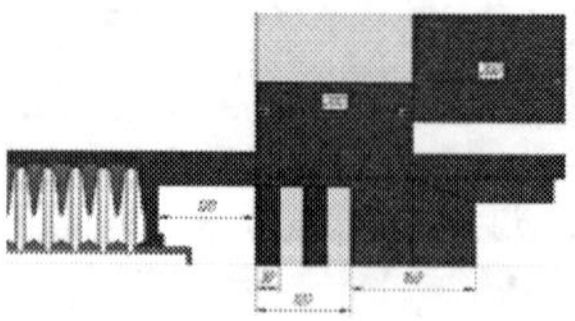

Fig. 10. The calculation model of the protection device.

To prevent the action of the flying fragments on the equipment, the alternating aluminum and polyethylene gaskets with the plug between them.

Fig. 11 illustrates the operation of the protection system.

0 ms	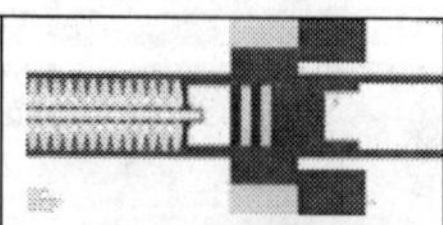	1,5 ms	
1 ms	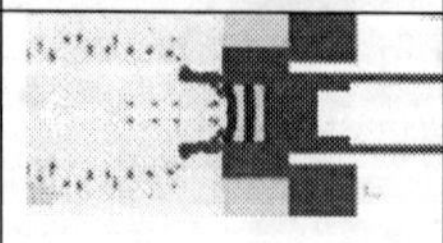	9 ms	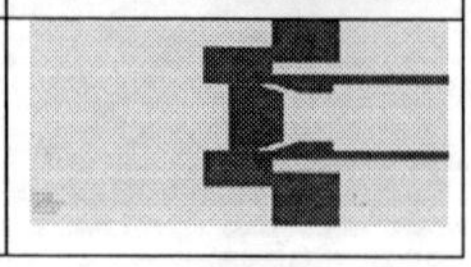

Fig. 11 Protection system operation process.

The results of the calculations demonstrate that the selected design of the protection device can provide safety of the recording equipment during the experiments. The HE mass in the first experiment is expected to be ~ 60 kg.

5. Determination of a possibility to X-ray the liner during acceleration by DEMG

A two-layered cylindrical liner was manufactured to check the capability of X-radiography of the liners in the process of turbulent mixing (see Fig. 12).

Fig. 12. Liner external view

The outer shell of the liner is made of copper, the inner shell is made of aluminum; there is no space between the shells. The aluminum shell imitated the turbulent mixing zone. Location of aluminum inside creates more rigid conditions for X-radiography than location of aluminum outside.

The liner was X-rayed by the unit "Argument-700". Similar units are planned to be used in the experiments to study turbulent mixing. The minimum focal distance (L_f=300 mm) was taken in order to get contrast and sharp image, at this the distance from the liner to the X-ray film was 50 mm. In addition, a barrier from sheet aluminum 5, 10 and 15 mm was placed in front of the liner (see Fig. 13a).

There is a plan to place a layer of polyethylene 50 mm thick between the X-ray units and the liner. The zone of turbulent mixing will appear in the process of liner deceleration against polyethylene. X-raying goes perpendicular to the surface of aluminum sheets. Fig. 13b shows the obtained liner image.

a) b)

Fig. 13 Two-layered liner with aluminum sheets: a) before X-raying; b) X-ray image of the liner.

In the image one can clearly see the outer and inner boundaries of the liner; copper-aluminum interface is not seen. Processing of the X-ray image made it possible to get sharper image of the two-layered liner (see Fig. 14a and 14b). The boundary between the layers in the image is clearly seen even in case the liner is screened by 15-mm aluminum.

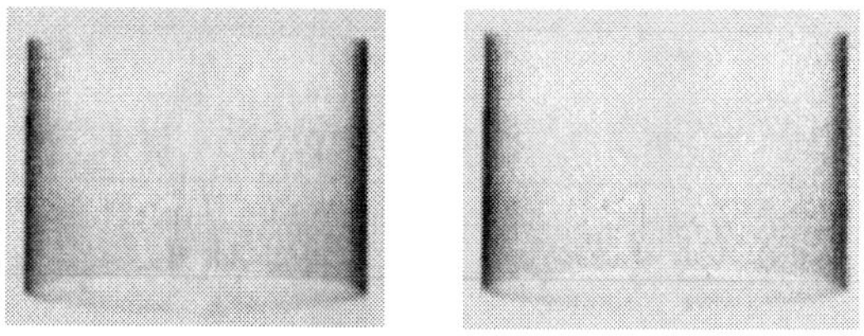

a) б)

Fig. 14. Liner image after processing:

a) filter 1; b) filter 2

By this is meant that units "Argument-700" can be used in the experiment to determine the zone of turbulent mixing in the process of electromagnetic drive of copper liner 1,5 mm thick to the velocity of ~10 km/s.

Conclusion

Design of the experiments to explore the process of turbulent mixing at the boundary of impact of high-velocity bodies has been determined. It has been demonstrated that the application of the pulsed power source on the basis of a small-size disk explosive magnetic generator to drive copper liners to velocities of 10 km/s and higher holds much promise.

The results of testing of the source model comprising 10 disk elements ∅ 250 mm and LECOS have been presented.

Gas-dynamic calculations of the multi-element DEMG ∅ 250 mm with low-inductive electrically exploded current opening switch (LECOS) designed to drive the copper shell 1.5 mm thick to a velocity of ~10 km/s have been performed. The protection device providing safety of the diagnostic equipment has been calculated. A possibility to record a layer of turbulent mixing has been demonstrated.

References

[1] A.V. Ivanovsky, High-velocity flat impactors driven by explosive magnetic generator'current, in Digest of papers of VII Khariton Scientific talks, Sarov, VNIIEF, 2005, pp. 506-511.

[2] V.K. Chernyshev and V.V. Vakhrushev, Disk EMG for fusion target compression study on models, Megagauss Magnetic Field Generation and Pulsed Power Applications, New York, USA, 1994, pp. 547-549.

[3] V. K. Chernyshev, V.V. Vakhrushev., V.B. Kudelkin, V.I. Shpagin, Small class (250mm) disk EMG testing result, 13th IEEE Int. Pulsed Power Conf. and 28 th IEEE Int. Conf. Plasma Science, Las Vegas, Nevada, USA, 2001, pp. 971-973.

[4] E.I. Bochkov, B.E. Grinevich, P.V. Duday, B.T. Egorychev, A.V. Ivanovsky K.N. Klimushkin, A.I. Kraev, V.B. Kudelkin, V.I. Mamyshev, Yu.I. Mattsev, E.V.Shapovalov, On possibilities of application of small-class DEMG to study implosion of solid liners // International Conference XVIII Khariton's Topical Readings. Problems of high-energy density physics. Book of Abstracts, Sarov, RFNC-VNIIEF, 2016, p. 187.

[5] Journal "RFNC-VNIIEF Main Achievements in 2007". Publ.House of RFNC-VNIIEF, Sarov, 2008, pp. 37-38.

[6] M.Yu. Aryutkin, B.E. Grinevich and et al., Experimental Test Bench on the basis of disk EMG ∅ 250 mm to study properties of materials under conditions of loading by cylindrical liners, MG-XIII, July 2010, China, pp. 593-599.

[7] M.Yu. Aryutkin, B.E. Grinevich, B.T. Egorychev, V.B. Kudelkin, A.I. Kraev, K.N. Klimushkin, V.I. Mamyshev, Yu.I. Mattsev, Family of disk EMG with flat disk elements, in RFNC-VNIIEF Digest of technical papers, Research Edition, Issue 18, Sarov.

Formation of Current Pulses with the Amplitude to 10 MA and Rise Time ~100 ns in Experiments with Explosive Magnetic Generator.

A.A. Bazanov, P.V. Duday, A.A. Zimenkov, A.V. Ivanovskiy, G.V. Karpov, A.I. Krayev, V.B. Kudelkin, V.I. Mamyshev, S.M. Polyushko, E.V. Shapovalov.

Russian Federal Nuclear Center - VNIIEF

607190, Sarov, Russia

Abstract - **The paper will present a circuit diagram of the device. The results of model experiments on electrophysical facility CASCADE and of an explosive experiment with a helical explosive magnetic generator (EMG) will be described. The program of works to test an explosive magnetic energy source able to form megampere current pulses with rise front ~ 100 ns will be given.**

***Keywords* -**

INTRODUCTION

The explosive magnetic generators (EMG) are able to produce the currents with the amplitude of several hundred megaamperes for the time of dozens of microseconds [1-4]. At this the level of magnetic energy delivered to the load can be hundreds of megajoules. Generation of pulses with such energy is not possible by any other means. The application of EMG turns out to be promising for the research at high energy densities.

Realization of such studies requires high energy level and small area of its localization, determined by the generation time. To peak the DEMG current pulse to ~ 1 μs, the foil current opening switches (FOS) – explosive (with the use of explosives (HE)) [5] and electrically exploded [6] - have been developed at VNIIEF. The principle of operation of the electrically exploded FOS is based on the phenomenon of increase (by a factor of ~100 and more) of resistance of the metal foil at the electrical explosion.

Further advance to the area of high energy densities is possible in case of current pulse peaking to ~100 ns. Such devices can be realized with two cascades of current peaking.

With short (~100 ns) current pulses the range pressures attainable with the use of EMG will become close to the range realized on laser facilities. This will allow posing the problem of achieving thermonuclear ignition with the use of EMG [7].

The paper is devoted to a solution of the problem of EMG current peaking to ~100 ns.

1. SCHEMATIC CIRCUIT OF THE DEVICE

To form the pulses of current with the amplitude to 10 MA and rise front ~100 ns, it is suggested that a small-class thirty-element disk EMG [2,4] with two cascades of current peaking [7] be used. Both cascades include an electrically exploded FOS in the form of a "serpentine" [8]. The proposed facility diagram is shown in Fig.1.

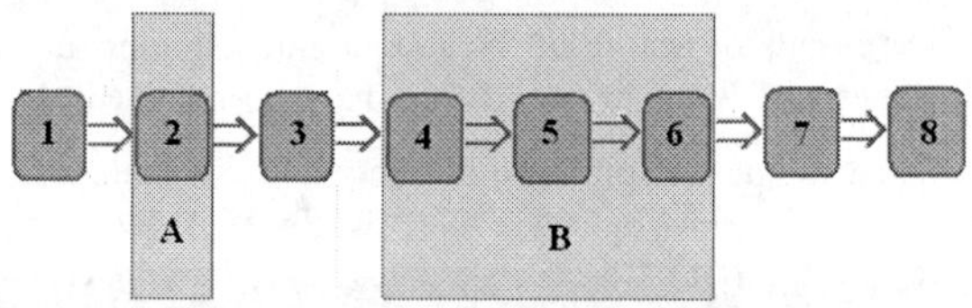

Fig. 1. Facility diagram: 1 – small-class disk EMG; 2,4 – current switches; 3 – inductive storage; 5 – bushing insulator; 6 – vacuum discharger; 7 – vacuum wave line; 8 – load (A and B – first and second cascades of current peaking).

The scheme of the first cascade of current peaking is shown in Fig.2.

Helical generator (1) produces the initial current in the DEMG (2) circuit. All HE charges are initiated simultaneously. The neighbor disks collapse under the effect of the explosion products thus bringing the magnetic flux to the wave line (3). The rise of current in the DEMG circuit results in an electrical explosion of the foil (4) and operation of explosive switch K1. At this the magnetic energy is delivered to the inductive storage (5).

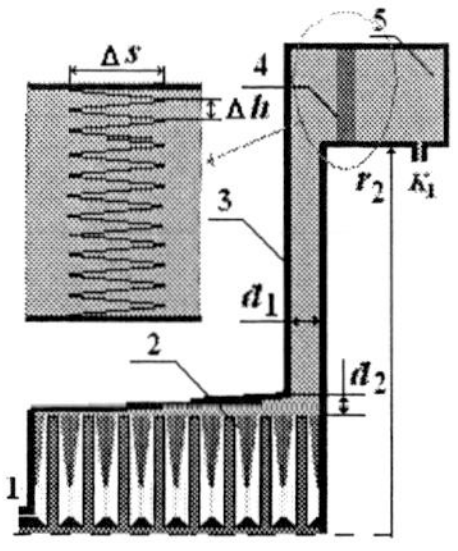

Fig. 2. Scheme of the first cascade of current peaking: 1. – helical EMG; 2. –30-element disk EMG; 3. – wave line; 4. -foil; 5. – inductive storage; K_1-explosive closing switch.

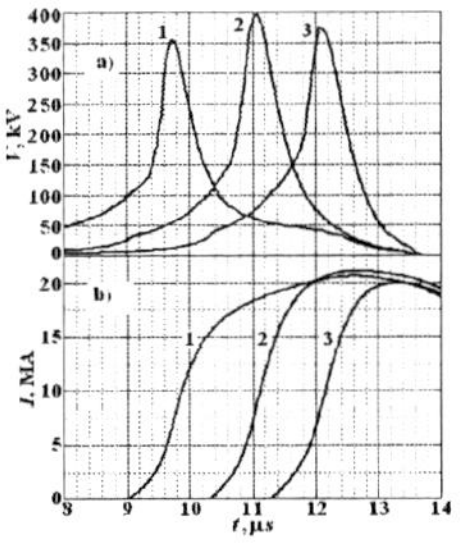

Fig. 3. Time dependences of voltage on the foil –a) and current in the inductive storage – б) at different foil thicknesses:
1. δ=30 µm; 2. δ=40 µm; 3. δ–50 µm.

The radius of foil location has been taken to be equal to r_2=50 cm, inductances L_W and L_{ind} of the wave line and inductive storage 10 and 15 nH, respectively. Length ℓ_f and thickness δ of the foil have been varied.

Fig. 3 presents the calculated time dependences of voltage on the foil and of current in the inductive storage at the length ℓ_f =1 m for several thicknesses δ. The maximum current in the inductive storage and voltage on the foil are realized at δ =40 µm. this foil thickness is considered to be optimum. The calculations demonstrated that the optimum thickness of the foil does not depend on its length ℓ_f, and the voltage on the foil is maximum at ℓ_f=1 m.

The inductance of the wave line L_W =10 nH taken in the calculations has been provided by a selection of thicknesses d_1 and d_2 (Fig. 2). At this the maximum intensity of the electric field in the wave line' dielectric has been ≤50 kV/mm.

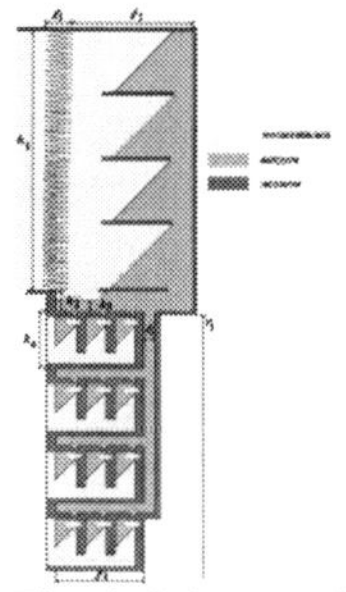

Fig. 4. Scheme of second cascade of current pulse peaking in a small-class disk EMG.

The proposed scheme of the assembly' second cascade for peaking the current of a small-class disk EMG is presented in Fig.4.

The scheme is based on a series connection of four bushing insulators and of four three-gap vacuum dischargers. The vacuum low-inductive dischargers, able to produce the current at megavolt voltages for the time less than 10 ns, have been developed under BAIKAL program.

At the selected parameter: ℓ_1 =25 mm; ℓ_2 =90 mm; ℓ_3 =10 mm; ℓ_4 =75 mm; h_1 =240 mm; h_2 =20 mm; h_3 =10 mm; h_4 =50 mm; r_1 =490 mm, the inductance in the vacuum cavity is L~10 nH.

The calculation optimization of the foil parameters has been carried out during the operation of the current source to the vacuum inductive load L_V =10 nH. The electric explosion of the foil in the second cascade is described by one-dimensional MHD equations. At this the magnetic fields on the foil surfaces have been determined by the currents in the circuit of the inductive storages L_{ind} and L_V, taken from the electrotechnical equations. Resistance of the vacuum discharger R_{disch} has been taken to be infinite and has been zeroed at achieving the breakdown value V_{br}.

The current in the vacuum inductive load is estimated from the ratio $I_p \sim I_{ind} \cdot L_{ind}/(L_{ind}+L_V) \sim 12$ MA. To transfer this current to the load for the time Δt ~100 ns, it is necessary to have a discharger with breakdown voltage $V_{br} \sim L_V \cdot I_p/\Delta t \sim 1.2$ MV.

Fig. 5 shows the calculated dependences of voltage on the foil and current in the vacuum wave line at the foil length ℓ_f =2.5 m, providing the maximum voltage and the minimum current pulse front, and different thicknesses δ. It is seen that the foil thickness δ =8 µm gives the maximum voltage and the minimum time of current rise. The proposed explosive magnetic device is able to form the current pulse with the amplitude 11.5 MA for the time ~100 ns in the inductive load 10 nH.

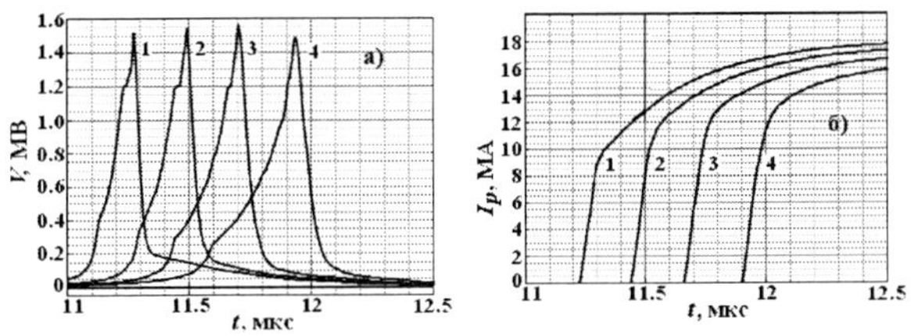

Fig. 5. Time dependence of voltage on the foil –a) and current in the vacuum wave line – b) at different foil thicknesses: 1. δ=4 µm; 2. δ=6 µm; 3. δ=8 µm; 4. δ=10 µm.

2. FIRST EXPERIMENTS

Before realization of a full-scale experiment it was necessary to make sure of operating capacity of separate elements of the explosive magnetic device. An explosive experiment has been carried out with the current source on the basis of a helical EMG ∅240 mm and with an explosive current opening switch (EOS). The sketch of the load comprising the FOS in the form of a "serpentine", a bushing insulator, a vacuum cavity and a three-gap vacuum discharger (V_{br}=300 kV), is presented n Fig.6.

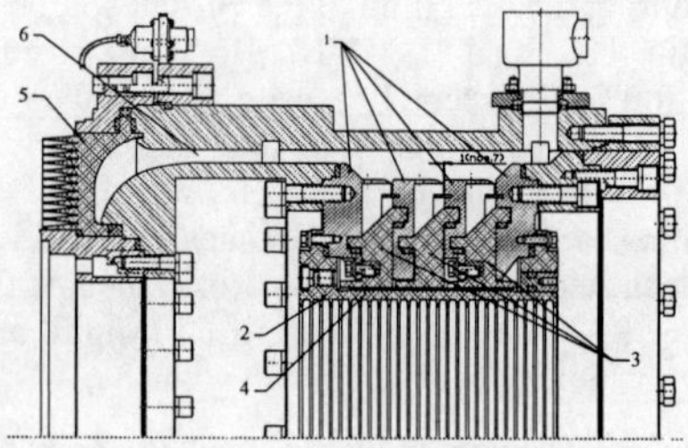

Fig. 6. Sketch of the load with a three-gap vacuum discharger.

1 – four annular steel electrodes forming a discharger; 2 – dielectric base; 3 – annular base insulators from plexiglass; 4 – volume filled with transformer oil; 5 – bushing insulator; 6 – vacuum cavity.

In the experiment the pulsed power source operated as expected. Comparison of the recorded and calculated current derivatives and currents in the source circuit after the EOS testifies to this fact (see Fig. 7, 8). Current measurements in the load circuit, recorded by the sensors located in solid dielectric, are presented in Fig. 7, 8. The current amplitude of 5 MA was achieved for the time of ~250 ns. The sensors in the vacuum cavity did not record the signals.

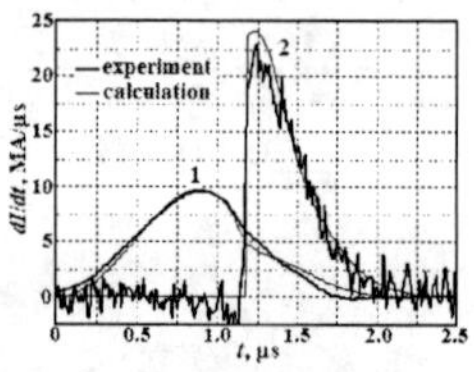

Fig. 7. Experimental and calculated current derivatives vs. time:
1 – source circuit;
2 – load circuit.

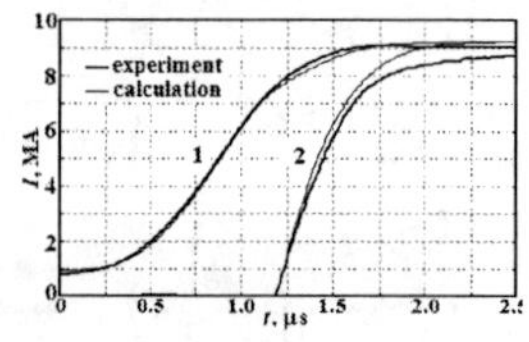

Fig. 8. Current vs. time
1 – source circuit;
2 – load circuit.

The electrical explosion of the foil happened, but the discharger did not operate. It means that there was a breakdown in the load circuit with a formation of the current contour before the place of discharger location.

The place and time of breakdown formation was found due to comparison of experimental and calculated current pulses. The results of the numerical simulations demonstrated that the agreement with the experiment (see Fig.7, 8) is observed at L_{br}~3 nH and corresponding V_{br}~70 kV. The conclusion was made that the breakdown happened along the surface of the bushing insulator.

To reveal the reasons of low electric strength of the bushing insulator, a series of experiments was started on the CASCADE capacitor bank [10]. The device on the basis of FOS in the form of a "serpentine" has been the same as in the explosive experiment (Fig.6). Copper foil 60 cm long has been placed in the form of 15 strips 28 μm thick (instead of 15 μm in the explosive experiment) and 10 mm wide. Vacuum discharger (1) was removed from the design. The FOS worked for the interrupted circuit ("infinite" load). The vacuum cavity (6) was pumped to the pressure ~ $5x10^{-4}$ torr. The current before the FOS and in the circuit of the "infinite" load was recorded by the inductive probes, the voltage at the output of the CASCADE facility was recorded by the conducting bridge.

There was a breakdown in the load in the experiment at the voltage U_f ~ 22 kV on the foil. After the experiment the traces of erosion were found on the surfaces of the electrodes in the vacuum cavity. It means that the nature of breakdown initiation was similar to breakdown in the explosive experiment.

In the second experiment only the operation of the FOS in the form of a "serpentine" to the "infinite' load was studied. The FOS had 36 stripes of copper foil 28 μm thick, 5 mm wide and 60 cm long. The inner electrode in the vacuum cavity was replaced by a dielectric flange, the other parts were the same. To provide the electric strength of the load elements, the inner volume of the cavity was filled with glycerol.

Fig. 9 shows the experimental (blue waveform) and calculated (red waveform) voltage on the FOS. The waveforms agree well, and it means that, as in the explosive experiment, the FOS functioned as expected.

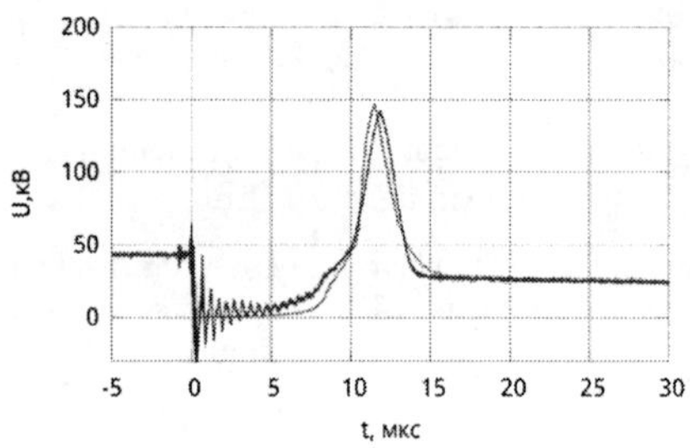

Fig. 9. Experimental and calculated voltage on FOS.

Usually the voltage in the vacuum cavity behind the bushing insulator appears simultaneously with the current and the regime of magnetic self-isolation is formed. In the scheme under consideration before the operation of the discharger the voltage between the electrodes in the vacuum cavity with the amplitude higher than the explosive emission threshold exists without current flowing. At this the presence of the leakage currents between the electrodes may lead both to potential limitation and to lighting of the bushing insulator with breakdown on its surface. This should be taken into account in the process of designing of the current peaking unit.

As the experiments demonstrated similar nature of breakdowns in the explosive experiment and in the experiments on CASCADE facility, it is expedient to finalize the design combining the laboratory and explosive experiments with EMG. In addition attention should be paid to:

- possible effect of the electrical explosion of the foil on stability of the bushing insulator by lighting the surface through a layer of polyethylene or by generating non-uniform electromagnetic field on it;

- impermissibility of occurrence of electric fields with intensity higher than the explosive emission threshold prior to the time of dischargers operation.

Further work will be divided into several stages.

The CASCADE facility will be used to continue the testing of the "serpentine" FOS, the electric strength of the design "FOS – bushing insulator" will be determined by testing with voltage formed by the FOS during the operation to open circuit.

At the second stage the explosive experiments will be realized with the pulsed power source on the basis of a helical EMG Ø240 mm in diameter. The objective of the experiments is to shape the current pulse with the amplitude ~5 MA for the time of 100 ns.

At the last stage the experiments will be conducted to create an explosive magnetic pulsed power source able to shape the current pulse with the amplitude to 10 MA during 100 ns.

CONCLUSION

The device shaping the submegampere-level EMG current pulses with the rise time of ~100 ns has been presented. The first experiments testing this device have been described. The problems have been revealed.

It is expected that these works will allow to create a current source on the basis of a helical EMG, making it possible to realize an explosive analogue of the Russian facility ANGARA (5 MA, 100 ns), and a source on the basis of a small-class disk EMG with the current parameters (10 MA, 100 ns) close to those realized on the capacitor PTS (IFP CAEP, China).

REFERENCES

[1] V.K. Chernyshev, M.S. Protasov, V.A. Shevtsov, P.N. Piskarev, E.I. Zharinov, G.I. Volkov, V.V. Vakhrushev, V.A. Ivanov, V.A. Demidov, S.V. Pak, N.P. Bidylo, B.E. Grinevich, V.P. Pogorelov, A.A. Petrukhin, A.I. Kuzyaev, V.B. Yakubov, V.I. Shpagin Explosive magnetic generators of "Potok" family//Questions of atomic science and technology (VANT). Ser. Mathematical modeling of physical processes. 1992, №4, pp. 33-41.

[2] V.K. Chernyshev, V.V. Vakhrushev, B.E. Grinevich, B.T. Egorychev, V.A. Demidov. High-speed disk explosive magnetic generator with flat disks//MEGAGAUSS-6, 1992, Albuquerque, USA.

[3] V.K. Chernyshev, B.E. Grinevich, V.V. Vakhrushev, V.I. Mamyshev. Scaling image of 90 MJ explosive magnetic generator//In Megagauss fields and pulse power system. Nova Science Publishers. New York.1990, p. 347.

[4] V.K. Chernyshev, A.V. Ivanovsky, V.B. Kudelkin, V.V. Vakhrushev Small-size explosive magnetic source of energy on the basis of second-generation disk EMG 250 mm in diameter//VIII Khariton's Topical Scientific Readings. Sarov, Russia, 2006.

[5] V.K. Chernyshev, V.V. Vakhrushev, G.I. Volkov Study of basic regularities of multi-MA-current pulses with short rise time by EMG circuit interruption//Megagauss Physics and Technology (Edited by P.J.Turchi) N.Y., L.:Plenum Press, 1980, p.663.

[6] V.K. Chernyshev, A.M. Buyko, V.N. Kostukov, A.I. Kuzyaev, A.A. Kulagin, V.I. Mamyshev, A.B. Mezhevov, A.I. Nechaev, A.A. Petrukhin, M.S. Protasov, V.A. Shevtsov and V.B. Yakubov Investigation of electrically exploded large area foil for current switching// Megagauss Field and Pulsed Power Systems (Edited by V.M. Titov and G.A. Shvetsov) N.Y., L.: Nova Science Publishers, 1990, p.465

[7] S.G. Garanin, A.V. Ivanovsky and L.S. Mkhitariyan. An ICF system based on Z-pinch radiation produced by an explosive magnetic generator//Nuclear Fusion., 2011, V.51, N10 (15pp).

[8] V.K. Chernyshev, A.I. Kucherov, A.I. Mezhevov, V.V. Vakhrushev. Electroexplosive foil 500 kV current opening switch characteristics research//11th IEEE Int. Pulsed Power Conf., Baltimore, Maryland, USA, pp.1208-1212 (1997).

[9] G.I. Dolgachev, D.D. Maslennikov, A.G. Ushakov, et al. Study of electric strength of vacuum gap at breakdown in non-uniform electric field for high-power pulses switching//PTE, №2, pp. 1-11(2007).

[10] V.E. Ablesimov, A.V. Andrianov, A.A. Bazanov, A.M. Glybin, Yu.N. Dolin, P.V. Duday, A.A. Zimenkov, A.V. Ivanovskiy, A.E. Kalinychev, G.V. Karpov, A.I. Kraev, S.S. Lomtev, S.V. Pak, N.I. Pozdov, S.M. Polyushko, A.F. Rybakov Development of plasma focus neutron source powered by an explosive magnetic generator//PMTF, 2015, V.56, №1, pp. 94-103.

Analysis of RF Generation Using NLTLs Driven by Pump Input Pulses of Different Shapes*

J.O. Rossi, E.G.L. Rangel, J.J. Barroso, F.S. Yamasaki
Space Research National Inst.- INPE
Sao Jose dos Campos, Brazil

L.P. Silva Neto
Federal University of Sao Paulo- UNIFESP
Sao Jose dos Campos, Brazil

E. Schamiloglu
Electrical and Computer Eng. Dept.
Univ. of New Mexico- UNM
Albuquerque, USA

***Abstract*— Nonlinear transmission lines (NLTLs) provide an alternative to generate RF without using vacuum electronic tubes, which normally requires heating filament and bias power supplies. In this work, the response of a NLTL driven by three pump input pulses of different shapes (rectangular, Gaussian and half sine) has been investigated. The performance of the NLTL was analyzed by using the Fast Fourier Transform (FFT) of the RF signal sampled on 29th section of a 30-section low voltage capacitive NLTL using BB809 diode varactors. The experiments demonstrated the NLTL's capability to shift energy content to higher frequencies while transforming a wide pulse into a burst of narrow solitons. The rectangular pulse shape generated oscillations of higher frequencies, around 30 MHz. The NLTL response is affected by the increase of the input pulse width leading to an increase of the average value of the amplitude of the RF oscillations with higher number of cycles, which follow a damped sine wave due to the line resistive losses. The maximum RF conversion corresponds to the 150 ns pulse width under a duty cycle of 33 % for all three waveforms, from those of which the rectangular pulse yields the highest efficiency of 16.4 %.**

Keywords—nonlinear transmission line; NLTL; soliton; RF generation; varactor

I. Introduction

Several works on nonlinear transmission lines (NLTLs) have demonstrated their suitability for application in a variety of high speed and wide bandwidth systems such as pulse compression [1-2], frequency translator [3], phase shifter [4-5], measurement circuitry [6], pulsed radar and battlefield communication disruption [7-8] and high power RF generation in remote sensing [9-10].

Research on RF generation using NLTs has been driven to generate RF without using vacuum electronic tubes, which normally requires heating filament and bias power supplies. Suitable applications in high power microwaves (HPM) could include RF transmitting systems of defense platforms, space vehicles and pulsed radars used for monitoring remote sensing.

In a NLTL, soliton generation occurs due to nonlinear and dispersive media obtained by the discreteness of the LC sections with nonlinear components (capacitors and/or inductors), as shown in Fig. 1. Because of the nonlinearity, pulse portions in the leading edge closer to the peak of the current or voltage waveforms will have a faster propagation (phase) velocity and will produce a shockwave front. On the other hand, dispersion occurs due to the discreteness of the NLTL and causes the waveform to spread out.

In a lumped capacitive NLTL constructed with varactors (Fig. 1) the nonlinearity is achieved by the capacitance changes due to the voltage variation, while its dispersion arises from its structural periodicity. When this type of NLTL is fed with a rectangular input pulse, it can be observed the forming of an electromagnetic shock wave superposed by oscillations generated with harmonics of higher order. The number of sections, the fractional change in the diode capacitance, the frequency dependent losses and the characteristics of the input signal (amplitude, frequency, duty cycle, and rise time) establish the signal propagation behavior along the line.

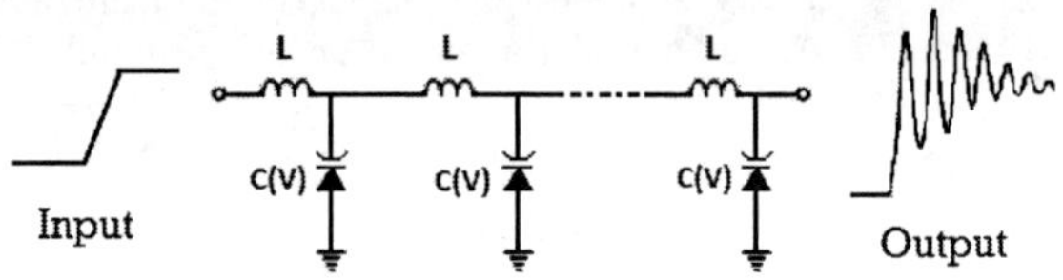

Fig. 1. Soliton generation produced in a capacitive NLTL.

The investigation of the response of a capacitive NLTL driven by three input pulses of different shapes (rectangular, Gaussian and half sine) is presented in this work. The performance of the capacitive NLTL was analyzed by using the Fast Fourier Transform (FFT) of the RF signal sampled on 29th section of a 30-section low voltage capacitive NLTL using BB809 diode varactors .

II. Basic Theory of NLTL

Fig. 1 shows a schematic diagram of a capacitive NLTL, which consists of a ladder network, built with nonlinear capacitors (varactors) and fixed inductors arranged as a series of LC section low-pass filters (Fig. 1). The combination of the dispersion (variation in phase velocity with frequency) and the nonlinearity of the capacitor allow the propagation of the output oscillations at a velocity V_p:

*Work supported by the Southern Office of Aerospace Research and Development (SOARD – USAF) under contract number FA9550-14-1-0133

$$V_p = 1/\sqrt{LC(V)} \quad (1)$$

with a cutoff (Bragg) frequency (f_c) given by[11]:

$$f_C = 1/\pi\sqrt{LC(V_{\max})} \quad (2)$$

where C(V) is the variable capacitance per section as a function of the voltage applied, and L is the linear inductance of each section, and $V_{\max}$ is the maximum voltage across the capacitor.

The characteristic impedance is a parameter that varies with the nonlinearity of the varactor. Disregarding any dissipative effects such as dielectric leakage and conductor resistance, the characteristic impedance (Z_0) is:

$$Z_0 = \sqrt{L/C(V)} \quad (3)$$

The NLTL can be used to transform an electrical pulse into a RF signal. The sharpening of the front of an electrical pulse is caused by the amplitude dependence of the phase velocity of electrical signals propagating along the ladder. The reduction in the rise time of a pulse injected into the NLTL is estimated as [11]:

$$\Delta T = n(\sqrt{LC_0}) - \sqrt{LC(V_{\max})}) \quad (4)$$

where n is the number of sections of the line, C_0 the unbiased capacitance, and $C(V_{max})$ the capacitance at the maximum amplitude of the propagating pulse. The sharpening occurs if the rise time of the input signal is larger than the pulse rise reduction (ΔT). When the rise time of the input signal is smaller than the pulse rise reduction the propagating pulse is broken into a train of solitons [12].

From Rayleigh's theorem, the correspondence between the energy of a nonperiodic signal in time and frequency domains is [13]:

$$E = \int_{-\infty}^{+\infty} v(t)^2 dt = \int_{-\infty}^{+\infty} V(f)^2 df \quad (5)$$

where $V(f)$ is the Fourier transform of $v(t)$.The efficiency of the NLTL can be evaluated by the percentage ratio of output to input energy:

$$\eta = E_{out} / E_{in} \quad (6)$$

III. EXPERIMENTAL RESULTS

A capacitive NLTL built using a printed circuit board (PCB) loaded with inductors of 2.7 μH, varactor diodes (BB809) as the nonlinear elements and a resistive load of 390 ohms was used in the experiments.

The BB809 C×V curve obtained from the manufacturer datasheet [14] is shown in Fig. 2. This curve can be modeled by a function given by [15]:

$$C(V) = \frac{C_{j0}}{(1 + V/V_j)^m} \quad (7)$$

where V_j is the diode junction potential, m is the nonlinearity factor and C_{j0} is the unbiased diode junction capacitance. These parameters are obtained from the datasheet: V_j= 0.7 V and m = 0.5. By extrapolation, for a voltage variation from zero to 10 V, the capacitance varies from 56 pF to 13 pF.

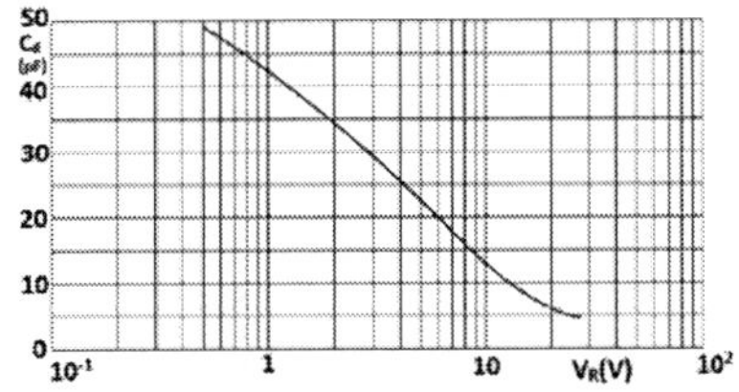

Fig. 2. Diode capacitance as function of the reverse applied voltage [14].

For the generation of the rectangular, Gaussian and half sine pulse shapes an arbitrary pulse generator (Agilent 81150A) was set using the software Agilent 33503A - Benchlink Waveform Builder. The generated signals had pulse widths varying between 100 and 300 ns, which was achieved by adjusting the frequency of the waveforms from 3.33 to 1.11 MHz, respectively.

The time and frequency domain analysis was performed with a digital oscilloscope Agilent MSO 9254A. The response of the NLTL was evaluated from the signal sampled with a clip on probe at the 29th section to avoid the signal distortion caused by impedance mismatch at the load.

For evaluating the RF conversion efficiency and considering that RF output energy includes higher-frequency spectral components not present in the input signal, a cutoff frequency of 20 MHz is chosen. According to (6) E_{in} and E_{out} are calculated by integrating the square of the continuous spectral density over the frequency ranges of 0-40 MHz and 20-40 MHz, respectively, on the corresponding FFT spectra for each signal.

The input signal used to check the performance of the NLTL was from an arbitrary function generator with constant duty cycle (33 %) so that the repetition frequency of the input pulse increased as its width decreased. Therefore, to avoid errors the estimations of the input and output energies (6) were computed by the areas under the envelope curves of the FFT spectra obtained for the pulses at the line input and at 29th section under single shot. This was made using a graphic software (called Origin).

A. Results

The responses of the capacitive NLTL driven by rectangular, Gaussian and half sine waveforms displayed from Figs. 3(a) to 11(a) show the time domain response of the sampled signal at the input and at 29th section for pulse widths of 100, 200 and 300 ns. The frequency domain responses from Figs. 3(b) to 11(b) show the corresponding sampled signal FFT directly windowed from the scope screen for the signals at the input and at 29th section of the NLTL.

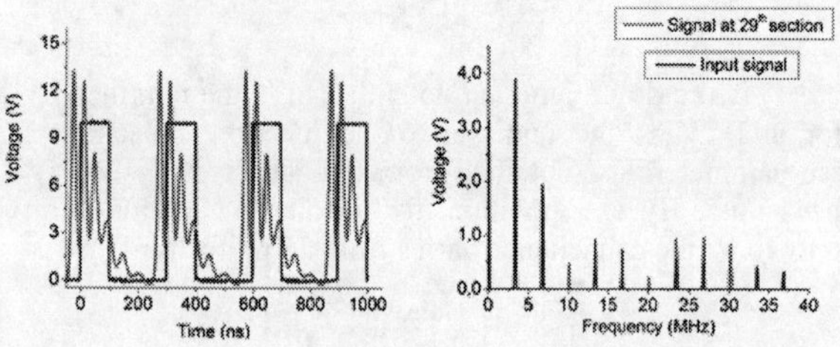

(a) (b)

Fig. 3. The response of the capacitive NLTL to the rectangular pulse train with input pulse width of 100 ns: (a) time domain and (b) frequency domain.

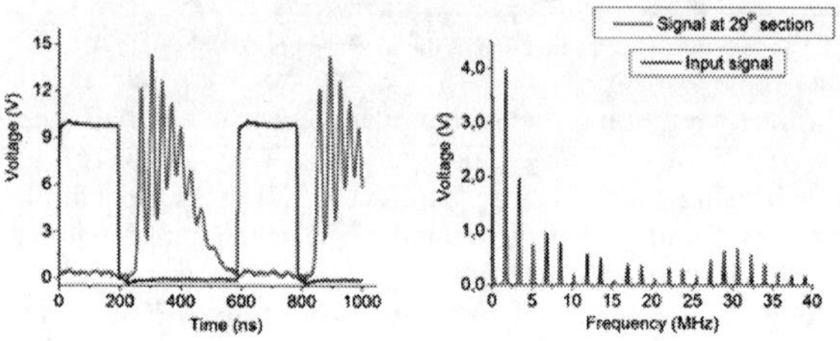

(a) (b)

Fig. 4. The response of the capacitive NLTL to the rectangular pulse train with input pulse width of 200 ns: (a) time domain and (b) frequency domain.

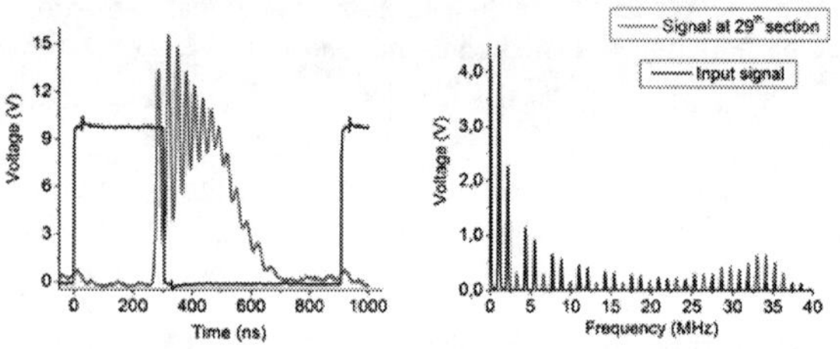

(a) (b)

Fig. 5. The response of the capacitive NLTL to the rectangular pulse train with input pulse width of 300 ns: (a) time domain and (b) frequency domain.

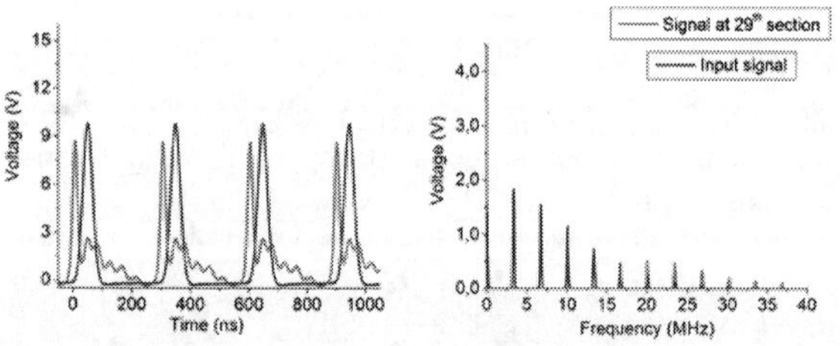

(a) (b)

Fig. 6. The response of the capacitive NLTL to the Gaussian pulse train with input pulse width of 100 ns: (a) time domain and (b) frequency domain.

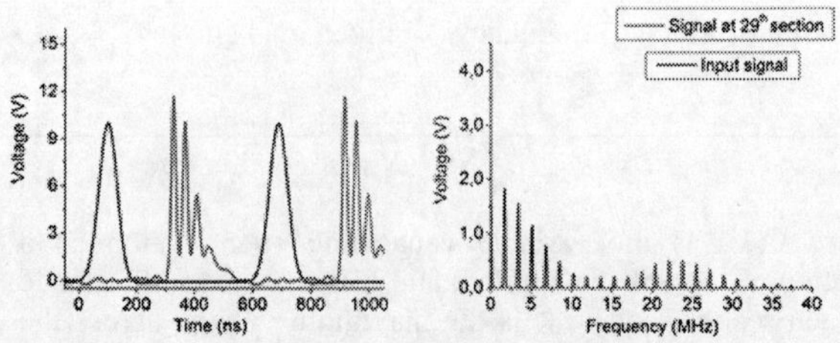

(a) (b)

Fig. 7. The response of the capacitive NLTL to the Gaussian pulse train with input pulse width of 200 ns: (a) time domain and (b) frequency domain.

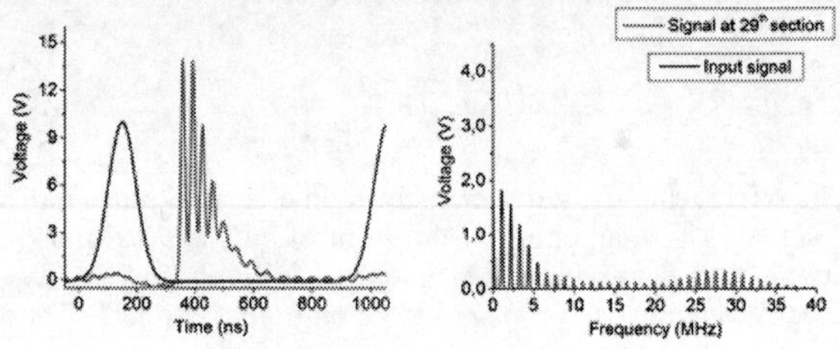

(a) (b)

Fig. 8. The response of the capacitive NLTL to the Gaussian pulse train with input pulse width of 300 ns: (a) time domain and (b) frequency domain.

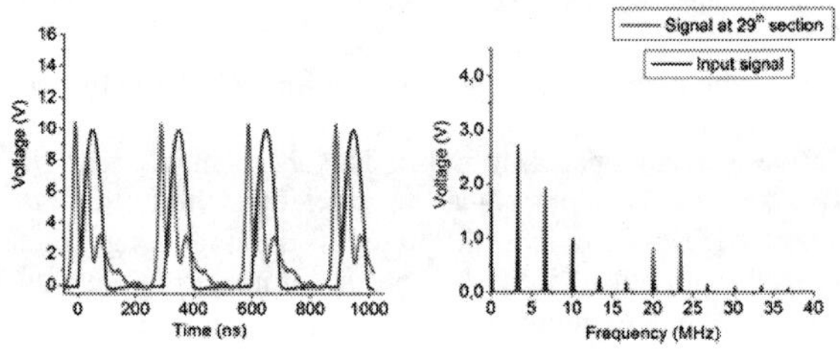

(a) (b)

Fig. 9. The response of the capacitive NLTL to the half sine wave pulse train with input pulse width of 100 ns: (a) time domain and (b) frequency domain.

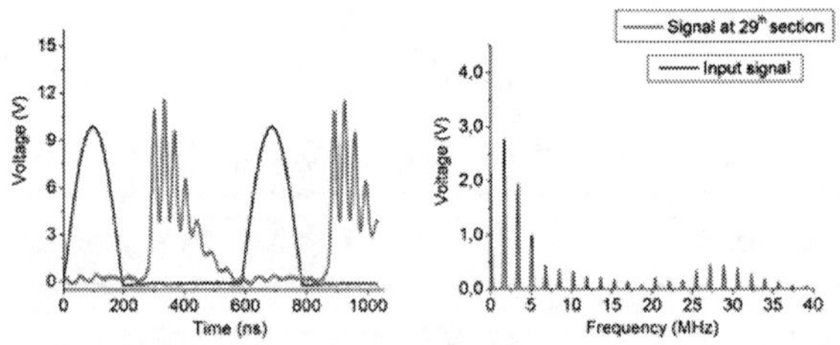

(a) (b)

Fig. 10. The response of the capacitive NLTL to the half sine wave pulse train with input pulse width of 200 ns: (a) time domain and (b) frequency domain.

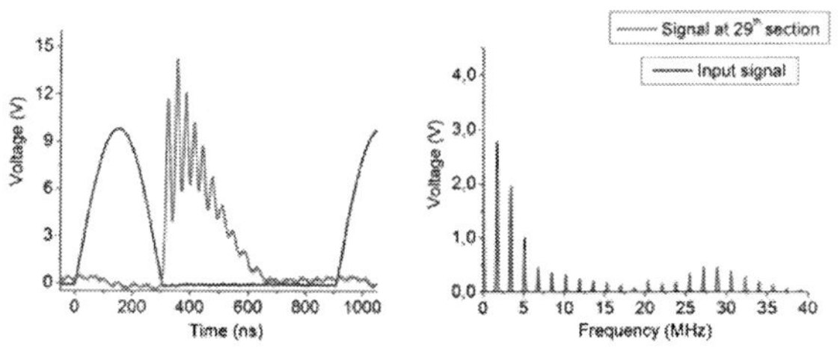

(a) (b)

Fig. 11. The response of the capacitive NLTL to the half sine wave pulse train with input pulse width of 300 ns: (a) time domain and (b) frequency domain.

B. Discussions

The results in section III.A show that the input signal with wider pulse width leads to an increase in the average value of the amplitude of the RF oscillations with higher number of cycles, which follow a damped sine wave due to the line resistive losses. This is so because as the voltage increases the capacitance decreases (Fig. 2), leading to an increase in the output frequency, which is nearly given by the half of the line cutoff frequency given by (2).

Fig. 12(a) shows that the frequency of spectral lines with the largest amplitudes increases with the pulse width. Higher frequency oscillations, around 30 MHz, were generated by the rectangular pulse wave train. Fig. 12(b) presents a comparison of the NLTL efficiency when driven by rectangular, Gaussian and half sine wave pulses for a pulse width range of 100 to 300 ns. It is seen that the rectangular pulse train presented the maximum efficiency (16.4 %) for a 150 ns pulse width.

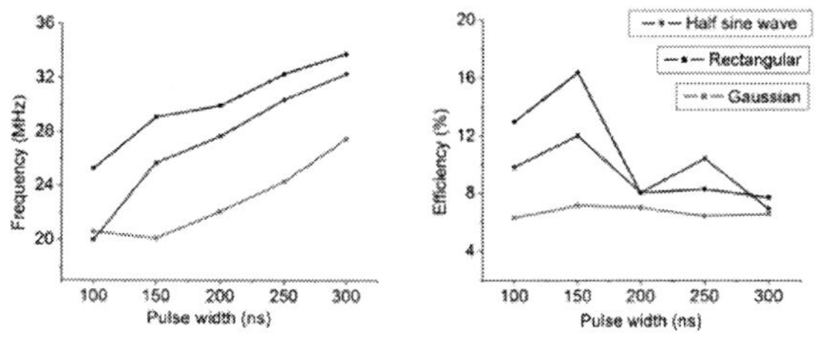

(a) (b)

Fig. 12. Comparison among the responses of the capacitive NLTL to the rectangular, Gaussian and half sine wave pulse trains: (a) frequency spectral component with maximum amplitude and (b) efficiency of RF generation in the range of 20 to 40 MHz.

IV. CONCLUSIONS

The performance of the capacitive NLTL was investigated when driven by three different pulse shapes: rectangular, Gaussian and half sine wave. The experiments demonstrated the NLTL's capability to shift energy content to higher frequencies while transforming a wide pulse into a burst of narrow solitons. The NLTL response is affected by the increase in the input pulse width, which leads to a higher average value of the amplitude of the RF oscillations with increased number of cycles, where the pulse envelope curve follows a damped sine wave due to the line resistive losses. This is because an increase in the DC level decreases the capacitance of the varactor diode, leading to a frequency increase in oscillations. The maximum RF conversion corresponds to the 150 ns pulse width under a duty cycle of 33 % for the three waveforms, from those of which the rectangular pulse yields the highest efficiency of 16.4 %.

REFERENCES

[1] E. Afshari and A. Hajimiri, "Nonlinear transmission lines for pulse shaping in silicon," *IEEE Journal of Solid-State Circuits,* , vol.40, no.3, pp.744-752, March 2005.

[2] L. P. Silva Neto, J. O. Rossi and J. J. Barroso, "Pulse sharpening and RF generation using nonlinear transmission lines," Microwave and Optoelectronics Conference (IMOC), 2015 SBMO/IEEE MTT-S International, Porto de Galinhas, Brazil, Nov. 2015, pp. 1-6.

[3] M. Pontón and A. Suárez, "Analysis of Two Coupled NLTL-Based Oscillators," in IEEE Transactions on Microwave Theory and Techniques, vol. 62, no. 12, pp. 3485-3499, Dec. 2014.

[4] Hongjoon Kim, Sung-Jin Ho, Chih-Chuan Yen, Kae-Oh Sun and D. W. van der Weide, "Balanced distributed-element phase shifter," in IEEE Microwave and Wireless Components Letters, vol. 15, no. 3, pp. 147-149, March 2005.

[5] A. S. Nagra and R. A. York, "Distributed analog phase shifters with low insertion loss," in IEEE Transactions on Microwave Theory and Techniques, vol. 47, no. 9, pp. 1705-1711, Sep 1999.

[6] Wohlgemuth, M. J. W. Rodwell, R. Reuter, J. Braunstein and M. Schlechtweg, "Active probes for 2-port network analysis within 70-230 GHz," Microwave Symposium Digest, 1999 IEEE MTT-S International, Anaheim, CA, USA, 1999, pp. 1635-1638 vol.4.

[7] M. J. W. Rodwell et al., "Active and nonlinear wave propagation devices in ultrafast electronics and optoelectronics [and prolog]," in Proceedings of the IEEE, vol. 82, no. 7, pp. 1037-1059, Jul 1994..

[8] S. Ibuka, K. Abe, T. Miyazawa, A. Ishii and S. Ishii, "Fast high-voltage pulse generator with nonlinear transmission line for high repetition rate operation," in IEEE Transactions on Plasma Science, vol. 25, no. 2, pp. 266-271, Apr 1997.

[9] D. V. Reale, J. M. Parson, A. A. Neuber, J. C. Dickens, and J. J. Mankowski, "Investigation of a stripline transmission line structure for gyromagnetic nonlinear transmission line high power microwave sources," Rev. Sci. Instrum., vol. 87, no. 3, pp. 034706-1–034706-5, Mar. 2016.

[10] D. V. Reale, J. M. Parson, A. A. Neuber, J. C. Dickens, and J. J. Mankowski, " Investigation of a stripline transmission line structure high power microwave sources", Rev. Sci. Instrum., vol. 87, no. 3, pp. 0347061-0347065, March 2016.

[11] P. W. Smith, "Transient electronics – pulsed circuit technology", John Wiley & Sons, West Sussex, England, 2002.

[12] N. S. Kuek, A. C. Liew, E. Schamiloglu, and J. O. Rossi, "Circuit modeling of nonlinear lumped element transmission lines including hybrid lines," IEEE Trans. Plasma Sci., vol. 40, no. 10, pp. 2523-2534, Oct. 2012.

[13] A. B. Carlson, Communication Systems. Tokyo: McGraw-Hill Kogakusha, 1968, pp. 22-30.

[14] Data sheet BB809 VHF variable capacitance diode. Available:http://www.datasheet4u.com/datasheet/B/B/8/BB809_Philips Semiconductors.pdf.html. Accessed: August, 9, 2016.

[15] F. S. Yamasaki, L. P. Silva Neto, J. O. Rossi, and J. J. Barroso "Soliton Generation Using Nonlinear Transmission Lines," IEEE Trans. Plasma Sci., vol. 42 , no. 11, pp. 3471-3477, Oct. 2014.

Development of Parallel Switching Circuit Using Wide Band-Gap Semiconductor Power Devices

Takashi Sakugawa, Naoto Yamashita, Kengo Tasaki, Koichiro Yamashita and Hidenori Akiyama
Kumamoto University
Kumamoto, Japan
sakugawa@cs.kumamoto-u.ac.jp

Tomoyuki Ohkubo, Hiroshi Umeda, Takashi Matsunaga and Hakaru Mizoguchi
Gigaphoton Inc.
Oyama, Japan
web_info@gigaphoton.com

***Abstract*— There are various methods of pulsed power generation using semiconductor switch. Semiconductor switching device is indispensable to realize a high repetitive operation of the pulsed power modulator. High repetitive operation of the modulator is contributed to industrial application of pulsed power. However the conventional semiconductor switching device is hard to generate fast rising pulse high voltage directly such as an excimer laser exciter. Generally, the switching speed of large-switching-capacity semiconductor switch is low, and the switching capacity of high speed semiconductor is small. In this study, we have used a new generation power switching semiconductor device as Silicon-Carbide based Metal Oxide Semiconductor Field Effect Transistor (SiC-MOSFET) for primary switching circuit. Primary switching module are using 6 parallel discrete SiC-MOSFETs per unit. Maximum switching voltage and current of the module is 1200V and 1500A respectively. We have developed multi connecting modules. And we have succeeded to multi parallel synchronized switching of modules. A pulsed power modulator consist of multi parallel modules, pulse transformer and 2 stage magnetic pulse compression (MPC) circuit. We have been generate a fast rising high voltage pulse using the pulsed power modulator. And we have obtained output voltage of several 10kV with fast rise time by the modulator.**

Keywords*— silicon carbide; MOSFET; parallel switching; magnetic pulse complession

I. Introduction

In recent years, research on practical industrial applications of high-repetition-rate pulsed power generated by a magnetic pulse compression circuit (MPC) and semiconductor switching devices has been progressing. Such research has long focused on development of photolithography light sources. In particular, the excimer laser that is used as a microlithographic light source in semiconductor fabrication requires a high repetition rate, high stability, and a long lifetime. Therefore, most of the excimer lasers for microlithography use a semiconductor switch and MPC for the exciter [1, 2]. In addition, applications to environmental fields involving the breakdown of harmful gases by utilizing a discharge plasma or the generation of ozone have been studied [3]. In this case too, high repetitively operation is possible and a long lifetime pulsed power generator is indispensable. In 1951, Melville (UK) described an MPC that performs pulse compression by using the nonlinear permeability of ferromagnetic materials [4]. In recent years, ferromagnetic materials with good characteristics have been developed, allowing the making of magnetic switches (MS) with high squareness-ratio(Br/Bs) and very low loss, which have found practical applications in solid-state pulsed power generators using a semiconductor switch and MPC. The semiconductor switch does not have the high voltage and high-speed properties of electric discharge switches. Generally, the switching speed of large-switching-capacity semiconductor switches is slow, and the switching capacity of high-speed semiconductor switches is small. Therefore, if we use a semiconductor switch of large capacity, many magnetic switches becomes necessary for the control of laser excitation and ozone generation, because an effective high-speed pulse is not obtained. Conversely, if we use a semiconductor switch with high-speed but small capacity, we must use a switching module that includes many parallel and series connections. In either case, the use of a conventional semiconductor switch is composited in a large and complicated circuit, hindering practical use.

A metal oxide semiconductor field effect transistor (MOSFET) is one type of high-speed semiconductor switch. Although the semiconductor switch module by using a MOSFET is able to high-speed switching, the switching power-capacity has been very small. Recently, development of a switching power devices using wide band-gap semiconductor material has been actively conducted. Silicon carbide (SiC) is a kind of wide band-gap semiconductor material has excellent physical properties as a power device than conventional silicon (Si). As a result, to use SiC based MOSFET (SiC-MOSFET) for power switching, in which the switching circuit design was focused on the turn-on characteristic, yielding high current pulse characteristic, were considered. We have developed a switch module using 6 parallel SiC-MOSFETs. In this study, we present the results of research on a high-speed, high-repetition-rate pulsed power generator using SiC-MOSFET modules and 2-stage MPC for pulsed discharge plasma generation. We describe experimental results of parallel switching, MPC and present data on the performance of a high-repetition-rate pulsed power generation circuit using SiC-MOSFET module.

II. Switching Module

A. Parameters of Switching Devices

We made a respective switch module using two kinds of discrete power devices. The parameters of the power devices used in each of the switch module shown in Table 1. SiC-MOSFET used in this study, another one is an IGBT using silicon semiconductor for comparison of switching characteristics.

TABLE I. Parameters of Switching Devices

Parameter	SiC-MOSFET (C2M0025120D,CREE)	IGBT (IRGPS60B120KDP,IR)
Breakdown Voltage : V_{DS} , V_{CE}	**1200V**	**1200V**
Current (pulse) : I_{DSp}, I_{CMp}	**250A**	**240A**
On-State Resistance : $R_{DS(on)}$	**25mΩ**	—
Rise Time : t_r	**31.6ns**	**32ns**
Turn-On Delay Time : $t_{d(on)}$	**14.4ns**	**72ns**
Fall Time : t_f	**28.4ns**	**45ns**
Turn-Off Delay Time : $t_{d(off)}$	**28.8ns**	**366ns**
Gate Source(Emitter) Voltage : V_{GS} , V_{GE}	**-10/+25V**	**-20/+20V**
Package Outline Ref. [5]Data sheet of C2M0025120D, CREE Inc. [6]Data sheet of IRGPS60B120KDP, International Rectifier	TO-247-3 [5] 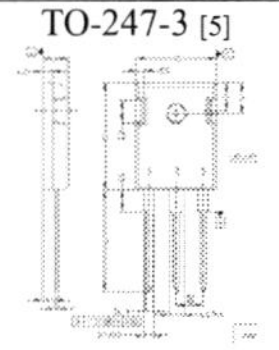	Super-247 [6] 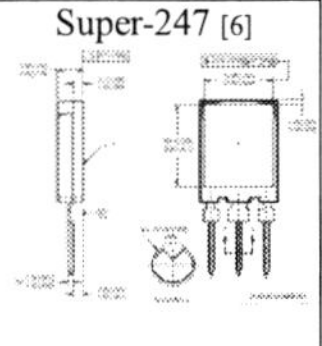

B. Switching Module Using 6 parallel Power Devices

The switching module are located 6 parallel power devices on the printed circuit board. The size of the switching module is (L)140 mm × (W)180 mm × (H)40 mm. It shows the photo of the switching module are shown in Fig. 1. A maximum switching voltage is 1000 V. A capacitance of storage capacitor is 2500 nF. Although each of the gate trigger is controlled by a common TTL signals, each power device has a gate circuit independent in order to obtain a sufficient input gate charge capacity.

Fig. 1. Switching Modules (Left: SiC-MOSFET, Right: IGBT)

C. Switching Waveforms of the Modules

The test circuit of the switching module is shown in Fig. 2. The switching module is connected to the primary side of the pulse transformer (PT). The switching test results of SiC-MOSFET module and IGBT module shown in Fig. 3. SiC-MOSFET module that the resonance current flows when the voltage drops to nearly 0. It can be confirmed from Fig.3. On the other hand, the IGBT module has large loss since the current flows in the course of the voltage drop is distorted wavcform. SiC-MOSFET module has an ideal switching waveform as compared with the IGBT module. The rate of loss calculated from a value obtained by dividing the input energy SiC-MOSFETs switching losses and IGBTs shown in Fig. 4.

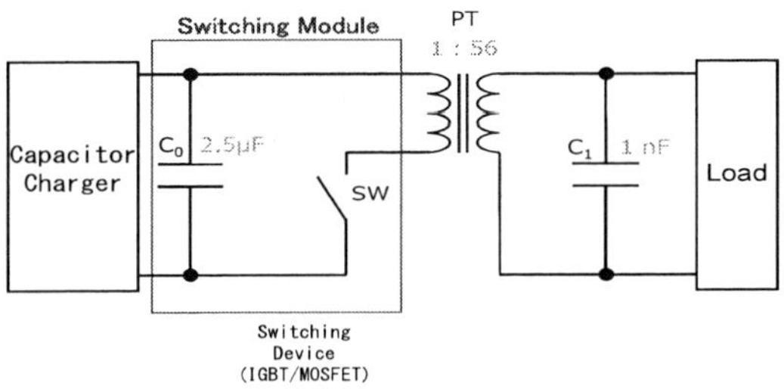

Fig. 2. Switching Test Circuit of the Module

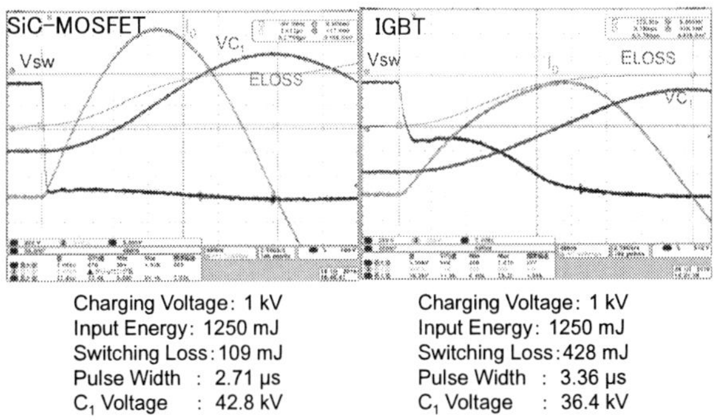

Fig. 3. Switching Waveforms of the Modules

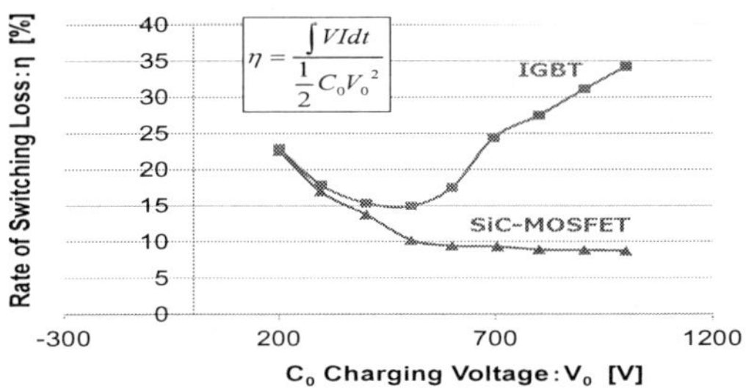

Fig. 4. Rate of Switching Loss of Switching Modules

III. PARALLEL CONNECTED SWITCHING MODULES

It has been demonstrated that experimental results of the SiC-MOSFET module is suitable for the switching circuit with low loss. We have experimentally investigated the switching characteristics are connected in parallel a plurality of SiC-MOSFET modules to the primary side of the pulse transformer.

A. Synchronous Parallel Switching of Modules

It performed synchronous turn-on operation is connected in parallel a plurality of switching modules in the primary side of the pulse transformer. The test circuit of a plurality parallel synchronous switching of the module shown in Fig. 5. A photo of 10 parallel connected modules to the primary side of the pulse transformer is shown in Fig. 6. Trigger signal is output ten TTL signals synchronized by the signal distributor to the gate circuit on each module. The switching waveforms of the 10 parallel connected modules to pulse transformer is shown in Fig. 7. It shows substantially the same waveforms as the switching waveforms of a single module. The relationship between the number of parallel and the pulse width of the switching module is shown in Fig. 8. The value of inductance in Fig. 8 is estimated from the pulse width.

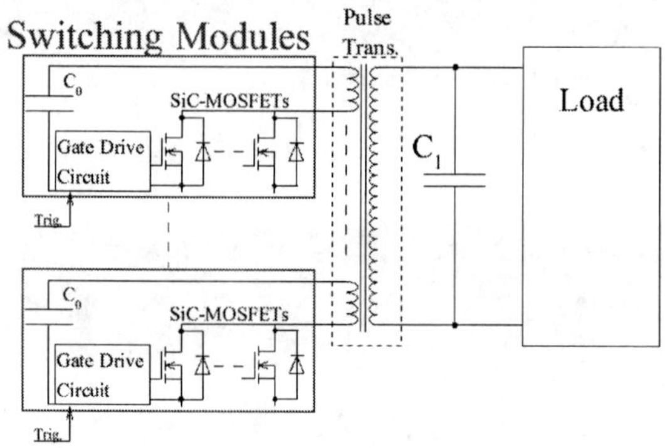

Fig. 5. Synchronous Parallel Switching Test Circuit

Fig. 6. 10 SiC-MOSFET Switching Modules Connected in Parallel

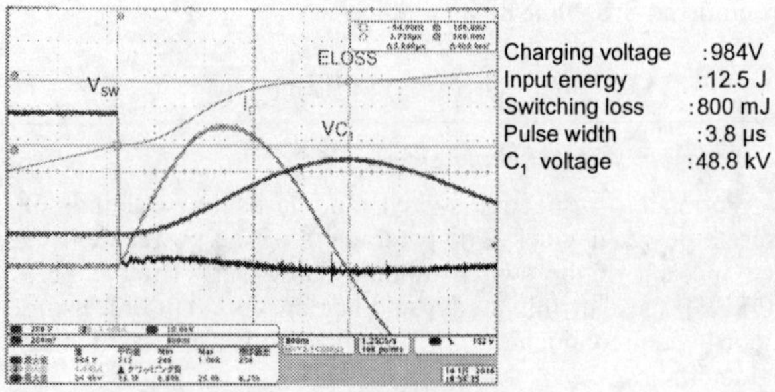

Fig. 7. Synchronous Parallel Switching Waveforms

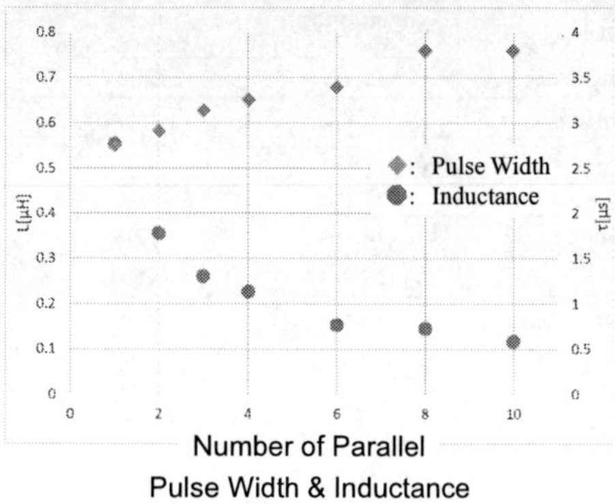

Fig. 8. Pulse Width Dependence of Number of Parallel

B. 2 Stage Magnetic Pulse Compression Circuit

It has connected the two-stage magnetic pulse compression circuit on the secondary side of the pulse transformer to generate a shorter pulse (ns level). A plurality of parallel modules and two-stage magnetic pulse compression (MPC) circuit shown in Fig. 9.

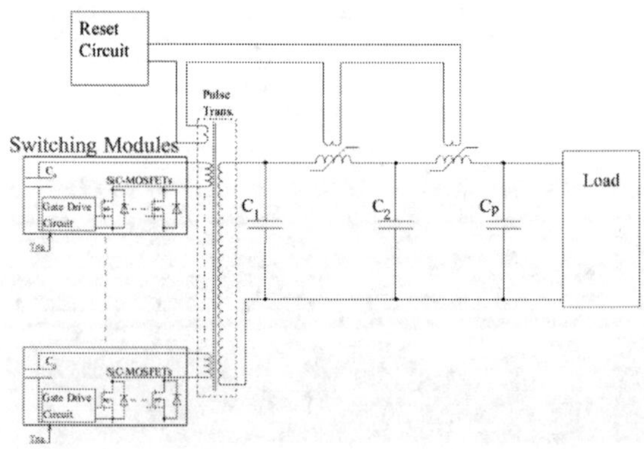

Fig. 9. Multi Paralleled Switching Modules and 2 Stage MPC Circuit

Each part of the voltage waveform of Fig. 9 is shown in Fig. 10. The number of parallel connections of the module is 10. Ratio of the number of windings of the pulse transformer is 1-to-56. And a capacitor of the capacitance of the MPC is each 8nF. When switching modules charging voltage of about 900V, the output voltage of the MPC -46kV is obtained. Fall time of the output voltage (V_{Cp}) is less than 100ns (10-90%).

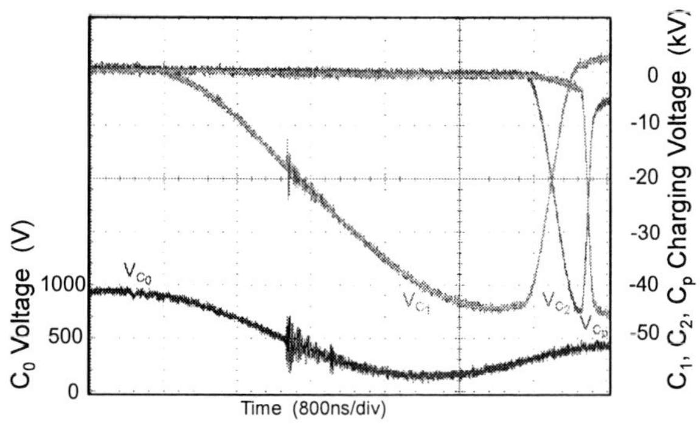

Fig. 10. Voltage Waveforms of the MPC and Paralleled Switching Modules

IV. Summary

We have developed a switching module in parallel the wide band gap semiconductor switch (SiC-MOSFET). The summarized as follows from the experimental results in combination with 2 stage magnetic pulse compression circuit.

a) We have developed multi connecting modules using SiC-MOSFETs. And we have succeeded to multi-parallel synchronized switching of modules.

b) A new developed pulsed power modulator consist of multi-parallel SiC-MOSFET modules, pulse transformer and 2 stage magnetic pulse compression (MPC) circuit.

c) We have been generate a fast rising high voltage pulse using the new developed pulsed power modulator. And we have obtained output voltage of several 10kV with fast rise time by the modulator.

We are expecting that the pulsed power modulator using a wide band gap semiconductor switch is to contribute to industrial applications.

Acknowledgment

This work was supported by Gigaphoton Inc.

References

[1] W. Partlo, R.Sandstrom, I. Fomenkov, "A low cost of ownership KrF excimer laser using a novel pulse power and chamber configuration", Proceedings of SPIE No. 2440, 90, (1995).

[2] H. Mizoguchi, O. Wakabayashi, T. Aruga, T. Sakugawa, T. Koganezawa, "High power KrF excimer laser with a solid state switch for microlithography", Proceedings of SPIE No. 2726, 831, (1996).

[3] T. Sakugawa, D. Wang, K. Shinozaki,T. Namihira, S. Katsuki and H. Akiyama, "Repetitive Short-Pulsed Generator Using MPC and Blumlein Line", Proceedings of 14th IEEE International Pulsed Power Conference, pp.657-660 (2003).

[4] WS. Melville, "The use of saturable reactors as discharge devices for pulse generators", Proceedings of IEE Radio Section ;98:185, (1951).

[5] Data sheet of C2M0025120D, CREE Inc..

[6] Data sheet of IRGPS60B120KDP, International Rectifier .

Analysis of PEF treatment using a monoaxial and a biaxial treatment chamber for food processing

[1]H.Bermaki, [1]A.Semmak, [1]Y. Bellebna, [2]M.Kaddour, [3]H.Belhassaini, [1]A. Tilmatine
[1]APELEC Laboratory, Djillali Liabes University of Sidi Bel-Abbes, Algeria
[2]LSTE Laboratory, University of Mascara, Algeria
[3]BVCV Laboratory, Djillali Liabes University of Sidi Bel-Abbes, Algeria

***Abstract*— Except the "Co-field" electrodes, the majority of PEF treatment chambers, used in food processing experiments, are constituted of two parallel metal electrodes energized by a pulsed voltage. The electric field lines which are perpendicular to the electrodes are monoaxial and oriented almost in only one direction. The objective of the present work is an experimental comparative analysis between such treatment chamber (TC1) to a similar one but comprising four identical metal plates placed on the side walls of a square treatment chamber (TC2). For this latter, each pair of the adjacent metal plates form one electrode, the electric field lines being biaxial and oriented in two directions. The chamber made of Plexiglas has a square paralelipedic shape of dimensions 4x4x10 cm3, in which are placed either two (TC1) or four (TC2) vertical and parallel stainless steel electrodes, of dimensions 10x4 cm2. The experimental analysis was made using a classic pulse voltage generator, by measuring the mass of PEF pretreated extracted juice from beet and the amount of betanine using a spectrophotometer. The obtained results, using the methodology of experimental designs, have shown that the TC2 model is more efficient, because higher quantities of juice and Betanine were obtained up to 25% with an energy saving of nearly 90%.**

Keywords— pulsed electric field, High Voltage, treatment chamber, electode.

I. INTRODUCTION

Pressing and pulsed electric field (PEF) treatments are widely combined for extraction of juice. Several studies confirmed the positive effect of PEF treatment on both juice yield and its quality **[1]**. Moreover, inactivation of microorganisms and acceleration of the drying process using PEF technique were demonstrated. Furthermore, the extraction rates of many compounds were increased in the food industry **[2]**.

Food product is placed in the treatment chamber, where two electrodes are connected together with a nonconductive material to avoid electrical flow from one to the other. There are two types of treatment chambers: static and dynamic. In both cases, high voltage electrical pulses are applied to the electrodes, which then conduct the high intensity electrical pulse to the product placed between the two electrodes, to achieve membrane electroporation **[3-6]**.

Static chamber where only a given volume can be processed at once, are for experimental applications. Dynamic chamber enable continuous processing, in accordance with the requirements of industrial application **[7]**.

While the parallel plate treatment chamber consists of two stainless steel plate electrodes and an insulator, the cofield chambers in which two stainless steel tubes are separated by an insulator, the electric field and the food flow concurrently. The parallel plate treatment chamber allows the uniformity of the electric field, and thus a homogeneous treatment, while co-field electric chambers which are more commonly used provide better fluid dynamics characteristics. The coaxial chamber is basically composed of an inner cylinder surrounded by an outer annular cylindrical electrode that allows food to flow between them. Furthermore, although coaxial treatment chambers are more simple, they have a smaller cross section available for the flow, that can be easily blocked by particulate products. The co-field treatment chamber (for the same electric field strength) allows treating large particle and is not blocked by particulate products **[8-10]**.

For classic PEF treatment chambers constituted of two parallel metal electrodes, the electric field lines which are perpendicular to the electrodes are monoaxial and oriented in only one direction. The objective of the present work is an experimental comparative analysis between such treatment chamber to a similar one but comprising four identical metal plates placed on the side walls of a square treatment chamber. For this latter, each pair of the adjacent metal plates form one electrode.

II. MATERIAL AND METHODS

Beets were crushed with a household robot to obtain a homogeneous leg. The sample was held in a closed container to prevent evaporation before use. A beet paw sample of mass 80 g was used for each experiment. After PEF treatment, an extraction step was achieved using an extraction chamber and a hydraulic pressing machine (Mega, 15 tons). The PEF treated extracted juice was then analyzed by measuring both its mass using an electronic balance of 0.1 mg precision and the betanine amount using a spectrophotometer (Optizen *200 plus*) for λ= 530 nm.

All experiments were performed while maintaining following factors at constant values: pulse repetition frequency

f = 1 Hz, extraction pressure P = 50 kg/cm^2, total pressing duration t = 300 s and the inter-electrodes gap d= 60 mm.

The experimental setup used in the present work is composed of a number of components, comprising a high DC voltage source, an energy storage capacitor, a spark gap switch and a treatment chamber (Fig.1). The PEF generator works in two consecutive phases: charging and discharging the capacitors. A DC high voltage (Spellman 40 kV, 9 mA) charges the bank of capacitors until producing the spark gap's breakdown, causing an abrupt voltage (shock) applied to the load (treatment chamber where the sample is disposed). The storage element is composed of three sets of five series capacitors (2 µf, 2 kV), with the possibility to reach a maximum voltage of 10 kV and a total capacitance of 1.2 µF (Fig.2).

Fig.1. The pulse generator.

a) Descriptive schematic of the setup; b) The photography of the setup

1- HV DC power supply, 2-Set of capacitors, 3- Spark gap switch, 4- Treatment chamber

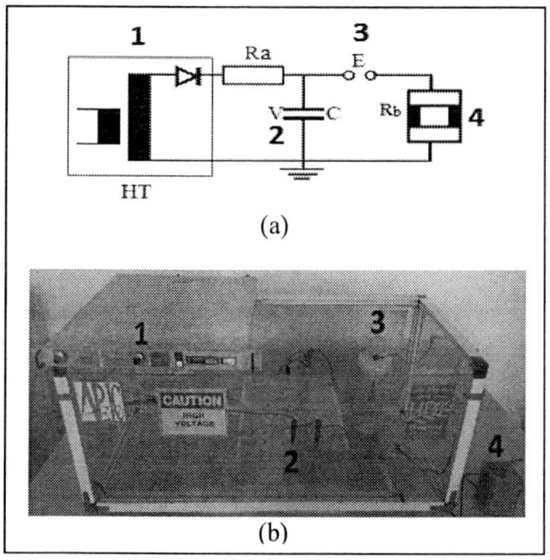

Fig.2 . The bank of capacitors.

a) Descriptive schematic; b) The photography of the capacitors bank

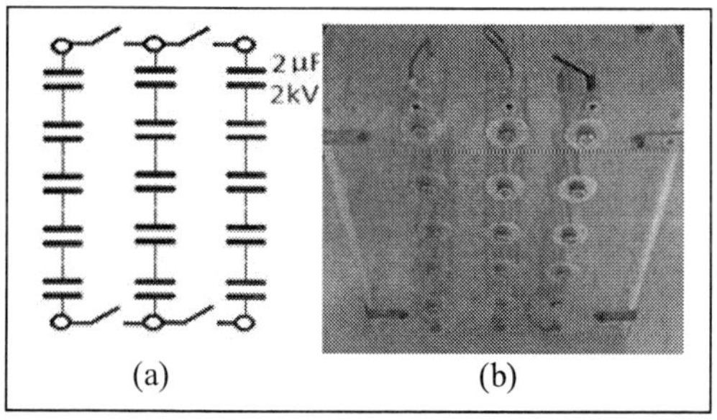

Two square paralelipedic treatment chambers made of Plexiglas, of dimensions 6x6x10 cm^3, in which are placed vertical stainless steel electrodes, were used in this work. The model TC1 comprises two parallel and opposite electrodes of dimensions 6x10 cm^2, while the model TC2 is constituted of four similar electrodes (Fig.3). For this latter, each pair of the adjacent metal plates form one electrode. The volume of both treatment chambers is 192.3 ml.

The methodology of the experimental designs makes it possible to determine the number of experiments to be achieved according to a well-defined objective, to analyze several factors simultaneously, to reduce dispersion related to measurements, to appreciate the effects of coupling between factors and finally to evaluate the respective influence of the factors and their interactions. Before starting the experiments, we need to opt for the most suitable design which can model the process with maximum precision. The Composite Centered Faces design (CCF), which gives quadratic models, was adopted in the present work.

Fig.3: The treatment chambers TC1 and TC2

1: electrodes 2: insulating. 3: Plexiglas

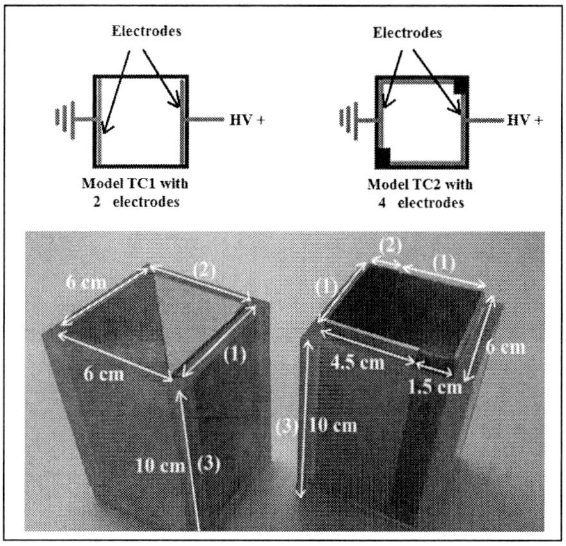

III. RESULTS AND DISCUSSION

The three following factors were considered in this work: the applied voltage V (kV), the number of pulses (n) and the pulse duration T (µs). The pulse duration is determined by the corresponding value of the capacitance, as shown in table 1. The pulse duration was deduced from the corresponding oscillogram for a determined value of the capacitance (Fig.4).

Moreover, the mass of extracted juice m (g), the amount of Betatine expressed in terms of Absorbance Abs (%) and the energy $W = 1/2\, n\, C\, V^2$ were considered significant to be considered as the response of the model.

TABLE1: Values of the pulse duration according to corresponding capacitance

C (µF)	0.2	0.4	0.5	0.8	1.2
T (µs)	8	20	24	40	60

A set of three preliminary experiments was performed. For each experiment, a factor was varied while the others were kept constant. The purpose of these preliminary experiments is to determine the variation domain for each factor in view of the following step by performing a CCF experimental design, which should enable the modeling and optimization of the PEF treated extraction process.

Fig.4 . Pulse shape of duration T= 40 µs obtained for C=0.8 µF

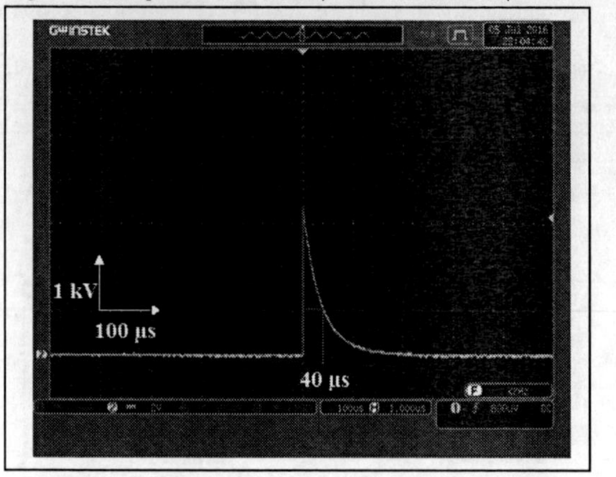

The obtained results of the preliminary experiments, for both treatment chambers, are plotted in Figs.5-7, representing the variation of the extracted juice mass *m* as function of the voltage *V*, the pulses number *n* and the pulse duration *T* respectively. Moreover, the variation of the absorbance is shown in Figs.8-10 for both treatment chambers.

Fig.5 . Evolution of the extracted beet juice mass according to the voltage (n=50,T=40µs)

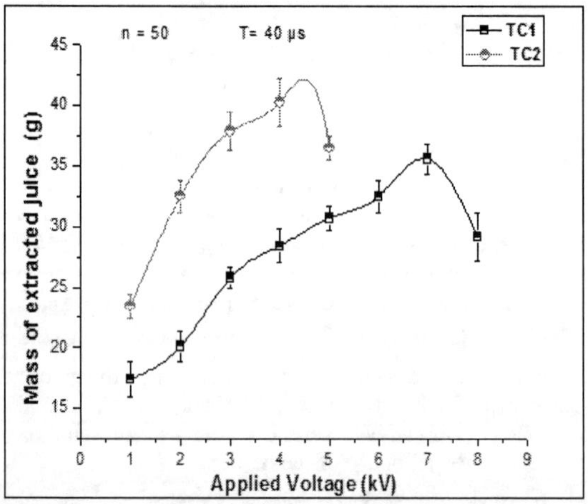

According to the obtained results of the preliminary experiments, following variation domains of each factor were selected, corresponding to the best limits giving maximum outcome:

Model TC1: V_{min}= 6 kV & V_{max}= 8 kV; n_{min}= 80 & n_{max}= 120; T_{min}= 20 µs & T_{max}= 60 µs

Model TC2: V_{min}= 3 kV & V_{max}= 5 kV; n_{min}= 60 & n_{max}= 100; T_{min}= 8 µs & T_{max}= 40 µs

Fig.6 . Evolution of extracted beet juice mass according to the pulses number (V=3kV, T= 40 µs)

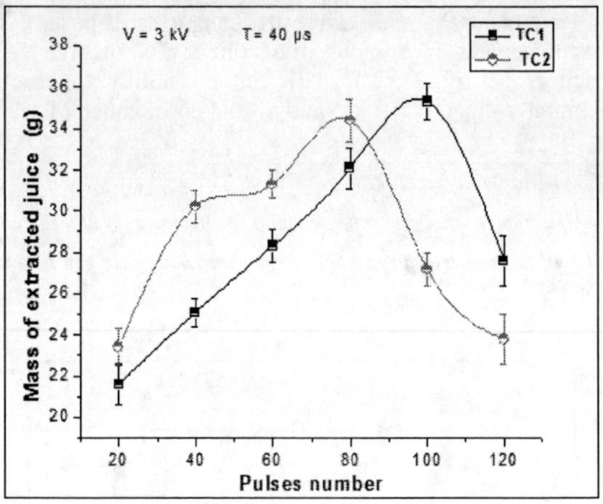

Fig.7 . Evolution of extracted beet juice mass according to the pulse duration (V=3kV, n=60)

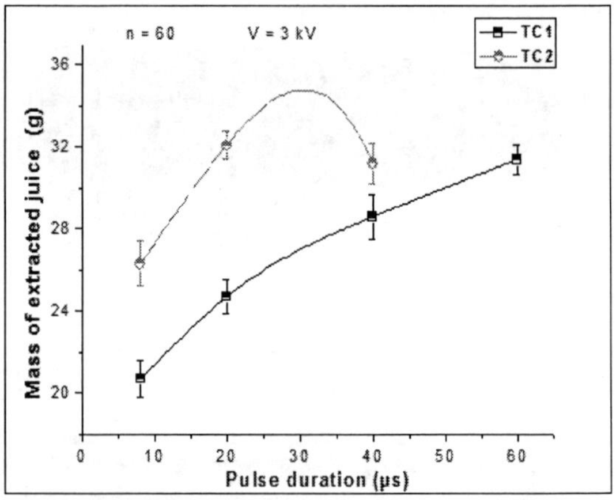

The following step is the modeling of the process using the design of experiments methodology, by performing two CCF designs for both TC1 and TC2 chambers (Tables 2 and 3).

Predictive and descriptive quality of the models are satisfactory since values of criteria $Q^2$and R^2 are close to the unit, leading to validated mathematical models for both chambers.

The mathematical model of the responses considered for optimization, which are the mass of extracted juice, the value of the absorbance and the consumed energy were obtained with MODDE 5.0, as shown in table 4.

Fig.8 . Evolution of the juice absorbance according to the voltage (n=50, T= 40 µs)

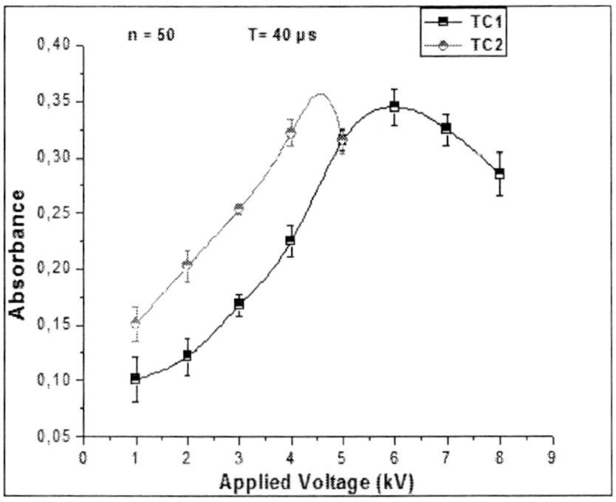

Fig.9 . Evolution of the juice absorbance according to the pulses number (V=3kV, T= 40 µs)

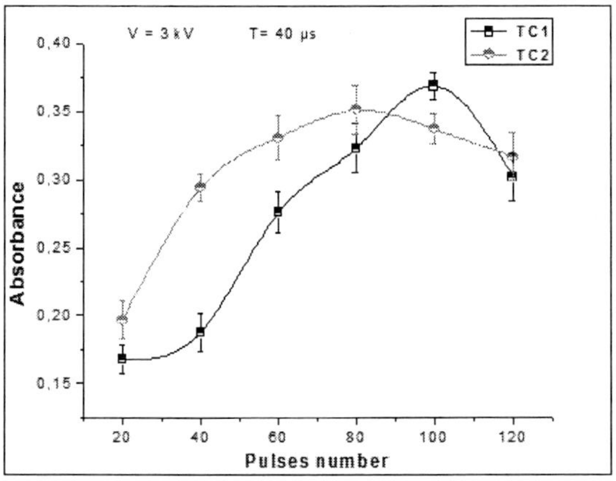

Fig.10. Evolution of the juice absorbance according to the pulse duration (V=3kV, n=60)

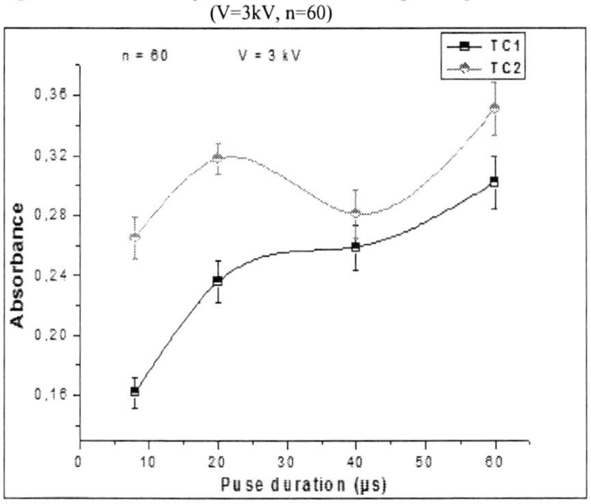

TABLE 2: Results of the CCF Design experiment (TC1)

Exp N°	V (kV)	n	T (µs)	m (g)	Abs	W (J)
01	6	80	20	21.3	0.317	576
02	8	80	20	29.3	0.384	1024
03	6	120	20	24.7	0.364	864
04	8	120	20	30.3	0.395	1536
05	6	80	60	27.4	0.389	1728
06	8	80	60	32.7	0.457	3072
07	6	120	60	28.5	0.377	2592
08	8	120	60	32.8	0.383	4608
09	6	100	40	25.2	0.386	1440
10	8	100	40	32.8	0.415	2560
11	7	80	40	31.3	0.371	1568
12	7	120	40	31.4	0.373	2352
13	7	100	20	26.6	0.355	980
14	7	100	60	29.9	0.388	2940
15	7	100	40	30.5	0.385	1960
16	7	100	40	30.5	0.382	1960
17	7	100	40	30.5	0.384	1960

TABLE 3: Results of the CCF Design experiment (TC2)

Exp N°	V (kV)	n	T (µs)	m (g)	Abs	W (J)
01	3	60	8	34.3	0.376	54
02	5	60	8	39.1	0.482	150
03	3	100	8	36.2	0.456	90
04	5	100	8	39.5	0.515	250
05	3	60	40	38.4	0.488	216
06	5	60	40	40.1	0.522	600
07	3	100	40	39.2	0.492	360
08	5	100	40	39.5	0.495	1000
09	3	80	24	37.7	0.474	180
10	5	80	24	41.5	0.543	500
11	4	60	24	35.9	0.404	240
12	4	100	24	36.8	0.422	400
13	4	80	8	37.7	0.53	128
14	4	80	40	39.8	0.577	512
15	4	80	24	37.6	0.449	320
16	4	80	24	37.2	0.471	320
17	4	80	24	37.4	0.459	320

The software offers the possibility to identify the optimal values of the factors which should give the highest amount of extracted juice m and Absorbance Abs for the smallest power consumption. It contains an optimization routine that is capable of simultaneously processing several responses, affected by different weighting coefficients. MODDE.05 has an optimizer tool, which proposes the optimal values of factors by maximizing m and Abs and minimizing the energy.

TABLE 4: Mathematical model of the responses

	TC1			TC2		
	Mass (g)	**Abs**	**W (J)**	**Mass (g)**	**Abs**	**W (J)**
Constant	30.3	0.382	1960	37.76	0.47	320
V	3.08	0.020	560	1.39	0.02	160
n	0.56	-0.00	398.4	0.34	0.01	84
T	1.91	0.01	996	1.02	0.02	201.6
V*V	-1.19	0.02	40	1.56	0.01	20
n*n	1.15	-0.01	0	-1.68	-0.07	0
T*T	-1.94	-0.01	0	0.71	0.06	0
V*n	-0.42	-0.01	112	-0.36	-0.01	40
V*T	-0.49	-0.09	280	-0.76	-0.01	96
n*T	-0.4	-0.01	200	-0.26	-0.01	51

Fig.11. Optimal values of the factors proposed by MODDE 5.0

	Response	Criteria	Weight	Min	Target	Max
1	Mass of juice	Maximize	1	32,5058	33,6688	
2	Absorbance	Maximize	1	0,443997	0,457647	
3	Energy consumption	Minimize	1		442,16	833,04

Iteration: 5001 Iteration slider:

	1	2	3	4	5	6	7	8
	Applied Voltage	Pulses number	Pulse duration	Mass of juice	Absorbance	Energy consumption	iter	log
1	6	120	20,0462	24,2611	0,3697	812,528	5000	1,5
2	8	80	20	29,4673	0,385	973,599	5000	1,1
3	8	80	35,8965	32,7684	0,4203	1828,83	5000	0,8
4	8	80	52,9693	33,5777	0,4448	2747,35	5001	1,0
5	7,9999	80	42,0506	33,3867	0,4308	2159,88	5001	0,8
6	7,6348	80,0044	20	28,7173	0,3605	888,425	5000	1,3
7	8	80	52,9639	33,5779	0,4448	2747,06	5001	1,0
8	8	80	52,9693	33,5777	0,4448	2747,35	5001	1,0

According to this model, the optimum of the process (i.e., maximizing the mass m and the absorbance Abs and minimizing the energy) should be obtained:

for model TC1 : V= 8 kV, n = 80 and T = 35.9 µs corresponding to m = 32.7 g, Abs= 0.48 and P =1828.8 J

and for Model TC2: V = 5 kV, n = 70 and T = 8 µs corresponding to m = 40.68 g, Abs= 0.5 and P =167.9 J

According to this model, the optimum of the process (i.e., the greatest amount of juice, of absorbance) for a minimum of energy consumed should be obtained for voltage V_0 = 5 kV, number of pulses n_0 = 70 and pulse duration T_0 = 8 µs.

Fig.12. Optimal values of the factors proposed by MODDE 5.0

	Response	Criteria	Weight	Min	Target	Max
1	Mass of juice	Maximize	1	40,3925	41,0424	
2	Absorbance	Maximize	1	0,552788	0,571042	
3	Energy consumption	Minimize	1		22,1401	112,66

Iteration: 5001 Iteration slider:

	1	2	3	4	5	6	7	8
	Applied Voltage	Pulses number	Pulse duration	Mass of juice	Absorbance	Energy consumption	iter	log(D)
1	4,95	71,4923	8	40,4992	0,5545	167,05	5001	0,1335
2	5	70,5589	8	40,6851	0,5546	167,94	5001	0,0921
3	3	79,3889	40	40,419	0,5689	282,697	5000	0,4876
4	4,7952	70,1153	38,2863	40,4748	0,5557	628,299	5001	1,1886
5	4,9559	71,3549	8	40,5186	0,5544	167,061	5001	0,1294
6	4,95	71,4923	8	40,4992	0,5545	167,05	5001	0,1335
7	4,9481	71,5119	8	40,4907	0,5545	166,956	5000	0,1363
8	4,7957	70,0222	38,3186	40,4733	0,5557	628,194	5001	1,1885

IV. CONCLUSION

The present paper describes an experimental comparative analysis between two square treatment chambers of same dimensions but having two or four metal electrodes placed on the side walls. For this latter, each pair of the adjacent metal plates form one electrode. The experimental analysis was made by measuring the mass of PEF pretreated extracted juice from beet and the amount of betanine using a spectrophotometer. The obtained results, using the methodology of experimental designs, have shown that the second model with four electrodes is more efficient, because higher quantities of juice and Betanine were obtained up to 25% with an energy saving of nearly 90%.

References

[1] Bazhal, M. I., Ngadi, M. O., & Raghavan, V. G. (2003). Influence of pulsed electroplasmolysis on the porous structure of apple tissue. *Biosystems engineering*, *86*(1), 51-57.

[2] El-Belghiti, K., Rabhi, Z., & Vorobiev, E. (2005). Kinetic model of sugar diffusion from sugar beet tissue treated by pulsed electric field. *Journal of the Science of Food and Agriculture*, *85*(2), 213-218.

[3] Maged E.A. Mohamed and Ayman H. Amer Eissa «Pulsed Electric Fields for Food Processing Technology» in Structure and Function of Food Engineering, Chapter 11, Intech, 2012.

[4] Sébastien GONZALEZ « Résolution numérique de l'équation de Lighthill par éléments finis et équations intégrales pour l'estimation du bruit rayonné par des écoulements en conduite ». doctoral thesis in Advanced mechanics. under the direction of Pr. Emmanuel Lefrançois. Compiégne. University of Technology, 2013.

[5] Humberto Vega-Mercado, M. Marcela Gongora-Nieto, Gustavo V. Barbosa-Canovas and Barry G. Swanson «Pulsed Electric Fields in Food Preservation» Handbook of Food Preservation, Second Edition, Taylor & Francis Group, LLC, 2007.

[6] Ratna Ika Putri, Ika Noer Syamsiana and La Choviya Hawa «Design of High Voltage Pulse Generator for Pasteurization by Pulse Electric Field (PEF)» International Journal of Computer and Electrical Engineering, Vol. 2, No. 5, October, 2010, pp 916-917.

[7] JEYAMKONDAN S., JAYAS D.S. et HOLLEY R.A. «Pulsed electric field processing of foods» : a review. J. Food Protect., 1999, 62, pp 1088-1096.

[8] Francesca De Vito, «Application of Pulsed Electric Field (PEF) Techniques in Food Processing »,Ph.D. Course in Chemical Engineering (IV Cycle-New Series), under the direction of Prof. Ernesto

REVERCHON, SALERNO, Department of Chemical and Food Engineering, 2006.

[9] Nicolas Meneses, Henry Jaeger and Dietrich Knorr «Understanding Enzyme Inactivation Mechanisms During Pulsed Electric Field Treatments» Department of Food Biotechnology and Food Process Engineering, Berlin University of Technology.

[10] Kambiz Shamsi and Frank Sherkat «Application of pulsed electric field in non-thermal processing of milk» Asian Journal of Food and Agro-Industry ISSN 1906-3040, 2009, pp 219-220.

Analysis of PEF treatment using a square multi-units chamber for food processing

[1]A.Semmak, [1]H.Bermaki, [1]Y. Bellebna, [2]M. Kaddour, [3]H.Belhassaini, [1]A. Tilmatine
[1]APELEC Laboratory, Djillali Liabes University of Sidi Bel-Abbes, Algeria
[2]LSTE Laboratory, University of Mascara, Algeria
[3]BVCV Laboratory, Djillali Liabes University of Sidi Bel-Abbes, Algeria

Abstract— **The objective of this paper is dedicated to the experimental analysis of a square treatment chamber comprising two and three parallel units constituted by three and five alternate HV-Ground electrodes respectively. The chamber made of Plexiglas has a square paralelipedic shape of dimensions 4x4x10 cm3, in which are placed vertical and parallel stainless steel electrodes, of dimensions 10x4 cm^2. This chamber has been compared to the classic one-unit constituted of two electrodes, one HV and one ground electrodes both placed on the side walls of the chamber. The experimental analysis was made using a soark-gap switch pulse generator, by measuring the mass of PEF pretreated extracted juice from beet m, the amount of Betanine and the consumed energy W. The obtained results have shown that the proposed models constituted of two and three units are better than the similar classic constituted of only one treatment unit. Higher values of m were obtained with the multi-units chamber with smaller energy.**

Keywords— Treatment chamber, electrodes, pulsed electric field, High Voltage, electroporation.

I. INTRODUCTION

During the last decade, the pulsed electric field (PEF)-treatment was found to be useful for enhancing the pressing, drying, extraction, and diffusion processes **[1-4]**. The PEF-treatment has also found application as a minimally invasive method for processing of plant tissues, allowing to avoid many undesirable changes in products, pigments, vitamins and flavouring agents, which are typical for other pre-treatment techniques, including thermal, chemical and enzymatic ones. Moreover, the PEF-treatment is also promising for purposes of microbial inactivation **[5-6]**. The PEF-treatment at moderate electric field strength E=500–1000 V/cm and treatment time within 10^{-4}–10^{-2} s allows to damage tissue effectively without any significant temperature increase **[7-8]**.

The unique property of PEF application is related to selective damage of the biological membranes. An external electric field E induces a transmembrane potential on a membrane. When the transmembrane potential exceeds some threshold value (typically about 0.2–1.0 V), the electric field causes a temporary loss of semi-permeability by the cell membranes (electro-permeabilization) or their damage.

The treatment chamber, in which is placed the product to be PEF treated, usually consists of two electrodes; one is connected to the pulse generator and the other is connected to the ground. The two electrodes are separated by an electrical insulator **[9]**. The basic idea of the treatment chamber is to keep the treated product inside during pulsing, although the uniformity of the process is highly dependent on the characteristic design of the treatment chamber **[10]**. Treatment chambers can be static (batch) or continuous. Static chambers are mostly used in laboratory scale studies for PEF treatments of both liquid and solid products. Continuous chambers, instead, are more suitable for pilot plant and industrial scale applications **[11]**.

Co-field treatment chambers described by "Yin et al" have two hollow cylindrical electrodes separated by an insulator to form a tube through which the product flows. Field distribution in this type of chamber is not uniform **[12]**. In this configuration the product flow as well as the electric field lines are parallel **[11]**.

The major challenge today lies in the application of PEF in the industry; it requires an ever-increasing processing rate by seeking ways to continuously reduce energy consumption. In all treatment chambers (TC) models used in industry and in research laboratories, the product to be treated is placed in one volume unit delimited between two electrodes. The objective of this paper is to perform an experimental analysis of a new TC model comprising two or three PEF units in parallel, consisting of several "alternate ground-HV electrodes," in order to increase the processing flow rate and save more energy.

II. MATERIAL AND METHODS

The experimental setup used in the present work is composed of a number of components, comprising a high DC voltage source, an energy storage capacitor, a spark gap switch and a treatment chamber (Fig.1). A DC high voltage (Spellman 40 kV, 9 mA) charges the bank of capacitors until producing the spark gap's breakdown, causing an abrupt voltage (shock) applied to the load (treatment chamber where the sample is disposed). The storage element is composed of three sets of five series capacitors (2 μF, 2 kV), with the possibility to reach a maximum voltage of 10 kV and a total capacitance of 1.2 μF (Fig.2).

Three square paralelipedic treatment chambers of same dimensions 60x60x100 mm^3 comprising one, two and three treatment units were used, constituted of two, three and five identical parallel stainless steel electrodes respectively. Each chamber is made of Plexiglas in which are placed identical vertical and parallel stainless steel electrodes, of dimensions 60x100x2 mm^3.

Fig.1. The pulse generator.

a) Descriptive schematic of the setup; b) The photography of the setup

1- HV DC power supply, 2-Set of capacitors, 3- Spark gap switch, 4- Treatment chamber

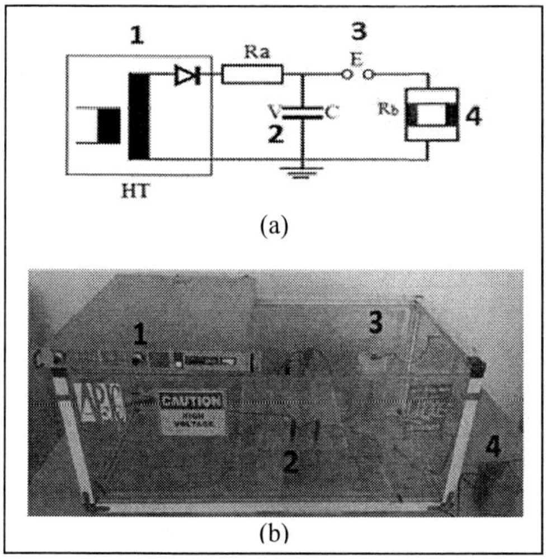

Fig.2 . The bank of capacitors.

a) Descriptive schematic; b) The photography of the capacitors bank

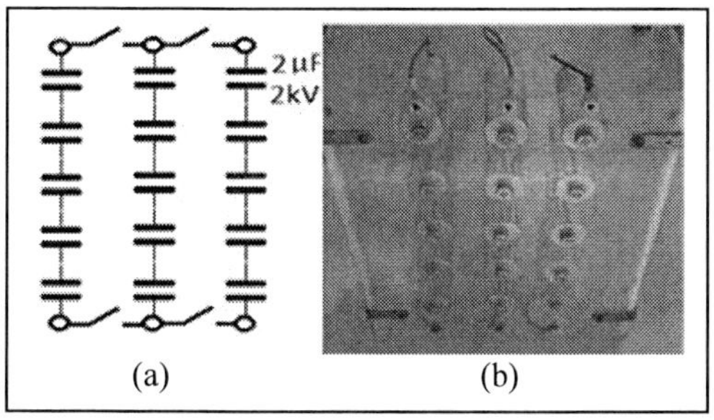

The three TC, having the same maximum volume of 360 ml, are described below (Fig.3):

- The model TC1 is constituted of one unit and comprises two electrodes: one HV electrode and one ground electrode both placed on the side walls of the chamber.

- The model TC2 is constituted of two units comprising three electrodes: one central HV electrode and two side ground electrodes.

- The model TC3 is constituted of three units comprising five alternate HV and ground electrodes: two HV electrodes and three ground electrodes.

Fresh beets, of average mass 60 g each one, were used. After sorting and cleaning operations, a homogenous mash was obtained using a domestic food processor. Before each experiment, the mash was properly mixed to obtain a homogenous mixture. Right after PEF treatment, a hydraulic pressing machine (Mega, 15 tons) was used for the juice extraction. The filled extraction chamber was pressed until a defined pressure of 100 kg/cm^2, and was then held at this pressure for 5 min. For all the experiments, the same sample mass m=80 g was used.

Fig.3: Multi-units treatment chambers

1: stainless steel electrodes; 2: Plexiglas (All dimensions in mm)

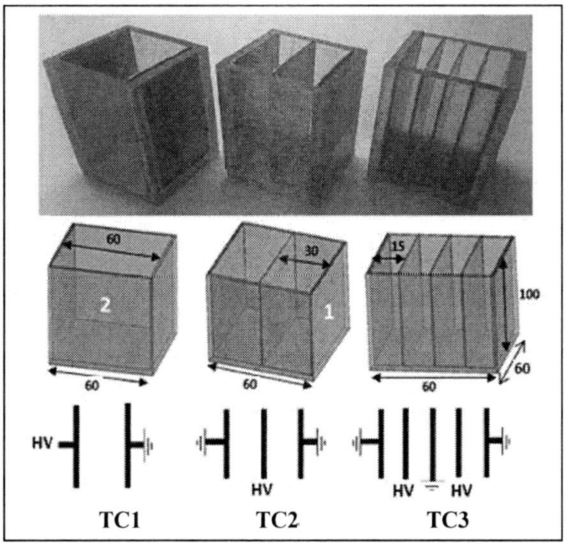

Extracted juice filtered through a stainless steel sieve, was collected in a plastic collector placed under the extraction chamber. A 0.1 mg precision electronic balance was used to weight the extracted mass M (g) of beet juice. All the experiments were carried out in stable climatic conditions of temperature (20-25 °C) and humidity (40-60 %).

An experimental investigation was performed to compare the PEF treatment efficiency between the three models. For each model, the influence of the applied voltage (V, kV), the pulses number (n) and the pulse duration (T, µs) was analyzed. The pulse duration which is determined by the corresponding value of the capacitance C (Table 1), is deduced from the corresponding oscillogram for a determined value of C (Fig.4).

Moreover, the mass of extracted juice m (g), the amount of Betatine expressed in terms of Absorbance Abs and the energy $W= 1/2\ n\ C\ V^2$ were considered significant to be considered for the evaluation of the PEF-treatment efficiency.

TABLE1: Values of the pulse duration according to corresponding capacitance

C (µF)	0.2	0.4	0.5	0.8	1.2
T (µs)	8	20	24	40	60

Fig.4 . Pulse shape obtained for C=1.2 μF.

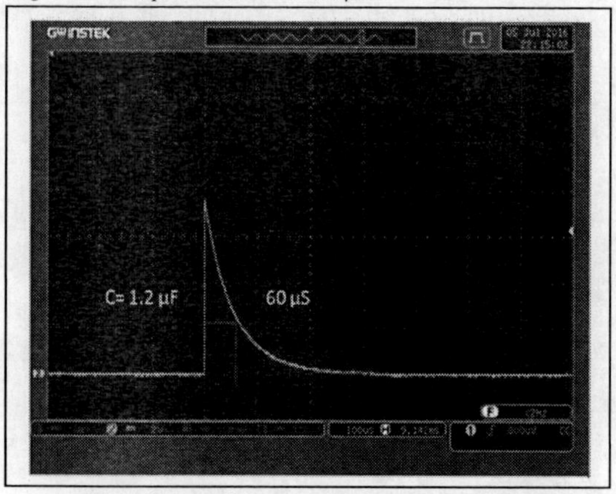

III. RESULTS AND DISCUSSION

For all the experiments carried out in this section, for each TC model, one factor was varied while the two other factors were kept constant. Thus, tables 2, 3 and 4 represent the variation of the PEF treatment efficiency, in terms of extracted juice mass *m*, absorbance *Abs* and Energy *W*, according to V, n and T, respectively.

TABLE 2: Extracted juice mass *m*, absorbance *Abs* and Energy *W* according to the applied voltage V (kV), for the three chambers (n=50, T= 40 μs)

	TC1		TC2		TC3		
V (kV)	M (g)	Abs	M (g)	Abs	M (g)	Abs	W (J)
1	16.6	0.237	17.2	0.346	19.3	0.390	20
2	19.3	0.253	21.5	0.381	25	0.432	80
3	21	0.302	27.7	0.4	30.4	0.453	180
4	25.4	0.346	28.2	0.427	33	0.470	320
5	28.0	0.356	30.5	0.446	33.4	0.474	500
6	30.1	0.382	34.2	0.452	30	0.386	720
7	31.9	0.394	29.8	0.362	28.4	0.332	980
8	29,8	0,345	28.2	0.325	19.3	0.390	1280

As expected, the mass of extracted juice and the quantity of Betanine obtained with a PEF treated sample increases according to the applied voltage, for the three chambers. Beyond a determined value of the voltage (5 kV for TC1, 4 kV for TC2 and 3 kV for TC3), the effect of the PEF treatment is inversed due to "oxidation" of the product. However, the treatment is more efficient for models TC3 and TC2 compared with the classic chamber TC1 comprising two electrodes. While for the model TC1, the optimal treatment was obtained for V = 7 kV (m= 31.9 g, Abs= 0.394 and W=980 J), for the multi-unit chambers greater values of m and Abs were obtained with smaller voltage. Thus, for models TC2 and TC3, the optimal treatment was obtained for V = 6 kV (m= 34.2 g, Abs= 0.452 and W=720 J), and V= 5 kV (m= 33.4 g, Abs= 0.474 and W=500 J) respectively.

Energy saving for treatment chambers TC2 and TC3 compared to the TC1 treatment chamber can be calculated by the following relation:

$$W_{saving} = (W_{TC1} - W_{TCi}) / W_{TC1} \quad (1)$$

W_{TC1}: Energy consumed by model TC1
W_{TCi}: Energy consumed by model TC2 or TC3

An energy saving of 26% and 48 % were achieved for models TC2 and TC3 respectively.

The advantage of the multi-units chambers is that for the same voltage the electric field is increased compared to the "one-unit" chamber comprising two electrodes. For example, when a voltage of 6 kV is applied, the electric field of the models TC1, TC2 and TC3 is equal to 1, 2 and 4 kV/cm respectively.

TABLE 3: Extracted juice mass m, absorbance Abs and Energy W according to the applied pulses number, for the three chambers (V = 3 kV, T = 40 μs)

	TC1		TC2		TC3		
N	m	Abs	m	Abs	m	Abs	W
20	24.6	0.247	28.9	0.291	28.8	0.324	72
40	28.1	0.331	30.8	0.336	31.8	0.333	144
60	29.6	0.327	32.4	0.382	36.2	0.606	216
80	32.0	0.328	36.2	0.476	31.4	0.418	288
100	36.7	0.359	28.6	0.450	28.5	0.450	360
120	28.6	0.304	27.1	0.321	25	0.414	432

As seen in Table 3, the mass of extracted juice and the quantity of Betanine obtained with a PEF treated sample increases with pulses number, for the three chambers. The treatment is more efficient for models TC2 and TC3 compared with the classic chamber TC1. While for the model TC1, the optimal treatment was obtained for n= 100 pulses (m= 36.7 g, Abs= 0.359 and W=360 J), for the multi-unit chambers greater values of m and Abs were obtained with less pulses. Thus, for models TC2 and TC3, the optimal treatment was obtained for n= 80 (m= 36.2 g, Abs= 0.476 and W=288 J), and n= 60 (m= 36.2 g, Abs= 0.606 and W=216 J) respectively. An energy saving of 20% and 40% were achieved for models TC2 and TC3 respectively.

On the other hand, the mass m and absorbance Abs obtained with a PEF treated sample increases with the pulse width, for the three chambers, and decreases beyond a determined value (Table 4). As for the previous factors, the treatment is more efficient for multi-units models compared with the classic chamber. For the model TC1, the optimal treatment was obtained for τ= 60 µS (m= 31.4 g & Abs= 0.302), however for the models TC2 and TC3 greater values were obtained with smaller pulse width. The optimal treatment was obtained for τ= 40µS with both models (m= 31.4 g & Abs= 0.331 with TC2; m= 32.7 g & Abs= 0.342 with TC3). For this latter, a more efficiency treatment was obtained for τ= 40 µs (m= 36.2 g & Abs= 0.496).

TABLE 4: Extracted juice mass m, absorbance Abs and Energy W according to the applied pulse width, for the three chambers (V=3 kV, n=60).

τ (µS)	C(µF)	TC1		TC2		TC3		
		m	Abs	m	Abs	m	Abs	W
8	0.2	21.7	0.256	23.5	0.263	24.6	0.297	54
20	0.4	24.7	0.264	29.4	0.266	32.7	0.342	108
40	0.8	27.6	0.269	31.4	0.331	36.2	0.496	216
60	1.2	31.4	0.302	28.1	0.288	30.8	0.441	324

IV. CONCLUSION

A comparative experimental analysis was performed between three different treatment chambers constituted of one (TC1), two (TC2) and three (TC3) parallel units. Although, the volume of the chambers is the same, however the one comprising three units has given the best results in terms of extracted juice mass, amount of Betanine and energy consumption. For the same voltage, thus for the same energy, the electric field of TC2 and TC3 models is two and three times higher respectively in comparison with the one unit model.

References

[1] BARSOTTI, L. AND CHEFTEL, J.C. (1998) Traitement des aliments par champs électriques pulses. Science des Aliments 18, 584–601.

[2] ANGERSBACH, A., HEINZ, V. AND KNORR, D. (2000) Effects of pulsed electric fields on cell membranes in real food systems. Innovative Food Science and Emerging Technologies 1(2), 135–149.

[3] VOROBIEV, E., JEMAI, A.B., BOUZRARA, H., LEBOVKA, N.I. and BAZHAL, M.I. (2005) Pulsed electric field assisted extraction of juice from food plants. In: G. BARBOSA-CANOVAS, M.S. Tapia and M.P. Cano (Eds.), Novel food processing technologies, CRC Press, New York, pp. 105–130.

[4] VOROBIEV, E. AND LEBOVKA, N.I. (2006) Extraction of intercellular components by pulsed electric fields. In: J. RASO AND V. HEINZ (Eds.), Pulsed electric field technology for the food industry. Fundamentals and applications. Springer, New York, pp. 153–194.

[5]] BARBOSA-CANOVAS, G. and VEGA-MERCADO, H. (1996) Dehydration of foods. Chapman & Hall, New York.

[6] TOEPFL, S., HEINZ, V. AND KNORR, D. (2007) High intensity pulsed electric fields applied for food preservation. Chemical Engineering and Processing 46(6), 537–546.

[7] FINCAN, M. AND DEJMEK, P. (2002) In situ visualization of the effect of a pulsed electric field on plant tissue. Journal of Food Engineering 55, 223–230.

[8] LEBOVKA, N.I., BAZHAL, M.I. AND VOROBIEV, E. (2002) Estimation of characteristic damage time of food materials in pulsed-electric fields. Journal of Food Engineering 54, 337–346.

[9] Nabil GRIMI. «VERS L'INTENSIFICATION DU PRESSAGE INDUSTRIEL DES AGRORESSOURCES PAR CHAMPS ÉLECTRIQUES PULSÉS : ÉTUDE MULTI-ÉCHELLES ». doctoral thesis in Industrial Process Engineering and Sustainable Development. under the direction of Pr. M. Eugène VOROBIEV . Compiegne. University of Technology, 2009.

[10] Maged E.A. Mohamed and Ayman H. Amer Eissa, «Pulsed Electric Fields for Food Processing Technology», Agricultural and Biological Sciences, Structure and Function of Food Engineering", book edited by Ayman Amer Eissa, ISBN 978-953-51-0695-1, Published: August 22, 2012 under CC BY 3.0 license, Chapter 11.

[11] Francesca De Vito, «Application of Pulsed Electric Field (PEF) Techniques in Food Processing »,Ph.D. Course in Chemical Engineering (IV Cycle-New Series), under the direction of Prof. Ernesto REVERCHON, SALERNO, Department of Chemical and Food Engineering.

[12] Paul TAKHISTOV , «pulsed electrical field in food processing and preservation». handbook of food science and engineering, volume 4, 126, CRC press 2005.

Dependence of Shock Strength on the Operating Conditions in Underwater Wire Explosion

K.J. Chung, K. Lee, Y.S. Hwang
Dep't. of Nuclear Engineering, Seoul National University
Seoul, Republic of Korea

D.K. Kim
Agency for Defense Development
Daejeon, Republic of Korea

***Abstract*—In this paper, we investigate the strength of cylindrical shock waves generated by exploding wires driven by pulsed capacitive discharges for various operating conditions such as wire dimension and circuit parameters. In the numerical model, we solve a circuit equation coupled with one-dimensional magneto-hydrodynamic equations with the help of appropriate wide-range equation of state and semi-empirical electrical conductivity for copper. The numerical model is validated by comparing the simulation results with the experimentally measured current, voltage, and pressure waveforms. It is found that the strength of shock wave generated by wire explosion is greatly influenced by wire dimension and the peak pressure is proportional to the current value at the time of voltage peak (explosion time) irrespective of operating conditions.**

Keywords—wire explosion; shock wave; magneto-hydrodynamic; equation of state; pulsed power

I. INTRODUCTION

Explosion of a thin metal wire with a rapidly-rising current pulse has been widely utilized as a simple and effective method to generate strong shock wave in water [1,2]. In this technique, phase transition of the metal wire from liquid state to gas or plasma state during the Joule heating is known to be responsible for the generation of such a strong shock wave. Therefore, knowledge on temporal evolution of the thermodynamic state of the heated metal on the 3-dimensional $p-\rho-T$ diagram is essential for understanding the origin of strong shock wave. An appropriate electrical conductivity model is also required to calculate the Joule heating exactly.

Recently, we have developed a time-transient simulation code to analyze the wire explosion system for generating strong shock wave in water [3]. The simulation code solves a circuit equation coupled with 1-dimensional magneto-hydrodynamic (MHD) equations with the help of wide-range equation of state (EOS) and semi-empirical electrical conductivity models for copper. It allowed us to calculate the temporal change in thermodynamic state of the metal wire heated by pulsed electrical current. The numerical model was validated by comparing the simulation results with the experimentally measured current, voltage, and pressure waveforms.

In this paper, we investigate the strength of cylindrical shock waves generated by exploding copper wires driven by pulsed capacitive discharges with our in-house 1-dimensional simulation code. The simulations are carried out for various operating conditions such as wire dimension and capacitor charging voltage and we compare the peak pressures calculated at the position 53 mm away from the wire.

II. MODEL

The numerical model couples the circuit equation for wire explosion system with the resistive MHD equations for an exploding wire as a time-transient, nonlinear load of the wire explosion system [3]. The calculation procedure of the numerical model for exploding wires is depicted in Fig. 1. For calculating the circuit equation, an appropriate calculation of the wire resistance (R_p) at each time is needed. Since the R_p is directly related to the radial distribution of electrical conductivity of the wire ($\sigma(r)$) depending on temperature and mass density, an appropriate model for electrical conductivity of the wire is essential. In addition, the shock pressure evolved in the surrounding medium (water) is closely related to the EOS of the wire (copper).

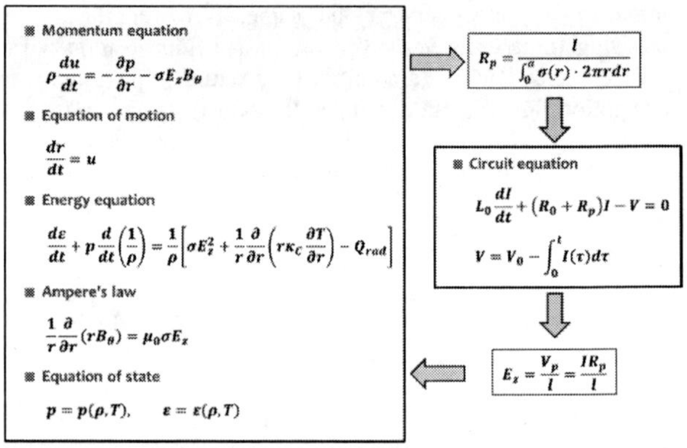

Fig. 1. Schematic description of the time-transient simulation code for calculation of the wire explosion system driven by pulsed capacitive discharge.

As described in our previous paper [3], we adopt the quotidian EOS (QEOS) for copper and Tait-type EOS for surrounding water. For the conductivity of copper, a semi-empirical form covering from condensed state to partially ionized plasma state is utilized. Since the conductivity of water is much lower than that of copper in all temperature and density range, electrical conductivity is assumed to be zero for water. Details for the simulation conditions and circuit parameters are described in elsewhere [3].

III. RESULTS AND DISCUSSION

In Fig. 2, we compare the voltage, current, and pressure waveforms calculated from the numerical model with the experimentally measured ones. One can find that the calculated current and voltage waveforms agree with the measured ones, except for slight difference at the current minimum after explosion. Here, the explosion time is defined as the time at the peak voltage, because the resistance change in the wire shows the maximum and the expansion speed of the wire also becomes the maximum, as shown in Fig. 3. Moreover, the Joule heating power is the maximum at the explosion time, indicating the maximum conversion of inductively stored energy in the circuit to resistive heating energy in the wire. It is also noted that the wire passes by near the critical point thermodynamically at the explosion time.

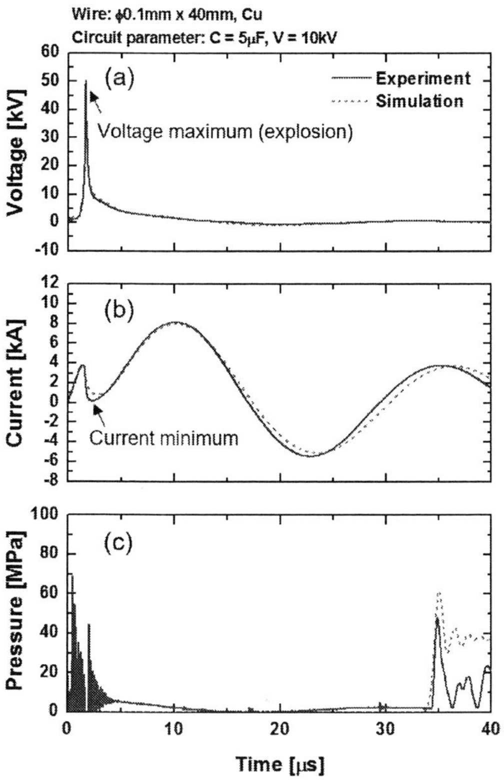

Fig. 2. Comparison of the voltage (a), current (b), and pressure (c) waveforms obtained from the simulation with those measured by experiment for the explosion of Cu wire with 100 μm in diameter. The capacitor charging voltage is 10 kV.

The peak pressure obtained from the numerical model shows a slightly larger value than that measured with a piezoelectric pressure sensor. Overall waveform is also quite different from the experiment. This discrepancy seems to be attributed to the imperfect EOS and electrical conductivity model. Nevertheless, we believe that the numerical simulation is applicable to the investigation of shock wave egenerated by wire explosion.

From Figs. 3(c) and 3(d), one can see that the radiation loss plays no role during the explosion, i.e. the process for generating shock wave, because the temperature of the wire is too low to emit strong radiation. It is also noticed that the Joule heating energy is consumed to increase the internal energy of the wire before the explosion. However, the internal energy does not further increase after explosion time. Instead, the difference between the Joule heating and the internal energy increases. This energy difference is nothing but the work done on the surrounding medium, thereby is closely related to the strength of shock wave generated in the medium.

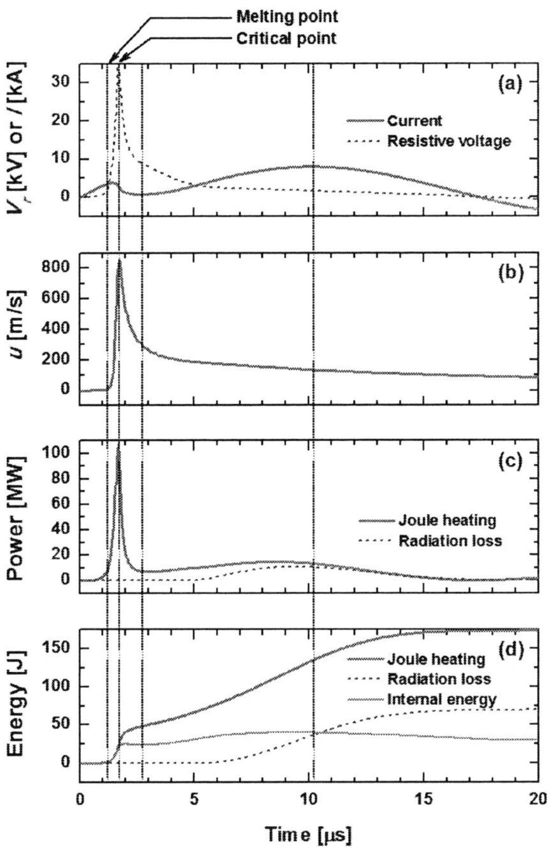

Fig. 3. Typical current and voltage waveforms (a) and corresponding expansion speed (b), power (c), and energy (d) for the explosion of Cu wire with 100 μm in diameter. The capacitor charging voltage is 10 kV.

We have performed a lot of simulations to find the correlation between the peak pressure and the operating conditions such as wire dimension and charging voltage. In view that the strength of shock pressure is closely related to the conversion of inductive energy stored in the circuit to resistive Joule heating energy during the explosion, we correlate the peak pressure with the current value at the time of wire explosion (peak voltage).

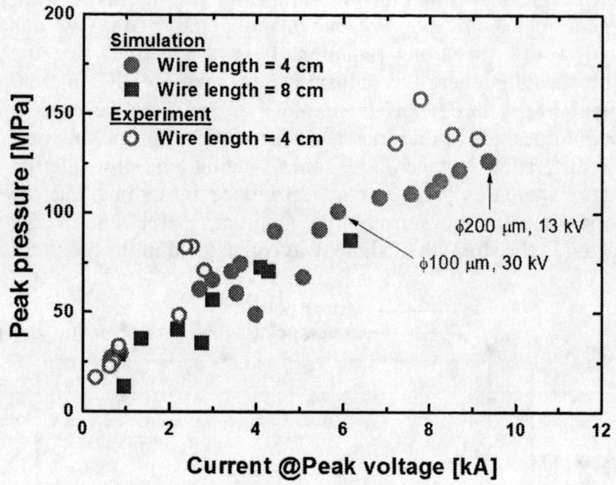

Fig. 4. Comparison of camera images before and after the well cleaning by underwater spark discharge.

The simulation result is shown in Fig. 4 for various wire diameters of raging from 25 to 300 μm and lengths of 4 and 8 cm. The capacitor charging voltage is varied from 10 to 30 kV. It is found that the peak pressure is well correlated with the current value at the time of wire explosion nearly irrespective of the wide-range operating conditions. It is clearly observed that the wire diameter is the most important factor to obtain the strong shock wave in the medium.

IV. Conclusion

In this paper, we show that the strength of shock wave generated by wire explosion is well correlated with the current value at the time of explosion. We find that optimizing the wire diameter is the most important in order to obtain the strongest shock wave in medium for given circuit parameters. This study is useful in optimizing the circuit parameters and wire dimension for utilization of shock waves generated by wire explosion in various applicaions.

Acknowledgment

This work was supported by the Defense Research Laboratory Program of the Defense Acquisition Program Administration and the Agency for Defense Development of Republic of Korea.

References

[1] W. G. Chase and K. H. Moore, Exploding Wires, Vol. 1, Plenum Press (1959).

[2] Y. E. Krasik, A. Grinenko, A. Sayapin, S. Efimov, A. Fedotov, V. T. Gurovich, and V. Oreshkin, "Underwater electrical wire explosion and its applications," IEEE Transactions on Plasma Science, vol. 36, no. 2, 2008, pp. 423-434.

[3] K. J. Chung, K. Lee, Y. S. Hwang, and D. K. Kim, "Numerical model for exploding copper wires in water," submitted to Physical Review E.

All solid-state Marx generator for exhaust gas treatment

K. Lee, K.J. Chung, and Y.S. Hwang
Department of Nuclear Engineering, Seoul National University
Seoul, Republic of Korea

Abstract—**A highly repetitive Marx generator based on solid-state devices has been designed for the controlled experiments on the exhaust gas treatment. The system consists of total 14-stack of Marx modules to be capable of achieving 56 kV at the reactor's anode. In order to organize the main discharge circuit, high voltage charging diodes and insulated-gate bipolar transistors (IGBTs) are used, and snubbers are also included to give additional protection to the IGBTs from the high voltage transient induced damage. The driver circuit is powered and isolated by ferrite core transformer whose primary side is excited by H-bridge oscillator. All the IGBT drivers are controlled by the optical links which enable us to modify the pulse width and repetition rate very flexibly. In this work, 14-stage solid-state Marx modulator has been built and successfully operated to generate 55 kV output pulses with pulse width of 1 μs, rise-time of 160 ns, and repetition rate of 10 kHz, as an initial test using resistive load. In this paper, we describe the design and construction of our device and present data obtained from the initial test.**

Keywords—Marx generator; solid-state switch; high voltage modulation, sub-microsecond streamer discharge, exhaust gas treatment

I. INTRODUCTION

Various type of pulsed power generators have been developed for the industrial applications [1,2] including exhaust gas treatment. In this application, the short duration of the pulse effectively prevent the transition from streamer corona to spark and produce nonthermal plasmas by selective heating of electrons [3-5]. Hence, the multistaged or triaxial Blumlein pulse-forming lines (PFL) were generally chosen for the nanosecond-scale pulsed streamer discharges [4,5]. However, their slow repetition rates and problems in impedance mismatching cannot deal with more wide range of experimental conditions. These limitations can be circumvented by the use of solid-state Marx generator. In virtue of recent technological advances in manufacturing solid-state switches, various concepts of solid-state high voltage modulators have been proposed [2,6-9]. Among them, the diode-directed structure, which is developed by Los Alamos National Laboratory (LANL), can be characterized by the diode charging and solid-state switching [6,7]. This structure has definite advantages in recharging efficiency, flexible shaping of waveform, and self-snubbing for switch protection.

The high voltage (HV) pulse modulator has been built in Seoul National University (SNU), based on the architecture proposed by LANL. In this article, we describe the design and construction of our solid-state Marx modulator, and the initial experimental results with resistive load.

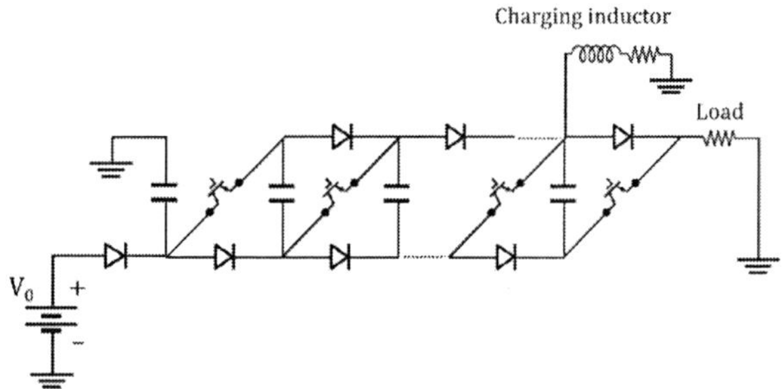

Fig. 1. Simplified circuit of solid-state Marx modulator.

II. DESIGN AND CONSTRUCTION

Figure 1 shows a simplified circuit of solid-state Marx modulator. As described in [6], the diodes provide a recharging path with low dissipation between pulses while effectively block the current leakage during the discharge. Also, insulated-gate bipolar transistor (IGBT) with high working voltage (≤ 4.5 kV) has been used to switch on and off the circuit. The charging resistor can be added to sustain flat-top pulse when the longer pulse width is needed, longer than 4 μs in our case, because the required dimension of charging inductor becomes unacceptably large. Since it reduces the charging efficiency, this charging resistance should be kept as low as possible. High voltage DC power supply, directly connected to the Marx board, continuously charges the capacitors between pulses in current limiting mode.

For the future experiments on exhaust gas treatment, the pulse modulator is required to generate 60 kV pulses at high repetition rate (≥1 kHz). Considering that the rise-time of single IGBT is about several hundred nanoseconds, only sub-microsecond pulsed discharges are available using this generator. The expected plasma impedance in streamer discharge is on the order of kiloohms [4]. Thus, the required discharge current is less than 100 A which can be safely sustained by IGBTs in use. For the modulation of 60 kV flat-top pulse, total 14 stages of solid-state Marx modules are assembled. As each stage is equipped with a charge storage of 330 nF Polyester film capacitor (SHV capacitors) which has a maximum charging voltage of 6 kV, the droop does not exceed 1 % even at the maximum pulse length.

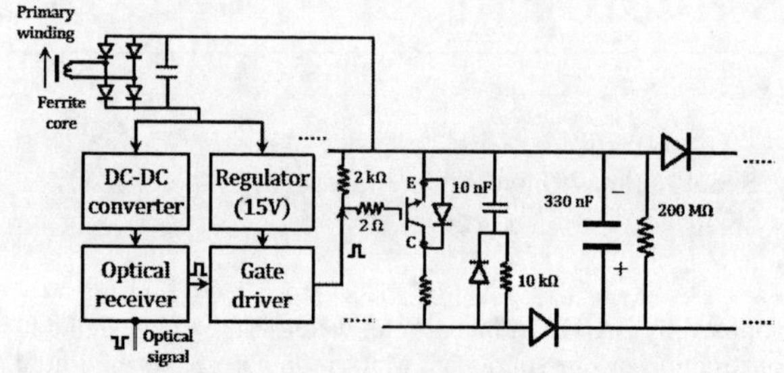

Fig. 2. Schematic diagram for a single Marx module.

Figure 2 illustrates the structure of single Marx module. There is minor modification on the design proposed by LANL [6]. Our choice for a solid-state switch is the IXYS model IXYX40N450HV because of its high working voltage and reasonable switching speed. The high voltage diode ED3085 of EDI is chosen to close the charging path. This topology has inherent advantage in switch protection from high voltage transient damage, so that the snubber plays minor role in this configuration. Each Marx module can be independently triggered by the external controller containing optical transmitters and multi-channel delay generator. For the safety, 200 MΩ resistor is added in parallel to the Marx capacitor to continuously dump the stored energy while the modulator is deactivated.

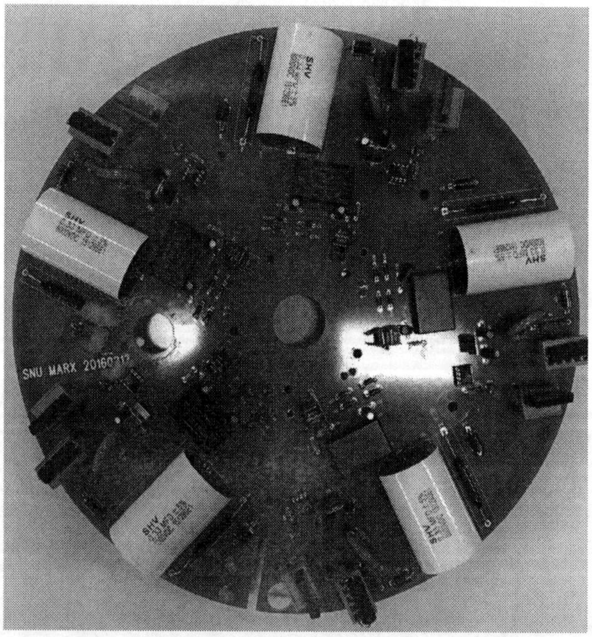

Fig. 3. Picture of a circular printed circuit board (PCB) with five Marx modules installed.

Power required to operate optical receiver and gate driver is delivered via ferrite core transformer. This point is most vulnerable due to the significant potential difference between primary and secondary windings when the entire Marx modulator is erected. From this reason, primary winding should be insulated very carefully to prevent arcing event-induced damage on the IGBT. In our modified design, five modules are distributed around the circular printed circuit board (PCB) as shown in figure 3. Each circuit board consists of five modules, and they share a single ferrite core located at the center. As total three circuit boards are linearly stacked up, the HV insulated cable passes through the center of ferrite cores to work as a primary winding. Half-bridge circuit supplies for primary side with oscillating frequency of 23 kHz. The oscillation frequency and voltage amplitudes at the primary side is optimized with proper turn ratio in order to guarantee that the delivered power is enough to operate all the electronics on each module. The circular layout of circuit much simplifies the construction of Marx modulator as well as its maintenance. The circuit has been drawn on the 2 mm thick substrate for its mechanical robustness. The outer diameter of circular PCB is 320 mm and each board is separated by three 185 mm-long supporting rods made of acetal. Fully assembled solid-state Marx modulator is shown in figure 4.

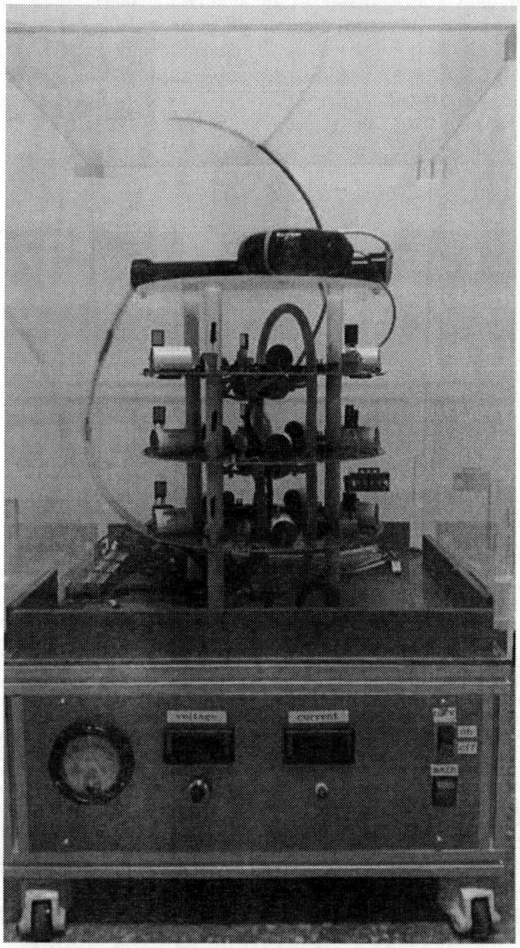

Fig. 4. 14-stage solid-state Marx modulator.

The polarity of output pulse can be easily changed by modifying the wire connection between adjacent boards and the use of positive DC power supply with exactly the same circuit composition. One can figure out that our circuit board is optimized to generate negative pulse. This is why the last module on the third floor is not being used in positive mode operation.

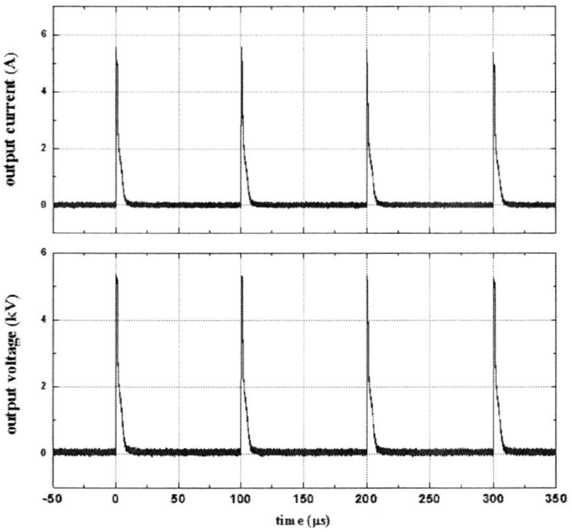

Fig. 5. Test with the resistive load 10 kΩ..

III. Experimental Results

Figure 5 shows experimental result in the initial test with the resistive load of 10 kΩ. In order to measure the output voltage, high voltage probe (VD-200, North Star High Voltage) has been used and its waveform was acquired by the digital oscilloscope (DPO 7104C, Tektronix) with the sampling rate of 2.5 GS/sec. All the optical transmitters are operated in parallel to produce the square pulses of 1 μs pulse width at repetition rate of 10 kHz. The rise-time of each pulse was about 160 ns. This rise-time of the output pulse is mainly governed by IGBT performance, Metal oxide semiconductor field effect transistor (MOSFET) can replace the IGBT if higher dv/dt with limited discharge current is required in the application. As more modules are stacked up for the higher output voltage generation, HV insulation becomes critical issue especially at the ferrite core transformer. The charging inductor also has definite limitations in the cases where the higher output voltage or the longer pulse width is required. One can replace the charging diode with another solid-state switch for the active switch charging [10-12]. However, charging inductor still effectively works for the intermediate level of pulse modulation with short pulse width (≤1 μs) based on much simpler circuit structure than the active switch charging scheme. The magnitude of output pulses gradually decreases due to limited performance of charging DC power supply. The elongation in the trailing edge of the waveforms is originated from the internal capacitance (24 pF) of the high voltage probe.

IV. Conclusions

All solid-state Marx modulator using IGBT and charging diode has been built at SNU for the research purpose on the exhaust gas treatment. It has been designed based on the basic architecture proposed by LANL [6] to handle wide range of experimental conditions such as the repetition rate, pulse width, and output voltages. Total 14-stages are assembled and tested with a 10 kΩ resistive load. The initial test result shows that the 55 kV output pulses with pulse width of 1 μs and rise-time of 160ns at the repetition rate of 10 kHz are successfully generated. Some minor modifications on the triggering circuit for the rapid turn-off of IGBTs are ongoing. This modulator is also can be used in other applications where well-defined flat top pulse with variable pulse width, low pulse current, and high repetition rate are required, such as plasma immersed ion implantation, micro pulse electrostatic precipitator, and pulsed beam extraction, etc.

Acknowledgment

This work was supported by the Defense Research Laboratory Program of the Defense Acquisition Program Administration and the Agency for Defense Development of Republic of Korea.

References

[1] H. Akiyama, T. Sakugawa, T. Namihira, K. Takaki, Y. Minamitani, and N. Shimomura, "Industrial applications of pulsed power technology," IEEE Transactions on Dielectrics and Electrical Insulation, vol. 14, pp. 1051-1064, 2007.

[2] B.M. Patel, S.H. Devmurari, and M.P. Rathod, "Development of high voltage solid state Marx generator for liquid applications", International Journal of Engineering Research and Technology, vol. 3, pp. 755-758, 2014.

[3] J.S. Chang and T. Yamamoto, "Corona discharge processes", IEEE Transactions on Plasma Science, vol. 10, no. 6, pp. 1152-1166, 1991.

[4] T. Namihira, D. Wang, T. Matsumoto, S. Okada, and H. Akiyama, "Introduction of nano-seconds pulsed discharge plasma and its applications", IEEJ Transactions on Fundamentals and Materials, vol. 129, no. 1, pp. 7-14, 2009.

[5] T. Matsumoto, D. Wang, T. Namihira, and H. Akiyama, "Exhaust gas treatment using nano seconds pulsed discharge", in 2009 IEEE Pulsed Power Conference, pp. 1035-1040, 2009.

[6] G.E. Dale, H.C. Kirbie, W.B. Haynes, C.E. Heath, T.A. Lopez, F.P. Romero, et al., "Design and application of a diode-directed solid-state Marx modulator," in 2005 IEEE Pulsed Power Conference, pp. 1211-1214, 2005.

[7] G.E. Dale, H.C. Kirbie, W.B. Haynes, C.E. Heath, T.A. Lopez, F.P. Romero, et al., "Performance of a diode-directed solid-state Marx modulator," in 2005 IEEE Pulsed Power Conference, pp. 1033-1036, 2005.

[8] S.C. Glidden and H.D. Sanders, "Solid state Marx generator", Conference Record of the 2006 Twenty-Seventh International Power Modulator Symposium, pp. 314-317, 2006.

[9] K. Liu, Y. Luo, and J. Qiu, "A repetitive high voltage pulse adder based on solid state switches", IEEE Transactions on Dielectrics and Electrical Insulation, vol. 16, pp. 1076-1080, 2009.

[10] A. Krasnykh, R. Akre, S. Gold, and R. Koontz, "A solid state Marx type modulator for driving TWT", in Power Modulator Symposium, 2000. Conference Record of the 2000 Twenty-Fourth International, pp. 209-211, 2000.

[11] G.E. Dale, R.M. Wheat, Jr., R. Aragonez, "Testing a scaled pulsed modulator for an IEC neutron source into a resistive load," in 2009 IEEE Pulsed Power Conference, pp. 1239-1243, 2009.

[12] H. Lim, D.H. Jeong, M. Lee, M. Lee, J. Yi, et al., “Solid-state pulse modulator using Marx generator for a medical linac electon-gun”, Journal of Instrumentation, vol. 11, P04003, 2016.

HV closing semiconductor switch for LHC beam dumping system

V. Senaj, L. Ducimetiere
CERN, TE/ABT
Geneva, Switzerland
viliam.senaj@cern.ch

Abstract*— *this document describes GTO like thyristors based HV closing switch used in Larger Hadron Collider beam dumping system

Keywords—GTO like thyristors, LHC beam dump, SEB

I. INTRODUCTION

Large Hadron Collider (LHC) is the world highest energy accelerator with two counter-rotating beams of protons or lead ions. Its maximum designed energy is 7 TeV with kinetic energy of up to 360 MJ per beam - high enough to provoke important damage to machine in case of incorrect or missing beam dump. Reliable operation of the beam dumping system is hence vital for the accelerator safety. Due to current level needed (up to 24 kA), pulse duration (> 89 μs) and required prefire-free operation, the only feasible switch technology was semiconductor based HV closing switch – in our case 10 series connected GTO like thyristors (GTO). LHC beam dumping system consists of 30 extraction and 20 dilution generators and serves to extract and sweep the LHC beam in order to deposit it safely on a graphite dumping block. It is designed with some redundancy so that one missing generator will not compromise machine safety. In order to keep the system reliability at a required level, system diagnostics is done after each pulse with the goal to keep it in "as good as new" condition. Surveillance systems observe important parameters like supply voltages and currents and create emergency dump in case of problem.

II. BEAM DUMPING SYSTEM DESCRIPTION

A. Extraction generators

Extraction of LHC beam at 7 TeV requires extraction magnet current of almost 19 kA lasting more than 89 μs and with a rise time of ~2.7 μs corresponding to extraction generator voltage of 29 kV. Generators have to be ready to operate anytime during the beam run lasting up to more than 36 hours. For reliability reasons, generators have two parallel redundant branches as can be seen on a Fig. 1 and one switch can conduct full current in case of not firing of the other one. Flattop ripple of less than 7.5% resulting from this topology is acceptably low to fit within the aperture of extraction path – Fig. 2. Energy storage capacitors are self-healing types and hence catastrophic internal failure is very unlikely. Generator internal current waveforms are captured and stored and are available for diagnostic purposes in case of need. In order to improve output current reproducibility the compartment with GTO stack is temperature stabilised – in such a way we can ensure long term current stability within ± 0.5 %. Thermal stabilisation is performed by Peltier module that provide heating or cooling according to the ambient temperature and internal voltage applied (power dissipation on internal HV dividers at maximum voltage is about 70W). Malfunctioning of more than one extraction generator out of 15 (per beam) can provoke important damage to downstream equipment and its reliable operation is hence of the highest importance..

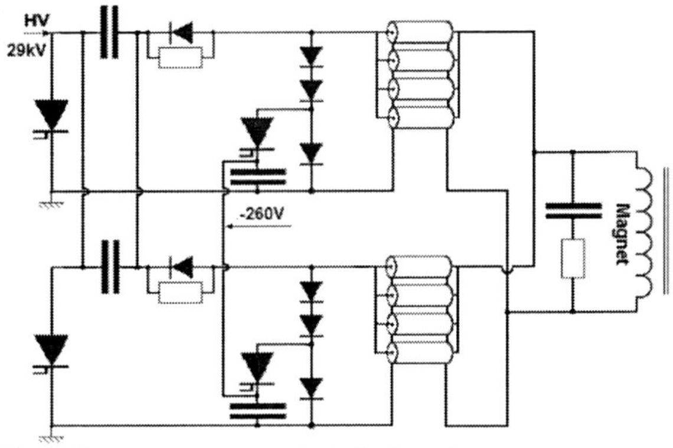

Fig. 1. Extraction generator principal schematics

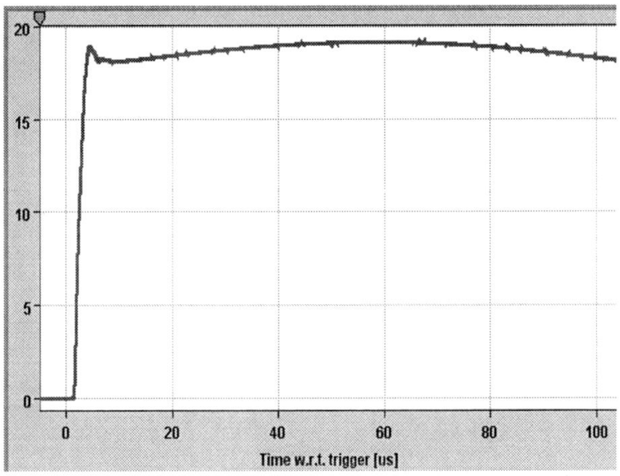

Fig. 2. Extraction generator current waveform

B. Dilution generators

Horizontal and vertical dilution generators schematics are on Fig. 3. They are based on the same HV switch topology – 10 series connected GTO. Since dilution is less critical for machine safety than extraction, the generators are not internally redundant. Horizontal and vertical dilution generators are operating at up 27 kV and 16 kV respectively with a peak current of 24 kA in both cases. They generate sinusoidal shape currents with frequency of ~14 kHz, horizontal diluter being phase delayed by 90 degrees (Fig. 4) in order to sweep a beam in e shape pattern (horizontal and vertical dilution magnet fields are mutually perpendicular). Generators are using self-healing type capacitors as well.

III. HV CLOSING SEMICONDUCTOR SWITCH – STACK OF 10 SERIES CONNECTED GTO LIKE THYRISTORS

View of 10 series connected GTO is on Fig. 5. Static voltage sharing within stack is ensured by 5 MΩ resistors in parallel with each GTO. Dynamic sharing is ensured by parallel snubber capacitors of 250 nF. Detection of eventual breakdown of individual GTO (resulting in its short circuit) is done by measurement of the voltage on the bottom most GTO. In case of one stage short circuit, the voltage on the bottom stage rises by 11% or fall to zero (in case of bottom stage short circuit). To avoid influence of GTO leakage current to the surveillance accuracy, GTO leakage currents are measured at 5 voltage levels within operational range at nominal temperature and GTO for position 1 is properly chosen. Two families of HV GTO with similar characteristics are presently in use. ABB made GTO 5STH20H4502 (wafer on left bottom) and Dynex made DG6478BH45-185. In both cases wafer diameter is ~ 60 mm, peak off state voltage is 4.5 kV, peak current is 80 kA and load integral is ~ 1e6 A2s. Due to parallel topology of extraction generator its HV closing switch is conducting roughly 9.5 kA during ~ 2.7 µs (magnet current rise time) then the current conduction is ensured by so called compensation part in the freewheel circuit operating at low voltage. Based on endurance test with 5e5 pulses, 16.5 kA peak current and pulse duration longer than in our application, the switch lifetime is estimated to ~ 1e6 pulses– well above expected number of pulses during beam dumping system lifetime.

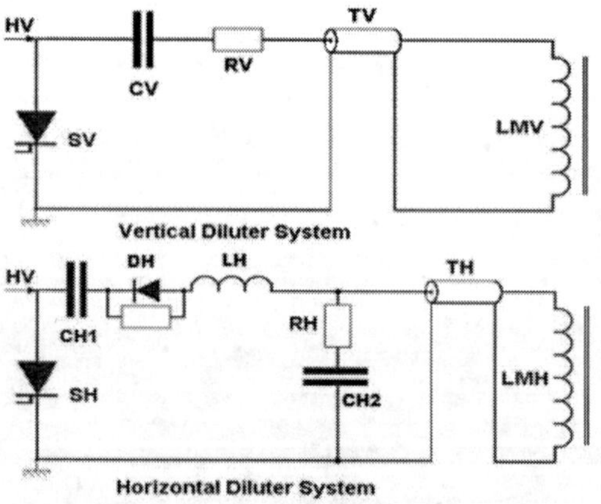

Fig. 3. Dilution generators principal schematics

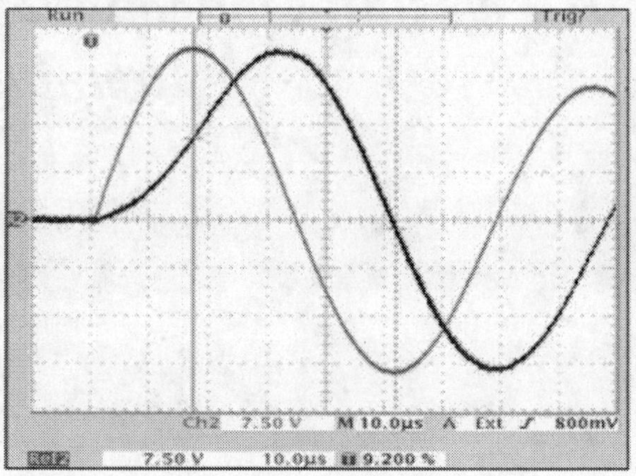

Fig. 4. Dilution generators current waveforms

IV. RADIATION TO ELECTRONICS EFFECTS

LHC beam dumping system is installed in underground galleries parallel to the LHC tunnel and connected with tunnel by cable ducts. High Energy Hadrons (HEH) penetrating from tunnel into the galleries via cable ducts present a potential risk of a Single Event Burnout (SEB) – a catastrophic failure mode in HV semiconductor. To estimate probability of this failure mode, the HV semiconductors – GTO and HV triggering IGBTs were tested in CERN irradiation facility using in house developed non-destructive method of SEB measurement [1]. SEB cross-section (SEBc-s) measurement for both GTO families is on Fig. 6. Based on SEBc-s, the mitigation measures were taken in order to keep the failure probability at acceptable level. Mitigation measures consisted of replacement of critical components (extraction generators are populated by less sensitive ABB GTO; HV triggering IGBT were replaced by higher voltage rating ones) and shielding of cable ducts. Based on SEBc-s measurement and simulated HEH fluency in galleries, the probability of SEB related failure is evaluated to ~ 0.1/year level.

Fig. 5. GTO wafers (left) and view of full GTO stack (right)

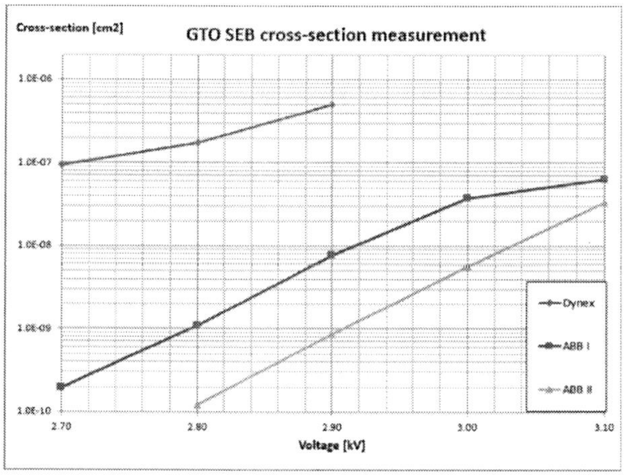

Fig. 6. SEBc-s measurement of 2 families of GTO used. ABB GTO consist of 2 subfamilies with different wafer passivation and with approximately order of magnitude difference in SEB sensitivity

V. GTO STACK TRIGGERING AND FUTURE DEVELOPMENT

GTO stack triggering is done by a triggering transformer presently with non-optimal performance; its leakage inductance is limiting achievable triggering current slew rate to some 400A/µs at triggering voltage of 3.5 kV. New triggering transformer with significantly reduced leakage inductance was developed recently and together with more powerful triggering system allows to increase triggering current slew rate to more than 1.5kA/µs. Significant reduction of the output current rise time (gain of 150 ns) and increased generator peak current (+ 1%) confirmed improved switch commutation performance with upgraded triggering circuits. In the past we observed sparking on the surface of the stack insulators sometimes leading to spontaneous firing of the stack. We found correlation with presence of dust or other foreign objects (fibres, hairs, insects...) in sensitive regions of HV insulators. Although the stack design was based on electrostatic field simulation with the goal to avoid excessive electrostatic field, it did not take into account possible presence of the foreign objects and their contribution to a surface charging they may contribute. First mitigation measures included meticulous cleaning of the exposed areas, modification of maintenance procedures (using of clean room protection clothes during maintenance) and modifications of the generators enclosure in order to avoid dust penetration. GTO stack with significantly reduced electrostatic fields was developed and is under test with the goal to gain safety margin and such reduce its fragility in case of future insulators surface pollution.

VI. CONCLUSION

LHC beam dumping system is employing 80 stacks of 10 series connected GTO working at up to 29 kV and 24 kA. Maintenance effort during last end year technical stop resulted in extremely reliable operation with zero hardware related fault during last year and with very good reproducibility within some 5e3 pulses at different energies. SEB related mitigation effort seems to work as well. Future development will be concentrated on improvement of triggering system performance and to gain immunity of the stacks to surface sparking in case of HV insulators surface pollution.

References

[1] V. Senaj, L. Ducimetiere, 'Non-destructive SEB test of HV GTO-like thyristors', in Single Event Effect Symposium, San Diego, CA, 2012.

Development of Relativistic Magnetrons*

Mikhail I. Fuks and Edl Schamiloglu
Department of Electrical and Computer Engineering
University of New Mexico, MSC01 1100
Albuquerque, NM 87131-0001, USA
fuchs@unm.edu, edls@unm.edu

Nikolay Kovalev
Department of Plasma Physics and High Power Electronics
Institute of Applied Physics, Russian Academy of Sciences
Nizhny Novgorod, 603950, Russia
kovalev@appl.sci-nnov.ru

***Abstract*— With the appearance of high current electron accelerators new life began for traditional microwave sources using relativistic intense electron beams. These are the O-type sources, such as BWOs, TWTs, orotrons, FELs, and others. However, for M-type sources, such as magnetrons, the availability of such accelerators gave the possibility to use high applied voltages, although the operating drift velocities of the electrons remained nonrelativistic. Since the author of the first high voltage magnetrons referred to them as "relativistic magnetrons" we follow this terminology in spite of the nonrelativistic electron motion. (Perhaps it would be more correct to refer to them as "high voltage magnetrons".) This paper reviews recent advances in the field.**

Keywords—relativistic magnetron, MDO, pulsed power, high power microwaves

I. INTRODUCTION

The design of the first relativistic magnetron [1] was similar to the design of conventional magnetrons, which is a reason for some of its limitations. For example, extraction of radiation from such magnetrons is achieved through a narrow slot from one (or several) of the cavities of the anode block (Fig. 1) that creates their asymmetric azimuthal design. Because of this, such magnetrons can operate only with non-degenerate modes, the π-mode or the 2π-mode. Any jump to other modes leads to splitting of the mode to two sub-modes: one has its node at the cavity with radiation output, while the other one has an antinode at the cavity, Fig. 1. Since the sub-mode with the node is almost decoupled with the output, the Q-factor of this sub-mode is higher and the magnetron will operate with this sub-mode while suppressing the slower sub-mode. As a result, the magnetron operates without any output, heating the magnetron electrodes until their ultimate disruption.

This paper describes advances in the relativistic magnetron that have been pioneered at the University of New Mexico for over the past decade.

II. CONCEPT OF MAGNETRON WITH AXIAL EXTRACTION OF RADIATION

Axial extraction of radiation through the horn antenna, in which all cavities of the anode block are continued up to a cross section that exceeds the dimenion that corresponds to the cutoff frequency for the radiated mode, is represented as the most suitable design for relativistic magnetrons. The first such relativistic magnetron that was named the "magnetron with diffraction output" (MDO) and was realized in 1977 [2] with gradually decreasing the depth of 8 equal cavities' continuations onto such a diffraction output (DO) - (Fig. 2).

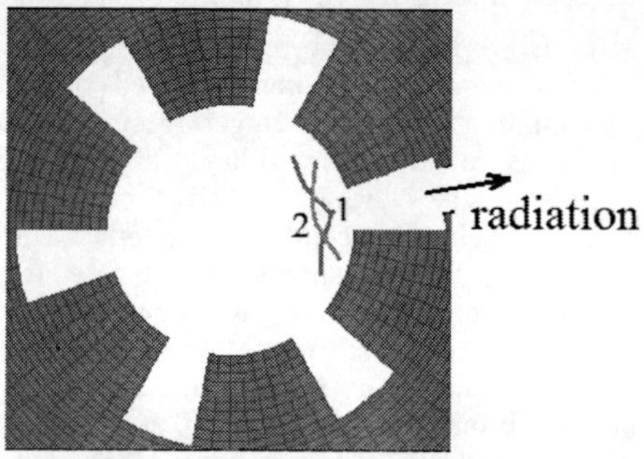

Fig. 1. Asymmetry anode block because of radiation being extracted through one of the cavities. 1 denotes a sub-mode with node near the cavity. 2 denotes an antinode.

Fig. 2. First relativistic magnetron [2] with axial extraction of radiation through the horn antenna with tapered cavity continuations.

The MDO can operate with any mode because symmetric loading of all cavities does not lead to splitting of degenerate modes into sub-modes with different Q-factors. Resistance to breakdown of MDOs is incomparably greater than that of conventional magnetrons.

Research at the University of New Mexico was sponsored by ONR Grants N00014-13-1-0565 and N00014-16-1-2352, AFOSR Grant FA9550-15-1-0094, and AFOSR DURIP Grant FA2386-12-1-3027.

Diffraction extraction of radiation is a selective element that gives an advantage to operating oscillations with the most uniform microwave field structure along the axis. In 2010 DO with gradually increased continuations of cavities onto the horn antenna (Fig. 3), which was suggested in [3], was optimized [4] using the particle-in-cell (PIC) code MAGIC [5], with optimized DO parameters: outer angle $\phi_O = 32°$ and inner angle $\phi_i = 17.5$ which resulted in electronic efficiency exceeding 70% (Figs. 4,5). In experiments [6,7] efficiency of this MDO was 63% prior to cathode endcap failure. It is pertinent to note that in the highly efficient MDO [4] a transparent cathode [8] was used.

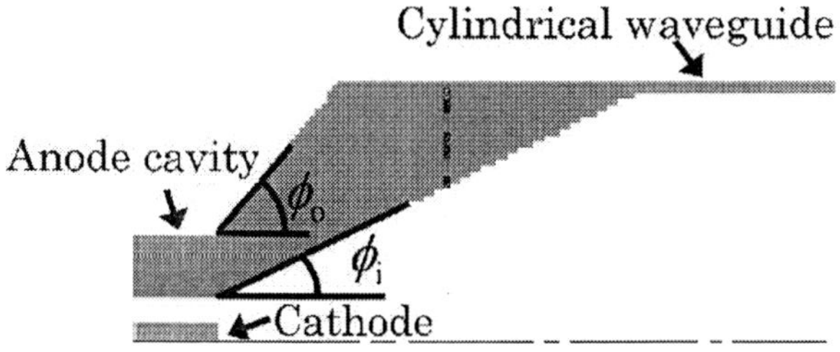

Fig. 3. Side view of the MDO with expanding continuations of cavities onto the horn antenna (with angle $\phi_O > \phi_i$).

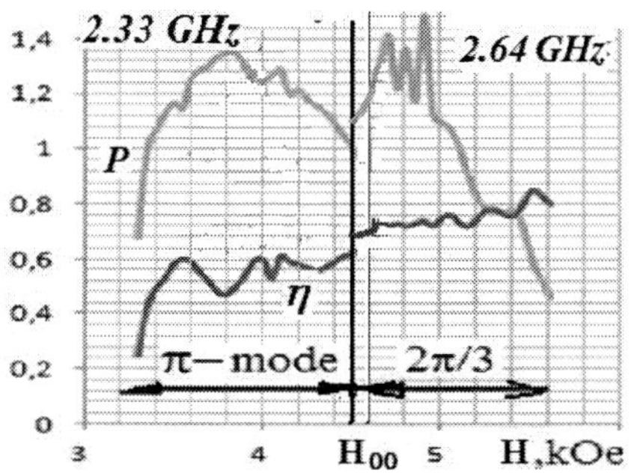

Fig. 4. Dependences of radiation power P and electronic efficiency η for the operating π-mode and for the (-1st) spatial harmonic of the $2\pi/3$-mode.

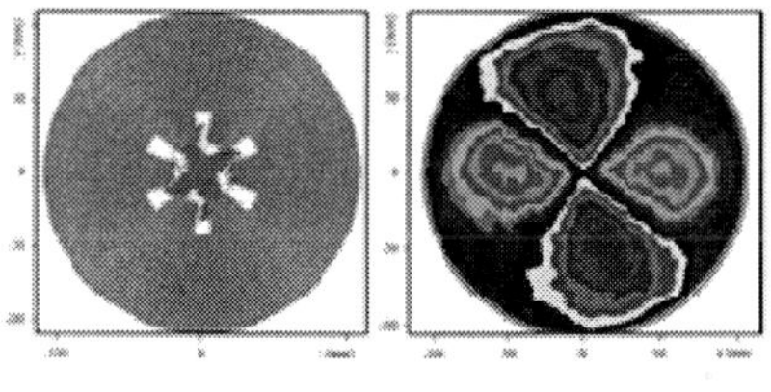

Fig. 5. Left – electron spokes; right – output field structure corresponding to the operating (-1st) spatial harmonic of the $2\pi/3$-mode.

III. Transparent Cathode in the MDO as a Means to Fast Start of Oscillations

For many applications, as for mutual phasing of magnetrons, it is important to provide a short front of magnetron radiation. To provide fast start of generation, we suggest to use a cathode [8] consisting of several individual emitters placed about the cathode radius (Fig. 6). The RF azimuthal electric field $\tilde{E}_\theta$ in such a cathode is transparent and goes to zero on axis. Therefore, this strong field in the electron flow drifting around the cathode provides a fast radial drift of electrons with velocity

$$v_r = c\frac{[\tilde{E}_\theta \times H_{0z}]}{H^2}, \quad (1)$$

which is the velocity to transfer electron potential energy to the electromagnetic fields. Here c is the light speed; H_{0z} is the applied axial magnetic field;

$$H^2 = H_{0z}^2 + H_\theta^2; H_\theta = \frac{2I_z}{R_c}, \quad (2)$$

and I_z is the axial current of the cathode. The field $\tilde{E}_\theta$ is the tangential field, which on the solid cathode must be zero. Thus, this field is small in the thin electron flow Δ for a solid cathode and much stronger for a transparent cathode (Fig. 7).

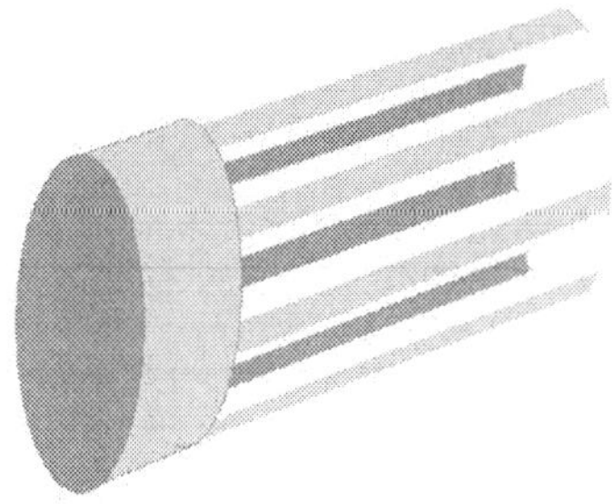

Fig. 6. A 6-emitter transparent cathode.

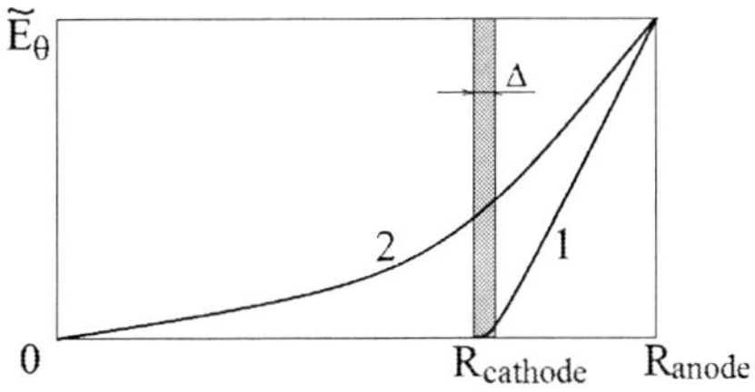

Fig. 7. Synchronous azimuthal electric field in an MDO with a solid cathode (1) and with a transparent cathode (2).

For the MDO [7] with a transparent cathode PIC simulations show a short rise time of MDO generation (Fig. 8) with such a cathode when the rise time of the applied voltage (400 kV) is 10 ns.

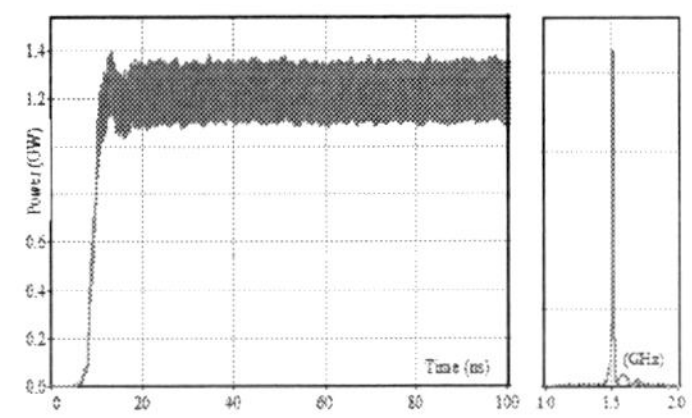

Fig. 8. Output power of MDO [7] with transparent cathode when the rise time of the applied 400 kV voltage is 10 ns (with a solid cathode, power saturation occurs in 40 ns).

IV. MDO with Virtual Cathode

In magnetrons cathodes are subjected to bombardment by electrons in unfavorable phases of the operating wave. In relativistic magnetrons cathodes with explosive electron emission are not only source of electrons, but also sources of plasma that expand across the axial magnetic field with velocity $v_p \approx 2 - 3$ cm/μs. This leads to a decreased effective gap spacing between the electrodes,

$$d_{\text{eff}} = d - v_p t \tag{3}$$

that decreases the electronic efficiency η via

$$\eta = 1 - \frac{\Delta}{d_{\text{eff}}}, \tag{4}$$

which leads to an unpredictable change of radiation frequency and to rapid pulse shortening. These reasons motivated us to remove such a cathode from the interaction space and to form a lengthy virtual cathode (VC) instead of a physical cathode. At the boundaries of the interaction space, we create two local VCs so that the direct electron flow from the first VC1 and reflected electron flow from the second VC2 form such a large space charge that its potential becomes close to the potential of a physical cathode. PIC simulations show (Fig. 9) that the existence of such a lengthy VC is possible [9].

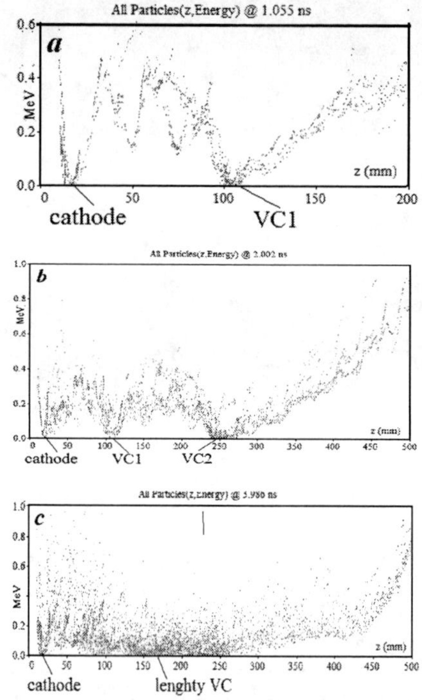

Fig. 9. Forming a lengthy VC. a): VC1; b): VC1 and VC2; c): the lengthy VC.

V. MDO as Mode Converter

Numerous PIC simulations of MDO operation show that when the number of cavities $N \leq 8$ realization of single mode operation is saved owing to good selection of transverse modes. Using this fact, we can use the DO for mode conversion when the operating mode in the MDO is the π-mode continuing onto the DO only those cavities whose polarization correspond to the polarization of the desired mode [10]. When each second cavity of the anode block is continued onto the DO the symmetric TE_{01} mode is radiated (Fig. 10). Fig. 11 shows as the operating π - mode converts to the TE_{11}- mode in a 6-cavity MDO.

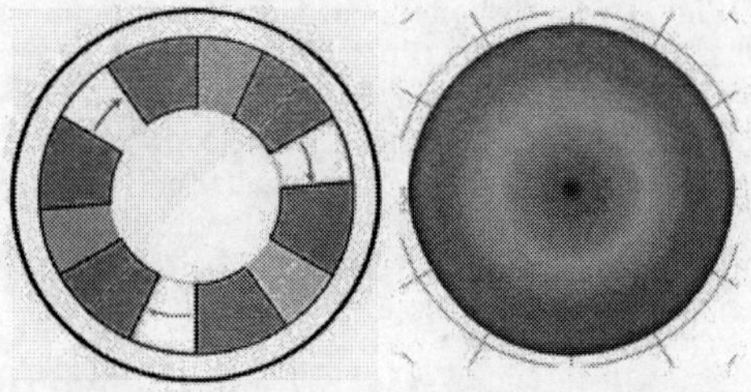

Fig. 10. Conversion to the radiated TM_{01}-mode.

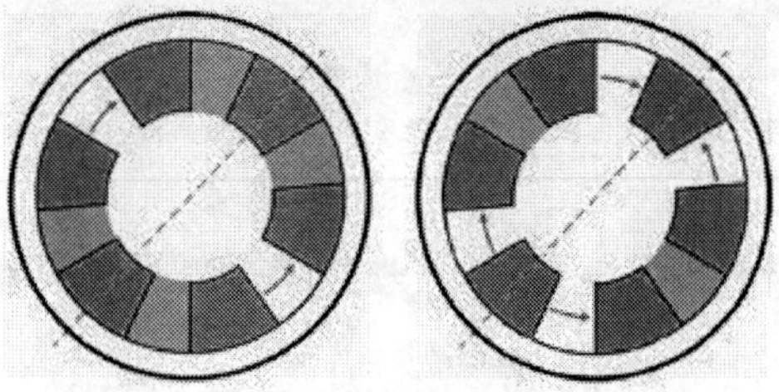

Fig. 11. Possible cavity continuations in a 6-cavity MDO for conversion to the radiated TM_{11}-mode.

VI. Compact MDO

As a rule, the radius corresponding to the cutoff frequency of the radiated TE_{11}-mode is less than the maximal radius of the cavities of the anode block. This gives a possibility to create a compact magnetron consisting of the anode block together with a mode converter of the operating π-mode to a Gaussian-like radiation pattern (TE_{11}-mode) with a minimal volume of the magnetic field-producing system [11], (Fig.12). The smooth horn antenna can be placed outside the magnetic system if it is necessary to increase the output window in order to avoid breakdown or to corrugate a surface of the horn antenna in order to add the TM_{11}-mode for formation of the real Gaussian microwave beam.

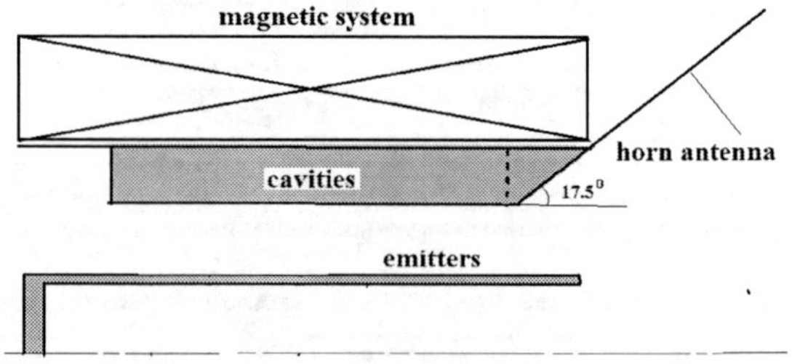

Fig. 12. The compact MDO.

VII. Conclusions

In this paper we have described basic improvements that, as we think, are inherent to relativistic magnetrons although many other problems were solved as well, such as elaboration of the theory of relativistic magnetrons, mode switching using an input signal, mode separation, reduction of leakage current and others. Here we shortly enumerate the main achievements that were presented above:

1. Symmetric axial extraction of radiation in an MDO that permits one to use any mode as the operating one; sufficiently increase resistance to electrical breakdown; improve longitudinal mode selectivity.
2. Optimization of diffraction output with result in achievement of electronic efficiency >70% in PIC simulations and 63% in experiments.
3. Transparent cathode provides fast start of generation.
4. Lengthy virtual cathode instead of a physical cathode in the interaction space – no electron bombardment, the possibility to use cathodes without appearance of plasma in the interaction space.
5. The possibility to use diffraction output as a mode converter to modes with simpler radiation pattern, including Gaussian microwave beam.
6. The possibility to realize a compact magnetron with Gaussian-like radiation.

[1] G. Bekefi and T.J. Orzechowski, "Giant microwave burst beam magnetron," Phys. Rev. Lett., vol. 37, pp. 379-382, 1976.

[2] N.F. Kovalev, B.D. Kol'chugin, V.E. Nechaev, M.M. Ofitserov, E.I. Soluyanov, and M.I. Fuks, "Relativistic magnetron with diffraction coupling," Sov. Tech. Phys. Lett., vol. 3, pp. 430-431, 1977.

[3] M. Daimon and W. Jiang, "Modified configuration of relativistic magnetron with diffraction output for efficiency improvement," Appl. Phys. Lett., vol. 91, pp. 191503-191505, 2007.

[4] M.I. Fuks and E. Schamiloglu, "70% efficient relativistic magnetron with axial extraction of radiation through a horn antenna," IEEE Trans. Plasma Sci., vol. 38, pp. 1302-1312, 2010.

[5] B. Goplen, L. Ludeking, D. Smithe, and G. Warren, "User-configurable MAGIC for electromagnetic PIC calculations," Comput. Phys. Commun., vol. 87, pp. 54-86, 1995.

[6] S. Prasad, M. Roybal, K. Prestwich, M. Fuks, C.J. Buchenauer, and E. Schamiloglu, "Magnetron experiments on the short-pulse 'SINUS-6' accelerator," *Proc. IVEC-2008* (Monterey, CA, 2008), pp. 441-442.

[7] C. Leach, S. Prasad, M. Fuks, and E. Schamiloglu, "Suppression of leakage current in a relativistic magnetron using a novel cathode endcap design," IEEE Trans. Plasma Sci., vol. 40, pp. 819-822, 2011.

[8] M.I. Fuks and E. Schamiloglu, "Rapid start of oscillations in a magnetron with a 'transparent' cathode," Phys. Rev. Lett., vol. 95, pp. 205101-1-4 2005.

[9] M. Fuks, S. Prasad, and E. Schamiloglu, "Efficient magnetron with a virtual cathode," IEEE Trans. Plasma Sci., vol. 44, pp. 129-1302, 2016.

[10] M.I. Fuks, N.F. Kovalev, A.D Andreev, and E. Schamiloglu, "Mode conversion in a magnetron with axial extraction of radiation," IEEE Trans. Plasma Sci., vol. 34, pp.620-626, 2006.

[11] C. Leach, S. Prasad, M.I. Fuks, and E. Schamiloglu, "Compact relativistic magnetron with Gaussian radiation pattern," IEEE Trans. Plasma Sci., vol. 40, pp. 3116-3120, 2012.

High Current Striplines for Neutrino Beams

T. Sipahi*[1], S.G. Biedron[1,2,3], J. Hylen[4], S.V. Milton[1,3], R. Zwaska[4]
[1]Department of Electrical and Computer Engineering, Colorado State University, Fort Collins, CO 80523 USA
[2]Faculty of Electrical Engineering, University of Ljubljana, Tržaška 25, SI-1000 Ljubljana, Slovenia
[3]Element Aero, Chicago, IL 60643 USA
[4]Fermi National Accelerator Laboratory, Batavia, IL 60555
*taylan@engr.colostate.edu

***Abstract*—To facilitate several genres of high-energy physics (HEP) experiments, protons that have been accelerated to a high energy collide with a fixed target material to generate new particles through nuclear interactions. Due to the manner in which they are generated, these new particles are emitted at large angles. To ensure these particles are useful for these HEP experiments, they must be carefully collected and directed toward the experimental stations. A very high, pulsed, magnetic field extending over the shape of a horn helps to harness a large fraction of these new particles. These devices focus the new particles back onto the forward beamline with the correct momentum. These devices are van der Meer magnets, referred colloquially simply as horns or magnetic horns, and are a key component of the HEP experiments at Fermilab outside of Chicago, USA at CERN in Geneva, Switzerland and at J-PARC in Ibraki, Japan. Achieving such a high, pulsed magnetic field is a challenge over the length of such a device. The high current required to power the magnetic device is delivered through striplines and this has led to several operational challenges. As the HEP experiments evolve, specifically the neutrino experiments at Fermilab, even higher currents will be required of the striplines. Here, we will discuss the investigations of the stresses on the magnetic horn striplines.**

Keywords*—high current striplines for magnetic horns; Van der Meer magnetic horns; neutrinos

I. Pulsed Magnetic Horns

A. Background on the Use of Pulsed Magnetic Horns

Fixed-target experiments comprise a significant portion of today's experimental high-energy physics (HEP) endeavors. The fixed-target is met by a high-energy proton beam generated in a particle accelerator, and upon interaction, generates myriad of new particles. Due to the physical configuration and the nuclear generation process of these new particles, the resultant particles are emitted at a wide angle and are of a wide spectrum of momentums. Since many of the effects/phenonenom/events in experimental HEP are of low probability of occuring, the goal in such experiments is to increase the flux – the number of particles per unit time for any given experiment. Since the experiments want to increase the ability to detect an event, by increasing the number of particles in the experimental beamline(s), the newly-generated particles need to be focused and directed toward the experimental stations. This is done through the use of a high, pulsed magnetic field in the shape of a horn. These magnetic horns, also known as Van der Meer magnets, focus the new particles back onto the forward beamline with the correct momentum. These devices are found in HEP experiments such as those at Fermilab in Batavia, Illinois, outside of Chicago, USA, at CERN in Geneva, Switzerland, and in J-PARC, Ibriki, Japan, as well as elsewhere.

B. Challenges in Magnetic Horns

Not only does there need to be a high, pulsed magnetic field at a high repetition rate over the length of the horn device, typically between 1-3 meters, but the entire device is also constantly subject to a very high radiation environment that contributes to issues for maintenance. The high current, typically several hunreds of kA, required to power the magnetic device is delivered through striplines and this has repeadedly led to several operational challenges worldwide [1, 2, 3].

The present and future generations of neutrino experiments apply a further challenge of precision to the horns. The experiments require that the neutrino flux vary no more than a few percent during the operational run (absolute precision on the produced flux is also required on the 10% level). The details of hadron production control the absolute precision, but the relative precision is determined by beam devices such as the target or horn. Time variations of the horn current or current distribution can lead to fluctuations in the neutrino flux and spoil the experiments. Horns and their stripline systems must be made to maintain their structural and electrical integrity with high fidelity during irradiation (where confirmatory measurements are not possible).

II. The MAgnetic Horns at Fermilab – Critical to Neutrino Experiment Success

A. Neutrino Experiment Background at Fermilab

Presently, there are several magnetic horn systems operational or either under development or construction at Fermilab for various neutrino experimental beamlines [4]. Both the NuMI (Neutrino at Main Injector) beamline that has been providing intense neutrino beams for several

This work is supported in part by the Visiting Scholars Program of the Universities Research Association (URA). Fermilab is operated by Fermi Research Alliance, LLC under Contract No. De-AC02-07CH11359 with the United States Department of Energy.

experiments (MINOS - Main Injector Neutrino Oscillation Search, MINERvA, NOvA - NuMI Off-axis ν_e Appearance Experiment) and the new Long Baseline Neutrino Facility (LBNF) beamline which plans to produce the highest power neutrino beam in the world for the Deep Underground Neutrino Experiment (DUNE) in South Dakota all require these pulsed magnetic horns to focus the pions (pi mesons) in the neutrino production process. The high current horn and its associated high current stripline design has been evolving as NuMI reconfigures for higher beam power and as LBNF produces designs for even higher beam powers.

B. Accelerator Based Neutrino Production

Neutrino beams are produced by first accelerating protons in an accelerator such as a synchrotron. The energy of the protons, along with the geometry of the beamline determines the energy of the neutrinos. The accelerated protons are extracted from the accelerator and directed on to a target where the protons interact with the target material, producing a large number of secondary pions among other particles. Shaped magnetic fields created by focusing horns are used to select out pions of the preferred charge, and focus them into a collimated beam. The horns are fed high current through high-current striplines as shown in Figure 2.

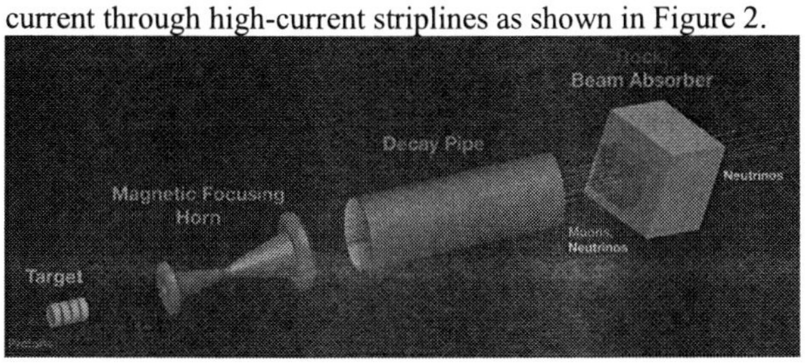

Fig. 1. Example of the major steps and components to neutrino production. Image courtesy of the Office of Communications, Fermilab.

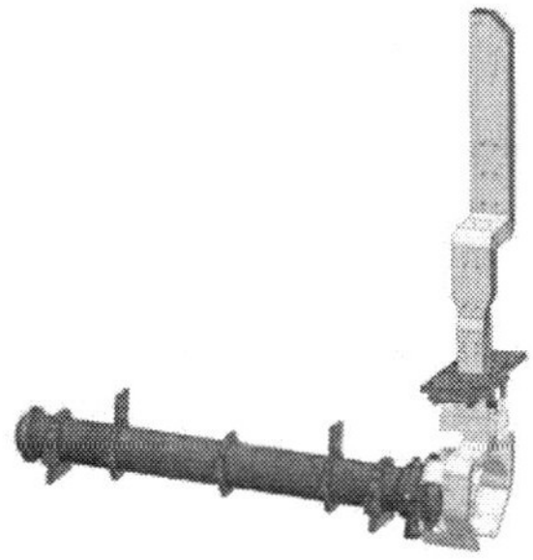

Fig. 2. 3D model of the attached magnetic horn (blue section) and the high current stripline (yellow section)

C. Recent Fermilab Horn System Challenges

Magnetic horns at Fermilab are designed to pulse 10-20 million times a year near the endurance fatigue limit. These horns fail after several years of operation, the goal being four years. Design of these horns always involve a compromise of various factors of performance or reliability, such as: maximum current, absorption by material, resistive heating, beam-radiation heating, radiation damage, electromechanical forces, thermomechanical forces, cooling, fatigue, erosion, and corrosion.

As an example: producing a horn with a thicker inner conductor will allow more force to be absorbed by the conductor and reduce the resistive losses; but it will increase the absorption of secondary particles (reducing performance) and increase the amount of beam-radiation heating without increasing the cooling capacity.

The stripline shown in Figure 2 has proven itself durable in fatigue resistance and overall robustness, however this stripline was modified, as shown in Figure 3, to a larger radius to the beam axis, reducing beam heating and increasing air cooling. Also they have been changed to chamfered plates thathas three bolt holes in the new design.

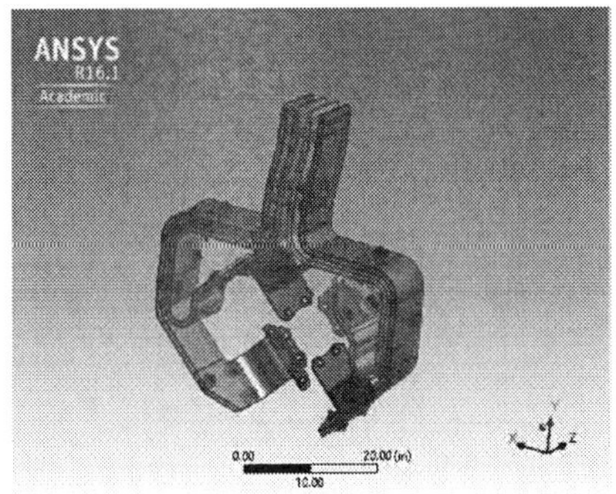

Fig. 3. Failed high-current stripline on th NuMI/NOvA Horn 1 (PH1-04). The aluminum is fractured due to fatigue of the eletromechanical forces (magnified by the corner) and extended vibration of the long span. This high-current stripline operated for 2 years (24 million pulses).

This updated design had a failure due to vibration-enhanced fatigue of an attached stripline. Electromechanical forces from the pulses excited a 424-Hz mode on a long span of the stripline. This excitation has since been observed to persist for 100s of pulses per impulse, greatly increasing the number of fatigue cycles to which the material is subjected.

Presently, to assist in the engineering design to reach these high currents, we are investigating the electromagnetic details of the magnetic horns and the high-current striplines. From these realistic simulations of the horns and striplines we hope to gain a better understanding of the complete system as well as find potential weaknesses, such as stress-induced structural damage to the striplines and/or the horn structure.

III. ELECTROMAGNETIC SIMULATIONS

We have performed simulations of the electromagnetic interaction of the stripline plates of the NoVA horn at critical stress points [4], using both the POISSON [5] and ANSYS Maxwell 3D [6] codes for a given applied current of 200 kA.

Four conductors are used to conduct the 200 kA into and out of the inner and outer conductor of the horn. The cross section of all the plates are 8 inches by 0.375 inches, and the distance between each plate is 0.375 inches. Each carry an applied current of 50 kA pulsed, however during our simulations direct current (DC) case is considered so far.

For example, two striplines carrying opposite currents generate magnetic fields that when coupled to the current generate a repulsive force between the two striplines. With

50-kA current each, the magnetic force between the two striplines is calculated as 6186 N/m from POISSON (Figure 4) and 6192 N/m as calculated by ANSYS Maxwell 3D (Figure 5).

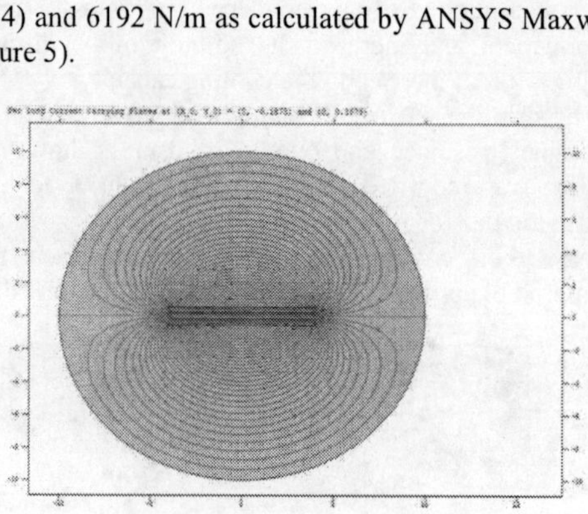

Fig. 4. Poisson result for the magnetic field map through a cross section of parallel plates in x-y planes with 50-kA DC in the opposite directions.

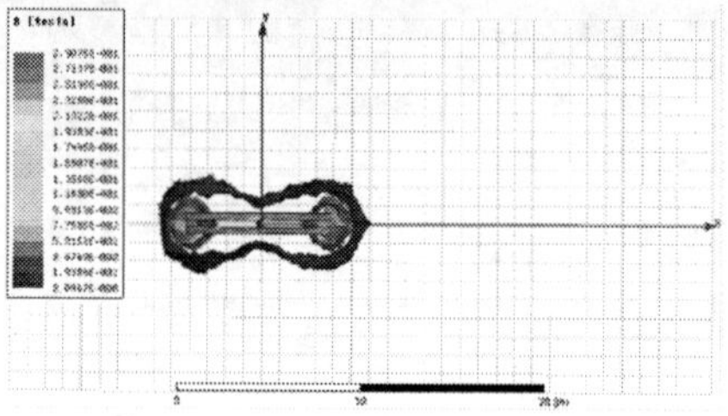

Fig. 5. ANSYS Maxwell 3D result for the magnetic field map of the two parallel plates of the stripline with 50-kA DC in the opposite directions.

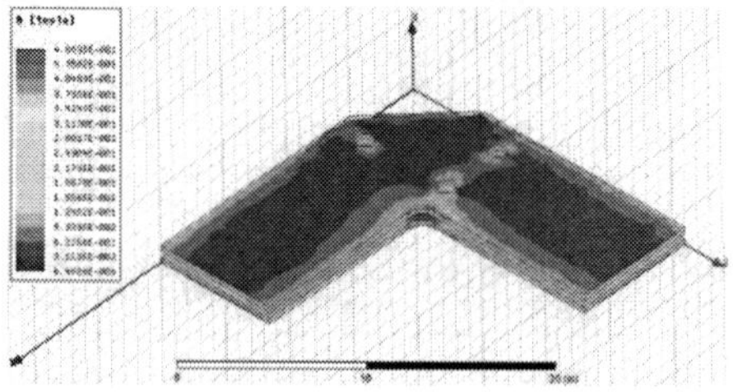

Fig. 6. Poisson result for the magnetic field map through a cross section of parallel plates in x-y planes each carrying 50 kA DC in the opposite directions.

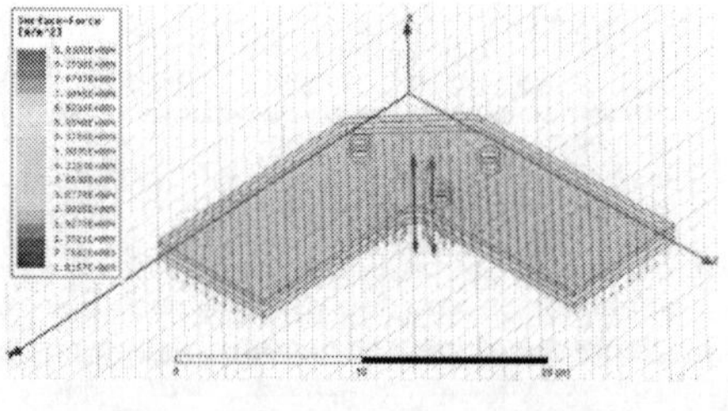

Fig. 7. ANSYS Maxwell 3D result for the pressure distribution of the flag plates of the high-current stripline each carrying 50-kA DC in the opposite directions.

After seeing that these two results are accurate enough for the straight sections of the stripline we continue to simulate using ANSYS Maxwell 3D for the stripline that failed in service at the 90 degree bend in the so-called "flag plate" (Figure 6). Simulations examined if the magnetic induced stresses were the primary source of the failures at this point.

We found that the magnetic field at the inner-most corner is larger than in the straights as shown in Figure 7. The average pressure on the flag plate is ~ 3.1 x 10^4 N/m^2; however, at the inner corner this pressure is ~ 8.8 x 10^4 N/m^2.

IV. CONCLUSION

The current densities and forces at the critical locations of the flag plate were evaluated and the inclusion of the chamfering and bolt holes affected the current density and forces, but not critically. The results for the magnetic field and force were calculated for the straight plates and agreed well with theory. Additional magnetic field related effects such as the local pressures and current densities were evaluated for real stripline flag plates. The magnetic field was found to be ~2 times larger at the inner corner of the plates as compared to the straight plates case. The pressure was ~2.5 times higher. These results alone do not cause the stress fractures but are likely related to repetitive heating and/or vibrational effects. Our results are being used in the upgrades of the magnetic horn stripline designs to handle increased electrical current and higher beam power. The corner sections need to be evaluated not only mechanically and thermally but also considering electromagnetic contributions. It is better if the inner corner radius of the plate will be enlarged for the higher current required in the next designs.

References

[1] Experiences from the antimatter decelerator at CERN http://cds.cern.ch/journal/CERNBulletin/2014/40/News%20Articles/1756355.

[2] T. Sekiguchi et al., "Development and operational experience of magnetic horn system for T2K experiment," Nuclear Instruments and Methods in Physics Research Section A: Accelerators, Spectrometers, Detectors and Associated Equipment, Volume 789, 21 July 2015, (pp. 57-80), http://dx.doi.org/10.1016/j.nima.2015.04.008.

[3] For a comprehensive summary of the neutrino experiments congoing and planned at Fermilab and its off-site detector locations, please visit https://www.fnal.gov/pub/science/particle-physics/experiments/neutrinos.html

[4] T. Sipahi *et al.*, "Simulations of High Current NuMI Magnetic Horn Striplines at FNAL," in *Proc. of the 7th International Particle Accelerator Conference (IPAC'16)*, Busan, Korea, May 8-13, 2016, paper TUPMR004, (pp. 1230-1233). JACOW, Geneva, Switzerland, 2016.

[5] J. Warren, et al., "POISSON/SUPERFISH Reference Manual," Los Alamos National Laboratory report LA-UR-87-126 (1987).

[6] ANSYS Academic Research, Release 16.

AUTHOR INDEX

AUTHOR INDEX

AUTHOR INDEX

AUTHOR INDEX

AUTHOR INDEX

AUTHOR INDEX